Alkane Functionalization

Alkane Functionalization

Edited by
Armando J. L. Pombeiro and M. Fátima C. Guedes da Silva
Instituto Superior Técnico
Universidade de Lisboa, Portugal

WILEY

This edition first published 2019

Registered Offices
John Wiley & Sons, Inc., 111 River Street, Hoboken, NJ 07030, USA
John Wiley & Sons Ltd, The Atrium, Southern Gate, Chichester, West Sussex, PO19 8SQ, UK

Editorial Office
9600 Garsington Road, Oxford, OX4 2DQ, UK

For details of our global editorial offices, customer services, and more information about Wiley products visit us at www.wiley.com.

Wiley also publishes its books in a variety of electronic formats and by print-on-demand. Some content that appears in standard print versions of this book may not be available in other formats.

Library of Congress Cataloging-in-Publication Data applied for

ISBN: 9781119378808

Cover image: Courtesy of Armando J. L. Pombeiro and M. Fátima C. Guedes da Silva
Cover design by Wiley

Set in 10/12pt WarnockPro by SPi Global, Chennai, India
Printed in Singapore by C.O.S. Printers Pte Ltd

10 9 8 7 6 5 4 3 2 1

Contents

List of Contributors

Mathieu Achard
Université de Rennes, CNRS, ISCR (Institut des Sciences Chimiques de Rennes)
UMR 6226
Rennes
France

Seihwan Ahn
Department of Chemistry
KAIST and Center for Catalytic Hydrocarbon Functionalizations
Daejeon
South Korea

and

Center for Catalytic Hydrocarbon Functionalizations
Institute for Basic Science (IBS)
Daejeon
South Korea

Elisabete C.B.A. Alegria
Centro de Química Estrutural
Instituto Superior Técnico
Universidade de Lisboa
Lisboa
Portugal

and

Chemical Engineering Department
Instituto Superior de Engenharia de Lisboa
Instituto Politécnico de Lisboa
Lisboa
Portugal

Mu-Hyun Baik
Department of Chemistry
KAIST and Center for Catalytic Hydrocarbon Functionalizations
Daejeon
South Korea

and

Center for Catalytic Hydrocarbon Functionalizations
Institute for Basic Science (IBS)
Daejeon
South Korea

Christian Bruneau
Université de Rennes, CNRS, ISCR (Institut des Sciences Chimiques de Rennes)
UMR 6226
Rennes
France

Konstantin P. Bryliakov
Novosibirsk State University and Boreskov Institute of Catalysis
Novosibirsk
Russian Federation

Fabrizio Cavani
Dipartimento di Chimica Industriale "Toso Montanari"
Università di Bologna
Bologna
Italy

and

Current address:
Strategy Development Research Division
Catalysis Refining and Base Chemicals Department
TOTAL Research & Technology
Seneffe
Belgium

Sunney I. Chan
Institute of Chemistry
Academia Sinica
Sec. 2, Nankang
Taipei
Taiwan

and

Department of Chemistry
National Taiwan University
Sec. 4, Taipei
Taiwan

Evgeniy G. Chepaikin
Institute of Structural Macrokinetics and Materials Science
Russian Academy of Sciences
Chernogolovka
Moscow
Russia

Alessandro Chieregato
Dipartimento di Chimica Industriale "Toso Montanari"
Università di Bologna
Bologna
Italy

Miquel Costas
Departament de Química I Institut de Química Computacional i Catàlisi
Facultat de Ciències
Campus de Montilivi
Girona
Catalonia
Spain

Robert H. Crabtree
Yale Chemistry Department and Energy Sciences Institute
Yale University
Yale West Campus
West Haven CT USA

Bruno Dominelli
Molecular Catalysis
Technische Universität München (TUM)
Department of Chemistry/Catalysis Research Center
8 Garching bei München
Germany

Ana M. Faisca Phillips
Centro de Química Estrutural
Instituto Superior Técnico
Universidade de Lisboa
Lisboa
Portugal

Huaquan Fang
State Key Laboratory of Organometallic Chemistry
Shanghai Institute of Organic Chemistry
Shanghai
China

Luís M. T. Frija
Centro de Química Estrutural
Instituto Superior Técnico
Universidade de Lisboa
Lisboa
Portugal

Christian R. Goldsmith
Department of Chemistry and Biochemistry
Auburn University
179 Chemistry Building
Auburn
AL
United States

M. Fátima C. Guedes da Silva
Centro de Química Estrutural
Instituto Superior Técnico
Universidade de Lisboa
Lisboa
Portugal.

Jianlin Han
School of Chemistry and Chemical Engineering
State Key Laboratory of Coordination Chemistry
Jiangsu Key Laboratory of Advanced Organic Materials
Nanjing University
Nanjing
People's Republic of China

Zheng Huang
State Key Laboratory of Organometallic Chemistry
Shanghai Institute of Organic Chemistry
Shanghai
China

Damodar Janmanchi
Institute of Chemistry
Academia Sinica
Sec. 2, Nankang
Taipei
Taiwan

Maximilian N. Kopylovich
Centro de Química Estrutural
Instituto Superior Técnico
Universidade de Lisboa
Lisboa
Portugal

Fritz E. Kühn
Molecular Catalysis
Technische Universität München (TUM)
Department of Chemistry/Catalysis Research Center
8 Garching bei München
Germany

Maxim L. Kuznetsov
Centro de Química Estrutural
Instituto Superior Técnico
Universidade de Lisboa
Lisboa
Portugal

Jay A. Labinger
Beckman Institute
California Institute of Technology
Pasadena
CA
USA

Osvaldo Lanzalunga
Dipartimento di Chimica
Università degli Studi di Roma "La Sapienza"
and Istituto CNR di Metodologie Chimiche
(IMC-CNR)
Sezione Meccanismi di Reazione
Rome
Italy

Johannes A. Lercher
Department of Chemistry
Technische Universität München
Garching
Germany

Anja C. Lindhorst
Molecular Catalysis
Technische Universität München (TUM)
Department of Chemistry/Catalysis Research
Center
8 Garching bei München
Germany

Chih-Cheng Liu
Institute of Chemistry
Academia Sinica
Sec. 2, Nankang
Taipei
Taiwan

Guixia Liu
State Key Laboratory of Organometallic
Chemistry
Shanghai Institute of Organic Chemistry
Shanghai
China

Oleg Y. Lyakin
Novosibirsk State University and Boreskov
Institute of Catalysis
Novosibirsk
Russian Federation

Kamran T. Mahmudov
Centro de Química Estrutural
Instituto Superior Técnico
Universidade de Lisboa
Lisboa
Portugal

and

Department of Chemistry
Baku State University
Baku
Azerbaijan

Luísa M. D. R. S. Martins
Centro de Química Estrutural
Instituto Superior Técnico
Universidade de Lisboa
Lisboa
Portugal

Dan Meyerstein
Chemical Sciences Department
Ariel University
Ariel
Israel

and

Ben-Gurion University
Beer-Sheva
Israel

Jean-Marc M. Millet
Institut de Recherche sur la Catalyse et
l'Environnement de Lyon
UMR 5256 Universite Claude-Bernard Lyon 1
Villeurbanne Cedex
France

Daniel J. Mindiola
Department of Chemistry
University of Pennsylvania
Philadelphia PA
USA

Dmytro S. Nesterov
Centro de Química Estrutural
Instituto Superior Técnico
Universidade de Lisboa
Lisboa
Portugal

Oksana V. Nesterova
Centro de Química Estrutural
Instituto Superior Técnico
Universidade de Lisboa
Lisboa
Portugal

Jose Manuel Lopez Nieto
Instituto de Tecnologia Quimica
Universitat Politècnica de València-Consejo Superior de Investigaciones Científicas
Valencia
Spain

Giorgio Olivo
Dipartimento di Chimica
Università degli Studi di Roma "La Sapienza" and Istituto CNR di Metodologie Chimiche (IMC-CNR)
Sezione Meccanismi di Reazione
Rome
Italy

António Palavra
Centro de Química Estrutural
Instituto Superior Técnico
Universidade de Lisboa
Lisboa
Portugal

Armando J. L. Pombeiro
Centro de Química Estrutural
Instituto Superior Técnico
Universidade de Lisboa
Lisboa
Portugal

Ravirala Ramu
Institute of Chemistry
Academia Sinica
Sec. 2, Nankang
Taipei
Taiwan

Ana P.C. Ribeiro
Centro de Química Estrutural
Instituto Superior Técnico
Universidade de Lisboa
Lisboa
Portugal

Maricruz Sanchez-Sanchez Department of Chemistry
Technische Universität München
Garching
Germany

Georgiy B. Shul'pin
Semenov Institute of Chemical Physics
Russian Academy of Sciences
Moscow
Russia

and

Chair of Chemistry and Physics
Plekhanov Russian University of Economics
Moscow
Russia

Alexander B. Sorokin
Institut de Recherches sur la Catalyse et l'Environnement de Lyon
IRCELYON
UMR 5256
CNRS – Université Lyon I
Villeurbanne
France

Dieter Sorsche
Department of Chemistry
University of Pennsylvania
Philadelphia PA
USA

Stefano Di Stefano
Dipartimento di Chimica
Università degli Studi di Roma "La Sapienza"
and Istituto CNR di Metodologie Chimiche
(IMC-CNR)
Sezione Meccanismi di Reazione Piazzale
Rome
Italy

Manas Sutradhar
Centro de Química Estrutural
Instituto Superior Técnico
Universidade de Lisboa
Lisboa
Portugal.

Evgenii P. Talsi
Novosibirsk State University and Boreskov
Institute of Catalysis
Novosibirsk
Russian Federation

Yi-Fang Tsai
Institute of Chemistry
Academia Sinica
Sec. 2, Nankang
Taipei
Taiwan

Wondemagegn H. Wanna
Institute of Chemistry
Academia Sinica
Sec. 2, Nankang
Taipei
Taiwan

and

Sustainable Chemical Science and Technology
Taiwan International Graduate Program
(TIGP)
National Chiao-Tung University and
Academia Sinica
Taiwan

Steve S.-F. Yu
Institute of Chemistry
Academia Sinica
Sec. 2, Nankang
Taipei
Taiwan

and

Sustainable Chemical Science and Technology
Taiwan International Graduate Program
(TIGP)
National Chiao-Tung University and
Academia Sinica
Taiwan

Pavel Zatsepin
Department of Chemistry
University of Pennsylvania
Philadelphia PA
USA

Jincan Zhao
College of Chemistry & Environmental Science
Hebei University
Baoding
Hebei
People's Republic of China

Fedor I. Zubkov
Organic Chemistry Department
RUDN University
Moscow
Russian Federation

About the Editors

Armando J. L. Pombeiro is Full Professor at the Instituto Superior Técnico, Universidade de Lisboa, President of the Centro de Química Estrutural and Coordinator of its Synthesis and Catalysis thematic line, President of the College of Chemistry of the University of Lisbon, Director of the Catalysis and Sustainability (CATSUS) PhD program, Full Member of the Academy of Sciences of Lisbon and former President of the Portuguese Electrochemical Society. His research group investigates the activation of small molecules with industrial, environmental or biological significance, including metal-mediated synthesis and catalysis (e.g., functionalization of alkanes under mild conditions), crystal engineering of coordination compounds, design and self-assembly of polynuclear and supramolecular structures, non-covalent interactions in synthesis, molecular electrochemistry, and theoretical studies. He was Chairman of the 25th ICOMC and member of organizing/scientific committees of 40 international conferences or schools. He authored one book (plus five as editor), (co-)authored over 800 research publications, 40 patents, and presented 110 invited lectures at international conferences. His work has received over 20,000 citations, h-index over 60 (Web of Science). Among his honors, he was awarded the Prix Franco-Portugais from the French Chemical Society, the Madinabeitia-Lourenço Prize from the Spanish Royal Chemical Society, the Vanadis Award, the Scientific Prize of the Technical University of Lisbon and the Prizes of the Portuguese Chemical and Electrochemical Societies.
http://orcid.org/0000-0001-8323-888X
https://fenix.tecnico.ulisboa.pt/homepage/ist10897

M. Fátima C. Guedes da Silva is an Associate Professor at Instituto Superior Técnico, Lisbon, Portugal. She is the Coordinator of the research group of Coordination Chemistry and Catalysis of the Centro de Química Estrutural, member of the Coordination Commission of this Centre and of the Directive Board of the Catalysis and Sustainability (CATSUS) PhD program. Her research activity follows a general streamline that usually starts with the design and synthesis of novel coordination compounds, then evolving to their eventual application by exploring structure – properties (including reactivity) – function (application) relationships. Her main research interests include: structural determination, by X-ray diffraction analysis, of metal complexes and organic compounds, metal polynuclear assemblies and supramolecular structures; activation, by transition metal centres, of small molecules with biological, pharmacological, environmental or industrial significance; metal mediated synthesis and catalysis; molecular electrochemistry and electrocatalysis; mechanistic investigation of fast reactions mainly by digital simulation of cyclic voltammetry. She has co-authored over 300 research publications and 13 patents. Her work has received over 6000 citations, h index over 42 (Web of Science). She was awarded the Scientific Prize Universidade de Lisboa/Caixa Geral de Depósitos in 2017 and the Scientific Prize of the Portuguese Electrochemical Society in 2013.
http://orcid.org/0000-0003-4836-2409
https://fenix.tecnico.ulisboa.pt/homepage/ist90142

Preface

Alkanes, in spite of being the major components of natural gas and oil and a highly rich natural carbon source, have been applied mainly as nonrenewable fossil fuels where they are burnt and carbon is completely lost to the atmosphere (full oxidation to carbon dioxide) with environmental problems.

In fact, the inertness of alkanes has hampered their potential application as a feedstock for organic synthesis of functionalized products with an added value – namely, carboxylic acids, ketones, alcohols, amines, and amides, for example. Such an obvious, but yet virtually unexplored, synthetic approach based on alkanes as starting materials consists of their functionalization by which the alkane carbon framework becomes decorated with functional groups (namely bearing oxygen, nitrogen, carbon, boron, halo, or other elements) as a blooming tree branch with attractive blossoms in spring (Figure 1).

However, such an overall approach for organic synthesis concerns one of the greatest challenges in modern chemistry, since alkanes are rather inert.

Their inertia usually requires the use of harsh reaction conditions and results in low-product yields and selectivities. These limitations should be overcome by finding sufficiently active, selective, and sustainable catalytic systems involving the selective activation of some of the alkane carbon–hydrogen (or –carbon) bonds.

The development of sustainable and direct processes to achieve functionalized products from alkanes would be highly advantageous, even in terms of simplicity, in comparison with the current multistage and often energy-demanding processes used in industry for such organic products.

This book gathers highly recognized researchers on the modes of catalytic functionalization of alkanes, preferably under mild conditions, covering, in an authoritative way, the state of the art in the field, under its modern perspectives and approaches.

This book consists of 27 chapters grouped in five parts, following an introduction/overview of the theme based on the contents of the book:

I- C—O Bond Formation. Hydroxylation and Other Oxygenation Reactions
II- Bioinspired Alkane Functionalization
III- C—B, C—C, and C—N Bond Formation
IV- Dehydrogenation Reactions
V- Unconventional Systems

Part I (C—O Bond Formation) concerns the oxidative alkane functionalization upon C—O bond formation, typically via catalytic hydroxylation and other oxygenation reactions, leading mainly to organoperoxides, alcohols, ketones, etc. Types of catalysts and mechanisms, in both homogeneous and heterogeneous systems are addressed, for both light gaseous alkanes (including methane) and liquid ones.

Figure 1 (a) Alkane functionalization; and (b) blooming of an almond tree branch. (*See insert for color representation.*)

Industrial perspectives are presented, namely, via interviews with agents of the industrial sector.

Therefore, a diversity of metal catalytic systems is described (as in other parts), including mononuclear, multinuclear homo and heterometallic ones, coordination polymers, and metal organic frameworks (MOFs). The different types of mechanisms are addressed, namely involving electrophilic activation with heterolytic C—H bond cleavage, radical processes, etc.

Of particular significance for alkane functionalization is the strategies and systems inspired on (and eventually mimicking) biological catalytic systems, usually based on iron (cytochrome P-450 and soluble methane monooxygenase, sMMO) or copper (particulate methane monooxygenase, pMMO). They are assembled in *Part II* (Bioinspired Alkane Functionalization) and include a diversity of non-heme iron and multicopper catalysts, metalloproteins, and nanobiomimetics. The mechanisms involved, strategies to control selectivities, and comparisons with the biological systems are also discussed.

Alkane functionalization reactions via C—C, C—N or C—B bond formation provide relevant synthetic approaches and are collected in *Part III*. The first type (C—C bond formation) focuses on alkane carboxylation to carboxylic acids and esters, amides, imides, and other products, in some cases with high yields and turnover numbers (TONs), under relatively mild conditions. Alkane hydrocarboxylation to carboxylic acids where water plays the key role of hydroxylating agent is also discussed. These catalytic reactions are of a high simplicity (single-pot) and promising eventual industrial application, allowing the synthesis of such valuable functionalized products in a much simpler and easier way than those of the current multistage industrial processes.

Other types of C—C bond formation reactions are presented, namely the formal insertion of a carbene into an alkane C—H bond.

Functionalizations upon catalytic C—B or C—N bond formation are focused on catalytic borylation, amidation, amination, and related reactions, leading to various types of functionalized organic products (organo-boron, amides, imides, amines, carbamates, nitro, nitroso, and azide compounds) via simpler and more direct routes than those commonly followed, an approach with a high synthetic potential. Types of catalytic systems and mechanisms are included for the various types of reactions.

Catalytic dehydrogenation reactions of alkanes to yield unsaturated compounds also constitute promising routes for functionalization, and both oxidative and nonoxidative strategies, homogeneous and heterogeneous catalytic systems, are discussed in *Part IV*.

To overcome the difficulties associated to alkane functionalization, unconventional systems and conditions have been applied in this field, some of which with a sustainable significance, as described in *Part V*. For instance, the catalysts involved in alkane functionalization in the overwhelming majority of cases are based on transition metals, but some nontransition metal complexes have been found to be able to catalyze alkane functionalization reactions, namely oxidations, in spite of being non-redox active. Such systems are described herein, as well as those where the use of a metal catalyst is dispensable (metal-free systems).

Attention is also paid to the application of unconventional reaction conditions of "green" significance, including the use of an ionic liquid or of a supercritical fluid ($scCO_2$) as the reaction medium, and of microwave irradiation instead of the conventional heating method.

For the successful design of alkane functionalization systems, a better understanding of the modes of interaction of alkanes with metal centers, inorganic and organic species, will be quite useful. Identification of inorganic and organic associates with alkanes via noncovalent interactions and of coordination compounds with alkane ligands (of expected instability and lability) are also covered in *Part V*, as an unconventional approach toward the establishment of the bases of functionalization.

Approaches for alkane functionalizations can be inspired not only on biological systems but also on selected types of functionalization of sp^3 C—H bonds pertaining to substrates other than alkanes. They can show prospects to be extended to alkane functionalization, as illustrated at the end of the book.

A more detailed description of the contents of the book, including its chapters, is provided in Chapter 1.

We hope the readers will enjoy reading the chapters of this book as much as we did when editing it, and will consider it as a source of inspiration for further studies toward the establishment of alkanes as alternative feedstocks for organic synthesis.

Can we already perceive the advent of an "Alkane Era" for synthesis?

We leave the answer to the readers by inviting them to navigate across this book and think about the potential of the variety of already-achieved functionalization reactions of alkanes to produce valuable functionalized organic products.

We acknowledge the authors of the various chapters for their worthful contributions. Thanks are also due to the Fundação para a Ciência e Tecnologia (FCT), Portugal, for its support (Project PTDC/QEQ-QIN/3967/2014).

Centro de Química Estrutural
Instituto Superior Técnico, Universidade de Lisboa

Armando J. L. Pombeiro
M. Fátima C. Guedes da Silva

List of Abbreviations

1,2-DMCH	1,2-dimethylcyclohexane
7-N-Etppz	3,3′-(1,4-diazepane-1,4-diyl)bis[1-(4-ethylpiperazine-1-yl)propan-2-ol
AA	acrylic acid
ACCH	1-aminocyclopropane-1-carboxylic acid oxidase
ACH	acetone-cyanohydrin process
adam	adamantane
AIBN	2,2′-azobis(isobutyronitrile)
AlkB	alkane hydroxylase
BDE	bond dissociation energy
BEA	beta polymorph A zeolite
BMOH	iron-containing hydroxylase
BNNTs	boron nitride nanotubes
BP	British Petroleum
cat	catecholate
CDC	cyclohexane-1,4-dicarboxylate
CHD	1,4-cyclohexadiene
CHP	*p*-cymene hydroperoxide
CIDNP	chemically induced dynamic nuclear polarization
CmOOH	cumyl hydroperoxide
CNT	carbon nanotube
COD	1,5-cycloctadiene
COE	cyclooctene
CP	coordination polymer
CpdI	compound I , $Fe^{IV}(O)$ porphyrin radical
CTF	covalent triazine-based framework
CuMMO	copper membrane monooxygenase
CyOH	cyclohexanol
Cyt P450	cytochromes P450
DET	direct electron transfer
DFT	density functional theory
DH	non-oxidative dehydrogenation

DHA	9,10-dihydroanthracene
DMCH	dimethylcyclohexane
dmpe	1,2-bis(dimethylphosphino)ethane
DMPO	5,5-dimethyl-1-pyrroline-N-oxide
EMT	hexagonal Y
EPR	electron paramagnetic ressonance
ESI-MS	electrospray ionization - mass spectrometry
ET-PT	electron-transfer-proton-transfer process
EXAFS	extended X-ray absorption fine structure
FAP	tris(perfluoroalkyl)trifluorophosphate anion
FAU	faujasite, siliceous cubic Y
FER	ferrierite
GLC	gas-liquid chromatography
GC–MS	gas chromatography-mass spectrometry
HAT	hydrogen atom transfer
h-BN	hexagonal boron nitride
HOTFA	trifluoroacetic acid
HPC	heteropolyacids
IL	Ionic liquid
IRC	intrinsic reaction coordinate
KIE	kinetic isotope effects
LM	α-$La_2Mo_2O_9$
LMV	β-$La_2Mo_{1.9}V_{0.1}O_{8.95}$
MA	methacrolein
MA	maleic anhydride
MAA	methacrylic acid
MAC	metal/azolate carboxylate compound
MCD	magnetic circular dichroism
m-CPBA	*m*-chloroperbenzoic acid
m-CPBA	*m*-chloroperoxybenzoic acid
MCPBA or mCPBA	3-chloroperoxybenzoic acid, *m*-chloroperbenzoic acid
MD	molecular dynamics
Me_3TACN	1,4,7-trimethyl-1,4,7-triazacyclononane
MFI	Silicalite-1
MOF	metal organic framework
MOR	mordenite
MO_x	transition metal oxides
MWCNT	multi-walled carbon nanotubes
NADH	dihydro nicotinamide adenine dinucleotide
NADPH	Nicotinamide Adenine Dinucleotide Phosphate Hydrogen

NHC	*N*-heterocyclic carbene
NPG	*N*-palmitoylglycine
OCM	oxidative coupling of methane
ODH	oxidative dehydrogenation
PBN	N-tertiary-butyl nitrone
PCA	pyrazine-2-carboxylic acid
PDP	pyrrolidin-1-yl]methyl]pyridine
PET	polyethylene terephthalate
pin	pinacolate
pMMO	particulate methane monooxygenase
pmtz	5-(2-pyrimidyl)tetrazolate
POM	Polyoxometalate
Pytz	5-(2-pyridyl)tetrazolate
pz	pyrazolyl
QM/MM	quantum-mechanical/molecular mechanics
RNS	reactive nitrogen species
ROS	reactive oxygen species
RTIL	room temperature ionic liquids
sBMO	soluble butane monooxygenase
$scCO_2$	supercritical carbon dioxide
SDIMO	soluble diiron monooxygenase
sMMO	soluble methane monooxygenase
SPCE	screen-printed carbon electrode
SSNMR	solid state nuclear magnetic resonance
TACN	1,4,7-triazacyclononane
TauD	taurine:aKG dioxygenase
TBHP	*tert*-butyl hydroperoxide
TBPB	*tert*-butyl peroxybenzoate
TBP_8Cz	octakis(4-tertbutylphenyl)corrolazinato
TCQ	tetrachlorobenzoquinone
TFA	trifluoroacetic acid
TFPPBrs	octabromotetra(pentafluorophenyl)porphyrin
TMC	tetraazamacrocycle
Tmphen	3,4,7,8-tetramethyl-1,10-phenanthroline
TMSP	transition-metal-substituted polyoxometalate
TOF	turnover frequency or time of flight (mass spectrometry)
TON	turnover number
TP	3-(trihydroxysilyl)-propylmethyl-phosphonate
TPA	tripicolylamine
TSR	two-state reactivity concept

VPO	vanadyl pyrophosphate
VPP	vanadyl pyrophosphate
VT-MS	variable-temperature mass-spectrometry
WGSR	water gas shift reaction
XANES	X-ray absorption near edge structure
XAS	X-ray absorption spectroscopy
ZVI	zero valent iron
HPC	heteropolyacid

1

Alkane Functionalization: Introduction and Overview

Armando J. L. Pombeiro

Centro de Química Estrutural, Instituto Superior Técnico, Universidade de Lisboa, Lisboa, Portugal

1.1 Why Alkane Functionalization?

Alkanes, as main components of natural gas and oil, constitute a huge reserve of carbon, which remains virtually unexplored as a feedstock for chemical synthesis. In fact, nowadays, the use of alkanes has focused on their application as an energy source upon their combustion.

This is a rather unsustainable approach since it depletes the Earth from carbon, which is lost to the atmosphere as carbon dioxide with recognized deleterious environmental effects, namely of the greenhouse type. This broad use of alkanes contributes greatly to the exhaust of these nonrenewable fossil fuels.

Moreover, such application is very disappointing and embarrassing from the chemical point of view, resembling the burning of undesired organic waste, although in the latter case the energy profit is often not associated.

In contrast, we can potentiate a sustainable vision to the use of alkanes by attempting to shift their current major application as nonrenewable fossil fuel to that as carbon feedstock for synthesis of functionalized organic compounds with an added value (e.g. alcohols, ketones, carboxylic acids, amides, amines, etc.). This approach concerns alkane functionalization in which the alkane provides the carbon framework that bears the desired functional groups in the final products. Its viability is supported by (i) the action of enzymatic systems that can catalyze alkane oxidation and (ii) the still rare, but already possible, industrial use of alkanes, typically cyclohexane to the production of Nylon intermediates, although in an energy demanding process with a high complexity and low selectivity, requires further developments.

A further encouragement for alkane functionalization results from the discovery of new natural gas and shale gas resources which potentiates light alkanes as promising new carbon resources for hydrocarbon conversion in the petrochemical industry.

1.2 Alkane Functionalization as a Challenge and Overcoming Approaches

The above overall aim of alkane functionalization constitutes a great challenge to chemistry since alkanes are rather *inert*, but nature shows that at least some transformations can be achieved under natural conditions, leading to alkane functionalization to alcohols. This has encouraged and inspired research in this field. Let us analyze further these points.

Alkane Functionalization, First Edition. Edited by Armando J. L. Pombeiro and M. Fátima C. Guedes da Silva.

The *inertia of alkanes* is well understood in terms of unavailability of lone pairs and of empty orbitals of their molecules, also in view of the low-energy lying HOMO (σ) and the high-energy lying LUMO (σ*). Hence, alkanes are difficult to react in general, e.g. to undergo oxidation and reduction.

However, catalysis can contribute greatly to the above aim since the use of a suitable catalyst can modify completely the reaction mechanism and accelerate the alkane reaction. Therefore, alkane functionalization commonly is achieved by using metal catalysts.

Metal catalysts can act in different modes, as illustrated in Schemes 1.1–1.3.

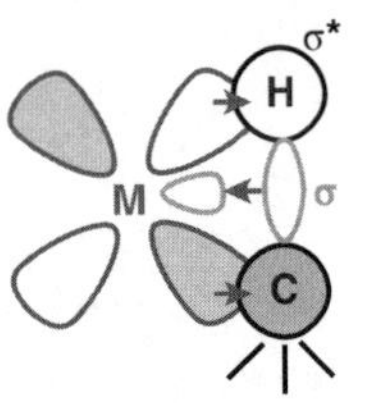

Scheme 1.1 Interaction of an alkane C—H bond with a metal (side-on mode). (*See insert for color representation.*)

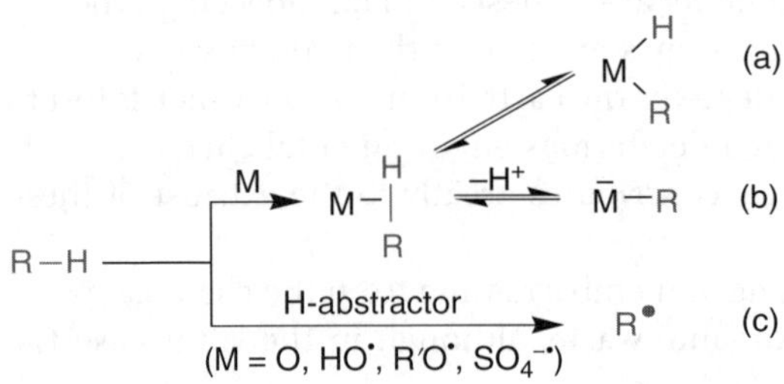

Scheme 1.2 Alkane C—H bond activation/cleavage: (a) oxidative addition to a metal; (b) heterolytic C—H bond cleavage (electrophilic activation); (c) homolytic C—H bond cleavage to form an alkyl radical. (*See insert for color representation.*)

$$RH + HO^{\bullet} \rightarrow R^{\bullet} + H_2O \qquad (a)$$

$$RH + R'O^{\bullet} \rightarrow R^{\bullet} + R'OH \qquad (b)$$

$$RH + SO_4^{-\bullet} \rightarrow R^{\bullet} + HSO_4^{-} \qquad (c)$$

$$RH + O{=}M^{n} \rightarrow R^{\bullet} + HO{-}M^{n-1} \rightarrow ROH + M^{n-2} \qquad (d)$$

Scheme 1.3 Alkane H-atom abstraction by (a) an hydroxyl; (b) an organoxyl; (c) a sulfate radical; (d) or a high-valent metal-oxido species, to form an alkyl radical (the formation of the derived alcohol upon rebound of the alkyl occurs in the last case).

In a direct one, upon *interaction of the metal with the alkan*e (RH), as shown in Scheme 1.1, which depicts the application to alkanes of the well-known Chatt-Dewar-Duncanson coordination model initially addressed to ethylene (olefins) and later to a diversity of unsaturated (carbon monoxide, dinitrogen, isocyanides, cyanide, nitriles, etc.) and even saturated (in particular dihydrogen and alkanes) molecules. In this model, the interaction (inner-sphere C—H activation) is accounted for by two components: σ-donation from a filled σ(C—H) orbital of the alkane to an empty metal orbital with a suitable local symmetry and π-backdonation from a filled π(d) orbital of the metal to the empty σ*(C—H) orbital of the alkane.

Such an *alkane-metal covalent interaction is difficult to achieve* on account of the low-lying energy of the filled σ(C—H) bonding orbital of the alkane and of the high energy of its empty σ*(C—H) orbital, usually at quite different energy levels of the empty and filled metal orbitals, respectively, with which they should combine.

Nevertheless, a few alkane coordination compounds are known and alkane metal complexes have been postulated as (transient) intermediates in alkane reactions (Chapter 27).

Moreover, alkane inorganic and organic associates via noncovalent interactions are documented (Chapter 25).

As a result of any or both components of the alkane-metal interaction (coordination), a weakening of the alkane C—H bond occurs with its eventual breaking to form a metal alkyl-hydride via an oxidative addition reaction (Scheme 1.2a). This reaction is *thermodynamically unfavored,* mainly in view of the low M—C bond energy and also in entropic terms. Typically that bond energy lies in the range of 30–45 kcal mol^{-1}, being unable to compensate, with the formed M—H bond energy, the energy of the alkane C—H bond (typical 90–110 kcal mol^{-1} range). Accordingly, the alkyl-hydride product is thermodynamically unstable toward alkane elimination and the reaction has a very low equilibrium constant. Nevertheless, it can become a synthetic value upon shifting the equilibrium toward the alkyl-hydride by inducing its further reaction, e.g. dehydrogenation of the alkyl ligand (to form an olefin) and insertion of carbon monoxide into the M—C(alkyl) bond (to form an acyl ligand), etc.

If the metal behaves as a strong electrophile (Lewis acid) toward the alkane C—H bond, i.e. if the σ-component of the metal-(C—H) bond interaction dominates, the *heterolytic cleavage* of the C—H bond can occur, leading to a ligated carbanion (R^-) (M—R complex) with proton liberation, without changing the metal oxidation state (Scheme 1.2b). The reaction can be an electrophilic substitution or a σ-bond metathesis at a starting M—X or M—R′ species, leading to M—R (and HX or R′H, respectively). Such a type of alkane C—H bond activation is well represented, namely by palladium catalysts and by the pioneering Shilov platinum system.

In other cases, the cleavage of the alkane C—H bond can be *homolytic,* forming a reactive alkyl radical (Scheme 1.2c). This requires the presence of an abstractor of the alkane hydrogen-atom, i.e. a suitable oxidant such as a peroxide derived species (e.g. the hydroxyl radical $HO^\bullet$ from hydrogen peroxide, an alkoxyl radical $R'O^\bullet$ from an organoperoxide R′OOH, the sulfate radical $SO_4^{-\bullet}$ derived from peroxydisulfate $S_2O_8^{2-}$) (Scheme 1.3a–c, respectively), or a high-valent metal-oxido species (M=O) (Scheme 1.3d). The metal catalyst can activate the oxidant to generate the active metal-oxido or oxyl-radical species.

The alkane C—H activation can involve a 1,2-insertion or addition to an unsaturated metal–ligand bond M=X to afford a metal-alkyl M(R)(XH) species.

The approach of route (c) (outer-sphere C—H activation) in Scheme 1.2 and of route (d) in Scheme 1.3 is that believed to be followed by cytochrome P-450 and methane monooxygenase (MMO), iron enzymes (the former with a mononuclear heme center and the latter with a dinuclear non-heme site) that catalyze the oxidation of alkanes (RH) to the corresponding alcohols (ROH) (Scheme 1.4). Cytochrome P-450 is not active for methane, the most inert alkane, the

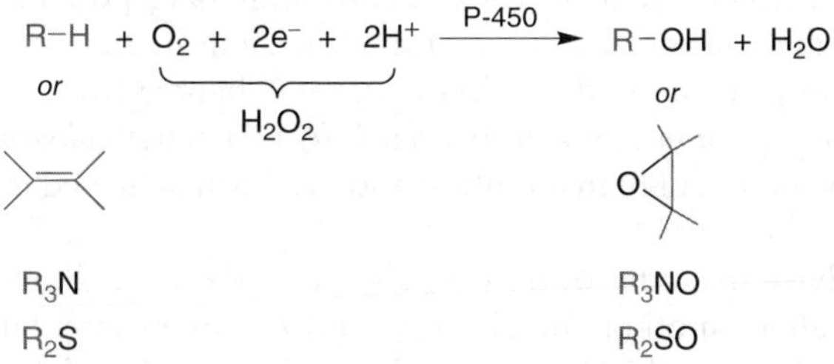

Scheme 1.4 Oxidation of alkanes and other substrates catalyzed by cytochrome P-450 (not active for methane) or by methane monooxygenase (MMO). (*See insert for color representation.*)

oxidation of which requires the action of MMO. However, it catalyzes the oxidation of other substrates (such as olefins, amines, and organosulfides) to the corresponding oxides (epoxides, aminooxides, and organosulfoxides, respectively) (Scheme 1.4).

Dioxygen is the natural oxidant but its combination with two electrons and two protons ($O_2 + 2e + 2H^+$), as shown in Scheme 1.4, is formally equivalent to hydrogen peroxide (H_2O_2), which is a common oxidant used in the laboratory.

The biological approach is quite clever in the sense that the enzyme activates the oxidant rather than the unreactive alkane, thus overcoming the difficulty associated with the inertia of the latter and taking advantage of the higher reactivity of the former reagent. This strategy is quite inspiring and has been followed by many laboratory catalytic systems.

With a capacity to catalyze alkane oxidation to alcohol, apart from the above iron monooxygenases, it is also known as a copper-based enzyme, particulate methane monooxygenase (pMMO). It is a membrane multicopper enzyme, active in the metabolic pathway of methanotrophs, which catalyzes C1–C5 alkanes oxidation. Although its metal content is still controversial, it appears to be a trimer with each subunit containing a mononuclear and a dinuclear copper center.

This has inspired the design and synthesis of multinuclear copper catalysts, including coordination polymers (or metal organic frameworks, MOFs), which have shown a high chemoselectivity and activity for alkane oxidation (see, e.g. Chapters 7 and 16).

Metal-based oxidants (M=O) are usually searched for mimetic systems and, in comparison with other H-atom abstractors, namely the hydroxyl radical, have the advantage of a higher selectivity (see, e.g. Chapters 2, 10–15). The final reaction of Scheme 1.3d (rebound of the short-lived radical $R^\bullet$, generated close to the metal, to the close-by M—OH moiety) is very fast and leads to the alcohol (ROH), selectively. It is even faster than the epimerization of $R^\bullet$ and, thus, stereoselectivity is also expected.

This contrasts with the cases where the radical $R^\bullet$ (long-lived) is formed upon H-abstraction from the alkane by the highly reactive hydroxyl radical, where usually the *selectivity toward the alcohol is low* (see, e.g. Chapters 3, 7, and 16). The corresponding ketone is also produced, the stereoselectivity also decreases (the long-lived alkyl radical can undergo epimerization) and, since the reactive hydroxyl radical is not considerably selective toward the various types of alkane carbon atoms, low regio- and bond-selectivities are commonly observed.

Nevertheless, the chemoselectivity of the alkane oxidation toward the mixture of alcohol (ROH) and corresponding ketone can be quite high, and very selective and active systems operating via radical mechanisms where the hydroxyl radical is generated in situ from H_2O_2 have already been achieved.

We have already illustrated a number of *challenges* associated to the functionalization of alkanes, starting with the *inertia of the alkanes* themselves, followed e.g. by the common *low selectivity* of their reactions.

Other challenges concern the *search for sustainable conditions*, which, namely, should involve environmentally benign and mild protocols, with "green" oxidants, water as solvent and, depending on the type of reaction (e.g. in alkane hydrocarboxylation), water as a possible reagent as well. Water was found to be able to act also as a promoter/catalyst, but this interesting behavior was recognized so far only in a limited number of cases (Chapter 16).

The *replacement of catalysts with expensive and noble transition metals by others with cheap metals* (namely, of 3D type) has also been the object of high attention and has been achieved in a considerable number of systems.

Moreover, *nontransition metal catalysts* have been discovered for alkane oxidations, allowing the replacement of the transition metal by another metal with a lower environmental impact and which, moreover, is nonredox active. This has been observed in cases where the coordination sphere of the metal has a redox active ligand that avoids the change of the oxidation state of the redox inactive non-transition metal (Chapter 22). Such a *ligand–metal cooperative behavior* has also been recognized in a few transition-metal catalysts, opening widely the possibilities of design of new types of catalysts.

The *development of solvent-free and/or metal-free catalytic systems* is demanding and appreciated in view of the advantages associated to the elimination of such components. Single-pot processes are also preferable on account of their simplicity.

The use of unconventional conditions is also promising toward the promotion of activity or selectivity, catalyst recycling, elimination of organic solvents, etc. They can involve the application of ionic liquids or supercritical fluids as solvents, or of microwave irradiation.

Naturally, the search for *insights into the mechanisms* involved is of a high relevance, since their understanding should provide a high contribution to the establishment of sustainable alkane functionalization systems.

These and other issues are covered in this book as indicated in the following sections, based on the topics of its various chapters, which the reader is encouraged to read.

1.3 Types of Alkane Functionalization

The functionalization of alkanes (RH) usually involves the conversion of an akane saturated $C(sp^3)—H$ bond into a C—X (X=O, B, C, N, or halogen) bond or of an alkane saturated $C(sp^3)—C(sp^3)$ bond into an olefinic $C(sp^2)=C(sp^2)$ one.

A variety of types of added-value organic products derived from alkane functionalization has already been achieved (Scheme 1.5), namely as follows:

- Alcohols (ROH), ketones, organoperoxide (ROOH) intermediates, carboxylic acids or esters $R—OCOR^1$ (upon oxidation with formation of a C—O bond, Scheme 1.5a, Chapters 2–16);
- Carboxylic acids RCOOH with an additional C (upon oxidative carboxylation or oxidative hydrocarboxylation upon formation of a C-CO bond, Scheme 1.5(b.1), Chapter 18);
- Esters $RC(R^1)COOR^2$ upon carbene insertion into an alkane C—H bond, usually by reaction of a diazoketone $NN=C(R^1)COOR^2$ (e.g. ethyldiazoacetate) as the carbene source (Scheme 1.5(b.2), Chapter 18); and esters upon alkane alkoxycarbonylation with CO and an alcohol (Scheme 1.5(b.3), Chapter 18), in both cases involving C—CO bond formation;
- Amides and aliphatic imides by reaction of CO with an amine or an amide (Scheme 1.5(b.4) and (b.5), respectively, Chapter 18), also with C—CO bond coupling;
- Amides, imides, carbamates, nitroalkanes, and nitroso compounds (upon amidation, imidation, carbamation, nitration, nitrosation, and azidation, with formation of a C—N bond, Scheme 1.5(c.1–c.7, respectively, Chapter 19));
- Organoboranes upon formation of a C—B bond (Scheme 1.5d, Chapter 17);
- Organohalides upon formation of a C-halogen bond (Scheme 1.5e, Chapters 2 and 8);
- Olefins upon dehydrogenation of alkanes, which can be either oxidative (ODH) or nonoxidative (NODH) (not shown in the Scheme) (Chapters 20 and 21).

Reactions leading to the formation of C bonds other than those mentioned above, such as C—S and C—Se, have also been reported, although they have usually been less studied.

The alkane-derived products can have immediate applications or can undergo further reactions, converting into other types of valuable compounds, widening the synthetic strategy based on alkanes as feedstocks for organic synthesis.

The synthetic potential of alkane functionalization is thus huge and, just to exemplify for one of the less-known types, *alkane borylation* (replacement of an alkane C—H bond by

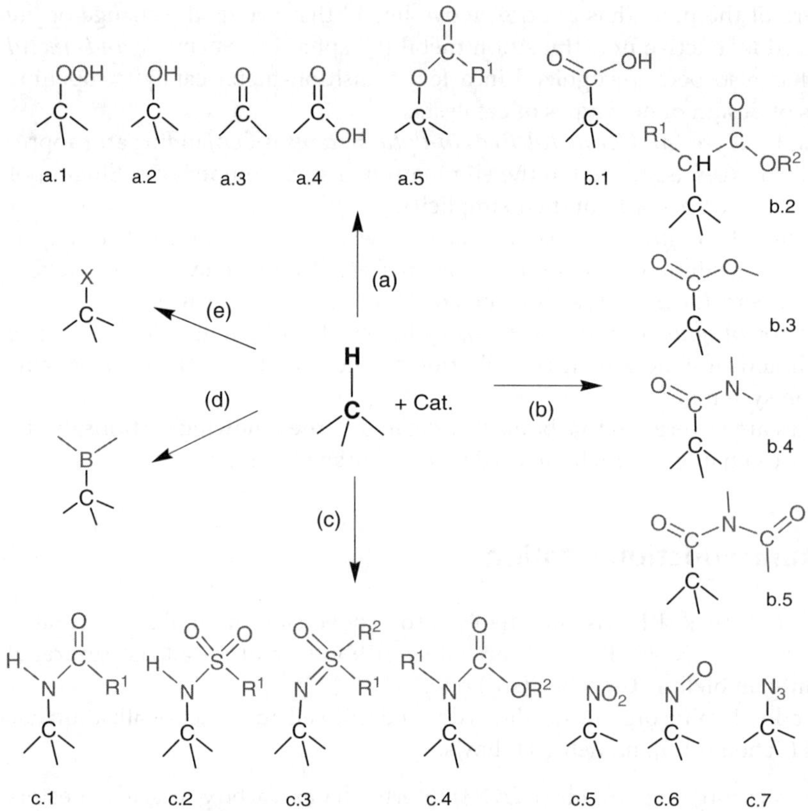

Scheme 1.5 General types of alkane functionalizations with formation of a (a) C—O; (b) C—C; (c) C—N; (d) C—B; or (e) C-halogen (X) bond. (*See insert for color representation.*)

a C—B bond) (Chapter 17), we can note that the presence of a boryl group in the product can allow a diversity of subsequent functionalizations, e.g. formation of a C—C bond upon Suzuki–Miyaura cross-coupling reaction, halogenation, hydroxylation, alkoxylation, or amination.

Catalytic *carbonylation* of alkanes is another general type of functionalization that deserves particular attention in view of the diversity and relevance of the derived products. It is illustrated in Schemes 1.5 (products b.1, b.3–b.5) and 1.6, where a C—CO bond is formed, and is discussed mainly in Chapter 18.

Carbon monoxide is the common carbonyl source, but other sources are possible, such as methanol, trifluoro acetic acid solvent (Scheme 1.6(a2)), or methane itself (Scheme 1.6c).

The possibility of avoiding CO is of environmental significance. For carbonylations in trifluoro acetic acid, the mixed anhydride is an intermediate step to the final carboxylic acid (Scheme 1.6(a2)). There are high expectations for the use of methane as the carbonyl source (Scheme 1.6c), as achieved in the oxidative condensation of methane (Periana system). However, further developments are required to achieve operation under milder experimental conditions, with higher yield and higher catalytic activity.

Replacement of trifluorocetic acid (a common solvent in carboxylation reactions) by a milder solvent (water/acetonitrile) and using water as the hydroxylating agent (hydrocarboxylation system of Pombeiro's group, Scheme 1.6b) also constitute promising developments toward carboxylic acids under sustainable conditions, using convenient catalysts or even proceeding under

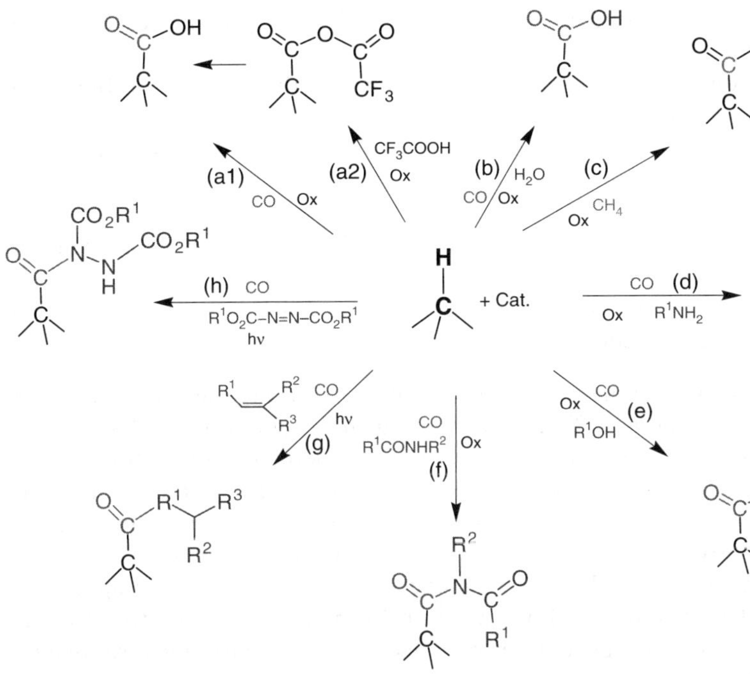

Scheme 1.6 Catalytic carbonylation of alkanes: (a) carboxylation with CO (a1) or with trifluoroacetic acid solvent (a2) to carboxylic acid; (b) hydrocarboxylation to carboxylic acid; (c) Oxidative condensation of methane (also as source of carboxylic carbon); (d) amidation with CO and primary amine; (e) alkoxycarbonylation with alcohol; (f) imidation with CO and amide; (g) photocatalytic carbonylation with CO and an electron-deficient olefin (R^3 = electron withdrawing group); (h) photocatalytic carbonylation with CO and an azodicarboxylate to an acyl hydrazide. Ox = oxidant (e.g. $K_2S_2O_8$, $^tBuOO^tBu$, H_2SO_4, O_2). (*See insert for color representation.*)

metal-free conditions. However, a more convenient oxidant than peroxydisulfate should be searched for.

The *combined use of CO and another reagent* can lead to different functionalized products, according to the involved catalytic functionalization reactions such as:

- Amidation with CO and a primary amine to afford an amide (Scheme 1.6d);
- Alkoxycarbonylation with CO and an alcohol to give an ester (Scheme1.6e);
- Imidation with CO and an amide to form an imide (Scheme 1.6f);
- Carbonylation with CO and an electron-deficient olefin to yield a ketone (Scheme 1.6g);
- Carbonylation with CO and an azocarboxylate to produce an acyl hydrazide (Scheme 1.6h).

Some of these types of functionalization reactions (namely the last two) proceed by photocatalysis.

Less studied catalytic alkane (RH) functionalizations also deserve attention, as exemplified by the variety of reactions leading to the *oxidative conversion of an alkane C—H bond into a C—N one*, in the presence of an oxidant (namely an organoperoxide or a hypervalent iodine reagent), as reviewed in Chapter 19:

- Amidation to $RNHC(=O)R^1$, with an amide $NH_2C(=O)R^1$ or with acetonitrile (R^1 = Me) as the N source, in the presence of $^tBuOO^tBu$ or NHPI/HIO_3 (Scheme 1.5, product c.1);
- Sulfonamidation to $RNHSO_2R^1$, via insertion of a metal-bound nitrene using a hypervalent iodine reagent, e.g. PhI=NR^1 (R^1 = tosyl, etc.) or via reaction of a sulfonamide with a hypervalent iodine-mediated generation of nitrene (Scheme 1.5, product c.2);

- Imidation to $RN{=}SOR^1R^2$, with a sulfoximine $SO({=}NH)R^1R^2$, in the presence of $^tBuOO^tBu$ (Scheme 1.5, product c.3);
- Carbamation to $RN(R^1)COOR^2$, using isocyanate $R^1N{=}C{=}O$ in the presence of $^tBuOO^tBu$ (Scheme 1.5, product c.4);
- Nitration to R-NO_2, using NO_2 or ^tBuONO (*tert*-butyl nitrite) as the nitrating agent, in the presence of NHPI (N-hydroxyphthalimide) as source of the phthalimide N-oxyl (PINO) radical (Scheme 1.5, product c.5);
- Nitrosation to the nitroso compound RN=O, e.g. with *tert*-butyl nitrite in the presence of NHPI (Scheme 1.5, product c.6);
- Azidation to RN_3, with NaN_3/PhIO (or a peroxide oxidant) or with a hypervalent iodine reagent containing an azide unit (Scheme 1.5, product c.7).

However, in spite of the synthetic potential of the oxidative functionalization of alkanes, it is worthwhile mentioning the general difficulty concerning the common limited selectivity, namely due to overoxidation, which can even lead to CO_2, the product of combustion.

Dehydrogenation of alkanes (Chapters 20 and 21) constitutes another important functionalization strategy, leading to their conversion into olefins with a high synthetic value. It can be achieved by oxidative or nonoxidative catalytic routes (ODH or NODH, respectively), either heterogeneous or homogeneous ones. The heterogeneous NODH can already compete with other methods to light olefins. Coupling with other transformations can help to overcome the unfavorable thermodynamics, as expected for other alkane functionalizations. Among the molecular catalysts, those based on noble metals (e.g. Ir and Rh) are the most effective ones, and their replacement by catalysts with more accessible metals should be researched.

Although the functionalization of alkanes typically proceeds via transition metal catalysts, as discussed above, the use of *systems without a transition metal* should be economically and environmentally significant, avoiding probably expensive and toxic catalysts.

An interesting step in this direction concerns the recognition of *nontransition metal catalysts* in the peroxidative oxidation of alkanes, and the disclosure of the involved mechanisms via a metal–ligand cooperation. Although the metal is non-redox active, the ligand is redox active and allows the preservation of the metal oxidation state along the catalytic process (Chapter 22). This approach, which involves the formation of a C—O bond, has already been followed typically for a number of group 13 and 15 metals, and deserves to be extended.

Further developments concern the use of metal-free systems (Chapter 23), which are already known for a variety of alkane functionalizations involving namely the formation of C—C, C—N, C—S, and C—Se bonds, and leading to different types of alkylated products. However, the use of a strong oxidant, high temperature, and low selectivity are drawbacks that still need to be overcome.

Concerning the alkanes investigated, *cyclohexane* (CyH) has been the most frequently used as a model substrate, in view of its convenient structure (with only one type of C—H and C—C bonds, i.e. simply one set of equivalent C—H bonds, what facilitates the selectivity) and properties (e.g. liquid and soluble in acetonitrile, a common solvent used in oxidation catalysis), easy handling and detection of its most common oxidation products, its application in industry for the production of oxidation products of a high added value and the need to improve the industrial catalytic system toward more sustainable conditions. Concerning the last point, it is noteworthy to mention that typical conditions in the industrial process of cyclohexane oxidation to cyclohexanol and cyclohexanone concern a considerably high temperature (c. 150 °C); moreover, the product yield is rather low (c. 5%) to achieve a good selectivity (c. 80%).

Cyclohexane oxidation (Scheme 1.7) typically leads to the mixture (KA oil) of the corresponding alcohol and ketone (i.e. cyclohexanol CyOH and cyclohexanone $Cy_{-H}{=}O$, respectively),

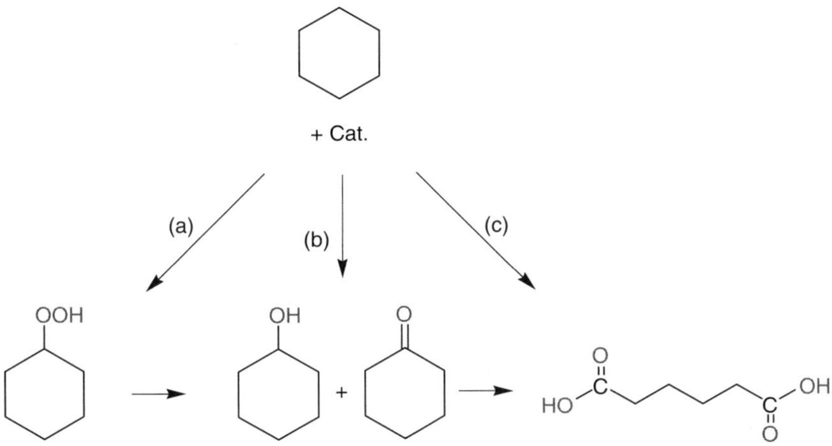

Scheme 1.7 Catalytic cyclohexane oxidations to (a) cyclohexylhydroperoxide; (b) cyclohexanol and cyclohexanone (KA oil); and (c) adipic acid. (*See insert for color representation.*)

materials for Nylon-6,6, whereas the involvement of the organoperoxide (CyOOH) has been proved, at least in a few cases, as a key intermediate (Chapters 3, 6, and 16).

In this context, it is also noteworthy the industrial application of cyclohexane to the production of adipic acid ($HOOC(CH_2)_4COOH$) which is a precursor for the manufacture of Nylon-6,6 and of polyester and polyurethane resins. In this multistage process, cyclohexane is firstly oxidized with dioxygen, in the presence of a Co^{2+} or Mn^{2+} catalyst, to the above-mentioned mixture of cyclohexanol and cyclohexanone. In a subsequent stage, the mixture of these products (KA oil) is further oxidized with nitric acid, in the presence of a Cu^{2+}/Mn^{2+} (or vanadium/copper) catalyst, to give adipic acid and the greenhouse gas N_2O.

Hence, it is evident that there is a need for single-pot selective catalytic systems, with a much higher activity and atom economy, and which can operate under mild conditions, what has been recently achieved (Chapters 6 and 18).

Another alkane of great interest is *methane*. It is particularly challenging (with a high C—H bond strength, low polarizability, and high inertia) and also particularly important as the main component of natural gas. It can undergo industrial conversion to synthesis gas (mixture of CO and H_2, by steam reforming), which can be subsequently converted to methanol (hydrogenation of CO) or to hydrocarbons (Fischer–Tropsch synthesis). These processes proceed via high temperature heterogeneous catalysis and require high-energy consumption.

The thus produced methanol can undergo further carbonylation to acetic acid (a valuable commodity product), by using a Rh or Ir catalyst (Monsanto or Cativa process, respectively) under homogeneous conditions, which, nevertheless, still require a considerably high temperature.

Therefore, the overall industrial production of acetic acid, based on the oxidative functionalization of methane, is a complex process involving three separate stages (methane to synthesis gas, the latter to methanol and this to acetic acid) and a high-energy expenditure. The search for simpler catalytic processes, which can operate under milder and more sustainable conditions, has been a relevant aim in this field: direct oxidation of methane to methanol (Scheme 1.8a) (Chapters 2 and 10 for homogeneous catalytic systems and Chapter 8 for heterogeneous ones) or, even more appealing, the direct conversion of methane into acetic acid by oxidative condensation (Periana system) where methane is the carbon source for both the methyl and carboxylic groups (Scheme 1.6c and Scheme 1.8b) (Chapters 2 and 18). Direct methane carboxylation to

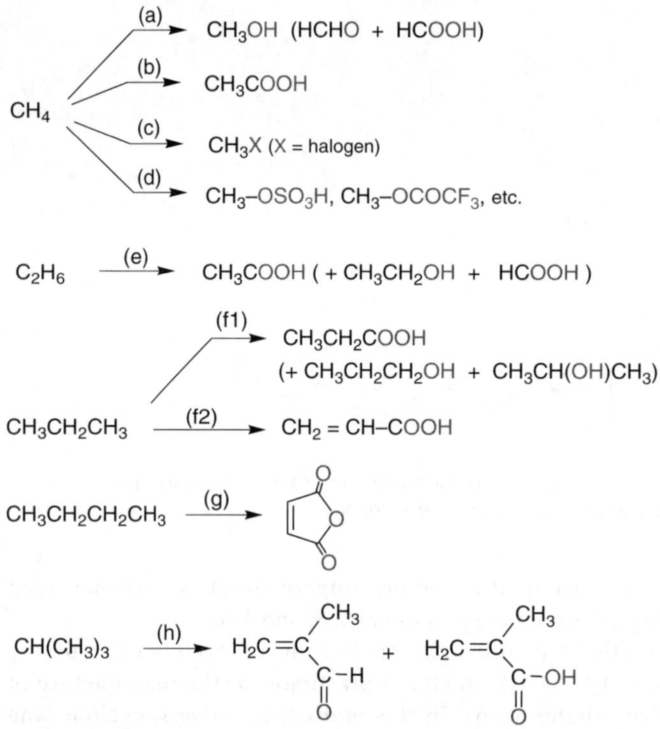

Scheme 1.8 Methane and other light alkanes oxidations: (a) methane oxidation to methanol (and formaldehyde); (b) methane oxidative condensation to acetic acid; (c) methane halogenation; (d) methane esterification; (e) ethane oxidation to acetic acid; (f) propane oxidation to acrylic acid; (g) butane oxidation to maleic anhydride; (h) *iso*-butane oxidation to methacrolein and methacrylic acid. (*See insert for color representation.*)

acetic acid is also a very promising alternative (see above, Scheme 1.6a,b, Fujiwara and Pombeiro systems) (Chapter 18).

Considering dioxygen as the oxidant, such methane oxidation reactions could be represented by the prospective equations of Scheme 1.9a,b, respectively, but the special precautions to avoid explosive mixtures with methane should be added to the inherent chemical challenge of achieving those reactions.

$$CH_4 + 1/2O_2 \xrightarrow{\text{Cat.}} CH_3OH \qquad \text{(a)}$$

$$2CH_4 + 2O_2 \xrightarrow{\text{Cat.}} CH_3COOH + 2H_2O \qquad \text{(b)}$$

Scheme 1.9 Prospective catalytic oxidations of methane with dioxygen to (a) methanol and (b) acetic acid.

Other possible catalytic *methane functionalizations* with a high synthetic potential (eventually with industrial significance) include its (i) halogenation to halomethane upon formation of a C-halogen bond (Scheme 1.8c, Chapters 2 and 8), (ii) esterification to a methyl ester (with formation of a C—O bond) (Scheme 1.8d, Chapters 2 and 5), which, upon further hydrolysis, can ultimately lead to methanol and the acid that concerns the ester, (iii) oxidative coupling to ethylene (Chapters 8 and 17), and (iv) borylation (Scheme 1.5d, Chapter 17).

Functionalization of *other light alkanes* with an industrial objective has also raised high interest.

Ethane oxidation to acetic acid (Scheme 1.8e, Chapters 10 and 18) is of a high significance toward the development of an alternative and advantageous preparative process to this acid in comparison with the current ones. This is also true for *propane oxidation* to propionic acid (Scheme 1.8(f1), Chapter 10). However, other oxygenated products are also obtained, namely, ethanol and formic acid from ethane, and propanol, *iso*-propanol and C_2 or C_1 products (ethanol, acetic acid, and formic acid) from propane (Chapter 10), which poses selectivity issues in spite of the interest of such functionalized compounds.

Oxidative dehydrogenation of butane to maleic anhydride (catalytic heterogeneous process) is already used in industry (Scheme 1.8g, Chapter 9), and the development of related catalytic systems applied to propane or *iso*-butane leading to acrylic acid or mathacrolein and methacrylic acid (Scheme 1.8(f2),h, respectively, Chapter 9) is also being researched.

1.4 Structure of the Book

This book comprises five sections, addressing the following main subjects of alkane functionalization: C—O bond formation (Part I); Bioinspired functionalization (Part II); C—B, C—C, and C—N bond formation (Part III); Dehydrogenation (Part IV); and Unconventional systems (Part V).

Part I concerns the oxygenation reactions of alkanes (leading typically to alcohols and ketones), which are those more studied, and covers a wide variety of metal catalysts and types of both radical and nonradical mechanisms, in homogeneous and heterogeneous systems. Noble metal (Pd, Pt, and Rh) molecular catalysts are described by Chepaikin (Chapter 2, also dealing with carbonylation reactions) and various other group metals (e.g. vanadium and iron) by Shul'pin (Chapter 3). Both chapters also describe the mechanisms involved in the homogeneous systems, as does Chapter 4 by Meyerstein on radical reactions in aqueous solutions. The mechanisms of activation of the oxidant are also discussed, namely of dioxygen (Chapter 2) and of peroxides (Chapter 3), as well as the methods of mechanism investigation (Chapter 3). Routes to the formation of alkyl radicals in water, their chemical properties, and mechanisms of reaction in solution and at the surface of immersed metals and of suspended nano-particles are addressed in Chapter 4.

Other particularly effective or promising types of homogeneous catalytic systems for the oxidation of alkanes to alcohols and ketones are treated in the following chapters: metal-based carbene catalysts (typically *N*-heterocyclic carbene, NHC, complexes of iron, iridium, and palladium, in a few cases also able to catalyze methane esterification, by Kuhn et al., Chapter 5), metal-based scorpionate catalysts (mainly vanadium or iron complexes with C-based tris(pyrazolyl)methane ligands) (Martins, Chapter 6), and multinuclear heterometallic catalysts usually prepared by self-assembly (Nesterov, Pombeiro, et al., Chapter 7).

The use of ozone (apart from the usual hydrogen peroxide) as an oxidant is addressed in Chapter 6, in which the single-pot oxidation of cyclohexane to adipic acid is also discussed (Scheme 1.7c), besides the common cyclohexane oxidation to cyclohexanol and cyclohexanone (KA oil) (Scheme 1.7a,b). The same chapter addresses the immobilization of the scorpionate catalysts on zeolites and carbon materials (supported catalysts).

The application of multinuclear heterometallic catalysts with a diversity of structures/compositions (including coordination polymers or MOFs) not only to oxygenation but also to hydrocarboxylation of alkanes (RH), where they are converted to carboxylic acids (RCOOH) upon reaction with carbon monoxide, water, and peroxydisulfate (Scheme 1.6b), is presented in Chapter 7, which also discusses selectivity and isotopic labeling studies, namely with $^{18}O_2$ and $H_2{}^{18}O$.

Heterogeneous catalytic systems and their potential industrial significance are dealt with by Sanchez-Sanchez and Lercher (Chapter 8) and by Cavani, Chieregato, Nieto, and Millet (Chapter 9) (Scheme 1.8a,c,f–h).

Chapter 8 focus on strategies for direct oxidation (with dioxygen) of methane to methanol with iron- and copper-zeolites, although discussing also halogenations and oxyhalogenation (to methyl halide which can be subsequently converted to methanol upon hydrolysis) and oxidative coupling of methane (to ethane and ethylene). Chapter 9 is complementary, dealing with light C_2-C_4 alkanes: their gas-phase oxidation (of isobutane to methacrylic acid, of *n*-butane to maleic anhydride, of propane to acrylic acid), and gas-phase ammoxidation of propane to acrylonitrile, apart from oxidative dehydrogenation to light olefins. The main sections of this chapter end with discussions of industrial perspectives through interviews with people from the industrial sector.

Part II on bioinspired functionalization systems extends the previous Part (I) to such catalytic systems of biological significance, often based on non-heme iron, manganese, copper, and vanadium catalysts. The iron catalysts are somehow inspired on soluble methane monooxygenase (sMMO) or cytochrome P450, whereas the copper catalysts can be inspired on pMMO. Mimicking the enzyme function is a major issue, and thus the selectivity and the possible involvement of high-valent oxido-metal intermediates are the object of high attention, although free radical mechanisms are also possible. The chapters are written by Sorokin (10), Goldsmith (11), Di Stefano et al. (12), Costas (13), Bryliakov et al. (14), Yu et al. (15), and Sutradhar, Pombeiro et al. (16).

Chapter 10 concerns the oxidation of methane, ethane, and propane, namely with di- or trinuclear iron and copper catalytic sites, which generate high-valent metal-oxo species using H_2O_2 and O_2 oxidants (Scheme 1.8a,e,(f1)). They include, e.g. μ-nitrido diiron phthalocyanine and porphyrin complexes, and iron- and copper-containing zeolites, apart from copper models of pMMO.

Chapter 11 addresses strategies to control the site-selectivity of the C—H activation promoted by non-heme iron complexes, based on controlling the weakening of the metal-based oxidant and increasing the steric properties of the ligand. The ligands in the catalysts are commonly neutral N-donors, consisting of derivatives of tripicolylamine, *N,N′*-bis(2-pyridylmethyl)-1,2-ethanediamine or 1,4,7-triazacyclononane. Adamantane and substituted cyclohexanes are the common substrates, as in the next chapters.

Chapter 12 revises in a systematic way the variety of imine-based iron and manganese catalysts and their mechanisms of action. It also describes the common tools to distinguish a metal-based from a radical chain oxidation. Stabilization by a high ligand denticity promotes the metal-based oxidation mechanism. Oxygenation of $C(sp^3)$-H bonds in compounds other than alkanes is also included and the simple preparation of such catalysts is highlighted.

Mononuclear iron catalysts with nitrogen ligands based on the triazacyclononane backbone are discussed in Chapter 13. The robustness of this type of ligands stabilizes high-valent iron-oxo species, which drive the hydroxylation mechanism of the aliphatic C—H bonds. The methods of type of mechanism assignment are also discussed, as well as the mechanism of incorporation into products of the oxygen atom from water.

Iron complexes with tetradentate N_4-donor aminopyridine ligands are addressed in Chapter 14 as catalysts for both alkane hydroxylation and olefin enantioselective epoxidation with hydrogen peroxide or a peroxycarboxylic acid as oxidant, in the presence of a carboxylic acid as additive. The involvement of oxo- or peroxo-iron intermediates is discussed.

However, free radical mechanisms are believed to account for the catalytic activity of a variety of copper and vanadium catalysts (Chapter 16) on the peroxidative oxidation of alkanes under mild conditions. The catalytic activity of aryl- (or aroyl)-hydrazone, tetrazolato or scorpionate and phenanthroline copper complexes is discussed, including multinuclear ones, MOFs, or coordination polymers, which can be more active than the mononuclear ones.

Although no vanadium enzyme is known to catalyze alkane oxidation, V enzymes such as haloperoxidases are able to catalyze related reactions of other substrates, thus also inspiring the use of inorganic mimetic V catalytic systems. Accordingly, oxido-V complexes containing azine fragment (C—N=N—C) ligands are quite efficient, and a ligand–metal cooperation can occur with preservation of the V oxidation state. Other V catalysts are based on pyrazole or scorpionate ligands. V complexes anchored to chemically modified carbamated silica gel can catalyze alkane oxidation with dioxygen. The possible promoting role of water by stabilization of a six-membered V transition state is also analyzed (Chapter 16).

A further biologically oriented approach is followed in Chapter 15, by designing the substrate binding pockets of alkane hydroxylase (AlkB) and cytochrome P450 BM3 using a heterogeneous *Escherichia coli* recombinant system. The two redesigned proteins can achieve efficient hydroxylations of *n*-butane at its C1 and C2 positions, respectively. Advantages in the use of nanoparticles as nano-biomimetics of metallo-monooxygenases are discussed.

Functionalization of alkanes upon formation of other types of carbon-X bonds, i.e. where X = B, C, or N, is also of a high synthetic significance, and this topic is treated in *Part III* of the book. The chapters herein are authored by Mindiola et al. (17), Phillips and Pombeiro (18), and Nesterov, Pombeiro et al. (19).

Catalytic methane borylation (with *C—B bond formation*) is addressed in Chapter 17, where a predictive use of theoretical DFT calculations allowed the design of a better Ir catalyst upon replacement of a 1,10-phenanthroline-based ligand by a more polarizable diphosphine with a promoting effect on the relevant oxidative addition step. A suitable borylating agent is the diboron reagent bis(pinacolato)diboron B_2pin_2 (pin = pinacolate), which leads to CH_3-Bpin (and H-Bpin). The chapter revises the field and related methane functionalizations.

C—C bond formation where an alkyl group is provided by an alkane can be involved in various important reactions, namely alkane carbonylation and carbene insertion into an alkane C—H bond, fields that are reviewed in Chapter 18.

Palladium, vanadium, or copper catalysts can promote (one-pot) carbonylation reactions of alkanes (including methane), leading to carboxylic acids, esters, amides, imides, and other products, in some cases in high yields and turnover numbers (TONs), under relatively mild conditions. Even the single-pot carbonylation of methane to acetic acid and of ethane to propionic acid can be achieved very effectively with very high TONs, using bioinspired vanadium catalysts such as *Amavadin* (a natural V complex present in some amanita toadstools) and models. The hydrocarboxylation can proceed in a metal-free process, under quite mild conditions. Some other alkane C—C bond formation reactions can also occur without a transition metal catalyst, although in the presence of a superelectrophile or superacid.

Insertion reactions of carbenes (typically derived from an ester functionalized diazo compound, commonly a diazoacetate) into an alkane C—H bond, usually with a rhodium or silver catalyst, can produce esters in high yields, in some cases with enantioselectivity.

C—N bond formation leading to an alkylated nitrogen compound where the alkyl group is furnished by an alkane can account for the synthesis of an array of products, including amides, imides, amines, carbamates, nitro, nitroso, and azide compounds. The functionalization is oxidative, requiring the presence of an oxidizing agent and commonly proceeding via a free radical mechanism. The topic is reviewed in Chapter 19, with indication, for each type of reaction (amidation, imidation, amination, carbamation, nitration, nitrosation, and azidation), of relevant metal catalytic systems and proposed mechanisms. Copper catalysts are common, but other metal catalysts (e.g. based on ruthenium, silver, rhodium, or iron) have also been found. Moreover, cases where the catalyst is metal-free are also discussed, namely involving N-hydroxyphthalimide (NHPI) and its derived PINO radical.

Part IV presents another approach for functionalization of alkanes by *dehydrogenation* to afford unsaturated products (olefins). The chapters are written by Labinger (20) and Huang (21).

They describe both oxidative and nonoxidative routes (ODH and NODH, respectively), using heterogeneous and molecular catalysts. Although heterogeneous catalytic nonoxidative processes dominate, homogeneous ones are also of interest, and the various types of catalytic alkane dehydrogenations are illustrated and their scopes are compared.

Thermochemical and photochemical dehydrogenations are discussed, as well as strategies to overcome the unfavorable thermodynamics of NODH, such as coupling to other transformations (e.g. alkane and alkene metathesis).

The last section of the book (*Part V*) collects a number of *unconventional approaches* toward alkane functionalization that do not fall within those covered in the previous chapters. The chapters of this part are authored by Kuznetsov (22), Zhao and Han (23), Ribeiro, Pombeiro et al. (24), Mahmudov, Guedes da Silva, Pombeiro et al. (25), Crabtree (26), and Bruneau (27).

The replacement of transition metal catalysts by *nontransition metal* ones in oxidative functionalization of alkanes is of scientific and practical interest, namely in view of the nonredox character of the latter. Theoretical calculations disclosed a different type of mechanism involved in the catalytic generation of the hydroxyl radical from hydrogen peroxide, when a suitable redox active ligand is available. This has been achieved with nontransition metals in periodic groups 13 (Al, Ga, In) and 15 (Bi), and their catalytic behavior was predicted by theoretical DFT calculations that drove the experimental studies. The topic is reviewed in Chapter 22.

A metal catalyst can even be avoided in alkane functionalization, and *metal-free systems* are mostly discussed in Chapter 23 for a variety of systems involving the formation of C—C, C—N, C—S, and C—Se bonds. The chapter revises different approaches, such as: cross-dehydrogenative-coupling, decarboxylative cross-coupling, conjugate addition, radical rearrangement, radical cascade reactions, thermally induced C—H (for C—N bond formation); thiolation with diarylsulfides (for C—S bond formation); and cross-couplings with disulfides or diselenides (for C—S or C—Se bond formation). Advantages and limitations of the metal-free systems are discussed.

The application of unconventional conditions in catalytic alkane functionalization concerning the performance of the reaction in an *ionic liquid* or in a *supercritical fluid* (carbon dioxide), or under *microwave irradiation*, is addressed in Chapter 24, namely for the oxidation of cyclohexane and for light alkanes, commonly using transition metal catalysts. Such conditions can allow the catalyst recycling (use of an ionic liquid) and avoid an organic solvent, which can be of significance toward the development of sustainable systems. A comparison with the conventional conditions is presented.

The recognition of *noncovalent interactions* of alkanes is meaningful toward understanding their properties, interactions with other species, and involvement in reactions. Such types of interactions can weaken C—H and C—C bonds of alkanes and thus assist their eventual functionalization. They can fall, e.g. in the C_{alkane}—H···H—C, C_{alkane}—H···π, C_{alkane}—H···O or C_{alkane}—H···metal center types, and occur in organic and inorganic alkane associates. The topic is covered in Chapter 25.

In continuation of this line, *complexation* of alkanes to transition metals can constitute a relevant step for their functionalization, and either transient or stable alkane complexes are known. Moreover, they have been proposed as transient reaction intermediates in a number of cases. Such systems and their relation to agostic and other sigma bond complexes are addressed in Chapter 26, which also highlights the importance of computational studies to identify those unconventional alkane compounds. Moreover, the general strategies for alkane complexes are revised.

Alkane C—H bond activation and functionalization can eventually be inspired on related C—H activations of other more reactive substrates. The final chapter (27) addresses the functionalization of alicyclic amines, namely with rhodium and ruthenium catalysts, leading to the formation of C—C bonds based on oxidation, hydrogen-transfer, or hydrogen-borrowing.

1.5 Final Remarks

Chemistry and, in particular, catalysis, should contribute to a better use of natural carbon resources and development of alternative raw materials with benefits relatively to the established processes, in terms of easier availability, higher simplicity, energy saving, environmental concerns, and sustainability.

This chapter addresses the bases of activation of alkanes and promising strategies and approaches followed in this book to their functionalization toward shifting their current main application as fuels to carbon feedstocks in organic synthesis, a challenge to modern chemistry of a synthetic high significance.

The field is promising, with high prospects, and we are already witnessing (and contributing to) the beginning of a new era in organic synthesis, the "alkane era," following the current "olefin era" and the past "acetylene era."

Acknowledgments

The financial support of the Fundação para a Ciência e Tecnologia (Project PTDC/QEQ-QIN/3967/2014), Portugal, is acknowledged.

Part I

C—O Bond Formation. Hydroxylation and Other Oxygenation Reactions

2

Activation and Oxidative Functionalization of Alkanes with Noble-Metal Catalysts: Molecular Mechanisms

Evgeniy G. Chepaikin

Institute of Structural Macrokinetics and Materials Science, Russian Academy of Sciences, Chernogolovka, Moscow, Russia

2.1 Introduction

Industrial-scale production of alcohols, aldehydes, ketones, carboxylic acids, and other products of organic and oil-and-gas synthesis via oxidation of lower alkanes (C_1–C_4) still remains to be in focus of current studies in the field of catalysis. It has been multiply declared that the oxidation of methane into methanol with dioxygen is a great challenge or even the Holy Grail of modern chemistry. The gas-phase oxidation of methane over heterogeneous catalysts has shown low selectivity since it is accompanied by transformation of methane into carbon dioxide at elevated temperatures required by the process [1, 2]. In this context, great expectations are associated with homogeneous catalysis that is operative in much milder conditions. The reactions proceeding by molecular mechanisms exhibit higher selectivity than radical and chain-radical reactions.

In 1965, Chatt and Davidson [3] reported on oxidative addition of naphthalene to Ru complex with 1,2-(dimethylphosphino)ethane. Later, Halpern [4] declared the exceeding importance of the activation of the C—H bond in alkanes as for their involvement into selective catalytic reactions in mild conditions. Shilov and co-workers [5] were the first who carried out such a process in laboratory. To date, the examples of C—H bond activation in alkanes are numerous. Meanwhile, it has become clear that a much more difficult problem is the functionalization of an activated C—H bond.

A number of monographs [6–15] and vast amounts of reviews and original papers have been devoted to the activation of alkanes and dioxygen and also to oxidation of alkanes and other substrates, including special issues of leading journals such as *Chem. Rev.* 2010, vol. 110, no. 2; *J. Organomet. Chem.* 2015, vol. 793; and *J. Mol. Catal. A: Chem.* 2017, vol. 426. Worth mentioning are excellent reviews [16–31]. The author of this chapter has also published three review papers [32–34].

2.2 Activation of C—H Bond in Alkanes

As is known, the activation of C—H bond in alkanes may proceed by so-called outer-sphere (**A**) or inner-sphere (**B**) mechanisms as shown in Scheme 2.1.

Alkane Functionalization, First Edition. Edited by Armando J. L. Pombeiro and M. Fátima C. Guedes da Silva.

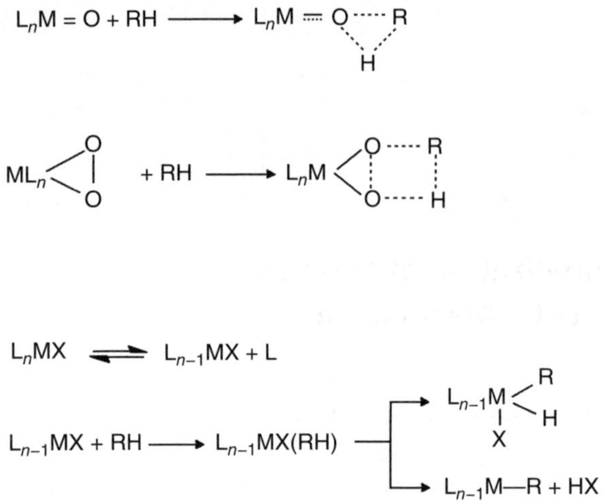

Scheme 2.1 Outher-sphere (**A**) and inner-sphere (**B**) mechanisms for activation of C—H bond in alkanes RH.

Mechanism **A** postulates that the activation of alkanes takes place with participation of oxygen atoms from various oxygenases and some model compounds [11], so that the activation proceeds on oxo- or peroxo-metal complexes. To date, there exist only indirect evidences for the formation of alkane associates with the oxo complex of Fe(IV), a model of methane monooxygenase (MMO) [35]. Outer-sphere electron transfer is known to happen in various reactions of metal complexes [36, 37] such as the interaction of $[Fe(IV)(H_2O)_5O]^{2+}$ oxo complex with different substrates [36]. Outer-sphere mechanism for oxidation of ethane under the action of other Fe(IV) oxo species was suggested in [38].

According to mechanism **B**, the reaction of alkanes with coordination-unsaturated complexes yields σ-complexes. This is followed by oxidative addition or electrophilic substitution resulting in formation of metal alkyls. Mechanism **B** is regarded as the most likely for anaerobic biooxydation of methane [30]. In this case, the active center may contain Ni, while sulfate or nitrate ions and some other compounds play the role of oxidant. In this sense, both the Shilov system ($PtCl_6^{2-}$ as oxidant) [31] and the Periana system (SO_3 as oxidant) [23, 24] can be regarded as suitable models for anaerobic oxidation of alkanes.

There is no doubt that the activation of alkanes by mechanism **B** starts with the formation of σ-complexes that were identified by IR and NMR spectra [39–41]. Later, σ-complexes of methane with pincer compounds of Rh and Ir were prepared [42–44] upon interaction of methyl complexes with $HB(Ar_f)_4$ at a low temperature, as illustrated in Scheme 2.2.

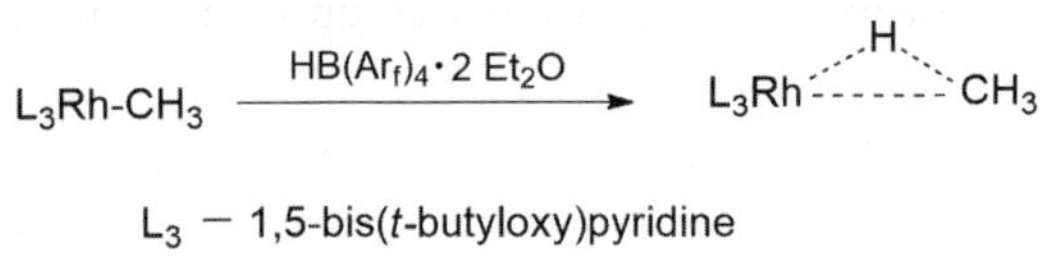

Scheme 2.2 Synthesis of σ-complex of methane with Rh.

Recently, the low-temperature hydrogenation of 1,3-pentadiene ligand was found to yield the σ-complex of pentane with diphosphine compound of Rh (Scheme 2.3). Investigation of single-crystal X-ray diffraction (150 K) and solid state nuclear magnetic resonance (SSNMR) (158 K) have shown that pentane is coordinated with Rh via two hydrogen atoms in positions 2, 4 [45].

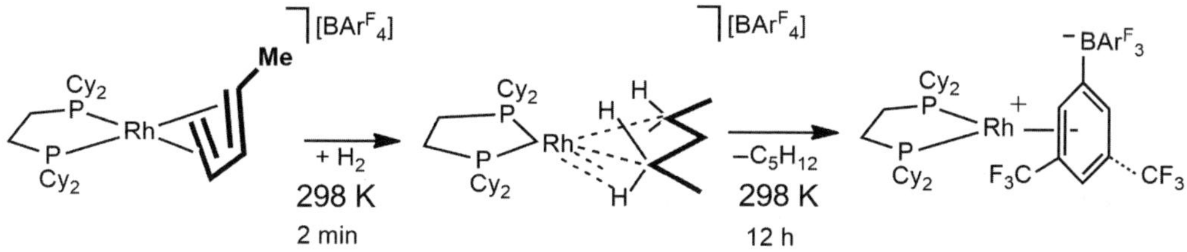

Scheme 2.3 Synthesis and destruction of pentane σ-complex with Rh [45].

A next stage in mechanism **B** is the formation of alkyl intermediates. Meanwhile, Shteinman [26] suggests that it is worthwhile to seek the reaction routes that would ensure the transformation of σ-complexes directly into functionalized products. Oxidative addition of alkanes at C—H bond was attained upon UV photolysis of rhodium and iridium carbonyls in alkanes [46–48]. In case of neopentane and cyclohexane, oxidative addition to a photolytically produced unsaturated complex of Ir(I) was found to yield alkyl hydrides. Upon interaction with CCl_4, these hydrides transform into alkyl chlorides [48]. Reactivity of nucleophilic Pd–carbene complexes toward C—H bond was investigated in [49]. Upon interaction with *n*-pentane and *n*-octane, pincer diacetate complex of Ir(III) yielded respective *n*-alkyl compounds [50].

2.3 Activation of Dioxygen

Organic peroxides, peroxy acids, ArIO, $S_2O_8^{2-}$, etc. seem reasonable for use in fine organic synthesis (pharmacology, perfumery, etc.). In addition, the use of H_2O_2 in synthesis of methanol was recognized as unprofitable [51]. Large-scale production of methanol from methane or propanol and acetone from propane can become profitable only if dioxygen or air will be used as an oxidant. At ambient conditions, dioxygen is rather inert and requires activation. Activation of dioxygen means both (i) cleavage over the first molecular bond resulting in formation of peroxides and hydroperoxides and (ii) complete bond cleavage with formation of oxo complexes.

Capture of electron by O_2 molecule is characterized by a negative potential of −0.32 V. This explains the kinetic inertness of dioxygen. Moreover, a dioxygen molecule is present largely in its triplet state while alkanes are present in their singlet state. Since the reactions of O_2 with alkanes are spin-forbidden, they exhibit high activation energies. Biological catalysts for activation of dioxygen (MMO, cytochrome P450, etc.) normally use an external reducing agent, dihydro nicotinamide adenine dinucleotide (NADH). Electron transfer from NADH to a reactive site in enzymes occurs through a number of intermediates in a way that excludes direct contact of O_2 with a reducing agent, thus ensuring high efficiency of the latter.

2.3.1 Peroxo Complexes

Peroxo complexes of Ni, Pd, and Pt can be readily formed [52] upon interaction of unsaturated phosphine or carbene metal complexes with O_2: see reaction (1) in Scheme 2.4.

The synthesis of peroxo complexes of Ru, Rh, and Ir was reported in [53–56]. In acidic media, peroxo complexes are transformed into hydroperoxides (reaction 2 in Scheme 2.4). Combining easy reduction of Pd(II) to Pd(0) under the action of CO (reaction 3) and reactions (1) and (2), the catalytic synthesis of H_2O_2 was carried out [57]. Later, this approach was applied to synthesize H_2O_2 in a two-phase system [58, 59]. Best results were obtained with 2,9-dimethyl-4,7-diphenyl-phenanthroline as a ligand [59]. Recently [60], hydrogen peroxide was prepared by hydrogenation of dioxygen in the presence of Pd nanoparticles stabilized with $C_{20}H_{46}NO_5P$ (HHDMA) as a ligand.

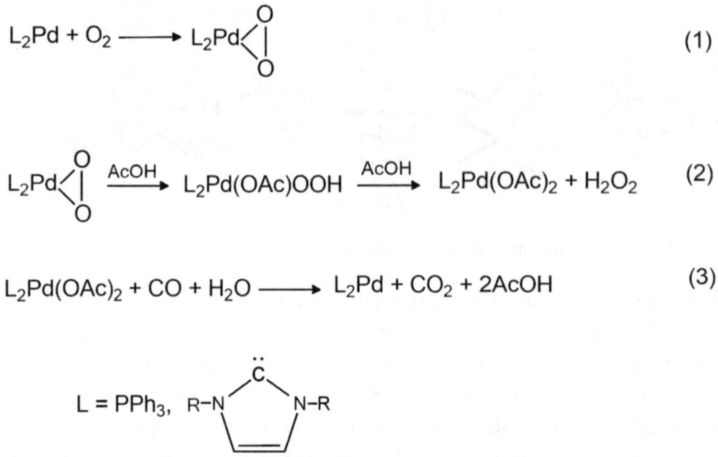

Scheme 2.4 Preparation and properties of peroxo complexes.

Of key importance seems to be the synthesis and characterization of binuclear complexes $[ClPy_4Rh–O–O–RhPy_4Cl]^{2+}$ with a bridging peroxo group [61]. Worth mentioning are binuclear copper compounds [62] and the formation of marked amounts of H_2O_2 upon interaction of Cu(I) with dioxygen [63]. Also studied was the kinetics for reactions of Cu(I) [64, 65] and Fe(II) [66] with dioxygen. We used the above property of Cu(I) and Fe(II) in designing catalytic systems for oxidation of alkanes [32–34]. Among the latest publications, worth mentioning is the synthesis of Pd complex with phosphine ligands in which phenyl groups at P atoms are linked by a hydroquinone or quinone fragment [67]. The starting complex is capable of absorbing 0.5 dioxygen equivalent to produce one water equivalent; at that, hydroquinone is oxidized to quinone. The UV, IR as well as ^{1}H and ^{31}P NMR spectra taken at low temperatures clearly demonstrated the formation of Pd(II) peroxo complex (Scheme 2.5).

An interesting effect of ligand switch was reported in [68] for heteronuclear $Ni(S)_2Ru$ complex. Replacement of η^6-hexamethylbenzene at Ru(0) by η^5-pentamethylcyclopentadienyl resulted (Scheme 2.6) in switching the complex from activating dihydrogen (reaction a) to activating dioxygen (reaction b). In addition, reaction (b) yielded the peroxo complex of Ru (Scheme 2.6).

According to [69], the reduction of Rh peroxo complex with ammonium formate leads (Scheme 2.7) to the formation of H_2O_2.

Reaction mechanisms for interaction of dioxygen with noble metals were discussed in [70].

2.3.2 Hydroperoxo Complexes

Hydroperoxo complexes can be prepared through (a) protonation of peroxo complexes or (b) reaction of O_2 with metal hydrides. In Scheme 2.4, we illustrate the transformation of Pd peroxo complex into hydroperoxo derivatives. The protonation of peroxo complexes was also reported in [71].

The insertion of dioxygen into a metal–hydride bond also results in the formation of Co, Rh, Pd, Ir, and Pt hydroperoxo complexes [7]. The transformation of Rh peroxo complexes into hydroperoxo complexes (R=H) and trimethyl silyl complexes [$R=Si(CH_3)_3$] was found [72] to proceed as shown in Scheme 2.8.

Mechanism of interaction between Pd(II) hydride complexes and dioxygen has been discussed in the literature. In the case of bi- and tridentate ligands, the process involves [73] peroxo

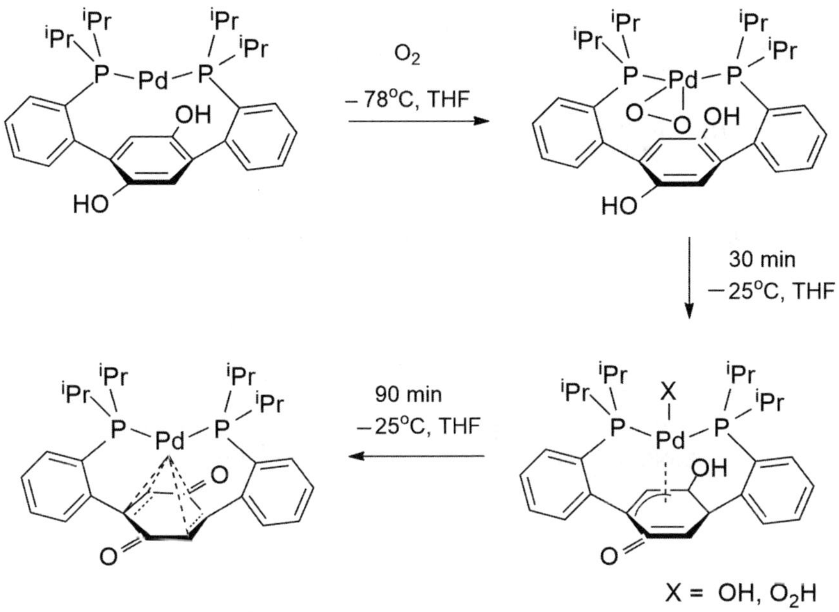

Scheme 2.5 Proposed mechanism for activation of dioxygen with Pd(II) diphosphine complex. Source: Adapted with permission from [67]. Copyright © 2016 American Chemical Society.

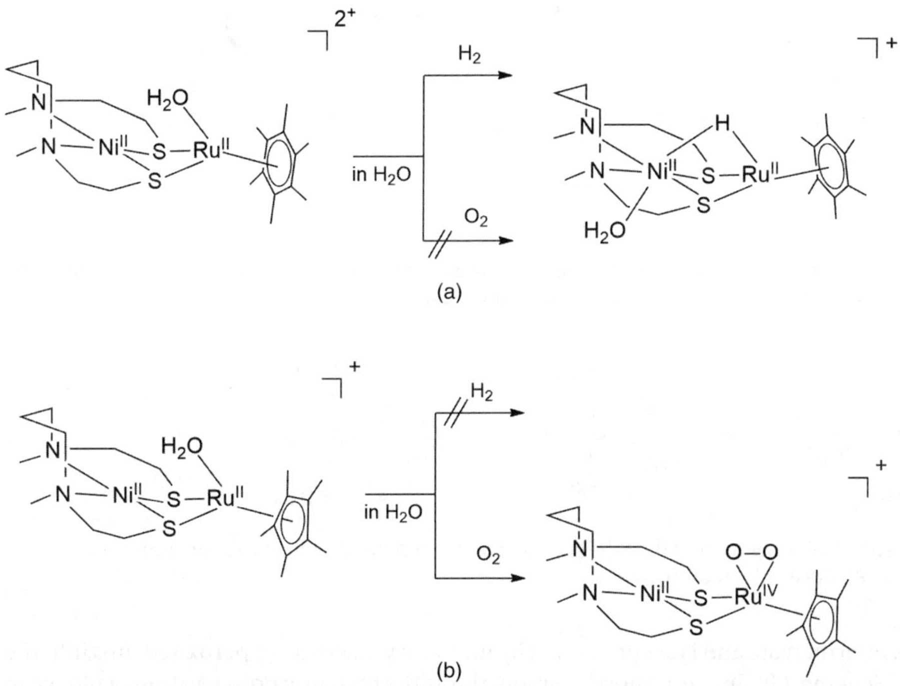

Scheme 2.6 Ligand switch effect for Ni–Ru complex. Source: Reprinted with permission from [68]. Copyright © 2012 Wiley-VCH Verlag GmbH & Co. KGaA, Weinheim.

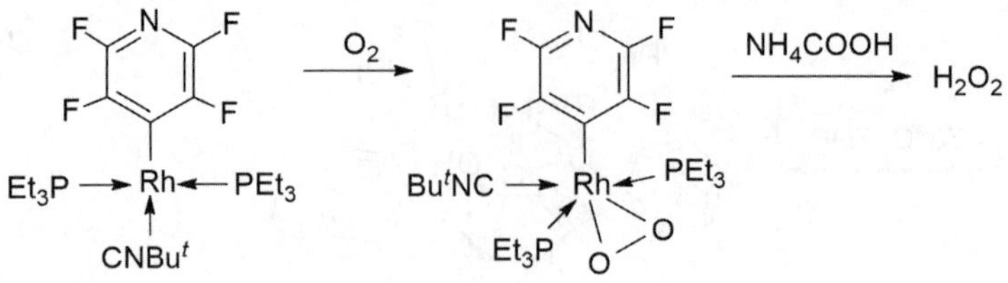

Scheme 2.7 Synthesis of H_2O_2 by reduction of Rh peroxo complex with ammonium formate.

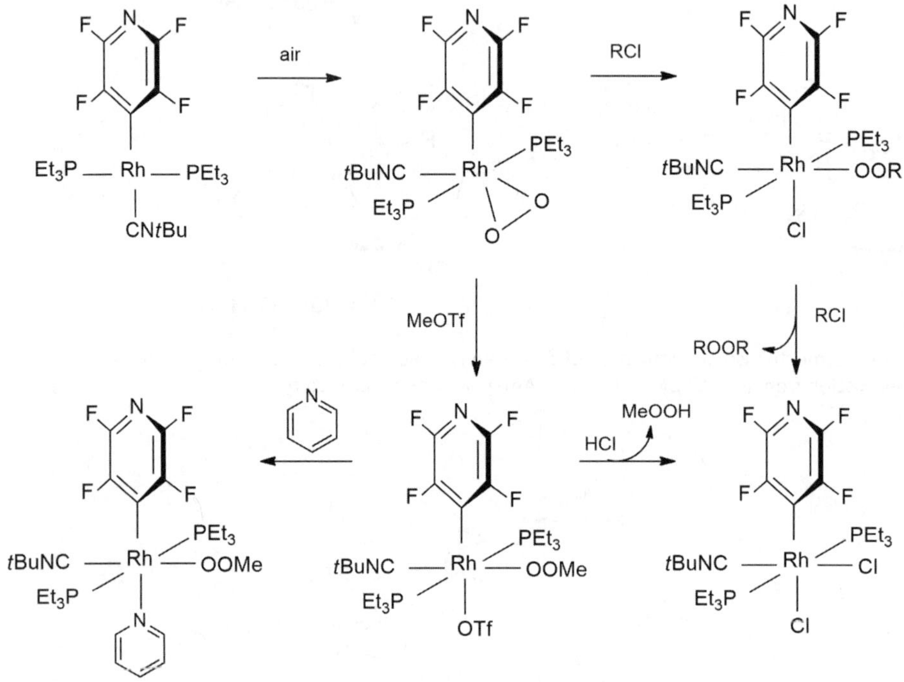

Scheme 2.8 Synthesis and reactivity of rhodium peroxo compounds. Source: Reprinted with permission from [72]. Copyright © 2005 Wiley-VCH Verlag GmbH & Co. KGaA, Weinheim.

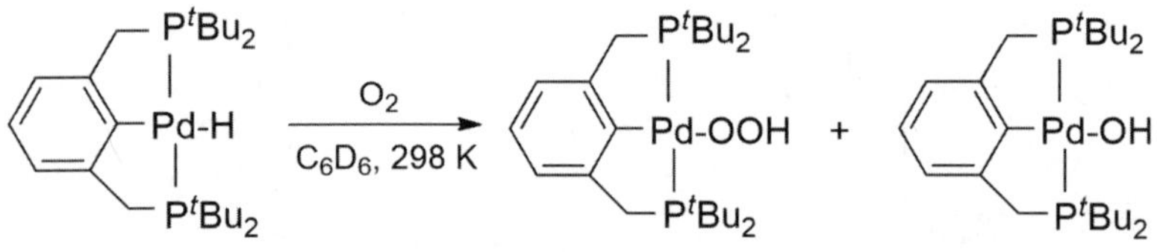

Scheme 2.9 Insertion of dioxygen into the Pd—H bond. Source: Reprinted with permission from [73]. Copyright © 2006 American Chemical Society.

complexes as intermediate and is accompanied by migratory insertion of peroxo groups into the Pd—H bond (Scheme 2.9). In the authors' opinion, this is the first direct observation of dioxygen insertion into the Pd—H bond.

The reaction of dioxygen with a hydride Pd complex with carbene ligands proceeds [74] as reductive elimination of benzoic acid (reaction 1 in Scheme 2.10) followed by formation of peroxo complex and its protonation to yield hydroperoxo complex.

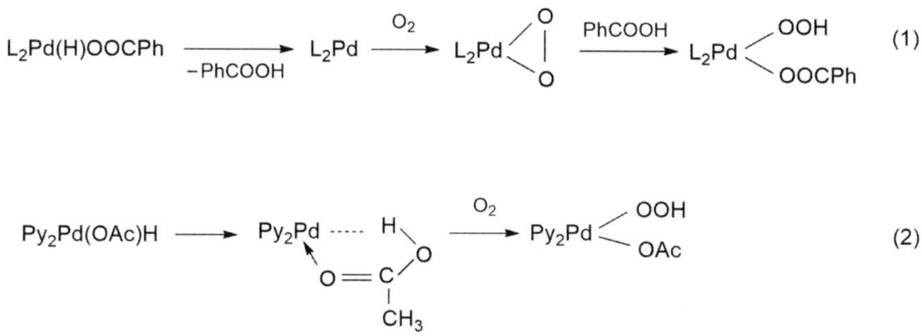

Scheme 2.10 Hydroperoxo complexes of Pd with (1) bidentate and (2) labile ligands.

In case of labile ligands, the interaction of Pd with O_2 occurs [75, 76] via reductive elimination of RCOOH coordinated with Pd(0) (reaction 2 in Scheme 2.10).

The reactivity of Co(III) hydroperoxo complexes was characterized recently [77]. Hydroperoxo complex of Ru(III) was identified in equilibrium reaction of $[(EDTA)Ru^{3+}(H_2O)]^-$ with H_2O_2 [78].

2.3.3 Oxo Complexes

Stability of oxo complexes is known [79] to decrease going from left to right across the Periodic table. Being a strong π-donor, an oxo group is stabilized by π-accepting ligands. A good example is the $(Me_3)_3Ir{=}O$ complex in which the interaction of *d*-electrons with oxo group is minimized due to tetrahedral configuration of the complex and acceptor properties of mesityl ligands. Among platinum metals, oxo complexes are typical of Ru and Os; these exist even as tetroxides, RuO_4 and OsO_4.

The reactivity of Ru oxo complexes was characterized in detail recently [80]. Osmium tetroxide is capable of oxidizing isobutane into *t*-butanol (reaction 1 in Scheme 2.11). Cyclohexane and cyclopentane are oxidized [81] through cleavage over C—C bonds (reactions 2, 3 in Scheme 2.11). Decalin is oxidized into alcohol, presumably by a concerted [3 + 2] mechanism.

Iron-oxo complexes play an exceedingly important role in biological oxidation of alkanes. In this context, of current importance seems to be the synthesis and XRD analysis of $[LFe{=}O(MeCN)](OTf)_2$ complex (L = 1,4,8,11-tetramethyl-1,4,8,11-tetraazacyclotetradecane) [82]. The pincer oxo complex of Pt was synthesized [83] as shown in Scheme 2.12.

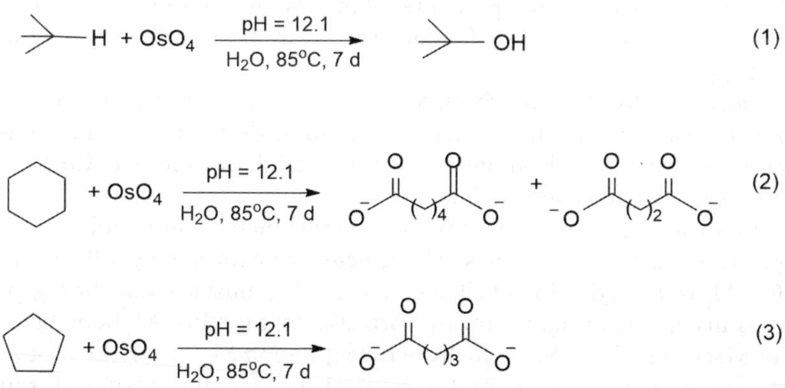

Scheme 2.11 Oxidative properties of OsO_4. Source: Adapted with permission from [81]. Copyright © 2005 American Chemical Society.

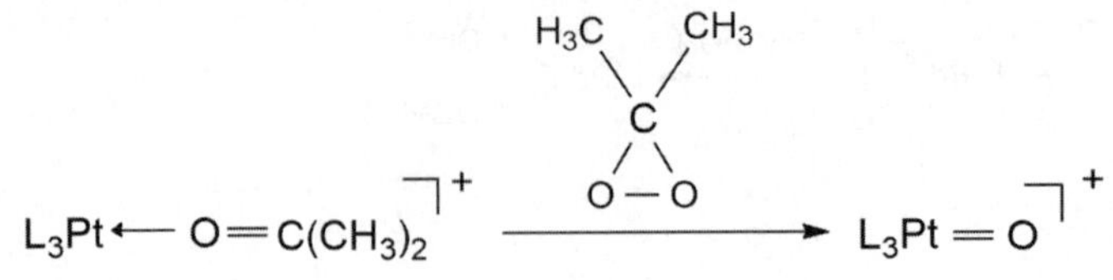

L_3 = 2(*t*-butylphosphino)methyl-6-(2-dimethyl-amino)ethylphenyl

Scheme 2.12 Synthesis of pincer oxo complex of Pt.

Stahl [25] considered in detail palladium oxidase catalysis in reactions of selective oxidation of various substrates (others than alkanes) with dioxygen.

2.4 Design of Catalytic Systems for Oxidation of Alkanes

In the design of catalytic systems, of key importance are the stoichiometric reactions previously analyzed in [32]. Provided that the oxidized form is regenerated by the oxidant (Scheme 2.13), a stoichiometric reaction can be transformed into a catalytic one. The pairs NO/NO_2 or SO_2/SO_3 can act as oxygen carriers in Scheme 2.13.

$$L_nM^{m+2}X_2 + RH \longrightarrow L_nM^m + RX + HX$$

$$RX + H_2O \longrightarrow ROH + HX$$

$$L_nM^m + Red{=}O + 2HX \longrightarrow L_nM^{m+2}X_2 + Red + H_2O$$

$$Red + 1/2\ O_2 \longrightarrow Red{=}O$$

$$RH + 1/2\ O_2 \longrightarrow ROH$$

Scheme 2.13 Design of catalytic systems containing no external reducing agent.

Generally, some organic or inorganic compounds that are prone to easy redox transitions can be selected for use as co-catalysts. Interaction of reduced co-catalysts with dioxygen yields either hydrogen peroxide or hydroperoxides and peroxides of metals. In terms of the so-called biomimetic approach (Scheme 2.14), regeneration of reduced co-catalysts takes place under the action of external reducing agent.

In the process of bio-oxidation, NADH plays the role of external reducing agent. For the role of reducing agent in chemical systems, most suitable seem to be dihydrogen and carbon monoxide, while for that of co-catalyst, hydroquinone/quinone, iodide/hypoiodide, Cu(I, II) salts, Fe(II, III) salts, and V(IV)/V(V) polyoxo metalates.

As follows from Scheme 2.14, a catalytic system must contain two metal compounds. Metal M_a must activate CO in protic media and hence it has to be selected from the metals of Pt group. But the main destination of M_a is the activation of alkane, whereas M_b must activate dioxygen. M_a and M_b admit electron transfer over bridging groups. Formation of the alkyl–M_a bond takes place due to transfer of two electrons from M_a to hydroperoxo group in M_bOOH. The proton from RH goes away to the medium, whereas the proton from M_b surrounding participates in the formation of water. Regeneration of low-valent M_b takes place at the expense of the CO reduction catalyzed by M_a. We applied the above approach to the design of catalytic systems. Some examples will be discussed as follows.

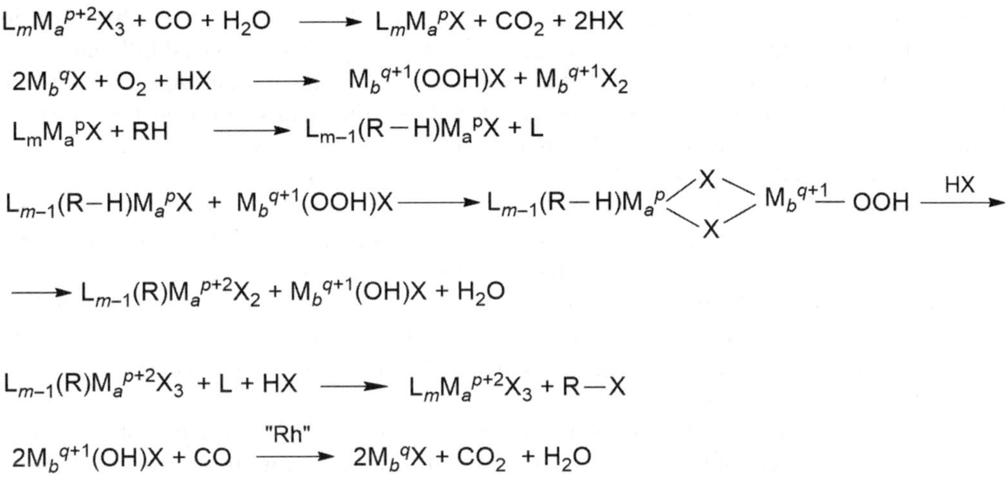

Scheme 2.14 Design of catalytic systems with external reducing agent.

2.5 Oxidation of Alkanes with Metal Complexes in Solution

A key point in oxidation of alkanes is higher reactivity of primary products, alcohols, as compared with that of starting alkanes, being estimated as 6/1for the CH_3OH/CH_4 pair [84]. The presence of acids provides some protection of alcohol from further oxidation due to formation of esters.

2.5.1 Shilov's System and Other Pt Catalysts

Kinetics and mechanism of alkanes' oxidation in the presence of $PtCl_4^{2-}/PtCl_6^{2-}$ system had been a subject of numerous studies [12, 18, 19, 21, 26, 31]. There is no doubt that the reaction mechanism is molecular. Detailed isotopic exchange studies provided strict evidence for the formation of alkyl complexes of Pt(II) and Pt(IV). The mechanism of the Shilov reaction is illustrated in Scheme 2.15 [31]. Since alkanes are weak nucleophiles, the activation of C—H bond cannot be associated with electrophilic substitution. Prof. Shilov believed that the activation of methane got started as oxidative addition and finished as the synchronous formation

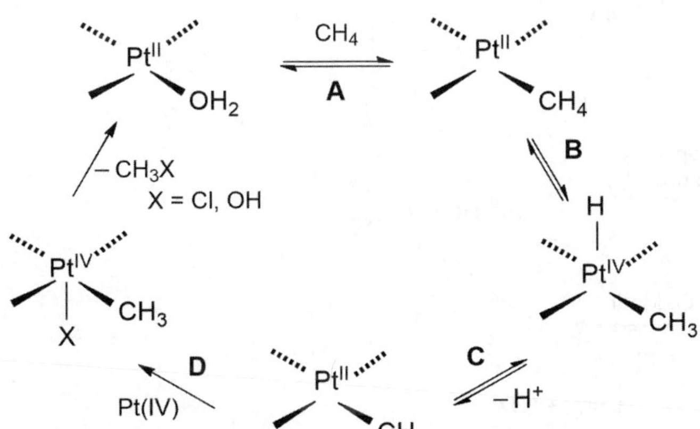

Scheme 2.15 Schematic diagram of Shilov chemistry [31].

of CH_3—Pt(II) fragment and release of proton and Cl^-. The formation of Pt methyl hydride (stage **B**) is thermodynamically limited and can hardly happen without external influence.

Although the Shilov system is unsuitable for practical implementation, its discovery gave strong impetus to rapid development of research in the field of homogeneous catalysis. An attempt was made to replace H_2PtCl_6 by dioxygen in the presence of $CuCl_2$ [85]. The reaction was conducted in acidified (with H_2SO_4) aqueous solution at 150°C and $P(O_2) = 8.26$ MPa. The substrates – $CH_3{-}SO_3H$, $C_2H_5{-}SO_3H$, *n*-$C_3H_7{-}SO_3H$, $C_2H_5{-}PO_3H_2$, and C_2H_5COOH – oxidized into respective hydroxy compounds. The suggested two-electron molecular mechanism involved isotopic exchange with environment and preemptive oxidation of C_2 and C_3 substrates at the methyl group. Competitive oxidation of ethane and $C_2H_5SO_3H$ gave products in the same relative amounts as in the $PtCl_4^{2-}/PtCl_6^{2-}$ system. Therefore, the Shilov catalytic system should be regarded as operating by inner-sphere molecular mechanism involving alkyl–Pt intermediates.

The process of methane oxidation in oleum in the presence of $HgSO_4$ was patented back in 1950 [86]. The overall yield of products (methyl bisulfate and methane sulfonic acid) attained a value of 44%. Rudakov et al. [12] systematically explored the behavior of alkanes in H_2SO_4 and found [87] that, in the presence of Hg salts, the intermediate species was R–Hg–SO_3H [87]. Later, Periana et al. [88–90] patented the process for methane oxidation with oleum in the presence of Hg and Pt salts. Results of basic studies on the catalytic action of $HgSO_4$ and Pt(bpym)Cl_2 (bpym = *bis*-1,2-dipyrimidyl) can be found in [91–94]. The above investigation involved quantum-chemical calculations and discussion of practical aspects. A detailed theoretical study on the mechanism and energetics of methane to methanol conversion in Periana Catalytica process was also conducted independently [95]. The most likely mechanism of catalytic action is presented in Scheme 2.16 [24].

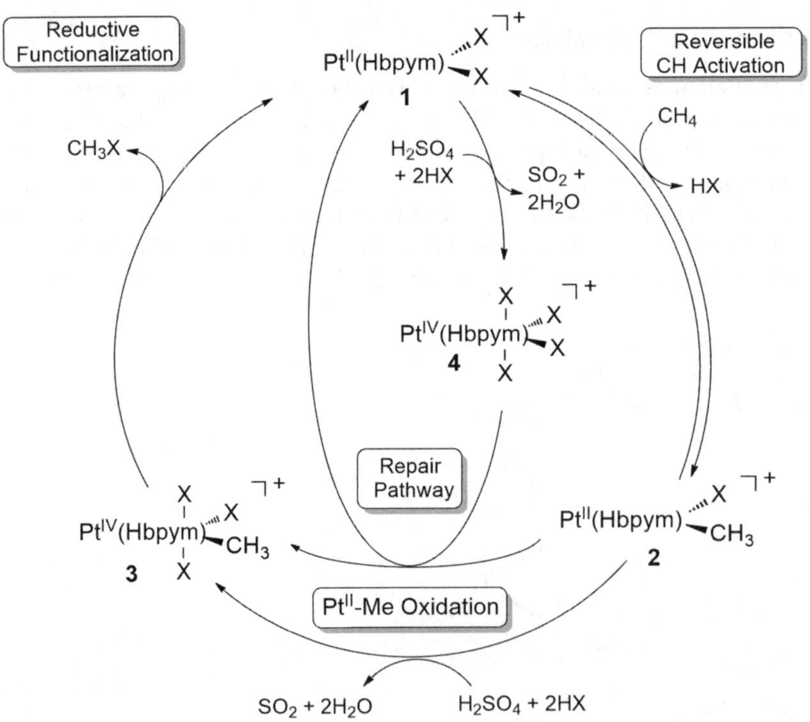

Scheme 2.16 Mechanism of methane oxidation with Periana catalyst. Source: Adapted with permission from [24]. Copyright © 2014 Wiley-VCH Verlag GmbH & Co. KGaA, Weinheim.

The activation of methane proceeds via electrophilic route **1** → **2**. CH_3–Pt(II) species **2** is oxidized to CH_3–Pt(IV) species **3** by SO_3. Compound **3** may also form upon interaction of intermediates **2** and **4**. The reaction product, methyl bisulfate, is obtained by reductive elimination.

Subsequent studies [96, 97] have shown that high catalytic activity is also exhibited by Pt(II) chlorides in the absence of bpym. A main drawback of Pt(bpym)Cl_2 is its sensitivity to water; and the attempt was made to reduce this drawback by addition of some ionic liquids [96]. In 92–98% H_2SO_4 containing 0.3 M 1-methylpyrazolium bisulfate, the activity of $PtCl_2$ exceeded that of Pt(bpym)Cl_2. But in 102% oleum, the situation reversed. Investigated was the immobilization of Pt(II) on porous solid substrates [98, 99]. For this to achieve, synthesized was a polymer based on 2,6-dicyanopyridine, CTF (covalent triazine-based framework) and Pt(CTF)Cl_2 complex [98]. The activity of Pt(CTF)Cl_2 catalyst turned close to that of Pt(bpym)Cl_2 but the former can be recirculated without loss of activity. Suggested was also another support, the activated coal produced from the exoskeleton of lobster (ExLob) [99]. This coal (specific surface 400 m^2 g^{-1}) is believed to contain nitrogen atoms that are capable of forming chelate complexes with $PtCl_2$. The activity of Pt(ExLob)Cl_2 was higher than that of Pt(bpym)Cl_2: TOF = 1200–1800 h^{-1} at 200–220 °C and $P(CH_4)$ = 65 bar in 20% oleum.

According to [97], the catalytic activity of $(NH_4)_2PtCl_4$, $PtCl_2$, Pt black, Pt(Acac)$_2$, K_2PtCl_4, and K_2PtCl_6 is exceedingly high. TOF strongly grows with decreasing catalyst concentration and conversion, which is quite reasonable. But the value of TOF > 25 000 h^{-1} looks too overestimated. As mentioned by the authors [97], even in the Cativa synthesis of AcOH via carbonylation of methanol the TOF value was 2000 h^{-1}. Note also that methanol (methyl iodide) is much more reactive than methane. It is important that such high activity is observed with a narrow range of [SO_3]. Since the oxidation of methane is accompanied by consumption of SO_3 and liberation of water, it is problematic to sustain a required value of [SO_3].

According to [97], commercialization of the process may be restrained by difficulties in extraction of methyl bisulfate and in recirculation of SO_3. Here, recirculation is understood as (i) the evacuation of SO_2 along with unreacted SO_3 from reaction medium and (ii) the oxidation of SO_2 to SO_3 and (iii) return of the latter into a technological process. Operations (i) and (ii) still remain to be worked out. In addition, we have to mention the environmental threat of the reactor (a 50-m^3 volume containing oleum at 215°C and 30–70 bar) In view of unprofitability of methane oxidation with H_2O_2 [51], the use of oleum also looks doubtful. The above circumstances cast serious doubt as concerning large-scale implementation of the Periana Catalytica process. The quest for new catalytic systems should be continued even with detriment of catalytic activity in favor of environmental friendliness.

The oxidation kinetics for methane in oleum was studied at temperatures 140–180°C, pressures 30–68 bar, and [SO_3] = 0.40–5.95 M with $PtCl_4$ as a precursor [100]. The suggested kinetic equation is applicable to calculating reaction rate and time evolution of $MeOSO_3H$ concentration.

It turned out [101] that the reaction can be carried out at 1 atm of methane in a counter-flow reactor filled with glass globules. $PtCl_4$ solution in oleum was fed from the top while methane, from the bottom. In case of 25-% oleum at 180°C, TOF attained a value of $1.7 \cdot 10^{-5}$ s^{-1}.

In an attempt to modify Pt(bpym)Cl_2 catalysts, one of nitrogen atoms not coordinated with Pt was quarternized. The catalytic activity of N-methylated Pt(bpym–CH_3)$^+X_2$ in isotopic exchange between C—H and CF_3COOD was found to be even smaller than that of standard Pt(bpym)X_2 [102]. In view of this, the quarternization of bpym ligand cannot be expected to have any influence on catalytic properties.

Another way to improve the activity of Pt catalysts was suggested and checked by Periana et al. [103]. A key point of their suggestion is in suppressing the negative effect of water, which may coordinate with Pt and thus raise the energy barrier for methane activation. This can be achieved by increasing the electron-donating properties of organic ligand, as shown in

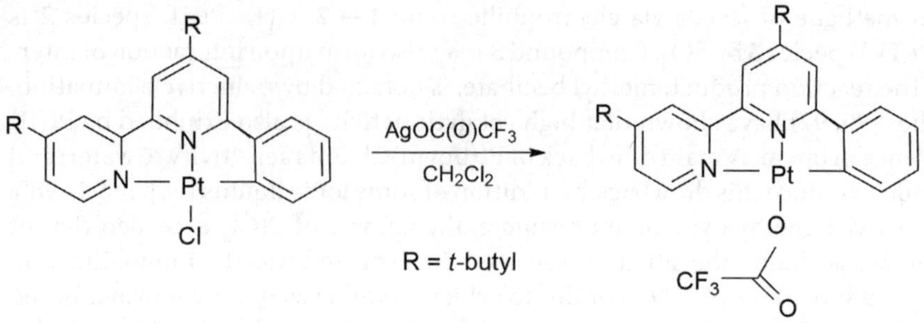

Scheme 2.17 Synthesis of Pt(N,N,C)(OAc$_f$). Reprinted with permission from [103]. Copyright © 2006 American Chemical Society.

Scheme 2.17 for Pt(II) complex with tridentate ligand Pt(N,N,C)(OAc$_f$). These hopes also were in vain: the activity of Pt(N,N,C)(OAc$_f$) turned out to be smaller than that of Pt(bpym)X_2.

Still another procedure for utilization of N-methylated bpym ligand was suggested in [104]. The authors synthesized a hybrid catalyst via ionic exchange between $[Pt(Me–bpym)Cl_2]^+$ $[HSO_4]^-$ and polyoxometalate $H_5PV_2Mo_{10}O_{40}$ (POM) and immobilized it on SiO_2. The oxidation of methane in acidified aqueous solution at 50°C, $P(CH_4) = 30$ bar, $P(O_2) = 2$ bar yielded methanol and, unexpectedly, acetaldehyde, with turnover number (TON) = 31–33 in 4 hours. The suggested reaction mechanism (Scheme 2.18) was similar to that of Shilov's chemistry. A main difference is that the oxidation of Pt(II) to Pt(IV) takes place under the action of dioxygen. Atom ^{18}O comes to methanol from $H_2{}^{18}O$. Apparently, the oxidation of Pt(II) to Pt(IV) involves some active POM intermediate such as V–OO–V. POM is capable of replacing

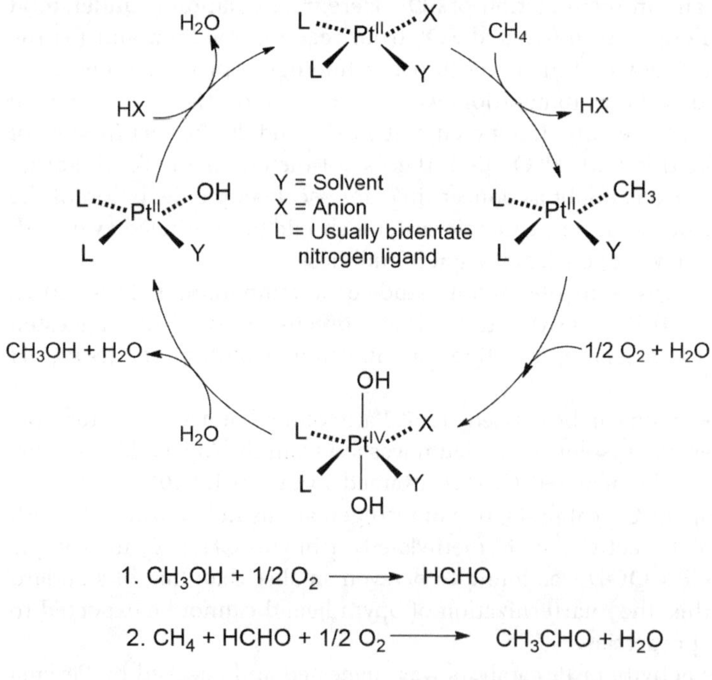

Scheme 2.18 Catalytic cycle of methane oxidation on $[Pt(Me–bpym)Cl_2][H_4PV_2Mo_{10}O_{40}]/SiO_2$ catalyst. Source: Adapted with permission from [104]. Copyright © 2004 American Chemical Society.

its ^{16}O by ^{18}O from water. Experiments with ^{13}C-labeled compounds have demonstrated that reactions 1 and 2 in Scheme 2.18 are most likely for the formation of acetaldehyde.

Activation of C—H bond with cationic complexes of Ni, Pd, and Pt with α,α-bipy, close analog of bpym, was explored in [105]. Complexes $[Ni(\alpha,\alpha\text{-bipy})CH_3]^+$ and $[Pd(\alpha,\alpha\text{-bipy})CH_3]^+$ turned active only with respect to the C—H bonds of aromatic compounds while $[Pt(\alpha,\alpha\text{-bipy})CH_3]^+$, to the C—H bonds in benzyl group of toluene. Chloride analog of the Pt complex was also active to the C—H bonds in benzyl group of toluene, whereas those of Ni and Pd complexes were inactive.

The reactivity of Pt(II, IV) complexes with another N,N-bidentate ligand, di(2-pyridyl) methane sulfonate (dpms), was studied in [106]. The reaction of $Pt(dpms)CH_3(OH)$ complex with dioxygen in water is pH dependent. For pH < 10, the intermediate species with CH_3–Pt(IV)–OOH fragment reacts with the starting complex to give two equivalents of $Pt^{IV}(dpms)(CH_3)(OH)_2$. For pH > 12, the reductive elimination of methanol is observed. It is stated that the dpms ligand imparts a capability of reacting with dioxygen to Pt(II).

Generally, influence of organic ligand on the catalytic activity of complexes in activation of C—H bond is essential. So the replacement of bpym by N,O-bidentate ligand–anion of picolinic acid increases the catalytic activity in isotopic exchange between benzene and Ac_fOH: from $TOF = 6.4{\cdot}10^{-5}\ s^{-1}$ to $1.2{\cdot}10^{-2}\ s^{-1}$ at 70°C [107].

Oxidative carbonylation of methane in the presence of $PtCl_2$ in fuming or concentrated sulfuric acid – at 189 °C, $P(CH_4)$ = 27 bar, and $P(CO)$ = 27 bar – was investigated in [108]. The catalytic activity grew upon addition of $CuCl_2$ and dioxygen.

It should be summarized in conclusion that platinum compounds are effective in catalyzing the activation of C—H bonds. Pt(IV) compounds also play an important role as intermediates in the functionalization of alkanes. The examples of using dioxygen as an ultimate oxidant are not numerous.

2.5.2 Oxidation of Alkanes with Pd Compounds

Compared to Pt complexes, the complexes of Pd are more labile in typical reactions within the coordination sphere: substitution, oxidative addition, reductive elimination, insertion, etc. For this reason, one can expect higher catalytic activity of Pd complexes. But in redox reactions with involvement of Pd(0) as an intermediate, it is necessary to prevent the formation of metallic palladium. This process can be suppressed by strongly π-accepting ligands. An alternative route can be the addition of an oxidant – such as Cu(II), Fe(III), quinones, etc – that would rapidly transform Pd(0) to Pd(II).

Systematic studies by Fujiwara and co-workers [109] revealed that carbonylation of hexane in Ac_fOH in the presence of $Pd(OAc)_2$ and $K_2S_2O_8$ at 80 °C and $P(CO)$ = 20 bar gave $C_6H_{11}COOH$ in a very low yield. But the above yield strongly increased upon addition of $Cu(OAc)_2$ as a co-catalyst [109]. The carbonylation of propane under similar conditions yielded *i*-butyric and *n*-butyric acids at the mole ratio $i/n \approx 3.4$–4.4. It is important that the yield of the above products increased for $Cu(OAc)_2/Pd(OAc)_2 \geq 2$. Interaction between the catalyst components was confirmed by UV–VIS absorption spectra and by other methods (ESCA, ^{19}F NMR, XRD). Methane is also subjected to carbonylation at 80 °C, $P(CH_4)$ = 20 bar, $P(CO)$ = 15 bar. Instead of $K_2S_2O_8$, the authors used dioxygen (15 bar), which increased the yield of acetic acid by a factor of two. The reaction mechanism is shown in Scheme 2.19.

The C—H bond in alkane is activated by $[Pd(Ac_f)]^+$ to give a Pd–alkyl species followed by activation of CO and insertion into the Pd–alkyl bond. Scheme 2.19 contains no Cu(II). Apparently it is the action of heteronuclear Pd–Cu complex, while Cu(II) promotes the oxidation of Pd(0) into Pd(II). Moreover, withdrawal of electrons from Pd(II) to Cu(II) occurs, thus increasing

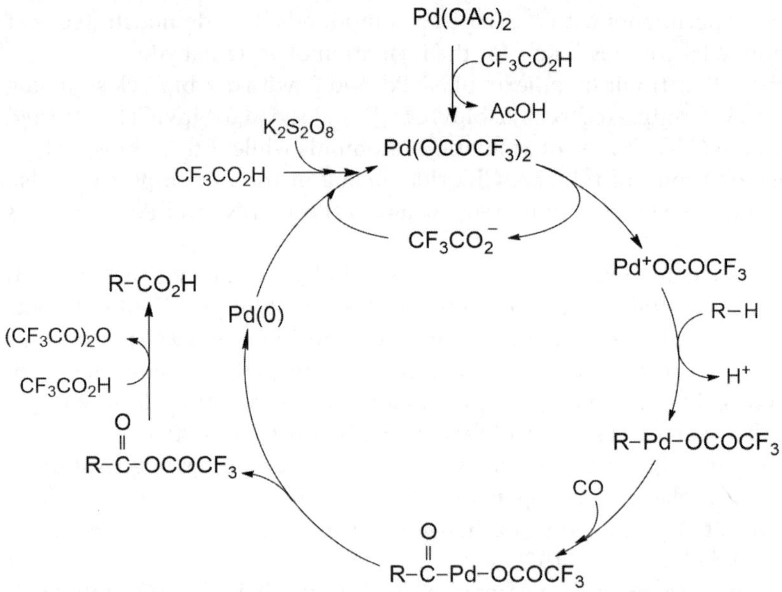

Scheme 2.19 Fujiwara mechanism for synthesis of carboxylic acids. Source: Adapted with permission from [109]. Copyright © 1996, Rights managed by Georg Thieme Verlag KG Stuttgart–New York.

the electrophility of $[PdOAc_f]^+$. Fujiwara et al. [17] reviewed some other important reactions involving the activation of C—H bond.

As demonstrated by Sen et al. [110, 111], methane and other alkanes underwent trifluoroacetoxylation in $(Ac_f)_2O$ and H_2O_2 in rather mild conditions (90 °C, 61 bar CH_4). The reaction proceeds via the outer-sphere activation of alkane with the anion of peroxy trifluoroacetic acid coordinated with Pd(II) (Scheme 2.20).

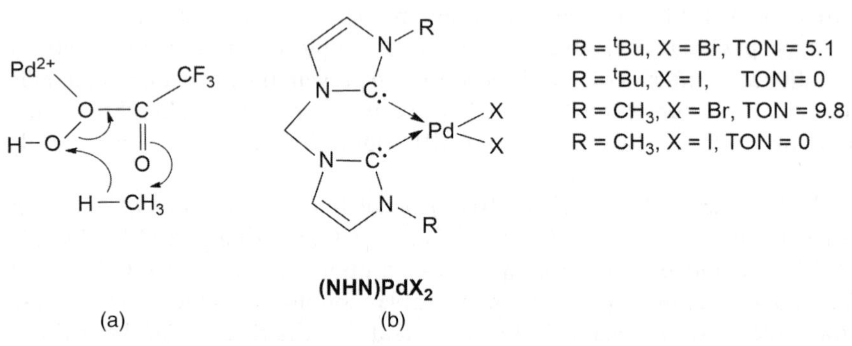

Scheme 2.20 (a) Sen intermediate. Reprinted with permission from [111]. Source: Copyright (1991) American Chemical Society; and (b) Pd(II) complexes with chelating biscarbene ligands. Source: Adapted with permission from [112]. Copyright © 2002 Wiley–VCH Verlag GmbH & Co. KGaA, Weinheim.

The reaction is thought to be electrophilic since: (*i*) the attack on the aromatic ring in *p*-xylene is two orders of magnitude stronger than that on the benzyl hydrogen atoms and (*ii*) upon oxidation, *cis*- and *trans*-1,2-dimethyl cyclohexane retain their configuration. Unfortunately, the initial accumulation of Ac_fOCH_3 is then followed by its oxidation. In authors' opinion, the final reaction product is CO_2.

Similar mechanism was suggested [113] for oxidation of methane with H_2O_2 in the presence of $Pd(hfacac)_2$ and $Cu((hfacac)_2$ in $(Ac_fO)_2O$, where hfacac = hexafluoroacetyl acetonate. The yield of Ac_fOCH_3 as a function of time passes through a maximum. Just as in [111], it is believed that Ac_fOCH_3 is oxidized to CO_2, despite good screening of CH_3O– group in this ester. Strong radical scavenger – 2,6-di-*tert*-butyl-4-methylphenol – produces little or no influence on the process of oxidation, which is indicative of its molecular mechanism. The activity of $Pd(hfacac)_2$ (TOF ≈ 20 h^{-1}) is greater than that of $Pd(OAc_f)_2$ (TOF ≈ 2 h^{-1}) even under slightly milder conditions. Upon recycling, $Pd(hfacac)_2$ was found to retain its structure. But as for the formation of CO_2, it has not been confirmed experimentally neither in [111] nor in [113]. Meanwhile, we have demonstrated [32–34] that the oxidation of methane in Ac_fOH/H_2O yields Ac_fOCH_3 and HCOOH but not CO_2.

The oxidation of alkanes with $K_2S_2O_8$ in the presence of Pd complexes with NHN ligands (Scheme 2.20b) in $Ac_fOH/(Ac_fO)_2O$ was studied by Strassner and co-workers [29, 112, 114–117]. The oxidation of methane at 90°C, $P(CH_4)$ = 30 bar for 14 hours gave 5 moles of Ac_fOCH_3 per 1 mol of Pd [112].

Later, the oxidation of propane was found effective in the presence of $(NHN)PdBr_2$ [118–120]. Assuming possible involvement of Br_2 in oxidation of alkanes, a new strategy was suggested for designing catalytic systems based on the use of dioxygen as a terminal oxidant [29, 119, 120]. A key point in this strategy is the use of "redox mediator" such as the $2Br^-/Br_2$ pair. In this case, the catalytic process consists of two cycles linked by redox mediator (Scheme 2.21). Br^- is oxidized to Br_2 with V(V), whereas the formed V(IV) is returned to V(V) by dioxygen. Therefore, the left cycle is responsible for the activation of alkane on Pd(II) and its functionalization by Pd(IV) while the right cycle, for the activation of dioxygen. The oxidation of propane at 90°C, $P(C_3H_8)$ = 6 bar, $P(O_2)$ = 4 bar in $Ac_fOH/(Ac_fO)_2O$ yielded Ac_fOPr^i as a main product (TON = 22 in 17 hours). The oxidation of propane with $K_2S_2O_8$ at 60°C gave the following mixture: Ac_fOPr^i (90%), Ac_fOPr^n (4%), $Ac_fO—CH_2—CH_2{=}CH_2$ (1%), acetone (1%), and $CH_3—CH(OAc_f)—CH_2—OAc_f$ (4%).

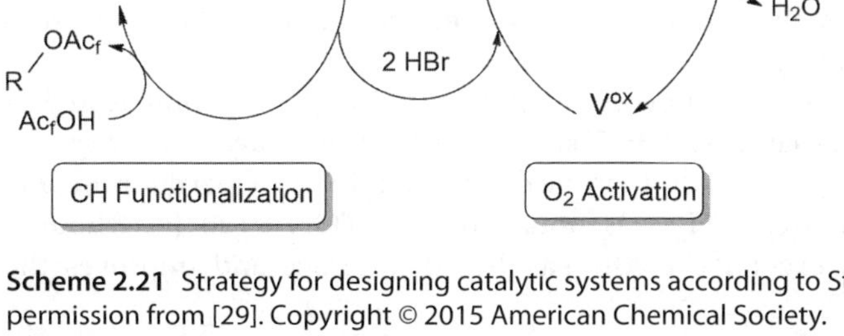

Scheme 2.21 Strategy for designing catalytic systems according to Strassner and Munz. Source: Adapted with permission from [29]. Copyright © 2015 American Chemical Society.

The strategy suggested by Strassner and Munz seems fruitful and promising. The catalytic activity is low, but it should be noted that the process is carried out with dioxygen and without external reducing agent.

The oxidation of alkyl aromatics by dioxygen in the presence of $Pd(o\text{-phen})(OAc)_2$ at 120–200 °C and $P(O_2)$ = 25 bar was reported in [121]. Radical scavenger had no influence on the process. Toluene oxidized to benzyl alcohol, benzaldehyde, benzoic acid, and benzyl benzoate as a main product with overall TOF = $8\,h^{-1}$.

The acetoxylation of alkylpyridines was carried out [122] in the presence of $Pd(OAc)_2$ and CuI (or CuCl) in AcOH at 100–120 °C and $P(O_2) = 8$ bar. The reaction occurs largely at the alkyl group in α-position. 2-Ethylpyridine oxidizes to 2-(α-acetoxyethyl) pyridine at 100 °C in 71-% yield and almost 100-% selectivity. The role of Cu(I) halides was not discussed, while Cu(II) salts were ineffective. In terms of our strategy (Scheme 2.14), Cu(I) activates dioxygen and acts as a co-reductant.

Chinese researchers conducted a number of interesting works. Designed was the catalytic system $Pd(OAc)_2$–Q–$NaNO_2$ (Q = benzoquinone) for oxidation of methane in Ac_fOH at $P(O_2) = 1$ bar and $P(CH_4) = 54$ bar [123]. Under optimal conditions, TOF was $0.5\,h^{-1}$. In the authors' opinion, the process proceeds as shown in Scheme 2.22 (**A**).

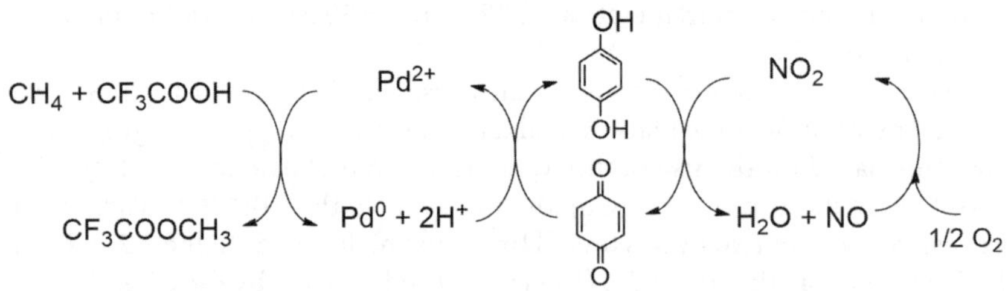

(**A**)

$$Pd(OAc)_2 + Ac_fOH \not\longrightarrow Ac_fOCH_3 \quad (1)$$

$$2Pd(OAc_f)(NO_2) + CH_4 + 3Ac_fOH \longrightarrow 2Pd(OAc_f)_2 + Ac_fOCH_3 + 2NO + 2H_2O \quad (2)$$

$$2Pd(OAc_f)_2 + 2NO + NO_2 + H_2O \longrightarrow 2Pd(OAc_f)_2(NO_2) + 2Ac_fOH \quad (3)$$

Scheme 2.22 (**A**) Mechanism of methane oxidation in the Bao system. Source: Adapted with permission from [123]. Copyright © 2006 American Chemical Society.

According to [124, 125], $Pd(OAc_f)_2$ does not oxidize CH_4 even at 180 °C and $P(CH_4)$ = 30–40 bar (reaction 1). In view of this, Scheme 2.22 seems doubtful. Tentatively, a true reason for oxidation of methane can be the nitrite complex of Pd(II) (reaction 2, Scheme 2.22). Regeneration of the nitrite complex may proceed through reaction 3 (Scheme 2.22). In this context, the role of quinone remains unclear. It can be assumed that quinone stabilizes Pd(0) and prevents the formation of Pd metal.

Others have [126] replaced NO_x in the above scheme by $H_3PMo_{10}V_2O_{40}$ polyoxometalate (POM). At 80 °C, $P(CH_4) = 25$ bar, $P(O_2) = 5$ bar in Ac_fOH, the catalytic activity was very low (TON = 3 in 8 hours) but markedly increased upon addition of perfluorooctane C_8F_{18}. At $C_8F_{18}/Ac_fOH = 1$ (v/v), Ac_fOCH_3 was obtained with TON ≈ 120 (in 8 hours). Supposedly, C_8F_{18} facilitates mass exchange at the gas–liquid interface and improves the solubility of O_2.

Later, it was suggested [127] that tetrachlorobenzoquinone (TCQ) be used as a redox co-catalyst. Palladium was used in the form of Pd/C and methane oxidation with O_2/CO was conducted in aqueous Ac_fOH. Under the assumption of water gas shift reaction (WGSR), the evolved dihydrogen is oxidized with dioxygen to yield H_2O_2, and the latter acts as an oxidant. The reaction yielded Ac_fOCH_3 and HCOOH. At 140 °C, $P(CH_4) = 18$ bar, $P(CO) = 10$ bar, and $P(O_2) = 5$ bar, the catalytic activity attained a value of TOF = $120\,h^{-1}$. According to [127], the reaction mechanism can be represented as shown in Scheme 2.23.

1. $CO + H_2O \xrightarrow{Pd} CO_2 + H_2$
2. $TCQ + H_2 \longrightarrow TCQ\text{-}H_2$
3. $TCQ\text{-}H_2 + O_2 \longrightarrow TCQ + H_2O_2$
4. $Pd + H_2O_2 + 2Ac_fOH \longrightarrow Pd(OAc_f)_2 + Ac_fOH$
5. $Pd(OAc_f)_2 + CH_4 \longrightarrow Pd(0) + Ac_fOCH_3 + Ac_fOH$

Scheme 2.23 Reaction scheme for oxidation of methane in the presence of TCQ as a co-catalyst.

As mentioned above, reaction 5 Scheme 2.23 seems unlikely. A reactive intermediate can be expected to form in the Pd(II)–H_2O_2 reaction. Also note that the yield of Ac_fOCH_3 as a function of [TCQ] passes through a maximum and the yield of HCOOH decreases. In order to activate O_2, an external reducing agent (CO) was added. It is noteworthy to mention that the addition of H_2 or CO for generation of H_2O_2 was suggested by Sen in 1998 [16].

The oxidation of methane in the presence of Pd compounds and POM as co-catalysts was found to occur even in the absence of quinone [128]. The K_2PdCl_4–NaCl mixture (NaCl/K_2PdCl_4 = 50) was used as a precursor. In Ac_fOH at 80°C, $P(CH_4)$ = 30 bar, $P(O_2)$ = 15 bar, Ac_fOCH_3 was obtained with TON = 80 (in 8 hours). Unexpectedly, the main reaction product was acetic acid (TON = 2000 in 8 hours). It was assumed that CH_3^+ was a key intermediate that eliminated –COOH from Ac_fOH to give AcOH or attached to Ac_fO^- to yield Ac_fOCH_3. The occurrence of oxidative condensation was not considered and no experiments with labeled $^{13}CH_4$ were conducted. But, if indeed acetic acid is formed from Ac_fOH, such a synthesis can hardly be regarded as interesting.

The oxidation of methane in the presence of Pd(bipy)Cl_2–benzoquinone–$H_5PMo_{10}V_2O_{40}$ system in aqueous Ac_fOH was explored in [129]. At H_2O/Ac_fOH = 1/5, the yield of Ac_fOCH_3 was maximal (TON = 13.3 8 h^{-1}). The yield of Ac_fOCH_3 increased threefold upon the addition of C_8F_{18} (1 ml per 5 ml Ac_fOH) and with an increase in $[H_5PMo_{10}V_2O_{40}]$. The reaction was first-order with respect to $P(CH_4)$ and $P(O_2)$. The suggested reaction mechanism [129] is illustrated in Scheme 2.24.

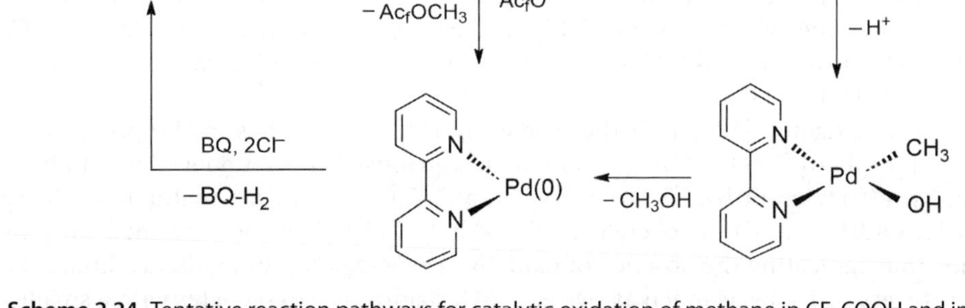

Scheme 2.24 Tentative reaction pathways for catalytic oxidation of methane in CF_3COOH and in CF_3COOH/H_2O. Source: Adapted with permission from [129]. Copyright © 2013, Springer Science + Business Media New York.

The first stage is the electrophilic substitution yielding Pd–CH_3 species; this is followed by the acceptation of two electrons by Pd(II) and formation of Ac_fOCH_3. Benzoquinone (BQ) regenerates Pd(0) into Pd(II). Another pathway includes the replacement of Cl^- by H_2O, followed by deprotonation of Pd(CH_3)(OH) and reductive elimination of CH_3OH. According to our results [130], Pd(bipy)Cl_2 does not catalyze isotopic exchange between methane and medium even at 95°C, $P(CH_4)$ = 60 bar, and Ac_fOD/D_2O = 2.85 (v/v). In this context, reversibility of the stage yielding Pd–CH_3 seems doubtful. Scheme 2.24 does not reflect the role of POM. Moreover, in Scheme 2.24 BQ is shown as a methane oxidant, although this is thermodynamically forbidden. It can be imagined that the role of oxidant is played by some species formed with involvement of dioxygen.

Apparently, Sen was one of the first who suggested [16, 131–133] using an external reducing agent (CO, H_2, etc.) in order to transform O_2 into active two-electron oxidants. For instance, Pd/C catalyst in D_2O–DCl was found [131] effective in oxidation of C_2H_6 at 70–110°C and, $P(C_2H_6)$ = 34 bar, $P(CO)$ = 6.8 bar, and $P(O_2)$ = 6.8 bar, by the mechanism depicted in Scheme 2.25.

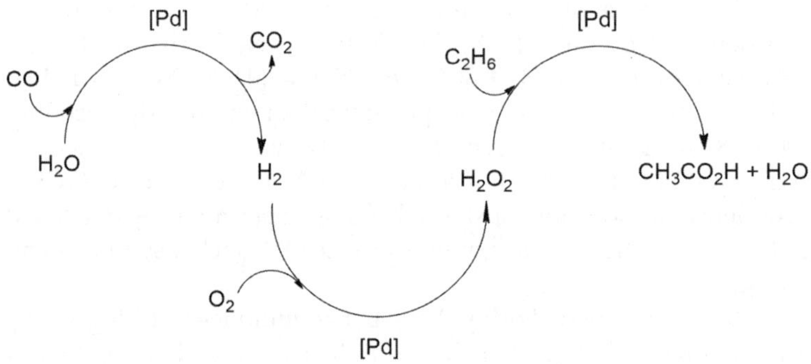

Scheme 2.25 Oxidation of methane in Pd/C–O_2–CO system. Source: Reprinted with permission from [131]. Copyright © 1992 American Chemical Society.

Under similar conditions, methane is oxidized to formic acid but at much lower efficiency [132]. The catalyzates were found to contain low amounts of H_2O_2.

Later, Sen et al. [133] made a step in proper direction: they added a redox co-catalyst into catalytic system. In aqueous Ac_fOH (Ac_fOH/H_2O = 3/1, v/v) at 85–95°C, $P(CH_4)$ = 61 bar, $P(CO)$ = 13.6 bar, and $P(O_2)$ = 6.8 bar, the catalytic Pd/C–$CuCl_2$ system showed the value of TOF = 20–25 h^{-1}. An important role was assigned to Cl^-: the use of $Cu(OAc_f)_2$ instead of $CuCl_2$ gave a negative result. A number of evidences were provided in favor of two-electron mechanism for oxidation of alkanes.

It also turned out [134] that dihydrogen can play the role of external reducing agent. Oxidation of cyclohexane with Pd/Al_2O_3–V(Acac)$_3$ system in glacial acetic acid at 38°C, $P(H_2)$ = 6.8 bar, $P(N_2)$ = 68 bar, and $P(O_2)$ = 6.8 bar yielded cyclohexanol/cyclohexanone = 2.7 (mol/mol) in overall TOF = 38 h^{-1}.

Previously, we investigated [130, 135] the oxidation of C_1–C_4 alkanes in the presence of Pd(bipy)Cl_2 or [Pd(phen)Cl_2]–$Cu(OAc_f)_2$ system in aqueous Ac_fOH under the action of O_2/CO. At 80°C, $P(CH_4)$ = 60 bar, $P(O_2)$ = 5.6 bar, and $P(CO)$ = 16 bar, methane oxidized to Ac_fOCH_3/HCOOH ≈ 3.5/1 in overall TOF ≈ 44 h^{-1}. CH_3COOH was formed only in trace amounts, thus indicating the absence of oxidative carbonylation of methane. Similar Pt compounds were less active by as much as twice. C_2–C_4 oxidized to oxygenates with a smaller number of carbon atoms. So the oxidation of ethane gave Ac_fOCH_3, $Ac_fOC_2H_5$, and AcOH. Propane oxidized to Ac_fOR (R = CH_3, C_2H_5, *i*-C_3H_7, *n*-C_3H_7) and acetone while *n*-butane, to Ac_fOR (R = CH_3, C_2H_5, *n*-C_3H_7, *sec*-Bu, *n*-Bu), methyl ethyl ketone, and AcOH. Addition of ionic liquids can be used to regulate the selectivity of propane oxidation [130]. With increasing $[PyH]^+OAc_f^-$ concentration (up to 4 M), the yield of acetone increased by a factor of 2.0, while that of Ac_fO^nPr and Ac_fO^iPr decreased by a factor of 1.5–2.

It follows that the catalytic activity of Pd compounds shows nothing special in comparison with that of Pt compounds.

2.5.3 Rh-Containing Catalysts and Catalytic Systems

First let us mention an interesting publication [136] concerning the oxidation of toluene to benzaldehyde in the presence of $Rh_2(Ac_fO)_4$ in $Ac_fOH/(Ac_fO)_2O$ under the action of Selectfluor, highly selective oxidant [136]. It is stated that the reaction does not proceed in the presence of other oxidants. It was established that Ac_fOCH_2–Ph was not an intermediate. The kinetic isotope effect (Ph–CH_3/Ph–CD_3) $k_H/k_D = 2.0–2.3$ is indicative of slow activation of C—H bond. According to results of subsequent density functional theory (DFT) studies, the reaction may proceed via the carbene intermediate Ph–CH=Rh(IV). Suggested was a catalytic cycle with low energy barriers.

Some Pt metals are known to act as catalysts for oxidation of alkanes with hydrogen peroxide in Ac_fOH [137], the highest activity being exhibited by $[Rh_3O(OAc)_6(H_2O)_2]OAc$ complex. Cyclohexane is oxidized to $C_6H_{11}OH$ and Ac_fO-C_6H_{11} in a ratio of 40 : 1. The oxidation of *n*-octane gives all of four isomeric hexyl trifluoroacetates. The suggested reaction mechanism involves the formation of Rh oxo complex with participation of peroxo trifluoroacetic acid.

Lin and Sen [138] reported on the oxidation and oxidative carbonylation of methane in aqueous solution containing $RhCl_3$ and HI. At 95 °C, $P(CH_4) = 54$ bar, $P(CO) = 13.6$ bar, and $P(O_2) = 6.8$ bar, the main reaction product was CH_3COOH, along with HCOOH and trace amounts of CH_3OH. Despite low activity (TON = 27.5 in 350 hours), this system seems rather promising, since the replacement of water by organic solvent can be expected to markedly improve the activity. Indeed, Sen and co-workers showed [139] that such an event took place in a Pr_fCOOH–H_2O medium. Later, we successfully used such media as CD_3COOD–D_2O [140, 141] and Ac_fOH–H_2O [142].

The oxidation and oxidative carbonylation of methane in Pr_fCOOH/D_2O (6/1 vol) gave Pr_fCOOCH_3 and AcOH in yields of 0.9 and 0.2 M for 80 hours and $[RhCl_3] = 0.005$ M. C_2–C_4 alkanes were more reactive than CH_4 and oxidized not only across C—H bonds but with cleavage over C—C bonds [139].

In reactions of methane, the carbonyl group comes to acetic acid from carbon monoxide (^{13}CO). The carbon dioxide detected in the gas phase originates from CO rather than from CH_4. Carbonylation of $^{13}CH_4$ and $^{12}CH_3OH$ showed that the reactant was methane. Hence, the reaction mechanism differs from that involving oxidative addition of CH_3I to Rh(I) [143]. Convincing evidences were provided in favor of two-electron catalytic process involving Rh–alkyl intermediate. The role of iodide co-catalyst was not discussed. For C_2H_6 and C_2D_6, $k_H/k_D = 2.65$. The role of oxidant was played by some oxygen-containing species formed upon oxidation of dihydrogen generated in WGSR. The latter statement seems erroneous.

In [144], we have shown that WGSR does not occur during catalytic oxidation of methane in the $RhCl_3$–KI–NaCl system. In the Monsanto process, this reaction happens because of a high concentration of I^- [143] and hence the role of CO as external reducing agent is interpreted reasonably. Its role of iodide promoter remains obscure.

In view of this, we extensively studied the $RhCl_3$–KI–NaCl system in $Ac_fOH–H_2O$ with special emphasis on the role of iodide co-catalyst [144]. The results (Scheme 2.26) can be summarized as follows: (i) reaction rates r_{CH_4} and r_{CO_2} pass through a maximum at $m_{H_2O} = 0.6–0.7$; (ii) for formation of acetic acid in $Ac_fOH–H_2O/Ac_fOD–D_2O$ pair, $(k_H/k_D)_{solv} = 2$, and for HCOOH in Ac_fOCH_3 and CO_2, $k_H/k_D \approx 1.7$; (iii) for methane, $(k_H/k_D)_{CH_4} \approx 4$; (iv) the efficiency of Ti(0) oxidation (test on H_2O_2) is maximal at $m_{H_2O} = 0.6–0.7$; (v) reaction rate has a maximum at $[KI] \approx 2\,M$, there is no reaction at $[KI] = 0$ и $[KI] \geq 4 \cdot 10^{-2}$; and (*vi*) radical scavengers have no influence on the process. In optimal conditions: $(TOF)_{AcOH} = 71\ h^{-1}$, $(TOF)_{AcfOCH_3} = 9\ h^{-1}$, $(TOF)_{HCOOH} = 12\ h^{-1}$, and $(TOF)_{CO_2} = 1300\ h^{-1}$.

1. $CH_4 + 2CO + O_2 \longrightarrow CH_3COOH + CO_2$

2. $CH_4 + CO + O_2 + Ac_fOH \longrightarrow Ac_fOCH_3 + CO_2 + H_2O$

3. $Ac_fOCH_3 + H_2O \rightleftharpoons Ac_fOH + CH_3OH$

4. $CH_3OH + O_2 \longrightarrow HCOOH + H_2O$

5. $2HI + 1/2O_2 \longrightarrow I_2 + H_2O$

6. $I_2 + CO + H_2O \longrightarrow 2HI + CO_2$

Scheme 2.26 Reactions of methane oxidation taking place in the presence of $RhCl_3$–KI–NaCl system in $Ac_fOH–H_2O$.

The above results agree with the two-electron mechanism for activation and oxidative transformations of methane. The oxidation of Ti(0) (in conditions of methane oxidation) is associated with the presence of two-electron oxidants such as H_2O_2 and HOI [145]. Hypoiodic acid may participate in the formation of acetic acid via oxidative carbonylation of methane from thermodynamic considerations ($\Delta G = -5$ kcal/mol).

Therefore, the role of iodide co-catalyst is the generation of two-electron oxidants. Such a catalyst is effective but inconvenient because of the formation of molecular iodine. Moreover, I_2 may escape from the system via sublimation. There are some other compounds – Cu(I) [62, 63], Fe(II) [64], hydroquinones [7], etc. – that are capable of reacting in their reduced form with dioxygen to yield H_2O_2 or its equivalents. Co-catalysts must also exhibit a redox potential admitting the reduction of their oxidized form with external reducing agent (CO, H_2). We established that Cu(I, II) [146, 147] and Fe(II, III) [148] are effective as co-catalysts for oxidation of methane in $RhCl_3$-containing systems. Fe(II, III) salts should be added in larger concentrations compared to Cu(I, II) salts.

Copper salts were studied to a greater extent. With increasing $[Cl^-]$, the oxidation rate attains a maximum at $[Cl^-] \approx 0.020–0.025$ g-ion/l. In $RhCl_3–CuCl_2$, Cl^- ions are excessive. In view of this, copper salts were added in the form of Cu_2O, CuO or $Cu(OAc_f)_2$. Cu_2O and CuO are well soluble in $Ac_fOH–H_2O$ at reaction temperature. Upon consumption of O_2 (taken in shortage from safety considerations), a precipitate of Cu_2Cl_2 formed in the reaction mixture. An important advantage of $RhCl_3$–Cu(I, II)–NaCl system in comparison with $RhCl_3$–KI–NaCl one is its tolerance to water content. In experimental conditions, Ti(0) also dissolves, which can be regarded as evidence for formation of two-electron oxidants – such as H_2O_2, Cl–Cu–O–O–Cu–Cl or Cl–Cu–O–O–H – from dioxygen. The mechanisms of action for I- and Cu-containing catalytic systems are generally similar as confirmed by isotopic studies [149]. The Rh–Cu–Cl system is more active as compared to the Rh–I–Cl one (TOF around $150\ h^{-1}$ against $100\ h^{-1}$).

The k_H/k_D values for methane in formation of Ac_fOCH_3 were 4.34 ± 0.16 for the Rh–I–Cl system and 3.93 ± 0.18 for the Rh–Cu–Cl one. They correspond to the mechanism of outer-sphere activation of methane on oxygen-containing species like peroxo complexes of Rh [12]. Isotopic studies with $^{18}O_2$, $H_2{}^{18}O$, and $Ac_f{}^{18}O$ confirmed [149] the insertion of O atom from dioxygen, except for acetic acid. All sets of the results comply with Scheme 2.27 involving heteronuclear Rh–O–O–Cu peroxo intermediates.

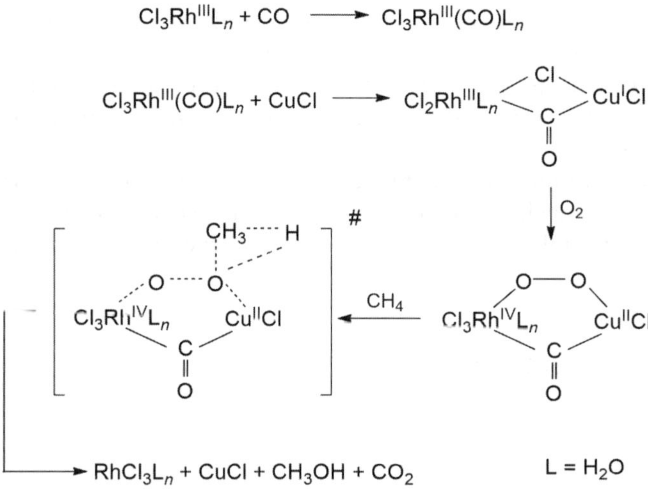

Scheme 2.27 Oxidation of methane via heteronuclear Rh–O–O–Cu complexes [34].

Carbonyl chloride complex of Rh(III) and CuCl forms a heteronuclear compound with chloride and carbonyl bridging groups, and this compound activates dioxygen. Methane is activated on one of O atoms in the peroxo complex. After some transformations, the activated complex (transition state) decomposes to yield methanol, CO_2, and this leads to regeneration of starting Rh and Cu.

The above mechanism is supported by the following observations: (i) dihydrogen is ineffective as external reducing agent; (ii) heteronuclear Cu–Ni complexes with a bridging peroxo group were reported [150]; (iii) the existence of Rh(IV) is permissible [151]; (iv) substitution of Cu(II) and CO for Cu(I) is inefficient; and (v) heteronuclear complexes Pd–Co, Pd–Ni, and Pd–Cu are known [152–154].

The synthesis of acetic acid in both systems (Rh–I–Cl and Rh–Cu–Cl) must nevertheless involve $Rh–CH_3$ intermediate. According to [139], the relative amount of Ac_fOCH_3 or AcOH in reaction products depends on reaction rate (Scheme 2.28).

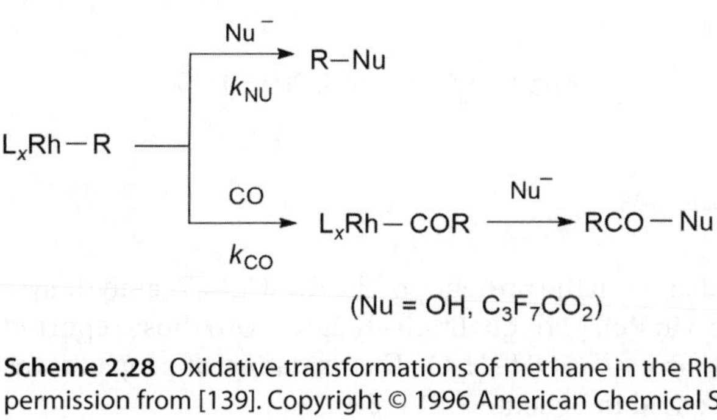

Scheme 2.28 Oxidative transformations of methane in the $RhCl_3$–KI–NaCl system. Source: Reprinted with permission from [139]. Copyright © 1996 American Chemical Society.

The mechanism suggested by Periana et al. [24] casts some serious doubts. This scheme is based largely on the results of DFT studies [155]. It assumes the electrophilic substitution in methane followed by reductive elimination of CH_3X (formation of methanol). $[Rh(CO)_2(CH_3)I]^-$ complex may undergo carbonylation to produce acyl intermediate whose hydrolysis leads to AcOH. Oxygen plays the role of oxidant for rhodium hydride. The destiny of newly formed highly reactive OOH^- (i.e. H_2O_2) remains unclear. Sen reasonably believed that neither I_2, nor $I^•$ can act as oxidants but nevertheless suggested some oxygen-containing species (most likely H_2O_2) as an oxidant. In the Periana scheme, I^- only plays the role of ligand, while the presence of free I^- is neglected, as well as the oxidation of I^- to I_2 observed in [142, 144]. In terms of the Periana mechanism (kind of Shilov chemistry), one should observe isotopic exchange between methane and medium. The occurrence of such an exchange has not been confirmed experimentally [144].

A key point of our strategy is the observation that a metal–alkyl intermediate in the systems under consideration is formed not only upon electrophilic substitution but also due to involvement of oxidant. In particular, in case of catalytic system Rh–Cl–I, we believe that OI^- acts as shown in Scheme 2.29 [34].

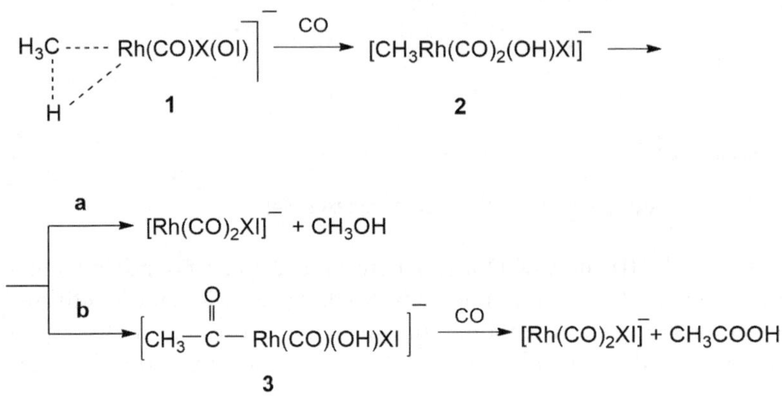

Scheme 2.29 Activation and functionalization of methane in the Rh–I–Cl system ($X = Cl^-, I^-$) [34].

Upon oxidation, methane σ-complex **1** transforms into methyl complex **2**. Upon further transformation, it may produce CH_3OH (route **a**). But complex **2** will sooner transform into acyl **3** with subsequent reductive elimination of AcOH. DFT studies [34] supported the feasibility of route **b**. Tentatively, the formation of CH_3–Rh(III) intermediate in the Rh–Cu–Cl system happens as shown in Scheme 2.30, where CuCl(OOH) fragment acts as an oxidant.

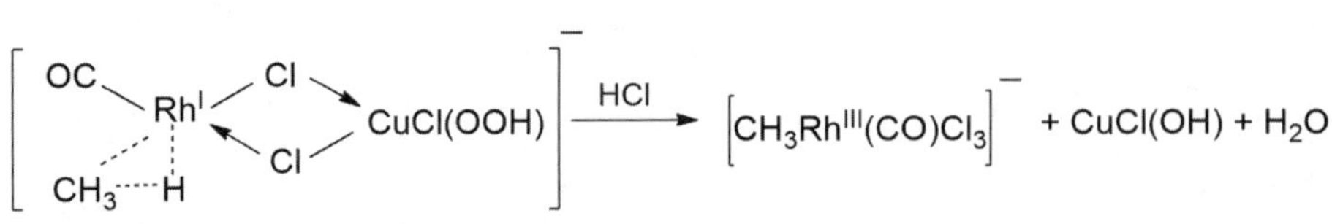

Scheme 2.30 Formation of CH_3–Rh(III) intermediate [34].

We also studied the oxidation of C_2–C_4 alkanes in the presence of the Rh–Cu–Cl catalytic system [130, 135, 156, 157]. The results (Table 3 in Ref. [34]) qualitatively agree with those reported by Sen et al. [139] for the Rh–I–Cl system in Pr_fCOOH–H_2O. The oxidation of propane was studied in more detail; the results are presented in Scheme 2.31 [150].

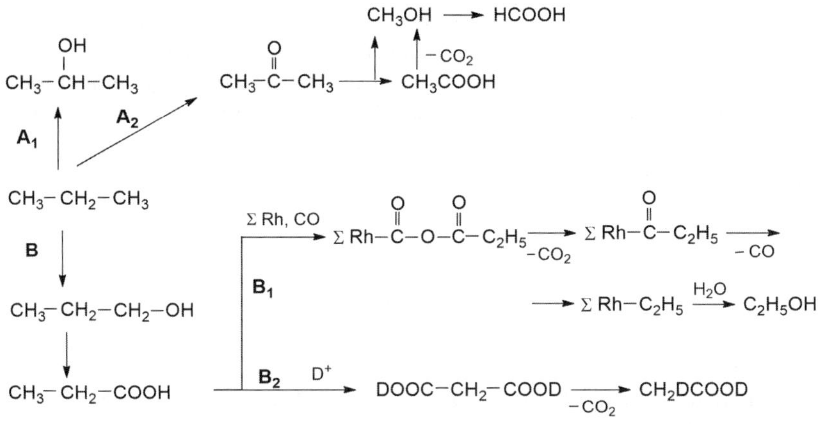

Scheme 2.31 Oxidation of propane in the Rh–Cu–Cl system [34]. Here Σ Rh stands for a catalytically active complex. Esters are in equilibrium with respective alcohols. For the sake of simplicity, oxidizers are not specified.

The process develops in two routes: reactions $\mathbf{A}_1$ yield *i*-propanol (ester), reaction $\mathbf{A}_2$ yields acetone while reactions **B**, *n*-propanol. Special experiments [130] showed that joint oxidation of C_3H_8 and $Ac_f{}^{18}OPr^i$ with $^{16}O_2$ gave CH_3–C(^{16}O)–CH_3 as a single product [130]. It means that, in competition between propane and *i*-propanol for formation of acetone, propane strongly prevails.

Scheme 2.31 is based on the results of our previous studies on oxidation of propane [130, 135, 150, 156]. So the oxidation of acetone yields Ac_fOCH_3 and AcOH, whereas the oxidation of AcOH results in its destruction to Ac_fOCH_3, and the oxidation of Ac_fOCH_3 gives HCOOH. The oxidation of propane in route **B** is less effective. The formed propionic acid reacts in routes $\mathbf{B}_1$ and $\mathbf{B}_2$ to give $Ac_fOC_2H_5$ and AcOH, respectively, via malonic acid as an intermediate (confirmed by experiments in deuterated solvent). Therefore, in conditions of alkanes oxidation in Ac_fOH–H_2O, acetic and propionic acids undergo oxidative destruction. This process was studied in detail [158–160] and the use of these acids as solvents for oxidative reactions was recommended, with precautions.

Therefore, the use of Rh compounds in catalytic systems looks promising, especially in the presence of thriftily consumed CO as a reducing agent. For justice's sake, we have to mention that, in the Monsanto process [143], 30–40% CO is spent for the side WGSR preventing the precipitation of RhI_3.

2.6 Final Comments

Over 50 years have passed since the pioneering work of Shilov's group [5], but no industrial-scale process for transformation of lower alkanes into alcohols or other oxygenates has been elaborated so far. To date, vast amounts of experimental and theoretical information on activation of C—H bond in alkanes, activation of dioxygen, and functionalization of alkanes have been accumulated in the literature. The most effective catalysts and catalytic systems are based on Pt group metals. However, the competition between homogeneously catalyzed synthesis and synthesis through syngas for production of methanol from methane (gross production 62 Mtons a year) still does not favor homogeneous catalysis. One of most effective catalysts, Pt(bpym)Cl_2 complex, is operative in rather harsh conditions and is not environmentally friendly. The rather effective rhodium–copper–chloride system requires unproductive expenditure of carbon monoxide.

It seems reasonable that further studies in the field be focused on: (i) the quest of effective catalysts among the complexes of Pt group metals, (ii) improving the catalytic activity, and (iii) lowering the consumption of precious reagents. More attention should be given to modeling the action of biocatalysts. Homogeneous metallocomplex catalysis has serious problems with recirculation of catalysts and large-volume solvents. From this standpoint, the design of hybrid homogeneous–heterogeneous catalysts operating in the gas-and-steam phase looks very promising.

References

1 Krylov, O.V. (2004). *Heterogeneous Catalysis*. Moscow: Akademkniga.
2 Chorkendorff, I. and Neimantsverdriet, J.W. (2007). *Concepts of Modern Catalysis and Kinetics*. Weinheim: Wiley–VCH.
3 Chatt, J. and Davidson, J.M. (1965). *J. Chem. Soc.* 843.
4 Halpern, J. (1968). *Discuss. Faraday Soc.* 46: 7.
5 Goldshleger, N.F., Tyabin, M.B., Shilov, A.E. et al. (1969). *Zh. Fiz. Khim.* 43: 2174.
6 Shilov, A.E. (1984). *Activation of Saturated Hydrocarbons by Transition Metal Complexes*. Dordrecht: D. Reidel.
7 Strukul, G. (1992). *Catalytic Oxidation with Hydrogen Peroxide as Oxidant*. Dordrecht: Kluwer.
8 Simandi, L.I. (ed.) (2003). *Advances in Catalytic Activation of Dioxygen by Metal Complexes*. New York, Boston, Dordrecht, London, Moscow: Kluwer Academic Publishers.
9 Hill, C.L. (ed.) (1989). *Activation and Functionalization of Alkanes*. Wiley–VCH.
10 Shilov, A.E. and Shul'pin, G.B. (1995). *Aktivatsiya i Kataliticheskie Reaktsii Uglevodorodov (Activation and Catalytic Reactions of Hydrocarbons)*. Moscow: Nauka.
11 Shilov, A.E. (1997). *Metal Complexes in Biomimetic Chemical Reactions*. New York: CRC Press.
12 Rudakov, E.S. (1985). *Reaktsii Alkanov s Okislitelyami, Metallokompleksami i Radikalami v Rastvorakh (Reactions of Alkanes with Oxidants, Metal Complexes and Radicals in Solution)*. Kiev: Naukova Dumka.
13 Sheldon, R.A. (1994). *Metalloporphyrins in Catalytic Oxidation*. New York: Marcel Dekker.
14 Shilov, A.E. and Shul'pin, G.B. (2000). *Activation and Catalytic Reactions of Saturated Hydrocarbons in the Presence of Metal Complexes*. New York, Boston, Dordrecht, Moscow: Kluwer Academic Publishers.
15 Henrici-Olive, G. and Olive, S. (1980). *Coordination and Catalysis*. Weinheim, New York: Verlag Chemie.
16 Sen, A. (1998). *Acc. Chem. Res.* 31: 550.
17 Jia, C.G., Kitamura, T., and Fujiwara, Y. (2001). *Acc. Chem. Res.* 34: 633.
18 Crabtree, R.H. (2001). *J. Chem. Soc. Dalton Trans.* 2437.
19 Crabtree, R.H. (2004). *J. Organomet. Chem.* 689: 24.
20 Neumann, R. and Khenkin, A.M. (2006). *Chem. Commun.* 2529.
21 Labinger, J.A. and Bercaw, J.E. (2011). *Top. Organomet. Chem.* 35: 29.
22 Sorokin, A.B. (2013). *Chem. Rev.* 113: 8152.
23 Hashiguchi, B.G., Bischof, S.M., Konnick, M.M. et al. (2012). *Acc. Chem. Res.* 45: 885.
24 Gunsalus, N.J., Konnick, M.M., Hashiguchi, B.G. et al. (2014). *Isr. J. Chem.* 54: 1467.
25 Stahl, S.S. (2004). *Angew. Chem. Int. Ed.* 43: 3400.
26 Shteinman, A.A. (2016). *Russ. Chem. Bull. (Engl. Transl.)* 65: 1930.
27 Shul'pin, G.B. (2016). *Catalysts* 50: 2.

28 Armentrout, B.P. (2017). *Chem. Eur. J.* 23: 10.
29 Munz, D. and Strassner, T. (2015). *Inorg. Chem.* 54: 5043.
30 Shilov, A.E. and Shteinman, A.A. (2012). *Russ. Chem. Rev. (Engl. Transl.)* 81: 291.
31 Shteinman, A.A. (2015). *J. Organomet. Chem.* 793: 34.
32 Chepaikin, E.G. (2011). *Russ. Chem. Rev. (Engl. Transl.)* 80: 363.
33 Chepaikin, E.G. (2014). *J. Mol. Catal. A: Chem.* 385: 160.
34 Chepaikin, E.G. and Borshch, V.N. (2015). *J. Organomet. Chem.* 793: 78.
35 Mitra, M., Nimir, H., Hrovat, D.A. et al. (2017). *J. Mol. Catal. A: Chem.* 426: 350.
36 Bataineh, H., Pestovsky, O., and Bakac, A. (2016). *Inorg. Chem.* 55: 6719.
37 Hasanayn, F., Al-Assi, L.M., Moussawi, R.N. et al. (2016). *Inorg. Chem.* 55: 7886.
38 Verma, P., Vogiatzis, K.D., Planas, N. et al. (2015). *J. Am. Chem. Soc.* 137: 5770.
39 Cowan, A.J. and George, M.W. (2008). *Coord. Chem. Rew.* 252: 2504.
40 Lawes, D.J., Darwish, T.A., Clark, T. et al. (2006). *Angew. Chem. Int. Ed.* 45: 4486.
41 Young, R.D., Lawes, D.J., Hill, A.F. et al. (2012). *J. Am. Chem. Soc.* 134: 8294.
42 Bernskoetter, W.H., Schauer, C.K., Goldberg, K.I. et al. (2009). *Science* 326: 553.
43 Campos, J., Kundu, S., Pahls, D.R. et al. (2013). *J. Am. Chem. Soc.* 135: 1217.
44 Walter, M.D., White, P.S., Schauer, C.K. et al. (2011). *New J. Chem.* 35: 2884.
45 Chadwick, F.M., Rees, N.H., Weller, A.S. et al. (2016). *Angew. Chem. Int. Ed.* 55: 3677.
46 Janowicz, A.H. and Bergman, R.G. (1982). *J. Am. Chem. Soc.* 104: 352.
47 Bromberg, S.E., Yang, H., Asplund, M.C. et al. (1997). *Science* 278: 260.
48 Hoyano, J.K. and Graham, W.A.G. (1982). *J. Am. Chem. Soc.* 104: 3723.
49 Comanescu, C.C. and Iluc, V.M. (2015). *Organometallics* 34: 4684.
50 Ito, J., Kaneda, T., and Nishiyama, H. (2012). *Organometallics* 31: 4442.
51 Hutchings, G.J. (2016). *Top. Catal.* 59: 658.
52 Wilke, G., Schott, H., and Heimbach, P. (1967). *Angew. Chem. Int. Ed.* 6: 92.
53 Graham, B.W., Laing, K.R., O'Connor, C.J. et al. (1972). *J. Chem. Soc., Dalton Trans.* 1237.
54 Vigalok, A., Shimon, L.J.W., and Milstein, D. (1996). *J. Chem. Soc., Chem. Commun.* 1673.
55 Bruin, B., Peters, T.P.J., Wilting, J.B.M. et al. (2002). *Eur. J. Inorg. Chem.* 2671.
56 Vaska, L. (1963). *Science* 140: 809.
57 Zudin, V.N., Likholobov, V.A., and Ermakov, Y.I. (1979). *Kinet. Katal.* 20: 1599.
58 Bianchi, D., Bortolo, R., D'Aloisio, R. et al. (1999). *Angew. Chem. Int. Ed.* 38: 706.
59 Bianchi, D., Bortolo, R., D'Aloisio, R. et al. (1999). *J. Mol. Catal. A: Chem.* 150: 87.
60 Lari, G.M., Puértolas, B., Shahrokhi, M. et al. (2017). *Angew. Chem. Int. Ed.* 56: 1775.
61 Addison, A.W. and Gillard, R.D. (1970). *J. Chem. Soc. A* 2523.
62 Matsumoto, T., Ohkubo, K., Honda, K. et al. (2009). *J. Am. Chem. Soc.* 131: 9258.
63 Zuberbuhler, A. (1967). *Helv. Chim. Acta* 50: 466.
64 Gray, R.D. (1969). *J. Am. Chem. Soc.* 91: 59.
65 Henry, P.M. (1966). *Inorg. Chem.* 5: 688.
66 Huffman, R.E. and Davidson, N. (1956). *J. Am. Chem. Soc* 78: 4836.
67 Horak, K.T. and Agapie, T. (2016). *J. Am. Chem. Soc.* 138: 3443.
68 Kim, K., Matsumoto, T., Robertson, A. et al. (2012). *Chem. Asian J.* 7: 1394.
69 Meier, G. and Braun, T. (2012). *Angew. Chem. Int. Ed.* 51: 12564.
70 Boisvert, L. and Goldberg, K.I. (2012). *Acc. Chem. Res.* 45: 899.
71 Jacobson, S.E., Mares, F., and Zambri, P.M. (1979). *J. Am. Chem. Soc.* 101: 6938.
72 Anijado, M., Braun, T., Noveski, D. et al. (2005). *Angew. Chem. Int. Ed.* 44: 6947.
73 Denney, M.C., Smythe, N.A., Kara Cetto, L. et al. (2006). *J. Am. Chem. Soc.* 128: 2508.
74 Konnick, M.M. and Stahl, S.S. (2008). *J. Am. Chem. Soc.* 130: 5753.
75 Keith, J.M. and Goddard, W.A. III, (2009). *J. Am. Chem. Soc.* 131: 1416.
76 Popp, B.V. and Stahl, S.S. (2009). *Chem. Eur. J.* 15: 2915.

77 Shin, B., Sutherlin, K.D., Ohta, T. et al. (2016). *Inorg. Chem.* 55: 12391.
78 Chatterjee, D., Mitra, A., and van Eldic, R. (2007). *Dalton Trans.* 943.
79 Limberg, C. (2009). *Angew. Chem. Int. Ed.* 48: 2270.
80 Ishizuka, T., Kotani, H., and Kojima, T. (2016). *Dalton Trans.* 45: 16727.
81 Bales, B.C., Brown, P., Dehestani, A. et al. (2005). *J. Am. Chem. Soc.* 127: 2832.
82 Rohde, J.-U., In, J.-H., Lim, M.H. et al. (2003). *Science* 299: 1037.
83 Poverenov, E., Efremenko, I., Frenkel, A.I. et al. (2008). *Nature (London)* 455: 1093.
84 Sen, A., Benevenuto, M.A., Lin, M. et al. (1994). *J. Am. Chem. Soc.* 116: 998.
85 Lin, M., Shen, C., Garcia-Zayas, E.A. et al. (2001). *J. Am. Chem. Soc.* 123: 1000.
86 US Pat. 2 493 038 (Chem. Abstr. 1950, 44, 20151).
87 Rudakov, E.S. (1987). *Zh. Fiz. Khim.* 61: 289.
88 R.A. Periana, D.J. Taube, E.R. Evit,"Process for converting lower alkanes to esters", US Grant US5233113A, 1993. 1991.
89 R.A. Periana, D.J. Taube, E.R. Evit, Catalytic process for converting hydrocarbonaceous feeds, particularly lower alkanes to esters, alcohols, and optionally to hydrocarbons", WO Application WO1992014738A1, 1992.
90 R.A. Periana, D.J. Taube, H. Taube, et al. "Ligated platinum group metal catalyst complex and improved process for catalytically converting alkanes to esters and derivatives thereof", WO Application, WO1998050333A1, 1998.
91 Periana, R.A., Taube, D.J., Evit, E.R. et al. (1993). *Science* 259: 340.
92 Periana, R.A., Taube, D.J., Gamble, S. et al. (1998). *Science* 280: 560.
93 Periana, R.A., Bhalla, G., Tenn, W.J. III, et al. (2004). *J. Mol. Catal. A: Chem.* 220: 7.
94 Conley, B.L., Tenn, W.J. III, Young, K.J.H. et al. (2006). *J. Mol. Catal. A: Chem.* 251: 8.
95 Paul, A. and Musgrave, C.B. (2007). *Organometallics* 26: 793.
96 Cheng, J., Li, Z., Haught, M. et al. (2006). *Chem. Commun.* 4617.
97 Zimmermann, T., Soorholtz, M., Bilke, M. et al. (2016). *J. Am. Chem. Soc.* 138: 12395.
98 Palkovits, R., Antonietti, M., Kuhn, P. et al. (2009). *Angew. Chem. Int. Ed.* 48: 6909.
99 Soorholtz, M., White, R.J., Zimmermann, T. et al. (2013). *Chem. Commun.* 49: 240.
100 Michalkiewicz, B. (2006). *Appl. Catal. A: General* 307: 270.
101 Michalkiewicz, B. and Kosowski, P. (2007). *Catal. Commun.* 8: 1939.
102 Villalobas, J.M., Hickman, A.J., and Sanford, M.S. (2010). *Organometallics* 29: 257.
103 Joung, K.J.H., Meier, S.K., Gonzales, J.M. et al. (2006). *Organometallics* 25: 4734.
104 Bar-Nahum, I., Khenkin, A.M., and Neumann, R. (2004). *J. Am. Chem. Soc.* 126: 10236.
105 Butschke, B. and Schwarz, H. (2011). *Organometallics* 30: 1588.
106 Liu, W.-G., Sberegaeva, A.V., Nielsen, R.J. et al. (2014). *J. Am. Chem. Soc.* 136: 2335.
107 Ziatdinov, V.R., Oxgaard, J., Mironov, O.A. et al. (2006). *J. Am. Chem. Soc.* 128: 7404.
108 Zerella, M., Kahros, A., and Bell, A.T. (2006). *J. Catal.* 237: 111.
109 Fujiwara, Y., Taniguchi, Y., and Takaki, T. (1996). *SYNLETT* 591.
110 Gretz, E., Oliver, T.F., and Sen, A. (1987). *J. Am. Chem. Soc.* 109: 8109.
111 Kao, L.-H., Hutson, A.C., and Sen, A. (1991). *J. Am. Chem. Soc.* 113: 700.
112 Muehlhofer, M., Strassner, T., and Herrmann, W.A. (2002). *Angew. Chem. Int. Ed.* 41: 1745.
113 Ingrosso, G. and Midolini, N. (2003). *J. Mol. Catal. A: Chem.* 204-205: 425.
114 Strassner, T., Muehlhofer, M., Zeller, A. et al. (2004). *J. Organomet. Chem.* 689: 1418.
115 Ahrens, S., Zeller, A., Taige, M. et al. (2006). *Organometallics* 25: 5409.
116 Meyer, D., Taige, M.A., Zeller, A. et al. (2009). *Organometallics* 28: 2142.
117 Munz, D. and Strassner, T. (2014). *Top. Catal.* 57: 1372.
118 Munz, D., Poethig, A., Tronnier, A. et al. (2013). *Dalton Trans.* 42: 7297.
119 Munz, D. and Strassner, T. (2014). *Angew. Chem. Int. Ed.* 53: 2485.

120 Munz, D. and Strassner, T. (2014). *Chem. Eur. J.* 20: 14872.
121 George, S.D., Sherman, S.C., Iretski, A.V. et al. (2000). *Catal. Lett.* 65: 181.
122 Jiang, H., Chen, H., Wang, A. et al. (2010). *Chem. Commun.* 46: 7259.
123 An, Z., Pan, X., Liu, X. et al. (2006). *J. Am. Chem. Soc.* 128: 16028.
124 Vargaftic, M.N., Stolyarov, I.P., and Moiseev, I.I. (1990). *J. Chem. Soc. Chem. Commun.* 1049.
125 Gol'dshleger, N.F., Kresova, E.N., and Moravskii, A.P. (1991). *Kinet. Catal. (Engl. Transl.)* 32: 923.
126 Yuan, Y., Wang, L., and Wang, Y. (2011). *Ind. Ing. Chem. Res.* 50: 6513.
127 Fan, Y., An, Z., Pan, X. et al. (2009). *Chem. Commun.* 7488.
128 Yuan, J., Liu, L., Wang, L. et al. (2013). *Catal. Lett.* 143: 126.
129 Yuan, J., Wang, Y., and Hao, C. (2013). *Catal. Lett.* 143: 610.
130 Chepaikin, E.G., Bezruchenko, A.P., Menchikova, G.N. et al. (2014). *Petrol. Chem. (Engl. Transl.)* 54: 374.
131 Lin, M. and Sen, A. (1992). *J. Am. Chem. Soc.* 114: 7307.
132 Sen, A. and Lin, M. US Pat. 5 393 922, 1995.
133 Lin, M., Hogan, T., and Sen, A. (1997). *J. Am. Chem. Soc.* 119: 6048.
134 Renias, J.E. and Sen, A. (2002). *J. Mol. Catal. A: Chem.* 189: 33.
135 Chepaikin, E.G., Bezruchenko, A.P., Menchikova, G.N. et al. (2010). *Petrol. Chem. (Engl. Transl.)* 50: 332.
136 Lin, Y., Zhu, L., Lan, Y. et al. (2015). *Chem. Eur. J.* 21: 14937.
137 Nomura, K. and Uemura, S. (1994). *J. Chem. Soc. Chem. Commun.* 129.
138 Lin, M. and Sen, A. (1994). *Nature* 368: 613.
139 Lin, M., Hogan, T., and Sen, A. (1996). *J. Am. Chem. Soc.* 118: 4574.
140 Chepaikin, E.G., Boiko, G.N., Bezruchenko, A.P. et al. (1997). *Dokl. Phys. Chem. (Engl. Transl.)* 353: 120.
141 Chepaikin, E.G., Boiko, G.N., Bezruchenko, A.P. et al. (1998). *J. Mol. Catal. A: Chem.* 129: 15.
142 Bezruchenko, A.P., Boiko, G.N., Grigoryan, E.H. et al. (1998). *Dokl. Phys. Chem. (Engl. Transl.)* 363: 396.
143 Maitlis, P.M., Haynes, A., Sunley, G.J. et al. (1996). *J. Chem. Soc. Dalton Trans.* 2187.
144 Chepaikin, E.G., Bezruchenko, A.P., Leshcheva, A.A. et al. (2001). *J. Mol. Catal A Chem.* 169: 89.
145 Luchinskii, G.P. (1971). *Khimia Titana (Chemistry of Titanium).* Moscow: Khimia.
146 Chepaikin, E.G., Bezruchenko, A.P., Leshcheva, A.A. et al. (2000). *Dokl. Phys. Chem. (Engl. Transl.)* 373: 105.
147 Chepaikin, E.G., Bezruchenko, A.P., and Leshcheva, A.A. (2002). *Kinet. Catal. (Engl. Transl.)* 43: 507.
148 Chepaikin, E.G., Bezruchenko, A.P., Boiko, G.N. et al. (2003). *Petrol. Chem. (Engl. Transl.)* 43: 395.
149 Chepaikin, E.G., Bezruchenko, A.P., Boiko, G.N. et al. (2006). *Kinet. Catal. (Engl. Transl.)* 47: 12.
150 Kundu, S., Pfaff, F.F., Miceli, E. et al. (2013). *Angew. Chem. Int. Ed.* 52: 1.
151 Bakac, A. (2006). *Dalton Trans.* 1589.
152 Brandon, R.W. and Claridge, D.V. (1968). *Chem. Commun.* 677.
153 Kozitsina, N.Y., Nefedov, S.E., Cherkashina, N.V. et al. (2005). *Russ. Chem. Bul. Int. Ed.* 54: 2215.
154 Hosokawa, T., Nomura, T., and Murahashi, S.-I. (1998). *J. Organomet. Chem.* 551: 387.
155 Hristov, I.H. and Ziegler, T. (2003). *Organometallics* 22: 3513.

156 Chepaikin, E.G., Bezruchenko, A.P., Menchikova, G.N. et al. (2010). *Kinet. Catal. (Engl.Transl.)* 51: 666.

157 Chepaikin, E.G., Bezruchenko, A.P., Menchikova, G.N. et al. (2017). *J. Mol. Catal., A: Chem.* 426: 389.

158 Chepaikin, E.G., Bezruchenko, A.P., Menchikova, G.N. et al. (2011). *Dokl. Phys. Chem. (Engl. Transl.)* 437 (part 1): 41.

159 Chepaikin, E.G., Bezruchenko, A.P., Menchikova, G.N. et al. (2011). *Russ. Chem. Bull. Int. Ed.* 60: 49.

160 Chepaikin, E.G., Bezruchenko, A.P., Menchikova, G.N. et al. (2011). *Petrol. Chem. (Eng. Transl.)* 51: 133.

3

Alkane-Oxidizing Systems Based on Metal Complexes. Radical Versus Nonradical Mechanisms[1]

Georgiy B. Shul'pin [1,2]

[1] *Semenov Institute of Chemical Physics, Russian Academy of Sciences, Moscow, Russia*
[2] *Chair of Chemistry and Physics, Plekhanov Russian University of Economics, Moscow, Russia*

3.1 Introduction

Functionalization of organic compounds containing C—H bonds is replacing these C—H bonds with any other functional groups. In a broader sense, the C—H fragment can be replaced by the ≥C—H fragment and halogen atom can be entered instead of hydrogen substituent [1]. This chapter deals with main principles of oxidation of saturated C—H compounds with inorganic and organic peroxides as well as molecular oxygen. Reactions catalyzed by metal complexes that have been reported in recent years are described here.

Organic and inorganic peroxides and especially hydrogen peroxide, as well as molecular oxygen, O_2, are widely employed for the oxidation of various organic compounds. Alkanes are among these organic substrates, although saturated hydrocarbons exhibit very high resistance in reactions in solutions under mild conditions. In recent decades, many metal complexes have been described in the literature as catalysts in alkane oxygenation with dioxygen or peroxides [2]. Simple metal salts are usually poor catalysts, but their activity becomes much higher if certain co-catalysts are added to the reaction solution [3]. The works by the author and his groups will be described here in more detail.

3.2 Main Features of Oxidation Processes

The main pathways of alkane or alcohol functionalization (oxygenation) via C—H bond activation are shown in Figure 3.1.

3.2.1 *Radical* Reactions, But Not Radical-*Chain* Reactions

In many cases, the reaction of alkanes (as well as alcohols containing C—H bonds) occur with intermediate formation of free O- and C-centered radicals. The latter easily react with molecular oxygen if the reaction is carried out in air. Finally, less stable (alkyl hydroperoxides) or stable (ketones) derivatives are produced as shown in Figure 3.2. The role of a metal complex in such transformations consists in generating active radicals that attack C—H bonds as well as the participation in various redox processes leading finally to the formation of stable products.

1 This work was presented in part at the 41st International Conference on Coordination Chemistry, Singapore, 2014, Lecture D4S-08(I), and 16th International Congress on Catalysis, Beijing, China, 2016, Lecture OF24.

Alkane Functionalization, First Edition. Edited by Armando J. L. Pombeiro and M. Fátima C. Guedes da Silva.

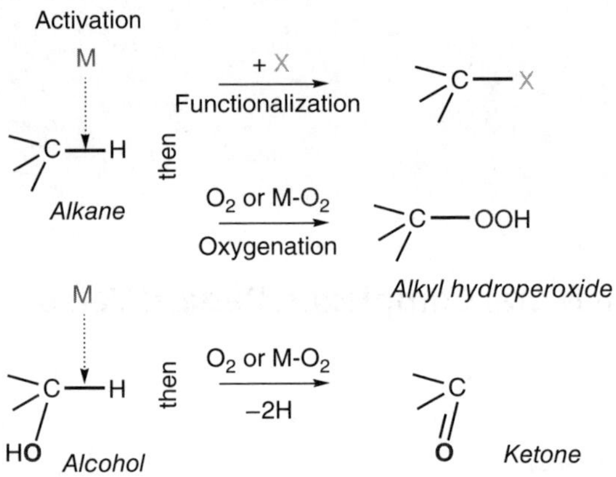

Figure 3.1 Main routes of alkane or alcohol functionalization (oxygenation) via C—H bond activation and further splitting with the participation of metal derivative M.

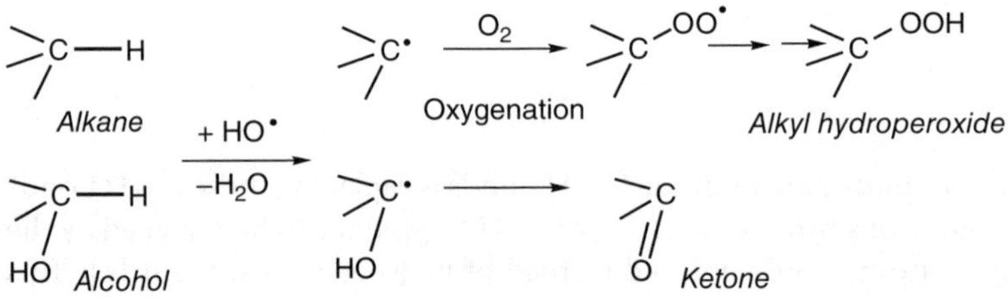

Figure 3.2 Radical and nonradical mechanisms of alkane or alcohol functionalization.

Autoxidation [1a, b, 4] of saturated hydrocarbons in the liquid phase is usually a branched chain process with a so-called "degenerate" chain branching. This means that branching of each chain happens much later than its termination and is caused by the formation of a rather stable intermediate. Usually, this intermediate is nevertheless chemically more active than the initial hydrocarbon and can form free radicals at a greater rate than that of the chain initiation process. Autoxidation of C—H compounds is of great importance for living organisms. Hydroperoxides are the intermediates in liquid phase oxidation. The following classical scheme presents a typical mechanism of liquid-phase hydrocarbon oxidation [5].

Chain initiation:

$$RH + O_2 \xrightarrow{k_A} R^\bullet + HOO^\bullet$$

Chain propagation:

$$R^\bullet + O_2 \xrightarrow{k_B} ROO^\bullet$$

$$ROO^\bullet + RH \xrightarrow{k_C} ROOH + R^\bullet$$

Chain branching:

$$ROOH \xrightarrow{k_D} RO^\bullet + HO^\bullet$$

or

$$2\,ROOH \xrightarrow{k_D} RO^\bullet + ROO^\bullet + H_2O$$

Chain termination:

$$R^\bullet + R^\bullet \longrightarrow R\text{–}R$$

$$ROO^{\bullet} + R^{\bullet} \longrightarrow ROOR$$

$$ROO^{\bullet} + ROO^{\bullet} \xrightarrow{k_G} ROH + R'COR'' + O_2$$

It should be noted that radicals $RO^{\bullet}$ and $HO^{\bullet}$ can take part in the fast following steps:

$$HO^{\bullet} + RH \longrightarrow H_2O + R^{\bullet}$$

$$RO^{\bullet} + RH \longrightarrow ROH + R^{\bullet}$$

Relatively unstable chemical initiators can be used to initiate the reaction generating alkyl radicals, $R^{\bullet}$. For example, in the case of azobis(isobutyronitrile) (AIBN, In—N=N—In), the chain initiation step

$$In\text{–}N = N\text{–}In \longrightarrow 2\,In^{\bullet} + N_2$$

$$In^{\bullet} + RH \longrightarrow InH + R^{\bullet}$$

is more efficient in comparison with stage

$$RH + O_2 \xrightarrow{k_A} R^{\bullet} + HOO^{\bullet}.$$

An alkane oxidation can be also initiated by any other free radicals, $X^{\bullet}$, which are capable of abstracting the hydrogen atom from an alkane:

$$X^{\bullet} + RH \longrightarrow XH + R^{\bullet}$$

Very reactive hydroxyl and alkoxyl radicals are among potential initiators of alkane oxidations. Molecular oxygen from air could be the oxidant while radicals-initiators can be produced by the decomposition of H_2O_2 or *tert*-BuOOH under the action of light or (much more efficiently) of a metal complex catalyst:

$$H_2O_2 \longrightarrow HO^{\bullet} + \ldots$$

$$\text{tert-BuOOH} \longrightarrow \textit{tert}\text{-BuO}^{\bullet} + \ldots$$

It should be emphasized that the radical-chain mechanism can operate only in the reactions at relatively high temperatures and/or with substrates containing weak C—H bonds (e.g. tetralin, decalin, cumene). At low temperatures, substrates containing strong C—H bonds (for example, methane, ethane) do not enter into the main route typical for radical-*chain* reactions:

$$RH + ROO^{\bullet} \longrightarrow R^{\bullet} + ROOH$$

Indeed, if the 10% conversion time at 100 °C for tetralin and cumene is only 0.2 hour, the same parameter for cyclohexane and methane is 6.5 and 30 days, respectively. The time necessary for the methane conversion for 10% at 30 °C with involvement solely of a radical-chain mechanism has been estimated to be 22 000 years (Table 3.1 [5a]). The parameters used in this table are as follows: k_p is the rate constant of the interaction between peroxyl radical and RH, k_t is the rate constant of the chain termination, the chain length of the process, ν equals the ratio of the ROOH formation rate and the chain initiation rate W_i:

$$W_{ROOH} = \frac{k_p}{\sqrt{2k_t}}\sqrt{W_i}\,[RH]$$

$$\nu = \frac{W_{ROOH}}{W_i}$$

Table 3.1 Highest possible rates of the hydrocarbon radical-chain low-temperature oxidation (chain length $\nu = 1$ and $W^0_{ROOH} = W_i$) and the minimum possible times $\tau_{0.1}{}^0$ for the transformation of these hydrocarbons to the extent of 10%.[a)]

		Rate $W_{ROOH}{}^0$ (M s^{-1})		Time $\tau_{0.1}{}^0$	
No	**Hydrocarbon**	**30 °C**	**100 °C**	**30 °C**	**100 °C**
1	Tetralin	9.1×10^{-6}	1.0×10^{-4}	1.7 hr	0.2 hr
2	Cyclohexene	6.3×10^{-6}	5.1×10^{-5}	2.5 hr	0.3 hr
3	Ethylbenzene	8.3×10^{-8}	7.3×10^{-6}	7.5 d	2.1 hr
4	Toluene	2.3×10^{-10}	1.0×10^{-7}	7.5 yr	6.5 d
5	Cumene	2.0×10^{-6}	1.2×10^{-4}	7.5 hr	0.2 hr
6	Cyclohexane	5.5×10^{-11}	1.0×10^{-7}	32.5 yr	6.5 d
7	Ethane	3.2×10^{-12}	1.0×10^{-7}	550 yr	6.5 d
8	Methane	7.9×10^{-14}	2.1×10^{-8}	22 000 yr	30 d

a) The parameters have been calculated in assumption of the hydrocarbon concentration of 1.0 M (which is often lower in reality). Note that at $W_{ROOH} > W^0_{ROOH}$ the chain length $\nu < 1$, which corresponds to a non-chain process, and at $W_{ROOH} < W^0_{ROOH}$ parameter $\nu > 1$ and $\tau_{0.1} > \tau_{0.1}{}^0$.

The condition $W_{ROOH} = W_i$ evaluates the rate of RH oxidation, W^0_{ROOH}, via a radical-chain mechanism with $\nu = 1$:

$$W^0_{ROOH} = \left(\frac{k_p}{\sqrt{2k_t}}\right)^2 [RH]^2$$

We can make the following important conclusion: at $W_{ROOH} < W^0_{ROOH}$ a radical-chain oxidation process is possible ($\nu > 1$), and at $W_{ROOH} > W^0_{ROOH}$ only the radical oxidation of RH with $\nu < 1$ is possible.

Figure 3.3 demonstrates the difference between a reaction that occurs via the classical radical-chain mechanism and a process that involves free radicals (hydroxyl, alkyl, and alkoxyl). It is important to emphasize that in a laboratory, under mild conditions, alkane oxidation does not include the propagation step, and we can say about such a reaction that it is free-radical process but by no means is a radical-chain one. In contrast, alkenes can be easily functionalized via their allylic hydrogens. Thus, an iron(III) complex with a macrocyclic ligand "is an efficient catalyst (or initiator?) for the allylic oxidation of cyclohexene with O_2 to produce cyclohex-2-enone predominantly" [5b].

3.2.2 Hydrogen Peroxide Decomposition to Generate Hydroxyl Radicals

One of the first and one of the most "famous" systems oxidizing with the participation of hydroxyl radicals is the combination of H_2O_2 with the iron salt. In the case of iron(II) salt, this system is called "the Fenton reagent" [6a–c]. Copper ions [6d] as well as ions of numerous transition metals in combination with hydrogen peroxide give Fenton or Fenton-like oxidizing systems.

In the course of the hydrogen peroxide oxidation by Fe^{3+} ion according to Eq. (A), the stoichiometric Fenton system, "Fe(II)–H_2O_2," is transformed gradually into the catalytic system "Fe(III)–Fe(II)–H_2O_2" ("Fenton-like" or "Fenton-type" system)

$$H_2O_2 + Fe^{3+} \rightarrow HOO^\bullet + H^+ + Fe^{2+} \qquad (A)$$

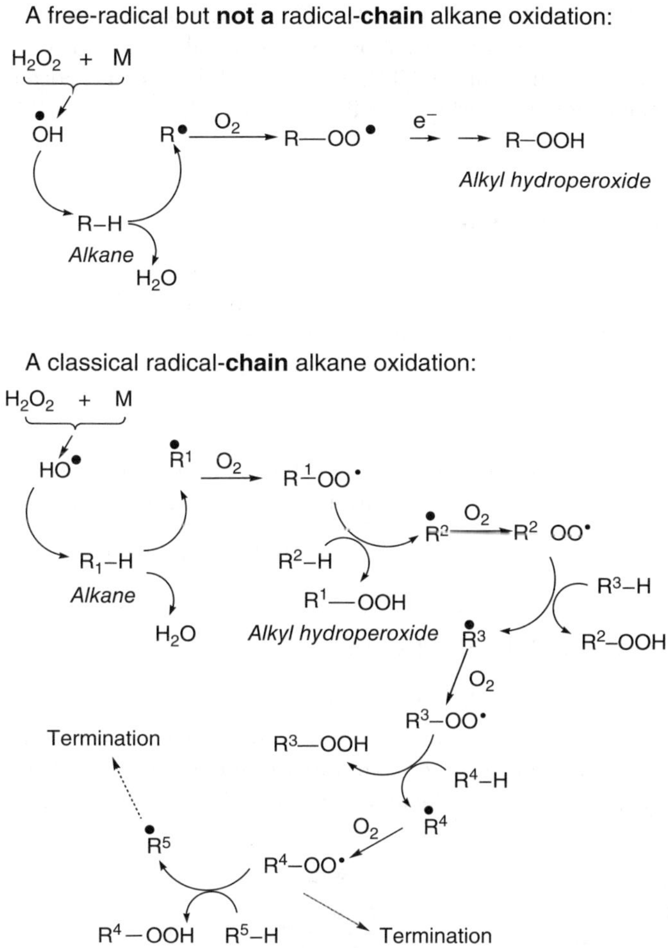

Figure 3.3 Two types of catalyzed alkane radical oxygenation with H_2O_2: radical but not radical-chain (which does not include chains) and radical-chain reaction (which involves chains).

with the effective rate constant $k_A = 2 \times 10^{-4}\ M^{-1}s^{-1}$ at 25 °C and $[H^+] = 1.0 \times 10^{-2}$ M. Fenton reaction proceeds in accord with the equation when hydrogen peroxide is reduced by Fe^{2+} ion:

$$H_2O_2 + Fe^{2+} \rightarrow Fe^{3+} + HO^{\bullet} + HO^{-} \qquad (B)$$

Earlier we discovered and studied [7] a new reagent based on a vanadium derivative in combination with pyrazine-2-carboxylic acid (PCA), which efficiently oxidizes alkanes in acetonitrile solution at low (20–70 °C) temperatures. Two crucial stages A and B (see above) in catalytic oxidation with the $H_2O_2 + Fe^{II} + Fe^{III}$ are analogous to the steps A′ and B′ proposed for the oxidation with the "$O_2 + H_2O_2 + n\text{-}Bu_4NVO_3 + PCA$" reagent:

$$V^{V} + H_2O_2 \rightarrow V^{IV} + HOO^{\bullet} + H^{+} \qquad (A')$$

$$V^{IV} + H_2O_2 \rightarrow V^{V} + HO^{\bullet} + HO^{-} \qquad (B')$$

Here, ions V^V and V^{IV} correspond, respectively, to Fe^{III} and Fe^{II}. It was proposed that the important stage of the reaction is proton transfer from coordinated H_2O_2 molecule to the oxo (=O) or peroxo (O—O) or hydroxy (OH) ligand at vanadium ion. One route is a pca-assisted mechanism ("robot's arm mechanism; pca is the anion of PCA [7b, c, 8a]) and the second pathway is a water-assisted mechanism [7e]. Both routes shown schematically in Figures 3.4 and 3.5 were confirmed with density functional theory (DFT) calculations. It has been reported recently that the reaction between the peroxide $VO(O_2)$(pyridine-2-carboxylate) and Fe(II) is not a Fenton-like reaction [8b].

3.2.3 Possible Mechanisms of H_2O_2 Decomposition

In an early work, Kozlov and Purmal [9] experimentally studied decomposition of hydrogen peroxide using the inhibitor tetranitromethane, $C(NO_2)_4$. It was concluded that at $[H_2O_2]_0 < 1$ M this process proceeds via radical-chain mechanism. The initiation rate W_i is

$$W_i = n\frac{[C(NO_2)_4]_0}{\tau} = n\frac{d[\overset{\ominus}{C}(NO_2)_3]}{dt}$$

Here τ is the kinetic induction period, n is the number of chains terminated by one tetranitromethane molecule. The initiation rate W_i at low concentration of H_2O_2 is expressed as

$$W_i = W_{i1} + W_{i2} = \varkappa_1[Fe^{3+}]_0[H_2O_2]/([H^+] + K_I) + \varkappa_2[Fe^{3+}]_0^2[H_2O_2]/([H^+] + K_I)^2$$

and at low $[Fe^{3+}]$ and high $[H_2O_2]$ as

$$W_i = W_{i1} + W_{i3} = \varkappa_1[Fe^{3+}]_0[H_2O_2]/([H^+] + K_{II}[H_2O_2]) + \varkappa_3[Fe^{3+}]_0[H_2O_2]^2/([H^+] + K_{II}[H_2O_2])$$

Here W_{i1}, W_{i2}, and W_{i3} are initiation rates of three initiation pathways, respectively; K_I and K_{II} are shown below in Eq. (i_1) equilibrium constants of the formation of hydroxy and peroxy iron complexes $\varkappa_1$, $\varkappa_2$, and $\varkappa_3$ are coefficients that are the combinations of

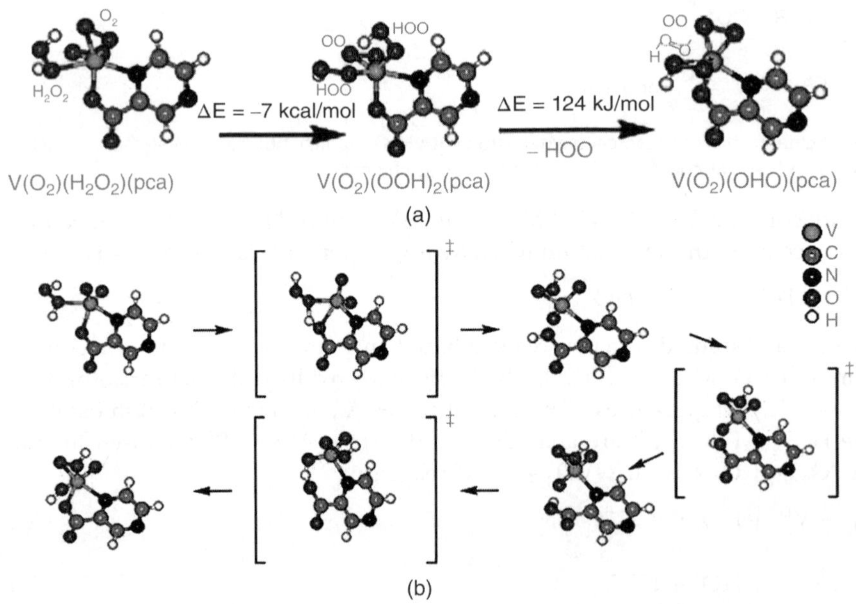

Figure 3.4 (a) Proton transfer from coordinated H_2O_2 to (O_2)-ligand followed by elimination of HOO• radical); (b) "Robot's arm mechanism" of the proton transfer. Source: Reproduced by permission of The Royal Society of Chemistry [7b] and American Chemical Society [8a]. (*See insert for color representation.*)

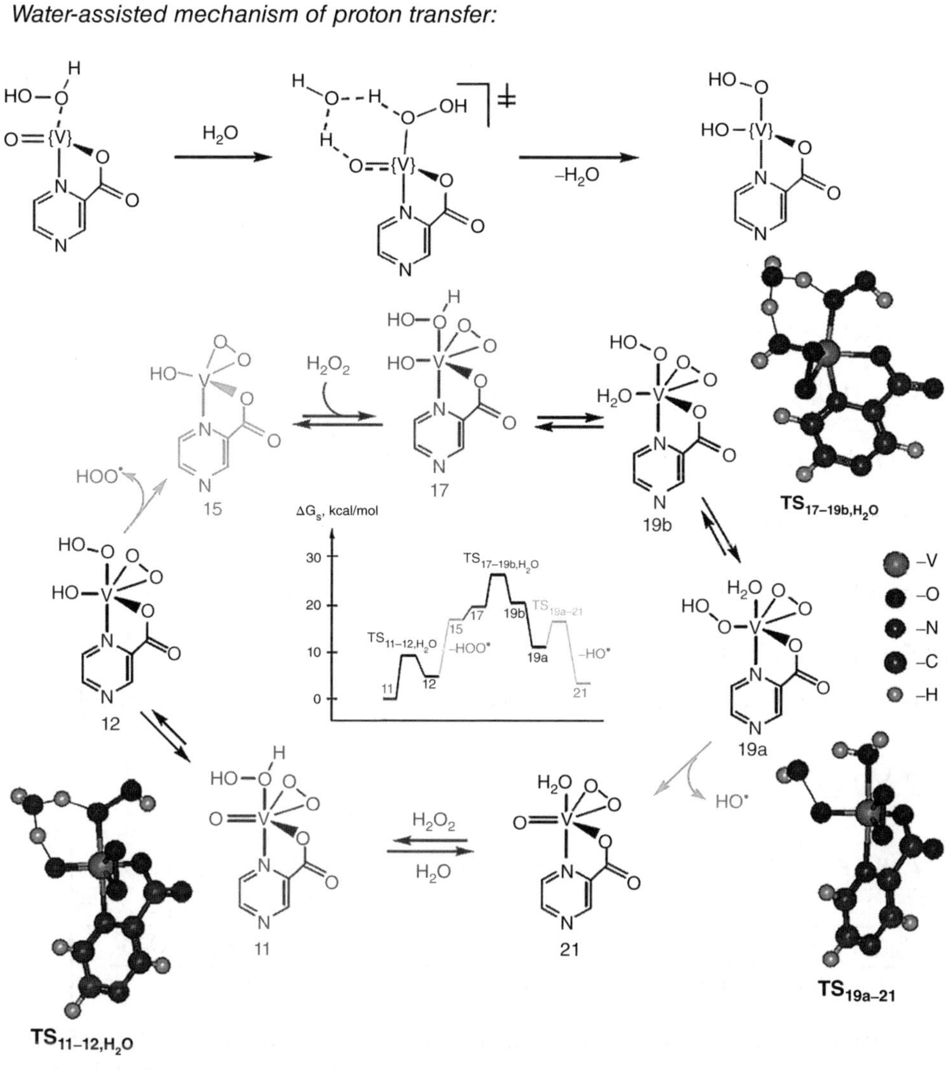

Figure 3.5 Generation of radicals by the "$O_2 + H_2O_2 + n\text{-}Bu_4NVO_3 + PCA$" reagent. Water-assisted mechanism of proton transfer from coordinated H_2O_2 to O-atom of the COOH group and then to (O_2)-ligand followed by elimination of HOO• radical. Enumeration of species is as in the original paper. Color code: red: formation of H_2O_2-adducts, blue: H-transfers, green: radical generation. Source: Adapted from Ref. [7e] with permission of Elsevier. (*See insert for color representation.*)

corresponding constants of the initiation rate reactions and equilibrium constants (the product of corresponding initiation rate constants and corresponding equilibrium constants). The last kinetic expression indicates the existence of one more initiation reaction involving one Fe^{3+} and two hydrogen peroxide molecules.

The initiation mechanism can be presented as a consequence of the following reactions:

$$Fe^{3+} + H_2O \xrightleftharpoons{K_I} FeOH^{2+} + H^+$$

$$Fe^{3+} + H_2O_2 \xrightleftharpoons{K_{II}} Fe(HOO)^{2+} + H^+$$

$$Fe(HOO^{\bullet})^{2+} \rightarrow Fe^{2+} + HOO^{\bullet}; Fe(OH)^{2+} \rightleftarrows Fe^{3+} + HO^{-} \quad (i_1)$$

$$Fe(HOO^{\bullet})^{2+} + FeOH^{2+} \rightarrow 2Fe^{2+} + H_2O + O_2 \quad (i_2)$$

$$Fe(HOO^{\bullet})^{2+} + H_2O_2 \ (H_2O_2)Fe(HOO^{\bullet})^{2+} \rightarrow Fe^{2+} + H_2O + O_2 + HO^{\bullet} \quad (i_3)$$

The kinetic scheme of the whole process of H_2O_2 decomposition should be supplemented by known terms of the chain propagation and termination:

$$Fe^{2+} + H_2O_2 \rightarrow Fe^{3+} + HO^{\bullet} + HO^{-} \quad \text{propagation (Fenton reaction)} \ k_1 = 68\ M^{-1}\,s^{-1}$$

$$HO^{\bullet} + H_2O_2 \rightarrow H_2O + HOO^{\bullet} \quad \text{propagation} \ k_2 = 1.7 \times 10^{7}\ M^{-1}\,s^{-1}$$

$$HOO^{\bullet} \rightarrow H^{+} + O_2^{-\bullet} \quad K_{III} = 1.3 \times 10^{-5}\ M$$

$$Fe^{3+} + O_2^{-\bullet} \rightarrow Fe^{2+} + O_2 \quad \text{propagation} \ k_3 = 4 \times 10^{8}\ M^{-1}\,s^{-1}$$

$$HOO^{\bullet} + Fe^{2+} \rightarrow HOO^{-} + Fe^{3+} \quad \text{termination} \ k_4 = 2.1 \times 10^{6}\ M^{-1}\,s^{-1}$$

$$Fe^{2+} + HO^{\bullet} \rightarrow Fe^{3+} + HO^{-} \quad \text{termination} \ k_5 = 5.1 \times 10^{8}\ M^{-1}\,s^{-1}$$

$$HOO^{\bullet} + HOO^{\bullet} \rightarrow H_2O_2 + O_2 \quad \text{termination} \ k_6 = 2.4 \times 10^{6}\ M^{-1}\,s^{-1}$$

The stationary rate of hydrogen peroxide decomposition can be expressed by the following equation:

$$-\frac{d\,[H_2O_2]}{dt} = 2\left\{\frac{\dfrac{k_{i1}K_{II}[Fe^{3+}]_0[H_2O_2]}{[H^+]+K_I+K_{II}[H_2O_2]} + \dfrac{k_{i2}K_IK_{II}[Fe^{3+}]_0^{\,2}[H_2O_2]}{([H^+]+K_I+K_{II}[H_2O_2])^2} + \dfrac{k_{i3}K_{II}[Fe^{3+}]_0^{\,2}[H_2O_2]^2}{[H^+]+K_I+K_{II}[H_2O_2]}}{\dfrac{k_4([H^+]+K_I+K_{II}[H_2O_2])}{K_{III}k_1k_3[Fe^{3+}]_0[H_2O_2]} + \dfrac{k_5}{k_1k_2[H_2O_2]^2} + \dfrac{k_6([H^+]+K_I+K_{II}[H_2O_2])^2}{K^2_{III}k_3^{\,2}[Fe^{3+}]_0^{\,2}}}\right\}^{1/2}$$

The authors using new ideas on the initiation mechanism and well-known reactions of propagation and termination were able to quantitatively describe numerous experimental data on the rates of catalytic H_2O_2 decomposition with iron perchlorate.

Under conditions when $[Fe^{3+}]_0/[H^+] \ll 1$ and $[Fe^{3+}]_0/[H_2O_2] \ll 1$ the main initiation reaction is (i_1) and the termination occurs via the reaction with k_4. In this case, the reaction rate is

$$-\frac{d\,[H_2O_2]}{dt} = 2\,\frac{[Fe^{3+}]_0[H_2O_2]}{([H^+]+K_I)}\sqrt{\frac{k_{i1}K_{II}k_1k_3K_{III}}{k_4}}$$

When the $[Fe^{3+}]_0/[H^+]$ ratio is relatively high and $[Fe^{3+}]_0/[H_2O_2] \geq 1$ the main initiation reaction is i_2, and the termination occurs via reaction above with k_5. In this case, we have

$$-\frac{d\,[H_2O_2]}{dt} \sim \frac{[Fe^{3+}]_0[H_2O_2]^{3/2}}{[H^+]+K_I}$$

Alkane oxidation with hydroperoxides (especially with *m*-chloroperoxybenzoic acid (MCPBA)) can proceed via concerted mechanism though with the formation of radical-like species. Xia et al. [10] found that compound Cp_2VCl_2 (**3.1**; Figure 3.6) catalyzes benzylic

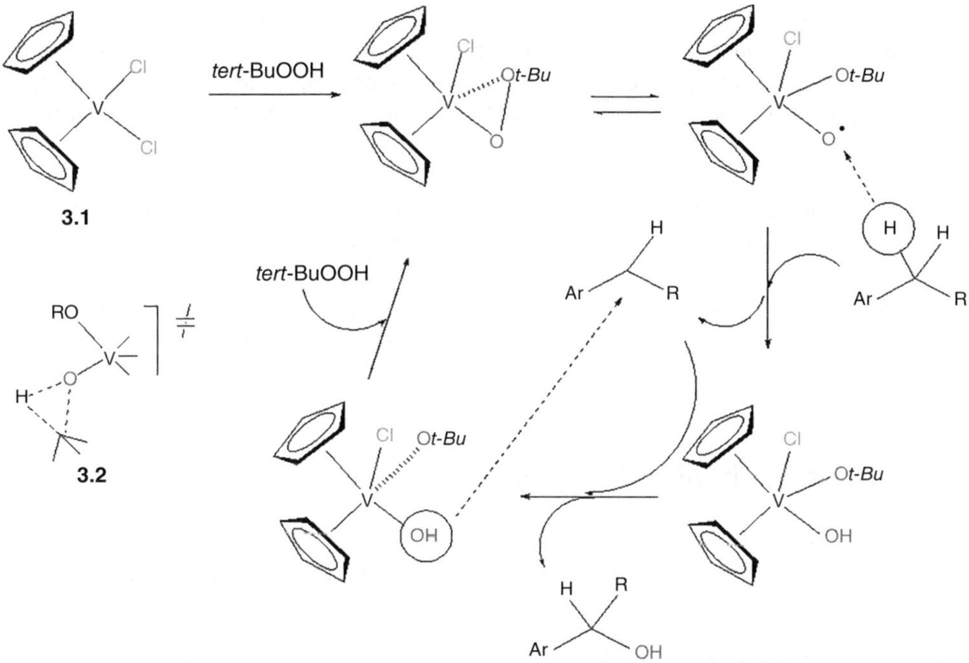

Figure 3.6 A catalytic cycle proposed for the benzyl hydroxylation catalyzed by vanadium complex **3.1**. Source: Adapted from Ref. [10a] (open access). This is a corrected Figure 1.11 reproduced from Ref. [10b].

C—H oxidation selectively and effectively, giving no aromatic oxidation products. The authors assumed that unlike the vanadate–PCA–H_2O_2 reagent [7b, e, 8a], the oxidation by the system based on Cp_2VCl_2 and *tert*-butyl hydroperoxide (TBHP) is not mediated by *t*BuO$^•$ or *t*BuOO$^•$, the Haber–Weiss decomposition products of TBHP. Six arguments have been reported to support this assumption. The rebound of V—O species and the benzylic radical is relatively fast and possibly occurs in the solvent cage. This C—H abstraction/rebound may proceed in a nearly concerted asynchronous manner with a transition state **3.2**, as shown in Figure 3.6.

3.2.4 Activated Adducts (Transition States) Containing Simultaneously Two H_2O_2 Molecules

For the stages i_2 and i_3 the activated adducts **A** and **B**, respectively, have been proposed [9]. The main remarkable feature of intermediate **B** is the simultaneous coordination of two peroxide fragments in a coordination sphere of iron. The first fragment is a hydroperoxy ligand and the second species is a whole H_2O_2 molecule ligated to the iron ion.

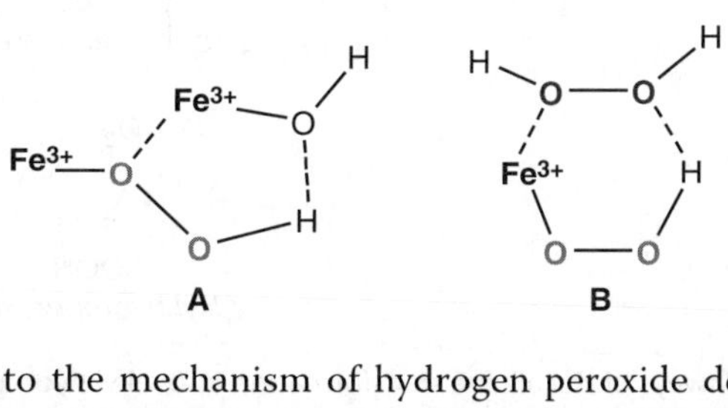

This idea was applied to the mechanism of hydrogen peroxide decomposition on an aluminum(III) ion [11a]. It was stated above that in a Fenton reaction with Fe(II) or a Fenton-like reaction (for example with Fe^{III}), a transition metal can change its oxidation state ($Fe^{II} \rightleftarrows Fe^{III}$) in the course of the hydrogen peroxide decomposition. Aluminum cation is unable to change its valency. To explain the catalytic action of the aluminum cation, we assumed [11b] that, in the case of Al^{III}, not one but simultaneously two H_2O_2 molecules interact in the cation's coordination sphere. The first H_2O_2 molecule – as a model of the electron-rich Fe^{II} ion – delivers one electron (like in the Fenton mechanism) to the system, and this leads to the splitting of the O—O bond in the second H_2O_2 molecule. The concept of the simultaneous participation of two hydrogen peroxide molecules in the hydroxyl radical generation was used in calculations of reactions catalyzed by other nontransition metals (Al, La, Y, In, Sc, Ga, Be, Zn, Cd, Bi) [11c–g]. Further development of this approach was made recently by Meyerstein and co-workers who studied kinetic and DFT methods generation of hydroxyl radical on ions of molybdenum [12a], cobalt [12b, c], iron, and cobalt [12d].

3.3 Methods of Mechanism Investigation

In the previous section, a kinetic method including a radical trap was described that allows us to propose stages of a catalytic cycle. Some other methods will be considered in this section.

3.3.1 Formation of Alkyl Hydroperoxides

The oxidations promoted by non-heme amine-based iron complexes commonly proceed via a metal-based mechanism (Figure 3.7). Olivo, Lanzalunga, and Di Stefano [13] noted that "a high level of variability and/or uncertainty between a free-radical and a metal-based mechanism is found for non-heme imine-based iron-catalyzed oxidations."

Di Stefano et al. [13] divided imine-based catalysts into two classes: (i) salen-Fe complexes and (ii) imine-Fe complexes. In some cases, clues for free-radical oxidation mechanisms have been reported, while in other cases evidence for metal-based mechanisms has been collected. The

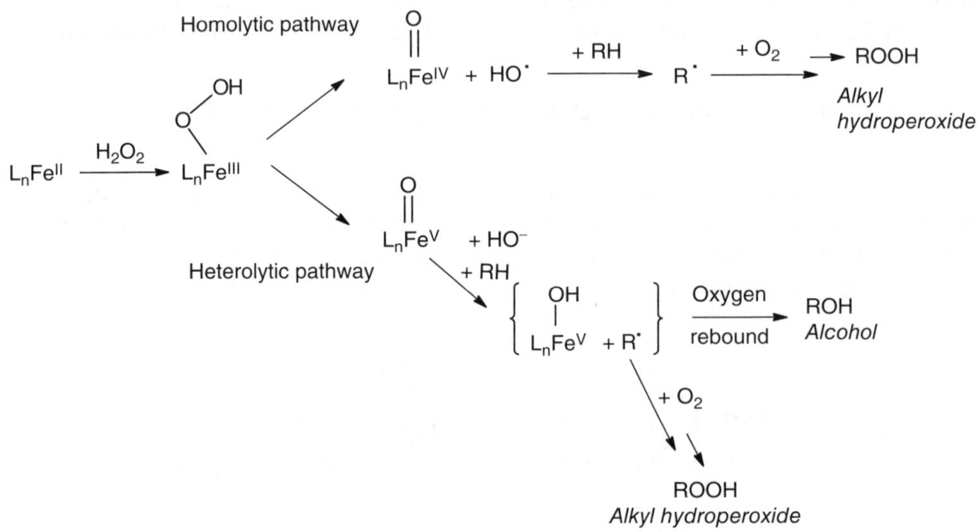

Figure 3.7 Two oxidation pathways of the alkane transformation into alkyl hydroperoxide or/and alcohol. Source: adapted from Ref. [13] with permission of Wiley.

preferred mechanistic pathway is shown to be a function of catalyst structures and features. Talsi and Bryliakov [14] and Costas and Norlander [15] recently carried out comprehensive studies focused on the evidence for nonradical mechanisms in alkane oxidations with H_2O_2 catalyzed by iron and manganese complexes.

Hydroxyl, alkyl radicals, and alkyl peroxyl radicals are unstable intermediates. Di Stefano et al. [13] described 10 tools to distinguish between metal-based and free-radical oxidation mechanisms: (i) regioselectivity, (ii) alcohol/ketone ratio (A/K), (iii) kinetic isotope effects (KIEs), (iv) reactions under argon/air, (v) Shul'pin test for alkyl hydroperoxides, (vi) epimerization, (vii) chirality transfer, (viii) labeling studies, (ix) use of radical traps, and (x) cyclohexene oxidation. Let us discuss here some of these methods.

3.3.2 Regioselectivity

The nature of the alkane oxidizing species (radical or nonradical) can be evaluated if we measure the selectivity parameters in oxidations of certain alkanes. Table 3.2 demonstrates these parameters for the systems generating hydroxyl radicals (entries 1–5). They are noticeably lower than parameters determined for the systems oxidizing without the participation of reactive free radicals (entries 6–9).

3.3.3 Competitive Oxidation of an Alkane and Solvent

A kinetic study of competitive simultaneous oxidation of an alkane and solvent can give a valuable information for understanding the nature of reaction under investigation. Let us consider an example. Stirring a solution of cyclohexane, CyH, in acetonitrile at 30 °C with 50% aqueous

Table 3.2 Selectivity parameters in H_2O_2 oxidations of alkanes by certain catalytic systems.[a)]

		C(1):C(2):C(3):C(4)	1°:2°:3°		*trans/cis*	
Entry	**Catalytic system**	***n*-Heptane**	**MCH**	**Isooctane**	***cis*-1,2 DMCH**	***trans*-1,2-DMCH**
1	hν	1:10:7:6	1:2:6	0.9	1.0	
2	$FeSO_4$	1:5:5:4.5	1:3:6	1.3	1.2	
3	PCA/VO_3^-	1:6:7:5	1:9:37	1:4:9	0.75	0.8
4	$Al(NO_3)_3$	1:5:5:5	1:6:23	1:4:7	0.8	0.8
5	Ti-MMM-2	1:9:7:6.5	1:6:113	1:4:30	0.9	0.9
6	$NaAuCl_4$	1:35:25:23	1:116:255	1:10:240		
7	$[Mn_2L_2(O)_3]^{2+}/MeCO_2H$	1:46:35:34	1:26:200	1:5:50	0.34	4.1
8	$[Mn_2L_2(O)_3]^{2+}/(COOH)_2$	1:14:13:12	1:0.3:06	1:0.4:32	0.2	7.3
9	$FeCl_3/L^3/m$-CPBA	1:29:30:27	1:21:211	0.25	3.0	
10	TS-1	1:80:193:100	n/p	n/p	n/p	n/p

a) Parameter C(1):C(2):C(3):C(4) is relative normalized (calculated taking into account the number of hydrogen atoms at each carbon) reactivities of hydrogen atoms at carbons 1, 2, 3, and 4 of the chain of *n*-heptane. Parameter 1°:2°:3° is relative normalized reactivities of hydrogen atoms at primary, secondary, and tertiary carbons of methylcyclohexane and isooctane. Parameter *trans/cis* is the ratio of isomers of *tert*-alcohols with mutual *trans*- and *cis*-orientation of the methyl groups formed in the oxidation of *cis*- and *trans*-1,2-dimethylcyclohexane. All parameters were measured after reduction of the reaction mixtures with triphenylphosphine before gas chromatography (GC) analysis and calculated based on the ratios of isomeric alcohols. hν means UV irradiation. n/p – no products detected. *L* is 1,4,7-trimethyl-1,4,7-triazacyclononane. L^3 is tetradentate amine *N,N'*-bis(2-pyridylmethylene)-1,4-diaminodiphenyl ether. *m*-CPBA is metachloroperoxybenzoic acid.

hydrogen peroxide in the presence of compound **3.3** and nitric acid leads to oxygenation of the alkane. Dependence of the oxidation rate W_0 on initial concentration of cyclohexane catalyzed by complex **3.3** is shown in Figure 3.8.

An assumption about the participation of hydroxyl radicals in the reaction under discussion is supported by the results shown in Figure 3.8: the curve of dependence of W_0 on the initial concentration of cyclohexane is approaching a plateau at $[CyH]_0 > 0.4$ M. The analysis [16a] of this mode of dependence of initial reaction rate $W_0 = W_{ROOH}$ on initial concentration of cyclohexane (approaching a plateau at high concentration of cyclohexane) indicates a competitive interaction of an oxidizing species with the alkane and acetonitrile. Assuming that this species is hydroxyl radical we can propose the following kinetic scheme for $HO^\bullet$-induced oxidation of the alkane, RH:

$$\mathbf{3.3} + H_2O_2 \rightarrow HO^\bullet \qquad \text{(i)}$$

$$HO^\bullet + CH_3CN \rightarrow \text{products} \qquad (1)$$

$$HO^\bullet + RH \rightarrow \text{products} \qquad (2)$$

where (i) is the reaction of $HO^\bullet$ generation in the H_2O_2 decomposition with the effective rate constant k_i and with rate W_i. Reactions (1) and (2) are the sequences of transformations under study. Their limiting stages are characterized by constants k_1 and k_2, respectively. The analysis of the proposed kinetic scheme in quasi-stationary approximation leads to the following equation

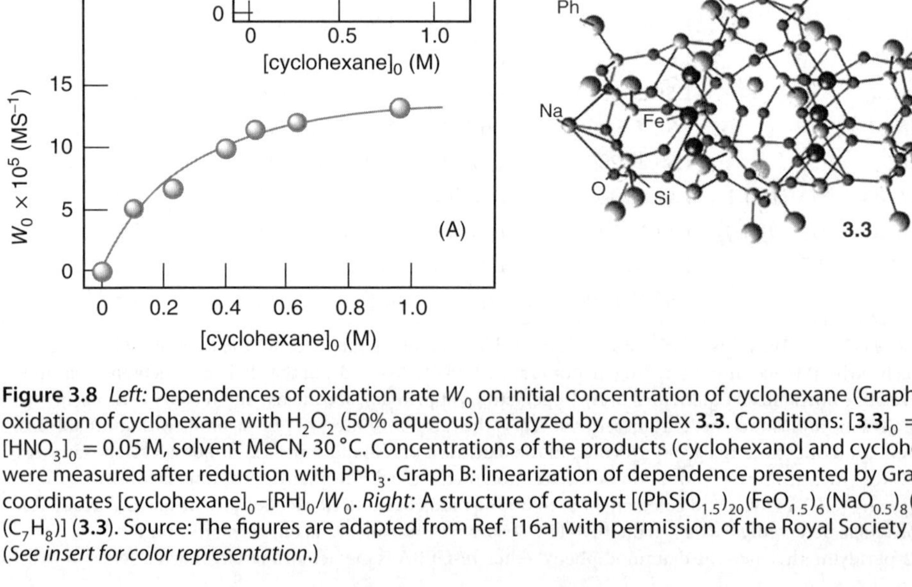

Figure 3.8 *Left:* Dependences of oxidation rate W_0 on initial concentration of cyclohexane (Graph A) in the oxidation of cyclohexane with H_2O_2 (50% aqueous) catalyzed by complex **3.3**. Conditions: $[\mathbf{3.3}]_0 = 2 \times 10^{-4}$ M, $[HNO_3]_0 = 0.05$ M, solvent MeCN, 30 °C. Concentrations of the products (cyclohexanol and cyclohexanone) were measured after reduction with PPh_3. Graph B: linearization of dependence presented by Graph A using coordinates $[\text{cyclohexane}]_0$–$[RH]_0/W_0$. *Right:* A structure of catalyst $[(PhSiO_{1.5})_{20}(FeO_{1.5})_6(NaO_{0.5})_8(n\text{-}BuOH)_{9.6}(C_7H_8)]$ (**3.3**). Source: The figures are adapted from Ref. [16a] with permission of the Royal Society of Chemistry. (*See insert for color representation.*)

for the rate of alkyl hydroperoxide, ROOH, formation:

$$W_{\mathrm{ROOH}} = \frac{d[\mathrm{ROOH}]}{dt} = \frac{k_2[\mathrm{RH}]W_{\mathrm{i}}}{k_1[\mathrm{CH_3CN}] + k_2[\mathrm{RH}]} \quad (3)$$

This expression can be easily transformed into Eq. (4):

$$\frac{[\mathrm{RH}]}{W_0} = \frac{1}{W_{\mathrm{i}}}\left\{\frac{k_1[\mathrm{CH_3CN}]}{k_2} + [\mathrm{RH}]\right\} \quad (4)$$

In accord with Eq. (4), the dependence of $[\mathrm{RH}]/W_0$ on [RH] should be linear, and the slope of this line should be $1/W_{\mathrm{i}}$, whereas the intercept with the Y-axis should be $(1/W_{\mathrm{i}})(k_1[\mathrm{CH_3CN}]/k_2)$. Treating the experimental dependence shown in Figure 3.8, Graph A using Eq. (4) gives a linear dependence of $[\mathrm{RH}]/W_0$ on [RH] (Figure 3.8, Graph B). This indicates that the experimental data are in agreement with Eq. (4). The following parameters have been calculated: $(1/W_{\mathrm{i}}) = 7\times10^3\ \mathrm{M^{-1}\,s}$ and $(1/W_{\mathrm{i}})(k_1[\mathrm{CH_3CN}]/k_2) = 10^3\ \mathrm{s}$. The ratio of the intercept to the slope of the line presented in Figure 3.8, Graph B is equal to $(k_1[\mathrm{CH_3CN}]/k_2) = 0.14\ \mathrm{M}$. This value is in agreement with an assumption that k_1 and k_2 are rate constants of the interaction between hydroxyl radicals with acetonitrile and cyclohexane, respectively (Table 3.3). Thus, we may conclude that the kinetic data support our proposal on the participation of hydroxyl radicals in the alkane oxidation.

It is important to emphasize that approaching the plateau in the dependence of W_0 on $[\mathrm{alkane}]_0$ (like demonstrated in Figure 3.8) has not been noticed in the cyclohexane oxidation with MCPBA catalyzed by $\mathrm{FeCl_3{\cdot}6H_2O}$ in the presence of tetradentate amine *N,N′-bis*(2-pyridylmethylene)-1,4-diaminodiphenyl ether (compound **3.4**, Figure 3.9). It can be concluded that hydroxyl radicals are not generated in this system.

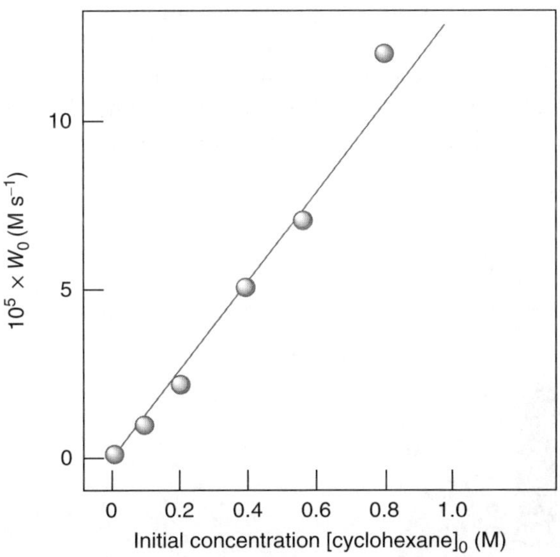

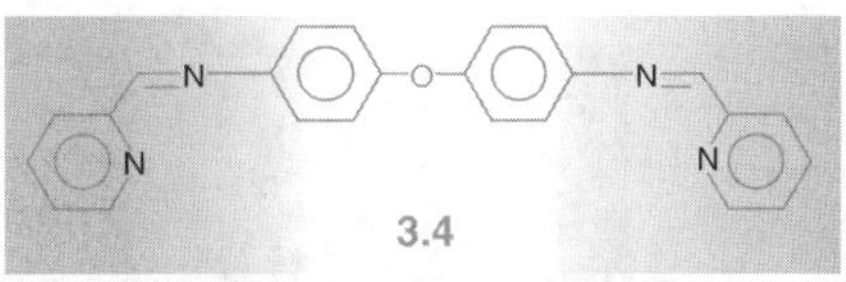

3.4

Figure 3.9 Dependence of initial oxidation rate W_0 on the concentration of cyclohexane in its oxidation by MCPBA (0.5 M) catalyzed by $\mathrm{FeCl_3{\cdot}6H_2O}$ (2×10^{-4} M) and tetradentate amine *N,N′-bis*(2-pyridylmethylene)-1,4-diaminodiphenyl ether (compound **3.4**; 16×10^{-4} M) in MeCN at 25 °C. Source: Adapted from Ref. [16c] with permission from Elsevier.

Table 3.3 Kinetic parameters for the competitive oxidation of cyclohexane and acetonitrile with various systems based on hydrogen peroxide.[a)]

Entry	System	Ratio $k_1[CH_3CN]/k_2$ (M)	Ratio k_1/k_2
1	$H_2O_2/O_2/(n\text{-}Bu_4N)VO_3$/PCA	0.14	0.008
2	H_2O_2/O_2/"Cu_4"/CF_3COOH	0.20	0.012
3	H_2O_2/O_2/"Cu_4"/HCl	0.10	0.006
4	$H_2O_2/O_2/[Co_4Fe_2OSae_8]/HNO_3$	0.14	0.008
5	$H_2O_2/O_2/Cp^*{}_2Os$/py	0.09÷0.19	0.0055÷0.011
6	H_2O_2/O_2/ Cp_2Fe/Py/PCA	0.19	0.011
7	H_2O_2/O_2/"Fe_2(TACN)"/PCA	0.19	0.011
8	H_2O_2/O_2/"Mn_2"/carboxylic acid	4	0.24

a) Concentration $[CH_3CN]_0$ was assumed to be 17 M. Abbreviations: PCA is pyrazine-2-carboxylic acid. "Cu_4" is tetracopper(II) triethanolaminate complex $[O\subset Cu_4\{N(CH_2CH_2O)_3\}_4(BOH)_4][BF_4]_2$. Complex $[Co_4Fe_2OSae_8]\cdot 4DMF\cdot H_2O$, where H_2Sae = salicylidene-2-ethanolamine. $Cp^*{}_2Os$ is decamethylosmocene. Cp_2Fe is ferrocene. "Fe_2(TACN)" is an iron(III) complex with 1,4,7-triazacyclononane. "Mn_2" is the dinuclear manganese(IV) complex $[LMn(O)_3MnL](PF_6)_2$ where L is 1,4,7-trimethyl-1,4,7-triazacyclononane, TMTACN (see Figure 3.19 and Ref. [16b]).

3.3.4 Shul'pin Test for Alkyl Hydroperoxides

It is reasonable to discuss here in more details the test for alkyl hydroperoxides. Moderately stable alkyl hydroperoxides, ROOH, are usually the main primary products of alkane oxidation with peroxides and especially molecular oxygen. Typically, concentration of fairly stable ROOH passes through a maximum in the time reaction profile. Determination of an alkyl hydroperoxide and measuring its concentration at different time periods allows us to deeply understand the reaction mechanism. Nevertheless, many chemists [17] prefer to inject the reaction solution directly into the chromatograph where alkyl hydroperoxides completely decompose.

Earlier we proposed and developed [18] a simple method based on the GC analysis of the reaction solution and involving PPh_3 (or thiourea). Alkyl hydroperoxides formed in various oxidations of alkanes are decomposed in the chromatograph to afford corresponding alcohols and ketones (aldehydes) (Figure 3.10). The alcohol/ketone (aldehyde) ratio is usually around

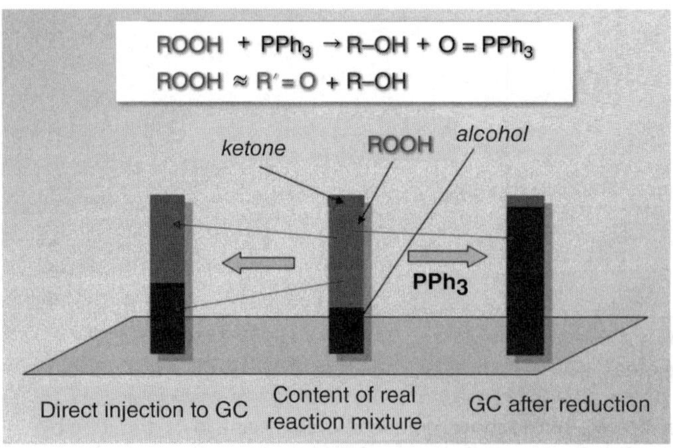

Figure 3.10 Determination of alkyl hydroperoxide, alcohol, and ketone using the GC. (*See insert for color representation.*)

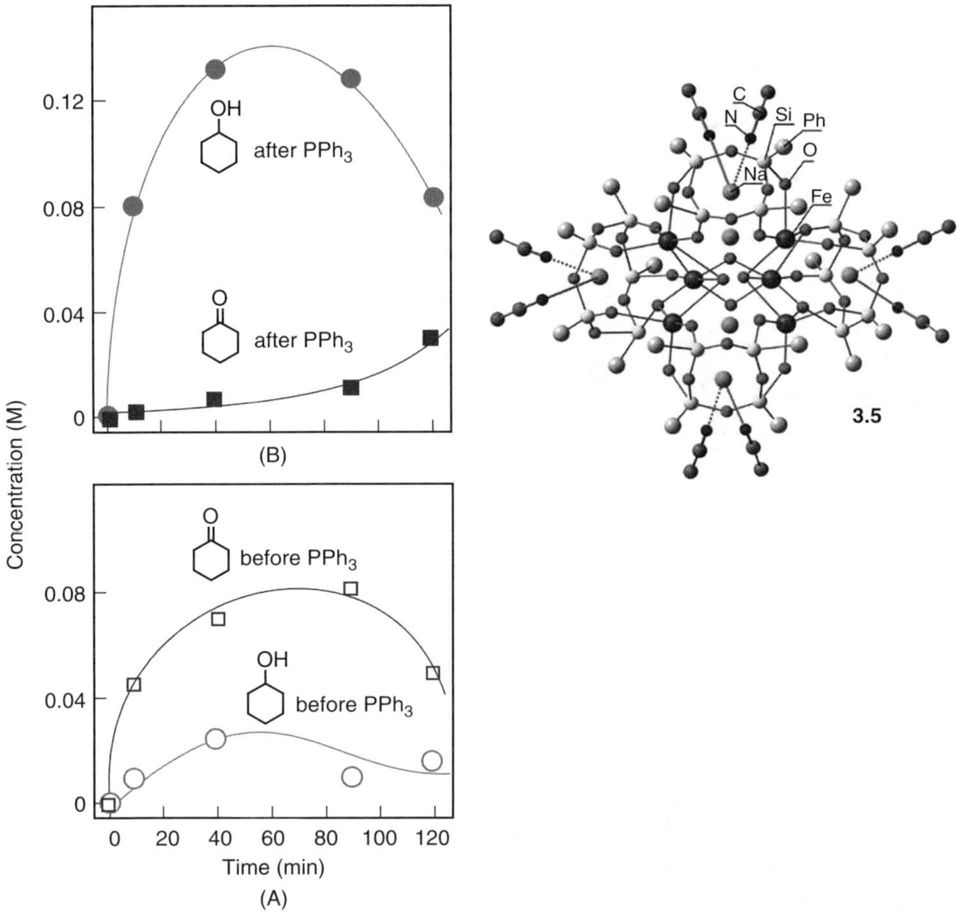

Figure 3.11 *Left:* Accumulation of cyclohexanol and cyclohexanone with time in the cyclohexane (0.46 M) oxidation with H_2O_2 (50%, aqueous, 1.5 M) catalyzed by the heterometallic cage silsesquioxane complex $[Ph_5Si_5O_{10}]_2[Ph_4Si_4O_8]_2Fe_6Na_6(O^{2-})_3$ $(CH_3CN)_{8.5}(H_2O)_{8.44}$ (**3.5**; 5×10^{-4} M), CF_3COOH (0.05 M), CH_3CN, 40 °C. Concentrations measured before (Graph A) and after (Graph B) reduction of the samples with PPh_3 are shown. *Right:* the crystal structure of catalyst **3.5**. Source: The data are adapted from Ref. [20a]. (*See insert for color representation.*)

1 : 1. The exact value of this ratio depends on the construction of the injector, its material, temperature of the injector, etc. It depends also on the nature and amount of the catalyst used in the oxidation reaction as this catalyst is injected into the instrument. Many reactions between the components can occur in the injector.

If an excess of solid PPh_3 (or thiourea) is added to a sample of the reaction solution ca. 10 minutes before GC analysis, the alkyl hydroperoxide present in the reaction mixture is completely reduced to the corresponding alcohol. In many cases, the sum of concentrations alcohol + ketone before the reduction is approximately equal to the same sum obtained after the reduction. The analysis after the reduction gives more precise quantification of products because PPh_3 reduces ROOH to ROH quantitatively.

We proposed [18d] formulae for the *real* concentrations of the three products, [ROOH], [ROH] and [R′C(=O)R″], expressed via concentrations of the alcohol (y_{before} and y_{after}) and

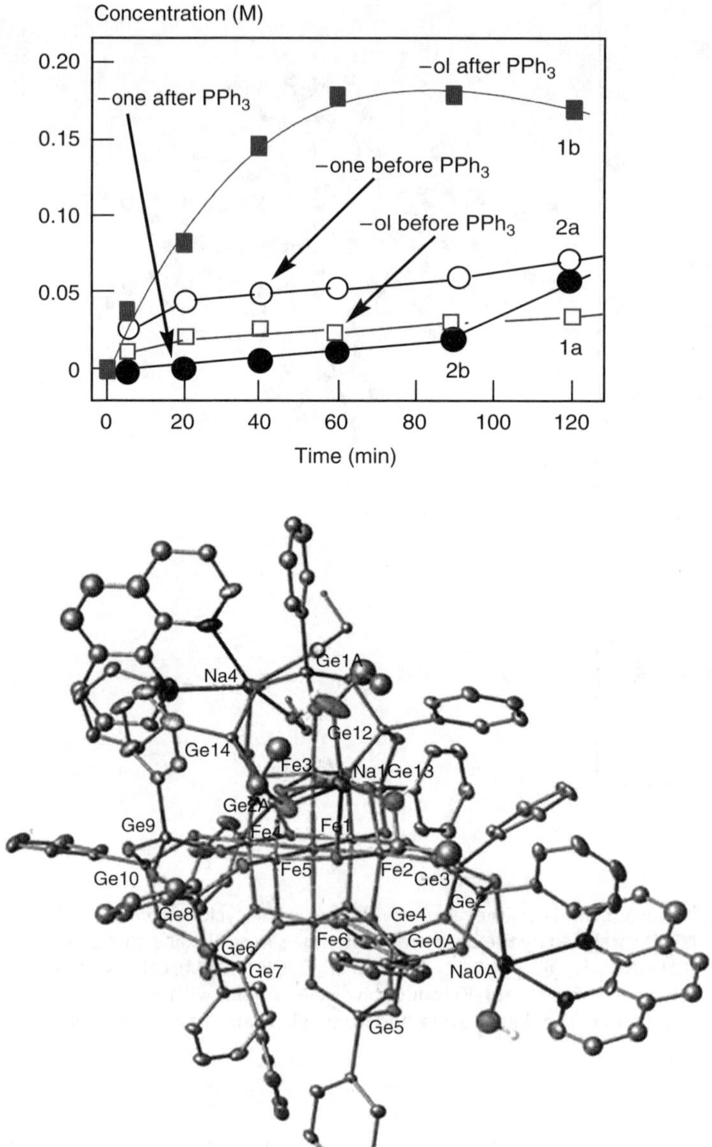

3.6

Figure 3.12 *Left:* Accumulation of oxygenates (cyclohexanol, curves 1, and cyclohexanone, curves 2) with time in the cyclohexane (0.46 M) oxidation with H_2O_2 (50%, aqueous, 1.5 M) catalyzed by the cage-like Fe, Na-germsesquioxane $(Ph_5Ge_5O_{10})_3Fe_6(OH)_3(O)Na_2(C_{12}H_8N_2)(EtOH)_4(H_2O)_{0.5}$ (**3.6**; 5×10^{-4} M) in the presence of CF_3COOH (0.05 M). Solvent was acetonitrile (total volume of the reaction solution was 5 ml); 40 °C. Concentrations of cyclohexanone and cyclohexanol were determined before (curves a) and after (curves b) reduction of the aliquots with solid PPh_3. Yield of oxygenates was 41% after 90 minutes. Right: the molecular structure of catalyst **3.6**. Source: Figures are adapted from Ref. [20b] with permission of Wiley.

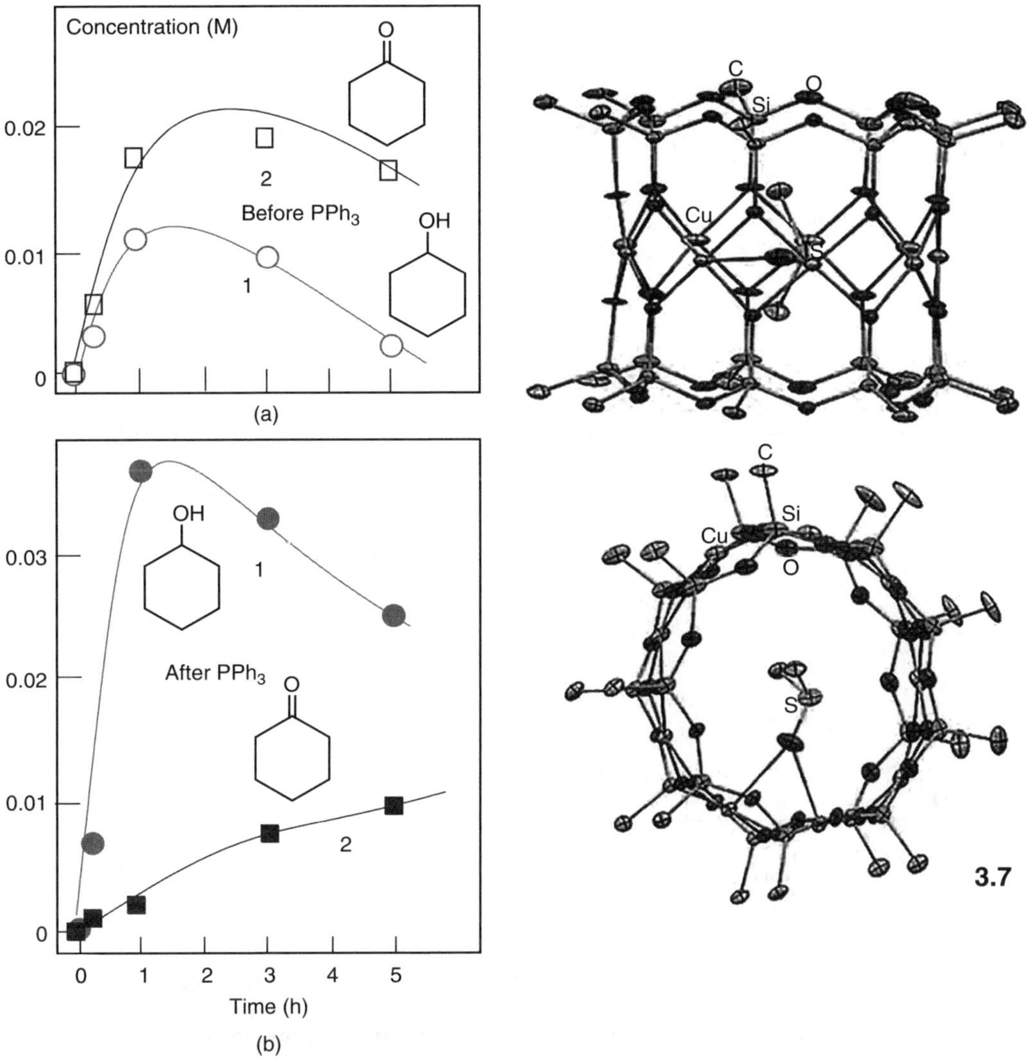

Figure 3.13 *Left:* Oxidation of cyclohexane (0.23 M) with H_2O_2 (50%, 2 M) catalyzed by the copper cage-like coppersilsesquioxanes nonanuclear $[(MeSiO_2)_{18}Cu_9]$ (**3.7**; 5×10^{-4} M) in the presence of HNO_3 (0.05 M). Solvent was acetonitrile (total volume of the reaction solution was 5 ml); 50 °C. Concentrations of cyclohexanol (curves 1) and cyclohexanone (curves 2) were measured both before and after reduction with PPh_3. *Right:* Molecular structure (two views) of complex **3.7**). Source: Adapted from Ref. [20c] with permission of American Chemical Society. (*See insert for color representation.*)

ketone ($z_{\text{before}} = z_{\text{after}}$) measured before and after reduction with PPh_3:

$$[\text{ROOH}] = x_{\text{real}} = m(z_{\text{before}} - z_{\text{after}}) + (y_{\text{after}} - y_{\text{before}}) \tag{3.1}$$

$$[\text{ROH}] = y_{\text{real}} = y_{\text{before}} - m(z_{\text{before}} - z_{\text{after}}) \tag{3.2}$$

$$[\text{R}'\text{C}(=\text{O})\text{R}''] = z_{\text{real}} = z_{\text{after}} \tag{3.3}$$

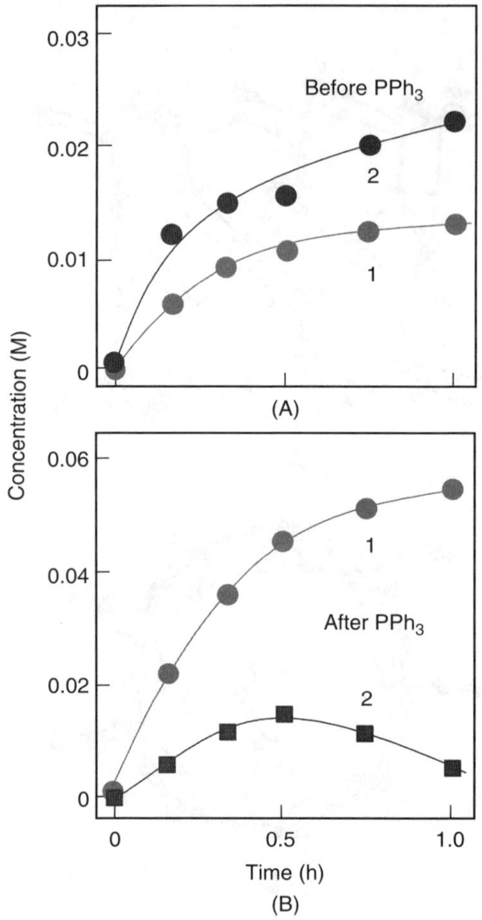

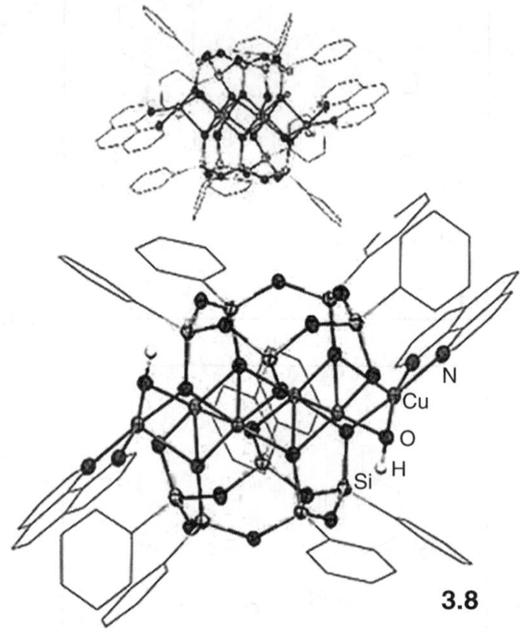

3.8

Figure 3.14 *Left:* Oxidation of cyclohexane (0.46 M) with H_2O_2 (50%, 1.5 M) catalyzed by hexacoppersilsesquioxane containing phenanthroline **3.8** (5×10^{-4} M) in the presence of HNO_3 (0.05 M) at 60 °C. Concentrations of cyclohexanol (curves 1) and cyclohexanone (curves 2) were measured both before (A) and after (B) reduction with PPh_3. *Right:* Two projections of the structure of cage hexacoppersilsesquioxane containing N-ligand phenanthroline $(PhSi)_{10}O_{20}(OH)_2Cu_6(phen)_2$ (complex **3.8**). Source: Adapted from Ref. [20d] with permission of American Chemical Society. (*See insert for color representation.*)

Here x_{real}, y_{real}, and z_{real} are real concentrations of the alkyl hydroperoxide, alcohol, and ketone, respectively, in the reaction solution, m is the y_{add}/z_{add} ratio (usually ~1) and $y_{add} = y_{before} - y_{real}$, $z_{add} = z_{before} - z_{real}$.

The [alcohol]/[ketone] ratio (presented in many publication as the A/K ratio) has no sense and gives valuable information only if [alkyl hydroperoxide] = 0 [19a]. It would be more correctly to write "the A/K ratio determined after reduction of the sample with PPh_3" [7e, 19b]. In this case, a rough parameter [A] = [alcohol] + [alkyl hydroperoxide] [19c] gives us absolutely no information on existence or nonexistence of ROOH in the solution and the information on the amount of alkyl hydroperoxide in the reaction mixture. Direct determination of relatively unstable alkyl hydroperoxides was possible when quartz-reinforced injectors and quartz columns were used. The use of metallic (steel or copper) injector and columns most often results in full decomposition of alkyl hydroperoxides with the formation of the corresponding ketones and alcohols (often in a ratio close to 1 : 1). Peaks corresponding to isomeric alkyl hydroperoxides were observed on the chromatograms of the reaction mixture (before reduction with PPh_3) along with less intensive peaks of isomeric alcohols and ketones. After reduction of the sample with PPh_3, the peaks of alkyl hydroperoxides completely disappear, and a great increase in the intensity of all the peaks corresponding to the isomeric alcohols is simultaneously observed. The intensity of the peaks attributed to ketones decreases, and this indicates a slight (10–15%) decomposition of the hydroperoxides in the chromatograph.

The examples collected below demonstrate the formation of alkyl hydroperoxides due to substantial difference between alcohol and ketone concentrations measured before and after treating with PPh_3 (Figures 3.11–3.16). Accumulation of both products with time in oxidation with H_2O_2 is given for the reactions catalyzed by silsesquioxane iron (Figure 3.11) and germsesquioxane iron (Figure 3.12) complexes. Various silsesquioxane copper derivatives turned out to be also efficient in alkane oxidations (Figures 3.13–3.16).

Complex **3.9** containing three 2,2′-dipyridil ligands (Figure 3.15) exhibited high activity in the cyclohexane oxidation with hydrogen peroxide in the absence of any added acid. Thus, the reaction afforded after 2 hours at 50 C oxygenates in 20% yield and turnover number (TON) = 190 (Figure 3.15). A comparison of chromatograms obtained before and after addition of PPh_3 to the sample indicated that the main product of the reaction cyclohexyl hydroperoxide was formed in high concentration because the alcohol/ketone ratio dramatically growth after the reduction. It should be noted that unlike many oxidations under different conditions this reaction gave much lower product concentration before PPh_3 than after reduction. This can be explained by the fact

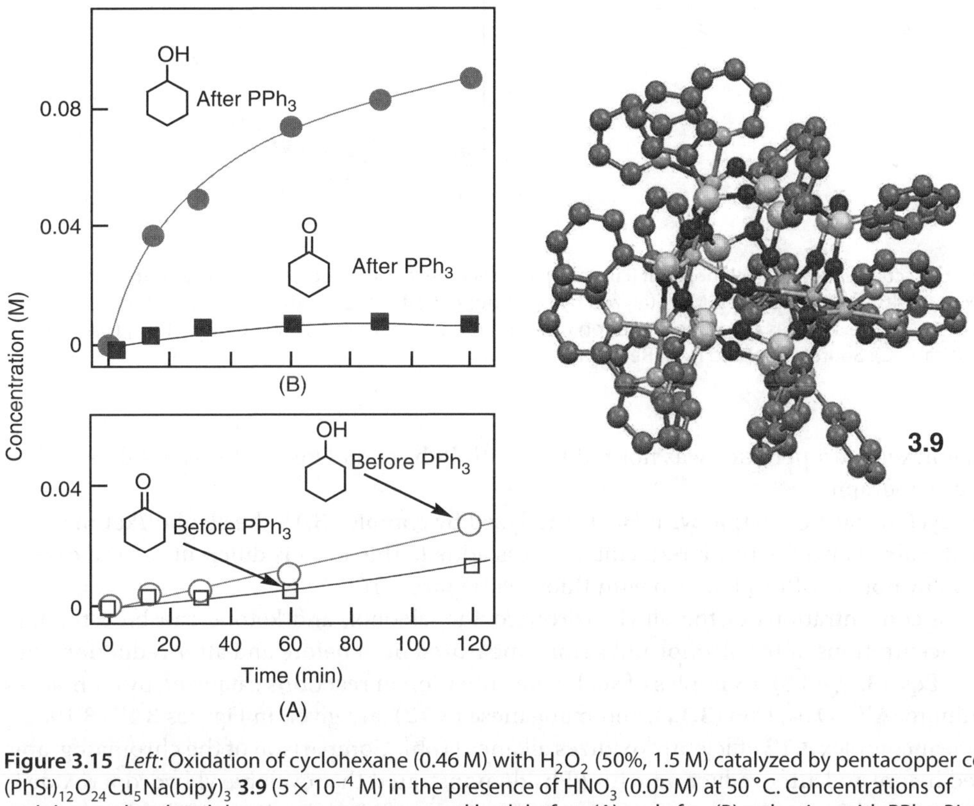

Figure 3.15 *Left:* Oxidation of cyclohexane (0.46 M) with H_2O_2 (50%, 1.5 M) catalyzed by pentacopper complex $(PhSi)_{12}O_{24}Cu_5Na(bipy)_3$ **3.9** (5×10^{-4} M) in the presence of HNO_3 (0.05 M) at 50 °C. Concentrations of cyclohexanol and cyclohexanone were measured both before (A) and after (B) reduction with PPh_3. *Right:* pentacoppersilsesquioxane containing N-ligand 2,2′-bipyridil). Source: Adapted from Ref. [20e].

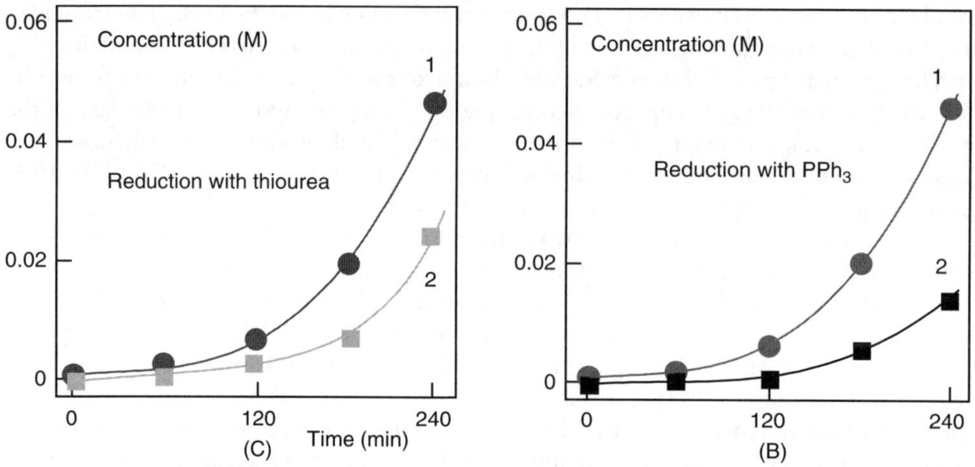

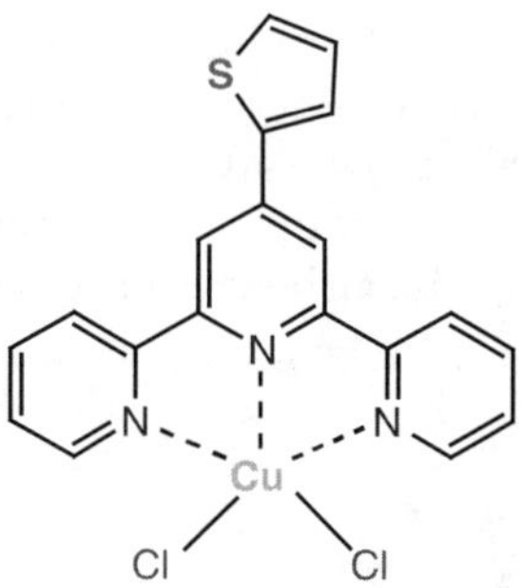

3.10

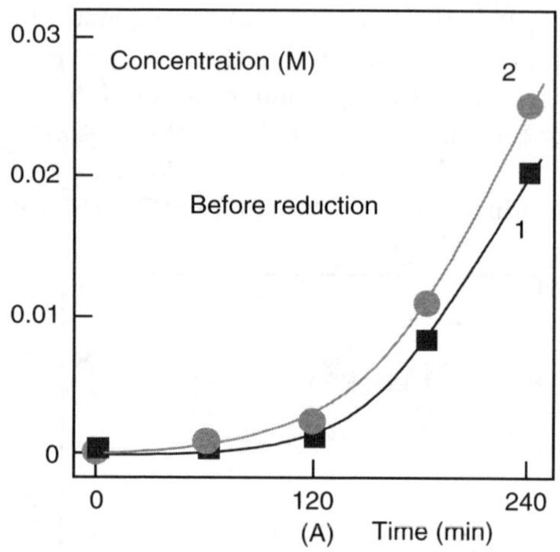

Figure 3.16 Accumulation of cyclohexanol (curves 1) and cyclohexanone (curves 2) in the oxidation of cyclohexane (0.46 M) with H_2O_2 (1.5 M) catalyzed by compound **3.10** (concentrations 5×10^{-4} M) in acetonitrile solution at 60 °C. The samples of reaction solution were not reduced (Graph A), reduced with PPh_3 (Graph B), or thiourea (Graph C). Source: Adapted from Ref. [20f].

that cyclohexyl hydroperoxide was not reduced with PPh_3 completely but was not detected by our chromatograph.

In the cyclohexane oxidation with H_2O_2 catalyzed by complex **3.10** the alcohol/ketone ratio is dramatically changed when a reducing agent is added. This ratio is different for the case of reduction by not only PPh_3 but also with thiourea (Figure 3.16).

The real concentrations of the alkyl hydroperoxide, alcohol, and ketone can be estimated using concentrations of the alcohol and ketone measured both before and after reduction with PPh_3; see Eqs. (3.1)–(3.3). Examples of such determination in reactions catalyzed by complexes of aluminum, $Al(NO_3)_3$, iron (**3.11**), and manganese (**3.12**), are given in Figures 3.17–3.19.

Binuclear complex **3.12** efficiently oxidizes alkanes [16b]. Comparison of the chromatograms obtained before and after reduction with PPh_3 demonstrated that cyclohexyl hydroperoxide is accumulated in this reaction in sufficient concentration and after that it gradually decomposes (Figure 3.19). However, the selectivity parameters clearly demonstrate that hydroxyl radicals do

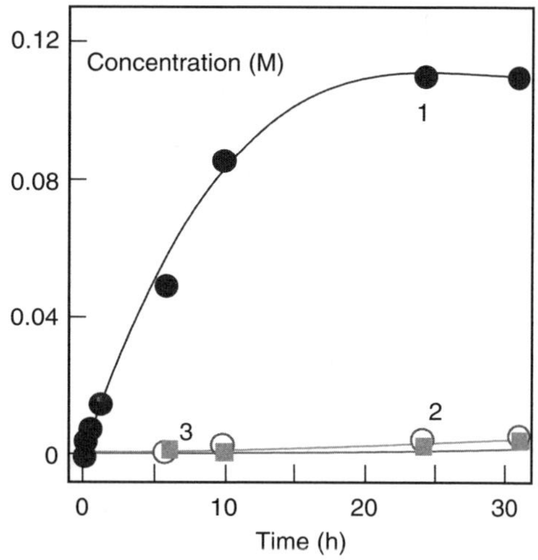

Figure 3.17 Accumulation of the products (cyclooctyl hydroperoxide, curve 1; cyclooctanol, curve 2; cyclooctanone, curve 3) in oxidation of cyclooctane (0.5 M) with hydrogen peroxide (70% aqueous; 1.0 M; the total concentration of water in the reaction mixture was 1.3 M) catalyzed by $Al(NO_3)_3$ (1.0 mM) in MeCN at 70 °C. Concentrations of cyclooctyl hydroperoxide, cyclooctanone, and cyclooctanol were estimated using double measuring concentration of cyclooctanone and cyclooctanol, before and after reduction of the aliquots with solid PPh_3. Source: Adapted from Ref. [11a] with permission of Elsevier.

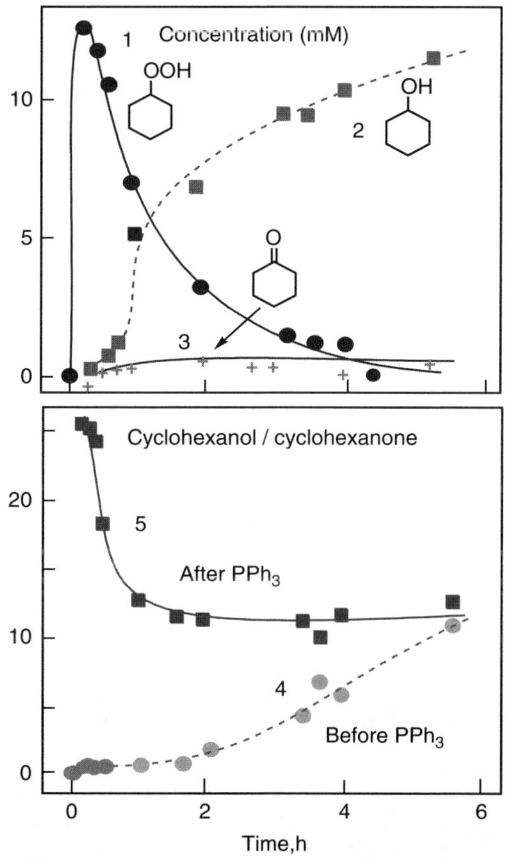

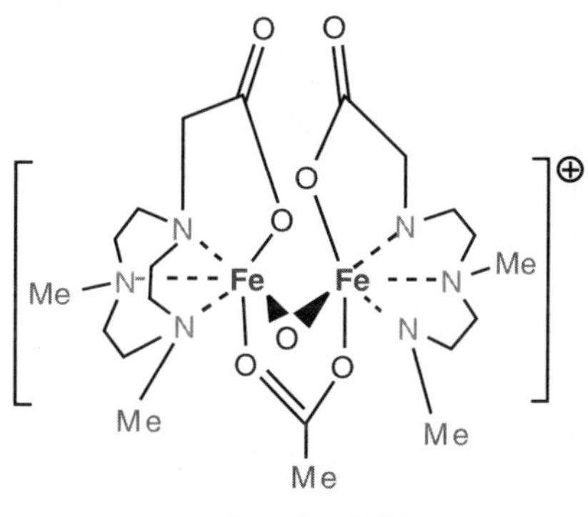

Complex **3.11**

Figure 3.18 Oxidation of cyclohexane (0.46 M) with H_2O_2 (0.6 M) in MeCN at 25 °C catalyzed by 1 mM of complex $[Fe_2(N_3O\text{-}L)_2(\mu\text{-}O)(\mu\text{-}OOCCH_3)]^+$ (**3.11**), where N_3O-L is 1-carboxymethyl-4,7-dimethyl-1,4,7-triazacyclononane; the following curves are shown: accumulation of cyclohexyl hydroperoxide (curve 1), cyclohexanol (2), cyclohexanone (3) as well as the alcohol/ketone ratio measured before (4) and after (5) the reduction of the samples with triphenylphosphine. Source: Adapted from Ref. [21] with permission of American Chemical Society.

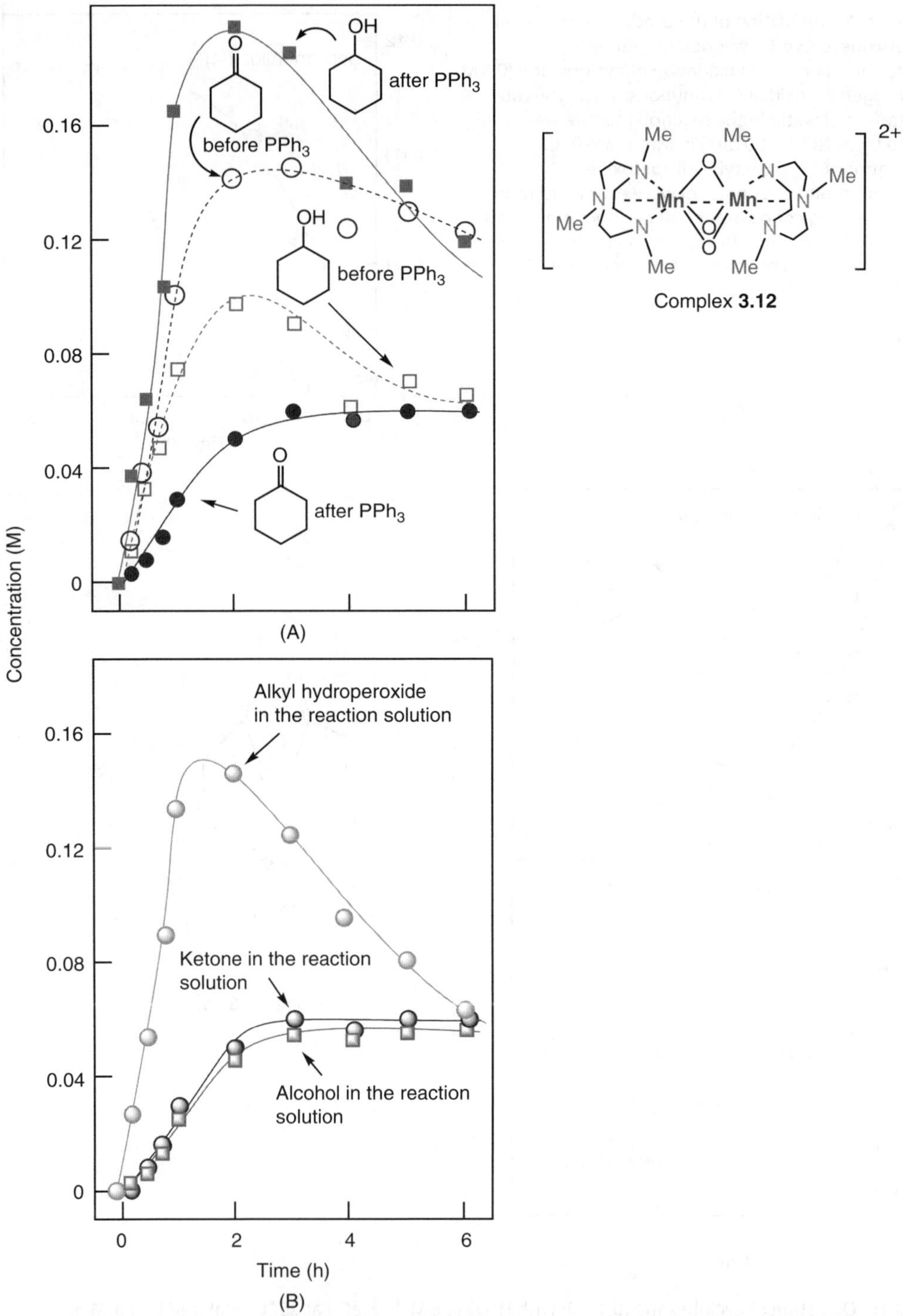

Figure 3.19 Graph A: accumulation of cyclohexanol and cyclohexanone in the oxidation of cyclohexane (0.46 M) with hydrogen peroxide (1 M; 50%) catalyzed by complex **3.12** (1×10^{-4} M) and oxalic acid (0.025 M) in acetonitrile at 30 °C. Graph B: real concentrations of cyclohexyl hydroperoxide, cyclohexanol, and cyclohexanone in the reaction solution estimated using data presented on Graph A. Source: Adapted from Ref. [16b] with permission from Elsevier.

not attack hydrocarbons. It was assumed that the transformation of an alkane into the corresponding alkyl hydroperoxide proceeds via generation of alkyl radicals R•, which rapidly react with atmospheric molecular oxygen.

Some authors wrote that they used the reduction with PPh_3 in order *only* to remove peroxides (starting H_2O_2, *tert*-BuOOH, produced ROOH, etc.) from the reaction mixtures [22a–d]. Although treatment with PPh_3 can be used in order *only* to remove from the reaction mixtures peroxides (starting H_2O_2, *tert*-BuOOH, produced ROOH, etc.) the main aim of the method is the estimation of real concentrations of ROOH, alcohol, and ketone formed in the reaction. Chromatograms obtained *only* after reduction of the reaction mixture *cannot* be used for determination of *real* concentration of each oxygenate. Nevertheless, such chromatograms give valuable information on the total concentration of produced oxygenates (Figure 3.20, [22e]).

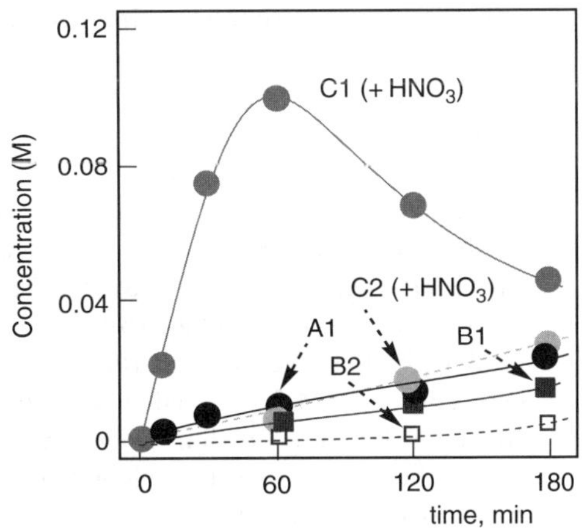

Figure 3.20 The accumulation of cyclohexanol (curve 1) and cyclohexanone (curve 2) in the course of cyclohexane oxidation with H_2O_2 catalyzed by the complex **3.13** (Figure 3.21) in acetonitrile in air. Conditions. Catalyst, 5×10^{-4} M; cyclohexane, 0.46 M; H_2O_2 (50% aqueous, 2 M). A: at 20 °C; in the presence of HNO_3, 0.05 M; maximum yield of oxygenates is 20% after 1 hour; TON = 182. B: at 60 °C; in the presence of HNO_3, 0.1 M. C: at 60 °C; in the absence of HNO_3. In all cases, concentrations of the oxygenates were measured after addition of PPh_3 to the reaction sample. C: at 60 °C; in the absence of HNO_3. In all cases, concentrations of the oxygenates were measured after addition of PPh_3 to the reaction sample.

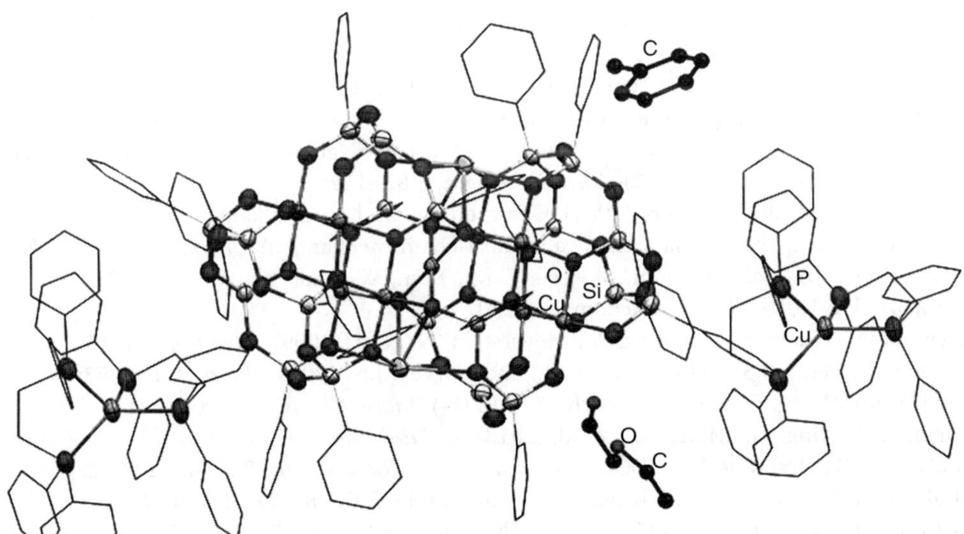

3.13

Figure 3.21 The molecular structure of catalyst $[(PhSiO_{1.5})_{12}(PhSiO_{1.5})_8(CuO)_8\ (NaO_{0.5})_4\{Cu(O_{0.5})_4\}][Cu(dppe)_2]_2$ (**3.13**; solvating molecules of water and EtOH are omitted in the sake of clarity). Source: Adapted from Ref. [22e].

3.4 Conclusions

The formation of oxygen derivatives of C—H compounds can proceed with the intermediate generation of free radicals (C- and O-centered). Alternatively, free radicals do not take part in such processes. Although usually the pathways with the involvement of free radicals give rise to a bit low selectivity of the reaction, radical processes are more efficient (exhibiting higher yields and TONs).

References

1 (a) Shilov, A.E. and Shul'pin, G.B. (1997). *Chem. Rev.* 97: 2879–2932. (b) Shilov, A.E. and Shul'pin, G.B. (2000). Activation and catalytic reactions of saturated hydrocarbons in the presence of metal complexes. In: *Kluwer Academic Publishers: Dordrecht/Boston/London*, 548. New York, Boston, Dordrecht, London, Moscow: Kluwer Academic Publishers. (c) Godula, K. and Sames, D. (2006). *Science* 312: 67–72. (d) Cernak, T., Dykstra, K.D., Tyagarajan, S. et al. (2016). *Chem. Soc. Rev.* 45: 546–576. (e) Roudesly, F., Oble, J., and Poli, G. (2017). *J. Mol. Catal. Chem.* 426: 275–296. (f) Gini, A., Brandhofer, T., and Mancheña, O.G. (2017). *Org. Biomol. Chem.* 15: 1294–1312.

2 (a) Shul'pin, G.B. (2004). Oxidations of C–H compounds catalyzed by metal complexes. In: *Transition Metals for Organic Synthesis*, 2e, vol. 2 (ed. M. Beller and C. Bolm), 215–242. Weinheim, New York: Wiley. (b) Shul'pin, G.B. (2009). *Mini-Rev. Org. Chem.* 6: 95–104. (c) Shul'pin, G.B. (2010). *Org. Biomol. Chem.* 8: 4217–4228. (d) Shul'pin, G.B. (2013). Selectivity in C–H functionalizations. In: *Comprehensive Inorganic Chemistry II*, 2e, vol. 6, Chapter 6.04 (ed. J. Reedijk, K. Poeppelmeier and L. Casella), 79–104. Elsevier. (e) Shul'pin, G.B. (2016). *Catalysts* 6 (4), Paper No. 50. (f) Ciriminna, R., Albanese, L., Meneguzzo, F. et al. (2016). *ChemSusChem* 9: 3374–3381. (g) Oszajca, M., Brindell, M., Orzeł, Ł. et al. (2016). *Coord. Chem. Rev.* 327-328: 143–165. (h) Wójtowicz-Młochowska, H. (2017). *Arkivoc* (part ii): 12–58.

3 (a) Shul'pin, G.B.J. (2002). *Mol. Catal. A: Chem.* 189: 39–66. (b) Shul'pin, G.B. (2013). *Dalton Trans.* 42: 12794–12818.

4 (a) Emanuel', N.M., Denisov, E.T., and Maizus, Z.K. (1965). *Chain Reactions of Hydrocarbon Oxidation in the Liquid Phase* (in Russian). Moscow: Nauka. (b) Shul'pin, G.B. (1988). *Organic Reactions Catalysed by Metal Complexes.* Moscow: Nauka (in Russian). (c) Roduner, E., Kaim, W., Sarkar, S. et al. (2013). *ChemCatChem* 5: 82–112.

5 (a) Kozlov, Y.N. and Shul'pin, G.B. (2001). Can methane and other alkanes be oxidized in solutions at low temperature via a classical radical-chain mechanism? *Chem. Preprint Arch.* 2001 (6): 57–65. (b) Sankaralingam, M., Lee, Y.-M., Nam, W., and Fukuzumi, S. (2017). *Inorg. Chem.* doi: 10.1021/acs.inorgchem.7b00220.

6 (a) Rachmilovich-Calis, S., Masarwa, A., Meyerstein, N. et al. (2009). *Chem. Eur. J.* 15: 8303–8309. (b) Petit, A.S., Pebbifold, R.C.R., and Harvey, J.N. (2014). *Inorg. Chem.* 53: 6473–6481. (c) Zhang, Q. and Goldsmith, C.R. (2014). *Inorg. Chem.* 53: 5206–5211. (d) Pham, A.N., Xing, G., Miller, C.J. et al. (2013). *J. Catal.* 301: 54–64.

7 (a) Shul'pin, G.B., Drago, R.S., and Gonzalez, M. (1996). *Russ. Chem. Bull.* 45: 2386–2388. (b) Shul'pin, G.B., Kozlov, Y.N., Nizova, G.V. et al. (2001). *J. Chem. Soc. Perkin Trans.* 2: 1351–1371. (c) de la Cruz, M.H.C., Kozlov, Y.N., Lachter, E.R. et al. (2003). *New J. Chem.* 27: 634–638. (d) Kozlov, Y.N., Romakh, V.B., Kitaygorodskiy, A. et al. (2007). *J. Phys. Chem. A* 111: 7736–7752. (e) Kirillova, M.V., Kuznetsov, M.L., Romakh, V.B. et al. (2009). *J. Catal.* 267: 140–157. (f) Gusevskaya, E.V., Menini, L., Parreira, L.A. et al. (2012). *J. Mol. Catal. A:*

Chem. 363–364: 140–147. (g) Sutradhar, M., Shvydkiy, N.V., da Silva, M.F.C.G. et al. (2013). *Dalton Trans.* 42: 11791–11803.

8 (a) Khaliullin, R.Z., Bell, A.T., and Head-Gordon, M. (2005). *J. Phys. Chem. B* 109: 17984–17992. (b) Burg, A., Fastovesky, E., Shamir, D. et al. (2016). *J. Coord. Chem.* 69: 1722–1729.

9 Kozlov, Y.N., Nadezhdin, A.D., and Purmal', A.P. (1973). *Kinetika Kataliz* (in Russian) 14: 141–148.

10 (a) Xia, J.-B., Cormier, K.W., and Chen, C. (2012). *Chem. Sci.* 3: 2240–2245. (b) Shul'pin, G.B. (2014). Organometallic complexes as catalysts in oxidation of C–H compounds, Chapter 1. In: *Advances in Organometallic Chemistry and Catalysis* (ed. A.J.L. Pombeiro), 3–14. Wiley.

11 (a) Mandelli, D., Chiacchio, K.C., Kozlov, Y.N. et al. (2008). *Tetrahedron Lett.* 49: 6693–6697. (b) Kuznetsov, M.L., Kozlov, Y.N., Mandelli, D. et al. (2011). *Inorg. Chem.* 50: 3996–4005. (c) Novikov, A.S., Kuznetsov, M.L., Pombeiro, A.J.L. et al. (2013). *ACS Catal.* 3: 1195–1208. (d) Novikov, A.S., Kuznetsov, M.L., Rocha, B.G.M. et al. (2016). *Catal. Sci. Technol.* 6: 1343–1356. (e) Kuznetsov, M.L., Teixeira, F.A., Bokach, N.A. et al. (2014). *J. Catal.* 313: 135–148. (f) Rocha, B.G.M., Kuznetsov, M.L., Kozlov, Y.N. et al. (2015). *Catal. Sci. Technol.* 5: 2174–2187. (g) Kuznetsov, M.L., Rocha, B.G.M., Pombeiro, A.J.L. et al. (2015). *ACS Catal.* 5: 3823–3835.

12 (a) Popivker, I., Zilbermann, I., Maimon, E. et al. (2013). *Dalton. Trans.* 42: 16666–16668. (b) Burg, A., Shamir, D., Shusterman, I. et al. (2014). *Chem. Commun.* 50: 13096–13099. (c) Burg, A., Shusterman, I., Kornweitz, H. et al. (2014). *Dalton Trans.* 43: 9111–9115. (d) Kornweitz, H., Burg, A., and Meyerstein, D. (2015). *J. Phys. Chem. A* 119: 4200–4206.

13 Olivo, G., Lanzalunga, O., and Di Stefano, S. (2016). *Adv. Synth. Catal.* 358: 843–863.

14 (a) Talsi, E.P. and Bryliakov, K.P. (2012). *Coord. Chem. Rev.* 256: 1418–1434. (b) Bryliakov, K.P. and Talsi, E.P. (2014). *Coord. Chem. Rev.* 276: 73–96. (c) Ottenbacher, R.V., Talsi, E.P., and Bryliakov, K.P. (2015). *ACS Catal.* 5: 39–44. (d) Talsi, E.P., Ottenbacher, R.V., and Bryliakov, K.P. (2015). *J. Organomet. Chem.* 793: 102–107. (e) Lyakin, O.Y., Zima, A.M., Samsonenko, D.G. et al. (2015). *ACS Catal.* 5: 2702–2707. (f) Ottenbacher, R.V., Talsi, E.P., and Bryliakov, K.P. (2016). *Molecules* 21: 1454.

15 (a) Canta, M., Font, D., Gómez, L. et al. (2014). *Adv. Synth. Catal.* 356: 818–830. (b) Codola, Z., Lloret-Fillol, J., Mitra, M. et al. (2017). *J. Mol. Catal. A: Chem.* 426: 350–356. (c) Milan, M., Birtti, M., and Costas, M. (2017). *ACS Cent. Sci.* 3: 196–204.

16 (a) Bilyachenko, A.N., Levitsky, M.M., Yalymov, A.I. et al. (2016). *RSC Adv.* 6: 48165–48180. (b) Shul'pin, G.B., Nesterov, D.S., Shul'pina, L.S. et al. (2017). *Inorg. Chim. Acta* 455 (P2): 666–676. (c) Shul'pin, G.B., Stoeckli-Evans, H., Mandelli, D. et al (2004). *J. Mol. Catal. A-Chem.* **219**: 255–264.

17 (a) Matin-Caballero, J., Wéry, A.S.J., Reinoso, S. et al. (2016). *Inorg. Chem.* 55: 4970–4979. (b) Fang, Y., Gong, W., Liu, L. et al. (2016). *Inorg. Chem.* 55: 10102–10105. (c) Yang, D., Wu, T., Chen, C. et al. (2017). *Green Chem.* 19: 311–318. (d)Kadwa, I., Friedrich, H.B., and Bala, M.D. (2017). *Inorg. Chim. Acta* 463: 112–117.

18 (a) Shul'pin, G.B. and Druzhinina, A.N. (1992). *React. Kinet. Catal. Lett.* 47: 207–211. (b) Shul'pin, G.B. (2003). *C. R. Chim.* 6: 163–178. (c) Shul'pin, G.B., Kozlov, Y.N., Shul'pina, L.S. et al. (2009). *Inorg. Chem.* 48: 10480–10482. (d) Shul'pin, G.B., Kozlov, Y.N., Shul'pina, L.S. et al. (2010). *Appl. Organomet. Chem.* 24: 464–472. (e) Garcia-Bosch, I. and Siegel, M.A. (2016). *Angew. Chem. Int. Ed.* 128: 13065–13068. (f) Sutradhar, M., Alegria, E.C.B.A., Barman, T.R. et al. (2016). *Polyhedron* 117: 666–671.

19 (a) Das, B., Al-Hunaiti, A., Haukka, M. et al. (2015). *Eur. J. Inorg. Chem.* 3590–3601. (b) Contaldi, S., Di Nicola, C., Garau, F. et al. (2009). *Dalton Trans.* 4928–4941. (c) Haslinger, S., Raba, A., Cokoja, M. et al. (2015). *J. Catal.* 331: 147–153.

20 (a) Yalymov, A.I., Bilyachenko, A.N., Levitsky, M.M. et al. (2017). *Catalysts* 7 (4), paper No. 101. (b) Bilyachenko, A.N., Levitsky, M.M., Yalymov, A.I. et al. (2016). *Angew. Chem. Int. Ed.* 55: 15360–15363. (c) Bilyachenko, A.N., Kulakova, A.N., Levitsky, M.M. et al. (2017). *Inorg. Chem.* 56: 4093–4103. (d) Kulakova, A.N., Bilyachenko, A.N., Levitsky, M.M. et al. (2017). *Inorg. Chem.* 56 (24): 15026–15040. (e) Bilyachenko, A.N., Kulakova, A.N., Korlyukov, A.A. et al. (2017). High catalytic activity of heterometallic (Fe_6Na_7 and Fe_6Na_6) cage silsesquioxanes in oxidations with peroxides. *Preprints* 2017030039. (f) Czerwińska, K., Machura, B., Kula, S. et al. (2017). *Dalton Trans.* 46: 9591–9604.

21 Romakh, V.B., Therrien, B., Süss-Fink, G. et al. (2007). *Inorg. Chem.* 46: 3166–3175.

22 (a) Xue, S., Chen, G., Long, Z. et al. (2015). *RSC Adv.* 5: 19306–19314. (b) Dias, S.P., Kirillova, M.V., André, V. et al. (2015). *Inorg. Chem. Front.* 2: 525–537. (c) Mncube, S.G. and Bala, M.D. (2016). *J. Mol. Liquids* 215: 396–401. (d) Fernandes, T.A., André, V., Kirillov, A.M. et al. (2017). *J. Mol. Catal. A: Chem.* 426: 357–367. (e) Bilyachenko, A.N., Kulakova, A.N., Levitsky, M.M. et al. (2017). *ChemCatChem* 9: 4437.

4

Reactions of Alkyl Radicals in Aqueous Solutions*

Dan Meyerstein

Ariel University, Chemical Sciences Department, Ariel, Israel
Ben-Gurion University, Chemistry of Department, Beer-Sheva, Israel

4.1 Introduction

Alkyl radicals are key intermediates in a variety of important processes in aqueous solutions, e.g. biological processes [1–6] (some of these processes are needed to regulate physiological processes [1, 3, 5] and/or are key steps in enzymatic processes, [7–11] whereas others induce deleterious biological processes [3, 6, 12–20]), advanced oxidation processes [21–37], catalytic processes [38–41], photochemical processes [42–53] of importance in solar energy conversion [54], environmental processes [27, 28, 30, 32, 33], and electrochemical processes [29, 55–59]. The aim of this chapter is to outline the main routes of the formation of alkyl radicals in aqueous media and their general mechanisms of reaction. As the field is very large, this chapter is limited to simple small alkyl radicals, and even for those, only typical, recent reactions are discussed with no aim to give a full literature survey.

4.2 Production of Alkyl Radicals in Aqueous Solutions

As most covalent bonds have high activation energy for dissociation, only several compounds yield radicals in aqueous solutions within a reasonable period to initiate radical processes. The best known is 2,2′-Azobis(isobutyronitrile) (AIBN), which is oil soluble, has a half-life of 10 hours at 65 °C, and decomposes into $N_2 + 2^{\bullet}C(CH_3)_2CN$. Analogous water-soluble di-azo-compounds include 2,2′-Azobis[2-(2-imidazolin-2-yl)propane]-dihydrochloride with a half-life of 10 hours at 44 °C and 2,2′-Azobis[2-methyl-*N*-(2-hydroxyethyl)propionamide] with a half-life of 10 hours at 86 °C. The alkyl radicals formed in these decompositions initiate radical processes. In principle, peroxides such as H_2O_2 also decompose upon heating, as the RO—OR is weak, but the process is far too slow to be practical for initiating radical processes. Therefore, radical reactions have to be initiated by redox processes or by energy absorption by the sample. Some of the more important such processes are discussed in the next sections.

Quite a few chemical reactions lead to the formation of alkyl radicals in aqueous solution. In the next sections, the major classes of such processes are discussed.

* This chapter was submitted in April 2017.

Alkane Functionalization, First Edition. Edited by Armando J. L. Pombeiro and M. Fátima C. Guedes da Silva.

4.2.1 The Fenton and Fenton-like Reactions

Probably the most common reactions leading to the formation of alkyl radicals in aqueous solutions are the Fenton reaction or Fenton-like reactions. In 1894, Fenton observed that when $Fe(H_2O)_6^{2+}$ is added to solutions containing tartaric acid and H_2O_2, they catalyze the oxidation of the tartaric acid by the peroxide. Since then, the reaction of $Fe(H_2O)_6^{2+}$ with H_2O_2 has been called the Fenton reaction, and the reactions of other Fe(II) complexes or other low-valent transition metal complexes with peroxides, often with other peroxides, have been called Fenton-like reactions. Fenton did not suggest a mechanism for this catalytic process. In 1932, two parallel mechanisms to the Fenton reaction were proposed:

1. Haber and Weiss proposed that hydroxyl radicals formed via Eq. (4.1) are the active oxidizing agent formed in the Fenton reaction [60, 61].

$$\mathrm{Fe(H_2O)_6}^{2+} + \mathrm{H_2O_2} \rightarrow \mathrm{Fe(H_2O)_6}^{3+} + \mathrm{OH^{\bullet}} + \mathrm{OH^-} \tag{4.1}$$

2. Bray and Gorin proposed that $Fe^{IV}{=}O^{2+}{}_{aq}$ formed via Eq. (4.2) is the active oxidizing agent formed in the Fenton reaction [62].

$$\mathrm{Fe(H_2O)_6}^{2+} + \mathrm{H_2O_2} \rightarrow \mathrm{Fe^{IV}{=}O^{2+}}_{aq} + \mathrm{H_2O} \tag{4.2}$$

The debate about which of these mechanisms is the correct one has lasted for over 50 years [63–67]. However, in the literature Eq. (4.1) is often referred to as the Fenton reaction.

The formation of $OH^{\bullet}$ radicals in the presence of alkyl compounds will be followed by one of the following reactions [68]:

$$\mathrm{OH^{\bullet}} + \mathrm{HCR^1R^2R^3} \rightarrow {}^{\bullet}\mathrm{CR^1R^2R^3} + \mathrm{H_2O} \tag{4.3}$$

$$\mathrm{OH^{\bullet}} + \mathrm{R^1R^2C{=}CR^3R^4} \rightarrow {}^{\bullet}\mathrm{CR^1R^2(C(OH)R^3R^4)} \tag{4.4}$$

All these reactions are very fast [68]; however, the relative rates of hydrogen atoms abstraction via reaction (4.3) differ considerably [69]. In some cases, $OH^{\bullet}$ radicals form alkyl radicals via more complex routes, e.g. methyl radicals are commonly formed via [70]:

$$\mathrm{OH^{\bullet}} + \mathrm{(CH_3)_2SO} \rightarrow \mathrm{(CH_3)_2S(O)(OH)^{\bullet}} \qquad k = 7.0 \times 10^9\ \mathrm{M^{-1}s^{-1}} \tag{4.5}$$

$$\mathrm{(CH_3)_2S(O)(OH)^{\bullet}} \rightarrow {}^{\bullet}\mathrm{CH_3} + \mathrm{(CH_3)S(O)(OH)} \qquad k = 1.5 \times 10^7\ \mathrm{s^{-1}} \tag{4.6}$$

It should be noted that the reactions of $Fe^{IV}{=}O^{2+}{}_{aq}$ with the same substrates often yield the same radicals, as shown below.

Later, it was realized that the Fenton and Fenton-like reactions usually proceed via (4.7) followed by (4.8) [71, 72].

$$\mathrm{M(H_2O)_6}^{2+} + \mathrm{H_2O_2} \leftrightarrow \mathrm{M(OOH)(H_2O)_5}^{+} + \mathrm{H_3O^+}\ \mathit{or}\ \mathrm{M^{II}(H_2O_2)(H_2O)_5}^{2+} + \mathrm{H_2O} \tag{4.7}$$

$$\mathrm{M^n(H_2O_2)(H_2O)_5}^{2+} \xrightarrow{\mathrm{H_3O^+}} \mathrm{M^{n+2}{=}O^{2+}}_{aq} + 2\mathrm{H_2O} \qquad \text{(b)}$$

$$\mathrm{M^n(H_2O_2)(H_2O)_5}^{2+} \xrightarrow{\mathrm{H_2O}} \mathrm{M^{n+1}(H_2O)_6}^{3+} + \mathrm{OH^{\bullet}} + \mathrm{OH^-} \qquad \text{(a)} \tag{4.8}$$

$$\mathrm{M^n(H_2O_2)(H_2O)_5}^{2+} \xrightarrow{\mathrm{RH,\ H_3O^+}} \mathrm{M^{n+1}(H_2O)_6}^{3+} + \mathrm{R^{\bullet}} + 2\mathrm{H_2O} \qquad \text{(c)}$$

Which route is followed depends on the following:

- The nature of the central cation if other central cations are used
- The nature of its ligands if other ligands replace water ligands

- The pH
- The nature of RH and its concentration etc. [71, 72].

Indeed, results point out that the reaction $Fe(H_2O)_6{}^{2+} + H_2O_2$ yields $OH^•$ radicals in acidic solutions and $Fe^{IV}{=}O^{2+}{}_{aq}$ in neutral solutions [73]. Later it was reported that at least for M = Fe, the following reaction must also be considered [74]:

$$Fe^{III}(H_2O)_6{}^{3+} + H_2O_2 \leftrightarrow Fe(OOH)(H_2O)_5{}^{2+} + H_3O^+ \tag{4.9}$$

$$Fe^{II}(H_2O)_6{}^{2+} + Fe(OOH)(H_2O)_5{}^{2+} \xrightarrow{H_3O+} Fe^{IV}{=}O^{2+}{}_{aq} + Fe^{III}(H_2O)_6{}^{3+} \text{ or}$$
$$2Fe^{III}(H_2O)_6{}^{3+} + OH^• \qquad k = 7.7 \times 10^5\ M^{-1}s^{-1} \tag{4.10}$$

The latter result clearly points out that the nature of the oxidizing agent depends also on $[H_2O_2]/[Fe^{II}(H_2O)_6{}^{2+}]$ [75] .

Recent results point out that $OH^•$ radicals and $HO_2{}^•$ radicals are also formed in the presence of cations that are not redox active [76, 77], e.g. $Al(H_2O)_6{}^{3+}$, and cations that have high redox potentials, e.g. $Co(H_2O)_6{}^{2+}$ [78], via:

$$(H_2O)_{6\text{-}l\text{-}1}M^n(OOH)_l(H_2O_2) \rightarrow (H_2O)_{6\text{-}l\text{-}1}M^n({}^•OOH)(OOH)_{l\text{-}1}(OH) + {}^•OH \tag{4.11}$$
$$(H_2O)_{6\text{-}l\text{-}1}M^n(OOH)_l(H_2O_2) \rightarrow (H_2O)_{6\text{-}l\text{-}1}M^{n+1}({}^•OOH)(OOH)_{l\text{-}1}(OH) + OH^- \tag{4.12}$$

Recently, it was shown that the catalytic oxidations initiated by $Co(H_2O)_6{}^{2+}$ and H_2O_2 in the presence of bicarbonate occur via the formation of $(CO_4)Co^{II}(OOH)(H_2O)^-$, which decomposes via [79]:

$$(CO_4)Co^{II}(OOH)(H_2O)^- \rightarrow Co^{II}({}^•OOH)(H_2O)(OH)_2 + CO_3{}^{•-} \tag{4.13}$$

That is, the reactive oxidizing species formed under these conditions is the carbonate anion radical. Finally, DFT calculations suggest that $CO_3{}^{•-}$ is also formed in the reaction of $Fe(H_2O)_6{}^{2+}$ with H_2O_2 in the presence of bicarbonate [80]. Furthermore, the DFT results also suggest that the reactions $Fe(H_2O)_6{}^{2+} + H_2O_2$ and $Co(H_2O)_6{}^{2+} + H_2O_2$ in the presence of acetate yield methyl radicals [80].

To sum up this section, Fenton and Fenton-like reactions often yield strong oxidizing agents that lead to the formation of alkyl radicals in the presence of suitable organic substrates. However, the nature of these strong oxidizing agents depends on the nature of the system studied and often differs from $OH^•$ radicals.

4.2.2 Other Oxidation Reactions

The major other oxidizing agents are high-valent transition metal complexes oxo complexes. The most important such complexes are the active sites in the heme-enzyme cytochrome P450 [81] and the nonheme enzyme taurine:aKG dioxygenase, TauD [82]. The mechanisms of oxidation by these enzymes are often initiated by a hydrogen atom abstraction forming an alkyl radical [41, 81–83]:

$$L(porphyrin^{•+})Fe^{IV}{=}O + RH \rightarrow L(porphyrin)Fe^{IV}{-}OH + R^• \tag{4.14a}$$

$$LFe^{IV}{=}O + RH \rightarrow LFe^{III}{-}OH + R^• \tag{4.14b}$$

In most cases, this reaction is followed by a rebound mechanism [41, 80–82]:

$$L(porphyrin)Fe^{IV}{-}OH + R^• \rightarrow L(porphyrin)Fe^{III} + ROH \tag{4.15a}$$

$$LFe^{III}{-}OH + R^• \rightarrow LFe^{II} + ROH \tag{4.15b}$$

However, in some enzymatic processes the radical $R^\bullet$ escapes the cage in which it is formed and either reacts with another substrate present in the solution or rearranges and then rebounds [83]. Because of the enzyme structure, for nearly all enzymatic processes of this type, the rebound mechanism is the mechanism taking place. However, for oxidations by synthetic $LM^{n+2}{=}O$ complexes, $M^{n+2} = Fe^{V}$; Fe^{IV}; Mn^{IV}; Cr^{IV}; Ru^{IV}, the radical usually escapes the cage and then reacts in the solution [41].

Alkyl radicals might also be formed by strong $M^{III}L$ and $M^{IV}L$ oxidizing agents: e.g. by Mn^{III} [84, 85], Cu^{III} [86, 87], and Ce^{IV} [88] complexes. As these are less-active oxidants, the radicals are often formed by oxidation of the ligand L, e.g. by decarboxylation of the ligand.

$$LM^{n+1}OC(O)R \rightarrow M^{n}L + RCO_2^{\bullet} \rightarrow LM^{n} + R^{\bullet} + CO_2 \quad (4.16)$$

Naturally even weaker oxidants can oxidize stronger reducing agents and form substituted alkyl radicals. Thus, for example, $Cu^{2+}{}_{aq}$ [89] and $Fe^{III}{}_{aq}$ [90] are known to oxidize ascorbate, HAA^-, to form the anion radical $AA^{\bullet-}$ in which there is a considerable spin density on the carbon atoms [91].

4.2.3 Reduction Reactions

The most common process for the formation of alkyl radicals using reducing agents is via Eq. (4.17) where X = Cl, Br, or I.

$$M^{n}L + RX \rightarrow LM^{n+1}X + R^{\bullet} \quad (4.17)$$

Many low-valent transition metal complexes were/are used for this process, including, for example: $Cu^{I}L$ [92–96], $Co^{I}L$ [97, 98], $Co^{II}L$ [92, 99], $Cr^{II}L$ [92], $Ni^{I}L$ [100–114]. The observation that the rate constants of these reactions depend on the bond strength of the R-X suggest that their mechanisms involve inner-sphere processes [92, 99, 100].

An analogous reaction is the reduction of alkyl-halides on the surface of zero valent metals (Eq. (4.18)):

$$(M^0)_{(s)} + RX \rightarrow (M^0)^{+}{}_{(s)} + X^- + R^{\bullet} \quad (4.18)$$

This reaction was observed in aqueous solutions for M = Fe, Mn, Co, Ni, and Zn [115]. This reaction is of major importance for M = Fe, as zero valent iron (ZVI) is used for remediation of groundwater polluted with halogenated compounds [116–124]. Recently, the use of nanoparticles ZVI is becoming a major environmental procedure [124–126]. It should be noted that as these systems are aerated, the ZVI also reacts with the dioxygen forming $O_2^{\bullet-}$, which is transformed into H_2O_2 and therefore Fenton processes also take place [125, 126]. The detailed mechanism of reaction (4.18) was not clarified.

Recently, it was shown that Au^0-NPs entrapped in sol-gel matrices catalyze the de-halogenation of halo-organic compounds by BH_4^- [127]. The results clearly point out that substituted alkyl radicals are formed via the reaction of $(Au^0\text{-NPs})\text{-}H_n^{m-}$ with the RX compounds. The mechanism of this reaction is still under study.

Though alkyl halides are the common starting materials for the formation of alkyl radicals by reducing species, other aliphatic compounds can also be reduced by strong single-electron reducing agents forming alkyl radicals as transients in the process. Thus, for example, $Ni^{I}L^+$ and $Co^{I}L^+$, L = tetra-aza-macrocyclic ligand, were shown to reduce alkenes, as shown in Scheme 4.1 [128–131]. The mechanism of reduction depends on the redox potential of the complex, steric hindrance by the ligand, the properties of the alkene, and the tendency of the central cation to be oxidized into the trivalent oxidation state [130, 131]. Radicals are formed via reaction I.9. A discussion of the properties of the transients with metal carbon σ bonds is given in a separate section in this chapter.

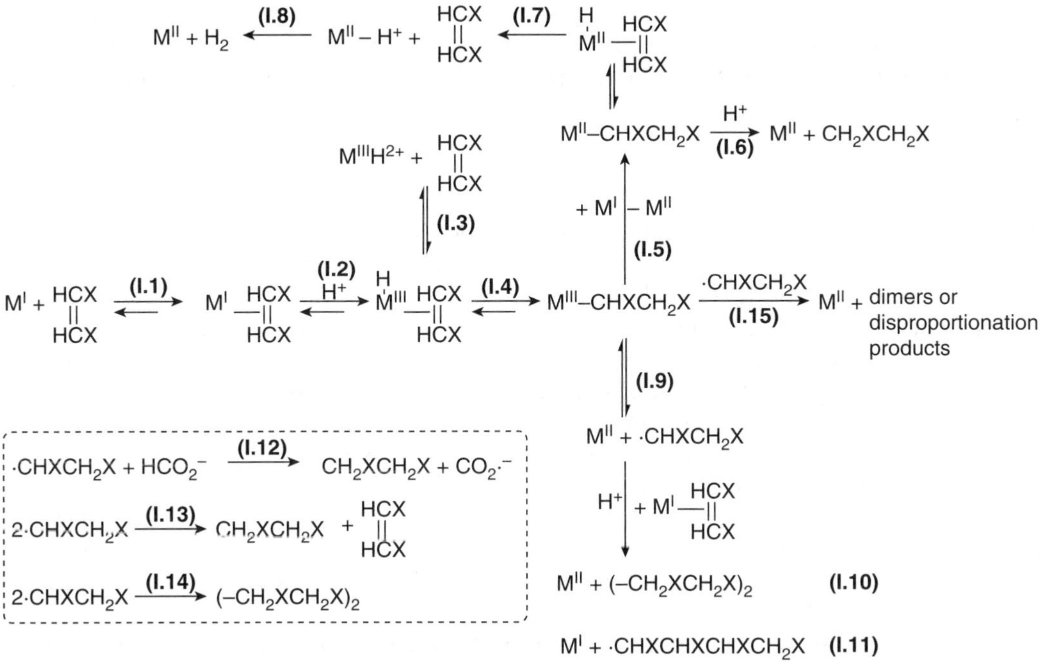

Scheme 4.1 Mechanism of reduction of alkenes by mono-valent transition metal complexes.

4.2.4 Reactions with Other Radicals

4.2.4.1 Inorganic Radicals

A large variety of inorganic radicals can be formed and studied by various techniques: $H^\bullet$, $O_2^{\bullet-}$, $HO_2^\bullet$, $O_3^{\bullet-}$, $CO_2^{\bullet-}$, $CO_3^{\bullet-}$, $CN^\bullet$, $N_3^\bullet$, $NH_2^\bullet$, $NO^\bullet$, $NO_2^\bullet$, $NO_3^\bullet$, $NCO^\bullet$, $PO_3^{\bullet 2-}$, $PO_4^{\bullet 2-}$, $HS^\bullet$, $S^{\bullet-}$, $HSS^{\bullet 2-}$, $RSSR^{\bullet-}$, $SO_2^{\bullet-}$, $SO_3^{\bullet-}$, $SO_4^{\bullet-}$, $SO_5^{\bullet-}$, $(SCN)_2^{\bullet-}$, $SeO_3^{\bullet-}$, $HSeO_4^{\bullet-}$, $(SeCN)_2^{\bullet-}$, $Cl^\bullet$, $Br^\bullet$, $I^\bullet$, $Cl^\bullet$, $Cl_2^{\bullet-}$, $Br_2^{\bullet-}$, $I_2^{\bullet-}$, $ClO^\bullet$, $BrO^\bullet$, $IO^\bullet$, $ClO_2^\bullet$, $BrO_2^\bullet$, $IO_2^\bullet$, $ClO_3^\bullet$ [132, 133]. The redox potentials of most of these radicals are known [134]. These radicals can react with different organic compounds via hydrogen atom abstraction in reactions analogous to reaction (4.3) as long as the R–H, where R is the inorganic radical, bond strength is larger than the C—H bond in $HCR^1R^2R^3$. Alternatively, most of these radicals can add to the unsaturated bond in alkenes or alkynes in reactions analogous to reaction (4.4). As $OH^\bullet$ is a stronger oxidizing agent than most of these radicals, their reactions are usually considerably slower and more selective than those of $OH^\bullet$ radicals [132, 133].

Some of these radicals, e.g.: $H^\bullet$, $O_2^{\bullet-}$, and $CO_2^{\bullet-}$, are also strong reducing agents and can form alkyl radicals via reduction processes. An example for this is the catalytic reduction of alkyl halides via reaction (4.19) [135, 136] followed by reaction (4.20), or, alternatively [137], via reaction (4.21), followed by (4.22) and then by reaction (4.20).

$$CO_2^{\bullet-} + RX \rightarrow R^\bullet + CO_2 + X^- \quad (4.19)$$

$$R^\bullet + HCO_2^- \rightarrow RH + CO_2^{\bullet-} \quad (4.20)$$

$$CO_2^{\bullet-} + Ni^{II}L^{2+} \rightarrow Ni^IL^+ + CO_2 \quad (4.21)$$

$$Ni^IL^+ + RX \rightarrow R^\bullet + Ni^{II}L^{2+} + X^- \quad (4.22)$$

4.2.4.2 Organic Radicals

Reactions of alkyl radicals, aryl radicals, or cyclohexadienyl radicals, $R^\bullet$ with aliphatic compounds, $HCR^1R^2R^3$, form new alkyl radicals via reaction (4.23) as long as the bond strength R–H is larger than that of $H\text{-}CR^1R^2R^3$.

$$R^\bullet + HCR^1R^2R^3 \rightarrow {}^\bullet CR^1R^2R^3 + RH \tag{4.23}$$

Thus, for example, methyl radicals react with nearly all aliphatic compounds, as the bond strength of CH_3-H is stronger than that of $H\text{-}CR^1R^2R^3$. The rate constants of many such reactions were reported [133]. The selectivity of these reactions increases with the decrease in the R–H bond strength. The $R^\bullet$ radicals react also often with alkenes and alkynes in reactions analogous to reaction (4.4). It should be noted that as in reactions (4.3) and (4.23), hydrogen abstraction might occur at several sites in the substrate often more than one radical is obtained in these reactions. The relative selectivity of different hydrogen atoms in aliphatic compounds toward reaction (4.3) was reported [69].

One should note that even in the presence of only one substrate in the solution the composition of the radicals formed is often expected to change with time. Thus, for example, $OH^\bullet$ radicals react with methanol to yield mainly ${}^\bullet CH_2OH$ but also some $CH_3CO^\bullet$. The latter radical isomerized with time into ${}^\bullet CH_2OH$, the more stable radical [138, 139]. It was also shown that the composition of the radical mixture obtained when $OH^\bullet$ radicals react with deoxyribose changes with time [140].

Thiyl radicals are often formed by abstraction of a hydrogen atom from the R′S-H substituent. However, these processes involve C—H and S—H bonds with similar bond strength and are therefore often equilibrium processes:

$$R^\bullet + R'SH \leftrightarrow RH + R'S^\bullet \tag{4.24}$$

The thiyl radicals are of importance in a wide range of synthetic processes, including the formation of alkyl radicals and in biological processes [141].

4.3 Electrochemistry

As most electron transfer reactions, including those on electrodes, involve single electron transfer processes, radicals are the primary products in these processes. In principle, these processes can involve direct redox reactions on the electrode or electro-catalytic processes in which the catalyst is dissolved in the solution or bound/adsorbed to the electrode.

4.3.1 Radicals as the Products of the Direct Oxidations of Substrates on the Electrode

The best-known process involving the formation of radicals in oxidation processes on electrodes is the Kolbe reaction (4.25) [141–145].

$$RCO_2^- - e^- \rightarrow RCO_2^\bullet \rightarrow R^\bullet + CO_2 \tag{4.25}$$

Another example is the oxidation of cysteine, CySH on NiOOH, which forms CyS-SyC via $CyS^\bullet$ [146].

4.3.2 Radicals as the Products of the Direct Reductions of Substrates on the Electrode

The electrochemical reductions of halo-organic compounds [147], diazonium compounds [148], and other organic substrates [149] yield alkyl radicals in aqueous solutions.

4.3.3 Radicals as the Products of Electro-Catalytic Oxidations

The electro-chemical oxidation can produce an oxidizing species that reacts with the aliphatic substrate present in the solution to form the alkyl radical. These processes involve between others the electro-chemical formation of $OH^{\bullet}$, $SO_4^{\bullet -}$, $CO_3^{\bullet -}$, and $H_2PO_4^{\bullet}$ as the initial oxidizing agents [55]. More complex processes include photo-electro-Fenton processes where the Fenton reaction in neutral solutions (4.1b) is followed by (4.26), thus inhibiting the precipitation of $Fe^{III}{}_{aq}$ [55].

$$Fe(H_2O)_6{}^{2+} + H_2O_2 \rightarrow Fe(OH)(H_2O)_5{}^{2+} + OH^{\bullet} + H_2O \tag{4.1b}$$

$$Fe(OH)(H_2O)_5{}^{2+} + h\nu \rightarrow Fe(H_2O)_6{}^{2+} + OH^{\bullet} \tag{4.26}$$

4.3.4 Radicals as the Products of Electro-Catalytic Reductions

Surprisingly, the most important electro-catalytic reduction process leading to the formation of alkyl radicals is the electro-catalytic Fenton reaction that is an oxidation process. This occurs via reactions (4.27)–(4.30), followed by the Fenton reaction [55].

$$O_2 + e^- \rightarrow O_2{}^{\bullet -} \tag{4.27}$$

$$O_2{}^{\bullet -} + H_3O^+ \rightarrow HO_2{}^{\bullet} + H_2O \tag{4.28}$$

$$O_2{}^{\bullet -} + HO_2{}^{\bullet} \rightarrow HO_2{}^{-} + O_2 \tag{4.29}$$

$$HO_2{}^{-} + H_3O^+ \rightarrow H_2O_2 + H_2O \tag{4.30}$$

Another important process leading to the formation of alkyl radicals is the electro-catalytic de-halogenation process catalyzed by a variety of $Ni^{II}L^{2+}$complexes [12–114, 150]:

$$Ni^{II}L^{2+} + e^- \rightarrow Ni^{I}L^{+} \tag{4.31}$$

$$Ni^{I}L^{+} + RX \rightarrow Ni^{II}L^{2+} + R^{\bullet} + X^{-} \tag{4.32}$$

Similar processes are catalyzed in the presence of other transition metal complexes that can be reduced electrochemically to form strong reducing agents, e.g. B_{12r} that can be reduced to B_{12s} [97].

4.4 Photochemistry

Four major processes to form alkyl radicals in aqueous solutions have to be considered:

4.4.1 Light Absorption that Leads Directly to the Dissociation of an R—X Bond

The example of choice for this process in aqueous solutions is the photo-homolysis of metal-carbon σ bonds:

$$LM^{n+1} - R + h\nu \rightarrow M^{n}L + R^{\bullet} \tag{4.33}$$

The most important reactions of this type are for $M = Co^{III}$ [102, 151]. The advantage of this process is that one can get a desired alkyl radical by synthesizing its cobalt complex.

4.4.2 Photolytic Formation of Active Species that React to Form Alkyl Radicals

The most important processes in this class are [152]:

$$H_2O_2 + h\nu \rightarrow 2OH^{\bullet} \tag{4.34}$$

$$S_2O_8{}^{2-} + h\nu \rightarrow 2SO_4{}^{\bullet -} \tag{4.35}$$

In the photo-Fenton reaction, the formation of Fe^{III} precipitates formed in the Fenton reaction is inhibited/diminished by the reaction (4.36), thus also increasing the $OH^{\bullet}$ radical yield [21, 153–155].

$$Fe(OH)(H_2O)_5^{2+} + h\nu \rightarrow Fe(H_2O)_6^{2+} + OH^{\bullet} \tag{4.36}$$

There are many other photochemical reactions that form an active redox reagent capable of forming alkyl radicals, see, for example, reaction (4.37) followed by (4.38) [156].

$$Cr(H_2O)_6^{2+} + h\nu \rightarrow Cr^{III}{}_{aq} + e^-{}_{aq} \tag{4.37}$$

$$e^-{}_{aq} + RX \rightarrow X^- + R^{\bullet} \tag{4.38}$$

4.4.3 Photosensitized Processes

One of the most important photo-sensitizers in inorganic chemistry is $Ru^{II}(bpy)_3^{2+}$, and analogous complexes. The excited state of this complex has a long lifetime that enables its reactions with sacrificial reducing [45, 114, 157–159] or oxidizing agents [45, 157–160].

$$Ru^{II}(bpy)_3^{2+} + h\nu \rightarrow Ru^{II}(bpy)_3^{2+*} \tag{4.39}$$

$$Ru^{II}(bpy)_3^{2+*} + N(CH_2CH_2OH)_3 \rightarrow Ru^{I}(bpy)_3^{+} + N(CH_2CH_2OH)_3^{\bullet +} \tag{4.40}$$

$$Ru^{II}(bpy)_3^{2+*} + S_2O_8^{2-} \rightarrow Ru^{III}(bpy)_3^{3+} + SO_4^{\bullet -} + SO_4^{2-} \tag{4.41}$$

The $Ru^{I}(bpy)_3^{+}$ thus formed is a powerful reducing agent and the $Ru^{III}(bpy)_3^{3+}$ is a powerful oxidizing agent that can react with appropriate substrates to yield alkyl radicals [114, 157]. Other central cations can be Ir^{III} [157], Cu^{I} [158] with analogous ligands. Other photosensitizers are often various aromatic compounds that have long-lived excited states [45] though these often are used in nonaqueous media.

4.4.4 Photocatalytic Processes Initiated by Light Absorption in Semiconductors

Light absorbed in semiconductors, e.g. in TiO_2, transfers an electron into the conduction band leaving a hole, h^+, in the valence band [161–163]. The electron in the conduction band can reduce O_2 into $O_2^{\bullet -}$ that disproportionates yielding H_2O_2 that is then reduced to form $OH^{\bullet}$ radicals, or if H_2O_2 is added to the system the $OH^{\bullet}$ radicals are formed directly [163]. The h^+ react with water to yield $OH^{\bullet}$ radicals or react directly with the substrate yielding the alkyl radical [163].

4.5 Ionizing Radiation

When ionizing radiation is absorbed in dilute aqueous solutions process (4.42) occurs [164, 165].

$$H_2O \xrightarrow{\gamma,\, e-} e^-{}_{aq}\,(2.65);\, OH^{\bullet}\,(2.65);\, H^{\bullet}\,(0.60);\, H_2\,(0.45);\, H_2O_2\,(0.75);\, H_3O^+\,(2.65) \tag{4.42}$$

where the numbers in the parenthesis are the G values, i.e. the relative yields of each product (G values are defined as the number of molecules/species of a product formed/100 eV of energy absorbed in the solution).

The problem is that $e^-{}_{aq}$ is a very strong reducing agent, $E^0 = -2.88$ V vs. SHE, [134] and $OH^\bullet$ is a very strong oxidizing agent, $E^0 = 2.73$ V vs. SHE at pH 0.0 [134]. However, one can overcome this problem by either converting the hydrated electron into $OH^\bullet$ radicals via (4.43) or by converting the $OH^\bullet$ radicals into strong reducing radicals via (4.44) [68].

$$e^-{}_{aq} + N_2O \rightarrow OH^\bullet + N_2 + OH^- \qquad k = 9.6 \times 10^9\ M^{-1}s^{-1} \tag{4.43}$$

$$OH^\bullet + HCO_2{}^- \rightarrow H_2O + CO_2{}^{\bullet -} \qquad k = 3.2 \times 10^9\ M^{-1}s^{-1} \tag{4.44}$$

The $CO_2{}^{\bullet -}$ radical anion is a very strong reducing radical, $E^0 = -1.90$ V vs. SHE [133]. Or by converting the $OH^\bullet$ radicals into $^\bullet C(CH_3)_2OH/^\bullet C(CH_3)_2O^-$, depending on the pH, via (4.45) [68].

$$\begin{aligned} &OH^\bullet + HC(CH_3)_2OH \rightarrow H_2O + {}^\bullet C(CH_3)_2OH \ \textit{or} \\ &{}^\bullet C(CH_3)_2O^- + H_3O^+ \qquad k = 1.9 \times 10^9\ M^{-1}s^{-1} \end{aligned} \tag{4.45}$$

The $^\bullet C(CH_3)_2OH/(^\bullet C(CH_3)_2O^-$ radicals are also very strong, reducing radicals $E^0 \sim\!-1.800/\!\sim\!-2.500$ V vs. SHE, respectively [166].

In analogous procedures, the primary radicals formed in reaction (4.42) can be transformed into all the radicals mentioned in Section 4.2.4.1 and into a very large variety of aliphatic and aromatic radicals [68] that have a variety of redox potentials [133, 166].

Radiation chemistry, pulse radiolysis, has an advantage over photochemistry in the study of the properties of different radicals as the energy is absorbed in the solvent [164, 165], and therefore, the addition of different substrates to the solution does not affect the nature of the primary products.

4.6 Sono-Chemistry

The sono-chemical processes are due to gaseous bubble cavitation induced by the ultrasound wave [36, 153, 167–173]. During the collapse of these bubbles, very high temperatures are formed in them, > 5000 °C. At these temperatures, the water molecules dissociate, yielding $OH^\bullet$ and $H^\bullet$ radicals. The latter react with each other, forming H_2 and H_2O_2, which are dispersed into the solution with some $OH^\bullet$ radicals that are responsible for the initiation of most sono-chemical processes. It should be pointed out that volatile organics present in the bubbles pyrolize totally and must therefore be avoided. Further alkyl radicals can be formed in the warm interface between the hot gaseous bubble and the liquid medium by thermal decomposition [170]. In some cases, combined ultrasound and electrochemical or photochemical technologies were shown to have interesting synergistic advantages [153, 169].

4.7 Mechano-Chemistry

Mechano-chemistry is usually performed in the solid phase by grinding two or more solid compounds. Sometimes some solvents are added to the mill, but as those are usually organic they are beyond the scope of this chapter. A special case is the liquid jets created when the gaseous bubbles formed in sono-chemistry collapse. These jets reach pressures of up to 2000 atm [171–173] and rupture polymers into two radicals. The rupture occurs at the weakest point in the polymer structure and thus two specific alkyl radicals are formed that can lead to desired chemical processes [171–173].

4.8 Microwave Chemistry

Many microwave-assisted organic synthetic process involve radical processes [173–175].

4.9 Production of Alkyl Radicals in Biological Systems

Two major types of radical processes occur in biological systems, as follows.

4.9.1 Enzymatic Processes

There are quite a few enzymatic processes involving alkyl radical as intermediates. Here only the major enzymatic processes involving such radicals are discussed.

Cytochrome P450: This family of enzymes is involved in oxidation processes: hydroxylations and epoxidations [81, 176]. In these oxidations the active species is $(\text{porphyrin}^{\bullet+})\text{Fe}^{\text{IV}}{=}\text{O}$ that oxidizes the substrate via the rebound mechanism, discussed above. Many analogous processes are of synthetic importance [176].

Taurine:aKG dioxygenase, TauD: [82, 177] This enzyme is also an oxidizing agent consisting of a non-heme $\text{Fe}^{\text{IV}}{=}\text{O}$ active center. Also, here an alkyl radical is formed as a transient in the oxidation process.

Adenosylcobalamin enzymes: In all these enzymes the catalytic cycle consists of the following steps [178, 179]:

$$\text{LCo}^{\text{III}}\text{–CH}_2\text{Ado(AdoCbl)} \leftrightarrow \text{LCo}^{\text{II}} + {}^{\bullet}\text{CH}_2\text{Ado} \tag{4.46}$$

(This homolysis occurs only after the substrate is bound to the enzyme.)

$$\text{RH} + {}^{\bullet}\text{CH}_2\text{Ado} \rightarrow \text{R}^{\bullet} + \text{CH}_3\text{Ado} \tag{4.47}$$

$$\text{R}^{\bullet} \rightarrow \text{R}'^{\bullet} \tag{4.48}$$

$$\text{R}'^{\bullet} + \text{CH}_3\text{Ado} \rightarrow \text{R}'\text{H} + {}^{\bullet}\text{CHAdo} \tag{4.49}$$

$${}^{\bullet}\text{CH}_2\text{Ado} + \text{LCo}^{\text{II}} \rightarrow \text{LCo}^{\text{III}}\text{–CH}_2\text{Ado(AdoCbl)} \tag{4.50}$$

Thus, the catalytic step is initiated by the homolysis of the $\text{LCo}^{\text{III}}\text{—C}$ bond. This step occurs only after binding of the substrate RH to the enzyme and is coupled with the next step, reaction (4.47) as the bond strength of the $\text{Co}^{\text{III}}\text{—C}$ bond in the enzyme is ca. $30\,\text{kcal mol}^{-1}$. Reaction (4.48) represents the enzymatic process in which the substrate R is transformed into the product R′.

There are several classes of adenosylcobalamin enzymes with different roles:

1. Carbon-skeleton mutases that catalyze the interchange of a hydrogen atom on one carbon with a carbon-containing group on an adjacent carbon. This group contains the methylmalonyl-CoA mutase, the glutamate mutase and the isobutyrl-CoA mutase.
2. The eliminases that catalyze 1,2-rearrangements of either 1,2-diols to 1,1-diols or 1,2-amino-alcohols to 1,1-amino-alcohols, that then undergoes enzyme-catalyzed elimination of water or ammonia to yield aldehydes as the final product. These enzymes include diol dehydrase, glycerol dehydrase, and ethanolamine ammonia-lyase.
3. Aminomutases catalyze 1,2-migrations of amino groups. These enzymes require in addition a pyridoxal phosphate as an additional cofactor. The enzymes belonging to this group are ornithine-4,5-aminomutase and lysine-5,6-aminomutase [178].

4. Finally, the class II ribonucleotide reductase enzyme also belongs to the adenosylcobalamin enzymes, though it catalyzes the reduction of the ribose unit in the nucleotide into the deoxyribose unit.

SAM enzymes: A somewhat related family of enzymes is the SAM enzymes. In these enzymes, a 5′-deoxyadenosyl radical is formed by the reduction of S-adenosylmethionine that is bound to a Fe_4S_4 cluster. The radical thus formed initiates the enzymatic reactions that include a large variety, e.g. class-III ribonucleotide reductase; pyruvate formate lyase activase; lysine 2,3-aminomutase; radical-mediated methylation of ribosomal RNA; FeFe-hydrogenase maturase; biotin synthase, and many others. All these enzymes operate via a hydrogen abstraction from the substrate by the 5′-deoxyadenosyl radical [178–182].

Class-III Ribonucleotide reductase: This class is the acting enzyme in mammals operating via a similar mechanism to class-II and class-III already discussed. In this enzyme, the initial H-abstraction is induced by a tyrosyl radical [183].

Glycine radical enzymes: There are quite a few enzymes in which glycyl radicals initiate the catalytic process, e.g. pyruvate formate-lyase, B_{12}-independent glycerol dehydratase, 4-Hydroxyphenylacetate decarboxylase, benzylsuccinate synthase and CutC (choline tri-methylamine-lyase) [184].

4.9.2 Processes Initiated by ROS/RNS

Reactive oxygen species, ROS, and reactive nitrogen species, RNS, include between others $OH^•$, $O_2^{•-}$, H_2O_2, $CO_3^{•-}$, $NO^•$, $NO_2^•$, NO_2^-, $ONOO^-$, and $ONOOCO_2^-$. All these are inorganic species and therefore are out of the scope of this chapter. However, in biological systems their reactions induce the formation of alkyl radicals, and that is the reason that they are mentioned here.

It should be stressed that ROS and RNS are essential to the biological systems as long as their concentration is not too high [185]. ROS and RNS have important roles in fighting viral infections, in cellular signaling pathways, and in immune response [185, 186]. However, when present in too-high concentrations ROS and RNS cause oxidative stress that has been claimed to cause aging and over 70 diseases. At present the aging claim is in debate [187–189]. The role of ROS and RNS in some diseases is now questioned, but clearly they are involved in one way or another in many health issues [186, 190–196]. It should be pointed out that all these roles of ROS and RNS are induced via their interaction with the organic components of the cells, thus forming a variety of alkyl radicals that in unknown reactions mechanisms influence health. It is of interest to note that nitroxides, which are alkyl radicals although the unpaired electron is located mainly on the oxygen and nitrogen atoms, can be used as antioxidants in biological systems and have the potential to be used as a tool to examine ROS-related processes [2].

4.10 General Comments on Radical Reactions

Though radicals differ significantly from each other, they have several similar properties, which are discussed in this section.

4.10.1 Nearly All Radicals Are Both Oxidizing and Reducing Agents

As a radical is a species with one electron in the HOMO orbital, it usually can react either by accepting another electron into this orbital, as it is clearly not too high, i.e. react as an oxidizing agent, or donate the unpaired electron, as it is in a relatively high orbital, thus acting as a reducing agent. Thus, nearly all radicals are relatively strong oxidizing and reducing agents. The best example is the hydrogen atoms that have the following redox potentials:

$$H^{\bullet} + e^- + H^+ \rightarrow H_2 \qquad E^0 = +2.31\ V\ vs.\ SHE \tag{4.51}$$

$$H^+ + e^- \rightarrow H^{\bullet} \qquad E^0 = -2.31\ V\ vs.\ SHE \tag{4.52}$$

Thus, for example, the $CO_2^{\bullet-}$ radical anion is considered a very strong reducing agent (4.53) but is also a relatively strong oxidizing agent (4.54).

$$CO_2 + e^- \rightarrow CO_2^{\bullet-} \qquad E^0 = -1.90\ V\ vs.\ SHE \tag{4.53}$$

$$CO_2^{\bullet-} + e^- + H^+ \rightarrow HCO_2^- \qquad E^0 = +1.52\ V\ vs.\ SHE \tag{4.54}$$

Furthermore, the $CO_2^{\bullet-}$ radical anion disproportionates at $pH < 3.8$ [197].

$$2CO_2^{\bullet-} \rightarrow CO_2 + HCO_2^- \tag{4.55}$$

There are only few exceptions to this: the clear-cut example is e^-_{aq} that is only a reducing agent. Usually, it is claimed that $OH^{\bullet}$ radicals are only oxidizing agents; however, the reaction (4.56) [198] that is a disproportionation reaction occurs when the two $OH^{\bullet}$ radicals encounter each other in the triplet state configuration [198].

$$2OH^{\bullet} \rightarrow H_2O + {}^3O \tag{4.56}$$

The redox potentials of methyl radicals are not known. However, as the bond dissociation energy of H_3C-H is 431 kJ mol^{-1} and of H—H is 436 kJ mol^{-1}, the oxidation potential of methyl radicals should be very similar to that of the hydrogen atom. On the other hand, the bond dissociation energy of H_3C—OH is 377 kJ mol^{-1}, which is considerably smaller than that of HO—H at 498 kJ mol^{-1}. Therefore, methyl radicals are considerably weaker reducing agents than hydrogen atoms.

4.10.2 Nearly All Redox Processes Involving Alkyl Radicals Proceed Via the Bridged Mechanism and Not Via the Outer-Sphere Mechanism

As stated above, alkyl radicals are good oxidizing and reducing agents. However, one should note that the redox potentials are always given for the final products, e.g. for methyl radicals the final products are methane or methanol, respectively. If the redox reactions would proceed via the outer-sphere mechanism, the initial products would be CH_3^- and CH_3^+, respectively. Clearly, totally different redox potentials apply for the formation of these transients. Furthermore, the CH_3^- and CH_3^+ would be formed with the solvation shell of the $CH_3^{\bullet}$ radical that is clearly considerably smaller than that required by the transient ions. Therefore, the redox reactions of alkyl radicals proceed via the bridged mechanism, the inner-sphere mechanism, and not via the outer-sphere mechanism. In this aspect the alkyl radicals behave similarly to $OH^{\bullet}$ radicals and $H^{\bullet}$ atoms [199].

This is also true for reductions by $CO_2^{\bullet-}$ [200], probably due to the different geometric structures of $CO_2^{\bullet-}$ and CO_2.

4.10.3 Three Electron-2 Center, Hemi-, Bonds

The nature of radicals in aqueous solutions is not fully understood; thus, for example, it was recently reported that $OH^{\bullet}$ radicals in aqueous media are bonded to a water molecule by a three-electron two-center bond, 3e-2c, often also called hemibond [201]. This type of bonds was first observed when R_2S compounds were oxidized by $OH^{\bullet}$ radicals, forming $R_2S\therefore SR_2^{+}$ transients [202–205]. Similarly $R_2S\therefore X$, X a halide or pseudo halide were observed [204]. Later, analogous compounds e.g. $Se\therefore N$ [205], were reported. Similarly, $S\therefore Se^{+}$ [206], $Se\therefore Se^{+}$ [206], $S\therefore Se^{+}$ [206], $S\therefore Te^{+}$ [206], $Te\therefore Te^{+}$ [206], $N\therefore N^{+}$ [206], $S\therefore O^{+}$ [207], $N\therefore Cl$ [208], $Se\therefore N^{+}$ [209], $P\therefore^{+}$ [210], were observed. $S\therefore S^{-}$ and their transition metal complexes were observed [211, 212]. Also $H_2S\therefore F$, $H_2S\therefore Cl$, $H_2S\therefore Br$ and $H_2S\therefore OH$ were observed [213]. Finally, it was suggested that 3e—2c bonds act as intermediates in several radical reactions [214, 215]. Thus, the question of whether this is not the case in many other radical reactions has to be considered.

4.10.4 Radical—Radical Reactions

The reaction between two radicals leads to colligation and/or to disproportionation. These reactions for alkyl radicals are usually very fast, approaching the diffusion-controlled limit [68]. It should, however, be noted that colligation, and often disproportionation, are only possible if the two radicals encounter each other in the singlet state configuration due to the spin conservation law. If they encounter each other in the triplet state configuration, then no reaction is expected. This phenomenon expresses itself in the CIDNP effect observed for the colligation reactions of several alkyl radicals in aqueous solutions [216–218].

The colligation reactions of unsubstituted alkyl radicals in aqueous solutions are not well studied. However, the reaction between two ethyl radicals was studied in detail [219], and the results point out that k(disproportionation)/k(colligation) = 0.35 in aqueous solutions. Theoretical calculations, in the gas phase, for the reaction $CH_3^{\bullet} + C_2H_5^{\bullet}$ yield similar values [220, 221]. As the disproportionation is not expected to proceed via an electron transfer process, see previous paragraph, it has to proceed via hydrogen atom abstraction, as confirmed by the theoretical studies [220, 221]. The rate constant of the reaction (4.57) [222] is as expected, approaching the diffusion controlled limit.

$$2C_2H_5^{\bullet} \rightarrow \begin{cases} C_4H_{10} \\ C_2H_4 + C_2H_{10} \end{cases} \qquad k = 7 \times 10^8\ M^{-1}s^{-1} \tag{4.57}$$

It is of interest to note that the disproportionation reaction is orders of magnitude faster than the hydrogen abstraction reactions from alkanes by alkyl radicals [68]. Consider the following reactions (4.58) [22], (4.59) [223, 224] and (4.60) [223].

$$2CH_2OH^{\bullet} \rightarrow (-CH_2OH)_2 \qquad k = 1.5 \times 10^9\ M^{-1}s^{-1} \tag{4.58}$$

$$2CH_3CHOH^{\bullet} \rightarrow \begin{cases} C_4H_{10}O_2 \quad 32\% \\ CH_3CHO + CH_3CH_2OH \quad 68\% \end{cases} \qquad k = 1.2 \times 10^9\ M^{-1}s^{-1} \tag{4.59}$$

$$2(CH_3)_2COH^{\bullet} \rightarrow (CH_3)_2CO + (CH_3)_2CHOH \qquad k = 6.5 \times 10^8\ M^{-1}s^{-1} \tag{4.60}$$

Other simple colligation reactions worth mentioning are (4.61), where the relative yields of the two routes are pH dependent [197], and reaction (4.62), where the relative yields are not known [224].

$$2COOH^{\bullet} \rightarrow (-COOH)_2$$
$$2COOH^{\bullet} \rightarrow CO_2 + HCOOH \qquad (4.61)$$

$$2HCO^{\bullet} \rightarrow (-CHO)_2$$
$$2HCO^{\bullet} \rightarrow CO + HCHO \qquad (4.62)$$

Naturally, alkyl radicals react in very fast reactions also with inorganic radicals, e.g. $OH^{\bullet}$, $H^{\bullet}$, $O_2^{\bullet-}$, [225] $NO^{\bullet}$, [226] $NO_2^{\bullet}$ [227] and with other organic radicals, e.g. nitroxides [228].

4.10.5 Measuring the Lifetime of Very Short-Lived Radicals

In many processes that presumably involve the formation of a radical as a short-lived intermediate in very low steady-state concentrations, it is difficult to prove the formation of a radical intermediate. This is correct, for example, in the rebound mechanism discussed in Section 4.2.2 and in many catalytic processes. To prove the formation of a radical as a transient in these systems, often radical clocks are used [40, 229, 230]. In principle, a radical clock involves the formation of a radical that isomerizes intra-molecularly with a known rate. Thus, for example, if a radical $R^{\bullet}$ rearranges into the $R'^{\bullet}$ radical within a known rate constant (4.63) and if $R^{\bullet}$ and $R'^{\bullet}$ react with YX via (4.64), then if R′X is the sole product obtained, the radical $R^{\bullet}$ was formed and its lifetime in the system is long enough to enable reaction (4.63) to proceed to completion.

$$R^{\bullet} \rightarrow R'^{\bullet} \qquad t_{1/2} = \text{A (a known value)} \qquad (4.63)$$

$$R^{\bullet} + YX \rightarrow RX + Y^{\bullet} : R'^{\bullet} + YX \rightarrow R'X + Y^{\bullet} \qquad (4.64)$$

If no R′X is formed, then there is no proof for the formation of the radical $R^{\bullet}$ and another probe for which $t_{1/2}$ is shorter has to be applied. If a mixture of RX and R′X is obtained, there is proof of the formation of the radical in the process and its lifetime can be determined. There are today known radicals R· for which k_{63} ranges between 3×10^2 s^{-1} and 5×10^{11} s^{-1}, [229] not all of them soluble in aqueous solutions.

4.11 Unimolecular Reactions

4.11.1 Isomerization Reactions

Many radicals isomerize to different structures, these include all the radical clocks discussed in the previous section i.e. intra-molecular addition of a radical to an alkene or a small ring opening [229, 230]. Also important are the carbon skeleton isomerizations catalyzed by B_{12} enzymes: e.g. glutamate mutase that catalyzes the reaction (4.65) [231] and the 2-methyleneglutarate mutase; methylmalonyl-CoA mutase; isobutyryl-CoA mutase [231, 232]. Analogous reactions are electrochemically catalyzed by B_{12} analogous complexes [233].

$$^{\bullet}CH(CO_2H)-CH_2CH(NH_2)(CO_2H) \rightarrow {}^{\bullet}CH_2-CH(CO_2H)(CH(NH_2)(CO_2H) \qquad (4.65)$$

A totally different fast isomerization reaction is (4.66) [234]

$$R^1R^2HCO^{\bullet} \rightarrow R^1R^2COH^{\bullet} \qquad (4.66)$$

4.11.2 β-Eliminations

When a radical is formed on a carbon α to a hydroxyl in a vicinal diol, an elimination of a hydroxyl occurs, e.g. (4.67) this reaction is acid and base catalyzed [235].

$$^{\bullet}CH(OH)CH_2OH \rightarrow {}^{\bullet}CH_2CHO + H_2O \tag{4.67}$$

This reaction is also catalyzed by co-enzyme B_{12} elimination reactions [236]. Furthermore this reaction is a key step in single strand brakes of DNA and RNA when hydroxyl radicals react with the deoxyribose/ribose and a phosphate ester is cleaved [237, 238].

4.11.3 Methyl Radical Eliminations

Reaction (4.6) is such a reaction. Another example is reaction (4.68) [239].

$$C(CH_3)_3O^{\bullet} \rightarrow C(CH_3)_2O + {}^{\bullet}CH_3 \tag{4.68}$$

4.12 Reactions of Alkyl Radicals with Different Substrates

It should be noted that when a radical reacts with a molecule that is not a radical, one of the products has to be another radical that is active on its own. Therefore, mechanistic studies are always complicated and stop only after termination reactions that include radical–radical reactions or radical transition–metal complexes reactions.

4.12.1 Reactions with Radicals

These reactions were discussed in Section 4.10.4. It should be added here that the reactions with dioxygen, a bi-radical, and with metals and semiconductors discussed below should in principle be included in this section as the unpaired electrons in the conduction bands behave as unpaired electrons in radicals.

4.12.2 Reactions with Organic Substrates

The width of radical reactions with organic substrates inhibits a full coverage of them. Therefore, only the major types of reaction will be mentioned.

Abstraction reactions: The most common abstraction reaction is the hydrogen atom abstraction:

$$R^{\bullet} + R'H \rightarrow RH + R'^{\bullet} \tag{4.69}$$

This reaction takes place in all systems with preference to the formation of the most stable radical [68]. Thus, for example, reactions of this type include antioxidants, e.g. ascorbate; tocopherol, and RSH compounds. Some of these reactions are relatively slow, e.g. (4.70) [240].

$$CH_3^{\bullet} + (CH_3)_2SO \rightarrow CH_4 + {}^{\bullet}CH_2(CH_3)SO \qquad k = 1.0 \times 10^2\ M^{-1}s^{-1} \tag{4.70}$$

However, other abstraction reactions are also prevalent, e.g. halogen atom abstraction (4.71) where R′ can be an organic compound, and (4.72), whereas the analogous reaction (4.73) proceeds via a proton-coupled electron transfer mechanism [241].

$$R^{\bullet} + R'X \rightarrow RX + R'^{\bullet} \tag{4.71}$$

$$C(CH_3)_2OH^\bullet + ICH_2C(O)NH_2/CCl_4 \rightarrow$$
$$\rightarrow C(CH_3)_2O + {}^\bullet CH_2C(O)NH_2/CCl_3{}^\bullet + I^-/Cl^- + H^+ \quad (4.72)$$

$$C(CH_3)_2OH^\bullet + ICH_2CO_2{}^- \rightarrow C(CH_3)_2O + CH_2CO_2{}^{\bullet-} + I^- + H^+ \quad (4.73)$$

R′ can also be an inorganic species as in (4.74) [242].

$$CH_3{}^\bullet + IrCl_6{}^{2-} \rightarrow IrCl_5(H_2O)^{2-} + CH_3Cl \qquad k = 1.2 \times 10^9\ M^{-1}s^{-1} \quad (4.74)$$

Another important abstraction reaction (4.75) [68] will be discussed in Section 4.12.5.

$$L_5M^{n+1} - CH_3 + CH_3{}^\bullet \rightarrow M^nL_5(H_2O) + C_2H_6 \quad (4.75)$$

Addition to alkenes, alkynes, and aromatic compounds: The general reaction (4.76) is of major importance.

$$R^1R^2C{=}CR^3R^4 + R^\bullet \rightarrow {}^\bullet C(CRR^1R^2)R^3R^4 \quad (4.76)$$

See, for example, dozens of such reactions for methyl radicals in reference [68]. Analogous examples are reactions (4.77)–(4.80) [68].

$$CH_3CH_2CH{=}NO_2{}^- + {}^\bullet CH_3 \rightarrow CH_3CH_2CH(CH_3)NO_2{}^{\bullet-} \quad k = 1.4 \times 10^7\ M^{-1}s^{-1} \quad (4.77)$$

$$(CH_3)_2C{=}NO_2{}^- + {}^\bullet CH_3 \rightarrow [(CH_3)_3CNO_2]^{\bullet-} \qquad k = 2.4 \times 10^6\ M^{-1}s^{-1} \quad (4.78)$$

$$(CH_3)_3C(NO) + {}^\bullet CH_2CO_2{}^- \rightarrow (CH_3)_3CN(O)CH_2CO_2{}^- \quad k = 7 \times 10^6\ M^{-1}s^{-1} \quad (4.79)$$

$$CO + {}^\bullet CH_3 \rightarrow CH_3CO \qquad k = 1.8 \times 10^6\ M^{-1}s^{-1} \quad (4.80)$$

The following reaction (4.81) proceeds via addition of the radical to the ring at the ortho position to the bromo substituent, followed by consecutive elimination of acetone and bromide [240].

$${}^\bullet C(CH_3)_2OH + \text{5-bromo-uracil} \rightarrow C(CH_3)_2O + \text{uracil}^\bullet + Br^- + H^+ \quad (4.81)$$

On the other hand, the reaction of 8-bromoguanine derivatives with $R^1R^2COH^\bullet$ radicals proceeds via addition of the radical at the 8 position followed by bromide elimination [243]. Of special interest are intra-molecular addition reactions that form macro-cyclic products [244].

Spin traps: A specially important radical addition reaction is that for spin traps, S = T (4.82) [245] where the R-S-T$^\bullet$ radical is long lived and can therefore be detected by EPR, i.e. can be characterized.

$$R^\bullet + S{=}T \rightarrow \text{R-S-T}^\bullet \quad (4.82)$$

Commonly used spin traps are: 5,5-dimethyl-1-pyrroline-N-oxide (DMPO) and N-tertiary-butyl nitrone (PBN), among others.

4.12.3 Reduction or Oxidation Processes

Alkyl radicals are strong oxidizing agents and are involved in a variety of oxidation processes. Thus, for example, in reaction (4.67) one of the carbon atoms is oxidized. All the reactions of alkyl radicals [68], or nitroxides [246], with ethanol or 2-propanol that lead to the formation of ${}^\bullet CH(CH_3)OH$ and ${}^\bullet C(CH_3)_2OH$, respectively, and analogous reactions, are oxidation processes that lead to the formation of aldehydes or ketones respectively, reactions (4.59) and (4.60). Also the reactions of alkyl radicals with thiols-forming thiyl radicals are oxidation reactions:

$$\text{RS-H} + R'^\bullet \leftrightarrow RS^\bullet + R'H \quad (4.83)$$

These reactions are important both in biology and in organic synthesis [68, 140, 247]. Furthermore, see Sections 4.12.5 and 4.12.6, alkyl radicals were shown to oxidize low-valent metal complexes and metal surfaces. It was also shown that methyl and $^{\bullet}CR^1R^2OH$ radicals oxidize $PW_{11}O_{39}{}^{8-}$ in aqueous solutions [248].

All alkyl radicals are also reducing agents; thus, for example, methyl radicals are oxidized by copper metal, as described in Section 4.12.5. Also, all the radicals involved in the rebound mechanism (Section 4.2.2, are formally acting as reductants. However, the commonly alkyl radicals used in a variety of reduction processes are the $R^1R^2COH^{\bullet}$ and their alkaline forms $R^1R^2CO^{\bullet-}$. (This is correct if the $CO_2{}^{\bullet-}$ radical anions are not included as they are formally inorganic radicals.) These radicals are powerful reducing agents [166]; some examples of their reactions are:

1. De-halogenation processes:

$$RX + R^1R^2COH^{\bullet}/R^1R^2CO^{\bullet-} \rightarrow R^{\bullet} + X^- + (R^1R^2CO + H^+)/R^1R^2CO \quad (4.84)$$

Usually, the detailed mechanism of reaction (4.84) was not elucidated; however, in some cases, mainly when R is an aromatic residue, a transient was observed or suggested, indicating that the mechanism does not occur via the outer sphere mechanism [243]. In some cases, these processes are catalytic chain reductions as the $R^{\bullet}$ radical reacts via (4.85) [249, 250].

$$R^{\bullet} + R^1R^2CHOH \rightarrow RH + R^1R^2COH^{\bullet}/(R^1R^2CO^{\bullet-} + H^+) \quad (4.85)$$

2. Reduction of RNO_2, R often an aryl, compounds:

$$RNO_2 + R^1R^2COH^{\bullet} \rightarrow RNO_2{}^{\bullet-} + R^1R^2CO + H^+ \quad (4.86)$$

This reaction is often followed by an intramolecular reduction of R by the $NO_2{}^{\bullet-}$ substituent [251–253], including aryl de-halogenation processes [252]. In a similar way, the radicals $R^1R^2COH^{\bullet}$ reduce also pyrazinecarboxylate [253, 254], pyrimidinecarboxylate [253], methyl-pyridinium carboxylate [253], nicotinamide [255], and 8-hydroxyquinoline [256]. It is of interest to note that in the latter two systems, intermediates that are radical adducts to the aromatic ring are observed [255, 256]. $R^1R^2COH^{\bullet}$ also reduce grapheme-oxide suspended in aqueous solutions [257].

The $R^1R^2COH^{\bullet}/R^1R^2CO^{\bullet-}$ also reduce transition metal complexes; see, for example, reaction (4.74). The kinetics of reduction of $Co(NH_3)_5X^{2+}$ and $Ru(NH_3)_5X^{2+}$ were studied in detail [258]. Of special interest are the results for $Co(NH_3)_6{}^{3+}$ and $Ru(NH_3)_6{}^{3+}$ that prove that their reductions by the $R^1R^2COH^{\bullet}$ radicals do not proceed via the outer-sphere mechanism [258].

4.12.4 Reactions with Dioxygen

All alkyl radicals react in fast reactions, with rate constants approaching the diffusion controlled limit, to form alkyl-peroxyl radicals (4.87) [68, 259]. These radicals, in the absence of appropriate substrates, decompose via (4.88). The rate constants of reactions (4.88) are relatively small for bi-radical reactions [68, 259].

$$R^{\bullet} + O_2 \rightarrow RO_2{}^{\bullet} \quad (4.87)$$

$$2RO_2{}^{\bullet} \rightarrow ROOOOR \quad (4.88)$$

The transient products ROOOOR decompose into a variety of products. Transients of the type $R_2CHOOOOCHR_2$ decompose via (4.89); the detailed composition of the products depends on the nature of R, on the pH of the solution, and on the temperature [260–263].

$$\begin{aligned} R_2CHOOOOCHR_2 &\rightarrow R_2CO + R_2CHOH + O_2 \\ &\rightarrow 2R_2CO + H_2O_2 \\ &\rightarrow 2R_2CHO^\bullet + O_2 \\ 2R_2CHO^\bullet &\rightarrow R_2CHOOCHR_2 \\ &\rightarrow R_2CO + R_2CHOH \\ &\rightarrow 2R^\bullet + 2RCHO \qquad (2R^\bullet + 2O_2 \rightarrow 2RO_2{}^\bullet) \\ &\rightarrow 2R_2COH^\bullet \\ &(2R_2COH^\bullet + 2O_2 \rightarrow 2R_2C(OH)OO^\bullet \rightarrow 2R_2CO + 2HO_2{}^\bullet) \end{aligned} \tag{4.89}$$

The $RO_2{}^\bullet$ radicals are relatively strong oxidizing agents, $0.6 < E^0 < 1.1$ V vs. NHE at pH 7 depending on the substituents on the carbon atom bound to the oxygen [150, 166, 264, 265].

Probably the most important reaction of alkyl-peroxyl radicals in biological systems is (4.90) [266–272]. This is a chain reaction that results in oxidative damage at many sites and produces a hydro-peroxide in each step. Therefore, an important role of antioxidants is to cut this chain short by scavenging the alkyl-peroxyl radicals [267, 268, 273].

$$RO_2{}^\bullet + R'H \rightarrow RO_2H + R'^\bullet \tag{4.90}$$

The mechanism of oxidation by alkyl-peroxyl radicals was studied by several groups [274–278]. The results clearly point out that electron-withdrawing substituents on the α carbon enhance the rate constants of these reactions. The detailed oxidation mechanisms seems to proceed via the inner-sphere mechanism [274] and/or via the hydrogen atom transfer mechanism.

Another important reaction of alkyl-peroxyl radicals is the elimination of peroxide/hydroperoxyl radical when the radical formed has the O_2 and an OH or OR bound to the same carbon (4.91) [279, 280]. This reaction has a major effect on the yield of oxidized products in alcohol oxidation by the Fenton reaction in aerated solutions [281].

$$R^1R^2C(OO^\bullet)OH \rightarrow R^1R^2CO + HO_2{}^\bullet \qquad 400 < k > 2600\ s^{-1} \tag{4.91}$$

Another reaction that is of importance in the Fenton-catalyzed processes is the reaction of the alkyl-peroxyl radicals with the low-valent transition-metal cation involved (4.92) [282].

$$M^nL_l + ROO^\bullet \rightarrow L_{l-1}M^{n+1}\text{-}OOR + L \tag{4.92}$$

Reaction (4.92) was observed for $M^nL_l = Fe(H_2O)_6{}^{2+}$ [150], Cu^{II}(glycine)$_2$ [283], Cu^{II}(glycylglycylglycine) [283] and Co^{II}(2,3,9,10-tetramethyl-1,4,8,11-tetraazacyclotetradeca-1,3,-8,10-tetraene)$^{2+}$ ($Co^{II}TIM)^{2+}$) [284]. The mechanisms of decomposition of the $L_{l-1}M^{n+1}$—OOR complexes formed in reaction (4.92) differ for each complex. The $Fe^{III}(OOR)(H_2O)_5{}^{2+}$ complex is short-lived and decomposes via three routes: homolysis of the Fe^{III}—OO bond, heterolysis of the same bond yielding $Fe(H_2O)_6{}^{3+}$, and via the Fenton reaction, reaction (4.10) [150]. The Cu^{III}(OOH)(glycine)$_2$ and the Cu^{III}(OOR)(glycylglycylglycine) decompose first by oxidation of the ligands by the peroxyl, leaving a Cu^{III} complex that then oxidizes the remaining ligand [283]. The (TIM)Co^{III}-OOR^{2+} complexes are stable for several days; however, they react with $CH_3OO^\bullet$ that oxidizes the bound peroxide to form the transient (TIM)Co^{III}-($^\bullet$OOR)$^{3+}$ that then either decomposes via loss of the $^\bullet$OOR radical leaving a stable $Co^{III}(TIM)^{3+}$ complex or via internal electron transfer between the bound radical and the central cobalt ion. The choice between the latter mechanisms depends on the nature of R [284].

Alkyl-peroxyl radicals naturally react with other radicals in fast reactions; thus, for example, they react with $NO_2^{\bullet}$ to form $ROONO_2$, which is in equilibrium with the parent radicals [285]. They react with NO to form ROONO that decomposes into $RO^{\bullet} + NO_2^{\bullet}$ in a cage as a part of these radicals recombine to yield $RONO_2$ and a part diffuses into the solution [285].

Finally, it should be mentioned that alkyl-peroxyl radicals are of importance in environmental Fenton processes [286] and in the aqueous phase in the atmosphere [287].

4.12.5 Reactions with Low-Valent Transition-Metal Complexes Forming Complexes with Metal-Carbon σ Bonds

A very important reaction of alkyl radicals is that with low-valent transition metal complexes (4.93).

$$M^nL_l + R^{\bullet} \rightarrow L_{l-1}M^{n+1}\text{-}R + L \tag{4.93}$$

This reaction is of importance as low-valent transition metal complexes are involved in the formation of the radicals; see, for example, reactions (4.7) +(4.8); (4.17); Scheme 4.1; (4.22); (4.32); (4.33); (4.36); (4.37) and (4.46). Furthermore, these reactions might be of importance in biological systems [288, 289]. Reactions of this type were reported for: Ni^+_{aq} [290, 291], Zn^+_{aq} [291], Cu^+_{aq} [292–297], Cd^+_{aq} [291, 298], $Cd_2^{2+}{}_{aq}$ [291, 298], Cr^{2+}_{aq} [299–301], V^{2+}_{aq} [291], Fe^{2+}_{aq} [302], Cu^{II}_{aq} [303], Ti^{III}_{aq} [302] and a variety of complexes of Ni^I [102, 103, 304], Cu^I [305], Cr^{II} [306, 307], Mn^{II} [308], Fe^{II} [308–310], Co^{II} [308, 311–319], Ni^{II} [320–322], Cu^{II} [323–326] and Pt^{II} [327]. The kinetics of these reactions, where analyzed, fit all an inner sphere mechanism that is controlled by the rate of the ligand exchange on the central cation [328–330].

Reaction (4.93) is often an equilibrium process, i.e. the $L_{l-1}M^{n+1}$-R complex decomposes homolitically [156, 292, 294, 306–308, 316, 319, 321, 322, 329, 331–335]:

$$M^nL_l + R^{\bullet} \leftrightarrow L_{l-1}M^{n+1}\text{-}R + L \tag{4.94}$$

The observation that the mechanism of reaction (4.94) is a ligand exchange mechanism either via I_d or an I_a mechanism [328–330] points out that one cannot determine the bond strength of the M^{n+1}—C bond from $\Delta H^{\#}$ of k_{-94} as this process involves partially the formation of the M—OH_2 bond.

In the cases where the homolysis rate is observable, it can be used for the insertion of dioxygen into the M—C bond [316, 319, 321, 331].

Of special interest are the reactions [306, 308, 319, 321, 322, 333–335]:

$$L_{l-1}M^{n+1}\text{-}R + R^{\bullet} \rightarrow M^nL_l + R\text{-}R/(R_{-H} + RH) \tag{4.95}$$

The detailed mechanism of this reaction might involve one of the following mechanisms:

$$LM^{n+1}\text{–}CH_3 + \bullet CH_3 \longrightarrow LM^{n+1}\text{–}CH_3\cdots CH_3 \longrightarrow LM^{n+} + C_2H_6 \tag{4.96}$$

$$LM^{n+1}\text{–}CH_3 + \bullet CH_3 \longrightarrow LM^{n+2}(CH_3)\text{–}CH_3 \longrightarrow LM^{n+} + C_2H_6 \tag{4.97}$$

For $L_{l-1}M^{n+1}$-R = $(NH_3)_5Co^{III}$-CH_3 mechanism (4.96) was proven [315]. This mechanism is probably true also for $L_{l-1}M^{n+1}$-R = $L(H_2O)Ni^{III}$-R [304, 319], where L is a tetra-aza-macro-cyclic ligand, due to the fact that in this complex no *cis* position to R is open. On the other hand mechanism (4.97) is probably correct for $L_{l-1}M^{n+1}$-R = (nta)$(H_2O)Co^{III}$-R [321].

Also, the formation of R-R without homolysis of the M—C bond was observed for $Cu^{II}CH_3^+$. This reaction proceeds via the following mechanism [297]:

$$2LCu^{II}\text{–}CH_3^+ \longrightarrow \{LCu^{II}\text{---}C\text{---}C\text{---}Cu^{II}L\}^{2+} \longrightarrow 2LCu^+ + C_2H_6 \quad (4.98)$$

The major other route of decomposition of the $L_{l-1}M^{n+1}$-R complexes is heterolysis of the metal-carbon bond. The heterolysis in aqueous solutions results in the formation of $L_{l-1}M^{n+1}$-OH_2 for complexes where this product is stable [297, 307, 317, 332, 336, 337] or of $L_{l-1}M^{n-1}$-OH_2 where reduction of the central cation leads to a more stable product [297, 303, 317].

Another important route of decomposition of the $L_{l-1}M^{n+1}$-R complexes is via β-elimination that occurs only when the $L_{l-1}M^{n+1}$-OH_2 complex is relatively stable. Two types of β-elimination were reported: β-hydride shift (4.99) [129, 130, 291, 306, 314, 317, 327, 338],

$$L_{l-1}M^{n+1}\text{-}CR^1R^2CHR^3R^4 \rightarrow L_{l-1}M^{n+1}\text{-}H + R^1R^2C{=}CR^3R^4 \quad (4.99)$$

and β-elimination of good leaving groups, e.g. OH [292, 299, 303, 310, 311, 314, 339–342]; OR [311, 342, 343]; NH_3^+ [344, 345]; NHR^1R^2 [346]; OPO_3 [347]; etc. (4.100).

$$L_{l-1}M^{n+1}\text{-}CR^1R^2CXR^3R^4 \rightarrow L_{l-1}M^{n+1}\text{-}OH_2 + R^1R^2C{=}CR^3R^4 + X^- \quad (4.100)$$

In one system (4.101) also β-elimination of CO_2 from a carboxylate was observed [348].

$$Cu^{III}\text{-}CH_2C(CH_3)_2CO_2H^{2+}{}_{aq} \rightarrow Cu^+{}_{aq} + CO_2 + CH_2{=}C(CH_3)_2 \quad (4.101)$$

The $L_{l-1}M^{n+1}$-R complexes are considerably stronger reducing agents than the $L_{l-1}M^{n+1}$-OH_2 complexes [324, 332, 349].

The $L_{l-1}M^{n+1}$-R complexes often transfer the R group to another metal complex that is a stronger reducing agent than $L_{l-1}M^n$. These reactions often proceed without homolysis of the M—C bond, probably via a S_H2. Thus, for example, the reactions (4.102) [350] and (4.103) [319] were observed.

$$RCoL^1(H_2O)^{2+} + Cr^{2+}{}_{aq} \rightarrow R\text{-}Cr^{2+}{}_{aq} + CoL^1(H_2O)_2{}^{2+} \quad (4.102)$$

$$L(H_2O)Ni^{III}\text{-}CH_3{}^{2+} + Cr(H_2O)_6{}^{2+} \rightarrow LNi^{2+} + (H_2O)_5Cr^{III}\text{-}CH_3{}^{2+} \quad (4.103)$$

Also reaction (4.104), that proceeds via an S_E2 mechanism, was reported [301, 332].

$$(H_2O)_5Cr\text{-}c\text{-}C_5H_9{}^{2+} + Hg^{2+} \rightarrow Cr(H_2O)_6{}^{3+} + Hg\text{-}C_5H_9{}^+ \quad (4.104)$$

In some cases, interesting isomerizations of the R group in M-R complexes were observed. Thus, for example, the complex $NC(CH_2)_3CH_2Ni(tmc)^+$ is involved in reaction (4.105), where tmc represents the tetraazamacrocycle [102] and the complex 1-cuprio-2-hydroxycyclohexyl(1+) rearranges into 6-hydroxy-7λ^2-cupra-bicyclo[3.2.0]heptane. The latter intermediate decomposes via heterolysis of one of the copper(II)-carbon σ bonds to form (1-cupriocyclopentyl)methanol(1+) [351].

$$NC(CH_2)_3CH_2Ni(tmc)^+ + 2H_2O \rightarrow Ni(tmc)^{2+} + c\text{-}(CH_2)_4CO + NH_3 + OH^- \quad (4.105)$$

Finally, all the living radical polymerization catalyzed by transition metal complexes clearly involve alkyl radicals and transients with M—C σ bonds [352–355].

4.12.6 Reactions with Surfaces and Nanoparticles

Reactions of alkyl radicals at surfaces have been suggested to be involved in variety of catalytic [356–358], photo-catalytic [225, 359, 360], electrochemical [361, 362], synthetic [363, 364], geological [365], mechano-chemical [366], and environmental processes [360, 367]. Methyl radicals bound to catalytic surfaces are key intermediates in the catalytic utilization of methane, e.g. in the oxidation of methane to methanol and formaldehyde [368–370]. These observations induced the study of the reactions of alkyl radicals with metal powders immersed in aqueous solutions [114, 240, 371]. The results point out that indeed alkyl radicals react in fast processes with the metal powders. The reaction of methyl radicals proceeds via (4.106) for $M^0 = Cr^0$, Mn^0, Fe^0, Ni^0, and Zn^0 [240].

$$[M^0]_{powder} + 2CH_3^{\bullet} \rightarrow 2CH_4 + M^{2+}{}_{aq} + [M^0]_{powder} \tag{4.106}$$

However, for $M^0 = Cu^0$ the reaction product is CH_3OH and not methane [240]. These results clearly point out the fast formation of transients of the type $[M^0]_{powder}$-CH_3 [240]. Indeed, the reaction of $CCl_3^{\cdot}$ radicals with Fe^0 powders yields methane via the same mechanism [371]. Furthermore, the reactions of $^{\bullet}CR^1R^2CR^3R^4X$, X = OH or NH_3^+, with $M^0 = Mn^0$, Fe^0, Co^0, Ni^0, and Zn^0 proceed via the following reaction (4.107) [114]:

$$\begin{aligned} & \rightarrow [M^{\circ}]^{+}{}_{powder} + R^1R^2C{=}CR^3R^4 + X^- \\ [M^{\circ}]_{powder} + {}^{\bullet}CR^1R^2CR^3R^4X & \rightarrow [M^{\circ}]_{powder} - CR^1R^2CR^3R^4X \\ & \rightarrow [M^{\circ}]^{+}{}_{powder} + HCR^1R^2CR^3R^4X \end{aligned} \tag{4.107}$$

The formation of the alkene clearly proves that an intermediate of the type $[M^0]_{powder}$-$CR^1R^2CR^3R^4X$ is formed in the reaction [114]. Furthermore when a considerable negative potential is applied on the Fe^0 powders the mechanism of decomposition of the transient chemically-adsorbed species is affected dramatically [372, 373].

As the kinetics of reaction of the radicals with the metal powders immersed in aqueous solutions are difficult to analyze, the reactions of alkyl radicals [374–377] and alkyl-peroxyl radicals [378] with M^0-nanoparticles, (M^0-NPs), and with (TiO_2-NPs) suspended in aqueous solutions were studied. The results point out that for (Ag^0-NPs) [374], (Au^0-NPs) [374], and (TiO_2-NPs) [375], TiO_2 mainly rutile, their reactions with methyl radicals follow this mechanism:

$$(M^0\text{-NPs}) + nCH_3^{\bullet} \rightarrow (M^0\text{-NPs})\text{-}(CH_3)_n \qquad k_{108} > 10^7\ M^{-1}s^{-1} \tag{4.108}$$

$$(M^0\text{-NPs})\text{-}(CH_3)_n \rightarrow (M^0\text{-NPs})\text{-}(CH_3)_{n-2} + C_2H_6 \tag{4.109}$$

The lifetime of the transients (M^0-NPs)-$(CH_3)_{n-2}$ is relatively long 1 seconds $< t_{1/2} >$ 180 seconds [375]. On the other hand, the product of the analogous reaction with (Cu^0-NPs) is methane [377]. The reaction of methyls with (Pt^0-NPs) is more complex [376]:

$$\begin{aligned} & & & \rightarrow m(C_2H_4) \quad m = \text{traces} \\ (Pt^0\text{–NPs}) + nCH_3^{\bullet} & \rightarrow (Pt^0\text{–NPs})\text{–}(CH_3)_n & & \rightarrow (Pt^0\text{–NPs})\text{–}(CH_3)_{n-2m-2l} \\ & & & \rightarrow l(C_2H_6) \quad l = \text{traces} \end{aligned} \tag{4.110}$$

The product (Pt^0-NPs)-$(CH_3)_{n-2m-2l}$ is stable for at least a month and reacts with H_2 to release CH_4 [377].

Also the reactions (4.101) are very fast [378].

$$(M^0{-}NPs) + CH_3O_2^{\bullet} \rightarrow (M^0{-}NPs){-}OOCH_3 \qquad (M = \text{Ag or Au})$$
$$k_{111}(\text{Au}) = 3.5 \times 10^8 M^{-1}s^{-1}; k_{111}(\text{Ag}) = 1.9 \times 10^9 M^{-1}s^{-1} \qquad (4.111)$$

The transient complexes $(M^0{-}NPs){-}OOCH_3$ decompose via (4.112) [378].

$$(M^0{-}NPs){-}OOCH_3 + H^+ \longrightarrow (M^0{-}NPs) = O^+ + CH_3OH \qquad (4.112)$$

The observation that all the reactions of the radicals studied with (M^0-NPs) or semiconductors are fast is attributed to the fact that the electrons in the conduction bands of these NPs are unpaired and therefore are likely to react with radicals.

Finally, recent results point out that the Au^0-NPs catalyzed reduction of halo-acetic acids by BH_4^- involves the formation of $CH_nX_{2\text{-}n}CO_2^{\bullet -}$ radicals, n = 0, 1, or 2 [126]. The radicals thus formed react with the Au^0-NPs [126].

4.13 Concluding Remarks

This chapter outlines the major routes to the production of alkyl radicals in aqueous solutions and the major types of reaction of simple alkyl radicals. Clearly, alkyl radicals are also formed in more complex systems – see, e.g. the reactions of $OH^{\bullet}$ radicals with penicillin [379]. The problem in these studies is that the $OH^{\bullet}$ radical can attack at several sites in the substrate. The formation of these initial radicals probably is often followed by reactions like reaction (4.69) forming at the end a mixture of more stable radicals than those formed initially. As reactions (4.69) are often slow [68], their contribution in flash-photolysis and pulse-radiolysis experiments is considerably smaller than in samples photolyzed or γ irradiated due to the competition between reactions (4.69) and radical-radical reactions, e.g. reaction (4.57), that are usually very fast [68].

One can, in principle, prepare a desired radical also in complex systems by synthesis of a halo-substituted compound, followed by its reduction, e.g. reactions (4.17) or (4.19). Alternatively, the use of $CH_3^{\bullet}$ radicals as the oxidizing agents will produce a much smaller number of radicals because the $CH_3^{\bullet}$ radicals are considerably more selective in their reactions than the $OH^{\bullet}$ radicals. [68].

Finally, it should be pointed out that though alkyl radicals are key intermediates in a large variety of processes, their mode of reaction in many systems is still not clear.

Acknowledgments

I am indebted to all my students and collaborators whose work is cited in this chapter, as without them I would have never learned so much about radicals. Thanks are due to the Ariel University, the Ben-Gurion University, and the Nuclear Research Centre Negev, the institutes in which these studies were performed. Finally, thanks to the funding agencies that supported these studies.

References

1 Eid, R., Arab, N.T.T., and Greenwood, M.T. (2017). *Biochim. Bioph. Acta* 1864: 399–430.
2 Prescott, C. and Bottle, S.E. (2017). *Cell Biochem. Biophys.* 75: 227–240.

3 Pohjoismäki, J.L. and Goffart, S. (2017). *Free Radical Biol. Med.* 106: 345–354.
4 Sharova, E.I. and Medvedev, S.S. (2017). *Russian J. Plant Physiol.* 64: 1–14.
5 Bachi, A., Dalle-Donne, I., and Scaloni, A. (2013). *Chem. Rev.* 113: 596–698.
6 Victorino, V.J., Vanessa, J., Mencalha, A.L. et al. (2015). *Life Sci.* 129: 42–47.
7 Liu, J., Chakraborty, S., Hosseinzadeh, P. et al. (2014). *Chem. Rev.* 114: 4366–4469.
8 Poulos, T.L. (2014). *Chem. Rev.* 114: 3919–3962.
9 Lubitz, W., Ogata, H., Rüdiger, O. et al. (2014). *Chem. Rev.* 114: 4081–4148.
10 Migliore, A., Polizzi, N.F., Therien, M.J. et al. (2014). *Chem. Rev.* 114: 3381–3465.
11 Broderick, J.B., Duffus, B.R., Duschene, K.S. et al. (2014). *Chem. Rev.* 114: 4229–4317.
12 Schuch, A.P., Moreno, N.C., Schuch, N.J. et al. (2017). *Free Radical Biol. Med.*, Ahead of Print.
13 Gozzelino, R. and Arosio, P. (2016). *Int. J. Mol. Sci.* 17: 1–14.
14 del Rio, L.A. (2015). *J. Exp. Bot.* 66: 2827–2837.
15 Kaur, S.J., McKeown, S.R., Stephanie, R. et al. (2016). *Gene* 577: 109–118.
16 Goswami, S. and Haldar, C. (2015). *J. Photochem. Photobiol., B* 153: 281–288.
17 Ramis, M.R., Esteban, S., Miralles, A. et al. (2015). *Curr. Med. Chem.* 22: 2690–2711.
18 Steinbacher, P. and Eckl, P. (2015). *Biomolecules* 5: 356–377.
19 Weidinger, A. and Kozlov, A.V. (2015). *Biomolecules* 5: 472–484.
20 Richards, S.L., Wilkins, K.A., Swarbreck, S.M. et al. (2015). *J. Exp. Bot.* 66: 37–46.
21 Mirzaei, A., Chen, Z., Haghighat, F. et al. (2017). *Chemosphere* 174: 665–688.
22 Ochando-Pulido, J.M., Pimentel-Moral, S., Verardo, V. et al. (2017). *Sep. Purif. Technol.* 179: 161–174.
23 Divyapriya, G., Nambi, I.M., and Senthilnathan, J. (2016). *J. Bionanosci.* 10: 356–368.
24 Ab Aziz, N.A., Palaniandy, P., Abdul Aziz, A.H. et al. (2016). *J. Chem. Res.* 40: 704–712.
25 Zhang, H., Ma, D., Qiu, R. et al. (2017). *Chem. Eng. J.* 313: 157–170.
26 Ghanbari, F. and Moradi, M. (2017). *Chem. Eng. J.* 310: 41–62.
27 Mamaghani, A.H., Haghighat, F., and Lee, C.S. (2017). *Appl. Cat., B* 203: 247–269.
28 Bethi, B., Sonawane, S.H., Bhanvase, B.A. et al. (2016). *Chem. Eng. Process.* 109: 178–189.
29 Moreira, F.C., Boaventura, R.A.R., Brillas, E. et al. (2017). *Appl. Cat., B* 202: 217–261.
30 Singh, P., Ojha, A., Borthakur, A. et al. (2016). *Environ Sci. Pollut. Res.* 23: 22340–22364.
31 Garcia-Segura, S., Bellotindos, L.M., Huang, Y.H. et al. (2016). *J. Taiwan Inst. Chem. Eng.* 67: 211–225.
32 Petronella, F., Truppi, A., Ingrosso, C. et al. (2017). *Catal. Today* 281: 85–100.
33 Padmanaban, V.C., Giri, N.M.S., Madhangi, P.G. et al. (2016). *Int. J. Environ. Sci. Technol.* 13: 1591–1606.
34 Kawamura, K. and Bikkina, S. (2016). *Atmos. Res.* 170: 140–160.
35 Chapleski, R.C. Jr.,, Zhang, Y., Troya, D. et al. (2016). *Chem. Soc. Rev.* 45 (13): 3731–3746.
36 Sathishkumar, P., Mangalaraja, R.V. et al. (2016). *Renewable Sustainable Energy Rev.* 55: 426–454.
37 Thakare, S.R., Thakare, J., and Kosankar, P.T. (2015). *Res. J. Chem. Environ.* 19 (6): 77–84.
38 Lykakis, I.N., Evgenidou, E., and Orfanopoulos, M. (2012). *Curr. Org. Chem.* 16: 2400–2414.
39 Samanta, D., Kratz, K., and Emrick, T. (2011). *Green Polymerization Methods* (ed. R.T. Mathers and M.A.R. Meier), 129–142. Wiley-VCH Verlag GmbH & Co.
40 Huang, X. and Groves, J.T. (2017). *J. Biol. Inorg. Chem.* 22: 85–207.
41 Cho, K.B., Hirao, H., Shaik, S. et al. (2016). *Chem. Soc. Rev.* 45: 1197–1210.
42 Angnes, R.A., Li, Z., Correia, C.R.D. et al. (2015). *Org. Biomol. Chem.* 13: 9152–9167.
43 Puga, A.V. (2016). *Coord. Chem. Rev.* 315: 1–66.
44 Johnston, C.P., Smith, R.T., Allmendinger, S. et al. (2016). *Nature* 536: 322–325.

45 Dadashi-Silab, S., Doran, S., and Yagci, Y. (2016). *Chem. Rev.* 116: 10212–10275.
46 Öberg, K.I. (2016). *Chem. Rev.* 116: 9631–9663.
47 Kärkäs, M.D., Porco, J.A. Jr.,, and Stephenson, C.R.J. (2016). *Chem. Rev.* 116: 9683–9747.
48 Huo, H., Harms, K., and Meggers, E. (2016). *J. Am. Chem. Soc.* 138: 6936–6939.
49 Romero, N.A. and Nicewicz, D.A. (2016). *Chem. Rev.* 116: 10075–10166.
50 Chen, M., Zhong, M., and Johnson, J.A. (2016). *Chem. Rev.* 116: 10167–10211.
51 Takeda, K., Fujisawa, K., Nojima, H. et al. (2017). *J. Photochem. Photobiol.* 340: 8–14.
52 Skubi, K.L., Blum, T.R., and Yoon, T.P. (2016). *Chem. Rev.* 116: 10035–10074.
53 Ravelli, D., Protti, S., and Fagnoni, M. (2016). *Chem. Rev.* 116: 9850–9913.
54 Hoang, S. and Gao, P.X. (2016). *Adv. Energy Mater.* 6: 1–29.
55 Sires, I., Brillas, E., Oturan, M.A. et al. (2014). *Environ. Sci. Pollut. Res.* 21: 8336–8367.
56 Martin, E.T., McGuire, C.M., Mubarak, M.S. et al. (2016). *Chem. Rev.* 116: 15198–15234.
57 Ligia, F.A., Magyar, A., and Kispert, L.D. (2015). *Arch. Biochem. Biophys.* 572: 167–174.
58 Ciriminna, R., Palmisano, G., and Pagliaro, M. (2015). *ChemCatChem* 7: 552–558.
59 Dey, G.R. (2014). *Green carbon dioxide* (ed. G. Centi and S. Perathoner), 25–50.
60 Haber, F. and Weiss, J. (1932). *Naturwissenschaften* 20: 948–950.
61 Haber, F. and Weiss, J. (1934). *Proc. R. Soc. London, Ser. A* 147: 332–351.
62 Bray, W.C. and Gorin, M.H. (1932). *J. Am. Chem. Soc.* 54: 2124–2125.
63 Baxendale, J.H., Evans, M.G., and Park, G.S. (1946). *Trans. Faraday Soc.* 42: 155–169.
64 Merz, J.H. and Waters, W.A. (1947). *Discuss. Faraday Soc.* 2: 179–188.
65 Walling, C. (1975). *Acc. Chem. Res.* 8: 125–131.
66 Sutton, H.C. and Winterbourn, C.C. (1989). *Free Rad. Biol. Med.* 6: 53–60.
67 Goldstein, S. and Czapski, G. (1990). *Int. Rev. Exp. Pathol.* 31: 133–164.
68 NDRL/NIST solution kinetic data base, on the web.
69 Anbar, M., Meyerstein, D., and Neta, P. (1966). *J. Chem. Soc. B* 742–747.
70 Veltwisch, D., Janata, E., and Asmus, K.D. (1980). *J. Chem. Soc. Perkin Trans.* 2: 146–153.
71 Masarwa, M., Cohen, H., Meyerstein, D. et al. (1988). *J. Am. Chem. Soc.* 110: 4293–4297.
72 Goldstein, S., Czapski, G., and Meyerstein, D. (1993). *Free Rad. Biol. Med.* 15: 435–445.
73 Bataineh, H., Pestovsky, O., and Bakac, A. (2012). *Chem. Sci.* 3: 1594–1599.
74 Rachmilovich-Calis, S., Masarwa, A., Meyerstein, N. et al. (2009). *Chem. Eur. J.* 15: 8303–8309.
75 Mansano-Weiss, C., Cohen, H., and Meyerstein, D. (2002). *J. Inorg. Biochem.* 91: 199–204.
76 Kuznetsov, M.L., Teixeira, F.A., Bokach, N.A. et al. (2014). *J. Catal.* 313: 135–148.
77 Novikov, A.S., Kuznetsov, M.L., Rocha, B.R. et al. (2016). *Catal. Sci. Technol.* 6: 1343–1356.
78 Burg, A., Shusterman, I., Kornweitz, H. et al. (2014). *Dalton Trans.* 43: 9111–9115.
79 Burg, A., Shamir, D., Shusterman, I. et al. (2014). *Chem. Commun.* 50: 13096–13099.
80 Kornweitz, H., Burg, A., and Meyerstein, D. (2015). *J. Phys. Chem. A* 119: 4200–4206.
81 Guengerich, F.P. and Macdonald, T.L. (1984). *Acc. Chem. Res.* 17: 9–16.
82 Bollinger, J.M. Jr.,, Price, J.C., Hoffart, L.M. et al. (2005). *Eur. J. Inorg. Chem.* 4245–4254.
83 Groves, J.T. (1985). *J. Chem. Educ.* 62: 928–934.
84 Kalmár, J., Biri, B., Lente, G. et al. (2012). *Dalton Trans.* 41: 11875–11884.
85 Kumar, G.S., Krishnaveni, N., Umayavalli, M. et al. (2013). *Asian J. Chem.* 25: 1447–1450. and references therein.
86 Meyerstein, D. (1971). *Inorg. Chem.* 10: 2244–2249.
87 Goldstein, S., Czapski, G., Cohen, H. et al. (1994). *Inorg. Chem.* 33: 3255–3260.
88 Saha, S.S. and Chaudhuri, A.K. (1972). *J. Polym. Sci. A* 1, 10: 797–808.
89 Xu, J. and Jordan, R.B. (1990). *Inorg. Chem.* 29: 2933–2936.
90 Hsieh, Y.H.P. and Hsieh, Y.P. (2000). *J. Agric. Food Chem.* 48: 1569–1573.

91 O'Malley, P.J. (2001). *J. Phys. Chem. B* 105: 11290–11293.
92 Kochi, J.K. (1978). *Organometallic Mechanisms and Catalysis*. New York: Academic Press.
93 Cornils, B. and Herrmann, W.A. (1996). *Applied Homogeneous Catalysis with Organometallic Compounds*. Weinheim, Germany: VCH.
94 Brandsma, L., Vasilevsky, S.F., and Verkruijsse, H.D. (1998). *Application of Transition Metal Catalysis in Organic Synthesis*. Berlin: Springer.
95 Navon, N., Cohen, H., Paoletti, P. et al. (2000). *Ind. Eng. Chem. Res.* 39: 3536–3540.
96 Navon, N., Burg, A., Cohen, H. et al. (2002). *Eur. J. Inorg. Chem.* 423–429.
97 Gao, J., Rusling, J.F., and Zhou, D. (1996). *J. Org. Chem.* 61: 5972–5977.
98 Schrauzer, G.N. and Grate, J.H. (1981). *J. Am. Chem. Soc.* 103: 541–546.
99 Chock, P.B. and Halpern, J. (1969). *J. Am. Chem. Soc.* 91: 582–588. and references therein.
100 Sadler, N., Scott, S.L., Bakac, A. et al. (1989). *Inorg. Chem.* 28: 3951–3954.
101 Ram, M.S., Bakac, A., and Espenson, J.H. (1988). *Inorg. Chem.* 27: 4231–4235.
102 Kelley, D.G., Marchaj, A., Bakac, A. et al. (1991). *J. Am. Chem. Soc.* 113: 7583–7587.
103 Bakac, A. and Espenson, J.H. (1986). *J. Am. Chem. Soc.* 108: 5353–5354.
104 Ram, M.S., Bakac, A., and Espenson, J.H. (1986). *Inorg. Chem.* 25: 3267–3272.
105 Dunach, E., Esteves, A.P., Medeiros, M.J. et al. (2009). *C. R. Chim.* 12: 889–894.
106 Dunach, E., Esteves, A.P., Medeiros, M.J. et al. (2004). *Tetrahedron Lett.* 45: 7935–7937.
107 Dunach, E., Jose Medeiros, M., and Olivero, S. (2006). *New J. Chem.* 30: 1534–1548.
108 Dunach, E. and Medeiros, M.J. (2008). *Electrochim. Acta* 53: 4470–4477.
109 Dunach, E., Paula Esteves, A., Medeiros, M.J. et al. (2005). *New J. Chem.* 29: 633–636.
110 Esteves, A.P., Ferreira, E.C., and Medeiros, M.J. (2007). *Tetrahedron* 63: 3006–3009.
111 Medeiros, M.J., Neves, C.S.S., Pereira, A.R. et al. (2011). *Electrochim. Acta* 56: 4498–4503.
112 Shandalov, E., Zilbermann, I., Maimon, E. et al. (2003). *Eur. J. Inorg. Chem.* 4105–4109.
113 Esteves, A.P., Neves, C.S., Medeiros, M.J. et al. (2008). *J. Electroanal. Chem.* 614: 131–138.
114 Mochizuki, K. and Suzuki, M. (2011). *Inorg. Chem. Comm.* 14: 902–905.
115 Rusonik, I., Cohen, H., and Meyerstein, D. (2006). *Inorg. Chem.* 45: 7389–7396.
116 Gillham, R.W. and O'Hannesin, S.F. (1994). *Groundwater* 32: 958–967.
117 Matheson, L.J. and Tratnyek, P.G. (1994). *Environ. Sci. Technol.* 28: 2045–2053.
118 Boronina, T., Klabunde, K.J., and Sergeev, G. (1995). *Environ. Sci. Technol.* 29: 1511–1517.
119 Chuang, F.W., Larson, R.A., and Wessman, M.S. (1995). *Environ. Sci. Technol.* 29: 2460–2463.
120 Helland, B.R., Alvarez, P.J.J., and Schnoor, J.L. (1995). *Hazard. Mater. J.* 41: 205–216.
121 Roberts, A.L., Totten, L.A., Arnold, W.A. et al. (1996). *Environ. Sci. Technol.* 30: 2654–2659.
122 Campbell, T.J., Burris, D.R., Roberts, A.L. et al. (1997). *Environ. Toxicol. Chem.* 16: 625–630.
123 Fennelly, J.P. and Roberts, A.L. (1998). *Environ. Sci. Technol.* 32: 1980–1988.
124 Mao, X., Oremland, R.S., Liu, T. et al. (2017). *Environ. Sci. Technol.* 51: 2366–2372.
125 Guan, X., Sun, Y., Qin, H. et al. (2015). *Water Res.* 75: 224–248.
126 Xiea, Y., Donga, H., Zenga, G. et al. (2017). *J. Hazard. Mater.* 321: 390–407.
127 Adhikary, J., Meistelman, M., Burg, A. et al. (2017). *Eur. J. Inorg. Chem.* 1510–1515.
128 Rasnoshik, H., Masarwa, A., Cohen, H. et al. (2008). *J. Phys. Chem.* 112: 12769–12771.
129 Rasnoshik, H., Masarwa, A., Cohen, H. et al. (2010). *Dalton Trans.* 39: 823–833.
130 Schutz, O., Masarwa, A., Zilbermann, I. et al. (2010). *J. Coord. Chem.* 63: 2528–2541.
131 Schutz, O., Masarwa, A., Zilbermann, I. et al. (2014). *Eur. J. Inorg. Chem.* 932–940.
132 Neta, P., Huie, R.E., and Ross, A.B. (1988). *J. Phys. Chem. Ref. Data* 17: 1024–1284.
133 NDRL/NIST Solution Kinetics Database on the Web.
134 D.A. Armstrong, R.E. Huie, W.H. Koppenol, et al. (2015) 87, 1139–1150.

135 Schutz, O. and Meyerstein, D. (2006). *Tetrahedron Lett.* 47: 1093–1096.
136 Shandalov, E., Zilbermann, I., Maimon, E. et al. (2004). *Tetrahedron Lett.* 45: 989–992.
137 E. Shandalov, Ph.D. thesis, Ben-Gurion University, (2007).
138 Adams, G.F., Bartlett, R.J., and Purvis, G.D. (1982). *Chem. Phys. Lett.* 87: 311–314.
139 Burgers, P.C. and Ruttink, P.J.A. (2005). *Int. J. Mass Spectrom.* 242: 49–56.
140 Rachmilovich-Calis, S., Meyerstein, N., and Meyerstein, D. (2009). *Chem. Eur. J.* 15: 7717–7723.
141 Dénes, F., Pichowicz, M., Povie, G. et al. (2014). *Chem. Rev.* 114: 2587–2693.
142 Shono, T. (1984). Anodic oxidations. In: *Electroorganic Chemistry As a New Tool in Organic Synthesis*, 5–124. Berlin: Springer-Verlag.
143 Andrieux, C.P., Gonzalez, F., and Savéant, J.M. (2001). *J. Electroanal. Chem.* 498: 171–180.
144 Andreev, V.N., Grinberg, V.A., Dedov, A.G. et al. (2013). *Prot. Met. Phys. Chem. Surf.* 49: 32–39.
145 Niu, J., Li, Y., Shang, E. et al. (2016). *Chemosphere* 146: 526–538.
146 Miao, Y., Ouyang, L., Zhou, S. et al. (2014). *Biosens. Bioelectron.* 53: 428–439.
147 Li, T. and Farrell, J. (2000). *Environ. Sci. Technol.* 34: 173–179.
148 Pinson, J. (2012). *Aryl diazonium salts*, 1–35. Wiley-VCH Verlag GmbH &_CoKGaA.
149 Geneste, F., Cadoret, M., Moinet, C. et al. (2002). *New J. Chem.* 26: 1261–1266.
150 Fenton, H.J.H. (1894). *J. Chem. Soc.* 65: 899–910.
151 Bakac, A. and Espenson, J.H. (1987). *Inorg. Chem.* 26: 4353–4355.
152 Zhang, R., Yang, Y., Huang, C.H. et al. (2016). *Water Res.* 103: 283–292.
153 Panda, D. and Manickam, S. (2017). *Ultrason. Sonochem.* 36: 481–496.
154 Litter, M.I. and Slodowicz, M. (2017). *J. Adv. Oxid. Technol.* 19: 3–21.
155 Pouran, S.R., Aziz, A.R.A., and Daud, W.M.A.W. (2015). *J. Ind. Eng. Chem.* 21: 53–69.
156 Bakac, A., Janni, J.A., and Espenson, J.H. (1992). *Inorg. Chem.* 31: 1088–1090.
157 Wallentin, C.J., Nguyen, J.D., Finkbeiner, P. et al. (2012). *J. Am. Chem. Soc.* 134: 8875–8884.
158 Pirtsch, M., Paria, S., Matsuno, T. et al. (2012). *Chem. Eur. J.* 18: 7336–7340.
159 Zeitler, K. (2009). *Angew. Chem. Int. Ed.* 48: 9785–9789.
160 Tossi, A.B., Roerner, H.G., and Schulte-Frohlinde, D. (1989). *Photochem. Photobiol.* 50: 585–591.
161 Valenzuela, M.A., Flores, S.O., Rios-Berny, O. et al. (2012). *Molecular Photochemistry – Various Aspects* (ed. S. Saha), 25–40. IntechOpen.
162 Rajeshwar, K., Osugi, M.E., Chanmanee, W. et al. (2008). *J. Photochem. Photobiol. C: Photochem. Rev.* 9: 171–192.
163 Egerton, T.A. (2014). *Molecules* 19: 18192–18214.
164 Wishart, J.F. and Nocera, D.G. (eds.) (1998). *Photochemistry and Radiation Chemistry – Complementary Methods for the Study of Electron Transfer*. American Chemical Society.
165 Spinks, J.W.T. and Woods, R.J. (1990). *An Introduction to Radiation Chemistry*, 3e. Wiley.
166 Wardman, P. (1989). *J. Phys. Chem. Ref. Data* 18: 1637–1755.
167 Sancheti, S.V. and Gogate, P.R. (2017). *Ultrason. Sonochem.* 36: 527–543.
168 Banerjee, B. (2017). *Ultrason. Sonochem.* 35: 1–14.
169 Magdziarz, A. and Colmenares, J.C. (2017). *Molecules* 22: 216–237.
170 Rehman, M.U., Jawaid, P., Uchiyama, H. et al. (2016). *Arch. Bichem. Biophys.* 605: 19–25.
171 Xiuyuan, N., Yuefang, H., Bailin, L. et al. (2001). *Eur. Polymer J.* 37: 201–206.
172 Li, H., Goestl, R., Delgove, M. et al. (2016). *ACS Macro Lett.* 5: 995–998.
173 Cogate, P.R. and Prajapat, A.L. (2015). *Ultrason. Sonochem.* 27: 480–494.
174 Rathi, A.K., Gawande, M.B., Zboril, R. et al. (2015). *Coord. Chem. Rev.* 291: 68–94.
175 Johnson, A.E. and Perchyonok, V.T. (2009). *Curr. Org. Chem.* 13: 1705–1725.
176 Fasan, R. (2012). *ACS Catal.* 2: 647–666.

177 Puri, M. and Que, L. Jr., (2015). *Acc. Chem. Res.* 48: 2443–2452.
178 Marsh, E.N.G. and Meléndez, G.D.R. (2012). *Biochim. Biophys. Acta* 1824: 1154–1164.
179 Frey, F.A. (2014). *Acc. Chem. Res.* 47: 540–549.
180 Dowling, D.P., Vey, J.L., Croft, A.K. et al. (2012). *Biochim. et Biophys. Acta* 1824: 1178–1195.
181 Tang, M.C., Zou, Y., Watanabe, K. et al. (2017). *Chem. Rev.* 117: 5226–5333.
182 Landgraf, B.J., McCarthy, E.L., and Booker, S.J. (2016). *Annu. Rev. Biochem.* 85: 485–514.
183 Guarino, E., Salguero, I., and Kearsey, S.E. (2014). *Semin. Cell Dev. Biol.* 30: 97–103.
184 Shibata, N. and Toraya, T. (2015). *J. Biochem.* 158: 271–229.
185 Li, Z., Xu, X., Leng, X. et al. (2017). *Arch. Virol.* 162: 603–610.
186 Gobert, A.P. and Wilson, K.T. (2017). *Free Radical Biol. Med.* 105: 16–27.
187 Lei, Y., Wang, K., Deng, L. et al. (2015). *Med. Res. Rev.* 35: 306–340.
188 Pole, A., Dimri, M., and Dimri, G.P. (2016). *AIMS Mol. Sci.* 3: 300–324.
189 Edrey, Y.H. and Salmon, A.B. (2014). *Free Rad. Biol. Med.* 71: 368–378.
190 Spahis, S., Borys, J.M., and Levy, E. (2017). *Antioxid. Redox Signaling* 26: 445–461.
191 Wu, J., Xia, S., Kalionis, B. et al. (2014). *Bio. Med. Res. Inter.* 1–13.
192 Spahis, S., Delvin, E., Borys, J.M. et al. (2017). *Antioxid. Redox Signaling* 26: 519–541.
193 Gerber, P.A. and Rutter, G.A. (2017). *Antioxid. Redox Signaling* 26: 501–518.
194 Carrier, A. (2017). *Antioxid. Redox Signaling* 26: 429–431.
195 Domenico, F.D., Barone, E., Perluigi, M. et al. (2017). *Antioxid. Redox Signaling* 26: 364–387.
196 Herrera, A., Muñoz, P., Steinbusch, H.W.M. et al. (2017). *ACS Chem. Neurosci.* 8: 702–711.
197 Flyunt, R., Schuchmann, M.N., and von Sonntag, C. (2001). *Chem. Eur. J.* 7: 796–799.
198 (a) Codorniu-Hernández, E., Hall, K.W., Boese, A.D. et al. (2015). *J. Chem. Theory Comput.* 11: 4740–4748; (b) Codorniu-Hernández, E., Hall, K.W., Ziemianowicz, D. et al. (2014). *Phys. Chem. Chem. Phys.* 16: 26094–26102; (c) Kusalik, P.G. (2015). *J. Chem. Theory Comput.* 11: 4740–4748.
199 Lati, J. and Meyerstein, D. (1978). *J. Chem. Soc. Dalton Trans.* 1105–1118.
200 (a) Jubran, N., Ginzburg, G., Cohen, H. et al. (1985). *Inorg. Chem.* 24: 205–258; (b) Sutin, N., Creutz, C., and Fujita, E. (1997). *Comments Inorg. Chem.* 19: 67–92.
201 Chipman, D.M. (2016). *J. Phys. Chem. A* 120: 9618–9624.
202 Asmus, K.D. (1986). *Preprints – Am. Chem. Soc., Div. Petroleum Chem.* 31: 870–874.
203 Asmus, K.D., Bahnemann, D., Bonifacic, M. et al. (1977). *Faraday Dis. Chem. Soc.* 63: 213–225.
204 Hiller, K.O. and Asmus, K.D. (1981). *Int. J. Radiat. Biol. Relat. Stud. Phys., Chem. Med.* 40: 583–595.
205 Assmann, A., Bonifacic, M., Briviba, K. et al. (2000). *Free Rad. Res.* 32: 371–376.
206 Gleiter, R. and Haberhauer, G. (2017). *Coord. Chem. Rev.* 344: 263–298.
207 KÖchling, T., Morozova, O.B., Yurkovskaya, A.V. et al. (2016). *J. Phys. Chem. B* 120: 9277–9286.
208 Sutradhar, D., Zeegers-Huyskens, T. et al. (2015). *J. Mol. Phys.* 113: 3232–3240.
209 Singh, B.G., Thomas, E., Sawant, S.N. et al. (2013). *J. Phys. Chem. A* 117: 9259–9265.
210 Ji, L.F., Li, A.Y., Li, Z.Z. et al. (2016). *J. Mol. Model* 22: 1.
211 Wenska, G., Filipiak, P., Asmus, K.D. et al. (2008). *J. Phys. Chem. B* 112: 10045–10053.
212 Berry, J.F. (2016). *Acc. Chem. Res.* 49: 27–34.
213 Alday, B., Johnson, R., Li, J., and Guo, H. (2014). *Theor. Chem. Acc.* 133: 1540.
214 Stefanic, I., Ljubic, I., Bonifacic, M. et al. (2009). *Phys. Chem. Chem. Phys.* 11: 2256–2267.
215 Mondal, S., Gold, B., Mohamed, R.K. et al. (2014). *Chem. Eur. J.* 20: 8664–8669.
216 Trifunac, A.D. and Nelson, D.J. (1977). *J. Am. Chem. Soc.* 99: 1745–1752.

217 Trifunac, A.D. and Nelson, D.J. (1978). *J. Am. Chem. Soc.* 100: 5244–5246.
218 Nuttall, R.H.D. and Trifunac, A.D. (1982). *J. Phys. Chem.* 86: 3963–3969.
219 Bakac, A. and Espenson, J.H. (1986). *J. Phys. Chem.* 90: 325–327.
220 Howe, P.T., Fahr, A., and Laufer, A.H. (2004). *J. Phys. Chem. A* 108: 1638–1639.
221 Zhu, R.S., Xu, Z.F., and Lin, M.C. (2004). *J. Chem. Phys.* 120: 6566–6573.
222 Getoff, N. (1989). *Appl. Radiat. Isot.* 40: 585–594.
223 Neta, P., Grodkowski, J., and Ross, A.B. (1996). *J. Phys. Chem. Ref. Data* 25: 709–1050.
224 Park, H.R. and Getoff, N. (1992). *Z. Naturforsch.* 47a: 985–991.
225 Furukawa, S., Shishido, T., Teramura, K. et al. (2011). *J. Phys. Chem. C* 115: 19320–19327.
226 Goldstein, S., Czapski, G., and Meyerstein, D. (1997). *Inorg. Chem.* 36: 2893–2897.
227 Elliot, A.J. and Simsons, A.S. (1984). *Can. J. Chem.* 62: 1831–1834.
228 Soule, B.P., Hyodo, F., Matsumoto, K.I. et al. (2007). *Free Radic. Biol. Med.* 42: 1632–1650.
229 Newcomb, M. (2012). *Encyclopedia of Radicals in Chemistry, Biology and Materials*, 1–18. Wiley.
230 Newcomb, M. (1993). *Tetrahedron* 49: 1151–1176.
231 Marsh, E.N.G. (2000). *Bioorganic Chem.* 28: 176–189.
232 Gruber, K., Puffer, B., and Krautler, B. (2011). *Chem. Soc. Rev.* 40: 4346–4363.
233 Hisaeda, Y., Nishioka, T., Inoue, Y. et al. (2000). *Coord. Chem. Rev.* 198: 21–37.
234 Konya, K.G., Paul, T., Lin, S. et al. (2000). *J. Am. Chem. Soc.* 122: 7518–7527.
235 Jiang, D., Barata-Vallejo, S., Golding, B.T. et al. (2012). *Org. Biomol. Chem.* 10: 1102–1107. and references therein.
236 Toraya, T. (2003). *Chem. Rev.* 103: 2095–2127.
237 von Sonntag, C. (2006). *Free-Radical-Induced DNA Damage and its Repair: A Chemical Persective*. Berlin: Springer-Verlag.
238 Yurkova, I.L. (2012). *Russ. Chem. Rev.* 81: 175–190.
239 Johnson, D., Cassanelli, P., and Cox, R.A. (2004). *Atmos. Environ.* 38: 1755–1765.
240 Rusonik, I., Polat, H., Cohen, H. et al. (2003). *Eur. J. Inorg. Chem.* 4227–4233.
241 Ljubic, I., Sabljic, A., and Bonifacic, M. (2016). *J. Phys. Chem. B* 120: 11810–11820.
242 Steenken, S. and Neta, P. (1982). *J. Am. Chem. Soc.* 104: 1244–1248.
243 Chatgilialoglu, C., Caminal, C., and Mulazzani, Q.G. (2011). *Org. Biomol. Chem.* 9: 3494–3498.
244 Li, C.J. and Chen, L. (2006). *Chem. Soc. Rev.* 35: 68–82.
245 Mason, R.P. (2016). *Redox Biol.* 8: 422–429.
246 Htet, Y. and Tennyson, A.G. (2016). *Chem. Sci.* 7: 4052–4058.
247 Dey, G.R., Naik, D.B., Kishore, K. et al. (1999). *Rad. Phys. Chem.* 54: 19–27.
248 Popivker, I., Zilbermann, I., Maimon, E. et al. (2010). *Inorg. Chim. Acta* 363: 4202–4206.
249 Kim, D., Chen, J.R., and Yen, T.F. (2009). *J. Hazard. Mater.* 163: 231–238.
250 Matasović, B. and Bonifačić, M. (2011). *Radiat. Phys. Chem.* 80: 750–754.
251 Bays, J.P., Blumer, S.T., Baral-Tosh, S. et al. (1983). *J. Am. Chem. Soc.* 105: 320–324.
252 Cohen, H. and Meyerstein, D. (1975). *J. Chem. Soc., Dalton Trans.* 2477–2479.
253 Cohen, H., Nutkovich, M., Meyerstein, D. et al. (1982). *J. Chem. Soc. Dalton Trans.* 943–950.
254 Wieghardt, K., Cohen, H., and Meyerstein, D. (1978). *Angew. Chem.* 90: 632–633.
255 Dhiman, S.B. and Naik, D.B. (2012). *Radiat. Phys. Chem.* 81: 635–641.
256 Naik, D.B., Dey, G.R., Dwibedy, P. et al. (2001). *Radiat. Phys. Chem.* 60: 61–69.
257 Flyunt, R., Knolle, W., Kahnt, A. et al. (2014). *Int. J. Rad. Biol.* 90: 486–494.
258 Cohen, H. and Meyerstein, D. (1977). *J. Chem. Soc., Dalton Trans.* 1056–1062.
259 Neta, P., Huie, R.E., and Ross, A.B. (1990). *J. Phys. Chem. Ref. Data* 19: 413–497.

260 Al-Sheikhly, M.I., Hissung, A., Schuchmann, H.P. et al. (1984). *J. Chem. Soc. Perkin Trans. II* 601–608.
261 Cooper, W.J., Cramer, C.J., Martin, N.H. et al. (2009). *Chem. Rev.* 109: 1302–1345.
262 Von Sonntag, C. and Schuchmann, H.P. (1997). *Peroxyl Radicals in the Chemistry of Free Radicals* (ed. Z.B. Alfassi), 173. Chichester: Wiley.
263 Von Sonntag, C. and Schuchmann, H.P. (1991). *Angew. Chem.* 103: 1255–1279.
264 Jovanovic, S.V., Jankovic, I., and Josimovic, L. (1992). *J. Am. Chem. Soc.* 114: 9018–9021.
265 Nath Das, T., Dhanasekaran, T., Alfassi, Z.B. et al. (1998). *J. Phys. Chem. A* 102: 280–284.
266 Gebicki, J.M. (2016). *Arch. Biochem. Biophys.* 595: 33–39.
267 Rahman, T., Hosen, I., Islam, M.M.T. et al. (2012). *Adv. Biosci. Biotechnol.* 3: 997–1019.
268 Xu, P., Qiao, K., and Stephanopoulos, G. (2017). *Biotechnol. Bioeng.* 114: 1521–1530.
269 Bhattacharjee, S. (2014). *Curr. Sci.* 107: 1811–1823.
270 Wadhwa, N., Mathew, B.B., Jatawa, S.K. et al. (2012). *Int. J. Curr. Sci.* 3: 29–38.
271 Milligan, J.R., Ng, J.Y.Y., Wu, C.C.L. et al. (1996). *Radiat. Res.* 146: 436–443.
272 Adhikari, A., Bothe, E., Jain, V., and von Sonntag, C. (2000). *Int. J. Radiat. Biol.* 76: 1157–1166.
273 Meyerstein, D. (2014). *Israel J. Chem.* 53: 279–283.
274 Alberto, M.E., Russo, N., Grand, A. et al. (2013). *Phys. Chem. Chem. Phys.* 15: 4642–4650.
275 Galano, A. and Alvarez-Idaboy, J.R. (2011). *RSC Adv.* 1: 1763–1771.
276 Alfassi, Z.B., Huie, R.E., Kumar, M. et al. (1992). *J. Phys. Chem.* 96: 767–770.
277 Huie, R.E. and Neta, P. (1986). *Int. J. Chem. Kinet.* 18: 1185–1191.
278 Packer, J.E., Willson, R.L., Bahnemann, D. et al. (1980). *J.C.S. Perkin II* 296–299.
279 Bothe, E., Schuchmann, M.N., Schultefrohlinde, D. et al. (1978). *Photochem. Photobiol.* 28: 639–644.
280 Bothe, E., Schultefrohlinde, D., and von Sonntag, C. (1978). *J. Chem. Soc., Perkin Trans.* 2: 416–420.
281 Rachmilovich-Calis, S., Masarwa, A., Meyerstein, N. et al. (2005). *Eur. J. Inorg. Chem.* 2875–2880.
282 Masarwa, A., Rachmilovich-Calis, S., Meyerstein, N. et al. (2005). *Coord. Chem. Rev.* 249: 1937–1943.
283 Mansano-Weiss, C., Masarwa, A., Cohen, H. et al. (2005). *Inorg. Chim. Acta* 358: 2199–2206.
284 Solomon-Rapaport, E., Masarwa, A., Cohen, H. et al. (2002). *Eur. J. Inorg. Chem.* 2427–2432.
285 Goldstein, S., Lind, J., and Merenyi, G. (2004). *J. Phys. Chem. A* 108: 1719–1725.
286 Dhakshinamoorthy, A., Navalon, S., Alvaro, M. et al. (2012). *Chem. Sus. Chem.* 5: 46–64.
287 Rudzinski, K.J., Szmigielski, R., Kuznietsova, I. et al. (2016). *Atmos. Environ.* 130: 163–171.
288 Meyerstein, D. (1999). *Metal Ions in Biological Systems*, vol. 36 (ed. H. Sigel and A. Sigel), 41–77.
289 Masarwa, A. and Meyerstein, D. (2011). *Molecular and Supramolecular Bioinorganic Chemistry*, vol. 2 (ed. A.L.R. Merce, J. Felcman and M.A.L. Recio), 187–217. N.Y: Nova Science Publishers, Inc.
290 Kelm, M., Lilie, J., Henglein, A. et al. (1974). *J. Phys. Chem.* 78: 882–887.
291 Freiberg, M. and Meyerstein, D. (1977). *J. Chem. Soc. Trans. Faraday Soc. 1* 73: 622–631.
292 Freiberg, M. and Meyerstein, D. (1980). *J. Chem. Soc. Trans. Faraday Soc. 1* 76: 1838–1845.
293 Cohen, H. and Meyerstein, D. (1986). *Inorg. Chem.* 25: 1505–1506.
294 Cohen, H. and Meyerstein, D. (1987). *Inorg. Chem.* 26: 2342–2344.
295 Cohen, H. and Meyerstein, D. (1988). *J. Chem. Sac., Faraday Trans. I* 84: 4157–4160.
296 Szulc, A., Meyerstein, D., and Cohen, H. (1998). *Inorg. Chim. Acta* 270: 440–445.

297 Shaham, N., Cohen, H., van Eldik, R. et al. (2000). *J. Chem. Soc. Dalton Trans.* 3356–3359.
298 Kelm, M., Lilie, J., and Henglein, A. (1975). *J. Chem. Soc., Faraday Trans. 1* 71: 1132–1144.
299 Cohen, H. and Meyerstein, D. (1974). *Inorg. Chem.* 13: 2434–2443.
300 Schmidt, W., Swinehart, J.H., and Taube, H. (1971). *J. Am. Chem. Soc.* 93: 1117–1123.
301 Leslie, J.P. II, and Espenson, J.H. (1976). *J. Am. Chem. Soc.* 98: 4839–4844.
302 Behar, D., Samuni, A., and Fessenden, R.W. (1973). *J. Phys. Chem.* 77: 2055–2059.
303 Freiberg, M. and Meyerstein, D. (1980). *J. Chem. Soc. Trans. Faraday Soc. 1* 76: 1825–1837.
304 Bakac, A. and Espenson, J.H. (1986). *J. Am. Chem. Soc.* 108: 713–719.
305 Navon, N., Golub, G., Cohen, H. et al. (1995). *Organometallics* 14: 5670–5676.
306 Huston, P., Espenson, J.H., and Bakac, A. (1991). *Inorg. Chem.* 30: 4826–4830.
307 Shi, S., Espenson, J.H., and Bakac, A. (1990). *Inorg. Chem.* 29: 4318–4322.
308 Cohen, H. and Meyerstein, D. (1988). *Inorg. Chem.* 27: 3429–3431.
309 Cohen, H. and Meyerstein, D. (1985). *J. Chem. Soc., Faraday Trans. 1* 81: 233–239.
310 Sorek, Y., Cohen, H., and Meyerstein, D. (1986). *J. Chem. Soc., Faraday Trans. 1* 82: 3431–3438.
311 Schrauzer, G.N., Michaely, W.J. et al. (1976). *J. Am. Chem. Soc.* 95: 2024–2026.
312 Tait, A.M., Hoffman, M.Z., and Hayon, E. (1973). *J. Am. Chem. Soc.* 98: 86–93.
313 Roche, T.S. and Endicott, J.F. (1974). *Inorg. Chem.* 13: 1575–1580.
314 Elroi, H. and Meyerstein, D. (1978). *J. Am. Chem. Soc.* 100: 5540–5548.
315 Shaham, N., Masarwa, A., Matana, Y. et al. (2002). *Eur. J. Inorg. Chem.* 87–92.
316 Bakac, A. and Espenson, J.H. (1987). *Inorg. Chem.* 26: 4305–4307.
317 Mulac, W.A. and Meyerstein, D. (1982). *J. Am. Chem. Soc.* 104: 4124–4128.
318 Sorek, Y., Cohen, H., and Meyerstein, D. (1989). *J. Chem. Soc. Faraday Trans. l* 85: 1169–1179.
319 Sauer, A., Cohen, H., and Meyerstein, D. (1988). *Inorg. Chem.* 27: 4578–4581.
320 Kelley, D.G., Espenson, J.H., and Bakac, A. (1991). *J. Chem. Soc., Chem. Commun.* 546–547.
321 Sauer, A., Cohen, H., and Meyerstein, D. (1989). *Inorg. Chem.* 28: 2511–2512.
322 Kurzion-Zilbermann, T., Masarwa, A., Maimon, E. et al. (2007). *Dalton Trans.* 3959–3965.
323 Mulac, W.A. and Meyerstein, D. (1979). *J. Chem. Soc. Chem. Commun.* 893–895.
324 Masarwa, M., Cohen, H., and Meyerstein, D. (1986). *Inorg. Chem.* 25: 4897–4900.
325 Masarwa, M., Cohen, H., Glaser, R. et al. (1990). *Inorg. Chem.* 29: 5031–5035.
326 Mansano-Weiss, C., Epstein, D.M., Cohen, H. et al. (2002). *Inorg. Chim. Acta* 339: 283–291.
327 Chen, W.J., Espenson, J.H., and Bakac, A. (1993). *J. Chinese Chem. Soc.* 40: 329–335.
328 van Eldik, R., Gaede, W., Cohen, H. et al. (1992). *Inorg. Chem.* 31: 3695–3696.
329 van Eldik, R., Cohen, H., and Meyerstein, D. (1994). *Inorg. Chem.* 33: 1566–1568.
330 van Eldik, R. and Meyerstein, D. (2000). *Acc. Chem. Res.* 33: 207–214.
331 Ryan, D.A. and Espenson, J.H. (1979). *J. Am. Chem. Soc.* 101: 2488–2489.
332 Espenson, J.H., Connolly, P., Meyerstein, D. et al. (1983). *Inorg. Chem.* 22: 1009–1013.
333 Mulac, W.A., Cohen, H., and Meyerstein, D. (1982). *Inorg. Chem.* 21: 4016–4020.
334 Meyerstein, D. and Schwarz, H.A. (1988). *J. Chem. Soc. Faraday Trans. l* 84: 2933–2949.
335 van Eldik, R., Cohen, H., and Meyerstein, D. (1991). *Angew. Chem.* 30: 1158–1160.
336 Rotman, A., Cohen, H., and Meyerstein, D. (1985). *Inorg. Chem.* 24: 4158–4164.
337 Cohen, H. and Meyerstein, D. (1984). *Inorg. Chem.* 23: 84–87.
338 Shaham, N., Cohen, H., and Meyerstein, D. (1999). *J. Chem. Soc., Dalton Trans.* 3805–3808.
339 Ryan, D.A. and Espenson, J.H. (1982). *Inorg. Chem.* 21: 527–530.
340 Sorek, Y., Cohen, H., Mulac, W.A. et al. (1983). *Inorg. Chem.* 22: 3040–3046.
341 Cohen, H., Meyerstein, D., Shusterman, A.J. et al. (1984). *J. Am. Chem. Soc.* 106: 1876–1877.
342 Cohen, H., van Eldik, R., Gaede, W. et al. (1994). *Inorg. Chim. Acta* 227: 57–61.

343 Cohen, H. and Meyerstein, D. (1985). *Angew. Chim. Int. Ed. Engl.* 24: 779–781.
344 Goldstein, S., Czapski, G., Cohen, H. et al. (1992). *Inorg. Chem.* 31: 798–803.
345 Goldstein, S., Czapski, G., Cohen, H. et al. (1992). *Inorg. Chim. Acta* 192: 87–93.
346 Goldstein, S., Czapski, G., Cohen, H. et al. (1993). *J. Chem. Soc. Faraday Trans.* 89: 4045–4051.
347 Goldstein, S., Czapski, G., Cohen, H. et al. (1990). *Free Radicals Biol. Med.* 9: 371–379.
348 Masarwa, M., Cohen, H., Saar, J. et al. (1990). *Israel J. Chem.* 30: 361–368.
349 Shi, S., Espenson, J.H., Meyerstein, D. et al. (1991). *Inorg. Chem.* 30: 4468–4470.
350 Lee, S. and Ku, T.Y. (2007). *Polyhedron* 26: 2901–2905.
351 Masarwa, M., Cohen, H., and Meyerstein, D. (1991). *Inorg. Chem.* 30: 1849–1854.
352 Lligadas, G., Grama, S., and Percec, V. (2017). *Biomacromolecules* 18: 1039–1063.
353 Zetterlund, P.B., Thickett, S.C., Perrier, S. et al. (2015). *Chem. Rev.* 115: 9745–9800.
354 Zhang, H. (2013). *Eur. Polymer J.* 49: 579–600.
355 Anastasaki, A., Nikolaou, V., Alsubaie, F. et al. (2015). *ACS Symp. Ser.* 1188: 29–45.
356 Busca, G., Lorenzelli, V., Ramis, G. et al. (1991). *Mater. Chem. Phys.* 29: 175–189.
357 Bent, B.E. (1996). *Chem. Rev.* 96: 1361–1390.
358 Kopple, K., Meyerstein, D., and Meisel, D. (1980). *J. Phys. Chem.* 84: 870–875.
359 Hai, Z.B., El Kolli, N., Chen, J.F. et al. (2014). *New J. Chem.* 38: 5279–5286.
360 Lee, J.S. and Choi, W.Y. (2004). *Environ. Sci. Technol.* 38: 4026–4033.
361 Vasini, E.J. and Giordano, M.C. (1990). *Electrochim. Acta* 35: 1237–1243.
362 Poizot, P., Laffont-Dantras, L., and Simonet, J. (2008). *J. Electroanal. Chem.* 624: 52–58.
363 Kanoufi, F., Combellas, C., Hazimeh, H. et al. (2006). *J. Phys. Org. Chem.* 19: 847–866.
364 Yang, Q.Y., Johnson, A.D., Maynard, K.J. et al. (1989). *J. Am. Chem. Soc.* 111: 8748–8749.
365 Andersson, L.O. (2008). *Phys. Chem. Miner.* 35: 505–520.
366 Kaupp, G. (2011). *Crystengcomm* 13: 3108–3121.
367 Kamat, P.V. and Meisel, D. (2003). *C. R. Chim.* 6: 999–1007.
368 Baltrusaitis, J., Jansen, I., and Christus, J.D.S. (2014). *Catal. Sci. Technol.* 4: 2397–2411.
369 Vines, F., Lykhach, Y., Staudt, T. et al. (2010). *Chem. Eur. J.* 16: 6530–6539.
370 Liu, H.F., Liu, R.S., Liew, K.Y. et al. (1984). *J. Am. Chem. Soc.* 106: 4117–4121.
371 Rusonik, I., Cohen, H., and Meyerstein, D. (2005). *Eur. J. Inorg. Chem.* 1227–1229.
372 Rusonik, I., Cohen, H., Lugowskoy, A. et al. (2010). *Eur. J. Inorg. Chem.* 3252–3255.
373 Borojovich, E.J.C., Bar-Ziv, R., Oster-Golberg, O. et al. (2017). *Appl. Catal. B: Environ.* 210: 255–262.
374 Zidki, T., Cohen, H., and Meyerstein, D. (2006). *Phys. Chem. Chem. Phys.* 8: 3552–3556.
375 Bar-Ziv, R., Zilbermann, I., Oster-Golberg, O. et al. (2012). *Chem. Eur. J.* 18: 4699–4705.
376 Bar-Ziv, R., Zilbermann, I., Zidki, T. et al. (2012). *Chem. Eur. J.* 18: 6733–6736.
377 Zidki, T., Elisseev, A., Cohen, H. et al. (2014). *The Optimization of Composition, Structure and Properties of Metals, Oxides, Composites, Nano and Amorphous Materials* (ed. M. Zinigrad and L. Leontiev), 259–276. Yekaterinburg: Ural Branch Russian Academy of Sciences.
378 Bar-Ziv, R., Zilbermann, I., Zidki, T. et al. (2009). *J. Phys. Chem. C.* 113: 3281–3286.
379 Szabó, L., Tóth, T., Rácz, G. et al. (2016). *Free Radical Res.* 50: 26–38.

5

C—H Bond Oxidation with Transition-Metal-Based Carbene Complexes

Bruno Dominelli, Anja C. Lindhorst, and Fritz E. Kühn

Technische Universität München (TUM), Department of Chemistry/Catalysis Research Center, Garching bei München, Germany

5.1 Introduction

In view of the pioneering synthetic works of Wanzlick, Öfele (1968), and Arduengo (1991), and the Nobel Prize-winning research in catalysis by Fischer (1973), Schrock, Grubbs, and Chauvin (2005), transition metal-based carbene complexes have been shown to play a decisive role in the synthesis of large-scale and fine chemical compounds [1–4]. However, it is still one of the biggest challenges in catalysis to develop organometallic catalysts able to convert saturated hydrocarbons in economically and environmentally benign processes [5]. In recent years, considerable research has been performed in order to apply transition metal-based carbene complexes for this purpose. From all known carbene types, *N*-heterocyclic carbenes (NHCs) have emerged as the best option for the oxygenation of alkanes due to their potent σ-donor character, allowing the formation of very stable transition metal complexes even under harsh reaction conditions such as oxidizing or acidic media. Since their first synthesis in the late 1980s, NHCs have found numerous applications in catalysis and gained high attention, as they are highly flexible in terms of electronic, steric and physical properties. A detailed overview regarding their properties, synthesis and catalytic applications was recently given by Glorius *et al.* [6].

In addition to NHC-based catalysts Fischer-type carbenes, for example, are also active in alkane activation. However, their applicability is limited to the transfer of an electrophilic carbon atom into a C—H bond finally forming a C—C bond [7, 8]. In the context of C—O bond formation the electrophilic character of the carbene carbon atom is disadvantageous, since the used oxidants are usually nucleophilic. Furthermore, Schrock-type carbenes with nucleophilic character are known to be very air and moisture sensitive and therefore suffer from decomposition under typical conditions applied in oxidation catalysis.

This chapter presents examples of transition metal-based NHC complexes, which were shown to be active catalysts in the oxygenation of C—H bonds of saturated hydrocarbons. As model substrates for activity tests, typically cyclic hydrocarbons like cyclohexane or cyclooctane are used, bearing only chemically equivalent C—H bonds. In order to elucidate the selectivity of catalysts toward different types of C—H bonds, adamantane or linear hydrocarbons are popular test substrates [9].

Alkane Functionalization, First Edition. Edited by Armando J. L. Pombeiro and M. Fátima C. Guedes da Silva.

5.2 Iron-Catalyzed C—O Bond Formation

The iron-catalyzed oxidation of C—H bonds is, at least in part, inspired by well-known and highly selective bioactive enzymes (often termed *biocatalysts* or *bio-inorganic catalysts*) such as soluble methane monooxygenase (sMMO) and cytochromes P450 (Cyt P450) [9, 10]. Additionally, iron is readily available, as it is the most abundant transition metal, both on Earth and in the universe, and many of its compounds are nontoxic.

The use of iron in oxidation catalysis is historically best known from Fenton's reagent [11–13]. In this case, simple Fe(II)-salts catalytically decompose hydrogen peroxide (H_2O_2) forming hydroxyl radicals, which unselectively oxidize hydrocarbons to the corresponding alcohols and ketones in a mixture of 1 : 1. In the past decades, several organometallic iron compounds, mainly based on N-donor ligands, were introduced as C—H bond oxygenation catalysts following a more selective metal-centered reaction mechanism [9]. In this context, two main issues still remain unsolved, including the selectivity (simultaneous formation of alcohols and ketones) and a limited catalyst lifetime.

In 2012, our group presented an octahedral Fe(II) complex (**1a**) bearing a tetradentate bis(pyridyl-*N*-heterocyclic carbene) ligand in the equatorial plane and two solvent molecules in the axial positions [14]. This compound is a pre-catalyst for the catalytic hydroxylation of aromatic compounds (e.g. benzene and toluene) and the efficient mediation of olefin epoxidation reactions with hydrogen peroxide as the oxidant [15, 16]. As a catalytically active intermediate, a high-valent iron-oxo species was proposed. The same compound was recently further applied for the Wittig reaction and also the catalytic oxidation of saturated hydrocarbons like cyclohexane and 2,3-dimethylbutane (Scheme 5.1) [17, 18]. Bioinspired modifications of the axial ligands were performed in order to analyze their influence on the catalytic activity.

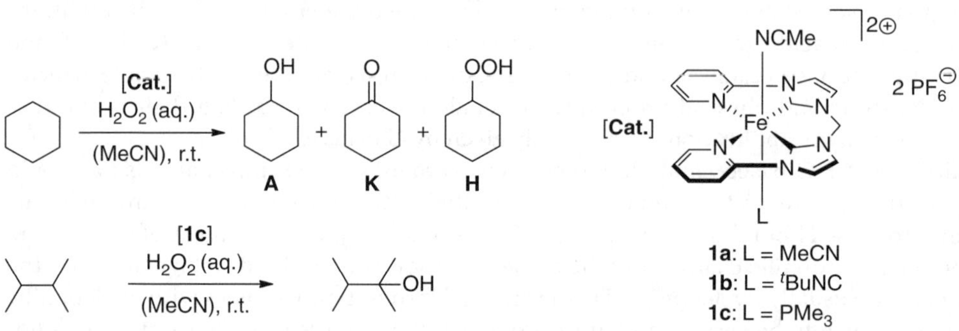

Scheme 5.1 Catalytic oxidation of cyclohexane and 2,3-dimethylbutane in presence of iron(II) complexes **1a–c** and hydrogen peroxide [17]. (**A** = alcohol, **K** = ketone, **H** = alkyl hydroperoxide).

Similar to the well-studied Cyt P450 enzyme, the nature of the axial ligands proved to be decisive for the stability and kinetic behavior of the catalyst. Under the same reaction conditions, the number of catalytic turnovers in cyclohexane oxidation is significantly higher for catalyst **1b** carrying a π-acceptor ligand (^{t}BuNC) compared to **1c** bearing a σ-donor ligand (PMe_3). However, substitution of the acetonitrile ligand by ^{t}BuNC also reduces the catalytic activity resulting in a slower conversion of cyclohexane. In line with this, an induction period becomes visible using catalyst **1b**, which can be explained by the π-acceptor properties of ^{t}BuNC disfavoring the dissociation of the acetonitrile ligand in *trans*-position. In all reactions, cyclohexyl hydroperoxide is the main oxidation product. It is known to be formed via a radical chain autoxidative pathway [19]. Subsequently, the desired cyclohexanol is mainly obtained after addition of triphenylphosphine as reductant.

Additionally, the most active catalyst **1b** was tested with 2,3-dimethylbutane as a substrate, showing 28 turnovers and a high selectivity toward the tertiary alcohol 2,3-dimethyl-2-butanol. The use of common organic peroxides like *tert*-butyl hydroperoxide (TBHP), *p*-cymene hydroperoxide (CHP), or *m*-chloroperoxybenzoic acid (*m*-CPBA) delivers poorer results compared to hydrogen peroxide.

5.3 Iridium-Catalyzed C—O Bond Formation

Another transition metal, which has been used to form catalysts for alkane oxidation reactions, is iridium in the oxidation state +III. In 2015, Sarkar and co-workers reported on the C—H bond oxygenation of cyclooctane to the corresponding ketone using several iridium(III)-based carbene complexes of the basic structure [IrCp*Cl(**L**)] (Cp* = pentamethylcyclopentadienyl). The bidentate ligands (**L** = **2a**–**2g**) contain either the click-chemistry derived 1,2,3-triazole motif or the corresponding 1,2,3-triazolydene (Scheme 5.2) [20]. In comparison to classical Arduengo-type imidazole-2-ylidenes triazolydenes possess stronger donor character due to their mesoionic properties, which is favorable for their application in oxidation catalysis [21].

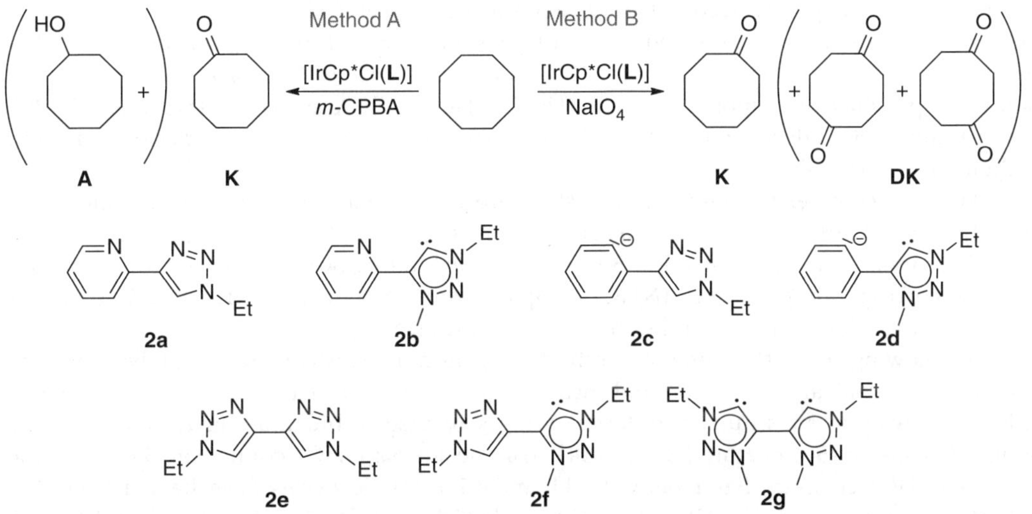

Scheme 5.2 Two methods for the catalytic oxidation of cyclooctane in presence of [IrCp*Cl(**L**)] and the oxidants *m*-CPBA (method A) or $NaIO_4$ (method B) [20].

As oxidants *m*-CPBA in dichloromethane and sodium periodate ($NaIO_4$) in aqueous solution were chosen. Depending on the oxidant, different reaction products were observed. While the reaction with *m*-CPBA yields cyclooctanol and cyclooctanone, a product mixture of cyclooctanone and diketones is observed upon oxidation with $NaIO_4$. Initially, all catalysts were tested in the catalytic oxidation of cyclooctane using both methods with 0.5 mol% of the catalyst at 45 °C. It was observed that those complexes bearing unsymmetrical ligands (**2a**–**2d**) are catalytically more active than those with symmetrical ligands. Furthermore, depending on the choice of oxidant, varying selectivities are found. With *m*-CPBA as oxidant (method A) a high percentage of cyclooctanol can be detected with ketone/alcohol ratios between 0.5–2.6. In direct comparison, the complexes bearing carbene ligands show higher selectivity for the ketone formation than their triazole analogues. Using $NaIO_4$ as an oxidant, no alcohol product is detected; instead, overoxidation to diketones becomes an issue.

For the two most active Ir-catalysts, **2a** and **2b**, the reaction conditions have been optimized. At a loading of 1 mol% both catalysts reach a conversion of approximately 70% in an acetone/water mixture of 4/1 (room temperature, method B). Recently, Crabtree and co-workers introduced several comparable Ir(III)-complexes, which also are able to catalytically oxidize cyclic hydrocarbons in presence of the oxidant ceric ammonium nitrate (CAN) [22–24]. However, they deliver lower yields in the range of 3–25%, showing high selectivity for cyclooctanone and up to 10 catalytic cycles (5 mol% catalyst and 9 equivalents CAN), which again underlines the high potential of NHC-based catalysts [23]. Based on experimental and computational studies, an iridium(V)-oxo complex is presumed to be the catalytically active species and mediate the reaction via a direct oxygen insertion pathway [23].

5.4 Palladium- and Platinum-Catalyzed C—O Bond Formation

Since Shilov *et al.* described the first example of C—H activation by a simple platinum salt in 1969, a lot of attention has been paid to palladium and platinum as active transition metals in hydrocarbon oxidation [25]. Several non–carbene-based organometallic catalysts such as the Periana- or Schüth-systems were introduced in this field of catalysis [26, 27].

In 2002, Straßner *et al.* introduced the first Pd(II) complexes (**3**) bearing bidentate NHC ligands, which exhibit catalytic activity for the oxidation of CH_4 to trifluoroacetic acid methyl ester in the presence of trifluoroacetic acid (HOTFA) and $K_2S_2O_8$ as oxidant (Scheme 5.3) [28]. The resulting methyl derivate is unlikely to be overoxidized and can be further converted to methanol via hydrolysis.

All Pd(NHC) complexes explored in this study are stable at elevated temperatures and under strongly acidic and oxidizing conditions. The highest yield (ca. 3%) of methanol can be obtained, according to this study, using catalyst **3e** (0.1 mol%) exhibiting 30 catalytic turnovers. In contrast, the corresponding Pt(II)-bis(NHC) complex is not catalytically active and decomposes after 20 hours under identical conditions to platinum(0) [28].

In the following years, the same research group continued studying Pd(II) catalysts bearing NHC ligands for alkane activation and introduced a plethora of palladium-based complexes with varied wingtip substituents, different bridges in case of bidentate ligands and diverse anionic ligands (**3–5**, Scheme 5.3) [29–34]. For the Pd-bis(NHC) complexes **3a–i** and the Pd-mono(NHC) catalysts **5a–g** between 11 and 41 catalytic cycles have been noted. The highest turnover number (TON) of 41 is obtained using complex **5a** having a $[(dmso)PdCl_3]^-$ metallate as counter ion, even though the catalyst loading has been reduced by 50% due to the homo-bimetallic structure of the complex. Sterically demanding, bulky wingtip substituents (e.g. complexes **3f** or **3 h**) have shown to reduce the catalytic activity of the catalyst. In case of bis(NHC) ligands the decomposition of the catalyst was traced to the acidic methylene bridge connecting the two NHC moieties. Indeed, elongation of the bridge (complexes **3b–3d**) affords slightly better or very similar catalytic activity. In addition, the anionic ligand in *trans*-position to the NHC moieties has an influence on the catalytic performance. Complexes **3g** and **3h** are not catalytically active, since the iodide ligands possess relatively low basicity compared to bromide and chloride and are therefore less readily protonated, which is a prerequisite in order to create free coordination sites and initiate the catalytic reaction. The resulting hydrogen halides are subsequently oxidized to elemental halogens by the oxidant ($K_2S_2O_8$). Moreover, addition of two to five equivalents of KBr was reported to inhibit the catalytic conversion of

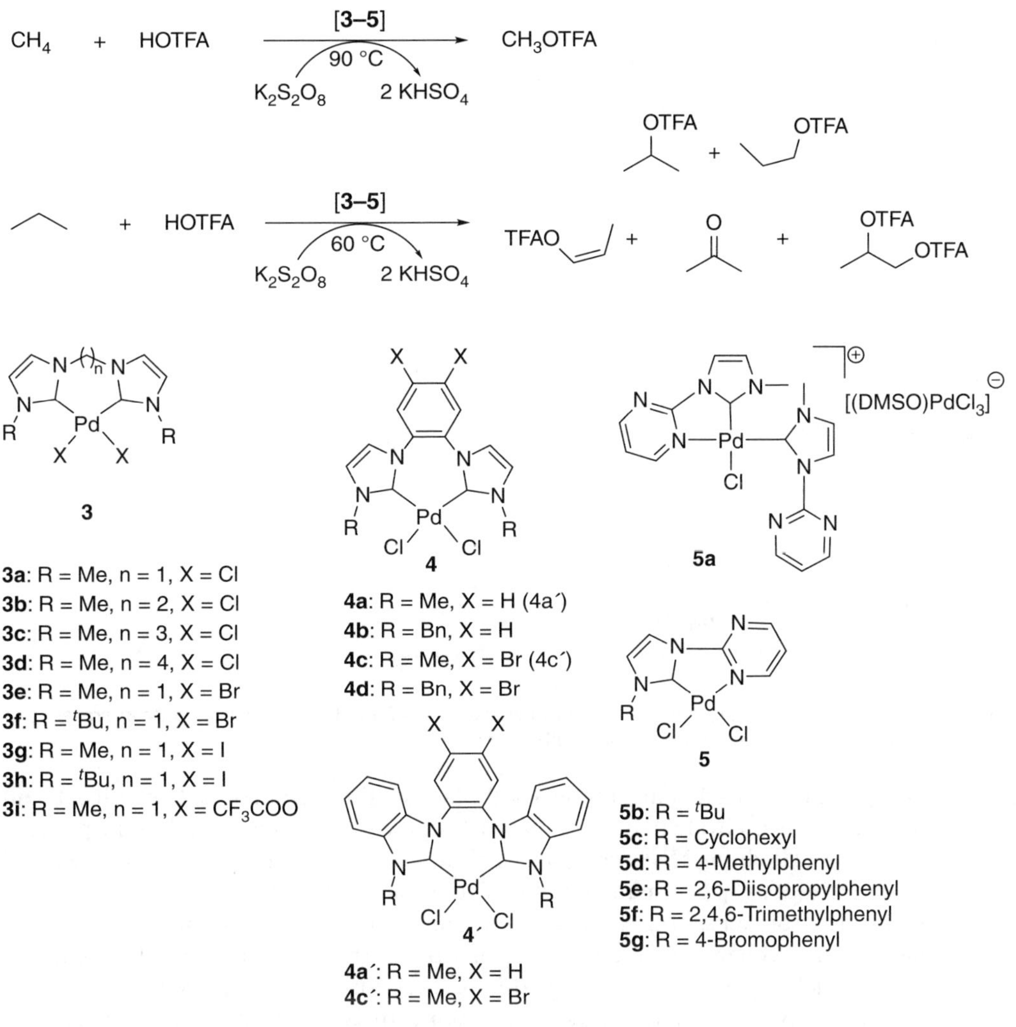

Scheme 5.3 Overview of Pd(II)-NHC complexes used in recent years for the catalytic oxidation of methane or propane in presence of $K_2S_2O_8$ and HOTFA [28–33].

both methane and propane. Most likely, additional bromide ions inhibit the creation of free coordination sites for the substrate.

Complexes **3a, 3e, 3g, 3i, 4a–4d, 4a′, and 4c′** were additionally examined with propane as substrate, which allows study of their selectivity toward secondary and primary C—H bonds. Expectedly, a high selectivity for isopropyl trifluoroacetate (90%) with TONs in the range between 6 and 30 could be observed.

Combining DFT calculations and previous experimental results, including the isolation of palladium(II) and –(IV) intermediates, a catalytic cycle has been proposed for the oxidation of aliphatic C—H bonds by catalyst **3e** (Scheme 5.4) [33]. The formation of a high oxidation state +IV of palladium complexes is presumed to be mediated by elemental bromine. Pd(IV)-bis(NHC) complexes with different anionic ligands (methyl, trifluoroacetate, halides) could be isolated and fully characterized, confirming the intermediary structures of the catalytic cycle [35]. The presence of palladium(0) could be excluded, since no metallic precipitate was observed under the strongly acidic milieu.

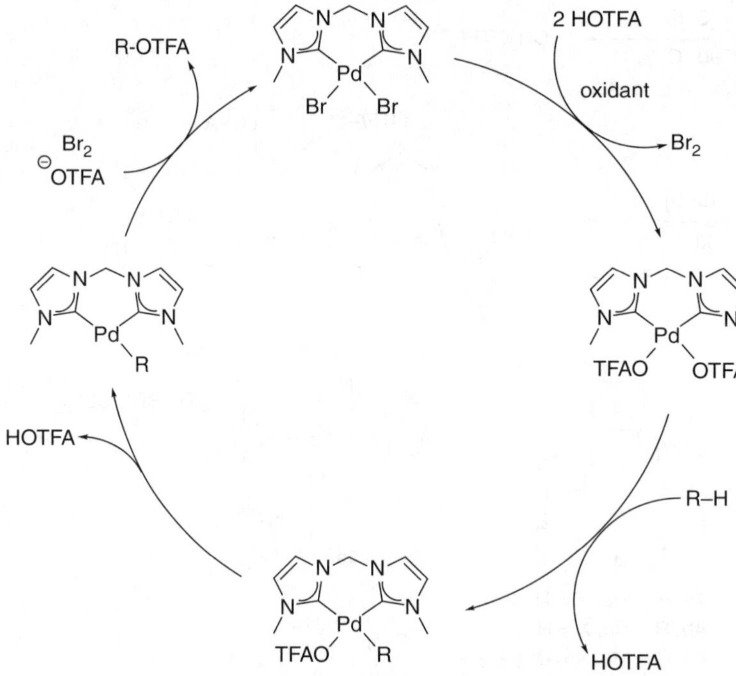

Scheme 5.4 Simplified catalytic cycle for the oxidation of alkanes by Pd(II)-bis(NHC) complex **3e** in presence of HOTFA and an oxidant excluding all intermediary Pd(IV) species [33].

Comparable platinum-based noncarbene catalysts like the Periana- ([Pt(bipy)]Cl_2) and Schüth-system (Pt with a covalent triazine-based framework), which served as inspiration for synthesis of the palladium NHC complexes, show TONs in the range of 158–355 and 201–246, respectively [26, 27]. Consequently, it is visible that the palladium(II)-based bis(NHC) complexes still suffer from a lower lifetime under the necessary reaction conditions.

In the context of green chemistry and economical aspects, the use of dioxygen as an oxidant for the oxidation of alkanes is an attractive goal [34]. Therefore, focus is placed on the development of catalysts capable of mediating such reactions. In 2014, Straßner *et al.* introduced a combination of Pd(II)-bis(NHC) complex **3e** and $NaVO_3$ as co-catalyst for the activation of dioxygen [33]. Approximatively 60 catalytic cycles can be reached for the oxidation of both, methane and propane, in the presence of HOTFA. In this case, bromine acts as a redox mediator, which is well known for the generation of transition metals complexes in high-oxidation states by a nonradical two-electron process.

5.5 Final Comments

In this chapter, examples of transition-metal-based NHC complexes with catalytic activity in the oxygenation of C—H bonds of saturated hydrocarbons were presented. Although NHCs are versatile ligands forming comparatively strong metal–carbon bonds, the corresponding complexes still suffer from limited stability under oxidizing conditions translating into a limited catalyst lifetime and low TONs. In the case of palladium complex **5a**, the homo-bimetallic nature has turned out to be very advantageous in alkane oxidation; thus, in the future, functionalized NHC moieties could be designed for the synthesis of homo- or even hetero-bimetallic complexes.

Acknowledgments

We gratefully acknowledge support and funding from the TUM Graduate School. A. C. L. thanks the Fonds der Chemischen Industrie for a grant.

References

1 Wanzlick, H.W. and Schönherr, H.J. (1968). *Angew. Chem.* 80: 154.
2 Öfele, K. (1968). *J. Organomet. Chem.* 12: P42–P43.
3 Arduengo, A.J., Harlow, R.L., and Kline, M. (1991). *J. Am. Chem. Soc.* 113: 361–363.
4 Cundari, T.R. and Prince, B.M. (2011). *J. Organomet. Chem.* 696: 3982–3986.
5 Arndtsen, B.A., Bergman, R.G., Mobley, T.A. et al. (1995). *Acc. Chem. Res.* 28: 154–162.
6 Hopkinson, M.N., Richter, C., Schedler, M. et al. (2014). *Nature* 510: 485–496.
7 Caballero, A., Diaz-Requejo, M.M. et al. (2015). *Dalton Trans.* 44: 20295–20307.
8 Doyle, M.P., Duffy, R., Ratnikov, M. et al. (2010). *Chem. Rev.* 110: 704–724.
9 Lindhorst, A.C., Haslinger, S., and Kühn, F.E. (2015). *Chem. Commun.* 51: 17193–17212.
10 Riener, K., Haslinger, S., Raba, A. et al. (2014). *Chem. Rev.* 114: 5215–5272.
11 Fenton, H.J.H. (1894). *J. Chem. Soc., Trans.* 65: 899–910.
12 Haber, F. and Weiss, J. (1934). *Proc. R. Soc. London, Ser. A* 147: 332–351.
13 Gozzo, F. (2001). *J. Mol. Catal. A: Chem.* 171: 1–22.
14 Raba, A., Cokoja, M., Ewald, S. et al. (2012). *Organometallics* 31: 2793–2800.
15 Raba, A., Cokoja, M., Herrmann, W.A. et al. (2014). *Chem. Commun.* 50: 11454–11457.
16 Kück, J.W., Raba, A., Markovits, I.I.E. et al. (2014). *ChemCatChem* 6: 1882–1886.
17 Haslinger, S., Raba, A., Cokoja, M. et al. (2015). *J. Catal.* 331: 147–153.
18 Karaca, Ö., Anneser, M.R., Kück, J.W. et al. (2016). *J. Catal.* 344: 213–220.
19 Shilov, A.E. and Shul'pin, G.B. (1997). *Chem. Rev.* 97: 2879–2932.
20 Hohloch, S., Kaiser, S., Duecker, F.L. et al. (2015). *Dalton Trans.* 44: 686–693.
21 Donnelly, K.F., Petronilho, A., and Albrecht, M. (2013). *Chem. Commun.* 49: 1145–1159.
22 Zhou, M., Schley, N.D., and Crabtree, R.H. (2010). *J. Am. Chem. Soc.* 132: 12550–12551.
23 Zhou, M., Balcells, D., Parent, A.R. et al. (2012). *ACS Catalysis* 2: 208–218.
24 Zhou, M., Hintermair, U., Hashiguchi, B.G. et al. (2013). *Organometallics* 32: 957–965.
25 Stahl, S.S., Labinger, J.A., and Bercaw, J.E. (1998). *Angew. Chem. Int. Ed.* 37: 2180–2192.
26 Palkovits, R., Antonietti, M., Kuhn, P. et al. (2009). *Angew. Chem. Int. Ed.* 48: 6909–6912.
27 Periana, R.A., Taube, D.J., Gamble, S. et al. (1998). *Science* 280: 560–564.
28 Mühlhofer, M., Straßner, T., and Herrmann, W.A. (2002). *Angew. Chem. Int. Ed.* 41: 1745–1747.
29 Straßner, T., Mühlhofer, M., Zeller, A. et al. (2004). *J. Organomet. Chem.* 689: 1418–1424.
30 Ahrens, S., Zeller, A., Taige, M. et al. (2006). *Organometallics* 25: 5409–5415.
31 Meyer, D., Taige, M.A., Zeller, A. et al. (2009). *Organometallics* 28: 2142–2149.
32 Munz, D., Pöthig, A., Tronnier, A. et al. (2013). *Dalton Trans.* 42: 7297–7304.
33 Munz, D. and Straßner, T. (2014). *Angew. Chem. Int. Ed.* 53: 2485–2488.
34 Munz, D. and Straßner, T. (2015). *Inorg. Chem.* 54: 5043–5052.
35 Meyer, D. and Straßner, T. (2016). *Beilstein J. Org. Chem.* 12: 1557–1565.

6

Alkane Oxidation with C-Scorpionate Metal Complexes

Luísa M. D. R. S. Martins

Instituto Superior Técnico, Universidade de Lisboa, Centro de Química Estrutural, Lisboa, Portugal

6.1 Introduction

C-scorpionate complexes (Figure 6.1) are one of the scarce mononuclear types of metal compounds able to catalyze the oxidation of alkanes. Their use as homo- or heterogeneous oxidation catalysts has experienced significant advances in the last years [1–4] in particular, in the challenging attempt to oxidize cyclohexane via an efficient and sustainable protocol.

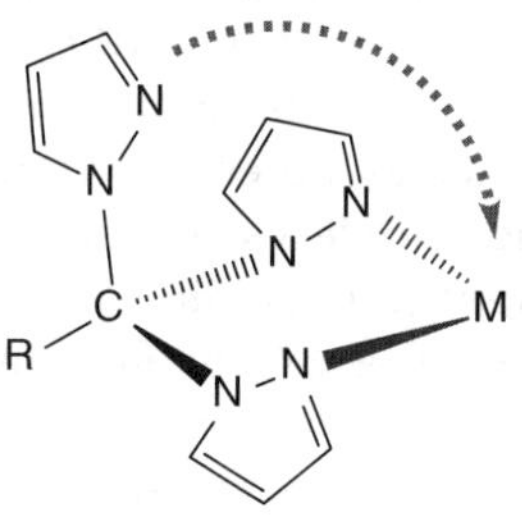

Figure 6.1 C-scorpionate tris(pyrazol-1-yl)methane compounds.

The researchers' preference for cyclohexane as a model substrate arises from the significance of the oxidized products (KA oil, i.e. cyclohexanol and cyclohexanone) in the manufacturing of adipic acid, a building block for several processes in the chemical, pharmaceutical and food industries. In fact, adipic acid is a relevant commodity chemical produced worldwide (over 3.5 millions of metric tons/year, growing ca. 5%/year) [5, 6] for the synthesis of nylon-6,6 polyamide (primary use) and is also widely used to produce polyester and polyurethane resins [5]. Moreover, adipic acid is industrially obtained by a process [5, 7] that shows great need for improvement (Scheme 6.1). It involves two steps: the inefficient oxidation of cyclohexane to KA oil (ca. 8% yield; 85% selectivity) [5], and then further oxidation with a huge excess of nitric acid, leading to an extensive liberation of the greenhouse gas N_2O (300 kg of N_2O per ton of adipic acid) [5, 8].

It is believed that the interchange between bi- and tridentate coordination modes of C-scorpionate ligands [1] as well as their stability toward hydrolysis [9, 10] are the main reasons for their significant catalytic activity that will be described in the following sections.

6.2 C-Scorpionate Complexes

C-scorpionate tris(pyrazol-1-yl)methane compounds (Figure 6.1) were first reported by Hückel et al. in 1937 [11]. However, their coordination chemistry remained dormant until 1966, when Jesson [12] described Co(II) and Ni(II) chelate systems of those ligands. Synthetic difficulties associated to the very low yields were responsible for such silent period. In fact, until Elguero's improved synthetic strategy (1984) [13] for the formation of hydrotris(pyrazol-1-yl)methane ($HC(pz)_3$, pz = pyrazol-1-yl) and functionalized derivatives, only a few reports appeared. More recently, the use of microwave-assisted synthesis [14] and breakthroughs in the formation

Alkane Functionalization, First Edition. Edited by Armando J. L. Pombeiro and M. Fátima C. Guedes da Silva.

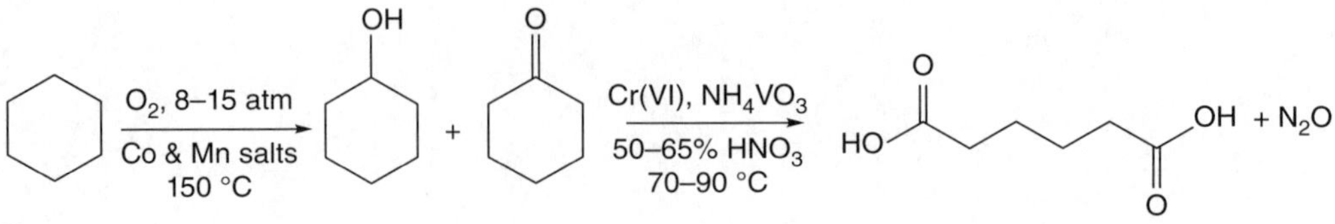

Scheme 6.1 Current industrial process of adipic acid production.

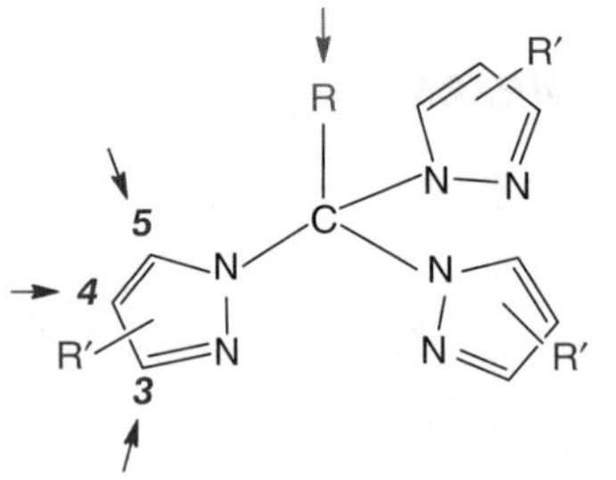

Figure 6.2 Schematic representation of a C-scorpionate with highlighted positions for functionalization.

of tris(pyrazol-1-yl)methanes [15–20] allowed developing this promising class of ligands. In fact, in the two last decades, coordination behavior and physicochemical properties of poly(pyrazol-1-yl)methanes have attracted considerable interest [1–4, 20–23] with applications in bio-inorganic, organic, and/or catalytic chemistry [1, 2, 24–29]. Tris(pyrazol-1-yl)methanes are used [30] to mimic the coordination chemistry of metalloproteins [e.g. Fe- (hemerythrin) Cu- (hemocyanin) or Mn-proteins (superoxide dismutase)], and a considerable number of water-soluble tris(pyrazol-1-yl)methane metal complexes proved to act as potent *in vitro* cytotoxic agents [2, 28]. In organic synthesis, tris(pyrazol-1-yl)methane complexes are able to catalyze the two main types of reactions, carbon–carbon bond formations (e.g. the nitro-aldol coupling) [31, 32] and functional group transformations (e.g. oxidations) [1–4, 33] involved in important industrial processes.

The electronic and structural versatility of scorpionate metal complexes results from the functionalization of carbon homoscorpionate ligands, which allowed the design of derivatives toward different and highly specific applications [10, 30]. The positions of the homoscorpionate molecule commonly recognized as significant for functionalization, the apical position and the heterocyclic ring, are depicted in Figure 6.2. Their modification leads to unique effects on the physical properties and coordination behavior of the corresponding compounds.

Moreover, although the most common coordination mode is the three-nitrogen heteroatoms of the corresponding pyrazol-1-yl rings occupying three facially adjacent vacant positions of the coordination sphere of a metal center (N_3-chelating mode), they can afford different coordination modes, adapting their geometry to the demands of the metal center, binding in a κ^2-mode (i.e. bidentate, Figure 6.3) or acting as a bridging ligand [1, 2, 10, 15, 34, 35, 37–41].

6.3 Oxidation of Cyclohexane to KA Oil

C-scorpionate transition metal complexes are good catalysts toward the mild oxidative functionalization of inert alkanes into valuable organic products. The existence of different oxidation states of transition metals appears to play a significant role for their catalytic properties (see below) [1–3, 38].

When used at similar operation conditions to those of the industrial process (typically at 15–19 atm of O_2, 140–150 °C, 8–18 hours reaction time [39, 40]), the tested C-scorpionate

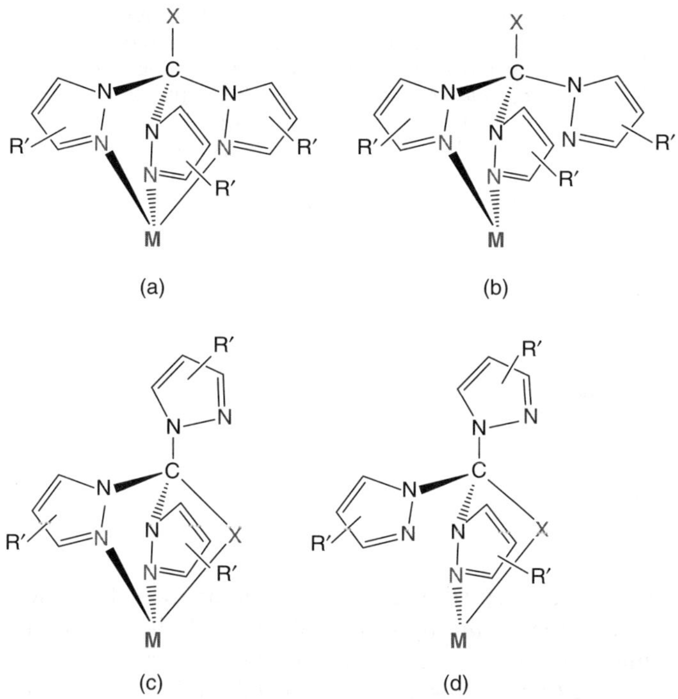

Figure 6.3 Schematic representation of the most common coordination modes of C-scorpionates: (*a*) tripodal N,N,N (κ^3-*N,N,N*), (*b*) tripodal N,N,X (κ^3-*N,N,X*), (*c*) bipodal N,N (κ^2-*N,N*), and (*d*) bipodal N,X (κ^2-*N,X*); X = donor atom from a ligand L.

metal complexes, either in homo- [39] or heterogeneous [40] conditions, are able to catalyze the oxidation of cyclohexane with improved yields relative to the industrial process [5]. One example is the vanadium(III or IV) complexes [$VCl_3\{\kappa^3$-$RC(pz)_3\}$] (R = H or SO_3) yielding cyclohexanol (the main product) and cyclohexanone with a 13% conversion in homogeneous conditions (Scheme 6.2). The reaction is slightly promoted up to 15% conversion by the presence of heteroaromatic carboxylic acids (e.g. pyrazinecarboxylic acid, Hpca) [39]. The Re(III) analogues [40], although inactive under aerobic homogeneous conditions, when immobilized on 3-aminopropyl functionalized silica, lead to up to 16% overall conversion toward KA oil. The supported C-scorpionate Re(III) complexes also catalyze the oxidation of other cycloalkanes (e.g. cyclo- pentane, heptane and octane) to the corresponding cycloalkanols and cycloalkanones, with conversions and selectivities comparable to those for cyclohexane oxidation [40].

The authors, supported by radical traps experiments, proposed radical mechanisms for the aerobic oxidation of cyclohexane [39, 40]) for both homo- and heterogeneous media.

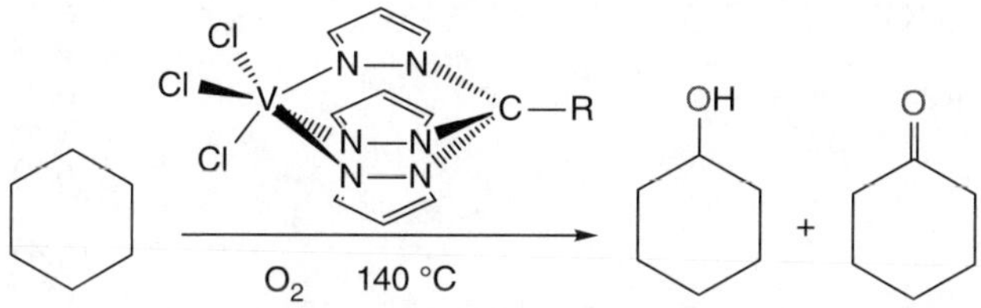

Scheme 6.2 Aerobic oxidation of cyclohexane to KA oil catalyzed by the vanadium(III or IV) complexes [$VCl_3\{\kappa^3$-$RC(pz)_3\}$] (R = H or SO_3).

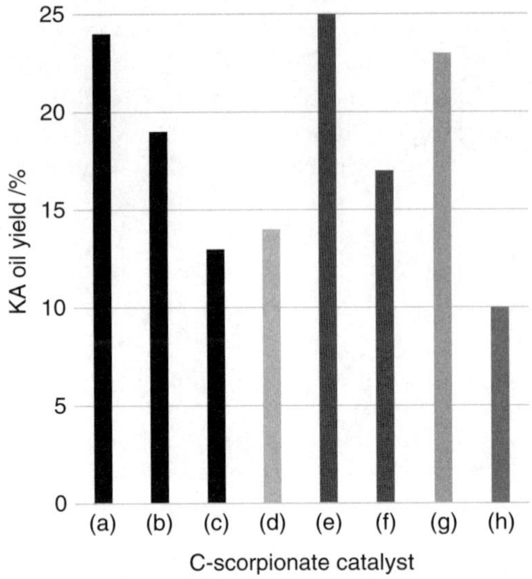

Figure 6.4 Comparison of KA oil yields obtained by room temperature oxidation of cyclohexane with hydrogen peroxide catalyzed by C-scorpionate complexes: (a) $[VO_2\{\kappa^3\text{-}HC(pz^{3,5Me2})_3\})]$, (b) $[VO_2\{\kappa^3\text{-}SO_3C(pz)_3\}]$, (c) $[VCl_3\{\kappa^3\text{-}HC(pz)_3\}]$, (d) $[ReO_3\{\kappa^3\text{-}SO_3C(pz)_3\}]$, (e) $Li[FeCl_2\{\kappa^3\text{-}SO_3C(pz)_3\}]$, (f) $[FeCl_2\{\kappa^3\text{-}HC(pz)_3\}]$, (g) $[CuCl_2\{\kappa^3\text{-}HOCH_2C(pz)_3\}]$, and (h) $[AuCl_2\{\kappa^3\text{-}HOCH_2C(pz)_3\}]Cl$.

The cyclohexyl radical ($Cy^•$) could be initially formed by the metal-promoted reaction (slow induction period) of dioxygen (O_2) with the alkane upon homolytic C—H bond rupture. The oxidation of $Cy^•$ with O_2 would form the cyclohexylperoxy radical ($CyOO^•$), which could lead to cyclohexyl hydroperoxide (CyOOH). This undergoes metal catalyzed homolytic decomposition to $CyO^•$ (O—O bond rupture) or $CyOO^•$ (O—H bond cleavage) that react further to form cyclohexanol and cyclohexanone. The Hpca co-catalyst [41, 42] could promote these bond cleavages by assisting the occurring hydrogen transfers. Cyclohexanol could be formed either upon H-abstraction from cyclohexane (CyH) by $CyO^•$ or upon decomposition of $CyOO^•$, leading to cyclohexanone [39].

The search for milder reaction conditions (e.g. operating at room temperature) led to the use of peroxides (mainly the clean H_2O_2) as oxidizing agents for the oxidation of cyclohexane catalyzed by C-scorpionate metal (e.g. V, Re, Fe, Co, Ni, Cu, or Au) complexes. The reactions are usually performed with aqueous hydrogen peroxide in acidic acetonitrile, yielding selectively KA oil up to 25% [1]. As depicted in Figure 6.4, the iron(II) and vanadium(V) scorpionate complexes were the most effective catalysts.

Detailed investigation of the effects of various experimental parameters in this reaction, the use of radical traps, and kinetic and selectivity studies complemented with theoretical calculations indicated [41, 43–45] the involvement of hydroxyl ($OH^•$) and cyclohexyl ($Cy^•$) radicals in a radical type mechanism. The proposed route for the metal-catalyzed decomposition of hydrogen peroxide (Haber–Weiss mechanism) and further reactions of hydroxyl ($HO^•$) radical are represented by the following key steps (1–9):

$$M^{n+} + H_2O_2 \rightarrow HO^• + M^{(n+1)+} + HO^- \quad (6.1)$$

$$M^{(n+1)+} + H_2O_2 \rightarrow HOO^• + H^+ + M^{n+} \quad (6.2)$$

$$HO^• + CyH \rightarrow H_2O + Cy^• \quad (6.3)$$

$$Cy^• + O_2 \rightarrow CyOO^• \quad (6.4)$$

$$CyOO^• + H_2O_2 \rightarrow CyOOH + HOO^• \quad (6.5)$$

$$CyOOH + M^{n+} \rightarrow CyO^• + HO^- + M^{(n+1)+} \quad (6.6)$$

$$CyOOH + M^{(n+1)+} \rightarrow CyOO^{\bullet} + H^{+} + M^{n+} \quad (6.7)$$

$$CyO^{\bullet} + CyH \rightarrow CyOH + Cy^{\bullet} \quad (6.8)$$

$$2\,CyOO^{\bullet} \rightarrow CyOH + Cy_{-H}{=}O + O_2 \quad (6.9)$$

6.3.1 Oxidation of Cyclohexane at Nonconventional Conditions

Several approaches have been developed [38, 46–48] to overcome safety and ecological drawbacks of current alkane oxidations aiming at operating under more sustainable conditions. Short reaction times are relevant for an economically feasible process. Thus, the use of microwave (MW) radiation as an alternative energy source is appealing (although unexplored) for its potential to improve product purity, selectivity, and yield, and because it is more energy-efficient compared to conventional (e.g. oil-bath) heating methods [49, 50].

MW irradiation was successfully applied to the cyclohexane oxidation by hydrogen peroxide catalyzed by the C-scorpionate complex [$FeCl_2$\{κ^3-HC(pz)$_3$\}], providing a highly selective and fast method to synthesize KA oil (28% yield after 1 hour reaction) [51]. This procedure allowed reaction time to be shortened by five hours, with the concomitant KA oil yield increase from 17% [52] to 28% when compared to the conventional oil-bath heating method.

Another important topic to consider in developing sustainable catalytic processes concerns the use of solvents. Room-temperature ionic liquids (RTILs) are suitable alternatives to volatile organic solvents (VOCs) in organic synthesis [53–56], although their use in alkane oxidations remains unexplored [57–59].

Recently, RTILs 1-butyl-3-methylimidazolium tetrafluoroborate [bmim][BF_4] and 1-butyl-3-methylimidazolium dicyanamide [bmim][$N(CN)_2$] (Figure 6.5) were tested as solvents for the oxidation of cyclohexane to KA oil catalyzed by [$FeCl_2$\{κ^3-HC(pz)$_3$\}] under the MW-assisted conditions. Significant advantages were observed when RTILs were used as solvents, such as the enhanced KA oil yield under additive-free conditions (4.8% in MeCN vs. 18.6% in [bmim][$N(CN)_2$] or 16.1% in [bmim][BF_4]) or the recycling and reuse of [$FeCl_2$\{κ^3-HC(pz)$_3$\}] up to nine consecutive catalytic cycles [51].

Moreover, an alcohol/ketone selectivity tuning dependent on the RTIL was found: high selectivity toward cyclohexanol production is obtained in [bmim][$N(CN)_2$] whereas in [bmim][BF_4] the formation of cyclohexanone is clearly favored. The selective oxidation of cyclohexane to cyclohexanone can be industrially applied as an improvement in the polyamide (PA) 6 production. In fact, in the current production process of PA 6, KA mixture obtained from cyclohexane oxidation is first converted to pure cyclohexanone (at high pressure, 250 °C, in the presence of copper and chromium oxides) and then used to make caprolactam (monomer for PA 6 synthesis) [5, 60]. Such a step can be avoided by choosing the suitable solvent. IL interactions with [$FeCl_2$\{κ^3-HC(pz)$_3$\}] were determined by DFT calculations.

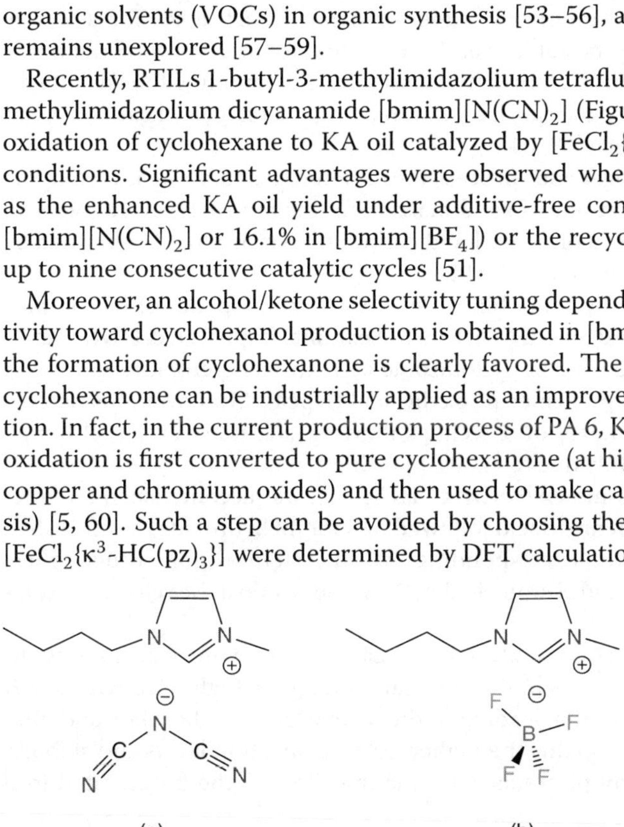

Figure 6.5 Schematic structure representation of (a) 1-butyl-3-methylimidazolium dicyanamide [bmim][$N(CN)_2$] and (b) 1-butyl-3-methylimidazolium tetrafluoroborate[bmim][BF_4].

The use of supercritical fluids (e.g. carbon dioxide, $scCO_2$) as safer solvent media for chemical reactions is also very important in sustainable chemistry design [61, 62]. The low environmental impact and the chemical inertness of $scCO_2$ under most catalytic process conditions are expected to promote safer reactions in this medium. Moreover, when combined with RTILs, i.e. RTIL/$scCO_2$ systems would facilitate recycling and reuse of catalysts [62].

The catalytic performance of the C-scorpionate iron(II) complexes [$FeCl_2${RC$(pz)_3$}] (R = H or CH_2OH) for the oxidation of cyclohexane to KA oil with *tert*-butyl hydroperoxide (TBHP, 70% aqueous solution), at 70 °C, was assessed [56] in 1-butyl-3-methylimidazolium hexafluorophosphate ([bmim][PF_6] and 1-butyl-3-methylimidazolium tris(pentafluoroethyl)trifluorophosphate, [bmim][(FAP)]) medium, in $scCO_2$ in acetonitrile or in $scCO_2$/[bmim][X] (X = PF_6 or FAP) mixtures. For both complexes in pure solvents, the following order of catalytic activity was found: [bmim][PF_6] > [bmim][FAP] > $scCO_2$ > CH_3CN. RTIL/$scCO_2$ (RTIL = [bmim][PF_6] or [bmim][FAP]) mixtures were found to enhance KA oil yield and selectivity toward cyclohexanone formation. Moreover, the catalysts can be recycled in those media, preserving their activity up to three consecutive cycles [56]. Therefore, the use of such alternative solvents led to promising efficient and selective protocols for the oxidation of cyclohexane.

6.3.2 Oxidation of Cyclohexane Catalyzed by Supported C-Scorpionate Complexes

The design of environmentally and economically superior catalytic processes for selective alkane oxidations also addresses catalysts stability and their reuse, aiming to overcome problems presented by current homogeneous systems. The possibility of combining the advantages of homogeneous and heterogeneous catalysts by immobilizing metal complexes on solid supports is therefore of great importance [63, 64].

A general strategy to convert a homogeneous process into a heterogeneous one is to support the soluble C-scorpionate metal catalyst onto a solid matrix. To this purpose, several types of materials have been used, in particular, zeolitic [65] and carbon materials [36, 66] obtained by different approaches most recently proposed in the literature [64].

The hybrid material resulting from the immobilization (by wet-impregnation) of complex [$FeCl_2${κ^3-HC$(pz)_3$}] on desilicated mordenite (MOR) zeolite exhibits improved catalytic activity (TON up to 2.90×10^3) for the oxidation of cyclohexane by hydrogen peroxide, with a KA oil yield of 38% [65] in the presence of pyrazinecarboxylic acid (Hpca). Moreover, this heterogeneous system offered practical advantages of recycling, allowing successive cycles, although with a considerable loss of activity mainly due to catalyst leaching [65].

A more efficient immobilization of complex [$FeCl_2${κ^3-HC$(pz)_3$}] was accomplished on functionalized (upon surface treatment with nitric acid followed by sodium hydroxide) multi-walled carbon nanotubes (MWCNTs) [66]. The corresponding hybrid system exhibited the highest catalytic activity (TON up to 5.6×10^3 and KA oil yield of 21%) [66] toward the selective oxidation of cyclohexane (Scheme 6.3).

Recovering the catalyst from the reaction mixture by an easy filtration and reusing it without loss of activity for five consecutive cycles [66] while maintaining the high selectivity to KA oil were also achieved. The functionalization, leading to the formation of carboxylate and phenolate groups on the MWCNT walls, explains the higher heterogenization levels and stability of [$FeCl_2${κ^3-HC$(pz)_3$}], since those groups act as coordinating sites for the unsaturated iron coordination sphere.

A similar behavior was found [36] for the immobilization of C-scorpionate gold(III) compounds [$AuCl_2${κ^2-RC$(R'pz)_3$}]Cl (R = R′ = H, R = CH_2OH and R′ = H, or R = H and

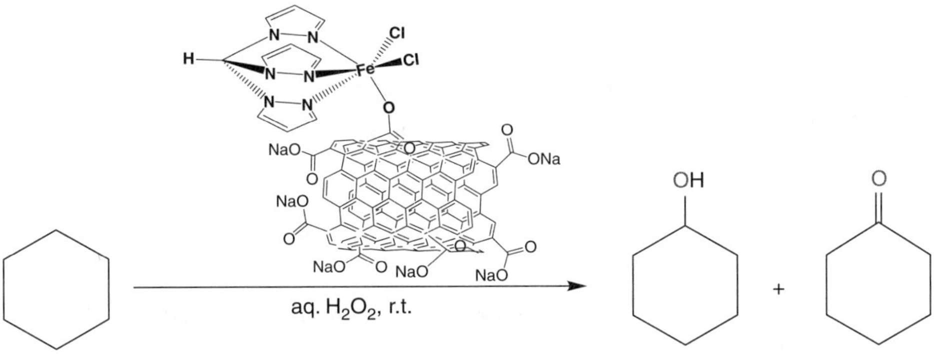

Scheme 6.3 Oxidation of cyclohexane to KA oil with hydrogen peroxide, catalyzed by $[FeCl_2\{\kappa^3\text{-}HC(pz)_3\}]$ supported at functionalized MWCNT. (*See insert for color representation.*)

R′ = 3,5-Me_2) and for the catalytic activity of the heterogenized species on carbon materials (activated carbon, carbon xerogel, and MWCNTs) when used for the oxidation of cyclohexane to cyclohexanol and cyclohexanone, with aqueous H_2O_2, under mild conditions.

6.4 Oxidation of Cyclohexane to Adipic Acid

Very recently, the direct oxidation of cyclohexane to adipic acid (96% yield, 98% selectivity, TONs up to 1.8×10^3) was achieved [67] by using ozone as oxidant and the catalyst $[FeCl_2\{\kappa^3\text{-}HC(pz)_3\}]$ (pz = pyrazol-1-yl) in a solvent-, heating-, radiation-, and N_2O-free new protocol (Scheme 6.4).

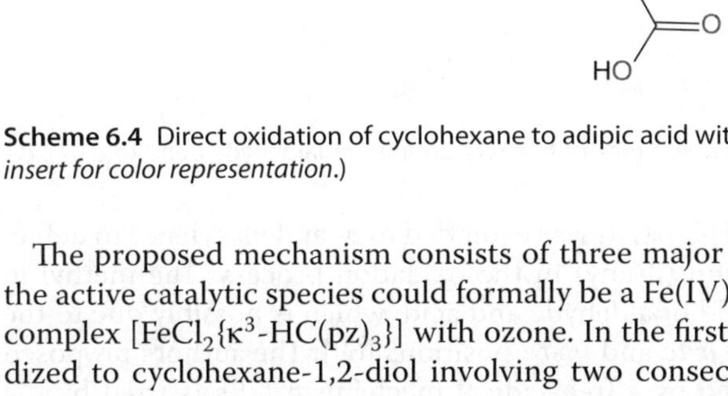

Scheme 6.4 Direct oxidation of cyclohexane to adipic acid with ozone catalyzed by $[FeCl_2\{\kappa^3\text{-}HC(pz)_3\}]$. (*See insert for color representation.*)

The proposed mechanism consists of three major oxidative steps (Scheme 6.5, [68]) where the active catalytic species could formally be a Fe(IV)=O one formed by oxidation of the Fe(II) complex $[FeCl_2\{\kappa^3\text{-}HC(pz)_3\}]$ with ozone. In the first step (*i*), Scheme 6.5), cyclohexane is oxidized to cyclohexane-1,2-diol involving two consecutive hydroxylation steps, each initiated by a hydrogen–atom transfer from the substrate to Fe(IV)=O). Follows (*ii*), Scheme 6.5) the cyclohexane-1,2-diol conversion to hexane-1,6-dial by intradiol C—C bond cleavage, and then (*iii*), Scheme 6.5) the hydroxylation of the carbonyl groups of hexane-1,6-dial to adipic acid [68].

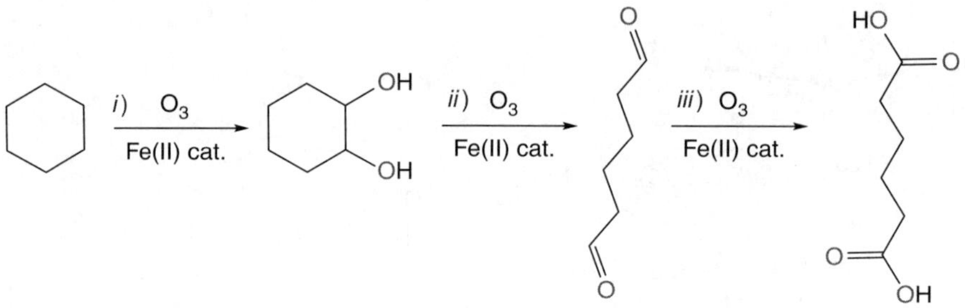

Scheme 6.5 The three major oxidative steps of the proposed reaction mechanism for the oxidation of cyclohexane with O_3.

6.5 Oxidation of Other Hydrocarbons

The oxidative catalytic activity of C-scorpionate metal complexes was successfully extended to other substrates, namely xylene hydrocarbons [69, 70], secondary alcohols [71], linear and cyclic ketones [4], and alkenes [72].

The oxidation of *para*-xylene to terephthalic acid is a very important process in the polyester industry [73, 74] due to the use of this acid in polyethylene terephthalate (PET) production, a commodity polymer for manufacturing fibers, films, and plastic goods. Greener catalytic processes than the current industrial one, the Amoco process (a homogeneous cobalt and manganese catalytic system in the presence of highly corrosive bromide dissolved in aqueous acetic acid), [73] have been reported [75–77] but these lack the necessary oxidation activity, and therefore the development of a halogen-free efficient catalytic procedure to oxidize xylenes is still needed.

Recently, *p*-xylene was successfully activated at one methyl group using the C-scorpionate iron(II) complex $[FeCl_2\{\kappa^3\text{-}HC(pz)_3\}]$ as a catalyst and aqueous hydrogen peroxide (considered as a clean oxidant) at 35 °C in acetonitrile, leading to the very fast (after 5 minutes of reaction) formation of 4-methyl benzyl alcohol, *p*-tolualdehyde and *p*-toluic acid (Scheme 6.6, overall oxygenates yield of 22%, TOF = $1.3 \times 10^2\ h^{-1}$) [69]. However, under the tested conditions, the second methyl group leading to terephthalic acid was not activated.

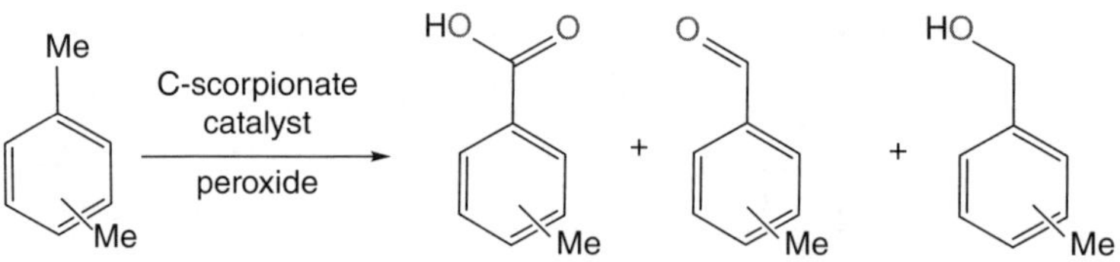

Scheme 6.6 Oxidation of xylenes to the corresponding methyl benzyl alcohols, tolualdehydes, and toluic acids.

The catalytic applicability of $[FeCl_2\{\kappa^3\text{-}HC(pz)_3\}]$ was extended to *o*- and *m*-xylenes to detect the influence of the methyl group position (if any) in the oxidation process. The methyl in the *para* position leads to the highest yield of aldehyde and acid, which is possibly due to the higher steric constrains associated to the *orto* and *meta* positions [69]. The authors proposed that the oxidation of xylenes is conducted by a free-radical mechanism [78] assisted by the C-scorpionate iron(II) catalyst, involving $HO^\bullet$ and $HOO^\bullet$ radicals produced in the iron promoted decomposition of hydrogen peroxide, according to equations (6.10)–(6.16):

$$Fe^{2+} + H_2O_2 \rightarrow HO^\bullet + Fe^{3+} + HO^- \tag{6.10}$$

$$Fe^{3+} + H_2O_2 \rightarrow HOO^{\bullet} + H^+ + Fe^{2+} \tag{6.11}$$

$$HO^{\bullet} + RCH_3 \rightarrow H_2O + RCH_2^{\bullet} \tag{6.12}$$

$$RCH_2^{\bullet} + O_2 \rightarrow RCH_2OO^{\bullet} \tag{6.13}$$

$$RCH_2OO^{\bullet} + RCH_3 \rightarrow RCH_3OOH + RCH_2^{\bullet} \tag{6.14}$$

$$RCH_2OO^{\bullet} + Fe^{2+} \rightarrow Fe^{3+} + RCHO + HO^- \tag{6.15}$$

$$2\,RCH_2OO^{\bullet} \rightarrow RCH_2OH + RCHO + O_2 \tag{6.16}$$

The oxidation of xylenes using vanadium complexes as catalysts, in homogeneous environment and heterogenized, has been recently [70] achieved for the first time by using C-scorpionate vanadium(IV) $[VO_xCl_{3-x}\{\kappa^3\text{-}RC(pz)_3\}]$ (x = 0, R = SO_3; x = 1, R = CH_2OH or CH_2OSO_2Me) complexes. The neat oxidation of *o*-, *m*- or *p*-xylene, with TBHP (70% aqueous solution), to the corresponding toluic acids (main products, yields up to 43%) [70], tolualdehydes and methyl benzyl alcohols (Scheme 6.6) was obtained in a simple microwave-assisted mild oxidation procedure, in the absence of any bromine source, catalyzed by the above V-complexes supported on functionalized MWCNTs [70]. The immobilization of C-scorpionate vanadium complexes on the functionalized carbon nanotubes is believed to occur via the oxygenated functional groups (carboxylate or phenolate) of the MWCNT, with possible formation of a covalent VO bond, upon displacement of a chloride ligand or by replacement of one of the coordinated pyrazolyl arms of the C-scorpionate ligands, which thus would change from the trihapto to a dihapto coordination mode (a known feature of these ligands [1–4]). Moreover, catalysts reuse with preservation of their activity was accomplished for up to six consecutive cycles.

Acknowledgments

The author gratefully acknowledges all the co-authors cited in the joint publications. Its work in this area has been partially supported by the Fundação para a Ciência e a Tecnologia: FCT (projects UID/QUI/00100/2013, PTDC/QEQ-ERQ/1648/2014 and PTDC/QEQ-QIN/3967/2014), Portugal.

References

1 Martins, L.M.D.R.S. and Pombeiro, A.J.L. (2014). *Coord. Chem. Rev.* 265: 74.
2 Martins, L.M.D.R.S. and Pombeiro, A.J.L. (2016). *Eur. J. Inorg. Chem.* 2236.
3 Martins, L.M.D.R.S. (2017). *Catalysts* 7: 12.
4 Martins, L.M.D.R.S. and Pombeiro, A.J.L. (2017). *Inorg. Chim. Acta* 455: 390.
5 (1999–2016). *Ullmann's Encyclopedia of Industrial Chemistry*, 6th, vol. 11, 41–49. Weinheim, Germany: Wiley-VCH.
6 (2016). *Adipic Acid (ADPA): 2016 World Market Outlook and Forecast up to 2020*. Birmingham, UK: Merchant Research and Consulting.
7 Hermans, I., Jacobs, P.A., and Peeters, J. (2006). *Chem. Eur. J.* 12: 4229.
8 Van de Vyver, S. and Román-Leshkov, Y. (2013). *Catal. Sci. Technol.* 3: 1465.
9 Smith, J.M. (2008). *Comments Inorg. Chem.* 29: 189.
10 Trofimenko, S. (1993). *Chem. Rev.* 93: 943.

11 Hückel, W. and Bretschneider, H. (1937). *Ber. Dtsch. Chem. Ges.* 70: 2024.
12 Jesson, J.P. (1966). *J. Chem. Phys.* 45: 1049.
13 Juliá, S., del Mazo, J.M., Avila, L. et al. (1984). *Org. Prep. Proced. Int.* 16: 299.
14 A.J.L. Pombeiro, L.M.D.R.S. Martins, E.C.B.A. Alegria, Use of microwaves in the synthesis of tris (pyrazolyl) methanes, PT103681; award date: 4/12/2009.
15 Alegria, E.C.B.A., Martins, L.M.D.R.S., Haukka, M. et al. (2006). *Dalton Trans.* 4954.
16 Reger, D.L., Grattan, T.C., Brown, K.J. et al. (2000). *J. Organomet. Chem.* 607: 120.
17 Reger, D.L., Collins, J.E., Rheingold, A.L. et al. (1996). *Organometallics* 15: 2029.
18 Wanke, R., Smoleński, P., Guedes da Silva, M.F.C. et al. (2008). *Inorg. Chem.* 47: 10158.
19 Silva, T.F.S., Mishra, G.S., Silva, M.F.G. et al. (2009). *Dalton Trans.* 42: 9207.
20 Wanke, R., Silva, M.F.C.G., Lancianesi, S. et al. (2010). *Inorg. Chem.* 49: 7941.
21 Bigmore, H.R., Lawrence, S.C., Mountford, P. et al. (2005). *Dalton Trans.* 635.
22 Pettinari, C. and Pettinari, R. (2005). *Coord. Chem. Rev.* 249: 525.
23 Otero, A., Fernández-Baeza, J., Lara-Sánchez, A. et al. (2013). *Coord. Chem. Rev.* 257: 1806.
24 Martins, L.M.D.R.S., Alegria, E.C.B.A., and Pombeiro, A.J.L. (2014, Chapter 4). *Ligands: Synthesis, Characterization and Role in Biotechnology* (ed. P. Smolenski and P. Gawryszewska), 117–140. New York: Nova Science Publishers, Inc.
25 Silva, T.F.S., Martins, L.M.D.R.S., Guedes da Silva, M.F.C. et al. (2012). *Dalton Trans.* 41: 12888.
26 Silva, T.F.S., Martins, L.M.D.R.S., Guedes da Silva, M.F.C. et al. (2014). *Chem. Asian J.* 9: 1132.
27 Pettinari, C., Marchetti, F., Lupidi, G. et al. (2011). *Inorg. Chem.* 50: 11173.
28 Niesel, J., Pinto, A., P N'Dongo, H.W. et al. (2008). *Chem. Commun.* 1798.
29 Martins, L.M.D.R.S. and Pombeiro, A.J.L. (2014, Chapter 22). Carbon-scorpionate complexes in oxidation catalysis. In: *Advances in Organometallic Chemistry and Catalysis, the Silver/Gold Jubilee ICOMC Celebratory Book* (ed. A.J.L. Pombeiro), 285. Hoboken, NJ: Wiley.
30 Pettinari, C. and Santini, C. (2003). *Comprehensive Coordination Chemistry II: From Biology to Nanotechnology*, vol. 1 (ed. J.A. McCleverty and T.J. Meyer), 159–172. Amsterdam: Elsevier Pergamon (APS).
31 Rocha, B.G.M., Mac Leod, T.C.O., Guedes da Silva, M.F.C. et al. (2014). *Dalton Trans.* 43: 15192.
32 Pettinari, C., Marchetti, F., Cerquetella, A. et al. (2011). *Organometallics* 30: 1616.
33 Ribeiro, A.P.C., Martins, L.M.D.R.S., and Pombeiro, A.J.L. (2017). *Green Chem.* 19: 4811–4815.
34 Silva, T.F.S., Luzyanin, K.V., Kirilova, M.V. et al. (2010). *Adv., Synth. Cat.* 352: 171.
35 Martins, L.M.D.R.S., Alegria, E.C.B.A., Smoleński, P. et al. (2013). *Inorg. Chem.* 52: 4534.
36 Almeida, M.P., Martins, L.M.D.R.S., Carabineiro, S.A.C. et al. (2013). *Catal. Sci. Technol.* 3: 3056.
37 Ellis, D.D., Jeffery, J.C., Jelliss, P.A. et al. (2001). *Inorg. Chem.* 40: 2041.
38 Shul'pin, G. (2016). *Catalysts* 6: 50.
39 Mishra, G.S., Silva, T.F.S., Martins, L.M.D.R.S. et al. (2009). *Pure Appl. Chem.* 81: 1217.
40 Mishra, G.S., Alegria, E.C.B.A., Martins, L.M.D.R.S. et al. (2008). *Cat. A: Chem.* 285: 92.
41 Shul'pin, G.B. (2013). *Dalton Trans.* 42: 12794.
42 Kirillova, M.V., Kuznetsov, M.L., Kozlov, Y.N. et al. (2011). *ACS Catal.* 1: 1511.
43 Shul'pin, G.B. (2003). *C. R. Chim.* 6: 163.
44 Kirillova, M.V., Kuznetsov, M.L., Romakh, V.B. et al. (2009). *J. Cat.* 267: 140.
45 Rachmilovich-Calis, S., Masarwa, A., Meyerstein, N. et al. (2009). *Chem. Eur. J.* 15: 8303.
46 Li, J.J. (ed.) (2015). *C-H Bond Activation in Organic Synthesis.* CRC Press.

47 Pérez, P.J. (ed.) (2012). *Alkane C-H Activation by Single-Site Metal Catalysis*. Springer.
48 Pombeiro, A.J.L. (2014, Chapter 1–3). *Advances in Organometallic Chemistry and Catalysis: The Silver/Gold Jubilee International Conference on Organometallic Chemistry Celebratory Book*, 22. Hoboken, USA: Wiley.
49 Dudley, G.B., Richert, R., and Stiegman, A.E. (2015). *Chem. Sci.* 6: 2144.
50 Varma, R.S. (2007). *Green Chem. Lett. Rev.* 1: 37.
51 Ribeiro, A.P.C., Martins, L.M.D.R.S., and Kuznetsov, M.L. (2017). A. J. L. Pombeiro. *Organomet.* 36: 192.
52 Silva, T.F.S., Alegria, E.C.B.A., Martins, L.M.D.R.S. et al. (2008). *Adv., Synth. Cat.* 350: 706.
53 Pârvulescu, V.I. and Hardacre, C. (2007). *Chem. Rev.* 107: 2615.
54 Wasserscheid, P. and Welton, T. (2007). *Ionic Liquids in Synthesis*. Weinheim: Wiley-VCH.
55 Plechkova, N.V. and Seddon, K.R. (2008). *Chem. Soc. Rev.* 37: 123.
56 Ribeiro, A.P.C., Martins, L.M.D.R.S., Alegria, E.C.B.A. et al. (2017). *Catalysts* 7: 230.
57 Jlassi, R., Ribeiro, A.P.C., da Silva, M.F.C.G. et al. (2014). *Eur. J. Inorg. Chem.* 4541.
58 Ribeiro, A.P.C., Martins, L.M.D.R.S., Hazra, S. et al. (2015). *C. R. Chim.* 18: 758.
59 Paul, A., Ribeiro, A.P.C., Karmakar, A. et al. (2016). *Dalton Trans.* 45: 12779.
60 Clerici, M.G., Ricci, M., and Strukul, G. (2006). *Metal-Catalysis in Industrial Organic* (ed. G.P. Processes and P.M.M. Chiusoli). Cambridge, UK: Royal Society of Chemistry.
61 Anastas, P. and Warner, J.C. (1998). *Green Chemistry – Theory and Practice*, 1e, 30. New York: Oxford University Press.
62 Muldoon, M. (2014). Catalysis in ionic liquid–supercritical CO_2. In: *Catalysis in Ionic Liquids: From Catalyst Synthesis to Application*, 1e, vol. 15, Chapter 4 (ed. C. Hardacre and V.I. Pârvulescu), 309–344. UK: Royal Society of Chemistry.
63 Carvalho, A.P., Nunes, N., and Martins, A. (2010). Hierarchical zeolites: preparation, properties and catalytic applications. In: *Comprehensive Guide for Mesoporous Materials*, vol. 3 (ed. M. Aliofkhazraei), 147–211. Nova Science Publishers, Inc.
64 Enterria, M. and Figueiredo, J.L. (2016). *Carbon* 108: 79.
65 Martins, L.M.D.R.S., Martins, A., Alegria, E.C.B.A. et al. (2013). *Appl. Catal. A:Gen.* 23: 464–465.
66 Martins, L.M.D.R.S., Peixoto de Almeida, M., Carabineiro, S.A.C. et al. (2013). *ChemCatChem* 5: 3847.
67 Ribeiro, A.P.C., Martins, L.M.D.R.S., and Pombeiro, A.J.L. (2017). *Green Chem.* 19: 1499.
68 Noack, H., Georgiev, V., Blomberg, M.R.A. et al. (2011). *Inorg. Chem.* 50: 1194.
69 Mendes, M., Ribeiro, A.P.C., Alegria, E.C.B.A. et al. (2017). *Polyhedron* 125: 151.
70 J. Wang, L. M. D. R. S. Martins, A. P. C. Ribeiro et al. *Chem. Asian J.*, 2017, 12, 1915.
71 Martins, L.M.D.R.S., Ribeiro, A.P.C., Carabineiro, S.A.C. et al. (2016). *Dalton Trans.* 45: 6816.
72 Neves, P., Gago, S., Balula, S.S. et al. (2011). *Inorg. Chem.* 50: 3490.
73 Tomás, R.A.F., Bordado, J.C.M., and Gomes, J.F.P. (2013). *Chem. Rev.* 113: 7421.
74 Sheehan, R.J. (2003). *Ullmann's Encyclopedia of Industrial Chemistry*, 6e, vol. 35, 639. Weinheim, Germany: Wiley-VCH.
75 Li, M., Ruddy, T., Fahey, D. et al. (2014). *ACS Sustainable Chem. Eng.* 2: 823.
76 Kim, D.S., Shin, Y.H., and Lee, Y.-W. (2014). *Chem. Eng. Comm.* 202: 78.
77 Osada, M. and Savage, P.E. (2009). *AIChE J.* 55: 1530.
78 Sheldon, R.A. and Kochi, J.K. (1981). *Metal Catalyzed Oxidation of Organic Compounds*. New York: Academic Press.

7

Alkane Oxidation with Multinuclear Heterometallic Catalysts

Dmytro S. Nesterov, Oksana V. Nesterova, and Armando J. L. Pombeiro

Centro de Química Estrutural, Instituto Superior Técnico, Universidade de Lisboa, Lisboa, Portugal

7.1 Introduction

Mild oxidation of inert alkanes (sp^3 C—H bonds, in a broader meaning) is a hot topic in current chemistry [1–5], for several reasons. Oxygen insertion into a C—H bond results in a hydroxyl group as a primary product [6–8]. This is a common terminal group in biosystems, and selective hydroxylation is of interest in fine chemistry for synthesis of chemicals with biological and pharmaceutical significance [9]. Moreover, the transformation of the hydroxyl group to the keto or carboxylic one is of great importance. There is also a potential interest in the large-scale industry: cyclohexane can be oxidized to cyclohexanol and cyclohexanone in a process with a number of limitations [10, 11], whereas the oxidation of light alkanes, such as methane, requires harsh conditions [12]. In contrast, natural enzymes, such as methane monooxygenases (MMOs) and cytochrome P450, are able to hydroxylate gaseous alkanes at ambient conditions [1, 6]. Nowadays, although significant achievements have already been reached in the field of mild alkanes oxidation, the enzymatic efficiency is still not reproduced under natural conditions with the artificial systems, and hence the search for novel catalytic systems continues.

The idea to use different metals within a single catalytic system can also be originated from nature. Many enzymes contain different metals in their active centres [13]. Examples include Fe/Mo or Fe/V in nitrogenase, Ni/Fe in hydrogenase, Ca/Mn in photosystem II, Cu/Zn in superoxide dismutase, and even the heterotrimetallic Fe/Cu/Ni core [14] in acetyl synthase/CO dehydrogenase. Despite the synthetic obstacles toward the synthesis of heterometallic coordination compounds (such as metal scrambling) [15], many heterometallic compounds have been already recognized as active catalysts in a wide range of chemical processes [16]. In this chapter, we focus on recent achievements in the field of sp^3 C—H bond oxidative transformations, first of all hydroxylation, employing heterometallic complexes as catalysts or as catalyst precursors.

7.2 Peroxidative Oxidation of Liquid Alkanes

Catalytic hydroxylation is one of the most often tested reactions in the field of alkanes functionalization. Cyclohexane (CyH) is a typical model for hydroxylation catalytic activity tests due to the following reasons: (i) cyclohexane has sufficiently high C—H bond dissociation energy of 99.3 kcal mol^{-1} [17], preventing its background oxidation under via a noncatalytic processes; (ii) cyclohexane, being a cyclic alkane with all positions equal, produces only one

Alkane Functionalization, First Edition. Edited by Armando J. L. Pombeiro and M. Fátima C. Guedes da Silva.

isomer of products; and (iii) among the other cycloalkanes, cyclohexane and its oxidation products (cyclohexanol and cyclohexanone) possess the most suitable range of boiling temperatures, what is important for catalytic experiments and analytical procedures (typically gas chromatography, GC). Notably, cyclohexane can be oxidized by some natural enzymes, such as sMMO (soluble methane monooxygenase), providing important comparative data [18].

In industry, cyclohexane is a raw material for the production of caprolactam, where its initial oxidation is performed by dioxygen in the presence of a soluble cobalt catalyst [10, 11]. This process shows only below 6% of cyclohexane conversion in order to achieve a good selectivity. Comparison to model catalytic systems should be done cautiously as it is based on the radical auto-oxidation mechanism, which is not followed by the enzymatic processes and their closer models.

The overview of cyclohexane oxidation is shown in Scheme 7.1.

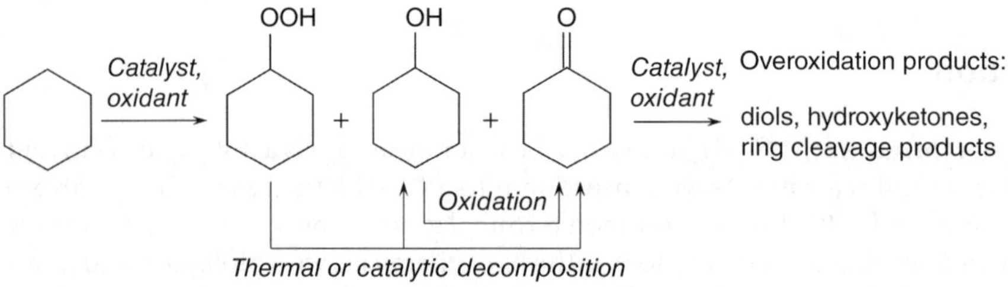

Scheme 7.1 Main products formed from the oxidation of cyclohexane.

Depending on the reaction mechanism, the main oxidation product can change from pure cyclohexyl hydroperoxide (CyOOH) to pure cyclohexanol (CyOH), but usually a mixture of these products with cyclohexanone (Cy=O) is obtained [19–22]. The product distribution provides a tool for tentative conclusions about the reaction mechanisms. A typical measured parameter is the alcohol/ketone (A/K) ratio. However, one has to consider that the presence of CyOOH in the reaction system may significantly influence the results of GC measurements: depending on the GC column used and on the experimental conditions, CyOOH may either spontaneously decompose to produce CyOH, Cy=O and other products, or can be seen as a single product (GC peak). The latter behaviour is typical for nonpolar GC columns, while polar ones promote decomposition of the hydroperoxide [23]. Hence, the determination of the final A/K ratio can involve treatment of the solution, at the end of reaction, with a mild reducing agent (PPh_3 or thiosulfate), quantitatively converting CyOOH to CyOH (and reducing other peroxides as well, e.g. the nonreacted oxidant) [20]. Comparison of the products analyses, with and without addition of the reducing agent, allows estimation of the amounts, at the end of the reaction, of CyOOH, CyOH, and Cy=O [20]. The overoxidation products can also be an indicator of the reaction mechanism, as they can show a well-recognizable pattern [24–27].

One of the first heterometallic complexes used for oxidation of cycloalkanes with H_2O_2 was the hexanuclear Cu/Co/Fe compound $[CuCoFe(Dea)_3(NCS)_2(MeOH)]_2{\cdot}3.2H_2O$ (**1**) (where H_2Dea = diethanolamine) (Figure 7.1), obtained via a direct synthesis [15] reaction using zerovalent copper as a starting material [28]. Compound **1** catalyses the oxidation of cyclohexane and cyclopentane with up to 44.7% and 35.7% of yield, respectively (with aqueous H_2O_2 in acetonitrile at room temperature) (Table 7.1, for the case of CyH oxidation). The presence of nitric acid promoter (20 equiv. excess to the catalyst) is essential for reaching a high yield of products. Such a high catalytic efficiency can be associated with the synergic effect of the three different metals (Cu, Co, and Fe). The A/K ratios, measured after reduction

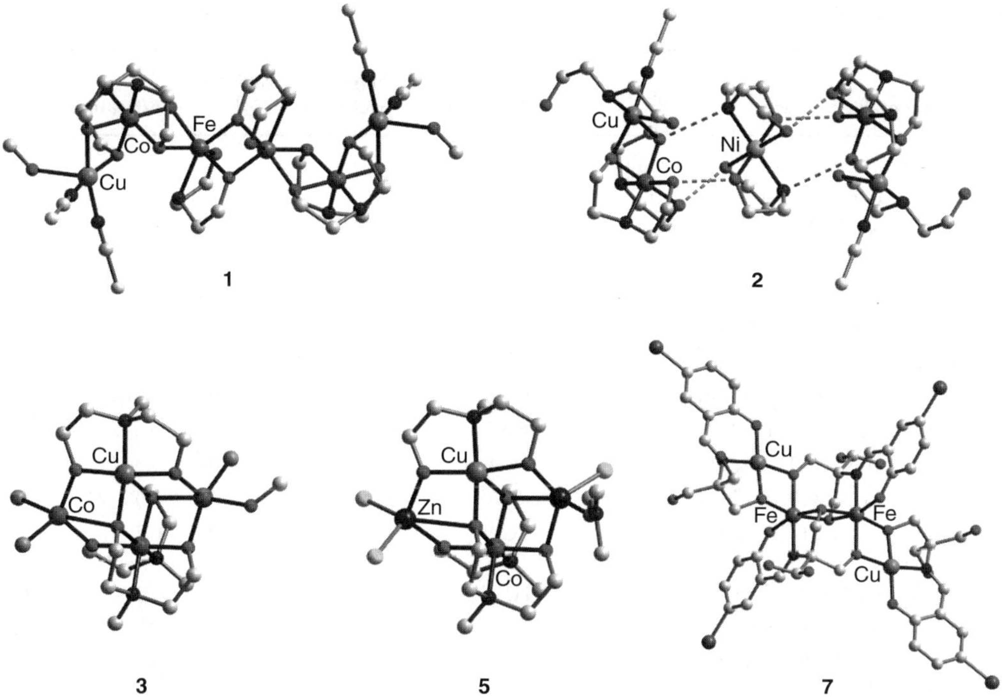

Figure 7.1 Molecular structures of heterometallic catalysts (the cores of **4** and **6** are identical to those of **5** and **7**, respectively, and thus are omitted for clarity) [28–31]. (*See insert for color representation.*)

with PPh_3, are in the range of 6.2 to 18.0 (for cyclohexane) and 3.4 to 6.0 (for cyclopentane), showing a pronounced prevalence of the alcohols.

The reaction mechanism is believed to involve free alkyl radicals, as evidenced by the use of radical trapping agents and the presence, although in low amounts, of cyclohexyl hydroperoxide (estimated from the measurements before and after addition of PPh_3 to the reaction solution).

In contrast to **1**, the heterotrimetallic Cu/Co/Ni compound (with Ni instead of Fe) $[Ni(H_2Dea)_2][CuCo(Dea)_2(H_2Dea)(NCS)]_2Br_2{\cdot}2H_2O$ (**2**) [29] (Fig. 7.1) does not show a notable activity providing the highest product yield of 8% for cyclohexane oxidation under the same conditions as **1** [28].

By replacing simple diethanolamine with the N-methyl one (H_2MeDea) and using Cu/Co and Cu/Co/Zn reaction mixtures, the respective heterometallic complexes have been synthesized, $[CuCoCo_2(MeDea)_3Cl_3(CH_3OH)]$ (**3**), $[CuCoZn_2(MeDea)_3Cl_3(CH_3OH)]$ (**4**), and $[CuCoZn_2(MeDea)_3Cl_3(DMF)]$ (**5**) (Figure 7.1) [30]. Under nearly the same catalytic conditions as those for **1** and **2**, the complex **3** leads to a good yield of products in cyclohexane oxidation (up to 22%). The effects of nitric acid promoter, oxidant, and water were studied, revealing the best conditions as $[HNO_3]_0/[\mathbf{3}]_0 = 20$ and $[H_2O_2]_0/[\mathbf{3}]_0 = 10^3$ and minimum possible amount of water. The A/K ratios range from 2 to 17. The zinc-containing complexes **4** and **5** exhibit a low activity (4% yield). Considering that the tetranuclear coordination cores of **3**–**5** are identical (Figure 7.1), the lower catalytic activity of **4** and **5** is conceivably due to the presence of zinc instead of cobalt in the tetranuclear structures. Although recently it was found that some metals possessing only one stable oxidation state (such as Zn^{2+}) can catalyze the peroxidative oxidation of alkanes (and olefins) by coordinating more than one oxidant

Table 7.1 Heterometallic homogeneous catalytic systems for oxidation of cyclohexane with H_2O_2.[a)]

No.	Heterometallic catalyst	Promoter	Highest yield,[b)]%	Highest TON[c)]	A/K ratios	Ref.
1	$[CuCoFe(Dea)_3(NCS)_2(MeOH)]_2 \cdot 3.2H_2O$	HNO_3	44.7	100	6.2–18.0	[28]
2	$[Ni(H_2Dea)_2][CuCo(Dea)_2(H_2Dea)(NCS)]_2Br_2 \cdot 2H_2O$	HNO_3	8	n.r.[d)]	n.r.	[28]
3	$[CuCoCo_2(MeDea)_3Cl_3(CH_3OH)]$	HNO_3	22	20	2.0–17.0	[30]
4	$[CuCoZn_2(MeDea)_3Cl_3(CH_3OH)]$	HNO_3	4	n.r.	n.r.	[30]
5	$[CuCoZn_2(MeDea)_3Cl_3(DMF)]$	HNO_3	4	n.r.	n.r.	[30]
6	$[Cu_2Fe_2(HL^1)_2(H_2L^1)_2] \cdot 10DMSO$	HNO_3	36	596	up to 10	[31]
7	$[Cu_2Fe_2(HL_2)_2(H_2L^2)_2] \cdot 2DMF$	HNO_3	44	1.1×10^3	up to 10	[31]
8	$[Co_4Fe_2O(Sae)_8] \cdot 4DMF \cdot H_2O$	HNO_3	38	3.6×10^3	0.4–15	[32]
9	$[(VO_2)_2(L^3)\{Na_2(\mu\text{-}H_2O)_2(H_2O)_2\}]_n$	PCA	15.2	1.4×10^3	0.8–1.5	[33]
10	$[\{V(\mu\text{-}O)_2\}_2(L^3)\{K_2(\mu\text{-}H_2O)_2(H_2O)_2\}]_n$	PCA	24.2	2.2×10^3	0.36–2.2	[33]
11	$[\{V(\mu\text{-}O)(\mu_3\text{-}O)\}_2(L^3)\{Cs_2(\mu\text{-}H_2O)_2(H_2O)_2\}]_n$	PCA	32.5	3.0×10^3	0.7–1.8	[33]
12	$[Cu_6(H_2Tea)_6Fe(CN)_6]_n(NO_3)_{2n} \cdot 6_nH_2O$	HNO_3	16.8	319	0.4–6.9	[34]
13	$[Cu_6(HMeDea)_6Fe(CN)_6]_n(NO_3)_{2n} \cdot 7_nH_2O$	HCl	21.5	467	0.3–9.8	[34]
15	$[(Ph_5Ge_5O_{10})_3Fe_6(OH)_3(O)Na_2(phen)] \cdot (EtOH)_4 \cdot (H_2O)_{0.5}$	TFA	44	405	n.r.	[35]
16	$[(PhSiO_{1.5})_{20}(FeO_{1.5})_6(NaO_{0.5})_8(n\text{-}BuOH)_{9.6}(C_7H_8)]$	HNO_3, TFA	32	750	n.r.	[26]
17	$[(PhSiO_{1.5})_{12}(CuO)_4(NaO_{0.5})_4]$	HNO_3	20	207	n.r.	[25]
18	$[(n\text{-}C_4H_9)_4N]_4[\gamma\text{-}HPV_2W_{10}O_{40}]$	$HClO_4$	1	36	49	[36]

a) The complex **14** is not included into the table because it was not tested in cyclohexane oxidation reaction.
b) Yield based on cyclohexane;
c) turnover number, moles of products per mol of catalyst;
d) not reported;

molecule [37–40], this possibility seems to be precluded by the absence of two labile ligands at Zn^{2+} in **4** and **5**.

The heterometallic compounds $[Cu_2Fe_2(HL^1)_2(H_2L^1)_2] \cdot 10DMSO$ (**6**) and $[Cu_2Fe_2(HL^2)_2(H_2L^2)_2] \cdot 2DMF$ (**7**) (where the ligands are products of condensation of salicylaldehyde (H_4L^1) or 5-bromo-salicylaldehyde (H_4L^2) and tris(hydroxymethyl)aminomethane), possessing similar structures (Figure 7.1), have been obtained using the same synthetic approach [15] as **1–5** and tested towards mild cyclohexane oxidation with H_2O_2 [31]. In general, they show a behaviour similar to that of **3**, but with much higher yields and turnover numbers (TONs) (Table 7.1). The A/K ratios are not reported for every entry, but noted to be up to 10 after reduction with PPh_3. In the oxidation of *n*-octane both **6** and **7** reveal a low averaged normalized regioselectivity of 8 : 1 between secondary and primary positions, with no preference between secondary carbon atoms in the chain. These parameters are typical for catalytic systems oxidizing C—H bonds with hydroxyl radical ($HO^\bullet$) [25, 27]. Oxidation of *cis*-1,2-dimethylcyclohexane was performed under conditions which could promote non-free radical mechanisms (low concentration of substrate and oxidant, gradual addition of the oxidant within 10 min time at 0 °C). Nevertheless, even under such conditions, both **6** and **7** did not reveal stereoselective oxidation, expected for systems abstracting hydrogen from the alkane with high-valent metal-oxo species [19].

The Co/Fe complex $[Co_4Fe_2O(Sae)_8]\cdot 4DMF\cdot H_2O$ (**8**) (Figure 7.2, left) bearing the Schiff base aminoalcohol ligand salicylidene-2-ethanolamine (H_2Sae) is one of the most active catalysts toward mild oxidation of alkanes with H_2O_2 [32]. Complex **8** exhibits a remarkable combination of parameters, as illustrated by the maximum yield/TON combinations of 26%/3.6×10^3 and 31%/1.6×10^3 (for cyclohexane) and 46%/2.3×10^3 (for cyclooctane), supported by maximum turnover frequencies (TOF) of 1.1×10^3 h^{-1} and initial reaction rate $W_0=3.5\times10^{-4}$ M s^{-1} (for cyclohexane). The highest yield of 38% (Table 7.1) was supported by lower TON of 543. A prominent feature of the catalytic system **8**/HNO_3/H_2O_2 is that the acute growth of the initial reaction rate (W_0) at $[\mathbf{8}]_0>4\times10^{-5}$ M (Figure 7.2, right) is associated to the presence of the heterometallic cationic species $[Co_2Fe(Sae)_4]^+$ in solution, as indicated by kinetic and electrospray ionization - mass spectrometry (ESI-MS) studies. At the lowest concentrations of **8** (c. 1×10^{-5} M) this species is almost absent, leading to a negligible catalytic activity in this region (Figure 7.2, right). The main pathway is assumed to proceed via a free radical mechanism with the hydroxyl radical as the main attacking species, as confirmed by the 7 : 1 secondary to primary averaged normalized regioselectivity in the oxidation of *n*-heptane and *n*-octane, with no selectivity among different secondary carbons. Also, oxidation of methylcyclohexane reveals the 21 : 7 : 1 averaged normalized bond selectivity, which is also in agreement with participation of the $HO^{\bullet}$ radical [25, 27]. An intriguing observation is the pronounced stereoselectivity in the *cis*-1,2-dimethylcyclohexane oxidation catalysed by **8** when hydrogen peroxide oxidant is replaced by *m*-chloroperbenzoic acid (*m*-CPBA). Such a behaviour is not expected for a free radical oxidation pathway and further investigation is under way.

A series of heterometallic V/M (M = Na, K, Cs) coordination polymers has been synthesized employing bis(salicylaldehyde)-oxaloyldihydrazone ligand H_4L^3: $[(VO_2)_2(L^3)\{Na_2(\mu\text{-}H_2O)_2(H_2O)_2\}]_n$ (**9**), $[\{V(\mu\text{-}O)_2\}_2(L^3)\{K_2(\mu\text{-}H_2O)_2(H_2O)_2\}]_n$ (**10**), and $[\{V(\mu\text{-}O)(\mu_3\text{-}O)\}_2(L^3)\{Cs_2(\mu\text{-}H_2O)_2(H_2O)_2\}]_n$ (**11**) (Figure 7.3) [33]. The complexes possess different structures, influenced by the nature of the s-metal M. They act as catalysts in the mild oxidation of cyclohexane with H_2O_2 in the presence of various promoters – HCl, H_2SO_4, trifluoroacetic acid (TFA), and

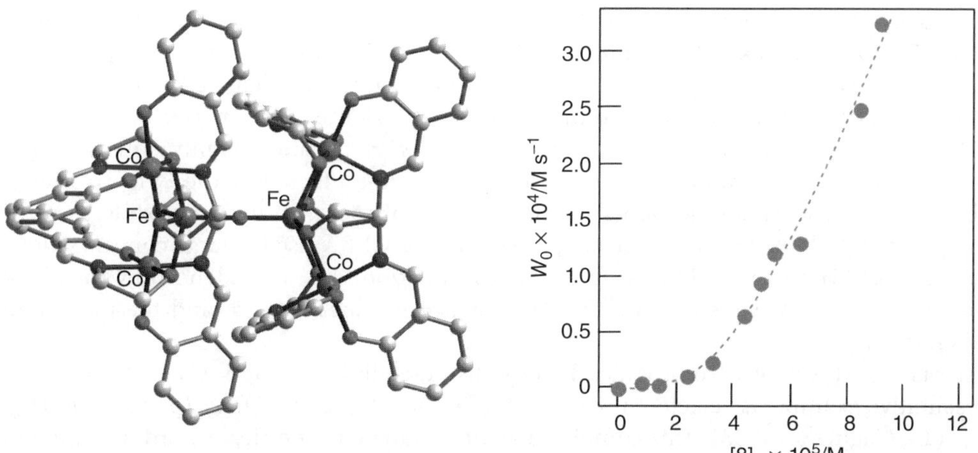

Figure 7.2 Left: molecular structure of the Co/Fe heterometallic complex **8**. Right: dependence of the initial reaction rate W_0 of cyclohexane oxidation to cyclohexanol and cyclohexanone on the concentration of catalyst **8** ($[C_6H_{12}]_0=0.37$ M, $[H_2O_2]_0=1$ M, $[HNO_3]_0=0.04$ M). Source: Adapted with permission from *Inorg. Chem.* 2012, *51*, 9110–9122. Copyright 2012 American Chemical Society [32]. (*See insert for color representation.*)

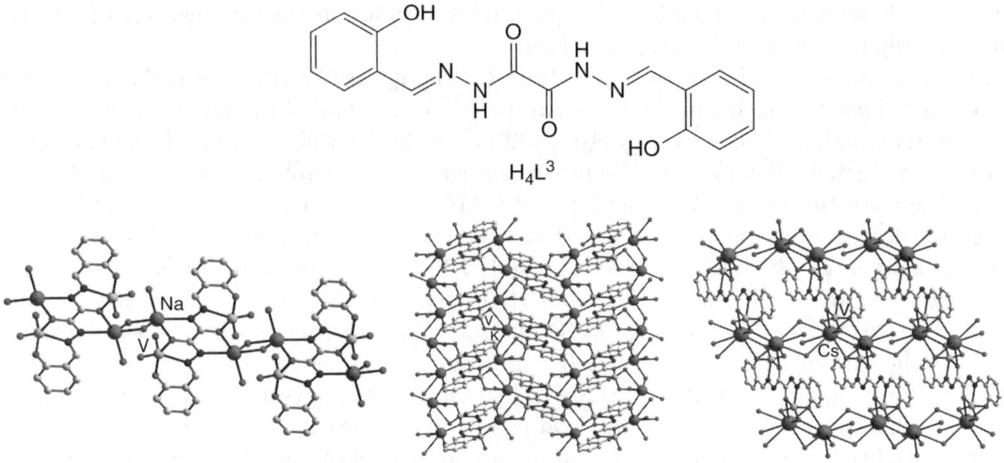

9 10 11

Figure 7.3 Top: ligand H_4L^3. Bottom: fragments of the structures of catalysts **9–11** [33].

pyrazinecarboxylic acid (PCA). The highest yields and TONs (Table 7.1) disclose a pronounced activity of **9–11** in the presence of a PCA promoter, while the other promoters lead to lower activities. The normalized regio- and bond selectivities for oxidation of *n*-heptane and methylcyclohexane are identical to those for the complex **8**, suggesting also a main free radical pathway.

A series of heterometallic Na/M (M = Co, Fe, Cu) complexes **14–17** based on macrocyclic silsesquioxane and germasesquioxane ligands have been reported [25, 26, 35, 41]. The Na/Co complex $[(PhSiO_{1.5})_{10}(CoO)_5Na(OH)]$ (**14**) [41] (Figure 7.4) was tested as a catalyst in a mild oxidation of *cis*- and *trans*-1,2-dimethylcyclohexanes with *m*-CPBA in the presence or absence of HNO_3 promoter (Scheme 7.2) (the activity of **14** toward cyclohexane oxidation was not reported).

The catalyst **14** exhibits a very high yield of tertiary alcohols up to 62% when oxidizing *cis*-1,2-dimethylcyclohexane (0.14 M) in the presence of HNO_3 (0.05 M), reaching TONs of 174 ($[\mathbf{1}]_0$ was 5×10^{-4} M and $[m\text{-CPBA}]_0 = 0.27$ M). Importantly, the system **14**/HNO_3/*m*-CPBA shows 91% retention of stereoconfiguration (RC), calculated as $(cis - trans)/(cis + trans) \times 100\%$. This corresponds to the *cis*/*trans* ratio of the products (tertiary alcohols) of 21. In the absence of a nitric acid promoter, the *cis*/*trans* ratio roughly equals 1 [41]. Furthermore, the composition of oxygenated products is not affected by the PPh_3 treatment (see above). This observation accounts for the absence of the alkyl hydroperoxide as main reaction product [20]. One should notice, however, that *m*-CPBA itself at large concentrations (greater than 0.1 M) may reveal metal-free alkanes oxidation [42], as evidenced by the blank test without catalyst **14,** which shows the best yield of products of 17% and the maximum *cis*/*trans* ratio of 8.

Other cases of this series are provided by cage-like metallagermsesquioxanes, from which a representative example is selected, namely $[(Ph_5Ge_5O_{10})_3Fe_6(OH)_3(O)Na_2(phen)]\cdot(EtOH)_4\cdot(H_2O)_{0.5}$ (**15**) (Figure 7.4) [35]. This complex exhibits catalytic properties toward oxidation of cyclohexane with H_2O_2 in the presence of TFA. The highest yield of oxygenates (sum of cyclohexanol and cyclohexanone) reaches 44%. The typical conditions used were $[CyH]_0 = 0.46$ M, $[TFA]_0 = 0.05$ M, $[H_2O_2]_0 = 1.5$ M, $[\mathbf{15}]_0 = 5 \times 10^{-4}$ M in acetonitrile at 40 °C. The main reaction mechanism is believed to proceed via a free radical pathway with formation of CyOOH, as evidenced by the low averaged normalized regio- (9 : 1) and bond- (33 : 10 : 1) selectivities in

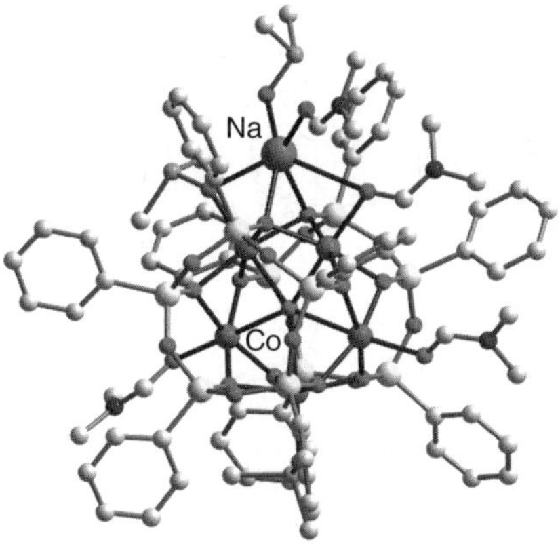

14

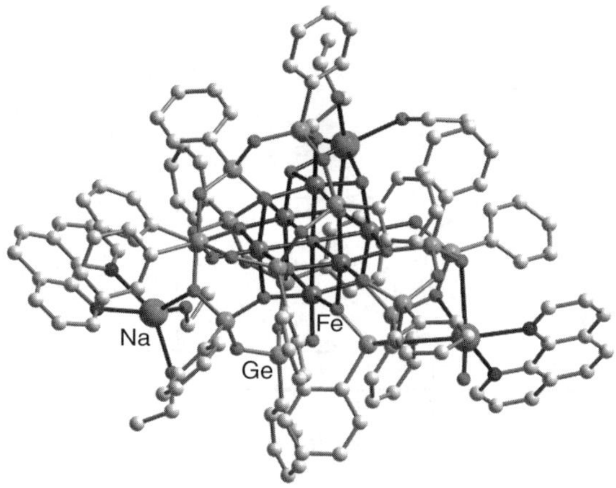

15

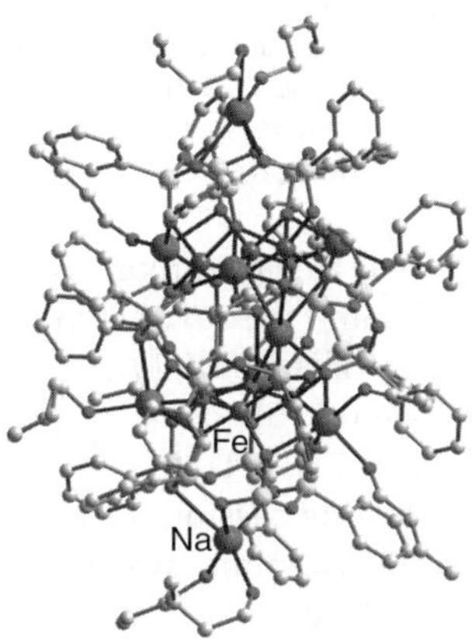

16

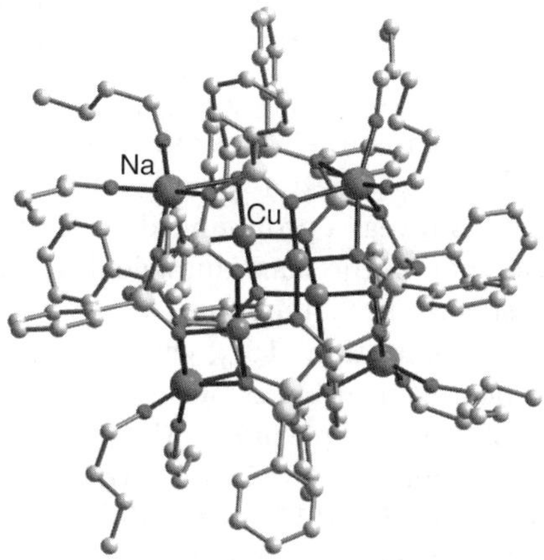

17

Figure 7.4 Structures of the Na/M complexes **14–17**, catalysts for cyclohexane and 1,2-dimethylcyclohexane oxidation [25, 26, 35, 41]. (*See insert for color representation.*)

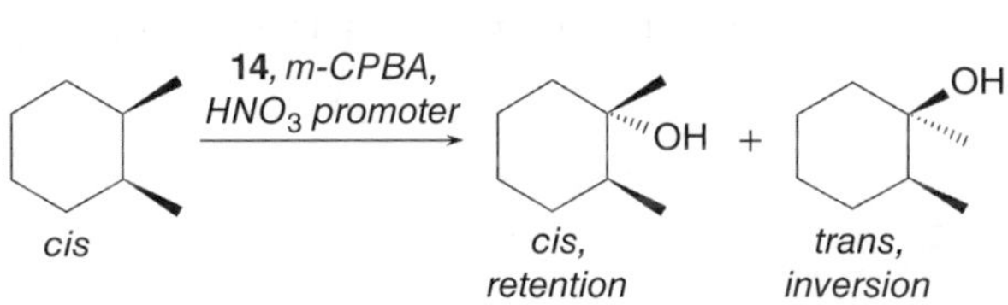

Scheme 7.2 Tertiary alcohols formed from the oxidation of *cis*-1,2-dimethylcyclohexane with *m*-CPBA, catalysed by **14**.

the oxidation of *n*-heptane and methylcyclohexane, respectively, where no preference among different secondary carbons of *n*-heptane is observed. The catalytic system exhibits a complex dependence of the initial reaction rate W_0 on $[\mathbf{15}]_0$, showing an S-shape curve [35]. As in the case of the catalyst **8** (Figure 7.2), such behaviour may account for a complex transformation of catalytic species in solution.

The heterometallic Fe/Na and Cu/Na coordination compounds $[(PhSiO_{1.5})_{20}(FeO_{1.5})_6(NaO_{0.5})_8(n\text{-}BuOH)_{9.6}(C_7H_8)]$ (**16**) [26] and $[(PhSiO_{1.5})_{12}(CuO)_4(NaO_{0.5})_4]$ (**17**) [25], respectively (Figure 7.4), also act as catalysts for the mild oxidation of cyclohexane with H_2O_2.

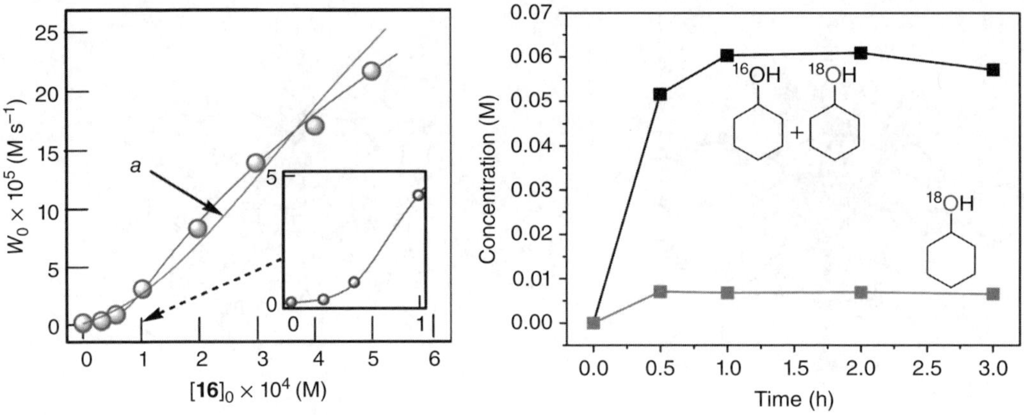

Figure 7.5 Left: dependence of the initial reaction rate W_0 of cyclohexane oxidation to cyclohexanol and cyclohexanone on the concentration of catalyst **16** ($[C_6H_{12}]_0 = 0.46$ M, $[H_2O_2]_0 = 1.5$ M, $[HNO_3]_0 = 0.05$ M). Curve *a* is a kinetical fit of the experimental data. Right: accumulation with time of cyclohexanol containing both partially labeled (the sum $^{16}O + ^{18}O$) and completely ^{18}O-labeled oxygenates when reaction is performed under $^{18}O_2$. The yields and isotopic abundances were measured after reduction of the reaction samples with PPh_3. Source: Adapted from Ref. [26] with permission from The Royal Society of Chemistry.

Both complexes reveal a pronounced activity when using nitric acid as a promoter. From the accumulation curves the maximum yields of products can be estimated approximately as 32% and 20%, respectively. Compound **16** shows a complex dependence of the initial reaction rate W_0 on the concentration $[\mathbf{16}]_0$, with negligible reaction rates when $[\mathbf{16}]_0 < 5 \times 10^{-5}$ M, and following a nearly linear behaviour for higher concentrations (Figure 7.5, left). In contrast, the complex **17** shows a simple linear W_0 vs. $[\mathbf{17}]_0$ dependence with the line passing through zero. The regio- and bond selectivites, as well as stereoselectivity in the 1,2-dimethylcyclohexane oxidation for both **16** and **17** suggest the presence of the hydroxyl radical as the C—H attacking species in a free radical reaction pathway.

Both **16** and **17** were subjected to tests in the presence of labelled oxygen sources, namely $^{18}O_2$ and $H_2{}^{18}O$. It is known [43] that dioxygen readily reacts with alkyl radicals and therefore, if the latter are present in the reaction mixture the formation of ^{18}O labelled alcohols is expected. The Cy–$^{18}O^{18}OH$ intermediate was directly trapped by the gas chromatography-mass spectrometry (GC–MS) methods [23] (Scheme 7.3 [a]).

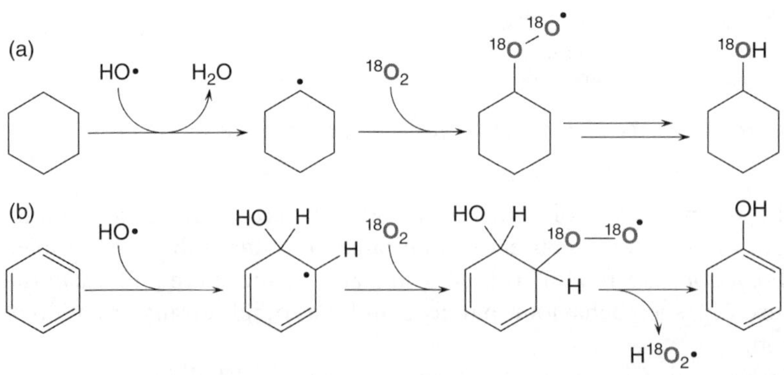

Scheme 7.3 Reaction of hydroxyl radical and dioxygen with a saturated hydrocarbon, cyclohexane (a), and with an aromatic one, benzene (b). (*See insert for color representation.*)

One should notice that the interaction of a hydroxyl radical with an aromatic hydrocarbon proceeds in a different way (Scheme 7.3b) and no (or lower) incorporation of ^{18}O from dioxygen is expected for the respective free radical process [44]. The results of reactions under $^{18}O_2$ atmosphere are similar for **16** (Fig. 7.5, right) and **17**, with the maximum incorporation of ^{18}O into cyclohexanol up to 14% and 66%, respectively. In both cases the highest level of ^{18}O was observed after a short reaction time, followed by a further decay, which can be explained by the gradual formation of unlabeled $^{16}O_2$ catalytically originating from $H_2{}^{16}O_2$. Initially, cyclohexanone also accumulated a significant portion of ^{18}O before the level fell to nearly zero. However, ketones can rapidly exchange oxygen with water [26] and therefore the measurement of $^{18}O/^{16}O$ composition of ketones can just point to a $H_2{}^{18}O/H_2{}^{16}O$ isotopic composition of the reaction mixture. The presence of ^{18}O in reaction byproducts (cyclohexanediols, hydrohycyclohexanones, and others) was also investigated in both **16** and **17** catalytic systems and is consistent with the same type of mechanisms as for the main oxidation to cyclohexanol and cyclohexanone. In the presence of $H_2{}^{18}O$, both systems show a noncatalytic labeling of ketones only, with no ^{18}O incorporation into alcohols. Water may exchange oxygen with high-valent metal-oxo species, in this way labeling the final product (alcohol) [43], but usually does not label the product in the case of a free radical oxidation mechanism, as expected for **16** and **17**. The complex **16**, however, shows an intriguing behaviour when oxidizing benzene in the presence of $H_2{}^{18}O$, giving 2.5% incorporation of ^{18}O into phenol.

A pronounced retention of stereoconfiguration of the substrate was achieved with the V/W polyoxometalate $[(n\text{-}C_4H_9)_4N]_4[\gamma\text{-}HPV_2W_{10}O_{40}]$ (**18**) (Figure 7.6) [36]. The carbon-free anion of **18** catalyzes the hydroxylation of various alkanes up to 98% yield based on the oxidant (H_2O_2) using $HClO_4$ as an acid promoter. The system operates in CH_3CN/tBuOH media, but can also work in 1,2-dichloroethane. Cyclohexane is oxidized mainly to cyclohexanol, with 49 A/K ratio (Table 7.1). However, the yields based on the substrate do not exceed 1% as a great excess of the alkane substrate (up to 200 relative to the oxidant) is used. These conditions could also explain the high A/K ratio. The competitive oxidation of normal and deuterated cyclohexane reveals a kinetic isotope effect value of 3.2. The system incorporates 90% of ^{18}O into the cyclohexanol in the course of cyclohexane oxidation with $H_2{}^{18}O_2$. The most prominent feature of the **18**/$HClO_4$/H_2O_2 system is its selectivity toward hydroxylation of secondary C—H bonds of some substrates in the presence of more reactive tertiary ones (Scheme 7.4).

Considering also that adamantane is oxidized mainly at tertiary carbon (as commonly expected) with 82% selectivity, one may suppose that the sterical hindrance of tertiary C—H

Figure 7.6 Structure of the anion core of **18** [36].

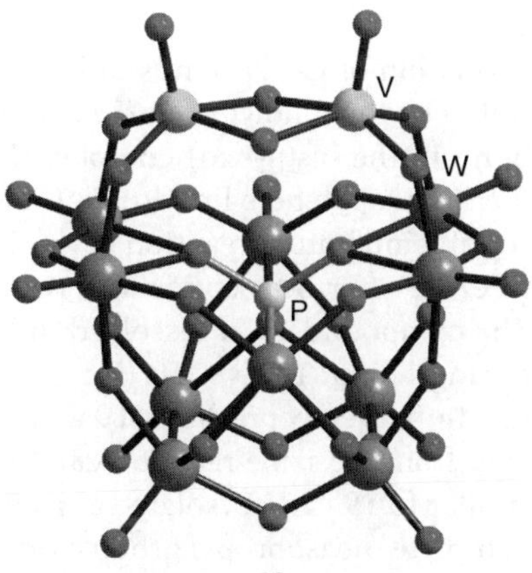

18

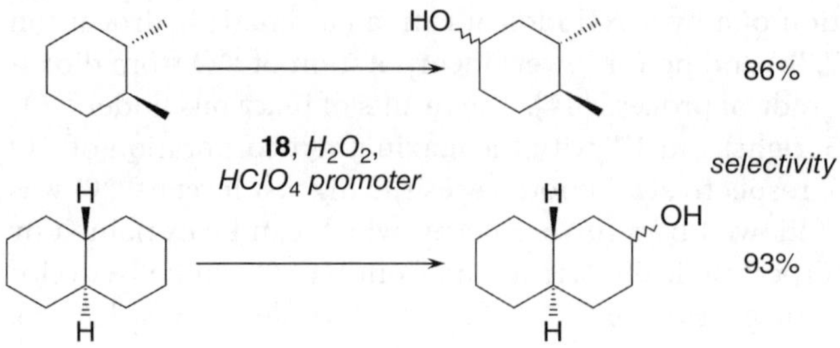

Scheme 7.4 Main products and respective selectivities in selected alkanes oxidation with H_2O_2 catalyzed by **18** [36].

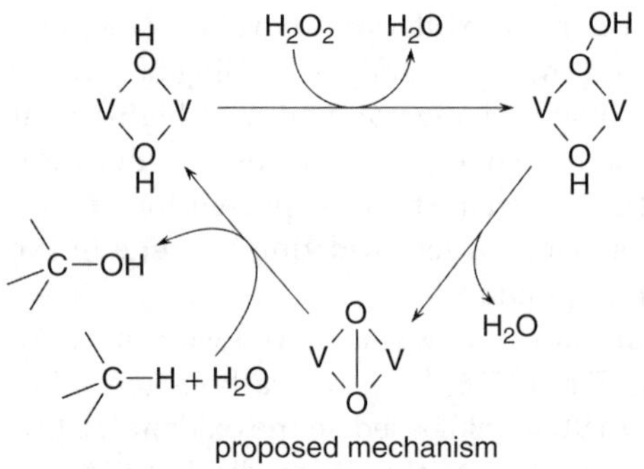

Scheme 7.5 Proposed alkane hydroxylation mechanism for the catalyst **18** [36].

bonds in dimethylcyclohexanes and decahydronaphthalenes (Scheme 7.4) is the main reason for their low reactivity with a bulky polyoxometalate core of **18**, which is assumed to be the C—H attacking species. A reaction mechanism (Scheme 7.5) is proposed on the basis of ESI-MS, kinetic and selectivity studies and is suggested to involve the participation of a binuclear vanadium fragment as the oxidant, while the remaining phosphotungstate core acts as a support and stabilizer [36].

7.3 Intramolecular C—H Oxygen Insertion

Oxidation of C—H bonds of ligands occurring within heterometallic complexes can be illustrated by the following Co/Fe example. It concerns the self-oxidation of the benzylic sp^3 C—H bond in the bis(μ-oxo) complex $[L^4Co(O)_2FeL^5]$ (where L^4 = 2,2,6,6-tetramethyl-3,5-bis(2,4,6-triisopropylphenylimido)hept-4-yl; L^5 = methyl analogue of L^4) (**19**) (Figure 7.7) [45]. The room temperature reaction of $[L^4Co(O)_2]$ with $[L^5Fe(C_6H_6)]$ mixed in benzene affords **19** upon freezing after 30 seconds and recrystallization from pentane/hexamethyldisiloxane mixture. The compound **19**, if dissolved in benzene at room temperature, leads to ligand self-oxidation during 4 hours to give **19a** (Fig. 7.7). Under the same procedure but without freezing, the above reaction directly produces **19a**. The presence of a different second metal accelerates the oxygen insertion: the same reaction can be performed using cobalt complexes only and a Co complex similar to **19** can be isolated at room temperature, but it requires 3 hours of heating at 80 °C to complete the isopropyl group oxidation, while no heating is required for the Co/Fe complex **19** in the same process [45].

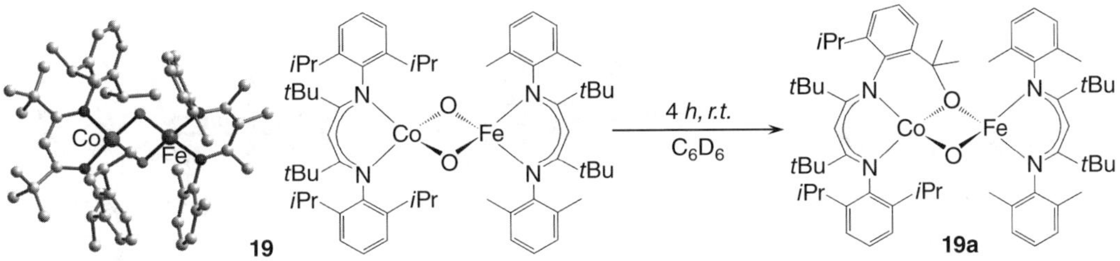

Figure 7.7 Left: structure of the complex **19**. Right: intramolecular C—H self-oxidation to form **19a** [45].

7.4 Peroxidative Hydrocarboxylation of Light and Liquid Alkanes

Direct transformation of alkanes to carboxylic acids, avoiding stepwise oxidation of alcohols and ketones, is of great potential interest for synthetic chemistry. Recently, it was demonstrated that alkanes (RH), including gaseous ones, can undergo direct hydrocarboxylation to yield the corresponding carboxylic acids (RCOOH) upon reaction with CO and H_2O in the presence of peroxydisulfate, typically $K_2S_2O_8$, in acetonitrile/water solution (Scheme 7.6) [46, 47]. The reaction is promoted by a suitable metal catalyst and proceeds under moderate pressure of CO and mild heating (typically up to 60 °C), exhibiting a high conversion [46, 47]. The advantages of this type of alkane carboxylation include the use of water as reagent and of an aqueous medium, the mild conditions, the absence of corrosive reagents (such as acid solvents), and the use of cheap source materials.

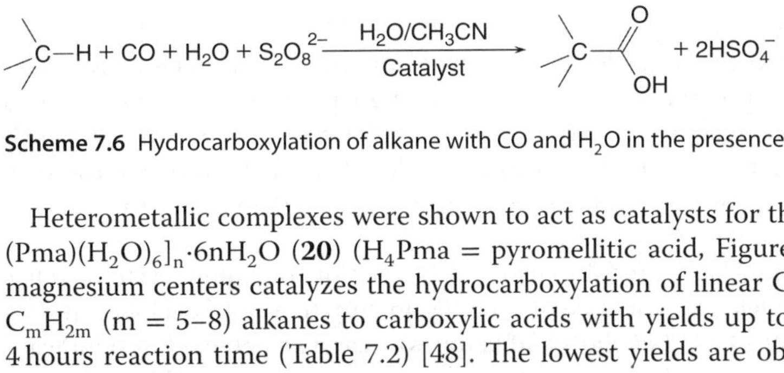

Scheme 7.6 Hydrocarboxylation of alkane with CO and H_2O in the presence of $K_2S_2O_8$ oxidant.

Heterometallic complexes were shown to act as catalysts for this reaction. $[Cu_2Mg_2(HTea)_2(Pma)(H_2O)_6]_n \cdot 6nH_2O$ (**20**) (H_4Pma = pyromellitic acid, Figure 7.8) containing copper and magnesium centers catalyzes the hydrocarboxylation of linear C_nH_{2n+2} (n = 5–9) and cyclic C_mH_{2m} (m = 5–8) alkanes to carboxylic acids with yields up to 44% (for cyclohexane) after 4 hours reaction time (Table 7.2) [48]. The lowest yields are observed for nonane (9%). The reactivity of alkanes lighter than pentane was not tested. The secondary carbons undergo carboxylation preferably to the primary ones, but no preference within the former (normalized to the same number of atoms) is observed (Table 7.2).

Three Cu/Na heterometallic complexes with the general formula $[Cu_4(HBes)_4(RC_6H_4COO)\{Na(H_2O)_x\}]_n \cdot 3ynH_2O$ (where H_3Bes = *N*,*N*-bis(2-hydroxyethyl)-2-aminoethanesulfonic acid; R = H, OH, OCH_3) are catalysts for the hydrocarboxylation of butane and ethane [49], the highest activity being exhibited by the complex with R = H (**21**). High conversions (90.5 and 70.6%) of butane or propane to 2-methylbutanoic acid or 2-methylpropanoic acid, respectively, are achieved (Table 7.2) [51]. The normalized selectivity toward secondary: primary carbon is identical for butane and propane (c. 28 : 1). Lower conversions of butane are observed for the other two Cu/Na catalysts (R = OH or OCH_3) [49] of this series (they are not included into the Table 7.2 and in the numeration).

Using the same ligand H_3Bes, but replacing sodium salts with lithium ones and changing the acid, three Cu/Li heterometallic compounds have been synthesized [50]. A representative example is $[Cu_4(HBes)_4(Hhba)\{Li(H_2O)_2\}]_n \cdot 3nH_2O$ (**22**) (where Hhba = 3-hydroxybenzoic

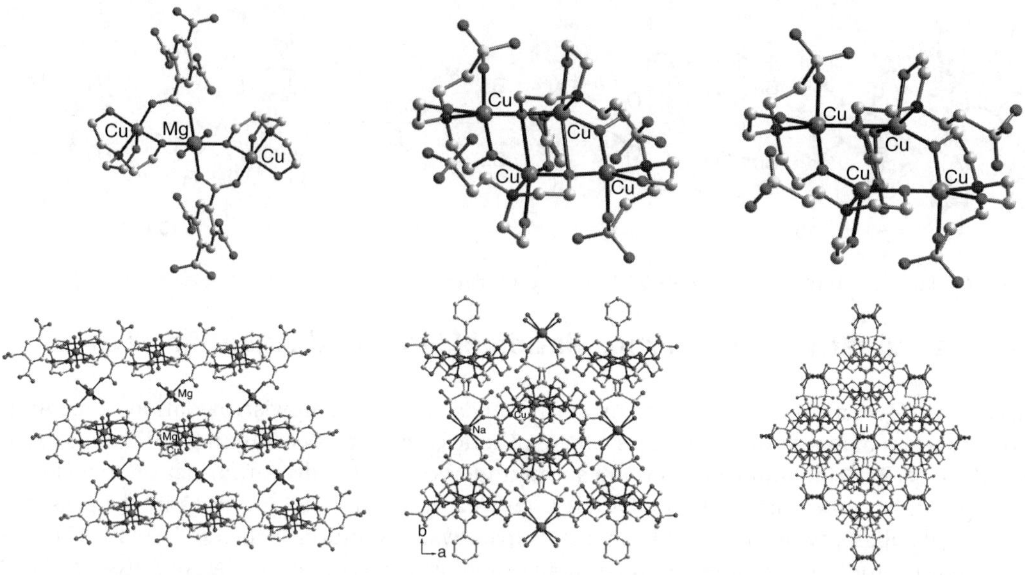

20 21 22

Figure 7.8 Cores (top) and packing (bottom) of Cu/Mg (**20**), Cu/Na (**21**) and Cu/Li (**22**) heterometallic complexes [48–50]. (*See insert for color representation.*)

acid). Although the structures of the polynuclear cores of **21** and **22** are close (Figure 7.8), their catalytic activities are different, the latter leading to lower butane and propane conversions (Table 7.2). The linear alkanes until nonane were tested, and higher conversions were obtained in comparison with **20**. The averaged normalized regioselectivities between secondary and primary carbon atoms for nonane are similar for both **20** and **22** (c. 20 : 1), showing equal reactivity of all secondary carbons in the chain. Attempts to hydrocarboxylate methane using the complex **22** as a catalyst were unsuccessful.

Complexes **20** and **22** were also tested for hydrocarboxylation of liquid cyclic alkanes (Table 7.2), showing the best yields for the case of cyclohexane (up to 44% of cyclohexanecarboxylic acid). Hydroxylation products (alcohols and ketones) were also detected in significant quantities (Table 7.2), while these products are rather minor (lower than 1% yield) in the case of hydrocarboxylation of linear alkanes.

The main reaction pathway is assumed to be common for all catalytic systems of this type, operating in H_2O/CH_3CN media, including **20–22** [52]. The proposed mechanism (Scheme 7.7) starts with homolytic splitting of peroxodisulfate anion to form sulfate radical, induced thermally or by the metal catalyst. Abstraction of H atom from the alkane by the sulfate radical affords the alkyl radical (which can be trapped by $CBrCl_3$) [52]. The acyl radical, formed by carboxylation of alkyl radical with CO, is oxidized with the metal-containing catalyst or with $S_2O_8^{2-}$ and, upon further hydrolysis, gives the carboxylic acid. The reaction performed in the presence of labeled water, $H_2{}^{18}O$, affords single-labeled acid $RCO^{18}OH$, thus being in agreement with the proposed role of water. In this mechanism, the alkane-attacking species is the sulfate radical, while the role of the metal catalyst is to promote further transformations. Hence, the presence of the catalyst does not have a marked influence on the overall selectivity of the system, in accordance with the similar regioselectivitites exhibited by the different catalysts (Table 7.2).

Table 7.2 Alkane hydrocarboxylation products (selected data) observed for catalysts **20–22** [48–50].[a)]

Linear alkanes	Hydrocarboxylation products			Cyclic alkanes	Hydrocarboxylation products	
 ethane	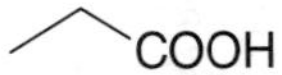 15.2% (**22**)			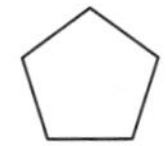 cyclopentane	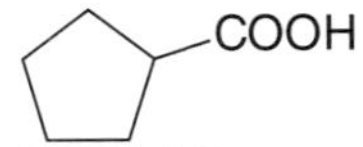 39.3% (**20**) 42.0% (**22**)	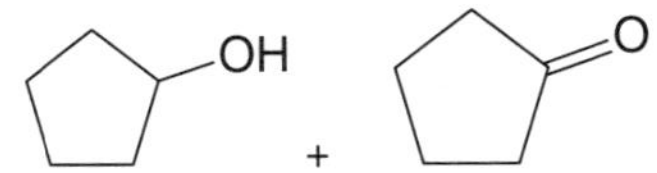 4.3% (**20**) 5.2% (**22**)
 propane	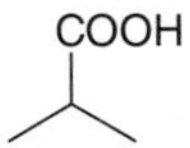 70.6% (**21**) 66.5% (**22**)	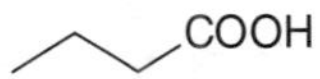 7.8% (**21**) 8.1% (**22**)		 cyclohexane	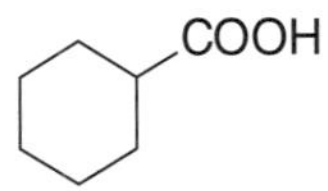 28.1% (**20**) 43.8% (**22**)	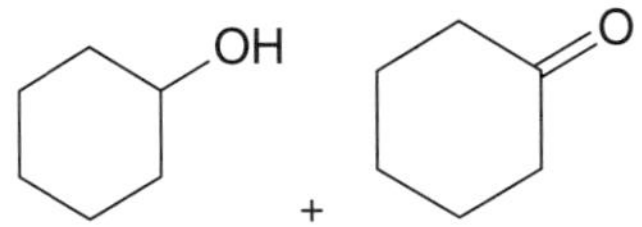 2.1% (**20**) 2.2% (**22**)
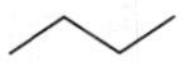 *n*-butane	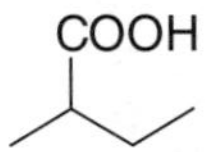 90.5% (**21**) 61.0% (**22**)	 4.7% (**21**) 4.4% (**22**)		 cycloheptane	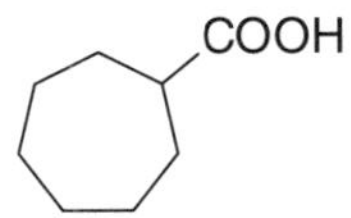 20.9% (**20**) 24.6% (**22**)	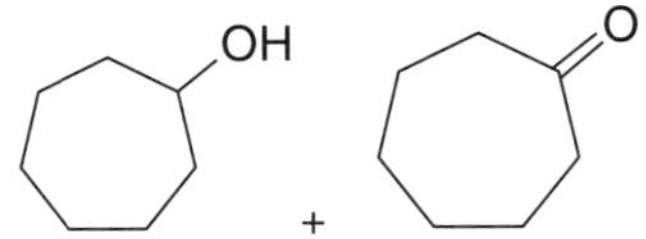 6.5% (**20**) 8.8% (**22**)
 *n*-pentane	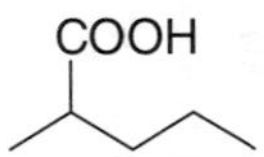 16.8% (**20**) 23.0% (**22**)	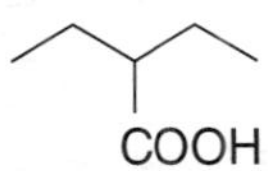 7.6% (**20**) 10.0% (**22**)	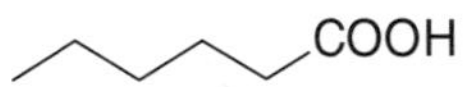 0.7% (**20**) 1.4% (**22**)	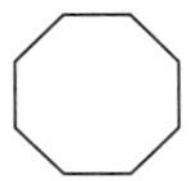 cyclopentane	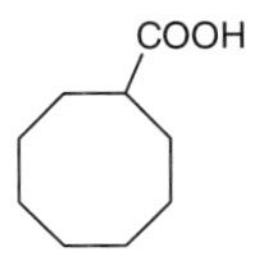 13.7% (**20**) 6.5% (**22**)	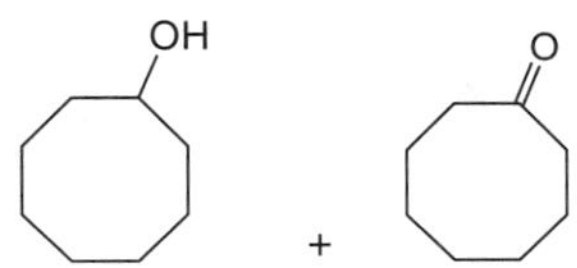 6.5% (**20**) 12.3% (**22**)
 *n*-hexane	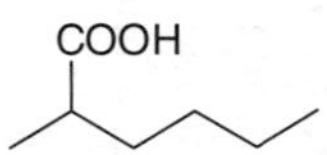 7.8% (**20**) 15.6% (**22**)	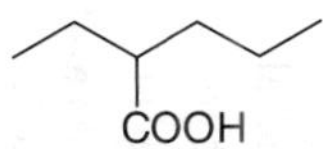 8.3% (**20**) 15.3% (**22**)	 0.5% (**20**) 1.2% (**22**)			

a) Typical reaction conditions: CO pressure 10–20 atm, $K_2S_2O_8$ oxidant, H_2O/CH_3CN medium, 60 °C temperature, 4–6 h reaction time.

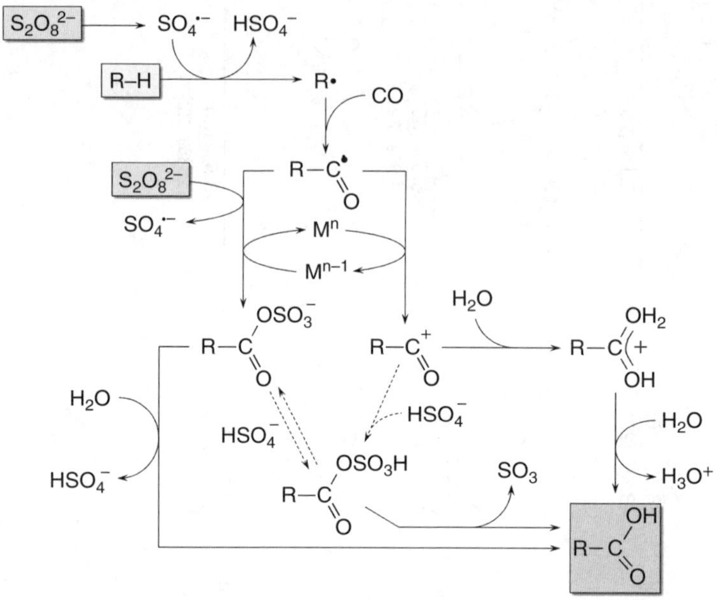

Scheme 7.7 Proposed reaction mechanism for peroxidative hydrocarboxylation of alkanes [52].

7.5 Concluding Remarks

Oxidation of inactive C—H bonds is a challenge in modern chemistry, and there are several issues when trying to design an efficient catalyst for this type of reactions, such as a proper choice of the metal and oxidant, the creation of specified environments around the metal center, and careful tuning of reaction conditions. The use of a second different metal is also an attractive approach, but despite the broad use of heterometallic systems in catalytic processes, including industrial ones [16], the field of alkanes oxidation using heterometallic catalysts is still very limited. Nevertheless, the use of different metals can expand the existing sets of efficient oxidation catalysts, as evidenced by the examples discussed in this chapter. They demonstrate that heterometallic systems may exhibit catalytic features among the best ones (such as the impressive turnover numbers for the Co/Fe complex **8**, Table 7.1). The role of a second metal may concern not only its participation in the catalytic cycle but also the stabilization of catalytically active species and reaction intermediates.

However, a better control of the design, structure, and behaviour of the catalyst is needed for understanding the processes, refining the parameters of the system toward the desired pathway and products, and establishing structure–properties relationships. Of particular significance will be the development of catalytic heterometallic systems with the potentially advantageous synergic effect associated with the presence of different metals.

Acknowledgments

This work has been supported by the Foundation for Science and Technology (FCT), Portugal (projects UID/QUI/00100/2013 and PTDC/QEQ-QIN/3967/2014, and fellowships SFRH/BPD/99533/2014 and SFRH/BPD/63710/2009).

References

1 Que, L. and Tolman, W.B. (2008). *Nature* 455: 333.
2 Godula, K. and Sames, D. (2006). *Science* 312: 67.
3 Shul'pin, G.B. (2016). *Catalysts* 6: 50.
4 Zhou, M. and Crabtree, R.H. (1875). *Chem. Soc. Rev.* 2011 (40).
5 Pombeiro, A.J.L. (2014). Toward functionalization of alkanes under environmentally benign conditions, Chapter 2. In: *Advances in Organometallic Chemistry and Catalysis* (ed. A.J.L. Pombeiro). Wiley.
6 Talsi, E.P. and Bryliakov, K.P. (1418). *Coord. Chem. Rev.* 2012, 256.
7 Gunay, A. and Theopold, K.H. (2010). *Chem. Rev.* 110: 1060.
8 Che, C.M., Lo, V.K.Y., Zhou, C.Y. et al. (1950). *Chem. Soc. Rev.* 2011, 40.
9 Yamaguchi, J., Yamaguchi, A.D., and Itami, K. (2012). *Angew. Chem. Int. Ed.* 51: 8960.
10 (2002). *Ullmann's Encyclopedia of Industrial Chemistry*. Weinheim, 6: Wiley-VCH.
11 Schuchardt, U., Cardoso, D., Sercheli, R. et al. (2001). *Appl. Catal. A* 211: 1.
12 Olah, G.A., Goeppert, A., and Prakash, G.K.S. (2006). *Beyond Oil and Gas: The Methanol Economy*. Weinheim: Wiley-VCH.
13 Que, L. Jr., and Tolman, W.B. (2004). Recurring structural motifs in bioinorganic chemistry. In: *Comprehensive Coordination Chemistry II*, From Biology to Nanotechnology, vol. 8 (ed. J.A. McCleverty and T.J. Meyer), 1. Elsevier.
14 Doukov, T.I., Iverson, T.M., Seravalli, J. et al. (2002). *Science* 298: 567.
15 Nesterov, D.S., Nesterova, O.V., Kokozay, V.N. et al. (2014). *Eur. J. Inorg. Chem.* 4496.
16 Buchwalter, P., Rose, J., and Braunstein, P. (2015). *Chem. Rev.* 115: 28.
17 Mitra, M., Nimir, H., Derneshko, S. et al. (2015). *Inorg. Chem.* 54: 7152.
18 Burrows, K.J., Cornish, A., Scott, D. et al. (1984). *J. Gen. Microbiol.* 130: 3327.
19 Olivo, G., Lanzalunga, O., and Di Stefano, S. (2016). *Adv. Synth. Catal.* 358: 843.
20 Shul'pin, G.B. (2002). *J. Mol. Catal. A* 189: 39.
21 Ribeiro, A.P.C., Martins, L.M.D.R.S., Hazra, S. et al. (2015). *Cr. Chim.* 18: 758.
22 Shul'pin, G.B. (2003). *Cr. Chim.* 6: 163.
23 Shul'pin, G.B., Nesterov, D.S., Shul'pina, L.S. et al. (2017). *Inorg. Chim. Acta.* 455: 666.
24 Nesterov, D.S., Nesterova, O.V., da Silva, M.F.C.G. et al. (2015). *Catal. Sci. Technol.* 5: 1801.
25 Vinogradov, M.M., Kozlov, Y.N., Bilyachenko, A.N. et al. (2015). *New. J. Chem.* 39: 187.
26 Bilyachenko, A.N., Levitsky, M.M., Yalymov, A.I. et al. (2016). *RSC Adv.* 6: 48165.
27 Vinogradov, M.M., Kozlov, Y.N., Nesterov, D.S. et al. (2014). *Catal. Sci. Technol.* 4: 3214.
28 Nesterov, D.S., Kokozay, V.N., Dyakonenko, V.V. et al. (2006). *Chem. Commun.* 4605.
29 Nesterov, D.S., Makhankova, V.G., Vassilyeva, O.Y. et al. (2004). *Inorg. Chem.* 43: 7868.
30 Nesterov, D.S., Kokozay, V.N., Jezierska, J. et al. (2011). *Inorg. Chem.* 50: 4401.
31 Nesterova, O.V., Chygorin, E.N., Kokozay, V.N. et al. (2013). *Dalton Trans.* 42: 16909.
32 Nesterov, D.S., Chygorin, E.N., Kokozay, V.N. et al. (2012). *Inorg. Chem.* 51: 9110.
33 Gupta, S., Kirillova, M.V., da Silva, M.F.C.G. et al. (2013). *Inorg. Chem.* 52: 8601.
34 Karabach, Y.Y., da Silva, M.F.C.G., Kopylovich, M.N. et al. (2010). *Inorg. Chem.* 49: 11096.
35 Bilyachenko, A.N., Levitsky, M.M., Yalymov, A.I. et al. (2016). *Angew. Chem. Int. Ed.* 55: 15360.
36 Kamata, K., Yonehara, K., Nakagawa, Y. et al. (2010). *Nat. Chem.* 2: 478.
37 Kuznetsov, M.L., Rocha, B.G.M., Pombeiro, A.J.L. et al. (2015). *ACS Catal.* 5: 3823.
38 Kuznetsov, M.L., Teixeira, F.A., Bokach, N.A. et al. (2014). *J. Catal.* 313: 135.
39 Rocha, B.G.M., Kuznetsov, M.L., Kozlov, Y.N. et al. (2015). *Catal. Sci. Technol.* 5: 2174.
40 Kuznetsov, M.L., Kozlov, Y.N., Mandelli, D. et al. (2011). *Inorg. Chem.* 50: 3996.
41 Bilyachenko, A.N., Yalymov, A.I., Levitsky, M.M. et al. (2016). *Dalton Trans.* 45: 13663.

42 Bravo, A., Bjorsvik, H.R., Fontana, F. et al. (1996). *J. Org. Chem.* 61: 9409.
43 Chen, K. and Que, L. (2001). *J. Am. Chem. Soc.* 123: 6327.
44 Pan, X.M., Schuchmann, M.N., and Vonsonntag, C. (1993). *J. Chem. Soc., Perkin Trans. 2* 289.
45 DeRosha, D.E., Mercado, B.Q., Lukat-Rodgers, G. et al. (2017). *Angew. Chem. Int. Ed.* 56: 3211.
46 Kirillova, M.V., Kirillov, A.M., and Pombeiro, A.J.L. (2010). *Chem. Eur. J.* 16: 9485.
47 Kirillova, M.V., Kirillov, A.M., and Pombeiro, A.J.L. (2009). *Adv. Synth. Catal.* 351: 2936.
48 Kirillov, A.M., Karabach, Y.Y., Kirillova, M.V. et al. (2012). *Cryst. Growth. Des.* 12: 1069.
49 Kirillov, A.M., Coelho, J.A.S., Kirillova, M.V. et al. (2010). *Inorg. Chem.* 49: 6390.
50 Kirillova, M.V., Kirillov, A.M., Martins, A.N.C. et al. (2012). *Inorg. Chem.* 51: 5224.
51 Kirillov, A.M., Kirillova, M.V., and Pombeiro, A.J.L. (2012). *Coord. Chem. Rev.* 256: 2741.
52 Kirillova, M.V., Kirillov, A.M., Kuznetsov, M.L. et al. (2009). *Chem. Commun.* 2353.

8

Oxidative Functionalization of Methane on Heterogeneous Catalysts

Maricruz Sanchez-Sanchez and Johannes A. Lercher

Technische Universität München, Department of Chemistry, Garching, Germany

8.1 Introduction

The discovery of new natural gas and shale gas resources has provided new carbon resources and has shifted the needs for hydrocarbon conversion processes in the petrochemical industry. The wide availability of light alkanes and particularly methane, has placed the focus on processes that can use hydrocarbons with high H/C ratios.

The current processes to convert methane involves reforming to synthesis gas, which is subsequently transformed via hydrogenation of CO to methanol [1, 2] or to hydrocarbons (Fischer-Tropsch synthesis) [3]. The steam reforming of methane to synthesis gas (a mixture of CO and H_2, syngas) is very well developed, but because of its endothermic nature, it is economically feasible only when applied on a large scale [4]. Thus, a significant fraction of methane in small streams or as side-product is currently still flared [5]. This is not only economically problematic, but also means a reduction of valuable carbon resources.

Overall, the high symmetry and low polarizability of methane, together with the high C—H bond strength (439 kJ mol^{-1}), make this molecule refractory and poses a challenge for chemical conversion.

Conceptually, partial oxidation of methane to products such as formaldehyde and methanol are highly rewarding conversion routes [6]. The selective oxidation of methane to methanol has been frequently addressed, but remained an elusive target of research for several decades. While the reaction is substantially exothermic, its enthalpy is only one-sixth that of total combustion (Eqs. 8.1 and 8.2):

$$CH_4 + \tfrac{1}{2} O_2 \rightarrow CH_3OH \qquad \Delta H = -126\ \text{kJ mol}^{-1} \tag{8.1}$$

$$CH_4 + \tfrac{1}{2} O_2 \rightarrow CO_2 + H_2O \qquad \Delta H = -802\ \text{kJ mol}^{-1} \tag{8.2}$$

This high difference in exothermicity points to the challenges in controlling the selectivity of the partial oxidation process [7]. These challenges are especially significant when higher reaction temperatures are used, as that opens the possibility of participation of gas phase oxidation reactions.

Alkane Functionalization, First Edition. Edited by Armando J. L. Pombeiro and M. Fátima C. Guedes da Silva.

8.2 Oxidative Routes for Methane Functionalization

8.2.1 High-Temperature Routes

8.2.1.1 Partial Oxidation of Methane to Methanol and Formaldehyde

A sizable number of studies appeared on heterogeneous catalysts for partial oxidation of methane to formaldehyde and methanol at temperatures above 500 °C. Otsuka and Wang [7] provided an extensive review on the oxides typically used and of their selectivities to methanol and formaldehyde. Mainly, Mo, V, W, Fe, Zn, and Cu oxides were studied for methane oxidation at temperatures between 600 and 750 °C. However, under the conditions applied, it is unclear whether the solid catalyst is beneficial at all when compared to conversion in an empty reactor or in a reactor filled with inert packing material [8, 9]. In general, it is believed that the gas phase radical reactions are likely controlling the process [10, 11]. It was proposed that the role of the solid catalyst is primarily to act as radical quencher, i.e. to block or reduce the self-acceleration of the gas phase radical-chain reaction and so to avoid overoxidation [11]. Overall, the interest in these processes is limited by the poor selectivities, even at low conversions, and the lack of their reproducibility, which is caused by reactor wall and gas phase reactions [7, 12].

8.2.1.2 Oxidative Coupling of Methane (OCM)

A more promising high-temperature route is the oxidative coupling of methane (OCM) to ethane and ethylene [13]. The reaction is thermodynamically limited at low temperatures, and it is necessary to work in the range 600–800 °C.

The activation of methane under OCM reaction conditions generates methyl radicals that react in the gas phase, producing ethane. Ethane subsequently undergoes (presumably oxidative) dehydrogenation to ethane. Even though the presence of heterogeneous catalysts is obviously important to achieve selective conversions, consensus exists that rates of radical generation and quenching on the solid catalyst and the gas phase reactions have to be balanced.

Besides the complex radical chemistry [14–16], it is obvious that higher concentrations of reactive intermediates in the gas phase will limit selectivity and will lead to the formation of high concentrations of CO and CO_2. As ethane and ethene are more reactive than methane and can easily be oxidized to CO_x [17], the overall conversion has to be limited in order to realize higher selectivities. The fact that total oxidation may occur via gas phase as well as surface catalyzed routes, makes it difficult to achieve significant selectivities to C_2 at high conversions [18].

The catalysts most frequently used for methane oxidative coupling are basic metal oxides, in combination with other oxide components to create oxygen vacancies [19, 20]. Multiple combinations in ternary and multiphase catalysts have been explored, with the most outstanding performance reported for MgO and La_2O_3 doped by alkali and alkaline-earth cations [21]. The role of the dopants is not only to create surface defects but also to adjust the electronic properties of the metal oxide, because charge transfer appears to be the critical element for the formation of active oxygen species [22].

By far, the most studied OCM catalyst is the Li-MgO mixed-oxide system [23–29]. Surprisingly, the nature of the active sites is still under debate [12, 30]. Active sites were described as a Li^+O^- ion pair that homolytically cleaves methane [28]. The presence of Li^+ promotes the formation of a surface O^- center with a strong radical character. More recent results of Schwarz et al. using gas phase metal-oxo clusters indicated that O^- species are able to abstract a hydrogen atom from methane, forming a $CH_3^\bullet$ [31].

The Lunsford mechanism has been debated because of difficulties to detect Li^+O^- sites under reaction conditions [30]. Kwapien et al. proposed an alternative route for methane activation in OCM on Li/MgO catalysts [27, 32], suggesting that corner-and-step defects of MgO crystals

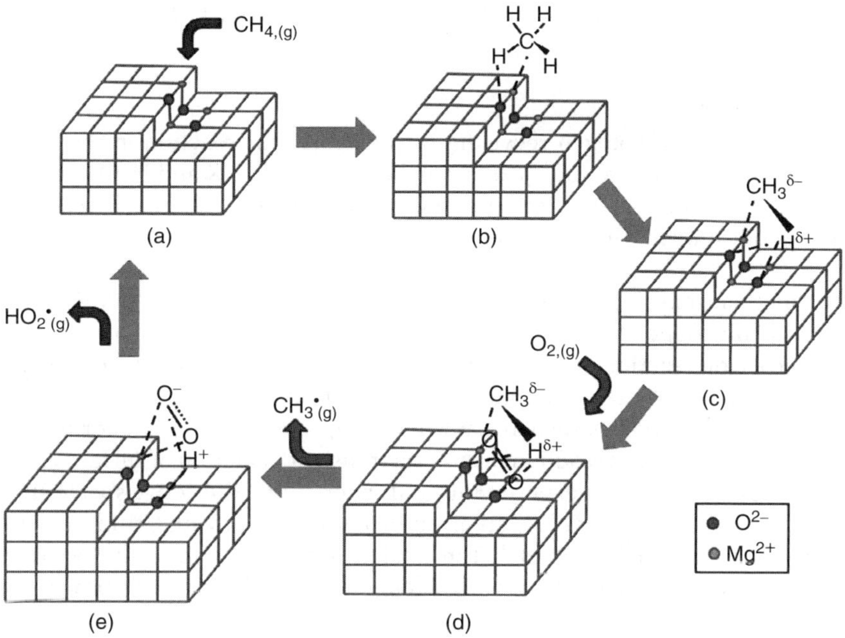

Figure 8.1 Mechanism of heterolytic C—H activation of methane on MgO defects. Source: Reproduced with permission from Schwach et al. [27]. (*See insert for color representation.*)

catalyze the heterolytic chemisorption of CH_4 (Figure 8.1), and Hu et al. [33] suggested the same mechanism based on dissociative adsorption of methane on Lewis acid–base pairs.

8.2.2 Low-Temperature Routes

8.2.2.1 Halogenation of Methane

Functionalization of methane via halogenation proceeds at moderate temperatures without a catalyst. In principle, methyl halides can be subsequently hydrolyzed to produce methanol and DME [34] or further functionalized, even substituting methanol in some processes like acid catalyzed methanol to olefins in zeolites [35–37].

The main drawback of this approach is the formation of higher halogenated products, which is difficult to avoid due to the radical nature of the reaction and the high reactivity of the intermediates. As HX is generated in all reaction steps, it must be considered the regeneration and recycling of the halogen via oxidation.

The highly corrosive character of FH and the low stability of CH_3I make Cl_2 and Br_2 the most adequate halogens for methane functionalization. Methane chlorination has been extensively studied [38–41]. To overcome the low selectivity to CH_3Cl, Olah et al. proposed using superacids to catalyze the electrophilic insertion [42, 43]. The reaction takes place at moderate temperatures (270 °C) over acid solids such as Al_2O_3-supported FeO_xCl_y, $TaOF_3$, $NbOF_3$ and $ZrOF_2$, or SbF_5 on graphite and TaF_5 on Nafion [38]. The Lewis acid sites of the catalyst are able to polarize the halogen leading to Cl^- and Cl^+ species. The activation of methane occurs via electrophilic insertion of Cl^+ into the C—H bond and the formation of a five-coordinated carbonium ion. This mechanism favors the formation of CH_3Cl as the barrier for the formation of carbonium ions increases with each additional halogen atom [38].

The reaction of methane with Br_2 to CH_3Br has also been considered as a feasible route, in spite of its higher cost in comparison to Cl_2 [44]. The gas phase chain-propagation in methane bromination occurs with lower rates and, therefore, selectivity to CH_3Br can be better controlled than for chlorination.

8.2.2.2 Oxyhalogenation of Methane

A way to overcome the problem of halogen recycling is to perform the reaction in the presence of oxygen. Methane reacts with HX and O_2 on a solid catalyst to CH_3X and H_2O [44, 45]. However, the exothermicity of the reaction, the presence of oxygen, and the higher reaction temperatures require a good control of heat transfer as well as highly corrosion resistant reactor materials [46].

The first attempts of methane oxychlorination were adaptations of the oxidative chlorination of ethane to vinyl chloride on copper-based catalysts [47, 48]. However, the use of a reducible metal also catalyzes the Deacon reaction, leading to formation of Cl_2. This not only reduces the availability of HCl but also leads to unselective multiple chlorination of methane.

The use of nonreducible oxide catalysts such as LaOCl [49, 50], synthesized via chlorination of La_2O_3, allowed higher selectivity. The reaction occurs via transient oxidation of surface chlorides, demonstrated by pressure transient experiments and theory [49]. The active species is a surface OCl^- in which Cl is positively charged. Methane is believed to react with Cl^+ in a mechanism similar to that proposed by Olah et al. [38]. The catalytic cycle is closed by the reaction of HCl with the transiently formed surface OH group, regenerating the chlorinated surface and producing H_2O.

Although the selectivity to CH_3Cl of LaOCl catalyst was nearly 100% at low methane conversions, the formation of CH_2Cl_2 occurs at considerable rates at higher conversions. The formation of multiple-chlorinated products suggests a reaction pathway involving radicals. Reducible oxides as dopants increase the selectivity. In the presence of Ni, the chlorination rate of CH_3Cl was reduced, leading to higher selectivities [51].

Methane oxybromination requires temperatures above 550 °C for Ru- and Rh-based catalysts [52, 53] and above 600 °C on $FePO_4$- [54], CeO_2- [55], and vanadium phosphate-based catalysts [56]. The catalytic mechanism is still unclear but likely involves redox pairs of metal cations [54].

8.2.2.3 Selective Oxidation of Methane

In early studies [57], the use of transition metal catalysis was considered nonsuitable for C—H activation of alkanes, because alkanes have neither lone electron pairs nor π orbitals to interact with the metal centers. Thus, the work by Shilov [58, 59] using Pt(II) salts in strong polar solvents to perform the electrophilic activation of methane meant an important breakthrough. The mechanism of C—H activation in the Shilov system is hypothesized to start with the substitution of a hydrogen of methane by the metal, yielding a methyl-Pt(II) organometallic complex [60]. Then, the reaction with $[PtCl_6]^{-2}$ generates an unstable methyl-Pt(IV) species that subsequently undergoes nucleophilic attack of the C—Pt bond, liberating CH_3X (being X either OH or Cl) and regenerating the Pt(II) complex [59]. Although the use of Pt(II) is catalytic, the necessity of using $[PtCl_6]^{2-}$ as stoichiometric oxidant is very impractical. Substitution of Pt(IV) salts by less expensive alternatives is possible, but the choice of oxidant is not trivial, because irreversible oxidation of the Pt(II) catalyst must be avoided. Quinones, Cu(II) and heteropolyacids are among the reported oxidants [61, 62]. Deactivation is also often observed due to reduction of Pt(II) with precipitation of Pt(0) [60].

Even though only a few turnover numbers (TONs) have been achieved by this type of catalysts, the Shilov chemistry has received much attention and has inspired many catalytic routes for oxidative activation alkanes, because it represents a nonradical mechanism in which the reactivity decreases with the degree of substitution. This, in turn, is expected to lead to higher selectivities toward alcohols or monohalogenated alkanes [63].

Based on the Shilov chemistry, Periana et al. explored the use of Hg(II) and Pt(II) catalysts for the selective oxidation of methane [64, 65]. By employing SO_3 as oxidant and a Hg(II) or Pt(II)-bidiazine catalyst in highly acidic media (concentrated sulfuric acid), methane was converted into methyl bisulfate at 200 °C with yields of 70%. In a subsequent step, methanol can be formed by reaction of methyl bisulfate with H_2O. In a variant, heterogeneous catalysts for this reaction were introduced, stabilizing Pt catalysts in covalent triazine frameworks [66] and unsupported Pt salts [67]. Catalytic activities were similar or higher than those with homogeneous analogs [66, 68].

The relatively high yields achieved by these processes are related to the protection of the methyl group against overoxidation via formation of a stable bisulfate [69]. The use of a strong oxidant such as SO_3 facilitates the reoxidation of the metal closing the catalytic cycle. However, the industrial realization is challenging because of product separation and corrosion problems with the SO_3/H_2SO_4 mixture.

The direct selective oxidation of methane with O_2 under mild conditions would, in contrast, allow small-scale conversion of methane into oxygenates. The principal feasibility of this route has been shown by enzymes, which are able to convert methane to methanol at ambient temperature [70]. However, a direct utilization has been proven to be challenging. Among other limitations, the reaction of methane on methane monooxygenase (MMO) requires the addition of reducing equivalents that cannot be regenerated out of the cells, increasing the costs of methanol production above the economic limit [71]. Thus, substantial activity to synthesize biochemical analogs has developed over the last years [72–75]. This activity will be explored in the following section.

8.3 Challenges in the Direct Oxidation of Methane to Methanol

It is generally agreed that only molecular O_2 is economically feasible for the selective oxidation of methane [76, 77]. However, its use as oxidant faces two significant challenges:

1. *The reaction is spin-forbidden.* O_2 is in a triplet ground state ($S = 1$), whereas substrate and products are in a singlet ground state. Thus, it will be necessary to use excited singlet states of O_2. The high reactivity of these species induces unselective oxidation [78]. Thus, it is necessary to use a metal catalyst able to facilitate the spin inversion at low temperatures and this way to generate the O-active species.
2. *Oxidation of methane to methanol requires the insertion of only 1 oxygen atom.* Thus, it is necessary not only to dissociate O_2 molecules but also to have a reducing agent (or a second site nearby) that consumes or stabilizes the second oxygen atom.

As a result, the formation of highly active oxygen species on the surface of the catalyst turns out to be one of the most relevant steps in the overall catalytic mechanism of methane oxidation to methanol [31, 79]. Early studies on partial oxidation found that methoxy species were already formed on some oxide surfaces at room temperature after exposure to methane [13, 80]. This indicates that, once the oxidizing species has been formed, the activation of methane has only a low enthalpic barrier. However, even at moderate temperatures the higher reactivity of methanol with respect to methane poses a major hurdle, because the highly active oxygen species that are able to oxidize methane also oxidize methanol. Thus, successful catalysts must protect methanol from further oxidation.

8.4 Strategies and Catalysts for the Selective Oxidation of Methane to Methanol

Two enzymes are capable of the efficient oxidation of methane to methanol at room temperature: the soluble and the particulate methane monooxygenase, sMMO and pMMO [81]. Their active sites contain Fe and Cu ions, respectively, constructing multinuclear metal-oxo structures that harness very active oxygen species to be inserted in one C—H bond to form the methanol [82, 83]. The regeneration of such active oxygen species after one catalytic cycle requires the presence of an electron donor, i.e. nicotinamide adenine dinucleotide (NADH) and NADPH cofactors, in order to remove the O atom left from the reaction with molecular O_2 [72]. The tertiary structure of the protein is believed to be essential to regulate the flow of protons, electrons, O_2, and hydrocarbon substrate to the active site [72]. The protection of the substrate from further oxidation is achieved by locating the active sites in hydrophobic pockets that allow methane and O_2 to enter but bind methanol weakly [73, 84].

The development of biomimetic catalysts has occurred almost in parallel to the advances in the structural determination of the Fe- and Cu- active sites of MMO enzymes. The active site of sMMO has been identified as a di-iron site [82]. Inspired by this structure, numerous studies have aimed to the synthesis of biochemical analogs [72] but only a few of these systems were active in oxidation of alkanes [85].

In contrast to the well-known structure of the active di-iron site of sMMO [82], the structure of the active species in the more reactive particulate MMO enzyme [86] has been the object of debate [87–90]. Either two or three Cu atoms are involved in the activity of pMMO, and, consequently, both dicopper and tricopper complexes have been investigated as biomimetic systems [73].

The heterogenization of these bioinspired catalysts has been realized by hosting Fe or Cu ions in microporous solids. The development of Cu- and Fe- exchanged zeolites have rendered catalysts active at low temperatures [91–103], although there are still significant limitations in comparison with the MMOs.

8.5 Zeolite-Based Catalysts for the Oxidation of Methane to Methanol

Zeolites are crystalline tectosilicates consisting of corner-sharing SiO_4 and AlO_4 tetrahedra. Substitution of a SiO_4 by an AlO_4 tetrahedron introduces one negative charge in the framework that is compensated by a metal cation or a proton. By ion exchange, metal cations with up to a 3+ charge can be introduced, leading to well-defined heterogeneous catalysts [104]. The zeolite crystal structure provides a controlled environment around the cations and stabilizes otherwise inaccessible sites. Thus, Cu- and Fe-oxo nanoclusters can be introduced via cation exchange.

Although Fe- and Cu-oxo species introduced, e.g. via ion exchange, are active in methane oxidation, catalysts based on such sites have significant limitations. First, the formation of the oxygen active species needs to be done in a separate step to avoid the overoxidation of methane. Second, the strong adsorption of methanol and other products limits the reaction to one turnover (reaction in stoichiometric mode) [74, 75, 91, 99, 105]. Methanol must be stripped from the catalyst by flowing H_2O and the active species needs to be regenerated prior a second catalytic cycle. Despite these limitations, such catalysts are fundamentally interesting as the variation of possible active structures may lead to better understanding of the sites in enzymes and, in turn, to designing better catalysts.

8.5.1 Fe-Zeolites

In the early 1990s, it was found that small concentrations of Fe in ZSM-5 could selectively oxidize methane to methanol in the presence of N_2O [75, 91]. The activity in methane oxidation of Fe-ZSM-5 catalysts was attributed to the formation of the highly active α-oxygen via decomposition of N_2O at $T < 300\,°C$ [106]. Studies with *in situ* Mössbauer spectroscopy identified the active site as a bis(−oxo)diiron cluster [107]. On the other hand, kinetic and adsorption studies showed a CH_4:Fe stoichiometry of 1 : 1, pointing to mono-iron species as the working active sites in Fe-ZSM-5 [107, 108]. This led to the hypothesis that the active Fe species are mononuclear Fe(II)α sites in a paired arrangement [109].

Snyder et al. used magnetic circular dichroism to identify the structure of active α-Fe species in Fe-BEA [110]. They concluded that this species is a mononuclear, high-spin, square planar Fe(II) site that, upon contact with N_2O, generates the active α-O species by forming a high-spin Fe(IV) = O species. The ability to activate methane was attributed to a constrained coordination geometry enforced by the β-6-MR of the BEA zeolite structure. It has been proposed that the zeolite lattice induces a geometric distortion of the Fe species, thus activating the site for the cleavage of the C—H bond.

The studies on Fe-zeolites mentioned above have encountered a common problem, i.e. that the reaction can only be performed in stoichiometric mode because the produced methanol remains strongly adsorbed on the catalysts.

In this respect, Stakoron et al. recently reported conversion under "quasi-catalytic" conditions. By contacting the Fe catalyst with a CH_4 stream containing low partial pressures of N_2O at 160 °C, it was possible to reach a TON of 3.6 [92]. Attempts to perform the reaction under fully catalytic mode at slightly higher temperatures (300 °C) resulted in a significant decrease in methanol selectivity [109].

In a different approach, Hutchings et al. [93, 111–114] reported that Fe in ZSM-5 selectively oxidized methane in presence of water at low temperatures, using H_2O_2 as an oxidating agent [93]. The reaction proceeds via a methylhydroxoperoxide that easily oxidizes to formic acid and CO_2 [93, 111, 115]. Thus, the mechanism of reaction and the Fe active species responsible for the oxidation of methane with H_2O_2 are proposed to be different from those reacting with N_2O. Resonance-enhanced Raman spectroscopy and DFT studies showed that the speciation of Fe ions on ZSM-5 is very complex and depends on the activation protocol [114]. This observation is in good agreement with the spectroscopic characterization of Fe species in ZSM-5 for different applications [116–118].

8.5.2 Cu-Zeolites

From a fundamental point of view, the studies on Fe-oxo species in zeolites have been crucial for the overall progress, exemplifying the oxidative mechanisms of methane activation and on the stabilization of ion clusters by the zeolite matrix. However, Fe-zeolites are only active in presence of N_2O or H_2O_2 as oxidants/reductants. Therefore, interest has shifted to pMMO-inspired Cu-zeolites, which are able to use molecular O_2 [119].

8.5.2.1 Structure of Active Sites

Most of the characterization studies are devoted to the two most active Cu-zeolites: Cu-ZSM-5 and Cu-mordenite (MOR) [99, 120–122]. The use of a wide variety of spectroscopic techniques in combination with DFT calculations have led to identification of two active clusters; namely, a $[Cu_2(\mu\text{-}O)]^{2+}$ dimer [120] and a $[Cu_3(\mu\text{-}O)_3]^{2+}$ trimer [99] (Figure 8.2). Their activity is attributed to the radical character of the O atom in μ-oxo bridges connecting Cu^{2+} ions

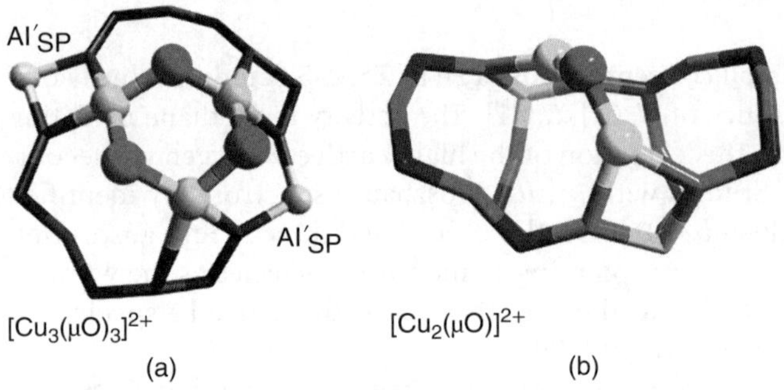

Figure 8.2 DFT model cluster of (a) $Cu_3(\mu\text{-}O)_3$ species located on the 8-MR of MOR [99], and (b) $Cu_2(\mu\text{-}O)$ species located on the γ cationic site of ZSM-5. Source: Adapted with permission from Li et al. [123]. (*See insert for color representation.*)

(Figure 8.2), which is able to abstract the first hydrogen from methane [31]. Recently, a third structure, a hydroxylated linear tri-Cu-oxo cluster, has been identified stabilized in a NU-1000 metal organic framework (MOF) structure [124].

The use of resonance Raman spectroscopy has demonstrated the presence of the dimer in Cu-ZSM-5 [120, 125] and Cu-MOR [122], whereas studies combining extended X-ray absorption fine structure (EXAFS) and IR spectroscopy have identified the trimer in Cu-ZSM-5 [100] and Cu-MOR [99]. Theoretical calculations have shown that both the dimeric and trimeric oxo-clusters are stable in these frameworks [99, 123, 126]. Under O_2-rich environments at high temperatures, the formation of the trimer is thermodynamically favored [99, 123].

Other proposals induced by theory include monomeric $[CuOH]^+$ sites in 8-MR in SSZ-13 [127], and $[Cu_4O_4]^{2+}$ and $[Cu_5O_5]^{2+}$ in MOR [128]. However, experimental evidence has not been found so far to support these theoretical models.

8.5.2.2 Activity of Explored Cu-Zeolites

Table 8.1 offers an overview of the most active Cu-zeolites. The majority of papers provide data from activity tests in stoichiometric mode (TON = 1). Therefore, comparisons of activities in Table 8.1 are made in terms of Cu efficiency of the material, defined as moles of methanol produced per mol of Cu in one catalytic cycle.

Table 8.1 shows that Cu efficiencies are in general low, with values well below 0.3. It should be noted that the expected efficiency for active sites $[Cu_2(\mu\text{-}O)]^{2+}$ and $[Cu_3(\mu\text{-}O)_3]^{2+}$ is 0.5 and 0.33, respectively. Even if an active site with nuclearity somewhat higher than 3 can be hypothesized, Cu_{eff} values lower than 0.1 point to large concentrations of inactive Cu species.

The Cu efficiencies vary widely, even for the same framework. These differences are attributed, in part, to differences in nonoptimal reaction parameters. For instance, the use of liquid phase solvents in the last step has been found to be less efficient in the extraction of products than the use of steam [95, 132].

Considering these aspects, two conclusions from the activity data reported in literature can be made:

i. *The zeolite topology has a strong effect on the activity.* Cu species hosted in large pore zeolites like faujasite (FAU), hexagonal Y (EMT), or beta polymorph A zeolite (BEA) showed very low activities. The most active catalysts were those with medium-pore zeolites MOR, ZSM-5, and ferrierite (FER). The small-pore zeolites SSZ-13, SSZ-39, SSZ-16 achieved similar Cu_{eff}, in the range of 0.05–0.10.
ii. *The presence of Na^+ cations adversely affects methane conversion.* This is particularly evident for ZSM-5- and MOR-based materials.

Table 8.1 Comparison of methanol yields and Cu efficiencies (expressed as mol MeOH/mol Cu) of different Cu zeolites for the oxidation of methane with O_2.

Catalyst (Si/Al)	Cu/Al	CH_3OH yield ($\mu mol\ g^{-1}$)	Cu_{eff}	References
Cu-Na-FAU (28)	0.34	<1[a)]	<0.005	Smeets et al. [94]
Cu-Na-EMT (4)	0.36	<1[a)]	<0.001	Smeets et al. [94]
Cu-Na-BEA (10)	0.50	4.2[a)]	0.0057	Smeets et al. [94]
Cu-Na-MOR (8,8)	0.50	16[a)]	0.020	Smeets et al. [94]
Cu-Na-MOR (11)	0.38	13[a)]	0.017	Alayon e al. [98]
Cu-Na-MOR (5)	0.34	31	0.04	Wulfers et al. [95]
Cu-Na-MOR (11)	0.32	82[b)]	0.19	Grundner et al. [101]
Cu-H-MOR (8.5)	0.30	23[a) c)]	0.047	Bozbag et al. [129]
Cu-H-MOR (20)	0.40	97[d)]	0.31	Kim et al. [130]
Cu-H-MOR (11)	0.31	115[b)]	0.27	Grundner et al. [101]
Cu-Na-ZSM-5 (14)	0.37	37[e)]	0.094	Narsimhan et al. [96]
Cu-H-ZSM-5 (14)	0.31	82[e)]	0.24	Narsimhan et al. [96]
Cu-H-ZSM-5 (14)	0.20	25[b)]	0.19	Markovits et al. [100]
Cu-Na-ZSM-5 (30)	0.47	2.7[a)]	0.011	Smeets et al. [94]
Cu-Na-ZSM-5 (11,5)	0.34	16	0.03	Wulfers et al. [95]
Cu-H-FER (26)	0.27	22[b)]	0.14	Grundner [131]
Cu-Na-SSZ-16 (6.5)	0.34	39	0.05	Wulfers et al. [95]
Cu-Na-SSZ-13 (12)	0.35	31	0.06	Wulfers et al. [95]
Cu-H-SSZ-13 (12)	0.40	45	0.09	Ipek et al. [97]
Cu-Na-SSZ-13 (12)	0.47	28[f)]	0.05	Ipek et al. [103]
Cu-Na-SSZ-39 (10)	0.48	23[f)]	0.03	Ipek et al. [103]
Cu-Na-SSZ-39 (10)	0.26	36	0.09	Wulfers et al. [95]

Unless otherwise stated, the three-stage reaction conditions are as follows: (i) activation in O_2 at 450–550 °C; (ii) reaction with CH_4 at 200 °C for 30–60 minutes; product desorption by steam treatment at 135–200 °C.

a) Methanol extraction with water or water/acetonitrile.
b) Methane reaction held for 4 hours.
c) Maximum yield achieved after 4 catalytic cycles.
d) Activation at 600 °C in O_2 and methane reaction at 150 °C.
e) Methanol extracted in the first 15 hours when flowing CH_4 (98.1 kPa), H_2O (3.2 kPa) and O_2 (0.0025 kPa) at 210 °C.
f) Methane reaction at 50 °C.

8.6 Approaches to Increase Methanol Yields on Cu-Zeolites

8.6.1 Synthetic Strategies for Highly Active Cu-Zeolites

Upon ion exchange, Cu^{2+} ions can attach to the framework compensating two negative charges of a Al-O-$(SiO_2)_n$-O-Al site (n = 1–3) when the two framework Al are in the same ring or in an accessible intersection [133]. Alternatively, Cu can form $[CuOH]^+$ ions and exchange on the zeolite, compensating the negative charge of any exchange site, regardless of the presence of a second exchange site in the vicinity. Grundner et al. [101] showed that a lower pH of the Cu^{2+} exchange solution favored the exchange of Cu^{2+} as divalent ions in Al pairs, whereas pH close to neutrality rendered a larger concentration of exchanged Cu. It has been concluded that this is related to the dissociation of H_2O, allowing for the formation of $[CuOH]^+$. During oxidative activation at temperatures above 400 °C, Cu ions migrate and/or combine with oxygen, forming multinuclear clusters. These Cu-oxo clusters will be located in the thermodynamically most stable positions, which are determined by the copper oxidation state and by the distribution of Al sites in the framework [133, 134].

Literature data compiled in Table 8.1 indicate that zeolite frameworks differ in their ability to stabilize Cu active species. IR spectroscopy studies showed that Cu-oxo species tend to be located in the 8-MR of the framework in the highly active CuMOR [99]. Structures such as FER, SSZ-13, SSZ-16, and SSZ-39 contain accessible 8-MR of similar size as MOR, and they have shown relatively high Cu efficiencies (Table 8.1). The second most efficient Cu-zeolite after MOR is Cu-ZSM-5. This zeolite framework (structure type MFI) lacks 8-MR, but its 10-MR can contain Al-$(SiO_2)_2$-Al pairs with an interatomic distance of 7.55 Å, similar to Al pairs in 8-MR in MOR (7.54 Å) [122], where Cu active species can be stabilized. However, it should be noted that the role of these structural features is not limited to stabilizing active sites but also to induce geometric constraints necessary to create or enhance the reactivity of oxygen with methane [100, 110]. For instance, when reaction barriers are calculated for $Cu_2(\mu\text{-}O)$ in different locations of MOR, it is found that the reaction of methane on the dimer in 8-MR positions is strongly disfavored with respect to the dimer in 12-MR [126].

Thus, it is clear that the first step for preparing a highly active Cu-zeolite catalysts is the choice of an adequate zeolite framework: Topologies containing 8-MR and 10-MR are able to hold the active Cu-oxo clusters [99, 122]. Furthermore, in order to minimize the concentration of Cu spectators in the catalyst, it is presumably important to avoid the presence of *d6r* building unit consisting in a double-six-member-ring prism, because it is known to strongly stabilize Cu^{2+} monomers [135, 136].

The presence of alkali and alkaline-earth exchanged cations has been found to be detrimental for the formation of active Cu species (Figure 8.3) [101]. As can be seen in Table 8.1, Cu-zeolites

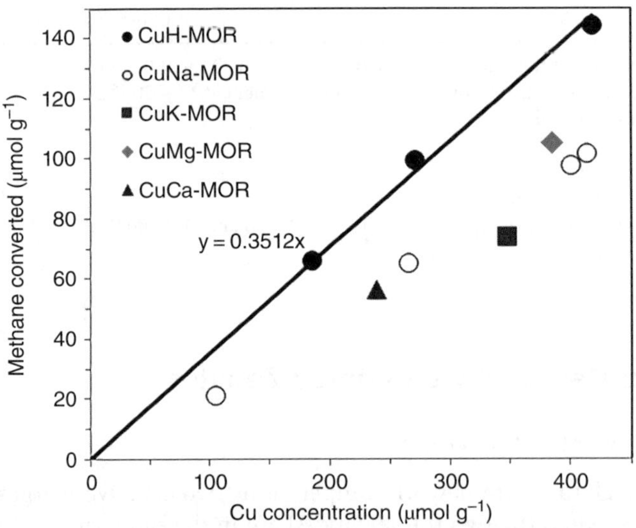

Figure 8.3 Methane converted versus Cu concentration for Cu-exchanged MOR with various co-cations. * The slope of 0.35 indicates that c. 3 Cu atoms are involved in the oxidation of one methane molecule. Source: Adapted from Grundner et al. [101]. (*See insert for color representation.*)

prepared from H-form have significantly higher efficiencies than their Na-form counterparts. The negative effect of co-cations is attributed to competition for specific exchange sites [101] and the change in the polarization of the Cu-μ-oxo species.

The active Cu-oxo clusters are charged species and are hydrolyzed during extraction of products in the presence of water [99], with Cu ions migrating to other positions that are more stable under wet conditions [137]. Therefore, the regeneration of the active cluster via oxidative treatment of the catalyst must combine the dehydration of Cu ions, the mobilization to adequate framework positions and the replenishment of O-active species, completing this way the reconstruction of the active cluster. This is likely the reason why the three-stage reaction of methane on Cu-zeolites only achieves significant yields to methanol when high temperatures (above 400 °C) are used in the activation/regeneration step [102, 130, 131].

8.6.2 Methanol Production over Cu-Zeolites in Catalytic Mode

The high-temperature activation of Cu-zeolites poses a major obstacle for the development of a process in continuous (catalytic) mode, because it is necessary to reduce the temperature below 250 °C before contacting the solid with methane to avoid unselective oxidation. CuMOR catalysts activated in O_2 at temperatures below 400 °C have shown some activity in the stepped reaction of methane to methanol [97, 102, 130, 131, 138]. However, the significantly lower methanol yields indicate that only a minority of Cu ions are forming active species under these conditions, and it is unclear whether the structure of such active sites is the same as the ones formed at $T > 400$ °C.

Narsimhan et al. [96] achieved the catalytic production of methanol on several Cu-zeolites by adding only a small concentration of O_2 to a wet stream of CH_4 at 210 °C (Figure 8.4). After a first desorption of products from the sites created during oxidative activation of the CuMOR catalyst at high temperatures (stoichiometric yield), the catalyst reaches steady-state production of methanol. Highest catalytic methanol yields were ca. 0.005 mol CH_3OH/mol Cu·h. The activity of Cu-zeolites decreased in the order Na-SSZ-13 > H-ZSM-5 > H-MOR > H-FER > H-BEA > Na-ZSM-5 > Na-MOR. This activity trend with zeolite frameworks is significantly different from the stepwise reaction. Thus, it is reasonable to speculate that a different type of Cu active site exists under wet conditions [96, 139].

8.6.3 Alternative Catalyst Components

The idea of using other transition metal ions in zeolites instead of Cu has also been explored. Co [140, 141] and Ni [142] exchanged in ZSM-5 were found to be active, but their activities were one order of magnitude lower than for the Cu counterparts.

A more interesting approach is replacing zeolites for MOF to host the active Cu-oxo clusters. The enormous structural and chemical versatility of MOFs offers the opportunity to mimic not only the active core of an enzyme but also its surroundings [143, 144]. For the case of methane activation over Cu species, the main advantage of using MOFs is the lower affinity of the organic support for the oxidation products. Ikuno et al. showed methane oxidation at 150 °C on Cu-oxo species hosted on NU-1000 [124]. X-ray absorption spectroscopy (XAS) characterization and DFT calculation showed that, in this material, the Cu-oxo active site has a trimeric Cu-hydroxide-like structure that bridges two nodes across the c-pore of the MOF. Although the selectivity to methanol was moderate (45–60%), and a significant fraction of the Cu atoms were inactive ($Cu_{eff} \approx 0.011$ mol CH_3OH/mol Cu), the results demonstrate the principal possibility of hydroxylated Cu-μ-oxo species for methane activation.

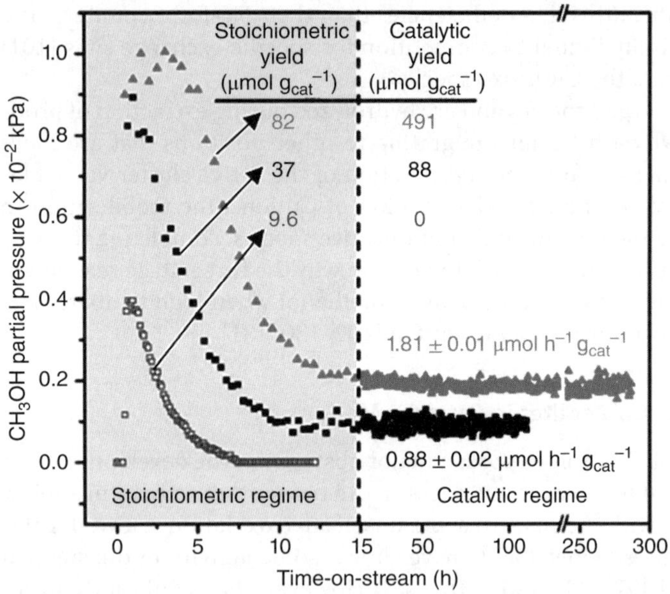

Figure 8.4 Methanol yield under stoichiometric and catalytic regime for Cu-Na-ZSM-5 (full squares), Cu-H-ZSM-5 (triangles) when treated at 210 °C with a gas stream of CH_4 (98.1 kPa), H_2O (3.2 kPa) and O_2 (0.0025). The curve with open squares is a Cu-Na-ZSM-5 in same mixture but substituting CH_4 by He. Prior to contact with the wet stream, the samples were contacted with CH_4 at 210 °C for 30 minutes. Source: Reproduced from Narsimhan et al. [96]. (*See insert for color representation.*)

8.7 Outlook

To date, all direct conversions do not measure up to the very-well-developed indirect route via methane-reforming and subsequent hydrogenation to methanol or hydrocarbons. Based on economics, a feasible direct route to methanol from methane must achieve high methanol selectivity between 80% and 90% at a methane conversion of at least 20–30%, respectively [77].

OCM requires quite high temperatures and a catalyst/reactor combination able to control the high reactivity of methyl radicals and C_2 products with O_2, in order to achieve reasonable yields. Despite efforts to optimize OCM catalysts to control the heterogeneous pathways, the highest ethylene yield achieved to date is 21% [21], which is still below the economic feasibility (yields >30%) [145]. A better understanding of the interplay between the catalyst-induced formation and quenching of radicals appears to be critical at present for a breakthrough. It should be noted that Siluria has recently completed a demonstration plant for OCM using a proprietary oxide catalyst in the form of whiskers [146].

Halogenation of methane can be achieved at temperatures below 270 °C, but the reaction tends to be unselective. The most successful strategy so far to a high selectivity to CH_3X is the electrophilic activation of methane by X^+ species [38]. McFarland and Stucky have published several integrated multistep processes for methane conversion to oxygenates and olefins via bromination [147–150]. In this context, a report by Li et al. [151] describes a multifunctional $MoO_2Br_2(H_2O)_2$ catalyst able to perform both methane bromination and the regeneration of bromine via metathesis. However, the disadvantages of these functionalization routes using halogens still outweigh the advantages. Aside from the difficulties in the regeneration of the halogen, the use of halides requires materials able to stand the corrosion, increasing the cost of a potential industrial application [44].

The performance of all the known catalysts able to oxidize methane with O_2 at low temperatures is restricted to very low methanol yields and in, most cases, the production in stoichiometric mode. A possibility to increase yields would be to use membrane technologies to separate methanol and avoid its overoxidation. However, the use of membranes has so far been reported only in biological applications [152]. The use of solid dense membranes for the transport of oxygen has been proposed in methane to syngas technology [153, 154] and could be of interest for direct methane oxidation applications.

Zeolites are excellent supports for transition metal ions, and they have been found to stabilize different active and inactive Fe and Cu species, including a type that can catalytically convert methane into methanol with molecular O_2 in the presence of H_2O at mild temperatures.

The speciation of transition metal ions in zeolites is rather complex and the spectroscopic identification of active species is often obscured by the presence of large concentrations of spectators [101]. Zeolite properties as well as protocols of ion exchange and thermal activation have a strong impact on the final siting and structure of the metal ion species in the framework [101, 117, 129, 135, 155]. Thus, a more detailed understanding of the speciation of transition metal ions in zeolites is required to design synthetic approaches. Research efforts should, in particular, aim to study important synthetic factors, such as the effect of Al distribution in the formation of active clusters, which until now have been only partially described for particular systems [99, 100, 118, 133–135].

Given the complexity of species present in Cu-zeolite catalysts, the use of *in situ* spectroscopies during O_2 and CH_4 activation together with DFT cluster models are very important tools for an unambiguous determination of the active site. Further insight is also necessary on the mechanistic requirements for harnessing highly reactive oxygen species on metal-oxo clusters. It has been seen in the study of methane activation on Fe sites that a strong geometric distortion forced by coordination of the metal ion to the β-6MR of BEA allows the activation of oxygen. Thus, it must be examined how the framework geometric constrains and confinement of the micropores influence the activity of Cu-oxo clusters hosted by zeolites. An analogous approach should be undertaken to explore the activation of methane on Cu-MOFs.

Summarizing, it becomes evident that impressive insight has been gained on the different mechanisms of C—H activation [13, 31, 33, 38, 156]. However, the findings for a variety of catalysts and reaction systems have not led yet to a scalable process for oxidative functionalization of methane. Further research is necessary not only toward new catalysts capable of effective alkane activation and control of selectivity, but also on associated aspects ranging from thermodynamic constraints to engineering and economic requirements of the process.

Acknowledgments

The authors are thankful to the Inorganometallic Catalyst Design Center, an EFRC funded by the U.S. Department of Energy (DOE), Office of Science, Office of Basic Energy Sciences (DE-SC0012702).

References

1 Huber, G.W., Iborra, S., and Corma, A. (2006). *Chem. Rev.* 106: 4044.
2 Caballero, A. and Perez, P.J. (2013). *Chem. Soc. Rev.* 42: 8809.
3 Dry, M.E. (2004). *Appl. Catal., A* 276: 1.
4 Rostrup-Nielsen, J.R. (1993). *Catal. Today* 18: 305.

5 Kondratenko, E.V., Peppel, T., Seeburg, D. et al. (2017). *Catal. Sci. Technol.* 7: 366.
6 Gradassi, M.J. and Wayne Green, N. (1995). *Fuel Process. Technol.* 42: 65.
7 Otsuka, K. and Wang, Y. (2001). *Appl. Catal., A* 222: 145.
8 Burch, R., Squire, G.D., and Tsang, S.C. (1989). *J. Chem. Soc., Faraday Trans. 1* 85: 3561.
9 Walker, G.S., Lapszewicz, J.A., and Foulds, G.A. (1994). *Catal. Today* 21: 519.
10 Hutchings, G.J. and Taylor, S.H. (1999). *Catal. Today* 49: 105.
11 Zhang, Q., He, D., Han, Z. et al. (2002). *Fuel* 81: 1599.
12 Horn, R. and Schlögl, R. (2015). *Catal. Lett.* 145: 23.
13 Lunsford, J.H. (2000). *Catal. Today* 63: 165.
14 Iwamoto, M. and Lunsford, J.H. (1980). *J. Phys. Chem.* 84: 3079.
15 Zavitsas, A.A. and Melikian, A.A. (1975). *J. Am. Chem. Soc.* 97: 2757.
16 Bohme, D.K. and Fehsenfeld, F.C. (1969). *Can. J. Chem.* 47: 2717.
17 Sinev, M.Y., Fattakhova, Z.T., Lomonosov, V.I. et al. (2009). *J. Nat. Gas Chem.* 18: 273.
18 Hammond, C., Conrad, S., and Hermans, I. (2012). *ChemSusChem* 5: 1668.
19 Derk, A.R., Li, B., Sharma, S. et al. (2013). *Catal. Lett.* 143: 406.
20 Borchert, H. and Baerns, M. (1997). *J. Catal.* 168: 315.
21 Zavyalova, U., Holena, M., Schlögl, R. et al. (2011). *ChemCatChem* 3: 1935.
22 Zhang, Z., Verykios, X.E., and Baerns, M. (1994). *Catal. Rev. Sci. Techol.* 36: 507.
23 Myrach, P., Nilius, N., Levchenko, S.V. et al. (2010). *ChemCatChem* 2: 854.
24 Johnson, M.A., Stefanovich, E.V., and Truong, T.N. (1997). *J. Phys. Chem. B* 101: 3196.
25 Lunsford, J.H., Hinson, P.G., Rosynek, M.P. et al. (1994). *J. Catal.* 147: 301.
26 Schwach, P., Willinger, M.G., Trunschke, A. et al. (2013). *Angew. Chem. Int. Ed.* 52: 11381.
27 Schwach, P., Hamilton, N., Eichelbaum, M. et al. (2015). *J. Catal.* 329: 574.
28 Ito, T., Wang, J., Lin, C.H. et al. (1985). *J. Am. Chem. Soc.* 107: 5062.
29 Leveles, L., Seshan, K., Lercher, J.A. et al. (2003). *J. Catal.* 218: 307.
30 Arndt, S., Laugel, G., Levchenko, S. et al. (2011). *Catal. Rev. Sci. Technol.* 53: 424.
31 Dietl, N., Schlangen, M., and Schwarz, H. (2012). *Angew. Chem. Int. Ed.* 51: 5544.
32 Kwapien, K., Paier, J., Sauer, J. et al. (2014). *Angew. Chem. Int. Ed.* 53: 8774.
33 Hu, Z., Li, B., Sun, X. et al. (2011). *J. Phys. Chem. C* 115: 3065.
34 Olah, G.A. (1987). *Acc. Chem. Res.* 20: 422.
35 Lersch, P. and Bandermann, F. (1991). *Appl. Catal.* 75: 133.
36 White, C.M., Douglas, L.J., Hackett, J.P. et al. (1992). *Energy Fuels* 6: 76.
37 Zhang, A., Sun, S., Komon, Z.J.A. et al. (2011). *PCCP* 13: 2550.
38 Olah, G.A., Gupta, B., Felberg, J.D. et al. (1985). *J. Am. Chem. Soc.* 107: 7097.
39 Bucsi, I. and Olah, G.A. (1992). *Catal. Lett.* 16: 27.
40 Batamack, P., Bucsi, I., Molnár, Á. et al. (1994). *Catal. Lett.* 25: 11.
41 Gorin, E., Fontana, C.M., and Kidder, G.A. (1948). *Ind. Eng. Chem.* 40: 2128.
42 Olah, G.A. and Mo, Y.K. (1972). *J. Am. Chem. Soc.* 94: 6864.
43 Olah, G.A., Renner, R., Schilling, P. et al. (1973). *J. Am. Chem. Soc.* 95: 7686.
44 McFarland, E. (2012). *Science* 338: 340.
45 Lange, J.P. and Tijm, P.J.A. (1996). *Chem. Eng. Sci.* 51: 2379.
46 Lin, R., Amrute, A.P., and Pérez-Ramírez, J. (2017). *Chem. Rev.* 117: 4182.
47 Ohtsuka, Y. and Tamai, Y. (1978). *J. Catal.* 51: 169.
48 Pieters, W.J.M., Conner, W.C., and Carlson, E.J. (1984). *Appl. Catal.* 11: 35.
49 Podkolzin, S.G., Stangland, E.E., Jones, M.E. et al. (2007). *J. Am. Chem. Soc.* 129: 2569.
50 A.E.J. Schweizer, Jones, M. E.; Hickman, D. A., 2002 U.S. Patent. 6, 452, 058.
51 Peringer, E., Salzinger, M., Hutt, M. et al. (2009). *Top. Catal.* 52: 1220.
52 Liu, Z., Huang, L., Li, W.S. et al. (2007). *J. Mol. Catal. A: Chem.* 273: 14.
53 Wang, K.X., Xu, H., F., Li, W.S. et al. (2006). *Appl. Catal. A* 304: 168.

54 Lin, R., Ding, Y., Gong, L. et al. (2010). *J. Catal.* 272: 65.
55 He, J., Xu, T., Wang, Z. et al. (2012). *Angew. Chem. Int. Ed.* 51: 2438.
56 Paunović, V., Zichittella, G., Moser, M. et al. (2016). *Nat. Chem.* 8: 803.
57 Halpern, J. (1968). *Disc. Faraday Soc.* 46: 7.
58 Shilov, A.E. and Shteinman, A.A. (1977). *Coord. Chem. Rev.* 24: 97.
59 Kushch, L.A., Lavrushko, V.V., Misharin, Y.S. et al. (1983). *Nouv. J. Chim.* 12: 729.
60 Luinstra, G.A., Wang, L., Stahl, S.S. et al. (1995). *J. Organomet. Chem.* 504: 75.
61 Freund, M.S., Labinger, J.A., Lewis, N.S. et al. (1994). *J. Mol. Catal.* 87: L11.
62 Geletii, Y.V. and Shilov, A.E. (1983). *Kinet. Katal.* 24: 486.
63 Labinger, J.A., Herring, A.M., Lyon, D.K. et al. (1993). *Organometallics* 12: 895.
64 Periana, R.A., Taube, D.J., Evitt, E.R. et al. (1993). *Science* 259: 340.
65 Periana, R.A., Taube, D.J., Gamble, S. et al. (1998). *Science* 280: 560.
66 Palkovits, R., Antonietti, M., Kuhn, P. et al. (2009). *Angew. Chem. Int. Ed.* 48: 6909.
67 Zimmermann, T., Soorholtz, M., Bilke, M. et al. (2016). *J. Am. Chem. Soc.* 138: 12395.
68 Soorholtz, M., White, R.J., Zimmermann, T. et al. (2013). *Chem. Commun.* 49: 240.
69 Ahlquist, M., Nielsen, R.J., Periana, R.A. et al. (2009). *J. Am. Chem. Soc.* 131: 17110.
70 Hanson, R.S. and Hanson, T. (1996). *Microbiol. Rev.* 60: 439.
71 Shiemke, A.K., Cook, S.A., Miley, T., and Singleton, P. (1995). *Arch. Biochem. Biophys.* 321: 421.
72 Friedle, S., Reisner, E., and Lippard, S.J. (2010). *Chem. Soc. Rev.* 39: 2768.
73 Wang, V.C.C., Maji, S., Chen, P.P.Y. et al. (2017). *Chem. Rev.* 117: 8574.
74 Groothaert, M.H., Smeets, P.J., Sels, B.F. et al. (2005). *J. Am. Chem. Soc.* 127: 1394.
75 Panov, G.I., Sobolev, V.I., Dubkov, K.A. et al. (1996). *Stud. Surf. Sci. Catal.* 101: 493.
76 Edwards, J.H. and Foster, N.R. (1986). *Fuel Sci. Technol. Int.* 4: 365.
77 Geerts, J.W.M.H., Hoebink, J.H.B.J., and van der Wiele, K. (1990). *Catal. Today* 6: 613.
78 Liu, X., Ryabenkova, Y., and Conte, M. (2015). *PCCP* 17: 715.
79 Ding, X.-L., Wu, X.-N., Zhao, Y.-X. et al. (2012). *Acc. Chem. Res.* 45: 382.
80 Kawabe, T., Tabata, K., Suzuki, E. et al. (2001). *Catal. Today* 71: 21.
81 Sirajuddin, S. and Rosenzweig, A.C. (2015). *Biochemistry* 54: 2283.
82 Merkx, M., Kopp, D.A., Sazinsky, M.H. et al. (2001). *Angew. Chem. Int. Ed.* 40: 2782.
83 Lieberman, R.L. and Rosenzweig, A.C. (2005). *Nature* 434: 177.
84 Whittington, D.A., Rosenzweig, A.C., Frederick, C.A. et al. (2001). *Biochemistry* 40: 3476.
85 Kudrik, E.V., Afanasiev, P., Alvarez, L.X. et al. (2012). *Nat. Chem.* 4: 1024.
86 Semrau, J.D., DiSpirito, A.A., and Yoon, S. (2010). *FEMS Microbiol. Rev.* 34: 496.
87 Chan, S.I. and Yu, S.S.F. (2008). *Acc. Chem. Res.* 41: 969.
88 Chan, S.I., Wang, V.C., Lai, J.C. et al. (2007). *Angew. Chem. Int. Ed.* 46: 1992.
89 Culpepper, M.A., Cutsail, G.E., Hoffman, B.M. et al. (2012). *J. Am. Chem. Soc.* 134: 7640.
90 Balasubramanian, R., Smith, S.M., Rawat, S. et al. (2010). *Nature* 465: 115.
91 Sobolev, V.I., Kharitonov, A.S., Panna, O.V. et al. (1995). *Stud. Surf. Sci. Catal.* 98: 159.
92 Starokon, E.V., Parfenov, M.V., Arzumanov, S.S. et al. (2013). *J. Catal.* 300: 47.
93 Hammond, C., Forde, M.M., Ab Rahim, M.H. et al. (2012). *Angew. Chem. Int. Ed.* 51: 5129.
94 Smeets, P.J., Groothaert, M.H., and Schoonheydt, R.A. (2005). *Catal. Today* 110: 303.
95 Wulfers, M.J., Teketel, S., Ipek, B., and Lobo, R.F. (2015). *Chem. Commun.* 51: 4447.
96 Narsimhan, K., Iyoki, K., Dinh, K., and Román-Leshkov, Y. (2016). *ACS Cent. Sci.* 2: 424.
97 Ipek, B. and Lobo, R.F. (2016). *Chem. Commun.* 52: 13401.
98 Alayon, E.M., Nachtegaal, M., Ranocchiari, M. et al. (2012). *Chem. Commun.* 48: 404.
99 Grundner, S., Markovits, M.A.C., Li, G. et al. (2015). *Nat. Commun.* 6: 7546.
100 Markovits, M.A.C., Jentys, A., Tromp, M. et al. (2016). *Top. Catal.* 59: 1554.
101 Grundner, S., Luo, W., Sanchez-Sanchez, M. et al. (2016). *Chem. Commun.* 52: 2553.

102 Tomkins, P., Mansouri, A., Bozbag, S.E. et al. (2016). *Angew. Chem. Int. Ed.* 55: 5467.
103 Ipek, B., Wulfers, M.J., Kim, H. et al. (2017). *ACS Catal.* 7: 4291.
104 Klier, K. (1988). *Langmuir* 4: 13.
105 Alayon, E.M.C., Nachtegaal, M., Kleymenov, E. et al. (2013). *Microporous Mesoporous Mater.* 166: 131.
106 Panov, G.I. (2000). *CATTECH* 4: 18.
107 Dubkov, K.A., Ovanesyan, N.S., Shteinman, A.A. et al. (2002). *J. Catal.* 207: 341.
108 Panov, G.I., Dubkov, K.A., and Starokon, E.V. (2006). *Catal. Today* 117: 148.
109 Parfenov, M.V., Starokon, E.V., Pirutko, L.V. et al. (2014). *J. Catal.* 318: 14.
110 Snyder, B.E.R., Vanelderen, P., Bols, M.L. et al. (2016). *Nature* 536: 317.
111 Hammond, C., Dimitratos, N., Jenkins, R.L. et al. (2013). *ACS Catal.* 3: 689.
112 Hammond, C., Dimitratos, N., Lopez-Sanchez, J.A. et al. (2013). *ACS Catal.* 3: 1835.
113 Hammond, C., Jenkins, R.L., Dimitratos, N. et al. (2012). *Chemistry* 18: 15735.
114 Hammond, C., Hermans, I., and Dimitratos, N. (2015). *ChemCatChem* 7: 434.
115 Rahman, A.K.M.L., Kumashiro, M., and Ishihara, T. (2011). *Catal. Commun.* 12: 1198.
116 Joyner, R.W. and Stockenhuber, M. (1997). *Catal. Lett.* 45: 15.
117 Joyner, R. and Stockenhuber, M. (1999). *J. Phys. Chem. B* 103: 5963.
118 Sazama, P., Wichterlová, B., Tábor, E. et al. (2014). *J. Catal.* 312: 123.
119 Vanelderen, P., Hadt, R.G., Smeets, P.J. et al. (2011). *J. Catal.* 284: 157.
120 Woertink, J.S., Smeets, P.J., and Groothaert, M.H. (2009). *Proc. Natl. Acad. Sci. U.S.A.* 106: 18908.
121 Vanelderen, P., Vancauwenbergh, J., and Tsai, M.-L. (2014). *ChemPhysChem* 15: 91.
122 Vanelderen, P., Snyder, B.E.R., and Tsai, M.-L. (2015). *J. Am. Chem. Soc.* 137: 6383.
123 Li, G., Vassilev, P., Sanchez-Sanchez, M. et al. (2016). *J. Catal.* 338: 305.
124 Ikuno, T., Zheng, J., Vjunov, A. et al. (2017). *J. Am. Chem. Soc.* 139: 10294.
125 Smeets, P.J., Hadt, R.G., Woertink, J.S. et al. (2010). *J. Am. Chem. Soc.* 132: 14736.
126 Zhao, Z.-J., Kulkarni, A., Vilella, L. et al. (2016). *ACS Catal.* 6: 3760.
127 Kulkarni, A.R., Zhao, Z.-J., Siahrostami, S. et al. (2016). *ACS Catal.* 6: 6531.
128 Palagin, D., Knorpp, A.J., Pinar, A.B. et al. (2017). *Nanoscale* 9: 1144.
129 Bozbag, S.E., Alayon, E.M.C., Pechacek, J. et al. (2016). *Catal. Sci. Technol.* 6: 5011.
130 Kim, Y., Kim, T.Y., Lee, H. et al. (2017). *Chem. Commun.* 53: 4116.
131 S. Grundner, PhD Dissertation 2016.
132 Starokon, E.V., Parfenov, M.V., Pirutko, L.V. et al. (2011). *J. Phys. Chem. C* 115: 2155.
133 Dědeček, J., Sobalík, Z., and Wichterlová, B. (2012). *Catal. Rev. Sci. Technol.* 54: 135.
134 Vanelderen, P., Vancauwenbergh, J., Sels, B.F. et al. (2013). *Coord. Chem. Rev.* 257: 483.
135 Paolucci, C., Parekh, A.A., Khurana, I. et al. (2016). *J. Am. Chem. Soc.* 138: 6028.
136 Fickel, D.W. and Lobo, R.F. (2010). *J. Phys. Chem. C* 114: 1633.
137 Smeets, P.J., Sels, B.F., van Teeffelen, R.M. et al. (2008). *J. Catal.* 256: 183.
138 Tomkins, P., Ranocchiari, M., and van Bokhoven, J.A. (2017). *Acc. Chem. Res.* 50: 418.
139 Alayon, E.M.C., Nachtegaal, M., Bodi, A. et al. (2014). *ACS Catal.* 4: 16.
140 Beznis, N.V., van Laak, A.N.C., Weckhuysen, B.M. et al. (2011). *Microporous Mesoporous Mater.* 138: 176.
141 Beznis, N.V., Weckhuysen, B.M., and Bitter, J.H. (2009). *Catal. Lett.* 136: 52.
142 Shan, J., Huang, W., Nguyen, L. et al. (2014). *Langmuir* 30: 8558.
143 Gascon, J., Corma, A., Kapteijn, F. et al. (2014). *ACS Catal.* 4: 361.
144 Olivos-Suarez, A.I., Szécsényi, À., Hensen, E.J.M. et al. (2016). *ACS Catal.* 6: 2965.
145 Labinger, J.A. (1988). *Catal. Lett.* 1: 371.
146 http://siluria.com/Technology/Oxidative_Coupling_of_Methane.
147 Lorkovic, I., Noy, M., Weiss, M. et al. (2004). *Chem. Commun.* 566.

148 Lorkovic, I.M., Noy, M.L., Schenck, W.A. et al. (2004). *Catal. Today* 98: 589.
149 Upham, D.C., Snodgrass, Z.R., Zavareh, M.T. et al. (2017). *Chem. Eng. Sci.* 160: 245.
150 Breed, A., Doherty, M.F., Gadewar, S. et al. (2005). *Catal. Today* 106: 301.
151 Li, F. and Yuan, G. (2006). *Angew. Chem. Int. Ed.* 45: 6541.
152 Reddy, P.V.L., Kim, K.-H., and Song, H. (2013). *Renew. Sustain. Energy Rev* 24: 578.
153 Li, Q., Zhu, X., He, Y., and Yang, W. (2010). *Catal. Today* 149: 185.
154 Balachandran, U., Dusek, J.T., Maiya, P.S. et al. (1997). *Stud. Surf. Sci. Catal.* 107: 409.
155 Kwak, J.H., Varga, T., Peden, C.H.F. et al. (2014). *J. Catal.* 314: 83.
156 Shilov, A.E. and Shul'pin, G.B. (1997). *Chem. Rev.* 97: 2879.

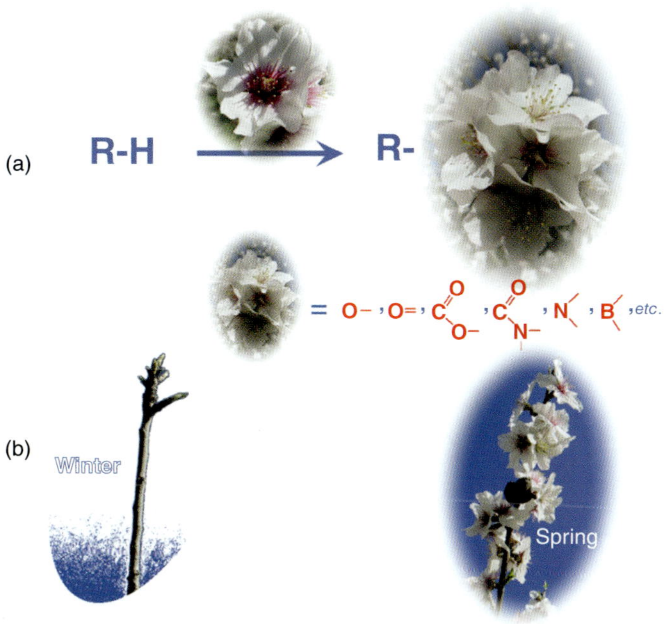

Figure 1 (a) Alkane functionalization; and (b) blooming of an almond tree branch.

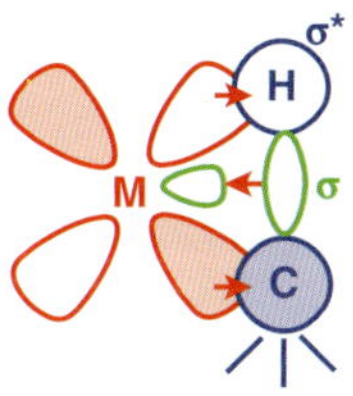

Scheme 1.1 Interaction of an alkane C—H bond with a metal (side-on mode).

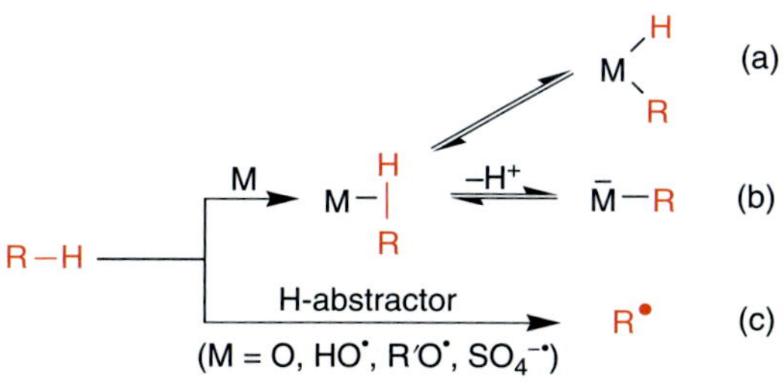

Scheme 1.2 Alkane C—H bond activation/cleavage: (a) oxidative addition to a metal; (b) heterolytic C—H bond cleavage (electrophilic activation); (c) homolytic C—H bond cleavage to form an alkyl radical.

Alkane Functionalization, First Edition. Edited by Armando J. L. Pombeiro and M. Fátima C. Guedes da Silva.

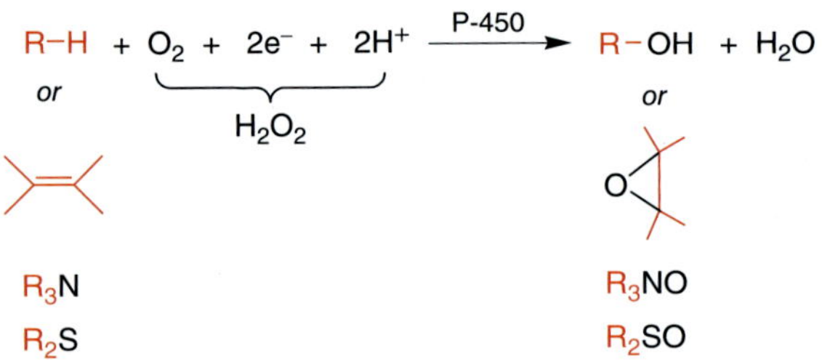

Scheme 1.4 Oxidation of alkanes and other substrates catalyzed by cytochrome P-450 (not active for methane) or by methane monooxygenase (MMO).

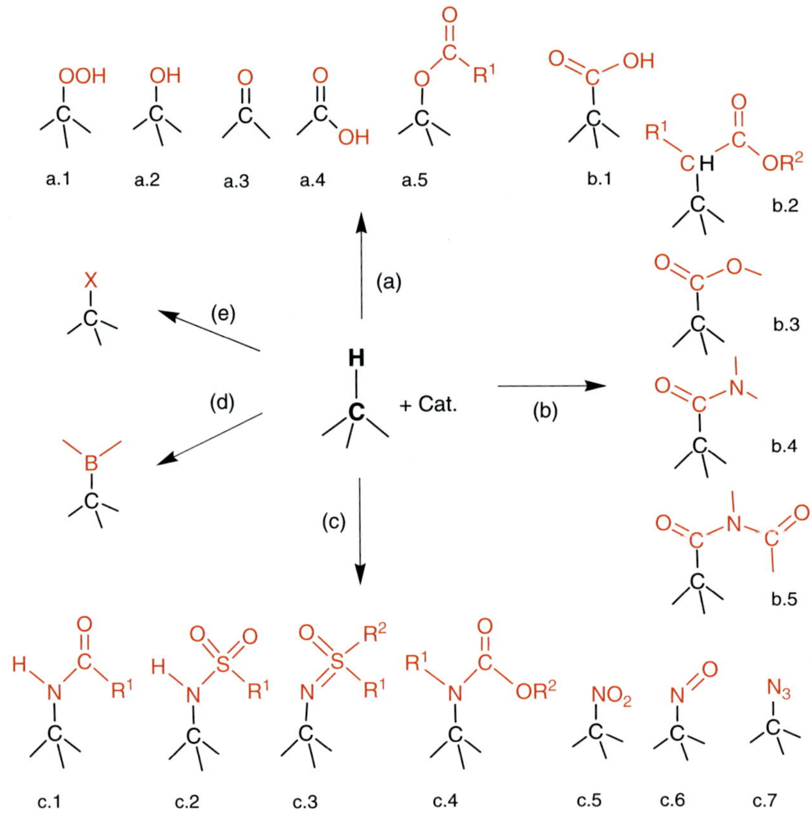

Scheme 1.5 General types of alkane functionalizations with formation of a (a) C—O; (b) C—C; (c) C—N; (d) C—B; or (e) C-halogen (X) bond.

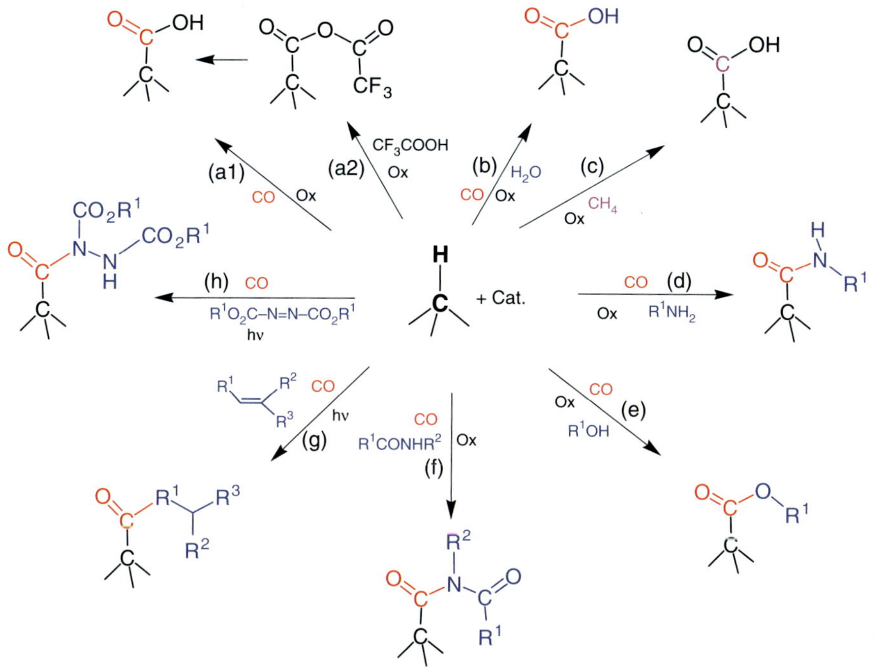

Scheme 1.6 Catalytic carbonylation of alkanes: (a) carboxylation with CO (a1) or with trifluoroacetic acid solvent (a2) to carboxylic acid; (b) hydrocarboxylation to carboxylic acid; (c) Oxidative condensation of methane (also as source of carboxylic carbon); (d) amidation with CO and primary amine; (e) alkoxycarbonylation with alcohol; (f) imidation with CO and amide; (g) photocatalytic carbonylation with CO and an electron-deficient olefin (R^3 = electron withdrawing group); (h) photocatalytic carbonylation with CO and an azodicarboxylate to an acyl hydrazide. Ox = oxidant (e.g. $K_2S_2O_8$, $^tBuOO^tBu$, H_2SO_4, O_2).

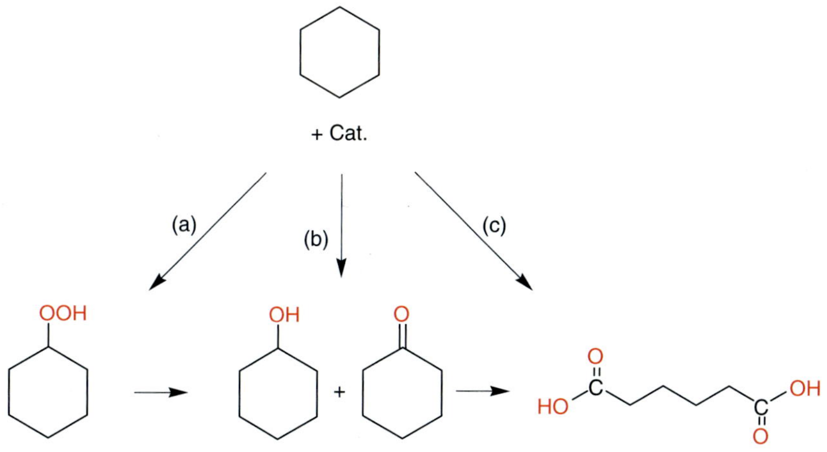

Scheme 1.7 Catalytic cyclohexane oxidations to (a) cyclohexylhydroperoxide; (b) cyclohexanol and cyclohexanone (KA oil); and (c) adipic acid.

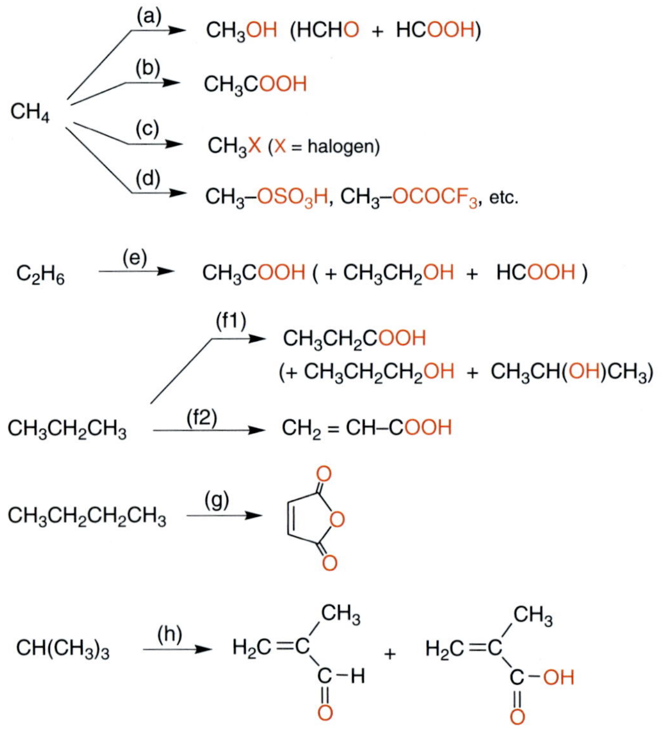

Scheme 1.8 Methane and other light alkanes oxidations: (a) methane oxidation to methanol (and formaldehyde); (b) methane oxidative condensation to acetic acid; (c) methane halogenation; (d) methane esterification; (e) ethane oxidation to acetic acid; (f) propane oxidation to acrylic acid; (g) butane oxidation to maleic anhydride; (h) *iso*-butane oxidation to methacrolein and methacrylic acid.

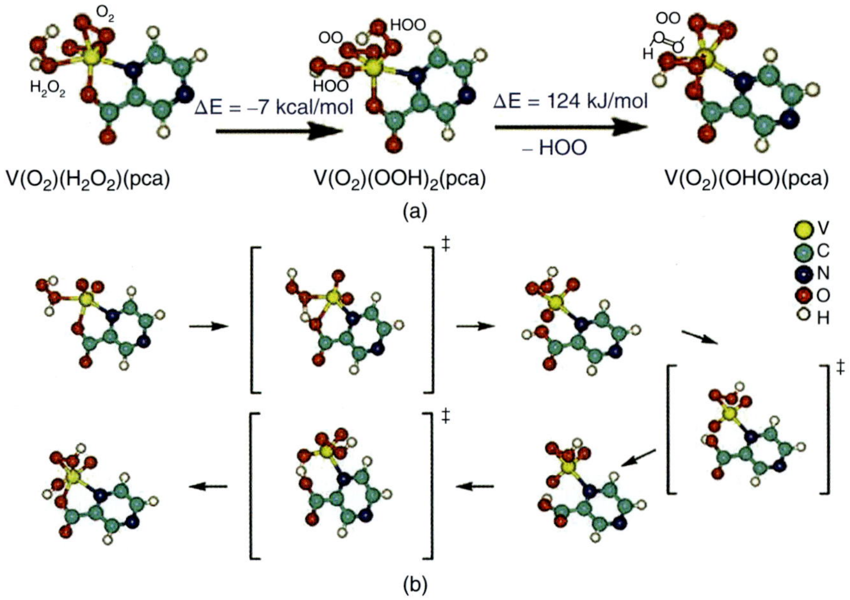

Figure 3.4 (a) Proton transfer from coordinated H_2O_2 to (O_2)-ligand followed by elimination of $HOO^{\bullet}$ radical); (b) "Robot's arm mechanism" of the proton transfer. Source: Reproduced by permission of The Royal Society of Chemistry [7b] and American Chemical Society [8a].

Water-assisted mechanism of proton transfer:

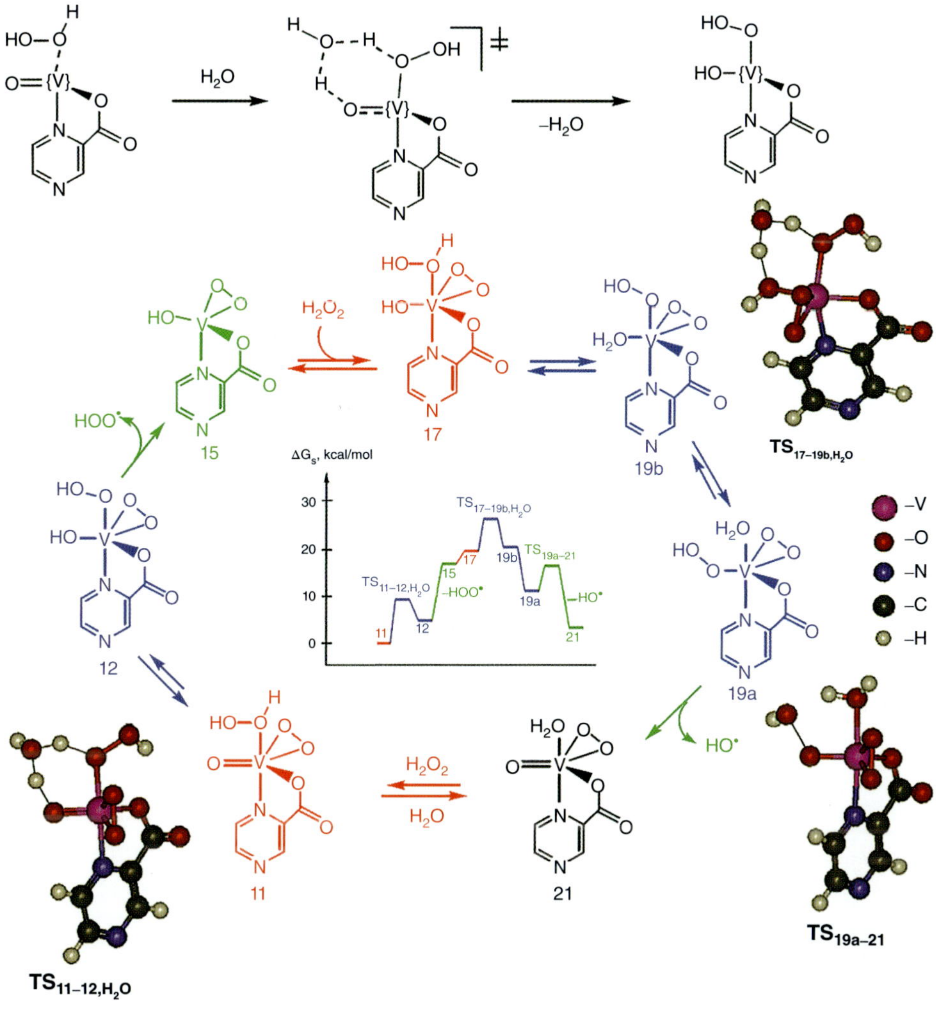

Figure 3.5 Generation of radicals by the "$O_2 + H_2O_2 + n\text{-}Bu_4NVO_3$ + PCA" reagent. Water-assisted mechanism of proton transfer from coordinated H_2O_2 to O-atom of the COOH group and then to (O_2)-ligand followed by elimination of $HOO^\bullet$ radical. Enumeration of species is as in the original paper. Color code: red: formation of H_2O_2-adducts, blue: H-transfers, green: radical generation. Source: Adapted from Ref. [7e] with permission of Elsevier.

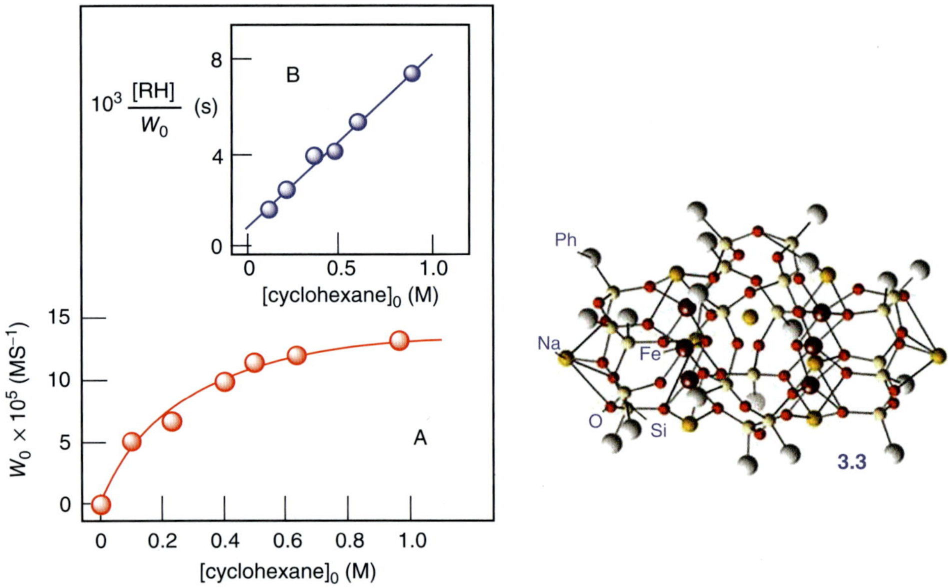

Figure 3.8 Left: Dependences of oxidation rate W_0 on initial concentration of cyclohexane (Graph A) in the oxidation of cyclohexane with H_2O_2 (50% aqueous) catalyzed by complex 3.3. Conditions: $[3.3]_0 = 2 \times 10^{-4}$ M, $[HNO_3]_0 = 0.05$ M, solvent MeCN, 30 °C. Concentrations of the products (cyclohexanol and cyclohexanone) were measured after reduction with PPh_3. Graph B: linearization of dependence presented by Graph A using coordinates $[\text{cyclohexane}]_0$–$[RH]_0/W_0$. Right: A structure of catalyst $[(PhSiO_{1.5})_{20}(FeO_{1.5})_6(NaO_{0.5})_8(n\text{-BuOH})_{9.6}(C_7H_8)]$ (**3.3**). Source: The figures are adapted from Ref. [16a] with permission of the Royal Society of Chemistry.

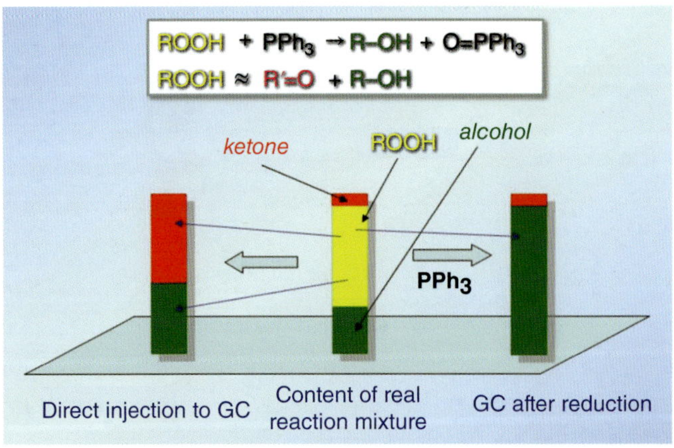

Figure 3.10 Determination of alkyl hydroperoxide, alcohol, and ketone using the GC.

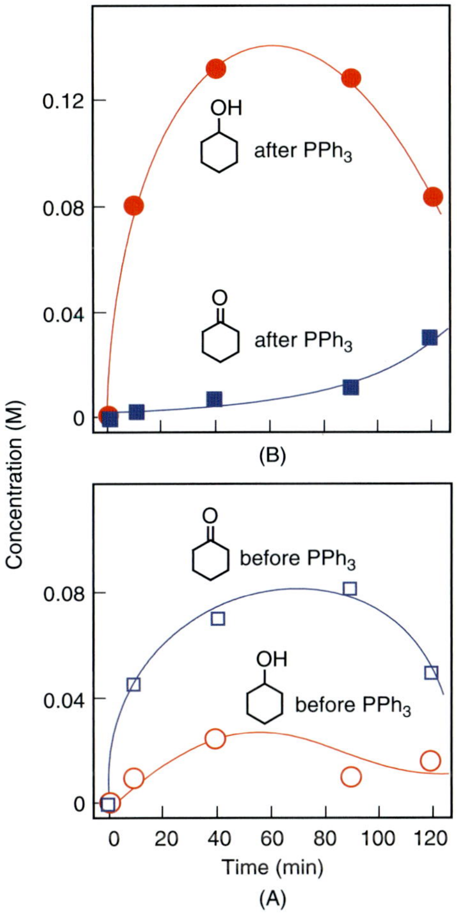

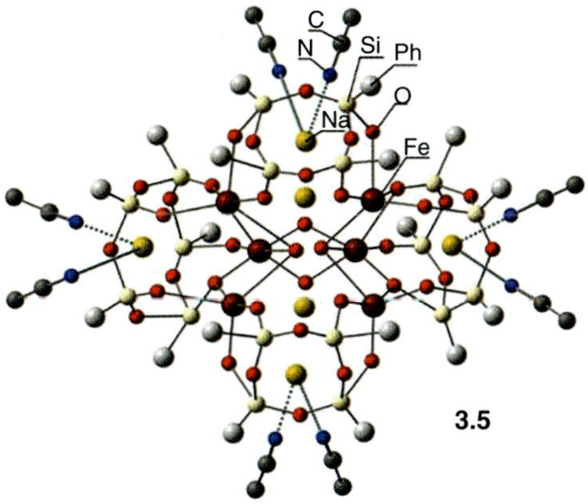

3.5

Figure 3.11 Left: Accumulation of cyclohexanol and cyclohexanone with time in the cyclohexane (0.46 M) oxidation with H_2O_2 (50%, aqueous, 1.5 M) catalyzed by the heterometallic cage silsesquioxane complex $[Ph_5Si_5O_{10}]_2[Ph_4Si_4O_8]_2Fe_6Na_6(O^{2-})_3$ $(CH_3CN)_{8.5}(H_2O)8.44$ (**3.5**; 5×10^{-4} M), CF_3COOH (0.05 M), CH_3CN, 40 °C. Concentrations measured before (Graph A) and after (Graph B) reduction of the samples with PPh_3 are shown. Right: the crystal structure of catalyst **3.5**. Source: The data are adapted from Ref. [20a].

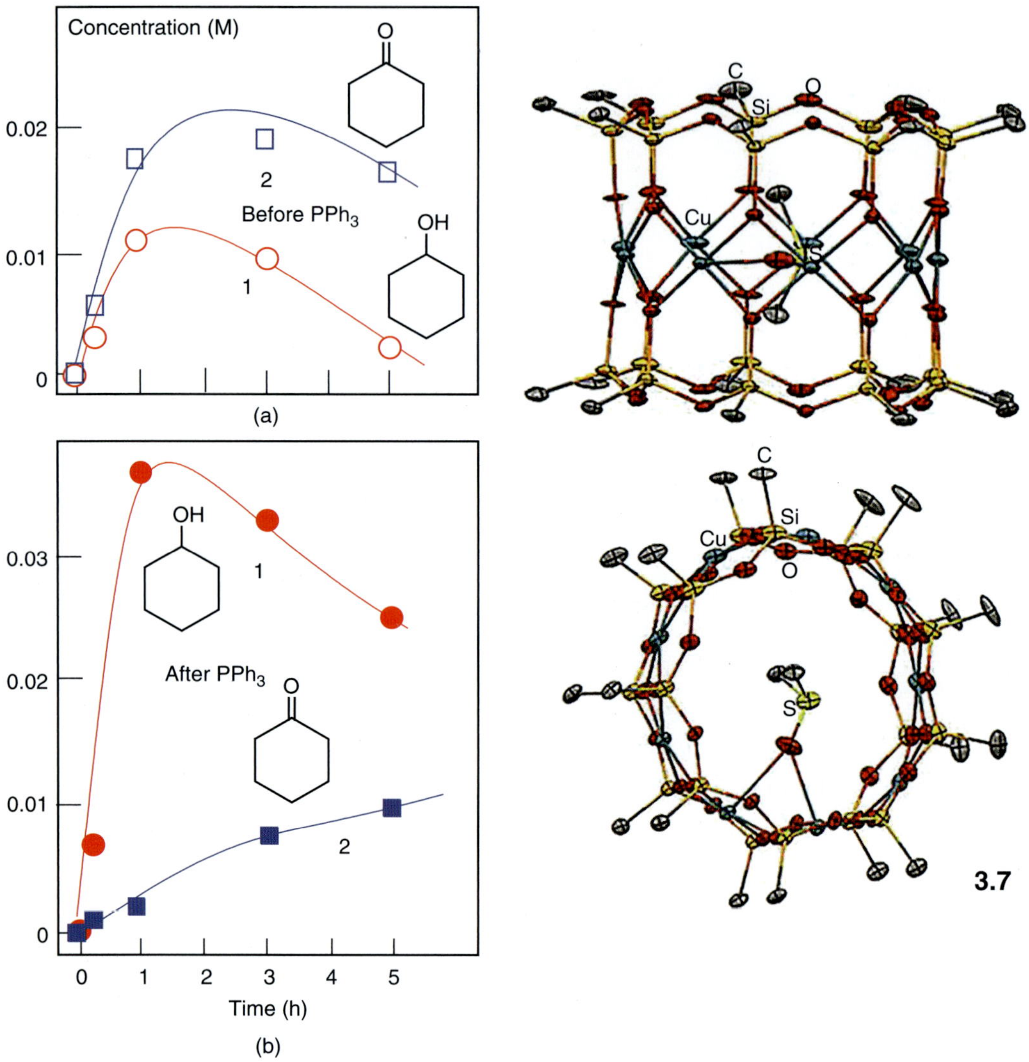

Figure 3.13 Left: Oxidation of cyclohexane (0.23 M) with H_2O_2 (50%, 2 M) catalyzed by the copper cage-like coppersilsesquioxanes nonanuclear $[(MeSiO_2)_{18}Cu_9]$ (**3.7**; 5×10^{-4} M) in the presence of HNO_3 (0.05 M). Solvent was acetonitrile (total volume of the reaction solution was 5 ml); 50 °C. Concentrations of cyclohexanol (curves 1) and cyclohexanone (curves 2) were measured both before and after reduction with PPh_3. Right: Molecular structure (two views) of complex **3.7**). Source: Adapted from Ref. [20c] with permission of American Chemical Society.

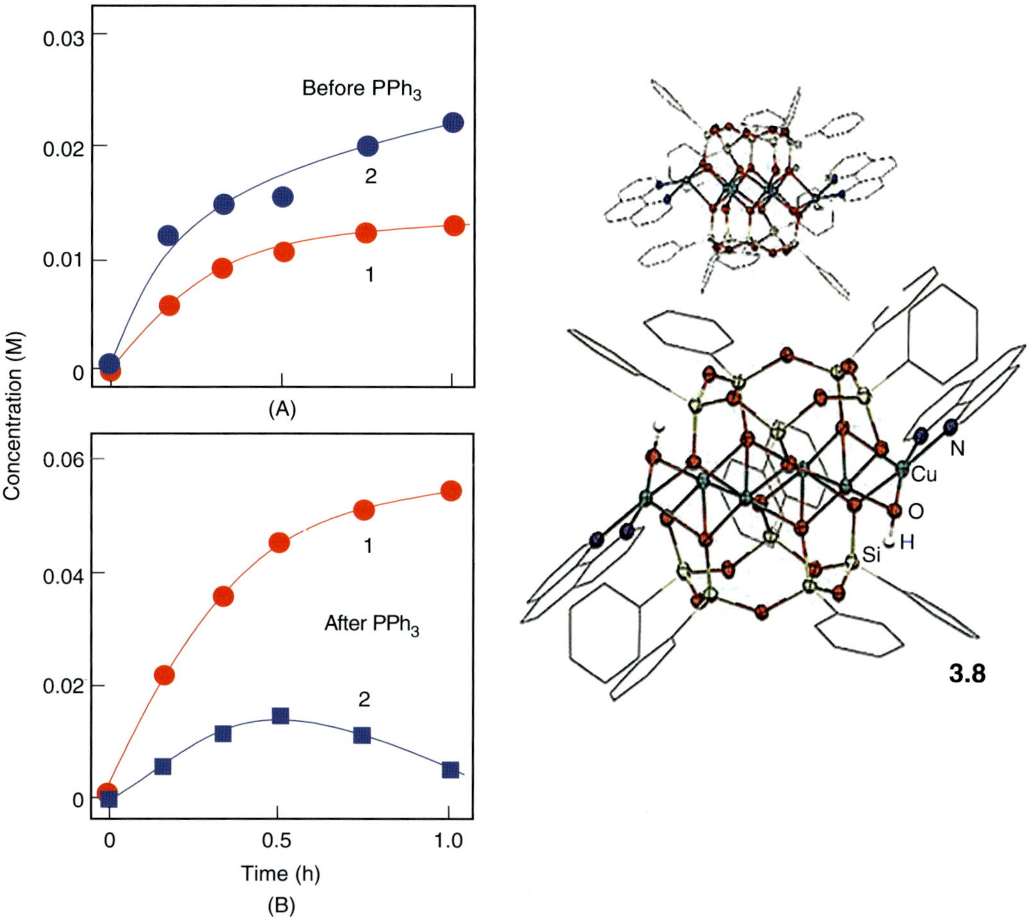

Figure 3.14 Left: Oxidation of cyclohexane (0.46 M) with H_2O_2 (50%, 1.5 M) catalyzed by hexacoppersilsesquioxane containing phenanthroline **3.8** (5×10^{-4} M) in the presence of HNO_3 (0.05 M) at 60 °C. Concentrations of cyclohexanol (curves 1) and cyclohexanone (curves 2) were measured both before (A) and after (B) reduction with PPh_3. Right: Two projections of the structure of cage hexacoppersilsesquioxane containing N-ligand phenanthroline $(PhSi)_{10}O_{20}(OH)_2Cu_6(phen)_2$ (complex **3.8**). Source: Adapted from Ref. [20d] with permission of American Chemical Society.

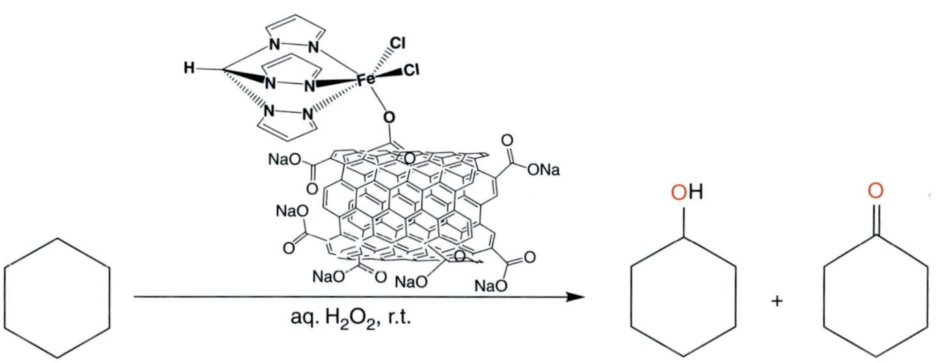

Scheme 6.3 Oxidation of cyclohexane to KA oil with hydrogen peroxide, catalyzed by $[FeCl_2\{\kappa^3\text{-}HC(pz)_3\}]$ supported at functionalized MWCNT.

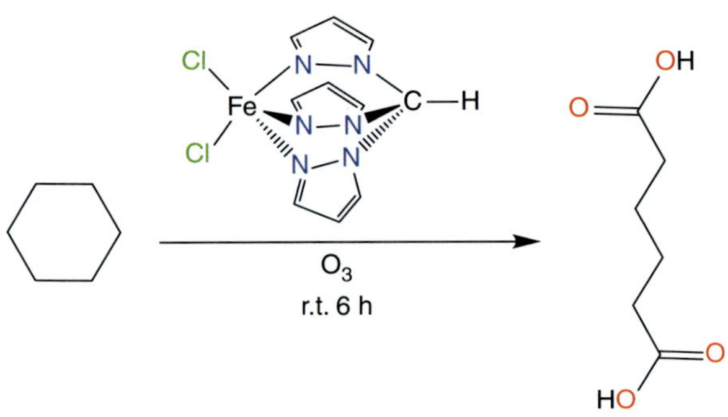

Scheme 6.4 Direct oxidation of cyclohexane to adipic acid with ozone catalyzed by $[FeCl_2\{\kappa^3\text{-}HC(pz)_3\}]$.

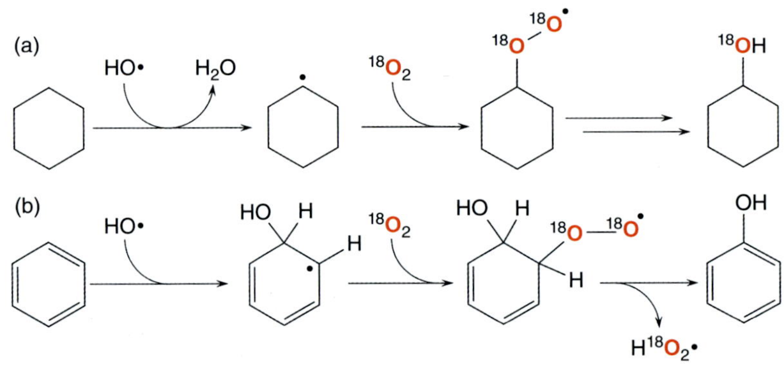

Scheme 7.3 Reaction of hydroxyl radical and dioxygen with a saturated hydrocarbon, cyclohexane (a), and with an aromatic one, benzene (b).

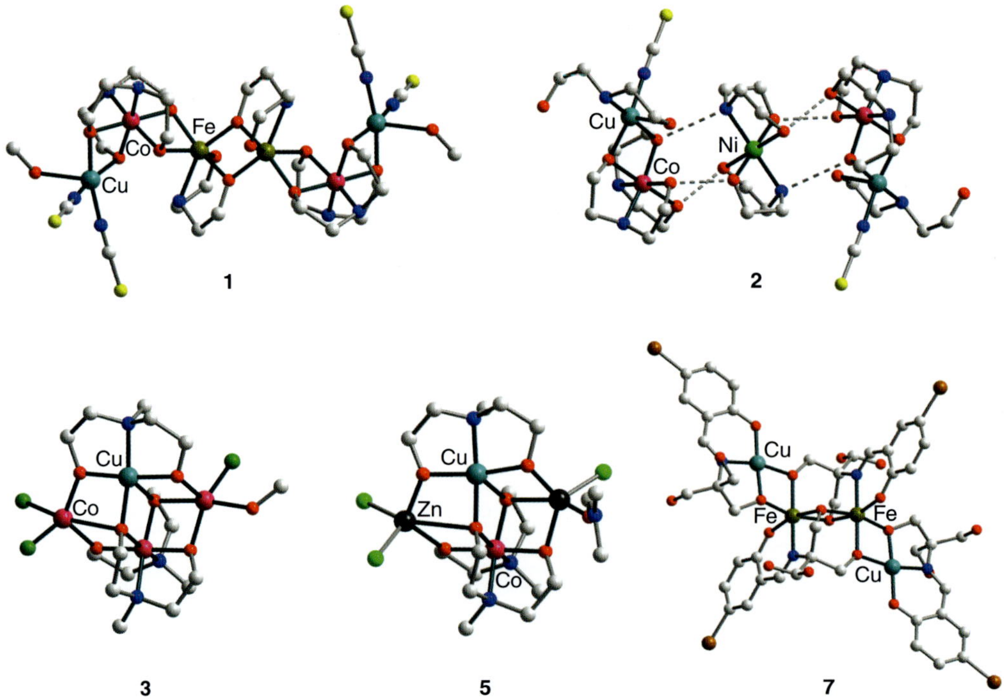

Figure 7.1 Molecular structures of heterometallic catalysts (the cores of **4** and **6** are identical to those of **5** and **7**, respectively, and thus are omitted for clarity) [28–31].

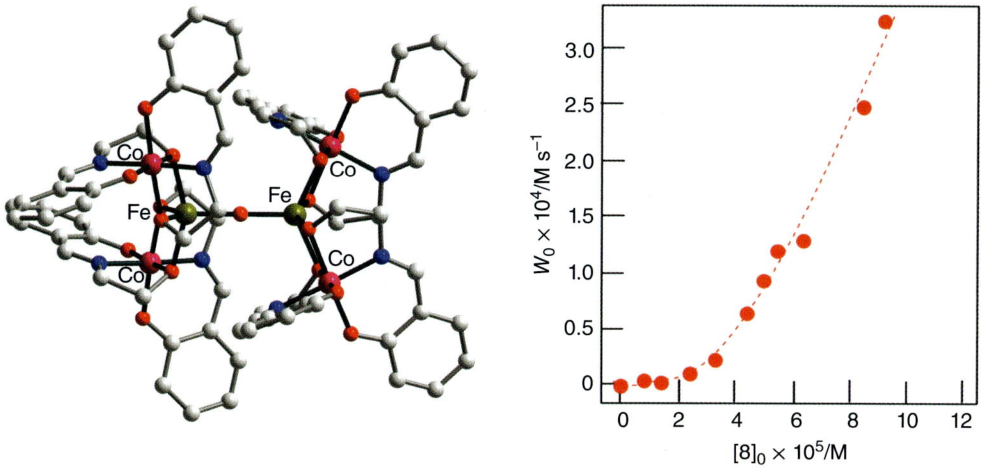

Figure 7.2 Left: molecular structure of the Co/Fe heterometallic complex **8**. Right: dependence of the initial reaction rate W_0 of cyclohexane oxidation to cyclohexanol and cyclohexanone on the concentration of catalyst **8** ($[C_6H_{12}]_0 = 0.37$ M, $[H_2O_2]_0 = 1$ M, $[HNO_3]_0 = 0.04$ M). Source: Adapted with permission from *Inorg. Chem.* 2012, *51*, 9110–9122. Copyright 2012 American Chemical Society [32].

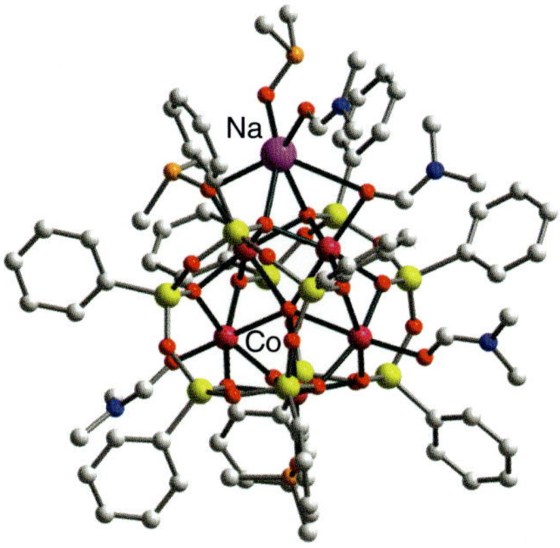

14

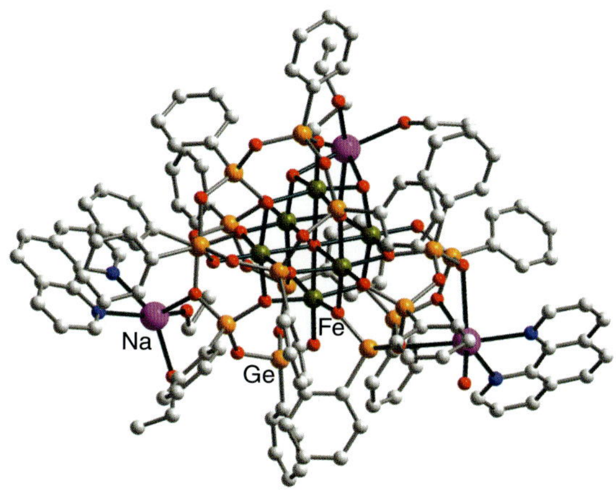

15

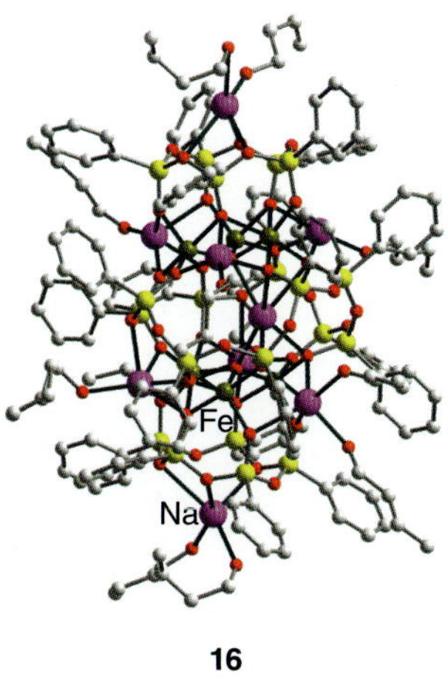

16

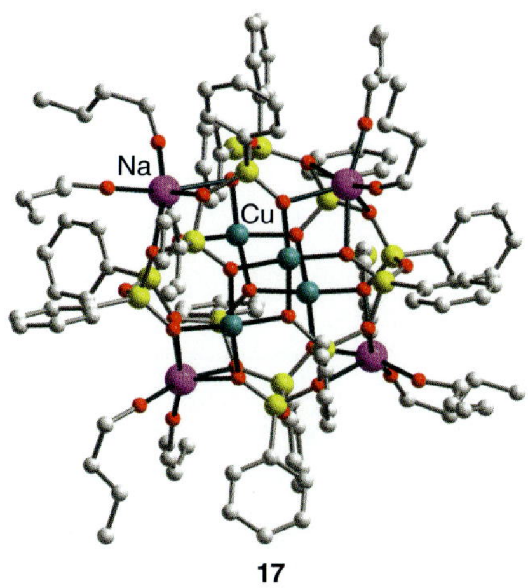

17

Figure 7.4 Structures of the Na/M complexes **14–17**, catalysts for cyclohexane and 1,2-dimethylcyclohexane oxidation [25, 26, 35, 41].

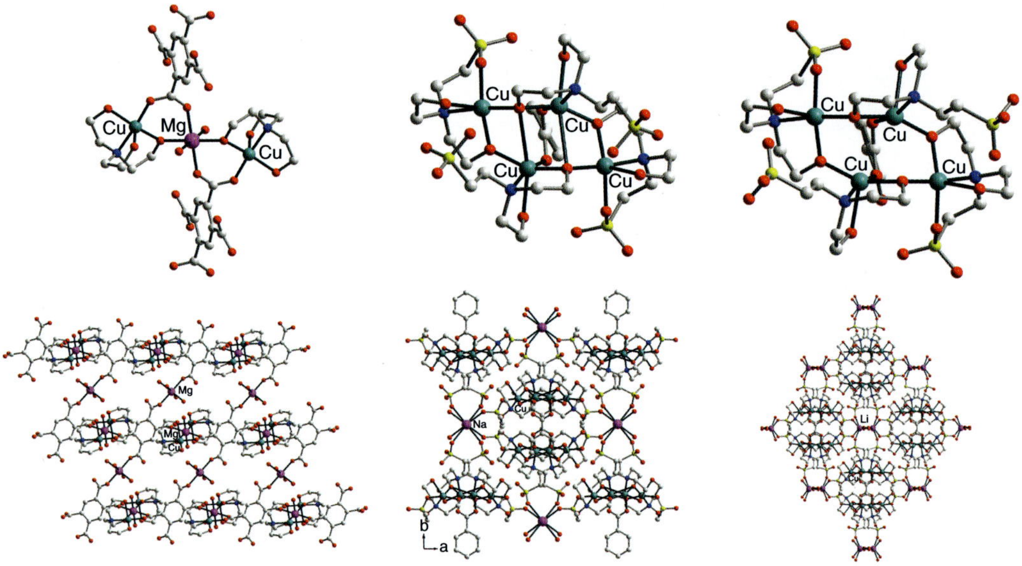

20 **21** **22**

Figure 7.8 Cores (top) and packing (bottom) of Cu/Mg (**20**), Cu/Na (**21**) and Cu/Li (**22**) heterometallic complexes [48–50].

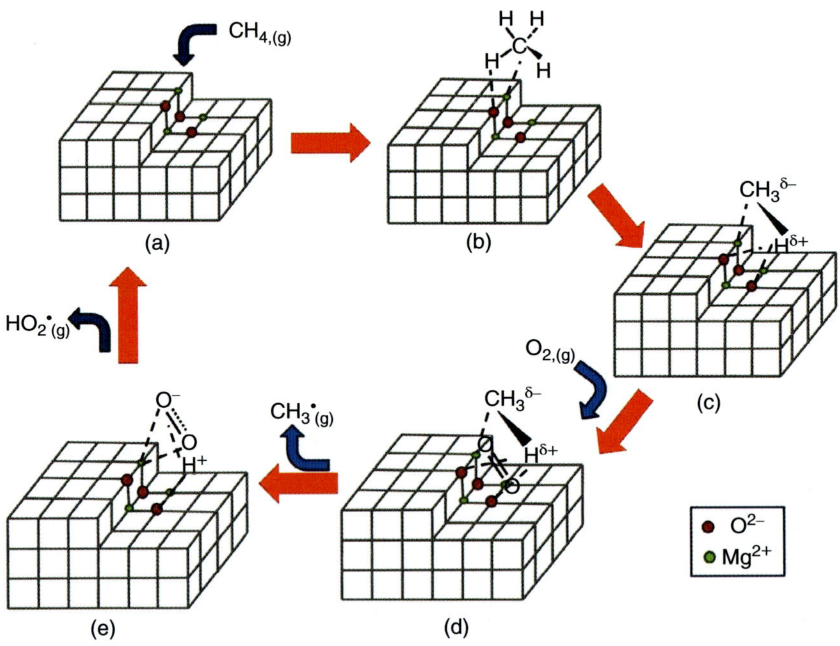

Figure 8.1 Mechanism of heterolytic C—H activation of methane on MgO defects. Source: Reproduced with permission from Schwach et al. [27].

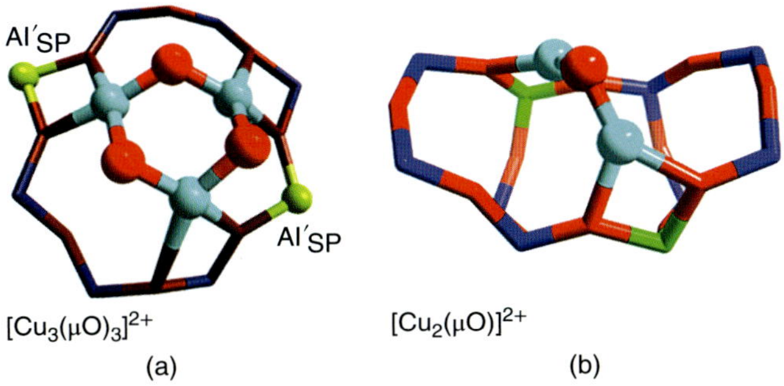

Figure 8.2 DFT model cluster of (a) $Cu_3(\mu\text{-}O)_3$ species located on the 8-MR of MOR [99], and (b) $Cu_2(\mu\text{-}O)$ species located on the γ cationic site of ZSM-5. Source: Adapted with permission from Li et al. [123].

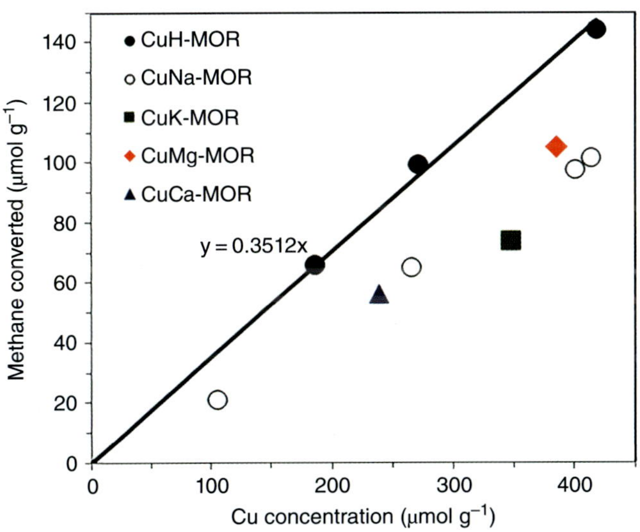

Figure 8.3 Methane converted versus Cu concentration for Cu-exchanged MOR with various co-cations. * The slope of 0.35 indicates that c. 3 Cu atoms are involved in the oxidation of one methane molecule. Source: Adapted from Grundner et al. [101].

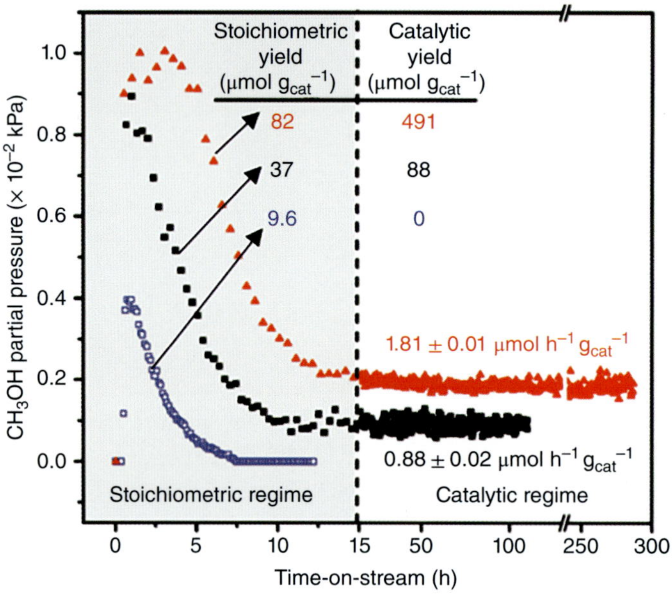

Figure 8.4 Methanol yield under stoichiometric and catalytic regime for Cu-Na-ZSM-5 (full squares), Cu-H-ZSM-5 (triangles) when treated at 210 °C with a gas stream of CH_4 (98.1 kPa), H_2O (3.2 kPa) and O_2 (0.0025). The curve with open squares is a CuNa-ZSM-5 in same mixture but substituting CH_4 by He. Prior to contact with the wet stream, the samples were contacted with CH_4 at 210 °C for 30 minutes. Source: Reproduced from Narsimhan et al. [96].

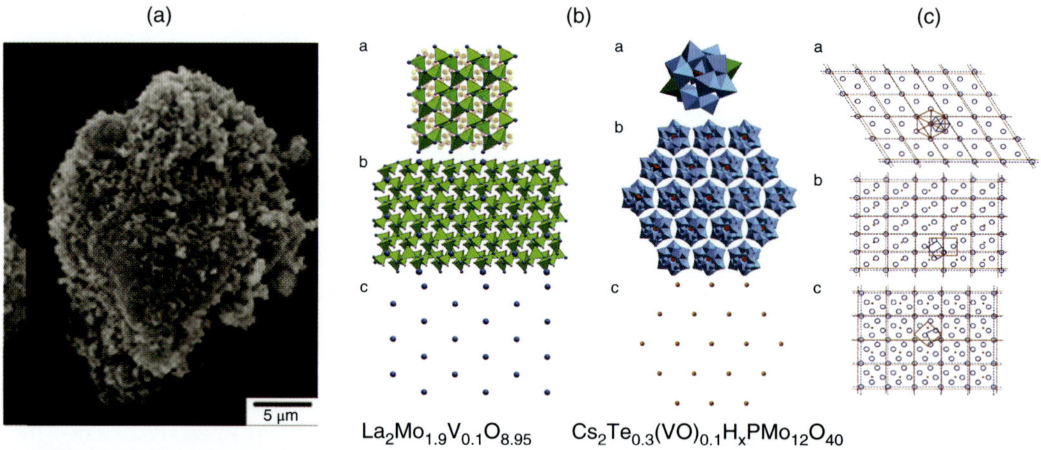

Figure 9.3 Study of the $Cs_2Te_{0.3}(VO)_{0.1}H_xPMo_{12}O_{40}$/$La_2Mo_{1.9}V_{0.1}O_{8.95}$ (CTV/LMV) catalyst (a) MEB image; (b) projection of the cationic arrangement of LMV along (111) with isolated La hexagonal array and anionic arrangement of CTV in the (111) plane with isolated P hexagonal array; (c) fits between the two arrays along different cubic directions of both lattices (from [22]).

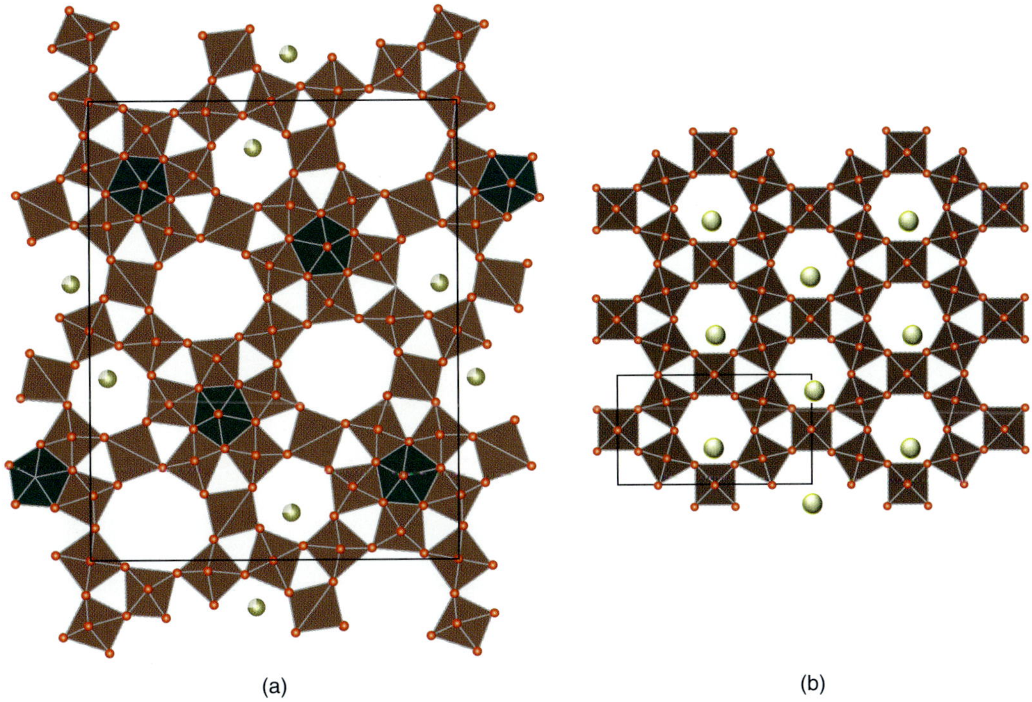

Figure 9.6 The structure of the (a) M1 and (b) M2 phase. Adapted from [125, 128].

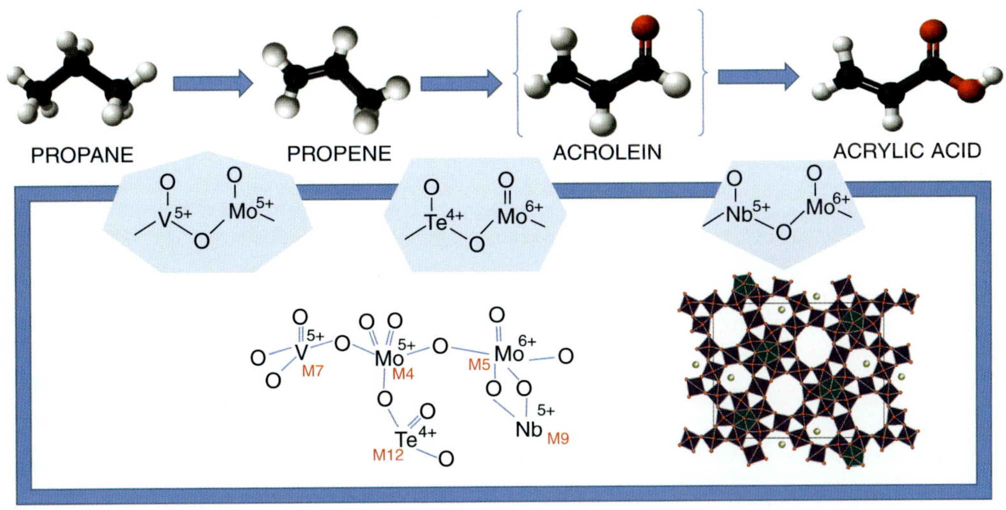

Figure 9.7 The mechanism proposed for propane to acrylic acid.

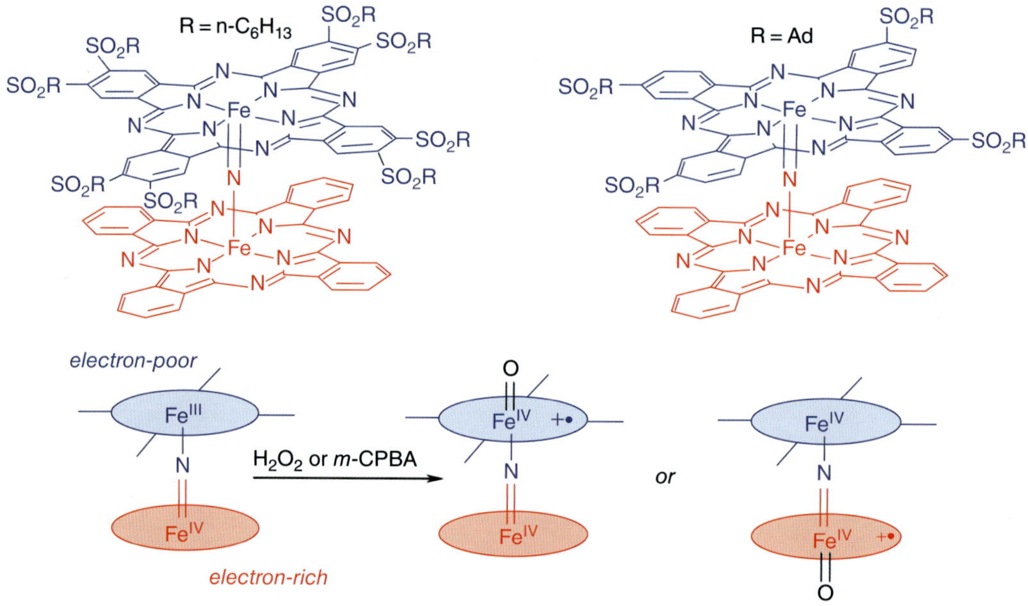

Figure 10.5 Formation of isomeric-oxo species using heteroleptic μ-nitrido diiron phthalocyanine platform.

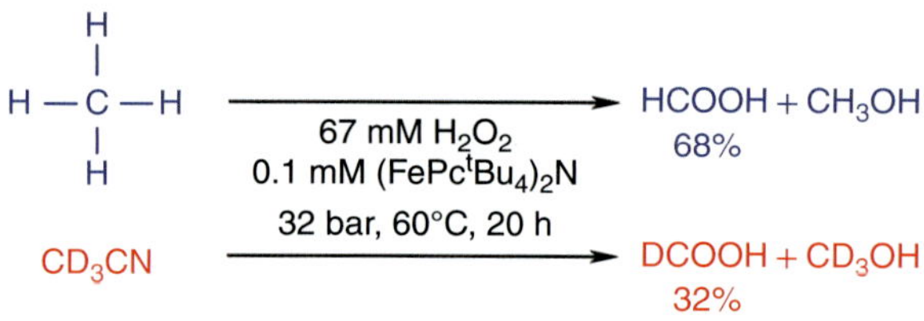

Scheme 10.1 Oxidation of CH_4 by $(FePc^tBu_4)_2N - H_2O_2$ system in CD_3CN.

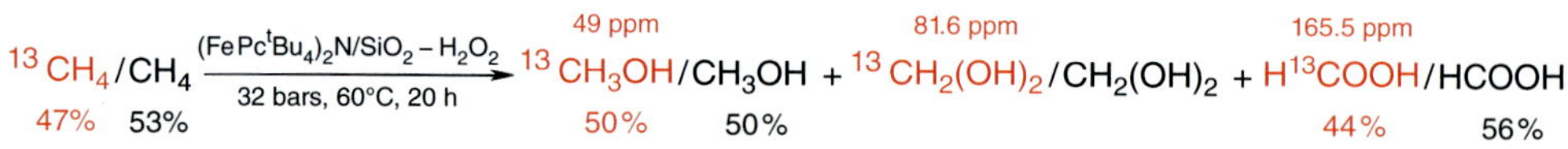

Scheme 10.2 Labeling study of the heterogeneous oxidation of methane by $(FePc^tBu_4)_2N/SiO_2—H_2O_2$ system.

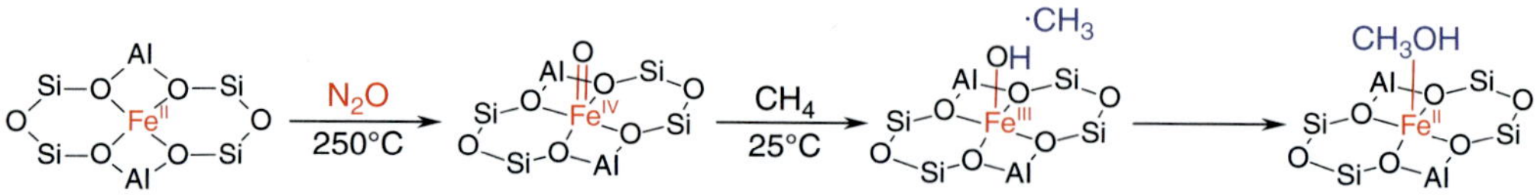

Scheme 10.3 Proposed structures of the zeolitic α-Fe sites and active α-Fe^{IV}=O site capable of reacting with methane.

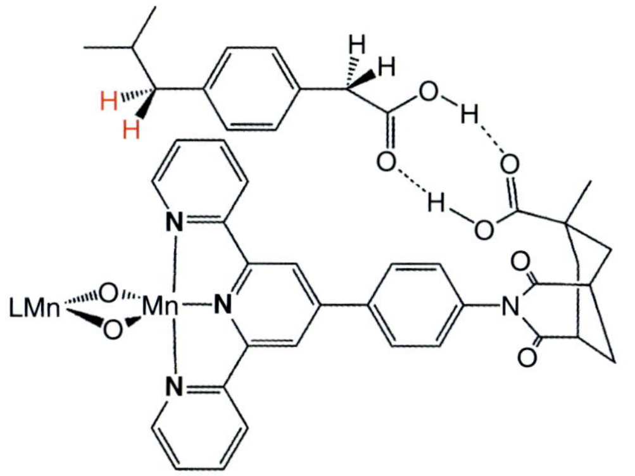

Scheme 11.12 Hydrogen bond interactions between a remote site on the ligand and the ibuprofen substrate direct manganese-catalyzed oxidation toward the highlighted benzylic hydrogens. L represents a second equivalent of the ligand with the targeting domain.

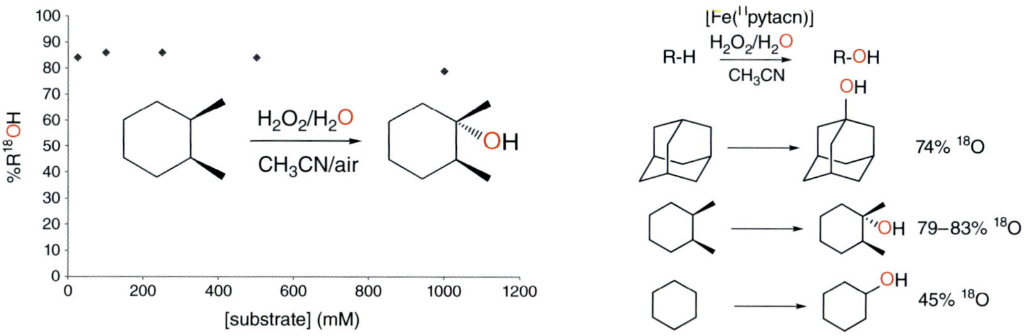

Scheme 13.6 Left, lack of dependence between substrate concentration and ^{18}O incorporation from $H_2{}^{18}O$ in catalytic oxidations with [Fe(^{H}pytacn)]. Right, ^{18}O incorporation for different substrates.

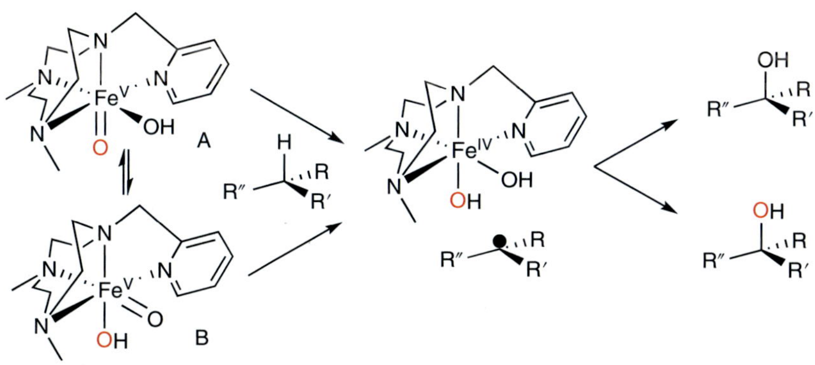

Scheme 13.7 Alternative rebound mechanism for iron catalyst [Fe(^{H}pytacn)], but that can be extended to iron catalysts containing two labile sites.

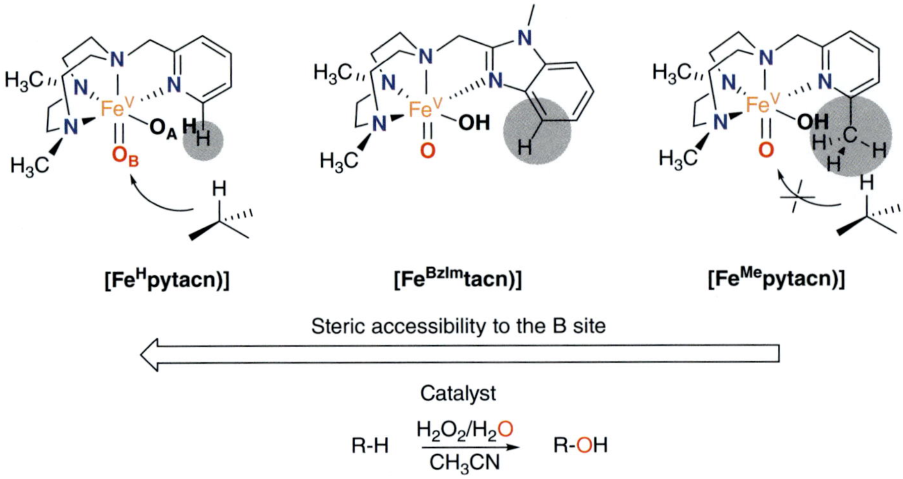

Substrate	[Fe^{BzIm}tacn)]	[Fe^{H}pytacn)]	[Fe^{Me}pytacn)]
c-Hexane	48	45	11
c-Hexane-d$_{12}$	48	40	–
c-Octane	41	44	–
cis-DMCH	26	79	2
Adamantane	28	74	3

Scheme 13.9 Impact of sterics in the extend of water incorporation into hydroxylated products.

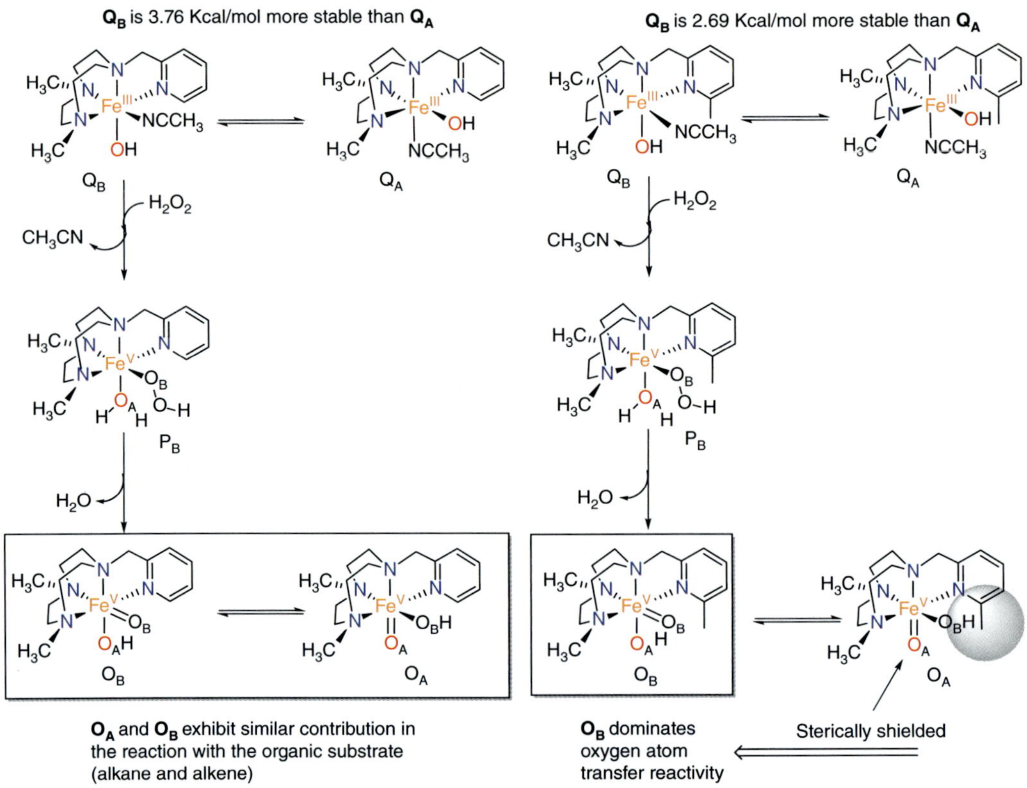

Scheme 13.10 Schematic diagram of the formation of the Fe^{V}(O)(OH) species in [Fe(^{x}pytacn)] catalysts.

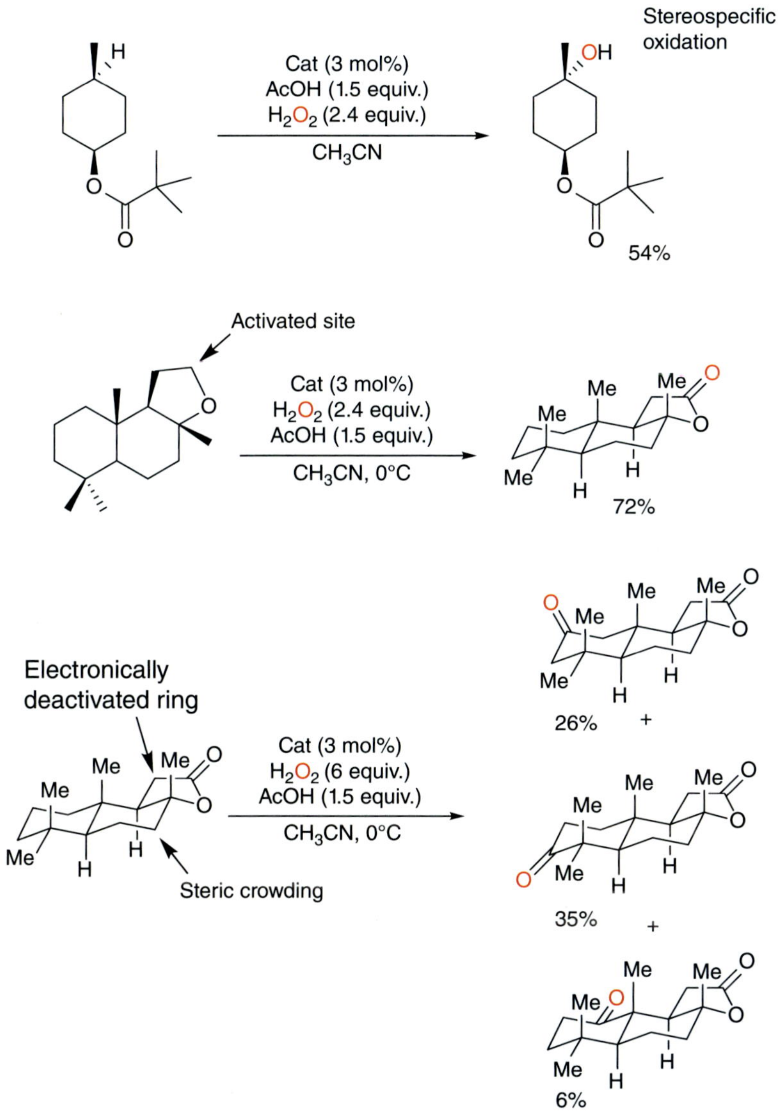

Scheme 13.12 Selected C—H oxidation reactions with the $[Fe(OTf)_2(^{Me}pytacn)]$ catalyst under substrate limiting conditions.

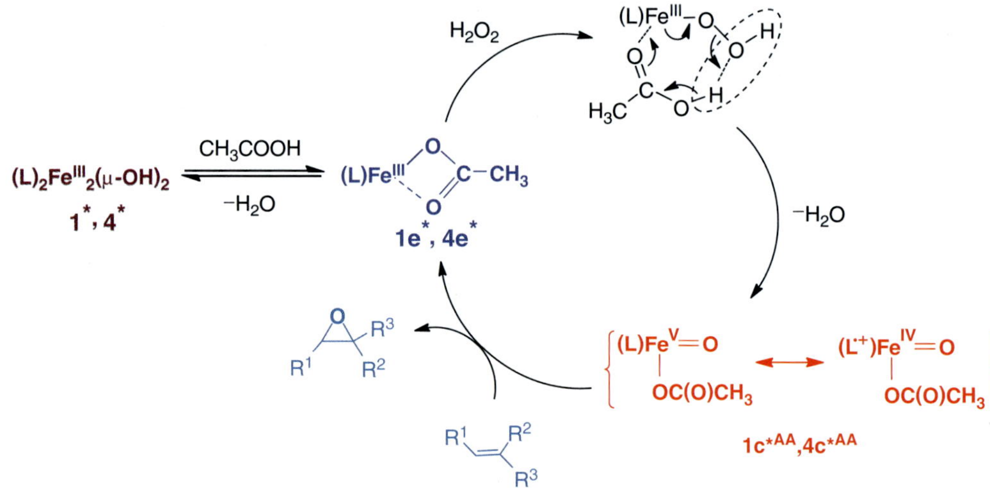

Scheme 14.1 Proposed mechanism of olefin epoxidation by catalyst systems **1***/H_2O_2/CH_3COOH and **4***/H_2O_2/CH_3COOH.

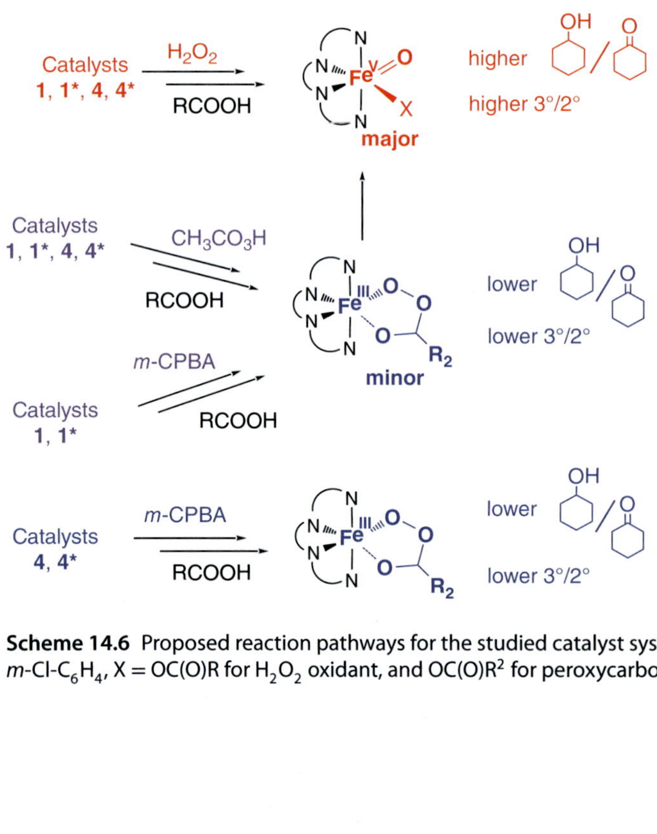

Scheme 14.6 Proposed reaction pathways for the studied catalyst systems (RCOOH = AA or EHA, R^2 = CH_3 or *m*-Cl-C_6H_4, X = OC(O)R for H_2O_2 oxidant, and OC(O)R^2 for peroxycarboxylic acid oxidant).

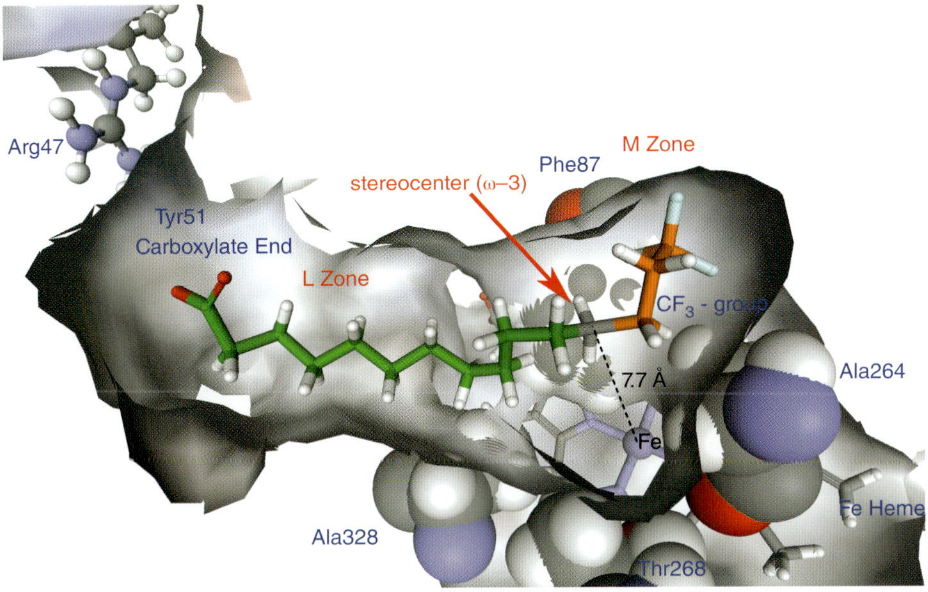

Figure 15.2 Three-dimensional modeling depicting the substrate binding in the active site of cytochrome P450 BM3 (PDB ID: 1FAG, modified by Accelrys®2.0). This model illustrates the encapsulation of the substrate, 15,15,15-trifluoropentadecanoic acid, instead of palmitoleic acid in the original structure. Source: Reproduced with permission from Ref. [56]. Copyright 2013 John Wiley & Sons.

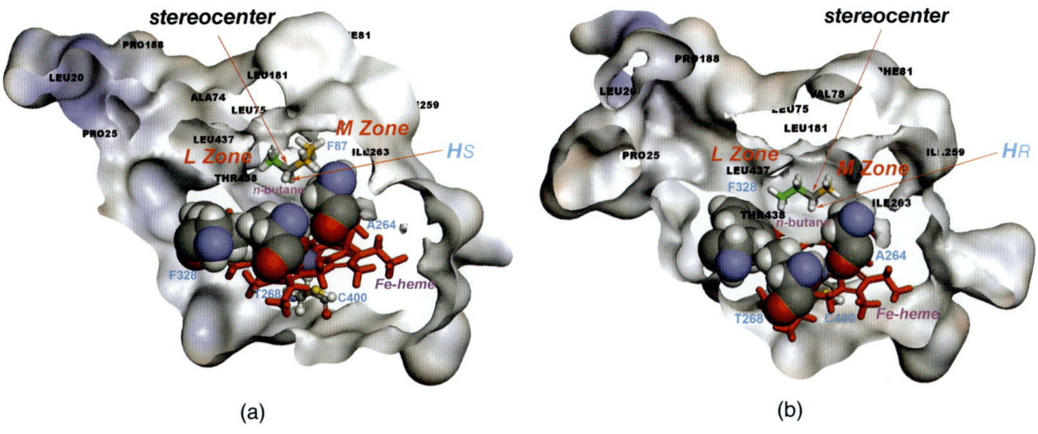

Figure 15.5 Two proposed conformational states of *n*-butane in the hydrophobic pocket of **BM3 PF** are indicated to lead to the oxidation of (a) C2-H_S and (b) C2-H_R bonds, respectively, by the iron heme center (reproduction of PDB: 1JPZ [53] by Discovery Studio 4.5, Dassault Systèmes BIOVIA). Source: Reproduced with permission from Ref. [97]. Copyright 2017 John Wiley & Sons.

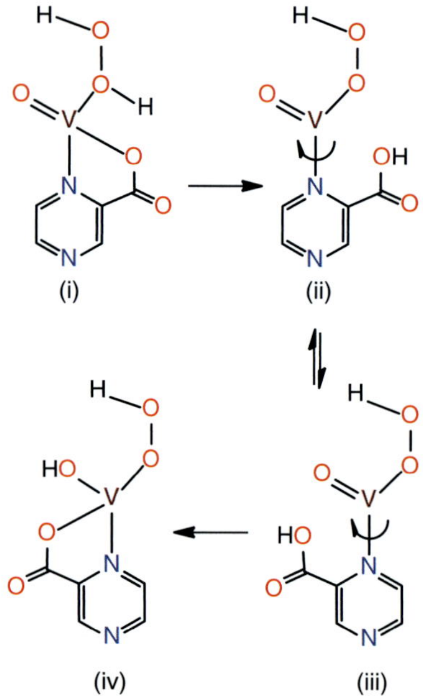

Scheme 16.2 PCA-assisted proton migration from a coordinated H_2O_2 molecule to an oxido ligand ("robot's arm mechanism") [16, 18, 22, 24, 32, 33].

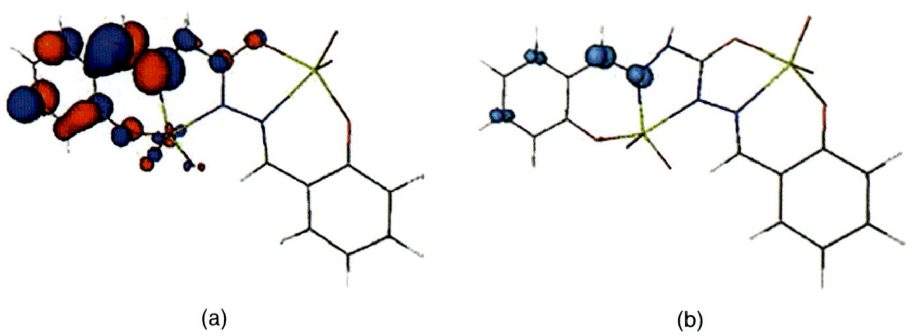

Figure 16.1 (a) LUMO of ^{1}[**7**]$^-$ in vacuum; (b) spin density of ^{2}[**7**]$^{2-}$ in vacuum [37]. With permission of ACS (http://pubs.acs.org/doi/abs/10.1021%2Facs.inorgchem.6b01011).

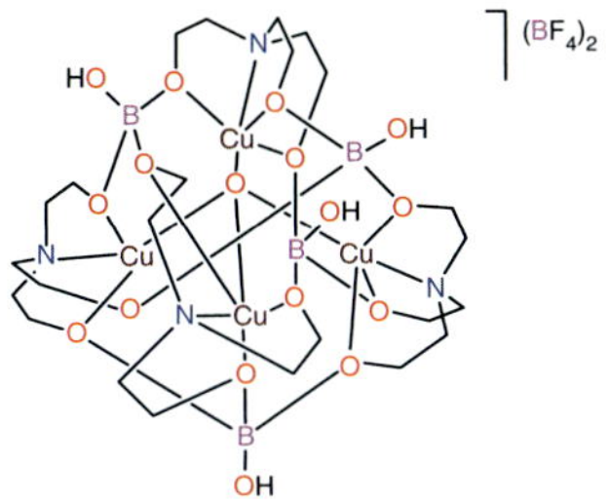

34

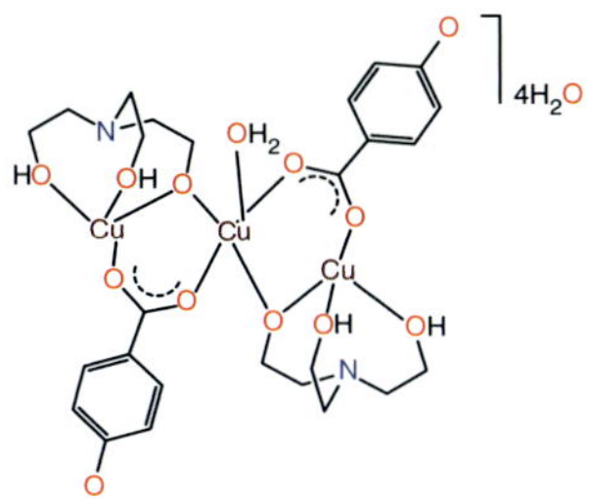

35

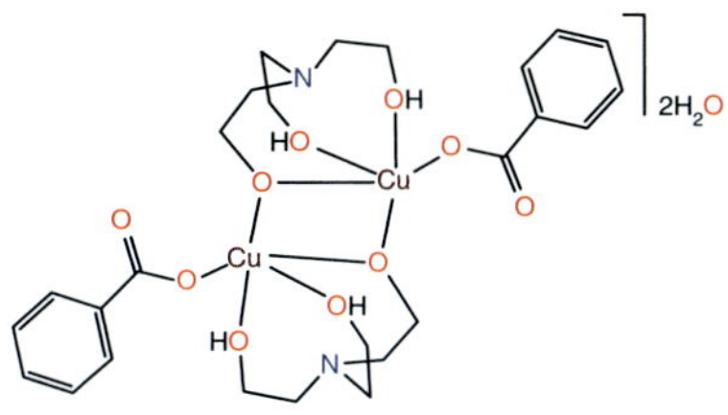

36

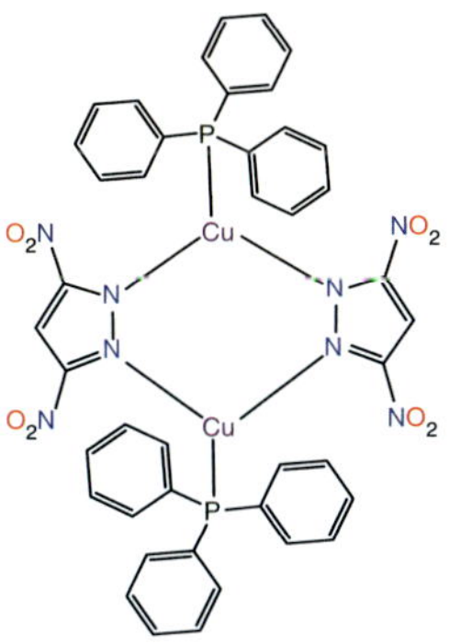

38

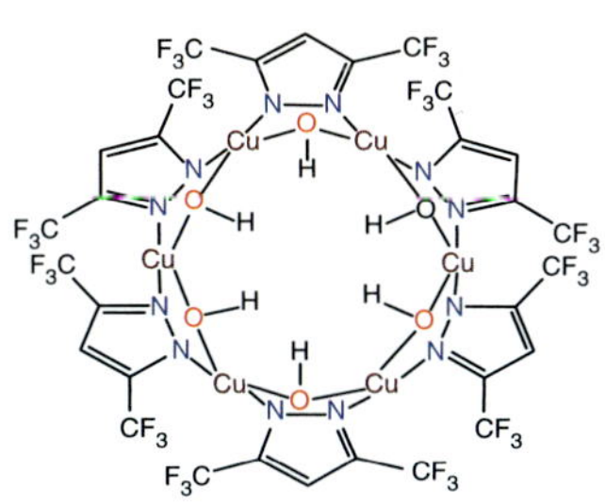

39

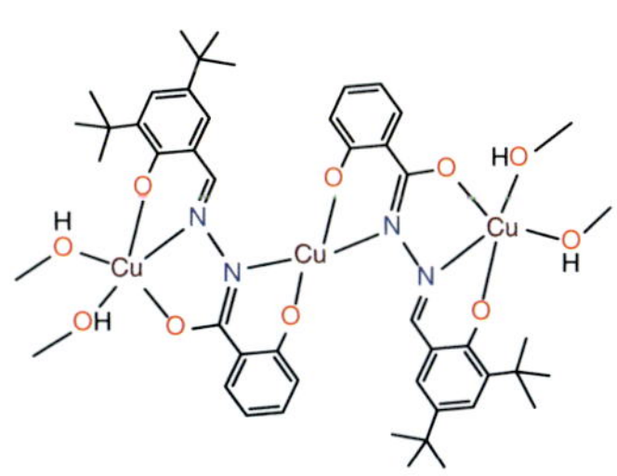

43

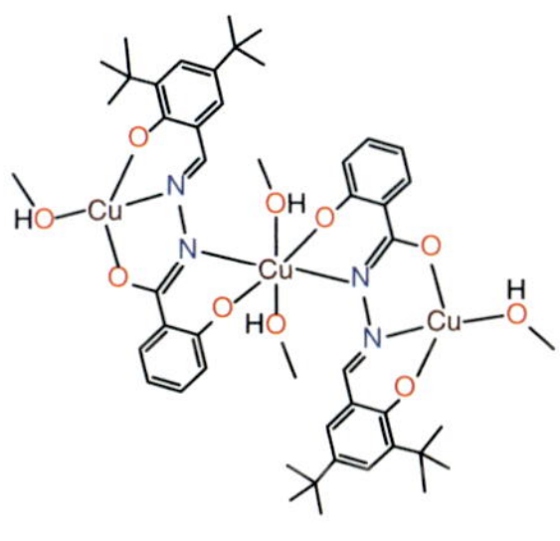

44

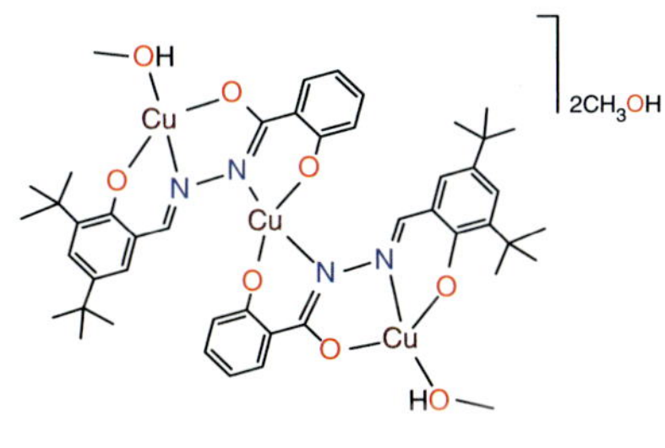

45

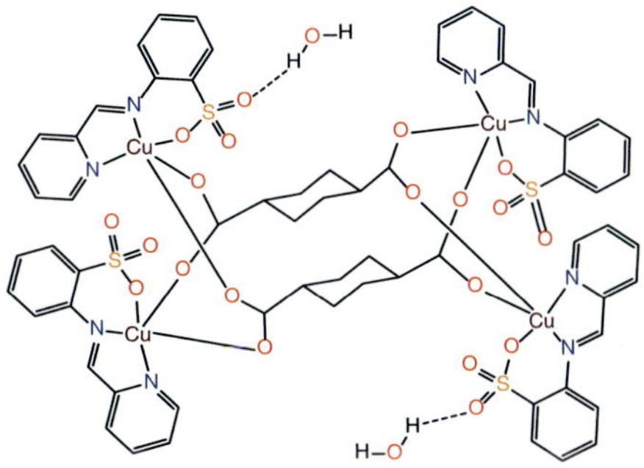

46

Chart 16.4

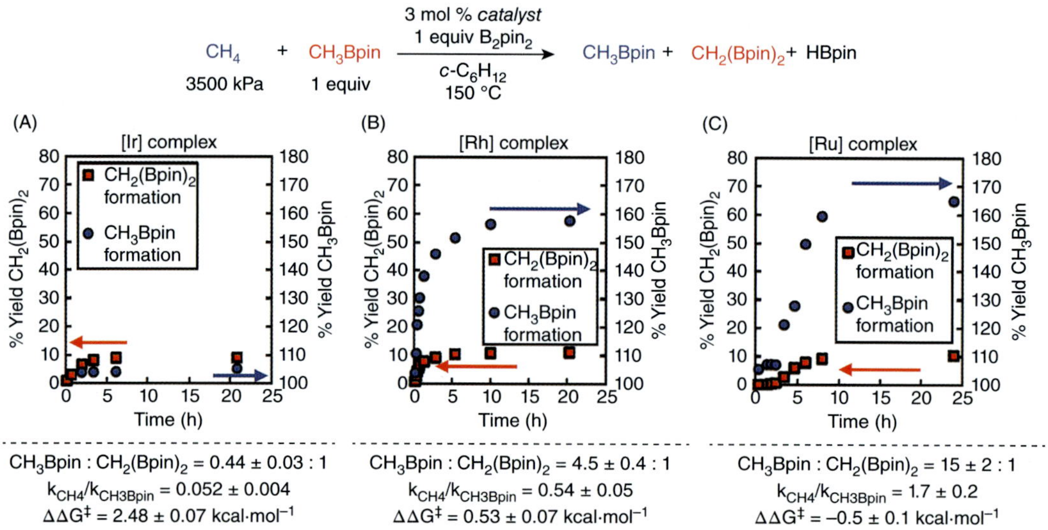

Figure 17.1 Reaction time profiles for three catalyst Ir (A), Rh (B), and Ru (C). The reaction time profiles show the formation of monoborylated **1** (red squares) and diborylated **2** (blue circles), from which $kCH_4/k\mathbf{1}$ and $\Delta\Delta G^{\ddagger}$ values can be inferred [108].

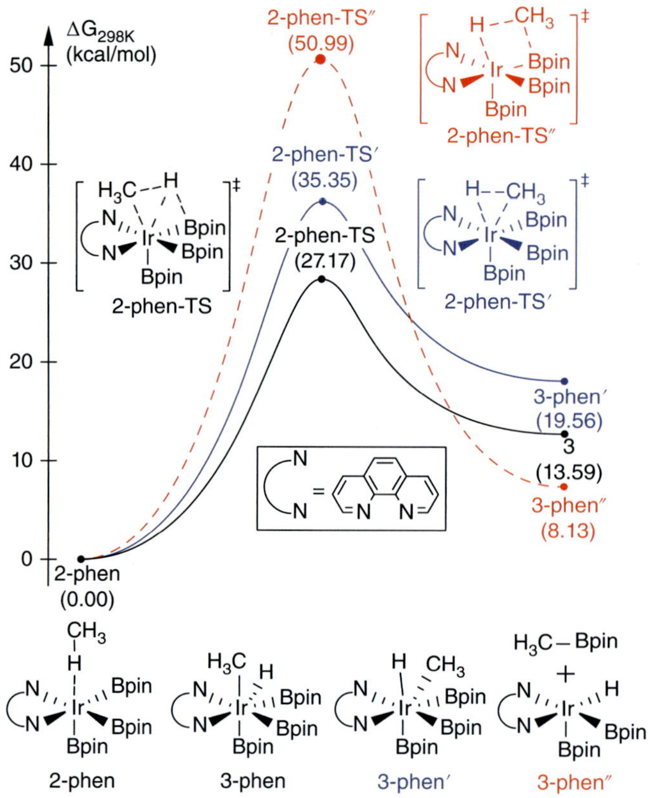

Figure 17.3 The calculated energy profile at the C—H bond cleavage step.

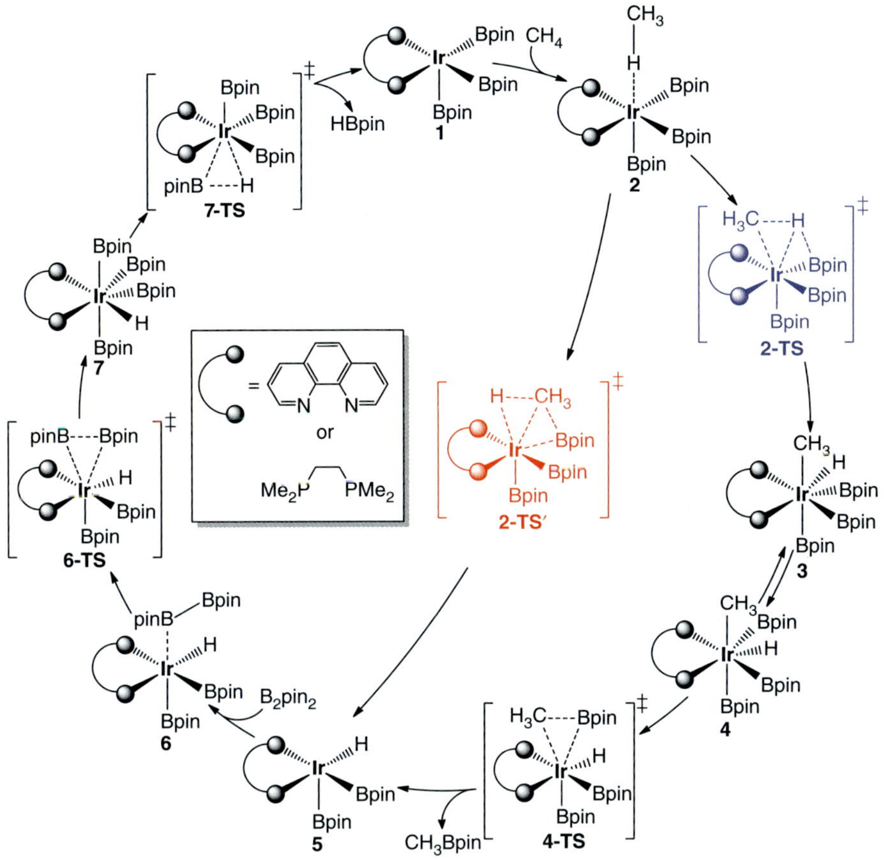

Figure 17.5 The proposed mechanism.

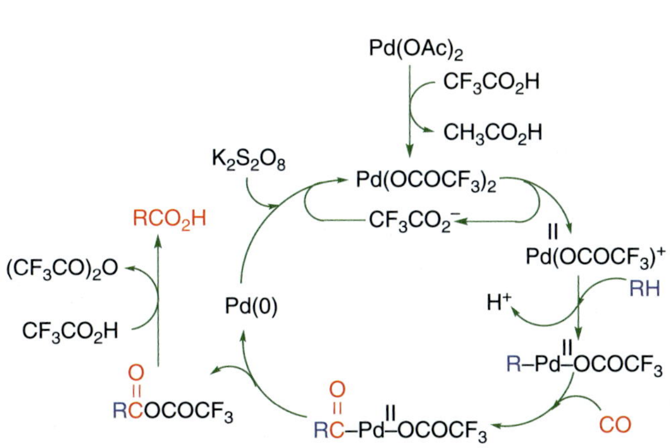

Scheme 18.16 Mechanism proposed for the Pd-catalyzed carbonylation of alkanes [46].

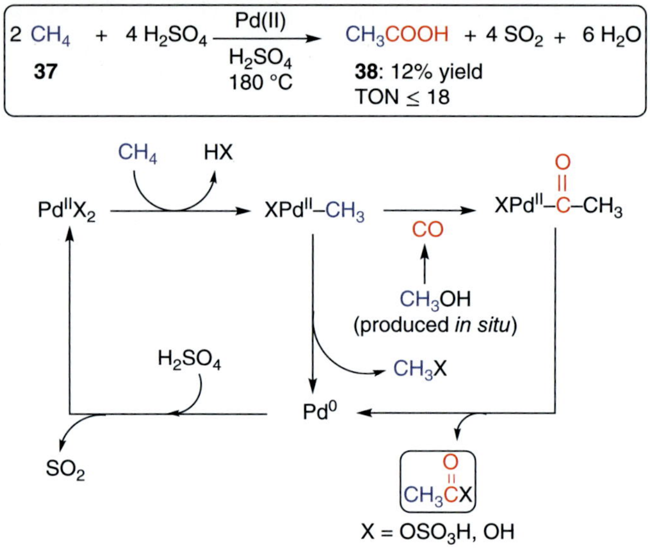

Scheme 18.17 Palladium-catalyzed oxidative condensation of methane and the mechanism proposed [48].

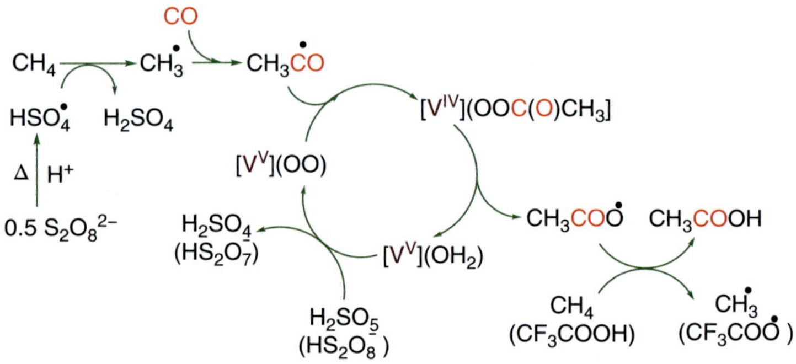

Scheme 18.24 Proposed mechanism for the vanadium-catalyzed carboxylation of methane [61].

$$RH + CO + H_2O + S_2O_8^{2-} \xrightarrow[\text{Catalyst}]{H_2O\,/\,MeCN} RCOOH + 2\,HSO_4^-$$

Scheme 18.28 The hydrocarboxylation of alkanes in a water-acetonitrile solution [69].

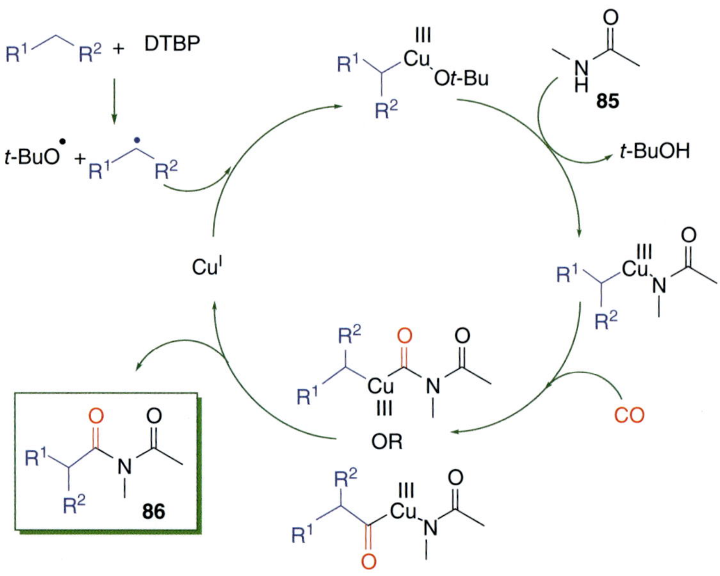

Scheme 18.34 Plausible mechanism for the copper-catalyzed carbonylative synthesis of aliphatic imides from alkanes and primary amines [82].

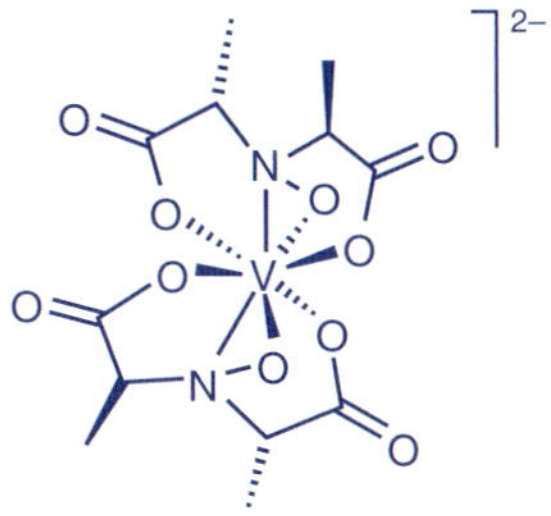

54: Amavadin

Figure 18.1 Amavadin (**54**), a non-oxido vanadium(IV) complex found in the fungus *Amanita muscaria*, seen in the figure, a poisonous mushroom with psychoactive and insecticidal properties. Amavadin catalyzes a few reactions of alkanes, including carboxylation and it has been used as a model for the development of other carboxylation catalysts.

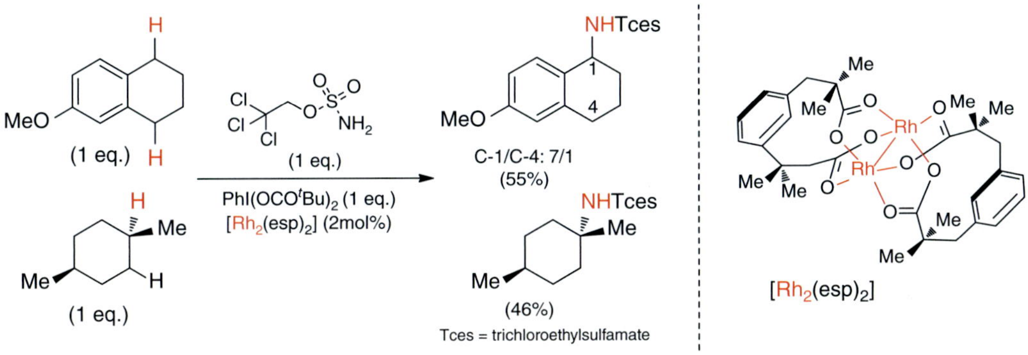

Scheme 19.17 Rhodium-catalyzed intermolecular C—H amination .[45,46]

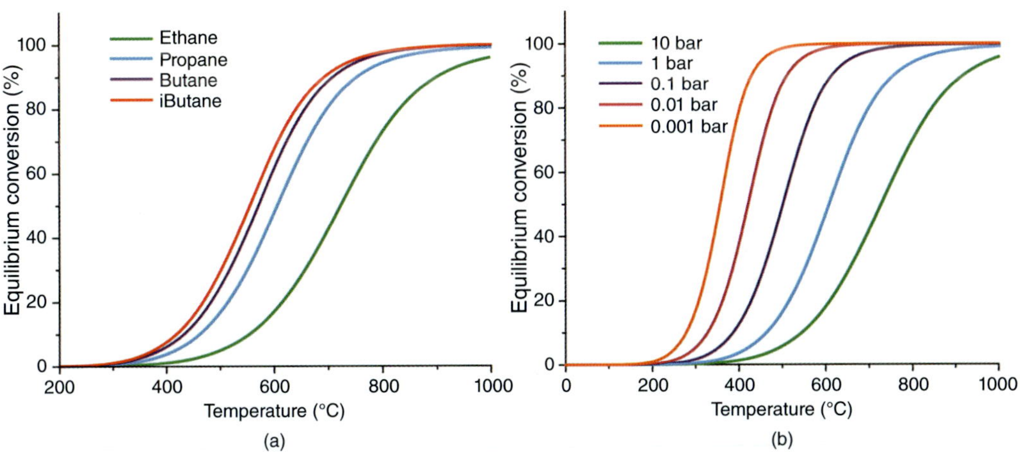

Figure 20.1 (a) Equilibrium conversion of ethane, propane, *n*-butane, and *i*-butane to the corresponding alkenes plus hydrogen as a function of temperature; the degree of conversion is greater for more substituted alkanes. (b) Equilibrium conversion of propane to propene plus hydrogen as a function of temperature and pressure; the degree of conversion is greater at lower pressures. Source: Reprinted with permission from reference [4], copyright 2014, American Chemical Society.

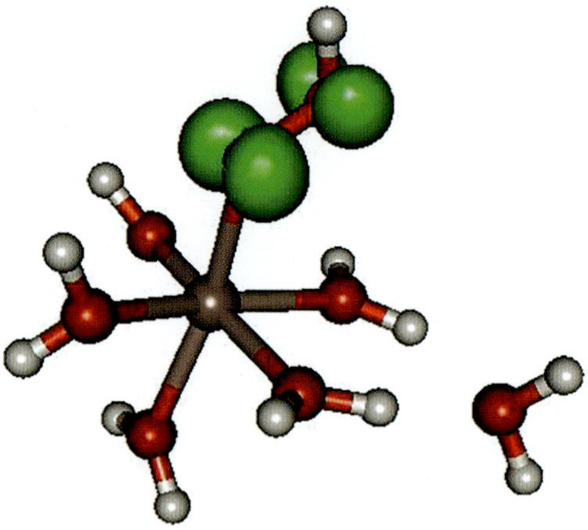

Figure 22.4 Spin density distribution in complex *cis*-$[Ga(H_2O)_4(OOH)(OH)](H_2O)^{2+}$.

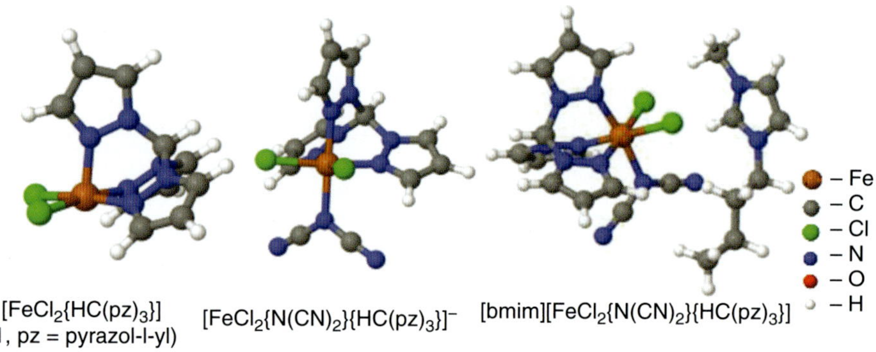

Figure 24.7 Calculated equilibrium structures for $[FeCl_2\{HC(pz)_3\}]$ **1** (pz = pyrazol-1-yl), $[FeCl_2\{N(CN)_2\}\{HC(pz)_3\}]^-$ and [bmim]$[FeCl_2\{N(CN)_2\}\{HC(pz)_3\}]$. Source: Reproduced with permission from ref. [79].

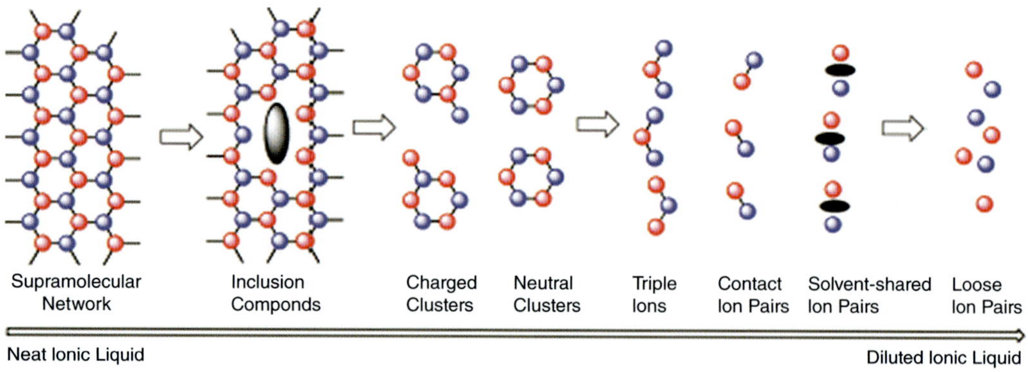

Scheme 24.2 Simplified 2D representation of the structure of a neat ionic liquid to its infinite dilution in the presence of other organic compounds. Most of these structures have been experimentally already observed (red spheres = anions, blue spheres = cations, black spots = solvent molecules; the lines represent the hydrogen bonds and/or other weaker interactions). Source: Reproduced with permission from ref. [50a].

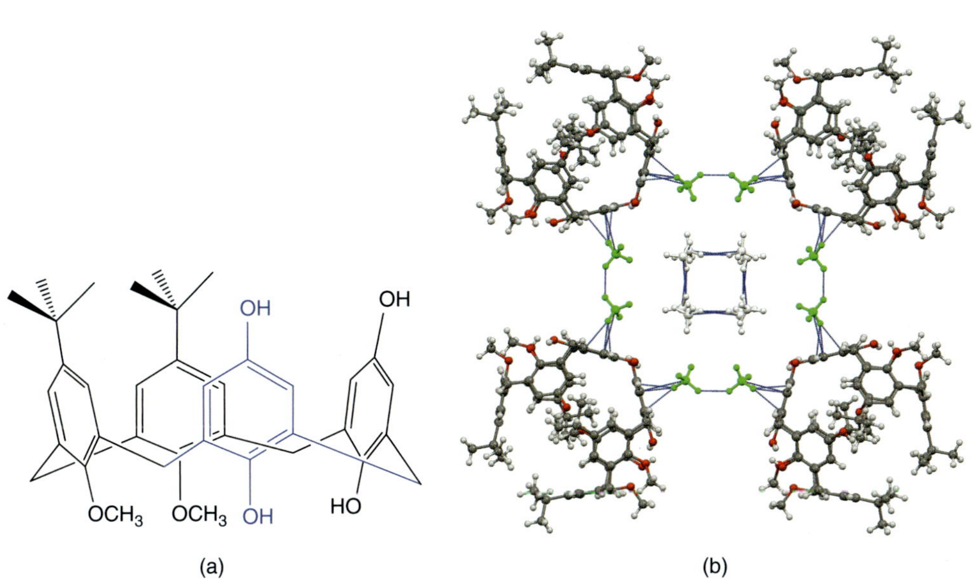

Figure 25.1 1,2-Dimethoxy-*p-tert*-butylcalix[4]dihydroquinone (a) and its associate with methane (b) [55].

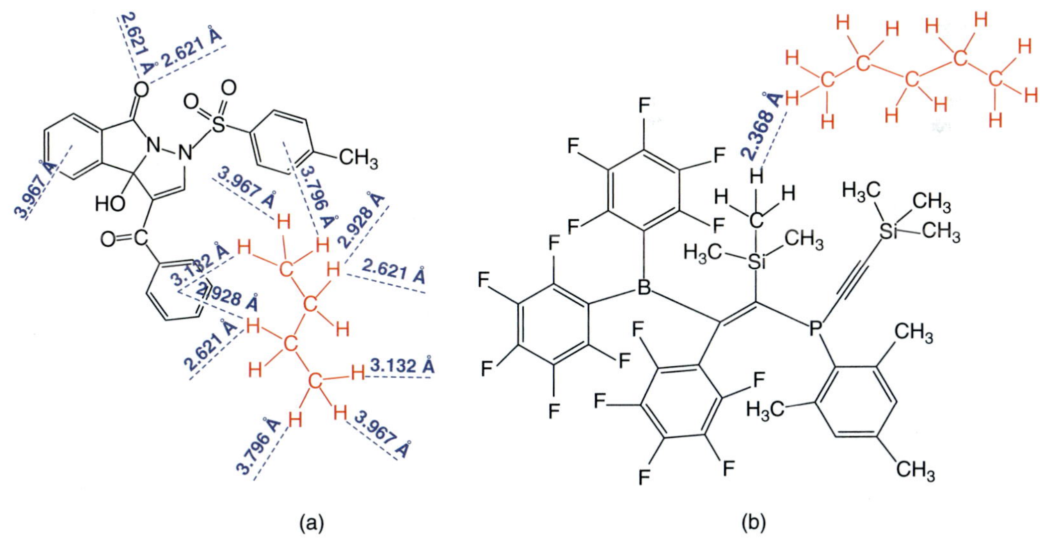

Figure 25.3 (a) Multiple hydrogen bonds between 3-benzoyl-3a-hydroxy-1-tosyl-1H-pyrazolo[5,1-a]isoindol-8(3aH)-one and butane [58]. (b) Dihydrogen van der Waals interactions in the associate of (*Z*)-(2-(*bis*(perfluorophenyl)boryl)-2-(perfluoro-phenyl)-1-(trimethylsilyl)vinyl)(mesityl)((tri-methylsilyl)ethynyl) phosphine with pentane [59].

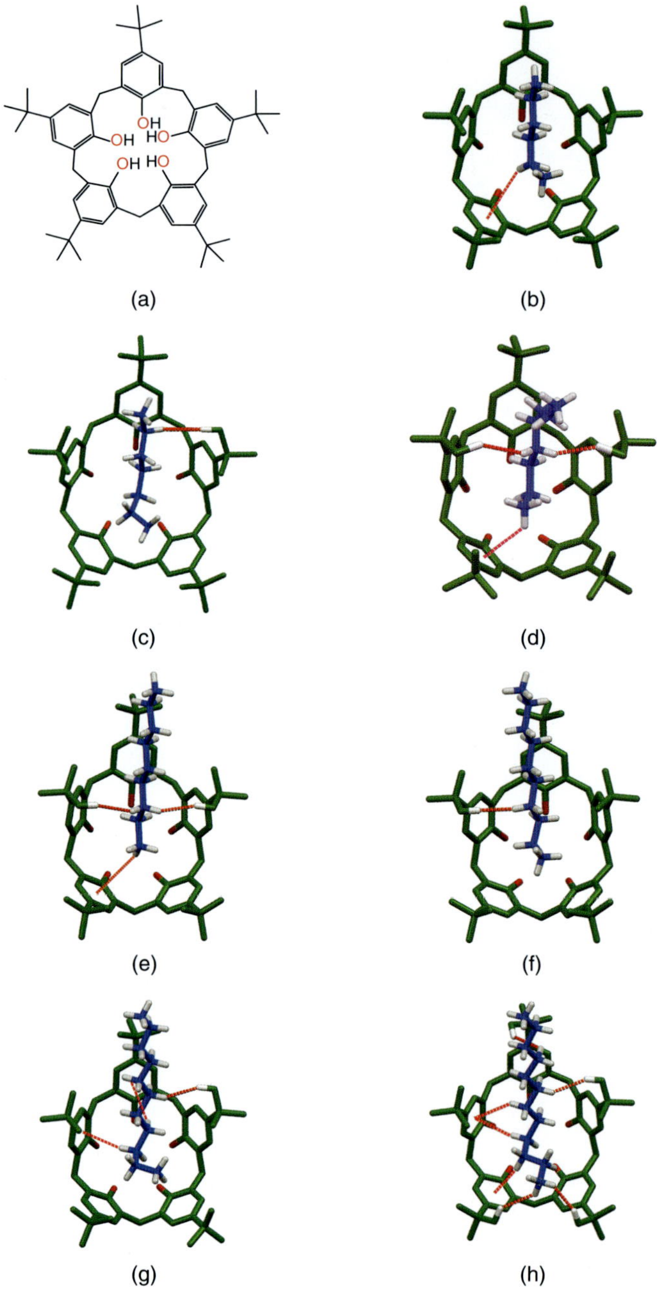

Figure 25.4 (a) *p-tert*-butylcalix[5]arene: (b) Host-guest complexes of *p-tert*-butylcalix[5]arene with hexane; (c) heptane; (d) octane; (e) nonane; (f) decane; (g) undecane; and (h) dodecane [60].

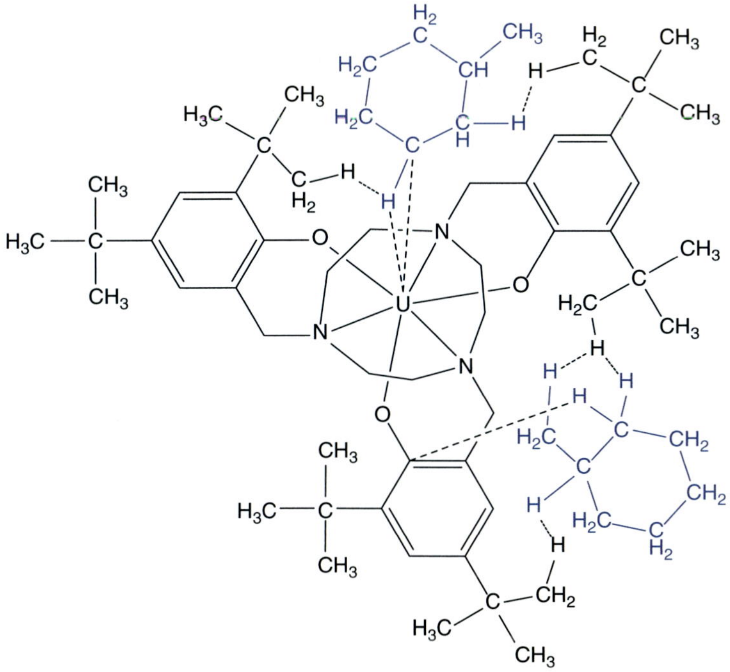

Figure 25.10 Noncovalent interactions in [U((ArO)$_3$tacn)(C$_6$H$_{11}$Me)]·(C$_6$H$_{11}$Me) [76].

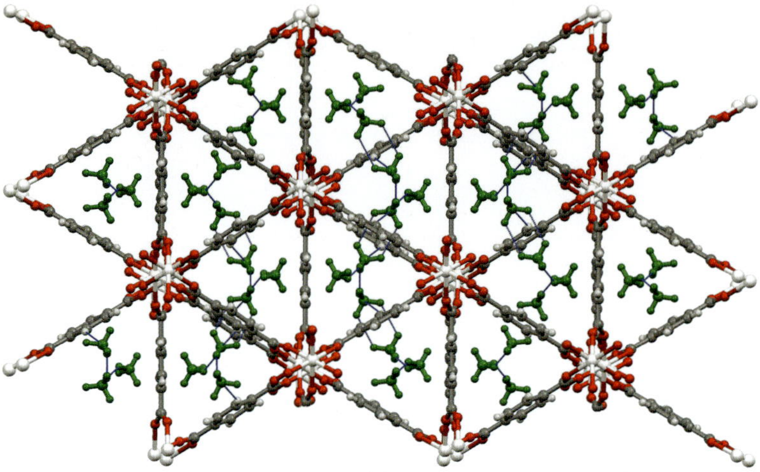

(a)

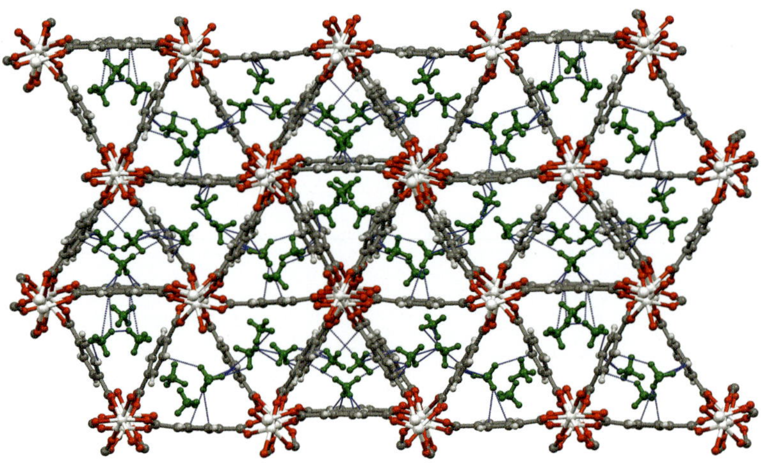

(b)

Figure 25.12 Noncovalent interactions in inorganic associate of $[Sc_2(BDC)_3]$ with methane at 3 (a) or 13 (b) kbar pressures [82].

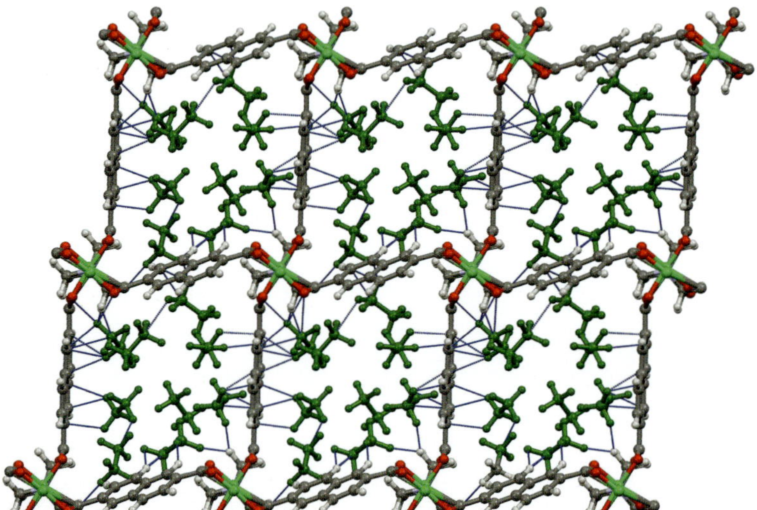

Figure 25.13 Inorganic associate of $Ni_2(naphthalene\text{-}2,6\text{-}dicarboxylate)_2$ 1,4-diazabicyclo[2.2.2] octane with ethane [84].

9

Gas-Phase Oxidation of Alkanes

Fabrizio Cavani[1], Alessandro Chieregato[1,2,3], Jose M. López Nieto[3], and Jean-Marc M. Millet[4]

[1] *Università di Bologna, Dipartimento di Chimica Industriale "Toso Montanari", Bologna, Italy*
[2] *TOTAL Research & Technology, Strategy Development Research Division, Catalysis Refining and Base Chemicals Department, Seneffe, Belgium**
[3] *Instituto de Tecnologia Quimica, Universitat Politècnica de València-Consejo Superior de Investigaciones Científicas, Valencia, Spain*
[4] *Institut de Recherche sur la Catalyse et l'Environnement de Lyon, UMR 5256 Universite Claude-Bernard Lyon 1, Villeurbanne Cedex, France*

9.1 The Gas-Phase Oxidation of Alkanes

This chapter deals with the oxidation of light C_3-C_4 alkanes to oxygenated compounds, and the oxidative dehydrogenation (ODH) of light alkanes to the corresponding olefins. Recent advancements and perspectives are discussed, with contributions from experts from industrial companies who kindly agreed to be interviewed.

9.2 Oxidation of Isobutane to Methacrylic Acid

9.2.1 Polyoxometalates Catalysts

Following the success achieved in the early 1980s with the oxidation of *n*-butane on vanadium phosphates, the oxidation of isobutane to methacrylic acid (MAA) was the reaction in which the highest hopes were placed for the development of an industrial process. The first patent related to this process was granted in 1981, and claimed highly interesting results. This represented a significant advance in the field of MAA production. Although it was not clearly reported as such, the catalysts contained phosphomolybdic heteropolyacid [1]. Very rapidly, polyoxometalates established themselves as the most suitable catalysts for this reaction, and although improvements have been made over the years, they are still the most efficient catalysts known for this reaction [2]. It was also very quickly recognized that optimal productivity could be attained with gas feeds having a high isobutane concentration (up to approximately 65%) [3]. One systematic outcome with this reaction was a relatively high selectivity to methacrolein (MA), with the MAA-to-methacrolein ratio ranging from 4:1 to 1:1. The best results correspond to isobutane conversions between 15% and 20%, with a selectivity to MAA and methacrolein as high as 70%. If higher conversions are sought, this is achieved at the cost of a considerably lower selectivity. In the context of an industrial application, it is thus necessary to recycle the isobutane, in order for the process to be economically viable. This recycling requires the prior separation of isobutane from the reaction products, which can be achieved using organic

* Current address

Alkane Functionalization, First Edition. Edited by Armando J. L. Pombeiro and M. Fátima C. Guedes da Silva.

solvents, such as a mixture of decanes, to adsorb the isobutane and methacrolein from the off-gas [3]. MAA is adsorbed with the other byproducts, from which it must then be separated.

In most situations, the unique physicochemical properties of polyoxometalates make these compounds efficient catalysts for heterogeneous oxidation reactions. First, their discrete anions constitute mixed-oxide clusters with a limited number of atoms, isolated from each other, into which electron transfer is possible but remains limited. This feature, which has already been emphasized for the oxidation with other catalytic systems, plays an important role in limiting the oxidation of alkanes to aldehydes or acids [4]. Furthermore, their structure permits multiple substitutions that can be used to tune the catalytic properties. The most active polyoxometalates are generally complex, making it difficult to study them thoroughly. The best catalysts for isobutane oxidation have a formula of the type $P_{1-1.2}Mo_{12}A_{2-2.5}B_{0.08-0.30}O_n$ (**1**).

These catalysts contain $[PMo_{12}O_{40}]^{3-}$ anions with the well-known Keggin structure, comprising a central tetrahedral PO_4 surrounded by 12 edge-sharing metal-oxygen MoO_6 groups (Figure 9.1a). Although other metals such as arsenium, As, could be used to supply the central atom, phosphorus leads to the most stable anion, and is present in most catalysts.

It is often proposed to substitute Mo with V as the addenda atom, and this strategy appears to have a positive effect on the catalytic properties. However, the origin of the observed effect

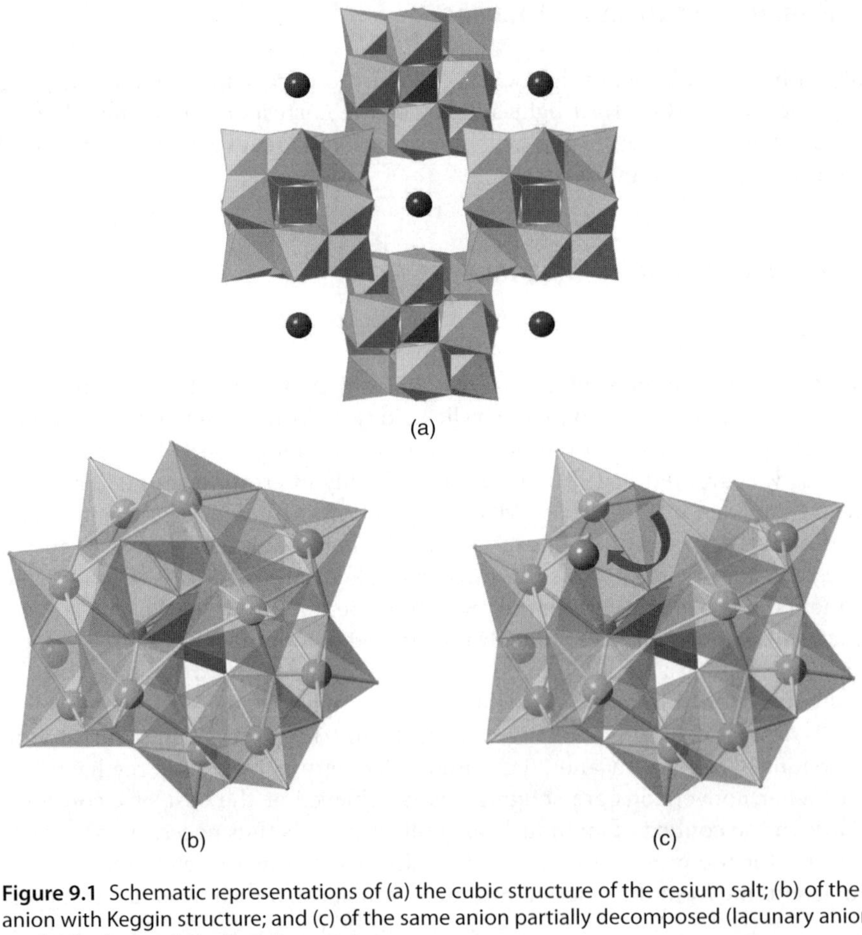

Figure 9.1 Schematic representations of (a) the cubic structure of the cesium salt; (b) of the $(PMo_{12}O_{40})^{3-}$ anion with Keggin structure; and (c) of the same anion partially decomposed (lacunary anion) proposed as a model for the catalytically active species.

is still debated. Partial migration of vanadium from the anion into the counter-cation position occurs systematically, concealing possible correlations between the position of the vanadium and its catalytic properties. Furthermore, a synergy between VO^{2+} and other cations in the counter-cation position (such as Cu^{2+}) has been established, even when its content is very low [5].

Element A in **1** corresponds to the counter-ions counterbalancing the negative charge of the anions. They are generally alkali or ammonium cations, which make it possible to form a salt with a stable cubic structure. It should be noted that, depending on the method of preparation, catalysts with more than two alkaline metals per Keggin unit are frequently nonhomogeneous, such that it can be preferable to consider them to be in a cubic acid phase that contains B cations covering the alkali or ammonium salt [6–8].

By virtue of their saline structure, polyoxometalates may accept multiple substitutions in the counter-cation position (B elements), even when their charge is greater than unity. The B elements may correspond to transition metals (Fe, Co, Ni, Cu) [8–16], or to elements such as tellurium, antimony, [17–19] rare earth elements, or even noble metal elements [11, 20]. When B = Fe, Ni, or Mn, the MAA and methacrolein (MA) yields increase, whereas when B = Co, Cu, Hg, Pt, or Pd the yields decrease. At the macroscopic level, none of these elements appears to modify the structural characteristics of the polyoxometalate, whereas they have a significant effect on its redox or acid-base properties, and under most circumstances positively modify either its catalytic activity, or its selectivity. They may also have an effect on the thermal stability of the catalyst, as has been shown in the case of rare earth elements such as Ce or La, and trivalent elements in general [21].

The oxidation of isobutane is a surface-type reaction, and the catalytic properties of polyoxometalates are controlled by both the oxidizing ability of the catalyst's surface, and its protonic acidity [11]. This property became apparent when, in the case of compounds with various cesium contents, but no transition metal cations, the intrinsic rate of catalytic oxidation of isobutane was found to vary linearly with the rate of reduction by CO. For any given transition metal cation content, the maximum level of activity depends not only on the catalyst's oxidizing ability, but also on its protonic acidity. The first of these properties increases with metal content, whereas the latter property decreases with proton substitution. These opposing trends make it possible to obtain optimal activity when the composition is correctly balanced. From these considerations, the two catalyst formulations proposed in the literature and patents, and having the most useful properties, i.e. $Mo_{12}V_{0.5}P_{1.5}As_{0.4}Cu_{0.3}Cs_{1.4}O_y$ and $Mo_{12}V_{0.1}P_1Te_{0.3}Cs_2O_x$, can be more readily explained. At this stage, it is important to note that these two formulations, which are the results of complicated optimizations taking full advantage of the unique flexibility of the polyoxometallate structure, lead to the same best yield of 9% to MAA, and to a methacrolein and MAA productivity of 6.4 mmol h^{-1} $g_{cat}{}^{-1}$ [18, 21, 22].

It has been shown that under reducing conditions, and at temperatures above approximately 300 °C, phosphomolybdic salt or acid $H_3[PMo_{12}O_{40}]$ and $Cs_2H[PMo_{12}O_{40}]$ were reduced and transformed with the migration of molybdenum from the Keggin ions onto the interstitial position, producing thermally stable, partially reduced lacunary Keggin ions [23] (Figure 9.1c). This migration, which can be that of vanadium if present in the polyanion [24], was shown to be correlated with catalytic activity in the case of the partial oxidation of propane and butane, as well as isobutane [25–28], and turns out to be a prerequisite for the polyoxometallate to become an active heterogeneous catalyst. Although the driving force for the formation of lacunary Keggin anions at elevated temperatures and under reducing conditions has not been exactly determined, it has been proposed that it may be linked to the relaxation of the Keggin structure after reduction, possibly accompanied by the formation of links between the Keggin ions involving the ejected molybdenum atoms [23]. It has been shown that these lacunary Keggin anions lead to catalysts with a high proton acidity, fast but localized electron transfer, and the ability to store reactive hydrogen species in anionic vacancies [29–31]. The strong acid sites formed would be able to abstract the H^- ion from isobutane, to form an i-$C_4H_9^+$ carbocation. This intermediate (I_0) is rapidly solvated, generating more stable alkoxy species (I_1) and a hydroxyl group with two transferred electrons (Figure 9.2). This explains why some authors have found recursive correlations between acidity and catalytic activity [11, 12, 32]. This mechanism is different from that proposed in the case of other well-known catalysts, such as MoVTeNb mixed oxides, on which an initial homolytic C—H bond is generally proposed. This may provide an explanation for heteropolyanions exhibiting a specific catalytic activity in the case of isobutane oxidation, and could possibly explain the weak and mixed effect of vanadium ions, which is generally described as an active species in the second type of activation. The alkoxy species would evolve toward the formation of dioxy-alkylidene species (I_2), which undergo either the dissociation of a C—O bond, leading to methacrolein, or hydration followed by a subsequent dehydrogenation, to form a final intermediate (I_3) leading to MAA [18, 33]. The ratio of the apparent rate constant values from TAP experiments, for the surface reaction of I_0–I_1 to the desorption rate of I_0 into isobutene, amounts to 100. This clearly shows that a surface route is favored to the gas phase route, and explains why isobutene is almost never observed in steady-state experiments. Furthermore, a ratio of 10 is observed between the respective adsorption rates of isobutane and isobutene. The isobutene formed in this reaction reacts rapidly.

Water is involved in the reaction mechanism and plays a positive role on both the activity and the selectivity of the catalysts. However, the results of the TAP experiments tend to show that its main role is to assist with the desorption of the reaction products, since the observation of methacrylic and acetic acid in these products depends on the addition of water in the pulses [18]. It is also certain that water plays a positive role in the stabilization of the lacunary Keggin anions, or in their regeneration.

Lacunary Keggin anions are relatively stable and may reconstruct after the loss of the addenda ions. However, they can also decompose into fragments of solely edge-sharing MoO_6 units, which can lead to the formation of molybdenum oxide [34]. After a certain time on stream, this decomposition leads to the systematic deactivation of the catalysts, which further decreases their efficiency, thereby barring the road to an industrial application. In recent years, the major industrial actors involved in research in this field have concentrated most of their efforts on regeneration of the catalysts [35, 36]. As described above, deactivation of the catalyst is due to a loss of specific surface area, and to the partial decomposition of the surface Keggin anions and formation of molybdenum oxide. The most commonly reported techniques used for catalyst regeneration consist in treating the deactivated catalyst in an aqueous solution containing nitrate and ammonium ions having a molar ratio 1 : 1.3, at a temperature of at least 100 °C. With this type of treatment, the surface area is only partially recovered, but the Kegging anions appear to be fully reconstructed.

Another commonly encountered, although less frequently investigated cause of deactivation of these catalysts arises from a high loss of specific surface area, during their activation under reaction conditions. This aspect of deactivation has been addressed in a paper proposing catalysts corresponding to mechanical-phase mixtures of an active phosphomolybdic cesium salt and α-$La_2Mo_2O_9$ (LM) or β-$La_2Mo_{1.9}V_{0.1}O_{8.95}$ (LMV), acting as a particle support and preventing sintering of the salt (Figure 9.3) [22]. The origin of these unconventional effects has been attributed to crystallographic fits between the two phases.

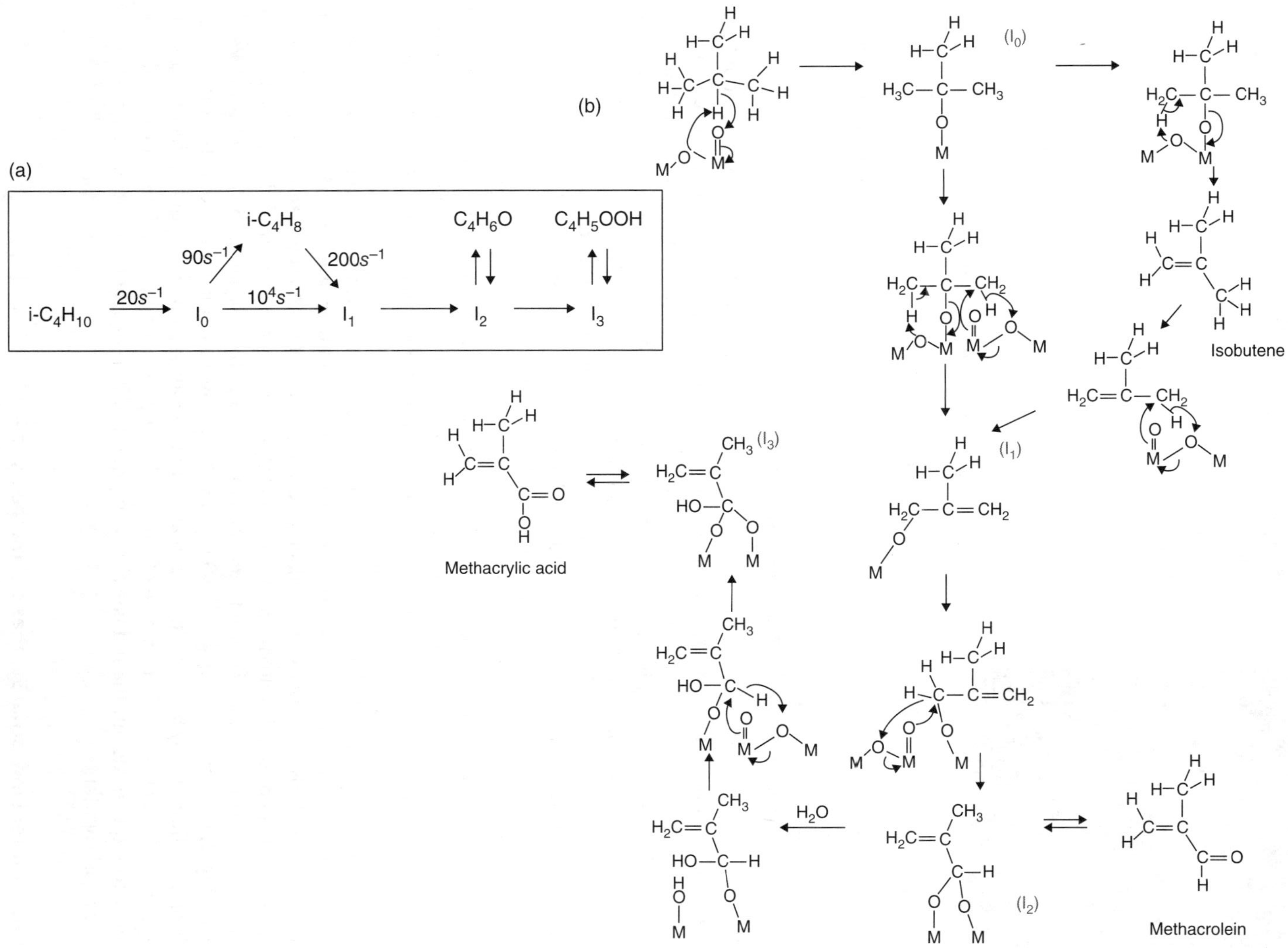

Figure 9.2 Reaction pathway (a) and proposed mechanism (b) for the formation of methacrolein and MAA on polyoxometalates based catalysts (from [18]).

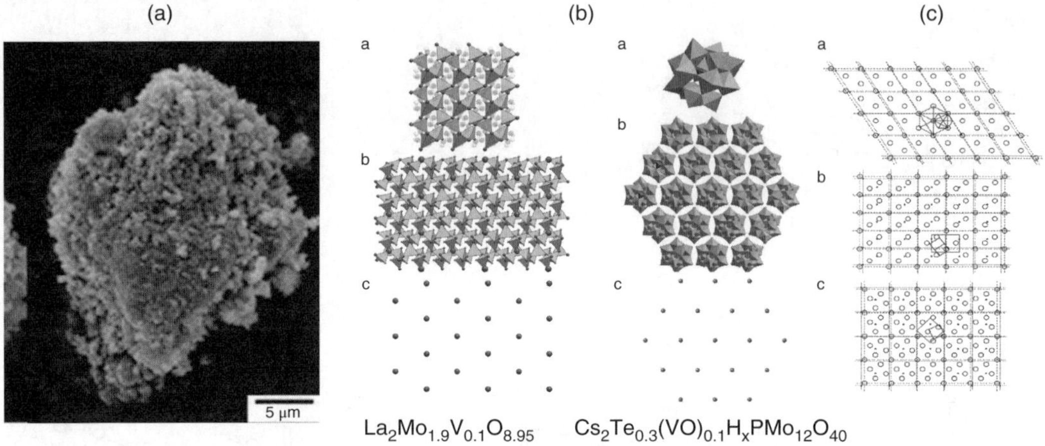

Figure 9.3 Study of the $Cs_2Te_{0.3}(VO)_{0.1}H_xPMo_{12}O_{40}/La_2Mo_{1.9}V_{0.1}O_{8.95}$ (CTV/LMV) catalyst (a) MEB image; (b) projection of the cationic arrangement of LMV along (111) with isolated La hexagonal array and anionic arrangement of CTV in the (111) plane with isolated P hexagonal array; (c) fits between the two arrays along different cubic directions of both lattices (from [22]). (*See insert for color representation.*)

9.2.2 Other Synthetic Pathways to Methacrylic Acid

The oxidation of isobutane to MAA and methacrolein is not the only way to oxidize isobutane. The ODH of isobutane to isobutene as well as its oxidation to *tert*-butanol has also been considered. Although numerous studies have proposed the use of catalysts for the ODH of isobutane, most of these exhibit a high level of selectivity, although this can only be achieved at a low rate of conversion, which ultimately leads to a low isobutene yield [37–42]. These catalysts are chromium oxide supported on various substrates, pyrophosphates and phosphates, or complex mixed oxides. Two noteworthy catalysts have recently been proposed, the first of which is chromium-based, supported on a novel ZnAlLaO mixed oxide, and the second makes use of boron nitride [43, 44]. The chromium-based catalyst has an isobutane conversion of 52.4% with a selectivity to isobutene of 84.0% at 577 °C, whereas the latter catalyst has a much lower conversion (5–15%), but a selectivity in excess of 90% at 460 °C. The chromium-based catalyst is reported to have a slow deactivation caused by coking; however, this effect can be recovered by means of simple regeneration treatment.

High-efficiency catalysts based on highly halogenated porphyrin iron complexes were proposed in the 1990s. One example of this type of catalyst is octabromotetra(pentafluorophenyl) porphyrin (TFPPBrs), used for the molecular O_2 oxidation of isobutane to *tert*-butanol. During an earlier stage of the development of these catalysts, it was hypothesized that the complexes function according to the accepted mechanism of alkane hydroxylation by cytochrome P450, involving the removal of H by a $Fe^{IV}{=}O$ species [45]. However, it is more likely that these complexes function mainly by initiating a radical chain mechanism, and catalytically decompose the organic peroxides [46, 47]. This type of catalyst is expensive, and its catalytic efficiency is currently too low to sustain the industrial development of hydrocarbon oxidation reactions such as that of isobutane.

9.2.3 Current Industrial Synthesis of Methacrylic Acid

It is not easy to assess the future prospects for the industrial oxidation of isobutane, since this will strongly depend on the price and availability of oil, as well as the cost and supply of basic chemical intermediates. Isobutane is still an accessible source, at a price that remains acceptable for large-scale industrial applications, with an annual production in excess of several million tons [48]. If oil becomes expensive, the price of gas and therefore that of isobutane will increase, and the energy value of isobutane may outweigh its value as a starting material for chemistry.

There are currently three MAA production processes used throughout the world. As these processes have been extensively described in various reviews, in the following we simply provide a brief outline of their principal characteristics: (i) the acetone-cyanohydrin process (ACH) uses acetone and hydrogen cyanide as starting materials; (ii) the process based on the carbonylation of ethylene to methylpropionate (Alpha process, from Lucite), or to propionaldehyde (BASF), together with condensation using formaldehyde; and (iii) the ASHAI process based on the selective oxidation of isobutylene or *tert*-butanol to methacrolein, followed by the oxidation of methacrolein to MAA. The latter process is currently being developed in Japan only, where the demand for isobutene is low. The oldest ACH process is generally integrated into the design of chemical production platforms, where it can take advantage of the acetone and hydrogen cyanide released as byproducts by other processes (synthesis of acrylonitrile and phenol). Unless these processes are supplanted by new techniques, leading to the disappearance of their associated byproducts, it will be difficult to impose a new process based on the oxidation of isobutane, even if it is intrinsically competitive. The second type of process of interest makes use of raw ethylene materials (CO, H_2, CH_3OH, H_2CO), which are readily available in modern refineries, in large quantities and at relatively low prices. In countries where such large industrial complexes have already been developed, or where ethylene is becoming more readily available (via steam cracking of shale gas ethane), it is unlikely that any competing technology will be introduced in the near future. Furthermore, when its CO_2 emissions are taken into account, the oxidation of isobutane is not an advantageous process. A recent calculation has shown that any new process based on the use of isobutane would require a minimum yield of 26%, in order to match the CO_2 emissions of the ACH process (within a life cycle assessment) [49].

9.2.4 Methacrylic Acid from Biomass

The current development of biomass technologies is bringing new processes to the forefront, which can compete with the oxidation of isobutane. For example, several companies have successfully developed the fermentative production of isobutene. Global Bioenergies and Clariant recently announced the first industrial, pilot-plant isobutene production process based on the use of a wheat straw hydrolyzate (http://www.chemengonline.com/global-bioenergies-and-clariant-begin-producing-bio-based-isobutene). Another process has been proposed, involving the bioconversion of 2-methyl-1,3-propanediol to 3-hydroxy-2-methylpropionic acid, and its catalytic dehydration, for the production of MAA [50]. This approach could also be competitive if an economically viable pathway to 2-methyl-1,3-propanediol could be developed. This compound is currently produced only as a byproduct, during the synthesis of 1,3-propanediol. Finally it can be noted that biomass yield in biosynthesis of 2-hydroxyisobutyric acid from renewable carbon is constantly increasing and the compound could also become a valuable precursor for MAA synthesis [51, 52].

9.2.5 Perspectives for Development

Although it may appear challenging to impose a process based on the oxidation of isobutane, there is still an enormous margin for improvement in catalyst performance, and it would be reasonable to expect that the current yields of the order of 9% per pass will be exceeded in the future. There are two main challenges in the development of polyoxometalate-based catalysts for the oxidation of isobutane: stabilizing the catalysts and improving their selectivity. It is difficult to propose a solution to the first problem, since the instability of the present catalysts is related to the structural instability of their lacunary polyoxometalates under reductive conditions. However, if this difficulty can be overcome, there are suitable regeneration processes, which could be implemented, and it could be relatively straightforward to improve the catalysts' selectivity, especially when there is a need for a substantial increase in selectivity to MAA, at the expense of selectivity to methacrolein. This would already represent a significant improvement.

From the fundamental point of view, there is still no suitable DFT model for the lacunary Keggin anion, with the ejected Mo cation still in interaction, or for its interaction with the isobutane molecule. This would allow us to fully understand how the first C—H bond breaking occurs. It would also be desirable to understand the multiple roles of the counter-cations and the texture (micro and mesoporosity). This would in practice be very challenging, due to the dynamic changes taking place in the catalyst under reaction conditions, and would certainly call for the use of *operando* characterization techniques.

Although one might hope for the discovery of a novel, efficient, non-polyoxometalate catalyst, it should be noted that none of the new catalysts developed for the oxidation of other alkenes [MoVO, MoVTeNbO, VPO, etc.) have allowed better results to be obtained than with polyoxomolybdates [53–58]. This tends to confirm the idea that polyoxometalates have specific redox and acid-base properties, allowing a different reaction network to take place [59].

The combination of efficient catalysts with new technological options might improve the viability of an isobutane-based MAA synthesis process. In this respect, transported bed reactors could lead to improved productivity, through the use of a concentrated isobutane flow, optimizing the re-oxidation and facilitating the implementation of a regeneration process, without stopping the catalytic reactor. Novel short contact time oxidation processes, combining highly active catalysts with an auto-thermal hybrid reactor, may also lead to a viable isobutene oxidation process with high selectivity to MAA.

9.2.6 An Industrial Perspective for Isobutane Oxidation: Interview with Dr. Jean-Luc Dubois, Arkema

1 Which are, in your opinion, the most important hurdles, which hinder the development and industrial implementation of direct isobutane oxidation?
JLD: Direct oxidation of isobutane would be the fourth technology to produce Methyl-Methacrylate, after the acetone cyanohydrin, the isobutene oxidation and the ethylene carbonylation route. The main advantage would be a "cheap" feedstock, isobutane being used as a fuel currently. But the advantage would be lost if the selectivity/yield is not high enough. Isobutane is not very reactive, so harsh conditions have to be used (high temperature and highly oxidative catalyst) so that it is difficult to keep a high selectivity at the same time.

2 Which is your vision about the future development of the oxidation of isobutane?
JLD: In the long term, if crude oil is expensive, so is isobutane, since it can also be used as energy source. So it is losing one of its advantages. And in the worst case, isobutane can be dehydrogenated to isobutylene with well-known established technologies. Cost differential between isobutane and isobutylene would be given by the cost for dehydrogenation and only slightly dependent on isobutane price (energy cost).
The industry is Risk averse, so in an equation where there is high risk – potential high reward (isobutane oxidation) – and low risk – high capital cost – known product quality – several

suppliers/technology providers, Top management of companies would favor the low-risk option.

3 Which are the most important scientific and technological challenges for the oxidation of isobutane?
JLD: One of the striking features in that reaction, is the lack of C_3 products. As expected there are C_4 (methacrolein and MAA) products, and C_2 and C_1 (acetic acid, CO, and CO_2), but there is a gap in the C_3. And because of the structure of the molecule, it is impossible to go from the C_4s (including isobutane and isobutylene) without forming in intermediate C_3 adsorbed on the surface, which degrades into C_1 and C_2. A better understanding of that mechanism, could lead to more selective catalysts.
Another challenge is linked to the heat of reaction. The process is cumulating in a single step 3 oxidation reactions (oxydehydrogenation, oxidation to methacrolein, and oxidation to MAA). In addition, there are all the nonselective oxidations, which are producing a lot of heat. The reactor design has to be able to cope with the heat produced, and keep a high productivity at the same time.

4 A general question: Which are the perspectives and trends for alkanes selective oxidation to chemicals?
JLD: For products from C1 to C3 and even C4, we have witnessed the development of alternative processes, which have reduced the pressure on the need to develop new technologies. For example, compared to 20 years ago, we now have methanol to olefins and methanol to propylene technologies, metathesis of ethylene and 2-butene to produce propylene, so the fear to lack of propylene has significantly reduced.
For methacrylic acid, there has been the development of the alpha process (Lucite), and even a second plant under construction, which further diversifies the possible starting materials.
Another general trend for oxygenated products has been to look for bio-based products, since sugars are already rich in oxygen. These routes have still to be implemented for the acrylic and methacrylic acid families, but have been done for 1,4-butanediol and succinic acid for example.
So a direct alkane conversion route would still be interesting, but has to face more challenges and more competition.

9.3 Oxidation of *n*-Butane to Maleic Anhydride

9.3.1 Introduction

The oxidation of *n*-butane is currently used by several industrial companies for the production of over 2 M tons of maleic anhydride (MA) per year [60–64]. Since the discovery of a catalyst for this direct – although multistep – transformation, this reaction has been one of the most investigated topics in the field of gas-phase selective oxidation. Indeed, some peculiarities make this reaction quite unique; in fact, the selective oxidation process involves 14 electrons and yet leads to quite a remarkable selectivity, despite a very complex mechanism.

The aim of the huge numbers of papers and patents published during the last decades of past century was mainly that one of investigating on the chemical-physical features of vanadyl pyrophosphate (VPP), $(VO)_2P_2O_7$, the catalyst used in industry for this reaction, and on developing efficient procedures for its synthesis. The role of promoters has also been studied, and in some cases specific promoters have been included in catalyst formulation, because of their ability to improve either activity or selectivity to MA, or both.

Over the latest years, research in this field has been mainly focused on a deeper understanding of the nature of the active phase that develops on the surface of the V/P/O catalyst. In fact, even though the main catalyst component is VPP, it has been ascertained that under reaction conditions surface domains of oxidized vanadium phosphate may form, which play an important role in both alkane activation and its selective oxidation [65–78]. Modeling of reactors and catalyst pore structure have also been investigated in few recent papers [79–84].

Here we report on most recent scientific achievements in *n*-butane selective oxidation, and set the point on future perspectives for the development of this technology by reporting two short interviews released by experts in this field, both working in industrial companies which supply the VPP catalyst and produce MA, i.e. Polynt SpA (Dr. Carlotta Cortelli) and Clariant (Dr. Gerhard Mestl).

9.3.2 $(VO)_2P_2O_7$ or $VOPO_4$ as the Active Species? (and the Role of "Excess P")

The role of the P/V atomic ratio in the VPP catalyst has been the subject of an intense debate, along with the understanding of the nature of the true active phase [60–64]. Indeed, even though in VPP (and in $VOPO_4$ as well) the stoichiometric P/V atomic ratio is 1.0, catalysts offering the best performance are those which are prepared with a slight excess of P, e.g. with a P/V ratio in the range 1.05–1.15. Moreover, during reaction it is necessary to co-feed to the reactor ppm levels of an organic phosphorus compound together with water in order to keep the optimal performance [69]. If the reactor is operated without P during prolonged periods, an increase in activity and selectivity to carbon oxides may drive the operation in a less stable region with the risk of runaway.

The importance of the P/V atomic ratio has been recently discussed in some papers; it was confirmed that the nature of the active layer that develops at the surface of VPP depends strongly on this composition parameter [65–70]. By means of *in-situ* Raman spectroscopy in a nonreactive atmosphere, and of steady-state and non-steady-state reactivity experiments, it was found that the nature of the VPP surface is affected by the P/V atomic ratio even in the case of small changes in this parameter. More specifically, with the catalyst having P/V = 1.0, a layer made of α_I-$VOPO_4$ developed in the temperature interval 340–400 °C in the presence of air over the VPP surface, but this catalyst then gave a low selectivity to MA when reaction temperatures in the range 340–400 °C were used. In the presence of steam, hydrolyzed $[VO_y + (PO_4)_n]$ domains also formed, and the relative amount of the two components was a function of the partial pressure of steam [65, 66]. However, at higher temperature, i.e. in the interval 400–440 °C, δ-$VOPO_4$ overlayers formed, and the catalyst became moderately active but selective to MA. Conversely, with the catalyst containing a slight excess of P (P/V = 1.05), δ-$VOPO_4$ was the prevailing phase over the entire temperature range investigated (340–440 °C). In the presence of steam the formation of $VOPO_4{\cdot}2H_2O$ was also observed, that dehydrated to δ-$VOPO_4$ when steam was removed. This catalyst showed the best catalytic performance with the highest selectivity to MA at both intermediate and high temperatures. Therefore, at high temperatures, i.e. above 400 °C, the active layer consists mainly of δ-$VOPO_4$ in both catalysts, and the catalytic behavior was similar regardless of the P/V ratio. Figure 9.4 summarizes these findings.

Recently, a slightly different picture was presented by Mestl and coworkers [69–72]. The authors agreed that water and phosphorus play a fundamental role in determining the nature of the true active layer under reaction conditions. The P loss from VPP catalyst during long-term operation has a strong effect on the temperature distribution in the industrial reactor, which, in turn, significantly affects the catalyst selectivity, and important changes in catalyst activity and reactor performance are induced by only minor variations in the feed

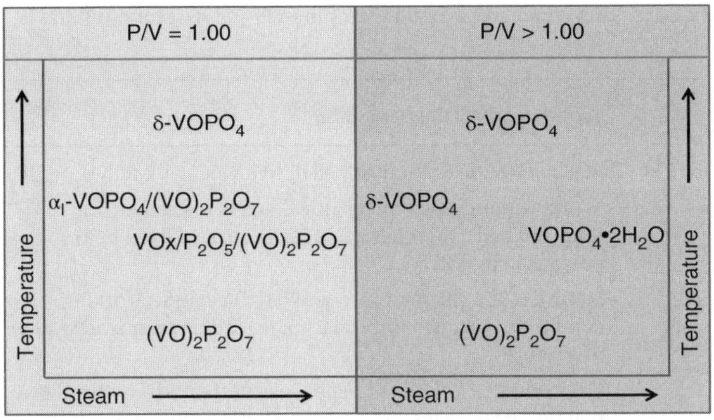

Figure 9.4 The nature of the VPP active layer under reaction conditions.

contents of phosphorus and water. According to the scheme proposed by Mestl et al., the active surface is a two-dimensional structure composed of various surface species, e.g. X_1-, δ-, α-, and β-$VOPO_4$, which may have various functions, and which are formed on VPP surface under reaction condition because of the presence of water. However, the composition of this active layer is determined by the P concentration at the catalyst surface.

The complexity of the picture regarding the nature of the active layer is confirmed by recent findings by Eichelbaum et al. [73–76]; it was found that the VPP catalyst acts like semiconducting gas sensors with a dynamic charge transfer between the bulk and the surface, as indicated by the gas-phase-dependent response of the VPP chemical-physical features, such as work function, and the surface potential barrier. The same authors also identified three general descriptors of selective and active oxidation catalysts: (i) a fast change of the conductivity (i.e. of surface V oxidation state) with the chemical potential of the gas phase; (ii) stability of the electronic structure under the reaction conditions; and (iii) the prerequisite that the conductivity under reaction conditions is as low as possible (compared to strongly reducing conditions for p-type or oxidizing conditions for n-type semiconductors). The latter descriptor is a physical representation of the "site-isolation" principle of selective oxidation catalysts, because a low amount of active charge carriers should prevent an overoxidation of the substrate.

9.3.3 Recent Patent Literature

A review by Grasselli and Trifirò [64] examines recent patent literature on *n*-butane selective oxidation. Table 9.1 summarizes the important aspects for the successful commercial operation of VPP catalysts, as reported in patent literature and highlighted by Grasselli and Trifirò.

For what concerns the synthesis of the VPP precursor, $VOHPO_4{\cdot}0.5H_2O$, one key aspect is the nature of the organic compound used for the reduction of the starting V oxide material, V_2O_5. Typically, isobutanol and benzyl alcohol are the preferred reactants, but the use of glycols (or other structure-directing agents) may affect the morphology of the crystalline precursor, and finally the chemical-physical features of the VPP obtained after the thermal treatment of $VOHPO_4{\cdot}0.5H_2O$ [85–88]. Another important aspect is the way the thermal treatment is carried out; in fact, depending on temperature, pressure, and composition of the gas phase (which can be modified by adding steam, thus developing hydrothermal-like conditions), the crystallinity of the VPP may considerably change [89, 90]. This affects the catalytic performance during the so-called equilibration stage.

Table 9.1 Most important aspects for the successful commercial operation of the VPP catalyst for *n*-butane oxidation to MA.

Process parameter	Tools to control the parameter
Pressure drop in multitubular fixed bed	Size and shape of catalyst pellet; bed voidage; catalyst density
Catalyst lifetime (including continuous make-up in fluidized bed)	Working conditions, mechanical-stress resistance (in fluidized bed), thermal-stress resistance (in fixed bed), poisons in inlet feed
"Equilibration time" (time needed to reach stable target performance)	Nature of the calcined (pre-activated) catalyst; working conditions during plant start-up; in-situ or ex-situ activation procedure
Temperature gradients and hot spots in fixed beds	Co-feeding of volatile phosphorus compounds; removal of reaction heat; mass and heat-transfer phenomena
Catalyst (VPP): manufacture and chemical-physical features	Tools to control chemical-physical features
V^{5+}/V^{4+} ratio under working conditions	Nature and amount of promoter (and procedure used to add it to VPP); nature of the catalyst before activation (main and spurious crystalline and amorphous phases)
VPP crystallinity (surface area)	Procedure used for the preparation of the precursor $(VO)HPO_4 0.5H_2O$
P/V atomic ratio	Recipe for catalyst preparation; continuous feed of volatile P compounds to the reactor

Several patents claim the importance of catalyst morphology and pellets shape [91, 92]. The importance of catalyst morphology has also been pointed out by Horn et al. [79]. Simulation results have demonstrated that the reactor performance is sensitive to the pore structure, especially to macro-pore porosity and mean micro-pore diameter; a bimodal catalyst pellet with bigger macro-pores and smaller micro-pores is preferred to achieve higher yields of MA.

9.3.4 The Role of Promoters

Promoters play an important role in the composition of the VPP catalyst [64, 93–96]. One of the most efficient promoters for VPP is Nb^{5+} [95, 96] In Nb-doped VPP catalyst, it was demonstrated that the presence of Nb is conducive to the development of a V^{5+} phosphate compound, not only in the precursor, but also in the calcined catalyst and in the "equilibrated" one. By means of *in-situ* Raman experiments, it was shown that during reaction a strongly dynamic behavior exists on the catalyst surface, with redox cycles between V^{4+} in VPP and V^{5+} in oxidized $VOPO_4$, with δ-$VOPO_4$ being the compound involved in the multielectron MA formation. The presence of Nb affected the redox behavior and catalytic performance considerably, depending on the V/Nb ratio used. For highly Nb-doped catalysts, the oxidized phases formed became more stable and less reducible, and undesired α_{II}-$VOPO_4$ was the predominant oxidized compound; this catalyst was the least selective to MA. On the other hand, in catalysts with a moderate amount of Nb (e.g. with V/Nb atomic ratio equal to 150–80), a discrete amount of δ-$VOPO_4$ developed, and the optimal V^{5+}/V^{4+} formed on the surface. This catalyst offered a performance improved over that of the undoped catalyst.

9.3.5 The Mechanism of the Reaction

Different mechanisms have been reported for the oxidation of *n*-butane catalyzed by VPP; one of the most successful one is the "olefin" mechanism (Figure 9.5). Indeed, the main difficulties

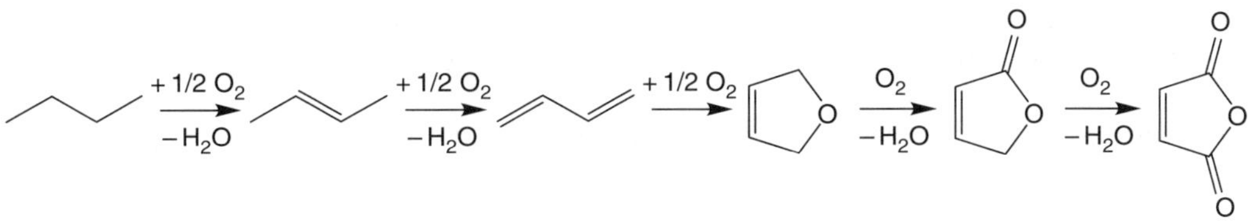

Figure 9.5 The mechanisms proposed for *n*-butane oxidation to MA.

met in finding convincing proofs for the reaction pathway are due to the lack of information of adsorbed intermediate species. This is due to both the relatively low number of active sites involved and the much greater reactivity of any partially oxidized intermediate compared to the starting alkane. An important implication of this observation is that the multi-electron oxidation can be completed on a limited number of active sites, which well fits with the principle of "site isolation," successfully applied to other partial oxidation reactions [97].

An important experimental finding is that when the reaction is carried out under alkane-rich conditions (that is, with a very high *n*-butane/O_2 inlet feed molar ratio, conditions necessarily leading to very low hydrocarbon conversion), butenes and butadiene intermediates are desorbed. In an alternative pathway, the so-called "alkoxide" mechanism, the alkane anchors on the surface and desorbs only when MA is formed, thus explaining the lack of partially oxidized intermediates under normal working conditions. In general, it is believed that the $(V{=}O)^{3+}$ species in $VOPO_4$ plays the major role.

However, recent quantum mechanical studies have disclosed an alternative option, with direct involvement of the O=P bond of the oxidized X1 phase of the $VOPO_4$ surface in cleaving the C—H bond of the alkane and extract the H atom; the electron is simultaneously transferred to the highly reducible V^{5+} next to the P=O bond (O=P—O—V=O unit), to form V^{4+} [77, 78]. This new mechanism decouples the proton transfer and electron transfer components of the H transfer reaction (reduction-coupled oxo-activation mechanism). The highest reaction barrier in the multistep process leading from *n*-butane to MA was found to be equal to 21.7 kcal mol^{-1}.

9.3.6 An Industrial Perspective for *n*-Butane Oxidation: Interview with to Dr. Gerhard Mestl, Clariant

1 Which are the most important scientific and technological challenges for the oxidation of *n*-butane to MA?
GM: Technologically, there is only one direction driven by the need of the MA producers to enhance their margins, to push the MA process to higher productivities, which means of course higher C4 loads and higher GHSV's – if possible, both together at the same time. This, in turn, means the catalyst has to be driven to higher yield maxima, i.e. higher reachable conversion, or current conversion levels but reached for higher C4 concentrations, with similar or, of course this would be better, improved MA selectivity.

2 Which are the most important hurdles which hinder the further improvement of current industrial technologies for *n*-butane-to-MA?
GM: The most important hurdle is, as almost always nowadays, an extremely careful risk management when new technologies are assessed in companies. New technologies are nowadays almost always considered as extremely risky so that even the investments into small demo units are hardly taken. This careful risk management on one side has its counterpart in the highly developed and for many decades proven standard technology, i.e. the salt bath cooled tubular bundle reactor, on which the very careful industry absolutely

relies. Even in this established case of reactors, it seems very difficult to realize new concepts, like the two-zone-reactor, or alternative cooling concepts.

3 Do you think that alternative reactor configurations for *n*-butane-to-MA might lead to improvements in performance?
GM: Two- or even better multiple-zone reactors would have a certain advantage as our reactor simulations have shown. Alternative cooling concepts also could improve the MA productivities. The largest performance step could be reached for reactor configurations allowing considerably improved heat transfer. It was shown that micro-reactors would lead to remarkable productivities and such stacks of micro-reactors would not have higher pressure drops, hence, current blower systems could be used. Moreover, the total reactor volume to MA productivity ratio would even be an advantage as compared to state-of-the-art 30.000 tube bundle reactors, i.e. such stacks would be considerably smaller in size for the same MA productivity.

9.3.7 An Industrial Perspective for *n*-Butane Oxidation: Interview with Dr. Carlotta Cortelli, Polynt SpA

1 Which are the most important scientific and technological challenges for the oxidation of *n*-butane to MA?
CC: The improvement of catalyst performances is the first main challenge for *n*-butane oxidation to MA, for both processes industrially used (fixed and fluidized bed).
In this case, scientific targets meet market needs, i.e. to have a more selective catalyst, able to run under severe conditions, particularly at high butane concentration and without impacting catalyst life. The design of an improved catalyst passes through the understanding of the complex surfaces interaction between the gas phase and the catalyst itself. Secondarily, the effective comprehension of fluidodynamic parameters (and therefore catalyst shape/dimension and reaction conditions) is an important topic that can have a very positive result in the reactor /catalyst performances. This is particularly true for fluidized bed reactors.

2 2. Which are the most important hurdles which hinder the further improvement of current industrial technologies for *n*-butane-to-MA?
CC: Maleic anhydride production in fixed bed reactors through *n*-butane oxidation is nowadays a well-consolidated process. Changes would need important investments in terms of costs, time, and resources, and therefore further improvement seems difficult to be foreseen. Fluidized bed processes are, on the contrary, relatively new technologies for MA (developed starting from the 1990s) and further improvement in reactor/equipment design to allow better productivity and use of resources seems easier to be introduced.
In all cases, safety aspects have to be extremely carefully considered, particularly for productivity improvement by butane concentration increase.
Modifications and improvement in MA recovery sections (continuous processes, new solvents, etc.) should be a chance to introduce new technologies.

3 Do you think that alternative reactor configurations for the *n*-butane-to-MA might lead to improvements in performance?
CC: We do not see in the near future the possibility of introducing new reactor configurations, even improvement in heat exchange could be beneficial for catalyst/reactor performances. In this view, an optimization of heat removal in ALMA fluidized bed reactor has been recently introduced, leading to a better catalyst performance.

9.4 Oxidation and Ammoxidation of Propane to Acrylic Acid and Acrylonitrile

9.4.1 Introduction

Acrylonitrile and acrylic acid are the most important products from propylene after polypropylene. Acrylonitrile is produced in a one-step propylene ammoxidation process using a multicomponent catalyst based on Mo oxide (in 2014, global production of acrylonitrile was around 6 million tons). Acrylic acid is, however, produced in a two-step propylene oxidation process, with acrolein as intermediate, using two different rectors and two type of catalysts: the first one related to the oxidation of propylene to acrolein similar to those used in propylene ammoxidation (in 2014, global production of acrylic acid was around 5.4 million tons). Moreover, during the last two decades, the number of studies on the production of both acrylic acid/acrylonitrile from propane has increased due to [59, 98–101]: (i) high influence of the cost of propylene in both processes (propane is cheaper and currently almost exclusively employed as fuel for cooking, industrial and domestic heating); (ii) one step is only required in both processes; and (iii) propane has a possible better sustainability.

The reaction in both oxidation and ammoxidation is carried out in propane/oxygen mixtures in addition to steam or ammonia, respectively, at similar reaction temperatures. Accordingly, therefore, one might think that the type of catalysts for both processes should be similar. However, strong differences in the type of catalysts have been mainly observed. Thus, Vanadium-antimonates of the $VSb_xM_yO_z$ type (M = W, Al, Nb, Sn, and also other such as Ti, Cu, or Bi), with a rutile structure, were proposed in the 1980s by British Petroleum (BP) for propane ammoxidation [102] and was intensively studied by several researchers [103–107]. But they show very low selectivity in partial propane oxidation.

In the case of propane oxidation to acrylic acid, V-containing bi-molybdate-based catalysts [108], Ni-molybdates [109], pure or modified vanadyl pyrophosphate (VPP) [110, 111], or materials based on heteropolyacids (HPC) and their salts [112–115] were reported, although yield of AA lower than 20% were reported. As an alternative, it has been also proposed the selective propane oxidation to acrylic acid by sequential ODH of propane (over Ni-based oxide catalyst) and selective oxidation to acrylic acid of propylene (over a mixed metal oxides catalyst) in one reactor loaded with a two-catalyst bed. Based on the use of an O_2-distributor, the authors observed a ~80% of acrylic acid selectivity at a ~19% of propane conversion [116]. But, in all of the last cases, they proved to show a low effectiveness in ammoxidation reaction.

During the 1990s, researchers from Mitsubishi proposed the most effective catalysts for both oxidation and ammoxidation, which are multicomponent mixed metal oxides, i.e. Mo,V-based catalysts promoted with Te and/or Sb and Nb (MoVTe(Sb)NbO catalysts). In fact, these catalysts present high activity and relatively high selectivity for both propane ammoxidation to acrylonitrile (with yield of ca. 60%) [117] and propane oxidation to acrylic acid (with yield of ca. 48%) [118], and as it will be indicated later, for the ODH of ethane to ethylene (with yield of ca. 75%) [119], which suggests a great effectiveness of these catalysts in the selective oxidative activation of C_2-C_4 alkanes [54].

9.4.2 On the Nature of Active and Selective Sites

In the 1990s, the multicomponent MoVTe(Sb)Nb mixed oxides, initially proposed by Mitsubishi, were shown to be highly selective in propane ammoxidation, with yields of acrylonitrile of 60% at reaction temperatures of 400–450 °C [117, 118, 120, 121]. These catalytic results are better than those achieved over V–Sb–O based catalysts, as a consequence of the higher

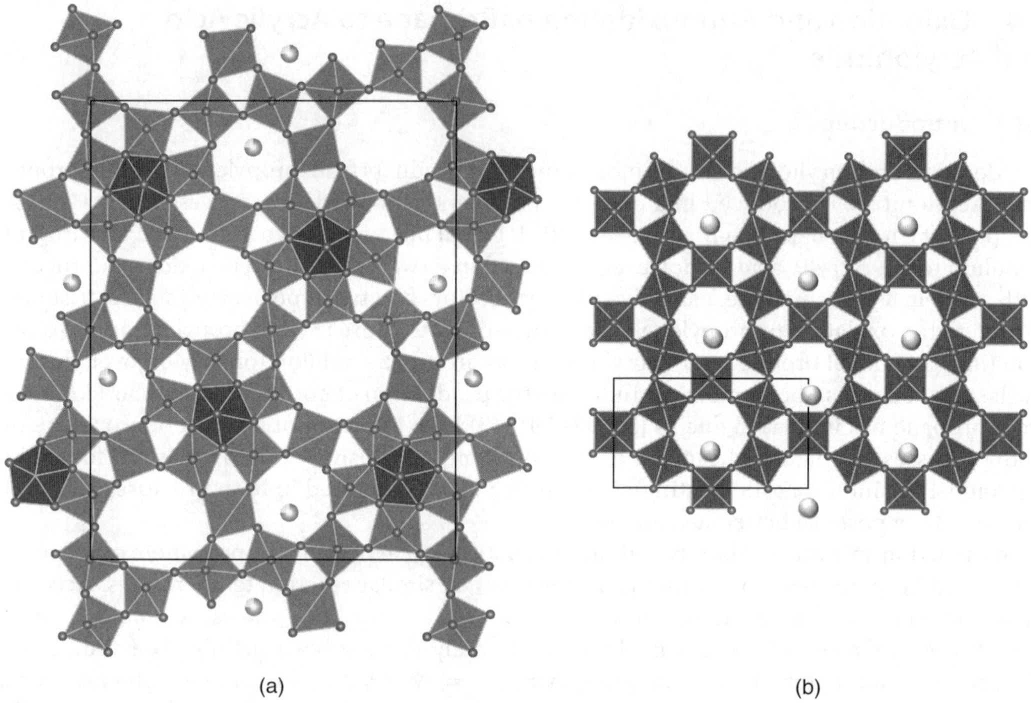

Figure 9.6 The structure of the (a) M1 and (b) M2 phase. Adapted from [125, 128]. (*See insert for color representation.*)

capability to activate ammonia [122]. But, in addition, they were also proposed as active and relatively selective catalysts for partial propane oxidation to acrylic acid with yield to acrylic acid of ca. 45% [123, 124]. Since these first results, an intense study on the synthesis and the characterization of these type of materials has been done in the last 15 years, which include not only an important number of patents but also some fundamental studies in order to understand the nature of active sites and the reaction mechanism.

Initially, it was proposed the existence of two crystalline phases in Mo-V-Te-Nb-O catalysts, called as M1 and as M2, which would be involved in the ODH of propane to propylene and the oxidation of propylene to acrylic acid, respectively [120]. It is now known that in the synthesis of these materials, more than one crystalline phase can be obtained, although only one of them is the active and selective phase: the Mo-bronze $Te_2M_{20}O_{57}$ (M = Mo, V, Nb) [125] (Figure 9.6a), which crystallizes in the orthorhombic space group Pba2, with a formula unit of $Mo_{7.8}V_{1.2}NbTe_{0.94}O_{28.9}$ [117], is able to transform by itself propane to acrylonitrile or to acrylic acid with good performance [126, 127]. This was finally named as the M1 phase.

However, in addition to this, other crystalline phases are observed, depending on the composition and the preparation procedure of catalysts. Of great importance is the presence of a $Te_{0.33}(Mo,V)O_{3.33}$ Mo-bronze [120, 128] (Figure 9.6b), which crystallizes in pseudo-hexagonal space group P*mm*2, with a formula unit of $Mo_{4.67}V_{1.33}Te_{1.82}O_{19.82}$ [129]. This was finally named as the M2 phase [127], because it is inactive in propane oxidation and active and selective in the partial oxidation of propylene to acrolein and/or acrylic acid [130]. However, it has been proposed that the M2 phase could act as a co-catalyst when working at high conversion levels [129] but a partial migration of tellurium from the M2 phase to the surface of M1 phase could also be responsible for a better catalytic behavior of catalysts containing mixtures of both crystalline phases [131].

In addition to these crystalline phases, other phases have been observed, as minorities, depending on the catalyst preparation procedure: $TeMo_5O_{16}$ or other Mo-bronzes such as $(Mo_{0.93}V_{0.07})_5O_{14}$, or $Nb_{0.09}Mo_{0.91}O_{2.80}$ [132], although their influence on activity or selectivity seems to be very low.

Subsequently, it was observed that the partial or total substitution of tellurium by antimony gave rise to the formation of similar crystalline phases, although the conditions of synthesis were in many cases different. Thus, in a similar way, Mo-V-Sb and Mo-V-Sb-Nb catalysts were also proposed [123, 124, 133–136], although with yield in propane oxidation lower than that of Mo-V-Te-based catalysts [99, 123, 124]. In this case, similar crystalline phases have been proposed: (i) an orthorhombic $(SbO_2)M_{20}O_{56}$ bronze [134]; and (ii) a pseudohexagonal $(Sb_2O)M_6O_{18}$ bronze [134], in which Sb can be as Sb^{3+} [137, 138] and/or Sb^{5+} [134], depending on the catalyst preparation procedure or the final catalyst activation method [139–141]. In this way, it was observed that under operando conditions at 380 °C, the oxidation states of Sb and V change, depending on the propane/oxygen ratio in the atmosphere [139].

In general, the selectivity to acrylic acid with many of Sb-containing catalysts could be a consequence of the higher formation of acetic acid, which could be favored on catalysts presenting Brønsted acid sites [137, 142]. In this way, it has been proposed that in both Mo-V-Sb-O [137, 141, 143] and Mo-V-Sb-Nb-O [144] catalysts, alkali-doped materials show a higher formation of acrylic acid. As will be discussed later, Brønsted acid sites in the unselective Sb-containing materials are removed when incorporating alkali metal in synthesis or in post-synthesis step. On the other hand, the use of post-synthesis methods for improving both the presence of M1 phase and the catalytic performance has been proposed [131], although in many cases, e.g. during the elimination of soluble phases, modification on the catalyst surface composition can be also observed [138].

An important factor is related to the synthesis of active and selective catalysts, especially the formation of the M1-phase. The most common preparation methods for these catalysts include the so-called dry-method [120, 121, 123–126, 145, 146], hydrothermal synthesis [135, 147–149], or other less conventional procedures [150]. Moreover, the mechanism for the formation of these catalysts seems to be similar [151], although the presence of each crystalline phase will require specific preparation conditions. The final results is the synthesis of Mo-based bronzes with some specific characteristics [152]: (i) needle-like crystals, growing along the crystallographic c direction; (ii) corner-sharing MO_6 octahedra in the crystallographic plane *ab* (M = Mo, V, W, or Nb), which position themselves in the three-dimensional space to form channels (pentagonal, hexagonal or heptagonal); (iii) except in the case of the hexagonal bronze, the presence of $[M_6O_{21}]^{6-}$ units, with M = Mo, V, W, or Nb, composed of a pentagonal bipyramidal MO_7 unit, which shares the five edges of the planar pentagon with five MO_6 octahedra; and (iv) the presence of different oxidation state for Mo and V, which can be modified on the surface of catalysts depending on the atmosphere.

This suggests that, in the case of multicomponent MoVTe(Sb)Nb mixed oxide catalysts [153], there would be greater flexibility to accommodate the average oxidation state of the elements on the catalyst surface, in reaction conditions, in both oxidation and ammoxidation reactions. This is an important feature in the explanation of the reaction mechanism [58], whereas the bulk should be taken into account as the support of a relatively low number of layers (which may change with the reaction conditions). A model of the M1 unit cell has been proposed [126, 154] (Figure 9.7), which can be used to explain the catalytic performance of catalysts presenting the so-called M1 phase.

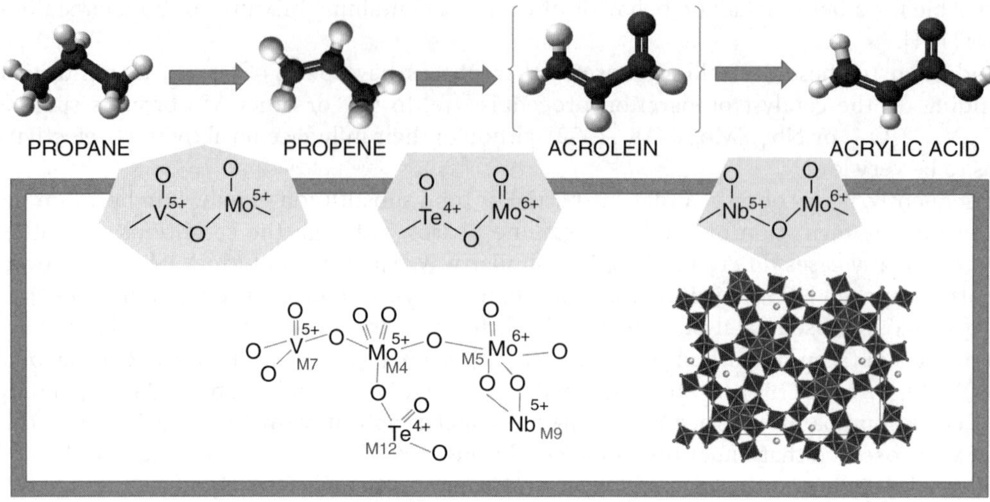

Figure 9.7 The mechanism proposed for propane to acrylic acid. (*See insert for color representation.*)

Most research on the subject suggests that acrylic acid is formed during propane oxidation, with propylene as intermediate [100, 129, 131, 132, 137, 155, 156] in a similar way to that proposed for acrylonitrile formation during the ammoxidation of propane [100, 129, 155, 156]. Thus, it has been proposed, a reaction network for selective oxidation/ammoxidation of propane over catalysts presenting M1 phase can be summarized by considering the following steps [100, 129, 147b]: (i) the ODH of propane to propene on V-O-Mo pairs (in the heptagonal channels, which is considered as the rate-determining step); (ii) propylene is oxidated/ammoxidated over Te^{4+}-O-Mo^{6+} or Sb^{3+}-O-Mo^{6+} pairs (with $O = Mo^{6+} = O$ entities), in the absence of acid sites; (iii) in the presence of acid sites, propylene is transformed to acetic acid and more easily to carbon oxides; and (iv) the presence of Nb deceases both the number of acid sites but also the deep oxidation of reaction products (acrylic acid or acrylonitrile), improving selectivity to partial oxidation products at higher propane conversion. In this way, it was initially proposed that the V^{5+}–O–Mo^{5+}(V^{4+}) moieties adjacent to Te^{4+} and Mo^{6+} sites in the [001] terminal plane provide a spatially isolated active site at which the selective ammoxidation of propane occurs [126]. In fact, it has been observed that V is the key element in the alkane activation [99, 100, 147, 155], although the incorporation of Nb has a positive effect in both activity and selectivity. On the other hand, and as proposed for propylene oxidation/ammoxidation [2, 3], Te^{4+} and Sb^{3+} are involved in the formation of the π-allylic intermediate, whereas the Mo-O-Mo pairs (with $O{=}Mo^{6+}{=}O$ entities) are responsible for the O- and N-insertion [155]. More recently, it has been also proposed that Te=O sites play an important role in the selective activation of propane [156].

It is generally accepted that acrylic acid is formed in the presence of water [157], although it has been also proposed that the reaction can occur in absence of water when using a fluidized bed reactor [158]. In any case, there is a limit of the maximum yield to acrylic acid (selectivity of about 70% at a propane conversion per pass of ca. 50%), as a consequence of the partial degradation of acrylic acid. Therefore, the objective of future studies must be to enhance the performance of catalysts in order to decrease the decomposition of the partial oxidation products [159].

On the other hand, the fact confirming that the same catalysts can be used in both propane oxidation and ammoxidation has been proposed to produce different acrylonitrile/acrylic acid ratios by adjusting the molar ratio of propane to ammonia from 2 to 10 and the molar ratio of oxygen to ammonia from 2 to 10 [160, 161]. In all cases, the partial oxidation of propane to acrylic acid continues to be of interest for industrial development. However, a comparison of different chemical processes from a life cycle perspective of acrylonitrile production indicates that the alternative routes starting from propane seem to have higher potential impact than the traditional SOHIO process, due to the lower catalysts activity [162]. In addition, several technical problems, such as the scale-up, the optimization of the catalysts composition, and energy efficiency, besides the stability of acrylic acid, must be solved. In this way, the propane ammoxidation to acrylonitrile seems to have a greater development than the partial oxidation process. Thus, *Asahi Kasei Chemicals Corp.* have developed a propane-process line at the Ulsan Plant of its *Korean subsidiary Tongsuh Petrochemical Corp.* for acrylonitrile production (coproducing also acetonitrile). *Asahi* and *Thailand's PTT Chemical* are building propane-based plant for acrylonitrile production in Malaysia, employing the same technology, whereas Saudi Basic Industries Corp. (SABIC), Asahi Kasei, and Mitsubishi Corp. could develop a new plant in Jubail (Saudi Arabia) using propane and coproducing sodium cyanide [101].

9.4.3 An Industrial Perspective for Propane Oxidation and Ammoxidation: Interview with Dr. Christian Walsdorff, BASF

CW: My answers will focus on gas-phase oxidations using air or oxygen as oxidant. Gas-phase technology allows to recuperate selectivity losses due to overoxidation in the form of valuable high-temperature energy. Oxidants more expensive than air or oxygen should severely compromise potential savings from using alkanes as feedstocks.

1 Which are the most important scientific and technological challenges for the oxidation of propane to acrylic acid (AA)?
CW: Three different reactions (dehydrogenation, aldehyde formation, and acid formation) have to be combined into one step.
Operation conditions (e.g. temperature or residence time) will have to be a compromise hardly reflecting optimum conditions for all of these reactions.
In order to achieve acceptable selectivity, single-pass conversion will probably have to be limited.
Acceptable overall yields should therefore require recycling of significant amounts of propane (and potential intermediate products such as propylene or acrolein).
Acrylic acid should be relatively easily separated from the product gas. However, propane, propylene, or acrolein would have to be separated from process purge streams (CO_2, CO, nitrogen,…) for recycling, and this will not be cheap.
Using oxygen as oxidant feed gas instead of air would probably be necessary to limit purge streams.

2 Which are the most important hurdles, which hinder the improvement of industrial technologies for propane to AA?
CW: Propane-to-AA oxidation technology has to compete with a troika of well-established technologies for propane-dehydrogenation /propylene-to-acrolein-oxidation/acrolein-to-AA-oxidation on the same raw material basis.
These technologies have been highly developed on a technology learning curve with huge global capacities in place.
In order to switch from propylene to propane, adding a separate propane dehydrogenation up front to propylene-based AA technology may be more attractive than propane oxidation.

Process integration between PDH and propylene-based AA technology may offer further synergies.
Regarding oxidation of propane to AA, concerns regarding impurities with respect to acrylic acid specification should be thoroughly addressed.

3 A general question: which are the perspectives and trends for alkanes selective oxidation to chemicals?
CW: Selective alkane oxidations will continue to face tough competition from alkylene-based technologies founded on dehydrogenation or steam cracking of alkanes and syngas-based technologies in case of methane. The well-established process for *n*-butane-based maleic anhydride is a special case. Maleic anhydride is a relatively stable product limiting consecutive overoxidation (total combustion) and thus allowing relatively high single-pass conversions with acceptable selectivity. Therefore, relatively high yields can be achieved without recycle of *n*-butane.
Maleic anhydride is readily separated from the product gas, and the remainder, including unconverted *n*-butane, can be incinerated to produce valuable high-temperature energy (steam). Air can be used as cheap oxidant (instead of oxygen), as there is no recycle from the product gas. Oxidation of higher alkanes will be generally less selective or at least offering lower carbon yields due to the higher number of potentially alternative carbon atoms for oxidation. Attractive products from the oxidation of lower alkanes will often be prone to overoxidation under the relatively harsh conditions required to convert the alkanes.
It remains to be seen whether relatively recent industrial processes for selective oxidation of acrylonitrile from propane or acetic acid from ethane will develop into a general threat for the respective alkylene- or syngas-based technologies.

9.5 Oxidative Dehydrogenation (ODH) of C_2-C_4 Alkanes

9.5.1 Introduction

For at least three decades, ODH of C_2-C_4 alkanes has been considered as the potential game-changer for production of base chemicals. Compared to conventional technologies like cracking or nonoxidative dehydrogenation (DH), ODH has the important advantage to work at significantly lower temperatures (as first approximation, 450 °C vs. 850 °C or 650 °C) while avoiding coke formation. Introduction of an oxidant (e.g. oxygen) in the reaction mixture changes the thermodynamics of the process, removing limitations that affect cracking and DH, and in turn eliminating the need of high temperatures to shift equilibrium toward olefins. Despite the potential energy save that characterizes alkanes ODH and the enormous scientific interest that it has found in the last 30 years [163, 164], production of light olefins in the petrochemical industry still largely relies on non-catalytic and relatively unselective cracking processes that are based on technologies developed in the early twentieth century. The reason for this failure lies behind important hurdles known to affect alkanes ODH. One above all is the consecutive total oxidation of the olefins, which combusts the valuable intermediates into carbon oxides. In order to limit consecutive total oxidation, hence increasing selectivity to olefins, conversion is typically kept low, creating the need to separate alkane from olefin(s) and recycle the alkane. Separation of an olefin from its parent alkane using conventional distillation technologies is very challenging, due to the small difference in boiling points. In turn, this translates into high distillation costs, both in terms of CAPEX (need of very tall distillation towers) and OPEX (need of high amounts of energy to manage distillation units). To reduce volumes of recycled gas (and avoid production of NO_x) as an oxidant, it would be ideal to use oxygen instead of air in order to avoid managing large amounts of nitrogen. To this end, a cryogenic distillation unit (or other competitive technologies) should be built close to the ODH reactor, further increasing investments and lowering margins compared to DH or cracking. Last but not least, safety concerns related to mixing oxygen with potentially flammable hydrocarbons are sometimes considered as important obstacles to the development of the oxidation processes in a type of industry, like the refining and petrochemical one, that is not always used to managing gaseous mixtures with oxidant and hydrocarbons present at the same time in the same environment. Overall, due to the lack of reactive allylic hydrogens, higher temperatures are required to activate ethane's C—H bond compared to C_3 and C_4 alkanes; nonetheless, for the very same reason, ethane ODH has found more success, since consecutive total oxidations are less favored; [165] indeed, ethylene yields higher than 75% at 400 °C were obtained [59, 119]. Much lower yields are instead obtained for C3 and C4 alkanes, the best values being generally below 30% yield [163, 166–168]. In any case, no alkane ODH process for production of olefins has so far attained industrialization.

About 10 years after it was first asked, "How far from implementation is alkanes ODH?" [163], it is worth asking today the same question and see if anything has really changed. Improvements on the process side, e.g. use of membranes, have been studied further and reported in literature [169, 170]. The next paragraph discusses novelties about catalytic active phases.

9.5.2 New Active Phases for Alkanes ODH

In the last 10 years or so, researchers have studied in depth the known active phases for C_2-C_4 alkanes ODH, such as combinations of more or less well-defined supported vanadium oxides [171], bulk mixed-oxide phases [172–175] or nonreducible metal oxides [176]. Catalytic performance of these phases were analyzed in detail and reviewed [163, 164, 167, 173, 174, 177, 178]. It is worth noting that some of these catalysts still represent the most active and selective materials for the titled reactions. Nonetheless, completely new active phases have found significant attention more recently, that is, carbon nanotubes (CNTs) and hexagonal boron nitride (h-BN). In chronological order, CNTs paved the way for use of metal-free materials as ODH catalysts [179]. Oxygen-functionalized terminations and defects of CNTs were found to be the active sites for the titled transformation [179, 180], with a redox cycle that involves ketonic carbonyl and hydroxyl pairs. Initially, it was thought that the redox process was similar to the one classically reported for metal-oxides. However, due to the nanoscopic nature of CNTs, a Mars-Van Krevelen-type mechanism is likely impossible to take place. More recently, a Langmuir-Hinshelwood mechanism was proposed; in this case, oxygen molecules were activated at defects sites (and terminations) of CNTs with diffusion of activated atomic oxygen happening on the basal plane of sp^2 carbon [181]. Further studies on model nanocarbon-catalysts support an intermediate view, where the entire catalytic process can be separated into a hydrogen abstraction step and a reoxidation step, which however should be distinguished from the typical Mars-van-Krevelen or Langmuir-Hinshelwood mechanisms because of the unique structure and chemical reactivity of the carbon materials, e.g. the absence of lattice oxygen species and relatively high activity in oxygen activation [182]. Several variations of CNTs as ODH catalysts have followed the pioneering work of J. Zhang et al. [179]; typical examples are graphene structures doped with heteroatoms, in general phosphorous, boron and nitrogen [183–185]. Role of these dopants is to modify alkane and oxygen activation and adsorption energies, eventually tuning both selectivity and conversion [180, 184]. As an example, CNTs showed propane conversion up to 15–20%, but propylene selectivity always below 25%, whereas boron or phosphorous-modified CNTs showed conversions always below 10% but selectivity between 90% and 40% [180]. Nitrogen doped CNTs outperformed boron and phosphorous-doped peers, showing propane selectivity between 60% and 30% at conversions up to 10–14% [184]. In any case, at conversions above 10%, selectivity to carbon oxides are always very high for all CNTs-based ODH catalysts, typically making carbon oxides the major oxidation products instead of the olefins.

Starting from late 2016, the inorganic equivalent of graphene, i.e. h-BN, was found to be an active and selective catalyst for C_2-C_4 alkanes ODH [186–189]. Using hydroxylated h-BN (BNOH), ethylene selectivity as high as 95% was observed at ethane conversion of 11% at 570 °C, with only small amounts of CO_2 (0.4%) formed. Higher ethylene yield, ca. 50%, was obtained at higher temperature (600 °C), the other major byproduct being CO. Worth of note, although very interesting, these values are not better than state-of-the-art M1 catalysts for the same reaction [190]. Compared to other types of active phases, h-BN showed a more interesting behavior for C_3 and C_4 alkanes. In a first report [188] at 14% propane conversion, 79% propene selectivity was obtained, whereas in a second publication 80% propylene selectivity was shown at 20% alkane conversion on BNOH [189]. The latter are among the highest selectivity values, at the mentioned level of conversion, so far ever reported for propane ODH. Moreover, contrary to what is typically observed with other catalysts, major byproducts were not carbon oxides but ethylene. Similarly, *n*-butane and isobutane ODH on h-BN showed, respectively, 70% and 60% selectivity to the corresponding C_4-olefin at 14% conversion [186]. As in the case of propane ODH, also for C_4-alkanes the major byproducts were lower olefin, i.e. propylene and ethylene, instead of CO_x. Indeed, the sum of C_4-C_2 olefins obtained from C_4-alkanes was 90–95% (at 14% alkane conversion). The fact that only a minor amount of carbon oxides were formed in favor of olefins is a first indication that alkanes ODH on BN follow a different reaction pathway compared to the vast majority of catalysts so far reported.

Postulated mechanisms derived from (*in situ*) characterization, kinetic analyses and computational studies [186–189], suggest that the active site for alkanes ODH on h-BN (and, similarly, on BNOH) is an inorganic peroxide species, >B—O—O—N<, that forms in presence of O_2 at high temperature on B—OH moieties present on h-BN terminations and defects. The peroxide site can activate a C—H bond in alkanes (probably on a secondary carbon for $C > 2$ alkanes), forming short-lived alkyl radicals that evolve in surface stabilized N—O—CH—$(R)_2$ intermediates ($(R)_2$ = aliphatic chains and/or H). Fast transformation of the alkyl radical into the more stable nitroxyl species could be the reason behind the high olefin selectivity observed and the limited formation of carbon oxides, the latter being typical consecutive products of the unstable alkyl radicals (as previously observed for nonreducible catalysts) [163]. For C_3-C_4 alkanes, formation of lower olefins as byproducts is synonym of the presence of a C—C cleavage reaction in competition with ODH. H-transfer from a B—OH moiety close to an alkyl radical previously formed on the catalyst's surface could lead to simultaneous breaking of the C—C bond, forming a B—O—CH_3 group and an intermediate alkyl-species that desorbs as olefin [186]. Overall, mechanistic hypotheses so far reported come from early-stage studies, which require further work to be validated.

An overview of the best ODH performance so far reported for CNTs and h-BN related materials is proposed in Table 9.2. Comparison with other state-of-the-art catalysts is presented. It can be noticed that, at low conversions, h-BN has a competitive performance compared to known active phases. However, the highest yields are still to be attributed to "conventional" catalytic systems.

9.5.3 Any Closer to Implementation?

From a scientific point of view, it is fair to say that the discovery which (modified) CNTs and h-BN can act as catalysts is among the most interesting novelties in the field of alkanes ODH and catalysis in general. However, apart from scientific curiosity, to date it remains unclear if a

Table 9.2 General overview of the catalytic performance of CNTs- and BN- related materials compared to reference active phases.

	ca. 15% alkane conversion							
Catalyst	**S C_2H_4 (%)**	**S C_3H_6 (%)**	**S C_4H_8 (%)**	**S CO_x (%)**	**Y (%)**	**T (°C)**	**Y_{max} (%) (T, °C)**	**Ref.**
C_2H_6 *ODH*								
MoVTeNbO (M1)	97	—	—	3	15	360	75 (400)	[190]
Li/Na/Mg/Dy/Cl	95	—	—	5	14	590	75(650)	[178]
V-CoAPO-18	80	—	—	20	12	600	34 (600)	[191]
o-CNTs	35	—	—	65	5	450	7 (450)	[192]
h-BN	95	—	—	5	15	570	50 (600)	[187]
C_3H_8 *ODH*								
V-MCF	2	92	—	6	14	550	28 (550)	[193]
VO_x/Al_2O_3	—	60	—	40	9	500	9 (500)	[194]
VO_x/TiO_2	—	43	—	57	7	500	7 (500)	[195]
N-CNTs	—	38	—	62	6	400	6 (400)	[184]
h-BN	10	76	—	14	12	500	12 (500)	[188]
C_4H_{10} (*n*-butane) *ODH*								
VMgO	—	—	60	40	9	500	20 (500)	[168]
VO_x/ Al_2O_3	—	1–3	63	34	10	600	14 (650)	[196]
P-o-CNTs	—	—	53	48	8	450	12 (450)	[179]
h-BN	14	18	65	1	10	460	10 (460)	[186]

Note: values reported in this table are indicative and rounded. Values that were not directly available at the mentioned level of conversion were interpolated.

net advantage is gained with these new phases compared to previous state-of-the-art catalysts. In the case of CNTs, at low conversion levels, reports of improved selectivity have increased, but olefin yields are still very low and large amounts of carbon oxides are formed. It is also likely that CNTs are not stable for long time on streams (longer than a few hundred hours), which are required for industrial operations: at high temperatures and in presence of oxygen, CNTs' fate is to burn into carbon oxides. To date, CNTs for alkanes ODH remain an interesting academic research, which, however, does not seem to have traction for large-scale industrial applications. In the case of boron nitride, if on one hand record olefin selectivity was obtained for C_3-C_4 alkanes at conversions below 20%, on the other hand no information is available at higher conversions, which seems to point to a relative inertness (that might not be too surprising) of the h-BN based materials so far studied. The mild slope with which olefin selectivities decrease in the range of conversions so far explored [186, 188] give hope for future promising results; however, to keep the process competitive with existing technologies, reaction temperatures should remain relatively low and productivities at least around 1 h^{-1}, i.e. typical industrial values. So far, olefins productivities remain rather low for propylene and butenes (<0.6 h^{-1}) except if boron nitride nanotubes (BNNTs) are used instead of conventional h-BN [188]. However, cost of BNNTs is currently far too high to think about loading ton-scale reactors with this material. Yields of the targeted olefins are also limited, but since the main byproducts obtained with h-BN are other valuable olefins rather than carbon oxides, an important economic advantage should be gained. Nonetheless, to further improve economics and maybe reach commercialization, total selectivity to the targeted olefin would be preferred to make ODH more competitive compared to steam cracking. Indeed, this well-established technology can produce (starting from naphtha) ethylene yields up to 30% and propylene/ethylene ratios up to 0.6 (although at much higher temperatures).

Concluding, about 10 years after first addressing the question of how far from implementation alkanes (C_2-C_4) ODH might be [163], it seems fair to say that no new catalyst has provided significantly better olefin yields compared to previously reported materials. The most promising ODH reaction remains ethane to ethylene, with the M1 phase reported to already allow shut-down economics in some cases [197, 198]. Nonetheless, interesting olefin selectivity and product distributions using h-BN-based materials have been recently reported. If alkanes conversions will be improved without affecting too much selectivity, the unique behavior of BN materials gives hope for future potential applications.

9.5.4 An Industrial Perspective for Alkanes ODH: Interview with Dr. Anne Gaffney, Idaho National Laboratory (Formerly ARCO, DuPont, Rohm and Haas, and Lummus)

1 Which are, in your opinion, the most important hurdles that hinder the development and industrial implementation of oxidative dehydrogenation of alkanes?
AG:
   a) First-of a-kind technology hurdles, risk aversion, capital already in the ground, are the responses we hear from U.S. manufacturers.
   b) High capital investment required if this technology is not coupled with membrane separations (i.e. need to replace distillation towers with alkane/alkene membrane separation systems). Therefore a modular approach should be investigated which includes the process intensified membrane system.
   c) Will require the commercialization of new catalytic materials which industry views as challenging. Idaho National Laboratory is filling this manufacturing need with its recently developed initiative, Catalytic Technologies Network.
   d) Need an integrated ODH membrane reactor system. Idaho National Laboratory is partnering with Compact Membrane Systems in commercializing an ODH membrane reactor technology that will eliminate the need for distillation towers.

2 Which is your vision about the future development of oxidative dehydrogenation of alkanes?
AG: The easiest path to commercialization is small, modular ODH membrane reactor systems, trademarked by INL as e-Zero™ that meet all 14 metrics set by the Department of Energy, Advanced Manufacturing Office, as outlined in the FY15 Quadrennial Technology Report. These systems will be located at disbursed, stranded, natural gas sites.

3 Which are the most important scientific and technological challenges for the ODH of alkanes?
AG: Catalyst scale-up is the most challenging according to industry. Therefore, Idaho National Laboratory has made catalyst scale-up a key component in the mission of its recently developed Catalytic Technologies Network. We will take new catalytic material from the laboratory and scale-up these new compositions for industrial applications.

4 Which are the perspectives and trends for alkane selective oxidation to chemicals (optional)?

AG: Working with academia and industry, Idaho National Laboratory has mapped out a new petroleum chemical roadmap which transforms light alkanes to higher-value consumer products. This is a paradigm shift from starting with heavy crude hydrocarbon resources. With the long term, abundant, cheap supply of light alkanes here in the U.S., we are poised to reshape the future of advanced manufacturing here in the U.S.

References

1 H. Krieger, L.S. Kirch, Process for the production of unsaturated acids, US Patent 4 260 822A, Röhm et Haas Co. 1981.
2 H. Imai, M. Nakatsuka, Mariko, A. Aoshima Production of methacrolein and methacrylic acid, from Jpn. Kokai Tokkyo Koho JP 62132832 A 19870616 1987.
3 S. Yamanatsu, T. Yamaguchi, (1993), Process for the preparation of methacrylic acid and methacrolein, US Patent 5191116 to Asahi Kasei Kogyo Kabushiki Kaisha.
4 Millet, J.M.M. and Védrine, J.C. (2001). *Top. Catal.* 15: 139.
5 Blouet-Crusson, E., Rigole, M., Fournier, M. et al. (1999). *Appl. Catal. A: Gen.* 178: 69.
6 Staroverova, I.N., Kutyrev, M., and Stakheev, A. (1992). *Kinetika i Kataliz* 33: 127.
7 Sultan, M., Paul, S., Fournier, M. et al. (2004). *Appl. Catal. A* 259: 141.
8 Langpape, M., Millet, J.M.M., Ozkan, U.S. et al. (1999). *J. Catal.* 182: 148.
9 Mizuno, N., Tateishi, M., and Iwamoto, M. (1994). *J. Chem. Soc. Chem. Commun.* 1411.
10 Cavani, F., Etienne, E., Favaro, M. et al. (1995). *Catal. Lett.* 32.
11 Mizuno, N., Tateishi, M., and Iwamoto, M. (1996). *J. Catal.* 163: 87.
12 Misono, N., Mizuno, N., Imumaru, K. et al. (1997). Third World Congress on Oxidation Catalysis. In: *Studies in Surface Science and Catalysis*, vol. 110 (ed. R.K. Grasseli, S.T. Oyama, A.M. Gafney and J.E. Lyons), 433. Amsterdam: Elsevier.
13 Albonetti, S., Cavani, F., and Trifirò, F. (1996). *Catal. Rev. Sci. Eng.* 38: 413.
14 Mizuno, N. and Yahiro, H. (1998). *J. Phys. Chem. B* 102: 437.
15 Langpape, M. and Millet, J.M.M. (2000). *Appl. Catal. A: Gen.* 200: 89.
16 Etienne, E., Cavani, F., Mezzogori, R. et al. (2003). *Appl. Catal. A: Gen.* 256: 275.
17 Cavani, F., Mezzogori, R., Pigamo, A. et al. (2001). *Chem. Eng. J.* 82: 33.
18 Huynh, Q., Shuurmann, Y., Delichere, O. et al. (2009). *J. Catal.* 261: 166.
19 Wu, S., Kan, Q., Ding, W. et al. (2012). *React. Kinet. Mech. Catal.* 106: 157.
20 Kendell, S., Alston, A., Ballam, N. et al. (2011). *Catal. Lett.* 141: 374.
21 Schindler, G.P., Knapp, C., Ui, T. et al. (2003). *Top. Catal.* 22: 117.
22 Huynh, Q., Selmi, A., Corbel, G. et al. (2009). *J. Catal.* 266: 64.
23 Wienold, J., Timpe, O., and Ressler, T. (2003). *Chem. Eur. J.* 9: 6007.
24 Brückner, A., Scholz, G., Heidemann, D. et al. (2007). *J. Catal.* 245: 369.
25 Li, W. and Ueda, W. (1997). 3rd world congress on oxidation. In: *Studies in surface science and catalysis*, vol. 110 (ed. R.K. Grasseli, S.T. Oyama, A.M. Gafney and J.E. Lyons), 433. Amsterdam: Elsevier.
26 Li, W. and Ueda, W. (1997). *Catal. Lett.* 46: 261.
27 Centi, G., Lena, V., Trifiro, F. et al. (1990). *J. Chem. Soc. Faraday* 86: 2775.
28 Cavani, F., Etienne, E., Mezzogori, R. et al. (2001). *Catal. Lett.* 75: 99.
29 Jalowwiecki-Duhamel, L., Monnier, A., Barbaux, Y. et al. (1996). *Catal. Today* 32: 237.
30 Cavani, F., Mezzogori, R., Pigamo, A. et al. (2003). *Top. Catal.* 23: 119.
31 Ushikubo, T. (2003). *Catal. Today* 78: 79.
32 Jing, F., Katryniok, B., Bordes-Richard, E. et al. (2013). *Catal. Today* 203: 32.
33 Busca, G., Cavani, F., Etienne, E. et al. (1996). *J. Mol. Catal. A* 114: 343.

34 Mestl, G., Ilkenhans, T., Spielbauer, D. et al. (2001). *Appl. Catal. A* 210: 13.
35 J. Shibata, E. Shiraishi, Assignee: Sumitomo Chemical Co., Ltd., Japan US 20090259070 A1 2009.
36 Y. Ohishi, J. Yoshizawa, E. Shiraishi, Method for regenerating catalyst for the production of methacrylic acid and process for preparing methacrylic acid, US 2009/0259071 A1 2010.
37 Al-Zahrani, S.M., Elbashir, N.O., Abasaeed, A.E. et al. (2001). *Ind. Eng. Chem. Res.* 40 (3): 781.
38 Hoang, M., Mathews, J.F., and Pratt, K.C. (1997). *J. Catal.* 171: 320.
39 Takita, Y., Sano, K., Kurosaki, K. et al. (1998). *Appl. Catal. A* 167: 49.
40 Takita, Y., Sano, K., Muraya, T. et al. (1998). *Appl. Catal. A* 170: 23.
41 Marcu, C., Millet, J.M.M., and Sandulescu, I. (2005). *J. Serb. Chem. Soc.* 70: 791.
42 Takita, Y., Kikutani, K., Xia, C. et al. (2005). *Appl. Catal. A: Gen.* 283: 209.
43 He, L., Fu, L., and Tang, Y. (2015). *Catal. Sci. Technol.* 5: 1115.
44 Hermans, I., Venegas, J., Grant, J.T. et al. (2017). *ChemCatChem* doi: 10.1002/cctc.201601686.
45 Ellis, P.E. and Lyons, J.E. (1989). *Catal. Lett.* 3: 389.
46 Labinger, J.A. (1994). *Catal. Lett.* 26: 95.
47 Moore, K.T., Horvath, I.T., and Therien, M.J. (2000). *Inorg. Chem.* 39: 3125.
48 2-Methyl-2-propen-1-ol (CAS 513-42-8) Market Research Report 2017.
49 Dubois, J.L. (2005). *Catal. Today* 99: 5.
50 Pyo, S.H., Dishisha, T., Dayankac, S. et al. (2012). *Green Chem.* 14: 1942.
51 Weichler, M.T., Kurteva-Yaneva, N., Przybylski, D. et al. (2015). *Appl. Environ. Microbiol.* 81: 4564.
52 Rohwerder, T. and Müller, R.H. (2017). *Appl. Environ. Microbiol.* doi: 10.1128/AEM.02622-16.
53 Centi, G., Trifiro, F., Ebner, J.R. et al. (1988). *Chem. Rev.* 88: 55.
54 Lopez Nieto, J.M., Solsona, B., Concepcion, P. et al. (2010). *Catal. Today* 157: 291.
55 Gulliants, V.V. and Holmes, S. (2001). *J. Mol. Catal. A: Chem.* 175: 227.
56 Guan, J., Xu, H., Song, K. et al. (2008). *Catal. Lett.* 126: 293.
57 Guan, J.Q., Wang, Z.Q., Xu, C. et al. (2009). *Catal. Lett.* 128: 356.
58 Shishido, T., Inoue, A., Konishi, T. et al. (2000). *Catal. Lett.* 68: 215.
59 Ivars, F. and Lopez Nieto, J.M. (2014). *Handbook of advanced methods and processes in oxidation catalysis* (ed. D. Duprez and F. Cavani), 767. London: Imperial College Press.
60 Ballarini, N., Cavani, F., Cortelli, C. et al. (2006). *Topics Catal.* 38: 147.
61 Dummer, N.F., Bartley, K.K., and Hutchings, G.J. (2011). *Adv. Catal.* 54: 189.
62 Bordes-Richard, E., Shekari, A., and Patience, G.S. (2014). *Handbook of advanced methods and processes in oxidation catalysis* (ed. D. Duprez and F. Cavani), 549. London: Imperial College Press.
63 Novelli, M., Leonardi, M., and Cortelli, C. (2014). *Handbook of advanced methods and processes in oxidation catalysis* (ed. D. Duprez and F. Cavani), 334. London: Imperial College Press.
64 Trifirò, F. and Grasselli, R.K. (2014). *Topics Catal.* 57: 1188.
65 Cavani, F., Luciani, S., Degli Esposti, E. et al. (2010). *Chem. Eur. J.* 16: 1646.
66 Cavani, F., De Santi, D., Luciani, S. et al. (2010). *Appl. Catal. A* 376: 66.
67 Caldarelli, A., Cavani, F., Folco, F. et al. (2010). *Catal. Today* 157: 204.
68 Ballarini, N., Cavani, F., Cortelli, C. et al. (2006). *Catal. Today* 117: 174.
69 Lesser, D., Mestl, G., and Turek, T. (2016). *Appl. Catal. A* 510: 1.
70 Mestl, G., Lesser, D., and Turek, T. (2016). *Topics Catal.* 59: 1533.

71 Mestl, G. (2014). *Handbook of advanced methods and processes in oxidation catalysis* (ed. D. Duprez and F. Cavani), 302. London: Imperial College Press.
72 Diedenhoven, J., Reitzmann, A., Mestl, G. et al. (2012). *Chem. Ing. Tech.* 84: 517.
73 Eichelbaum, M., Hävecker, M., Heine, C. et al. (2015). *Angew. Chem. Int. Ed.* 54: 2922.
74 Heine, C., Hävecker, M., Stotz, E. et al. (2014). *J. Phys. Chem. C* 118: 20405.
75 Eichelbaum, M., Glaum, R., Hävecker, M. et al. (2013). *ChemCatChem* 5: 2318.
76 Eichelbaum, M., Hävecker, M., Heine, C. et al. (2012). *Angew. Chem. Int. Ed.* 51: 6246.
77 Cheng, M.-J., Goddard, W.A. III,, and Fu, R. (2014). *Topics Catal.* 57: 1171.
78 Cheng, M.-J. and Goddard, W.A. III, (2013). *J. Am. Chem. Soc.* 135: 4600.
79 Dong, Y., Keil, F.J., Korup, O. et al. (2016). *Chem. Eng. Sci.* 142: 99.
80 Romano, A., Di Giuliano, A., Gallucci, K. et al. (2016). *Chem. Eng. Res. Des.* 114: 79.
81 Chaudhari, P. and Garg, S. (2017). *Comp. Chem. Eng.* 100: 198.
82 Wilkinson, S.K., Simmons, M.J.H., Stitt, E.H. et al. (2013). *J. Catal.* 299: 249.
83 Shekari, A. and Patience, G.S. (2012). *Int. J. Chem. React. Eng.* 10: A63.
84 Shekari, A. and Patience, G.S. (2013). *Canad. J. Chem. Eng.* 91: 291.
85 Fan, X.-B., Dummer, N.F., Taylor, S.H. et al. (2012). *Catal. Today* 183: 52.
86 Alhumaimess, M., Lin, Z., Dummer, N.F. et al. (2016). *Catal. Sci. Technol.* 6: 2910.
87 S. Ligi, F. Cavani, S. Albonetti, et al., 2005, EP 1183101 B1, assigned to Lonza SpA.
88 S. Altwasser, C.K. Dobner, H. Wiener, et al. 2012, US 20120149919 A1, assigned to BASF.
89 G. Mazzoni, G. Stefani, F. Cavani, 1997, Process for the tranformation of a vanadium/phosphorous mixed oxide catalyst precursor into the active catalyst for the production of maleic anhydride, EP 804963B1, Polynt SpA 2000.
90 Albonetti, S., Budi, F., Cavani, F. et al. (2001). *Stud. Surf. Sci. Catal.* 136: 141.
91 A. Reitzmann, W.M. Brandstaedter, L. Streitfinger, et al. 2012, Catalyst shaped body for flow-conducting fixed-bed reactors, WO 2012069481 A1, Süd-Chemie AG.
92 C.K. Dobner, S. Altwasser, W. Hagen, et al. 2012, Catalyst molded bodies and method for producing maleic acid anhydride, US 20110257414 A1, BASF.
93 C. Dobner, M. Duda, A. Raichle, et al. 2008, Catalyst and methods for producing maleic anhydride, US 20080227992 A1, assigned to BASF.
94 M.S. Haddad, R.A. Gustaferro, 2013, Catalyst for n-butane oxidation to maleic anhydride, patent US20130102455A1, INEOS Europe AG.
95 F. Cavani, F. Pierelli, F. Ghelfi, et al., 2010, Catalyst for n-butane oxidation to maleic anhydride, EP1663482B1, Polynt SpA.
96 Caldarelli, A., Bañares, M.A., Cortelli, C. et al. (2014). *Catal. Sci. Technol.* 4: 419.
97 Grasselli, R.K. (2014). *Catal. Today* 238: 10.
98 Vitry, D., Dubois, J.-L., and Ueda, W. (2004). *J. Mol. Catal. A: Chem.* 220: 67.
99 Lin, M.M. (2003). *Appl. Catal. A: Gen.* 250: 287.
100 Grasselli, R.K., Burrington, J.D., Buttrey, D.J. et al. (2003). *Top. Catal.* 23: 5.
101 Schlögl, R. (2016). *Top. Catal.* 59: 1461.
102 (a) A.T. Guttmann, R.K. Grasselli, J.F. Brazdil, Ammoxidation of paraffins and catalysts therefor, patent US4788317A, Standard Oil Co. (1988); (b) A.T. Guttmann, R.K. Grasselli, J.F. Brazdil, (1988), US Pat 4746641 A assigned Standard Oil Co., USA).
103 (a) Catani, R., Centi, G., Trifiro, F. et al. (1992). *Ind. Eng. Chem. Res.* 31: 107. (b) Centi, G., Perathoner, S., and Trifiro, F. (1997). *Appl. Catal. A: Gen.* 157: 143.
104 Nilsson, J., Landa-Canovas, A.R., Hansen, S.A. et al. (1999). *J. Catal.* 186: 442.
105 Zanthoff, H.W., Grunert, W., Buchholz, S. et al. (2000). *J. Mol. Catal. A: Chem.* 162: 443.
106 (a) Albonetti, S., Blanchard, G., Burattin, P. et al. (1998). *Catal. Lett.* 50: 17; (b) Cavani, F., Ballarini, N., Cimini, M. et al. (2006). *Catal. Today* 112: 12.

107 (a) Arcozzi, E., Ballarini, N., Cavani, F. et al. (2008). *Catal. Today* 138: 97. (b) Ballarini, N., Cavani, F., Marion, P. et al. (2009). *Catal. Today* 142: 170.
108 J.P. Bartek, A.M. Elmer, J.R. Brazdil, (1993), Process for oxidation of propane, US Pat 5,198,580, The Standard Oil Co.
109 (a) Kaddouri, A., Mazzocchia, C., and Tempesti, E. (1999). *Appl. Catal. A: Gen.* 180: 271. (b) Madeira, L.M., Portela, M.F., and Mazzochia, C. (2004). *Catal. Rev. Sci. Eng.* 46: 53.
110 Ai, M. (1992). *Catal. Today* 13: 679.
111 (a) Landi, G., Lisi, L., and Volta, J.C. (2004). *Catal. Today* 91–92: 275. (b) Landi, G., Lisi, L., and Volta, J.C. (2003). *Chem. Commun.* 492.
112 Mizuno, N., Tateishi, M., and Iwamoto, M. (1995). *Appl. Catal. A: Gen* 128: L165.
113 Li, W., Oshihara, K., and Ueda, W. (1999). *Appl. Catal. A: Gen.* 182: 357.
114 (a) Dillon, C.J., Holles, J.H., Davis, R.J. et al. (2003). *J. Catal.* 218: 54; (b) Holles, J.H., Dillon, C.J., Labinger, J.A., and Davis, M.E. (2003). *J. Catal.* 218: 42.
115 Dimitratos, N. and Vedrine, J.C. (2006). *J. Mol. Catal. A: Chem* 255: 184.
116 Fang, W., Ge, Q.-J., Yu, J.-F. et al. (2011). *Ind. Eng. Chem. Res.* 50: 1962.
117 T. Ushikubo, K. Oshima, A. Kayo, et al. (1992), Process for Producing Nitriles, patent EP0529853B1, Mitsubishi Chemical Corp.
118 T. Ushikubo, H. Nakamura, Y. Koyasu, et al. (1995), Method for producing an unsaturated carboxylic acid, patent US5380933A, Mitsubishi Chemical Corp.
119 J.M. Lopez Nieto, P. Botella, M.I. Vazquez, et al. (2003), Method for the oxidative dehydrogenation of ethane, patente WO2003064035B1, Consejo Superior De Investigaciones Cientificas, Universidad Politecnica De Valencia, UPV-CSIC.
120 Ushikubo, T., Oshima, K., Kayou, A. et al. (1997). Ammoxidation of propane over Mo-V-Nb-Te mixed oxide catalysts. In: *Spillover and Migration of Surface Species on Catalysts*, vol. 112 (ed. C. Li and Q. Xin), 473.
121 Ushikubo, T., Oshima, K., Kayou, A. et al. (1997). *J. Catal.* 169: 394.
122 Vaarkamp, M. and Ushikubo, T. (1998). *Appl. Catal. A: Gen.* 174: 99.
123 Tsuji, H. and Koyasu, Y. (2002). *J. Am. Chem. Soc.* 124: 5608.
124 Tsuji, H., Oshima, K., and Koyasu, Y. (2003). *Chem. Mater.* 15: 2112.
125 Millet, J.M.M., Roussel, H., Pigamo, A. et al. (2002). *Appl. Catal. A: Gen.* 232: 77.
126 (a) DeSanto, P., Buttrey, D.J., Grasselli, R.K. et al. (2004). *Z. Kristallogr.* 219: 152. (b) DeSanto, P. Jr.,, Buttrey, D.J., Grasselli, R.K. et al. (2003). *Top. Catal.* 23: 23.
127 Baca, M., Pigamo, A., Dubois, J.L. et al. (2003). *Top. Catal.* 23: 39.
128 Garcıa-Gonzalez, E., Lopez Nieto, J.M., Botella, P. et al. (2002). *Chem. Mater.* 14: 4416.
129 Grasselli, R.K., Buttrey, D.J., DeSanto, P. Jr., et al. (2004). *Catal. Today* 91–92: 251.
130 Botella, P., López Nieto, J.M., and Solsona, B. (2002). *Catal. Lett.* 78: 383.
131 Baca, M., Aouine, M., Dubois, J.L., and Millet, J.M.M. (2005). *J. Catal.* 233: 234.
132 Botella, P., Garcıa-Gonzalez, E., Lopez Nieto, J.M. et al. (2005). *Solid State Sci.* 7: 507.
133 (a) M. Hatano, K. Kayou, Procédé pour la production de nitriles, patent EP0318295B1, Mitsubishi Kasei Corp (1992); (b) H. Hinago, K. Hiroyuki, Verfahren zur Herstellung eines Oxidkatalysators für die Oxidation oder Ammonoxidation, patent DE10145958A1, Asahi Kasei Corp (2002).
134 Millet, J.M.M., Baca, M., Pigamo, A. et al. (2003). *Appl. Catal. A: Gen.* 244: 359.
135 Ueda, W. and Oshihara, K. (2000). *Appl. Catal. A: Gen.* 200: 135.
136 Vedrine, J.C., Novakova, E.M., and Derouane, E.G. (2003). *Catal. Today* 81: 27.
137 Blasco, T., Botella, P., Concepcion, P. et al. (2004). *J. Catal.* 228: 362.
138 Ivars, F., Solsona, B., Rodriguez-Castellon, E. et al. (2009). *J. Catal.* 262: 35.
139 Safonova, O.V., Deniau, B., and Millet, J.M.M. (2006). *J. Phys. Chem. B* 110: 23962.
140 Baca, M. and Millet, J.M.M. (2005). *Appl. Catal. A: Gen.* 279: 67.

141 Ivars-Barceló, F., Millet, J.M.M., Blasco, T. et al. (2014). *Catal. Today* 238: 41.
142 Baca, M., Pigamo, A., Dubois, J.L. et al. (2005). *Catal. Commun.* 6: 215.
143 Ueda, W., Endo, Y., and Watanabe, N. (2006). *Top. Catal.* 38: 261.
144 M. Takahashi, X. Tu, T. Hirose, et al. (1999), Process for producing acrylic acid, patent US5994580A, Toagosei Co Ltd.
145 Manhua, L. (2003). *Appl. Catal. A: Gen.* 250: 287.
146 Beato, P., Blume, A., Girgsdies, F. et al. (2006). *Appl. Catal. A: Gen.* 307: 137.
147 (a) Concepcion, P., Hernandez, S., and López Nieto, J.M. (2011). *Appl. Catal. A: Gen.* 391: 92. (b) Botella, P., López Nieto, J.M., Solsona, B. et al. (2002). *J. Catal.* 209: 445.
148 (a) Ueda, W., Vitry, D., and Kotau, T. (2005). *Catal. Today* 90: 43. (b) Pyrz, W.D., Blom, D.A., Sadakane, M. et al. (2010). *Chem. Mater.* 22: 2033.
149 Shiju, N.R., Rondinone, A.J., Mullins, D.R. et al. (2008). *Chem. Mater.* 120: 6611.
150 (a) Kolen'ko, Y.V., Zhang, W., d'Alnoncourt, R.N. et al. (2011). *ChemCatChem* 3: 1597. (b) Kolenko, Y.V., Amakawa, K., d'Alnoncourt, R.N. et al. (2012). *ChemCatChem* 4: 495. (c) Carreon, M.A. and Guliants, V.V. (2005). *Eur. J. Inorg. Chem.* 27. (d) Yang, X.-J., Feng, R.-M., Ji, W.-J., and Au, C.-T. (2008). *J. Catal.* 253: 57.
151 (a) Sadakane, M., Yamagata, K., Kodato, K. et al. (2009). *Angew. Chem. Int. Ed.* 121: 3840. (b) Canioni, R., Marchal-Roch, C., Leclerc-Laronze, N. et al. (2011). *Chem. Commun.* 47: 6413. c Sanchez Sanchez, M.C., Girgsdies, F., Jastak, M. et al. (2012). *Angew. Chem. Int. Ed.* 51: 7194. d Ishikawa, S., Tashiro, M., Murayama, T. et al. (2014). *Cryst. Growth Des.* 14: 4553.
152 Chieregato, A., López Nieto, J.M., and Cavani, F. (2015). *Coord. Chem. Rev.* 3: 301–302.
153 (a) Celaya Sanfiz, A., Hansen, T.W., Teschner, D. et al. (2010). *J. Phys. Chem. C* 114: 1912. (b) Govindasamy, A., Muthukumar, K., Yu, J. et al. (2010). *J. Phys. Chem. C* 114: 4544.
154 Hävecker, M., Wrabetz, S., Kröhnert, J. et al. (2012). *J. Catal.* 285: 48.
155 (a) Yu, J., Xu, Y., and Guliants, V.V. (2014). *Catal. Today* 238: 28. (b) Goddard, W.A. III, Chenoweth, K., Pudar, S. et al. (2008). *Top. Catal.* 50: 2. (c) Chenoweth, K., van Duin, A.C.T., and Goddard, W.A. III, (2009). *Angew. Chem. Int. Ed.* 48: 7630.
156 (a) Goddard, W.A. III,, Liu, L., Mueller, J.E. et al. (2011). *Top. Catal.* 54: 659. (b) Cheng, M.-J. and Goddard, W.A. III, (2016). *Top. Catal.* 59: 1506.
157 (a) d'Alnoncourt, R.N., Csepei, L.-I., Havecker, M. et al. (2014). *J. Catal.* 311: 369. (b) Heine, C., Hävecker, M., Trunschke, A. et al. (2015). *Phys. Chem. Chem. Phys.* 17: 8983.
158 J.L. Dubois, D. Garrait, A.L. Gall, et al. Method of preparing acrylic acid from propane in the absence, patent US20080139844A1 (2008).
159 Lintz, H.-G. and Müller, S.P. (2009). *Appl. Catal. A: Gen.* 357: 178.
160 A. Kayou, T. Ihara, (2000), Process for the simultaneous preparation of acrylonitrile and acrylic acid, patent US6166241A, Mitsubishi Chemical Corp.
161 Oliver, J.M., López Nieto, J.M., and Botella, P. (2004). *Catal. Today* 96: 241.
162 Cespi, D., Passarini, F., Cavani, F. et al. (2014). *Chem. Eng. Trans.* 36: 169.
163 Cavani, F., Ballarini, N., and Cericola, A. (2007). *Catal. Today* 127: 113.
164 Blasco, T. and Nieto, J.M.L. (1997). *Appl. Catal. A* 157: 117–142.
165 Kung, H.H. (1994). *Adv. Catal.* 40: 1.
166 Bashin, M.M. (2003). *Top. Catal.* 23: 145.
167 Vedrine, J.C. (2016). *J. Energy Chem.* 25: 936.
168 López Nieto, J.M., Concepción, P., Dejoz, A. et al. (2000). *J. Catal.* 189: 147.
169 Che-Galicia, G., Ruiz-Martínez, R.S., López-Isunza, F. et al. (2015). *Chem. Eng. J.* 280: 682.
170 Lobera, M.P., Escolástico, S., and Serra, J.M. (2011). *ChemCatChem* 3: 1503.
171 Carrero, C.A., Schlögl, R., Wachs, I.E. et al. (2014). *ACS Catal.* 4: 3357.

172 Zhu, H., Rosenfeld, D.C., Anjum, D.H. et al. (2015). *J. Catal.* 329: 291.
173 Zhai, Z., Wang, X., Licht, R. et al. (2015). *J. Catal.* 325: 87.
174 Nguyen, T.T., Deniau, B., Delichere, P. et al. (2014). *Top. Catal.* 57: 1152.
175 Li, X., Lunkenbein, T., Pfeifer, V. et al. (2016). *Angew. Chem.* 55: 4092.
176 Krasnobaeva, O.N., Belomestnykh, I.P., Nosova, T.A. et al. (2016). *Russ. J. Inorg. Chem.* 61: 1074.
177 Bulánek, R., Kalužová, A., Setnička, M. et al. (2012). *Catal. Today* 179: 149.
178 Kumar, C.P., Gaab, S., Müller, T.E. et al. (2008). *Top. Catal.* 50: 156.
179 Zhang, J., Liu, X., Blume, R. et al. (2008). *Science* 322: 73.
180 Frank, B., Zhang, J., Blume, R. et al. (2009). *Angew. Chem.* 48: 6913.
181 Qi, W. and Su, D. (2014). *ACS Catal.* 4: 3212.
182 Guo, X., Qi, W., Liu, W. et al. (2017). *ACS Catal.* 7: 1424.
183 Goyal, R., Sarkar, B., Bag, A. et al. (2016). *J. Mater. Chem. A* 4: 18559.
184 Chen, C., Zhang, J., Zhang, B. et al. (2013). *Chem. Commun.* 49: 8151.
185 de Jesús Díaz Velásquez, J., Suárez, L.M.C., and Figueiredo, J.L. (2006). *Appl. Catal. A* 311: 51.
186 Venegas, J.M., Grant, J.T., McDermott, W.P. et al. (2017). *ChemCatChem* doi: 10.1002/cctc.201601686.
187 Shi, L., Yan, B., Shao, D. et al. (2017). *Chin. J. Catal.* 38: 389.
188 Grant, J., Carrero, C., Goeltl, F. et al. (2016). *Science* 354: 1570.
189 Shi, L., Wang, D., Song, W. et al. (2017). *ChemCatChem* doi: 10.1002/cctc.201700004.
190 Lopez Nieto, J.M., Botella, P., Vazquez, M.I. et al. (2002). *Chem. Commun.* 1906.
191 Concepción, P., Blasco, T., López Nieto, J.M. et al. (2004). *Microporous Mesoporous Mater.* 67: 215.
192 Frank, B., Morassutto, M., Schomäcker, R. et al. (2010). *ChemCatChem* 2: 644.
193 Liu, Y., Feng, W., Li, T. et al. (2006). *J. Catal.* 239: 125.
194 Kondratenko, E.V. and Sinev, M.Y. (2007). *Appl. Catal. A* 325: 353.
195 Christodoulakis, A. (2004). *J. Catal.* 222: 293.
196 Madaan, N., Haufe, R., Shiju, N.R. et al. (2014). *Top. Catal.* 57: 1400.
197 Gaffney, A.M. and Mason, O.M. (2017). *Catal. Today* 285: 159.
198 Maffia, G.J., Gaffney, A.M., and Mason, O.M. (2016). *Top. Catal.* 59: 1573.

Part II

Bioinspired Alkane Functionalization

10

Recent Developments of Bioinspired Approaches to Functionalization of Light Alkanes

Alexander B. Sorokin

Institut de Recherches sur la Catalyse et l'Environnement de Lyon, IRCELYON, UMR 5256, Université Claude Bernard Lyon 1, CNRS, Villeurbanne, France

10.1 Introduction

Selective oxidation of light alkanes (methane, ethane, propane) under mild conditions is a challenging task from both fundamental and industrial perspectives. Methane is the principal component of the natural gas, accounting for 70–90% of the composition. It is one of the main sources of energy and raw materials for chemical industry. Shale gas also contains methane as a principal component. Depending on the origin, natural gas also contains 1–10% of ethane and propane.

Transformation of methane, the most abundant and the least reactive compound of natural gas, to valuable products remains one of the major challenges in chemistry and sustainable development in view of rational use of hydrocarbon feedstock and prevention of the accumulation of this greenhouse effect gas in atmosphere. The current industrial approach for conversion of methane is based on capital-intensive synthesis gas technologies operating at high temperatures (>600 °C). Direct oxidation processes require high temperatures, but conversion and selectivity of these processes are low [1]. Selective low-temperature oxidation of CH_4 is very difficult because of the inertness of the methane C—H bond, one of the least reactive bonds known (Bond Dissociation Energy, BDE = 104.9 kcal mol^{-1}). Only the most potent oxidizing species are capable of oxidizing CH_4. In particular, enzymes of aerobic methanotrophic bacteria catalyze the oxidation of CH_4 to CH_3OH by O_2 at physiological conditions. Principal methane oxidizing enzymes are soluble methane monooxygenase (sMMO) containing diiron non-heme active site and particulate methane monooxygenase (pMMO) having copper active site [2]. The chemistry of sMMO is relatively well-established whereas the structure(s) of the active site of pMMO and reaction mechanism(s) are still unclear. Native cytochrome P-450 enzymes catalyze oxidation of the strong C—H bonds but are not able to perform the oxidation of CH_4. Remarkable catalytic properties of MMO and P450 enzymes are associated with their capacity to form high-valent metal oxo species, which are powerful oxidants. In contrast to organometallic activation of CH_4 at Pt, Hg, Tl, Pd, Au centers in strong acids at elevated temperatures [3], MMO enzymes oxidize CH_4 to MeOH under physiological conditions (water, ambient temperature), thus indicating that such a difficult process is possible at mild conditions [4, 5]. MMOs operate using different strategy, by creating very strong oxidizing species that are able to attack C—H bond in CH_4. The chemistry of soluble MMO and particulate MMO has been intensively reviewed [2, 6, 7]. The state-of-the-art of these outstanding biocatalysts in the context of biological gas-to-liquids conversion was recently analyzed by Lawton and Rosenzweig [8].

Alkane Functionalization, First Edition. Edited by Armando J. L. Pombeiro and M. Fátima C. Guedes da Silva.

Extensive research has been performed to develop the chemical models of sMMO. Complexes mimicking a structural organization and spectral features of sMMO have been described [9], but functional chemical models able to oxidize CH_4 are still to be created [10]. Considerable efforts have been devoted to designing chemical bioinspired systems for selective catalytic oxidation of alkanes in mild and clean conditions. The principal efforts have been directed on the development of mononuclear metalloporphyrin catalysts as models of cytochromes P450 [11, 12] and dinuclear [9] or mononuclear [13, 14] iron non-heme complexes as models of sMMO and non-heme monooxygenases. Although a variety of efficient biomimetic catalysts for different oxidation reactions were elaborated, bioinspired chemical systems involving Earth-abundant metals capable of oxidizing gaseous alkanes under environmentally friendly conditions are still rare. This chapter describes the progress in the selective oxidation of methane, ethane, and propane using bioinspired catalysts made in the last 10 years. Mechanistically related inorganic catalytic systems operating in aqueous solutions at ambient and near-ambient temperatures are also discussed.

10.2 Bioinspired Oxidation of Methane

10.2.1 μ-Nitrido Diiron Phthalocyanine and Porphyrin Complexes

A novel bioinspired strategy for development of efficient catalysts for the oxidation of strong C—H bonds was recently proposed. μ-Nitrido diiron macrocyclic complexes associate the structural features of the active sites of two enzymes: a diiron site as in sMMO but situated in macrocyclic ligand environment as in cytochrome P-450 (Figure 10.1) [15].

This construction with two iron ions supported by phthalocyanine or porphyrin ligands and tightly connected by the nitrogen bridge is very suitable for the stabilization of the high iron oxidation states [16, 17]. These complexes bearing one unpaired electron are in the formal $Fe^{III}Fe^{IV}$ oxidation state, but iron sites of these symmetrical structures are equivalent, exhibiting one doublet in the Mössbauer spectrum with parameters compatible with $Fe^{+3.5}$ sites [16]. Available X-ray structures indicate equivalent Fe—N bonds with distances 1.65–1.68 Å [18]. Consequently, μ-nitrido diiron complexes are better described as delocalized $Fe^{+3.5}Fe^{+3.5}$ species. Despite their initial high oxidation state, $LFe^{+3.5}(\mu\text{-N})Fe^{+3.5}L$

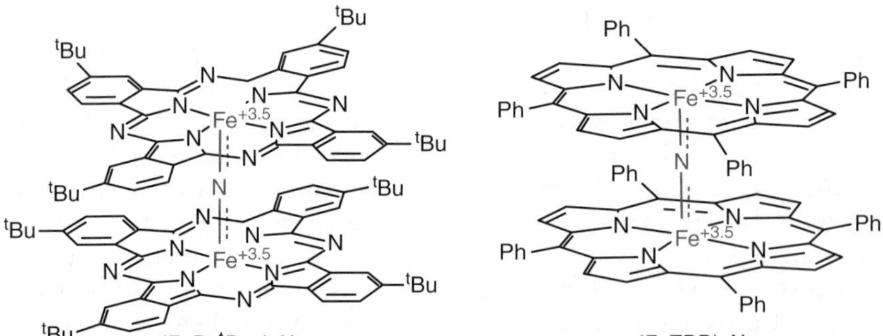

(FePc^{t}Bu$_4$)$_2$N

Fe-N$_{bridge}$ length:	1.65 Å
Fe-Fe distance:	3.33 Å
Interplane distance:	3.73 Å

(FeTPP)$_2$N

Fe-N$_{bridge}$ length:	1.68 Å
Fe-Fe distance:	3.36 Å
Interplane distance:	3.93 Å

Figure 10.1 Structure of μ-nitrido diiron phthalocyanine and porphyrin complexes and their principal structural parameters.

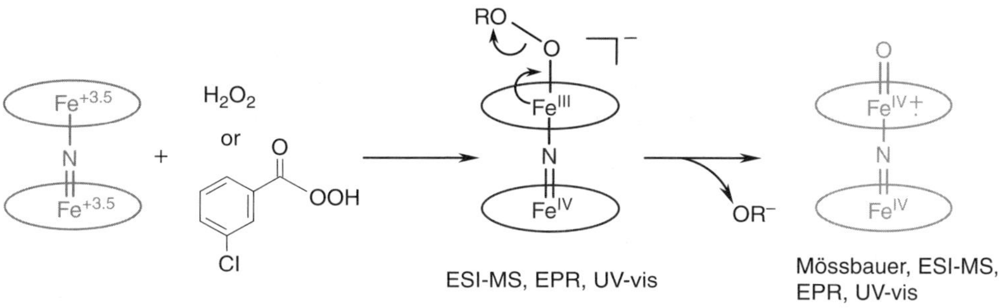

Figure 10.2 Formation of ultra-high-valent diiron-oxo species from μ-nitrido diiron porphyrin and phthalocyanine complexes and active oxygen donors. Oval stands for macrocyclic ligand.

complexes (L – macrocyclic ligand) react with oxidants like H_2O_2 and *m*-chloroperbenzoic acid (*m*-CPBA) to form peroxo complexes that undergo a heterolytic O—O bond cleavage to generate ultra-high-valent diiron oxo species (Figure 10.2) [15].

These complexes were detected at very low temperatures and characterized by spectroscopic methods [15, 19–21]. It should be pointed out that MMO and P450 enzymes and their biomimetic models activate O_2 or active oxygen donors like H_2O_2 and *m*-CPBA using Fe^{II}, Fe^{III}, or Cu^{I} sites, i.e. low metal oxidation states. Due to initial higher oxidation state of the $LFe^{+3.5}(\mu\text{-N})Fe^{+3.5}L$ complexes, their two-electron oxidation by H_2O_2 or *m*-CPBA formally generates $(L)Fe^{IV}(\mu\text{-N})Fe^{V}{=}O(L)$ entities, which can be better described as $Fe^{IV}Fe^{IV}$ species with cation-radical at macrocyclic ligand, $(L)Fe^{IV}(\mu\text{-N})Fe^{IV}{=}O(L^{+\cdot})$. The formation of these extremely electrophilic species should be facilitated by delocalization of the large positive charge at the two iron sites, two macrocyclic ligands and the nitrogen bridge. The $(L)Fe^{IV}(\mu\text{-N})Fe^{IV}{=}O(L^{+\cdot})$ species show much stronger oxidizing properties than their mononuclear counterparts [21]. Taking into account that enzymatic oxidation of CH_4 involves potent high-valent metal-oxo species it was suggested that $(L)Fe^{IV}(\mu\text{-N})Fe^{IV}{=}O(L^{+\cdot})$ might be competent for this challenging oxidation.

Catalytic tests confirmed the capacity of the homogeneous $(FePc^tBu_4)_2N$ – H_2O_2 system to oxidize CH_4 to provide HCOOH as principal oxidation product accompanied by minor amounts of MeOH and CH_2O. Since the CH_4 oxidation was performed in MeCN, a particular caution should be taken concerning the origin of the oxidation products [22]. Organic solvent containing C—H bonds can be also oxidized with the formation of C_1 oxidation products. This important issue cannot be excluded because the organic solvent is in a huge excess with respect to dissolved CH_4 (typically in mM concentration range) and BDE of the C—H bonds of organic solvent is smaller than the methane $BDE_{C\text{-}H}$. The origin of C_1 oxidation products can be determined in labeling experiments. When the oxidation of CH_4 was carried out in CD_3CN, the GC-MS analysis of the isotopic composition of formic acid obtained with turnover number (TON) of 60 showed the presence of 68% HCOOH and 32% DCOOH due to the oxidation of CH_4 and CD_3CN, respectively (Scheme 10.1) [22].

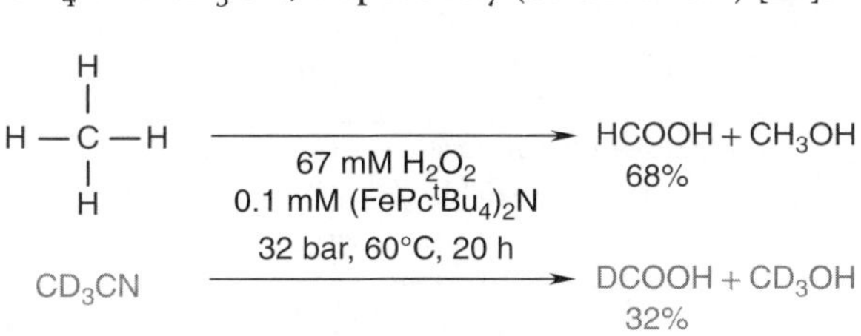

Scheme 10.1 Oxidation of CH_4 by $(FePc^tBu_4)_2N$ – H_2O_2 system in CD_3CN. (*See insert for color representation.*)

Consequently, the $(FePc^tBu_4)_2N - H_2O_2$ system does oxidize CH_4 with the effective TON of 41, but the parallel oxidation of MeCN to formic acid with TON of 19 leads to overestimated turnover numbers. This finding is not surprising, since if the oxidizing species is strong enough to oxidize CH_4, it should also oxidize organic solvent with weaker C—H bonds. In order to avoid this complication, the oxidation was performed in water. To this aim, $(FePc^tBu_4)_2N$ was supported onto silica (Degussa aerogel, $200\,m^2\,g^{-1}$) with the complex loading of 20–40 $\mu mol\,g^{-1}$. To further evidence that C_1 products originated from the oxidation of methane, the mixture of 53% of CH_4 and 47% of $^{13}CH_4$ was heterogeneously oxidized in water in the presence of $(FePc^tBu_4)_2N/SiO_2$ (Scheme 10.2) [19, 22].

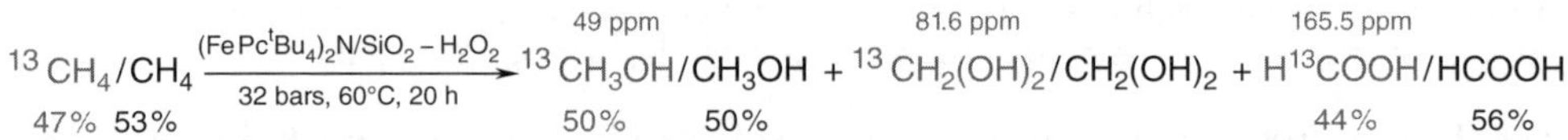

Scheme 10.2 Labeling study of the heterogeneous oxidation of methane by $(FePc^tBu_4)_2N/SiO_2$—H_2O_2 system. (*See insert for color representation.*)

The formation of ^{13}C-labeled products was confirmed by ^{13}C NMR and GC-MS data, which showed the similar isotopic compositions of the initial methane mixture and the oxidation products. Thus, the $(FePc^tBu_4)_2N/SiO_2 - H_2O_2$ system does oxidize CH_4 in water. When $(FePc^tBu_4)_2N$ was supported onto Al_2O_3 or graphite, the efficiency of the oxidation decreased. Although optimal reaction temperature was found to be 60 °C, CH_4 was oxidized in pure water at 25 °C with moderate TON of 13. Time course of the CH_4 oxidation showed that ~90% was formed after 8 hours, but the reaction time of 20 hours was typically used to obtain maximal yields of products [22]. The proposed reaction mechanism involves the formation of the active $(L)Fe^{IV}(\mu\text{-}N)Fe^{IV}{=}O(L^{+\cdot})$ species, which reacts with CH_4, from the hydroperoxo complex $(L)Fe^{IV}(\mu\text{-}N)Fe^{III}\text{-}OOH(L)$ (Figure 10.3).

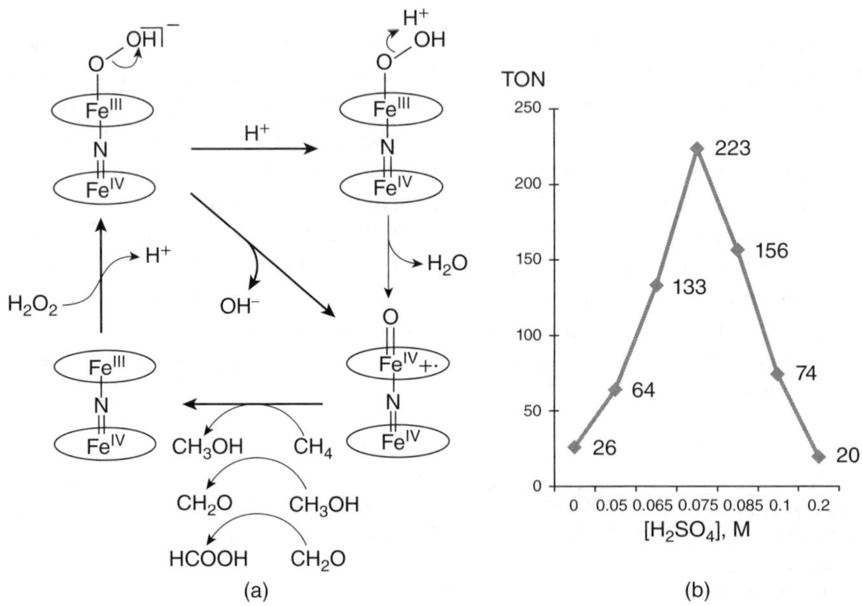

Figure 10.3 Proposed mechanism of the oxidation of methane by (a) $(FePc^tBu_4)_2N - H_2O_2$ system and (b) the dependence of the TON on the acidity of water solution.

This key step can be accelerated by protonation of the peroxide oxygen, which should facilitate the heterolytic cleavage of the O—O bond with release of water molecule, a better leaving group than OH^-. Indeed, the efficiency of the oxidation of CH_4 strongly increased in the presence of diluted H_2SO_4, showing a bell-shaped dependence on the H_2SO_4 concentration (Figure 10.3b). The highest TON of 223 was observed in 75 mM H_2SO_4 solution at 60 °C [22]. Comparison of the efficiency of CH_4 oxidation in pure water and in 75 mM H_2SO_4 solution at different temperatures shows a beneficial effect of a small amount of acid (Figure 10.4).

Control experiments showed that MeOH having a weaker C—H bond ($BDE_{H\text{-}CH2OH}$ = 96.1 kcal mol^{-1}) compared to CH_4 was readily oxidized by the $(FePc^tBu_4)_2N/SiO_2 - H_2O_2$ system to HCOOH. Thus, MeOH was the primary oxidation product, which is further converted to CH_2O and HCOOH under reaction conditions. Depending on the reaction conditions, the selectivity to HCOOH ranged from 88% to 97% and the product yields based on the H_2O_2 attained 50%. Importantly, the $(FePc^tBu_4)_2N/SiO_2$ catalyst was significantly more stable in 75 mM H_2SO_4 than in water. $(FePc^tBu_4)_2N/SiO_2$ can be used in three successive oxidations to provide 171, 161 and 160 turnovers per run to attain the total TON of 492 [15].

The powerful oxidizing properties of the $(FePc^tBu_4)_2N - H_2O_2$ system were associated with the formation of the high-valent $(Pc)Fe^{IV}(\mu\text{-}N)Fe^{IV}{=}O(Pc^{+\cdot})$ complex, which was detected by EPR and by ESI-MS using labeled $H_2{}^{18}O_2$ [19]. Additional evidence of the involvement of $(L)Fe^{IV}(\mu\text{-}N)Fe^{IV}{=}O(L^{+\cdot})$ came from the study of the oxidation of benzene. Similarly to enzymatic oxidation performed by high-valent oxo species, $(FePc^tBu_4)_2N - H_2O_2$ system oxidizes benzene with the intermediate formation of benzene epoxide and NIH shift [23].

The high-valent diiron-oxo species supported by tetraphenylporphyrin ligands (TPP) was prepared at −90 °C using *m*-CPBA and was characterized by low temperature UV-vis, EPR, cryospray MS, and Mössbauer methods as well as by reactivity studies [21]. A molecular ion cluster centered at m/z = 1366.0 in the cryospray mass spectrum corresponding to $(TPP)Fe^{IV}(\mu\text{-}N)Fe^{IV}{-}O(TPP^{+\cdot})$ disappeared upon addition of cyclohexene with regeneration of the starting $(FeTPP)_2N$. Only cyclohexene epoxide was detected without any allylic oxidation products which evidences a two-electron oxo-transfer mechanism rather than one-electron

Figure 10.4 Dependence of TONs of the oxidation of CH_4 in water and in 75 mM H_2SO_4 on the reaction temperature.

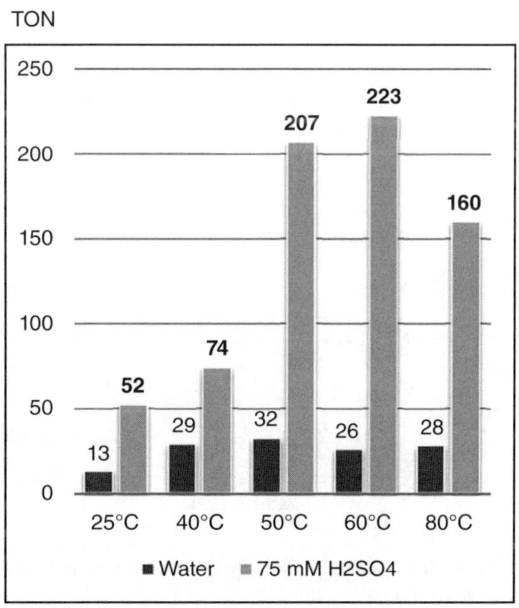

radical pathways. A strong narrow EPR signal at $g = 2.001$ unchanged in profile and position between 10 and 120 K indicated a well-isolated $S = 1/2$ ground state with one unpaired electron on a porphyrin ligand of (TPP)Fe^{IV}(μ-N)Fe^{IV}=O($TPP^{+\cdot}$). This green species showed a broad Q band between 600 and 650 nm, which was also indicative of the TPP cation-radical. The Mössbauer spectrum of the labeled (TPP)$^{57}Fe^{IV}$(μ-N)$^{57}Fe^{IV}$=O($TPP^{+\cdot}$) exhibited one double signal with $\delta = 0.001\ \text{mm s}^{-1}$ and $\Delta E_{Q1} = +0.752\ \text{mm s}^{-1}$ parameters indicating Fe^{IV} oxidation state. The applied magnetic field Mössbauer spectra showed a zero spin density on the Fe^{IV}(μ-N)Fe^{IV} fragment, thus representing a $S_{FeFe} = 0$ site weakly coupled to a porphyrin radical [21]. This complex obtained with *m*-CPBA was competent for the oxidation of CH_4 under catalytic conditions. Despite the concurrent oxidation of *m*-CPBA and its degradation products, HCOOH was formed with a 43.5% yield and TON of 13.7 [21].

The catalytic properties of μ-nitrido diiron complexes depend on the ligand structure. Two iron sites of μ-nitrido complex can be accommodated in different ligand environments. This affords a unique mechanistic possibility to probe the difference in their reactivity within the same molecule, i.e. in the *intra*-molecular approach. To this aim, two heteroleptic μ-nitrido diiron complexes bearing one phthalocyanine ligand with strong electron-withdrawing alkylsulfonyl groups and one more electron-rich unsubstituted phthalocyanine moiety were designed (Figure 10.5) [24].

Upon interaction with H_2O_2 or *m*-CPBA, two isomeric high-valent diiron-oxo species can be formed. These intermediates were trapped in the quadrupole part of the mass spectrometer and subjected to collision induced dissociation tandem MS/MS fragmentation. As was shown by the analysis of the fragmentation patterns and supported by DFT calculations, both heteroleptic complexes in combination with either H_2O_2 or *m*-CPBA preferentially bind oxo entity on the more electron-rich site [24]. Thus, the nature of macrocyclic ligand governs the formation of the active diiron species and therefore influences on their reactivity. Comparison of the catalytic properties of three μ-nitrido dimers showed that

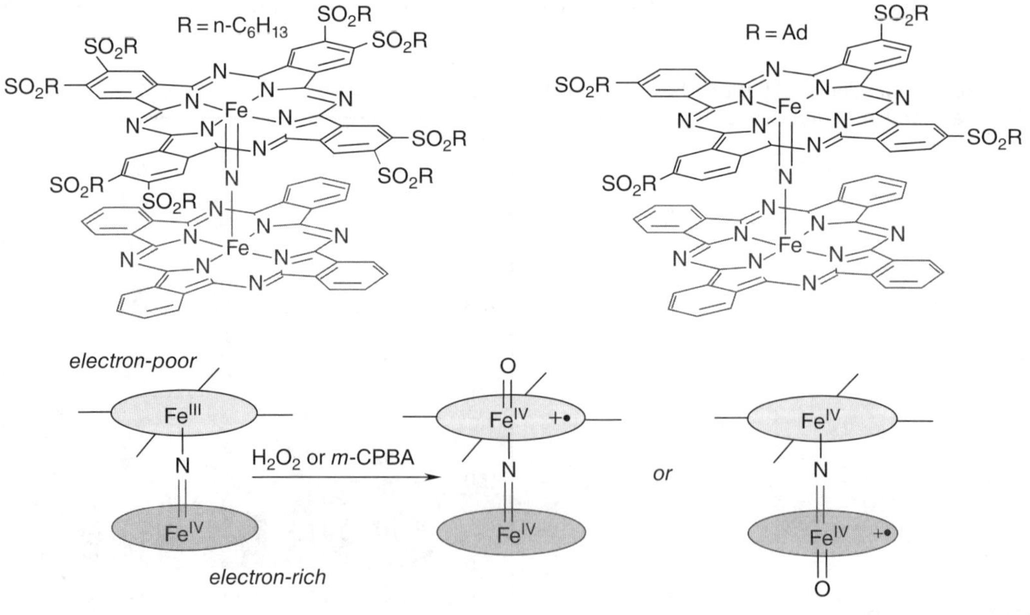

Figure 10.5 Formation of isomeric-oxo species using heteroleptic μ-nitrido diiron phthalocyanine platform. *(See insert for color representation.)*

the increase of the electron-donating character of phthalocyanine ligand along the series $[FePc(SO_2{}^tBu_4)]_2N \rightarrow (FePc)_2N \rightarrow (FePc^tBu_4)_2N$ resulted in the gain of the catalytic efficiency (Figure 10.6) [24].

It appears that μ-nitrido high-valent diiron-oxo complexes supported by porphyrin and phthalocyanine ligands are more potent oxidants than cytochrome P-450 active species. The DFT calculations confirmed that μ-nitrido diiron-oxo species readily react with CH_4 via oxygen atom transfer [25]. Free energy of activation of the reaction of μ-nitrido diiron-oxo species with CH_4 was calculated to be 15.7 kcal mol^{-1}, which is more than 10 kcal mol^{-1} lower than that determined for Compound I of cytochrome P450. Analysis of the electronic configuration of reactants and transition states showed that strong oxidizing properties of $(L)Fe^{IV}(\mu\text{-}N)Fe^{IV}{=}O(L^{+\cdot})$ are related to the orbital mixing patterns along the Fe—O axis. The axial Fe^{IV}-nitrido unit donates sufficient electron density to affect pK_a of the oxo entity and to increase the strength of the O—H bond formed after abstraction by $(L)Fe^{IV}(\mu\text{-}N)Fe^{IV}{=}O(L^{+\cdot})$ of the hydrogen atom from CH_4 [25]. Importantly, the μ-nitrido group plays an essential role in the reactivity. Neither related μ-oxo and μ-carbido dinuclear nor mononuclear complexes were active in the oxidation of CH_4. The DFT-calculated geometries and electronic structures of hypothetical high-valent oxo species of μ-oxo, μ-carbido, and μ-nitrido species were compared (Figure 10.7) [26].

Among these diiron-oxo species, the Fe=O bond is the longest in the μ-nitrido complex with considerable spin density on the oxygen atom. Thus, μ-nitrido diiron-oxo species can be considered as oxyl radical species with enhanced reactivity in the H-abstraction from the strong C—H bonds [26].

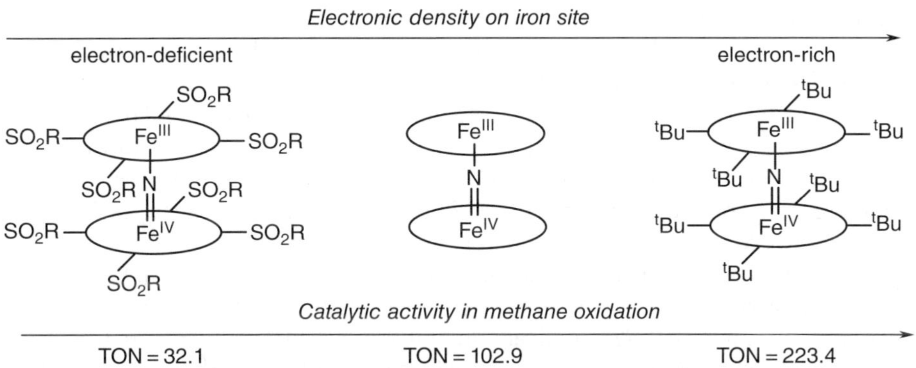

Figure 10.6 Influence of the electronic structure of μ-nitrido diiron phthalocyanines on the turnover numbers of the CH_4 oxidation. Oval stands for phthalocyanine ligand.

Figure 10.7 DFT-optimized structures of the hypothetical oxo species of μ-carbido, μ-nitrido, and μ-oxo diiron phthalocyanine complexes.

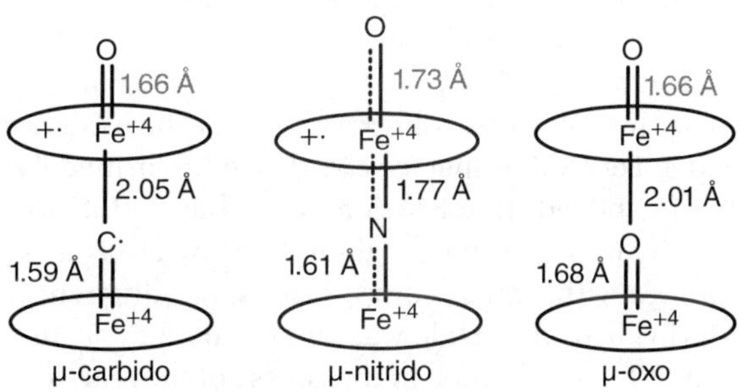

10.2.2 Fe- and Cu-Containing Zeolite Catalysts

The Fe- and Cu-containing zeolites are described in this chapter because they are often regarded as biomimetic inorganic models since the structures of putative active sites and some aspects of their reactivity are reminiscent of the bioinspired oxidation.

The group of Panov reported that calcinated Fe-containing ZSM-5 reacted with N_2O at 250 °C to generate co-called α-oxygen species capable of reacting with CH_4 at 25 °C [27–29]. Despite intensive research, the structure of the active site and the factors determining such a high reactivity were unclear until very recently [30]. Snyder et al. studied the Fe-BEA zeolite with the highest content of the α-Fe sites (~80%) by DR UV-vis spectroscopy, magnetic circular dichroism (MCD), including parameters' dependence on temperature and magnetic-field strength as well as by Mössbauer technique. On the basis of spectroscopic and computational data, previously proposed oxo-bridged and hydroxo-bridged diiron(II) entities were excluded and α-Fe site was defined as mononuclear, square planar, high-spin Fe(II) site embedded into zeolite lattice (Scheme 10.3).

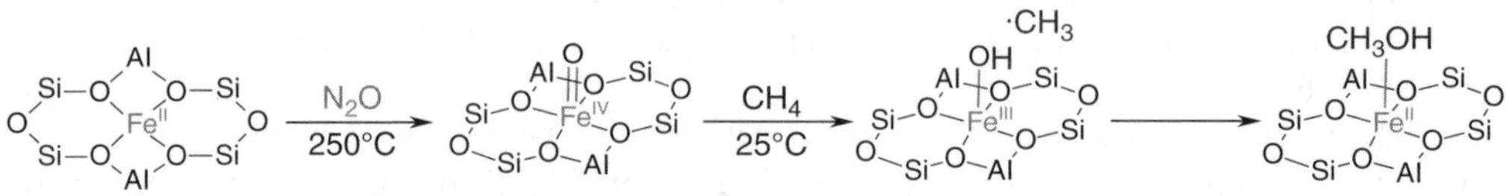

Scheme 10.3 Proposed structures of the zeolitic α-Fe sites and active α-Fe^{IV}=O site capable of reacting with methane. (*See insert for color representation.*)

Transfer of the oxygen atom from N_2O leads to the formation of a mononuclear $S = 2$ Fe^{IV}=O site with a very short, 1.59 Å Fe=O bond. The remarkable reactivity of the α-Fe^{IV}=O site originates from a constrained coordination geometry enforced by the zeolite lattice [30]. According to calculations, H-atom abstraction occurs with a very low activation barrier of 3.6 kcal mol^{-1} due to the exceptional strength of the forming O—H bond, which is similar to those of the methane C—H bond. In the transition state, Fe—O bond elongates to 1.72 Å, which indicates that Fe^{IV}=O bond acquires a radical character, corresponding to a very reactive Fe^{III}-$O^{+\bullet}$ species capable of abstracting hydrogen atom from methane [30].

However, this approach has a serious drawback from industrial perspective [31]. In fact, this strong oxidizing species carried out the selective oxidation of CH_4 only at 25 °C. Surface-bound methoxy species were obtained with 80% conversion and 90% selectivity. Attempts to desorb methanol from the surface at 250 °C led to deeper oxidation resulting in the high selectivity to CO_x [32]. Methanol can be extracted from the material using a suitable solvent [33], but it would add an additional regeneration cycle in the process already consisting of Fe-ZSM-5 activation – N_2O treatment at 250 °C – reaction with CH_4 at 25 °C – MeOH extraction with the solvent. In overall, the MeOH yield of 208 μmol g^{-1} of Fe-ZSM-5 containing 895 μmol Fe (TON ~ 0.25) was obtained [31].

The ability of the activated Fe-ZSM-5 to react with CH_4 prompted Hutchings and co-workers to further develop zeolite catalysts for the oxidation of CH_4 and C_2H_6 [34, 35]. They argued that active Fe-ZSM-5 species could also be formed by using other oxidants, in particular, aqueous H_2O_2. Indeed, Fe-ZSM-5 activated at 550 °C for three hours showed a remarkable catalytic activity in the oxidation of methane by H_2O_2 in aqueous medium at 50 °C. Interestingly, other zeolites with similar compositions, but different pore structures, were significantly less active. Methyl hydroperoxide was shown to be the primary oxidation product and MeOH, HCOOH, and CO_2 were formed in the course of the reaction. Formic acid was the principal final oxidation product obtained with >70% selectivity after two hours. Interestingly, a 30% selectivity toward HCOOH was obtained after two minutes suggesting that it may be formed from CH_4 in a parallel pathway. The conversion of CH_4 was 0.1–0.7% and typical product distribution of the methane oxidation is shown in Scheme 10.4 [34].

$$\underset{30.8\ \text{mmol}}{CH_4} \xrightarrow[\substack{30.5\ \text{bar, water (10 mL)} \\ 50\,°C,\ 30\ \text{min}}]{\substack{\text{Catalyst (27 mg)} \\ 0.5\ \text{M}\ H_2O_2}} CH_3OOH + CH_3OH + HCOOH + CO_2$$

Catalyst	CH_3OOH	CH_3OH	HCOOH	CO_2
ZSM-5 (0.014 wt.% Fe)	17.7	15.4	44.0	4.4
Fe-ZSM-5 (2.5 wt.% Fe)	1.8	22.3	164.0	38.6
Fe-silicate-1 (0.5 wt.% Fe)	7.5	15.4	60.5	5.8
Cu/Fe-ZSM-5 (2.5 wt.% Fe)	0.5	188.8	0.0	33.3
Fe-ZSM-5/$Cu(NO_3)_2$ (2.5 wt.% Fe)	1.1	140.1	0.0	18.8

Scheme 10.4 Distribution of the oxidation products of methane expressed in μmol obtained with different zeolites.

Using Fe-containing zeolites, the selectivity to MeOH attained 12–22% and the total selectivity to oxygenated products (MeOOH, MeOH, and HCOOH) was around 80–95%. Formaldehyde was not observed during the oxidation. In the very beginning of the reaction performed under strictly anaerobic conditions (>99.999% CH_4), MeOOH was obtained with a ~55% selectivity [34]. Usually, hydroperoxides can be formed in the free-radical aerobic oxidation of hydrocarbons at high temperatures in the reaction of alkyl radicals with O_2. Although the formation of O_2 from the decomposition of H_2O_2 cannot be excluded, such a high initial selectivity to MeOOH suggests a possible iron-centered reaction, which, however, requires the formation of methyl radicals. $CH_3^{\bullet}$ can be produced via the reaction with $OH^{\bullet}$ generated in the Fenton decomposition of H_2O_2. Although the Fenton system did oxidize CH_4, over 100 times more iron was needed to produce half the total amount of oxidation products obtained with Fe-ZSM-5 [35]. Consequently, Fe-ZSM-5 is much more efficient in the oxidation of CH_4 with respect to Fenton system, but some contribution of the Fenton mechanism leading to over-oxidation products cannot be excluded. Further evidence that iron sites in zeolite were responsible for the oxidation of CH_4 was obtained in the experiments with Na_2SO_3 (scavenger for $OH^{\bullet}$ radicals) when no decrease in the catalytic activity was observed [35].

UV-vis and FT-IR studies showed that upon treatment at 550 °C, iron ions migrated from the zeolite framework to extra-framework positions. Increase of the extra-framework iron component led to the increase of the catalytic activity. The nature of these extra-framework sites was calculated using extended X-ray absorption fine structure (EXAFS) and DFT. EXAFS data showed the Fe—Fe coordination number of 1.6 with average Fe—Fe and Fe—O distances of 2.97 ± 0.04 Å and 2.04 ± 0.02 Å, respectively [34]. Several possible diiron structures were considered, and the best match between EXAFS and DFT data was obtained for $[Fe_2(\mu_2\text{-}OH)_2(OH)_2(H_2O)_2]^{2+}$ with antiferromagnetically coupled high-spin octahedral Fe(III) sites (Figure 10.8) [34].

Hutchings and co-workers have proposed that this diiron site can bind two H_2O_2 molecules to form an active species capable of reacting with CH_4 to form MeOOH in the initial step. The brief mechanism proposed by Hutchings and co-workers [34] is completed in Figure 10.9 by adding the correct changes of the Fe oxidation states and changes in the coordination sphere of diiron complex.

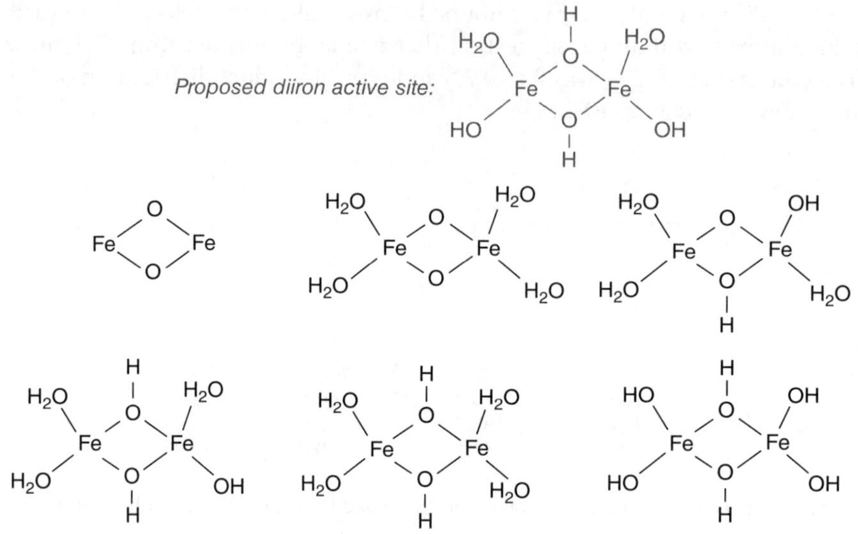

Figure 10.8 Proposed $[Fe_2(\mu_2\text{-}OH)_2(OH)_2(H_2O)_2]^{2+}$ active site and other diiron structures considered in DFT calculations [34].

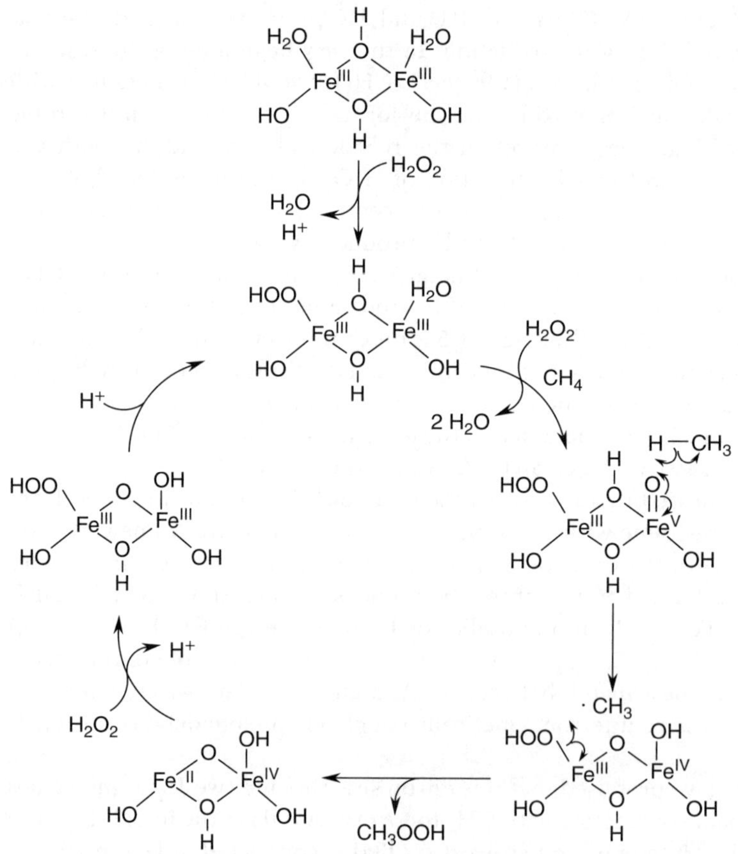

Figure 10.9 Mechanism of the oxidation of CH_4 to CH_3OOH by the Fe-ZSM-5 – H_2O_2 system proposed by Hutchings et al. in Ref. [34].

According to Ref. [34], the starting $(H_2O)Fe^{III}(\mu\text{-}OH)_2Fe^{III}(H_2O)$ (spectator OH ligands are omitted for clarity) exchanges H_2O with H_2O_2 to form peroxo complex HOO-$Fe^{III}(\mu\text{-}OH)_2$ $Fe^{III}(H_2O)$. Coordination of the second H_2O_2 molecule followed by the heterolytic cleavage of one peroxide O—O bond should provide a HOO-$Fe^{III}(\mu\text{-}OH)_2Fe^{V}$=O complex. This high-valent diiron complex should be able to abstract a hydrogen atom from CH_4 molecule to generate HOO-$Fe^{III}(\mu\text{-}OH)_2Fe^{IV}$-OH and $CH_3^{\bullet}$ in the cage. The MeOOH product might be formed via recombination of $CH_3^{\bullet}$ radical with $HOO^{\bullet}$ radical liberated at adjacent iron site. This unusual mechanism can therefore explain the initial formation of MeOOH under anaerobic conditions of the experiment, but classical H_2O_2 decomposition route and some involvement of Fenton chemistry cannot be definitely excluded. Indeed, EPR radical trapping studies indicated the formation of significant amounts of $HO^{\bullet}$ radicals, whereas $CH_3^{\bullet}$ was not detected. The primary MeOOH product undergoes further reactions to afford MeOH and HCOOH. Radical reaction pathways involving $HO^{\bullet}$ radicals lead to the formation of over-oxidation products including CO_2.

Importantly, the selectivity of the methane oxidation was changed in favor of MeOH formation in the presence of Cu additives (Scheme 10.4). The beneficial effect of Cu on the MeOH selectivity was observed (i) when Cu was added as homogeneous additive, (ii) in a physical mixture of heterogeneous Fe and Cu-containing materials, and (iii) using a bimetallic Fe/Cu-ZSM catalyst [34, 35]. EPR radical trapping experiments showed that the presence of copper suppressed the formation of $HO^{\bullet}$ radicals. This resulted in drastic reduction of the over-oxidation processes and in increasing of selectivity of MeOH formation up to 80–90%. In the presence of 2.5 wt% Cu/2.5 wt% Fe-ZSM-5, formic acid was not observed. However, even in the presence of Cu, CO_2 still formed with almost 20% selectivity. It was suggested that Cu may quench or convert $HO^{\bullet}$ radicals to O_2, H_2O_2, H_2O or $O_2^{-\bullet}$. An alternative explanation was that Cu ions can promote the heterolytic cleavage of O—O bond of MeOOH to furnish MeOH [36]. Consequently, additional studies are necessary to understand the origin of this important influence of Cu on the selectivity of CH_4 oxidation.

Significant improvement of the catalytic activity of Fe sites was achieved upon incorporation of Al^{3+} or Ga^{3+} cations into zeolite framework [37]. The Al^{3+} or Ga^{3+} cations do not constitute catalytic sites, but they promote the extraction of iron from the zeolite framework and formation of the active extra-framework Fe sites during thermal activation of the Fe-zeolite. In addition, these trivalent cations provide a negatively charged framework, which stabilizes and maintains the dispersion of the cationic extra-framework Fe species responsible for catalytic activity [37]. The highest catalytic activity in methane oxidation was obtained with zeolite containing 0.095 wt% Fe and 2.2 wt% Al. The turnover number accounting for all oxidation products achieved 761 [37].

The activity of Fe-containing zeolites is sensitive to the preparation methods since they may contain different iron sites. Some iron sites are selective to MeOH formation while others, obtained with high Fe loading, carry out unproductive H_2O_2 decomposition. Fe-ZSM-5 materials prepared by chemical vapor impregnation using $Fe(acac)_3$ showed a high catalytic activity in the oxidation of CH_4 to HCOOH [38]. Heating this material under a reducing atmosphere provided catalysts with changed selectivity profile. MeOH was obtained with up to 78% selectivity.

Despite extensive spectroscopic studies, the nature of Fe sites responsible for the selective oxidation of CH_4 by Fe-ZSM – H_2O_2 system is not yet definitely established. Although Hutchings and co-workers have initially proposed that the CH_4-oxidizing site in Fe-ZSM-5 is diiron-μ-oxo-hydroxo cluster [34], their further studies showed that the precise nature of the active Fe species in zeolites becomes more complex due to the possibility of speciation within the pores and on the external surface of zeolite upon iron loading [38]. Along with diiron species, reduced materials contain octahedral-isolated Fe(III)/Fe(II) irons with OH and/or H_2O ligands in the first coordination sphere, but a definitive model of this site could not be provided [38]. Taking into account that the Fe-ZSM-5 – H_2O_2 system is one of the most efficient and green systems for the mild selective oxidation of methane, the identification of well-defined iron species in Fe-ZSM-5 responsible for the catalytic activity is of primary importance for understanding their reaction mechanism in order to make possible the preparation of advanced materials with better catalytic properties.

10.2.3 Bioinspired Models of Particulate MMO

Although pMMO is the principal CH_4-oxidizing enzyme in Nature, its bioinspired models are much less developed because the structure and chemistry of pMMO active site is still less understood compared to sMMO. The nature of the pMMO active site has been the subject of extensive research [7]. The first structural data obtained by EXAFS revealed the dinuclear Cu site with Cu—Cu distance of 2.5 Å and mononuclear Cu site, both coordinated with histidine residues. Three available crystal structures of pMMO of different origin also contain dicopper active site while three others are best modeled with a single Cu ions [7]. In fact, all purified pMMO samples are heterogeneous in terms of the content of Cu ions and their catalytic activities are quite different. A tricopper cluster was also proposed as possible pMMO active site [7]. This structural uncertainty arises from strong dependence of Cu content on the pMMO isolation procedure, which makes difficult the correlation of the reactivity with Cu stoichiometry and development of pMMO chemical models. To achieve this goal, the structure of the pMMO active site and the mechanism of CH_4 oxidation should be better understood.

Despite extensive research on the bioinspired Cu complexes, examples of their use for the catalytic oxidation of light alkanes are extremely rare. Published to date model mono- and dinuclear Cu complexes are capable of oxidizing only relatively weak C—H bonds [39]. Recently, Chan et al. have published oxidation of CH_4 to MeOH in MeCN performed by tricopper complexes [40]. The $Cu^ICu^ICu^I$ complex supported by 3,3′-(1,4-diazepane-1,4-diyl)bis[1-(4-ethylpiperazine-1-yl)propan-2-ol] ligand (Figure 10.10) was shown to activate O_2 to form a $Cu^{II}Cu^{II}(\mu\text{-}O)_2Cu^{III}$ complex that reacts with CH_4 (2.44 atm) at room temperature to produce 0.92 equiv. of MeOH.

In the presence of a small amount of H_2O_2 (20 equiv. to Cu_3 complex), a catalytic oxidation of CH_4 with TON ~ 5.5 was obtained. This transformation of the single-turnover reaction to the catalytic oxidation was explained by the ability of $Cu^ICu^{II}(\mu\text{-}O)Cu^{II}$ complex to oxidize H_2O_2

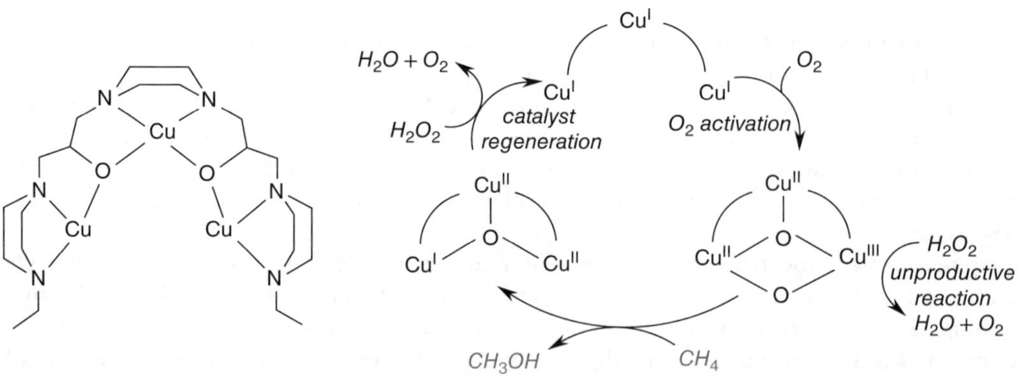

Figure 10.10 Structure of tricopper complex and proposed mechanism of CH_4 oxidation.

with regeneration of the initial $Cu^{I}Cu^{I}Cu^{I}$ form (Figure 10.10) [40]. The rate-limiting step of the catalytic cycle was proposed to be the reduction of the $Cu^{I}Cu^{II}(\mu\text{-}O)Cu^{II}$ form to the $Cu^{I}Cu^{I}Cu^{I}$ state rather than the reaction of the active $Cu^{II}Cu^{II}(\mu\text{-}O)_2Cu^{III}$ species with CH_4 [40]. The efficiency of the oxidation of CH_4 depended on the amount of H_2O_2 used. At higher H_2O_2 concentration, the principal reaction was unproductive oxidation of H_2O_2 by $Cu^{II}Cu^{II}(\mu\text{-}O)_2Cu^{III}$. The maximal TON of ~5.5 was obtained using 20 equiv. H_2O_2. Since O_2 consumed for the formation of the active $Cu^{II}Cu^{II}(\mu\text{-}O)_2Cu^{III}$ complex is again regenerated from H_2O_2 in the regeneration of the $Cu^{I}Cu^{I}Cu^{I}$ form, the total reaction is

$$CH_4 + H_2O_2 \rightarrow CH_3OH + H_2O$$

although the oxygen atom of MeOH originates from O_2. The tricopper complex with peptide was incorporated in the mesoporous carbon. The oxidation of CH_4 using this model catalyst in water was lower and the TON dropped to 0.12 [40].

Importantly, this high selectivity to alcohol without formation of further oxidation products was also observed in the oxidation of ethane and propane with the TON of 11 and 18.2, respectively, in the presence of 100 equiv. H_2O_2 [41]. The efficiency of CH_4 oxidation in MeCN was improved to ~17 cycles when 100 equiv. H_2O_2 were added in five portions each 10 minutes to limit unproductive H_2O_2 consumption. In the absence of CH_4, MeCN was oxidized to $HOCH_2CN$ [42]. It should be noted that the $Cu^{II}Cu^{II}(\mu\text{-}O)_2Cu^{III}$ complex was inactive for the oxidation of MeOH, EtOH, and 2-PrOH having much weaker C—H bonds compared with that of CH_4. The reason for this unusual behavior is unclear.

DFT calculations confirmed the superiority of the tricopper species $Cu^{II}Cu^{II}(\mu\text{-}O)_2Cu^{III}$ over dicopper models $Cu^{III}(\mu\text{-}O)_2Cu^{III}$ and $Cu^{II}(\mu\text{-}O)_2Cu^{III}$ in the oxidation of CH_4. The activation energy is significantly lower for $Cu^{II}Cu^{II}(\mu\text{-}O)_2Cu^{III}$ (15.0 kcal mol^{-1}) compared to $Cu^{III}(\mu\text{-}O)_2Cu^{III}$ (20.1 kcal mol^{-1}) and $Cu^{II}(\mu\text{-}O)_2Cu^{III}$ (19.1 kcal mol^{-1}) resulting in the reaction rate of several orders of magnitude faster [42].

Several trinucleating ligands were prepared to accommodate the tricopper cluster [42]. These complexes exhibited similar reactivity in the oxidation of cyclohexane to cyclohexanone/cyclohexanol (~2/1 ratio), styrene to benzaldehyde accompanied by minor amounts of acetophenone and styrene oxide and benzene to hydroquinone/phenol (2–3:1 ratio). The influence of the ligand structure of tricopper complex on the reactivity vis-à-vis CH_4 was not published.

The catalytic efficiency of tricopper complex in the oxidation of CH_4 in anhydrous MeCN was notably improved when this complex was supported onto mesoporous silica by electrostatic interaction between positively charged tricopper complexes and negatively charged silica surfaces [43]. Negative charges were provided either by grafting methylphosphonate groups or by doping of Al into the silica framework. While the homogeneous system was unable to produce MeOH in the presence of more than 40 equiv. H_2O_2, heterogeneous catalyst (complex loading: 204 μmol g^{-1}, specific surface: 759 m^2 g^{-1}, pore diameter: 2.6 nm) oxidized CH_4 to MeOH using 200 equiv. H_2O_2 with TON of 171.2. This high performance was explained by the higher solubility of CH_4 in MeCN confined in the mesopores of the support. Thus, productive oxidation of CH_4 prevailed over the unproductive reaction of the active species $Cu^{II}Cu^{II}(\mu\text{-}O)_2Cu^{III}$ with H_2O_2. The heterogeneous catalyst was used in three successive oxidation of CH_4 with turnover number of 166.5, 150.0, and 136.3, respectively. The gradual decrease of the catalytic activity was shown to be due to ~10% loss of the tricopper complex owing to its leaching to solution or decomposition [43].

The results obtained by Chan and co-workers can be considered as significant since (i) tricopper complex activates O_2 to form a very strong oxidizing species; (ii) this species reacts with

CH_4 at room temperature; and (iii) the oxidation of CH_4 resulted in the formation of MeOH without over-oxidation products.

Copper-containing zeolites have provided some mechanistic insight related to pMMO chemistry. Cu-ZSM-5 can be activated by O_2 to oxidize CH_4 at 100–200 °C to adsorbed CH_3O species [44] similarly to Fe-ZSM-5 activated by N_2O. The efficiency of the Cu-ZSM-5 is lower than that of Fe-ZSM-5 since only small fraction of the Cu sites reacts with CH_4. The resonance Raman, absorption, and DFT data are compatible with a bent Cu^{II}(μ-oxo)Cu^{II} site containing one μ-oxo group [45]. Due to the oxyl character of the bridging oxygen, the activation energy of CH_4 oxidation is only 15.7 kcal mol^{-1} and the strong O—H bond of $[Cu\text{-}OH\text{-}Cu]^{2+}$ intermediate formed after the abstraction of H atom contributes to lowering the reaction barrier. This reaction exhibits a kinetic isotope effect of 3.1 at 175 °C indicating C—H bond breaking in the rate-limiting step [45]. Catalytic studies of Fe- and Cu-containing zeolites often result in conflicting insights since active sites account for only a minor part of the total metal content. Recently Lercher and co-workers prepared Cu-exchanged zeolite with mordenite structure containing single-site tricopper clusters formulated as $[Cu_3(\mu\text{-}O)_3]^{2+}$ on the basis of in situ EXAFS, XANES, and UV-vis data, DFT modelization and reactivity studies [46]. In Cu-mordenite materials, they achieved a high concentration of the uniform Cu^{II} trinuclear species by controlled deposition of $Cu(OH)^+$ ions into zeolite matrix followed by calcinations at 450 °C in a O_2 flow. Treatment of the activated material with CH_4 at 200 °C followed by purging with water afforded MeOH and MeOMe with productivity up to 160 μmol of MeOH g^{-1}. The MeOH amount increased linearly with the Cu concentration and indicated that three Cu^{II} ions are necessary for the formation of one MeOH molecule. The CH_4 oxidation occurs via a homolytic C—H bond cleavage with the activation barrier of 17.7 kcal mol^{-1} followed by the radical rebound to form MeOH adsorbed to a reduced Cu cluster. Desorption of MeOH was performed by steam treatment at 135 °C. The Cu mordenite can be reactivated at 500 °C in the presence of O_2 and be used in eight cycles [46].

10.2.4 Miscellaneous Approaches

Several other catalysts were recently published to carry out the mild oxidation of methane. The 0.01 wt% Pd-ZSM-5 material containing PdO_4 single-sites anchored on the internal surface of micropores mediated oxidation of CH_4 (30 bar) by H_2O_2 in water phase at 50–95 °C to HCOOH (53.4%), MeOOH (35.3%), MeOH (6.6%), and CO_2 (3.8%) [47]. In line with the results obtained using Fe-ZSM-5, the selectivity to MeOH was improved to 86.4% when 2 wt% CuO was impregnated to 0.01 wt% Pd-ZSM-5. This material exhibited turnover frequency of 2.78 CH_4 molecules s^{-1} at 95 °C. Trimetallic AuPdCu/TiO_2 catalysts carried out the oxidation of CH_4 (30 bar) by aqueous H_2O_2 at 50 °C with 82.7% selectivity with TON of 0.7 [48].

Homogeneous dimeric and tetrameric μ-oxo iron complexes supported by 1-carboxymethyl-4,7-dimethyl-1,4,7-triazacyclononane ligands catalyzed the oxidation of CH_4 by H_2O_2 at 50–80 °C in aqueous solutions [49]. MeOH, MeOOH, and HCOOH were obtained with total TONs 18–24 after four hours. Tetranuclear copper triethanolamine complex with μ_4-O atom encapsulated in the cluster center catalyzed the oxidation of CH_4 (30 atm) by H_2O_2 in MeCN at 40 °C [50]. Using H_2O_2:HNO_3:catalyst ratio of 1000:20:1, MeOH was obtained with TON of 47.

10.3 Oxidation of Ethane

Selective oxidation of ethane to ethanol or directly to acetic acid represents a considerable interest since valuable products can be obtained from the abundant carbon feedstock. Current industrial processes are based on the steam cracking of ethane to ethylene which is used for the production of EtOH, CH_3CHO, CH_3COOH, ethylene epoxide, vinyl chloride, and polypropylene [51]. Direct oxidation of C_2H_6 is hampered by the strength of the ethane C—H bond (101.4 kcal mol^{-1}). Catalytic homogeneous and heterogeneous systems for the oxidation of CH_4 described above can be also applied for the direct oxidation of ethane.

Fe- and Cu-containing ZSM-5 catalysts mediated the oxidation of C_2H_6 by aqueous H_2O_2 to several C_2 and C_1 products generated via multiple reaction pathways [52]. Under optimal conditions, using 2.5 wt% Fe-ZSM-5 catalyst, a 56.4% C_2H_6 conversion and TON = 75.4 were achieved and the product distribution obtained is shown in Scheme 10.5.

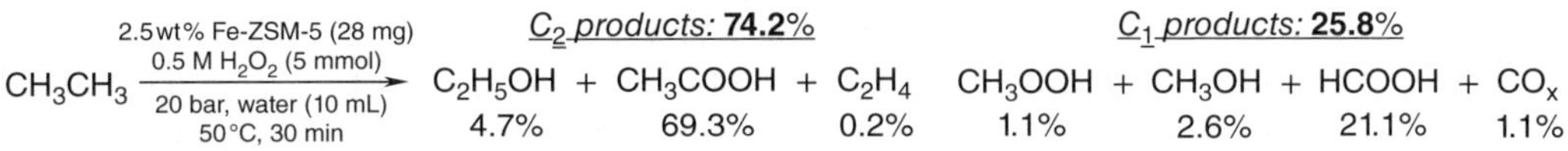

Scheme 10.5 Oxidation of ethane by H_2O_2 mediated by Fe-ZSM-5.

Acetic and formic acids were obtained with 90.4% selectivity in a 3.3:1 ratio and productivity attained 65 mol C_2H_6 kg$_{cat}^{-1}$ h^{-1} [52]. In contrast to the CH_4 oxidation mediated by the same catalyst, the addition of Cu doesn't notably change the selectivity of the C_2H_6 oxidation in favor of the alcohol formation. Another difference from the CH_4 oxidation is the radical mechanism of the C_2H_6 oxidation involving carbon-based radicals [52]. This system was applied for the oxidation of C_2H_6 in a continuous flow reactor [53]. A single pass C_2H_6 conversion was 22% and CH_3COOH was obtained with 16% yield (73% selectivity).

μ-Nitrido diiron phthalocyanine complexes supported onto SiO_2, graphite or nafion mediate the oxidation of ethane by H_2O_2 to CH_3COOH and HCOOH accompanied by the trace amounts of EtOH and CH_3CHO [54]. The better selectivity to CH_3COOH was 71% and yield on H_2O_2 attained 65%. In marked contrast to oxidation of CH_4, where the catalytic performance was significantly improved in 75 mM H_2SO_4 (TON = 223 vs TON$_{H2O}$ = 26), the oxidation of C_2H_6 was not much affected by the acid (Scheme 10.6).

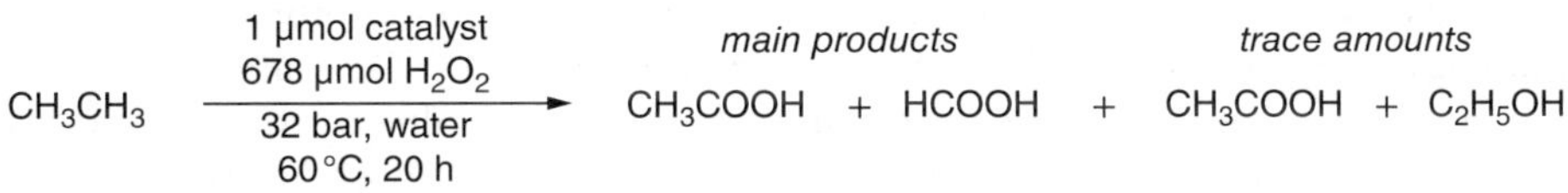

Catalyst (solvent)	TON_{AcOH}	TON_{HCOOH}	Selectivity to AcOH, %	Yield on H_2O_2, %
$(FePc^tBu_4)_2N/SiO_2$ (H_2O)	33	27	71	25
$(FePc^tBu_4)_2N/SiO_2$ (75 mM H_2SO_4)	41	44	65	36
$(FePc^tBu_4)_2N$/graphite (H_2O)	50	59	63	50
$(FePc^tBu_4)_2N$/nafion (H_2O)	48	84	53	65

Scheme 10.6 Oxidation of ethane by H_2O_2 catalyzed by supported $(FePc^tBu_4)_2N$.

The TON in acids only slightly increased from 60 in H_2O to 85 in 75 mM H_2SO_4. The $(FePc^tBu_4)_2N$/graphite was more efficient than $(FePc^tBu_4)_2N/SiO_2$ to afford CH_3COOH and HCOOH with TON of 50 and 59, respectively, compared to 33 and 27 obtained with $(FePc^tBu_4)_2N/SiO_2$. The $(FePc^tBu_4)_2N$/nafion catalyst having strong acid sites provided by support showed a higher total TON of 132 on the expense of predominant HCOOH formation [54]. Acid sites of support promote the cleavage of the ethane C—C bond. The nature of the support, the strength and loading of the acidic groups should be optimized to obtain more efficient heterogeneous catalysts.

Long and co-workers have prepared metal organic framework Fe_2(dobdc) (dobdc^{4-} – 2,5-dioxido-1,4-benzenedicarboxylate) and its Mg-diluted analogue $Fe_{0.1}Mg_{1.9}$(dobdc) [55]. Coordinatively unsaturated high-spin Fe^{II} sites of these materials react with N_2O at 60 °C in gas phase to form Fe^{III}-hydroxide species via putative short-lived Fe^{IV}=O intermediate. The treatment of Fe_2(dobdc) with N_2O:C_2H_6:Ar mixture (10:25:65) at 75 °C resulted in the formation of EtOH, CH_3CHO, Et_2O, and ether oligomers owing to close proximity of the active Fe sites. In contrast, diluted $Fe_{0.1}Mg_{1.9}$(dobdc) material with more isolated Fe sites afforded only EtOH and CH_3CHO in a 25:1 ratio and TON of 1.6 [55]. The design of other MOFs with unsaturated Fe sites can be envisaged within this interesting approach.

The [$Fe^{III}Me_3$tacn)(Cl-acac)Cl]Cl complex catalyzed the oxidation of C_2H_6 to EtOH (TON = 0.5) and CH_3COOH (TON = 2.9) by $KHSO_5$ in MeCN/H_2O medium at room temperature [56]. Low TON values were explained by the degradation of the complex during reaction. ESI-MS labeling and DFT data indicated the involvement of $(L^{+\cdot})Fe^{IV}$=O species. Tetranuclear Cu triethanolamine complex with centrosymmetric oxo group mediated the oxidation of C_2H_6 (30 atm) in MeCN by H_2O_2 at 40 °C [50]. Using H_2O_2:HNO_3:catalyst ratio of 1000:20:1, EtOH was obtained with TON of 23.

10.4 Oxidation of Propane

Because of the high BDE of the propane C—H bonds, 100.8 and 97.7 kcal mol^{-1} for primary and secondary bonds, respectively, current industrial processes of the valorization of C_3H_8 are energy demanding. Initially, C_3H_8 is converted to propene via steam cracking or dehydrogenation routes followed by oxidation to *i*-PrOH, acetone, acrolein, or acrylic acid [57]. Several catalysts for the direct oxidation of propane under mild and clean conditions have been published. The Fe-ZSM-5 catalysts mediate oxidation of propane by H_2O_2 at 50 °C in water [57]. This catalytic activity was attributed to extra-framework dimeric μ-oxo-hydroxo iron sites representing only small part of the total Fe content of zeolites. Removal of ~77% Fe by washing with acid decreased a propane conversion only by 17%. Comparison of different zeolites, SiO_2 and Al_2O_3 materials showed the highest efficiency of Fe-ZSM-5 due to a combination of MFI framework topology, presence of dimeric iron sites and availability of AlO_4^- cation exchange sites. Noteworthy, even commercial solids showed oxidizing activity due to the presence of trace amounts of Fe (0.007–0.038 wt%). Impregnation of ZSM-5 with 2.5 wt% Fe provided catalyst with a high productivity level of 23.5 mol C_3H_8 kg_{cat}^{-1} h^{-1}. This catalyst provided a 7.9% conversion of propane to the mixture of C_3, C_2, and C_1 oxidation products with 34.3%, 27.2%, and 38.2% selectivity, respectively, and total TON of 31.8 (Scheme 10.7) [57].

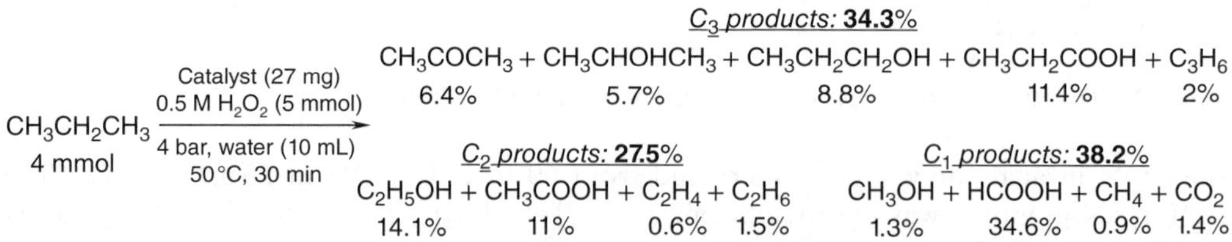

Scheme 10.7 Oxidation of propane in the presence of Fe-ZSM-5 catalyst.

The principal oxidation products were HCOOH (34.6%), EtOH (14.1%), propionic (11.4%), and acetic (11%) acids. Addition of a radical trap decreased conversion of C_3H_8, indicating at least a partial involvement of a free radical mechanism. The formation of numerous over-oxidation and C_1 and C_2 products is in agreement with this conclusion. Hutchings and co-workers concluded that iron sites of ZSM-5 behave as a heterogeneous Fenton's catalyst and acid sites in the hydrophobic pores of zeolite are responsible for the cracking of C_3 products [57].

The $(FePc^tBu_4)_2N/SiO_2$ – H_2O_2 system also mediated the oxidation of propane to propan-1-ol, propanol-2-ol, propionaldehyde, acetone, propionic, and acetic acids with TON of ~60 (Scheme 10.8) [58].

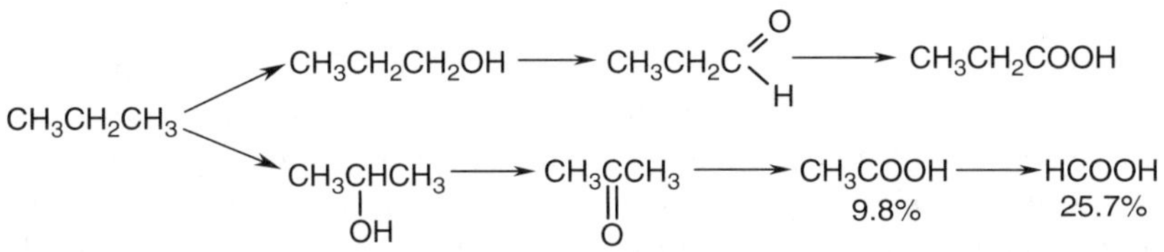

Scheme 10.8 Oxidation of propane by $(FePc^tBu_4)_2N/SiO_2 - H_2O_2$ system at 60 °C.

Acetic and formic acids were the main reaction products obtained with 9.8% and 25.7% selectivity, respectively.

The $[Fe^{III}Me_3tacn)(Cl\text{-}acac)Cl]Cl$ complex catalyzed the oxidation of propane to *i*-PrOH (TON = 4.9), acetone (TON = 14.2) and CH_3CH_2COOH (TON = 1.3) by $KHSO_5$ in MeCN/H_2O mixture at room temperature with the total TON of 20.4 [56]. Participation of $(L^{+\cdot})Fe^{IV}{=}O$ species was proposed on the basis of ESI-MS and DFT data. Complexes with hydrotris(1-pyrazolyl)-borate instead of Me_3tacn and with acac ligand instead of Cl-acac showed lower activity owing to their degradation during reaction.

Aforementioned studies show that the major problem in the direct oxidation of propane is the selectivity of the oxidation when a variety of the oxidation products are formed in comparable quantities. To achieve a good selectivity to target products, C—C cleavage must be controlled and over-oxidation must be prevented.

10.5 Final Comments

The Holy Grail of bioinspired studies in the catalytic oxidation of methane is the development of new efficient chemical systems based on Earth-abundant metals like iron, which would operate in the clean conditions with a low energy demand. This is a challenging task because of the chemical inertness of methane, ethane, and propane having high strengths of their C—H bonds. Another obstacle is the higher reactivity of the primary oxidation products leading to decrease of the selectivity to desired product because of over-oxidation. However, if such a technology could be developed, it should allow for the rational use of the huge natural gas feedstock and its implementation, even at remote locations without heavy investment.

Another important goal of the studies in bioinspired oxidation of methane is the understanding of the structure of active sites of enzymes and bioinspired catalysts, as well as the mechanisms of the oxidation of strong C—H bonds performed under mild conditions with high efficiency and selectivity. These studies might provide fundamentals for the development of catalysts for industrial processes.

Several bioinspired catalytic systems based on iron or copper catalysts have been developed for the oxidation of CH_4 with efficiencies attaining several hundred turnover numbers. These examples show that such a challenging reaction can be performed in aqueous solutions at near-ambient temperatures without the need for elevated temperatures and strong acids. The catalysts capable of oxidizing methane under mild conditions described in this chapter have the common feature. This activity is principally associated with di- or trinuclear iron and copper catalytic sites, which generate high-valent metal-oxo species using H_2O_2 and O_2 oxidants. They include μ-nitrido diiron phthalocyanine and porphyrin complexes with $(L)Fe^{+3.5}(\mu{-}N)Fe^{+3.5}(L)$ structure [15], dimeric μ-oxo iron triazacyclononane complex [49], extra-framework μ-hydroxo diiron sites $[Fe^{III}_2(\mu_2\text{-}OH)_2(OH)_2(H_2O)_2]^{2+}$ in Fe-ZSM-5 zeolites [34], bent $Cu^{II}(\mu\text{-}O)Cu^{II}$ sites in Cu-ZSM-5 [45], as well as tricopper cluster embedded into tri-piperazine ligand developed by Chan and co-workers as pMMO model [40], and tricopper $[Cu_3(\mu\text{-}O)_3]^{2+}$ cluster found in Cu-mordenite zeolites [46]. The only mononuclear α-iron site in Fe-BEA zeolite capable of reaction with methane after activation with N_2O was reported by Snyder et al. [30]. Upon activation of the oxidant (H_2O_2, N_2O, O_2), all these sites generate high-valent metal species, some of which were shown to exhibit the similar properties: (i) elongated iron-oxo and copper-oxo bonds developing oxyl radical character and (ii) high strength of the formed OH bond after H atom abstraction by high-valent oxo species. These features are responsible for the relatively low activation energy of the reaction of the oxo species with methane.

Although several promising catalytic systems for the mild catalytic oxidation of light alkanes have been developed, there are still unsolved issues concerning the oxidation selectivity and applicability of these reactions on the industrial scale. A major problem in achieving a high selectivity to target products, e.g. MeOH or EtOH in the oxidation of CH_4 and C_2H_6, respectively, is their higher reactivity compared to substrate alkanes leading to over-oxidation products. To achieve a direct selective oxidation of C_2H_6 to CH_3COOH, it is necessary to limit C—C bond cleavage reactions. Utilization of O_2 of air as a terminal oxidant for the oxidation of methane, ethane, and propane represents a key obstacle on the way to industrial implementation of these processes. Future studies directed to the development of more efficient catalysts based on accessible Earth-abundant metals for clean and energy-efficient processes for methane oxidation should also address these issues.

Acknowledgments

The author is grateful to all co-workers whose names are given in the references for their valuable contributions in this work. The research in the author's laboratory was supported by Agence National de Recherche (ANR, France, grants ANR-08-BLANC-0183-01 and ANR-16-CE29-0018-01), by CNRS (PICS project 6295) and by Région Rhône-Alpes.

References

1 Kondratenko, E.V., Peppel, T., Seeburg, D. et al. (2017). *Catal. Sci. Technol.* 7: 366.
2 Ross, M.O. and Rosenzweig, A.C. (2017). *J. Biol. Inorg. Chem.* 22: 307.
3 Hashiguchi, B.G., Bischof, S.M., Konnick, M.M. et al. (2012). *Acc. Chem. Res.* 45: 885.
4 Merkx, M., Kopp, D.A., Sazinsky, M.H. et al. (2001). *Angew. Chem. Int. Ed.* 40: 2782.
5 Baik, M.-H., Newcomb, M., Friesner, R.A. et al. (2003). *Chem. Rev.* 103: 2385.
6 Sazinsky, M.H. and Lippard, S.J. (2015). *Met. Ions Life Sci.* 15: 205.
7 Wang, V.C.-C., Maji, S., Chen, P.P.-Y. et al. (2017). *Chem. Rev.* 117: 8574 https://doi.org/10.1021/acs.chemrev.6b00624
8 Lawson, T.J. and Rosenzweig, A.C. (2016). *J. Am. Chem. Soc.* 138: 9327.
9 Tshuva, E.Y. and Lippard, S.J. (2004). *Chem. Rev.* 104: 987.

10 Friedle, S., Reisner, E., and Lippard, S.J. (2010). *Chem. Soc. Rev.* 39: 2768.
11 McLain, J.L., Lee, J., and Groves, J.T. (2000). *Biomimetic Oxygenations Related to Cytochrome P450: Metal-Oxo and Metal-Peroxo Intermediates* (ed. B. Meunier), 91–169. London: Imperial College Press.
12 Costas, M. (2011). *Coord. Chem. Rev.* 255: 2912.
13 Olivo, G., Cussó, O., Borrell, M. et al. (2017). *J. Biol. Inorg. Chem.* 22: 425.
14 Bryliakov, K.P. and Talsi, E.P. (2014). *Coord. Chem. Rev.* 276: 73.
15 Afanasiev, P. and Sorokin, A.B. (2016). *Acc. Chem. Res.* 49: 583.
16 Floris, B., Donzello, M.P., and Ercolani, C. (2003). Single-Atom Bridged Dinuclear Metal Complexes with Emphasis on Phthalocyanine Systems. In: *The Porphyrin Handbook*, vol. 18 (ed. K.M. Kadish, K.M. Smith and R. Guilard), 1–62. San Diego: Elsevier Science.
17 Afanasiev, P., Bouchu, D., Kudrik, E.V. et al. (2009). *Dalton Trans.* 9828.
18 Sorokin, A.B. (2017). *Adv. Inorg. Chem.* 70: 107–165.
19 Sorokin, A.B., Kudrik, E.V., and Bouchu, D. (2008). *Chem. Commun.* 2562.
20 Afanasiev, P., Kudrik, E.V., Millet, J.-M.M. et al. (2011). *Dalton Trans.* 40: 701.
21 Kudrik, E.V., Afanasiev, P., Alvarez, L.X. et al. (2012). *Nat. Chem.* 4: 1024.
22 Sorokin, A.B., Kudrik, E.V., Alvarez, L.X. et al. (2010). *Catal. Today* 157: 149.
23 Kudrik, E.V. and Sorokin, A.B. (2008). *Chem. Eur. J.* 14: 7123.
24 İşci, Ü., Faponle, A.S., Afanasiev, P. et al. (2015). *Chem. Sci.* 6: 5063.
25 Quesne, M.G., Senthilnathan, D., Singh, D. et al. (2016). *ACS Catal.* 6: 2230.
26 Colomban, C., Kudrik, E.V., Briois, V. et al. (2014). *Inorg. Chem.* 53: 11530.
27 Dubkov, K.A., Sobolev, V.I., and Panov, G.I. (1998). *Kinet. Catal.* 39: 72.
28 Ovanesyan, N.S., Shteinman, A.A., Dubkov, K.A. et al. (1998). *Kinet. Catal.* 39: 792.
29 Starokon, E.V., Parfenov, M.V., Pirutko, L.V. et al. (2011). *J. Phys. Chem. C* 115: 2155.
30 Snyder, B.E.R., Vanelderen, P., Bols, M.L. et al. (2016). *Nature* 536: 317.
31 Hammond, C., Conrad, S., and Hermans, I. (2012). *ChemSusChem* 5: 1668.
32 Zecchina, A., Rivallan, M., Berlier, G. et al. (2007). *Phys. Chem. Chem. Phys.* 9: 3483.
33 Panov, G.I., Sobolev, V.I., Dubkov, K.A. et al. (1997). *React. Kinet. Catal. Lett.* 61: 251.
34 Hammond, C., Forde, M.M., Ab Rahim, M.H. et al. (2012). *Angew. Chem. Int. Ed.* 51: 5129.
35 Hammond, C., Jenkins, R.L., Dimitratos, N. et al. (2012). *Chem. Eur. J.* 18: 15745.
36 Tano, T., Sugimoto, H., Fujieda, N. et al. (2012). *Eur. J. Inorg. Chem.* 4099.
37 Hammond, C., Dimitratos, N., Lopez-Sanchez, J.A. et al. (2013). *ACS Catal.* 3: 1835.
38 Forde, M.M., Armstrong, R.D., McVicker, R. et al. (2014). *Chem. Sci.* 5: 3603.
39 Itoh, S. (2015). *Acc. Chem. Res.* 48: 2066.
40 Chan, S.I., Lu, Y.-J., Nagababu, P. et al. (2013). *Angew. Chem. Int. Ed.* 52: 3731.
41 Nagababu, P., Yu, S.S.-F., Maji, S. et al. (2014). *Catal. Sci. Technol.* 4: 930.
42 Chen, P.P.-Y., Nagababu, P., Yu, S.S.-F. et al. (2014). *ChemCatChem* 6: 429.
43 Liu, C.-C., Mou, C.-Y., Yu, S.S.-F. et al. (2016). *Energy Environ. Sci.* 9: 1361.
44 Smeets, P.J., Hardt, R.G., Woertink, J.S. et al. (2010). *J. Am. Chem. Soc.* 132: 14736.
45 Woertink, J.S., Smeets, P.J., Groothaert, M.H. et al. (2009). *Proc. Natl. Acad. Sci. U.S.A* 106: 18908.
46 Grundner, S., Markovits, M.A.C., Li, G. et al. (2015). *Nat. Commun.* 6: 7546.
47 Huang, W., Zhang, S., Tang, Y. et al. (2016). *Angew. Chem. Int. Ed.* 55: 13441.
48 Ab Rahim, M.H., Armstrong, R.D., Hammond, C. et al. (2016). *Catal. Sci. Technol.* 6: 3410.
49 Romakh, V.B., Therrien, B., Süss-Fink, G. et al. (2007). *Inorg. Chem.* 46: 3166.
50 Kirillov, A.M., Kopylovich, M.N., Kirillova, M.V. et al. (2005). *Angew. Chem. Int. Ed.* 44: 4345.
51 Armstrong, R.D., Hutchings, G.J., and Taylor, S.H. (2016). *Catalysts* 6: 71.
52 Forde, M.M., Armstrong, R.D., Hammond, C. et al. (2013). *J. Am. Chem. Soc.* 135: 11087.

53 Armstrong, R.D., Freakley, S.J., Forde, M.M. et al. (2015). *J. Catal.* 330: 84.
54 Alvarez, L.X. and Sorokin, A.B. (2015). *J. Organomet. Chem.* 793: 139.
55 Xiao, D.J., Bloch, E.D., Mason, J.A. et al. (2014). *Nat. Chem.* 6: 590.
56 Tse, C.-W., Chow, T.W.-S., Guo, Z. et al. (2014). *Angew. Chem. Int. Ed.* 53: 798.
57 Peneau, V., Armstrong, R.D., Shaw, G. et al. (2017). *ChemCatChem* 9: 642.
58 Kudrik, E.V., Afanasiev, P., Bouchu, D. et al. (2008). *J. Porphyrins Phthalocyanines* 12: 1078.

11

Regioselectivity of Non-heme Iron Catalysts for C—H Activation

Christian R. Goldsmith

Auburn University, Department of Chemistry and Biochemistry, 179 Chemistry Building, Auburn, AL, USA

11.1 Introduction

Chemists have long sought the ability to activate C—H bonds in a controllable manner, that is, to take a complicated substrate and oxidize only one out of its many sorts of potentially reactive C—H bonds into a more useful functional group. When I use the term "regioselectivity" in the context of C—H activation, I refer to a reaction's preference for cleaving a C—H bond on one or more particular carbons on the molecular structure to yield a limited subset of constitutional isomers. This is also commonly referred to as "site-selectivity" [1, 2]. Taken to an extreme, only one out of several potential constitutional isomers is made (Scheme 11.1). The ability to stereo- and/or regioselectively activate C—H bonds would greatly streamline the production of fine chemicals from fossil fuel sources and enable more extensive late-stage synthetic modification of natural product pharmaceuticals. Consequently, selective C—H activation has long been considered a "Holy Grail" of synthetic chemistry [2–4].

The conversion of a C—H bond to a C—X group can be broken down into two fundamental steps: the cleavage of the C—H bond and the creation of the C—X bond. In non-heme iron chemistry, the C—H bond cleavage is typically homolytic. The metal-based oxidant abstracts a hydrogen atom from the substrate, yielding an organic radical that reacts quickly with a redox-active X donor, such as an Fe(III)-OH species or O_2, to generate the new functional group. A sample reaction between an alkane and an Fe(IV)-oxo species is illustrated in Scheme 11.2. In this reaction, the oxygen that accepts the hydrogen atom is subsequently transferred to the organic radical as a hydroxyl group; this is frequently referred to as an oxygen atom rebound [5]. Re-oxidation of the iron to the +4 oxidation state by a terminal oxidant enables the catalysis of C—H activation to continue.

There are two fundamental factors that determine how easily a C—H bond on an organic molecule can be activated: the strength of the bond and its accessibility to potential oxidants. There are several ways in which the strength of a bond can be gauged [2], but the measure most relevant to the non-heme iron oxidants discussed in this chapter is the bond dissociation energy (BDE), which is defined as the standard enthalpy change that results upon bond homolysis. The BDEs of C—H bonds are modulated by nearby electron-donating or -withdrawing groups. Substituents that stabilize the carbon-centered radical that results from hydrogen atom abstraction weaken the C—H bond and lower its BDE. There is a particularly strong effect for substituents with π bonds, such as phenyl groups or alkenes, which stabilize the generated radical by efficiently delocalizing the unpaired electron. Due to the greater thermodynamic stabilities of their corresponding radicals, benzylic, and allylic C—H bonds are consequently much weaker

Alkane Functionalization, First Edition. Edited by Armando J. L. Pombeiro and M. Fátima C. Guedes da Silva.

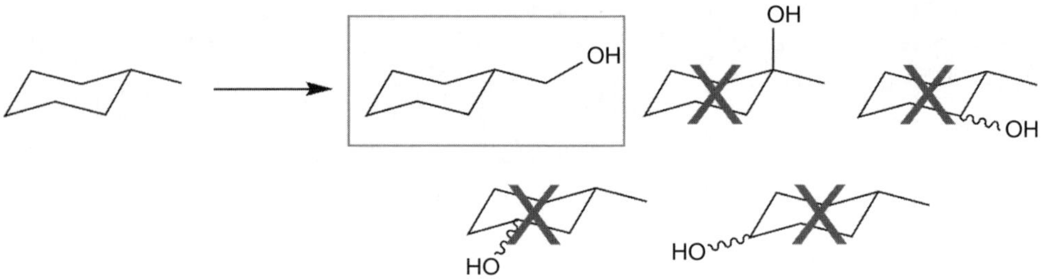

Scheme 11.1 Regioselective hydroxylation of methylcyclohexane at the primary carbon.

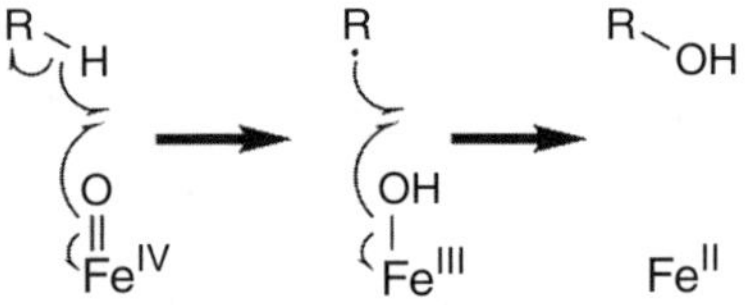

Scheme 11.2 C—H activation by a high-valent iron oxidant, depicting oxygen rebound.

and more readily oxidized than aliphatic ones [6–9]. Adjacent heteroatoms also substantially weaken C—H bonds. Consequently, many oxidants will preferentially or even exclusively target the C—H bonds near π-systems and heteroatoms [10–13]. C—H bonds weaken with increasing substitution on the carbon. The C—H bonds on tertiary carbons have lower BDEs than those on secondary carbons, which are, in turn, easier to activate than those on primary carbons. These trends hold as long as the substituents are comparable in terms of their ability to donate and delocalize electrons. As an alternative to considering the stabilities of the carbon-centered radicals that result from C—H homolysis, one can think about electron-donating groups rendering C—H bonds more susceptible to attack by electrophilic oxidants, as was done in a recent review by Newhouse and Baran [14]. The Hammett constants of electron-donating or -withdrawing substituents can potentially serve as quantitative measures to describe the strengths of nearby C—H bonds, but it can be difficult to properly account for the distance between the substituent and the C—H bonds when predicting which ones are more likely to react.

Accessibility is unfortunately much more difficult to quantify. The presence of sterically cumbersome substituents around a C—H functionality will tend to limit the access of bulkier oxidants. This can increase the kinetic barrier for C—H activation without the need for a large change in the exothermicity of the reaction. The two factors often oppose each other in that alkyl or aryl substituents will impede the approach of incoming oxidants while providing electron-donating substituents that lower the BDE of the target C—H bond. Although the accessibility of C—H bonds is generally accepted to follow the trend tertiary < secondary < primary, a quantitative scale suitable for making predictions has not yet been established. As a consequence, arguments concerning the accessibility of C—H bonds are most often made to explain an observed set of results rather than to predict the outcomes of new oxidation reactions.

Oxidants tend to preferentially activate the weakest C—H bonds that they can access. As the material in the remainder of this chapter will illustrate, there has been great success in directing oxidation toward tertiary and benzylic/allylic carbons. By limiting substrate access, one can also direct the oxidation away from tertiary carbons and toward secondary carbons for some, but not all, substrates. The regioselective activation of C—H bonds on aliphatic primary carbons, conversely, remains elusive, since these bonds are usually strong enough to completely nullify the benefits that their optimal accessibility can provide. The only system that preferentially and catalytically oxidizes aliphatic primary carbons is Shilov's system, which has the drawback of using an expensive platinum complex as the terminal oxidant [4, 15]. A number of other organometallic complexes can activate C—H bonds on terminal methyl C—H bonds, but these are stoichiometric reagents rather than catalysts [16].

Many researchers in the field of oxidative catalysis have been influenced by bioinorganic chemistry and, specifically, by oxidoreductase enzymes. Many of these enzymes can stereo- and regioselectively oxidize C—H bonds by O_2 under mild reaction conditions using common and inexpensive first-row transition metals, such as iron, manganese, or copper [17–22]. In certain cases, the generated oxidants can even activate the C—H bonds in methane. These systems have inspired decades of efforts to replicate this functionality. Although first-row transition metals have been used to activate C—H bonds under ambient conditions, sometimes even using O_2 as the terminal oxidant, directing that activity toward specific carbons on the substrates has proven to be much more difficult. The exquisite regioselectivity of certain oxidoreductase enzymes sometimes results from the presence of substrate-binding sites near the active sites [23]. Noncovalent interactions between the binding site and substrate orient the molecule in a such a manner that the active site oxidant can only reach and react with a certain site on the substrate. Alternatively, substrate channels may block reagents, or parts of reagents, with the incorrect size and/or shape from accessing the active site.

In this chapter, I will focus on efforts to control the regioselectivity of C—H activation by mononuclear non-heme iron catalysts. These catalysts are of high interest since they use naturally abundant iron as the catalytic metal, can oxidize C—H bonds with a variety of terminal oxidants under ambient conditions, and are readily modifiable. Throughout this discussion, I will refer to the organic ligand components of the catalysts using both a number specific for this chapter and the abbreviation used by the authors in their original report.

11.2 The Assessment of Regioselectivity

At first glance, systematically assessing regioselectivity appears to be trivial. One can take a single substrate with inequivalent C—H bonds, oxidize it, and then determine which C—H bonds are activated. The ratio of products is then analyzed to provide a measure of the oxidant's regioselectivity. In theory, any substrate with a set of inequivalent C—H bonds can be used for this purpose. The C—H bonds on primary carbons (1°) tend to be more accessible than those on secondary carbons (2°), which are in turn, more accessible than those on tertiary carbons (3°). This said, the identity of the substituents significantly impacts the accessibility to the extent that some 3° C—H bonds appear to be more accessible than the 2° C—H bonds in other molecules.

In order for the substrate to be useful, the different oxidation products must be readily distinguishable and identifiable by either nuclear magnetic resonance or gas chromatography. Substrates with commercially available oxidation products that can be used as standards are understandably extremely valuable for these sorts of assessments.

11.2.1 Commonly Used Substrates

The substrate adamantane is frequently used to study non-heme iron catalysis (Scheme 11.3). This substrate has both secondary and tertiary carbons. The ratio of 3°–2° C—H bonds is 4 : 12, and most reported 3°/2° oxidation ratios are corrected to account for the uneven numbers of bonds [24, 25]. The 3° C—H bonds appear to be unusually accessible relative to those in other hydrocarbon substrates, and non-heme iron-catalyzed oxidation is normally directed heavily toward these atoms, with the overwhelming majority of statistically adjusted 3°/2° oxidation ratios in excess of 10 : 1 (Table 11.1) [26–36].

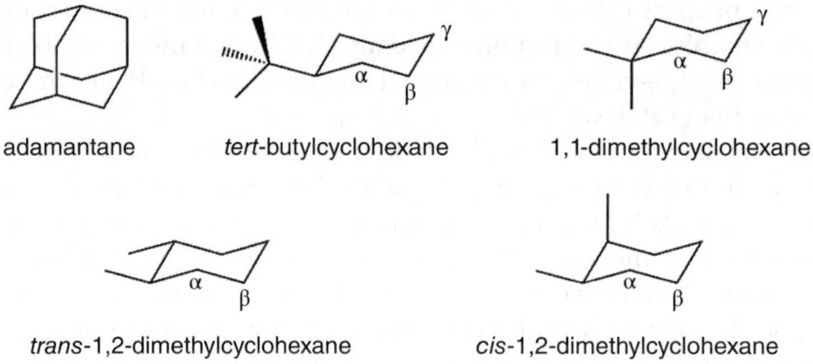

Scheme 11.3 Aliphatic substrates frequently used to assess the regioselectivity of non-heme iron oxidants.

Table 11.1 Summary of Fe(II) catalysts' site-selectivities with three common alkane substrates.

Catalyst	Adamantane 3°/2°[a)]	*cis*-1,2-DMCH 3°/2° (α/β)[b)]	*trans*-1,2-DMCH 3°/2° (α/β)[c)]	Reference(s)[d)]
$[Fe(\mathbf{12})(MeCN)_2]^{2+}$	17 : 1	3 : 1	1 : 1	[26, 27]
$[Fe(\mathbf{15})(MeCN)_2]^{2+}$	13 : 1	2.8 : 1 (1 : 1.4)	1 : 1.5 (1.6 : 1)	[28–30]
cis-α $[Fe(\mathbf{16})(MeCN)_2]^{2+}$	15 : 1	1.8 : 1 (1 : 1.6)	1 : 1.9 (1 : 1.2)	[28, 31]
cis-β $[Fe(\mathbf{16})(MeCN)_2]^{2+}$	17 : 1	Not tested	Not tested	[31]
$[Fe(\mathbf{17})(MeCN)]^{2+}$	**3.1 : 1**	No 2° oxidation noted	No 2° oxidation noted	[25]
$[Fe(\mathbf{18})(MeCN)_2]^{2+}$	30 : 1	No 2° oxidation noted	Not tested	[27]
$[Fe(\mathbf{19})(MeCN)_2]^{2+}$	33 : 1	No 2° oxidation noted	Not tested	[27]
$[Fe(\mathbf{20})(MeCN)_2]^{2+}$	15 : 1	No 2° oxidation noted	Not tested	[27]
$[Fe(\mathbf{23})(MeCN)_2]^{2+}$	21 : 1	No 2° oxidation noted	Not tested	[27]
$[Fe(\mathbf{26})(MeCN)_2]^{2+}$	27 : 1	No 2° oxidation noted	Not tested	[27]
$[Fe(\mathbf{28})(MeCN)_2]^{2+}$	15 : 1	No 2° oxidation noted	Not tested	[27]
$[Fe(\mathbf{29})(OTf)_2]$	Not tested	**5.1 : 1**[e)]	1 : 2.7 (1 : 1.7)[e)]	[32]
$[Fe(\mathbf{4})(MeCN)_2]^{2+}$	15 : 1	1.4 : 1 (1.4 : 1)	1 : 4.8 (1 : 1.2)	[28]
$[Fe(\mathbf{5})(OTf)_2]$	4.6 : 1	1 : 2.7 (5.7 : : 1)	1 : 3.6 (1 : 1.3)	[33]
$[Fe(\mathbf{30})(MeCN)_2]^{2+}$	Not tested	3.7 : 1 (1.5 : 1)	1 : 1.7 (1 : 1.3)	[34]
$[Fe(\mathbf{31})(MeCN)_2]^{2+}$	Not tested	Not tested	1 : 10 (1 : 1.7)	[35]
$[Fe(\mathbf{32})(OTf)_2]$	Not tested	Not tested	1 : 4 (1 : 2.1)	[36]
$[Fe(\mathbf{33})(OTf)_2]$	Not tested	Not tested	**1 : 13 (1 : 3.0)**	[36]

a) Statistically corrected for the uneven numbers of 3° and 2° C—H bonds.
3°/2° = 3 × (1-adamantanol)/(2-adamantanol + 2-adamantanone).

b) 3°/2° = (*trans*-1,2-dimethylcyclohexanol + *cis*-1,2-dimethylcyclohexanol)/(*cis*-2,3-dimethylcyclohexanone + *cis*-2,3-dimethylcyclohexanol + *cis*-3,4-dimethylcyclohexanone + *cis*-3,4-dimethylcyclohexanol).
α/β = (*cis*-2,3-dimethylcyclohexanone + *cis*-2,3-dimethylcyclohexanol)/(*cis*-3,4-dimethylcyclohexanone + *cis*-3,4-dimethylcyclohexanol).

c) 3°/2° = (*trans*-1,2-dimethylcyclohexanol + *cis*-1,2-dimethylcyclohexanol)/(*trans*-2,3-dimethylcyclohexanone + *trans*-2,3-dimethylcyclohexanol + *trans*-3,4-dimethylcyclohexanone + *trans*-3,4-dimethylcyclohexanol).
α/β = (*trans*-2,3-dimethylcyclohexanone + *trans*-2,3-dimethylcyclohexanol)/(*trans*-3,4-dimethylcyclohexanone + *trans*-3,4-dimethylcyclohexanol).

d) All reactions were run in MeCN and used H_2O_2 as the terminal oxidant. Note that the reaction conditions vary. Please see the provided references for further details.

e) Average for the two isolated enantiomers.

Two isomers of 1,2-dimethylcyclohexane (1,2-DMCH) have been used to assess regioselectivity with increasing frequency (Scheme 11.3) [14, 28, 33, 34]. *cis*-1,2-DMCH has been extensively used in non-heme iron catalysis in order to assess the degree of stereoretention upon the oxidation of the substrate [37], and most readers will be more familiar with this substrate in this capacity. Like adamantane, these substrates contain both 2° and 3° C—H bonds. The 1° C—H bonds on the methyl substituents could also be oxidized, but this has thus far not been observed, likely due to the much greater strengths of these bonds. The 3° C—H bonds appear to be less accessible than those on adamantane, and when either isomer of 1,2-DMCH is oxidized with the aid of a non-heme iron catalyst, a greater proportion of the oxidation occurs on the secondary carbons. More of the C—H bonds of 1,2-DMCH are found on secondary carbons relative to adamantane, and the 2 : 8 ratio of 3°–2° C—H bonds should statistically weight the oxidation slightly more toward the secondary carbons.

11.2.2 Substrates Used to Assess Longer-Range Impacts on Regioselectivity

More highly substituted carbons would be expected to hinder access to the C—H bonds on nearby carbons. In the aforementioned 1,2-DMCH, the α positions would be anticipated to be less accessible than the β positions since the former are immediately adjacent to the methylated carbons (Scheme 11.3). In practice, more oxidation is normally seen on the α carbons for the *cis* isomer and on the β carbons for the *trans* isomer [28, 33, 34]. Two other substrates that have been used to assess the longer-range impacts of sterics on the reactivity of individual C—H bonds are *tert*-butylcyclohexane and 1,1-dimethylcyclohexane (Scheme 11.3). The former substrate tends to be most strongly oxidized on the γ position, with some oxidation on the β and no oxidation of either the 3° or the α carbons [28, 33, 34]. The ratio of β : γ oxidation varies from catalyst to catalyst; statistically, activation of the β C—H bonds is favored since there are twice as many β sites. The two methyl groups in 1,1-dimethylcyclohexane exert a lesser, but still noticeable, effect [28, 33, 34]. The methyl groups are not large enough to entirely preclude oxidation of the α C—H bonds, but the β and γ carbons are noticeably more susceptible to oxidation. As with *tert*-butylcyclohexane, both the ratio of α : β : γ oxidation and indeed the preferred site of C—H activation depend strongly on the catalyst.

11.2.3 Commonly Encountered Complications in Determining the Regioselectivity of a Metal-Based Oxidant

One underappreciated complication with the assessment of regioselectivity is that patterns of oxidation do not scale perfectly with each other. One catalyst may have a higher 3°/2° oxidation ratio than another catalyst for one *but not all* hydrocarbon substrates. A stronger preference for oxidizing the secondary carbons on *cis*-1,2-DMCH, for instance, may not translate to a stronger preference for oxidizing the secondary carbons on either adamantane or *trans*-1,2-DMCH. Regrettably, many studies on non-heme iron oxidants demonstrate the site-selectivity of the promoted oxidation using a single substrate. This makes it difficult to determine which oxidants are intrinsically better or worse at accessing more sterically congested C—H bonds.

Another complication is the possible agency of multiple oxidants. As some examples in this chapter will illustrate, mixing a terminal oxidant and an Fe(II) precursor can generate a mixture of metal-containing and metal-free oxidants with discrete reactivity profiles and site-selectivities. It can be difficult to determine how each oxidant contributes to the overall product distribution. The most practical catalysts would therefore proceed through a single oxidant with a strong preference for a particular sort of C—H bond.

11.3 Non-heme Iron Oxidants with Strong Preferences for Weak C—H Bonds

Selectively activating the weakest C—H bonds on a hydrocarbon substrate has traditionally been easier to accomplish. Since rates of hydrogen atom abstraction generally have a strong free energy relationship with the enthalpy of the broken C—H bond at parity of oxidant [38–40], weaker C—H bonds break much more quickly than their stronger counterparts.

The selectivity can be further improved by using weak oxidants, which often decrease the rates of activation of stronger C—H bonds to negligible degrees. The major drawback to this strategy is that this necessarily slows the rate of the desired C—H activation reaction as well. This phenomenon is first encountered by students in the form of free radical halogenation. Chlorination is much less regioselective than bromination due to the much higher exothermicity of the reactions between C—H bonds and chlorine radicals (Scheme 11.4) [41]. 1°, 2°, and 3° C—H bonds are all chlorinated readily. Bromination occurs much more heavily on secondary, tertiary, and benzylic/allylic carbons [42]. If the oxidant is too weak, however, no reactivity will be observed. Free radical iodination of alkanes, for instance, doesn't occur since the I—H bond that forms upon hydrogen atom abstraction is not stable enough to make hydrogen atom abstraction from even the weakest of known C—H bonds thermodynamically favorable.

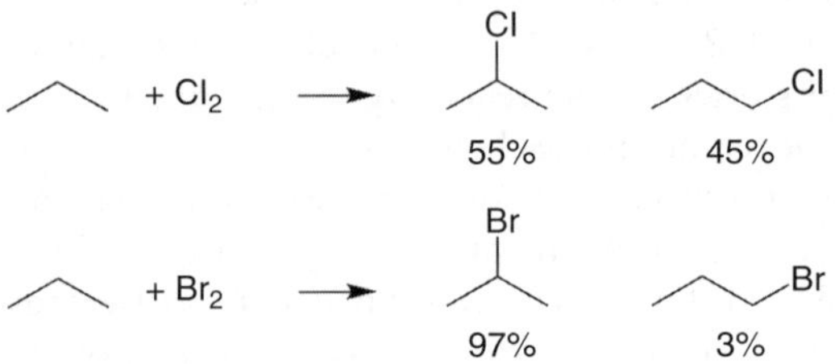

Scheme 11.4 Regioselectivities of free radical chlorination and bromination, using propane as the standard substrate.

In non-heme iron chemistry, certain mononuclear Fe(III)-containing species have been found to be relatively weak oxidants. As such, they tend to react exclusively with substrates with benzylic and allylic C—H bonds (Scheme 11.5). Reactivity with aliphatic carbons is extremely rare, and when found, it is both slow and limited to 3° C—H bonds. Most aliphatic C—H bonds are simply too strong to allow the transfer of hydrogen atoms to the following iron-containing oxidants. In the following cited manuscripts, the regioselectivity is rarely used to "sell" the content, but nonetheless, the use of these oxidants can direct oxidation toward specific sites on hydrocarbon substrates.

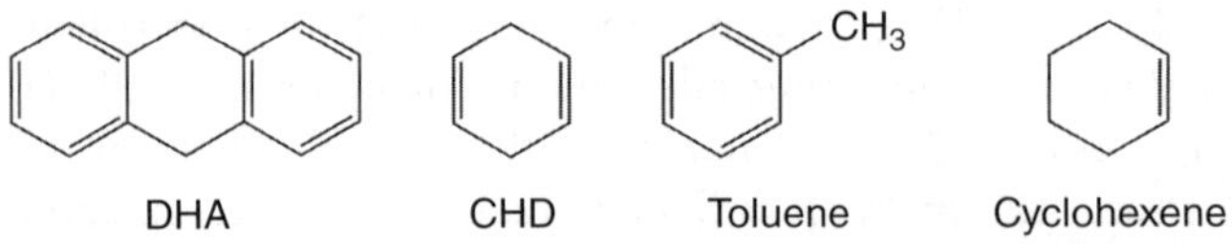

Scheme 11.5 Commonly used substrates with weak C—H bonds.

11.3.1 Fe(III)-Hydroxo and -Alkoxo Species

Fe(III)-OH and Fe(III)-OMe complexes with the pentadentate PY5 ligand (**1**, Scheme 11.6) were found to oxidize an assortment of allylic and benzylic C—H bonds, yielding Fe(II)-H_2O

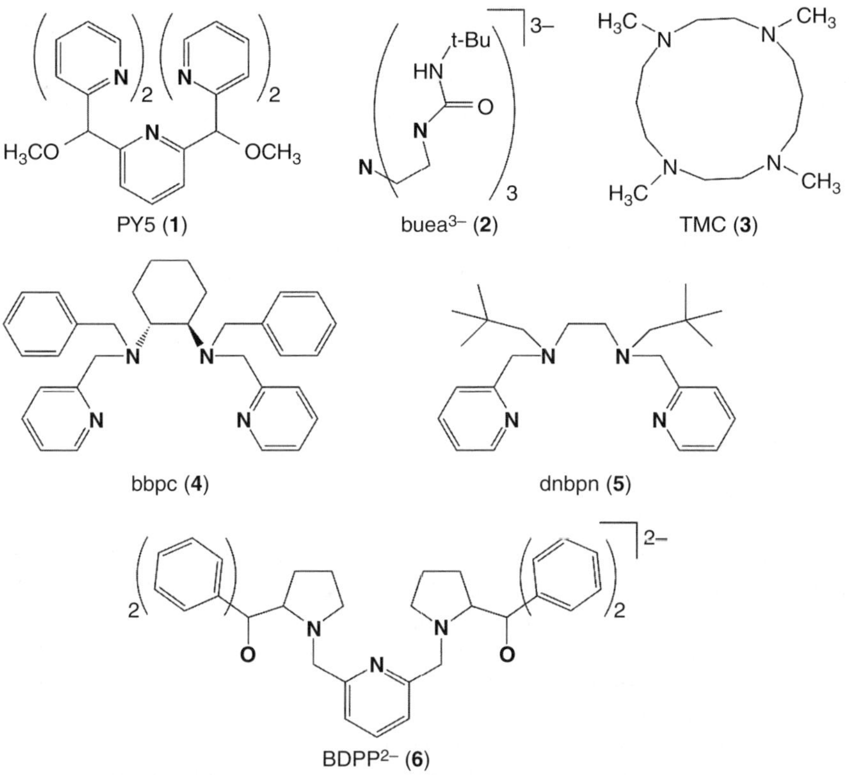

Scheme 11.6 Ligands used to generate observed and proposed Fe(III)-OH, Fe(III)-O, and Fe(III)-O_2 oxidants.

and Fe(II)-MeOH complexes, respectively, in addition to the organic products [43–45]. The active oxidants were designed to resemble Fe(III)-OH oxidants that are proposed to be generated in the active sites of lipoxygenase enzymes [46]. Thermodynamic cycles suggest that the new O—H bonds that form upon reduction of the iron have BDEs of 80 and 84 kcal mol^{-1} [43–45]. The Fe(III) complexes react readily with C—H bonds that are about the same strength or weaker. 9,10-Dihydroanthracene (DHA) and 1,4-cyclohexadiene (CHD) are quantitatively oxidized to anthracene and benzene when they react with two equiv. of $[Fe^{III}(\mathbf{1})(OMe)]^{2+}$ under a N_2 atmosphere [44, 45]. The reaction between $[Fe^{III}(\mathbf{1})(OMe)]^{2+}$ and toluene, which has a BDE of 88 kcal mol^{-1}, was found to be exceedingly slow in acetonitrile, with a k_2 of $5.7 \times 10^{-4}\ M^{-1}\ s^{-1}$ [44], and substrates with stronger C—H bonds do not react with either of the PY5 oxidants. Overall, $[Fe^{III}(\mathbf{1})(OH)]^{2+}$ and $[Fe^{III}(\mathbf{1})(OMe)]^{2+}$ react with substrates at approximately the same rates, despite the 4 kcal mol^{-1} greater driving force for the Fe(III)-OMe oxidant. This may suggest that the methyl group on the methoxide may be hindering substrate access to the reactive portion of $[Fe^{III}(\mathbf{1})(OMe)]^{2+}$. The Fe(II) products can be re-oxidized to the Fe(III) species through the addition of either H_2O_2 or iodosobenzene, but excesses of H_2O_2 predominantly yield an Fe(III)-OOH species that has a substantially different reactivity profile [47].

11.3.2 Fe(III)-Oxo Species

Borovik's group has found similar oxidation chemistry for Fe(III)-oxo complexes with tripodal anionic ligands, such as $buea^{3-}$ (**2**, Scheme 11.6) [48, 49]. The ligand was used to make $[Fe^{III}(\mathbf{2})(O)]^-$, which reacts with DHA and CHD to produce $[Fe^{II}(\mathbf{2})(OH)]^-$ and either anthracene (90% yield) or benzene (70% yield), respectively [48]. A thermodynamic cycle was used to estimate a BDE of $66\,kcal\,mol^{-1}$ for the O—H bond in the Fe(II)-OH product. Relative to **1**, **2** stabilizes the +3 oxidation state to a much greater extent, with a highly unfavorable Fe(III)/Fe(II) reduction potential of −1.79 V vs. $Fc^{+/0}$. Although the Fe(III) oxidant has an unfavorable reduction potential, its greater basicity enables it to activate C—H bonds. A subsequent computational study of the reactivity with DHA suggests that the reactivity proceeds through sequential proton and electron transfer steps as opposed to a concerted hydrogen atom transfer [49]. The 2003 study found that the O—H bond in $[Fe^{III}(\mathbf{2})(OH)]^-$ should have a BDE of approximately $115\,kcal\,mol^{-1}$; this would predict that the unobserved $[Fe^{IV}(\mathbf{2})(O)]^-$ would be a much more powerful oxidant. $[Fe^{III}(\mathbf{2})(OH)]^-$ itself does not react with either DHA or CHD.

11.3.3 Fe(III)-Superoxo Species

Within non-heme iron chemistry, there is mounting evidence that Fe(III)-superoxo complexes are likewise weak oxidants that specifically target mostly benzylic and allylic C—H bonds. These species are commonly proposed in both heme and non-heme iron enzymatic cycles, typically resulting from reactions between coordinatively unsaturated Fe(II) species and O_2 [50–52]. A computational study by Chung et al. concluded that non-heme Fe(III)-superoxo species were much weaker oxidants than Fe(IV)-oxo complexes due to their lower electrophilicity [53]. The BDEs of the O—H bonds formed upon the formation of the Fe(III)-OOH product were estimated to range from 64 to $75\,kcal\,mol^{-1}$.

Lee et al. found that an Fe(II) complex with the TMC ligand (**3**, Scheme 11.6) reacted with O_2 and cyclohexene to yield an Fe(IV)-oxo species [12]. Substrates with stronger C—H bonds led to slower rates of formation for the Fe(IV) product and no reactivity was observed with substrates with exclusively aliphatic C—H bonds. The authors speculated that the reactivity proceeded through an Fe(III)-superoxo complex, which would abstract a hydrogen atom from the cyclohexene to yield a Fe(III)-OOH species that subsequently converted to the Fe(IV) end-product. Although a Fe(III)-OOH complex was not directly observed, Nam's research group identified and subsequently isolated its conjugate base, an Fe(III)-peroxo complex, when the reaction was run with a base additive [12, 54].

My own lab found that the Fe(II) complexes with the sterically encumbered ligands bbpc (**4**) and dnbpn (**5**) react with O_2 to oxidize allylic and benzylic C—H bonds [33, 55, 56]. A kinetic study of the formation of an Fe(III)-OOH complex from $[Fe^{II}(\mathbf{4})(MeCN)_2]^{2+}$, O_2, and cyclohexene suggested the intermediacy of a hydrogen atom abstracting Fe(III)-superoxo species [55, 56]. In the absence of a hydrocarbon substrate with a weak C—H bond, the Fe(III)-OOH species does not form [56]. The Fe(III)-OOH species was also synthesized from a mixture of the Fe(II) precursor and H_2O_2 [28]. $[Fe^{II}(\mathbf{4})(MeCN)_2]^{2+}$ is notable in that it can catalyze the oxidation of substrates containing tertiary aliphatic C—H bonds by air. When *cis-* or *trans*-1,2-DMCH are used as substrates, oxidation of both secondary and tertiary carbons is observed [56]. The oxidation of the secondary carbons is attributed to the Fe(III)-OOH product, rather than the initially generated Fe(III)-superoxo oxidant. Substrates that lack benzylic/allylic and 3° C—H bonds, such as cyclohexane, are not oxidized under these conditions.

Chiang et al. were able to isolate a non-heme Fe(III)-superoxo species with the $BDPP^{2-}$ ligand (**6**, Scheme 11.6) at −80 °C, using Mössbauer and resonance Raman spectroscopy to make the identification [57]. A resonance Raman band at $1125\,cm^{-1}$ was assigned to an O—O stretch. The energy of the $^{16}O—^{16}O$ stretch is consistent with other metal-superoxo complexes. When ^{18}O-labeled dioxygen was used to prepare the compound the peak shifts by $63\,cm^{-1}$. In addition to confirming that O atoms are indeed involved in the vibrational feature, the magnitude of the shift is consistent with an end-on conformation for the superoxide. Fe(III)-superoxo complexes were not directly observed for the aforementioned TMC or bppc systems. The neutral N-donor ligands used in these studies favor lower oxidation states for the iron. In the absence of a weak substrate capable of donating a hydrogen atom to the Fe(III)-superoxo intermediate, the equilibrium position consists overwhelmingly of the Fe(II) precursor and O_2. The use of the dianionic **6** stabilizes the +3 oxidation state enough to allow observation and identification of the Fe(III)-superoxo species. Even though the greater stability would be anticipated to reduce the reactivity of the superoxide complex with external substrates, the compound still reacts with DHA at −70 °C, marking the first instance of an isolated non-heme Fe(III)-superoxo abstracting H atoms from a hydrocarbon.

Although non-heme Fe(III)-superoxo complexes themselves react exclusively with weaker C—H bonds and use nontoxic O_2 as the terminal oxidant, the Fe(III)-OOH products formed from the initial C—H activation reactions can subsequently either directly or indirectly oxidize much stronger C—H bonds [58]. For this reason, Fe(III)-superoxo complexes currently seem to be poor candidates for the active oxidants in site-selective C—H activation catalysis.

11.3.4 Other Weak Non-Heme Iron Oxidants

A few other Fe(III) complexes are rarely mentioned in non-heme iron literature due to their weaker resemblances to enzymatically relevant species but merit discussion here. In these, the abstracted hydrogen atom localizes on a N rather than an O on the iron-containing oxidant. Roth and Mayer reported that $[Fe^{III}(H_2bim)_2(Hbim)]^{2+}$ (H_2bim = 7, Scheme 11.7) reacts with hydrocarbons containing C—H bonds with BDEs less than 80 kcal mol^{-1} [59]. When performed under O_2, the oxidation of DHA is catalytic and yields a mixture of anthracene, anthraquinone, and $[Fe^{II}(7)_3]^{2+}$. The catalysis results partly from the reoxidation of the ferrous complex with 7 and partly from organic radical chain propagation. The N—H bond that forms upon reduction of the Fe(III) has a BDE of 76 kcal mol^{-1}; despite this being weaker than the O—H bond in $[Fe^{II}(\mathbf{1})(MeOH)]^{2+}$, $[Fe^{III}(7)_2(Hbim)]^{2+}$ and $[Fe^{III}(\mathbf{1})(OMe)]^{2+}$ react with hydrocarbon substrates at approximately the same rates [44, 45, 59, 60].

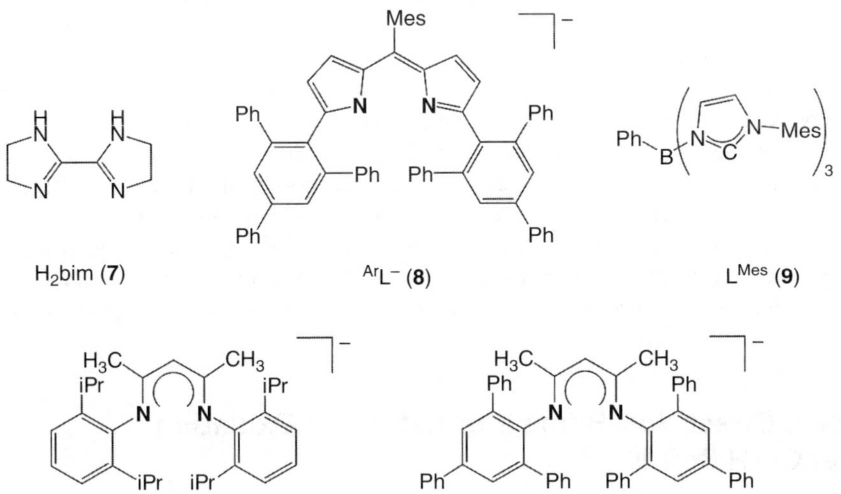

H_2bim (**7**) $^{Ar}L^-$ (**8**) L^{Mes} (**9**)

$(L^{Me,iPr2})^-$ (**10**) $(L^{Me,Ph3})^-$ (**11**)

Scheme 11.7 Ligands used to stabilize weak iron oxidants with imido, nitride, and iminyl radical ligands.

Fe(III) complexes with dipyrromethene ligands and iminyl radicals have also been reported to abstract hydrogen atoms from benzylic and allylic substrates [61, 62]. $[(^{Ar}L)FeCl(N(p\text{-}^tBuC_6H_4)]$ (^{Ar}L = **8**, Scheme 11.7) reacts with both toluene and CHD [61]. The latter reaction yields benzene, which is the anticipated product from a double hydrogen atom abstraction from CHD. The reaction with toluene yields $PhCH_2NH(C_6H_4\text{-}p\text{-}^tBu)$, which would result from a "nitrogen rebound" to an initially generated benzyl radical. This reaction is catalytic when excess alkyl azide – the source of the iminyl radical – is added, and 12 equiv. of aminated toluene per equiv. of iron are observed [61]. When mesityl groups are installed on the pyrrole rings instead of a *tert*-butyl or phenyl group, solely intramolecular oxidation is observed [63].

Smith's group has found that isoelectronic Fe(IV) imido species and Fe(IV) nitrido complexes can react with DHA and CHD [64, 65]. The first isolated complex, $[L^{Mes}Fe^{IV}(NAd)]^+$ (L^{Mes} = **9**, Ad = 1-adamantyl, Scheme 11.7), was generated by oxidizing $[Fe^{III}$(**9**)(NAd)] with $[Cp_2Fe](OTf)$ [64]. The authors were unable to directly observe the anticipated ferric product of hydrogen atom abstraction, $[Fe^{III}$(**9**)$(NHAd)]^+$, but estimated that the N—H bond formed upon hydrogen atom abstraction was 88 kcal mol^{-1} using density functional theory calculations and observations about the reactivity of $[Fe^{III}$(**9**)(NAd)] with a variety of weak acids to support their argument [64]. The Fe(IV) complex and its Fe(III) precursor undergo rapid electron self-exchange, with a second-order rate constant of $7.6 \times 10^7\ M^{-1}\ s^{-1}$. Despite the seemingly low reorganizational barrier for redox and the approximately 10 kcal mol^{-1} driving force, the oxidation of DHA requires heating the reaction at 100 °C for 24 hours and only results in a 25% yield of anthracene. The nitrido complex with the same ligand, $[Fe^{IV}$(**9**)N], reacts with CHD to yield benzene [65]. Other modes of reactivity aside from hydrogen atom abstraction, however, are evident since 1,3-cyclohexadiene and ethylene are also observed as organic products with this substrate.

Work from Holland's group found that ferric complexes with diketiminate ligands could also activate weak C—H bonds [66, 67]. In a 2006 study, a metal-based oxidant was generated by the addition of an alkyl azide to an Fe(I) precursor with $L^{Me,iPr2}$ (**10**, Scheme 11.7); the relatively short-lived complex was proposed to be an Fe(III) imido species "in an unusual intermediate-spin electronic configuration" [66]. The compound reacts with CHD and indene in deuterated toluene. Ligand **10** contains benzylic C—H bonds, and intramolecular oxidation is observed in the absence of either CHD or indene. Although DHA has weaker C—H bonds than indene, this substrate was not oxidized under the same conditions. This led the authors to speculate that steric repulsions between DHA and **10** hindered access of the molecule to the reactive Fe(III)=N group. A subsequent study used a diketiminate ligand, $(L^{Me,Ph3})^-$ (**11**, Scheme 11.7), that lacked benzylic C—H bonds [67]. This change provided a much more stable Fe(III) imido species by eliminating pathways for self-decomposition and allowed much faster reactivity with CHD. The authors also used computational methods to examine the accessibility of the Fe(III)=N group responsible for abstracting the hydrogen atom. Although the diketiminate ligands investigated had similar steric bulk, they found that of **11** was distributed in such a way to allow better substrate access to the oxidant generated in the core of the complex.

11.4 Attempts to Direct Non-Heme Iron-Catalyzed Oxidation Toward Stronger C—H Bonds

Tuning the selectivity away from the weakest C—H bonds on the substrate has proven to be a difficult challenge, and the successes thus far in non-heme iron chemistry have only been partial. Most of the following discrete iron complexes still direct oxidation toward tertiary aliphatic and benzylic/allylic carbons selectively. The stronger C—H bonds on secondary carbons can, in some cases, be oxidized preferentially over tertiary C—H bonds on the same molecule. The C—H bonds on primary carbons, however, have never been targeted over those on secondary carbons. Additionally, oxidation of the weakest C—H bonds in the molecule has never been fully eliminated, even with the catalysts most selective for 2° C—H bonds. In the following studies, arguments about site-selectivity are made with substrates containing exclusively aliphatic C—H bonds; all of the described catalysts would be highly likely to preferentially direct oxidation toward allylic or benzylic C—H bonds if they were present.

Mixtures of iron salts and H_2O_2 are capable of oxidizing a wide variety of substrates, including alkanes [68–70]. Discrete higher-valent non-heme iron oxidants are also known to be capable of readily activating aliphatic C—H bonds [51, 58]. These oxidants are commonly generated by reacting an Fe(II) precursor with H_2O_2 or an alkyl hydroperoxide. Fe(III)-OOH or Fe(III)-OOR intermediates are proposed to be initially generated, with the O—O bond breaking either homolytically or heterolytically to yield Fe(IV)-oxo or Fe(V)-oxo oxidants, respectively. The homolytic cleavage of the O—O bond would generate hydroxyl and alkoxyl radicals, which will themselves react with hydrocarbon substrates. Due to the strong BDEs of the O—H bonds that form upon abstracting hydrogen atoms, these small inorganic radicals are analogous to chlorine atom radicals in that they do not distinguish between different sorts of chemical environments. One must carefully consider the possible contribution of hydroxyl and/or alkoxyl radicals when determining the intrinsic regioselectivity of non-heme iron oxidants.

Most of the ligands that support non-heme iron catalysis of aliphatic C—H activation contain exclusively neutral N-donors and consist mostly of derivatives of tripicolylamine (TPA, **12**, Scheme 11.8), *N,N'*-bis(2-pyridylmethyl)-1,2-ethanediamine (bispicen, **13**), or 1,4,7-triazacyclononane (TACN, **14**).

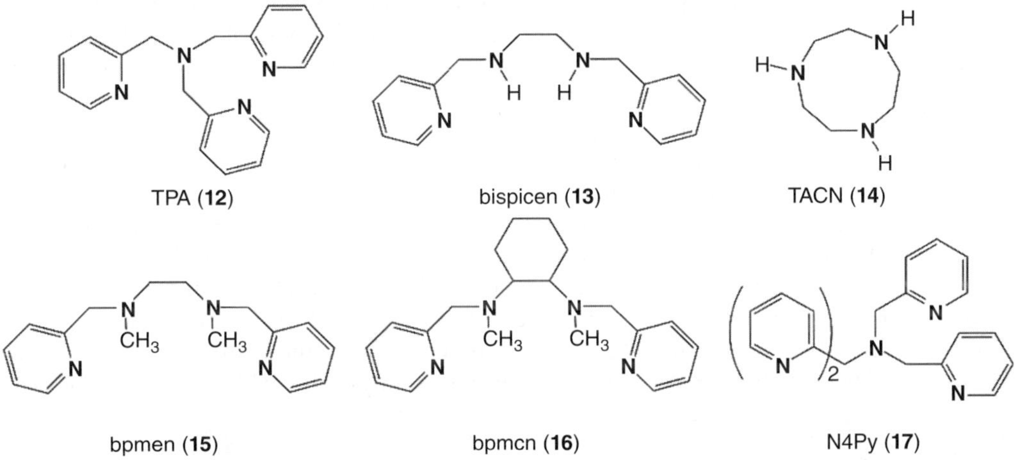

Scheme 11.8 Common ligands used for non-heme iron catalysis of the activation of aliphatic C—H bonds.

11.4.1 Catalysis with Fe(II) Complexes with Unmodified TPA

Iron complexes with TPA and its derivatives were some of the first discrete non-heme iron catalysts for alkane oxidation to be thoroughly explored. Que's group found that $[Fe^{II}(\mathbf{12})(MeCN)_2]^{2+}$ catalyzed the reaction between cyclohexane and H_2O_2 to yield 3.0 equiv. of cyclohexanol and 0.7 equiv. of cyclohexanone per equiv. of iron [26]. Overall, the $[Fe(\mathbf{12})(MeCN)_2]^{2+}$ complex directs alkane oxidation to the tertiary carbons of alkane substrates slightly more than iron complexes with other pyridylamine ligands (Table 11.1). Substituting cyclohexane with *cis*- and *trans*-1,2-DMCH yields 3°/2° oxidation ratios of 3 : 1 and 1 : 1, respectively. The distinction between α and β oxidation products was not made. This is understandable, however, given that the authors were rightly focused on highlighting the first instance of stereospecificity in the non-heme iron-catalyzed oxidation of the 3° C—H bonds; *cis*-1,2-dimethylcyclohexanol and *trans*-1,2-dimethylcyclohexanol are observed exclusively, amounting to >99% retention of the original stereochemistries of the substrates [26, 27]. The retention of stereochemistry is strong evidence against the participation of hydroxyl radicals in the hydrocarbon oxidation. Adamantane was explored as a substrate in a subsequent study; the statistically corrected 3°/2° oxidation ratio was found to be 17 : 1 [27].

11.4.2 Simple Pyridylamine Complexes

Iron complexes of *N,N′*-dimethylated bispicen, or bpmen (**15**, Scheme 11.8), have been frequently studied, and $[Fe^{II}(\mathbf{15})(MeCN)_2]^{2+}$ remains a widely used standard in non-heme iron chemistry due to its high catalytic activity [29, 30, 37]. When used as a catalyst for the oxidation of cyclohexane by 10 equiv. of H_2O_2, 5.6 equiv. of cyclohexanol, and 0.7 equiv. of cyclohexanone are produced [30]. The counteranion does not significantly impact the reactivity, and similar results are obtained when perchlorate is used in this capacity instead of triflate [71]. The yields of oxidized product and the specificity for the alcohol are both high for a non-heme iron catalyst. The catalyst preferentially directs oxidation to the tertiary carbons in substrates lacking allylic or benzylic C—H bonds. Adamantane is still oxidized primarily to 1-adamantol, but with a statistically corrected 3°/2° oxidation ratio of 13 : 1 that is lower than that seen for $[Fe^{II}(\mathbf{12})(MeCN)_2]^{2+}$ (Table 11.1) [29]. When used as a catalyst for the oxidation of *cis*-1,2-DMCH, only *cis*-1,2-dimethylcyclohexanol is observed as a product; the stereospecificity was again used as an argument against the involvement of hydroxyl radicals in the oxidation of the alkane [30]. No 2° oxidation was noted, but my own research group found a 3°/2° oxidation ratio of 2.8 : 1 when a different reaction protocol was followed (Table 11.1), with slightly more of the oxidation being directed toward the α position [28]. We also found that *trans*-1,2-DMCH was oxidized primarily on the secondary carbons, with a 3°/2° oxidation ratio of 1 : 1.5. The β position is oxidized to a slightly greater extent for this substrate.

Replacing the 1,2-ethanediamine linkage in bpmen with 1,2-cyclohexanediamine yields the bpmcn ligand (**16**, Scheme 11.8) [72]. The cyclohexane portion appears to stabilize transition metal complexes by fixing the amine donors in place, to the extent that two different conformational isomers can be isolated: *cis*-α and *cis*-β (Scheme 11.9) [31]. The two isomers have distinct reactivities, demonstrating that the ligand topology has a pronounced impact on non-heme iron catalysis. When the bpmcn complexes are used to catalyze the oxidation of cyclohexane

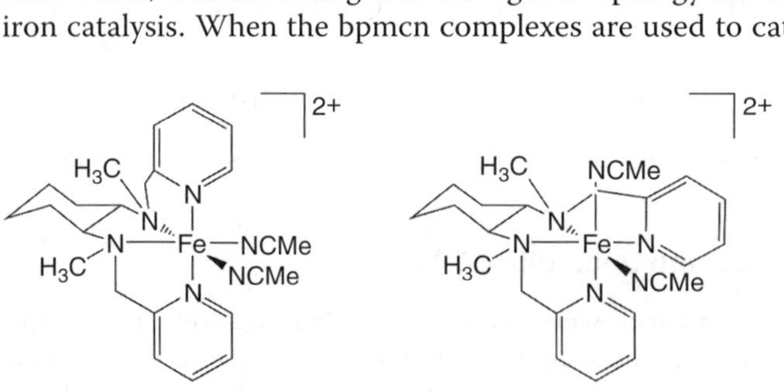

cis-α   *cis*-β

Scheme 11.9 Illustration of conformers observed for $[Fe(16)(MeCN)_2]^{2+}$.

by H_2O_2, for instance, the *cis*-α isomer is both more active, with 5.9 turnovers as opposed to 1.9 for the *cis*-β, and more selective for the cyclohexanol product (A : K = 9 for *cis*-α, 0.9 for *cis*-β). The cyclohexane portion would also be anticipated to render the ligand bulkier than the related bpmen. Despite this, both bpmcn complexes have stronger tendencies to oxidize the 3° C—H bonds of adamantane, with statistically corrected 3°/2° oxidation ratios of 15 : 1 and 17 : 1, respectively (Table 11.1) [31]. The greater predilection toward the oxidation of tertiary carbons, however, is not found in other substrates. The *cis*-α isomer yields lower 3°/2° oxidation ratios than the bpmen catalyst when either isomer of 1,2-DMCH is used as the substrate [28]. Slightly more oxidation is directed toward the β positions in both cases.

11.4.3 N4Py – An Illustration of the Effects of Multiple Oxidants on Site-Selectivity

The N4Py ligand (**17**, Scheme 11.8) differs from the aforementioned ligands in that it is pentadentate, rather than tetradentate [73]. Roelfes et al. found that the complex $[Fe^{II}(\mathbf{17})(MeCN)]^{2+}$ catalyzes the oxygenation of a variety of aliphatic, allylic, and benzylic substrates by H_2O_2 [25]. Cyclohexane, for instance, is converted to a mixture of cyclohexanol and cyclohexanone, with the ratio of products highly depending on the manner in which the terminal oxidant is added. Adamantane is oxidized to a mixture of 1-adamantol, 2-adamantol, and 2-adamantone. The statistically corrected 3°/2° oxidation ratio of 3.1 : 1 is much more heavily weighted toward the secondary carbons than for previously described catalysts (Table 11.1). *cis*- and *trans*-1,2-DMCH were investigated as substrates as well, but no oxidation of the 2° C—H bonds was noted. The relatively high incidence of 2° oxidation for adamantane is likely attributable to the generation of hydroxyl radicals. Three observations made by Roelfes et al. support this. First, *cis*- and *trans*-1,2-DMCH both show substantial losses of stereochemistry upon oxidation, and the retentions of configuration are low relative to those associated with other non-heme iron catalysts discussed in this chapter. *cis*-1,2-DMCH, for instance, was oxidized to a 1.75 : 1 mixture of *cis*- and *trans*-1,2-dimethylcyclohexanol. Second, mixtures of $[Fe^{II}(\mathbf{17})(MeCN)]^{2+}$ and H_2O_2 can also oxidize benzene to phenol. Discrete high-valent iron oxidants cannot normally do this, but hydroxyl radicals can [74]. Third, the results of the hydrocarbon oxidation reactions depend highly on the solvent, and the addition of radical traps decreased the activity and made the reaction more selective for ketone products. The low 3°/2° oxidation ratio reported for adamantane is therefore largely a consequence of the agency of hydroxyl radicals rather than a regioselective iron-containing oxidant.

11.4.4 Sterically Modified TPA and Pyridylamine Ligands

A commonly used strategy to targeting stronger C—H bonds has been to modify the organic ligand of the iron complex to engineer steric interactions between the substrate and catalyst. These can hinder the oxidation of more sterically congested C—H bonds. C—H bonds that are intrinsically harder to access are bypassed in favor of stronger C—H bonds that can be reached. The downside to this strategy is that it induces selectivity by slowing down competing reactions, as opposed to speeding up the desired one. This strategy was successfully applied to metalloporphyrin complexes in the mid-1980s [75, 76] and has subsequently been explored for heterogeneous catalysts for alkane oxygenation [77, 78].

A large number of TPA derivatives with methyl or methylesters installed on either the 3-, 5-, or 6-positions of the pyridine rings have been prepared (Scheme 11.10) [27]. The impact of these modifications on the regioselectivity of alkane oxidation, however, appears to be inconsistent. When adamantane is used as a substrate, the statistically corrected 3°/2° oxidation ratios range

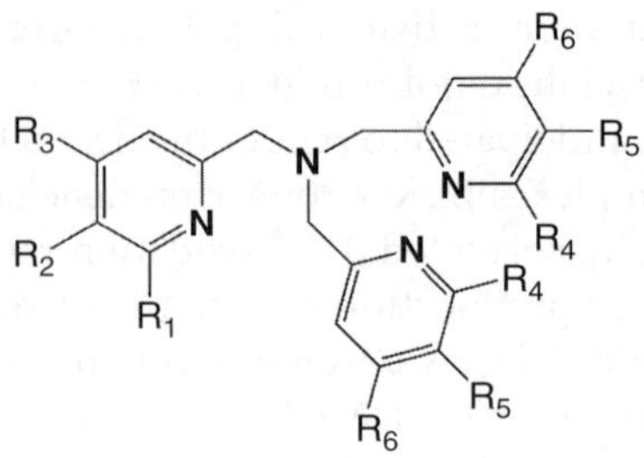

$R_{1\text{-}6}$ = H: TPA (**12**)
R_1 = Me; $R_{2\text{-}6}$ = H: 6-Me-TPA (**18**)
R_4 = Me; $R_{1\text{-}3,5,6}$ = H: 6-Me_2-TPA (**19**)
$R_{1,4}$ = Me; $R_{2,3,5,6}$ = H: 6-Me_3-TPA (**20**)
R_2 = Me; $R_{1,3\text{-}6}$ = H: 5-Me-TPA (**21**)
R_5 = Me; $R_{1\text{-}4,6}$ = H: 5-Me_2-TPA (**22**)
$R_{2,5}$ = Me; $R_{1,3,4,6}$ = H: 5-Me_3-TPA (**23**)
R_3 = Me; $R_{1,2,4\text{-}6}$ = H: 3-Me-TPA (**24**)
R_6 = Me; $R_{1\text{-}5}$ = H: 3-Me_2-TPA (**25**)
$R_{3,6}$ = Me; $R_{1,2,4,5}$ = H: 3-Me_3-TPA (**26**)
R_2 = CO_2Me; $R_{1,3\text{-}6}$ = H: 5-(MeOOC)-TPA (**27**)
R_5 = CO_2Me; $R_{1\text{-}4,6}$ = H: 5-$(MeOOC)_2$-TPA (**28**)

Scheme 11.10 Examples of modified TPA ligands.

from 15 to 30 (Table 11.1). Adding substituents to the 6-positions of TPA systematically lowers the retentions of configuration for the oxidation of *cis*-1,2-DMCH, but only the trimethylated ligand **20** displays a lower 3°/2° oxidation ratio for adamantane, and even this value of 15 : 1 is almost identical to the 17 : 1 ratio measured for $[Fe^{II}(\mathbf{12})(MeCN)_2]^{2+}$. Methylating the three 3- or 5-positions to yield **23** or **26** actually shifts the catalyzed oxidation more toward the 3° C—H bonds. Although the lesser stereoretention with *cis*-1,2-DMCH was associated with free radical generation in the case of the N4Py catalyst, the strong preferences for 3° C—H activation for all of the TPA derivatives suggest that these are not key agents in the observed oxidation.

Costas's group modified the bpmcn framework to include pinene rings on the 4- and 5-positions of the pyridine rings (**29**, Scheme 11.11) [32, 79, 80]. The addition of steric bulk to these positions does not appear to hinder the approach of either terminal oxidant or substrate to the bound metal ion. The installation of the pinenes increased the catalytic efficiency of the Fe(II) complex by reducing the decomposition of the catalyst, and the iron-catalyzed oxidation of cyclohexane by H_2O_2 provides cyclohexanone in up to 70% yield [80]. When *cis*-1,2-DMCH is used as the substrate, the 3°/2° oxidation ratios for the stereoisomers of the catalysts are approximately 5 : 1, representing a substantial shift toward 3° C—H oxidation relative to the related bpmcn complex [32]. *trans*-1,2-DMCH, conversely, is oxidized more heavily on the 2° positions than it is with the bpmcn catalyst, with 3°/2° oxidation ratios for the two stereoisomers averaging 1 : 2.7. This highlights one of the major conclusions of this chapter: that a shift in the site-selectivity for one substrate does not necessarily translate into a similar shift for other substrates. As Gómez et al. wrote in their 2013 work, "It is quite illustrative to remark that changes in regioselectivity dictated by the nature and chirality of the diamine or the orientation of the pinene correspond to very modest variations in the relative energies of the paths leading to oxidation of the different C—H groups. Therefore, interpretation and predictability needs caution" [32].

My own research group likewise modified the bpmcn ligand and installed benzyl groups in place of the methyl groups on the two amine nitrogens (Scheme 11.6), yielding the highly bulky **4**. $[Fe(\mathbf{4})(MeCN)_2]^{2+}$ catalyzes the oxidation of aliphatic C—H bonds by H_2O_2. Despite the added bulk of the benzyl groups, the reactivity between $[Fe(\mathbf{4})(MeCN)_2]^{2+}$, H_2O_2, and cyclohexane is typical for a non-heme iron catalyst with a tetradentate neutral N-donor ligand. When 10 equiv. of terminal oxidant are added to mixtures of cyclohexane and $[Fe(\mathbf{4})(MeCN)_2]^{2+}$, 4.7

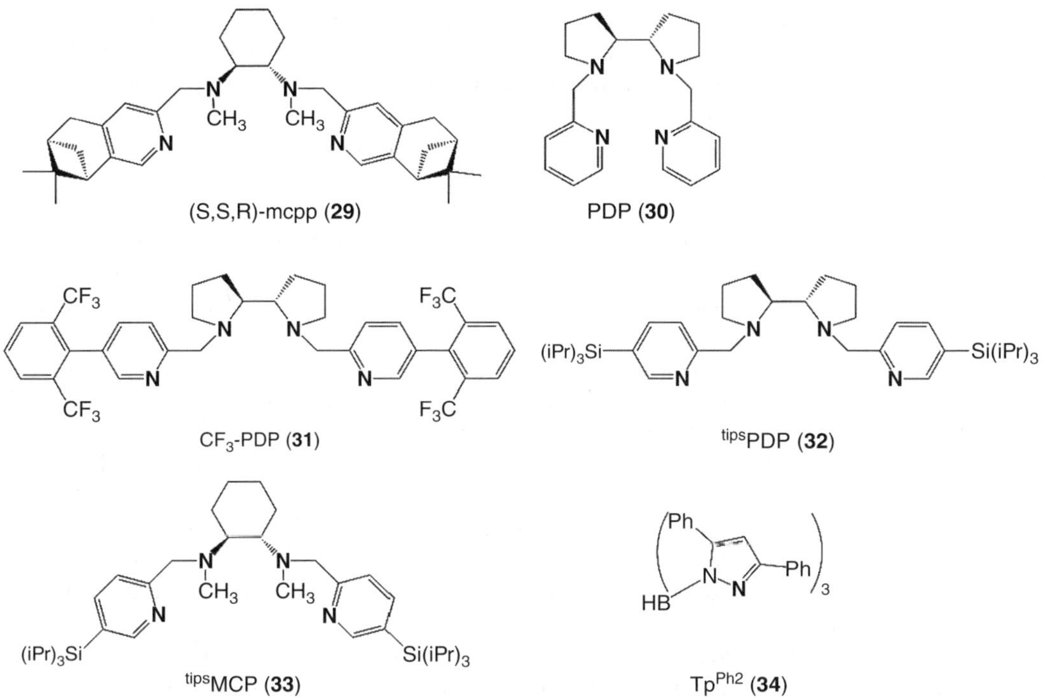

Scheme 11.11 Recently developed tetradentate N-donor ligands for non-heme iron catalysis of C—H activation.

equiv. of cyclohexanol and cyclohexanone are produced. When substrates with a mixture of tertiary and secondary carbons are used, however, much more oxidation tends to be seen on the secondary carbons. With *cis*-1,2-DMCH, tertiary oxidation is slightly favored, with a 3°/2° oxidation ratio of 1.4 : 1. *trans*-1,2-DMCH is oxidized overwhelmingly on the secondary carbons with a 3°/2° oxidation ratio of 1 : 4.8. The oxidation is never entirely shifted away from the tertiary carbons. With adamantane, the tertiary carbons are still preferentially oxidized with a statistically corrected 3°/2° oxidation ratio of 14.5 : 1. This ratio is relatively, but not drastically, low for a non-heme iron catalyst and slightly higher than that found for **15** (Table 11.1). The overall reactivity suggests that **4** is directing the catalyzed oxidation toward the secondary carbons relative to **16**, but it would be difficult to support that conclusion from just a single substrate.

Installing larger groups onto the amines is synthetically difficult, and it becomes increasingly more challenging to fully saturate the amines as the substituents become bulkier. My research group could not install two neopentyl groups onto the bpmcn framework, but we were able to do so with bispicen, which lacks the cyclohexane ring [33]. The resultant **5** (Scheme 11.6) is more comparable to **15**. The excessive bulk of the two neopentyl groups appears to destabilize the metal-based oxidant(s) formed from the reaction between the Fe(II) complex and H_2O_2; as a consequence, the catalytic activity of [Fe(**5**)$(OTf)_2$] is severely curtailed. Exceeding one turnover with cyclohexane requires 100 equiv. of the terminal oxidant H_2O_2. Although the reactivity with *trans* and *cis*-1,2-DMCH is slight, the oxidation is preferentially directed to the secondary carbons for both substrates, with 3°/2° oxidation ratios of 1 : 3.6 and 1 : 2.7, respectively. The reactivity of the *cis* isomer is notable in that the oxidation of tertiary carbons was preferred for all previously characterized non-heme iron catalysts. Adamantane is still oxidized preferentially on the tertiary carbons, but the statistically corrected 3°/2° oxidation ratio of 7.5 : 1 is low for a non-heme iron catalyst. *cis*-1,2-DMCH was oxidized to a 4.6 : 1 mixture of *trans*- and *cis*-1,2-dimethylcyclohexanol; the 82% retention of configuration and the lack of aromatic C—H oxidation suggest that hydroxyl radicals are not contributing to the reactivity as they are for the N4Py system.

White's research group has also tuned the regioselectivity of iron-catalyzed C—H activation using the ligand 2-[[2-(1-(pyridin-2-ylmethyl)pyrrolidin-2-yl)pyrrolidin-1-yl]methyl]pyridine (PDP) (**30**) and a derivative (**31**, Scheme 11.11) [34, 35, 81]. $[Fe(\mathbf{30})(MeCN)_2]^{2+}$ catalyzes the oxidation of aliphatic C—H bonds by H_2O_2 to a greater extent than normal and is efficient enough to use in natural product synthesis [34, 81]. The efficiency was attributed to the rigidity of the dipyrrolidine backbone [81]. Ligand **30** results in regioselectivity similar to that associated with the bpmen ligand. The reactivity with *cis*-1,2-DMCH occurs predominantly on the tertiary carbons, with a 3°/2° oxidation ratio of 3.7 : 1 whereas that for the *trans* isomer is 1 : 1.7 (Table 11.1) [34]. Modifying the 5-positions of the pyridines with 2,6-bis(trifluoromethyl)phenyl groups directs more of the catalyzed oxidation toward the secondary carbons of *trans*-1,2-DMCH, with a 3°/2° oxidation ratio of 1 : 10 [35]. This result is notable in that it is more selective for 2° C—H bonds than the catalysts containing either **4** or **5** while maintaining a high level of activity. Although *cis*-1,2-DMCH and adamantane were not investigated as substrates, the reactivity of other cyclic and acyclic hydrocarbons, such as *trans*-4-methylcyclohexyl acetate, supports the notion that **31** is indeed redirecting catalysis toward less sterically encumbered 2° C—H bonds.

Recently, Font et al. reported a catalyst that is even more selective for secondary carbons [36]. Much like White, they attach bulky substituents, specifically (isopropyl)silyl groups (tips), to the 5-positions of the pyridines of both PDP (**32**) and bpmcn (**33**, Scheme 11.11). When *trans*-1,2-DMCH is used as a substrate, the 3°/2° oxidation ratio is 1 : 4 for $[Fe^{II}(\mathbf{32})(OTf)_2]$ and a remarkable 1 : 13 for $[Fe^{II}(\mathbf{33})(OTf)_2]$ (Table 11.1). No *trans*-1,2-dimethylcyclohexanol is observed as a product, demonstrating the complete retention of stereochemistry. The lack of this product provides strong evidence against the participation of hydroxyl radicals in the hydrocarbon oxidation, and the regioselectivity can therefore be entirely attributed to the generated metal-based oxidant. The reactivity with *trans*-1,2-DMCH is also notable for the high amount of oxidation on the β positions. Isopropylcyclohexane and 1,1-dimethylcyclohexane are both oxidized predominantly on the β positions with both catalysts. Although ligand **33** is bulky enough to disfavor the oxidation of 3° C—H bonds, it does not appear to hinder the activation of 2° C—H bonds enough to favor the oxidation of 1° C—H bonds. n-Hexane, for instance, is still oxidized exclusively on the secondary carbons.

Mukherjee et al. reported unusual size-selective activity with a sterically encumbered tris(pyrazoyl)borate ligand, Tp^{Ph2}, (**34**, Scheme 11.11) [82]. In this ligand, phenyl groups are installed on both the 3- and 5-positions of the pyrazoyl rings. The Fe(II) complex with the ligand and benzoylformate (BF), $[Fe^{II}(\mathbf{34})(BF)]$, was previously observed to react with O_2 to generate $[Fe^{III}(\mathbf{34})(OBz)]$ [83]. In the 2009 work, Mukherjee et al. found that they could intercept a reactive intermediate using a variety of substrates with weak C—H bonds [82]. In the presence of 0.1 M DHA, for instance, anthracene is formed in 35% yield. Cyclohexene is oxidized to 1,3-cyclohexadiene in 40% yield. The dehydrogenation product is the only one observed; with most other non-heme iron catalysts, cyclohexene is instead oxidized to a mixture of 2-cyclohexenol and 2-cyclohexenone [27, 84, 85]. Cyclopentane and cyclooctane also react, but they appear to intercept the transient metal-containing oxidant at much lower efficiencies [82]. The poorer reactivities of the two cyclic alkanes was attributed to their stronger C—H bonds (96 kcal mol^{-1}). The most notable aspect of the reactivity is that certain benzylic substrates, specifically toluene, ethylbenzene, and diphenylmethane, do not react at all, despite their having C—H bonds with BDEs between 78 and 96 kcal mol^{-1}. The authors proposed that the greater size of these substrates was limiting their access to the reactive portion of the oxidant. This proposal was supported by additional experiments that demonstrated that the normally less readily oxidized cyclohexene was more efficient at accessing the active site than DHA, with the former substrate reacting even when present at only 1.0 mM. The authors concluded that the pocket formed from the phenyl groups favored molecules "with an oblate spheroidal shape" [86]. The lack of oxygenated products from cyclohexene suggests that the initially generated cyclohexenyl radical is both blocked from O_2 and positioned too far away from the generated Fe(III)-OH intermediate to allow an oxygen rebound step. The results suggest that the installed phenyl groups are effectively shielding the generated metal-based oxidant from bulkier substrates.

As the preceding results illustrate, care must be taken when determining where to make ligand modifications. Installing alkyl groups on either the metal-coordinating amine nitrogens or on the six-positions of pyridines can direct oxidation away from 3° aliphatic C—H bonds and toward 2° C—H bonds, but if the substituents are too large, they can largely eliminate the catalysis. Modifying the 5-positions of the pyridine rings impacts ligand binding to a much lesser extent while still maintaining the ability to restrict substrate access to the generated metal-containing oxidants. Installing large substituents on these sites can efficiently direct catalyzed oxidation toward 2° C—H bonds without compromising the metal-based oxidant.

11.5 Directing Oxidation Through Interactions Between Functional Groups on the Ligand and Substrate: A Currently Unexplored Route for Non-heme Iron Catalysis

Another possible approach to modulating regioselectivity is to incorporate functional groups on both the substrate and the catalyst that can link to each other transiently through either non-covalent or covalent bonds. The linkage orients a portion of the substrate near the oxidizing portion of the catalyst, thereby ensuring that only a select few bonds of the substrate can be activated. This approach is typified from an example from Das et al. in which interactions between carboxylic acids directed a manganese oxidant toward a specific pair of benzylic C—H bonds on ibuprofen (Scheme 11.12) [87]. A second benzylic carbon on the substrate resists oxidation but reacts with manganese catalysts that lack the targeting domain. Das et al.

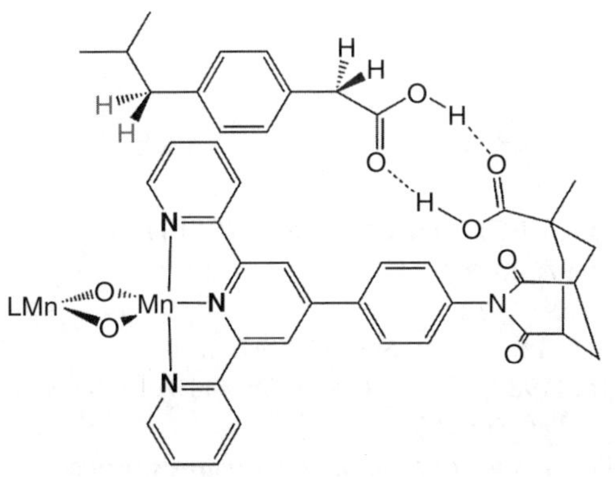

Scheme 11.12 Hydrogen bond interactions between a remote site on the ligand and the ibuprofen substrate direct manganese-catalyzed oxidation toward the highlighted benzylic hydrogens. L represents a second equivalent of the ligand with the targeting domain. (*See insert for color representation.*)

also demonstrated that their catalyst could distinguish between the two tertiary carbons on 4-methylcyclohexylacetic acid. The primary drawback to this strategy is that a specific functional group needs to be installed onto a specific site on the substrate. In practice, one normally doesn't have this much latitude with respect to the starting material, particularly in the later steps of a lengthy organic synthesis. A second disadvantage is that the linkage between the substrate and oxidant needs to be carefully designed. Any flexibility in the connection would be anticipated to disrupt the regioselectivity. Designing linkers to access different parts of the same substrate and obtain complementary products is difficult and labor-intensive. Although the ability to distinguish two sterically and electronically similar C—H bonds was and remains a fantastic accomplishment, this approach has not yet been adapted to non-heme iron catalysis.

11.6 Final Comments

Non-heme iron catalysts have proven to be attractive due to the low cost of the catalytic metal and their ability to accelerate the oxidation of C—H bonds under ambient conditions. Steps have been taken to direct the oxidation toward specific C—H bonds either by weakening the generated oxidant or by modifying the ligand to size-exclude either whole substrates or portions of substrates. The former approach has been more extensively explored, but the latter approach is attractive for its potential to generate products that are complementary to the first. In the past few years, there has been a strong interest in investigating second-sphere influences on small molecule reactivity. The installation of groups that can orient substrates to direct catalysis toward specific portions of the molecules certainly fit within these efforts. One complication in this field is that regioselectivity trends are not conserved between substrates. One can therefore obtain a misleading view of an oxidant's intrinsic site-selectivity if only a single substrate is used to assess this property. Using a standard set of reference substrates would provide a superior account of a non-heme iron oxidant's regioselectivity of this property and facilitate comparisons between different catalysts.

Acknowledgments

I am grateful to Ms. Angela Bell-Taylor for her assistance in proofreading this work.

References

1 Huang, Z. and Dong, G. (2017). *Acc. Chem. Res.* 50: 465.
2 Hartwig, J.F. (2017). *Acc. Chem. Res.* 50: 549.
3 Arndtsen, B.A., Bergman, R.G., Mobley, T.A. et al. (1995). *Acc. Chem. Res.* 28: 154.
4 Shilov, A.E. and Shul'pin, G.B. (1997). *Chem. Rev.* 97: 2879.
5 Groves, J.T. (1985). *J. Chem. Educ.* 62: 928.
6 Bordwell, F.G., Cheng, J.-P., Ji, G.-Z. et al. (1991). *J. Am. Chem. Soc.* 113: 9790.
7 Bordwell, F.G., Cheng, J.-P., and Harrelson, J.A. Jr., (1988). *J. Am. Chem. Soc.* 110: 1229.
8 Blanksby, S.J. and Ellison, G.B. (2003). *Acc. Chem. Res.* 36: 255.
9 Luo, R.-Y. (2003). *Handbook of Bond Dissociation Energies in Organic Compounds.* Boca Raton, FL: CRC Press LLC.
10 Davies, H.M.L., Dai, X., and Long, M.S. (2006). *J. Am. Chem. Soc.* 128: 2485.
11 Fraunhoffer, K.J. and White, M.C. (2007). *J. Am. Chem. Soc.* 129: 7274.

12 Lee, Y.-M., Hong, S., Morimoto, Y. et al. (2010). *J. Am. Chem. Soc.* 132: 10668.
13 Wender, P.A., Hilinski, M.K., and Mayweg, A.V.W. (2005). *Org. Lett.* 7: 79.
14 Newhouse, T. and Baran, P.S. (2011). *Angew. Chem. Int. Ed.* 50: 3362.
15 Gol'dshleger, N.F., Es'kova, V.V., Shilov, A.E. et al. (1972). *Zh. Fiz. Khim.* 46: 1353.
16 Jones, W.D. (1999). *Top. Organomet. Chem.* 3: 9.
17 Que, L. Jr., and Ho, R.Y.N. (1996). *Chem. Rev.* 96: 2607.
18 Solomon, E.I., Decker, A., and Lehnert, N. (2003). *Proc. Nat. Acad. Sci. U.S.A* 100: 3589.
19 Costas, M., Mehn, M.P., Jensen, M.P. et al. (2004). *Chem. Rev.* 104: 939.
20 Su, C. and Oliw, E. H. (1998). *J. Biol. Chem.* 273: 13072.
21 Wieghardt, K. (1989). *Angew. Chem. Int. Ed.* 28: 1153.
22 Mirica, L.M., Ottenwaelder, X., and Stack, T.D.P. (2004). *Chem. Rev.* 104: 1013.
23 Holm, R.H., Kennepohl, P., and Solomon, E.I. (1996). *Chem. Rev.* 96: 2239.
24 Tanase, S., Marques-Gallego, P., Browne, W.R. et al. (2008). *Dalton Trans.* 2026.
25 Roelfes, G., Lubben, M., Hage, R. et al. (2000). *Chem. Eur. J.* 6: 2152.
26 Kim, C., Chen, K., Kim, J. et al. (1997). *J. Am. Chem. Soc.* 119: 5964.
27 Chen, K. and Que, L. Jr., (2001). *J. Am. Chem. Soc.* 123: 6327.
28 He, Y., Gorden, J.D., and Goldsmith, C.R. (2011). *Inorg. Chem.* 50: 12651.
29 Britovsek, G.J.P., England, J., and White, A.J.P. (2005). *Inorg. Chem.* 44: 8125.
30 Chen, K. and Que, L. Jr., (1999). *Chem. Commun.* 1375.
31 Costas, M. and Que, L. Jr., (2002). *Angew. Chem. Int. Ed.* 41: 2179.
32 Gómez, L., Canta, M., Font, D. et al. (2013). *J. Org. Chem.* 78: 1421.
33 Zhang, Q., Gorden, J.D., and Goldsmith, C.R. (2013). *Inorg. Chem.* 52: 13546.
34 Chen, M.S. and White, M.C. (2010). *Science* 327: 566.
35 Gormisky, P.E. and White, M.C. (2013). *J. Am. Chem. Soc.* 135: 14052.
36 Font, D., Canta, M., Milan, M. et al. (2016). *Angew. Chem. Int. Ed.* 55: 5776.
37 Lindhorst, A.C., Haslinger, S., and Kühn, F.E. (2015). *Chem. Commun.* 51: 17193.
38 Evans, M.G. and Polanyi, M. (1938). *Trans. Faraday Soc.* 34: 11.
39 Mayer, J.M. (1998). *Acc. Chem. Res.* 31: 441.
40 Mayer, J.M. (2011). *Acc. Chem. Res.* 44: 36.
41 Hass, H.B., McBee, E.T., and Weber, P. (1936). *Ind. Eng. Chem.* 28: 333.
42 Russell, G.A. and Brown, H.C. (1955). *J. Am. Chem. Soc.* 77: 4025.
43 Goldsmith, C.R. and Stack, T.D.P. (2006). *Inorg. Chem.* 45: 6048.
44 Goldsmith, C.R., Jonas, R.T., and Stack, T.D.P. (2002). *J. Am. Chem. Soc.* 124: 83.
45 Jonas, R.T. and Stack, T.D.P. (1997). *J. Am. Chem. Soc.* 119: 8566.
46 Gardner, H.W. (1989). *Biochim. Biophys. Acta* 1001: 274.
47 de Vries, M.E., La Crois, R.M., Roelfes, G. et al. (1997). *Chem. Commun.* 1549.
48 Gupta, R. and Borovik, A.S. (2003). *J. Am. Chem. Soc.* 125: 13234.
49 Usharani, D., Lacy, D.C., Borovik, A.S. et al. (2013). *J. Am. Chem. Soc.* 135: 17090.
50 Que, L. Jr., and Tolman, W.B. (2008). *Nature* 455: 333.
51 Nam, W. (2007). *Acc. Chem. Res.* 40: 522.
52 Krebs, C., Fujimori, D.G., Walsh, C.T. et al. (2007). *Acc. Chem. Res.* 40: 484.
53 Chung, L.W., Li, X., Hirao, H. et al. (2011). *J. Am. Chem. Soc.* 133: 20076.
54 Cho, J., Jeon, S., Wilson, S.A. et al. (2011). *Nature* 478: 502.
55 Zhang, Q. and Goldsmith, C.R. (2014). *Inorg. Chem.* 53: 5206.
56 He, Y. and Goldsmith, C.R. (2012). *Chem. Commun.* 48: 10532.
57 Chiang, C.-W., Kleespies, S.T., Stout, H.D. et al. (2014). *J. Am. Chem. Soc.* 136: 10846.
58 Que, L. Jr., (2007). *Acc. Chem. Res.* 40: 493.
59 Roth, J.P. and Mayer, J.M. (1999). *Inorg. Chem.* 38: 2760.
60 Roth, J.P., Yoder, J.C., Won, T.-J. et al. (2001). *Science* 294: 2524.

61 King, E.R., Hennessy, E.T., and Betley, T.A. (2011). *J. Am. Chem. Soc.* 133: 4917.
62 Iovan, D.A. and Betley, T.A. (2016). *J. Am. Chem. Soc.* 138: 1983.
63 King, E.R. and Betley, T.A. (2009). *Inorg. Chem.* 48: 2361.
64 Nieto, I., Ding, F., Bontchev, R.P. et al. (2008). *J. Am. Chem. Soc.* 130: 2716.
65 Lee, W.-T., Juarez, R.A., Scepaniak, J.J. et al. (2014). *Inorg. Chem.* 53: 8425.
66 Eckert, N.A., Vaddadi, S., Stoian, S. et al. (2006). *Angew. Chem. Int. Ed.* 45: 6868.
67 Cowley, R.E. and Holland, P.L. (2012). *Inorg. Chem.* 51: 8352.
68 Sugimoto, H. and Sawyer, D.T. (1985). *J. Org. Chem.* 50: 1784.
69 Sawyer, D.T., Hage, J.P., and Sobkowiak, A. (1995). *J. Am. Chem. Soc.* 117: 106.
70 Fenton, H.J.H. (1894). *J. Chem. Soc. Trans.* 65: 899.
71 England, J., Gondhia, R., Bigorra-Lopez, L. et al. (2009). *Dalton Trans.* 5319.
72 Costas, M., Tipton, A.K., Chen, K. et al. (2001). *J. Am. Chem. Soc.* 123: 6722.
73 Lubben, M., Meetsma, A., Wilkinson, E.C. et al. (1995). *Angew. Chem. Int. Ed.* 34: 1512.
74 Pan, X.-M., Schuchmann, M.N., and von Sonntag, C. (1993). *J. Chem. Soc., Perkin Trans. 2* 289.
75 Cook, B.R., Reinert, T.J., and Suslick, K.S. (1986). *J. Am. Chem. Soc.* 108: 7281.
76 Costas, M. (2011). *Coord. Chem. Rev.* 255: 2912.
77 Thomas, J.M., Raja, R., Sankar, G., and Bell, R.G. (1999). *Nature* 398: 227.
78 Herron, N. (1988). *J. Coord. Chem.* 19: 25.
79 Zhang, C.X., Kaderli, S., Costas, M. et al. (2003). *Inorg. Chem.* 42: 1807.
80 Gómez, L., Garcia-Bosch, I., Company, A. et al. (2009). *Angew. Chem. Int. Ed.* 48: 5720.
81 Chen, M.S. and White, M.C. (2007). *Science* 318: 783.
82 Mukherjee, A., Martinho, M., Bominaar, E.L. et al. (2009). *Angew. Chem. Int. Ed.* 48: 1780.
83 Mehn, M.P., Fujisawa, K., Hegg, E.L. et al. (2003). *J. Am. Chem. Soc.* 125: 7828.
84 Kaizer, J., Klinker, E.J., Oh, N.Y. et al. (2004). *J. Am. Chem. Soc.* 126: 472.
85 Company, A., Gómez, L., Güell, M. et al. (2007). *J. Am. Chem. Soc.* 129: 15766.
86 Mukherjee, J., Lucas, R.L., Zart, M.K. et al. (2008). *Inorg. Chem.* 47: 5780.
87 Das, S., Incarvito, C.D., Crabtree, R.H. et al. (2006). *Science* 312: 1941.

12

Imine-based Iron and Manganese Complexes as Catalysts for Alkane Functionalization

Giorgio Olivo, Osvaldo Lanzalunga, and Stefano Di Stefano

Università degli Studi di Roma "La Sapienza" and Istituto CNR di Metodologie Chimiche (IMC-CNR), Dipartimento di Chimica, Sezione Meccanismi di Reazione, Rome, Italy

12.1 Introduction

The direct functionalization of C—H bonds is one of the most attractive features of natural biosynthetic routes. Replication of such reaction in organic synthesis has been long sought after, and yet it remains a challenging transformation from the synthetic chemist's point of view. In particular, selective C—H oxidation represents a powerful tool to introduce into the target backbone oxidized functionalities prone to further manipulations even at late stages avoiding many unnecessary synthetic steps [1–4], as demonstrated for instance in eudesmane synthesis [5]. The development of catalytic systems able to perform the oxidation of aliphatic C—H bonds (mainly hydroxylation) in such a selective manner represents a highly attractive goal for the scientific research. Given the rising environmental concerns in our society, such catalysts would be even more desirable if based on abundant and sustainable reagents, and if they minimize at the same time chemical waste production.

In the field of C—H oxidations, great advances have been achieved in recent years by the development of bioinspired non-heme aminopyridine iron or manganese complexes/H_2O_2 catalytic systems [6–17]. Such systems replicate the oxidation mechanism of the elaborated nonheme iron oxygenases with simple iron coordination complexes, showing a high potential as effective C—H hydroxylation reagents [18–26]. They present a number of key advantages in comparison with the limited existing C—H oxidation protocols. First, a selective, metal-based oxidation mechanism that avoids the formation of free diffusing radicals operates in most cases [25–27]. Such a mechanism ensures a good degree of selectivity in the reaction that has not been found in other synthetic systems [28–30]. Second, the synthesis of non-heme amine-based iron and manganese complexes is relatively straightforward, and significantly less demanding than metalloporphyrin heme model complexes [31–34]. Moreover, such compounds are rather resistant to air (albeit not for prolonged amounts of time, i.e. few days), and so quite easy to handle. Third, the modular nature of the ligands allows for a simple modification of the resulting complex structure, and this translates into a fine tuning of the oxidation selectivity exhibited by the catalyst [8, 22, 24, 35–38]. This is very hard to achieve with other C—H oxidation reagents, such as dioxiranes or polyoxometalates [39, 40]. Eventually, they meet the requirements of green chemistry, since H_2O_2 is used as the terminal oxidant producing H_2O as waste, and the abundant, cheap, and green iron (or the slightly less preferable manganese) is used as the metal center. This constitutes a significant advantage in comparison with several other catalysts, which rely on more polluting terminal oxidants or heavier and more toxic metals [1].

Alkane Functionalization, First Edition. Edited by Armando J. L. Pombeiro and M. Fátima C. Guedes da Silva.

Studies on non-heme amine-based iron catalysts have been paralleled in recent years by investigations on non-heme imine-based iron catalysts [21]. However, the latter have received less attention, probably due to the idea that imines (or Schiff bases) are not robust enough to tolerate the typical oxidation conditions. Actually, not only have many imine-based complexes been found to be stable enough to serve as oxidation catalysts, but, in many cases, the related imine ligands present the significant advantage of a much easier synthesis [21, 41–47]. Moreover, the reversible nature of the imine bond allows the self-assembly of both the ligand and the complex directly *in situ* from simple building blocks, eliminating the need of pre-synthesizing the catalyst [21, 44, 48]. This chapter describes imine-based iron and manganese catalysts for C—H oxidations in the hope of stimulating the interest toward this topic.

It has to be said that the combination of imine-based (or amine-based) iron or manganese complexes with oxidants can generate hydrocarbon oxidized products by means of either a selective, metal-based oxidation mechanism or a free radical pathway (see Scheme 12.1). In the case of aminopyridine complexes, certain ligand structures privilege the former mechanism [25, 26], but such a clear structure-activity relationship has not been observed in the related imine-based complexes, except for salen systems [21]. The reaction mechanism is of critical importance since a selective oxidation, and its related synthetic application, requires that the catalyst performs the reaction in a controlled manner without the involvement of unselective free radical intermediates. Therefore, the first part of the following discussion will be devoted to highlight the main mechanistic tools that allow discrimination between a metal-based and a free-radical mechanism.

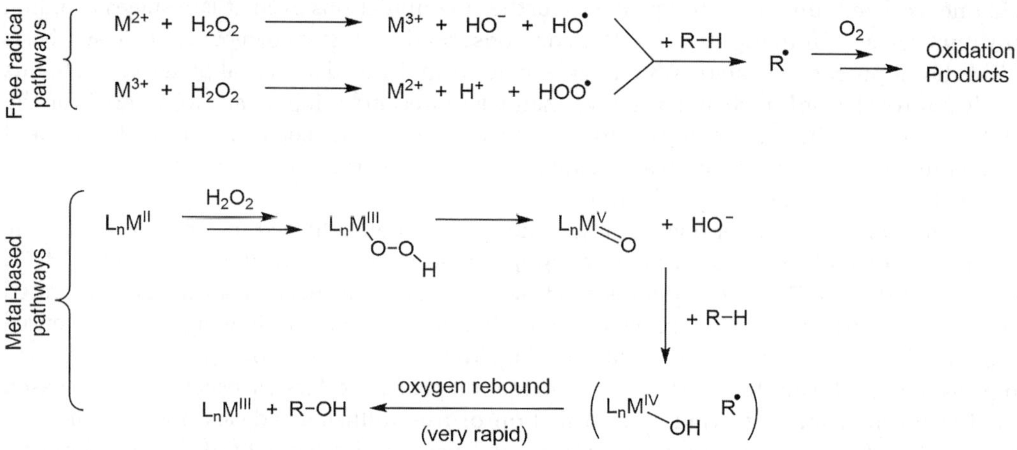

Scheme 12.1 Free-radical and metal-based oxidation pathways. This dichotomy is often invoked in the literature when iron is involved (M = Fe), but a similar mechanistic scenario is considered valid also in the case of manganese (M = Mn).

12.2 Free-Radical or Radical-Free Oxidation Mechanism?

Free-radical reactions differ from metal-based ones in several aspects (Scheme 12.1) [27, 49, 50]. First, they are usually performed by highly reactive, oxygen-centered radicals, while high-valent metal-oxo species are generally involved in the other mechanism. The high reactivity of free radicals translates into a poor selectivity among the tiny differences in the electronic and steric properties of C—H bonds. Second, hydrogen atom transfer (HAT) from the substrate C—H bond to the initial oxygen-centered radical forms a carbon centered radical, which can, in turn, rearrange, epimerize, or give radical coupling reactions beside the oxygenation itself [49]. All these secondary (and undesirable) pathways are usually switched off in metal-based oxidations, where oxygen rebound of the OH ligand from the metal to the carbon-centered radical is generally faster than cage escape and trigger of the other processes [50]. Therefore, the presence of reaction products deriving from these pathways is taken as a clue of a free radical mechanism. Some mechanistic probes adopted in order to discriminate between metal-based or a free-radical mechanisms are described here [50]. Of course, none of the following pieces of evidence on its own is a definitive proof in favor of a given mechanism, but if many of them converge to one of the two dichotomous mechanisms, the latter will be considered operating.

(i) *Alcohol/ketone ratio (A/K).* The measure of A/K ratio between alcohol and ketone molar equivalents in the oxidation products is a powerful test for radical-based oxidations [21, 50]. This test has to be carried out under air, with the substrate (usually cyclohexane or cyclooctane, see Scheme 12.2) added in large excess with respect to the oxidant (10 : 1 or 100 : 1). If formed, freely diffusing alkyl radicals are very rapidly trapped by molecular oxygen to produce alkylperoxyl radicals. Such intermediates eventually react according to Russell-type terminations [51] forming equimolar amounts of alcohol and ketone, no matter the initial substrate/oxidant ratio [51, 52]. Thus, close to 1 A/K ratios very often indicate a radical chain mechanism [27, 52]. Conversely, when a metal-based mechanism is operating, oxygen rebound is faster than O_2 trapping of the incipient radical [50, 52]. As a consequence, higher A/K values (>5) are observed in limiting oxidant conditions, being the ketone generated only by over-oxidation of the alcohol [27, 52].

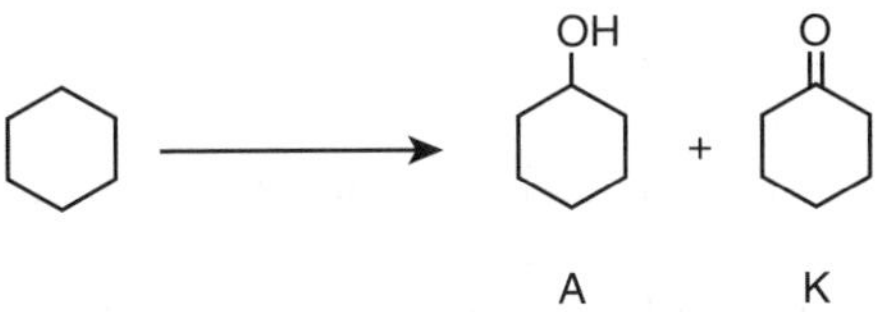

Scheme 12.2 Oxidation of cyclohexane to cyclohexanol to determine the A/K ratio.

(ii) *Reaction under argon/air.* As stated above, carbon-centered radicals can be very rapidly trapped by O_2 producing comparable amounts of alcohols and ketones. Therefore, if a free-radical mechanism is involved, the presence/absence of molecular oxygen has to exert a significant influence on the reaction outcome. For instance, the exclusion of O_2 from the reaction mixture will translate into a remarkable drop of the yields of the oxidized products. In contrast, if a metal-based mechanism is operating, negligible effects, if any, are expected when O_2 is excluded from the reaction mixture [28, 53].

(iii) *Shul'pin test for alkyl hydroperoxides.* This rapid test allows to detect the presence of alkyl hydroperoxides (e.g. cyclohexyl hydroperoxide) which are typical byproducts of radical chain oxidations, in the reaction mixture. Actually, hydroperoxide detection is not trivial due to its decomposition into alcohols and ketones on the gas chromatography (GC) injector. In the Shul'pin's procedure [54–56], the GC analysis of the product mixture is carried out before and after treatment with Ph_3P that fully reduces alkyl hydroperoxides to the corresponding alcohols (Scheme 12.3). If the GC chromatograms obtained before and after treatment with Ph_3P reveals an increased amount of alcohols, alkyl hydroperoxides are probably formed during the reaction, which has to proceed with a free radical chain mechanism.

(iv) *Regioselectivity.* The ratios among tertiary:secondary:primary C—H bond oxidation provide an indication on the nature of the species competent for the abstraction of the

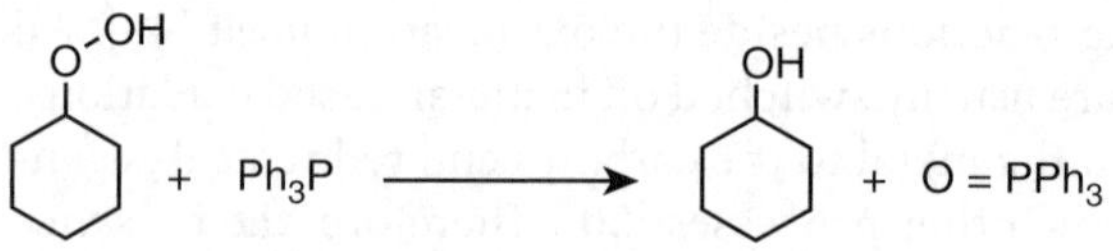

Scheme 12.3 Reduction of alkyl hydroperoxides in the Shul'pin's test.

hydrogen atom [50, 57]. In a typical adamantane oxidation (Scheme 12.4), the ratio of tertiary over secondary C—H bond oxidation products (3°/2°), normalized for statistical factors, is low when the highly reactive hydroxyl and alkoxyl radicals are involved (3°/2° < 6) [58]. Conversely, when the more stabilized active species of cytochrome P450 enzymes or of other heme catalysts are involved, the above ratio substantially increases (3°/2° > 13–15), in accordance with the reactivity/selectivity rule [50, 59]. Of course, such test can be extended also to other types of substrates. If the reaction is sensitive to the steric or electronic properties of the substrate, a metal-based oxidant is probably involved, while the absence of selectivity strongly suggests a free radical oxidizing species.

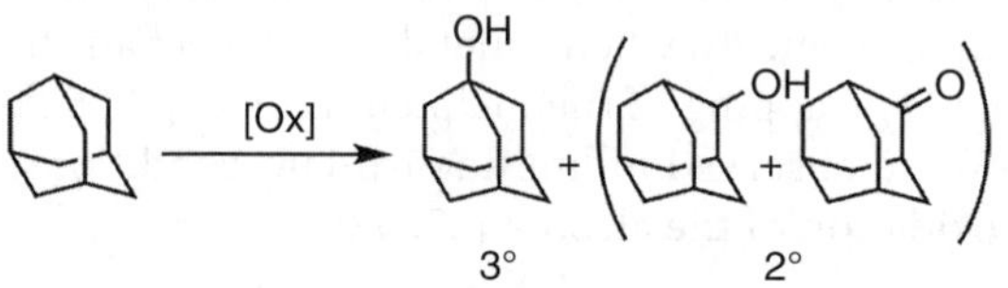

Scheme 12.4 Adamantane oxidation typically used to determine the 3°/2° ratio.

(v) *Epimerization.* The involvement of free-diffusing alkyl radical intermediates usually brings about the loss of the stereochemical information of a tertiary carbon center due to rapid inversion of configuration of the radical. On the other hand, metal-based oxidations afford high retention of configuration since the oxygen rebound step is generally more rapid than the epimerization of the radical carbon. Thus, retention of configuration or epimerization in the *cis*-1,2-dimethylcyclohexane oxidation (Scheme 12.5) is often taken as a convincing evidence in favor of one of the two mechanisms. Again, such a test could also be performed with other substrates that have a comparable low barrier for the inversion of configuration at the radical center [50].

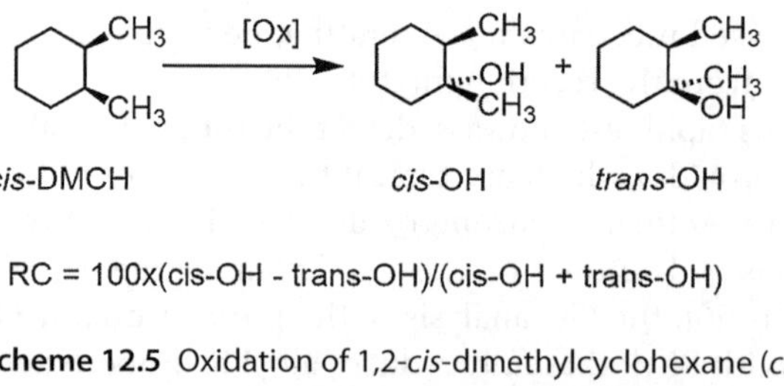

Scheme 12.5 Oxidation of 1,2-*cis*-dimethylcyclohexane (*cis*-DMCH).

(vi) *Chirality transfer*. When C—H oxidation catalyzed by a chiral catalyst preferentially forms one product enantiomer over the other, such a chirality transfer is only possible if a metal-based mechanism is operating. Of course, such a transfer cannot be observed in the case of oxidations in which an achiral oxygen-centered radical is the active species or a long-lived achiral carbon-centered radical is formed [15].

(vii) *Use of radical traps.* Involvement of free diffusing radical intermediates in the reaction mechanism can be evidenced by the use of radical traps that very rapidly react with such intermediates, significantly affecting the reaction outcome. However, radical traps may coordinate the metal catalyst and thus interfere in the oxidation mechanism. Therefore, a careful choice of the radical trap is usually required. $CBrCl_3$ (Scheme 12.6) is a particularly suitable radical scavenger capable of trapping free diffusing carbon radicals with high efficiency (rates at diffusion limits) forming alkyl bromides. A significant advantage of this radical trap is its inability to coordinate the metal complex and thus interfere with the reaction, if used at low concentrations [52].

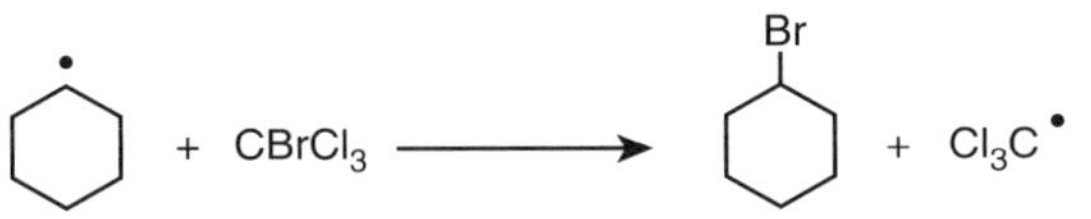

Scheme 12.6 Carbon-centered radical trapped by $CBrCl_3$.

(viii) *Labeling studies.* Isotope labeling is one of the most powerful tools to discriminate oxidation mechanisms promoted by metal-oxo from free radical species [27]. Free diffusing alkyl radicals are readily trapped by atmospheric O_2 during a radical-chain process. Consequently, a substantial part of the O-atom incorporated into the products derives from air. Conversely, when the oxidant is the exclusive O-atom source, a metal-based mechanism is operating [27, 33, 59]. However, a certain amount of the O-atom in the products can derive from water in oxidations mediated by high-valent metal-oxo species due to a rapid oxo-hydroxo tautomerism prior to the oxidation step (Scheme 12.7) [27, 60–62]. Actually, O-atom incorporation from water has been often used as an indirect evidence for the involvement of an iron-oxo species along the reaction pathway [60]. From an experimental point of view, two complementary experiments are required: an oxidation reaction performed in the presence of $H_2{}^{18}O$ and H_2O_2, and a second oxidation carried out in the presence of H_2O and $H_2{}^{18}O_2$. When a genuine metal-based mechanism is operating, the sum of the percentages of labeled oxygenated product obtained in the two experiments should amount to 100%. If this is not the case, the missing amount of unlabeled products should derive from atmospheric O_2, and it thus weights the parallel autoxidation reaction.

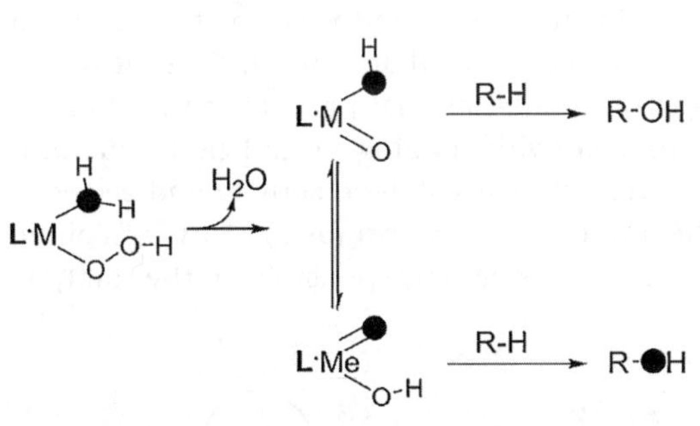

Scheme 12.7 Oxo-hydroxo tautomerism of high-valent metal-oxo species.

(ix) *Kinetic isotope effects (KIEs).* The intermolecular isotope effect is commonly measured in a competitive experiment in which a hydrocarbon (usually cyclohexane, C_6H_{12}) and its deuterated analogous (C_6D_{12}) are oxidized in the presence of the catalytic system for which the action mechanism has to be elucidated (Scheme 12.8). Most times, highly reactive oxygen-centered radicals do not discriminate between C—H and C—D bonds, and *KIE* values ranging from 1 to 2 are measured [63]. Conversely, definitely higher *KIE* values, from 3 up to 4–5, are obtained for metal-based oxidants that turn out to be more selective for the cleavage of the C—H over the C—D bond [11, 27, 50].

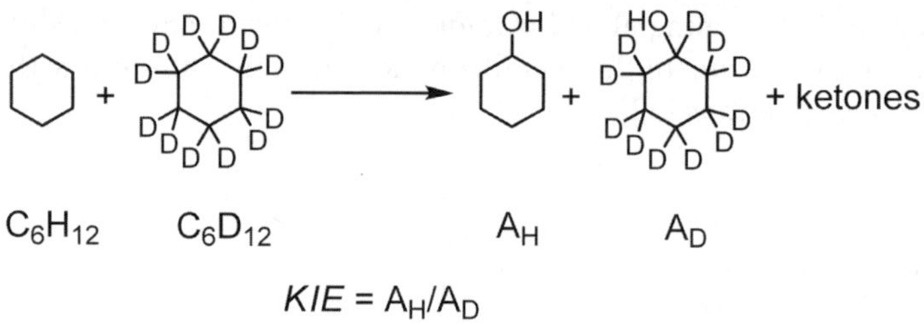

Scheme 12.8 Competitive oxidation of cyclohexane and cyclohexane-d_{12} to determine *KIE*.

12.3 Imine-Containing Complexes Competent for C—H Oxidations

The key structural features that enable a selective, metal-based reactivity in C—H oxidation at iron sites have been evidenced by several investigations on aminopyridine complexes. The presence of two *cis* labile sites, the use of a tetradentate, strong-field nitrogen ligand, and the formation of a stable complex with five-membered chelate rings have been highlighted as the main elements characteristic of such systems [27, 64–66], although some recent reports suggest that these structural requirements are not as strict as they seem to be. However, even in the case of a metal-based process, catalytic activity (measured, for instance by turnover number, TON) correlates with catalyst stability: the more stable the catalyst, the higher its activity [66]. This can be ascribed to the fact that the ligand is crucial to modulate the reactivity of the metal ion and to avoid a free radical mechanism. Ligand degradation weakens metal binding and eventually causes the release of the metal ion, with subsequent catalyst deactivation and trigger of radical reactions. Therefore, ligand stability is essential for a good catalytic performance.

Imines can be indeed subjected to two degradation pathways, i.e. hydrolysis in the mildly acidic media or oxidation to amides (Scheme 12.9). Hydrolysis leads to the loss of a ligand arm and eventual ligand fragmentation, while oxidation forms amides and weakens the ligand to metal binding since the latter functions are poorer Lewis bases than imines. Both degradation pathways have been observed in a vast array of amine-, aminal-, and imine-containing manganese catalytic systems by Browne's group [67]. Notwithstanding a significant structural diversity, all the catalysts investigated displayed a very similar catalytic activity and selectivity. A closer investigation revealed that all these ligands oxidatively degrade to Mn(II)/picolinic acid complexes in the reaction conditions, and this latter species is responsible for the catalytic activity observed.

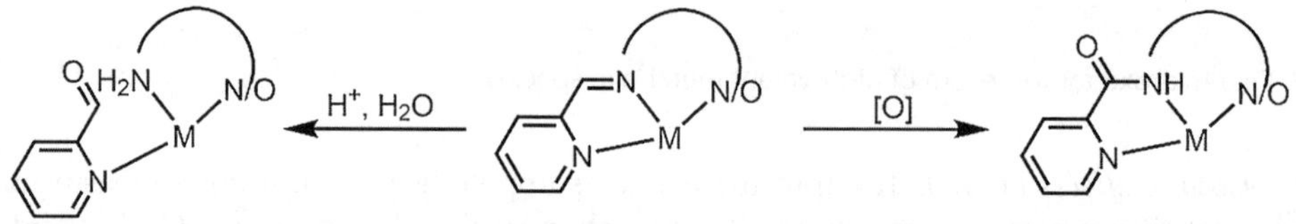

Scheme 12.9 Degradation pathways of imine-based ligands.

However, such a complete degradation is not always the case for both amine- and imine-containing ligands. Focusing on imine complexes, some strategies have been profitably developed to counterbalance the tendency of imine ligand systems to undergo degradation and ultimate metal release. The key feature of these strategies is reinforcement of the ligand to metal binding.

12.3.1 Salen-Based Catalysts

An example of this strategy is represented by the use of salen-type ligands. These are tetradentate ligands, with two negative charges and a delocalized π system, which form highly stable (and robust) complexes. This exceptional stability makes them suitable oxidation catalysts, and up to 500 turnovers have been reported for such systems [47].

Another main advantage is that salen ligands are easily tunable [68]. Chiral centers can be introduced on the diamine backbone (see Figure 12.1) enabling enantioselective transformations, while substituents on the aromatic ring allow to manipulate the electronic and steric properties of the ligand (substituents in *para* (R_1) or *ortho* (R_2) positions, respectively) [68–71]. Salen-Mn complexes are excellent catalysts for asymmetric olefin epoxidation [47, 68, 70], while salen-Fe complexes effectively promote enantioselective sulfoxidations [71].

Salen-Mn(III) complexes react with a variety of oxygen atom donor oxidants (such as ArIO, 2,6-dichloropyridine N-Oxide, peracids, etc.) to generate a high-valent (salen)Mn^V=O species responsible for the challenging C—H oxidation process (see Scheme 12.10) [72–76]. Direct evidence for such a reactive intermediate has been collected in few cases [77, 78], but the nature of the oxidizing species turned out to be highly dependent on the oxidant used and on the

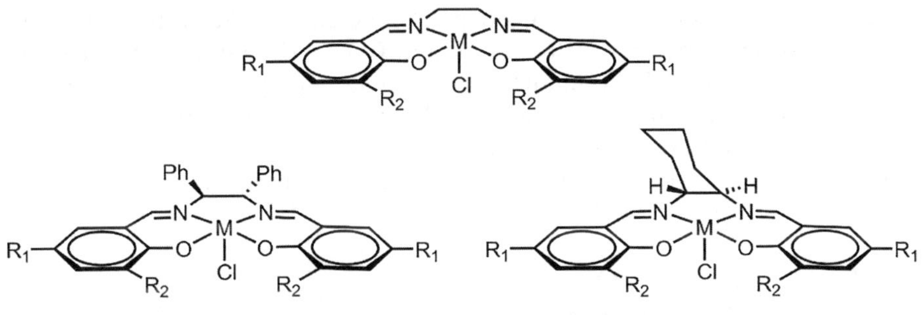

Figure 12.1 [(Salen)M(Cl)] complexes employed in oxidation reactions.

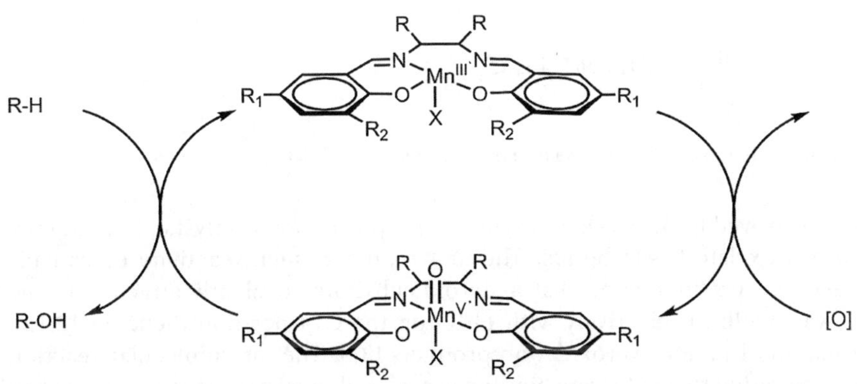

Scheme 12.10 Catalytic cycle for R-H oxidations catalyzed by Salen-Mn(III) complexes.

reaction conditions [74, 76]. Evidence in favor of "multiple oxidants" has been collected, and the terminal oxidant is often associated to some extent to the active species [75, 79]. After the oxidation step, the initial salen-Mn(III) complex is regenerated and enters again the catalytic cycle.

Also, the nature of the oxidizing species in salen-Fe catalysts is controversial. Initially, a (salen$^{+\bullet}$)Fe^{IV}(O)(X) intermediate was proposed as the active oxidant [80, 81]. However, strong evidence was provided a few years later for a (salen)Fe^{III}(PhIO)(X) active species generated with ArIO oxidants by ^{1}H-NMR and EPR spectroscopy [82]. This assignment was reinforced by the observation that the enantioselectivity of the sulfoxidation reaction was dependent on the steric properties of the ArIO terminal oxidant, suggesting that this is coordinated to the active species [83]. Electrochemical oxidation of the salen-Fe(III) complex [84], as well as chemical oxidation with mCPBA [82], was reported not to alter the oxidation state of the iron (which remained at +III), but the oxidation equivalents are stored on the aromatic π systems. However, a high-valent (salen$^{+\bullet}$)Fe^{IV}(O)(X) is still invoked with H_2O_2 as the oxidant [80, 85], although conclusive evidence on its assignment has not been yet collected.

The high stabilization of the +III oxidation state by the salen ligand suggests that strongly oxidizing, high-valent iron complexes are not accessible, and this was proposed to account for the low oxidizing activity of salen-Fe catalysts. The relatively high stability of both iron and manganese oxidizing species, due to the electron donation of the dianionic ligand to the electrophilic metal center, translates into a selective oxidation of more reactive substrates (such as olefins or sulfides), but transfer of such activity to the oxidation of alkanes proved to be more challenging. Indeed, most of the reports focused on the oxidation of relatively activated C—H bonds, such as those proximal to an aromatic moiety, an ethereal oxygen or an amide or amine function.

Kaizer reported the efficient conversion of a series of cyclic and acyclic amino acids into a carbonyl compound (or ethylene), HCN (or NH_3), and CO_2 with Fe-salen catalysts and several oxidants (see Scheme 12.11) [85], mimicking 1-aminocyclopropane-1-carboxylic acid (ACCH) oxidase. The products formed depended on the amino acid structure [86]. Initial coordination of the amino-acid to the iron center was proposed to be followed [85, 86] by an electron-transfer-proton-transfer (ET-PT) process, triggering the subsequent fragmentations [85, 86].

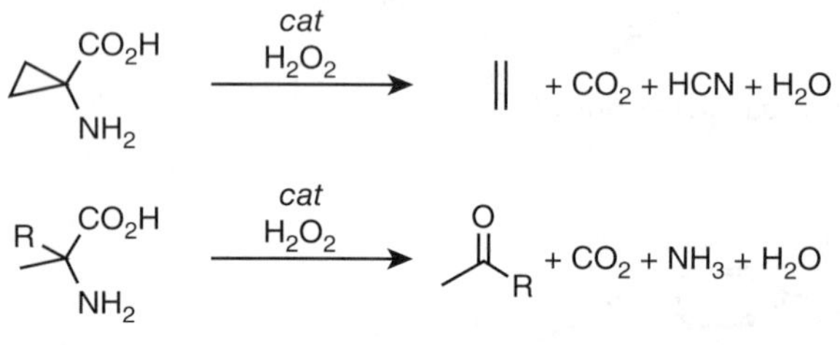

Scheme 12.11 Amino acid oxidation catalyzed by salen-Fe(X) complexes and H_2O_2.

Salen-Mn complexes proved to be excellent asymmetric epoxidation catalysts [47], but they were found also to hydroxylate C—H bonds. The first report of such reactivity dates back to 1993, when Grieco's group discovered that a steroid substrate covalently attached to the salen-Mn complex was oxidized selectively with PhIO on the C14 position (Scheme 12.12), along with the formation of minor overoxidation products [87]. The intermolecular reaction does not occur, demonstrating that substrate binding is required to overcome the intrinsic low reactivity of aliphatic C—H bonds.

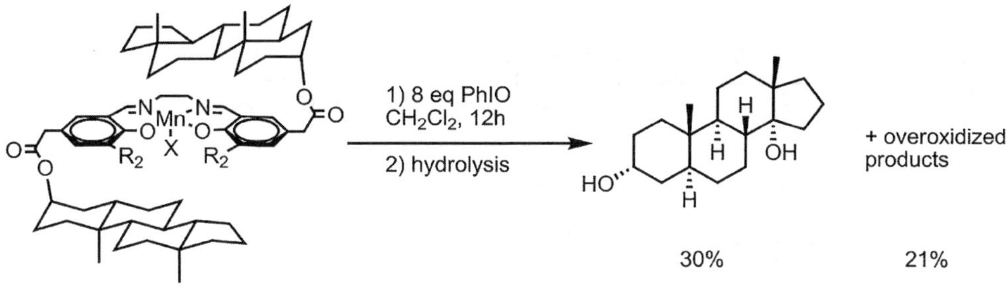

Scheme 12.12 Intramolecular oxidation of the steroid arms of a Salen-Mn complex.

Therefore, the attention turned toward the oxidation of more activated positions, on which an intermolecular pathway is feasible. In 1994, Jacobsen et al. observed that a salen-Mn complex is able not only to epoxidize 1,2-dihydronaphthalene substrates with NaOCl as oxidant but also to promote benzylic hydroxylation of the epoxides formed (Scheme 12.13, cat 1). Interestingly, this reaction favors one epoxide enantiomer over the other, enabling a kinetic resolution process [88]. Benzylic C—H oxidation with PhIO as the terminal oxidant, with isolated yields ranging from 22% to 92%, has been also reported [89]. Katsuki's group designed a new class of salen ligands, decorated with bulky, chiral biaryl derivatives (Scheme 12.13, cat 2 and cat 3), to perform the challenging asymmetric benzylic C—H hydroxylation of prochiral substrates. The reaction formed benzylic alcohols in low-to-moderate yield (19–29%) and, more remarkably, with enantiomeric excesses up to 64% [90, 91].

The use of oxidatively stable but viscous solvents (such as chlorobenzene) was crucial to achieve such enantioselectivity, preventing escape of the carbon-centered radical intermediate from the solvent cage. Later, the same group extended this asymmetric hydroxylation to other classes of enantiotopic, activated C—H bonds. In these cases, both higher yields and asymmetric induction were observed. The C—H bonds proximal to an ethereal oxygen (up to 59% yield and 82% ee) [92] or to a cyclic amide (up to 65% yield and 76% ee) were effectively hydroxylated with good stereoselectivity [93]. Murahashi and coworkers used a *t*-Bu salen-Mn catalyst to achieve enantiotopic C—H oxidations of cyclic alkylaromatic substrates (up to 15% yield and 70% ee) [94] or the desymmetrization of cyclic disilylethers [95]. Addition of an extra axially chiral center was also viable to achieve enantioselective C—H oxidations of benzylic positions (Scheme 12.13, cat 4) [96]. The replacement of PhIO oxidants with the more environmentally attractive H_2O_2 was later demonstrated to be viable in the oxidation of activated C—H bonds, with isolated yields up to 75%, but no studies on the asymmetric version of this transformation have been carried out. A large excess of the oxidant (10 mol eq with respect to the substrate) was required to overcome competitive hydrogen peroxide disproportionation. The oxidizing species was proposed to be the same formed with ArIO oxidants on the basis of UV-Vis experiments [97]. Later, tuning the alcohol/ketone ratio of the oxidized products was shown to be possible by simple modification of the ligand structure (replacement of cyclohexanediamine with different diamines and replacement of salicylaldehydes with 2-formylpyridines) [98].

Oxidation of fully nonactivated aliphatic C—H bonds proved to be quite challenging for salen-metal catalysts. Several investigations reported rather low conversions in cycloalkane oxidation with H_2O_2 or *t*-BuOOH as terminal oxidants and selectivity patterns close to the ones of Fenton-type reactions with both iron- [71, 81, 99, 100] and manganese-salen complexes [101–103]. The addition of an acid co-catalyst improves the reaction outcomes, increasing both the conversion and the selectivity of the process. Both selected inorganic acids (HNO_3) [104] and organic acids (pyrazine-carboxylic acid, oxalic acid) [101, 104] were proved to be suitable promoters. Increase of the electron-density on the aromatic rings of the ligand was found

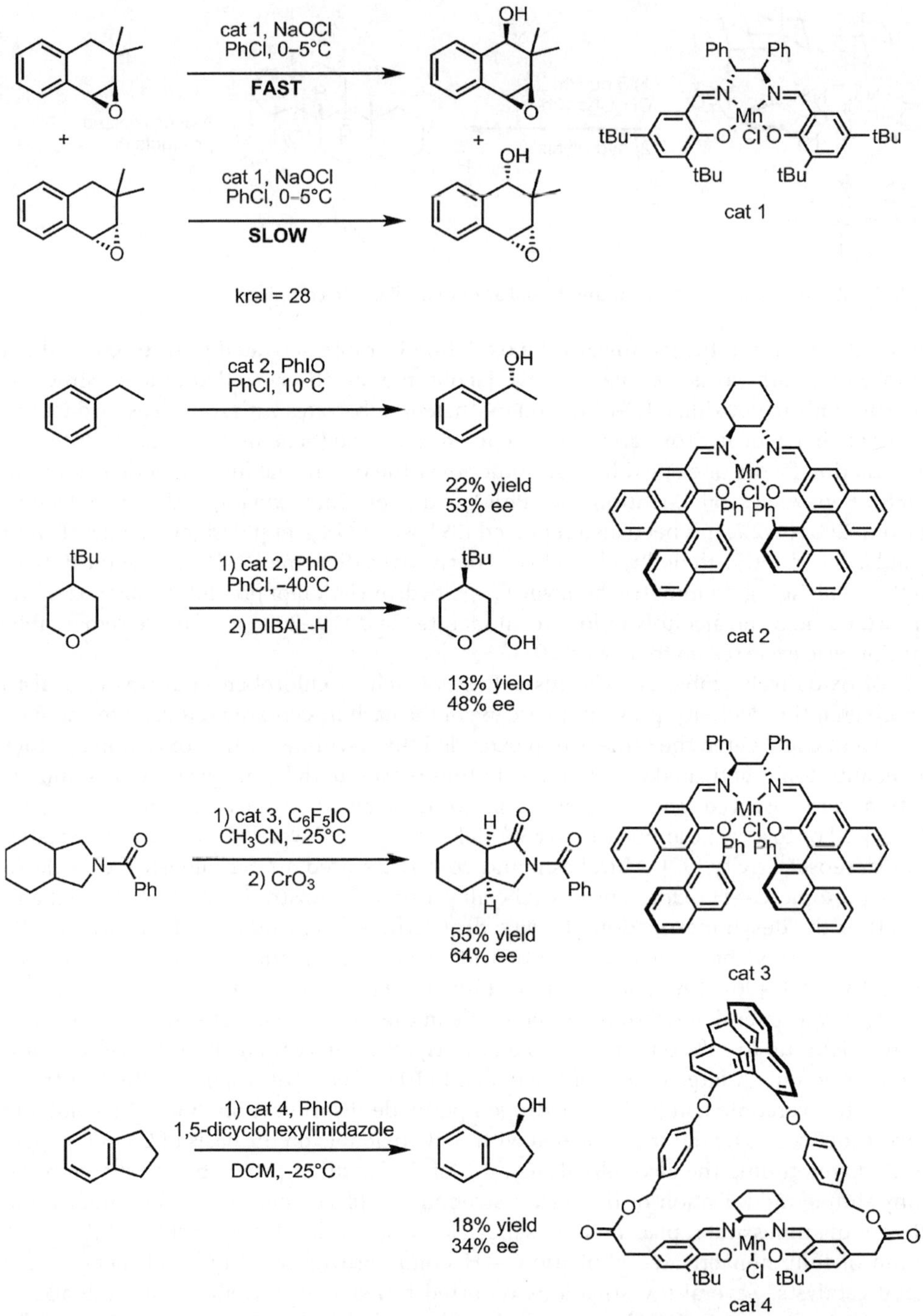
cat 1, NaOCl
PhCl, 0–5°C
FAST
OH
+
+
cat 1, NaOCl
PhCl, 0–5°C
SLOW
krel = 28
cat 1
cat 2, PhIO
PhCl, 10°C
22% yield
53% ee
1) cat 2, PhIO
PhCl, –40°C
2) DIBAL-H
13% yield
48% ee
cat 2
1) cat 3, C6F5IO
CH3CN, –25°C
2) CrO3
55% yield
64% ee
cat 3
1) cat 4, PhIO
1,5-dicyclohexylimidazole
DCM, –25°C
18% yield
34% ee
cat 4

Scheme 12.13

to exert a beneficial effect in alkane oxidation with salen-Fe catalysts (substitution on R1 in Figure 12.1) [104].

Oxo-bridged dimers of salen-Fe(III) complexes can also catalyze oxidation reactions. In 1980, Tabushi reported that a salen-Fe(III) μ-oxo dimer complex preferentially hydroxyates the secondary adamantane C—H bond oxidation with O_2 in the presence of a reducing agent (2-mercapto-ethanol) [105]. More recently [106], a dinulear μ-oxo salen$_2$-Fe$_2$^{III} complex was found to mediate effective C—H oxidation with *m*CPBA (TON up to 30). A rather high A/K ratio (12), declining over time, pointed to a metal-based oxidation mechanism, which is not operating with the mononuclear catalyst analog (A/K = 1.7).

However, the main strategy adopted to prevent ligand oxidative degradation and enhancement of the catalytic activity is immobilization of salen-metal complexes in a solid matrix. Such a heterogenization significantly facilitates separation processes and catalyst recycling at the end of the reaction. Several solid supports have been explored, such as zeolites [107, 108], diatomite [102], alumina [109, 110], or polymers [101, 111]. In some cases, cooperative effects between different metal ions are observed [112]. As in homogeneous systems, several clues point to a free-radical oxidation mechanism [101] (i.e. loss of the stereochemistry in the oxidation of tertiary carbons, A/K ratios close to 1 or the formation of alkyl hydroperoxide intermediates). Immobilization of the catalyst indeed allows to increase conversions; moreover, the catalyst can be successfully recycled for few cycles [102]. Interestingly, catalyst immobilization in a solid support can influence the hydroxylation regioselectivity. Bahramian et al. [102] reported an increase of the preference for tertiary over secondary C—H bonds in adamantane oxidation when the catalyst is grafted onto porous diatomite. When the salen-Mn complex is anchored onto polydimethylsiloxane and an hydrophobic environment surrounds the metal ion, a significant change in the distribution of products of linear alkanes oxidation was observed. In particular, selectivity for C2 over the more internal carbons increases if compared to the homogeneous system. This effect was ascribed to the different folding of linear alkanes in polar or hydrophobic environments, which exposes different sites to the attack of the reagent (in this case, presumably *t*-BuO• radical) [101].

12.3.2 Other Imine-Based Complexes

Besides salen systems, several different types of iron and manganese imine-containing complexes have been explored as C—H oxidation catalysts. In these cases, there are no common structural motifs, aside from the imine function, present in all the catalytic systems. This lack of structural homogeneity translates into a vast array of catalyst stabilities, mechanistic pathways, and oxidation outcomes. In this scenario, we are inclined to discuss these complexes on the basis of the denticity of the imine-containing ligands.

12.3.2.1 Bidentate Ligands

Given the preference of iron and manganese ions for the octahedral coordination geometry, the use of bidentate ligands usually results in the formation of complexes in which two or more molecules of ligand bind the central ion. This occurs in the case of imine-containing, bidentate ligands L1 and L2, which form complexes with iron in a 2 : 1 ligand to metal stoichiometry (see [(L1)$_2$Fe(OTf)$_2$] in Figure 12.2) [113, 114]. These complexes catalyze the oxidation of cyclohexane to cyclohexanol and cyclohexanone by H_2O_2 or t-BuOOH, although with limited yields. Low A/K ratios (0.37–0.61, or 0.19 in the case of *t*-BuOOH oxidant) and inhibition of the reaction in the presence 2,4,6-tri-*tert*-butylphenol used as a radical scavenger indicate that a radical mechanism is occurring with hydrogen peroxide. In the case of benzylic C—H oxidation, the use of *t*-BuOOH oxidant allows transformation to ketones in moderate-to-excellent

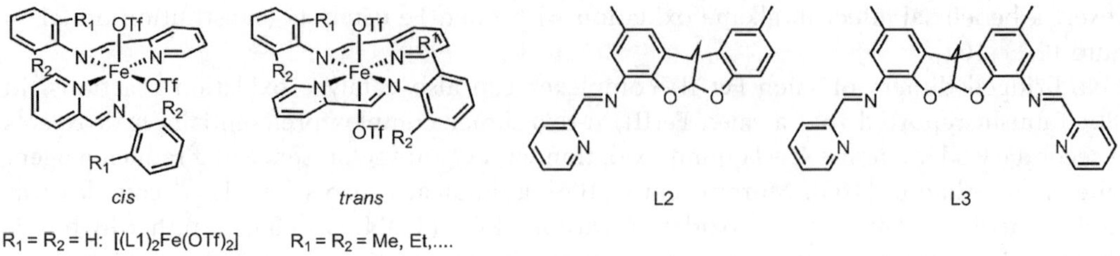

Figure 12.2

yields (22–91%), with no inhibition from radical scavengers. However, no further studies to ascertain the mechanistic pathway with *t*-BuOOH oxidant have been performed.

The related monotopic (L2) and ditopic (L3) ligands form iron complexes that are able to promote diphenylmethane oxidation with different oxidants, in moderate to good yields (41–73%) [115]. The active oxidants show an electrophilic character, but no further investigation on the oxidation mechanism has been provided.

12.3.2.2 Tridentate Ligands

The tridentate iminopyridine complexes [(^{R}BIP)Fe(CH_3CN)(OTf)]OTf (see Figure 12.3, OTf = triflate, R = CH_3— or $(CH_3)_2CH$—)) were employed by Britovsek in the H_2O_2 oxidation of cyclohexane to cyclohexanol [116]. In these complexes, the steric bulk of the aniline substituents prevents the formation of adducts with a 2 : 1 ligand to metal ratio. These complexes were found to sluggishly catalyze cyclohexane oxidation with a free radical mechanism, which was revealed by the low A/K ratios observed (comparable to those obtained in the sole presence of $Fe(OTf)_2$) and the detection of cyclohexyl hydroperoxide among the reaction products. The amine analog (see Figure 12.3) gives almost the same results [116]. Replacement of iron with manganese leads to fully inactive catalysts.

The use of monoanionic ligands was also considered. Fe(III) tridentate imine-based complex [(mqmp)Fe(CH_3OH)$(Cl)_2$] (Figure 12.3) was shown to fully oxidize cyclohexane within

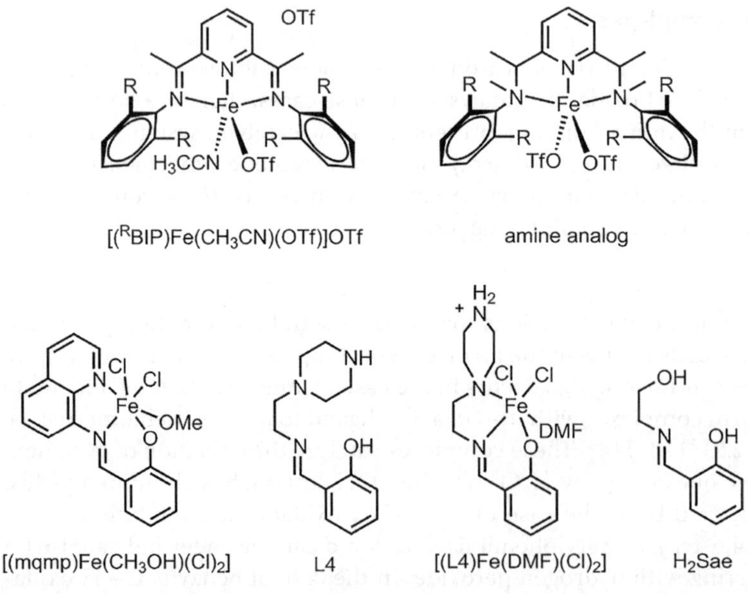

Figure 12.3

24 hours at 50 °C [117]. Cyclohexyl hydroperoxide was detected in the initial stages of the reaction, and hydroxyl radical scavengers, such as DMSO, significantly inhibited the reaction. These pieces of evidence point to a free-radical oxidation mechanism. However, a high valent iron-oxo species ([(mqmp)Fe(O)(CH_3OH)(Cl)]) was also detected by ESI-MS monitoring of the reaction.

The Fe(II) complex of the imine-based ligand L4 (Figure 1.3) is a very highly active catalyst for C—H oxidation [118]. This complex can be prepared in-situ, and results to be protonated on the distal nitrogen of the piperazine ring (see Figure 12.3) in the reaction conditions. Very low catalyst loadings (0.05–0.2 mol%) display an astonishing catalytic activity with 37% yield and 900 TONs reached in the cyclohexane oxidation with H_2O_2 and HNO_3 promoter. It seems that nitric acid causes a catalyst reorganization, that, in combination with H_2O_2, activates the catalyst. Low selectivity for tertiary over secondary C—H bonds, the absence of stereospecificity in 1,2-*cis*-dimethylcyclohexane oxidation and kinetic evidences definitely leads to the conclusion that a radical mechanism promoted by hydroxyl radicals is operating.

High catalytic activity in cyclohexane oxidation due to efficient formation of hydroxyl radicals was also found by Pombeiro and Kokozay with the heterometallic $[Co^{III}{}_4Fe^{III}{}_2(Sae)_8]$ imine-based catalyst (obtained with tridentate H_2Sae ligand depicted in Figure 12.3) [119]. H_2O_2 oxidant and HNO_3 transform this complex into the active $[Co_2Fe(Sae)_4]$ form, which shows a very high catalytic activity in cyclohexane oxidation (up to 3670 TONs and 26% yield). Mechanistic studies pointed to a Fenton-type oxidation process, with an extremely high efficient hydroxyl radical generation.

The use of less hindered tridentate ligands results in the formation of complexes with a 2 : 1 ligand:metal stoichiometry, as in the case of complex $[(L5)_2Fe](OTf)_2$ depicted in Figure 12.4. This simple Fe(II) complex has been shown to catalyze the H_2O_2 oxidation of aliphatic C—H bond with high efficiency (yields up to 47% and TONs in the range 25–50) [44]. This complex has the significant advantage of greatly easing preparation, since no other synthetic steps are required than addition in solution of the precursors, namely 2-picolyl aldehyde, 2-picolyl amine and $Fe(OTf)_2(CH_3CN)_2$ in the 2 : 2 : 1 molar ratio. In other words, complex $[(L5)_2Fe](OTf)_2$ can be rapidly and quantitatively assembled *in situ* just before substrate and H_2O_2 addition (see Figure 12.4) [44, 120, 121]. It displays a high catalytic activity (15–43% in aliphatic C—H oxidations), comparable to that of many more synthetically demanding and structurally more elaborated amine-based complexes [19, 20], and much higher than most of other imine-based catalysts appeared in the literature so far [21]. Several substrates are hydroxylated by this complex in good yields and with a catalyst loading as low as 1%.

However, its more remarkable property is the oxidation mechanism, which has been demonstrated to occur for the overwhelming part through a metal-based pathway [120]. The oxidations carried out on a series of mechanistic probes rule out the involvement of free radical species in the oxidation process (A/K ratio of 11, 3°/2° ratio of 13 in adamantane oxidation, *KIE* 3.3, 96% retention of configuration in *cis*-dimethylcyclohexane oxidation, absence of CyOOH) and point to the formation of a selective, metal-based oxidant [120]. This is in line

Figure 12.4

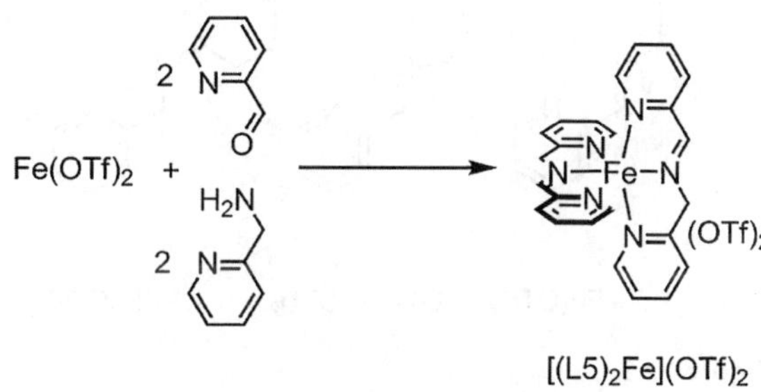

with isotope-labeling experiments, which demonstrated that the O-atom incorporated in the products largely comes from H_2O_2 (80–96%). No incorporation from H_2O was detected. This observation suggests that the H_2O_2 activation mechanism differs from the one commonly operating in tetradentate, aminopyridine iron complexes. This was further confirmed by the negligible effect exerted by acetic acid on catalytic activity, and by the lack of *cis*-diol products in olefin oxidation experiments. Moreover, ligand flexibility is required for catalytic activity. Rigid, hexadentate iron complexes are not able to promote oxidation reactions. All these clues point to the involvement of an oxidizing species that lacks a second labile coordination site close to the one occupied by hydrogen peroxide (or by some activated form of the latter reagent). On the basis of the above pieces of evidence, a mechanism was proposed that involves initial oxidation of the Fe(II) present in the $[(L5)_2Fe](OTf)_2$ complex to Fe(III), subsequent detaching of a pyridine arm, and H_2O_2 coordination and activation. However, the exact nature of catalytically active species has still to be identified.

12.3.2.3 Tetradentate Ligands

Few tetradentate imine-containing iron complexes have been tested as oxidation catalysts. This is probably due to the fact that the most investigated tetradentate ligand platforms are the salen compounds. The neutral N_4 type iron complex $[Fe(pyhd)(Cl_2)]$ [104] (see Figure 12.5) gave low yields in cyclohexane oxidation (17 TONs) with *t*-BuOOH, comparable to the ones obtained with salen systems (19–45 TONs). The use of its Mn analog, however, allows the retention of a good catalytic activity in the oxidation of activated, benzylic positions [98]. A related tetradentate N_2S_2 iron (III) complex [Fe(gma•) (Bu_3P)] (see Figure 12.5) was reported to catalyze aliphatic C—H oxidation with H_2O_2 in the presence of pyrazine carboxylic acid [122]. Cyclohexane is oxidized with high turnover numbers (up to 490), albeit in modest yield (up to 15%), likely through a free-radical oxidation pathway [122].

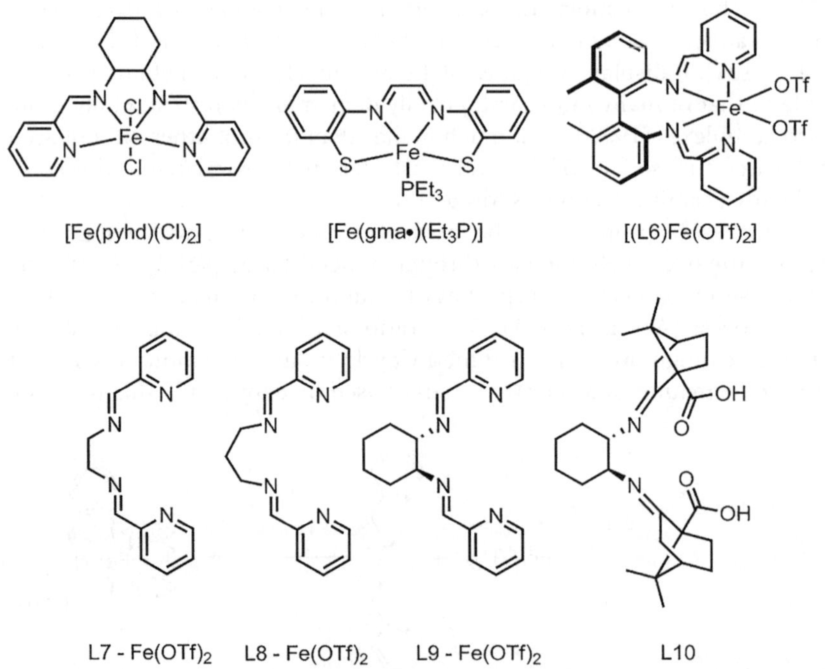

Figure 12.5

Britovsek prepared complex [(L6)Fe(OTf)$_2$] (Figure 12.5) [123], which does not assume a *trans* coordination, but is forced to adopt a *cis* one by the biaryl bridge on the diamine motif. This complex mediates the H_2O_2 oxidation of cyclohexane slightly more efficiently than the simple Fe(OTf)$_2$ salt, and the low A/K ratio as well as the formation of cyclohexyl hydroperoxide in the reaction mixture point to a free radical mechanism. Moreover, the use of higher amounts of H_2O_2 (from 1 : 100 to 1 : 10) significantly decrease the oxidation efficiency, due to probable catalyst decomposition.

A more thorough study of different tetradentate ligand structures has been performed by Feringa and coworkers [124]. They investigated a series of tri, tetra, and pentadentate amino and imino-pyridine iron complexes in the oxidation of ethylbenzene with H_2O_2 or O_2 as the terminal oxidant. With H_2O_2, the best results have been obtained with the Fe(II) complexes of iminopyridine ligands bearing a 1,2-diamine motif (L7, L9, Figure 12.5). As reported for aminopyridine complexes [65, 66] increase of the size of the spacer between the two aminic nitrogens (L8) translates into a lower catalytic activity. The A/K ratios (5.6–6.1) pointed to a metal-based reaction mechanism, and the oxidation yields are dependent on the aromatic substituents (an increase in the electron density enhances the activity). O_2 can also be used as the terminal oxidant, albeit at the expense of a certain efficiency and of a radical oxidation pathway (A/K ratios close to unity).

The tetradentate complex of Mn(OAc)$_2$·4H_2O and ligand L10 (Figure 12.5) was also found to catalyze activated C—H oxidation [125]. The O-donor of the ligand is a carboxylate anion derived from camphore, and the ligand is obtained by condensation of this ketoacid with 1,2-cyclohexane diamine. Good-to-high yields (52–91%) have been obtained in the oxidation of benzylic C—H bonds with *t*BuOOH as the terminal oxidant.

12.3.2.4 Pentadentate Ligands

Pentacoordinated complexes leave just one labile site on the metal to interact with the oxidant. The iminopyridine iron complex [(L11)Fe(OTf)$_2$] (Figure 12.6) is rapidly oxidized to the related μ-oxo Fe(III) dimer by molecular oxygen [126]. However, both the monomeric [(L11)Fe(OTf)$_2$] and the μ-oxo dimer act as C—H oxidation catalysts (cyclohexane and adamantane oxidation). In particular [(L11)Fe(OTf)$_2$] furnished an A/K value of 4.5 and a 3°/2° ratio of 4.6 for cyclohexane and adamantane oxidation, respectively. The μ-oxo dimer was even more selective (A/K value of 7.2 and a 3°/2° ratio of 9.4), albeit with a lower catalytic activity (2.9–4.6 TONs to be compared with 4.8–5.7 for [(L11)Fe(OTf)$_2$]). Unfortunately, increase of the oxidant/substrate ratio from 1 : 100 to 1 : 10 lowers the oxidation efficiency due to a pernicious catalase-like activity for both the monomer and the μ-oxo dimer.

Also the related complex of ligand dapab (Figure 12.6) with Fe(BF$_4$)$_2$ [127], in which the pyridine donors have been replaced by methoxy donors, is an efficient catalyst for cyclohexane oxidation with H_2O_2. 1 mol% of such complex completely fully converts cyclohexane into a 1 : 1 mixture of cyclohexanol and cyclohexanone and other byproducts. The reaction rate was found to be dependent on the presence of molecular oxygen, and inhibited by scavengers of

Figure 12.6

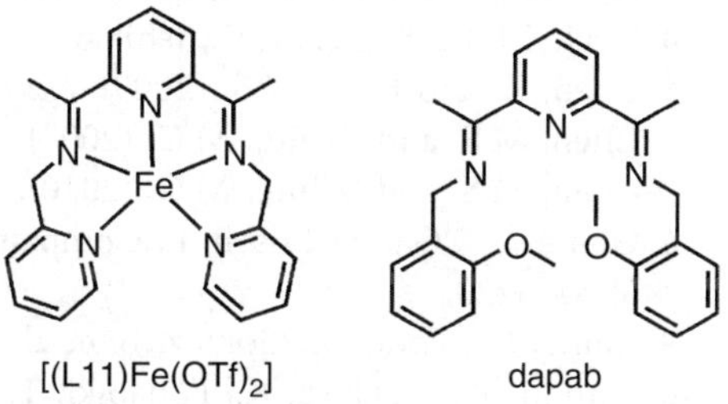

HO$^{\bullet}$ radicals like DMSO and acetone. These pieces of evidence point to a radical-based mechanism. However, comparable yields of oxygenated products have also been obtained with the simple iron salt, without the dapab ligand.

12.4 Conclusions

In this chapter, several imine-based iron and manganese complexes which can act as C—H oxidation catalysts have been described with a focus on their reaction mechanism. With the aim of introducing the following discussion, the common mechanistic tools used to distinguish a metal-based oxidation from a radical chain process have been presented in the first section. The high variability in the ligand and complex structure found in imine-based iron or manganese complexes translates into a large variety of catalytic efficiencies and oxidation pathways in C—H hydroxylation, going from negligible to quantitative substrate conversion. In a large number of cases such catalysts operate through a free-radical based mechanism. Nevertheless, when the complex is quite stable and polydentate (tetradentate or higher denticity systems), and the ligand is rather robust, the oxidation can occur through a metal-based mechanism. This is the case of salen complexes, which catalyze even enantioselective C—H oxidations with oxygen atom donor oxidants. However, the selected oxidant and the strength of the targeted C—H bond deeply affects the mechanistic pathway and thus the reaction outcome. Non-salen, polydentate complexes were also proved to mediate metal-based C—H oxidation, demonstrating that such a controlled reactivity could be successfully implemented not only with aminopyridine ligands, but also with imine-based ones. No doubt that the significant advantage offered by the imine-based catalysts is the possibility of an *in situ* catalyst self-assembly from simple, often commercial starting materials, which avoids expensive and highly demanding syntheses. This feature makes such catalysts highly attractive and, in perspective, competitive with other iron and manganese catalysts.

Acknowledgments

Thanks are due to Ministero dell'Istruzione, dell'Università e della Ricerca (MIUR, PRIN 2010CX2TLM). This work was also partially supported by Università di Roma La Sapienza (Progetti di Ricerca 2014).

References

1 Godula, K. and Sames, D. (2006). *Science* 312: 67–73.
2 White, M.C. (2012). *Science* 335: 807–809.
3 Newhouse, T. and Baran, P.S. (2011). *Angew. Chem. Int. Ed.* 50: 3362–3374.
4 Uckl, T.B.R., Baxter, R.D., Ishihara, Y., and Baran, P.S. (2012). *Acc. Chem. Res.* 45: 826–839.
5 Chen, K. and Baran, P.S. (2009). *Nature* 459: 824–828.
6 Chen, M.S. and White, M.C. (2007). *Science* 318: 783–787.
7 Chen, M.S. and White, M.C. (2010). *Science* 327: 566–571.
8 Gómez, L., Garcia-Bosch, I., Company, A. et al. (2009). *Angew. Chem. Int. Ed.* 48: 5720–5723.
9 Canta, M., Font, D., Gómez, L. et al. (2014). *Adv. Synth. Catal.* 356: 818–830.
10 Hitomi, Y., Arakawa, K., Funabiki, T. et al. (2012). *Angew. Chem. Int. Ed.* 51: 3448–3452.

11 Company, A., Gómez, L., Fontrodona, X. et al. (2008). *Chem. A Eur. J.* 14: 5727–5731.
12 Olivo, G., Lanzalunga, O., Mandolini, L. et al. (2013). *J. Org. Chem.* 78: 11508–11512.
13 Haslinger, S., Raba, A., Cokoja, M. et al. (2015). *J. Catal.* 331: 147–153.
14 Jana, S., Ghosh, M., Ambule, M. et al. (2017). *Org. Lett.* 19: 746–749.
15 Milan, M., Bietti, M., and Costas, M. (2017). *ACS Cent. Sci.* 3: 196–204.
16 Ottenbacher, R.V., Samsonenko, D.G., Talsi, E.P. et al. (2012). *Org. Lett.* 14: 4310–4313.
17 Dong, J.J., Unjaroen, D., Mecozzi, F. et al. (2013). *ChemSusChem* 6: 1774–1778.
18 Que, L. Jr. and Tolman, W.B. (2008). *Nature* 455: 333–340.
19 Talsi, E.P. and Bryliakov, K.P. (2012). *Coord. Chem. Rev.* 256: 1418–1434.
20 Lindhorst, A.C., Haslinger, S., and Kühn, F.E. (2015). *Chem. Commun.* 51: 17193–17212.
21 Olivo, G., Lanzalunga, O., and Di Stefano, S. (2016). *Adv. Synth. Catal.* 358: 843–863.
22 Cussó, O., Ribas, X., and Costas, M. (2015). *Chem. Commun.* 51: 14285–14298.
23 Nam, W., Lee, Y.-M., and Fukuzumi, S. (2014). *Acc. Chem. Res.* 47: 1146–1154.
24 Olivo, G., Cussò, O., and Costas, M. (2016). *Chem. Asian J.* 11: 3148–3158.
25 Oloo, W.N. and Que, L. Jr. (2015). *Acc. Chem. Res.* 48: 2612–2621.
26 Olivo, G., Cussó, O., Borrell, M. et al. (2017). *J. Biol. Inorg. Chem.* 22: 425–452.
27 Chen, K. and Que, L. Jr. (2001). *J. Am. Chem. Soc.* 123: 6327–6337.
28 Barton, D.H.R. and Doller, D. (1992). *Acc. Chem. Res.* 25: 504–512.
29 Kiani, S., Tapper, A., Staples, R.J. et al. (2000). *J. Am. Chem. Soc.* 122: 7503–7517.
30 Stavropoulos, P., Çelenligil-Çetin, R., and Tapper, A.E. (2001). *Acc. Chem. Res.* 34: 745–752.
31 Lu, H. and Zhang, X.P. (2011). *Chem. Soc. Rev.* 40: 1899–1909.
32 Che, C.-M., Lo, V.K., Zhou, C.-Y. et al. (2011). *Chem. Soc. Rev.* 40: 1950–1975.
33 Meunier, B., de Visser, S.P., and Shaik, S. (2004). *Chem. Rev.* 104: 3947–3980.
34 Meunier, B. (1992). *Chem. Rev.* 92: 1411–1456.
35 Gormisky, P.E. and White, M.C. (2013). *J. Am. Chem. Soc.* 135: 14052–14055.
36 Font, D., Canta, M., Milan, M. et al. (2016). *Angew. Chem. Ed.* 55: 5776–5779.
37 Gómez, L., Canta, M., Font, D. et al. (2013). *J. Org. Chem.* 78: 1421–1433.
38 Prat, I., Gómez, L., Canta, M. et al. (2013). *Chem. A Eur. J.* 19: 1908–1913.
39 Curci, R., D'Accolti, L., and Fusco, C. (2006). *Acc. Chem. Res.* 39: 1–9.
40 Kamata, K., Yonehara, K., Nakagawa, Y. et al. (2010). *Nat. Chem.* 2: 478–483.
41 Ciaccia, M., Cacciapaglia, R., Mencarelli, P. et al. (2013). *Chem. Sci.* 4: 2253–2261.
42 Ciaccia, M., Pilati, S., Cacciapaglia, R. et al. (2014). *Org. Biomol. Chem.* 12: 3282–3287.
43 Ciaccia, M. and Di Stefano, S. (2015). *Org. Biomol. Chem.* 646–654.
44 Olivo, G., Arancio, G., Mandolini, L. et al. (2014). *Catal. Sci. Technol.* 4: 2900–2904.
45 Kopylovich, M.N., Ribeiro, A.P.C., Alegria, E.C.B. et al. (2015). *Adv. Organomet. Chem.* 91–174.
46 Retcher, B., Costa, J.S., Tang, J. et al. (2008). *J. Mol. Catal. A: Chem.* 286: 1–5.
47 Katsuki, T. (1996). *J. Mol. Catal. A: Chem.* 113: 87–107.
48 Legros, J. and Bolm, C. (2003). *Angew. Chem. Int. Ed.* 42: 5487–5489.
49 Kochi, J.K. (ed.) (1973). *Free Radicals*. New York: Wiley-Interscience.
50 Costas, M., Chen, K., and Que, L. Jr. (2000). *Coord. Chem. Rev.* 200–202: 517–544.
51 Russell, G.A. (1957). *J. Am. Chem. Soc.* 3871–3877.
52 MacFaul, P.A., Ingold, K.U., Wayner, D.D.M. et al. (1997). *J. Am. Chem. Soc.* 119: 10594–10598.
53 Arends, I.W.C.E., Ingold, K.U., and Wayner, D.D.M. (1995). *J. Am. Chem. Soc.* 117: 4710–4711.
54 Shul'pin, G.B. (2002). *J. Mol. Catal. A: Chem.* 189: 39–66.
55 Shul'pin, G.B. and Nizova, G.V. (1992). *React. Kinet. Catal. Lett.* 48: 333–338.
56 Shul'pin, G.B. and Druzhinina, A.N. (1992). *React. Kinet. Catal. Lett.* 47: 207–211.

57 Baciocchi, E., D'Acunzo, F., Galli, C. et al. (1996). *J. Chem. Soc., Perkin Trans.* 2: 133–140.
58 Barton, D.H.R., Beck, A.H., and Taylor, D.K. (1995). *Tetrahedron* 51: 5245–5254.
59 Ortiz De Montellano, P.R. (2010). *Chem. Rev.* 110: 932–948.
60 Bernadou, J. and Meunier, B. (1998). *Chem. Commun.* 2167–2173.
61 Company, A., Gómez, L., Güell, M. et al. (2007). *J. Am. Chem. Soc.* 129: 15766–15767.
62 Puri, M., Company, A., Sabenya, G. et al. (2016). *Inorg. Chem.* 55: 5818–5827.
63 Buxton, G.V., Greenstock, C.L., Helman, W.P. et al. (1988). *J. Phys. Chem. Ref. Data* 17: 513–886.
64 Chen, K., Costas, M., and Que, L. Jr. (2002). *J. Chem. Soc., Dalton Trans.* 672–679.
65 Britovsek, G.J.P., England, J., and White, A.J.P. (2005). *Inorg. Chem.* 44: 8125–8134.
66 England, J., Davies, C.R., Banaru, M. et al. (2008). *Adv. Synth. Catal.* 350: 883–897.
67 Pijper, D., Saisaha, P., de Boer, J.W. et al. (2010). *Dalton Trans.* 10375–10381.
68 Katsuki, T. (2000). *Asymmetric Epoxidation Unfunctionalized Olefins and Related Reactions* (ed. I. Ojima), 287–325. New York: Wiley-VCH.
69 Baleizão, C. and Garcia, H. (2006). *Chem. Rev.* 106: 3987–4043.
70 McGarrigle, E.M. and Gilheany, D.G. (2005). *Chem. Rev.* 105: 1556–1602.
71 Gupta, K.C.C. and Sutar, A.K. (2008). *Coord. Chem. Rev.* 252: 1420–1450.
72 Bryliakov, K.P., Babushkin, D.E., and Talsi, E.P. (2000). *J. Mol. Catal.* 158: 19–35.
73 Hamada, T., Fukuda, T., Imanishi, H. et al. (1996). *Tetrahedron* 52: 515–530.
74 Dalton, C.T., Ryan, K.M., Wall, V.M. et al. (1998). *Top. Catal.* 5: 75–91.
75 Srinivasan, K., Perrier, S., and Kochi, J.K. (1986). *J. Mol. Catal.* 36: 297–317.
76 Cavallo, L. and Jacobsen, H. (2000). *Angew. Chem. Int. Ed.* 39: 589–592.
77 Feichtinger, D. and Plattner, D.A. (1997). *Angew. Chem. Int. Ed. Engl.* 36: 1718–1719.
78 Feichtinger, D. and Plattner, D.A. (2001). *Chem. Eur. J.* 7: 591–599.
79 Collman, J.P., Zeng, L., and Brauman, J.I. (2004). *Inorg. Chem.* 43: 2672–2679.
80 Jayaseeli, A.M.I. and Rajagopal, S. (2009). *J. Mol. Catal. A: Chem.* 309: 103–110.
81 Sivasubramanian, V.K., Ganesan, M., Rajagopal, S. et al. (2002). *J. Org. Chem.* 67: 1506–1514.
82 Bryliakov, K.P. and Talsi, E.P. (2004). *Angew. Chem. Int. Ed.* 43: 5228–5230.
83 Bryliakov, K.P. and Talsi, E.P. (2007). *Chem. Eur. J.* 13: 8045–8050.
84 Kurahashi, T., Kobayashi, Y., Nagatomo, S. et al. (2005). *Inorg. Chem.* 44: 8156–8166.
85 Baráth, G., Kaizer, J., Pap, J.S. et al. (2010). *Chem. Commun.* 46: 7391–7393.
86 Góger, S., Bogáth, D., Baráth, G. et al. (2013). *J. Inorg. Biochem.* 123: 46–52.
87 Kaufman, M.D., Grieco, P.A., and Bougie, D.W. (1993). *J. Am. Chem. Soc.* 115: 11648–11649.
88 Larrow, J.F. and Jacobsen, E.N. (1994). *J. Am. Chem. Soc.* 116: 12129–12130.
89 Ho Lee, N., Lee, C.-S., and Jung, D.-S. (1998). *Tetrahedron Lett.* 39: 1385–1388.
90 Hamachi, K., Irie, R., and Katsuki, T. (1996). *Tetrahedron Lett.* 37: 4979–4982.
91 Hamada, T., Irie, R., Mihara, J. et al. (1998). *Tetrahedron* 54: 10017–10028.
92 Miyafuji, A. and Katsuki, T. (1997). *Synlett* 7: 836–838.
93 Punniyamurthy, T. and Katsuki, T. (1999). *Tetrahedron* 55: 9439–9454.
94 Komiya, N., Noji, S., and Murahashi, S.I. (1998). *Tetrahedron Lett.* 39: 7921–7924.
95 Murahashi, S.-I., Noji, S., Hirabayashi, T. et al. (2004). *Synlett* 1739–1742.
96 Murahashi, S.-I., Noji, S., and Komiya, N. (2004). *Adv. Synth. Catal.* 346: 195–198.
97 Mardani, H.R. and Golchoubian, H. (2006). *J. Mol. Catal. A: Chem.* 259: 197–200.
98 Golchoubian, H. and Ghasemi, N. (2011). *J. Chin. Chem. Soc.* 58: 470–473.
99 Salomão, G.C., Olsen, M.H.N., Drago, V. et al. (2007). *Catal. Commun.* 8: 69–72.
100 Biswas, A.N., Das, P., Kandar, S.K. et al. (2009). *Catal. Commun.* 10: 708–711.

101 Mac Leod, T.C.O., Kirillova, M.V., Pombeiro, A.J.L. et al. (2010). *Appl. Catal., A* 372: 191–198.
102 Bahramian, B., Doulati, F., Mirkhani, V., and Badii, K. (2008). *Appl. Catal. A* 345: 97–103.
103 Shul'pin, G.B. (2001). *Pet. Chem.* 41: 405–412.
104 Silva, A.R., Mourão, T., and Rocha, J. (2013). *Catal. Today* 203: 81–86.
105 Tabushi, I., Nakajima, T., and Seto, K. (1980). *Tetrahedron Lett.* 21: 2565–2568.
106 R. Mayilmurugan, H. Stoeckli-Evans, E. Suresh, et al. *Dalton Trans* 2009, 5101–5114.
107 Corrêa, R.J., Salomão, G.C., Olsen, M.H.N. et al. (2008). *Appl. Catal., A* 336: 35–39.
108 Kadwa, E., Bala, M.D., and Friedrich, H.B. (2014). *Appl. Clay Sci.* 95: 340–347.
109 Mirkhani, V., Moghadam, M., Tangestaninejad, S. et al. (2008). *Catal. Commun.* 9: 2171–2174.
110 Farzaneh, F., Poorkhosravani, M., and Ghandi, M. (2009). *J. Mol. Catal. A: Chem.* 308: 108–113.
111 Gupta, K.C., Kumar Sutar, A., and Lin, C.-C. (2009). *Coord. Chem. Rev.* 253: 1926–1946.
112 Fan, B., Li, H., Fan, W. et al. (2008). *Appl. Catal., A* 340: 67–75.
113 Shejwalkar, P., Rath, N.P., and Bauer, E.B. (2011). *Dalton Trans.* 40: 7617–7631.
114 Lenze, M., Martin, F.T., Rath, N.P. et al. (2013). *Chem. Plus Chem.* 78: 101–116.
115 Martínez-Ferraté, O., Britovsek, G.J.P., Claver, C. et al. (2015). *Inorg. Chim. Acta* 431: 156–160.
116 Britovsek, G.J.P., England, J., Spitzmesser, S.K. et al. (2005). *Dalton Trans.* 945–955.
117 Pevec, A., Roubeau, O., Dehnen, S. et al. (2010). *Polyhedron* 29: 2291–2296.
118 Nesterov, D.S., Nesterova, O.V., Guedes da Silva, M.F.C., and Pombeiro, A.J.L. (2015). *Catal. Sci. Technol.* 5: 1801–1812.
119 Nesterov, D.S., Chygorin, E.N., Kokozay, V.N. et al. (2012). *Inorg. Chem.* 51: 9110–9122.
120 Olivo, G., Nardi, M., Vìdal, D. et al. (2015). *Inorg. Chem.* 54: 10141–10152.
121 Olivo, G., Giosia, S., Barbieri, A. et al. (2016). *Org. Biomol. Chem.* 14: 10630–10635.
122 Fernandes, R.R., Kirillova, M.V., da Silva, J.A.L. et al. (2009). *Appl. Catal., A* 353: 107–112.
123 Britovsek, G.J.P., England, J., and White, A.J.P. (2006). *Dalton Trans.* 1399–1408.
124 Klopstra, M., Hage, R., Kellogg, R.M. et al. (2003). *Tetrahedron Lett.* 44: 4581–4584.
125 Pan, J.F. and Chen, K. (2001). *J. Mol. Catal. A: Chem.* 176: 19–22.
126 Tanase, S., Reedijk, J., Hage, R. et al. (2010). *Top. Catal.* 53: 1039–1044.
127 Tang, J., Gamez, P., and Reedijk, J. (2007). *Dalton Trans.* 4644–4646.

13

Alkane Oxidation with Biologically Inspired Nonheme Iron Catalysts Based in the Triazacyclononane Ligand Scaffold

Miquel Costas

Departament de Química I Institut de Química Computacional i Catàlisi, Facultat de Ciències, Campus de Montilivi, Girona, Catalonia, Spain

13.1 Introduction

13.1.1 C—H Oxidation in Biological Systems. The Biologically Inspired Approach

The selective functionalization of aliphatic alkyl moieties constitutes a challenge for modern synthetic chemistry, but its realization will have a tremendous impact [1–3]. Oxygenation of alkyl moieties is an important subclass of these reactions, which besides its technological relevance [4] also plays a major role in a number of metabolic processes taking place in aerobic organisms [5–7].

Conversion of an sp^3 C—H bond into a C—O bond in a selective manner is a very difficult problem; alkyl C—H bonds are poorly reactive and often considered inert, only accessible with highly reactive species. Owing to their inert character, they are often much less reactive than organic functional groups. Consequently, oxidation of C—H bonds is often incompatible with the presence of other functional groups in the substrate molecule. Moreover, since organic molecules often contain different C—H bonds, the target of a specific C—H bond can be a formidable problem in terms of selectivity.

High-valent metal-oxo species are powerful electrophilic reagents that can cleave and functionalize aliphatic C—H bonds. High-valent iron-oxos are used by nature in a number of enzymes, including heme and non-heme oxygenases. In P450 and related enzymes, the reaction of the high-valent $Fe^{IV}(O)$ porphyrin radical, known as Compound I, with aliphatic C—H bonds has been thoughtfully studied, and serves as a paradigm for this reaction [5, 8, 9]. There is current consensus that the reaction of CpdI with hydrocarbons occurs via the so-called oxygen rebound mechanism that entails an initial hydrogen atom transfer (HAT) reaction from the C—H bond to the metal-oxo species, forming a carbon-centered radical. This step is followed by a rapid hydroxyl ligand transfer to the incipient carbon-centered radical, forming a hydroxylated product. A remarkable consequence of this mechanism is that the lifetime of the radical is so short that epimerization does not take place before hydroxyl ligand transfer. Therefore, the overall hydroxylation reaction occurs with stereoretention. This is an important and characteristic feature because most of the known C—H cleaving reagents generate a carbon-centered radical that diffuses in solution and can engage in radical rearrangements, including epimerization.

Evidence has mounted over the last decade that a number of non-heme iron-dependent oxygenases oxygenate aliphatic C—H bonds via mechanisms and reactive species that resemble those operating in heme enzymes [10]. Non-heme oxoiron(IV) species have been established

Alkane Functionalization, First Edition. Edited by Armando J. L. Pombeiro and M. Fátima C. Guedes da Silva.

as the C—H cleaving species in α-ketoglutarate-dependent enzymes such as TauD [11], prolyl hydroxylase [12], and in pterin-dependent enzymes such as tyrosine hydroxylase [13] and phenylalanine hydroxylase [14], and also in halogenase enzymes such as CytC3 [15, 16] and SyrB2 [17]. Particularly interesting are Rieske oxygenases, which are implicated in an array of oxidative transformations that bypass those associated with the heme systems in terms of versatility, and include *inter alia* the hydroxylation of C—H bonds (Scheme 13.1) [18]. Electron inventory experiments performed in single turnover reactions of naphthalene-1,2-dioxygenase, a member of this family [19], suggests that Rieske oxygenases operate via a catalytic cycle where ferric hydroperoxide species are reaction intermediates, and crystallographic characterization for such species has been obtained [20]. It has been proposed that these species may evolve toward a $Fe^V(O)$ species that is the substrate oxidizing agent. Mechanistic studies in the oxidation of radical clock substrates has provided evidence for radical intermediates, and further suggests that reactions also proceed via a P450-like rebound mechanism (Scheme 13.1) [21].

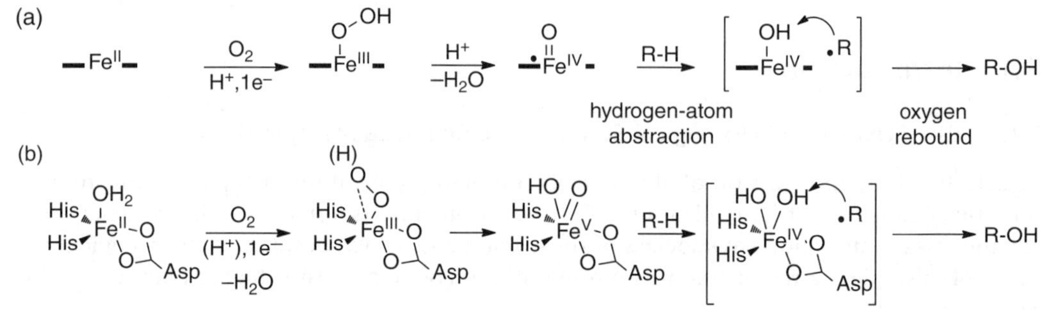

Scheme 13.1 Reaction mechanism for C—H hydroxylation at (a) P450 and (b) Rieske dioxygenases.

Unlike hemes, the iron center in non-heme iron-dependent enzymes is ligated to oxygen and nitrogen atoms from amino acid residues and water molecules. For example, a common ligand architecture is the so-called facial triad (two histidines and one carboxylate), which leaves at the metal center three other coordination positions, all of them *cis* to each other, available for interaction with the substrate and/or the oxidant [22, 23].

Oxidation reactions taking place at these enzymes present very attractive features from the point of view of synthetic organic chemistry; they often exhibit outstanding chemo and regioselectivities in reactions that are either difficult or unattainable with traditional reagents. Hydroxylation of C—H bonds constitute one of the most interesting examples. They rely on a readily available metal and employ nonharmful oxidants such as O_2 of H_2O_2, producing no waste. Because of these very attractive features, the structure and reactivity of non-heme iron enzymes have for long served as a motif of inspiration for synthetic bioinorganic chemistry and the design of oxidation catalysis [24–29]. On the one hand, the use of complexes with amine and carboxylate-based ligands have been considered as a simplistic approach toward the mimicking of the enzyme active center. On the other hand, it has been presumed that reaction of these complexes with oxidants, and most specially hydrogen peroxide, that can be seen as a 2e− reduced surrogate of O_2, may lead to iron-based species analogous to those operating in enzymatic active sites.

However, this has turned out to not be a straightforward approach, and extensive evidence exists that Fenton-type free-diffusing radical paths appear to be quite common when iron-based coordination compounds are used as oxidation catalysts in combination with peroxides [30–35]. However, several iron complexes react with H_2O_2, generating powerful high valent iron-oxo species that hydroxylate C—H bonds via enzyme-like mechanisms, without involvement of free radical species. These catalysts have started to be explored as synthetic tools in C—H oxidation reactions [2, 27].

13.1.2 Non-Heme Iron Catalysts that Oxidize Strong C—H Bonds with Stereoretention

Prototypical examples of iron complexes that catalyze the hydroxylation of aliphatic C—H bonds via a mechanism that departs from free diffusing Fenton processes are shown in Scheme 13.2 [36–38]. These complexes share common structural features; they contain a tetradentate ligand that leaves two sites at the iron center, *cis*- with respect to each other. Ligands all combine aliphatic amine and pyridine donors, and exert a relatively strong crystal field in the iron center. The two positions not occupied by the tetradentate ligand are fulfilled with labile ligands such as acetonitrile or triflate, but since reactions usually take place in acetonitrile, the bis-solvato dicationic species are always the dominant species in solution. Complex $[Fe(tpa)(CH_3CN)_2]^{2+}$ is the first example for which stereoretentive C—H hydroxylation was described [36]. $[Fe(mep)(CH_3CN)_2]^{2+}$ constitutes the prototypical example of a family of catalysts that have proven particularly efficient and that currently receive use in synthetic C—H oxidation protocols [37, 39, 40]. These two complexes preceded and inspired the design of $[Fe(^{x}pytacn)(CH_3CN)_2]^{2+}$ (x = H and CH_3) [38, 41]. We envisioned that the latter family of complexes could be a relevant contribution to this class of catalysts because it fulfilled all the required structural aspects. In addition, it is well-known that the tacn macrocycle forms very stable coordination complexes [42–44], and its highly basic nature was expected to be an important element for stabilizing high oxidation states.

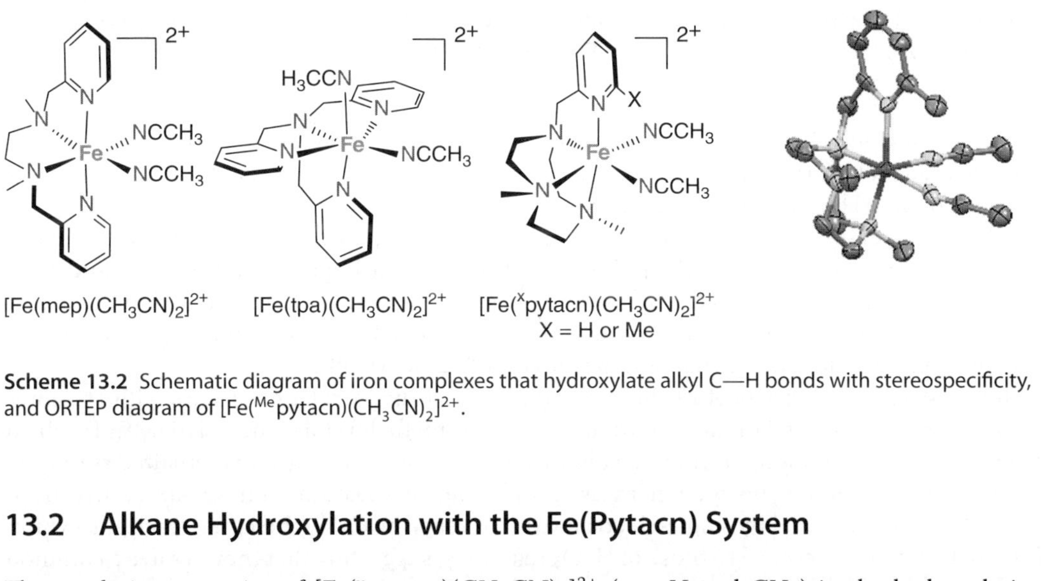

Scheme 13.2 Schematic diagram of iron complexes that hydroxylate alkyl C—H bonds with stereospecificity, and ORTEP diagram of $[Fe(^{Me}pytacn)(CH_3CN)_2]^{2+}$.

13.2 Alkane Hydroxylation with the Fe(Pytacn) System

The catalytic properties of $[Fe(^{x}pytacn)(CH_3CN)_2]^{2+}$ (x = H and CH_3) in the hydroxylation of alkanes with H_2O_2 are shown in Scheme 13.3 [38, 41]. Using a large excess of substrate (cyclohexane), the two complexes are efficient catalysts and convert up to 76% of H_2O_2 (10–100 equiv. delivered via syringe pump) into oxidized products, with a remarkably high A/K (alcohol/ketone) ratio. The latter is a first indication that the catalysts operate via a mechanism that does not entail long-lived carbon-centered radicals, since the latter, under aerobic conditions, are converted into equimolar mixtures of alcohol and ketones via a Russell-termination. The

Alkane hydroxylation reactions catalyzed by **[Fe(OTf)2(Hpytacn)]** and **[Fe(OTf)2(Mepytacn)]**.[a]

	Cyclohexane		adamantane	DMCH
catalyst	A+K[b] (A/K)	KIE[c]	3º/2º	RC(%)
Fe(Hpytacn)	6.5 (12.3)	4.3	30	93
Fe(Mepytacn)	7.6 (10.2)	3.4	20	94

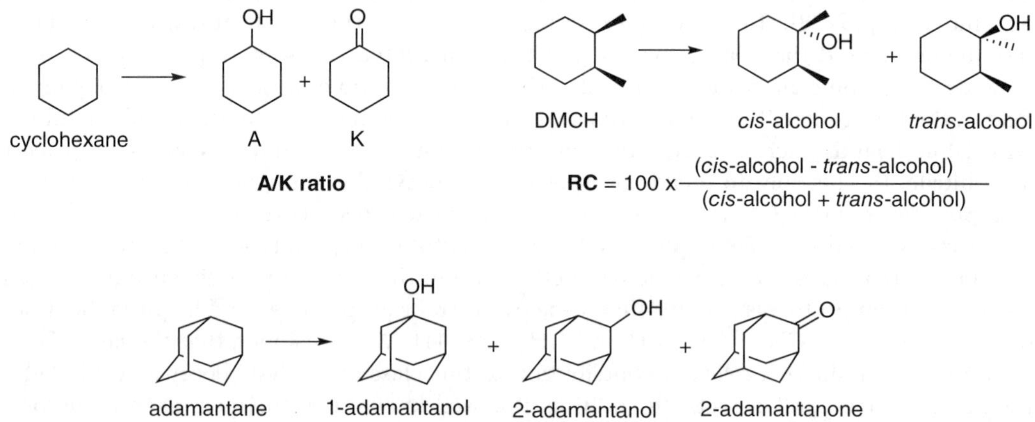

$$3°/2° = 3 \times \frac{\text{1-adamantanol}}{\text{(2-admantanol + 2-adamantanone)}}$$

Scheme 13.3 Mechanistic proves in the oxidation of C—H catalyzed by [Fe(OTf)$_2$(^{X}pytacn)] complexes. (a) Catalysts:H_2O_2:substrate ratio 1:10:1000 except for adamantane (10), (b) TN, and (c) for cyclohexanol formation.

excellent catalytic abilities of the complexes are maintained when larger amounts of peroxide are delivered. When 100 equiv. of peroxide are added, [Fe(^{X}pytacn)$(CH_3CN)_2]^{2+}$ produces 52 TN of alcohol (A) and 12 TN of ketone (K).

Catalytic oxidations of substrate probes demonstrate the involvement of highly selective metal-centered oxidants [38, 41, 45]. Thus, reactions exhibit large A/K ratios (12.3 and 10.2) and high selectivities, as indicated by the kinetic isotope effects (KIEs) evaluated in the oxidation of a 1:3 mixture of cyclohexane:d_{12}-cyclohexane (KIE ≈ 3–4) and the high $C_{tertiary}/C_{secondary}$ selectivity in the oxidation of adamantane (adam) (3°/2° ≈ 20–30).

The most remarkable aspect of these reactions is that hydroxylation occurs with stereoretention. This is evidenced in the oxidation of *cis*-1,2-dimethylcyclohexane (cis-DMCH), which provided the corresponding tertiary alcohol where the two methyl groups remain *cis*- to each other. The corresponding *trans* epimer was only obtained in trace amounts. Of notice, hydrogen peroxide can be replaced by single-electron oxidants such as Ce^{IV}. The latter reactions exhibit selectivity patterns identical to those of H_2O_2 reactions, suggesting that they involve a common reactive species [46].

Valuable information of the reaction mechanism was gained by isotopic labeling analyses. The origin of the oxygen atoms introduced into the cyclohexanol product was determined by performing the oxidation of cyclohexane using 10 equiv. $H_2{}^{18}O_2$ in the presence of 1000 equiv. H_2O. The complementary experiment involved the use of unlabeled H_2O_2 and 1000 equiv. $H_2{}^{18}O$. Interestingly, oxygen atoms from water were incorporated into cyclohexanol both for [Fe(^{H}pytacn)$(CH_3CN)_2]^{2+}$ (45%) and [Fe(^{Me}pytacn)$(CH_3CN)_2]^{2+}$ (11%). Oxygen atoms incorporated from $H_2{}^{18}O_2$ are 47% and 85%. Therefore, on the basis of the O-atom inventory, it was concluded that incorporation of O derived from air accounts for less than 10% in the oxidation reactions catalyzed by these two complexes. This is a mechanistically important observation because O_2 reacts with carbon-centered radicals at nearly diffusion controlled rates. Its nearly negligible participation in the reactions strongly argue in favor of a non-free-radical mechanism.

Overall, these mechanistic studies indicated that reactions involved a selective C—H cleaving reagent, that reactions proceed with stereoretention, and that the reaction involves reaction intermediates that can incorporate oxygen atoms from water into the product.

An additional important piece of experimental observation is the identification by ESI-MS of $[Fe(L)(OH)]^{2+}$, L = ^{x}pytacn as the resting state under catalytic conditions [45].

13.2.1 The Reaction Mechanism for the C—H Hydroxylation Reaction

The reactivity pattern observed for the $[Fe(OTf)_2({}^{x}pytacn)]$ family of complexes was highly reminiscent of that previously documented for $[Fe(tpa)(CH_3CN)_2]^{2+}$ and $[Fe(mep)(CH_3CN)_2]^{2+}$ complexes [47], and because of that, a common mechanistic scheme was derived [45]. This mechanism was first proposed by Que et al. for the latter systems and is shown in Scheme 13.4.

The reaction is initiated by the ferrous complex with H_2O_2 to generate ferric complex **Q**. In the case of $[Fe^{III}(OH)({}^{H}pytacn)]^{2+}$, this species is dominant in solution, as ascertained by ESI-MS analysis [45], but in the case of L = mep and tpa, these species are in equilibrium with oxo-bridged dimers [36, 48]. Reaction of **Q** with excess H_2O_2 results in the formation of ferric hydroperoxide **P**. For the specific case of $[Fe^{III}(OOH)(tpa)]^{2+}$, the intermediate can be spectroscopically characterized at low temperatures by EPR, resonance Raman, UV-Vis spectroscopy, and ESI-MS [47, 49]. Kinetic analyses demonstrate that **P** is not capable of attacking strong

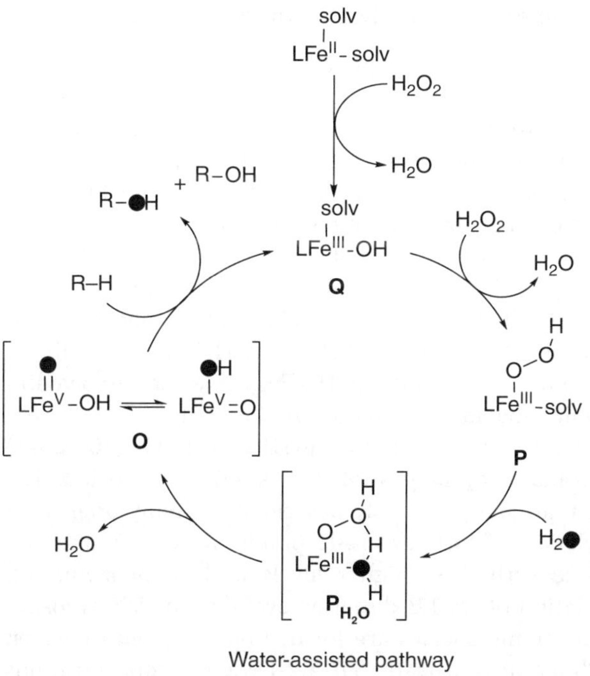

Scheme 13.4 Reaction mechanism proposed for the hydroxylation of aliphatic C—H bonds with iron complexes $[Fe^{II}(L)(CH_3CN)_2]^{2+}$, L = tpa, mep, and ^{x}pytacn.

C—H bonds, but instead it decays into a novel, spectroscopically undetected and highly reactive species [49]. For the specific case of putative $[Fe^{III}(OOH)(^{x}pytacn)]^{2+}$, it does not accumulate in solution, so direct spectroscopic evidence cannot be gained.

Species **O** are proposed to be formed via heterolytic cleavage of the O—O bond in **P**, assisted by a water molecule, which ends up as a ligand (species $\mathbf{P}_{\mathbf{H_2O}}$). Thus, **Q** contains one oxygen atoms that originates from H_2O_2 and a second oxygen atom that originates from a water molecule. Indeed, when this class of catalysts react with olefins via a *syn*-dihydroxylation reaction, the resulting diol contains one oxygen from water and one oxygen from H_2O_2, as predicted by the model [38, 48]. Evidence for **Q** was obtained by cryospray mass spectrometry [50]. On the other hand, the alternative homolytic cleavage was discarded because this reaction will produce a $Fe^{IV}O$ complex and a hydroxyl radical, which is an indiscriminate oxidant, incompatible with the observed selectivity.

Of notice, $[Fe^{V}(O)(OH)(^{x}pytacn)]^{2+}$, **O** can engage in a prototropic tautomerism by which the hydroxide ligand is transformed into an oxo, and the oxo is transformed into the hydroxide ligand (Scheme 13.5). This reaction, reminiscent of that occurring in hemes and metalloporphyrins [51], receives the name of oxo-hydroxo tautomerism, and provides a path for incorporating the oxygen atom from water into the oxidized products.

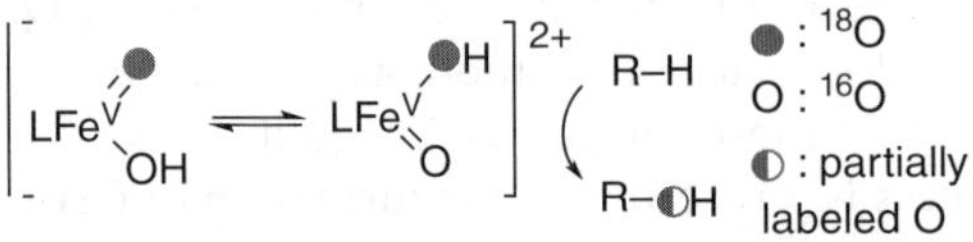

Scheme 13.5 Model that predicts competition between tautomerism and reaction with the substrate.

13.2.2 Rethinking the Oxo-Hydroxo Tautomerism at Catalysts with *cis*-Labile Sites

Following the precedents of heme systems and metalloporphyrin models, the tautomerism postulated for non-heme iron catalysts is in competition with the reaction with the substrate [36, 48]. Therefore, the faster reacting is the substrate, the less extended is the conversion of the initial hydroxide ligand (derived from water) into the terminal oxo, which is presumed to be the C—H cleaving species, and the oxygen atom that is transferred to the substrate. Since the initial reaction between the C—H bond and **O** is a hydrogen abstraction, the stronger the C—H bond, the larger percentage of incorporation of water must be observed. Considering that tertiary C—H bonds are weaker than secondary C—H bonds, the latter must be more extensively ^{18}O labeled when reacting with a mixture of $H_2O_2/H_2{}^{18}O$. Indeed, this was the trend observed in oxidation reactions with the $[Fe(tpa)(CH_3CN)_2]^{2+}$ and the $[Fe(mep)(CH_3CN)_2]^{2+}$ catalysts. However, $[Fe(OTf)_2(^{H}pytacn)]$ (which becomes $[Fe(^{H}pytacn)(CH_3CN)_2]^{2+}$ when dissolved in acetonitrile) exhibited a completely different behavior (Scheme 13.6) [38, 41, 45].

Oxidation of cyclohexane (1000 equiv. with respect to the catalyst) with H_2O_2 (10 equiv.) in the presence of $H_2{}^{18}O$ (1000 equiv.) leads to cyclohexanol that is 45% ^{18}O-labeled. The number is remarkably high, much larger than any previously described with any other iron catalyst. But most interestingly, hydroxylation of tertiary C—H bonds in adamantane and in *cis*-1,2-dimethylcyclohexane takes place with even higher levels of ^{18}O incorporation from water (74–83%). Of notice, hydroxylation of *cis*-1,2-dimethyl cyclohexane is stereospecific, ruling out the possibility that carbocationic species are formed and trapped by water. Furthermore, reactions performed at different substrate (*cis*-1,2-DMCH) concentrations show virtually the same level of ^{18}O incorporation from $H_2{}^{18}O$. Finally, cyclohexane and *cis*-DMCH were oxidized in competition with each other, and ^{18}O incorporation from water

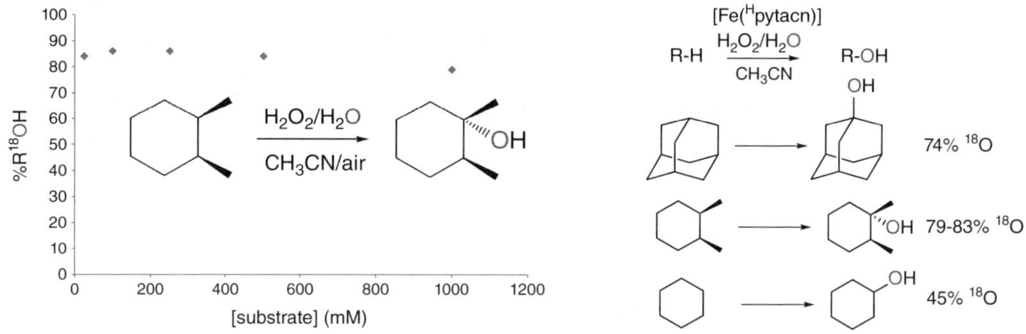

Scheme 13.6 Left, lack of dependence between substrate concentration and ^{18}O incorporation from $H_2{}^{18}O$ in catalytic oxidations with [Fe(^{H}pytacn)]. Right, ^{18}O incorporation for different substrates. (*See insert for color representation.*)

was determined for both substrates. *cis*-DMCH reacted much faster than cyclohexane, and so should have a shorter-lived oxidant. Nevertheless, the 3°-ol thus formed contained 74% label from water while cyclohexanol contained only 38%. Thus, the difference in label incorporation cannot be determined by how much ^{18}O is present in the common Fe^{V} oxidant, but instead must be determined by the nature of the substrate. These experiments show that the proposed model of the competition between tautomerism and reaction with the substrate needs to be reconsidered.

An alternative mechanism for explaining the experimental observations considers that the structure of the substrate (not the strength of the C—H bond) determines the course of C—O bond formation. It must be considered first that the two sites where the oxo and the hydroxo ligands can bind to the iron center are not equivalent. Therefore, there are two possible isomers for the [Fe^{V}(O)(OH)(^{H}pytacn)] species **O** (**O_A** and **O_B**, Scheme 13.7), in rapid equilibrium via a oxo-hydroxo tautomerism. The two Fe^{V}(O)(OH) isomers (**O_A** and **O_B**) are distinct because the ligands *trans* to the oxo and hydroxo groups are not chemically equivalent and the orientations of the pyridine ligand relative to the Fe=O bond are different. No matter which isomer abstracts the H-atom from substrate, a common $Fe^{IV}(OH)_2$ species is formed.

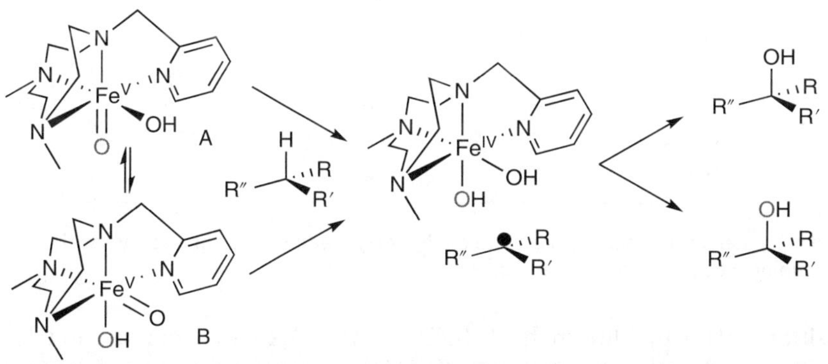

Scheme 13.7 Alternative rebound mechanism for iron catalyst [Fe(^{H}pytacn)], but that can be extended to iron catalysts containing two labile sites. (*See insert for color representation.*)

The alkyl radical can, in principle, rebound with either OH group, the outcome of which determines the $\%H_2{}^{18}O$ incorporation. Thus, the labeling results indicate that 2°-alkyl radicals do not discriminate between the two OH groups (45% ^{18}O incorporation), but 3°-alkyl radicals favor rebound with the OH group derived from water. An inspection of the crystal structure of the $[Fe(OTf)_2(^{H}pytacn)]$ catalyst suggests steric effects may provide a possible rationale for this preference, since the hemisphere surrounding the oxygen atom parallel to the pyridine ring is less sterically congested that the oxygen with the Fe—O vector perpendicular to this ring. This scenario predicts that the oxygen atom originating from water should be trans to the NCH_2py site, and that H_2O_2 must bind and cleave at the site *trans* to a NMe group. This prediction was further substantiated by computational analyses that will be described below.

13.2.3 Ligand Effects in C—H Hydroxylation

Modifications in the structure of the pytacn ligand provided a series of catalysts where the electronic and steric properties could be tuned in a systematic manner. These modifications provide further insight in the mechanism of C—H oxidation by this type of catalysts [45, 52].

The impact of steric and electronic properties of the catalysts in determining the extend of water incorporation into products was studied by systematically modifying positions α and γ of the pyridine of the ligand.

Modifications of position γ have a measurable and important impact in the electronic properties of the catalysts, as best evidenced by the 200 mV window of the Fe(III)/Fe(II) red-ox pair in the series of complexes $[FeCl_2L]$, upon going from X = NO_2 to X = NMe_2 (Scheme 13.8). However, the percentage of water incorporation into hydroxylated products obtained in the catalytic oxidation of cyclohexane and *cis*-DMCH does not respond in a systematic manner to the electronic properties of the catalysts. Hydroxylation of cyclohexane produces the corresponding alcohol 36–50% ^{18}O-labeled when reactions are conducted in the presence of $H_2{}^{18}O$ (1000 equiv.), while oxidation of cis-DMCH under analogous experimental conditions is stereospecific and provides alcohol 65–79% ^{18}O labeled.

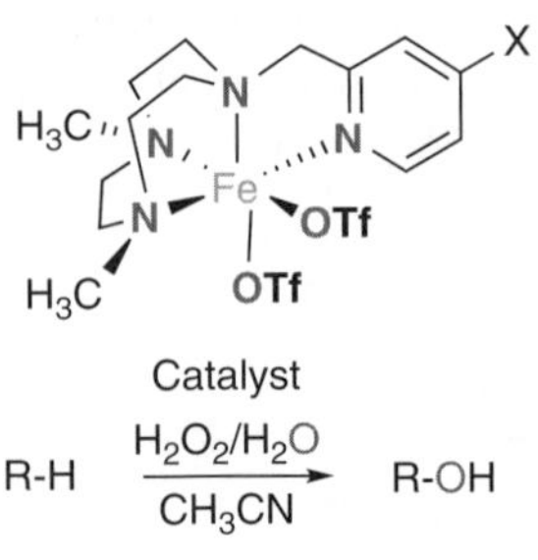

Percentage of ^{18}O incorporation from $H_2{}^{18}O$

	Substrate		
X	c-hexane	cis-DMCH	$E_{1/2}$ Fe(III)/Fe(II) (mV vs SCE)
NMe_2	36	68	11
MeO	44	72	44
Me	39	65	176
H	45	79	138
Cl	47	70	165
CO_2Et	48	72	171
NO_2	50	66	225

Scheme 13.8 Percentage of oxygen atoms incorporated into the hydroxylated products in reactions with [Fe(pytacn)] catalysts containing electronically tuned pyridine arms.

Modification of position α of the pyridine in the $[Fe(OTf)_2(^{x}pytacn)]$ catalyst, or replacing this heterocycle by a N-methylimidazole donor (catalyst $[Fe(OTf)_2(^{BzIm}pytacn)]$) has a much more drastic outcome (see Scheme 13.9) [45, 53].

Similar levels of ^{18}O-label incorporation from $H_2{}^{18}O$ (~40–45%) were observed in the case of the oxidation of secondary C—H bonds (cyclohexane, cyclooctane, and cyclohexane-d_{12}) when $[Fe(OTf)_2(^{H}pytacn)]$ and $[Fe(OTf)_2(^{BzIm}pytacn)]$ were used as catalysts. These levels of water incorporation are unusually high, and constitute strong evidence that $\mathbf{O_A}$ and $\mathbf{O_B}$ are roughly

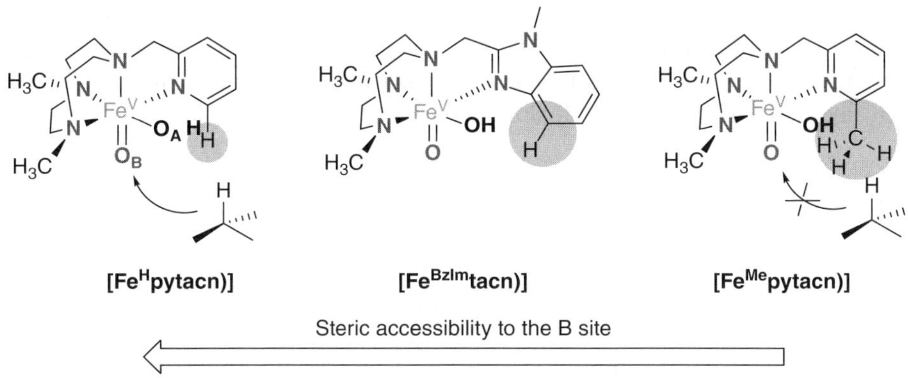

$$\text{R-H} \xrightarrow[\text{CH}_3\text{CN}]{\text{Catalyst} \;\; \text{H}_2\text{O}_2/\text{H}_2\text{O}} \text{R-OH}$$

Substrate	$[Fe^{BzIm}tacn)]$	$[Fe^{H}pytacn)]$	$[Fe^{Me}pytacn)]$
c-Hexane	48	45	11
c-Hexane-d_{12}	48	40	-
c-Octane	41	44	-
cis-DMCH	26	79	2
Adamantane	28	74	3

Scheme 13.9 Impact of sterics in the extend of water incorporation into hydroxylated products. (*See insert for color representation.*)

equally reactive against secondary C—H bonds. Most interestingly, when the substrates contain tertiary C—H bonds (e.g. DMCH and adamantane), the percentages of ^{18}O incorporation from $H_2{}^{18}O$ were found to be in the range 25–29%, indicating a preferential oxidation via $\mathbf{O_A}$.

On the other hand, hydroxylation of tertiary C—H bonds mediated by $[Fe(OTf)_2(^{H}pytacn)]$ is predominantly performed by $\mathbf{O_B}$ as shown by the large extent of oxygen atoms originating from water in the alcohol product (up to 79%, Scheme 13.9). In sharp contrast, hydroxylation catalyzed by $[Fe(OTf)_2(^{Me}pytacn)]$ exhibits a relatively small extent (~10%) of water incorporation in the hydroxylation of the secondary C—H bonds of cyclohexane, and negligible (<3%) incorporation in the hydroxylation of the tertiary C—H sites of cis-DMCH, indicating that hydroxylation is almost exclusively performed by $\mathbf{O_A}$.

Therefore, the relative reactivity of the two tautomeric forms of the $Fe^{V}(O)(OH)$ intermediate is finely tuned among the series of catalysts ($[Fe(OTf)_2(^{H}pytacn)]$, $[Fe(OTf)_2(^{BzIm}pytacn)]$, and $[Fe(OTf)_2(^{Me}pytacn)]$). These trends may be rationalized on the basis of steric effects. The benzylimidazole ring introduces steric bulk in the proximity of position B at the iron center that is intermediate between that set by pyridine and 6-Me-pyridine arms (Scheme 13.9). Accordingly, when secondary C—H bonds are hydroxylated, $[Fe(OTf)_2(^{H}pytacn)]$ behaves as $[Fe(OTf)_2(^{BzIm}pytacn)]$, i.e. tautomers $\mathbf{O_A}$ and $\mathbf{O_B}$ are equally implicated in the C—H oxidation reaction. On the other hand, oxidation of sterically more demanding tertiary C—H bonds yields levels of water incorporation that suggest predominant participation of $\mathbf{O_A}$, as in the case of $[Fe(OTf)_2(^{Me}pytacn)]$, although unlike in the latter case, implication of $\mathbf{O_B}$ remains significant (~25%) because steric hindrance at position B induced by the C—C-sp^2 benzylimidazole moiety is smaller than the one caused by a C-sp^3 methyl substituent.

Overall, the data strongly suggest that steric effects are at the basis of the relative reactivity of the two hydroxyl ligands in the $[Fe^{IV}(OH)_2(L)]$ intermediate that forms after hydrogen atom abstraction. In other words, the trajectories of the substrate carbon radical in the rebound process are basically dictated by the steric shielding of the hydroxyl ligands.

13.2.4 Computational Analysis on the C—H Oxidation Reactions

Mechanistic analyses in the catalytic C—H oxidation reactions of non-heme iron complexes are difficult because very reactive paramagnetic metal-based species are involved. Owing to their high reactivity, these very rarely accumulate in solution, and direct detection is very rare. Their paramagnetism also complicates characterization. Furthermore, quite often reactions are performed via syringe pump addition of H_2O_2, so oxidant delivery is the rate determining step. Overall, these factors make kinetic analysis very challenging. In this scenario, computational analyses have served as a powerful tool to validate mechanistic proposals derived from the study of the oxidation of mechanistic probes and isotopic labeling [45, 54, 55].

Computational analyses have provided valuable information to address the three most important aspects of these reactions. In the first place, they provide a theoretical description of the nature of the iron species that form upon reaction with H_2O_2, especially when the O—O bond is cleaved, and validate the paths for their formation. In the second place, they provide a basis to support the viability of tautomerically related pairs of reaction intermediates, arising from the unequivalent presence of two labile sites at the iron coordination sphere. Finally, computations have provided a description of the mechanism of C—H hydroxylation. The three aspects are discussed in Section 13.2.4.2.

13.2.4.1 Formation of the Active Species and Tautomerism

The reaction mechanism proposed on the basis of isotopic labeling analysis entail initial formation of a ferric hydroperoxide species **P** (S = 1/2) $[Fe^{III}(OOH)(^{x}pytacn)]^{2+}$ (Scheme 13.4) by reaction of $[Fe^{III}(OH)(^{x}pytacn)(S)]^{2+}$ (**Q**, S is an undefined ligand, probably CH_3CN, H_2O or OTf) with H_2O_2, that then evolves via a heterolytic O—O cleavage, assisted by a water molecule to form **O** (S = 3/2) $[Fe^{V}(O)(OH)(^{x}pytacn)]^{2+}$. Ferric hydroperoxide species have been spectroscopically characterized in reactions catalyzed by $[Fe(CH_3CN)_2(tpa)]^{2+}$ and $[Fe(CH_3CN)_2(mep)]^{2+}$ type of complexes [47, 56, 57], but not for the $[Fe(OTf)_2(pytacn)]$ family. Since the two labile positions at the coordination sphere of the iron site in $[Fe(CH_3CN)_2(pytacn)]$ complexes are not equivalent, a priori **Q**, **P**, and **O** can exist as two tautomerically related species (**A** and **B** in Scheme 13.10). DFT analysis indicates that in both $[Fe(^{H}pytacn)]$ and $[Fe(^{Me}pytacn)]$ $\mathbf{Q_B}$ is energetically (Gibbs energy) favored over $\mathbf{Q_A}$ by 3.76 and 2.69 kcal mol^{-1}, respectively, and because of that it was considered that this isomer dominates the reaction with H_2O_2.

Then, heterolytic O—O bond cleavage of $\mathbf{P_B}$ yields isomer $\mathbf{O_B}$ $[Fe^{V}(O)_B(OH)_A(L)]^{2+}$ (L = ^{H}pytacn and ^{Me}pytacn) in which the oxo group comes from H_2O_2 and the hydroxo ligand from water. Assistance of a water molecule, which acts as a sixth ligand, is crucial for the reaction to have an accessible activation barrier. In this situation, activation barrier for this transformation is ~21 kcal mol^{-1} and the reaction is slightly exoergonic (1–5 kcal mol^{-1}). That means that high-valent species **O** are slightly favored over the ferric hydroperoxide precursor **P**. Computations also indicate that isomers $\mathbf{O_A}$ and $\mathbf{O_B}$ are energetically very similar, with a Gibbs energy difference of <1 kcal mol^{-1} in favor of isomer $\mathbf{O_A}$. The Gibbs energy barrier of the tautomerism between $\mathbf{O_B}$ and $\mathbf{O_A}$ is relatively small, ~13 kcal mol^{-1} and is assisted by a water molecule. As a result, following the formation of $\mathbf{O_B}$, a rapid oxo-hydroxo tautomerism will allow the fast interconversion with $\mathbf{O_A}$, and the presence of both isomers in solution.

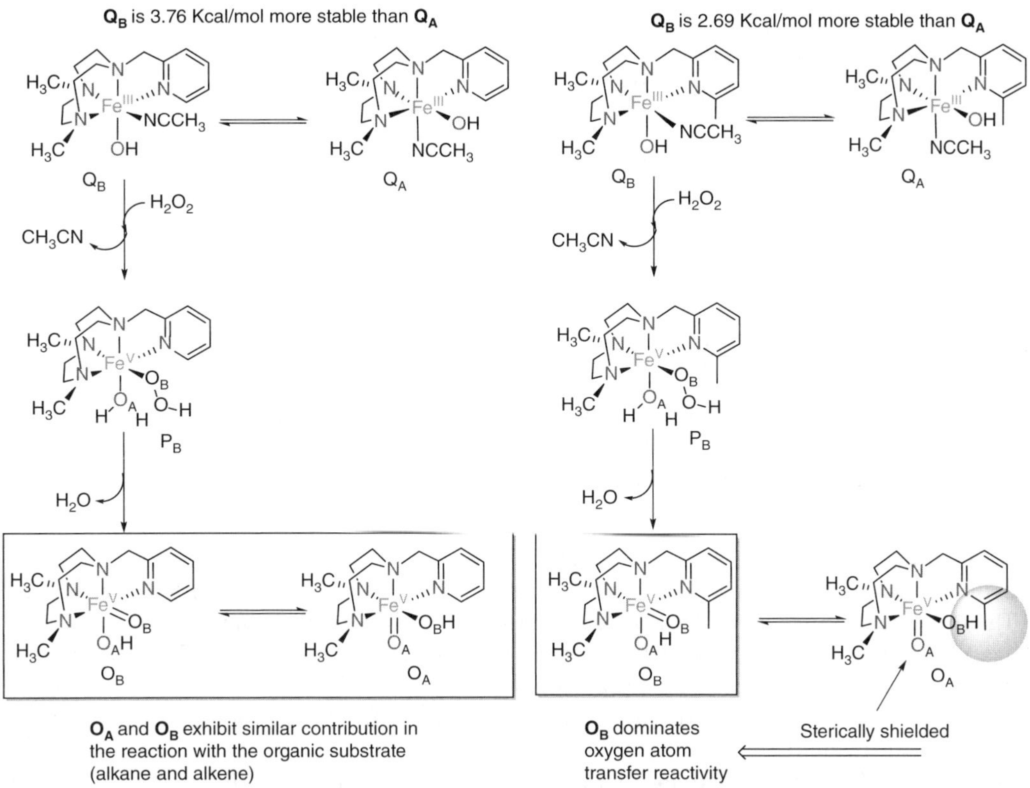

Scheme 13.10 Schematic diagram of the formation of the $Fe^{V}(O)(OH)$ species in [Fe(^{X}pytacn)] catalysts. (*See insert for color representation.*)

As originally expected on the basis of the isotopic labeling results, the hydroxylation reactivity of pairs of isomers $\mathbf{O_A}/\mathbf{O_B}$ are not equivalent. In the case of [Fe(^{H}pytacn)], they can be considered practically equally reactive because the hydroxylation of cyclohexane presents essentially the same activation barriers for hydrogen abstraction ($\Delta\Delta G° = 0.26\,kcal\,mol^{-1}$). Indeed, comparable amounts of ^{16}O-/^{18}O-labeled alcohols are obtained in the catalytic oxidation of cyclohexane conducted in the presence of $H_2{}^{18}O$ (see Scheme 13.9). Instead, for [Fe(^{Me}pytacn)] the hydroxylation of cyclohexane performed by $\mathbf{O_B}$ (in which the oxo ligand originates from H_2O_2) is slightly favored over $\mathbf{O_A}$ ($\Delta\Delta G° = 2.21\,kcal\,mol^{-1}$) [45, 54]. In nice agreement, hydroxylation with this catalyst proceeds with small incorporation of water into products (see Scheme 13.10) [45].

13.2.4.2 The Mechanism of C—H Hydroxylation

The mechanism of C—H hydroxylation by species **O** was also interrogated by computational methods. Three different possibilities were considered (Scheme 13.11) [54].

The first possibility is a rebound-type mechanism (path a in Scheme 13.11), initiated by a HAT from the alkane C—H bond to **O**, generating an alkyl radical and a $LFe^{IV}(OH)_2$ species. The mechanism is thus reminiscent to that of P450. A second proposed mechanism involves direct formation of the hydroxylated products through a concerted highly asynchronous mechanism (path b in Scheme 13.11). The transition state in this single-step process is given by the HAT, which is followed by the C—O bond formation without the generation of any intermediate. The third mechanism (path c in Scheme 13.11) involves an initial hydride transfer, generating

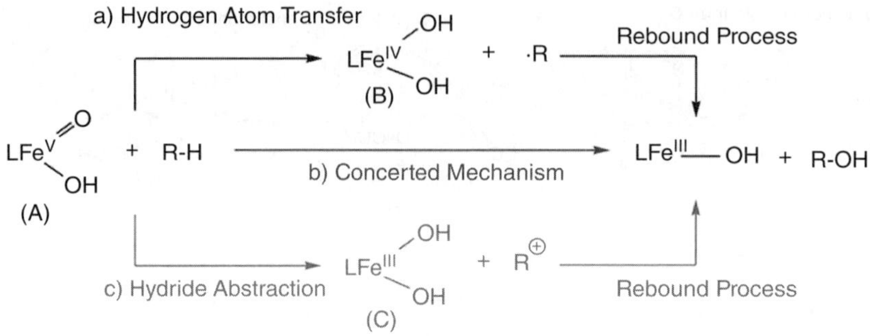

Scheme 13.11 Mechanisms considered for the hydroxylation of C—H bonds by species **O**.

a carbocation and a ferric bis-hydroxide species. Hydroxyl ligand transfer provides the alcohol and the ferric-monohydroxide species. The analysis of the three reaction mechanisms was performed for four substrates (methane, ethane, cyclohexane, and 2,3-dimethylbutane), employing $[Fe^{V}(O)(OH)(^{H}pytacn)]^{2+}$ as the oxidant. The study lead to the following conclusions:

1. In acetonitrile solution, a stepwise mechanism that starts with a HAT and follows with a hydroxyl radical rebound is determined. In all cases, $[Fe^{V}(O)(OH)(^{H}pytacn)]^{2+}$ is a quartet. The same mechanism is valid for the gas-phase hydroxylation of methane. On the contrary, gas-phase cyclohexane hydroxylation evolves through a concerted and highly asynchronous mechanism. The differences between acetonitrile solution and gas-phase cyclohexane mechanisms can be attributed to the important stabilization by the acetonitrile solvent of the $[Fe^{IV}(OH)_2(^{H}pytacn)]^{2+}/R\cdot$ intermediate.
2. With the exception of methane, structures resulting from a hydride transfer, namely $[Fe^{III}(OH)_2(^{H}pytacn)]^{+}/R^{+}$ are thermodynamically more stable than $[Fe^{IV}(OH)_2(^{H}pytacn)]^{2+}/R\cdot$ intermediates (resulting from a HAT). However, intrinsic reaction coordinate (IRC) calculations and Marcus theory prove that the former species are not involved in the hydroxylation mechanism. Electron transfer barriers of the evolution of $[Fe^{IV}(OH)_2(^{H}pytacn)]^{2+}/R\cdot$ to $[Fe^{III}(OH)_2(^{H}pytacn)]^{2+}/R^{+}$ are always far larger than OH radical rebound barriers. Thus, after the hydrogen abstraction, the iron-bishydroxo intermediate goes to the alcohol products instead of evolving via an ET process. The kinetics of the HAT step is always energetically more demanding that for the rebound step, and because of that the former is always the rate-determining step.
3. The rebound with the OH ligand *trans* to NCH_2 is always the most favorable one but the rebound with the hydroxyl *trans* to NCH_3 is only less than 1 kcal mol^{-1} higher in energy. Therefore, these results should be taken with some caution since energy differences between both rebounds are within the limits of precision of the computational methodology.
4. For the oxidation of cyclohexane with $[Fe^{V}(O)(OH)(^{H}pytacn)]^{2+}$, hydroxyl rebounds barriers for the two —OH groups are nearly identical ($\Delta\Delta G^{\ddagger} < 0.1$ kcal mol^{-1}) in nice agreement with the labeling results. Unfortunately, an analogous analysis of the $[Fe^{V}(O)(OH)(^{Me}pytacn)]^{2+}$ catalysts is still not available.

13.2.5 Use of the $[Fe(OTf)_2(^{Me}pytacn)]$ Catalyst in Preparative C—H Oxidation Reactions

Despite the last decade has witnessed a large number of studies devoted to iron-catalyzed C=H oxidations, catalysts that can provide synthetically useful yields in stereospecific oxidation have been very rare. The seminal studies of White and Chen provided the first examples of iron catalysts that oxidize complex organic molecules in synthetically valuable yields, and provided a basic set of selectivity rules that introduced predictability in the site selectivity of the reactions [39, 40]. An important aspect of White's work was the discovery that acetic acid plays a very positive role in enhancing product yields and selectivities in C—H oxidation reactions.

Taking advantage of White's findings, the family of [Fe(pytacn)] catalysts were tested in substrate limiting conditions. Previous reports showed that [Fe(OTf)$_2$(^{H}pytacn)] and [Fe(OTf)$_2$(^{Me}pytacn)] provided excellent C—H oxidation activity under conditions of large excess of substrate [38], and because of that they were considered as promising catalyst candidates. Interestingly, [Fe(OTf)$_2$(^{Me}pytacn)] proved to be an excellent catalyst, providing product yields comparable to those described by White et al., but using lower catalyst loadings [58]. Reactions were stereospecific, and C—H site selectivity in oxidation reactions basically follow the basic rules described by White that include sterics, electronics, and stereoelectronic factors. Application to relatively complex molecules was demonstrated (Scheme 13.12).

A particularly interesting aspect of this catalyst is that it exhibits an enhanced preference toward secondary over tertiary C—H bonds, despite the later are weaker. The origin of this unusual selectivity is the steric encumbrance of the catalyst active center, which favor oxidation of the spatially less demanding secondary C—H bonds. An illustrative example of the changes in C—H selectivity observed for this catalyst, in comparison with other literature examples, are shown in Scheme 13.13. Oxidation of neomenthyl esters with White catalyst [Fe(pdp)(CH$_3$CN)$_2$]$^{2+}$ provides comparable amounts of oxygenated products arising from oxidation at tertiary site C3 and at methylenic site C2. Use of the pinene derived catalyst [Fe(OTf)$_2$(mcpp)] concentrates oxidation at C3 [59], while [Fe(OTf)$_2$(^{Me}pytacn)] oxidizes preferentially C2 [58].

13.2.6 C—H Oxidation with a Well-Defined Fe(IV) Complex

Oxoiron(IV) species are aliphatic C—H cleaving agents in a number of nonheme iron dependent enzymes [10]. Interestingly, [Fe(OTf)$_2$(^{H}pytacn)] is an excellent platform to access oxoiron(IV) complexes bearing a *cis*-labile site adjacent to the terminal oxo ligand [60]. Despite oxoiron(IV) complexes having been extensively studied during the last 15 years, few examples exist where the oxo ligand is *cis*- to a labile site susceptible to binding additional ligands [61–63]. This actually constitutes an interesting structural aspect to include in model compounds because this site plays a major role in defining the reactivity of some iron enzymes, non-heme iron-dependent halogenases being a prominent example [64]. In the latter case, a halide atom is bound, and it is eventually transferred to the substrate during the halogenation reaction.

Reaction of [Fe^{II}(CF$_3$SO$_3$)$_2$(^{H}pytacn)] with 2 equiv. of peracetic acid in CH$_3$CN at 15 °C produces [Fe^{IV}(O)(S)(^{H}pytacn)]$^{2+}$, S = CH$_3$CN or H$_2$O, which is unusually stable in solution, ($t_{1/2}$ ~ of 2.4 hours at 15 °C). The nature of these species could be elucidated by a combination of Mössbauer, ^{1}H-NMR spectroscopy, resonance Raman, ESI-MS spectrometry and DFT calculations. These analysis show that the complex bears an S = 1 Fe^{IV} complex.

[Fe^{IV}(O)(S)(^{H}pytacn)]$^{2+}$ exhibits reactivity aspects characteristic of nonheme iron oxygenases (Scheme 13.14). It exchanges its oxygen atom with water, as shown by ESI-MS monitoring of the reaction of the complex with H$_2$^{18}O. It also probes to be a powerful oxygen atom transfer agent to sulfides, providing the corresponding sulfoxide and [Fe^{II}(CH$_3$CN)$_2$(^{H}pytacn)]$^{2+}$. Most interestingly, the complex reacts with aliphatic C—H bonds, producing their oxidation. Reaction with 9,10-dihydroanthracene (DHA) at 258 K affords anthracene as the only detected product, and use of its deuterated analogue (d_4-DHA) reveals that the reaction proceeds with a large KIE = 27. In this reaction [Fe^{IV}(O)(S)(^{H}pytacn)]$^{2+}$ acts as a 1e^{-} oxidant and forms

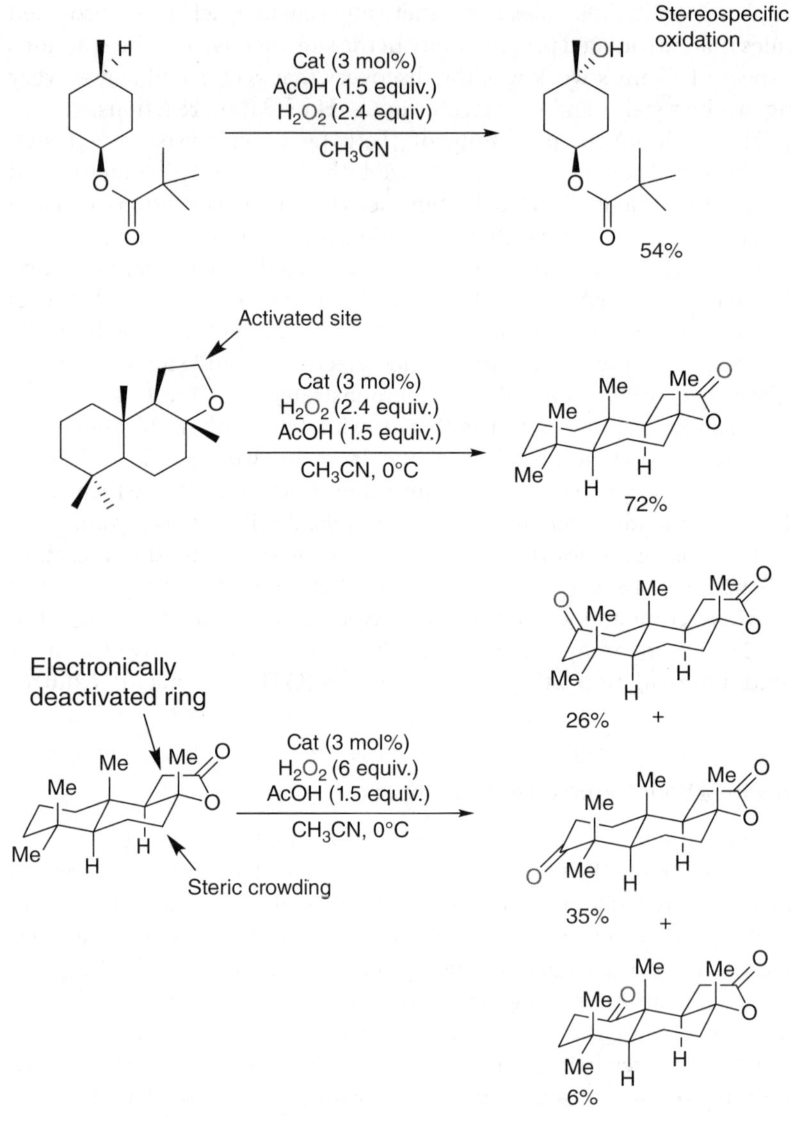

Scheme 13.12 Selected C—H oxidation reactions with the $[Fe(OTf)_2(^{Me}pytacn)]$ catalyst under substrate limiting conditions. (*See insert for color representation.*)

$[Fe^{III}(OH)(S)(^{H}pytacn)]^{2+}$. The complex can also oxidize substrates with stronger C—H bonds such as tetrahydrofuran (THF) (bond dissociation energy (BDE) = 93 kcal mol^{-1}) and even cyclohexane (BDE = 99.3 kcal mol^{-1}) although these reactions proceed at very slow rates and produce carbon-centered radicals that diffuse and are trapped by O_2.

Particularly interesting are also reactions with Cl^- and Br^-, leading to species $[Fe^{IV}(O)(X)(^{H}pytacn)]^+$, that can be regarded as models for non-heme iron-dependent halogenases [65] (Scheme 13.14). $[Fe^{IV}(O)(X)(^{H}pytacn)]^+$ performs hydrogen-atom abstraction of C—H bonds at reaction rates two to three times faster than the corresponding solvato dicationic species, thus modeling the first step in C—H functionalization taking place in natural halogenations. Reaction of these complexes with C—H bonds (R—H) entails an initial HAT, as also proposed for

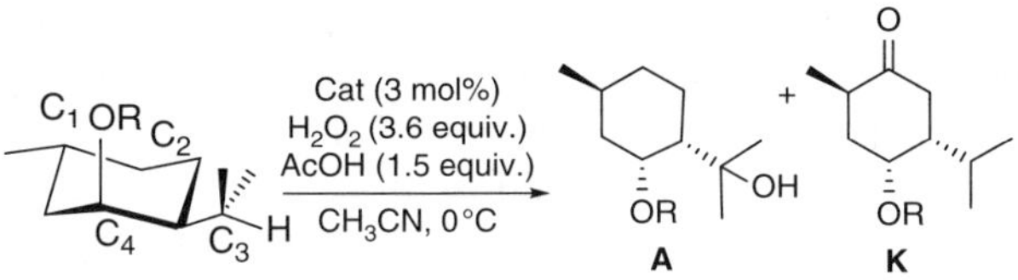

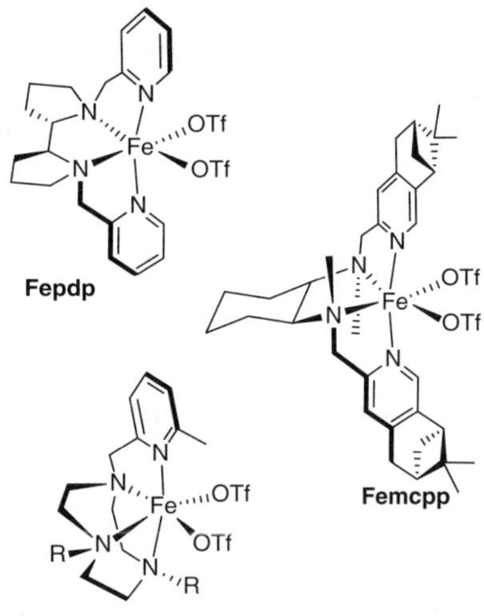

Catalyst	Conversion(%)	A(%)[c]	K(%)[c]	[A]:[K]
	R = Ac			
Fepdp[b]	66	22	22	**1 : 1**
Femcpp[b]	78	36	7	**5 : 1**
Fe^{Me}pytacn[a]	72 (15)[d]	15(14)[e]	33(32)[e]	**1 : 2**
	R = Piv			
Fepdp[b]	50	15	23	**1 : 1.5**
Femcpp[b]	71	43	11	**4 : 1**
Fe^{Me}pytacn[a]	62 (31)[d]	11 (11)[e]	30 (30)[e]	**1 : 3**
	R = $COCF_3$			
Fepdp[b]	61	13	20	**1 : 1.5**
Femcpp[b]	63	29	6	**5 : 1**
Fe^{Me}pytacn[a]	64	7	31	**1 : 4.5**

Fe^{Me}pytacn

[a] Cat:H_2O_2:substrate:AcOH 3:360:100:150. [b] Cat:H_2O_2:substrate:AcOH 3:200:100:150. [c] GC yield. [d] Recovered starting material. [e] Isolated yield.

Scheme 13.13 Regioselective C—H oxidation reactions catalyzed by different iron catalysts.

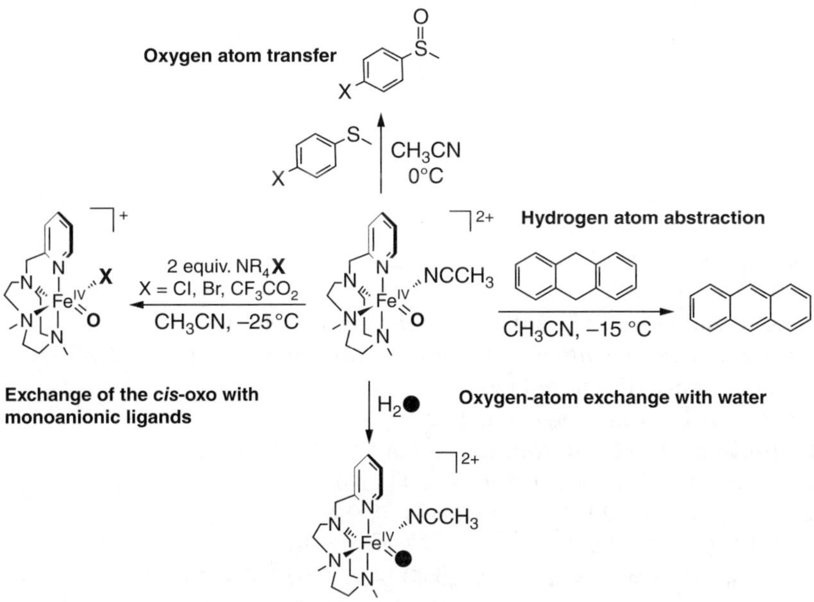

Scheme 13.14 Reactivity of $[Fe^{IV}(O)(pytacn)]^{2+}$.

non-heme iron-dependent halogenases. In this enzymatic reaction, a carbon-centered radical (R·) is generated in the substrate, concomitant with a $LFe^{III}(OH)(X)$ species. In halogenases, "rebound" of the substrate radical toward the Fe^{III}—X unit results in substrate halide (R—X) formation. However, $[Fe^{IV}(O)(X)(^{H}pytacn)]^{+}$ does not produce halogenated products. Rebound is not taking place, and carbon-centered radical intermediates diffuse in solution and react with other molecules such as adventitious O_2. Therefore, while $[Fe^{IV}(O)(X)(^{H}pytacn)]^{+}$ reproduce the key *cis*-Cl-Fe^{IV}=O unit of the active species of non-heme iron-dependent halogenases (albeit with a different spin state) and it can model the initial C—H bond activation step, subsequent halide transfer does not take place.

13.3 Concluding Remarks

Iron complexes bearing the tetradentate pytacn ligand are powerful C—H oxidation catalysts and have provided valuable mechanistic information to understand the mechanisms of stereospecific C—H hydroxylation. Their ability to stabilize iron centers in high oxidation states also constitutes a valuable characteristic of these ligands. Mechanistic analyses on this class of complexes have established that oxoiron(V) species are responsible for the stereoretentive C—H hydroxylation. The mechanism by which oxygen atoms from water are incorporated into products has been redefined on the basis of the experimental evidences that arise from the activity of [Fe(pytacn)] complexes. Water appears to be incorporated in the active species during the O—O lysis event. On the other hand, reactions with well-defined oxoiron(IV) complexes demonstrate a different mechanistic scenario, where an initial HAT is followed by diffusion of the carbon-centered radical.

Acknowledgments

We acknowledge financial support from MINECO of Spain (CTQ2015-70795-P) and the Catalan DIUE of the Generalitat de Catalunya (2009SGR637). M.C. thanks an ICREA-Academia award.

References

1 Newhouse, T. and Baran, P.S. (2011). *Angew. Chem. Int. Ed.* 50: 3362.
2 White, M.C. (2012). *Science* 335: 807.
3 McMurray, L., O'Hara, F., and Gaunt, M. (2011). *J. Chem. Soc. Rev.* 40: 1885.
4 Sheldon, R.A. (2000). *Biomimetic Oxidations Catalyzed by Transition Metal Complexes* (ed. B. Meunier), 613. London: Imperial College Press.
5 Ortiz de Montellano, P.R. (2010). *Chem. Rev.* 110: 932.
6 Kovaleva, E.G. and Lipscomb, J.D. (2008). *Nat. Chem. Biol.* 4: 186.
7 Tinberg, C.E. and Lippard, S.J. (2011). *Acc. Chem. Res.* 44: 280.
8 Shaik, S., Cohen, S., Wang, Y. et al. (2010). *Chem. Rev.* 110: 949.
9 Huang, X. and Groves, J.T. (2017). *JBIC J. Biol. Inorg. Chem.* 22: 185.
10 Krebs, C., Galonić Fujimori, D., Walsh, C.T. et al. (2007). *Acc. Chem. Res.* 40: 484.
11 Price, J.C., Barr, E.W., Tirupati, B. et al. (2003). *Biochemistry* 42: 7497.
12 Hoffart, L.M., Barr, E.W., Guyer, R.B. et al. (2006). *Proc. Natl. Acad. Sci. U.S.A.* 103: 14738.
13 Eser, B.E., Barr, E.W., Frantom, P.A. et al. (2007). *J. Am. Chem. Soc.* 129: 11334.

14 Panay, A.J., Lee, M., Krebs, C. et al. (2011). *Biochemistry* 50: 1928.
15 Fujimori, D.G., Barr, E.W., Matthews, M.L. et al. (2007). *J. Am. Chem. Soc.* 129: 13408.
16 Galonić, D.P., Barr, E.W., Walsh, C.T. et al. (2007). *Nat. Chem. Biol.* 3: 113.
17 Matthews, M.L., Neumann, C.S., Miles, L.A. et al. (2009). *Proc. Natl. Acad. Sci. U.S.A.* 106: 17723.
18 Barry, S.M. and Challis, G.L. (2013). *ACS Catal.* 3: 2362.
19 Wolfe, M.D., Parales, J.V., Gibson, D.T. et al. (2001). *J. Biol. Chem.* 276: 1945.
20 Karlsson, A., Parales, J.V., Parales, R.E. et al. (2003). *Science* 299: 1039.
21 Chakrabarty, S., Austin, R.N., Deng, D. et al. (2007). *J. Am. Chem. Soc.* 129: 3514.
22 Hegg, E.L. and Lawrence Que, Jr. (1997). *Eur. J. Biochem.* 250: 625.
23 Solomon, E.I., Brunold, T.C., Davis, M.I. et al. (2000). *Chem. Rev.* 100: 235.
24 Que, L., Jr. and Tolman, W.B. (2008). *Nature* 455: 8.
25 Olivo, G., Cussó, O., Borrell, M., and Costas, M. (2017). *JBIC J. Biol. Inorg. Chem.* 22: 425.
26 Olivo, G., Cusso, O., and Costas, M. (2016). *Chem. Asian J.* 11: 3148.
27 Oloo, W.N. and Que, L., Jr (2015). *Acc. Chem. Res.* 48: 2612.
28 Bryliakov, K.P. and Talsi, E.P. (2014). *Coord. Chem. Rev.* 276: 73.
29 Talsi, E.P. and Bryliakov, K.P. (2012). *Coord. Chem. Rev.* 256: 1418.
30 Costas, M., Chen, K., and Que, L. Jr. (2000). *Coord. Chem. Rev.* 200-202: 517.
31 Perkins, M. (1996). *J. Chem. Soc. Rev.* 25: 229.
32 Barton, D.H.R. and Doller, D. (1992). *Acc. Chem. Res.* 25: 504.
33 Walling, C. (1998). *Acc. Chem. Res.* 31: 155.
34 Ingold, K.U. and MacFaul, P.A. (2000). *Biomimetic Oxidations Catalyzed by Transition Metal Complexes* (ed. B. Meunier), 45. London: Imperial College Press.
35 Stavropoulos, P., Celenligil-Cetin, R., and Tapper, A.E. (2001). *Acc. Chem. Res.* 34: 745.
36 Chen, K. and Que, L. Jr. (2001). *J. Am. Chem. Soc.* 123: 6327.
37 Chen, K. and Que, L. Jr. (1999). *Chem. Commun.* 1375.
38 Company, A., Gómez, L., Fontrodona, X. et al. (2008). *Chem. Eur. J.* 14: 5727.
39 Chen, M.S. and White, M.C. (2010). *Science* 327: 566.
40 Chen, M.S. and White, M.C. (2007). *Science* 318: 783.
41 Company, A., Gómez, L., Güell, M. et al. (2007). *J. Am. Chem. Soc.* 129: 15766.
42 Wieghardt, K., Bossek, U., Nuber, B. et al. (1988). *J. Am. Chem. Soc.* 110: 7398.
43 Halfen, J.A., Mahapatra, S., Wilkinson, E.C. et al. (1996). *Science (New York, N.Y.)* 271: 1397.
44 Padamati, S.K., Angelone, D., Draksharapu, A. et al. (2017). *J. Am. Chem. Soc.* 139: 8718.
45 Prat, I., Company, A., Postils, V. et al. (2013). *Chem. Eur. J.* 19: 6724.
46 Garcia-Bosch, I., Codolà, Z., Prat, I. et al. (2012). *Chem. Eur. J.* 18: 13269.
47 Kim, C., Chen, K., Kim, J. et al. (1997). *J. Am. Chem. Soc.* 119: 5964.
48 Chen, K., Costas, M., Kim, J. et al. (2002). *J. Am. Chem. Soc.* 124: 3026.
49 Oloo, W.N., Fielding, A.J., and Que, L. Jr. (2013). *J. Am. Chem. Soc.* 6438.
50 Prat, I., Mathieson, J.S., Güell, M. et al. (2011). *Nat. Chem.* 3: 788.
51 Bernadou, J. and Meunier, B. (1998). *Chem. Commun.* 2167.
52 Prat, I., Company, A., Corona, T. et al. (2013). *Inorg. Chem.* 52: 9229–9244.
53 Mitra, M., Lloret-Fillol, J., Haukka, M. et al. (2014). *Chem. Commun.* 50: 1408.
54 Postils, V., Company, A., Solà, M. et al. (2015). *Inorg. Chem.* 54: 8223.
55 Bassan, A., Blomberg, M.R.A., Siegbahn, P.E.M. et al. (2002). *J. Am. Chem. Soc.* 124: 11056.
56 Mikhalyova, E.A., Makhlynets, O.V., Palluccio, T.D. et al. (2012). *Chem. Commun.* 48: 687.
57 Makhlynets, O.V. and Rybak-Akimova, E.V. (2010). *Chem. Eur. J.* 16: 13995.
58 Prat, I., Gomez, L., Canta, M. et al. (2013). *Chemistry* (Weinheim an der Bergstrasse, Germany) 19: 1908.
59 Gomez, L., Garcia-Bosch, I., Company, A. et al. (2009). *Angew. Chem. Int. Ed.* 48: 5720.

60 Company, A., Prat, I., Frisch, J.R. et al. (2011). *Chem. Eur. J.* 17: 1622.
61 Nam, W. (2015). *Acc. Chem. Res.* 48: 2169.
62 Ray, K., Pfaff, F.F., Wang, B. et al. (2014). *J. Am. Chem. Soc.* 136: 13942.
63 McDonald, A.R. and Que, L. Jr. (2013). *Coord. Chem. Rev.* 257: 414.
64 Wong, S.D., Srnec, M., Matthews, M.L. et al. (2013). *Nature* 499: 320.
65 Planas, O., Clemancey, M., Latour, J.-M. et al. (2014). *Chem. Commun.* 50: 10887.

14

The Nature of High-Valent Oxometal Intermediates of Iron-Aminopyridine Mediated Oxidations

Oleg Y. Lyakin[1,2], *Konstantin P. Bryliakov*[1,2], *and Evgenii P. Talsi*[1,2]

[1] *Novosibirsk State University, Novosibirsk, Russian Federation*
[2] *Boreskov Institute of Catalysis, Novosibirsk, Russian Federation*

14.1 Introduction

In nature, the oxidation of hydrocarbons is mediated by heme and non-heme metalloenzymes. The high selectivity of iron-containing enzymatic systems has inspired the search for synthetic Fe-based catalyst systems capable of conducting such transformations. Asymmetric epoxidation, which plays important role in modern organic synthesis because of the high synthetic value of chiral epoxides, is a particularly attractive transformation for this bioinspired oxidation strategy. In the last 20 years, significant progress has been achieved in the chemo-, regio-, and enantioselective oxidations of olefins, with H_2O_2 as a green terminal oxidant. Iron complexes with chiral N_4-donor aminopyridine ligands (of the type **1**–**5**), possessing two *cis*-α-oriented labile coordination sites, available for peroxide activation, constitute the most stereoselective iron-based family of epoxidation catalysts (Chart 14.1).

In the previous decade, Jacobsen, Que, and co-workers reported the first examples of the preparatively useful epoxidations of various olefins with up to 100% GC and 90% isolated epoxide yield, based on complexes **1** and **2** as catalysts, using H_2O_2 as the oxidant, and acetic acid as catalytic additive [1, 2]. Another milestone result was the synthesis and application of the first catalyst of Fe(PDP) family ($\mathbf{4^{SbF6}}$) in bioinspired oxidation catalysis by White and coworkers [3, 4]. The catalyst system $\mathbf{4^{SbF6}}/H_2O_2/CH_3COOH$ was shown to selectively oxidize sp^3 and methylene C—H groups of complex organic molecules and natural products with considerable level of predictability and in preparatively relevant yields.

Subsequently, Bryliakov, Talsi, and co-workers proposed catalyst system [((*S*,*S*)-PDP)Fe $(CF_3SO_3)_2$](**4**)/H_2O_2/RCOOH (Chart 14.1) capable of epoxidizing prochiral electron-deficient olefins with high enantioselectivity. The enantioselectivity was found to increase upon the use of branched carboxylic acids (such as 2-ethylhexanoic acid) instead of acetic acid [5]. As a complementary approach, Costas and co-workers showed that the yields and enantioselectivities of epoxidations, catalyzed by aminopyridine-iron complexes of the PDP family, could be improved by introducing electron-donating substituents at the pyridine rings of the PDP ligand [6].

Originally, both the approaches to the tuning of the epoxidation enantioselectivities have remained purely phenomenological. It is only recently that the mechanistic details of iron-aminopyridine-based catalyst systems for chemo-, regio-, and stereoselective C—H and C=C oxidations have been uncovered, thus providing the relationship between the nature of the oxidants, additives, and ligand substituents, on the one hand, and the reactivity and

Alkane Functionalization, First Edition. Edited by Armando J. L. Pombeiro and M. Fátima C. Guedes da Silva.

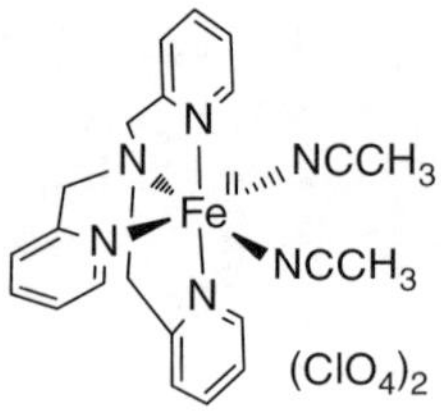

Fe^{II}(TPA)
1

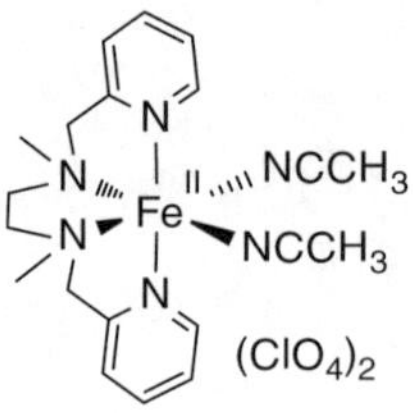

Fe^{II}(BPMEN)
2

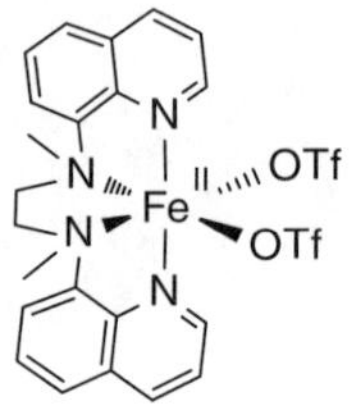

Fe^{II}(BQEN)
3

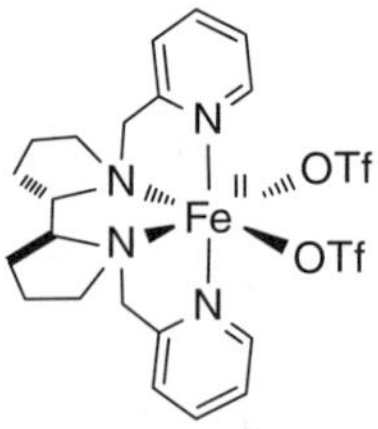

Fe^{II}((*S*,*S*)-PDP)
4

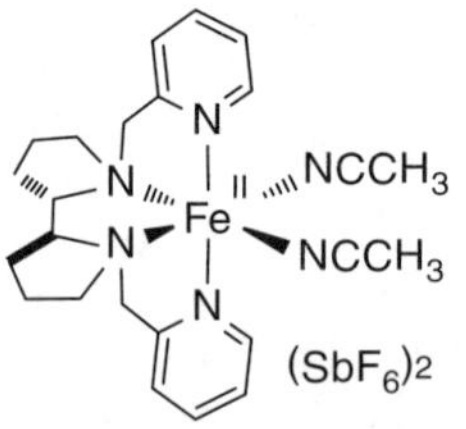

4^{SbF6}

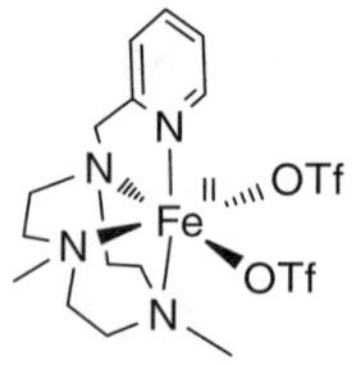

Fe^{II}(Me_2PyTACN)
5

Chart 14.1 Examples of non-heme iron complexes – efficient and selective catalysts for oxidation of alkenes and alkanes with H_2O_2.

stereoselectivity, on the other hand. The essential details of those studies are summarized in this chapter.

14.2 Oxoiron Active Species of Olefin Epoxidations

14.2.1 Oxoiron Active Species with High *g*-Factor Anisotropy Formed in the Catalyst Systems Based on Nonsubstituted Aminopyridine Iron Complexes

Selective oxidation of organic molecules is an important step of many biological and industrial processes. For heme enzymes such as cytochrome P450, formally $Fe^{V}{=}O$ species (($P^{\bullet+}$)$Fe^{IV}{=}O$, $P^{\bullet+}$ = porphyrin cation radical) has been accepted as the active intermediate of oxidation [7–11]. For non-heme iron enzymes such as *α*-KG-dependent dioxygenases, the high-spin ($S = 2$) $Fe^{IV}{=}O$ intermediate is considered to be responsible for the selective oxidation [12].

One can expect that the catalytically active species, responsible for the oxidation of organic substrates by synthetic biomimetic catalyst systems, can be similar to those of the metalloenzyme-based systems (i.e. $Fe^{IV}{=}O$ or $Fe^{V}{=}O$ species). However, until now, for the existing synthetic non-heme catalyst systems capable of preparative epoxidation of olefins with H_2O_2, only $Fe^{V}{=}O$ species have been considered as the active species of epoxidation, whereas $Fe^{IV}{=}O$ complexes were shown to be responsible for side reactions, deteriorating the epoxidation selectivity [13, 14].

The first synthesis of an oxoiron(V) complex [(TAML)$Fe^{V}{=}O$]$^-$ (**6c**, Chart 14.2) was reported in 2007 [15]. The reaction of [PPh_4][(TAML)Fe^{III}(H_2O)] with *m*-chloroperoxybenzoic acid (*m*-CPBA) in *n*-butyronitrile at −60 °C yielded a deep green complex identified as

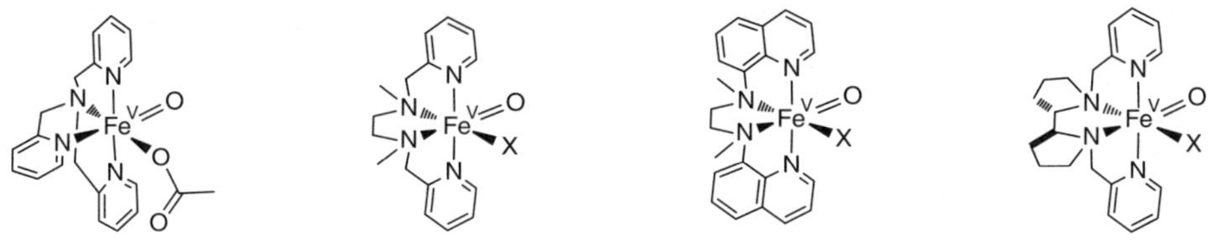

$[(TPA)Fe^V{=}O(OAc)]^{2+}$ **1c**

$[(BPMEN)Fe^V{=}O(X)]^{2+}$ **2c**

$[(BQEN)Fe^V{=}O(X)]^{2+}$ **3c**

$[((S,S)\text{-}PDP)Fe^V{=}O(X)]^{2+}$ **4c**

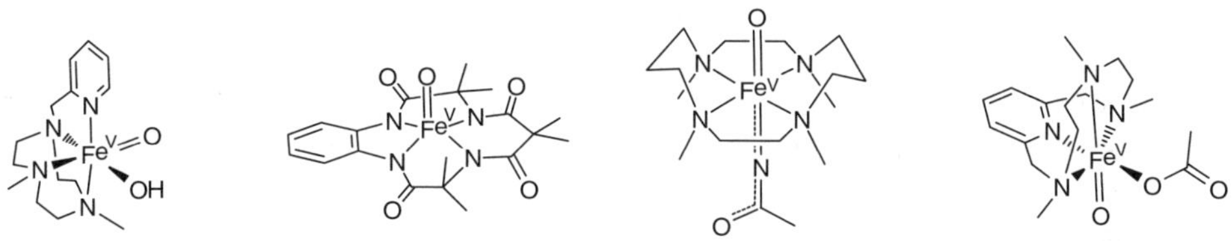

$[(Me_2PyTACN)Fe^V{=}O(OH)]^{2+}$ **5c**

$[(TAML)Fe^V{=}O]^-$ **6c**

$[(TMC)Fe^V{=}O(NC(O)CH_3)]^+$ **7c**

$[(PyNMe_3)Fe^V{=}O(OAc)]^{2+}$ **8c**

Chart 14.2 Putative iron-oxo intermediates **1c–4c** and the spectroscopically characterized iron-oxo species **5c–8c**.

$[(TAML)Fe^V{=}O]^-$ on the basis of Mössbauer, EPR, EXAFS, and ESI-MS studies. This low-spin ($S = 1/2$) complex had rhombically anisotropic EPR spectrum ($g_1 = 1.99$, $g_2 = 1.97$, $g_3 = 1.74$), and demonstrated the capability of oxidizing a variety of substrates, such as thioanisole, styrene, cyclooctene, ethylbenzene, 9,10-dihydroanthracene. Inspired by this discovery, we undertook the search for iron-oxygen intermediates in the catalyst systems **1–5**/H_2O_2. EPR spectroscopy was suitable for the detection of the $S = 1/2$ species, e.g. Fe^{III}—OOH, Fe^{III}—OH, and putative $Fe^V{=}O$ intermediates, and ^1H and ^2H NMR spectroscopy was used to observe the EPR silent $Fe^{IV}{=}O$ intermediates [5, 16–18].

Three types of unstable iron-oxygen intermediates (**2a**, **2b**, and **2c**) have been observed in the catalyst system **2**/H_2O_2 by EPR spectroscopy at low temperature (−70 °C to −60 °C) (Table 14.1). It is of note that the observed concentrations of **2a** and **2c** increase with an increase in the [H_2O_2]:[**2**] ratio.

The EPR spectrum of **2a** is a superposition of the spectra of two species: **2a-CH$_3$CN** ($g_1 = 2.218$, $g_2 = 2.175$, $g_3 = 1.966$) and **2a-H$_2$O** ($g_1 = 2.197$, $g_2 = 2.128$, $g_3 = 1.970$) (Table 14.1). **2a-CH$_3$CN** and **2a-H$_2$O** were assigned to hydroperoxo complexes $[(BPMEN)Fe^{III}{—}OOH(CH_3CN)]^{2+}$ and $[(BPMEN)Fe^{III}{—}OOH(H_2O)]^{2+}$, respectively [17]. ESI-MS data of Rybak-Akimova and Makhlynets have confirmed this assignment [24]. Complexes **2a-CH$_3$CN** and **2a-H$_2$O** do not directly react with cyclohexene. The rate of their self-decay at −70 °C does not change in the presence of cyclohexene. Complex **2b** ($g_1 = 2.43$, $g_2 = 2.21$, $g_3 = 1.91$) was assigned to ferric hydroxo complex $[(BPMEN)Fe^{III}(OH)(S)]^{2+}$ (S = H_2O or CH_3CN), since its EPR parameters are close to those of known ferric hydroxo complexes with N_4-donor ligands [25, 26].

Importantly, in contrast to **2a** and **2b**, complex **2c** reacted with cyclohexene at −70 °C, affording cyclohexene epoxide. A clear connection between the rate of decay of **2c** and the rate of cyclohexene epoxidation by the catalyst system **2**/H_2O_2 was documented [16]. So, it is reasonable to assign intermediate **2c** to the active species of epoxidation. Intermediate **1c** with similar EPR spectrum was observed in the catalyst system **1**/H_2O_2/CH_3COOH (Table 14.1). The rate of decay of **1c** increased with increasing cyclohexene concentration [16].

Table 14.1 EPR spectroscopic data for $S = 1/2$ iron-oxygen species with N_4-donor tetradentate ligands.

No	Species	g_1	g_2	g_3	Ref.
1	$[(BPMEN)Fe^{III}—OOH(CH_3CN)]^{2+}$ (**2a-CH$_3$CN**)	2.218	2.175	1.966	[17]
2	$[(BPMEN)Fe^{III}—OOH(H_2O)]^{2+}$ (**2a-H$_2$O**)	2.197	2.128	1.970	[17]
3	$[((S,S)\text{-}PDP)Fe^{III}—OOH(CH_3CN)]^{2+}$ (**4a-CH$_3$CN**)	2.206	2.171	1.955	[5]
4	$[((S,S)\text{-}PDP)Fe^{III}—OOH(H_2O)]^{2+}$ (**4a-H$_2$O**)	2.191	2.124	1.963	[5]
5	$[(TPA^*)Fe^{III}—OOH(CH_3CN)]^{2+}$ (**1a*-CH$_3$CN**)	2.201	2.163	1.960	[19]
6	$[(BPMEN)Fe^{III}—OH(S)]^{2+}$ (**2b**)$^{a)}$	2.43	2.21	1.91	[17]
7	$[((S,S)\text{-}PDP)Fe^{III}—OH(S)]^{2+}$ (**4b**)$^{a)}$	2.44	2.21	1.89	[5]
8	$[(TPA)Fe^{V}{=}O(OC(O)CH_3)]^{2+}$ (**1c**)	2.71	2.42	1.53	[16, 17]
9	$[(BPMEN)Fe^{V}{=}O(X)]^{2+}$ (**2c**)$^{b)}$	2.69	2.42	1.70	[16, 17]
10	$[(BQEN)Fe^{V}{=}O(X)]^{2+}$ (**3c**)$^{b)}$	2.57	2.41	1.76	[16]
11	$[((S,S)\text{-}PDP)Fe^{V}{=}O(X)]^{2+}$ (**4c**)$^{b)}$	2.66	2.42	1.71	[5]
12	$[(Me_2PyTACN)Fe^{V}{=}O(OH)]^{2+}$ (**5c**)	2.66	2.43	1.74	[18]
13	$[(TAML)Fe^{V}{=}O]^{-}$ (**6c**)	1.99	1.97	1.74	[15]
14	$[(TMC)Fe^{V}{=}O(NC(O)CH_3)]^{+}$ or $[(TMC)Fe^{IV}{=}O(^{+\bullet}NC(O)CH_3)]^{+}$ (**7c**)	2.053	2.010	1.971	[20]
15	$[(PyNMe_3)Fe^{V}{=}O(OC(O)CH_3)]^{2+}$ or $[(PyNMe_3)^{\bullet+}Fe^{IV}{=}O(OC(O)CH_3)]^{2+}$ (**8c**)	2.07	2.01	1.95	[21]
16	$[(TPA^*)^{\bullet+}Fe^{IV}{=}O(OC(O)CH_3)]^{2+}$ (**1c*^{AA}**)	2.070	2.005	1.956	[19]
17	$[((S,S)\text{-}PDP^*)^{\bullet+}Fe^{IV}{=}O(OC(O)CH_3)]^{2+}$ (**4c*^{AA}**)	2.071	2.008	1.960	[19]
18	$[(TPA^*)Fe^{III}(\eta^2\text{-}OOC(O)CH_3)]^{2+}$ (**X**)	2.58	2.38	1.72	[22]
19	$[(TPA^*)Fe^{III}(\eta^2\text{-}OC(O)CH_3)]^{2+}$ (**1e***)	2.58	2.38	1.72	[19]
20	$[((S,S)\text{-}PDP^*)Fe^{III}(\eta^2\text{-}OC(O)CH_3)]^{2+}$ (**4e***)	2.54	2.41	1.79	[19]
21	$[((S,S)\text{-}PDP^*)Fe^{III}(\eta^2\text{-}OOC(O)CH_3)]^{2+}$ (**4x***)	2.52	2.41	1.80	[23]
22	$[((S,S)\text{-}PDP^*)Fe^{III}(\eta^2\text{-}OOC(O)C_6H_4\text{–}3\text{-}Cl)]^{2+}$ (**4x****)	2.52	2.41	1.80	[23]

a) $S = CH_3CN$ or H_2O.
b) $X = OH$ or $OC(O)CH_3$.

In 2011, the detection of the elusive $HO—Fe^{V}{=}O$ intermediate in the catalyst system **5**/H_2O_2, was reported, using cryospray-assisted variable-temperature mass-spectrometry (VT-MS) [27]. The presence of the terminal metal-oxo moiety was confirmed by isotopic (^{18}O) labeling experiments, and the reactivity of the intermediate was confirmed by the *in-situ* observation of the reaction with cyclooctene. The same system was subsequently studied by EPR spectroscopy, and the reactive species was found to be a $S = 1/2$ complex **5c** with EPR parameters close to those of the intermediates **1c** and **2c** (Table 14.1) [18]. Intermediate **5c**, which directly reacted with cyclohexene at −60 °C, can be plausibly assigned to the $HO—Fe^{V}{=}O$ species observed by Costas, Cronin, and coworkers [27]. These data support the assignment of intermediates **1c**, **2c**, and **5c** to the low-spin oxoiron(V) species.

The catalyst system **2**/H_2O_2/CH_3COOH was the first highly efficient non-heme iron-based system, conducting olefin epoxidation with synthetically useful yields [1]. Another milestone result was the synthesis and application of the catalyst $\mathbf{4}^{\mathbf{SbF6}}$ (Chart 14.1). The catalyst system

$4^{SbF6}/H_2O_2/CH_3COOH$ is capable of regioselective and stereospecific oxidation of aliphatic C—H groups in complex organic molecules, with selectivities in some cases approaching those of enzymatic reactions [3, 4].

The EPR spectrum of the sample $4/H_2O_2 = 1:2.5$, frozen one minute after mixing the reagents at −70 °C, displays resonances of unstable hydroperoxo complexes **4a-CH$_3$CN** and **4a-H$_2$O** (*g*-values in the range 1.96–2.20), and resonances in the range of $g = 4.2$–8 from unidentified high-spin ($S = 5/2$) ferric species, stable up to room temperature (Figure 14.1a, Table 14.1). The increase of the [H_2O_2]:[**4**] ratio in the sample **4**/H_2O_2 results in the rise of the concentration of **4a** and in the appearance of resonances of an unstable low-spin ($S = 1/2$) iron-oxygen species **4c** (Figure 14.1b, Table 14.1). The EPR parameters of **4c** are very close to those of the active intermediate of epoxidation **2c**. The situation changes abruptly in the presence of acetic acid. The EPR spectrum of the sample **4**/H_2O_2/CH_3COOH = 1 : 2.5 : 15 (Figure 14.1c) shows predominantly the resonances of **4c**. Thus, one can conclude that acetic acid facilitates the conversion of iron(III)-hydroperoxo complex **4a** to the active species of epoxidation **4c** [5].

We have shown that chiral complex **4** catalyzes the chemo- and enantioselective epoxidation of chalcone with H_2O_2 at −30 °C. Remarkably, the high yields of epoxide (90–100%) were observed only in the presence of carboxylic acid, and the epoxidation enantioselectivity (i.e. the enantiomeric excess value, *ee*) increased monotonously from 65% to 86% in line with rising steric bulk of the carboxylic acid [5]. Apparently, the role of carboxylic acid additive is dual: it assists the formation of the active species and affects the epoxidation enantioselectivity; obviously, the acid molecule is incorporated into the active species at the enantioselectivity-determining step, presumably acting as an auxiliary ligand. The active species is most plausibly represented as $[((S,S)\text{-PDP})Fe^{V}{=}O(OC(O)R)]^{2+}$ (**4c**, Chart 14.2) [5]. A similar active species $[(TPA)Fe^{V}{=}O(OC(O)CH_3)]^{2+}$ (**1c**) was previously proposed for the catalyst system **1**/H_2O_2/CH_3COOH on the basis of the catalytic and ^{18}O-labeling studies [2].

We directly compared the reactivity of the putative $Fe^{V}{=}O$ and $Fe^{IV}{=}O$ intermediates in the catalyst systems studied. It was found that the catalyst system **2**/PhIO displays characteristic

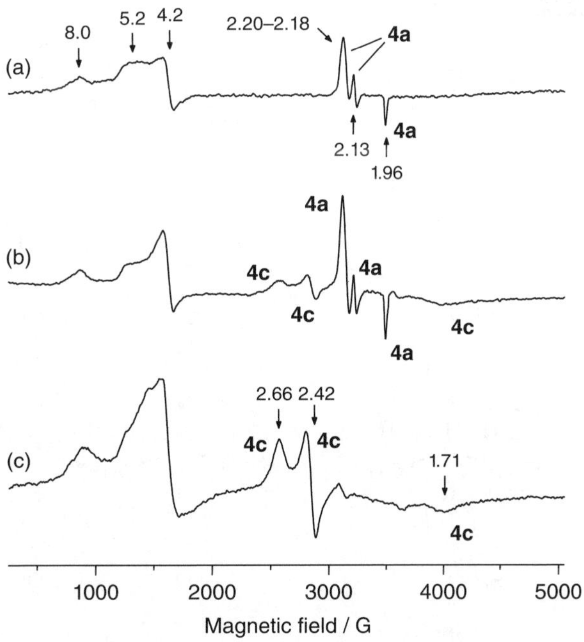

Figure 14.1 EPR spectra (−196 °C) of the samples (a) **4**/H_2O_2 ([**4**]:[H_2O_2] = 1 : 2.5), (b) **4**/H_2O_2 ([**4**]:[H_2O_2] = 1 : 10) and (c) **4**/H_2O_2/CH_3COOH ([**4**]:[H_2O_2]:[CH_3COOH] = 1 : 2.5 : 15) frozen one minute after mixing the reagents at −70 °C. [**4**] = 0.05 M. A 1.7 : 1 CH_2Cl_2/CH_3CN mixture was used as a solvent.

^{1}H NMR peaks of oxoiron(IV) intermediate [(BPMEN)Fe^{IV}=O]$^{2+}$ (**2d**) in the range from 50 to −200 ppm, and no EPR resonances from the putative oxoiron(V) intermediate **2c**. Contrariwise, catalyst system **2**/H_2O_2 showed resonances from **2c** but no signals from **2d** [17]. It was demonstrated that **2c** is much more reactive toward cyclohexene at −70 °C compared with **2d**. The half-life time of **2c** decreased from 20 to 2 minutes in the presence of 12 equiv. of cyclohexene at −70 °C, whereas the half-life time of **2d** (~2 hour) decreased to ~1 hour under the same conditions. Cyclohexene oxide was the major oxidation product of the catalyst system **2**/H_2O_2/C_6H_{10}, while mostly allylic oxidation products (i.e. 2-cyclohexen-1-one and 2-cyclohexen-1-ol) were observed in the catalysts system **2**/PhIO/C_6H_{10} [17]. We can conclude that **2c** is able to selectively epoxidize cyclohexene, whereas **2d** mainly drives its allylic oxidation. Reactivity studies of the well characterized intermediates [(TPA)Fe^{IV}=O]$^{2+}$ (**1d**) [2], [(BQEN)Fe^{IV}=O]$^{2+}$ (**3d**) [28] and [(Me_2PyTACN)Fe^{IV}=O]$^{2+}$ (**5d**) [29] also showed that the Fe^{IV}=O species cannot conduct the selective epoxidation of olefins. Therefore, oxoiron(V) intermediates are the likely active species of the catalyst systems based on complexes **1**–**5** and H_2O_2. We have assumed the highly unstable and reactive complexes **1c**–**5c** (Chart 14.2) to be these particular intermediates.

However, the EPR parameters of **1c**–**5c** drastically differ from those of oxoiron(V) complexes [(TAML)Fe^V=O]$^-$ (**6c**) and [(TMC)Fe^V=O(NC(O)CH_3)]$^+$ (**7c**), previously characterized by various spectroscopic techniques (Table 14.1, Chart 14.2) [15, 20]. Confused by this apparent discrepancy, Wang et al. hypothesized that **1c**–**5c** were ferric-peracetate complexes Fe^{III}—OOC(O)CH_3 rather than iron-oxo complexes [30]. In 2014, Oloo et al. reported the detection of the intermediate **X** (g_1 = 2.58, g_2 = 2.38, g_3 = 1.72) in the catalyst system **1**^{M*}/H_2O_2/CH_3COOH in acetonitrile at −40 °C [22]. In contrast to complex **1** with nonsubstituted TPA ligand, the TPA* ligand of mononuclear complex **1**^{M*} contains electron-donating substituents at the pyridine rings (Chart 14.3). By a combination of EPR, Mössbauer, resonance Raman and ESI-MS methods, **X** was identified as a low-spin (S = 1/2) acylperoxoiron(III)

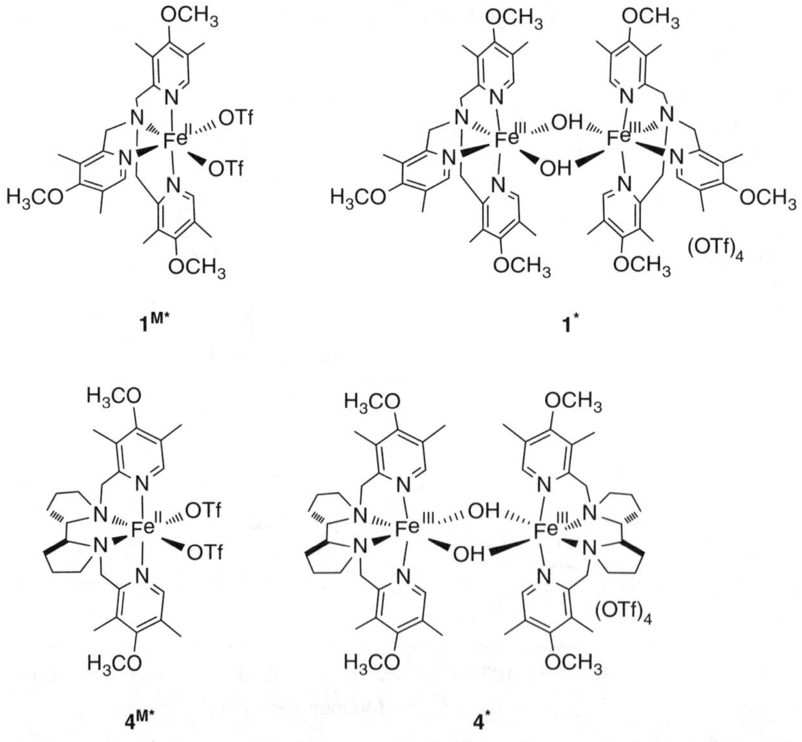

Chart 14.3 Mononuclear and dinuclear aminopyridine iron complexes with substituted pyridine rings.

species $[(TPA^*)Fe^{III}(\kappa^2\text{-}OOC(O)CH_3)]^{2+}$. The rate of decay of complex **X** at −40 °C was unaffected by the presence of cyclooctene, 1-octene, or 2-heptene in the reaction solution, ruling out the assignment of **X** to the actual oxygen-transferring species [22].

The *g*-values of **X** resemble those for complexes **1c**–**5c** (Table 14.1). However, the assignment of **1c**–**5c** to ferric-peracetate complexes $Fe^{III}\text{-}OOC(O)CH_3$ similar to **X** would be incorrect. Species **2c**, **4c**, and **5c** can also be observed in the catalyst systems **2**, **4**, **5**/H_2O_2, without any additives of CH_3COOH [5, 16–18], and thus cannot be iron(III)-acylperoxo species. The rate of decay of **2c** increased in the presence of cyclohexene at −70 °C [16], whereas the rate of decay for complex **X** at −40 °C was unaffected by the addition of olefins [22]. To shed light on the possible reason of the sharp difference of the EPR parameters of **1c**–**5c** and of well spectroscopically characterized $Fe^V{=}O$ species (Table 14.1), we have undertaken the search for the true active species in the catalyst systems **1***/H_2O_2/CH_3COOH and **4***/H_2O_2/CH_3COOH. More stable and oxygen-resistant dinuclear ferric complexes **1*** and **4*** can be used instead of the corresponding mononuclear ferrous complexes $\mathbf{1^{M*}}$ and $\mathbf{4^{M*}}$ as pre-catalysts (Chart 14.3), since the first 0.5 equiv. of H_2O_2 rapidly converts $\mathbf{1^{M*}}$ and $\mathbf{4^{M*}}$ into **1*** and **4***, respectively [19].

14.2.2 Oxoiron Active Species with Low *g*-Factor Anisotropy Formed in the Catalyst Systems Based on Substituted Aminopyridine Iron Complexes

Dinuclear complex **1*** is EPR silent. Just after the addition of 50 equiv. of CH_3COOH to the solution of **1*** at room temperature and freezing the sample at −196 °C, the EPR signal of a new ferric complex with proposed structure $[(TPA^*)Fe^{III}(\kappa^2\text{-}OC(O)CH_3)]^{2+}$ (**1e***) is observed (Figure 14.2a). This complex displays resonances at $g_1 = 2.58$, $g_2 = 2.38$, and $g_3 = 1.72$ [19]. Related complexes $[(TPA)Fe^{III}(acac)]^{2+}$ and $[(5\text{-}Me_3TPA)Fe^{III}(acac)]^{2+}$ (Hacac = acetylacetone)

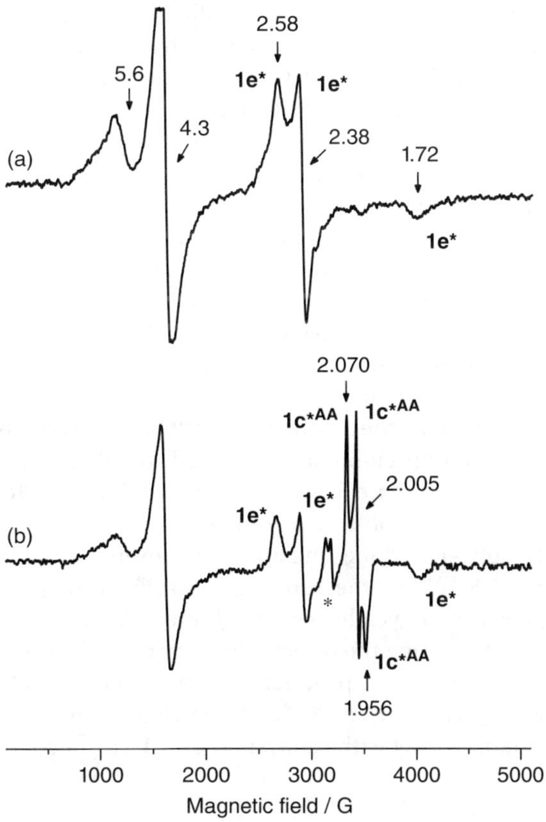

Figure 14.2 EPR spectra (−196 °C) of the sample (a) **1***/CH_3COOH ([CH_3COOH]:[**1***] = 50, [**1***] = 0.05 M) frozen one minute after mixing the reagents at room temperature in a 1.2 : 1 CH_2Cl_2/CH_3CN mixture, and the sample (b) **1***/H_2O_2/CH_3COOH ([**5**]:[H_2O_2]:[CH_3COOH] = 1 : 6 : 20, [**1***] = 0.02 M) frozen 1.5 minutes after mixing the reagents at −80 °C in a 1.8 : 1 CH_2Cl_2/CH_3CN mixture. Signals denoted by an asterisk belong to iron(III)-hydroperoxo complex $[(TPA^*)Fe^{III}\text{–}OOH(CH_3CN)]^{2+}$ (**1a***-CH_3CN, Table 14.1). Intense resonances at $g = 5.6$ and $g = 4.3$ belong to unidentified stable high-spin ($S = 5/2$) ferric species.

showed very similar EPR spectra ($g_1 = 2.57, g_2 = 2.35, g_3 = 1.71$) [31]. The EPR spectrum of **1e*** disappeared with the half-life time ($\tau_{1/2}$) of 2 minutes at room temperature. It is of note that EPR spectrum of **1e*** was virtually identical to that of the acylperoxo intermediate **X**, observed by Oloo et al. [22]. Apparently, it is difficult (perhaps impossible) to distinguish intermediates $[(L)Fe^{III}(\kappa^2\text{-}OC(O)CH_3)]^{2+}$ and $[(L)Fe^{III}(\kappa^2\text{-}OOC(O)CH_3)]^{2+}$ by EPR spectroscopy.

The EPR spectrum of the sample **1***/H_2O_2/CH_3COOH = 1 : 6 : 20 recorded 1.5 minutes after mixing the reagents at −80 °C exhibited resonances of residual **1e*** and an EPR signal $g_1 = 2.070$, $g_2 = 2.005, g_3 = 1.956$ from a new species **1c***^{AA} (Figure 14.2b, Table 14.1) [19]. The subscript AA is used to denote that acetic acid was used as an additive. **1c***^{AA} is much less temperature-stable ($\tau_{1/2}$ = 5 minutes at −85 °C) as compared to intermediate **X** with $\tau_{1/2}$ = 5 minutes at −40 °C [22] The evaluated maximum steady-state concentration of **1c***^{AA} was about 2–3% of the total Fe concentration in the sample.

To directly measure the reactivity of **1c***^{AA} toward olefin epoxidation, the rates of decay of **1c***^{AA} in the absence and in the presence of olefins were compared. It was shown that the addition of relatively small amounts of cyclohexene or *cis*-β-methylstyrene ([olefin]/[Fe] = 0.5) to the sample **1***/H_2O_2/CH_3COOH = 1 : 6 : 20 ([**1***] = 0.02 M) resulted in the drop of $\tau_{1/2}$ of **1c***^{AA} from 5 to <0.5 minutes at −85 °C, whereas electron-deficient olefins (1-acetyl-1-cyclohexene and cyclohexene-1-carbonitrile with an [olefin]/[Fe] ratio up to 10) did not appreciably affect the **1c***^{AA} decay. In the presence of 1-octene as a substrate, the pseudo-first-order rate constant versus the amount of added 1-octene was measured. A linear correlation was observed ($k_{obs} = k_1 + k_2 \cdot$[1-octene]), revealing the second-order rate constant $k_2 = 3.2 \times 10^{-3}\ M^{-1}\ s^{-1}$, which means that **1c***^{AA} is highly reactive toward electron-rich olefins even at −85 °C.

The *g*-values of **1c***^{AA} drastically differ from those of intermediates **1c**–**5c** (Table 14.1). Apparently, this difference is associated with the presence of electron-donating substituents at the pyridine rings of **1***. To verify this assumption, we undertook the search for the intermediate similar to **1c***^{AA} in a structurally related catalyst system **4***/H_2O_2/CH_3COOH. EPR spectroscopic study of this system revealed iron-oxygen species similar to those observed in the catalyst system **1***/H_2O_2/CH_3COOH. It was found that the EPR spectrum recorded just after the addition of 20 equiv. of CH_3COOH to the solution of complex **4*** at room temperature exhibits a rhombic EPR spectrum of a new complex **4c*** (Table 14.1). **4e*** is relatively stable and can be observed even 30 minutes after storing the sample at room temperature. By analogy with complex **1e***, **4e*** can be assigned to the ferric complex $[((S,S)\text{-PDP}^*)Fe^{III}(\eta^2\text{-}OC(O)CH_3)]^{2+}$. After the addition of 6 equiv. of H_2O_2 to the sample **4***/CH_3COOH at −75 °C and mixing the reagents during one minute, the EPR resonances of a new intermediate **4c***^{AA} appeared. The *g* values of **4c***^{AA} ($g_1 = 2.071, g_2 = 2.008, g_3 = 1.960$) are very close to those of **1c***^{AA} ($g_1 = 2.070, g_2 = 2.005,$ $g_3 = 1.956$). The half-life time of **4c***^{AA} (2 minutes at −80 °C) is rather close to that for **1c***^{AA} (5 minutes at −80 °C). The evaluated maximum concentration of **4c***^{AA} was c. 1% of the total iron concentration.

To evaluate the reactivity of **4c***^{AA} toward olefins, the sample **4***/H_2O_2/CH_3COOH = 1 : 6 : 20, containing preliminary generated **4c***^{AA}, was prepared. An olefin was added to this sample at −90 °C within 1 minute, and the decay of **4c***^{AA} was monitored at −85 °C by EPR spectroscopy. In the absence of olefins or in the presence of electron-withdrawing olefins, i.e. 1-acetyl-1-cyclohexene and cyclohexene-1-carbonitrile, **4c***^{AA} decayed with $\tau_{1/2}$ = 4 minutes at −85 °C. At the same time, **4c***^{AA} disappeared within <0.5 minutes at −85 °C in the presence of cyclohexene, *cis*-β-methylstyrene, and 1-octene with the [olefin]/[Fe] ratio of 0.5, demonstrating the direct reactivity of **4c***^{AA} toward electron-rich alkenes.

To quantify the products of cyclohexene oxidation by **1c***^{AA} and **4c***^{AA}, catalytic experiments were performed at −85 °C; 4.5 and 5.3 TN cyclohexene oxide were obtained 15 minutes after the reaction onset with catalysts **1*** and **4***, respectively (TN = catalyst turnover number = moles of product per mole of Fe) [19]. The sum of other cyclohexene oxidation products was less than 1% with respect to the epoxide. Thus, the true active species **1c**^{*AA} and **4c**^{*AA} oxidize cyclohexene even at −85 °C, demonstrating >99% epoxide selectivity. The reactivity of **1c**^{*AA} and **4c**^{*AA} resembles that of **2c**, which epoxidizes cyclohexene at −70 °C [16].

14.3 Electronic Structure of Iron-Oxo Intermediates with High and Low *g*-Factor Anisotropy

Taking into account the close reactivity patterns of **1c**^{*AA}, **4c**^{*AA}, and **2c** toward cyclohexene oxidation, one can assume that they are chemically similar intermediates. However, the EPR spectra of **1c**^{*AA} and **4c**^{*AA} exhibit a relatively small *g*-factor anisotropy ($g_{max} - g_{min} = 0.1$) and sharp resonances ($\Delta H_{1/2} = 15$–40 G), whereas species **1c**–**5c** display highly anisotropic EPR spectra ($g_{max} - g_{min} = 1.0$–1.2), with much broader resonances ($\Delta H_{1/2} = 100$–350 G) (Table 14.1).

The *g*-values and line widths of **1c**^{*AA} and **4c**^{*AA} are very similar to those of the spectroscopically well-characterized iron-oxo complex **7c** with TMC ligand (Table 14.1). The great difference in the EPR parameters of **1c**^{*AA}, **4c**^{*AA} and of **1c**–**5c** implies that the electronic structures of these intermediates significantly differ. It is well-known that paramagnetic metal complexes demonstrate larger *g*-factor anisotropy and broader EPR resonances than organic radicals, due to the occurrence of relatively low-lying excited states and stronger spin-orbit coupling [32]. One can expect that iron complexes with the unpaired spin mainly resting at the metal should display broader resonances and larger *g*-factor anisotropy than complexes with the unpaired spin mostly delocalized over the ligand. As an example, the EPR parameters of a complex with proposed structure $[(TBP_8Cz^{+\bullet})Fe^{IV}{=}O]$ ($g_1 = 2.09$, $g_2 = 2.05$, $g_3 = 2.02$, $g_{max} - g_{min} = 0.07$; antiferromagnetic coupling between $Fe^{IV}{=}O$ ($S = 1$) and $TBP_8Cz^{+\bullet}$ ($S = {}^1/_2$) is suggested) drastically differ from those of the $S = {}^1/_2$ complex $[(TBP_8Cz)Fe^{III}(Py)]$: $g_1 = 2.39$, $g_2 = 2.20$, $g_3 = 1.90$, $g_{max} - g_{min} = 0.49$ (TBP_8Cz = octakis(4-tertbutylphenyl)corrolazinato) [33]. It is worth mentioning that antiferromagnetic coupling between $Fe^{IV}{=}O$ ($S = 1$) and a ligand cation radical ($S = {}^1/_2$) is typical for compounds I of heme enzymes, for example, CYP119, the thermophilic cytochrome P450 from *Sulfolobus acidocaldarius*. The reaction of ferric CYP119 with an excess of *m*-chloroperoxybenzoic acid generates $S = {}^1/_2$ CYP119 compound I with EPR parameters $g_1 = 2.00$, $g_2 = 1.96$, and $g_3 = 1.86$ [34]. Recently, ferric complex with 1,4,7-trimethyl-1,4,7-triazacyclononane (Me_3TACN) has been used for the oxidation of alkanes with oxone, including cyclohexane, propane, and ethane. ESI-MS, EPR, UV–Vis spectroscopy and ^{18}O labeling experiments pointed to $[(Me_3TACN)Fe^{IV}{=}O(\{Cl\text{-}acac\}^{+\bullet})]^{2+}$ as the catalytically active species (Cl-acac = 3-chloro-acetylacetonate) [35]. EPR parameters of this complex ($g_1 = 1.97$, $g_2 = 1.93$, $g_3 = 1.91$) are rather close to those of CYP119 compound I.

On the basis of the data just presented, it seems reasonable to assign the intermediates **1c**^{*AA} and **4c**^{*AA} to the species of the type $[(L^{+\bullet})Fe^{IV}{=}O(OC(O)CH_3)]^{2+}$. The presence of strong electron donors in **1c**^{*AA} and **4c**^{*AA} may favor the intramolecular redistribution of electron density, leading to the reduction of iron to the formal +4 oxidation state, the positive charge being spread over the substituted pyridine rings. In these structures, the unpaired spin is mostly delocalized over the organic ligand framework, thus accounting for the low *g*-factor anisotropy. On the contrary, this scenario is unlikely for intermediates **1c**–**5c** due to the lack of electron density at their nonsubstituted pyridine rings. We conclude that **1c**^{*AA}, **4c**^{*AA} and probably **7c** may be better represented as $(L^{+\bullet})Fe^{IV}{=}O$ rather than $(L)Fe^{V}{=}O$. In contrast, **1c**–**5c** are most likely $(L)Fe^{V}{=}O$ species (Table 14.1).

Recently, the intermediate **8c** with EPR spectrum virtually identical to those of **1c***AA and **4c***AA has been observed in the catalyst system [(PyNMe$_3$)Fe^{II}(CF$_3$SO$_3$)$_2$]/CH$_3$CO$_3$H. On the basis of EPR, UV–VIS, mass spectroscopic data and reactivity studies, **8c** has been assigned to the iron-oxo complex with iron in the formal +5 oxidation state (Chart 14.2, Table 14.1). The reactivity of **8c** toward stoichiometric oxidation of various alkanes at −35 °C has been directly determined by the UV–Vis stopped-flow technique [21].

The data obtained clearly show that active iron-oxygen intermediates with EPR parameters close to those of known formally Fe^V=O species can be detected in bioinspired catalyst systems for chemo-, regio-, and stereoselective oxidation of C—H and C=C groups of various organic substrates with H_2O_2. On the basis of our data and milestone results of other research groups [36, 37], the following mechanism of alkene epoxidation by the catalyst systems studied was proposed (Scheme 14.1). At the initial stage, dimeric iron complexes **1*** and **4*** reversibly transform into monomeric ferric-acetate complexes **1e*** and **4e***, detectable by EPR even at room temperature. Subsequent interaction of **1e*** and **4e*** with H_2O_2, followed by heterolytic O—O bond cleavage with removal of a water molecule, generates the active species **1c***AA and **4c***AA, capable of epoxidizing olefins directly even at a very low temperature (−85 °C).

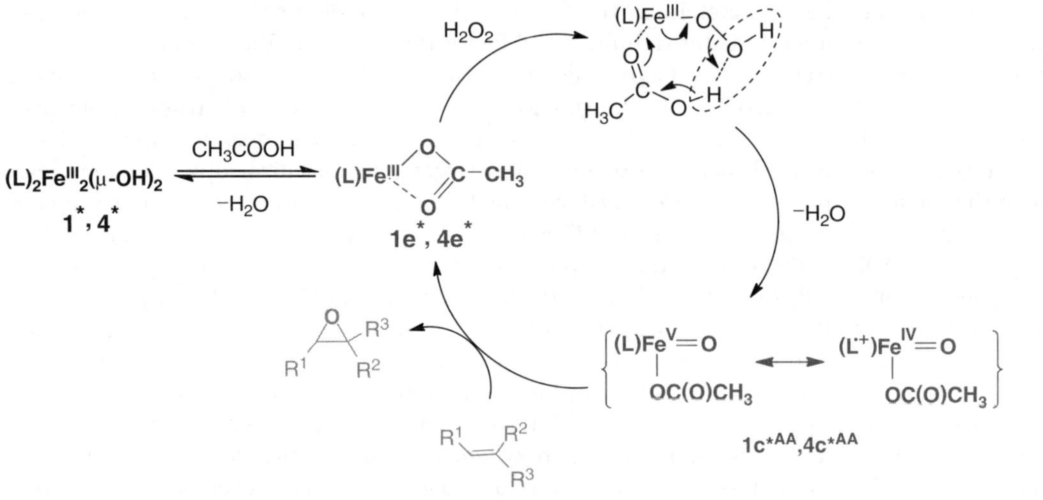

Scheme 14.1 Proposed mechanism of olefin epoxidation by catalyst systems **1***/H_2O_2/CH_3COOH and **4***/H_2O_2/CH_3COOH. (*See insert for color representation.*)

14.4 Switching the Electronic Structure of the Active Species by Selecting a Carboxylic Acid Additive

Another important factor, affecting the structure and properties of the catalytically active species, is the nature of the carboxylic acid, since the use of different carboxylic acids has been shown to significantly influence the epoxidation enantioselectivity observed in the catalyst systems based on complex **4** (Table 14.3) [38]. To get direct insight into the effect of the structure of the carboxylic acid on the active species of aminopyridine iron catalysts, we have undertaken the search for the active species in the catalyst systems **4**/H_2O_2/RCOOH with eight different carboxylic acids RCOOH (Chart 14.4). Spectacularly, it was found that the use of carboxylic acids with primary (AA) and secondary α-carbons (BA, CA) led to the iron-oxo

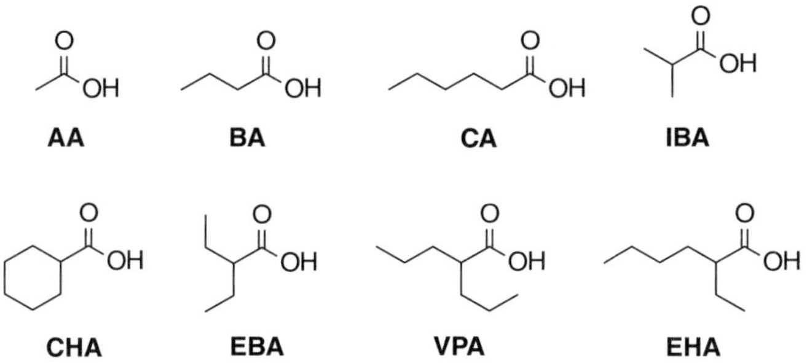

Chart 14.4 Carboxylic acids studied.

intermediates with large g-factor anisotropy of the type **4c**^{AA} (Figure 14.3a), while in the presence of some acids with tertiary α-carbon, intermediates with small g-factor anisotropy were detected (Table 14.2) [38]. In particular, the intermediates with small g-factor anisotropy of the type **4c**^{EHA} were observed for EBA, VPA, and EHA (Figure 14.3b), while IBA and CHA led to the intermediates with large g-factor anisotropy (Table 14.2).

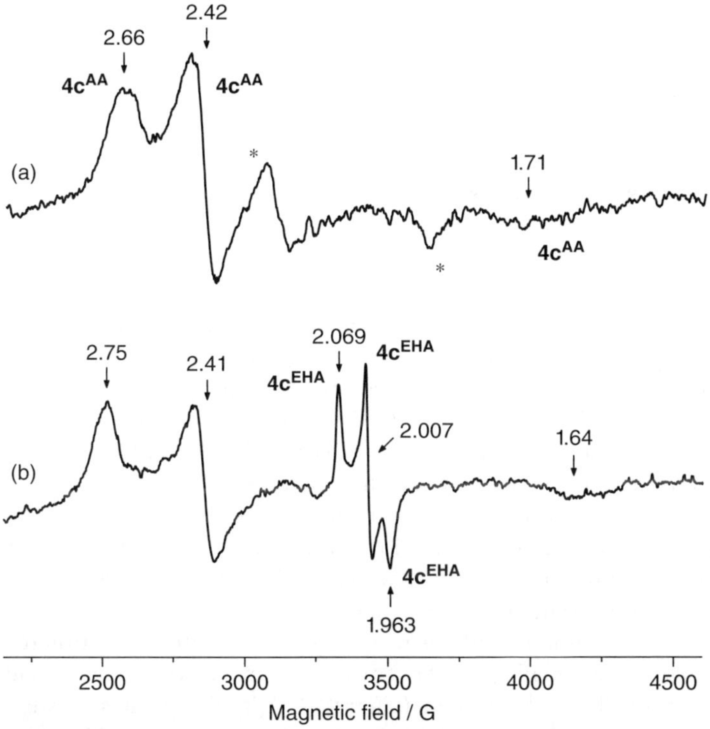

Figure 14.3 EPR spectra (−196 °C) of the sample (a) **4**/H_2O_2/CH_3COOH ([**4**]:[H_2O_2]:[CH_3COOH] = 1 : 3 : 10, [**4**] = 0.04 M) frozen one minute after mixing the reagents at −75 °C and storing the sample at −85 °C for one minute, and the sample (b) **4**/H_2O_2/EHA ([**4**]:[H_2O_2]:[EHA] = 1 : 3 : 10, [**4**] = 0.04 M) frozen 2.5 minutes after mixing the reagents at −65 °C. Signals denoted by an asterisk belong to iron(III)-hydroxo complex [((*S,S*)-PDP)Fe^{III}–OH(S)]$^{2+}$ (**4b**, Table 14.1). Signals in "B" at g_1 = 2.75, g_2 = 2.41, g_3 = 1.64 belong to ferric complex [((*S,S*)-PDP)Fe^{III}(η^1-OC(O)R)(RCOOH)]$^{2+}$ stable at room temperature (RCOOH = EHA). A 1.8 : 1 CH_2Cl_2/CH_3CN mixture was used as a solvent.

Table 14.2 Effect of the carboxylic acid on the EPR parameters of iron-oxygen species formed in the systems catalyst/H_2O_2/RCOOH.

Species	g_1	g_2	g_3
$[((S,S)\text{-PDP})Fe^{V}{=}O(OC(O)R)]^{2+}$, type $\mathbf{4c}^{AA}$:			
$\mathbf{4c}^{AA} = \mathbf{4c}$	2.66	2.42	1.71
$\mathbf{4c}^{BA}$	2.70	2.42	1.67
$\mathbf{4c}^{CA}$	2.70	2.42	1.66
$\mathbf{4c}^{IBA}$	2.72	2.42	1.66
$\mathbf{4c}^{CHA}$	2.69	2.42	1.67
$[((S,S)\text{-PDP})Fe^{IV}{=}O(^{\bullet}OC(O)R)]^{2+}$, type $\mathbf{4c}^{EHA}$:			
$\mathbf{4c}^{EBA}$	2.069	2.007	1.961
$\mathbf{4c}^{VPA}$	2.069	2.007	1.962
$\mathbf{4c}^{EHA}$	2.069	2.007	1.963
$[((S,S)\text{-PDP*})^{\bullet+}Fe^{IV}{=}O(OC(O)R)]^{2+}$, type $\mathbf{4c}^{*AA}$:			
$\mathbf{4c}^{*AA}$	2.071	2.008	1.960
$\mathbf{4c}^{*BA}$	2.070	2.008	1.958
$\mathbf{4c}^{*CA}$	2.069	2.008	1.957
$\mathbf{4c}^{*IBA}$	2.069	2.007	1.957
$\mathbf{4c}^{*CHA}$	2.070	2.008	1.957
$\mathbf{4c}^{*EBA}$	2.069	2.008	1.957
$\mathbf{4c}^{*VPA}$	2.069	2.007	1.958
$\mathbf{4c}^{*EHA}$	2.070	2.008	1.958

Taking into account the explanation of the reason of the small g-factor anisotropy in paragraph Section 14.3, it is logical to expect that in the intermediates of the type $\mathbf{4c}^{EHA}$, the unpaired spin mostly resides on the coordinated carboxylic moiety (Scheme 14.2) [38]. Apparently, the active oxidizing species formed in the presence of EBA, VPA, and EHA may be better represented as $[((S,S)\text{-PDP})Fe^{IV}{=}O(^{\bullet}OC(O)R)]^{2+}$. Higher enantioselectivity of the catalyst systems **4**/H_2O_2/RCOOH, where RCOOH = EHA, VPA, EBA, (Table 14.3) is plausibly explained by the additional stabilization of the corresponding active species $[((S,S)\text{-PDP})Fe^{IV}{=}O(^{\bullet}OC(O)R)]^{2+}$ due to the unpaired electron delocalization over the carboxylic moiety, which leads to more product-like transition states and thus ensures tighter interactions between the substrate and the iron-oxo species within the transition state, eventually resulting in higher enantioselectivity [38].

Iron complexes **1***, **4*** with aminopyridine ligands bearing electron-donating substituents generated iron-oxo intermediates with small g-factor anisotropy even with acetic acid [19]. In effect, the use of all carboxylic acids studied (Chart 14.4) in the catalyst system **4***/H_2O_2/RCOOH afforded active species with virtually the same g-factor values (Table 14.2). At the same time, the use of carboxylic acid with tertiary α-carbon still enhances the chalcone epoxidation enantioselectivity by the systems **4***/H_2O_2/RCOOH, yet to a smaller extent as compared with the systems **4**/H_2O_2/RCOOH (Table 14.3). Remarkably, the enantioselectivities well correlated with the observed maximum steady-state concentrations of the active species evaluated by EPR spectroscopy (Table 14.3). The obtained results provides a spectacular illustration for the widely-used Hammond-Leffler postulate, which for the present catalyst

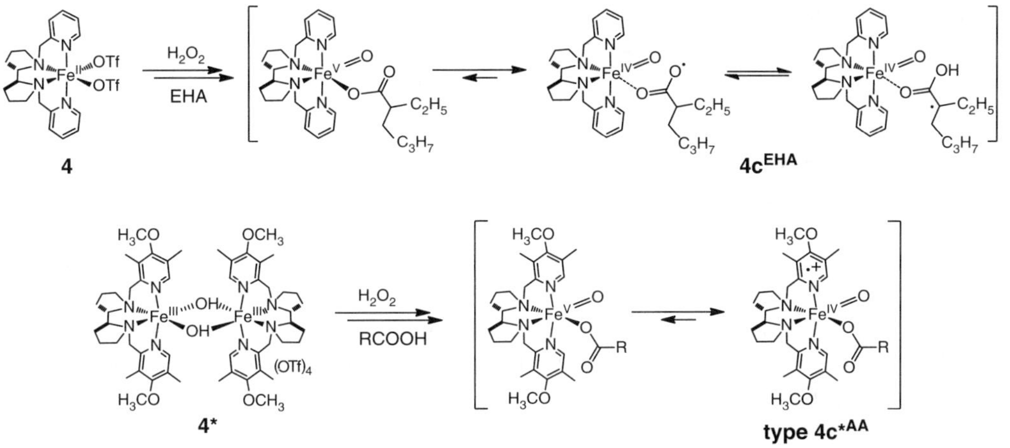

Scheme 14.2 Plausible mechanistic scenarios for the formation of the active species with small *g*-factor anisotropy. For **4c**^{EHA}, possible tautomeric forms are presented to illustrate the additional stabilization caused by the carboxylic acid with tertiary α-carbon.

Table 14.3 Effect of carboxylic acid structure on the asymmetric epoxidation of chalcone with hydrogen peroxide in the presence of complexes **4** and **4***.

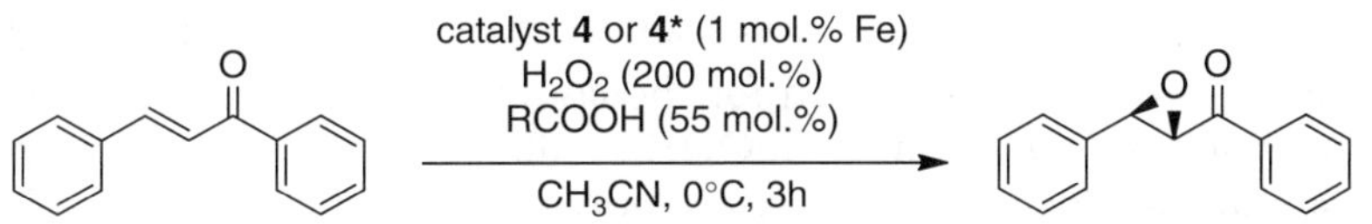

RCOOH additive	Catalyst **4** Conversion (%)/epoxide yield (%)	*ee* (%)[a)]	Catalyst **4*** Conversion (%)/epoxide yield (%)	*ee* (%)[a)]	Concentration of the active species[b)]
AA	77/75	62	74/73	69	1.0
CA	65/63	64	48/46	70	1.0
BA	68/65	64	61/59	70	1.4
IBA	85/85	73	92/92	78	2.2
CHA	91/90	73	72/71	78	2.2
EBA	89/89	82	54/54	81	2.0
VPA	90/89	83	69/69	82	2.6
EHA	92/92	82	73/73	83	3.0

Reaction conditions: at 0 °C, solvent CH_3CN, chalcone (100 μmol), H_2O_2 (200 μmol), catalyst load 1 mol% Fe, oxidant was added by a syringe pump over 30 min, and the mixture was stirred for 2.5 h, followed by LC analysis.

a) Enantiomeric excess value. The predominant enantiomer was (2*R*,3*S*).

b) Maximum observed concentration of the iron-oxo active species (for catalyst **4***) calculated from EPR spectra, expressed as the percentage of total iron concentration in the sample.

system may be expressed as follows: *the less reactive (i.e. those approaching higher steady-state concentration) oxygen-transferring species lead to more product-like transition states*, resulting in higher epoxidation stereoselectivity [38].

14.5 Active Species of Olefin Epoxidations in the Catalyst Systems with Different Oxidants

Obviously, environmentally benign hydrogen peroxide, having 47% active oxygen content, is the oxidant of choice for the asymmetric olefin epoxidations in the presence of bioinspired iron complexes. At the same time, other oxidants, e.g. peracetic acid, *m*-chloroperoxybenzoic acid, alkyl, and aryl hydroperoxides can also be used for the asymmetric olefin epoxidations in the presence of non-heme iron complexes, affording epoxides with comparably high enantioselectivities [17, 37, 39–41].

We have undertaken the combined EPR spectroscopic, enantioselectivity, and stereoselectivity, Hammett and ^{18}O labeling studies of the olefin epoxidations catalyzed by complexes **4** and **4*** in the presence of carboxylic acid additive RCOOH (acetic, AA or 2-ethylhexanoic acid, EHA). The following oxidants have been used: H_2O_2, *t*-butyl hydroperoxide (*t*-BuOOH), cumyl hydroperoxide (CmOOH), peracetic acid (CH_3CO_3H), and *m*-chloroperoxybenzoic acid (*m*-CPBA) [23].

Asymmetric epoxidation of chalcone with *t*-BuOOH and CmOOH, catalyzed by complex **4**, in the absence of RCOOH additive demonstrated lower *ee*'s (52% and 56%) and much lower epoxide yields (17% and 22%) as compared with the same experiments in the presence of AA (64% and 65% *ee*; 70% and 42% yield) or EHA (81% and 80% *ee*; 76% and 44% yield). The same picture was documented for the *cis*-stilbene epoxidation stereoselectivity. Catalyst systems **4**/*t*-BuOOH and **4**/CmOOH exhibited several times lower value of [*cis*-stilbene oxide]:[*trans*-stilbene oxide] (i.e. *cis*:*trans* ratio = 10% and 5%, Table 14.4, entries 4 and 7) than the catalyst systems **4**/*t*-BuOOH/AA (61%), **4**/*t*-BuOOH/EHA (63%), **4**/CmOOH/AA (48%), and **4**/CmOOH/EHA (59%) (Table 14.4, entries 5, 6, 8, 9). The search for unstable iron-oxygen intermediates in the systems **4**/*t*-BuOOH and **4**/CmOOH by EPR revealed only $[((S,S)\text{-PDP})Fe^{III}(OO\textit{t}\text{-Bu})(CH_3CN)]^{2+}$ (g_1 = 2.19, g_2 = 2.14, g_3 = 1.96) and $[((S,S)\text{-PDP})Fe^{III}(OOCm)(CH_3CN)]^{2+}$ species (g_1 = 2.18, g_2 = 2.13, g_3 = 1.96) with virtually the same EPR parameters. On the basis of obtained data, it was assumed that the epoxidations by catalyst systems **4**/*t*-BuOOH and **4**/CmOOH proceed via the general radical mechanism presented in Scheme 14.3 [23].

In contrast to systems without carboxylic acid additives, catalyst systems **4**/H_2O_2/EHA and **4**/R_1OOH/EHA (R_1 = *t*-Bu or Cm) generate iron-oxo species $\mathbf{4c^{EHA}}$, highly reactive in alkene epoxidation, the former system demonstrating the maximum concentration of the active species. In accordance with the EPR spectroscopic results, the catalyst system **4**/H_2O_2/EHA exhibited higher epoxide yield (90%) in asymmetric chalcone epoxidation, in comparison with those for the systems **4**/*t*-BuOOH/EHA (76%) and **4**/CmOOH/EHA (44%). At the same time, three systems with different oxidants demonstrated virtually identical *ee*'s (80–82%).

It is not surprising that the addition of EHA to systems **4**/*t*-BuOOH and **4**/CmOOH, which is shown to result in the formation of highly reactive and selective iron-oxo intermediate $\mathbf{4c^{EHA}}$, improves both the epoxide yield (from 14% and 21% to 89% and 80%) and the stereoselectivity in *cis*-stilbene epoxidation (*cis*:*trans* from 10 and 5 to 63 and 59, Table 14.4, entries 4 vs. 6 and 7 vs. 9). The latter values are very close to that for the catalyst system **4**/H_2O_2/EHA (*cis*:*trans* = 65, Table 14.4 entry 3). In turn, for the catalyst systems **4**/*t*-BuOOH/AA and **4**/CmOOH/AA,

Table 14.4 Epoxidation of *cis*-stilbene with various oxidants.

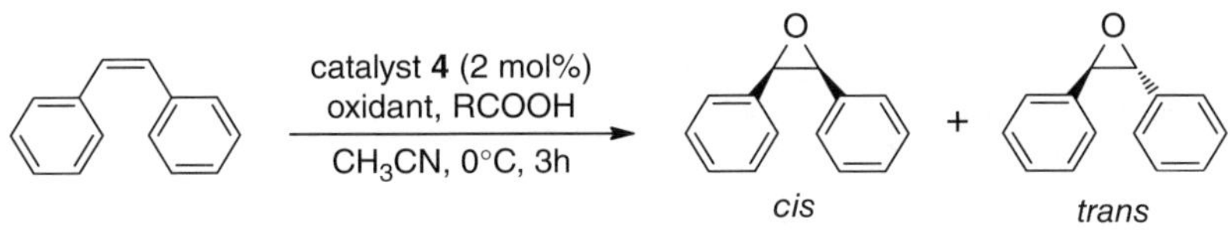

Entry	Catalyst	Oxidant	Additive	Conversion (%)/epoxide yield (%)	*cis*: *trans*
1	4	H_2O_2	—	87/75	18
2	4	H_2O_2	AA	99/84	49
3	4	H_2O_2	EHA	99/91	65
4	4	*t*-BuOOH	—	18/14	10
5	4	*t*-BuOOH	AA	84/69	61
6	4	*t*-BuOOH	EHA	100/89	63
7	4	CmOOH	—	25/21	5
8	4	CmOOH	AA	76/63	48
9	4	CmOOH	EHA	90/80	59
10	4	CH_3CO_3H	—	60/50	83
11	4	CH_3CO_3H	AA	57/48	82
12	4	CH_3CO_3H	EHA	63/54	81
13	—	CH_3CO_3H	—	8/7	50
14	4	*m*-CPBA	—	47/38	34
15	4	*m*-CPBA	AA	46/37	33
16	4	*m*-CPBA	EHA	54/44	37
17	—	*m*-CPBA	—	41/40	30

Reaction conditions: at 0 °C, *cis*-stilbene (100 μmol), oxidant, H_2O_2 (200 μmol), *t*-BuOOH, CmOOH, CH_3CO_3H, *m*-CPBA (110 μmol), carboxylic acid additive (55 μmol), catalyst load 2 mol%, oxidant was added by a syringe pump over 30 min, and the mixture was stirred for an additional 2.5 h followed by ^1H NMR analysis.

cis:*trans* ratios of 61 and 48 have been documented (Table 14.4 entries 5 and 8), that are close to the value of 49 for the system $4/H_2O_2/AA$ (Table 14.4 entry 2).

The above data point to the common active species operating in the catalyst systems $4/H_2O_2/RCOOH$ and $4/R^1OOH/RCOOH$ (R^1=*t*-butyl or cumyl, RCOOH=AA or EHA). The following mechanism for the epoxidation by these catalyst systems has been

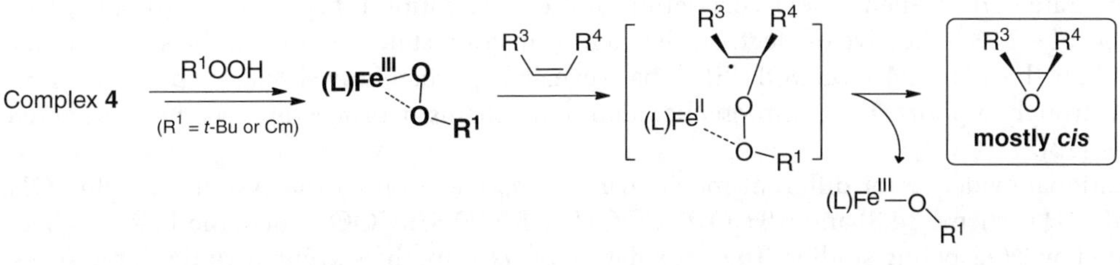

Scheme 14.3 Proposed reaction pathway for the asymmetric epoxidation by catalyst systems **4**/alkyl or aryl hydroperoxide.

suggested (Scheme 14.4) [23]. The same catalytic pattern obtained for the catalyst systems **4***/H_2O_2/RCOOH and **4***/R^1OOH/RCOOH implies a similar reaction sequence.

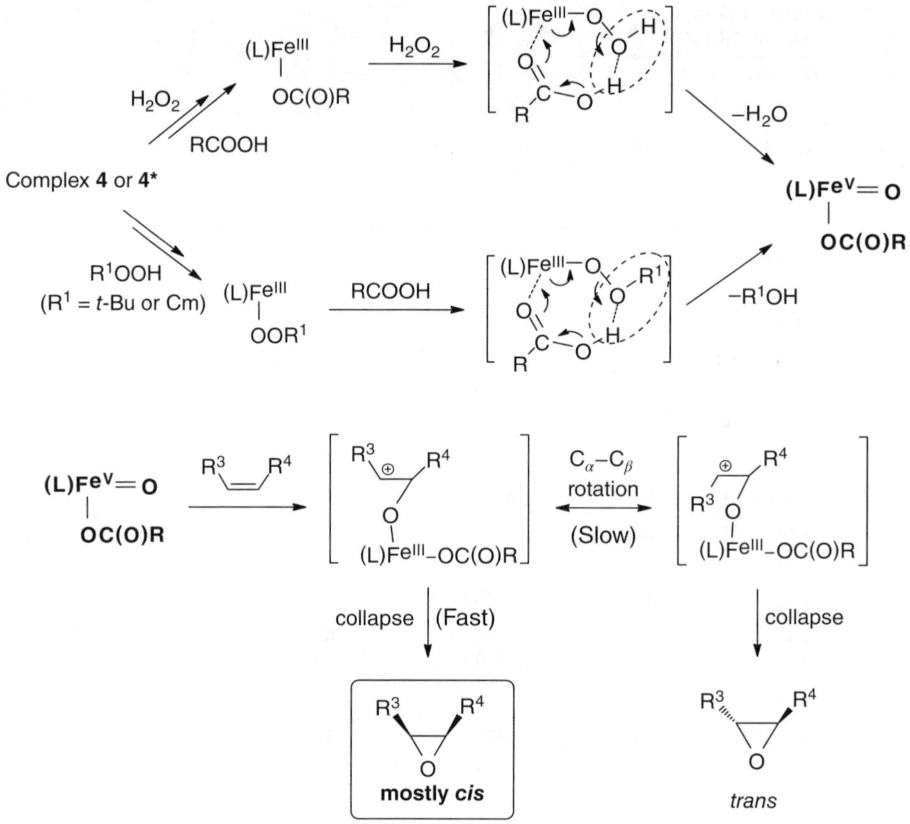

Scheme 14.4 Proposed reaction sequences for the epoxidations by the catalyst systems **4**/H_2O_2/RCOOH, **4**/R^1OOH/RCOOH, **4***/H_2O_2/RCOOH and **4***/R^1OOH/RCOOH (R^1 = *t*-butyl or cumyl).

In Scheme 14.4, the catalytic epoxidation is assumed to proceed via acyclic intermediate capable of rotating around the single C_α—C_β bond. This mechanism is similar to that proposed for the epoxidation by the related catalyst systems based on structurally similar Mn aminopyridine complex, with the evidence in favor of carbocationic nature of the acyclic intermediate obtained from the observed linear correlation between the relative epoxidation rates of *para*-substituted chalcones and the Brown-Okamoto parameter σ_p^+ [42, 43]. A similar Brown-Okamoto analysis applied to the epoxidation of *para*-substituted chalcones with the system **4**/H_2O_2/EHA has demonstrated an excellent linear correlation of the epoxidation rates vs. σ_p^+, with a negative slope ρ^+ of −2.83 indicative of electron-deficient transition state. Moreover, the same analysis applied for the system **4**/*t*-BuOOH/EHA has revealed spectacularly similar slope $\rho^+ = -2.84$, which strongly supports the common epoxidation mechanism, proceeding via the same active species [23].

Additional evidence for different mechanisms of epoxidation by the systems **4**/*t*-BuOOH, **4**/CmOOH (Scheme 14.3) and **4**/H_2O_2/RCOOH, **4**/R^1OOH/RCOOH (Scheme 14.4) has been obtained by ^{18}O labeling studies. The epoxidation of *cis*-*β*-methylstyrene with the catalyst system **4**/$H_2{}^{16}O_2$/$H_2{}^{18}O$ has resulted in a 15% incorporation of ^{18}O into the epoxide, characteristic of the presence of $[(L)Fe^V{=}O(OH)]^{2+}$ active species, capable of exchanging its oxygen atom with that of ^{18}O-enriched water [27]. On the contrary, the epoxidation with the system **4**/*t*-BuOOH/$H_2{}^{18}O$ has demonstrated a much smaller ^{18}O incorporation of 4%, indicating that the major epoxidation pathway for this system does not involve an iron-oxo intermediate, apparently proceeding with the participation of a less reactive and less selective ferric oxidant.

Further, peroxycarboxylic acids were studied as oxidants. Catalyst systems **4**/CH_3CO_3H, **4**/CH_3CO_3H/AA and **4**/CH_3CO_3H/EHA displayed identical enantioselectivities in chalcone epoxidation (67% *ee*). The enantioselectivity of the catalyst system **4**/CH_3CO_3H/EHA did not change upon the six-fold increase of the EHA concentration. The systems **4**/*m*-CPBA, **4**/*m*-CPBA/AA, and **4**/*m*-CPBA/EHA also demonstrated equal epoxidation enantioselectivities (51–52% *ee*). This enantioselectivity level is noticeably different from the value of 67% *ee* for the epoxidations with peracetic acid, and from those observed in the systems with hydrogen peroxide: **4**/H_2O_2/EHA (82% *ee*) and **4**/H_2O_2/AA (63% *ee*). A similar picture has been documented for the catalyst **4***.

For the *cis*-stilbene epoxidation with *m*-CPBA in the presence of complex **4**, the *cis:trans* selectivity was 33–37 (Table 14.4, entries 14–16), which is close to the value of 30 for uncatalyzed Prilezhaev epoxidation with *m*-CPBA (Table 14.4 entry 17), known to proceed via concerted mechanism [44]. In turn, for the epoxidations with the catalyst systems **4**/CH_3CO_3H, **4**/CH_3CO_3H/AA, and **4**/CH_3CO_3H/EHA, the *cis:trans* ratio was 82 ± 1 (Table 14.4, entries 10–12), irrespective of the presence of added carboxylic acid and of its structure. For comparison, the uncatalyzed Prilezhaev epoxidation of *cis*-stilbene with CH_3CO_3H showed a *cis:trans* ratio of 50 (Table 14.4 entry 13).

The above data indicate that the epoxidations with catalyst systems **4,4***/peracid and **4,4***/peracid/RCOOH demonstrate substantial mechanistic distinctions from the catalyst systems **4,4***/H_2O_2/RCOOH, that is (i) different levels of enantioselectivity, (ii) independence of the presence and structure of added carboxylic acid, and (iii) different *cis*-stilbene epoxidation stereoselectivity pattern. It is logical to explain these differences by the occurrence of different oxygen transfer mechanisms with different oxidizing species.

To directly access the unstable iron-oxygen intermediates, catalyst systems with peracids have been monitored by in situ EPR spectroscopy. The systems with complex **4*** as catalyst and EHA as catalytic additive appeared to be the most informative. It has been shown that the interaction of complex **4*** with both CH_3CO_3H and *m*-CPBA generates iron(III)-acylperoxo complexes of the type $[((S,S)\text{-PDP*})Fe^{III}(\eta^2\text{-OOC(O)}R^2)]^{2+}$ ($R^2{=}CH_3$ or 3-Cl-C_6H_4) with EPR parameters $g_1 = 2.52, g_2 = 2.41, g_3 = 1.80$. In contrast to the system **4***/*m*-CPBA/EHA, iron-oxo species $[((S,S)\text{-PDP*})Fe^V{=}O(OC(O)R)]^{2+}$ (**4c**^{*EHA}, R = EHA) has also been observed in the catalyst system **4***/CH_3CO_3H/EHA. On the basis of catalytic data, we have proposed that both iron(III)-acylperoxo and iron-oxo species can participate in oxygen atom transfer to olefin in the catalyst systems with complex **4*** and peracetic acid.

Interestingly, the epoxidation of *cis*-β-methylstyrene by the catalyst system **4**/*m*-CPBA in the presence of ^{18}O-labeled water showed zero incorporation of ^{18}O into the epoxide. In addition, competitive epoxidations of *para*-substituted chalcones have exhibited the Hammett slope ρ of −2.64, which is evidently different from the Brown-Okamoto slope ρ^+ of −2.83, observed for the epoxidations conducted by the $[(L)Fe^V{=}O(OH)]^{2+}$ active species [23]. These facts, as well as EPR spectroscopic data, provide evidence in favor of the epoxidation mechanism different from that involving the iron-oxo active species. In combination, the observations by EPR, the unique *cis:trans* stilbene oxide ratio pattern, ^{18}O-labelling and Hammett data can be reasonably interpreted in favor of the concerted mechanism of the epoxidations by the catalyst systems with *m*-CPBA as the oxidant (Scheme 14.5).

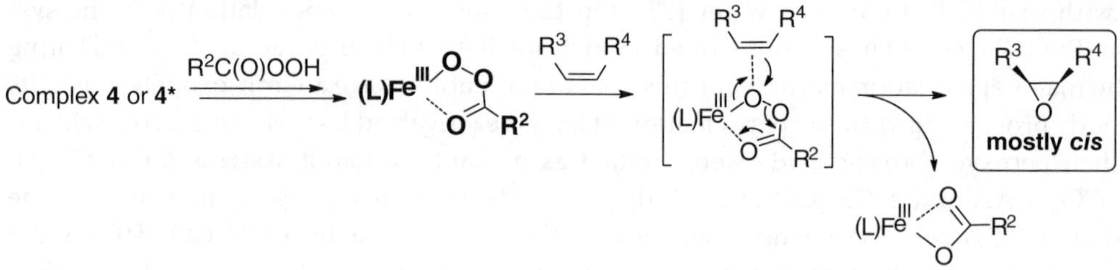

Scheme 14.5 Proposed reaction sequence for the asymmetric olefin epoxidations with peroxycarboxylic acids catalyzed by complexes **4** or **4***.

14.6 Iron-oxo Intermediates of C—H Oxidations

The regio- and stereospecific oxidation of nonactivated aliphatic C—H groups of hydrocarbons and complex organic molecules is a highly challenging problem that has attracted great attention in the last decade. In several cases, preparatively useful yields have been reported for stereospecific C—H oxidations with catalyst systems based on aminopyridine iron(II) complexes (e.g. **4**, $\mathbf{4^{SbF6}}$, and **5**) as catalysts, hydrogen peroxide as the oxidant, and acetic acid as the catalytic additive [3, 4, 27, 45, 46]. The selectivity of the most prominent bioinspired catalyst systems based on complexes **4**, $\mathbf{4^{SbF6}}$, and **5** (Chart 14.1) toward the oxidation of particular C—H group in complex organic molecules approaches those for iron oxygenases, thus inspiring interest in the nature and reactivity of the key metal-oxygen active species responsible for the selective C—H oxidation.

We have undertaken the systematic *in situ* examination of the alkane oxidation reactivity of the active species formed in the catalyst systems based on complexes **1**, **1***, **4**, and **4*** (Charts 14.1 and 14.3), various oxidants (H_2O_2, CH_3CO_3H, or *m*-CPBA), and catalytic additives (AA or EHA). Additionally, the efficiency and chemo- and regioselectivity of the same catalyst systems in the oxidation of cyclohexane and adamantane have been systematically examined [47].

Cyclohexane is widely used as "test substrate" for catalytic C—H oxidations, since the oxidation products ratio (cyclohexanol/cyclohexanone, A/K ratio) allows discrimination between free-radical-driven and metal-based oxidation mechanisms, and characterizes the reactivity of the metal-based intermediate. Under the high-substrate-excess conditions, in the case of free-radical involvement, the A/K ratio is close to 1, while for the latter mechanism, the A/K ratio is much higher; higher A/K ratios reflect lower sensitivity of the metal-based intermediates to the substrate bond-dissociation energy (BDE) and hence their higher reactivity. Another widely used mechanistic probe is the regioselectivity pattern of adamantane oxidation, metal-based oxidation mechanisms ensuring much higher selectivity toward the oxidation of tertiary C—H groups, compared with the free-radical-driven mechanisms [48].

The oxidation of adamantane with the studied catalyst systems afforded 1-adamantanol as the main product, with byproducts 2-adamantanol and 2-adamantanone. In the absence of carboxylic acid, the observed conversion and the 3°/2° ratio with the catalyst system $\mathbf{4}/H_2O_2$ are rather low (Table 14.5, entry 1). The addition of AA or EHA dramatically increases both the conversion and the 3°/2° ratio (Table 14.5 entries 2 and 3). The 3°/2° ratios for the systems $\mathbf{4}/H_2O_2$/AA and $\mathbf{4}/H_2O_2$/EHA are similar (41 and 38). So, the iron-oxo intermediates $\mathbf{4c^{AA}}$ and $\mathbf{4c^{EHA}}$, operating in these systems, display similar regioselectivities in the adamantane C—H oxidation, irrespective of differing EPR parameters (Table 14.2).

The oxidations of adamantane by the catalyst systems with peracetic acid ($\mathbf{4}/CH_3CO_3H$/AA and $\mathbf{4}/CH_3CO_3H$/EHA) exhibited close conversions and regioselectivities, as compared to the similar H_2O_2-based systems (3°/2° of 39, 38 vs. 41, 38, Table 14.5 entries 5, 6 and 2, 3), reflecting

Table 14.5 Regioselective oxidation of adamantane catalyzed by complexes **4** and **4***.

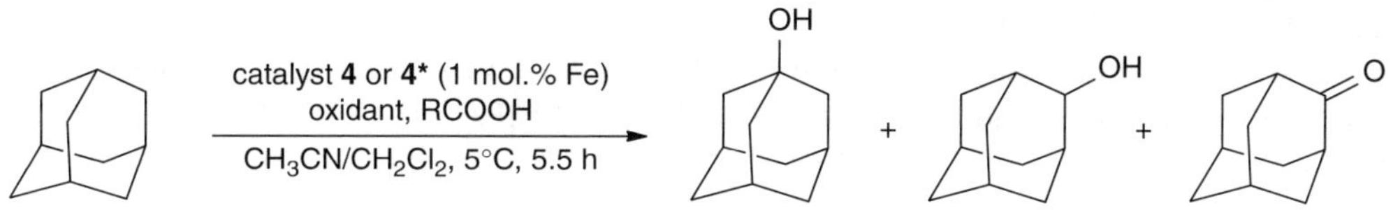

Entry	Catalyst	Oxidant	Carboxylic acid additive	Conversion (%)[a] (= TON)	3°/2°[b]
1	**4**	H_2O_2	—	6	9
2	**4**	H_2O_2	AA	57	41
3	**4**	H_2O_2	EHA	57	38
4	**4**	CH_3CO_3H	—	44	30
5	**4**	CH_3CO_3H	AA	42	39
6	**4**	CH_3CO_3H	EHA	50	38
7	**4**	*m*-CPBA	—	12	22
8	**4**	*m*-CPBA	AA	9	27
9	**4**	*m*-CPBA	EHA	17	20
10	**4***	H_2O_2	—	6	9
11	**4***	H_2O_2	AA	46	40
12	**4***	H_2O_2	EHA	53	41
13	**4***	CH_3CO_3H	—	34	28
14	**4***	CH_3CO_3H	AA	67	36
15	**4***	CH_3CO_3H	EHA	64	34
16	**4***	*m*-CPBA	—	10	13
17	**4***	*m*-CPBA	AA	7	15
18	**4***	*m*-CPBA	EHA	7	13

Reaction conditions: at 5 °C, solvent CH_3CN/CH_2Cl_2 (v/v = 3 : 4), adamantane (100 μmol), oxidants, H_2O_2 (200 μmol) CH_3CO_3H, *m*-CPBA (120 μmol), carboxylic acid (110 μmol), catalyst load 1 μmol Fe, oxidant was added by a syringe pump over 30 min, and the mixture was stirred for 5 h, followed by GC analysis.
a) Based on substrate.
b) 3°/2° = 3 × [1-adamantanol]/([2-adamantanol] + [2-adamantanone]).

close reactivities of the intermediates operating in these systems. So, the contribution of the putative iron(III)-acylperoxo species to the oxidation of adamantane by the catalyst systems **4**/CH_3CO_3H/AA and **4**/CH_3CO_3H/EHA cannot be significant, the reaction being predominantly conducted by iron-oxo intermediates.

The catalyst systems **4**/*m*-CPBA/AA and **4**/*m*-CPBA/EHA display much smaller conversions than the systems **4**/H_2O_2/AA and **4**/H_2O_2/EHA (9–17% vs. 57%, Table 14.5, entries 8, 9 and 2, 3). The 3°/2° ratios for the former two systems are also smaller (20, 27 vs. 41, 38, Table 14.5, entries 8, 9 and 2, 3), implicating the likely participation of the iron(III)-*m*-chlorobenzoylperoxo intermediates in the oxidation of adamantane.

Like the system **4**/H_2O_2, the system **4***/H_2O_2 exhibits low conversion and 3°/2° ratio (Table 14.5, entry 10); the addition of carboxylic acids improves the yield of the oxygenated products and enhances the 3°/2° ratio (entries 11, 12). The regioselectivities of the systems

$4^*/H_2O_2/AA$ and $4^*/H_2O_2/EHA$ were virtually identical (3°/2° of 40 vs. 41). This agrees with close electronic structures of the corresponding active species $\mathbf{4c}^{*AA}$ and $\mathbf{4c}^{*EHA}$, displaying identical EPR spectra (Table 14.2).

The oxidation of adamantane by the catalyst systems **4***/*m*-CPBA/AA and **4***/*m*-CPBA/EHA, exhibiting EPR active intermediates of the type $[((S,S)\text{-PDP}^*)\text{Fe}^{III}(\eta^2\text{-OOC(O)C}_6\text{H}_4\text{-3-Cl)}])]^{2+}$ (**4x****), has shown low conversions and low 3°/2° ratios (13, 15, Table 14.5, entries 17 and 18), as compared to the systems based on terminal oxidant H_2O_2 (40, 41, Table 14.5, entries 11 and 12). This indicates that different active species may predominantly operate in the H_2O_2- and *m*-CPBA-based catalyst systems (iron-oxo and iron-*m*-chlorobenzoylperoxo intermediates, respectively).

The catalyst systems $4^*/CH_3CO_3H/AA$ and $4^*/CH_3CO_3H/EHA$ display EPR spectra of two potentially active intermediates (**4x*** and $\mathbf{4c}^{*AA}$, **4x*** and $\mathbf{4c}^{*EHA}$, respectively). The conversions and 3°/2° ratios for the $4^*/CH_3CO_3H/AA$ and $4^*/CH_3CO_3H/EHA$ systems are rather close to those for the $4^*/H_2O_2/AA$ and $4^*/H_2O_2/EHA$ systems (3°/2° of 36, 34 vs. 40, 41, Table 14.5, entries 14, 15 and 11, 12). Therefore, the iron-oxo intermediates are most likely the major active species, responsible for adamantane oxidation in both catalyst systems $4^*/CH_3CO_3H/AA$ and $4^*/CH_3CO_3H/EHA$.

The presented data show that for catalysts **4** and **4***, the iron(III)-acylperoxo intermediates can strongly affect the oxidation of adamantane only in the case of *m*-CPBA as oxidant. For H_2O_2 and CH_3CO_3H oxidants, the major active species of adamantane oxidation are the high-valent iron-oxo intermediates.

A similar study of the adamantane oxidation, catalyzed by complexes **1** and **1***, has revealed that the iron-oxo intermediates are the key active species of regioselective oxidation of adamantane, irrespective of the nature of the oxidant (hydrogen peroxide or peroxycarboxylic acid) (Scheme 14.6). The mechanistic landscape in Scheme 14.6 is corroborated by the chemoselectivity data, i.e. the A/K ratio pattern, and kinetic isotope effect values obtained for the cyclohexane (cyclohexane/cyclohexane-d_{12}) oxidation by catalyst systems with different complexes (**1**, **1***, **4**, and **4***) and different oxidants (H_2O_2, CH_3CO_3H and *m*-CPBA) [47].

To confirm the key role of the iron-oxo intermediates in the chemo- and regioselective oxidation of alkanes, the direct data on their reactivity toward alkanes have been obtained. The catalyst systems based on complex **4*** have appeared to be the most suitable for the detailed study on the reactivity of the iron-oxo species toward C—H oxidations, since these systems generate iron-oxo active species in higher concentrations than the systems based on other iron complexes used in our study.

The system $4^*/CH_3CO_3H/AA$ displays intense EPR signal of the intermediate $\mathbf{4c}^{*AA}$, assigned to the iron-oxo species (Tables 14.1 and 14.2) The self-decay of $\mathbf{4c}^{*AA}$ at −70 °C follows first-order kinetics, with the rate constant $k_1 = 1.8 \times 10^{-3}\ s^{-1}$. In the presence of cyclohexane, the decay of $\mathbf{4c}^{*AA}$ is dramatically accelerated, and is described by the effective rate constant $k_{obs} = k_1 + k_2 \cdot [C_6H_{12}]$, where $k_2 = 2.0 \times 10^{-3}\ M^{-1}\ s^{-1}$ (Figure 14.4). Therefore, $\mathbf{4c}^{*AA}$ directly oxidizes cyclohexane at −70 °C [47].

Cyclooctane has a weaker C—H bond than cyclohexane (BDE = 95.7 vs. 99.5 kcal mol^{-1}). Therefore, one can expect larger k_2 value for cyclooctane oxidation by $\mathbf{4c}^{*AA}$ than for cyclohexane. Indeed, the value for cyclooctane oxidation, $k_2 = 1.8 \times 10^{-2}\ M^{-1}\ s^{-1}$ has been found by an order of magnitude larger than the corresponding value for cyclohexane ($k_2 = 2.0 \times 10^{-3}\ M^{-1}\ s^{-1}$). The decay of $\mathbf{4c}^{*AA}$ at −70 °C in the presence of compounds with weaker benzylic C—H bonds (ethylbenzene, BDE = 87 kcal mol^{-1}, or cumene, BDE = 84.8 kcal mol^{-1}) was too fast ($k_2 > 0.1\ M^{-1}\ s^{-1}$) for measurements by the *in situ* EPR technique.

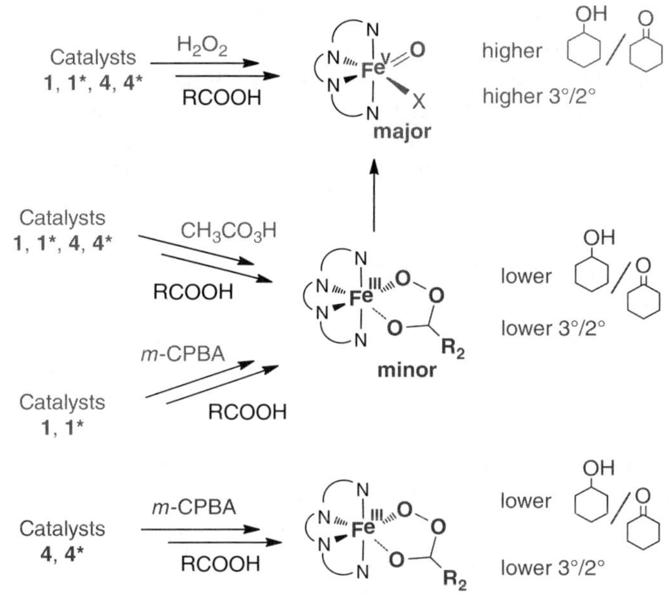

Scheme 14.6 Proposed reaction pathways for the studied catalyst systems (RCOOH = AA or EHA, $R^2 = CH_3$ or m-Cl-C_6H_4, X = OC(O)R for H_2O_2 oxidant, and OC(O)R^2 for peroxycarboxylic acid oxidant). (*See insert for color representation.*)

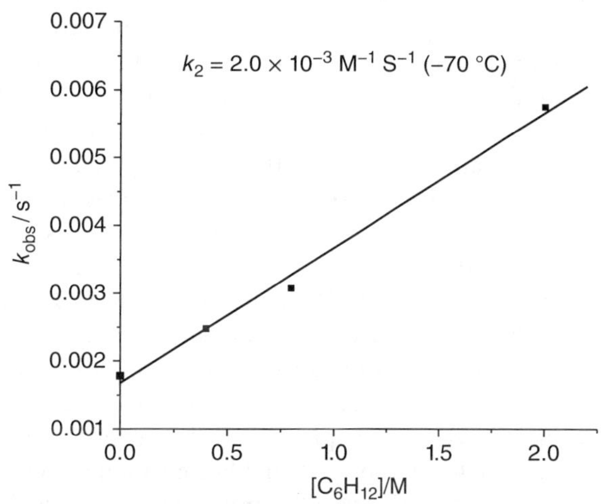

Figure 14.4 Kinetic data for the reaction of **4c*^{AA}** with cyclohexane at −70 °C, to evaluate the second-order rate constant.

The self-decay of **4c*^{EHA}** at −70 °C in the catalyst system **4***/H_2O_2/EHA ([**4***]:[H_2O_2]:[EHA] = 1 : 6 : 20, [**4***] = 0.02 M) obeys first-order kinetics, with the rate constant $k_1 = 3.3 \times 10^{-3}$ s^{-1}. In the presence of cyclohexane, the decay of **4c*^{EHA}** at −70 °C is described by the effective rate constant $k_{obs} = k_1 + k_2 \cdot [C_6H_{12}]$, with $k_2 = 2.6 \times 10^{-3}$ M^{-1} s^{-1}.

14.7 Final Comments

This chapter summarizes our recent findings related to the nature of active species and mechanisms of asymmetric epoxidation of olefins and C—H oxidations in the presence of bioinspired chiral iron aminopyridine complexes. The most important result is the direct EPR spectroscopic detection and identification of the metal-oxo intermediates in the catalyst systems (L)Fe/H_2O_2/RCOOH: intriguingly, two types of oxoperferryl complexes, differing in the electronic/spin state, have been detected. Intermediates of the first type, assigned to $[(L)Fe^{V}{=}O(OC(O)R)]^{2+}$ complexes, have large g-factor anisotropy ($g_1 \sim 2.7, g_2 \sim 2.4, g_3 \sim 1.7$), while those best represented as $[(L^{+\bullet})Fe^{IV}{=}O(OC(O)CH_3)]^{2+}$ and $[(L)Fe^{IV}{=}O(^{\bullet}OC(O)R)]^{2+}$ have small g-factor anisotropy ($g_1 = 2.07$, $g_2 = 2.01$, $g_3 = 1.96$). Intermediates of both types are reactive toward olefins even at low temperatures (−70…−85 °C). The catalyst systems displaying intermediates with small g-factor anisotropy exhibit superior epoxidation enantioselectivity, which has been explained in terms of additional stabilization via the unpaired electron delocalization over either the organic ligand or the carboxylic moiety. The key structural peculiarities, facilitating the formation of the intermediates with small g-factor anisotropy, are the presence of electron-donors at the pyridine ring, or the use of carboxylic acid with tertiary α-carbon as an additive.

A remarkable feature of aminopyridine iron-based catalysts is the capability of conducting olefin epoxidations with different oxidants, such as peroxycarboxylic acids and alkylhydroperoxides, with comparably high enantioselectivities. Combined catalytic, spectroscopic, kinetic, stereoselectivity, and isotopic labeling studies provide evidence in favor of different catalytically active species, and different mechanisms of oxygen transfer in these catalyst systems. In particular, the epoxidation by catalyst systems (L)Fe/R^1OOH/RCOOH (R^1 = H, t-butyl or cumyl) is predominantly conducted by the oxoperferryl intermediates, via stepwise mechanism, with formation of short-lived acyclic intermediate. At the same time, in the systems (L)Fe/R^2CO_3H(/RCOOH) ($R^2{=}CH_3$ or 3-Cl-C_6H_4), ferric acylperoxo complexes may have significant (even predominant) contribution to the stereoselective oxidation; in this case, the oxygen transfer to the olefinic double bond occurs in a concerted fashion.

We notice that the intermediates $[(L^{+\bullet})Fe^{IV}{=}O(OC(O)CH_3)]^{2+}$ are the first reported artificial analogues of species I in the catalytic cycle of cytochrome P450; it would thus be logical to systematically examine the reactivity and selectivity patterns of such intermediates in the C—H hydroxylation reactions. Not surprisingly, it has been found that the same oxoperferryl intermediates (both those with large and small g-factor anisotropy) are the predominant species, responsible for the C—H oxidation of cyclohexane and adamantane in the catalyst systems (L)Fe/H_2O_2/RCOOH and (L)Fe/CH_3CO_3H/RCOOH. For some of the iron-oxo species, their direct reactivity toward alkanes has been evaluated *in situ* at low temperature (−70 °C).

A notable exception is the catalyst systems **4**/*m*-CPBA/RCOOH and **4***/*m*-CPBA/RCOOH (RCOOH = AA or EHA), using terminal oxidant *m*-CPBA. The EPR and catalytic data obtained for the above systems provide evidence for the major role of iron(III)-*m*-chlorobenzoylperoxo intermediates in the C—H oxidations mediated by these systems.

We hope that the results summarized above will be a useful contribution to the fundamental understanding of oxidations mediated by biomimetic iron complexes, and will encourage the design of green and efficient asymmetric catalyst systems for practical applications. The mechanistic studies aimed at broadening the scope of iron complexes with different aminopyridine ligands are currently underway in our laboratory.

Acknowledgments

The authors acknowledge the financial support from the Russian Science Foundation (#17-13-01117).

References

1 White, M.C., Doyle, A.G., and Jacobsen, E.N. (2001). *J. Am. Chem. Soc.* 123: 7194.
2 Mas-Ballesté, R. and Que, L. Jr. (2007). *J. Am. Chem. Soc.* 129: 15964.
3 Chen, M.S. and White, M.C. (2007). *Science* 318: 783.
4 Chen, M.S. and White, M.C. (2010). *Science* 327: 566.
5 Lyakin, O.Y., Ottenbacher, R.V., Bryliakov, K.P. et al. (2012). *ACS Catal.* 2: 1196.
6 Cussó, O., Garcia-Bosch, I., Ribas, X. et al. (2013). *J. Am. Chem. Soc.* 135: 14871.
7 Groves, J.T. (2006). *J. Inorg. Biochem.* 100: 434.
8 Meunier, B., de Visser, S.R., and Shaik, S. (2004). *Chem. Rev.* 104: 3947.
9 Mansuy, D. (1993). *Coord. Chem. Rev.* 125: 129.
10 Nam, W. (2007). *Acc. Chem. Res.* 40: 522.
11 Costas, M. (2011). *Coord. Chem. Rev.* 255: 2912.
12 Krebs, C., Fujimori, D.G., Walsh, C.T. et al. (2007). *Acc. Chem. Res.* 40: 484.
13 Bryliakov, K.P. and Talsi, E.P. (2014). *Coord. Chem. Rev.* 256: 1418.
14 Lyakin, O.Y., Ottenbacher, R.V., Bryliakov, K.P. et al. (2013). *Top. Catal.* 56: 939.
15 De Oliveira, F.T., Chanda, A., Banerjee, D. et al. (2007). *Science* 315: 835.
16 Lyakin, O.Y., Bryliakov, K.P., Britovsek, G.J.P. et al. (2009). *J. Am. Chem. Soc.* 131: 10798.
17 Lyakin, O.Y., Bryliakov, K.P., and Talsi, E.P. (2011). *Inorg. Chem.* 50: 5526.
18 Lyakin, O.Y., Prat, I., Bryliakov, K.P. et al. (2012). *Catal. Commun.* 29: 105.
19 Lyakin, O.Y., Zima, A.M., Samsonenko, D.G. et al. (2015). *ACS Catal.* 5: 2702.
20 Van Heuvelen, K.M., Fiedler, A.T., Shan, X. et al. (2012). *Proc. Natl. Acad. Sci. U.S.A.* 109: 11933.
21 Serrano-Plana, J., Oloo, W.N., Acosta-Rueda, L. et al. (2015). *J. Am. Chem. Soc.* 137: 15833.
22 Oloo, W.N., Meier, K.K., Wang, Y. et al. (2014). *Nat. Commun.* 5: 3046.
23 Zima, A.M., Lyakin, O.Y., Ottenbacher, R.V. et al. (2017). *ACS Catal.* 7: 60.
24 Makhlynets, O.V. and Rybak-Akimova, E.V. (2010). *Chem. Eur. J.* 16: 13995.
25 Roelfes, G., Lubben, M., Chen, K. et al. (1999). *Inorg. Chem.* 38: 1929.
26 Duelund, L., Hazell, R., McKenzie, C.L. et al. (2001). *J. Chem. Soc., Dalton Trans.* 152.
27 Prat, I., Mathieson, J.S., Güell, M. et al. (2011). *Nat. Chem.* 3: 788.
28 Yoon, J., Wilson, S.A., Jang, Y.K. et al. (2009). *Angew. Chem. Int. Ed.* 48: 1257.
29 Company, A., Prat, I., Frisch, J.R. et al. (2011). *Chem. Eur. J.* 17: 1622.
30 Wang, Y., Janardanan, D., Usharani, D. et al. (2013). *ACS Catal.* 3: 1334.
31 Zang, X., Kim, J., Dong, Y. et al. (1997). *J. Am. Chem. Soc.* 119: 5964.
32 Weil, J.A. and Bolton, J.R. (2007). *Electron Paramagnetic Resonance: Elementary Theory and Practical Applications*, 2e, 85–117. Hoboken, NJ: Wiley.
33 McGown, A.J., Kerber, W.D., Fujii, H. et al. (2009). *J. Am. Chem. Soc.* 131: 8040.
34 Rittle, J. and Green, M.T. (2010). *Science* 330: 933.
35 Tse, C.-W., Show, T.W.-S., Guo, Z. et al. (2014). *Angew. Chem. Int. Ed.* 53: 798.
36 Oloo, W.N. and Que, L. Jr. (2015). *Acc. Chem. Res.* 48: 2612.
37 Cussó, O., Ribas, X., and Costas M. (2015). *Chem. Commun.* 51: 14285.
38 Zima, A.M., Lyakin, O.Y., Ottenbacher, R.V. et al. (2016). *ACS Catal.* 6: 5399.
39 Wu, M., Miao, C.-X., Wang, S. et al. (2011). *Adv. Synth. Catal.* 353: 3014.
40 Nishikawa, Y. and Yamamoto, H. (2011). *J. Am. Chem. Soc.* 133: 8432.
41 Dai, W., Li, G., Chen, B. et al. (2015). *Org. Lett.* 17: 904.
42 Ottenbacher, R.V., Samsonenko, D.G., Talsi, E.P. et al. (2014). *ACS Catal.* 4: 1599.
43 Ottenbacher, R.V., Samsonenko, D.G., Talsi, E.P. et al. (2016). *ACS Catal.* 6: 979.

44 Kim, C., Traylor, T.G., and Perrin, C.L. (1998). *J. Am. Chem. Soc.* 120: 9513.
45 Gómez, L., Garcia-Bosch, I., Company, A. et al. (2009). *Angew. Chem. Int. Ed.* 48: 5720.
46 Prat, I., Gómez, L., Canta, M. et al. (2013). *Chem. Eur. J.* 19: 1908.
47 Zima, A.M., Lyakin, O.Y., Bryliakov, K.P. et al. (2018). *Mol. Catal.* 455: 6.
48 Chen, K. and Que, L. Jr. (2001). *J. Am. Chem. Soc.* 123: 6327.

15

Selective Oxidation of Alkanes by Metallo-Monooxygenases and Their Nanobiomimetics

Chih-Cheng Liu[1], Yi-Fang Tsai[1], Wondemagegn H. Wanna[1], Ravirala Ramu[1], Damodar Janmanchi[1], Sunney I. Chan[1,2], and Steve S.-F. Yu[1]

[1] *Academia Sinica, Institute of Chemistry, Sec. 2, Nankang, Taipei, Taiwan*
[2] *National Taiwan University, Department of Chemistry, Sec. 4, Taipei, Taiwan*

15.1 Introduction

Selective oxy-functionalization of normal alkanes including methane, ethane, propane, and n-C_4-C_{12} alkanes is a difficult process. The C—H bond-dissociation energies for methane, and for primary CH_3—, secondary —CH_2—, and tertiary —CH— groups have been reported to be 105, 101, 99 to 96 kcal mol^{-1}, respectively [1–3], and the reactivity of the C—H bond for oxy-functionalization is expected to decrease as the bond strength increases. As a result, the activation of methane is a holy grail for researchers in organic chemistry. On the other hand, once a C—H bond does become activated, the bond strength of the α-C—H or C—C bond would subsequently decrease, facilitating a second oxidation to form a ketone or aldehyde as well as a third oxidation to form a carboxylic acid. If the products of these reactions exhibit even higher reactivity than the parent substrates toward other reactive channels, then it would not be surprising to observe lower oxygenated hydrocarbons arising from the rupture of C—C bonds.

Various mechanisms have been proposed for the catalysis of oxyfunctionalization of various hydrocarbons by biological metalloproteins and their biomimetics in their homogeneous or heterogeneous platforms. Most studies of these metalloproteins involve the iron and copper monooxygenases [4–6]. The pathways used by these metalloenzymes in the oxyfunctionalization invoke either a radical mechanism, including Fenton's chemistry, or electrophilic oxygenation by metal-oxygen species [7–11]. In the case of iron-based catalysts or enzyme systems, the oxygenated metal species have been shown to be Fe^{IV}=O, $Fe^{IV}(\mu\text{-}O)_2Fe^{IV}$, and Fe^{III}—OOH. For copper, the corresponding metal-oxygenated species are Cu^{III}=O, $Cu^{II}(\mu\text{-}(\eta^2{:}\eta^2)\text{-peroxo})Cu^{II}$ and $Cu^{III}(\mu\text{-}O)_2Cu^{III}$, and Cu^{II}—OOH. Recently, it has been also shown that a tricopper cluster can be exploited to harness a highly reactive "singlet oxene" for facile O-atom insertion across a C—C or C—H bond of an organic molecule at room temperature [12]. Theoretical calculations indicate that introducing one additional Cu^I ion into the $Cu^{II}(\mu\text{-}(\eta^2{:}\eta^2)\text{-peroxo})Cu^{II}$ and $Cu^{III}(\mu\text{-}O)_2Cu^{III}$ system leads to the formation of the $Cu^{II}Cu^{II}(\mu\text{-}O)_2Cu^{III}$ species [13].

The reactivity of each of the aforementioned oxygenated iron and copper species is strongly dependent on its electronic structure, including the spin state. Thus, when a number of electronic states are thermally accessible, the reactivity of these activated metal-based catalysts can become highly complex. In the case of the $Cu^{II}(\mu\text{-}(\eta^2{:}\eta^2)$ peroxo)Cu^{II} and $Cu^{III}(\mu\text{-}O)_2Cu^{III}$ system, the ground electronic state is a singlet due to strong through-bond "super-exchange"

Alkane Functionalization, First Edition. Edited by Armando J. L. Pombeiro and M. Fátima C. Guedes da Silva.

interaction between the Cu^{II} ions in the cluster [14, 15]. Typically, the excited triplet state is not thermally accessible in these activated dicopper complexes under ambient conditions. This simple scenario is not obtained with the $Fe^{IV}(\mu\text{-}O)_2Fe^{IV}$ complex, and the overall reactivity of this catalyst represents a composite of the intrinsic reactivity associated with each of the accessible electronic states of the spin ladder.

Maintaining the electronic spin state in the overall singlet state simplifies the chemistry of oxidation of the "inert" C—H bonds [16]. Thus, the mechanism of oxidation of a C—H bond mediated by the $Cu^{II}(\mu\text{-}(\eta^2{:}\eta^2)\text{-peroxo})Cu^{II}$ or $Cu^{III}(\mu\text{-}O)_2Cu^{III}$ catalyst involves either hydrogen-abstraction followed by geminal radical rebound or simple direct "singlet oxene" insertion across the C—H bond. Introducing one additional Cu^{I} ion into the $Cu^{II}(\mu\text{-}(\eta^2{:}\eta^2)\text{-peroxo})Cu^{II}$ and $Cu^{III}(\mu\text{-}O)_2Cu^{III}$ system to form the $Cu^{II}Cu^{II}(\mu\text{-}O)_2Cu^{III}$ species not only maintains the electronic state of the catalytic system in the ground singlet state but also facilitates tuning of the hardness of the O-atom to be transferred to the C—H bond in the O-atom transfer. The "singlet oxene" is highly reactive in this activated tricopper cluster and is responsible for the efficient methane oxidation in the particulate methane monooxygenase (pMMO) from methanotrophic bacteria [17]. It has been demonstrated that the O-atom transfer from the dioxygen-activated tricopper cluster to the C—H bond of the CH_4 is concerted [4, 18, 19]. A series of trinuclear model complexes based on this chemistry has been developed and shown to achieve efficient oxidation of linear alkanes to alcohols and/or alkanones under ambient conditions [17, 20].

15.2 Oxyfunctionalization of Hydrocarbons in Microbial Systems

Several bacteria have been isolated for their ability to utilize gaseous or liquid alkanes as growth substrates. These alkane-oxidizing bacteria have been conveniently categorized into three groups based on the lengths of the linear alkanes used as supplements for the bacterial growth media [21]. The three groups include methanotrophic bacteria that oxidize methane (C_1), and bacteria that oxidize other gaseous alkanes (C_2–C_4), and those grown on the liquid alkanes (C_5–C_{12}).

Methane-oxidizing bacteria have been shown to use methane monooxygenases (MMOs) to convert methane to methanol for their feedstock. In general, methanotrophic bacteria differentially express two forms of MMOs under conditions of different ratios of copper to biomass. One form is found in the cytoplasm of the cell, and is referred to as soluble methane monooxygenase (sMMO) [22–24]. sMMO is an iron enzyme, and the hydroxylase component is an $(\alpha\beta\gamma)_2$ dimer containing two non-heme binuclear iron clusters that activate O_2 and mediate the hydroxylation chemistry. pMMO is the other MMO, and is a membrane-bound protein [18, 25–28] and a multi-copper enzyme [4, 19, 29, 30]. Whereas pMMO has been found in all methanotrophic bacteria, sMMO has been isolated only from certain strains of methanotrophs.

Several Gram-positive bacteria mainly exploit the short-chain gaseous alkanes (C_2–C_4) for growth. The Corynebacterium-Nocardia-Mycobacterium-Rhodococcus complex can oxidize propane to 1- and 2-propanol [31]. Soluble diiron monooxygenases (SDIMOs) are known to be essential for the growth of the Gram-positive propanotrophs *Gordonia* sp. TY-5 and *Rhodococcus* sp. RHA-1 on propane [32–34]. In addition, the copper membrane monooxygenase (CuMMO) and the canonical diiron alkane monooxygenase (AlkB) from *Pseudomonas putida* GPo1 have been shown to be able to catalyze the oxidation of C_3 and C_4 *n*-alkanes in microbial organisms [35–38].

The Gram-negative *Pseudomonas spp.* (ATCC 43655) or *Thauera butanovorans* is reported to grow on short-chain alkanes other than methane [21, 39, 40]. Soluble butane monooxygenases (sBMOs) have been identified in these bacteria and have been characterized superficially. The sBMO complex consists of an iron-containing hydroxylase (BMOH), an NADH-oxidoreductase with flavin and Fe—S cofactors, and a small regulatory component protein, just like sMMO. O_2 is required for the degradation of *n*-butane, and 1-butanol is found to be the major product formed during the *n*-butane degradation.

Among the group of bacteria that oxidize longer-chain liquid alkanes using the AlkB system is *Pseudomonas oleovorans* (ATCC 29347) or *P. putida* GPo1, which can grow on C_3–C_{12} *n*-alkanes. The substrate range of this system includes straight-chain, branched, and cyclic alkanes as well as fatty acids, but not arenes. The AlkB system from *P. putida* GPo1 is a powerful enzymatic machine that introduces a terminal hydroxyl to alkanes of medium chain length (C_5–C_{12}) [41–43]. The hydroxylase in this system, AlkB, is an integral membrane non-heme diiron oxygenase [44, 45]. Together with its two redox components, AlkT (rubredoxin reductase) and AlkG (rubredoxin), AlkB converts *n*-alkanes to primary alcohols in a regioselective and stereoselective fashion with a high turnover frequency (TOF). However, detailed studies on the AlkB have been hampered by its membrane association and its multi-component nature.

Topological analysis and growth selection experiments [46, 47] have indicated that residue Trp 55 of AlkB controls the chain length of the hydrocarbon entering the active site of the AlkB enzyme. When this residue is mutated to a smaller yet still relatively hydrophobic residue such as serine, alanine, valine, leucine, or isoleucine, the AlkB system is shown to be able to oxidize alkanes longer than C_{13}. We surmise that the interaction of the terminal methyl group of the hydrocarbon substrate with the aromatic π-system of Trp 55 (10 π electrons) situated at the bottom of the active-site pocket of the AlkB system could be one of the key factors influencing the observed substrate specificity and regioselectivity, as well as stabilizing certain substrate orientations (Figure 15.1). Since earlier studies have suggested C—H…π and C—F…π interactions

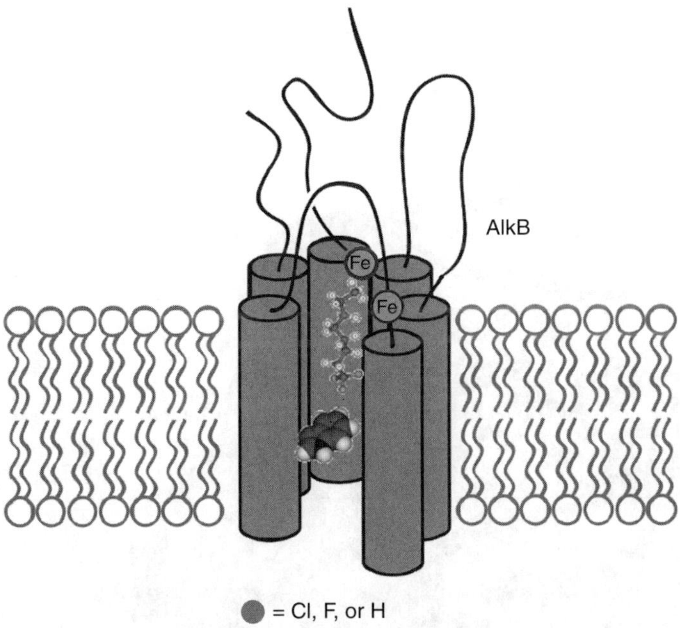

Figure 15.1 The topology model predicted from the hydropathy profile suggests that the hydrophobic pocket comprising Trp 55 could encapsulate medium-sized straight-chain substrates and exploit distal London dispersive forces or van der Waals interactions between C—H, C—X, or XC—H (X = F, Cl) of the substrate and the aromatic π-system of the tryptophan to tighten up the active site and promote activation of the omega terminal C-1. Source: Adapted with permission from Ref. [5]. Copyright 2014 Elsevier B.V.

to be quite comparable [18, 48, 49], it is, in principle, possible to test various C—X (or X) and C—H......π interactions (X = F, Cl, and Br) to tune the reactivity of the enzyme [50–52].

The native substrates of cytochrome P450 BM3 are C_{12}–C_{16} fatty acids. It has been suggested from crystallographic studies that the stereochemical controls on the substrate are exerted by this protein at a site within the heme pocket some 6–8 Å above (distal) the iron porphyrin center [53–55]. According to crystal structures in PDB ID: 1JPZ [53] and 1FAG [54], this asymmetric pocket can be divided into large and medium sub-pockets, labeled as "**L** zone," and "**M** zone," respectively, in Figure 15.2. The solvent-exposed polar residues Arg 47 and Tyr 51 of the protein, together with the water network, are shown to serve as hydrogen-bonding acceptors or donors to anchor the carboxylate end of the native substrate fatty acid, and in this manner exert distal control over the oxidations of C_{12}–C_{16} fatty acids at the ω-x positions ($x = 1$–3) [57–60].

However, in addition to the hydrogen bonding of the carboxylate end of the fatty acid with the distal water networks, van der Waals interactions between the acyl chain and the wall of the hydrophobic pocket are observed to occur in the **M** zone. Chiang et al. [56] have shown that these van der Waals interactions or dispersions (London dispersion forces) can be substantially enhanced by the introduction of fluorinated substituents into the alkyl chain in the vicinity of the **M** zone of the hydrophobic pocket. For example, substituting one, two, and especially all three of the hydrogens of the terminal methyl group of lauric acid with fluorines significantly improves the regioselectivity of the oxidation reaction, with the trifluorinated lauric acid yielding a single product and the hydroxylation occurring at the ω-3 position.

Long-chain perfluorocarboxylic acids have been exploited as rigid chemical space fillers or "dummy atoms" to compress the size of the substrate-binding pocket of cytochrome P450 BM3 for the hydroxylation of small gaseous alkanes. With this strategy, the wild-type cytochrome P450 BM3 has been shown to mediate the hydroxylation of n-butane and propane and allow

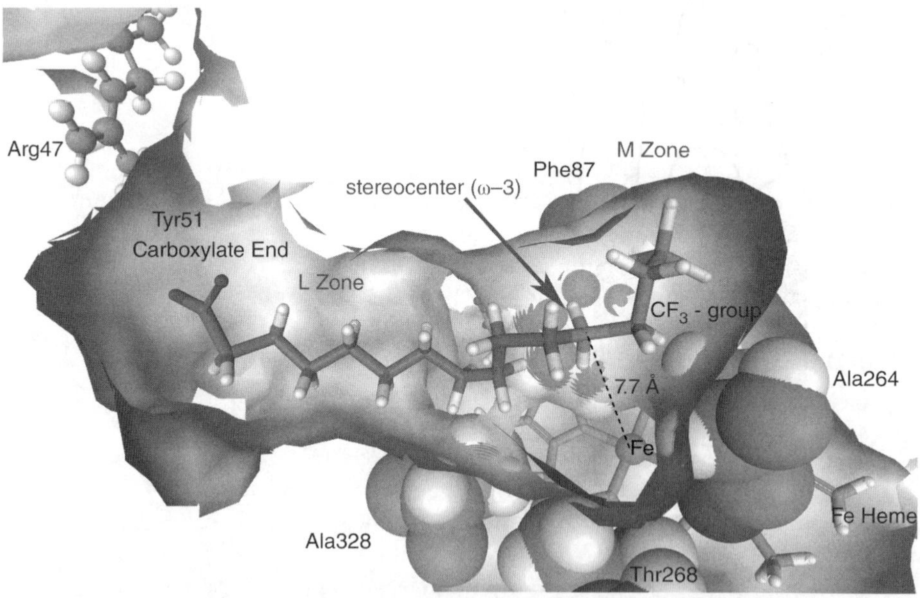

Figure 15.2 Three-dimensional modeling depicting the substrate binding in the active site of cytochrome P450 BM3 (PDB ID: 1FAG, modified by Accelrys®2.0). This model illustrates the encapsulation of the substrate, 15,15,15-trifluoropentadecanoic acid, instead of palmitoleic acid in the original structure. Source: Reproduced with permission from Ref. [56]. Copyright 2013 John Wiley & Sons. (*See insert for color representation.*)

these gaseous alkanes to be converted into secondary alcohols with reasonable turnover numbers (TONs) [61]. More excitingly, this enzyme has been shown under ambient and high pressures to oxidize ethane and propane [57, 62, 63].

15.3 Metal Active Site Features of Metallo-Monooxygenases for Inert Aliphatic C—H Bond Oxidation

All the metallo-monooxygenase systems under discussion here, pMMO, AlkB, sMMO, and cytochrome P450, are capable of oxidizing strong aliphatic C—H bonds. The metal active sites of these enzymes are found to contain at least one transition metal ion, which can be manipulated and tuned to achieve the difficult C—H bond oxidations. In the case of cytochrome P450, the active site consists of a heme iron porphyrin. The active site in the hydroxylase of sMMO as well as AlkB comprises a non-heme diiron core. In pMMO, it is either a dinuclear or trinuclear copper cluster [17, 19, 25, 29, 64]. To initiate catalytic turnover, these metal ions must be reduced before they can be activated by O_2 to form the oxygenated metal species participating in the oxidation chemistry.

15.3.1 pMMO

The catalytic conversion of methane into methanol by pMMO has been proposed to occur at either a dinuclear (site B) or trinuclear copper active site (site D). Site D, consisting of a $Cu^{I}Cu^{I}Cu^{I}$ tricopper-cluster center, can be activated by dioxygen to produce a mixed-valence $[Cu^{II}Cu^{II}(\mu\text{-}O)_2Cu^{III}]$ intermediate that promotes facile concerted insertion of a highly reactive "singlet oxene" into one of the C—H bonds of methane as well as C_2–C_5 straight-chain alkanes [4, 13, 17, 19]. In this chemistry, the products of C—H bond oxidation have been shown to display full retention of configuration, with the reactions proceeding in a concerted manner. The pMMO consistently displays H/D kinetic isotopic effects (KIEs) of 5.2–5.5 in experiments using tailored chiral $[1\text{-}^2H_1,1\text{-}^3H_1]$ethanes and $[2\text{-}^2H_1,3\text{-}^2H_1]$butanes as substrates [65, 66]. These observations have led to the conclusion that during methane oxidation the copper monooxygenase generates a "singlet oxene" intermediate harnessed by the tricopper cluster. The pMMO is capable of efficiently oxidizing methane to methanol, ethane to ethanol, propane and *n*-butane to 2-propanol and 2-butanol, respectively [4, 65, 67]. It also epoxidizes propene to propylene oxide, as well as related alkenes. Consistent with the "singlet oxene" mechanism of O-atom transfer to the substrate, the C=C bond epoxidation proceeds by electrophilic syn addition without epimerization [48].

The H/D KIEs obtained for pMMO are inconsistent with a strong KIE (15–25) predicted from the radical mechanism for a dicopper system [13, 68]. The result, however, is close to the theoretically calculated KIE value of 4.3 for oxidation of methane by the dioxygen-activated tricopper cluster. Moreover, when nuclear tunneling effects are included in the calculations of the rate constants, the KIE is 5.2. This agreement between theory and experiment offers strong, if not compelling, evidence that the active site of the copper monooxygenase is a trinuclear copper cluster [13].

15.3.2 sMMO

For the reaction of CH_4 with O_2 catalyzed by sMMO, a non-heme diiron monooxygenase, the key reaction intermediate is the $Fe^{IV}(\mu\text{-}O)_2Fe^{IV}$ species, often denoted compound **Q**, which has been shown to have a "diamond core" structure [69–72]. Both of the irons within the active site are found to have similar geometrical and electronic environments, with a short 2.46 Å distance between the two metal ions according to X-ray absorption spectroscopic data. But theoretical studies and representative diiron model complexes show that the distance could be longer in such a diiron active species and argue that the structure of **Q** could be more open, with the catalytic site functioning as individual/terminal $Fe^{IV}{=}O$ units [73]. Moreover, in a recent study, it has been shown that a synthetic valence-delocalized $[Fe^{3.5}(\mu\text{-}O)_2Fe^{3.5}]^{3+}$ diamond core is readily converted into a valence-localized $[HO{-}Fe^{III}{-}O{-}Fe^{IV}{=}O]^{2+}$ open core to efficiently cleave C—H bonds over a million-fold faster than its parent compound in the case of the C—H bond oxidation of 9,10-dihydroanthracene (C—H bond strength = 78 kcal mol^{-1}) [74]. In the case of sMMO, it has been suggested that the conversion of the terminal $Fe^{IV}{=}O$ species into $Fe^{IV}{=}O_{LS}$ and $Fe^{IV}{=}O_{HS}$ centers (where LS is low spin and HS is high spin) may be important for the catalytic activity of the active site [73]. On the other hand, it is also possible that redistribution of the electron density in the intermediate **Q** to form the more highly reactive Fe^{V}/Fe^{III} diiron center may be essential for the enzyme to catalyze a facile transfer of the O-atom to CH_4 [4]. Regardless, the intrinsic reactivity of Compound **Q** toward C—H bond oxidation is complicated because super-exchange between the two terminal $Fe^{IV}{=}O$ units leads to a spin ladder of thermally accessible electronic states, possibly, each with different intrinsic reactivity toward O-atom transfer to an alkane substrate. At this juncture, it appears that the mechanism of CH_4 oxidation in sMMO remains an open issue.

15.3.3 AlkB

AlkB is a diiron monooxygenase [45] with an anti-ferromagnetically coupled high-spin Fe^{III}-Fe^{III} active site in its resting state, as evidenced by Mössbauer studies. Similar to the mammalian integral membrane protein stearoyl-CoA desaturase (SCD), AlkB contains four histidine-containing motifs (two HXXHH motifs, one HXXXH motif, and one NXXH motif) that presumably comprise the ligands to coordinate the bimetallic center. In a recently determined SCD crystal structure of a variant in which Zn^{II} ions have been substituted for the two Fe^{III} ions, the distance between the two metal (Zn^{2+}) ions is observed to be 6.4 Å, with each metal ion associated with four to five histidine residues including the TM4 motif ($N^{261}XXXH^{265}$) [75]. An X-ray absorption spectroscopic study suggests that the rising K-edge of normalized Fe X-ray absorption near-edge spectroscopy (XANES) spectra of a purified recombinant AlkB [38] is consistent with the oxidized form of the non-heme diiron hydroxylase in sMMO, namely $MMOH_{ox}$, and its complex with the regulatory component (MMOD) in the $MMOH_{ox}$ + MMOD protein–protein complex [76]. This observation implies that the AlkB is purified as an iron protein in the ferric state. Pre-edge data fittings [38] indicate that these spectroscopic features of AlkB are also similar to those of a series of non-heme diiron complexes without a short oxo-bridged active site, and in particular, are comparable to those of the oxidized form of the hydroxylase subunit ($MMOH_{ox}$) in sMMO. The integrated area of the pre-edge peak obtained from diferric AlkB is 11.2 units, suggesting that the irons in AlkB exist either in a 5-coordinate or distorted 6-coordinate environment [77]. Further, the 1s→3d transition is resolved into two peaks at 7114.1 and 7115.9 eV [78] with integrated intensities of 6.7 and 4.5 units, respectively. This pre-edge profile of the purified AlkB is reminiscent of distinct pre-edge signals observed in the case of diiron model complexes such as $[FeOH(H_2O)Chel]_2(H_2O)_4$ and $[FeOH(H_2O)Dipic]_2$ with an intense feature at low energy but a shoulder at high energy [79]. The pre-edge profiles of $MMOH_{ox}$ and the $MMOH_{ox}$ + MMOD complex in sMMO also exhibit two resolved peaks with intensities of 7.7 ± 1.3 and 8.3 ± 0.4 units at 7114.0 eV as well as 1.2 ± 0.1 and 2.3 ± 0.8 units at 7115.5 eV, respectively [76, 78]. Thus, the pre-edge data on the purified AlkB offer compelling evidence,

Table 15.1 The Fe EXAFS data of AlkB, MMO_{ox} of sMMO, and stearoyl-acyl carrier Δ^9-desaturase indicate that these catalysts of aliphatic C—H bond oxidation all exhibit shorter Fe—O/N (first shell) and Fe—Fe (second shell) bonds than those displayed by $metHrN_3$, a dioxygen carrier, and RNR $R2_{met}$, an enzyme that converts nucleoside diphosphate into deoxynucleoside diphosphate.

Non-heme diiron enzymes	Fe—O′		Fe—O/N		Fe—C		Fe—Fe		Fe—C/N	
	N	*R* (Å)	*N*	*R* (Å)	*N*	*R* (Å)	*N*	*R* (Å)	*N*	*R* (Å)
AlkB	—	—	5	2.02			1	3.08		
$sMMO_{ox}$	—	—	5	1.99–2.01			1	3.00–3.03		
$metHrN_3$	1	1.80	5	2.13	4.2	3.05	1	3.19	5.1	4.33
RNR $R2_{met}$	1	1.80	5	2.06	3.1	3.03	1	3.22	3.8	4.30
Δ^9-desaturase	0.7–1	1.79	5	2.03	3.1	2.97	1	3.13		

albeit indirect, for the presence of a high-spin non-heme diferric core center in this system. Interestingly, the integrated area of the pre-edge peak obtained from AlkB is also in agreement with the values obtained from another study of MMO_{ox} (~11.6 units), its complex with MMOB (~11.2 units), [80] hemerythrin (~10.4 units), soluble stearoyl-acyl carrier protein Δ^9–desaturase (10.8 or 11.5 units) and RNR $R2_{met}$ (~10.1 units; RNR R2: ribonucleotide reductase subunit R2) [81, 82].

Fe K-edge extended X-ray absorption fine structure (EXAFS) spectroscopy has provided local structural information on the metal active site of AlkB as purified. The data indicate that the active site of AlkB exists in its oxidized form and consists of 5 × Fe—O/N with bond lengths of 2.02–2.03 Å (first-shell) and one Fe-Fe with a distance of about 3.07–3.08 Å (second-shell) [38]. These determined distances are shorter than the corresponding ones in the two non-heme diiron proteins $metHrN_3$ and RNR $R2_{met}$, which are known to display μ-oxo bridges (Table 15.1). These results suggest that the active site of AlkB is in a more congested coordination environment than those found for the active sites of $metHrN_3$ and RNR $R2_{met}$, and presumably, similar to the active site of MMOH, which has been reported to exhibit a somewhat shorter Fe—Fe bond (c. 3.0 Å) [76, 81, 83, 84]. A comparison of the structural features between the active sites of AlkB and MMOH is expected to be helpful toward elucidating how AlkB activates the C—H bonds of linear alkanes. As in sMMO, the catalytic intermediate is produced upon reduction of the two iron centers to the ferrous state and activation by O_2. However, the subsequent chemistry remains to be elucidated.

15.3.4 Cytochrome P450 BM3

A catalytic cycle of alkane hydroxylation by cytochrome P450 BM3 has been proposed. According to this proposal, the resting state of P450 BM3 contains a 6-coordinate ferric iron heme with a water molecule as an axial ligand (Figure 15.3). It is suggested that the hydrophobic interaction between the aliphatic substrate and P450 BM3 expels water molecules from the hydrophobic pocket and leads to the formation of a 5-coordinate ferric iron heme (step *a* in Figure 15.3). It appears that the latter process is largely controlled by desolvation [53, 85–87]. The resulting high-spin ferric intermediate has a high redox potential and is readily reduced to the ferrous heme by accepting electron(s) from NADPH mediated by the FAD/FMN cofactors in the reductase domain of the BM3 protein (BMR) (step *b* in Figure 15.3). The iron heme center is then, according to this proposal, poised to bind O_2 and to receive additional electrons, culminating in the formation of the high-valence ferryl oxo porphyrin cation radical species, namely compound I (Cpd I) [9, 59, 88, 89, 106, 107] (step *c* in Figure 15.3). The formation of Cpd I

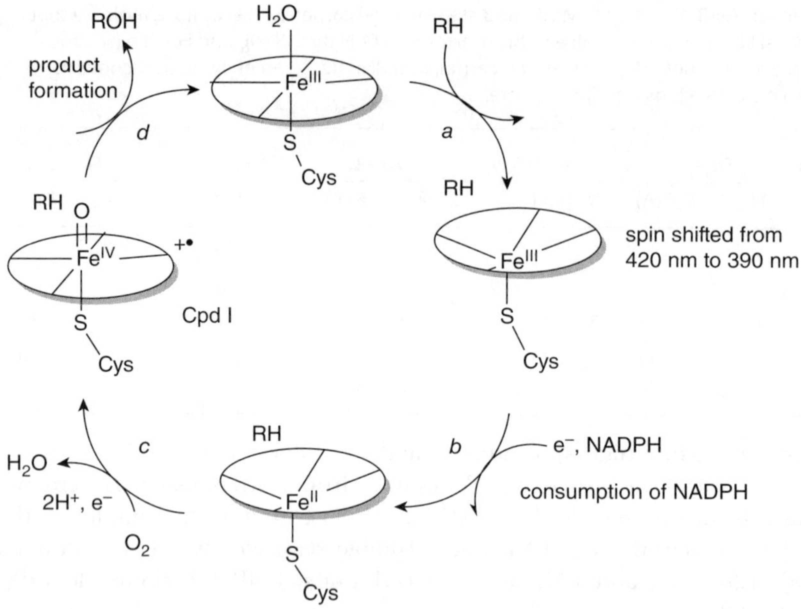

Figure 15.3 The catalytic cycle of alkane hydroxylation for cytochrome P450 BM3 (CYP102A1). Source: Reproduced with permission from Ref. [56]. Copyright 2013 John Wiley & Sons.

here plays a critical role in the hydrogen atom abstraction/radical rebound reaction or O-atom insertion mechanism for the controlled C—H activation in the catalysis (step *d* in Figure 15.3) [8, 9, 88, 90–92]. Regardless of the reaction mechanism, for the metalloprotein to display its compound selectivity, regioselectivity, and stereoselectivity, there must be forces between the hydrocarbon molecule to be oxidized and the protein pocket to direct the substrate toward the oxygenated iron heme and orient it within the active site after entering from the solution following de-solvation of the hydrophobic pocket(s). These interactions include roughness of the protein inner surface within the active site, hydrophobic interactions, and nonbonding interactions, including electrostatic and van der Waals/London dispersive forces.

Cpd I, the oxyferryl/porphyrin cation radical species in cytochrome P450, possesses essentially degenerate doublet (LS) and quartet (HS) spin states, which may give rise to different reaction features based on the two-state reactivity (TSR) concept [93]. There will be two distinct transition states for the two electronic states in the chemical reaction mediated by the oxygenated iron heme. In the LS state, the kinetic barrier is expected to be low so that hydrogen tunneling should contribute relatively little to the H-abstraction process. In this scenario, it is possible that the O-atom transfer can proceed in a predominantly concerted barrier-free manner. In contrast, the kinetic barrier for hydrogen abstraction should be substantially higher in the HS state, namely the quartet spin state of Cpd I. Accordingly, the substrate oxidation mediated by the HS state should follow a kinetic pathway that resembles more closely to the H-abstraction radical rebound mechanism. Experimentally, different H/D KIEs (k_H/k_D) of 3.4–4.5 and ca. 5.5 for the $S = 1/2$ and $S = 3/2$ states, respectively, have been reported for a series of *N*,*N*-dimethylaniline derivatives, providing the first example that different spin states may give rise to different KIE values [94]. It should be noted, however, that the computed KIE_{LS} values for the *N*,*N*-dimethylaniline substrates are substantially smaller than the computed KIE_{HS} values: only the KIE_{LS} values fit the experimental KIE. In the case of Cpd I of P450, the HS and LS states involve two energetically similar spin-state surfaces and earlier calculations show tunneling to modestly affect the H/D KIE values only [95]. Here, changes in the H/D KIE values due to tunneling corrections have been estimated to be less than 20% for the two electronic spin states. In principle, since the tunneling corrections are expected to be higher for the HS state than the LS state, the H/D KIE could serve as a probe for the reactive electronic state [96].

Recently, two tailored dideuterated enantiomers of butane, specifically (2*R*,3*R*)- and (2*S*,3*S*)-[2-$^2H_1$,3-$^2H_1$]butanes, have been subjected to oxidation by cytochrome P450 BM3 variants with the reactivity toward the C2—H or C2—D bonds delineated [97]. Two distinct H/D KIE values, 7.3–7.8 and 18–25, have been obtained for these two butanes, respectively. These experimental findings are highly significant. First, the heterogeneity in the KIEs for the deuterated butane substrates offer compelling support for the proposed TSR in the catalysis of the alkane oxidation by the cytochrome P450 system. Second, it is interesting that H/D tunneling contributes so significantly to the activation of the (2*S*,3*S*)-[2-$^2H_1$,3-$^2H_1$]butane C-H_R bond, given the rather similar spin-state surfaces predicted by theory. Typically, TSR gives higher H/D KIE values for weaker C—H bonds. For example, the C—H bond activation by the non-heme enzyme, TauD, has been reported to proceed in the quintet state with an observed H/D KIE above 30 [98]. At this juncture, we know of limited cases with higher H/D KIE values for oxidation of the stronger C—H bonds in small alkanes [96, 99, 162]. In any case, the above study offers the first evidence for H/D KIE tunneling observed for secondary C—H bond oxidation mediated by one of the most efficient cytochrome P450 systems known to date. The implication is that the chemical reaction mediated by the iron heme P450 system is mediated by more than one distinct transition state. However, the outcome is dramatically different between the two deuterated *n*-butane isotopomers examined, with the oxidation of one of the isotopomers proceeding in the LS state with tunneling contributing relatively little to the H-abstraction process, thus proceeding predominantly in a concerted barrier-free manner, while the oxidation of the other isomer proceeding mostly in the HS state with substantial H/D tunneling to form the 2-butanol according to the hydrogen abstraction radical rebound mechanism (Figure 15.4).

15.4 Engineering Cytochrome P450 BM3 and Alkane Hydroxylase (AlkB) for Gaseous Alkane Oxidation

15.4.1 Cytochrome P450 BM3

The hydroxylation of fatty acid substrates by P450 BM3 has been shown to occur at the ω-1, ω-2 or ω-3 position of the fatty acid, with the regioselective and stereoselective activation of the C—H bonds thought to occur at a site within a pocket 6–8 Å above (*distal*) the iron porphyrin [5, 53, 55]. The spatial orientation of the substrate in this pocket together with the asymmetric environment of this pocket resulting from it having distinct zones (**M** zone versus **L** zone) has been shown to yield enantiomeric excess (*ee*) values for the hydroxylation of C_{12}, C_{14}, and C_{16} fatty acids at the ω-1, ω-2, and ω-3 positions of 86–99%, 86–96%, and 44–74%, respectively, favoring the formation of secondary alcohols in the *R* configuration [5, 55, 101–103]. Recent molecular dynamics simulations (MD) and quantum-mechanical/molecular mechanics (QM/MM) studies indicate residue F87 to be essential in the control of the regioselectivity by juxtaposing the CH_2 positions ω-1/ω-2/ω-3 of *N*-palmitoylglycine (NPG) with Cpd I [104]. MD trajectory analysis shows the *pro-R* C—H bond to be closer to Cpd I for both the ω-1 and ω-2 sites than the *pro-S* C—H bond, and indicates that F87 enantioselectively blocks the *pro-S*

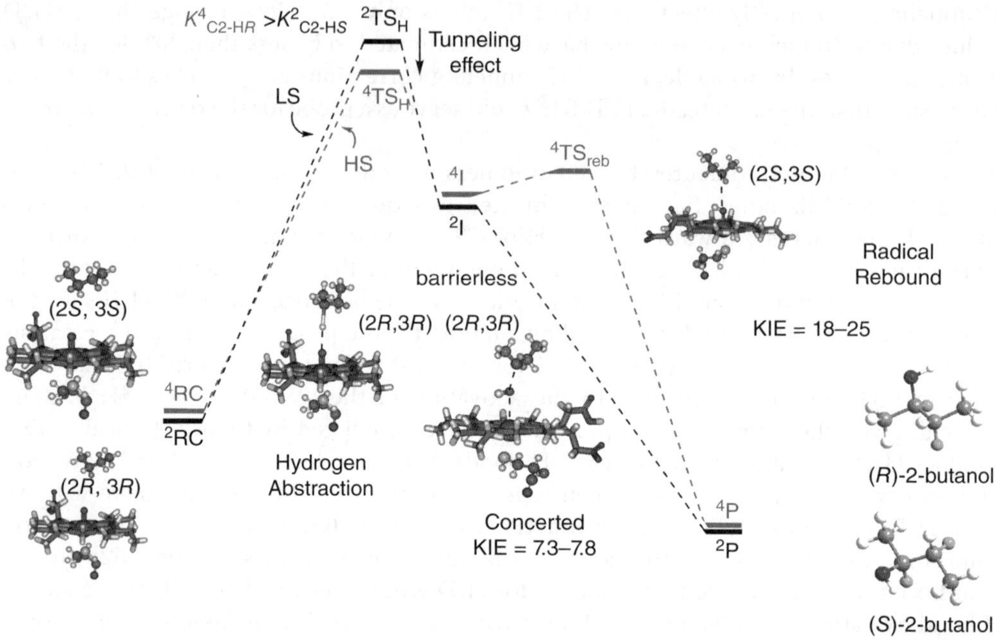

Figure 15.4 Reproduced two-state reactivity diagram [100] showing the initially degenerate HS and LS spin states of Cpd I in cytochrome P450 BM3 variants leading to two distinct KIEs obtained from butane substrates, with the oxidation of the C—H_R bond in **(2S,3S)-[2-$^2H_1$,3-$^2H_1$]butane** predominantly proceeding from the HS state after tunneling, and forming (*R*)-2-butanol according to the radical rebound reaction mechanism [95, 100]. RC: reactant complexes; TS_H: transition state for H-abstraction; I: intermediate; TS_{reb}: transition state for rebound species; P: product. Source: Reproduced with permission from Ref. [97]. Copyright 2017 John Wiley & Sons.

C—H bond, features resulting in a more facile abstraction of the H from the *pro-R* C—H bonds and a preference for *R* hydroxylation.

The hydroxylation of *n*-octane to form a 2-, 3-, or 4-ol has been carried out by an engineered cytochrome P450 BM3 variant containing the mutations A74G, F87V, and L188Q referred to as the 3mt protein or **BM3 GVQ** [105]. In these experiments, the *ee* values for the (2*R*)-octan-2-ol and (3*R*)-octan-3-ol products are only 20% and 39%, respectively [55]. However, after introducing *gem*-difluoromethylene substituents onto the *n*-octanes at their C-1 to C-4 positions, respectively, to tighten up the substrate-binding pocket, a significant improvement is observed in the stereoselective hydroxylation by the oxyferryl species/Cpd I to give the the *gem*-difluorinated 2-ol and 3-ol products, with the *ee* values increasing, respectively, to 52–67% and 65–80% in favor of the *R* configuration [55]. Enhancement in the coupling efficiency between the transfer of reducing equivalents from NADPH and product formation is observed as well. It is evident that the enhanced dispersive forces resulting from the fluorine substituents on the aliphatics substantially affect the free-energy landscape of the binding of the substrate to the enzyme. This change in the interaction of the hydrocarbon molecule with the enzyme influences the orientation and/or a specific conformer adopted by the substrate when bound in the hydrophobic pocket of the enzyme, leading to the superior regioselectivity and/or enantioselectivity in the hydroxylation.

An A328F variant of cytochrome P450 BM3 has also been shown to exclusively catalyze the hydroxylation of C-2 (or ω-1) position on *n*-octane derivatives, including the *gem*-fluorine-substituted *n*-octanes. In this mutant, the linear *n*-octane molecule presumably

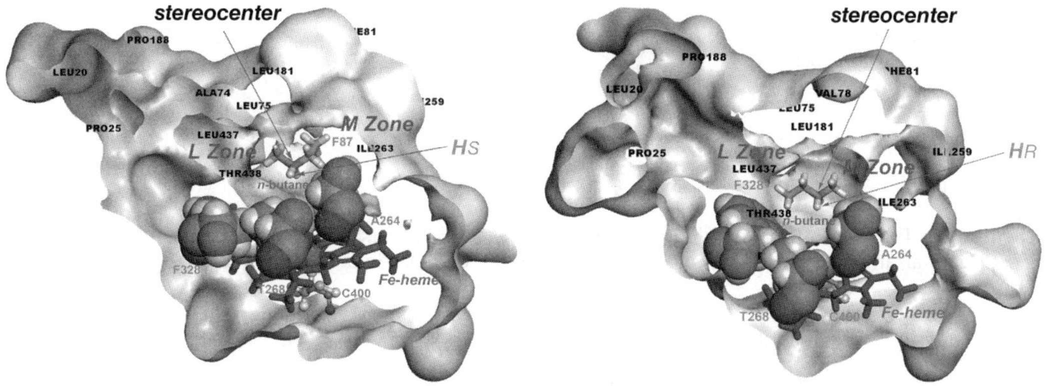

(a) (b)

Figure 15.5 Two proposed conformational states of *n*-butane in the hydrophobic pocket of **BM3 PF** are indicated to lead to the oxidation of (a) C2—H_S and (b) C2—H_R bonds, respectively, by the iron heme center (reproduction of PDB: 1JPZ [53] by Discovery Studio 4.5, Dassault Systèmes BIOVIA). Source: Reproduced with permission from Ref. [97]. Copyright 2017 John Wiley & Sons. (*See insert for color representation.*)

lodges in the substrate-binding pocket of this particular mutant in a way that allows the oxyferryl species to directly contact the ω-1 carbon atom.

To enable the activation of *n*-butane by P450 BM3, the P450 BM3 variants **BM3 PF** (L188P A328F) and **BM3 EPF** (A74E L188P A328F) have been prepared, expressed in *E. coli*, and purified to homogeneity. With these two variant enzymes, *n*-butane is converted into 2-butanol as the sole product. A P450 BM3 A74E L188P variant has also been reported to assist the activation of smaller alkanes [108]. In this mutant, with the conformation of the butanes shown in Figure 15.5, one can surmise that *n*-butane (or propane) is the smallest linear alkane that can be tightly bound to the substrate-binding pocket. However, with an *n*-butane substrate confined in the "**M**" zone, the dispersion forces are stronger than in the "L" zone, and this should be manifested in the *ee* of the product butanols. Accordingly, we expect the dominant conformer positions itself in the binding site with one of the terminal and the adjacent secondary carbon atoms of the *n*-butane in the "**M**" zone and the adjacent C3—H center points toward the iron heme center. In this configuration, the oxidation would prefer to take place at the C3—H_S bond (Figure 15.5). Experimentally, *de* (or *ee*) values of −52% and −56% in favor of the *S* configuration are observed for the conversion of *n*-butane into 2-butanol mediated by the cytochrome P450 **BM3 PF** and **EPF** mutants, respectively, in accordance with prediction.

15.4.2 AlkB

The fluorinated substituents in fluorinated *n*-octanes have been shown to exert a much stronger control when these C—F groups were embedded within the substrate pocket of AlkB. The terminal hydroxylation of the *gem*-difluorinated octanes, including trifluorinated octane by recombinant AlkB (Scheme 15.1) from *P. putida* GPo1, exhibits high selectivity for the formation of fluorinated octan-1-ols at the distal methyl group [52]. Interestingly, even with 3,3- and 4,4-difluorooctanes as substrates, the major products still give C—H activation at the omega position (terminal methyl group) (Scheme 15.2). However, minor products are also formed resulting from hydroxylation of the more proximal terminal methyl group. Apparently, the fluorine substituent at the C-3 or C-4 position of *n*-octane can guide the reaction to occur at the more proximal position, albeit to a low extent. Since only omega C—H activation has been

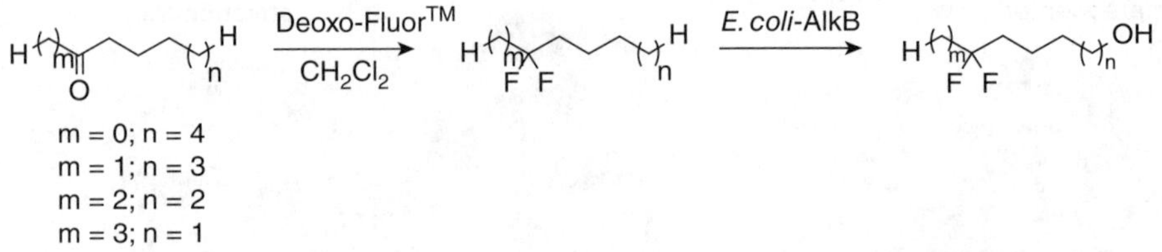

Scheme 15.1 The synthesis of *gem*-di-fluorinated octanes and their regioselective oxidations mediated by a recombinant alkane hydroxylase system (from *alkBGT* genes) of *P. putida* GPo1.

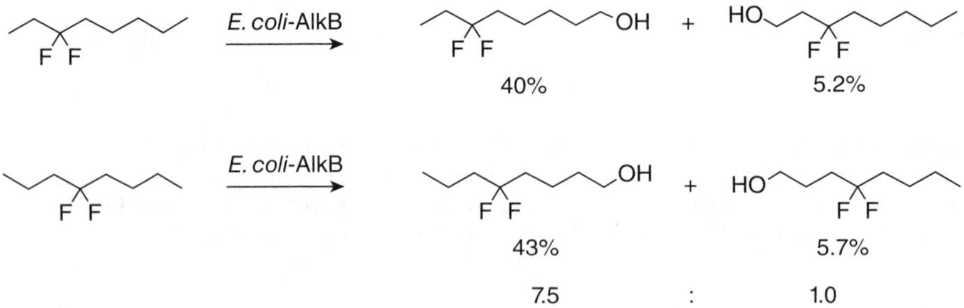

Scheme 15.2 Regioselective oxidation of 3,3- and 4,4-difluoro-octane substrates mediated by *Escherichia coli* whole cells with the recombinant alkane hydroxylase system (from *alkBGT* genes) of *P. putida* GPo1. Source: Adapted with permission from Ref. [5]. Copyright 2014 Elsevier B.V.

observed with 1,1- and 2,2-difluorooctanes, we surmise that the control must have been exerted by van der Waals interactions of the —CF_2— group with Trp 55 within the hydrophobic pocket of the alkane hydroxylase [47]. The shape of the pocket presumably coerces the linear alkanes into the all trans conformation and hence exerts the distal control of the diiron active site toward primary C—H bond oxidation (Figure 15.1).

A series of XANES experiments have shown that the fully reduced AlkG can efficiently reduce AlkB [38]. Based on the redox experiments in these XANES studies, a strategy has been developed to catalytically oxidize linear alkanes to primarily alcohols by directly supplying electrons to $AlkG_{oxidized}$ using an electrochemical approach to shuttle electrons to $AlkB_{oxidized}$ in the AlkG-AlkB protein-protein complex. Fully reduced AlkG has been prepared electrochemically through a direct electron transfer (DET) process using AlkG immobilized on a disposable screen-printed carbon electrode (SPCE) with a surface roughness (Ra) of 2.02 nm. The ratio of the effective surface coverage of the AlkG proteins on the SPCE based on the AlkG added to the SPCE is within 0.5–2% [109]. The catalytic conversion of medium-chain-length (C_5–C_{12}) normal alkanes into their corresponding primary alcohols mediated by the SPCE electrodes exhibits apparent TOFs of between 250 and 1000 min^{-1} per unit AlkG when the concentration ratio of AlkB to AlkG is about 2 (TOF is determined based on the amount of AlkG placed on top of the SPCE instead of the electroactive AlkG, i.e. the effective surface coverage ratio of AlkG) [38] (Figure 15.6). The specific activities (expressed as TOF) for the conversion of C_6, C_7, and C_9 *n*-alkanes into their corresponding primary alcohols are measured to be between 500 and 1000 min^{-1}, and those for the oxidation of C_5, C_8, C_{10}–C_{12} *n*-alkanes between 250 and 400 min^{-1}. Note here that there is no obvious correlation between the specific activity and the chain lengths of the alkanes. This lack of a correlation may be due to possible heterogeneous expression of the recombinant AlkB in the *E. coli* membranes, which would lead to promiscuity in the selection of substrates [110].

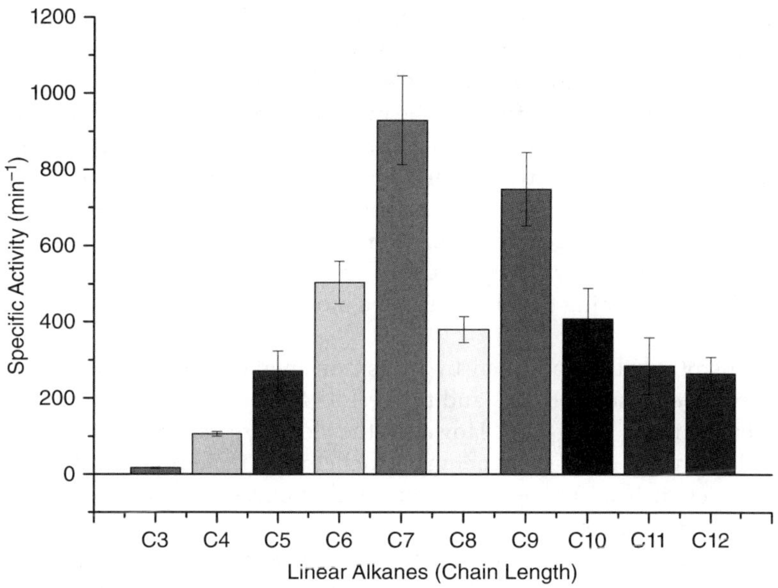

Figure 15.6 Specific activity in terms of turnover frequency (TOF) for the electrochemical conversion of medium-chain-length (C_3–C_{12}) normal alkanes to primary alcohols mediated by an AlkB-enriched membrane bound to AlkG immobilized on an SPCE. Source: Reproduced from Ref. [38].

Although the morphology and three-dimensional fold of the AlkB and its hydrophobic pocket have been effectively probed by using a variety of fluorinated *n*-octanes [5, 52], adsorption of the AlkBG protein complex on a large surface area of the SPCE, with its intrinsic properties of surface cavity, hydrophobicity, and roughness, may alter the shape of the hydrophobic pocket in AlkB. Accordingly, the relationship between the reaction activity and the chain length observed for the C_5–C_{12} conversion on the SPCE would be expected to differ from those observed for the wild-type AlkBGT proteins in the membrane interface. For example, to date, there has been no report of efficient oxidation of C_3 and C_4 *n*-alkanes to their corresponding primary alcohols by an enzymatic assay. Using the SPCE, electro-catalytic conversions of propane (C_3) and *n*-butane (C_4) to 1-propanol and 1-butanol, respectively, have been observed with unprecedented TOFs of 16 ± 2 and 106 ± 6 min^{-1}, respectively, for these two gas-to-liquid conversions [38]. The catalytic process presented here is environmentally benign and, in principle, can be combined with solar energy/photovoltaic cells for the sustainable conversion of either natural gas or LPG to the corresponding primary alcohols.

15.5 Homogeneous Catalysis of Oxyfunctionalization of Alkanes: Examples of Biomimetics of Metallo-Monooxygenases

A series of oxidations of alkanes including natural gas of methane (90 bar), ethane (27 bar), propane (7 bar) and isobutane (5 bar) as well as higher liquid alkanes including cyclohexane, 2,2,4-trimethylpentane, methylcyclohexane, 3-methylhexane, and *cis*- and *trans*-decalin catalyzed by a series of Fe complexes including $Fe(ClO_4)_3$ (naked iron cation), $FeCl_3$, $Fe(CH_3COO)_3$ as well as a binuclear Fe^{III} complex of 1,4,7-triazacyclononane, in a H_2O_2-H_2O-CH_3CN system, have been investigated and demonstrated [111]. TONs (the number of moles of all products per mole of a catalyst) as high as 30 and 70 have been reported for the oxidations of methane and ethane, respectively. The oxidation of methane produced methanol, formaldehyde, and formic acid, and that of ethane produced ethanol and acetaldehyde. The yields of the gaseous alkane oxidations were 1–6% based on the initial concentration of H_2O_2. Tertiary carbons of alkanes have been found to be most susceptible to being oxidized, and the oxidation selectivity presented in the order of $3° > 2° > 1°$ carbon of alkanes. The H_2O_2-CH_3CN system appears to be very sensitive toward almost any experimental variable, in particular solvent, ligand, water content in the organic solvent, and the oxidation state of the metal ion, all of which can be rapidly changed at the beginning of the experiment [112]. The authors propose that alkane oxidation catalyzed by $Fe(ClO_4)_3$ can proceed with the formation of ferryl ion species $(Fe^{IV}{=}O)^{2+}$, suggesting $Fe(ClO_4)_3$ to be a biomimetic of alkane-oxidizing metalloenzymes [111].

Many oxyfunctionalizations of alkanes catalyzed by transition-metal-based systems using peroxides, H_2O_2 and O_2 have been also reported, and the results for chemical selectivity and diversity of the catalysts are promising [113–119]. However, the yields are usually only moderate, and the substrates liquid alkanes, most of them being petrochemicals including cyclohexane and *n*-alkanes. However, advances in ligand design have indicated promise for the further development of these systems as selective catalysts [120].

A series of multinuclear copper complexes mimicking the active sites of copper monooxygenases have been developed for their catalytic oxidative conversions of C_3–C_9 *n*-alkanes, cyclohexane derivatives, methane and ethane using H_2O_2 or O_2 in a CH_3CN solution [121–123]. For such conversions of gaseous methane to methanol and ethane to ethanol carried out at 40 °C and at alkane pressures of 30 atm, the TONs are measured to be, respectively, 47 and 23 with yields of 2.2% and 2.1%. The TONs and the maximum yields for H_2O_2 oxidation of cyclohexane to cyclohexanol and cyclohexanone are 360–380 and 32%, respectively [121].

The $[Cu^ICu^ICu^I$(**7-N-Etppz**)$]^{1+}$ catalyst system, where **7-N-Etppz** stands for the organic ligand 3,3′-(1,4-diazepane-1,4-diyl)bis[1-(4-ethylpiperazine-1-yl)propan-2-ol], has been designed as a mimic of the trinuclear copper cluster site of pMMO, and developed to promote efficient oxidation of CH_4 as well as other small alkanes with H_2O_2 as the oxidant/reductant in CH_3CN at room temperature [17, 20]. The oxidation is found to be regiospecific, and the TON is limited only by the amount of H_2O_2 used to drive the process under controlled conditions. Only methanol, ethanol, and 2-propanol are produced, respectively, in the catalytic oxidations of methane, ethane, and propane, as mediated by the tricopper cluster [20]. There is no evidence for overoxidation of these hydrocarbons to form their corresponding aldehydes or ketones. In the case of *n*-C_4–*n*-C_6 alkanes, however, the formation of 2-alkanol and 2-alkanone is observed. With cyclohexane, cyclohexanol, and cyclohexanone are produced in roughly equal proportions. For cyclohexene, the oxidation yields the epoxide, cyclohexanol, and cyclohexanone. The catalytic efficiency is calculated based on the amounts of the $[Cu^ICu^ICu^I$(**7-N-Etppz**)$]^{1+}$ complex as the catalyst of the hydrocarbon oxidation and on the amounts of H_2O_2 used to drive the turnover. Since three molecules of H_2O_2 are consumed per productive cycle to form product(s), the formation of each equivalent of alkanol and alkanone would require three and six equivalents of H_2O_2, respectively, when the catalytic turnover is driven by H_2O_2 alone. The catalytic efficiencies of the oxidative conversion of the C_1–C_6 linear alkanes (i.e. methane to *n*-hexane) and of cyclohexane are 40–98% for the H_2O_2 consumption during the course of a one-hour experiment.

An alternative homogeneous approach for the catalytic oxidation of alkanes is to exploit polyoxometalates (POMs) [124]. POMs are metal complexes consisting of transition metal oxides (MO_x) with multiple metal centers. Because of the heterogeneity in the compositions of these metal complexes, these systems present a high variety of charge densities, surface roughness levels and surface areas for the catalysis, as well as complex redox chemistry. Due to their unique compositions and electronic structures, POMs can actually be considered as the alternative MO_x solid-state assembly or soluble metal oxides for homogeneous catalysis. In fact, it has been quite easy to form various transition-metal-substituted polyoxometalates (TMSPs) by incorporating specific transition metals into the POM structure. Such incorporation has been shown to stabilize the POM structure and to tune its properties by engendering certain electrostatic interactions, offering a useful strategy to yield improved catalysis, similar to the improvement resulting from doping alternative elements for bulk MO_x in heterogeneous catalysis. Due to the advantages offered by POMs, several robust inorganic catalysts [125] featuring well-defined multinuclear active sites have been developed for their sustainable and efficient oxidation of inert C—H bonds of alkanes using environmentally benign O_2 or H_2O_2 as the oxidants; these inorganic catalysts have been shown to be very good biomimetics of metallo-monooxygenases [126].

Shilov and Shul'pin have reported an electrophilic activation of the strong C—H bonds of methane by deploying a Pt^{II} complex [127]. The resulting Pt^{IV} intermediate is shown to be essential for the oxidation of methane to methanol when using Pt. A hybrid catalytic complex of bipyrimidinylplatinum-polyoxometalate ($H_5PV_2Mo_{10}O_{40}$) grafted onto a silica substrate has been developed to carry out mild and aqueous oxidation of methane to methanol and formaldehyde, and then partially further to acetaldehyde via cross coupling reaction on the POM surface [128]. In the presence of CH_4 (30 bar) and O_2 (2 bar) at 50 °C, the TONs (moles of products per mole of catalyst) here have been shown to be as high as 33. The $H_5PV_2Mo_{10}O_{40}$ POM acts as a counter-anion of the platinum complex and hence beneficially mediates the oxidation of Pt^{II} to Pt^{IV}. The POM moiety is found to be easily re-oxidized with the formation of water to drive the catalytic conversion of alkane oxidation.

Finally, a catalyst consisting of phosphotungstate with high-valence divanadium catalyst has been suggested to generate a strong electrophilic oxidative species, $V^V(\mu\text{-}O)_2V^V$, in the presence of $HClO_4$ and H_2O_2. This catalyst has been used for selective oxidations of cyclohexane, cyclooctane, cyclodecane, propane, *n*-butane, and *n*-hexane to form alcohols with high selectivity and reasonable yields (17–94%) [129, 130]. The use of this catalyst usually does not lead to wasteful decomposition; moreover, this catalyst seldom overoxidizes the target compound and shows significant recyclability.

15.6 Heterogeneous Nanobiomimetics for Oxyfunctionalization of Normal Alkanes

A series of metal-organic frameworks (MOFs) and polymers coordinating transition metals have been employed for alkane oxidation based on their biomimetic properties including multimetallic centers and unique porosity [131–136]. These catalysts, in reaction mixtures of CH_3CN/H_2O, showed high catalytic activity and recyclability levels, as well as reasonable product yields for the conversion of cyclohexane to cyclohexanol and cyclohexanone when H_2O_2 is employed as the oxidant, namely for peroxidative oxidation. The good catalytic activity is attributed to the high surface area, many active sites on the surface, and the stability of the materials [132].

Zeolites with molecularly sized microporous/mesoporous networks have been used to stabilize certain features of embedded metal complexes and hence allowed these features to be manifested [137]. One of the most revealing such examples is the use of ZSM-5 matrix to stabilize the α-site of iron complexes to promote the generation of α-oxygen or α-O via the decomposition of N_2O by the α-site [138, 139]. The corresponding specific oxygenated species can efficiently activate C—H bonds of organic molecules including methane and benzene in a highly selective and controlled manner for the formation of methanol (90–100% yields) and phenol (85–93% yields) [138]. Mössbauer spectral features of the α-O are highly reminiscent of the Mössbauer spectral data reported for the sMMO diiron active site, seemingly favoring a binuclear structure for the α-O site. However, more recent spectroscopic results obtained employing "site selective" methods reveal the α-Fe^{II} to be a mononuclear, high-spin, square-planar Fe^{II} site, and magnetic circular dichroism (MCD) measurements suggest that the reactive intermediate, α-O, is associated with a mononuclear high-spin Fe^{IV}=O species [140]. These findings indicate that the vacant *trans* axial position of α-O, which is stabilized by the zeolite lattice, might be providing a highly confined space for the reaction with CH_4 to form methanol.

Iron-containing zeolites or silicates have been developed as oxidation catalysts and shown to be capable of oxidizing about 10% of methane in $H_2O_{2(aq)}$ at 50 °C under a high-pressure supply of methane gas at 30.5 bar, but only 8% of the products formed is methanol [141]. The rest of the products are formic acid, methyl hydroperoxide, and carbon dioxide, and under some conditions, ~83% of the products is formic acid. However, the production of methanol can be significantly improved by adding Cu species to the zeolites via hydrothermal as well as solid state ion-exchange processes. Under the same operating conditions (temperature and methane pressure) indicated earlier, 10% conversion of methane with 93% of the products being methanol has been achieved. An Fe K-edge XANES study indicates that an increased concentration of octahedral iron is required for this catalytic activity. From analysis of the Fe EXAFS study, average Fe—Fe and Fe—O backscattering distances of 2.97 and 2.04 Å have been determined. These results suggest that the extra-framework diiron species is an anti-ferromagnetically coupled high-spin octahedral Fe^{III} centers, sharing features with sMMO and AlkB.

The iron- and copper-modified ZSMs have also been used to catalyze the oxidation of ethane and propane, each at a pressure of 20 bar, by H_2O_2 at 50 °C. Here, a relatively complicated profile of oxygenated products is observed [142, 143]. Just as the oxidation of methane preferentially leads to the formation of formic acid, the major products of the oxidation of ethane and propane catalyzed by Fe/ZSM (30) using H_2O_2 are acetic acid and propanoic acid, although C_1 and/or C_2 oxygenates as well as carbon dioxide are also produced. The oxidation of ethane catalyzed by Cu/ZSM (30) and Fe-Cu/ZSM (30) also produces considerable amounts of acetic acid and ethanol, as well as C_1 oxygenates including formic acid, presumably from C—C bond scission processes. A significant amount of ethylene is also observed in the gas phase, probably due to the further elimination of the ethane substrate from the zeolite acid sites. In the case of propane as the substrate, propanoic acid, and acetaldehyde are the major products when it is oxidized by Fe-based ZSMs in H_2O_2, with other C_1/C_2 oxygenates also observed. Apparently, C—C bond scissions are also detected, with these scissions attributed by the authors to the supported Fe/ZSM-5 catalyst behaving as a heterogeneous Fenton's type catalyst.

Upon substituting some of the Al^{III} ions of aluminophosphate molecular sieves, AlPO-18 and 36, with four-coordinated Co^{III} and Mn^{III}, these porous materials catalyze the oxidation of linear alkanes by molecular oxygen [144, 145]. Here, terminal methyl group oxidation is observed and attributed to the end-on approach of the linear alkanes to the transition metal active site. The reaction pathway of this oxidation is proposed to involve a classical free-radical chain autooxidation mechanism [146, 147] since the addition of a free-radical initiator (*t*-butylhydroperoxide) is found to reduce the induction period for catalysis, and the introduction of a scavenger (hydroquinone) reduces the reaction rate. The TON and conversion yield (mol%) values of a 24-hour CoAlPO/MnAlPO-mediated oxidation of n-C_5H_{12} are 109–175 and 5.5–8.8%, respectively, and those of n-C_6H_{14} are 41–149 and 2.4–8.7%, respectively, better than those of n-C_8H_{18} (72–83 and 5.8–6.8%, respectively). The primary index, defined as the ratio of the concentration of primary products to the concentration of secondary/tertiary products normalized to the respective number of hydrogens in the alkane molecule, is greater for the longer linear alkanes, i.e. the primary index of n-C_8H_{18} is found greater than that of n-C_6H_{14}, which is, in turn, greater than that of n-C_5H_{12}. The size, shape, morphology, and pore diameters of the AlPO molecular sieves influence their catalytic efficiencies and selectivity for primary versus secondary/tertiary C—H bonds as revealed by the primary index data. The pore aperture diameters of AlPO-18 and 36 have been reported to be 3.8 Å and 6.5 × 7.5 Å, respectively.

The [$Cu^ICu^ICu^I$(**7-N-Etppz**)]$^{1+}$ complex has been previously shown to mediate efficient oxidations of methane and other small alkanes to alkanols and alkanones in homogeneous solutions when the catalytic turnover is driven by H_2O_2 in CH_3CN [20] (Scheme 15.3). However, with the tricopper complex as a homogeneous catalyst, the system suffers from low conversion yields due to the inherent low solubility of O_2 and alkane gases in the solvent. Chemistry is increasingly conducted in confined spaces to overcome some of the limitations of homogeneous catalysis. For instance, mesoporous silica nanoparticles (MSNs) possess a rigid framework that can be exploited to immobilize small biomimetic model complexes for many applications, including catalysis. MSNs exhibit some unique properties, such as a large surface area, controllable geometries, and a modifiable surface to accommodate the types and sizes of molecules that can penetrate them. Accordingly, the [$Cu^ICu^ICu^I$(**7-N-Etppz**)]$^{1+}$ complex has been reformulated as a heterogeneous catalytic system for selective methane and cyclohexane oxidation by immobilizing the tricopper cluster complex into MSNs with a particle size of ~ 80 nm [148, 149]. In order to anchor the positively charged tricopper complex, the surface of the MSN nano-channels has been functionalized with the anionic 3-trihydroxysilylpropyl-methylphosponate (TP) to generate a negatively charged surface (MSN-TP) and a pore-expanded MSN obtained by adding decane [150], and increasing the surface charge density and acidity by doping Al into the framework (AlMSN30-ex) [151] (Figure 15.7). Expansion of the silica pore and increasing the surface charge density is shown to further increase the loading of the immobilized complexes, which, in turn, yields a higher conversion of methane. The expanded-pore AlMSN30-ex sample has a pore diameter of 4.80 ± 0.57 nm, and shows cleaner mesopores than does the MSN-TP sample, which has a pore diameter of 2.80 ± 0.14 nm.

Significantly, the mesopores of the functionalized MSNs overcome some of the shortcomings of the homogeneous catalytic system by having the capability to accommodate high CH_4 and O_2 solubilities. The interactions of the gas molecules with the nanoporous solid or with the solvent are considerably weaker than the interactions of the solvent molecules with the walls of the confining host framework. The stronger solvent–solid interactions create regions of low solvent density in the confined solvent and enhance the uptake of CH_4 and O_2. With the "oversolubility" of methane in the liquid confined in the nanoporous materials, the catalytic efficiencies and

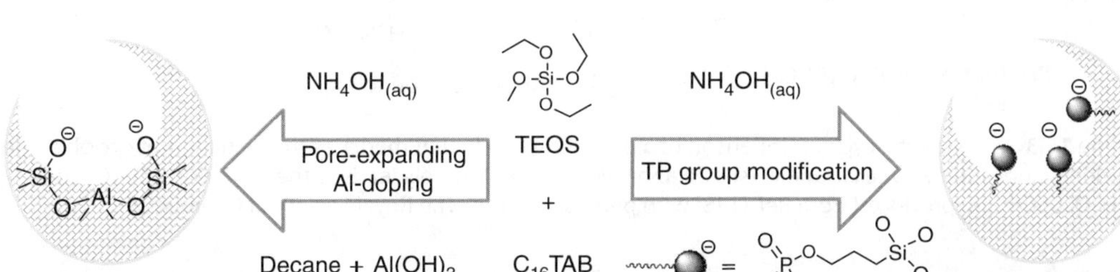

Figure 15.7 Two mesoporous silica nanoparticles (MSNs) prepared by modification of the anionic TP on the surface of the nano-channels and Al doping (Si/Al ratio = 30) introduced into the framework of pore-expanded MSN. Source: Reproduced from Ref. [148] with permission from The Royal Society of Chemistry.

TON of methane oxidation are no longer limited by abortive cycling. In addition, in the presence of O_2, the tricopper cluster is activated by O_2 instead of H_2O_2, and each catalytic cycle consumes only one H_2O_2 molecule to regenerate the catalyst after O-atom transfer from the activated tricopper complex to the methane [148] (Scheme 15.3). With this strategy, the TON of the methane oxidation reaction mediated by the **Cu_3(7-N-Etppz)**@AlMSN-ex catalyst shows significant improvement and approaches to 171 (moles of methanol produced per mole of complex in catalyst) with high overall catalytic efficiency (86%) when the catalytic turnover is initiated by 200 equiv. of H_2O_2. By comparison, as a homogeneous catalyst $[Cu^ICu^ICu^I$(**7-N-Etppz**)$]^{1+}$ without the silica support, only a maximum TON of 6.5 is observed when the catalytic turnover is initiated by 20 equiv. of H_2O_2, and essentially no methanol is detected with 200 equiv. of H_2O_2 due to abortion of the activated tricopper complex by reaction with H_2O_2 (producing O_2 and H_2O) at this high concentration of H_2O_2 used to drive the catalytic turnover. In the typical organic solvent, with its low CH_4 solubility, methane oxidation, i.e. productive cycling of the catalyst, cannot compete well with abortive cycling of the catalyst (Table 15.2). Regarding its recyclability, its activity decreased by 10% per cycle for each of three tested cycles.

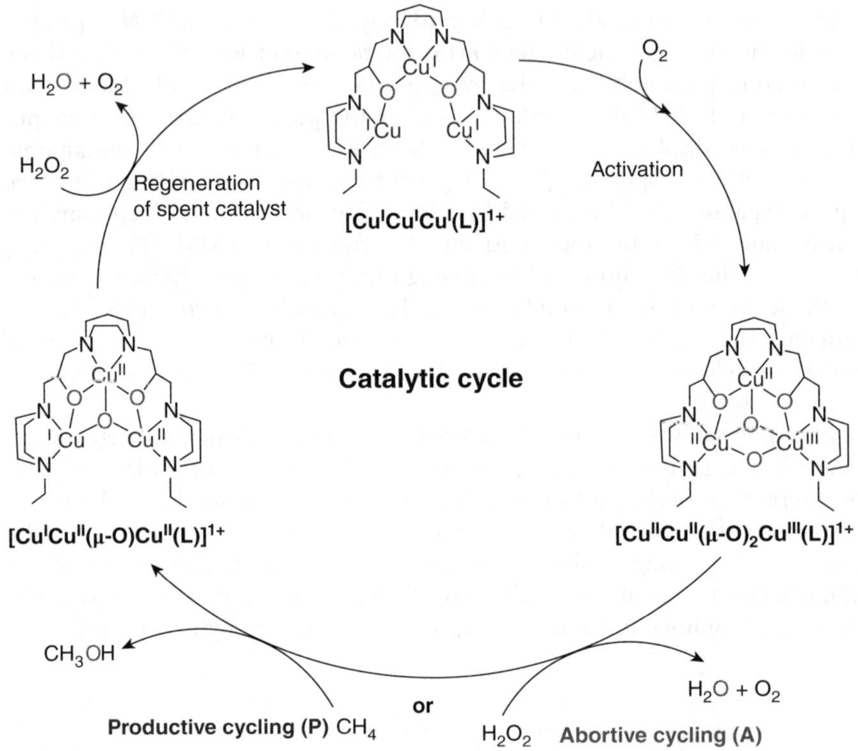

Scheme 15.3 The catalytic cycle: comparison of a productive cycle that mediates the oxidation of cyclohexane to cyclohexanol and cyclohexanone versus an abortive cycle that merely leads to the oxidation of H_2O_2 to O_2 and H_2O. Source: Reproduced from Ref. [148] with permission from The Royal Society of Chemistry.

A similar strategy has been applied for the selective catalytic oxidation of cyclohexane (C_6H_{12}), and this oxidation at room temperature shows good TONs, excellent catalytic efficiencies, high conversion yields and high product selectivity, when the catalysis is mediated by a tricopper cluster complex immobilized in the nanochannels of MSNs [149]. The anionic 3-(trihydroxysilyl)-propylmethyl-phosphonate (TP) has been also employed to generate a

Table 15.2 Comparison of the catalytic performance (limiting TONs) of the Cu_3**(7-N-Etppz)** complex as a homogeneous catalyst and its reformulation **Cu_3(7-N-Etppz)**@AlMSN30-ex as a heterogeneous catalyst for methane oxidation at room temperature.

Cu_3(7-N-Etppz)			Cu_3(7-N-Etppz)@AlMSN30-ex		
Homogeneous			Heterogeneous		
H_2O_2 (equiv.)	TON_{CH_3OH}	OCE[a)]	H_2O_2 (equiv.)	TON_{CH_3OH}	OCE[a)]
0	1	—	0	1.0	—
20	6.5	0.33	20	15.9	0.80
40	~0	~0	50	42.5	0.85
100	~0	~0	100	81.1	0.81
180	~0	~0	200	171.2	0.86

a) OCE denotes the overall catalytic efficiency, as estimated by the ratio of the CH_3OH produced (N_{CH_3OH}) in the experiment to the amount of H_2O_2 used to initiate the turnover (N_o), namely, N_{CH_3OH}/N_o.

Source: Taken from Ref. [148] with permission from The Royal Society of Chemistry.

negatively charged surface of the MSN channels to anchor the positively charged tricopper complexes. The average particle diameter of the MSN-TP is about 70 nm (71.4 ± 7.8 nm). The pore diameter is 2.8 ± 0.14 nm, which is large enough to accommodate the tricopper complexes inside the nanochannels of the MSN-TP to form a confined space for selective catalytic oxidation of cyclohexane. Again, the $Cu^ICu^ICu^I$ tricopper cluster is activated by O_2 to mediate the oxidation of cyclohexane to cyclohexanol ($C_6H_{12}O$) and cyclohexanone ($C_6H_{10}O$) by a direct O-atom transfer mechanism and the turnover of the catalyst is driven by H_2O_2. "Chemistry in confined space" is exploited to control the access or availability of different co-substrates to the catalyst. The MSN-TP system restricts the H_2O_2 to the bulk solution suspending the MSNs, and the H_2O_2 is used efficiently, almost totally to drive the turnover. When the H_2O_2 is in large excess, the conversion of C_6H_{12} and selectivity to $C_6H_{10}O$ can reach close to 100% (Figure 15.8). Finally, different tricopper complexes with different binding affinities for the hydrocarbon substrate have also been studied. The results obtained underscore the importance of substrate recognition and binding in the control of substrate specificity and selectivity.

To produce biomimetics of the copper active sites of pMMO (the pMMO from methanotrophic bacteria), many other heterogeneous Cu-based catalysts have been explored and developed for the conversion of methane to methanol [152–157]. Copper-modified zeolites have been developed to stoichiometrically convert methane to methanol at low temperatures. The dinuclear copper center Cu—O—Cu in Cu-ZSM-5 has been proposed to efficiently convert methane to methanol [153]. One interesting recent development is the use of water at a high temperature, i.e. in the vapor form, to anaerobically reconstitute the active species in the Cu-containing zeolite and to release methanol and hydrogen (H_2) from the catalysts [158].

Recent copper-exchanged zeolites having the mordenite structure and accommodating highly stable trinuclear copper-oxo clusters [$Cu_3(\mu\text{-}O)_3$], have been shown to efficiently convert methane to methanol at 200 °C [155]. The formed methanol and dimethyl ether products can be released by purging the zeolite with a water stream at 135 °C. The materials are reactivated in O_2 at 500 °C and recycled eight times. A recently developed facile solid-state ion exchange of the NH_4^+ form of mordenite with $Cu(acac)_2$ has also been shown to efficiently

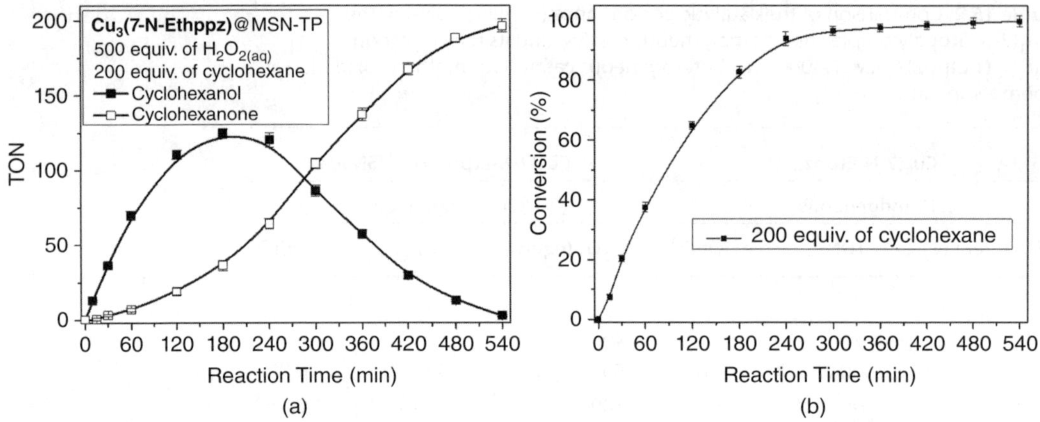

Figure 15.8 Time course study of (a) the TONs for the cyclohexane oxidation reaction catalyzed by **Cu_3(7-N-Ethppz)**@MSN-TP at room temperature for 9 hours; and (b) the corresponding conversion yields. The amount of cyclohexane (200 equiv.) available for oxidation is significantly less than the amount of H_2O_2 (500 equiv.) used in this reaction; thus, H_2O_2 is sufficient to drive the catalytic turnover completely, converting nearly 100% of cyclohexane to cyclohexanone with near 100% selectivity. Source: Reproduced from Ref. [149] with permission from The Royal Society of Chemistry.

oxidize methane to methanol with reasonable recyclability [159]. With the discovery of these heterogeneous platforms of biomimetics for gaseous alkane oxidation, the development of even more sophisticated catalysts for selective oxidation of higher alkanes can be expected by further engineering the hydrophobicity as well as porosity of the nanomaterials.

15.7 Final Comments

Much has been learned about how to use metalloenzyme systems to control the chemoselectivity, regioselectivity, and stereoselectivity in the oxidation of alkanes. Aside from the selection of the transition metal ion or metal cluster to activate O_2 (or H_2O_2) to form the necessary oxygenated metal intermediate with sufficient reactivity to accomplish the difficult C—H bond oxidations of the aliphatics, the active site of the enzyme must possess certain properties to provide a significant driving force to accommodate these chemically inert substrates after desolvation of the hydrophobic pocket. In addition to high Michaelis–Menten affinities (low K_m and optimal k_{cat}) for these substrates, the polar alcohol products must dissociate with high rates from the enzyme, namely, high k_{off}, to attain the high efficacy of these systems. The roughness of the surface of the pocket, and the nonbonding interactions exerted within the pocket, including van der Waals interactions, hydrogen bonding, and dispersive forces, are essential to position and orient the inert alkane substrates to ensure the desired regioselectivity and stereoselectivity during the oxidation. Also important for the sustainable oxidation of hydrocarbons within the cell is the supply of electrons from reduced cofactors, such as the NADH or NADPH, and redox partners of the metalloenzymes, to reduce the metal active site for subsequent activation by O_2. For optimal catalytic efficiencies, the rate of electron input must be matched as closely as possible to the rate of product formation to avert futile cycles. Unfortunately, most of these proteins are relatively unstable outside of the cell and would show poor recyclability in industrial chemical transformation applications [160]. Nevertheless, the significant insights provided by these molecular machines have already made metalloproteins a very useful example to follow for the design and construction of future efficient biomimetics.

Many novel homogeneous and heterogeneous catalysts based on the metal active sites of metalloenzymes have been developed for hydrocarbon oxidation during the past several decades. Within the context of the present review, selective or controlled oxidation is often a challenge, not mentioning the regioselectivity and stereoselectivity. Typically, there are many additional products resulting from the overoxidation of alkanes to ketones, aldehydes, and/or acids as well as C—C bond ruptures, and even the elimination processes that form alkenes. With metalloenzymes, these processes are tightly controlled. In future work, we should learn to design new metallocatalysts in order to achieve efficient single, double or triple oxidations of hydrocarbons by increasing the k_{off} around the metal active site or the substrate-anchoring pocket, like what has been achieved with the tricopper cluster catalyst in methane oxidation [17, 20, 148]. To achieve high catalytic efficiencies, strategies must be also developed to minimize and, if possible, eliminate abortive cycling, and improve the rate of regeneration of the "spent" catalyst [4, 19, 161].

Gaining an understanding of the complexity of the free-energy landscape associated with chemical transformations and finding conditions to carry out the catalytic oxidation of alkanes under more controlled ambient conditions is certainly an important area of scientific endeavor in future studies. The nanostructures of the current catalysts cannot in most cases be well controlled, and the corresponding processes are not well understood either. The development of well-characterized nanoparticles grafted with rationally designed catalysts tailored to specific reaction partners may lessen these difficulties. Besides controlling the pore sizes and other characteristics of the nanoparticles, including their shape, spatial distribution, surface composition, and electronic structure, we need to endow the surfaces of the nanoparticles with increased selectivity, stability, catalytic efficiency, and reactivity under more ambient conditions to reduce the amount of energy consumed and to reduce side reactions, and, hence, to create more efficient catalysis.

Acknowledgments

This work was supported by Academia Sinica, including the National Taiwan University and Academia Sinica Innovative Joint Program (NTU-SINICA-104R104502 and NTU-SINICA-105R104502) and the Academia Sinica Research Program on Nanoscience and Nanotechnology (2393-105-0200), and grants from the Research Program of "Taiwan's Deep Decarbonization Pathways toward a Sustainable Society" (106-0210-02-11-05) and Ministry of Science and Technology, R.O.C. (MOST 101-2113-M-001-007-MY3 and 104-2113-M-001-013-MY3) and the Taiwan International Graduate Program (TIGP), Sustainable Chemical Science and Technology, in a consortium collaboration between the National Chiao-Tung University and Academia Sinica.

References

1 Lide, D.R. (ed.) (2010). *CRC Handbook of Chemistry and Physics*. Boca Raton, FL: CRC Press/Taylor and Francis.
2 Cherkasov, A. and Jonsson, M. (2000). *J. Chem. Inf. Comput. Sci.* 40: 1222–1226.
3 Xue, X.-S., Ji, P., Zhou, B. et al. (2017). *Chem. Rev.* 117: 8622–648. doi: 10.1021/acs.chemrev.1026b00664.
4 Wang, V.C.C., Maji, S., Chen, P.P.Y. et al. (2017). *Chem. Rev.* 117: 8574–8621. doi: 10.1021/acs.chemrev.1026b00624.

5 Chen, Y.S., Luo, W.I., Yang, C.L. et al. (2014). *J. Inorg. Biochem.* 134: 118–133.
6 Sirajuddin, S. and Rosenzweig, A.C. (2015). *Biochemistry* 54: 2283–2294.
7 Holland, H.L. and Weber, H.K. (2000). *Curr. Opin. Biotechnol.* 11: 547–553.
8 Meunier, B., de Visser, S.P., and Shaik, S. (2004). *Chem. Rev.* 104: 3947–3980.
9 Ortiz de Montellano, P.R. (2010). *Chem. Rev.* 110: 932–948.
10 Sawyer, D.T., Sobkowiak, A., and Matsushita, T. (1996). *Acc. Chem. Res.* 29: 409–416.
11 Groves, J.T. (1985). *J. Chem. Educ.* 62: 928.
12 Chen, P.P.-Y., Yang, R.B.-G., Lee, J.C.-M. et al. (2007). *Proc. Natl. Acad. Sci. U.S.A.* 104: 14570–14575.
13 Chen, P.P.-Y. and Chan, S.I. (2006). *J. Inorg. Biochem.* 100: 801–809.
14 Allendorf, M.D., Spira, D.J., and Solomon, E.I. (1985). *Proc. Natl. Acad. Sci. U.S.A.* 82: 3063–3067.
15 Anderson, P.W. (1959). *Phys. Rev.* 115: 2–13.
16 Lewis, E.A. and Tolman, W.B. (2004). *Chem. Rev.* 104: 1047–1076.
17 Chan, S.I., Lu, Y.-J., Nagababu, P. et al. (2013). *Angew. Chem. Int. Ed.* 52: 3731–3735.
18 Chan, S.I. and Yu, S.S.-F. (2008). *Acc. Chem. Res.* 41: 969–979.
19 Chen, P.P.-Y., Nagababu, P., Yu, S.S.-F. et al. (2014). *ChemCatChem* 6: 428–437.
20 Nagababu, P., Yu, S.S.F., Maji, S. et al. (2014). *Catal. Sci. Technol.* 4: 930–935.
21 Hamamura, N., Storfa, R.T., Semprini, L. et al. (1999). *Appl. Environ. Microbiol.* 65: 4586–4593.
22 Tinberg, C.E. and Lippard, S.J. (2011). *Acc. Chem. Res.* 44: 280–288.
23 Sazinsky, M.H. and Lippard, S.J. (2006). *Acc. Chem. Res.* 39: 558–566.
24 Wallar, B.J. and Lipscomb, J.D. (1996). *Chem. Rev.* 96: 2625–2657.
25 Balasubramanian, R., Smith, S.M., Rawat, S. et al. (2010). *Nature* 465: 115–119.
26 Smith, S.M., Balasubramanian, R., and Rosenzweig, A.C. (2011). Metal reconstitution of particulate methane monooxygenase and heterologous expression of the pmoB subunit. In: *Methods in Methane Metabolism, Part B: Methanotrophy* (ed. A.C. Rosenzweig and S.W. Ragsdale), 195–210.
27 Balasubramanian, R. and Rosenzweig, A.C. (2007). *Acc. Chem. Res.* 40: 573–580.
28 Lieberman, R.L. and Rosenzweig, A.C. (2005). *Nature* 434: 177–182.
29 Chan, S.I., Wang, V.C.-C., Lai, J.C.-H. et al. (2007). *Angew. Chem. Int. Ed.* 46: 1992–1994.
30 Chan, S.I., Chen, K.H.-C., Yu, S.S.-F. et al. (2004). *Biochemistry* 43: 4421–4430.
31 Ashraf, W., Mihdhir, A., and Murrell, J.C. (1994). *FEMS Microbiol. Lett.* 122: 1–6.
32 Kotani, T., Yamamoto, T., Yurimoto, H. et al. (2003). *J. Bacteriol.* 185: 7120–7128.
33 Kotani, T., Kawashima, Y., Yurimoto, H. et al. (2006). *J. Biosci. Bioeng.* 102: 184–192.
34 Sharp, J.O., Sales, C.M., LeBlanc, J.C. et al. (2007). *Appl. Environ. Microbiol.* 73: 6930–6938.
35 Coleman, N.V., Le, N.B., Ly, M.A. et al. (2012). *ISME J.* 6: 171–182.
36 Johnson, E.L. and Hyman, M.R. (2006). *Appl. Environ. Microbiol.* 72: 950–952.
37 Masuda, H., McClay, K., Steffan, R.J. et al. (2012). *Lett. Appl. Microbiol.* 55: 175–181.
38 Tsai, Y.-F., Luo, W.-I., Chang, J.-L. et al. (2017). *Sci. Rep.* 7: 8369.
39 Dubbels, B.L., Sayavedra-Soto, L.A., and Arp, D.J. (2007). *Microbiology* 153: 1808–1816.
40 Cooley, R.B., Dubbels, B.L., Sayavedra-Soto, L.A. et al. (2009). *Microbiology* 155: 2086–2096.
41 van Beilen, J.B. and Funhoff, E.G. (2007). *Appl. Microbiol. Biotechnol.* 74: 13–21.
42 Austin, R.N. and Groves, J.T. (2011). *Metallomics* 3: 775–787.
43 Ji, Y.R., Mao, G.N., Wang, Y.Y. et al. (2013). *Front. Microbiol.* 4: 1–13.
44 Alonso, H., Kleifeld, O., Yeheskel, A. et al. (2014). *Biochem. J* 460: 283–293.

45 Shanklin, J., Achim, C., Schmidt, H. et al. (1997). *Proc. Natl. Acad. Sci. U.S.A.* 94: 2981–2986.
46 van Beilen, J.B., Penninga, D., and Witholt, B. (1992). *J. Biol. Chem.* 267: 9194–9201.
47 van Beilen, J.B., Smits, T.H., Roos, F.F. et al. (2005). *J. Bacteriol.* 187: 85–91.
48 Ng, K.Y., Tu, L.-C., Wang, Y.-S. et al. (2008). *ChemBioChem* 9: 1116–1123.
49 Nakagawa, Y., Irie, K., Yanagite, R.C. et al. (2005). *J. Am. Chem. Soc.* 127: 5746–5747.
50 Matter, H., Nazaré, M., Güssregen, S. et al. (2009). *Angew. Chem. Int. Ed.* 48: 2911.
51 Fujii, A., Shibasaki, K., Kazama, T. et al. (2008). *Phys. Chem. Chem. Phys.* 10: 2836–2843.
52 Ramu, R., Chang, C.-W., Chou, H.-H. et al. (2011). *Tetrahedron Lett.* 52: 2950.
53 Haines, D.C., Tomchick, D.R., Machius, M. et al. (2001). *Biochemistry* 40: 13456–13465.
54 Li, H. and Poulos, T.L. (1997). *Nat. Struct. Mol. Biol.* 4: 140–146.
55 Wu, L.-L., Yang, C.-L., Lo, F.-C. et al. (2011). *Chem. Eur. J.* 17: 4774–4787.
56 Chiang, C.-H., Ramu, R., Tu, Y.-J. et al. (2013). *Chem. Eur. J.* 19: 13680–13691.
57 Zilly, F.E., Acevedo, J.P., Augustyniak, W. et al. (2011). *Angew. Chem. Int. Ed.* 50: 2720–2724.
58 Hegde, A., Haines, D.C., Bondlela, M. et al. (2007). *Biochemistry* 46: 14010–14017.
59 Whitehouse, C.J.C., Bell, S.G., and Wong, L.-L. (2012). *Chem. Soc. Rev.* 41: 1218–1260.
60 Li, C. and Shaik, S. (2013). *RSC Adv.* 3: 2995–3005.
61 Kawakami, N., Shoji, O., and Watanabe, Y. (2011). *Angew. Chem. Int. Ed.* 50: 5315–5318.
62 Kawakami, N., Shoji, O., and Watanabe, Y. (2013). *Chem. Sci.* 4: 2344–2348.
63 Cong, Z.Q., Shoji, O., Kasai, C. et al. (2015). *ACS Catal.* 5: 150–156.
64 Culpepper, M.A., Cutsail, G.E., Hoffman, B.M. et al. (2012). *J. Am. Chem. Soc.* 134: 7640–7643.
65 Wilkinson, B., Zhu, M., Priestley, N.D. et al. (1996). *J. Am. Chem. Soc.* 118: 921–922.
66 Yu, S.S.-F., Wu, L.-Y., Chen, K.H.-C. et al. (2003). *J. Biol. Chem.* 278: 40658–40669.
67 Elliott, S.J., Zhu, M., Tso, L. et al. (1997). *J. Am. Chem. Soc.* 119: 9949–9955.
68 Yoshizawa, K. (2002). *Coord. Chem. Rev.* 226: 251–259.
69 Lee, S.K., Fox, B.G., Froland, W.A. et al. (1993). *J. Am. Chem. Soc.* 115: 6450–6451.
70 Tinberg, C.E. and Lippard, S.J. (2009). *Biochemistry* 48: 12145–12158.
71 Tinberg, C.E. and Lippard, S.J. (2010). *Biochemistry* 49: 7902–7912.
72 Shu, L.J., Nesheim, J.C., Kauffmann, K. et al. (1997). *Science* 275: 515–518.
73 Shaik, S. (2010). *Nat. Chem.* 2: 347–349.
74 Xue, G.Q., De Hont, R., Munck, E. et al. (2010). *Nat. Chem.* 2: 400–405.
75 Bai, Y.H., McCoy, J.G., Levin, E.J. et al. (2015). *Nature* 524: 252–256.
76 Rudd, D.J., Sazinsky, M.H., Merkx, M. et al. (2004). *Inorg. Chem.* 43: 4579–4589.
77 Roe, A.L., Schneider, D.J., Mayer, R.J. et al. (1984). *J. Am. Chem. Soc.* 106: 1676–1681.
78 The energy calibration of Fe K-edge absorption energy was set at 7112 eV for Fe foil.
79 Westre, T.E., Kennepohl, P., DeWitt, J.G. et al. (1997). *J. Am. Chem. Soc.* 119: 6297–6314.
80 Shu, L., Liu, Y., Lipscomb, J.D. et al. (1996). *J. Biol. Inorg. Chem.* 1: 297–304.
81 Shu, L., Broadwater, J.A., Achim, C. et al. (1998). *J. Biol. Inorg. Chem.* 3: 392–400.
82 True, A.E., Scarrow, R.C., Randall, C.R. et al. (1993). *J. Am. Chem. Soc.* 115: 4246–4255.
83 Whittington, D.A. and Lippard, S.J. (2001). *J. Am. Chem. Soc.* 123: 827–838.
84 Elango, N., Radhakrishnan, R., Froland, W.A. et al. (1997). *Protein Sci.* 6: 556–568.
85 Lewis, D.F.V., Jacobs, M.N., and Dickins, M. (2004). *Drug Discovery Today* 9: 530–537.
86 Fishelovitch, D., Shaik, S., Wolfson, H.J. et al. (2010). *J. Phys. Chem. B* 114: 5964–5970.
87 Shaik, S., Kumar, D., de Visser, S.P. et al. (2005). *Chem. Rev.* 105: 2279–2328.
88 Shaik, S., Cohen, S., Wang, Y. et al. (2010). *Chem. Rev.* 110: 949–1017.
89 Feenstra, K.A., Starikov, E.B., Urlacher, V.B. et al. (2007). *Protein Sci.* 16: 420–431.
90 Jung, C. (2011). *Biochim. Biophys. Acta* 1814: 46–57.

91 Sono, M., Roach, M.P., Coulter, E.D. et al. (1996). *Chem. Rev.* 96: 2841–2887.
92 J. T. Groves, K. Shalyaev and J. Lee in Vol. 4 Eds.: K. M. Kadish, K. M. Smith and R. Guilard), Academic Press, New York, 2000, pp. 17–40.
93 Shaik, S., Filatov, M., Schroder, D. et al. (1998). *Chem. Eur. J.* 4: 193–199.
94 Wang, Y., Kumar, D., Yang, C. et al. (2007). *J. Phys. Chem. B* 111: 7700–7710.
95 Ogliaro, F., Filatov, M., and Shaik, S. (2000). *Eur. J. Inorg. Chem.* 12: 2455–2458.
96 Shaik, S., Hirao, H., and Kumar, D. (2007). *Acc. Chem. Res.* 40: 532–542.
97 Yang, C.-L., Lin, C.-H., Luo, W.-I. et al. (2017). *Chem. Eur. J.* 23: 2571–2582.
98 Price, J.C., Barr, E.W., Glass, T.E. et al. (2003). *J. Am. Chem. Soc.* 125: 13008–13009.
99 Hirao, H., Kumar, D., Que, L. et al. (2006). *J. Am. Chem. Soc.* 128: 8590–8606.
100 Shaik, S., Lai, W.Z., Chen, H. et al. (2010). *Acc. Chem. Res.* 43: 1154–1165.
101 Truan, G., Komandla, M.R., Falck, J.R. et al. (1999). *Arch. Biochem. Biophys.* 366: 192–198.
102 Cryle, M.J. and De Voss, J.J. (2007). *Tetrahedron: Asymmetry* 18: 547–551.
103 Cryle, M.J., Matovic, N.J., and De Voss, J.J. (2007). *Tetrahedron Lett.* 48: 133–136.
104 Dubey, K.D., Wang, B.J., and Shaik, S. (2016). *J. Am. Chem. Soc.* 138: 837–845.
105 Li, Q.S., Schwaneberg, U., Fischer, P. et al. (2000). *Chem. Eur. J.* 6: 1531–1536.
106 Rittle, J. and Green, M.T. (2010). *Science* 330: 933–937.
107 Dawson, J.H. (1988). *Science* 240: 433–439.
108 Fasan, R., Meharenna, Y.T., Snow, C.D. et al. (2008). *J. Mol. Biol.* 383: 1069–1080.
109 Prodromidis, M.I., Florou, A.B., Tzouwara-Karayanni, S.M. et al. (2000). *Electroanalysis* 12: 1498–1501.
110 Staijen, I.E., van Beilen, J.B., and Witholt, B. (2000). *Eur. J. Biochem.* 267: 1957–1965.
111 Shul'pin, G.B., Nizova, G.V., Kozlov, Y.N. et al. (2004). *Adv. Synth. Catal.* 346: 317–332.
112 Pacześniak, T. and Sobkowiak, A. (2003). *J. Mol. Catal. A: Chem.* 194: 1–11.
113 Shul'pin, G.B. (2002). *J. Mol. Catal. A: Chem.* 189: 39–66.
114 Kirillov, A.M. and Shul'pin, G.B. (2013). *Coord. Chem. Rev.* 257: 732–754.
115 Shul'pina, L.S., Kudinov, A.R., Mandelli, D. et al. (2015). *J. Organomet. Chem.* 793: 217–231.
116 Sutradhar, M., Martins, L.M.D.R.S., Guedes da Silva, M.F.C. et al. (2015). *Coord. Chem. Rev.* 301–302: 200–239.
117 G. B. Shul'pin in 6.04 - Selectivity in C–H Functionalizations A2 - Reedijk, Jan, (Ed. K. Poeppelmeier), Elsevier, Amsterdam, 2013, pp. 79–104.
118 Shul'pin, G.B. (2013). *Dalton Trans.* 42: 12794–12818.
119 Shul'pin, G. (2016). *Catalysts* 6: 50.
120 Doble, M.V., Ward, A.C.C., Deuss, P.J. et al. (2014). *Bioorg. Med. Chem.* 22: 5657–5677.
121 Kirillov, A.M., Kopylovich, M.N., Kirillova, M.V. et al. (2005). *Angew. Chem. Int. Ed.* 44: 4345–4349.
122 Kirillova, M.V., Kozlov, Y.N., Shul'pina, L.S. et al. (2009). *J. Catal.* 268: 26–38.
123 Kirillov, A.M., Kirillova, M.V., and Pombeiro, A.J.L. (2013). Chapter one – Homogeneous multicopper catalysts for oxidation and hydrocarboxylation of alkanes. In: *Advances in Inorganic Chemistry*, vol. 65 (ed. R. van Eldik and D.H. Colin), 1–31. Academic Press.
124 Carraro, M., Sartorel, A., Ibrahim, M. et al. (2012). Polyoxometalatese as homogeneous oxidation catalysts. In: , 1–25. Innovative Catalysis in Organic Synthesis – Oxidation, Hydrogenation, and C-X bond forming Reactions, P. G. Andersson Wiley-VCH.
125 Wang, S.-S. and Yang, G.-Y. (2015). *Chem. Rev.* 115: 4893–4962.
126 Bordeaux, M., Galarneau, A., and Drone, J. (2012). *Angew. Chem. Int. Ed.* 51: 10712–10723.
127 Shilov, A.E. and Shul'pin, G.B. (1987). *Russ. Chem. Rev.* 56: 442–464.
128 Bar-Nahum, I., Khenkin, A.M., and Neumann, R. (2004). *J. Am. Chem. Soc.* 126: 10236–10237.

129 Mizuno, N., Kamata, K., and Yamaguchi, K. (2012). *Catal. Today* 185: 157–161.
130 Kamata, K., Yonehara, K., Nakagawa, Y. et al. (2010). *Nat. Chem.* 2: 478–483.
131 Gupta, S., Kirillova, M.V., Guedes da Silva, M.F.C. et al. (2013). *Inorg. Chem.* 52: 8601–8611.
132 Das, S.K., Mukherjee, S., Lopes, L.M.F. et al. (2014). *Dalton Trans.* 43: 3215–3226.
133 Dias, S.S.P., Kirillova, M.V., Andre, V. et al. (2015). *Inorg. Chem. Front.* 2: 525–537.
134 Martins, L., Nasani, R., Saha, M. et al. (2015). *Molecules* 20: 19203.
135 Timokhin, I., Pettinari, C., Marchetti, F. et al. (2015). *Cryst. Growth Des.* 15: 2303–2317.
136 Fernandes, T.A., Santos, C.I.M., André, V. et al. (2016). *Inorg. Chem.* 55: 125–135.
137 Rodkin, M.A., Sobolev, V.I., Dubkov, K.A. et al. (2000). *Stud. Surf. Sci. Catal.* 130: 875–880.
138 Panov, G.I., Uriarte, A.K., Rodkin, M.A. et al. (1998). *Catal. Today* 41: 365–385.
139 Panov, G.I., Dubkov, K.A., and Starokon, E.V. (2006). *Catal. Today* 117: 148–155.
140 Snyder, B.E.R., Vanelderen, P., Bols, M.L. et al. (2016). *Nature* 536: 317–321.
141 Hammond, C., Forde, M.M., Ab Rahim, M.H. et al. (2012). *Angew. Chem. Int. Ed.* 51: 5129–5133.
142 Forde, M.M., Armstrong, R.D., Hammond, C. et al. (2013). *J. Am. Chem. Soc.* 135: 11087–11099.
143 Peneau, V., Armstrong, R.D., Shaw, G. et al. (2017). *ChemCatChem* 9: 642–650.
144 Thomas, J.M., Raja, R., Sankar, G. et al. (1999). *Nature* 398: 227–230.
145 Thomas, J.M., Raja, R., Sankar, G. et al. (2001). *Acc. Chem. Res.* 34: 191–200.
146 Gómez-Hortigüela, L., Corà, F., Sankar, G. et al. (2010). *Chem. Eur. J.* 16: 13638–13645.
147 Zhan, B.-Z., Modén, B., Dakka, J. et al. (2007). *J. Catal.* 245: 316–325.
148 Liu, C.C., Mou, C.Y., Yu, S.S.F. et al. (2016). *Energy Environ. Sci.* 9: 1361–1374.
149 Liu, C.-C., Ramu, R., Chan, S.I. et al. (2016). *Catal. Sci. Technol.* 6: 7623–7630.
150 Kao, K.-C. and Mou, C.-Y. (2013). *Microporous Mesoporous Mater.* 169: 7–15.
151 Liu, C.-C., Lin, T.-S., Chan, S.I. et al. (2015). *J. Catal.* 322: 139–151.
152 Groothaert, M.H., Smeets, P.J., Sels, B.F. et al. (2005). *J. Am. Chem. Soc.* 127: 1394–1395.
153 Woertink, J.S., Smeets, P.J., Groothaert, M.H. et al. (2009). *Proc. Natl. Acad. Sci. U.S.A.* 106: 18908–18913.
154 Vanelderen, P., Hadt, R.G., Smeets, P.J. et al. (2011). *J. Catal.* 284: 157–164.
155 Grundner, S., Markovits, M.A.C., Li, G. et al. (2015). *Nat. Commun.* 6.
156 Solomon, E.I., Heppner, D.E., Johnston, E.M. et al. (2014). *Chem. Rev.* 114: 3659–3853.
157 Tomkins, P., Ranocchiari, M., and van Bokhoven, J.A. (2017). *Acc. Chem. Res.* 50: 418–425.
158 Sushkevich, V.L., Palagin, D., Ranocchiari, M. et al. (2017). *Science* 356: 523–527.
159 Le, H.V., Parishan, S., Sagaltchik, A. et al. (2017). *ACS Catal.* 7: 1403–1412.
160 Duetz, W.A., van Beilen, J.B., and Witholt, B. (2001). *Curr. Opin. Biotechnol.* 12: 419–425.
161 Chan, S.I., Chien, C.Y.-C., Yu, C.S.-C. et al. (2012). *J. Catal.* 293: 186–194.
162 Nesheim, J. C., Lipscomb, J. D., (1996) *Biochemistry*, 35: 10240–10247.

16

Alkane Oxidation with Vanadium and Copper Catalysts

Manas Sutradhar, Luísa M. D. R. S. Martins, M. Fátima C. Guedes da Silva, and Armando J. L. Pombeiro

Universidade de Lisboa, Centro de Química Estrutural, Instituto Superior Técnico, Lisboa, Portugal

16.1 Introduction

Transition metal complexes, namely, of vanadium and copper, can display a good catalytic activity toward the mild oxidative functionalization of inert alkanes into valuable organic products. The use of hydrogen peroxide or molecular oxygen as oxidant, under mild conditions, constitutes a convenient tool from a sustainable point of view. The existence of different metal oxidation states plays a key role for the catalytic properties of vanadium and copper complexes and their use as catalysts or catalyst precursors for the oxidation of liquid and gaseous alkanes has been the object of current and growing interest from both biological and chemical perspective [1–12].

The primary source of liquid alkanes is petroleum, which, however, are mainly used as fuel source. Alternatively, petrochemical feedstocks, i.e. chemical feedstocks derived from refined or partially refined petroleum fractions, are principally used in manufacturing chemicals, synthetic rubber, and plastics. Alkanes are a potentially rich carbon source of great interest for the synthesis of added value functionalized organic products, and thus there is high interest in researching their functionalization [1–19].

However, there are limitations of the usage of alkanes as raw materials in view of their inert character, such as common harsh reaction conditions, low product yield, and/or low selectivity. Therefore, the development of efficient catalytic systems under environmentally benign conditions for selective oxidation of alkanes is highly demanding.

Several vanadium and copper (mononuclear and polynuclear) complexes, including metal-organic frameworks (MOFs), have raised significant interest in this field. This chapter focuses on recent advances toward the efficient oxidation of light alkanes, in particular of cyclohexane (CyH) in view of its relevance as a model substrate for research and of its industrial significance.

16.2 Vanadium-Catalyzed Oxidation of Alkanes

The use of vanadium catalysts (or catalyst precursors) can provide a significant improvement in the oxidation of alkanes toward their eventual application as raw materials for organic synthesis. For example, the oxidation of CyH to cyclohexanol and cyclohexanone, which are important reagents to produce adipic acid and caprolactam applied in the manufacture of nylon. Among

Alkane Functionalization, First Edition. Edited by Armando J. L. Pombeiro and M. Fátima C. Guedes da Silva.

the various oxidants that have been used in vanadium-catalyzed oxidation, aqueous hydrogen peroxide has been the most attractive one, as it is cheap and environmentally benign. In the oxidation of alkanes, an acid promoter can also play an important role. In the following section, a general overview of vanadium-catalyzed oxidation of alkanes is unfolded.

16.2.1 Liquid Alkanes

CyH, a model substrate, is widely used for studying the functionalization of liquid alkanes and H_2O_2 as a common oxidant under mild conditions (Scheme 16.1). The usual oxidation products, cyclohexanol and cyclohexanone, are often obtained via cyclohexyl hydroperoxide.

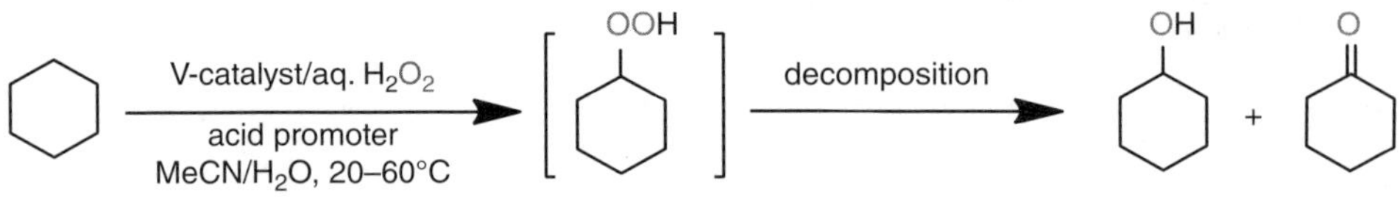

Scheme 16.1 Vanadium-catalyzed peroxidative oxidation of cyclohexane [1–3, 7–12].

Addition of a small amount of an acid as promoter, such as 2-pyrazinecarboxylic acid (PCA) [20–25], nitric [26], sulfuric [27, 28] or oxalic [27, 28] acid, can dramatically improve the activity. It was first discovered by Shul'pin and co-workers in 1993 that PCA can enhance the catalytic efficiency of vanadium systems for the oxidation of organic compounds (alkanes, arenes, alcohols) with hydrogen peroxide under mild conditions (20–60 °C, in acetonitrile solution) [29–31]. Several subsequent studies have confirmed that PCA has an important role toward the vanadium-catalyzed peroxidative oxidation of both liquid and gaseous alkanes under mild conditions [1–3, 7–12, 20, 21, 32]. As a promoter, PCA usually has a higher efficiency than those of other acids in the vanadium-catalyzed oxidation of alkanes. It has been assumed that PCA as a co-catalyst/promoter plays a significant role to accelerate the proton transfer from a coordinated hydrogen peroxide molecule to an oxido, hydroxido, or peroxido ligand via the so-called "robot's arm mechanism" (Scheme 16.2) [16, 18, 22, 24, 32, 33]. The importance of such a proton shift will be discussed below, being involved in the generation of the hydroxyl radical from H_2O_2.

In combination with PCA as a promoter, oxidovanadium(V) complexes having *azine fragment (C=N—N=C) ligands* exhibit an efficient catalytic activity in NCMe with aq. H_2O_2 under mild conditions. They are hemi-labile ligands that stabilize vanadium complexes in various coordination forms and oxidation states [20, 21, 34–37] and can enhance the catalytic activity of the vanadium centers, e.g. by promoting proton-shift steps and also by participating in redox steps (they are redox active) (see below).

Examples (see Chart 16.1) of such catalysts are provided by the dinuclear oxidovanadium(V) complexes $[\{VO(OEt)(EtOH)\}_2L]$ (**1**, H_4L = bis(2-hydroxybenzylidene)terephthalohydrazide; **2**, H_4L = bis(2-hydroxybenzylidene)oxalohydrazonic acid), the hexanuclear mixed-valence oxidovanadium(IV/V) complex $[V_3O_3(OEt)(ashz)_2(\mu{-}OEt)]_2$ (**3**, H_3ashz = N′-acetylsalicylhydrazide), and the heterometallic dioxidovanadium(V)/alkali metal coordination polymers (CPs) $[(VO_2)_2(\mu_4\text{-}L)\{Na_2(\mu\text{-}H_2O)_2(H_2O)_2\}]_n$ (**4**, H_4L as in **2**), $[\{V(\mu\text{-}O)_2\}_2(\mu_4\text{-}L)\{K_2(\mu\text{-}H_2O)_2(H_2O)_2\}]_n$ (**5**, H_4L as in **2**), $[\{V(\mu\text{-}O)(\mu_3\text{-}O)\}_2(\mu_8\text{-}L)\{Cs_2(\mu\text{-}H_2O)_2(H_2O)_2\}]_n$ (**6**, H_4L as in **2**) [20, 21, 36].

The set (Chart 16.1) of dinuclear oxidovanadium complexes $NH_4[(VO_2)_2(^{H}LH)]$ (**7**), $NH_4[(VO_2)_2(^{t\text{-}Bu}LH)]$ (**8**), $[(VO_2)(VO)(^{H}LH)(CH_3O)]$ (**9**), $[(VO_2)(VO)(^{t\text{-}Bu}LH)(C_2H_5O)]$ (**10**) and $[(VO_2)(VO)(^{Cl}LH)(CH_3O)\text{-}(CH_3OH/H_2O)]$ (**11**) (where $^{H}LH_4$ = 1,5-bis(2-hydroxybenzal

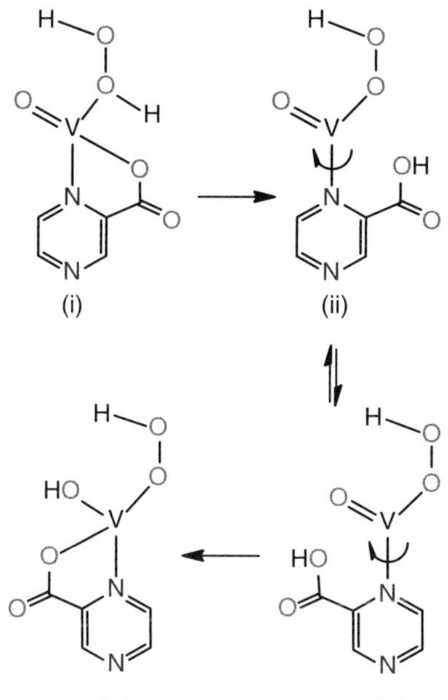

Scheme 16.2 PCA-assisted proton migration from a coordinated H_2O_2 molecule to an oxido ligand ("robot's arm mechanism") [16, 18, 22, 24, 32, 33]. (*See insert for color representation.*)

dehyde)carbohydrazone, $^{t\text{-Bu}}LH_4$ = 1,5-bis(3,5-di-*tert*-butyl-2-hydroxybenzaldehyde)carbohydrazone, and $^{Cl}LH_4$ = 1,5-bis(3,5-dichloro-2-hydroxybenzaldehyde)carbohydrazone), with the triply deprotonated ligand (using two different binding sites, ONN and ONO, separated by a diazine unit) show good catalytic activity toward the oxidation of CyH to cyclohexanol and cyclohexanone under MW-assisted and solvent-free peroxidative (with aqueous *tert*-butyl hydroperoxide, TBHP) conditions [37]. A cyclic voltammetric study shows that these complexes are reducible and DFT calculations indicate that the redox active carbohydrazone ligand plays an important role in the formation of the reduced form of the catalyst with preservation of the vanadium +5 oxidation state [37]. The first reduction in the ligand framework is indicated by the shape of LUMO for singlet $^1[7]^-$ under vacuum (Figure 16.1a). Calculations also show that the spin density of the paramagnetic doublet $^2[7]^{2-}$ is located predominantly on the ligand (Figure 16.1b), which is further supported by electrochemical and spectroelectrochemical studies [37]. Hence, a metal-ligand cooperation is believed to occur in the alkane oxidation with such vanadium catalysts.

The highest yields (above 25%) of cyclohexanol and cyclohexanone are obtained by the dinuclear oxidovanadium(V) complex **2**, the mixed-valence oxidovanadium(IV/V) complex **3**, and the heterometallic V^V/Cs coordination polymer **6** for the oxidation of CyH in the presence of PCA under mild conditions [20, 21, 36]. The highest turnover number (TON) of 4.4×10^4 was obtained for the **1**/PCA/H_2O_2 system at 50 °C, with a low catalytic amount (1×10^{-6} M) of **1**. The aqua-soluble coordination polymers **4–6** having similar $[(VO_2)_2(\mu_{4/8}\text{-}L)]^{2-}$ (H_4L = bis(salicylaldehyde)oxaloyldihydrazone) cores follow the trend **4** < **5** < **6** in the presence of PCA under optimized conditions, which parallels that of the ionic radius and the coordination number of the alkali cation [36].

The dioxidovanadium(V) complex $[VO_2(pca)(hmpa)]$ (**12**, Chart 16.1) (hmpa = hexamethylphosphoramide and pcaH = pyrazine-2-carboxylic acid) contains a deprotonated PCA ligand that was also applied to the oxidation of CyH under mild conditions [38]. Addition of PCA as a

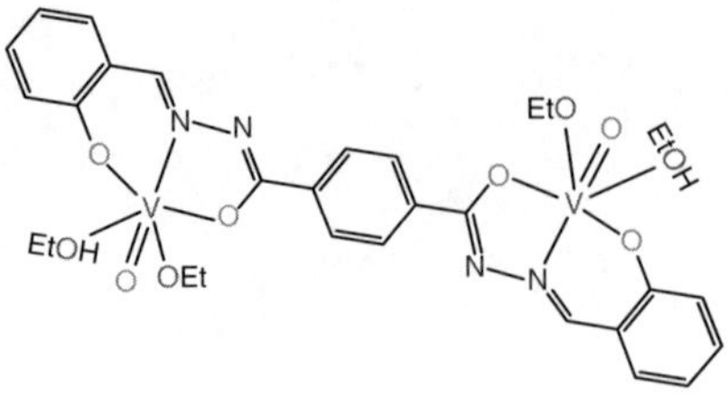

1

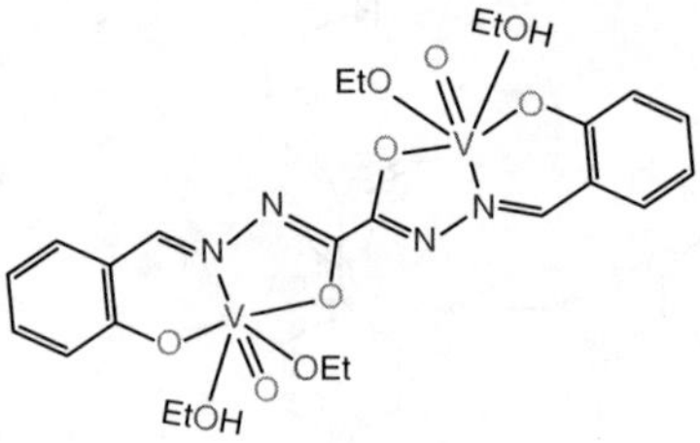

2

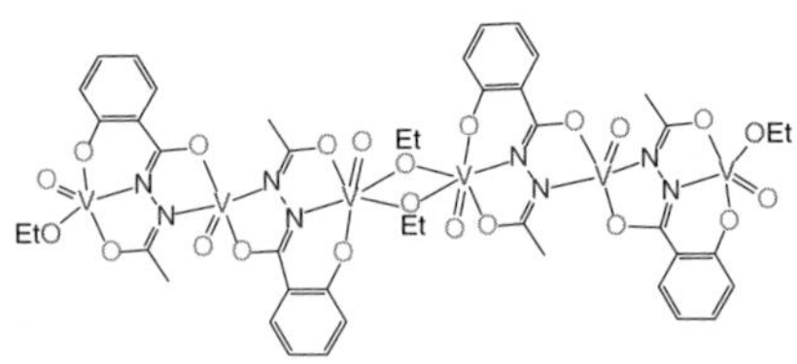

3

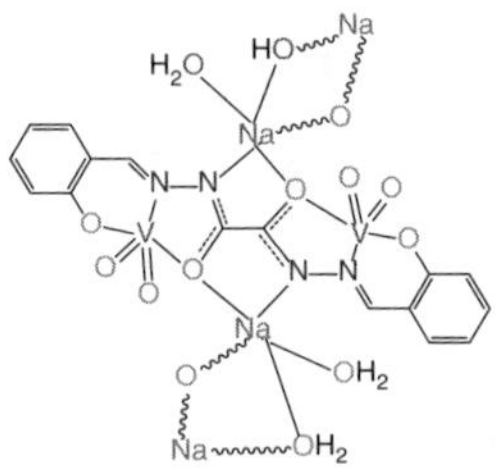

4

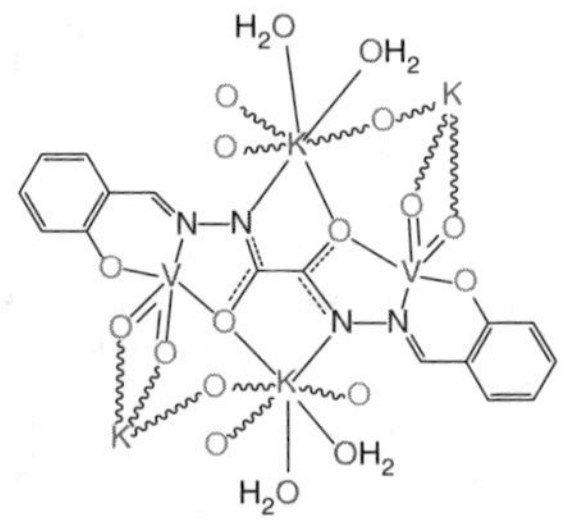

5

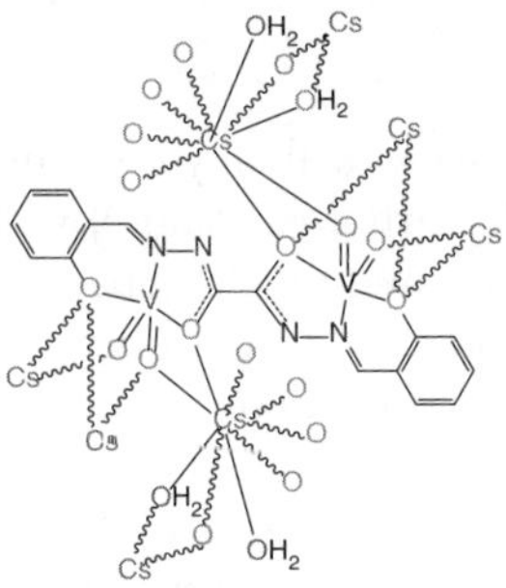

6

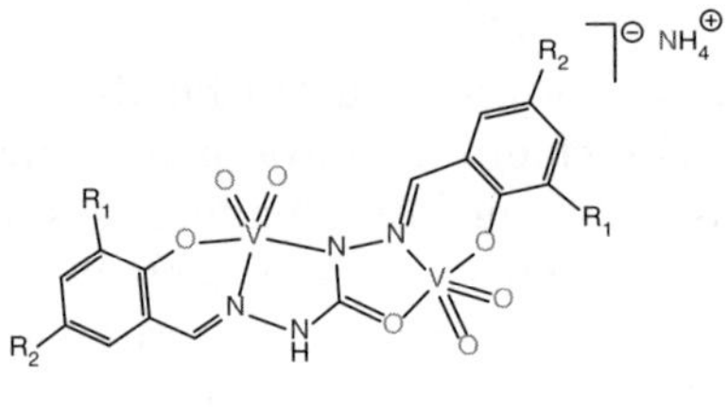

7: $R_1 = R_2 = H$
8: $R_1 = R_2 =$ *tert*-Bu

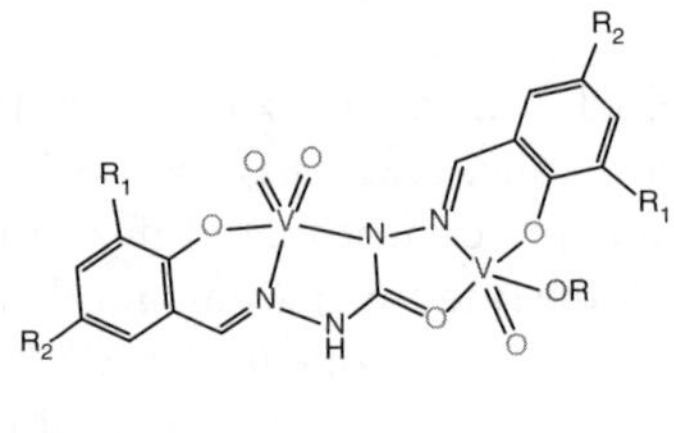

9: $R_1 = R_2 = H$, $R = Me$
10: $R_1 = R_2 =$ *tert*-Bu, $R = Et$

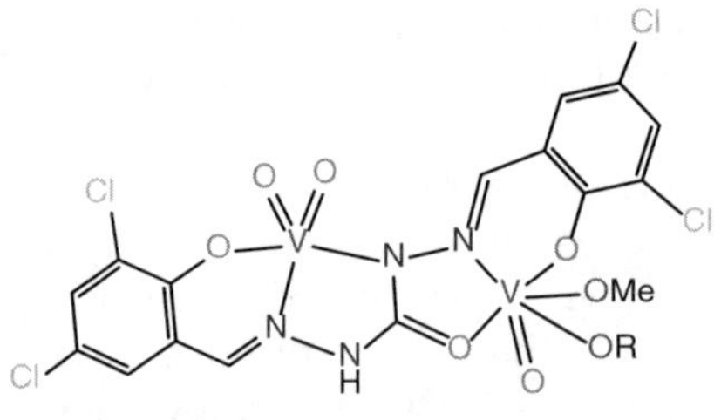

11: R = Me or H

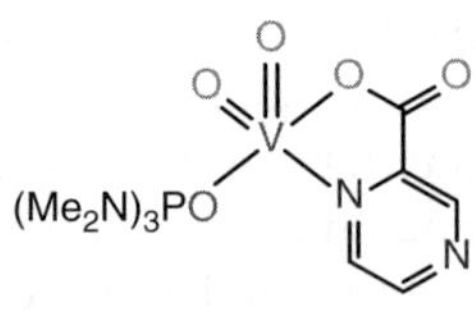

12

Chart 16.1

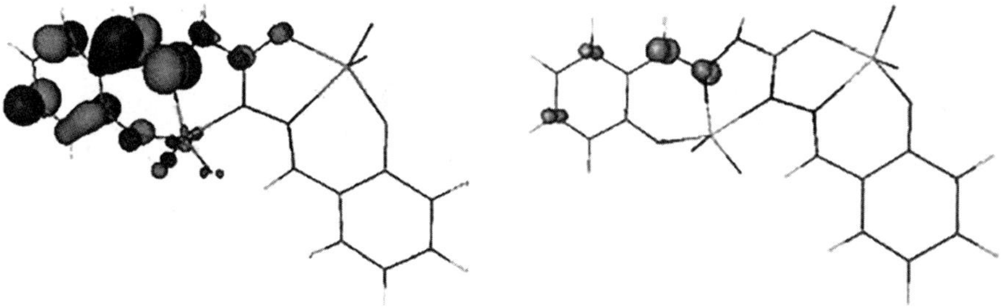

Figure 16.1 (a) LUMO of $^1[\mathbf{7}]^-$ in vacuum; (b) spin density of $^2[\mathbf{7}]^{2-}$ in vacuum [37]. With permission of ACS (http://pubs.acs.org/doi/abs/10.1021%2Facs.inorgchem.6b01011). (*See insert for color representation.*)

co-catalyst (V/PCA ratio 1:4) also accelerates the reaction. CyH oxidation proceeds with a high initial rate and a high TON (1370) in 24 hours [38].

Vanadium complexes bearing *pyrazole or scorpionate C-based or B-based trispyrazolyl ligands* also catalyze the peroxidative oxidation (in water/acetonitrile) of liquid alkanes [27, 39, 40]. Instead of PCA, the HNO_3 was used as a promoter (for typical reaction conditions, n(HNO_3)/n(catalyst) = 100) for this system. For example (see Chart 16.2), the dioxidovanadium(V) $[VO_2(3,5\text{-}Me_2Hpz)_3][BF_4]$ (**13**) (pz = pyrazolyl), $[VO_2\{SO_3C(pz)_3\}]$ (**14**), the oxidovanadium(IV) $[VO\{HB(pz)_3\}\{H_2B(pz)_2\}]$ (**15**), $[VO(acac)_2(Hpz)]\cdot HC(pz)_3$ (**16**), the non-oxidovanadium(IV) $[VCl_3\{HC(pz)_3\}]$ (**17**) and vanadium(III) $[VCl_3\{SO_3C(pz)_3\}]$ (**18**) complexes act as catalysts for oxidation with e.g. H_2O_2 (in water/acetonitrile) of liquid alkanes (CyH and cyclopentane) to the corresponding alcohols and ketones in the presence of HNO_3. A good catalytic activity in the oxidation of CyH, at 20 °C in 6 hours was obtained by the dioxidovanadium(V) complexes **13** (TON 113) and **14** (TON 117), with dimethylpyrazolyl or tris(pyrazolyl)methanesulfonate ligands, respectively [39]. The highest TON value (ca. 410) [40], was obtained by the oxidovanadium(IV) complexes **16**, using a low catalyst amount.

Different oxidants, such as, benzoyl peroxide (BPO), TBHP, *m*-chloroperoxybenzoic acid (mCPBA) or the urea-hydrogen peroxide adduct (UHP), were also studied with the catalysts **17–18** [27]. mCPBA was the most effective one in the oxidation of CyH, in accord with its strongest oxidizing character [27].

Molecular oxygen (cheaper and greener than the other oxidants) can also act as an oxidant for the oxidation of alkanes with oxidovanadium complexes.

The bis(maltolato)oxovanadium complexes (see Chart 16.2) *cis*-$[VO(OCH_3)(ma)_2]$ (**19**), *cis*-$[VO(OC_2H_5)(ma)_2]$ (**20**), $[VO(ma)_2]$ (**21**) and $[VO(py)(ma)_2]$ (**22**) anchored to chemically modified carbamated silica gel act as supported catalysts for the oxidation of *n*-pentane and *n*-hexane, with molecular oxygen under relatively mild reaction conditions, to the corresponding ketones (with good selectivities) and alcohols (in smaller amounts) [41]. The **19**-supported catalyst showed the highest activity and selectivity at 150 °C, p(O_2) = 15 atm for *n*-pentane oxidation to 2-pentanone (84% selectivity) with a small amount of 2-pentanol.

These supported catalysts were also studied for cyclopentane and cyclooctane oxidation [42]. At 150 °C and 6 hours reaction time with p(O_2) ca. 14 atm, the highest conversion and selectivity toward cyclooctanone are observed. However, the highest selectivity toward cyclopentanone was achieved at the lower pressure of 7 atm, at 160 °C.

The effect of some heteroaromatic acids, such as picolinic acid, 2,6-pyrazinedicarboxylic acid, pyrazine-2,6-dicarboxylic acid, 3-aminopyrazine-2-carboxylic acid, and PCA as co-catalysts, was also examined and PCA was found to be the most effective one for both linear and cycloalkanes [41, 42]. The catalysts can be reused without separation and reactivation,

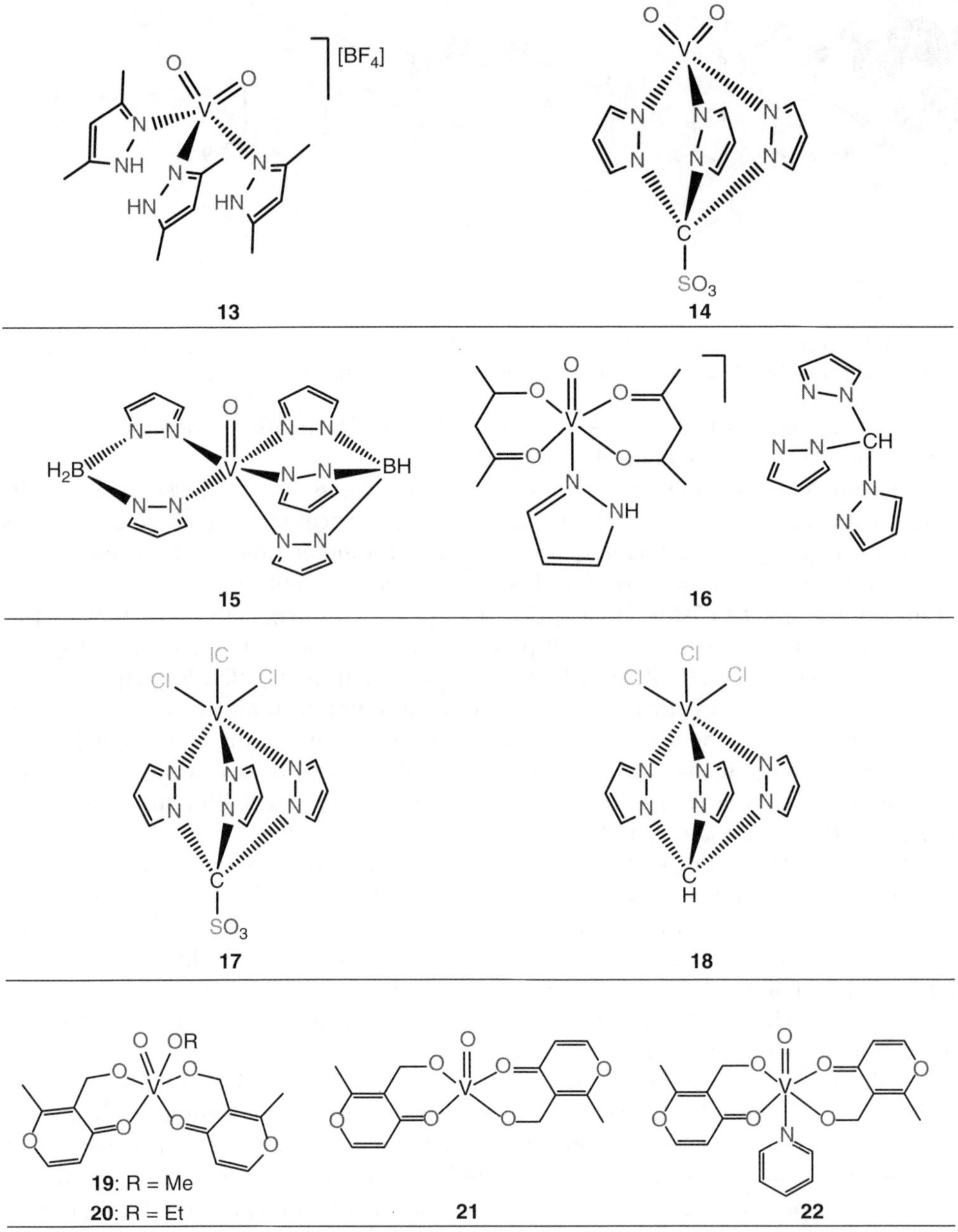

Chart 16.2

by simple admission of further molecular oxygen to the reaction system, which is a good advantage over the homogeneous catalysts.

16.2.2 Gaseous Alkanes

Oxidation of gaseous alkanes (methane, ethane, propane, and *n*-butane) is less studied compared to liquid alkanes. Peroxidative oxidation of such alkanes produces the corresponding oxidized compounds (alcohols, aldehydes, ketones, and/or carboxylic acids).

The oxido-V-cat/PCA/aq.H_2O_2/CH_3CN systems using complexes **1** and **2** efficiently catalyze the oxidation of gaseous alkanes at 50 °C [13, 35]. Methane is less reactive and undergoes oxidation to methanol with a much lower yield than the oxidation of the other alkanes. The oxidation of ethane produces acetaldehyde as a main product and ethanol, in addition to a smaller amount of acetic acid eventually formed due to the oxidation of the acetaldehyde [13, 35]. Ethane and propane oxidations reached a very high TON (in the 1×10^3–3×10^3 range) [13, 35], with propane being more reactive due to the presence of the secondary carbon atom. Acetone and isopropanol are the main products for the oxidation of propane, along with some minor amounts of propanal, propanol, and propionic acid [13, 35]. The major products for the oxidation of *n*-butane are butanone-2 and butanol-2 and the minor products are butanal, 1-butanol and butyric acid [13, 35].

Sodium metavanadate ($NaVO_3$) with aqueous hydrogen peroxide as an oxidant can also oxygenate light alkanes in the presence of H_2SO_4 as a promoter. The highest TON (200) was achieved for propane and the highest yield (19%) was obtained in the case of *n*-butane [28].

16.2.3 Mechanism of the Vanadium-Catalyzed Peroxidative Oxidation of Alkanes

Various vanadium catalytic systems, such as V-cat/PCA/aq.H_2O_2/CH_3CN [13, 14, 35] and V-cat/HNO_3/aq.H_2O_2/CH_3CN, [39] were used to determine the selectivity parameters for the oxidations of different linear and branched alkanes. These parameters are important to elucidate the type of mechanism for the oxidation of alkanes. They showed very low regioselectivities for *n*-heptane oxidation and for methylcyclohexane oxidation. The oxidation of the *cis* and *trans* isomers of 1,2-dimethylcyclohexane was non-stereoselective. The low selectivity features are indicative of the involvement of the highly reactive hydroxyl radical during the course of reaction.

In general, the selectivity, kinetic, and theoretical studies suggest that the oxidation of alkanes proceeds via formation of the hydroxyl radical (HO•) upon metal catalyzed decomposition of H_2O_2 [22–25, 33]. HO• acts as a hydrogen atom abstractor from the alkane (RH) to form the alkyl radical (R•). Reaction of R• with O_2 (e.g. from air) yields the alkyl peroxyl radical $ROO^•$, which leads to the formation of the alkyl hydroperoxide ROOH as a primary product. Then the metal-assisted decomposition of ROOH yields the corresponding alcohol (ROH) and ketone.

Equations (16.1) and (16.2) represent the overall reactions leading to the formation of HO• starting from a V^V catalyst, and Scheme 16.3 displays a possible catalytic mechanism for these reactions [22–25].

$$V^V + H_2O_2 \rightarrow V^{IV} + HOO\bullet + H^+ \tag{16.1}$$

$$V^{IV} + H_2O_2 \rightarrow V^V + HO^- + HO\bullet \tag{16.2}$$

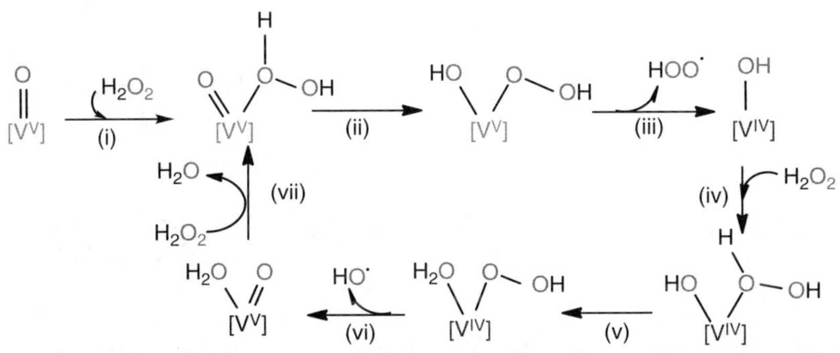

Scheme 16.3 Proposed catalytic cycle for vanadium-catalyzed hydroxyl radical generation [22–25].

For the proton transfer reactions from H_2O_2 to oxido or hydroxido ligands, different possible transition states can be envisaged as shown in Scheme 16.4 [23–25, 33, 43]. The involvement of H_2O [25] or an oxido-vanadium group [33] in the transition state [Scheme 16.4c or d, respectively,] forming a six-membered metallacycle markedly lowers the energy of the transition state relative to that for the direct proton-transfer [Scheme 16.4a] or even when assisted by PCA [Scheme 16.4b], thus effectively promoting the reaction, as indicated by DFT calculations [25, 33]. Hence, water in a controlled amount can play a relevant promoting role, acting as a catalyst for proton-transfer steps.

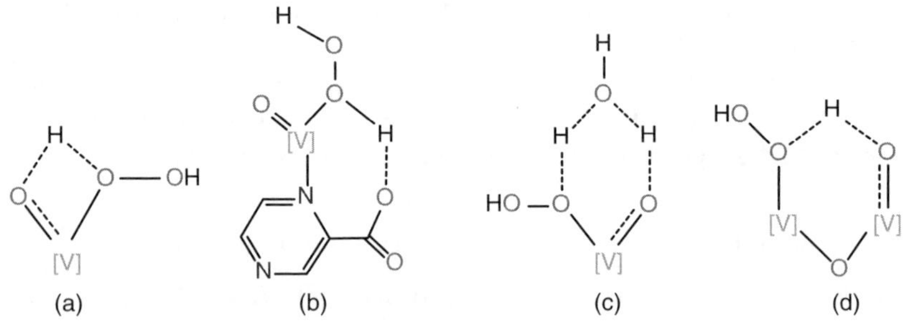

Scheme 16.4 Possible transition states involved in proton-transfer steps: direct proton-transfer (a), assisted by PCA (b), assisted by H_2O (c), and assisted by a second V center (d).

The dependences of initial oxidation rates W_0 on the initial concentrations of the various components of the reaction mixture have been determined, e.g. for the **1**/H_2O_2/PCA system. An estimate of rate constants of the pseudo-first order reactions of HO• with PCA, H_2O_2 and acetonitrile indicates that acetonitrile is the main competitor for CyH in the interaction with hydroxyl radical. The kinetic data support the assumption about the generation of hydroxyl radicals in the crucial stage [13, 20].

Theoretical DFT calculations also support the generation of the HO• radical in the case of oligovanadates (which are formed in acidic medium from monovanadate), as shown in Scheme 16.5 for an oxidodivanadate catalyst [33]. The second vanadium fragment plays the role of a stabilizer of the key transition state for step (iv) (Scheme 16.5) due to the formation of a six-membered cyclic structure [Scheme 16.4d] [33].

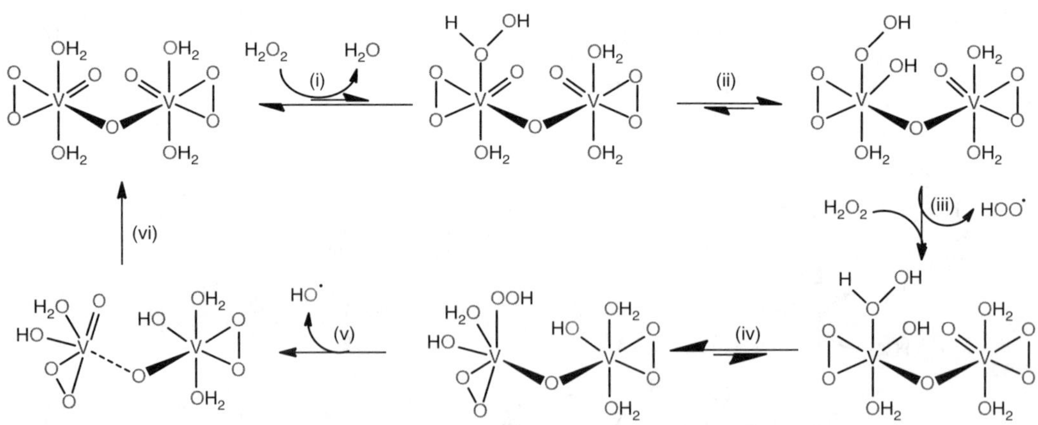

Scheme 16.5 Proposed catalytic cycle for the hydroxyl radical generation based on an oxidodivanadate [33].

16.3 Copper-Catalyzed Oxidation of Alkanes

16.3.1 Mononuclear Complexes

The use of mononuclear copper catalysts for the oxidation of alkanes is still relatively scarce. Among them, the complexes derived from aryl- (or aroyl)-hydrazone, tetrazolato, or scorpionate ligands are in the frontline. In fact, the application of complexes with C-scorpionate poly(pyrazol-1-yl)methane ligands as oxidation catalysts is experiencing a marked development [9]. It is believed that the interchange between bidentate and tridentate coordination modes of such hemi-labile ligands (consistent with the usual scorpionate designation) is at the core of the structural and chemical versatility of their metal complexes and is essential for their catalytic activity. Mononuclear tris(pyrazol-1-yl)methane copper complexes (e.g. **23** and **24**, see Chart 16.3) have been successfully applied as catalysts or catalyst precursors for relevant alkane oxidation reactions, under mild conditions, namely with peroxides (usually hydrogen peroxide) to give alcohols and ketones [27, 44]. Complexes derived from functionalized scorpionates with an ionic or polar group (e.g. -SO_3Na, -CH_2OH or -$OCH_2SO_2CH_3$) are water soluble, with the advantage of promoting the use of aqueous media, and thus of performing the catalytic reactions under greener conditions.

Recently, the water-soluble full-sandwich compound $[Cu\{CH_3SO_2OCH_2C(pz)_3\}_2](OTf)_2$ (**25**, pz = pyrazolyl) was synthesized [45] and shown to catalyze the single-pot cyclooctane oxidation under mild conditions with H_2O_2, which is promoted in acidic medium reaching yields (based on the alkane) up to ca. 27%. Hence, a coordinately saturated full-sandwich complex can also present a high catalytic activity, in view of the hemi-lability of the scorpionate ligands to form an active species.

Detailed information on the use of scorpionate complexes in alkane functionalization is given in Chapter 6.

The mononuclear aroylhydrazone Cu(II) complexes $[Cu(H_2L)(NO_3)(H_2O)]$ (**26**) and $[Cu(H_2L)Cl]\cdot 2MeOH$ (**27**), with H_2L^- in the *keto* form (Chart 16.3), were prepared by reaction of the Schiff base (3,5-di-*tert*-butyl-2-hydroxybenzylidene)-2-hydroxybenzohydrazide (H_3L) with a suitable copper(II) salt which played a key role for stabilizing the *keto* form over the *enol* one [46].

Complexes **26** and **27** act as efficient catalysts toward the peroxidative oxidation of CyH to cyclohexyl hydroperoxide, cyclohexanol, and cyclohexanone in NCMe/H_2O media, under mild conditions and in the absence of any additive. Yields may be enhanced by prolonging the reaction time achieving e.g. for **26** a maximum overall yield of 30% and TON 294 after 24 hours (at room temperature and without any additive), or by adding an acid co-catalyst (e.g. trifluoroacetic acid, TFA) [46].

The *ortho*-substituted arylhydrazone of barbituric acid 5-(2-(2-hydroxyphenyl)hydrazono) pyrimidine-2,4,6(1*H*,3*H*,5*H*)-trione (H_4L^1) was synthesized and used for the preparation of the water-soluble Cu(II) complex $[Cu(H_2L^1)(H_2O)(im)]\cdot 3H_2O$ (**28**) (im = imidazole) [47], which turned out to be an active catalyst for the peroxidative (with H_2O_2) oxidation of CyH at room temperature (yields up to 21% and TON up to 213). The ease of **28** to undergo redox processes, as found by cyclic voltammetry, probably favors the free-radical CyH oxidation.

Metal mediated [2 + 3] cycloaddition reaction between a copper-bound azide polymer and different organonitriles was used to give the mononuclear copper(II) tetrazolato complexes $[Cu(pmtz)_2(phen)]$ (**29**) and $[Cu(pytz)_2(phen)]$ (**30**) (pmtz = 5-(2-pyrimidyl)tetrazolate; pytz = 5-(2-pyridyl)tetrazolate) [48]. These complexes were explored as homogeneous catalysts for the selective oxidation of CyH to cyclohexanol and cyclohexanone with TBHP under solvent- and additive-free mild conditions using microwave (MW) irradiation. Although the

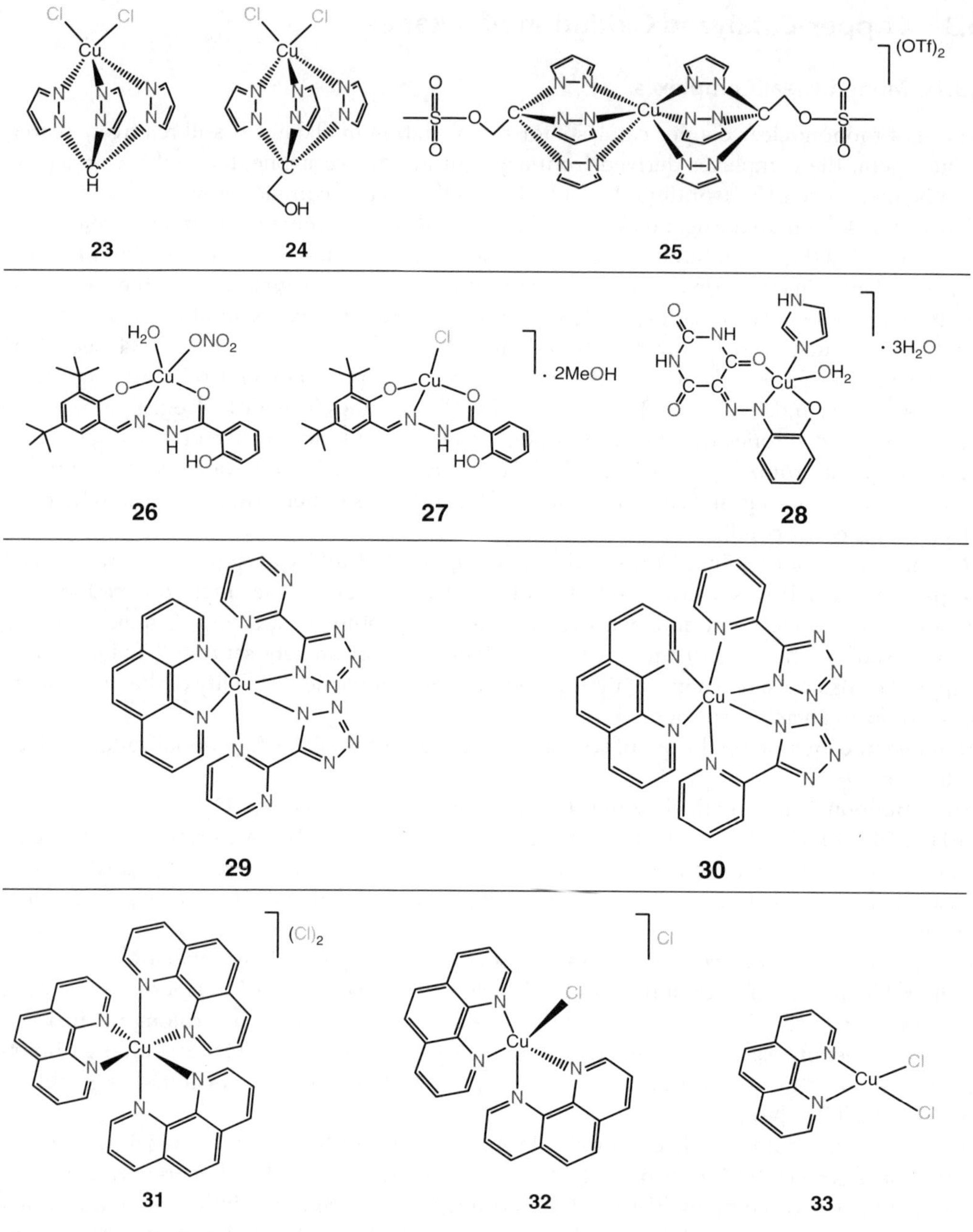

Chart 16.3

obtained yields (up to ca. 21%) are comparable to those of other catalysts, these features are of significance in terms of sustainability.

The mononuclear Cu(II) complexes $[Cu(phen)_3]Cl_2{\cdot}7H_2O$ (**31**), $[CuCl(phen)_2]Cl{\cdot}5H_2O$ (**32**), $[CuCl_2(phen)]$ (**33**), with different numbers of phenanthroline ligands, have also been examined in the oxidation of CyH using hydrogen peroxide as oxidant, in acetonitrile-water solution at temperatures ranging from 25 to 70 °C [49]. A high conversion with a total yield of 67%

(TON = 737) was obtained with **32** at 70 °C. At a higher temperature range (50–70 °C), a small percentage of adipic acid was also formed.

16.3.2 Polynuclear Complexes

Multinuclear copper complexes are of significance in terms of mimicking the action of particulate methane monooxygenase (pMMO), a multi-copper enzyme that catalyzes the oxidation of alkanes. The pioneering work [50] within this approach concerned the use of the tetranuclear μ-oxido complex $[O\subset Cu_4(tea)_4(BOH)_4][BF_4]_2$ (**34**, H_3tea = triethanolamine), the trinuclear $[Cu_3(H_2tea)_2(4\text{-}OC_6H_4COO)_2(H_2O)]\cdot 4H_2O$ (**35**), the dinuclear $[Cu_2(H_2tea)_2(C_6H_5COO)_2]\cdot 2H_2O$ (**36**), and the coordination polymer $[Cu_2(H_2tea)_2\{\mu\text{-}C_6H_4(COO)_2\text{-}1,4\}]_n\cdot 2nH_2O$ (**37**) (Chart 16.4). In acetonitrile, at room temperature, they form highly selective and active catalytic system for the oxidation of alkanes (namely CyH, methane, and ethane) [50].

The tetra and trinuclear complexes **34** and **35** exhibit the highest activity with overall yields (based on cycloalkane) up to 32% for a typical acid-to-catalyst molar ratio of 10:1. TONs of ca. 360–380 can be reached for a sufficiently high ratio of alkane to catalyst [50]. The multicopper catalysts can exhibit a higher catalytic activity in comparison with mononuclear ones [50–52].

Recently, copper(I) and copper(II) pyrazolate compounds have been reported [53]. For 3,5-NO_2-pyrazole ligands, the dinuclear cyclic Cu(I) complex $[Cu_2(\mu\text{-}N,N\text{-}3,5\text{-}(NO_2)_2pz)_2(PPh_3)_2]$ (**38**) was obtained, while in the case of 3,5-bis(trifluoromethyl)-pyrazole, a spontaneous rearrangement to the hexanuclear cyclic $[trans\text{-}Cu_6(\mu\text{-}OH)_6(\mu\text{-}3,5\text{-}(CF_3)_2pz)_6]$ (**39**) was observed.

Compounds **38** and **39** were successfully tested as catalysts for the microwave-assisted neat oxidation (with TBHP) of CyH to the cyclohexanol and cyclohexanone mixture, affording considerably high yields (up to 58%) in rather short reaction times (30 minutes, for **39**). The obtained yields are higher than those reported (although under considerably different conditions) for copper complexes with related pyrazole ligands [54, 55]. Moreover, this catalytic protocol (TBHP/MW) is very fast, solvent-free, and uses a green energy source.

A convenient route to metallasilsesquioxanes was disclosed, the simplest compound being the binuclear cage-like coppersilsesquioxane $[(PhSiO_{1.5})_{10}(CuO)_2(NaO_{0.5})_2(EtOH)_4]$ (**40**) [56]. Moreover, isomers containing four [57] and six [58] copper ions, $[(PhSiO_{1.5})_{12}(CuO)_4(NaO_{0.5})_4]$ (**41**) and $[(PhSiO_{1.5})_{12}(CuO)_6(NaCl)(C_4H_8O_2)_{12}(H_2O)_2]$ (**42**), respectively, have been isolated. All compounds exhibited high catalytic activity in oxidation of alkanes with peroxides [59]. Compound **40** is a good pre-catalyst for the alkane oxygenation with hydrogen peroxide in air and in acetonitrile solution. Moreover, the oxidation of *trans*-1,2-dimethylCyH with TBHP catalyzed by complex **40** proceeds stereoselectively with the inversion of configuration. Compound **41** is stable under oxidation (TBHP) conditions and is an efficient catalyst in the absence of any acid. The oxidation of saturated hydrocarbons with TBHP catalyzed by complex **40** proceeds more selectively in comparison with the oxidation using H_2O_2. This difference can be explained tentatively by considering the steric hindrance created by the bulky silsesquioxane ligands at the copper reactive centers.

In the isomeric trinuclear copper complexes $[Cu_3(L)_2(MeOH)_4]$ (**43** or **44**) and $[Cu_3(L)_2(MeOH)_2]\cdot 2MeOH$ (**45**) (H_3L = 3,5-di-*tert*-butyl-2-hydroxybenzylidene-2-hydroxybenzohydrazide), the L^{3-} ligand exhibits trianionic pentadentate $1\kappa O,O',N{:}2\kappa N',O''$ chelation modes [60]. All compounds act as efficient catalyst precursors, in aqueous-acetonitrile medium, at room temperature, for the peroxidative oxidation (with H_2O_2) of CyH to cyclohexyl hydroperoxide, cyclohexanol, and cyclohexanone, leading to overall yields (based on the alkane) up

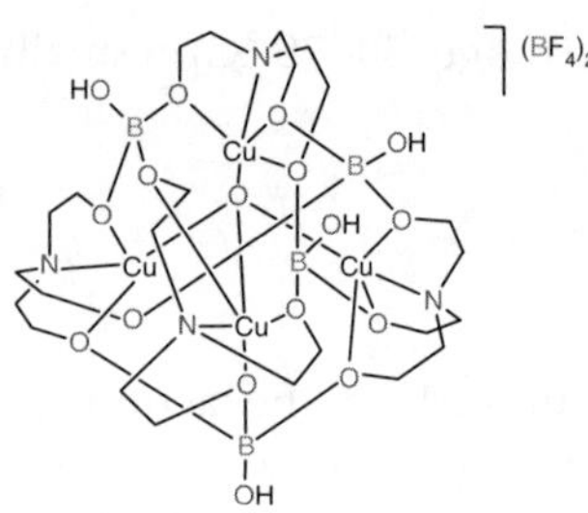

34

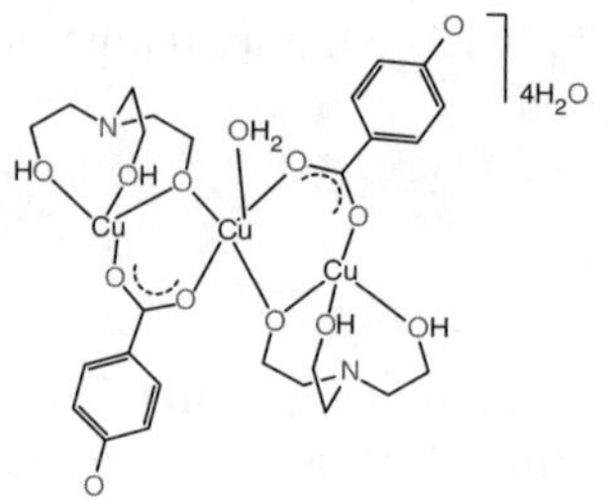

35

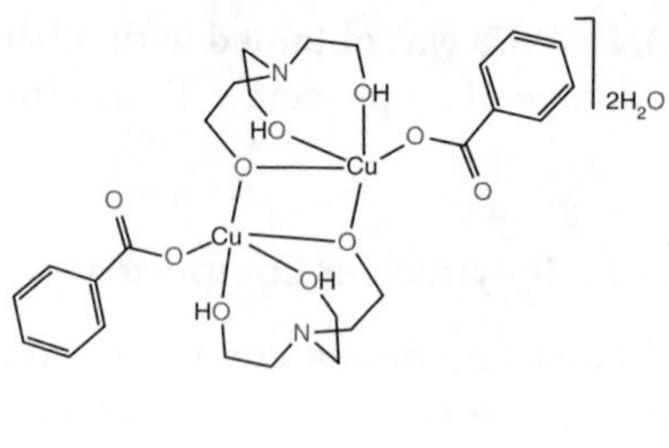

36

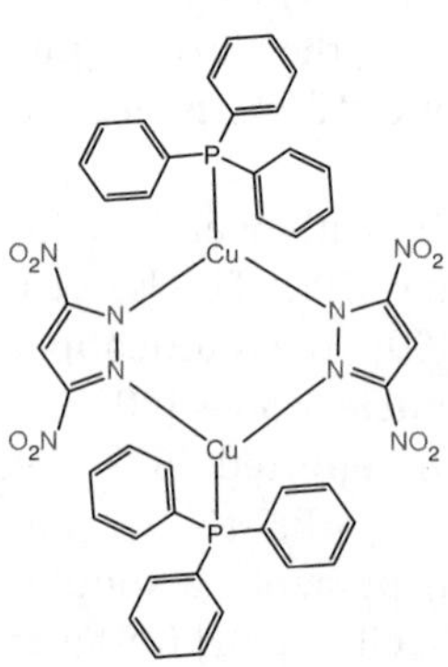

38

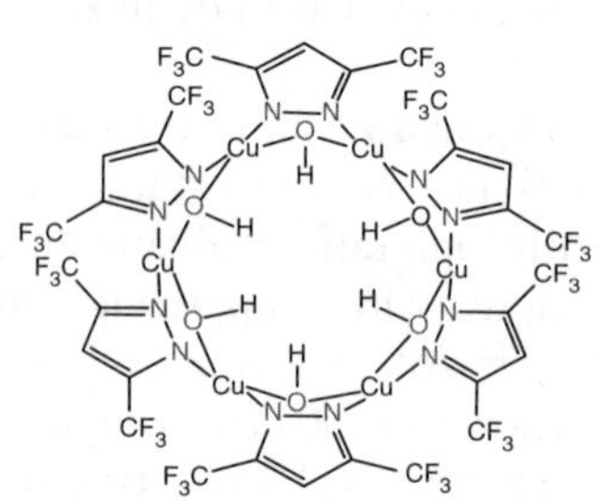

39

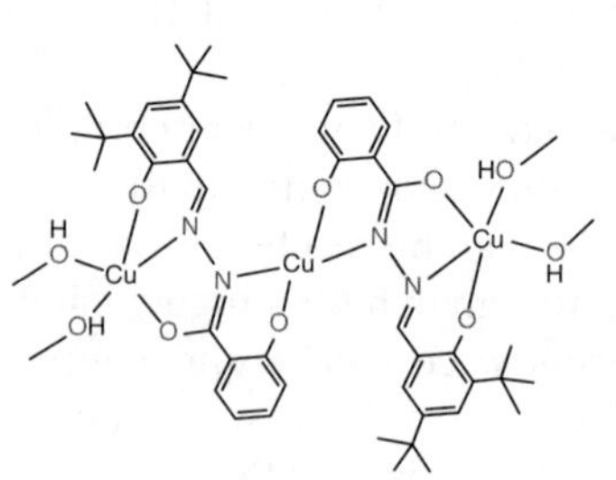

43

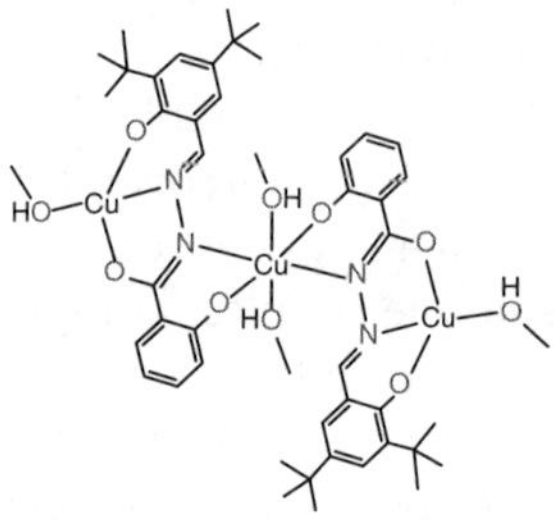

44

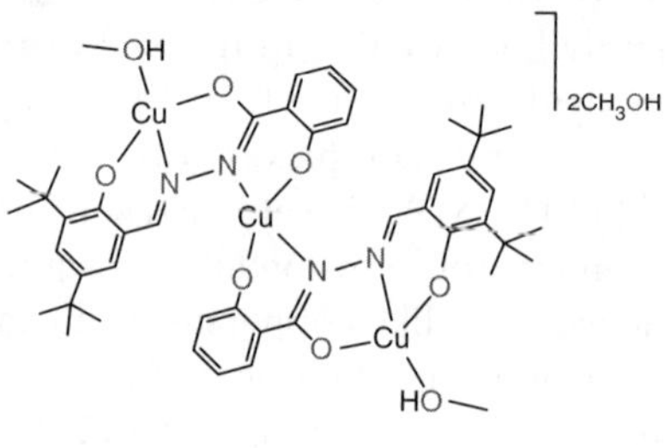

45

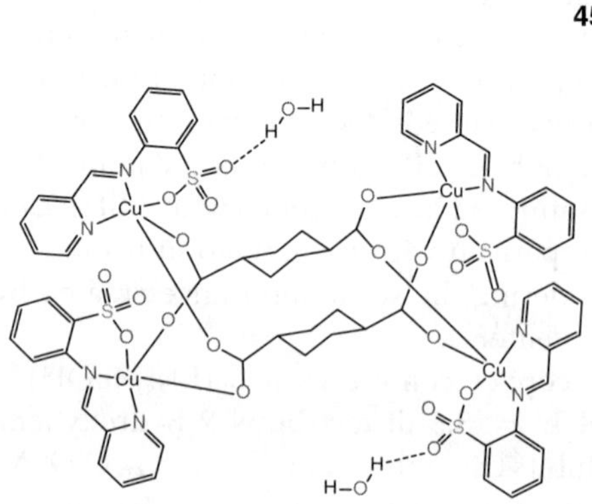

46

Chart 16.4 (*See insert for color representation.*)

to 31% (TON = 1.55×10^3) after 6 hours, in the presence of pyrazine carboxylic acid as promoter [60].

The catalytic oxidation with H_2O_2 of CyH in the ionic liquid (IL) 1-butyl-3-methylimidazolium hexafluorophosphate [bmim][PF_6] using the tetracopper(II) complex $[(CuL)_2(\mu_4\text{-}O,O',O'',O'''\text{-CDC})]_2{\cdot}2H_2O$ (**46**) [HL = 2-(2-pyridylmethyleneamino) benzenesulfonic acid, CDC = cyclohexane-1,4-dicarboxylate] as catalyst was also reported [61].

Significant improvements on the catalytic performance, in terms of product yield (up to 36%), TON (up to 529), reaction time, selectivity toward cyclohexanone, and easy recycling (negligible loss in activity after three consecutive runs), are observed using the [bmim][PF_6] IL instead of a molecular organic solvent including the commonly used acetonitrile [61]. The possible interactions between ILs and the catalyst, substrate, oxidant, and even reaction intermediates can turn ILs to act as multifunctional solvents for such catalytic oxidation reactions.

16.3.3 Copper MOF Catalysts

Copper MOFs or coordination polymers (CPs) can display a high catalytic activity in alkane oxidations with peroxides [51].

An example of MOF catalysts (complex **37**) [50] was already mentioned, and a recent study concerns the water-soluble 2D copper(II) CPs $[\{Cu_2(\mu_2\text{-dmea})_2(H_2O)\}_2(\mu_4\text{-pma})]_n{\cdot}4nH_2O$ (**47**), $[\{Cu_2(\mu_2\text{-Hedea})_2\}_2(\mu_4\text{-pma})]_n{\cdot}4nH_2O$ (**48**) and $[\{Cu(bea)(Hbea)\}_4(\mu_4\text{-pma})]_n{\cdot}2nH_2O$ (**49**) (H_4pma = pyromellitic acid, Hdmea = N,N-dimethylethanolamine, H_2edea = N-ethyldieth anolamine, Hbea = N-benzylethanolamine) [62]. These compounds are precatalysts for the oxidation, under the usual conditions, with aqueous H_2O_2 in acidic MeCN/H_2O medium, of various cycloalkanes to the corresponding alcohols and ketones with an overall product yield up to 45% (based on cycloalkane). A promoting behavior of H_2O was detected when using **47** [62], as previously reported in a few other catalytic systems with rhenium, vanadium, or copper catalysts [25, 33], and discussed above for vanadium systems.

The copper(I)-containing CPs $[Cu_2(L^1)]_n$ (**50**, H_2L^1 = bis(3,5-dimethyl-4-pyrazolyl)methane) (Figure 16.2) and $[Cu_2(L^2)]_n$ (**51**, H_2L^2 = 1,4-bis-4′-(3′,5′-dimethyl)-pyrazolylbenzene) [at high-temperature, reaction between copper(II) acetate and bis(pyrazolato)-based ligand reduces Cu(II) to Cu(I)] were tested [63] as catalysts in the peroxidative oxidation of CyH to

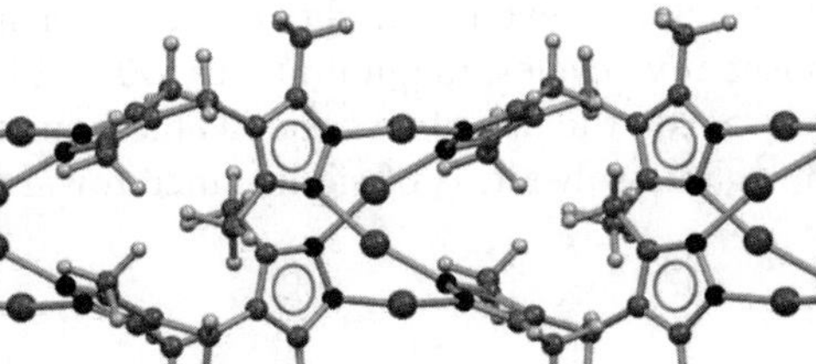

50

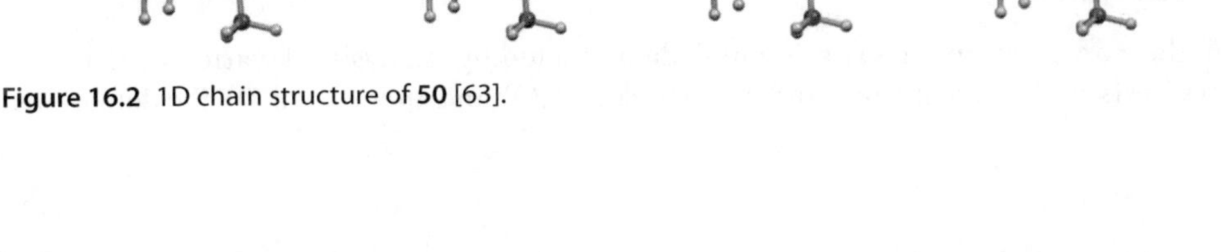

Figure 16.2 1D chain structure of **50** [63].

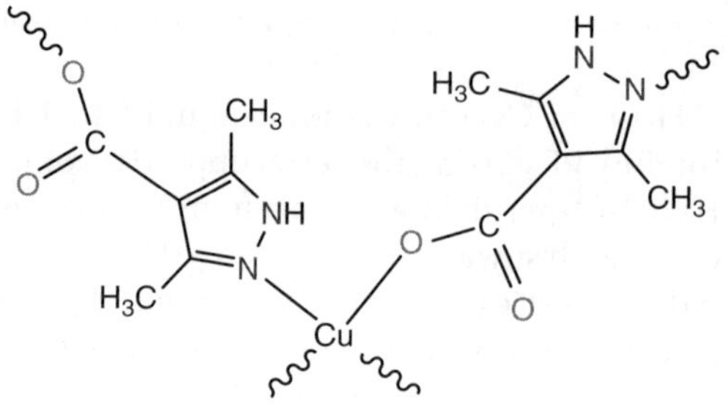

52

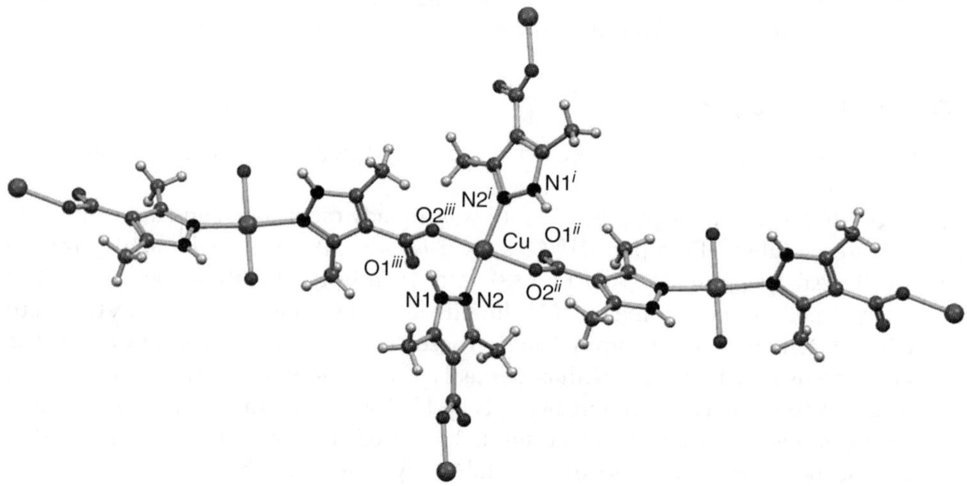

Figure 16.3 X-ray structure of **52** [64].

cyclohexanol and cyclohexanone with H_2O_2 in acetonitrile and found to exhibit high selectivity leading to a maximum product yield of 11%.

The copper azolate carboxylate compound $[Cu(Hdmpzc)_2]_n$ (**52**, H_2dmpzc = 3,5-dimethyl-1*H*-pyrazol-4-carboxylic acid) (Figure 16.3) is a nonporous framework that catalyzes the CyH oxidation yielding 12% of cyclohexanol and cyclohexanone after 9 hours, with TBHP in acetonitrile [64].

Microwave-assisted synthesis of the Cu(I) MOF $[Cu(\mu_4\text{-}4\text{-ptz})]_n$ [4-ptz = 5-(4-pyridyl) tetrazolate] (**53**) (Figure 16.4) has been performed by employing a relatively easy method within a short period of time [65]. It is insoluble in water and revealed a high catalytic activity (maximum yield of 35%, TON = 35) in the heterogeneous oxidation of cyclic alkanes (cyclo-pentane, -hexane and -octane) with aqueous hydrogen peroxide, under mild conditions (at room temperature), without any added solvent or additive. Moreover, it was easily recovered and reused, at least for four consecutive cycles, maintaining ca. 90% of the initial activity and concomitant high selectivity. Moreover, the use of an aqueous medium at room temperature, without the requirement of an organic solvent, is of significance toward the development of green catalytic systems.

16.3.4 Mechanism

Based on the combination of experimental data (including analysis of various selectivity parameters, tests with radical traps and ^{18}O-labeled H_2O) and theoretical DFT calculations,

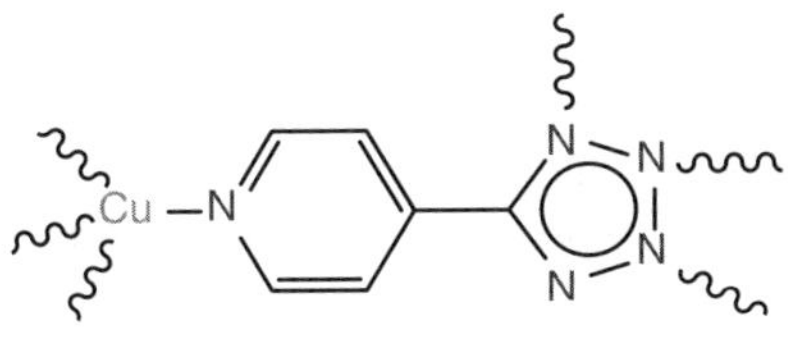

53

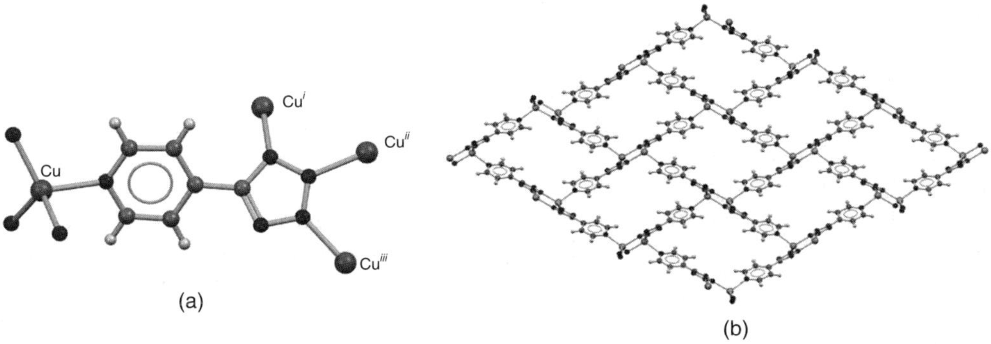

Figure 16.4 X-ray structure of **53**: asymmetric unit (a), 2D structure (b) [65]. (*See insert for color representation.*)

the mechanism is believed to usually involve the formation of free radicals and is illustrated for the CyH oxidation to cyclohexanol (CyOH) and cyclohexanone (C=O), with tert-butyl hydroperoxide (THBP), by the following equations [48–50]:

$$Cu^{2+} + t\text{-BuOOH} \rightarrow t\text{-BuOO}\bullet + H^+ + Cu^+$$

$$Cu^+ + t\text{-BuOOH} \rightarrow t\text{-BuO}\bullet + Cu^{2+} + OH^-$$

$$t\text{-BuO}\bullet + CyH \rightarrow t\text{-BuOH} + Cy\bullet$$

$$Cy\bullet + O_2 \rightarrow CyOO\bullet$$

$$CyOO\bullet + t\text{-BuOOH} \rightarrow CyOOH + t\text{-BuOO}\bullet$$

$$CyOOH + Cu^+ \rightarrow CyO\bullet + Cu^{2+} + OH^-$$

$$CyOOH + Cu^{2+} \rightarrow CyOO\bullet + H^+ + Cu^+$$

$$CyO\bullet + CyH \rightarrow CyOH + Cy\bullet$$

$$2CyOO\bullet \rightarrow CyOH + Cy = O + O_2$$

16.4 Conclusions

Vanadium and copper complexes play an important role in the metal-catalyzed peroxidative oxidation of alkanes to the corresponding alcohols and ketones, under mild conditions, and the field has markedly developed, as illustrated above.

Oxidovanadium(IV) and (V) complexes containing azine fragment (C—N=N—C) ligands are quite efficient. A recent development on vanadium compounds with azine ligands shows that the redox active carbohydrazone ligand can play an important role in the formation of the reduced form of the catalyst with preservation of the vanadium +5 oxidation state [37]. Other vanadium complexes based on pyrazole or scorpionate ligands have also shown a high activity.

The bis(maltolato)oxidovanadium complexes anchored to chemically modified carbamated silica gel can also catalyze the oxidition with molecular oxygen under relatively mild reaction conditions, to the corresponding ketones and alcohols.

The efficiency of the catalytic system can be promoted in the presence of a suitable acid, and PCA is often the most active one.

Mononuclear copper catalysts have been less used for oxidation of alkanes. Aryl- (or aroyl)-hydrazone, tetrazolato, or scorpionate and phenanthroline copper complexes are promising for such an oxidation reaction, but multinuclear copper complexes can exhibit a higher catalytic activity in comparison with the mononuclear ones. The multicopper-based systems are of significance to mimic the multi-copper pMMO. Copper compounds can exhibit a high catalytic activity in this field, either in homo- or heterogeneous medium, under mild conditions. A growing attention has been paid to copper-containing MOFs or coordination polymers (CPs), which can display a good catalytic activity.

In spite of the developments already reported, various challenges are still evident, namely toward more effective and selective catalysts, greener methods, further exploration of the metal-ligand cooperation with redox active ligands, of the possible promoting role of water, of catalytic recyclability, and extensive kinetic and theoretical studies for mechanistic elucidation for the various types of catalytic systems.

Acknowledgment

Authors are grateful to the Fundação para a Ciência e a Tecnologia: FCT (projects UID/QUI/00100/2013, PTDC/QEQ-QIN/3967/2014 and PTDC/QEQ-ERQ/1648/2014), Portugal, for financial support. M.S. acknowledges the FCT, Portugal, for a postdoctoral fellowship (SFRH/BPD/86067/2012).

References

1 Sutradhar, M., Martins, L.M.D.R.S., Guedes da Silva, M.F.C., and Pombeiro, A.J.L. (2015). *Coord. Chem. Rev.* 301–302: 200.
2 da Silva, J.A.L., Frausto da Silva, J.J.R., and Pombeiro, A.J.L. (2011). *Coord. Chem. Rev.* 255: 2232.
3 Shul'pin, G. (2016). New trends in oxidative functionalization of carbon-hydrogen bonds: a review. *Catalysts* 6: 50.
4 Butler, A. (1999). *Coord. Chem. Rev.* 187: 17.
5 Hirao, T. (1997). *Chem. Rev.* 97: 2707.
6 Conte, V., Coletti, A., Floris, B. et al. (2011). *Coord. Chem. Rev.* 255: 2165.
7 Sutradhar, M. and Pombeiro, A.J.L. (2017). Vanadium complexes in catalytic oxidations. In: *Elsevier Reference Module in Chemistry, Molecular Sciences and Chemical Engineering* (ed. J. Reedijk). Waltham, MA: Elsevier http://doi.org/10.1016/B978-0-12-409547-2.13525-5.
8 da Silva, J.A.L., Fraústo da Silva, J.J.R., and Pombeiro, A.J.L. (2013). *Coord. Chem. Rev.* 255: 2388.
9 Martins, L.M.D.R.S. and Pombeiro, A.J.L. (2014). *Coord. Chem. Rev.* 265: 74.
10 Martins, L.M.D.R.S. and Pombeiro, A.J.L. (2016). *Eur. J. Inog. Chem.* 2236.
11 Pombeiro, A.J.L. (2013). Toward functionalization of alkanes under environmentally benign conditions. In: *Advances in Organometallic Chemistry and Catalysis* (ed. A.J.L. Pombeiro), 15–25. Weinheim: Wiley-VCH, Ch. 2.

12 Kirillov, A.M., Kirillova, M.V., and Pombeiro, A.J.L. (2013). *Adv. Inorg.Chem.* 65: 1.
13 Shilov, A.E. and Shul'pin, G.B. (2000). *Activation and Catalytic Reactions of Saturated Hydrocarbons in the Presence of Metal Complexes.* Dordrecht: Kluwer Academic Publishers.
14 Backvall, J.-E. (2004). *Modern Oxidation Methods.* Weinheim: Wiley-VCH.
15 Pombeiro, A.J.L. (ed.) (2014). *Advances in Organometallic Chemistry and Catalysis, the Silver/Gold Jubilee ICOMC Celebratory Book.* Wiley.
16 G. B. Shul'pin, in *Transition Metals for Organic Synthesis,* ed.: M. Beller, C. Bolm, Vol. 2, 2nd edn, Wiley-VCR, New York, 2004 , Chapter 2 pp. 215.
17 Shilov, A.E. and Shul'pin, G.B. (1997). Activation of C–H bonds by metal complexes. *Chem. Rev.* 97: 2879.
18 Shul'pin, G.B. (2009). Hydrocarbon oxygenations with peroxides catalyzed by metal compounds. *Mini-Rev. Org. Chem.* 6: 95.
19 Crabtree, R.H. (2001). Alkane C–H activation and functionalization with homogeneous transition metal catalysts: a century of progress—a new millennium in prospect. *J. Chem. Soc., Dalton Trans.* 2437.
20 Sutradhar, M., Shvydkiy, N.V., Guedes da Silva, M.F.C. et al. (2013). *Dalton Trans.* 42: 11791.
21 Sutradhar, M., Kirillova, M.V., Guedes da Silva, M.F.C. et al. (2012). *Inorg. Chem.* 51: 11229.
22 Shul'pin, G.B., Kozlov, Y.N., Nizova, G.V. et al. (2001). *J. Chem. Soc. Perkin Trans.* 2: 1351.
23 Kirillova, M.V., Kuznetsov, M.L., Romakh, V.B. et al. (2009). *J. Catal.* 267: 140.
24 Shul'pin, G.B. (2013). *Dalton Trans.* 42: 12794.
25 Kuznetsov, M.L. and Pombeiro, A.J.L. (2009). *Inorg. Chem.* 48: 307.
26 Reis, P.M., Silva, J.A.L., Frausto da Silva, J.J.R. et al. (2000). *Chem. Commun.* 1845.
27 Silva, T.F.S., Alegria, E.C.B.A., Martins, L.M.D.R.S. et al. (2008). *Adv. Synth. Catal.* 350: 706.
28 Shul'pina, L.S., Kirillova, M.V., Pombeiro, A.J.L. et al. (2009). *Tetrahedron* 65: 2424.
29 Shul'pin, G.B., Attanasio, D., and Suber, L. (1993). *J. Catal.* 142: 147.
30 Shul'pin, G.B., Attanasio, D., and Suber, L. (1993). *Russ. Chem. Bull.* 42: 55.
31 Shul'pin, G.B., Druzhinina, A.N., and Nizova, G.V. (1993). *Russ. Chem. Bull.* 42: 1327.
32 Khaliullin, R.Z., Bell, A.T., and Head-Gordon, M. (2005). *J. Phys. Chem. B.* 109: 17984.
33 Kirillova, M.V., Kuznetsov, M.L., Kozlov, Y.N. et al. (2011). *ACS Catal.* 1: 1511.
34 Sutradhar, M. and Pombeiro, A.J.L. (2014). *Cord. Chem. Rev.* 265: 89.
35 Gupta, S., Kirillova, M.V., Guedes da Silva, M.F. et al. (2013). *Appl. Catal. A: General* 460–461: 82.
36 Gupta, S., Kirillova, M.V., Guedes da Silva, M.F. et al. (2013). *Inorg. Chem.* 52: 8601.
37 Dragancea, D., Talmaci, N., Shova, S. et al. (2016). *Inorg. Chem.* 55: 9187.
38 Süss-Fink, G., Cuervo, L.G., Therrien, B. et al. (2004). *Inorg. Chim. Acta* 357: 475.
39 Silva, T.F.S., Luzyanin, K.V., Kirillova, M.V. et al. (2010). *Adv. Synth. Catal.* 352: 171.
40 Silva, T.F.S., Mac Leod, T.C.O., Martins, L.M.D.R.S. et al. (2013). *J. Mol. Catal. A: Chem.* 367: 52.
41 Mishra, G.S. and Pombeiro, A.J.L. (2006). *Appl. Catal., A* 304: 185.
42 Mishra, G.S., Fraústo da Silva, J.J.R., and Pombeiro, A.J.L. (2007). *J. Mol. Catal. A: Chem.* 265: 59.
43 Sheldon, R.A., Arends, I., and Hanefeld, U. (2007). *Green Chemistry and Catalysis.* Weinheim, Germany: Wiley-VCH.
44 Silva, T.F.S., Mishra, G.S., Silva, M.F.G. et al. (2009). *Dalton Trans.* 42: 9207.
45 Silva, T.F.S., Rocha, B.G.M., Guedes da Silva, M.F.C. et al. (2016). *New J. Chem.* 40: 528.
46 Sutradhar, M., Alegria, E.C.B.A., Guedes da Silva, M.F.C. et al. (2016). *Molecules* 21: 425.
47 Palmucci, J., Mahmudov, K.T., Guedes da Silva, M.F.C. et al. (2015). *RSC Adv.* 5: 84142.
48 Saha, M., Vyas, K.M., Martins, L.M.D.R.S. et al. (2017). *Polyhedron* 132: 53.

49 Detoni, C., Carvalho, N.M.F., Aranda, D.A.G. et al. (2009). *Appl. Catal. A: General* 365: 281.
50 Kirillov, A.M., Kopylovich, M.N., Kirillova, M.V. et al. (2005). *Angew. Chem. Int. Ed.* 44: 4345.
51 Kirillov, A.M., Kirillova, M.V., and Pombeiro, A.J.L. (2012). *Coord. Chem. Rev.* 256: 2741.
52 Adams, R.D., Rassolov, V., and Wong, Y.O. (2016). *Angew. Chem. Int. Ed.* 55: 1324.
53 Galassi, R., Simon, O.C., Burini, A. et al. (2017). *Polyhedron* 137: 143.
54 Silva, T.F.S., Silva, M.F.C.G., Mishra, G.S. et al. (2011). *Organomet. Chem.* 696: 1310.
55 Contaldi, S., Di Nicola, C., Garau, F. et al. (2009). *Dalton Trans.* 4928.
56 Bilyachenko, A.N., Dronova, M.S., Yalymov, A.I. et al. (2013). *Eur. J. Inorg. Chem.* 5240.
57 Dronova, M.S., Bilyachenko, A.N., Yalymov, A.I. et al. (2014). *Dalton Trans.* 43: 872.
58 Bilyachenko, A.N., Dronova, M.S., Yalymov, A.I. et al. (2015). *Chem. Eur. J.* 21: 8758.
59 Vinogradov, M.M., Kozlov, Y.N., Bilyachenko, A.N. et al. (2015). *New J. Chem.* 39: 187.
60 Sutradhar, M., Martins, L.M.D.R.S., Guedes da Silva, M.F.C. et al. (2015). *Eur. J. Inorg. Chem.* 3959.
61 Ribeiro, A.P.C., Martins, L.M.D.R.S., Hazra, S. et al. (2015). *C.R. Chim.* 18: 758.
62 Fernandes, T.A., Santos, C.I.M., André, V. et al. (2016). *Inorg. Chem.* 55: 25.
63 Timokhin, I., Pettinari, C., Marchetti, F. et al. (2015). *Cryst. Growth Des.* 15: 2303.
64 Tăbăcaru, A., Xhaferaj, N., Martins, L.M.D.R.S. et al. (2016). *Inorg. Chem.* 55: 5804.
65 Martins, L.M.D.R.S., Nasani, R., Saha, M. et al. (2015). *Molecules* 20: 19203.

Part III

C—B, C—C and C—N Bond Formation

17

Catalytic Borylation of Methane: Combining Computational and High-Throughput Screening Approaches to Discover a New Catalyst

Pavel Zatsepin[1], *Dieter Sorsche*[1], *Seihwan Ahn*[2,3], *Mu-Hyun Baik*[2,3], *and Daniel J. Mindiola*[1]

[1] *University of Pennsylvania, Department of Chemistry, Philadelphia PA, USA*
[2] *Korea Advanced Institute of Science and Technology (KAIST), Department of Chemistry, Daejeon, Republic of Korea*
[3] *Institute for Basic Science (IBS), Center for Catalytic Hydrocarbon Functionalizations, Daejeon, Republic of Korea*

17.1 Introduction

Methane is the simplest and most abundant of all alkanes, accounting for 70–90% of the vast reserves of natural gas accessible today [1]. When pure, methane has a higher heat of combustion than any other hydrocarbon (~56 MJ kg^{-1}) and produces energy cleaner than alternative fossil fuels such as coal, diesel, or gasoline. Hence, it is not surprising that methane continues to play an important role in the world energy economy. From a chemical perspective, there are several notable reactions besides its simple combustion that make methane important in industrial processes such as steam reforming, which accounts for more than 95% of H_2 production in the USA alone, and translating to approximately nine million metric tons of hydrogen produced annually (Eqs. (17.1) and (17.2) with overall balanced Eq. (17.3)) [2, 3]. Worldwide, this process produces over 500 billion metric tons of hydrogen each year. Methane is a potent greenhouse gas, and its large-scale utilization with potential emissions into the atmosphere will greatly affect the global climate change [4, 5].

$$CH_4(g) + H_2O(g) \xrightarrow[700\text{–}1100\,°C]{\text{Ni catalyst}} CO + 3H_2 \qquad \Delta H_f = 206\ \text{kJ/mol} \tag{17.1}$$

$$CO\ (g) + H_2O\ (g) \xrightarrow[300\text{–}450\,°C]{\text{Fe catalyst}} CO_2 + H_2 \qquad \Delta H_f = -41\ \text{kJ/mol} \tag{17.2}$$

$$CH_4(g) + 2H_2O\ (g) \longrightarrow CO_2 + 4H_2 \qquad \text{overall reaction} \tag{17.3}$$

Thus, technologies are widely accepted and available for industrial scale utilization of methane to generate heat/energy or to obtain feedstock reagents such as H_2 or CO. Unfortunately, the steam-reforming process requires enormous amounts of energy with unavoidable formation of carbon dioxide via a secondary transformation that is referred to as the water–gas shift reaction (Eq. (17.2)) [6]. H_2 production from CH_4 and steam is best described as a two-step process involving first a highly endergonic step, that is the methane steam reforming (Eq. (17.1)). This step is perhaps the most critical and requires a Ni catalyst producing the greatest amount of H_2, but this process requires vast amount of energy (700–1100 °C) [7], unfortunately. The formation of CO and H_2 (syngas) is quite useful, since it can be converted to more valuable energy carriers such as methanol [8] or to diesel fuels by means of the

Alkane Functionalization, First Edition. Edited by Armando J. L. Pombeiro and M. Fátima C. Guedes da Silva.

Fischer–Tropsch process [8, 9]. The formation of syngas for methanol production can be improved by combining Eq. (17.1) with a partial oxidation step involving CH_4 and $\frac{1}{2}O_2$ to CO and $2H_2$, resulting in an ideal stoichiometry of 1 : 2 for CO : H_2. This reaction, although convenient for the Fischer-Tropsch process of forming methanol, is more costly. Alternatively, syngas can be produced by a bi-reforming process involving met-gas, an ideal feedstock for the preparation of methanol, allowing for producing greater equivalents of syngas without introducing the water–gas shift step. Instead, Eq. (17.1) is combined with a second reaction referred to as dry-reforming, where CH_4 and CO_2 are combined to make 2CO and $2H_2$ [10]. A recent comprehensive review by Caballero and Perez further discusses the use of methane for the production of more value-added chemicals by direct and indirect routes [11]. As noted above, steam reforming contains a second step (Eq. (17.2)), which is thermodynamically favored and is referred to as the water–gas shift process [6]. In Eq. (17.2), the use of an Fe catalyst results in an additional equivalent of H_2 being produced in addition to CO_2. This process, although much less energy intensive, is highly sensitive to temperature (higher temperatures favoring the equilibrium process to reactants), with the ideal temperature range for operation being 300–450 °C.

One of the main technical uses of H_2 derived from steam reforming is the catalytic reduction of N_2 to NH_3. In the United States alone, over 109 million metric tons of H_2 are utilized for the industrial production of NH_3 from its elemental constituents [12]. Without this reaction, known as the Haber–Bosch process [13], the feeding of a rapidly growing world population would have reached a critical point at the start of the twentieth century since most NH_4^+ and NO_3^- based salts are derived from natural and non-renewable resources such as guano (ammonium oxalates) and saltpeter (KNO_3) [14]. Other energy-intensive processes such as Birkeland–Eyde [15] and Frank–Caro (cyanamide) [16] were not efficient enough to produce NH_3 or derivatives thereof on an industrial scale. It is now estimated that the combination of these two processes (H_2 production and usage) consumes approximately ~2% of the world's energy supply [17].

Although natural gas reforming is the most economically viable process among all hydrogen production pathways, there is also interest in directly utilizing methane as a C_1 reagent for the preparation of liquid fuels such as methanol or other fine chemicals with much higher storage capacity, unlike the parent alkane. This is precisely the reason why methane is not practical as an energy carrier of hydrogen; it is a gas at room temperature and poses many other obstacles as a reagent. In addition to having a very strong C—H bond (~104 kcal mol^{-1}) [18], this gas has limited solubility in polar and nonpolar solvents, has very high ionization energies [19, 20], and has very low triple, boiling, and flashing points. Here, we discuss several examples of well-defined systems that can catalyze the activation and functionalization of methane. In this chapter we have chosen to overlook most non-organometallic, heterogeneous systems and radical-based chain processes that have been shown to convert methane to partially oxidized products. A recent comprehensive review on methane activation and functionalization by most elements in the periodic table was reported by Periana and co-workers [21].

17.2 Homogeneous Catalysts for Methane Activation and Functionalization

17.2.1 Electrophilic Methane Activation and Functionalization

Well-defined homogeneous systems capable of catalytically converting methane to liquid fuels are known, and such systems can even operate at relatively low temperatures. However, these transformations are generally confined to highly electrophilic conditions and in some cases require the use of superacids and/or strong oxidants or demand strongly oxidizing conditions. Examples of such systems include earlier studies by Shilov (Scheme 17.1) involving a platinum catalyst that converts methane to methylchloride and methanol in acidic/aqueous conditions. Constituting an electrophilic C—H activation reaction, the Shilov system is catalytic in Pt(II) but requires a stoichiometric Pt(IV) as an oxidant to subsequently promote C—X bond formation. A generic cycle for the Shilov reaction converting CH_4 to CH_3X (X = Cl or OH) is shown below. In 1972, Shilov's seminal paper described how CH_4 could be catalytically converted to a mixture of CH_3X species by introducing a Pt(IV) oxidant [22]. Earlier studies showed how $[PtCl_4]^{2-}$ could exchange H for D in CH_4 using a D_2O/DOAc medium [23]. Such exchange implicated a halide for methyl equilibrium to produce $[Pt(II)X_3(CH_3)]$, where X may be any generic ligand such as Cl, OH, or H_2O. This step is debatable in that C—H activation could involve an oxidative addition step followed by elimination or by a concerted σ-bond metathesis step. Oxidation by two electrons with $[PtCl_6]^{2-}$ or a Pt(IV) species may then form a 6-coordinate intermediate $Pt(IV)X_5(CH_3)$ or quite possibly a 5-coordinate species based on studies by Goldberg and co-workers [24]. In the presence of a nucleophile such as OH^- or Cl^-, reductive elimination ensues by forming H_3CCl (in HCl) or H_3COH (if done in H_2O) which would then close the cycle by reforming a square planar $Pt(II)X_4$ (Scheme 17.1). Scheme 17.1 depicts a generic cycle where the X ligand in question could be monoanionic or neutral, hence formal charges on the Pt intermediates have been omitted. Instead, each intermediate is represented with the formal oxidation state on Pt giving the overall balanced reaction shown in Eq. (17.4).

$$CH_4 + H_2O + [PtCl_6]^{2-} \xrightarrow[\text{catalyst}]{\text{Pt(II)}} H_3COH + 2H^+ + [PtCl_4]^{2-} + 2Cl^- \tag{17.4}$$

What is notable about this remarkable cycle is how the Pt(II) center kinetically imparts chemoselectivity for strong C—H bonds over weaker ones. As a result, the C—H bonds on the product H_3CX are not further activated (or oxidized). Efforts into replacing the oxidant

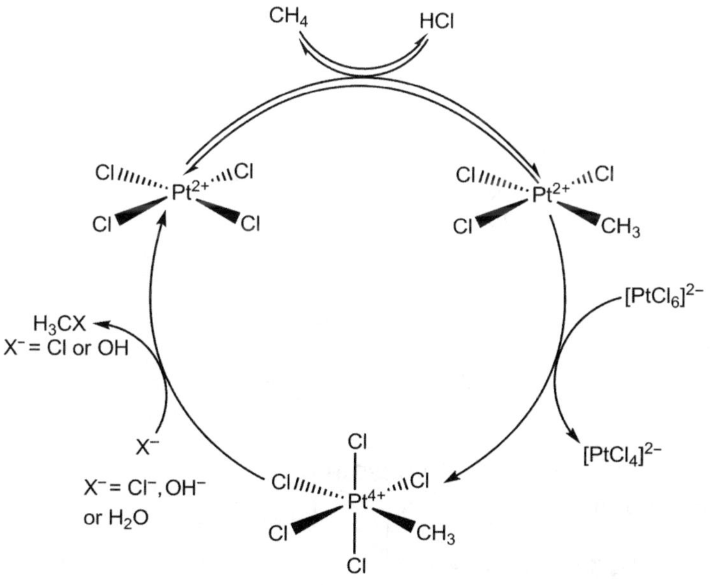

Scheme 17.1 Proposed Shilov cycle for the oxidation of methane to methylchloride or methanol using a generic Pt^{2+} precursor in addition to stoichiometric Pt^{4+} oxidant.

Pt(IV) with cheaper and more environmentally benign oxidants such as O_2 have been pursued [25, 26].

Most likely inspired by the electrophilic activation of methane, Periana and co-workers reported a Hg(II) catalyst in 1993, namely $Hg(OSO_3H)_2$, that could convert with good yields and high turnover frequencies (TOFs) methane to methylbisulfate (H_3COSO_3H) under acidic conditions using 100% H_2SO_4 at 180 °C [27]. Hg^{2+} in the form of $Hg(OAc)_2$ is a well-known reagent, perhaps even one of the oldest reported reagents, to display the ability to functionalize C—H bonds in benzene in what is now known as mercuration of arenes [28]. A balanced reaction is shown in Eq. (17.5). This reaction was much more selective than earlier studies in 1950 using $Hg(SO_4)$ in SO_3 and CH_4 in temperature ranges from 100 to 400 °C [29].

$$CH_4 + 2H_2SO_4 \xrightarrow[\substack{180\,°C \\ 100\%H_2SO_4}]{\substack{Hg(OSO_3H)_2 \\ \text{catalyst}}} H_3COSO_3H + 2H_2O + SO_2 \tag{17.5}$$

The mechanism of this remarkable transformation [27] was speculated to be similar in several aspects to the mechanism proposed in the Shilov cycle outlined in Scheme 17.1, but with H_2SO_4 being the oxidant. The binuclear two-electron process $2Hg^{2+} + 2e^- \rightarrow Hg_2{}^{2+}$ allows for functionalization of the methyl moiety by H_2SO_4 as well as reoxidation of Hg(I) to Hg(II) by the strong acid. Scheme 17.2 shows a proposed cycle for this reaction and illustrates the multifaceted role of the acid. Neat H_2SO_4 not only promotes electrophilic activation of methane by forming $Hg(OSO_3H)(CH_3)$ presumably via a highly electrophilic species such as $[Hg(OSO_3H)]^+$ in a reversible process, but also participates in the functionalization of the methyl moiety by forming H_3COSO_3H and the reduced Hg(I) species $Hg_2(OSO_3H)_2$. H_2SO_4 also re-oxidizes the latter intermediate to two equiv of $Hg(OSO_3H)_2$. In the process of oxidizing the methyl moiety in $Hg(OSO_3H)(CH_3)$ and the Hg(I) in $Hg_2(OSO_3H)_2$, both SO_2 and H_2O are formed as side products from H_2SO_4 reduction. H_3COSO_3H, the product resulting from methyl oxidation by H_2SO_4, gradually converts to H_3COH in the presence of H_2O. In principle, the cycle shown in

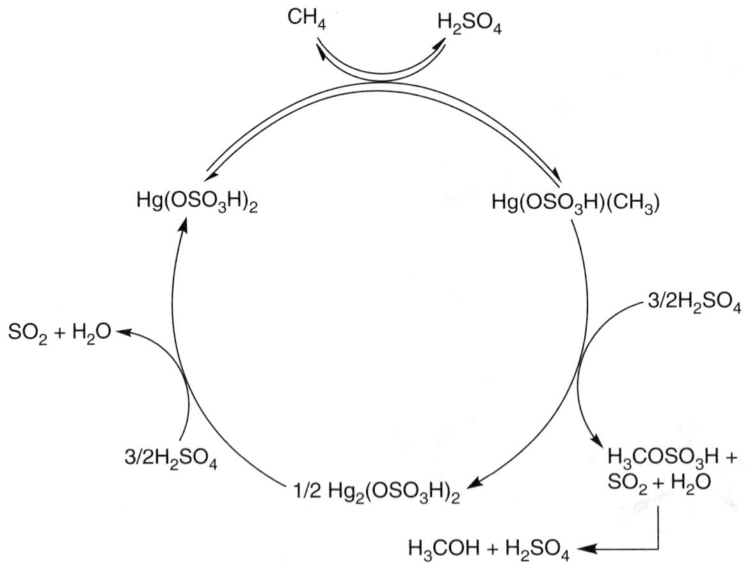

Scheme 17.2 Proposed cycle for the oxidation of methane to methylbisulfate and methanol using a Hg^{2+} catalyst in 100% sulfuric acid.

Scheme 17.2 could be further improved by coupling several cycles involving the oxidation of SO_2 with $1/2O_2$ to SO_3 and hydrolysis to form H_2SO_3, as well as the complete hydrolysis of H_3COSO_3H to H_3COH, as shown below in Eqs. (17.6) and (17.7), respectively.

$$SO_2 + 1/2O_2 + H_2O \rightarrow H_2SO_4 \tag{17.6}$$

$$H_3COSO_3H + H_2O \rightleftharpoons H_3COH + H_2SO_4 \tag{17.7}$$

Combining all these steps, one could in principle envision an aerobic oxidation reaction involving the conversion of methane to methanol, akin to how methane monooxygenase bacteria do it [30], but with higher TOF (10^{-3} s^{-1}). It is important to mention that strong acids (such as H_2SO_4 or HOTf) are needed for this reaction to operate efficiently since a weaker acid such as $HOCOCH_3$ is a good ligand under both basic or acidic conditions.

Utilizing a chelating ligand, Periana and co-workers optimized conditions for the Shilov process discussed in Scheme 17.1. It was found that simple reagents such as *cis-* or *trans-* $Pt(NH_3)_2Cl_2$ could catalyze the conversion of methane to methylbisulfate in concentrated H_2SO_4 at 180 °C with comparable TOF to those observed for the Hg^{2+} catalyst [31]. Although the catalyst was active, it was short- lived in 100% H_2SO_4 due to demetallation of the NH_3 ligands by forming NH_4^+ salts. For this reason, a new ligand scaffold was used that could be stable to protonolysis, oxidizing conditions, and which was less amenable to demetallation. Accordingly, the chelating ligand 2,2-bipyrimidyl was used to prepare the Pt(II) catalyst (2,2-bipyrimidyl)$PtCl_2$. In 102% H_2SO_4, using ~500 psi of methane, this Pt(II) complex readily promoted the activation and functionalization of the parent alkane to H_3COSO_3H with higher yield and greater selectivity but with slightly lower activity when compared to the Hg^{2+} system (TOF = 10^{-2} s^{-1}) [31]. The mechanism to formation of H_3COSO_3H is again quite similar to that of the Shilov cycle, and a simplified set of reactions are shown in Scheme 17.3. In 102%

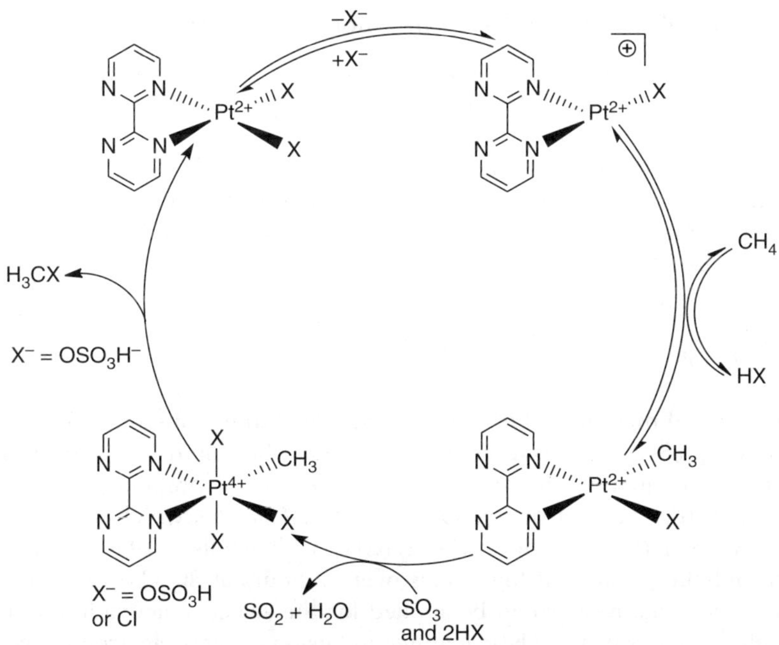

Scheme 17.3 Proposed cycle for the oxidation of methane to methylbisulfate and methanol using a Pt^{2+} catalyst in 102% sulfuric acid. For clarity, other cycles involving the protonation of the 2,2-bipyrimidyl ligand on Pt are not shown.

H_2SO_4, precursor (2,2-bipyrimidyl)$PtCl_2$ most likely transforms to (2,2-bipyrimidyl)PtX_2 where X^- represents Cl^- or OSO_3H^-. Dissociation of one X^- would open up a coordination site via putative $[(2,2\text{-bipyrimidyl})PtX]^+$ where methane could then be activated by producing (2,2-bipyrimidyl)$PtX(CH_3)$ and HX. SO_3 with HX (or alternatively, H_2SO_4) would then oxidize the latter complex to form (2,2-bipyrimidyl)$PtX_3(CH_3)$ concomitant with formation of SO_2 and H_2O. Akin to the Shilov cycle in Scheme 17.1 (*vide supra*), reductive elimination in (2,2-bipyrimidyl)$PtX_3(CH_3)$, most likely promoted by X^-, would then result in H_3CX ($X^- = OSO_3H$) formation, which would then hydrolyze to form H_3COH. More detailed and revised mechanistic studies have implicated protonation of 2,2-bipyrimidyl to be relevant in the cycle for C—H activation and functionalization, which we neglect here for the purpose of clarity [21, 32–34].

Using superacid conditions (in some cases using oleum, which is 30% SO_3 in H_2SO_4), examples of methane conversion to carbocationic H_3CX forms were extended to elemental Pd [35]. What is notable for this type of reaction is that the cycle most likely involves a Pd(0)–Pd(II) couple, where the catalyst was reported to be recycled up to 10 times without a decrease of catalytic activity! Periana and co-workers also reported that $Pd(SO_4)_2$ could be used in the oxidative coupling of two methane molecules to form HOAc using 96% sulfuric acid with a turnover number (TON) of 18, given the tendency to form Pd black [36]. For more details about these reactions, refer to an excellent comprehensive review by Periana and co-workers on homogeneous systems that promote methane activation and functionalization [21]. We regretfully exclude from this chapter discussions of other interesting catalytic reactions using metals and nonmetals in superacid conditions, including seminal studies by Olah et al. using HF/SbF_5 mixtures [37] as well as optimized conditions by Periana and co-workers for methane oxidation by elemental iodine in oleum (H_2SO_4/SO_3) at 195 °C to H_3COSO_3H [38].

17.2.2 Activation and Functionalization of Methane via σ-Bond Metathesis

In 1983, Watson discovered how rare-earth metallocene complexes such as $Cp^*_2RE(CH_3)$ (RE = Lu and Y) undergo exchange with $^{13}CH_4$ to form $Cp^*_2RE(^{13}CH_3)$ and CH_4 at 70 °C [39]. This degenerate transformation, which involves a four-centered transition state, portrays how metal-alkyl σ-bonds can undergo swapping with an incoming C—H σ-bond from an alkane, hence the term σ-bond metathesis. Taking advantage of this type of transformation, Sadow and Tilley introduced a hydromethylation reaction using the precatalyst $Cp^*_2Sc(CH_2{}^tBu)$ with methane and propene (Eq. (17.8)) [40].

$$H_3C\text{-}H + H_2C{=}CHCH_3 \xrightarrow[80\,°C,\ C_6H_{12},\ 12\ h]{Sc\ Catalyst} H_3CCH(CH_3)_2 \tag{17.8}$$

Scheme 17.4 shows how σ-bond metathesis between $Cp^*_2Sc(CH_2{}^tBu)$ and methane results in formation of the catalyst $Cp^*_2Sc(CH_3)$ along with neopentane. The microscopic-reverse reaction is unlikely to occur for kinetic reasons. Migratory insertion of propene into the Sc—CH_3 bond results in formation of the isobutyl species $Cp^*_2Sc(CH_2{}^iPr)$, which can then undergo σ-bond metathesis with methane to reform $Cp^*_2Sc(CH_3)$ with expulsion of isobutane. Although this cycle could only be performed four times over 12 hours at 80 °C, this transformation showed how σ-bond metathesis could be applied for the selective activation and transformation of the smallest alkanes in a catalytic manner and using a highly electropositive metal ion such as Sc^{3+}. The system is also selective for aliphatic C—H bonds over aromatic

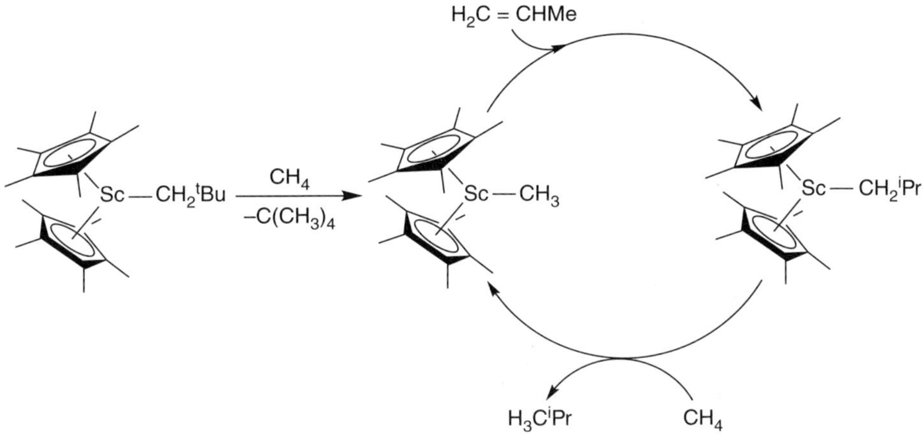

Scheme 17.4 Catalytic hydromethylation of propene by $Cp^*{}_2Sc(CH_3)$ via σ-bond metathesis and migratory insertion steps.

ones but unfortunately attempts to extend the hydromethylation cycle to other olefins did not work [40]. Soon after, it was found by the same authors that the ansa-bridged metallocene derivative $[Me_2Si(C_5Me_4)_2]Sc(CH_2{}^tBu)$ could undergo more facile σ-bond metathesis with CH_4, but with the migratory insertion reaction being compromised since the insertion product of propene, $[Me_2Si(C_5Me_4)_2]Sc(CH_2{}^iPr)$, was found to be more stable and hence resistant to σ-bond metathesis with the incoming CH_4 based on computational work [41].

Taking advantage of σ-bond metathesis but also the relatively weak nature of the Si—H bond, Sadow and Tilley cleverly developed an organometallic scandium catalyst that could also perform dehydrosilation of methane by making inherently stronger Si—C bonds at the expense of breaking weaker Si—H bonds [42]. The overall reaction is shown in Eq. (17.9) using a secondary silane such as Ph_2SiH_2 under various pressures of methane (0.5–150 atm) in cyclohexane (CyH). Increasing the pressure of methane improved the yield of methylsilane product (from 0.4 to 5 TON) at 80 °C over 7 days.

$$H_3C\text{-}H + H\text{-}SiHPh_2 \xrightarrow[80^\circ C,\, C_6H_{12}\; 7\ days]{Sc\ Catalyst} H_3C\text{-}SiHPh_2 + H\text{-}H \qquad (17.9)$$

Scheme 17.5 displays how the monohydride complex $Cp^*{}_2ScH$ can activate methane by σ-bond metathesis to form $Cp^*{}_2Sc(CH_3)$ and H_2. This step is in pre-equilibrium since the microscopic reverse reaction of $Cp^*{}_2Sc(CH_3)$ with H_2 is quite fast to form $Cp^*{}_2ScH$. In addition, the activation of Si—H bonds in Ph_2SiH_2 by $Cp^*{}_2ScH$ would be degenerate, leading to pre-equilibrium processes that should not affect the overall transformation. A second σ-bond metathesis between $Cp^*{}_2Sc(CH_3)$ and the Si—H bond in Ph_2SiH_2 results in formation of the methyl substituted silane $H_3CSiHPh_2$ along with regeneration of the hydride $Cp^*{}_2ScH$. However, it is not possible to discard a secondary pathway involving cyclometallation in $Cp^*{}_2ScH$ to form $(\eta^1{:}\eta^5\text{-}C_5(CH_3)_4CH_2)Cp^*Sc$, which could also participate in the methane dehydrosilation cycle by undergoing σ-bond metathesis with methane to form $Cp^*{}_2Sc(CH_3)$ and H_2 [42, 43]. This is not unreasonable to propose, since strained metallacycles have been shown to engage in σ-bond metathesis with the C—H bond of methane [44, 45].

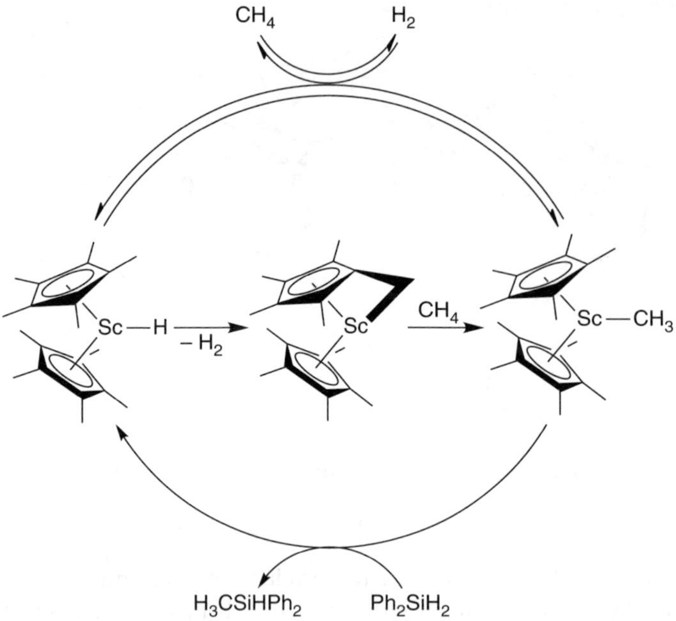

Scheme 17.5 Catalytic dehydrosilation of methane by $Cp^*{}_2ScH$ via σ-bond metathesis steps.

17.2.3 Electrophilic Insertion of Carbenes into the C—H Bond of Methane

The insertion of carbenes into C—H bonds of organic molecules (methylenation) is an old reaction stemming from the use of diazoalkanes with ketones to form homologated ketones in what is now known as the Büchner–Curtius–Schlotterbeck reaction [46]. Metal catalyzed carbene insertion into C—H bonds is quite a mature field [47], stemming from seminal studies by Teyssié and co-workers [48], but the insertion of a carbene into one of the C—H bonds of methane was unknown. Perez, Etienne, and co-workers found that tris(pyrazolyl)borate) ligands having electron withdrawing groups such as Br or F could coordinate to Ag^+ to form catalysts that would convert methane to an ethyl propionate by transferring and inserting the carbene moiety from ethyl diazoacetate into the methane C—H bond [49]. A balanced reaction is shown in Eq. (17.10). The key to this reaction was to use supercritical conditions (160 : 90 psi ratios of CH_4/CO_2) to dissolve the catalyst and reactants, therefore generating TON up to 478 at 40 °C over 14 hours [49]. Although the carbene source was restricted to ethyl diazoacetate, these catalysts can also functionalize other volatile hydrocarbons such as ethane and propane [11, 49] as well as insert the carbene into aromatic rings to generate cycloheptatrienes [50] or the C—H bond of aromatics [51]. Mindiola, Caulton, and co-workers also developed Ag^+ based catalysts using a trifluoromethyl substituted chelating ligand (2,2′-pyridyl-3,5-$(CF_3)_2$-pyrrolide) that could transfer carbene from ethyl diazoacetate to volatile alkanes such as ethane [52] as well as insert into the C=C bonds of aromatics [53].

$$H_3C\text{-}H + N_2CHCO_2Et \xrightarrow[\text{sc-CO}_2]{\text{Ag Catalyst}} H_3C\text{-}CH_2CO_2Et + N_2 \qquad (17.10)$$

A proposed cycle for carbene: $CHCO_2Et$ transfer and insertion into methane is shown in Scheme 17.6 using Ag^+ catalysts supported by two types of tris(pyrazolyl)borate ligands having perbrominated or perfluorinated rings. Carbene transfer to the Ag^+ complex most likely

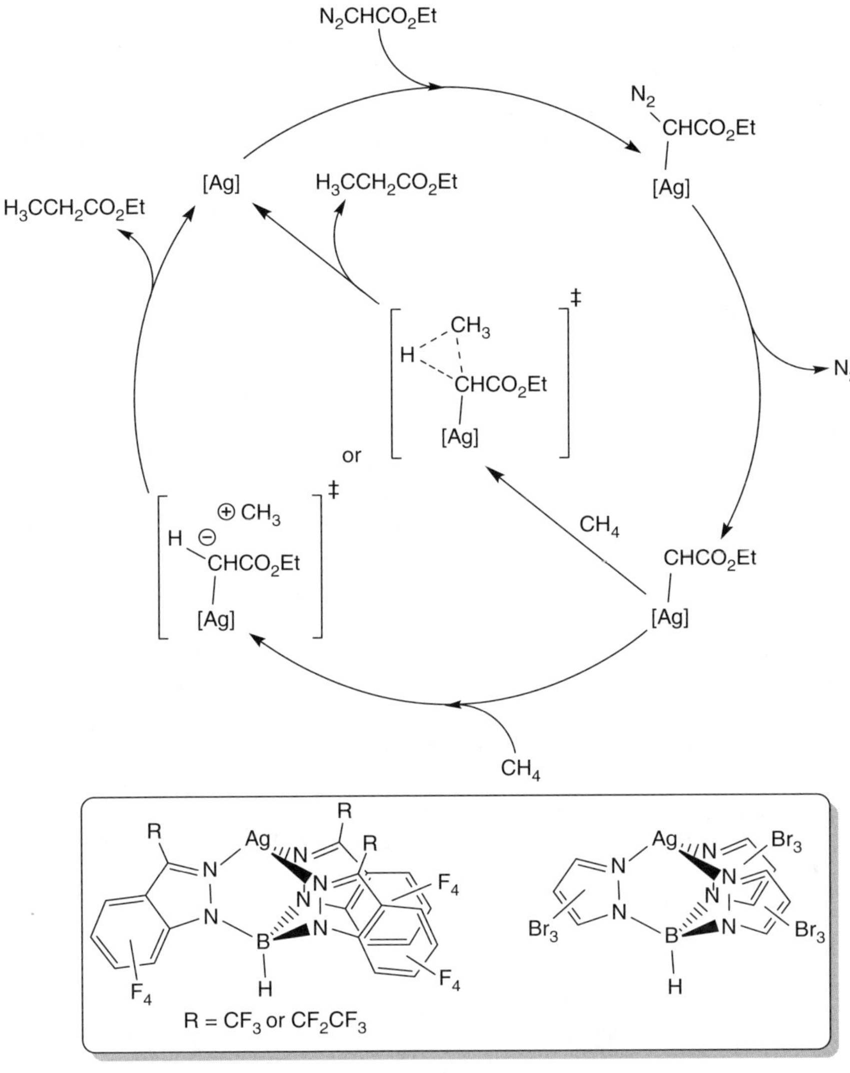

Scheme 17.6 Carbene insertion from ethyl diazoacetate to the C—H bond of methane using a Ag^+ catalyst.

occurs via coordination of the very reactive diazo reagent via the CH moiety. Rapid N_2 expulsion results in formation of a carbene species $(Tp_x)Ag(:CHCO_2Et)$ intermediate, which undergoes 1,2-H—C bond insertion of methane on the carbene carbon [11, 52]. Elimination of the ester would then close the cycle by reforming the resting state of the Ag^+ catalyst. Theoretical studies suggest the insertion process to occur in an asynchronous manner involving a three-centered transition state where the C—H bond is homolytically cleaved. The electrophilic nature of the carbenoid could also involve hydride transfer to form charge separated species, which could then recombine to form the C—C bond of the ester [54, 55]. Theoretical studies by Mindiola, Caulton, and co-workers using a lower-coordinate Ag^+ catalyst suggest that C—H insertion is overall rate-determining with significant charge flow from the alkane C—H bond to the electrophilic carbene carbon [52]. Since the Ag^+ catalysts serve as a carbene delivery vehicle, it is important that reactions be conducted at low concentration of catalyst and the carbene source ethyl diazoacetate to discourage carbene dimerization (homo coupling).

More recently, Perez and co-workers expanded their arsenal of catalysts for carbene insertion reactions into methane to more earth-abundant metals such as Copper [56]. Their conditions use a Cu^+ catalyst under 90 atm of supercritical CO_2 and 160 atm of methane at 40 °C leading to a 4% yield of the C—H insertion product. The catalyst is supported by halogenated Tp ligand, by forming the complex $[Tp(CF_3)_2\text{-}4\text{-}Br]Cu(NCMe)$ ($Tp(CF_3)_2\text{-}4\text{-}Br^-$ = hydrotris((3,5-bis(trifluoromethyl)-4-bromo)-pyrazol-1-yl)borate). It is quite likely that the same type of mechanistic scenario depicted in Scheme 17.6 takes place with Cu [56].

17.2.4 Dehydrocoupling of Methane to Ethane and Ethylene with Early-Transition Metals

Basset and co-workers have reported how well-defined organometallic complexes of early transition metals can be incorporated onto silicon oxide surfaces [57–59]. The thermal stability of these metal-ions on this surface allows for higher temperature reactions, therefore promoting transformations that are generally disfavored thermodynamically. One such reaction is the dehydrocoupling of methane to ethane (inset in Scheme 17.7). Although this reaction does not immediately seem like an important alkane homologation transformation, ethane could be viewed as an alkane with much more value-added properties than methane, given its ability to liquify at much higher temperatures and lower pressures. Likewise, the weaker C—C bond present in ethane is quite useful in the making of ethylene and α-olefins via the energy intensive steam cracking process [60]. Using solid-state multinuclear NMR spectroscopy, tantalum metal ions supported on a silicon oxide surface can be examined allowing for the detection of plausible intermediates involved in the C—H activation and functionalization [57]. Scheme 17.7 proposes how methane can be activated and dehydrocoupled to ethane as well as olefinated to ethylene using a Ta^{3+}/Ta^{5+} redox couple. From a trihydride tantalum species $(O_3{\equiv}SiO)_2Ta(H)_3$ generated from hydrogenation of an organometallic Ta(V) precursor, H_2

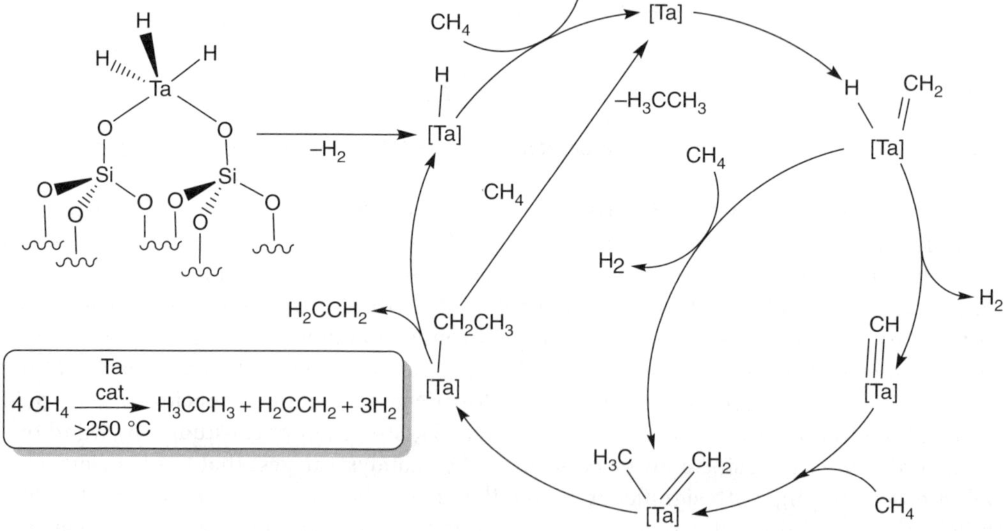

Scheme 17.7 Proposed mechanism to the dehydrocoupling of methane to ethane and ethylene by surface organometallic tantalum supported on silicon oxide. [Ta] represents a silicon oxide supported Ta surface.

is eliminated to form a putative Ta^{3+} hydride [Ta]—H ([Ta]=$(O_3{\equiv}SiO)_2Ta$). Through σ-bond metathesis, methane is activated to form [Ta]—CH_3, which then undergoes α-elimination to produce the methylidene-hydride [Ta]=CH_2(H). A second σ-bond metathesis of the latter complex can activate an additional equiv of methane to form [Ta]=$CH_2(CH_3)$, or alternatively, undergo α-hydrogen abstraction to form the methylidyne [Ta]≡CH, which would then activate methane by virtue of 1,2-CH bond addition across the Ta≡C multiple bond. Examples of α-elimination have been observed in low-valent metallocene complexes of Ta [61–63], whereas 1,2-CH bond addition and dehydrogenation of methane involving titanium alkylidynes have also been reported [64]. From [Ta]=$CH_2(CH_3)$, and perhaps what must be the bottleneck of the cycle, α-methyl migration would result in C—C bond formation to produce [Ta]—CH_2CH_3, which would then undergo σ-bond metathesis with methane to reform [Ta]—CH_3 and eject ethane. Intramolecular α-methyl migration to a methylidene is a rare phenomenon, although examples of such a transformation have been observed with metallocenes of Ta [65], Ir [66], Ru [67], and W [68] including the photolysis of explosive $W(CH_3)_6$ in the presence of phosphines [69]. It should also be noted that Scheme 17.7 portrays an additional side reaction where [Ta]—CH_2CH_3 undergoes β-hydride elimination to form [Ta]—H and ethylene, a second minor byproduct observed in the reaction mixture at temperatures >250 °C [57].

17.3 The Inspiration Behind Iridium: Early Studies

Seminal studies by Graham and co-workers showed how photolysis of $LIr(CO)_2$ (L^- = Cp or Cp*), under methane pressure (~8 atm) afforded the oxidative addition product $LIr(CO)(H)(CH_3)$, albeit in low yields (c. 20%) [70]. Soon after, Bergman and co-workers determined that thermolysis at 140–150 °C of $Cp^*Ir(PMe_3)(^cC_6H_{11})(H)$ under a pressure of CH_4 (20 atm) resulted in elimination of CyH, followed by formation of the methane C—H activation product $Cp^*Ir(PMe_3)(CH_3)(H)$ in higher yields (~60%) [71]. Scheme 17.8 depicts this early set of reactions. In both of these reactions the intermediacy of a coordinatively unsaturated Ir(I) [LIrX] or σ-complex [LIr(X)(σ-CH_4)] [72] could be invoked (X = CO or PMe_3), followed by oxidative addition of the methane C—H bond. As will be noted later, the Cp*Ir(III) systems studied by Bergman and co-workers would become the foundation for some of the first catalytic studies involving the borylation of arenes.

Ten years later, Bergman and co-workers found milder conditions to activate methane. Their strategy was to utilize an Ir(III) precursor instead that would possess an open coordination

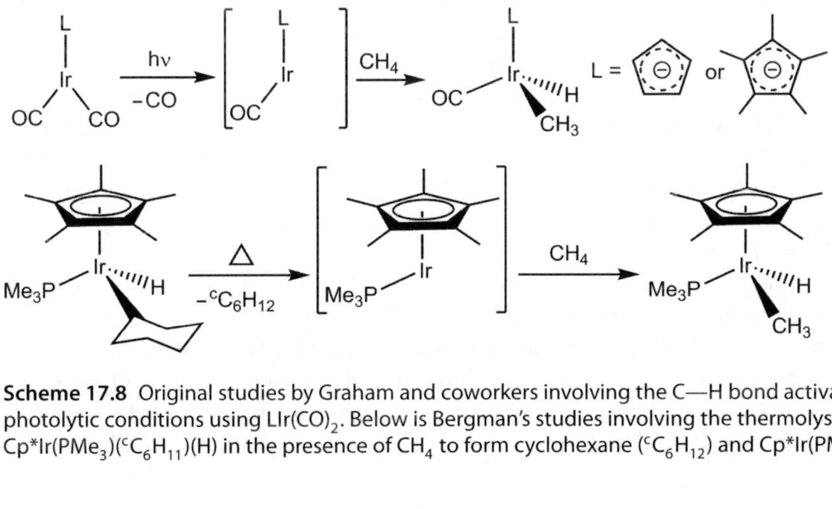

Scheme 17.8 Original studies by Graham and coworkers involving the C—H bond activation of CH_4 under photolytic conditions using $LIr(CO)_2$. Below is Bergman's studies involving the thermolysis of $Cp^*Ir(PMe_3)(^cC_6H_{11})(H)$ in the presence of CH_4 to form cyclohexane ($^cC_6H_{12}$) and $Cp^*Ir(PMe_3)(CH_3)(H)$.

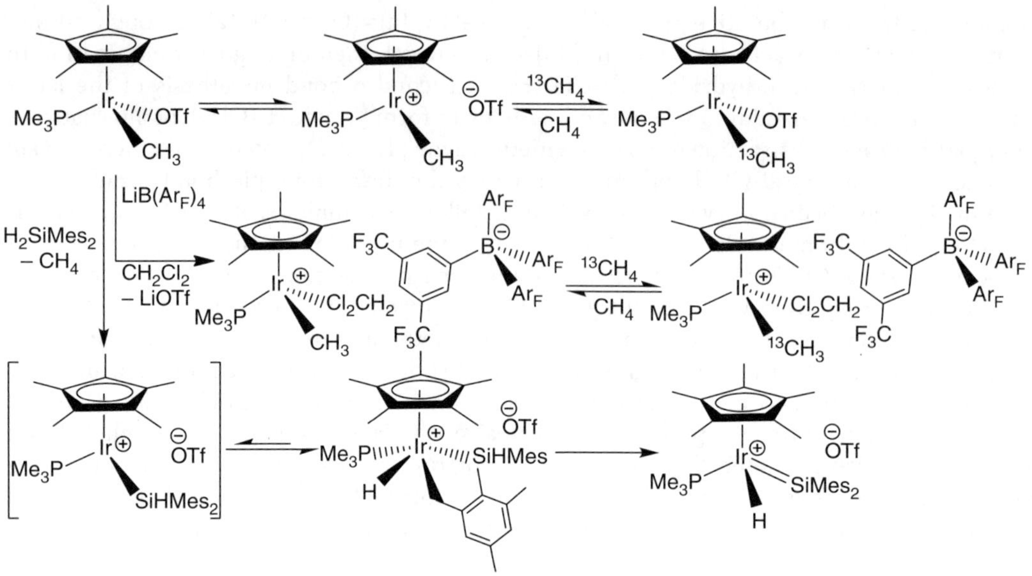

Scheme 17.9 Reactivity studies of Cp*Ir(PMe$_3$)(CH$_3$)(OTf) with $^{13}CH_4$, and replacement of the $^-$OTf with a robust borate anion $^-$B(ArF)$_4$ (ArF = 3,5-(CF$_3$)$_2$C$_6$H$_3$) in CH$_2$Cl$_2$ to generate a discrete salt [Cp*Ir(PMe$_3$)(CH$_3$)(Cl$_2$CH$_2$)][B(ArF)$_4$], which can then activate $^{13}CH_4$ at 10 °C. Also shown is the reactivity of Cp*Ir(PMe$_3$)(CH$_3$)(OTf) with H$_2$SiMes$_2$ and subsequent cyclometallation to form an Ir(V) species, which then converted to the silylene [Cp*Ir=SiMes$_2$(H)(PMe$_3$)][OTf].

site for methane binding. Accordingly, the complex Cp*Ir(PMe$_3$)(CH$_3$)(OTf), which likely serves as a synthon to [Cp*Ir(PMe$_3$)(CH$_3$)][OTf], activates methane at 45 °C according to $^{13}CH_4$ labeling studies (Scheme 17.9), whereas formation of a discrete salt [Cp*Ir(PMe$_3$)(CH$_3$)(Cl$_2$CH$_2$)][B(Ar$_F$)$_4$] (B(Ar$_F$)$_4^-$ = B(3,5-(CF$_3$)$_2$C$_6$H$_3$)$_4$) shown in Scheme 17.9 resulted in methane activation (also via a degenerate exchange using $^{13}CH_4$) at 10 °C [66, 73]! Hall and co-workers explored the mechanism of the latter reaction using the truncated model [CpIr(PH$_3$)(CH$_3$)]$^+$ and found that oxidative addition of methane by an Ir(V) four-legged piano-stool species [CpIr(PH$_3$)(CH$_3$)$_2$(H)]$^+$ was likely operative [74]. Although the role of the PMe$_3$ moiety was contested in the proposed mechanism on the grounds that such a ligand can undergo cyclometallation [75], subsequent studies by Bergman and co-workers revealed that indeed species such as [Cp*Ir(η^2-CH$_2$PMe$_3$)]$^+$ can readily activate alkane C—H bonds [76]. Regardless, the mechanism to C—H activation of methane by a species such as [Cp*Ir(PMe$_3$)(CH$_3$)]$^+$, either by σ-bond metathesis or oxidative addition, is debatable. In support of an oxidative mechanism pathway, Bergman and Tilley established in 2000 that an intramolecular oxidative addition (cyclometallation) in the Ir(III) silyl system [Cp*Ir(SiHMes$_2$)PMe$_3$)][OTf] resulted in the isolation of the Ir(V) complex [Cp*Ir{SiHMesC$_6$H$_2$Me$_2$CH$_2$}(H)PMe$_3$)][OTf] [77]. The latter species further rearranges to a silylene [Cp*Ir=SiMes$_2$(H)(PMe$_3$)][OTf] (Mes = 2,4,6-Me$_3$C$_6$H$_2$) by means of an α-hydrogen abstraction step (Scheme 17.9).

Iridium has been shown to catalyze, albeit with few TON (c. 6), the activation and controlled oxidation of methane to H$_3$CO$_2$CCF$_3$ [21, 78]. In a study by Goddard, Periana, and co-workers the iridium precatalyst (NNC)Ir(CH$_2$CH$_3$)(O$_2$CCF$_3$)(η^2-C$_2$H$_4$) (NNC$^-$ = 6-phenyl-4,4′-^{t}Bu$_2$-2,2′-bipyridine) can convert CH$_4$ to the ester H$_3$CO$_2$CCF$_3$ in strong acidic media (HO$_2$CCF$_3$) and in the presence of a strong oxidizer NaIO$_4$ under ~500 psi of the alkane at 180 °C (Eq. (17.11)) [78]. Using a weaker oxidant such as KIO$_3$ at slightly lower temperatures resulted in a TON of 2 whereas the usage of other oxidants such as $Na_2S_2O_8$, OsO_4, or $K_2Os(O)_2(OH)_4$ in HO_2CCF_3 led instead to stoichiometric conversion of methane to the ester (Eq. (17.11)). Although it is currently unclear how the Ir oxidizes the hydrocarbon, theoretical studies demonstrate that an Ir(III) species is likely involved in the C—H activation event. In acidic media, loss of ethylene and ethane in $(NNC)Ir(CH_2CH_3)(O_2CCF_3)(\eta^2\text{-}C_2H_4)$ likely forms the five-coordinate Ir(III) species $(NNC)Ir(O_2CCF_3)_2$, and subsequent C—H activation of methane would result in formation of $(NNC)Ir(O_2CCF_3)(CH_3)(HO_2CCF_3)$. In the C—H activation step and based on theoretical studies, both the $^-O_2CCF_3$ and the anionic aryl motif of the pincer ligand could be facilitating such a process.

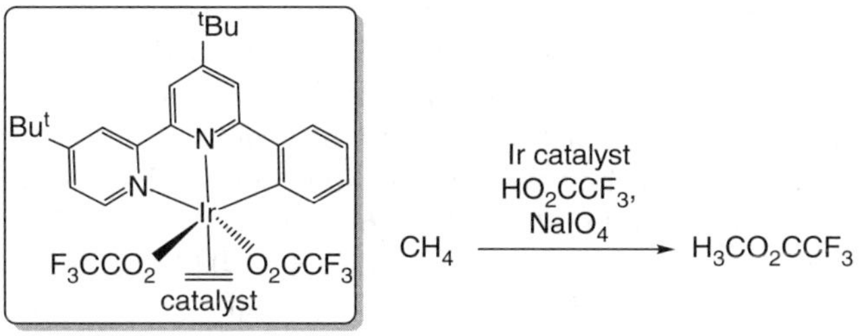

(17.11)

17.4 Catalytic Borylation of Aliphatic C—H Bonds

Replacing a C—H bond of an arene or an alkane with a boryl group represents an important transformation since the boryl group can essentially be used as a delivery vehicle or handle for numerous functionalizations such as the construction of C—C bonds by the Suzuki-Miyaura cross-coupling reaction. Carbon-boron bonds can also be utilized for cyanation, halogenation (with I, Br, or Cl), and can be hydroxylated/oxidized or alkoxylated, as well as aminated. Borylation of aromatic and aliphatic C—H bonds in the presence of other functional groups can also be achieved with some degree of regioselectivity and with a high degree of functional group tolerance. The boryl group can also be tuned for subsequent cross-coupling chemistries, such as being converted to borate (c. RBX_3^-) or boronic acids (c. $RB(OH)_2$). Comprehensive reviews detailing the background and significance of C—H bond borylation have been documented [79–81]. This reaction has received noteworthy attention because borylation of the C—H group does not require a directing group, and is generally regarded as a reaction that is primarily dominated by kinetics, which in principle could serve as a means to incorporate functional groups *contra* electronics. However, the site of borylation can be also quite modular and recent studies by Hartwig [82] and Smith [83, 84] have shown that anchoring groups such as silanes or amines (with an NH moiety) can help direct the boryl group through Si—H activation or outer sphere hydrogen bonding networks with the boryl oxygen, respectively. As a result, the boryl group can be placed nearby or away from a substituent, depending on what role it plays in the C—H activation and borylation cycle [79, 80]. Borylation of C—H bonds has therefore replaced, to a great degree, the need to perform classical hydroborations of alkenes, given the fact that alkanes, a feedstock to alkenes and alkynes, can now be directly used as the reactant.

The first documented case of borylation of a C—H bond of toluene can be found in the supporting information of a communication of Marder and co-workers involving the synthesis of an iridium trisboryl complex (arene)Ir(Bcatechol)$_3$ (arene = mesitylene or toluene) [85]. Although at that time it was not recognized, the significance of having traces of borylated "impurities," that is, the two isomeric borylated forms of toluene, their work catalogued for the first time that iridium could deliver the boryl group from HBcatechol (the reagent used to prepare the triboryl complex) to an arene. Soon after, pioneering work by Hartwig and co-workers demonstrated that $CpFe(CO)_2(Bcat)$ (cat = catecholate) could be used, under photolytic conditions (similar studies were conducted with $Mn(CO)_5(Bcat)$ and $Re(CO)_5(Bcat)$), as a stoichiometric borylation reagent to substrates such as arenes and terminal olefins [86]. Unfortunately, the propensity of the $CpFe(CO)_2$ to dimerize as well as the instability of the $CpFe(CO)_x(H)$ formed from C—H activation and C—B bond expulsion (under photolytic conditions and in the presence of H_2) prevented any turnover. Likewise, the vulnerability of the Cp ligand to becoming borylated led to decomposition. Although $CpFe(CO)_2(Bcat)$ was unable to borylate alkanes, they soon discovered the analogue $Cp^*Fe(CO)_2(Bcat)$ to be a stoichiometric reagent for the borylation of pentane at the primary C—H bond [87]. The yield of borylated pentane improved by switching to $Cp^*Ru(CO)_2(Bcat)$, but ultimately $Cp^*W(CO)_3(Bcat)$ was found to be the most active system that could photolytically borylate linear or branched alkanes to make alkylboronate esters, with predominant selectivity toward primary C—H bonds [87].

In 1999 Hartwig and co-workers reported that $Re(CO)_5(Bcat)$ could stoichiometrically borylate the ortho-CH bond of anisole [88]; however, it was soon found that $Cp^*Re(CO)_3$ could catalyze (under photolytic conditions and under a slight pressure of CO), the borylation of alkanes and ethers with primary selectivity for the least hindered C—H bond (Eq. (17.12)) [89]. This seminal paper is perhaps the first documented case of catalytic borylation of an alkane, and at room temperature! Scheme 17.10 shows the proposed mechanism to catalytic alkane C—H borylation by the photolysis of $Cp^*Re(CO)_3$. It was also here where the usage of

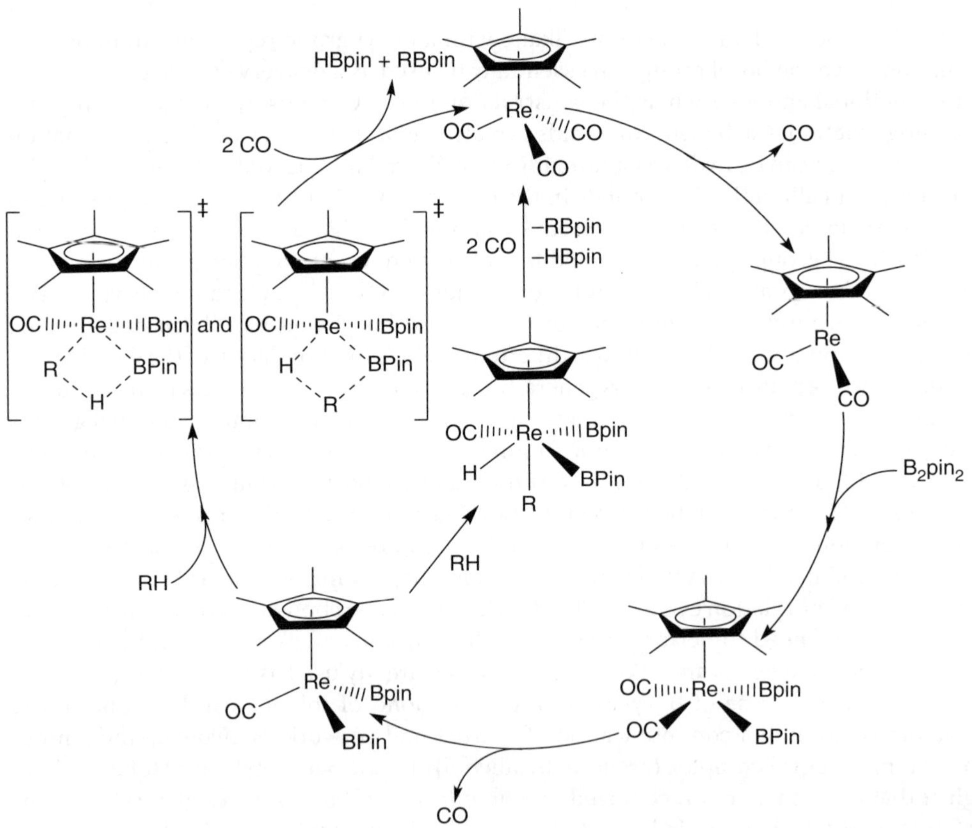

Scheme 17.10 Proposed mechanism for the photocatalytic borylation of alkanes with $Cp^*Re(CO)_3$ under CO atm and B_2pin_2.

the more reactive diboron reagent B_2pin_2 (pin = pinacolate) was introduced to the catalytic cycle. Shown in Scheme 17.12, photo-triggered decarbonylation of Cp*Re(CO)$_3$ results in formation of an unsaturated Cp*Re(CO)$_2$, which then undergoes oxidative addition of the diboron reagent B_2pin_2 to produce the four-legged piano-stool complex Cp*Re(CO)$_2$(Bpin)$_2$. Further decarbonylation to Cp*Re(CO)(Bpin)$_2$ and activation of the alkane C—H bond results in formation of either an oxidative-addition species Cp*Re(CO)(H)(R)(Bpin)$_2$, or the σ-bond metathesis transition-states [Cp*Re(CO)(HRBpin)(Bpin)]‡ or [Cp*Re(CO)(RHBpin)(Bpin)]‡, respectively. Dicarbonylation (not likely in a concerted way) would then close the cycle by reforming Cp*Re(CO)$_3$, with concurrent elimination of the alkylborane ester RBpin along with HBpin [89].

$$\text{R-H} + \text{B}_2\text{pin}_2 \xrightarrow[25^\circ\text{C}]{\text{Re catalyst, CO (2 atm), h}\nu} \text{H}_3\text{C-Bpin} + \text{H-Bpin} \qquad (17.12)$$

A few years later, Hartwig and Chen examined the catalytic borylation reaction of alkanes and arenes using Cp*Rh(η^4-C$_6$Me$_6$), but under thermal conditions at 150 °C [90]. The use of a hindered arene was introduced since it could represent a better leaving group and would eliminate the need for the pressurized CO. Likewise, it was found that this hindered arene was impervious to borylation. In a general reaction, 5 mol% of the Rh(I) catalysts could borylate ^{n}octane with B_2pin_2 to the ^{n}octylBpin in 88% yield. This work established that catalytic, regiospecific functionalization of alkanes could be conducted under thermal conditions, but it also revealed that two catalytic cycles, in tandem, were likely taking place in the alkane borylation reaction using B_2pin_2 given that HBpin, a byproduct from alkane borylation could also be used as a borylating agent under similar reaction conditions. Replacing B_2pin_2 with HBpin is a more economically viable reaction since diboron reagents require additional synthetic steps to prepare that generally involve strongly reducing conditions [91]. Other Rh and Ir catalysts were tested but were found to be less efficient. Shown in Scheme 17.11 are the proposed two cycles for the borylation of alkane using B_2pin_2, RH (alkane) and Cp*Rh(η^4-C$_6$Me$_6$). We chose oxidative addition of B_2pin_2 and HBpin to the unsaturated Cp*Rh(I) fragment as the first step in the reaction since recent studies by Hartwig, Smith, and co-workers showed that Ir(III) trisboryl (with Bpin or Bcat) systems are generally the resting state of the catalyst [92]. By analogy, we invoke a similar set of steps here, but one can easily propose oxidative addition of the alkane HR first, followed by addition of the B—B and/or H—B substrates. Scheme 17.11 clearly shows how two competitive cycles, one for B_2pin_2 (above) and HBpin (below), could be promoted by the Rh(I) catalyst. For the B_2pin_2 cycle, oxidative addition to the [Cp*Rh] fragment would result in Cp*Rh(Bpin)$_2$, followed by C—H activation of the alkane to form Cp*Rh(Bpin)$_2$(H)(R). Reductive elimination in the Rh(V) system would eliminate RBpin by forming Cp*Rh(Bpin)(H), which can then add another B—B bond from B_2pin_2 to form Cp*Rh(Bpin)$_3$(H). Reductive elimination of HBpin would close the cycle, at which point the other cycle could commence. Akin to the above cycle in Scheme 17.11, HBpin can oxidatively add to the Cp*Rh(I) fragment, followed by the steps oxidative addition of the alkane RH, reductive elimination of H_2, and oxidative addition of another HBpin to produce Cp*Rh(Bpin)$_2$(H)(R), also a common intermediate in the catalysis involving B_2pin_2. Elimination of RBpin from the latter would close the cycle by reforming Cp*Rh(Bpin)(H).

Then in 2006, Hartwig and co-workers established that catalysts such as [Cp*Ru(μ-Cl)(Cl)]$_2$ or [Cp*Ru(μ-Cl)]$_4$, could also cleanly borylate linear alkanes such as neat ^{n}octane using B_2pin_2 and without the need of photolysis, although reaction conditions did require high temperatures (150 °C), as shown in Eq. (17.13) [93]. Other derivatives of Ru(II) were tested as well and were found to be catalytically active. It is quite possible that B_2Pin_2 replaces the Cl ligands as well as

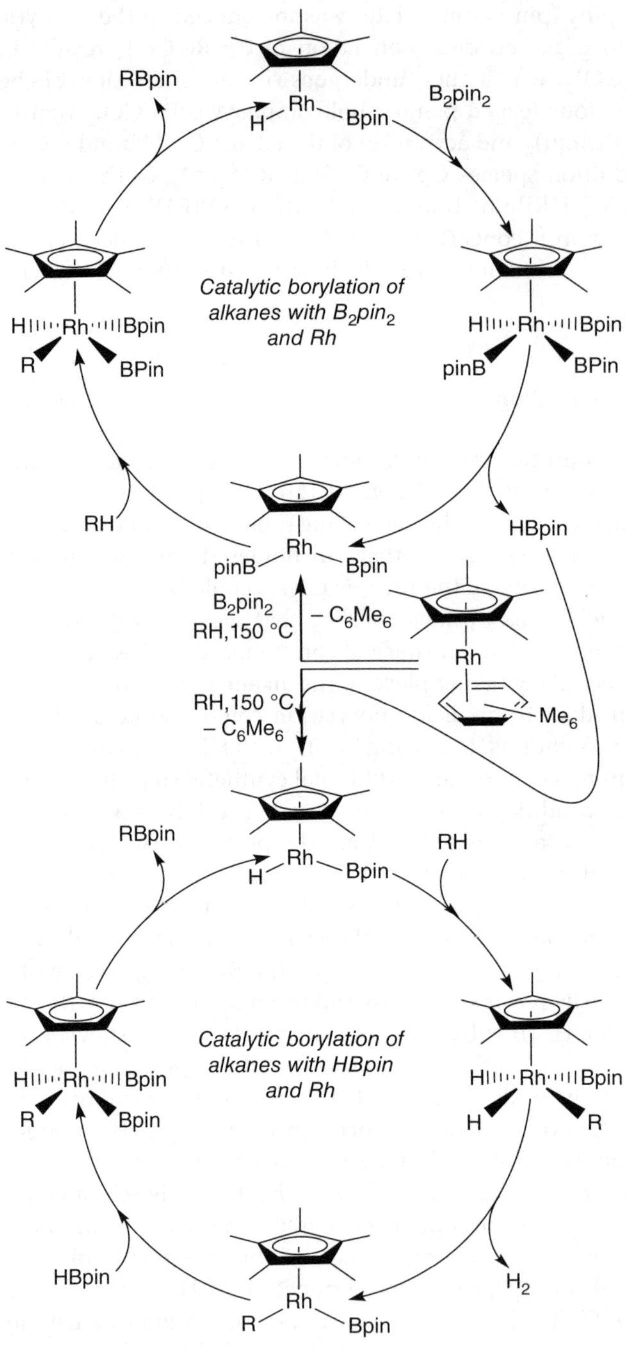

Scheme 17.11 Proposed mechanism for the borylation of the alkane (RH) by precatalyst Cp*Rh(η^4-C_6Me_6) with B_2pin_2 (top cycle), and HBpin (bottom cycle).

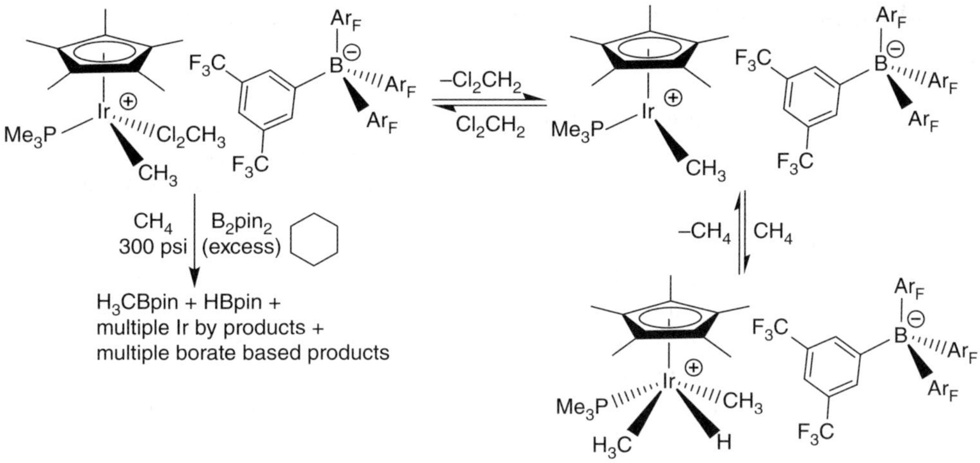

Scheme 17.12 Attempted catalysis with methane using $[Cp^*Ir(CH_3)(PMe_3)(ClCH_2Cl)][B(Ar_F)_4]$ in cyclohexane with excess B_2pin_2 and CH_4 (300 psi).

oxidatively adds to $[Cp^*Ru(\mu\text{-}Cl)]_4$ to form a putative species such as $Cp^*Ru(Bpin)_3$. C—H activation and borylation akin to that shown in the top of Scheme 17.11 would result in elimination of ^{n}octylBpin and HBpin. Although speculative at this point, it is likely that the Ru(III) catalyst $[Cp^*Ru(\mu\text{-}Cl)(Cl)]_2$ operates via a similar mechanism involving the formation $Cp^*Ru(Bpin)_3$, but where an additional "Bpin" equiv is converting Ru(III) to Ru(IV), or alternatively, reducing Ru(III) to Ru(II).

$$\text{R-H} + \text{B}_2\text{pin}_2 \xrightarrow[150\,^\circ\text{C}]{\text{Ru catalyst}} \text{R-Bpin} + \text{H-Bpin} \tag{17.13}$$

As noted before, Hartwig and co-workers reported Ir(I) or Ir(V) systems that could borylate hydrocarbons, albeit in lower yield when compared to the Rh catalysts [90]. Prior to this work, however, Smith and co-workers were exploring control reactions to establish the viability of Ir complexes for borylation catalysis. As such, they centered their attention on Ir systems known to activate alkane and arene C—H bonds. For example, it was found that Bergman's complex $Cp^*Ir(PMe_3)(^cC_6H_{11})(H)$ [71] or $Cp^*Ir(PMe_3)(H)(Ph)$ reacted with boranes such as HBR_2 ($R_2 = (C_6F_5)_2$ (dimer via bridging B—H—B bonds), catecholate, pinacolate) to form the alkyl boranes $^cC_6H_{11}BR_2$ or $PhBR_2$, respectively, along with $Cp^*Ir(PMe_3)(Y)(X)$ (X = H or BR_2), depending on the nature of the R_2 group on boron. For $R_2 = (C_6F_5)_2$, Y = X = H, whereas for R_2 = catecholate (cat), Y = X = Bcat. In the case of R_2 = pinacolate (pin), the Ir complex contained both X = H and Y = Bpin [94]. After considering both thermal and photochemical conditions, it was established that catalytic borylation of benzene could be achieved using $Cp^*Ir(PMe_3)(H)(Bpin)$ at 150 °C in benzene and in the presence of HBpin to form PhBpin in 53% yield over 5 days (Eq. (17.14)). These studies set the premise for Ir catalysts being likely candidates for the borylation of arenes, without the need of photolysis and by using more economically viable borylating agents such as HBpin. A follow-up manuscript showing catalytic reactions at 150 °C with $Cp^*Ir(PMe_3)(H)(Bpin)$ with HBpin revealed that borylation of various arenes (regioselectively by sterics) could be performed, albeit in low turnovers (20 mol% catalyst loadings) as shown in Eq. (17.14).

$$\text{PhH} + \text{HBpin} \xrightarrow[95\,^\circ\text{C}]{\text{Ir catalyst}} \text{PhBpin} + \text{H}_2 \tag{17.14}$$

17.5 Preliminary Studies on Methane Borylation Using *N*-Based Ligands

In 1996, Rablen and Hartwig reported computational studies examining the bond dissociation energy (BDE) in C—H, C—C, B—H, and B—C of various substrates using ab initio molecular orbital calculations at the G-2 and CBS-4 computational levels [95]. Among some of their conclusions were that low-coordinate boron reagents possess greater B—H bonds energies when compared to borate B—H bonds. It was also concluded that electronic effects play minor roles in the BDE in B—X bonds. However, the most notable finding was that B—C and B—H bond energies are stronger than those found for C—C and C—H bonds, whereas the average B—H and B—C bonds for organoboranes are lower than those observed in hydrocarbons. Taken all together, when one compares the computed BDE to those found for the C—H BDE of CH_4, then one can conclude that using a borane as a borylating reagent for these simple alkanes represents a reaction that is essentially isothermal, as noted by Smith and co-workers (Eq. (17.15)) [94]. Likewise, theoretical studies by Hall and co-workers using Cp*Rh based catalysts and CH_4 as their model alkane have also demonstrated this type of reaction to be thermodynamically feasible [96].

Shown in Eq. (17.15) are BDE for the borylation of methane [16] using both HBpin and HBcat, where the BDE for B—H and B—CH_3 bond having a pinacolate group has been truncated to cyclo-OCH_2CH_2O- [97]. Although the ΔH_o for these reactions are essentially isothermal, the reaction using HBcat and methane is found to be slightly more favored due to the inherent B—C bond strength in H_3CBcat (113.0 kcal mol^{-1}) when compared to H_3CBpin (111.6 kcal mol^{-1}).

	$R-H$	$+\ H-BR'_2$	$\rightarrow\ R-BR'_2$	$+\ H-H$	ΔH_o(kcal/mol)
$R = CH_3$	104.9	111.3 (cat)	113.0 (cat)	104.2	− 1.0
	104.9	110.8 (pin)^a	111.6(pin)^a	104.2	− 0.1
R = CH_3CH_2	101.1	111.3 (cat)	113.0(cat)^b	104.2	− 4.8
	101.1	110.8 (pin)^a	111.6(pin)^{a,b}	104.2	− 3.9

^{a}pin represents the truncated $^-OCH_2CH_2O^-$ group;

^{b}B-C BDE for $H_3CCH_2BR'_2$ and $H_3CRB'_2$ are assumed to be the same. (17.15)

Using the BDE for ethane, Eq. (17.15) reveals that borylating this alkane with HBcat or HBpin renders such a reaction slightly more favorable given the slightly weaker C—H in ethane versus methane (104.9 kcal mol^{-1}). However, this prediction is based on the assumption that B—C bond energies in H_3CH_2Bcat and H_3CCH_2Bpin are essentially similar to those for H_3CBcat (113.0 kcal mol^{-1}) and H_3CBpin (111.6 kcal mol^{-1}).

	$R-H$	$+\ B_2R'_4$	$\rightarrow\ R-BR'_2$	$+\ HBR'_2$	ΔH_o
$R = CH_3$	104.9	103.6^a	113.0 (cat)	111.3 (cat)	− 15.8
	104.9	103.6^a	111.6(pin)^a	110.8 (pin)^b	− 13.9
R = CH_3CH_2	101.1	103.6^a	113.0(cat)^c	111.3 (cat)	− 19.6
	101.1	103.6^a	111.6(pin)^{b,c}	110.8 (pin)^b	− 17.7

(17.16)

^{a}The B − B BDE for B_2OH_4 is used (R′ = OH);

^{b}Represents the truncated version $^-OCH_2CH_2O^-$ group;

^{c}B-C BDE for $H_3CCH_2BR'_2$ and $H_3CBR'_2$ are assumed to be the same.

In Eq. (17.16) we try to depict how switching to a diboron reagent results in the reaction being more thermodynamically favored. We estimate the ΔH_o for the reaction using the BDE for $B_2(OH)_4$, and must assume the B—C BDE H_3CH_2Bcat and H_3CCH_2Bpin to be essentially similar to those for H_3CBcat (113.0 kcal mol^{-1}) and H_3CBpin (111.6 kcal mol^{-1}).

As can be seen from Eq. (17.16), the use of a diboron reagent enthalpically facilitates the reaction by at least 10 kcal mol^{-1} for methane given the weaker B—B bond and stronger C—B bond as opposed to that found in H_2 (Eq. (17.15)). For ethane, the reaction is again, more favored given the slightly weaker BDE for ethane versus methane. Unfortunately, we cannot make any distinction between B_{cat} and B_{pin} BDE since we are estimating these using the same model, namely $B_2(OH)_4$ (Eq. (17.16)) [97]. The fact that this reaction is thermodynamically favored led us to pursue conditions using diboron reagents for the catalytic borylation of methane using diboron reagents as our borylating source. Although calculations for the borylation of methane were reported in the literature [78, 89, 90], we were surprised to see that experimental studies validating this interesting reaction had not been reported despite there being reports for very active alkane borylation catalysts [74, 94, 95]. Our starting point was to use Ir reagents to help catalyze the borylation of methane. We chose Ir since Graham and Bergman found Cp- and Cp*-based systems to be capable of activating CH_4 [69, 70]. In particular, studies by Smith and co-workers using complexes such as Cp*Ir(PMe_3)($^cC_6H_{11}$)(H) and Cp*Ir(PMe_3)(H)(Bpin) [94, 98] demonstrated that such systems could, in principle, serve as catalysts for the activation and borylation of aliphatic and aromatic C—H bonds given the formation of C—B bonds when treated with HBcat under thermolytic conditions [94].

In our initial attempt to explore catalytic borylation we explored iridium complexes that could activate methane under very mild conditions (low temperature and pressures). For this reason, we focused our attention in the salt [Cp*Ir(CH_3)(PMe_3)($ClCH_2Cl$)][B(Ar_F)$_4$] reported by Bergman (*vide supra*, Scheme 17.9). Complex [Cp*Ir(CH_3)(PMe_3)($ClCH_2Cl$)][B(Ar_F)$_4$] has been shown to activate methane, by exchange with $^{13}CH_4$, under low pressures at 10 °C [73]. For us, when this complex was exposed to methane at 300 psi and the diboron reagent B_2pin_2 in CyH for for several hours at 25 °C or above, examination of the mixture revealed formation of H_3CBpin, HBpin, in addition to other boron containing products resulting from degradation of the Ir component (based on ^{31}P NMR spectroscopy) and the counter anion B(Ar_F)$_4$ component (based on ^{11}B NMR spectroscopy). Although this reaction mixture suggested [Cp*Ir(CH_3)(PMe_3)($ClCH_2Cl$)][B(Ar_F)$_4$] to be unstable under these conditions, the transfer of the methyl ligand from the Ir complex to Bpin implied that methane could be conceivably activated and transferred to the boryl group since such systems readily exchange the methyl moiety with methane via a putative Ir(V) intermediate such as [Cp*Ir(PMe_3)(CH_3)$_2$(H)][B(Ar_F)$_4$] (Scheme 17.12). The highly reactive nature of Bergman's iridium borate salt (especially at room temperature or above) [73], poor solubility in CyH, as well as the ambiguity regarding the source of the methyl moiety in the formed H_3CBpin, led us to explore more stable Ir reagents.

Some of the most active borylation catalysts or precatalysts have been based on Ir [99, 100]. As a result, we focused our attention on commercially available iridium reagents such as [Ir(COD)(μ-Cl)]$_2$ (COD = 1,5-cyclooctadiene), [Ir(COD)(μ-OMe)]$_2$, and [Ir(COE)$_2$(μ-Cl)]$_2$ (COE = cyclooctene) since these are thermally robust and would not require photochemical means to promote C—H activation of the hydrocarbon. We also decided to explore more robust reagents of Ir, specifically the pre-catalyst [Ir(COD)(μ-OMe)]$_2$ with the ligand 4,4′-di-tert-butyl-2,2′-bipyridine (^{tBu}bipy) [100] or 3,4,7,8-tetramethyl-1,10-phenanthroline (tmphen) [101] under similar conditions. Our choice of ligand was based on previous studies by Hartwig and co-workers reporting a highly active Ir catalyst (^{tBu}bipy)(COE)Ir(Bpin)$_3$ [100]. Our solvent of choice was again CyH given its more kinetically hindered hydrogens. Although other borylation catalysts are known for Ir [102–105], the more active systems developed by Hartwig [100], Maleczka, Smith, and co-workers [101] were our original targets.

Using the Ir(I) precatalysts with nitrogen-based ligands, we exposed these for 16 hours at 120 °C to 300 psi of methane in the presence of B_2pin_2 and results indicate that the ligand tmphen resulted in a mixture of H_3CBpin (**1**), $H_2C[Bpin]_2$ (**2**), HBpin (**3**), and $H_3COBpin$ (**4**). We also observed the production of $O[Bpin]_2$ (**5**) (Eq. (17.17)) [106]. Based on some control experiments, formation of **5** is likely due to a separate cycle using B_2pin_2, which is metal catalyzed. Smith and co-workers observed and even isolated an Ir borylene species from ring-opening of pinacolborane using Ir borylation catalyst [107].

$$CH_4 \xrightarrow[\substack{150^\circ C,\ 16\ h \\ {}^cC_6H_{12}}]{\substack{[Ir(COD)OMe]_2 \\ 2\ tmphen \\ B_2pin_2}} H_3CBpin + HBpin + H_2C[Bpin]_2 + H_3COBpin + O[Bpin]_2 \qquad (17.17)$$

In these reactions, we did not observe production of tri- or tetra-borylated methyl products, $(H_{4\text{-}x}C)[B(pinacol)]_x$ ($x = 3$ or 4). Borylation of methane was not improved appreciatively when using higher pressures of methane ($300 \le p_{CH4} \le 1200$ psi), whereas a reaction was observed when B_2pin_2 is used instead of B_2cat_2. Given that borylation of methane under these conditions led essentially to stoichiometric conversions, we turned our attention to theoretical methods as a guiding tool to understand how we could improve catalyst performance by analyzing the most plausible reaction profile. Specifically, we wanted to understand the role of the supporting ligand during the C—H activation of methane and B—C bond forming reactions. During the same time, and independently, Sanford and co-workers explored Cp*Rh catalysts for methane borylation [108] since Hall and co-workers predicted these to activate the parent alkane with relatively low barriers [96]. Their studies established Cp*Rh(η^3-C_6Me_6) to catalyze the borylation of methane in CyH at 150 °C and in the presence of B_2pin_2 with up to 68 TON using 0.75 mol% loadings and with high selectively in forming monoborylated **1** versus diborylated product **2** (18 : 1), as shown in Eq. (17.18). The borylation of solvent, although observed, was also minimized. Ir and Ru catalysts also catalyzed the reaction but with lower TON. However, in the case of the Ru system, reactions were completed within one hour with high selectivity for **1** versus **2** [108].

$$CH_4 \xrightarrow[\substack{150^\circ C,14\ h \\ {}^cC_6H_{12}}]{\substack{Cp^*Rh(\eta^4\text{-}C_6Me_6) \\ catalyst \\ B_2pin_2}} H_3CBpin + HBpin + H_2C[Bpin]_2 \qquad (17.18)$$

17.6 Mechanistic Studies on the Borylation of Methane

17.6.1 Experimental Studies

As shown in Table 17.1, the use of chelating 1,10-phenanthroline based ligands such as **L1**, **L2**, and **L3** led to stoichiometric borylation of methane to form mixtures of **1** and **2** among other products. Bipy-based ligands also gave similar results. The best Ir systems found for this transformation were the Ir(I) precursor $[(COD)Ir(\mu_2\text{-}OMe)]_2$ and the Ir(III) system $(Mes)Ir(Bpin)_3$ in the presence of B_2pin_2 under 300 psi of CH_4. What is notable is the lack of borylation of the solvent, whether it be tetrahydrofuran (THF) or CyH [106].

Given the stoichiometric amount of monoborylated product **1** we turned our attention to using theoretical studies to probe why this catalysis did not work well. We considered two possible pathways for the catalytic cycle described in Scheme 17.13 using tmphen as the supporting ligand (calculations were based on a truncated model with 1,10-phenanthroline,

Table 17.1 1,10-phenanthroline based ligands used in the borylation of methane. The ligands were added in a 2 : 1 ratio relative to dimeric Ir reagent and in a 1 : 1 ratio relative to independently prepared (MesH)Ir(Bpin)$_3$. Solvent was either tetrahydrofuran (THF) or cyclohexane (CyH) [106].

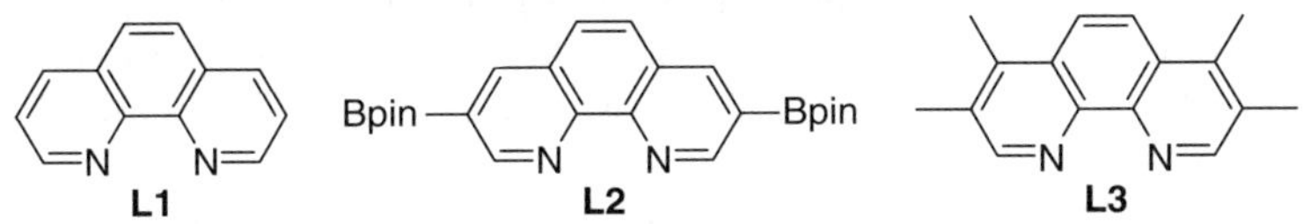

$$\underset{300\ \text{psi}}{CH_4} + B_2pin_2 \xrightarrow[\text{solvent, 16 hr, 120 °C}]{\text{Ir reagent, ligand } \mathbf{L}} \underset{\mathbf{1}}{H_3C\text{-}Bpin} + \text{other products}$$

Entry	Ir reagent	Loading (mol%)	Ligand	Solvent	% yield 1	Ratio 1 : 2
1	[Ir(COD)(μ-OMe)]$_2$	2.5	L3	THF	2.6	n/a
2	[Ir(COD)(μ-OMe)]$_2$	5.0	L1	THF	1.5	n/a
3	[Ir(COD)(μ-OMe)]$_2$	5.0	L2	THF	1.5	n/a
4	[Ir(COD)(μ-OMe)]$_2$	5.0	L3	THF	2.7	n/a
5	[Ir(COD)(μ-OMe)]$_2$	2.5	L3	CyH	2.0	3.9 : 1
6	[Ir(COD)(μ-OMe)]$_2$	5.0	L3	CyH	2.3	2.5 : 1
7	(MesH)Ir(Bpin)$_3$	2.5	L3	THF	4.1	n/a

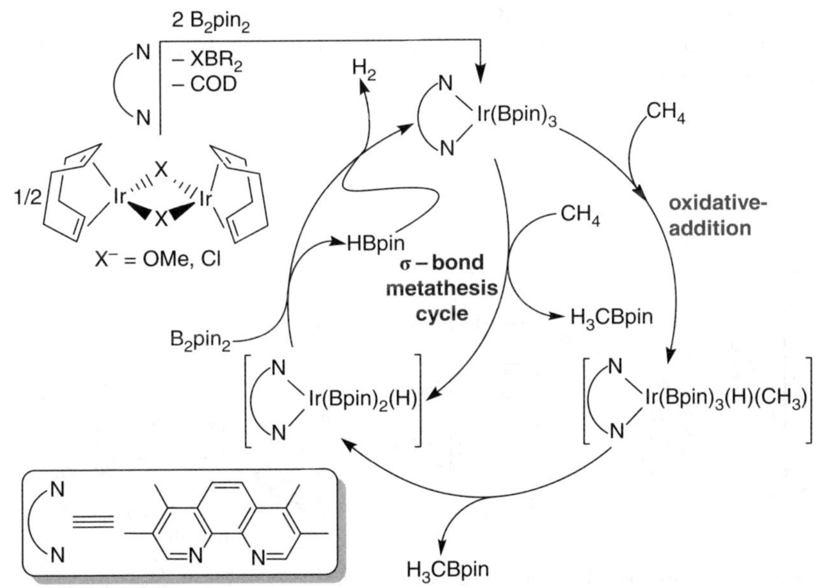

Scheme 17.13 Proposed catalytic reactions involving the borylation of methane by Ir and B$_2$pin$_2$. For clarity purposes, a separate cycle showing the borylation of methane with HBpin as well as the diborylation of methane to form **2** are not shown.

phen, as ligand). Once the resting state of the catalyst, (tmphen)Ir$(Bpin)_3$ is formed from COD displacement, X^- metathesis and oxidative addition of B_2pin_2, C—H activation could take place by either an oxidative addition via (tmphen)Ir$(Bpin)_3(H)(CH_3)$ or σ-bond metathesis to release **1** and form (tmphen)Ir$(Bpin)_2$(H). Scheme 17.13 shows how (tmphen)Ir$(Bpin)_2$(H) may then react with another equiv of B_2pin_2 to extrude HBpin and reform (tmphen)Ir$(Bpin)_3$. Scheme 17.13 also illustrates how HBpin could re-enter the cycle by reacting with (tmphen)Ir$(Bpin)_2$(H) to release H_2 akin to the two cycles shown in Scheme 17.11 (*vide supra*) [106].

At the time we were exploring this chemistry, the Sanford group also studied the catalytic borylation of methane, independently from our own work [108]. In their studies, three catalysts were found to activate methane. Although they obtained the monoborylated methane in reasonable yields using a Rh(I) pre-catalyst, superior selectivity for **1** versus **2** was achieved with the $[Cp^*RuCl(\mu_2\text{-}Cl)]_2$ (31 : 1 selectivity for **1** : **2**), a system which unfortunately did not display high TON (17). Their studies also included ethane borylation, and it was found that the Rh system displayed the highest activity and higher yield. Consistent with our findings, the Ir catalyst (tmphen)Ir$(Bpin)_3$, presumably formed from addition of tmphen ligand to (Mes)Ir$(Bpin)_3$, is not the most active borylation catalyst, with a TON of only 15 and is therefore comparable to the Ru catalyst in performance. Figure 17.1 presents kinetic studies by Sanford and co-workers where the reaction time profiles of three different catalysts were collected: Ir ((tmphen)Ir$(Bpin)_3$), Rh ($Cp^*Rh(\eta^4\text{-}C_6Me_6)$), and Ru ($[Cp^*RuCl(\mu_2\text{-}Cl)]_2$). During the course of the reaction the yields of **1** and **2** were obtained and it was found that the Ir catalyst showed the least selectivity for methane borylation to form **1** versus borylation of **1** to form **2**. This feature is reflected by the small $k\mathrm{CH_4}/k\mathbf{1}$ value (~0.05) for formation of **1** versus **2**, which translates to a small but positive $\Delta\Delta G^{\ddagger}$ (~2.5 kcal mol^{-1}) and which reflects faster borylation for **1** than for CH_4. In the case of Rh, there is essentially little selectivity for CH_4 versus **1** borylation ($k\mathrm{CH_4}/k\mathbf{1} \sim 0.5$, $\Delta\Delta G^{\ddagger}$ ~0.5 kcal mol^{-1}), whereas for Ru the selectivity is inverted showing greater selectivity for borylation of CH_4 over **1**, consistent with the observed greater ratios of **1** versus **2** ($k\mathrm{CH_4}/k\mathbf{1} \sim 1.7$, $\Delta\Delta G^{\ddagger} \sim -0.5$ kcal mol^{-1}). We therefore carried out studies to find a new catalyst to increase the yield and selectivity of the monoborylated methane, **1**, simultaneously.

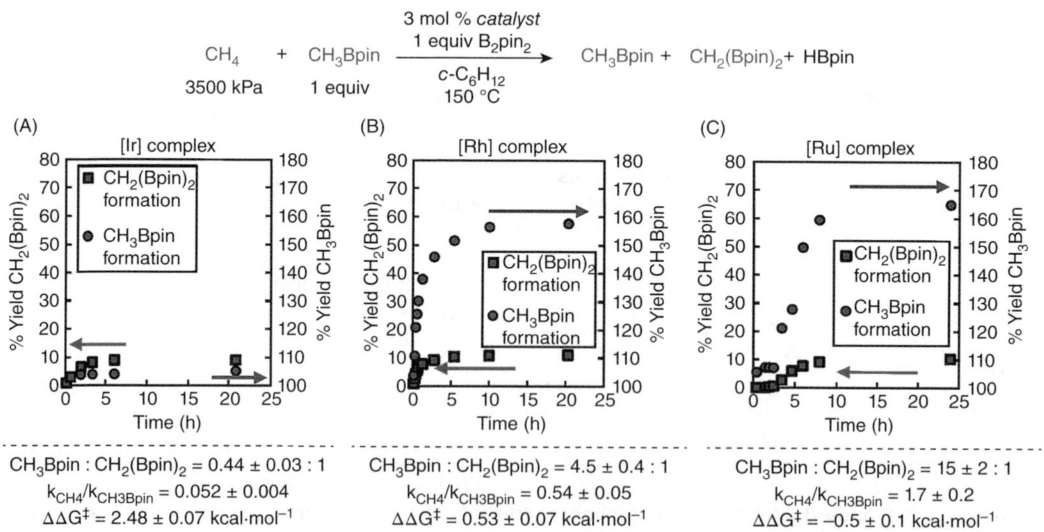

Figure 17.1 Reaction time profiles for three catalyst Ir (A), Rh (B), and Ru (C). The reaction time profiles show the formation of monoborylated **1** (red squares) and diborylated **2** (blue circles), from which $k\mathrm{CH_4}/k\mathbf{1}$ and $\Delta\Delta G^{\ddagger}$ values can be inferred [108]. (*See insert for color representation.*)

17.6.2 Computational Studies

The preliminary experimental studies provided an ideal foundation for the exploration of the mechanism using a computational molecular model. Figure 17.2 [109] displays the calculated energy profile of the borylation of methane mediated by the catalyst (phen)Ir(Bpin)$_3$, **1-phen**. The catalytic reaction consists of two parts. The first phase of the reaction is a formation of H_3CBpin by activating the C—H bond of methane. In the second phase, the catalyst is regenerated by producing hydroborane.

The catalytic cycle starts with methane approaching the Ir-center of the catalyst **1-phen** to afford intermediate **2-phen**. To activate the C—H bond of methane, **2-phen** may proceed via two possible pathways: (i) σ-bond metathesis, (ii) oxidative addition where the Lewis acidic boron may or may not provide assistance [110]. To confirm which pathway is most appropriate, we carried out the calculations specifically targeting these reaction scenarios, as summarized in Figure 17.3. The σ-bond metathesis pathway could be rejected quickly, as we were unable to locate any reasonable transition state despite significant efforts and we were able to estimate the barrier to be at least in the range of 50 kcal mol^{-1} or higher from manual scans of the potential energy surface. Our calculations indicated that oxidative addition is a much more likely pathway. The oxidative addition transition state **2-phen-TS** is lowest in energy with 27.2 kcal mol^{-1} compared to the alternatives **2-phen-TS′** and **2-phen-TS″**, where we considered lack of boron-assistance and inverted arrangement of the C—H bond toward the Ir—B fragment, respectively, as illustrated in Figure 17.3. Thus, our calculations indicate that the C—H bond of methane is activated via the B-assisted oxidative addition pathway. Because the formation of the methane-adduct is slightly uphill, the predicted barrier is 35.4 kcal mol^{-1}, as illustrated in Figure 17.2. After traversing the key transition state, a seven-coordinated intermediate **3-phen** is generated, which is formally an Ir(V)-methylhydride. Next, **3-phen** may produce HBpin via reductive elimination, but the *trans*-influence of Bpin ligand makes the **3-phen** intermediate disfavored and **3-phen** undergoes a structural rearrangement to afford **4-phen** by exchanging the position of the hydride with one of the Bpin ligands. Intermediate **4-phen** is located at a solution phase free energy of 18.1 kcal mol^{-1}. The methyl borane is

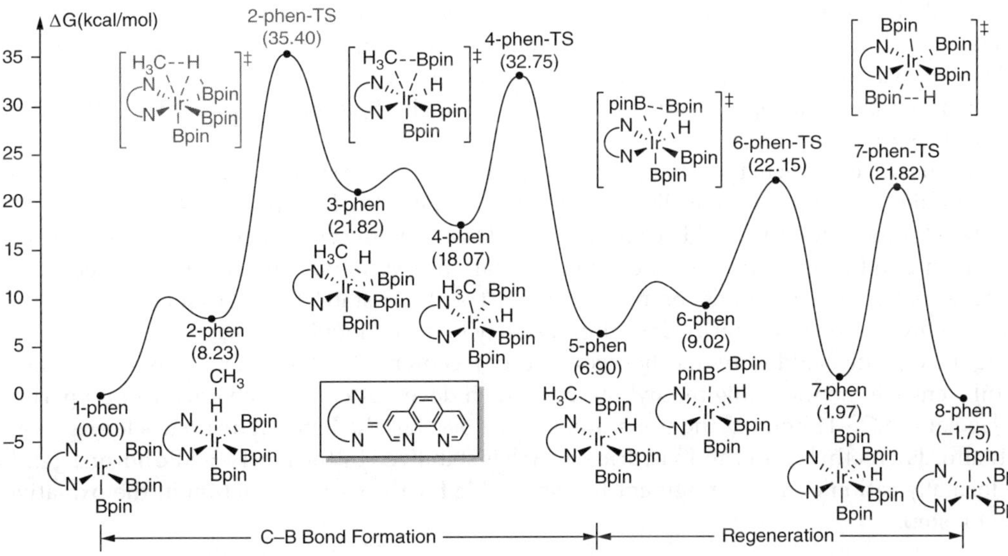

Figure 17.2 The calculated energy profile of 1,10-phenanthroline system.

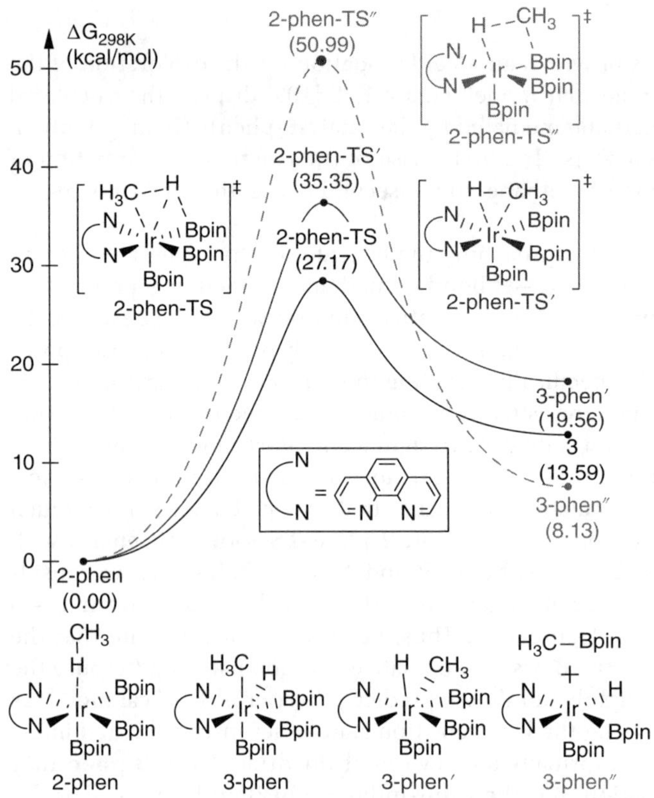

Figure 17.3 The calculated energy profile at the C—H bond cleavage step. *(See insert for color representation.)*

obtained via reductive elimination traversing **4-phen-TS** with a step barrier of 14.7 kcal mol^{-1}, which completes the first half of the catalytic cycle.

The catalytic cycle continues by consuming an equivalent of diborane via oxidative addition associated with **6-phen-TS** to produce intermediate **7-phen**, in which the hydride from the methane activation is still bound to the iridium center. Next, the active form of the catalyst is regenerated by releasing hydroborane via reductive elimination at **7-phen-TS** with a step barrier of 19.9 kcal mol^{-1}.

In conclusion, our calculations suggest that the Ir-catalyst carrying the 1,10-phenanthroline ligand should be able to catalytically borylate methane. But the high barrier of 35.4 kcal mol^{-1} indicates that the reaction should be inefficient and low-yielding even at elevated temperatures, which is in good agreement with the experimental observations. A more detailed electronic structure analysis and energy decomposition of the transition state allowed to speculate that the iridium center in the key intermediate **1-phen** may be too difficult to polarize the C–H bond during the oxidative addition. In other words, the Ir-center is too hard of a Lewis-base to successfully engage the incoming methyl moiety, which develops cationic character upon partial cleavage of the C—H bond at the transition state. Note that the heterolytic bond-cleavage formally affords a methyl cation and a hydride, which is stabilized by the Lewis-acidic boron ligand. The formally cationic methyl fragment is responsible for the oxidative action in the oxidative addition step.

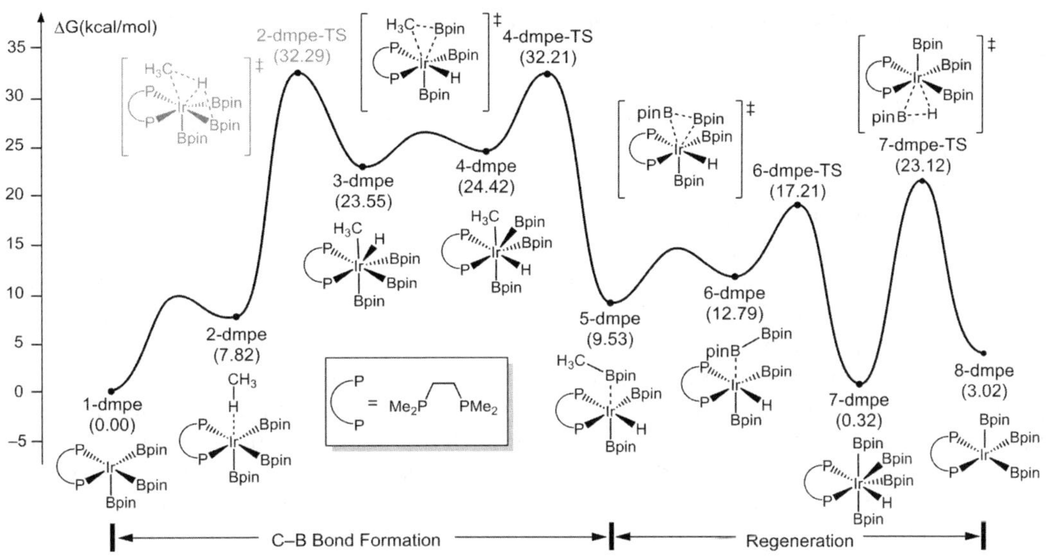

Figure 17.4 The calculated energy profile for a dmpe Ir-based system.

This conceptual conclusion from the computed energies and electron-density flows during the C—H activation suggests an obvious optimization strategy: To make the oxidative addition step more efficient, the dπ lone-pair orbital of the Ir-center must be made more polarizable, i.e. the metal center must become a softer Lewis-base. The relatively hard Lewis basic character of the Ir-center is of course a direct consequence of the relatively hard N-donor ligand. Therefore, a viable hypothesis is to replace the phenanthroline ligand by a more polarizable donor-ligand, such as phosphines. To test this conceptual hypothesis, we considered 1,2-bis(dimethylphosphino)ethane (dmpe), a chelating ligand providing a much softer Lewis basic phosphorus instead of pyridylic nitrogens, to afford the hypothetical catalyst **1-dmpe**.

The calculated energy profile of the methane activation catalyzed by (dmpe)Ir$(Bpin)_3$, **1-dmpe** is shown in Figure 17.4 [109]. Overall, the catalytic cycle is quite similar to that found for the catalyst **1-phen** and commences with methane approaching the catalyst **1-dmpe** to give the adduct **2-dmpe**. The oxidative addition is accomplished through transition state **2-dmpe-TS**. Interestingly, this transition state is associated with a barrier of 32.3 kcal mol^{-1}, which is notably lower by 3.1 kcal mol^{-1} than what was found with the phenanthroline ligand. The rest of the catalytic cycle follows the identical trajectory as found for the N-donor ligand with slightly different energies that were unremarkable. In summary, these calculations clearly suggested that replacing the N-donor ligands with P-donor ligands is a viable strategy for obtaining a more effective borylation catalyst. If the computed energy difference of 3 kcal mol^{-1} is taken at face value, the phosphine-carrying Ir-catalyst should display a catalytic performance that is four to five orders of magnitude faster and more efficient. Although reaction barriers computed with density functional calculations should be evaluated with some care, this decisive and chemically sensible difference was encouraging, and we decided to prepare and study the phosphine analogues of the initially used iridium catalyst.

17.7 Improving Catalyst Performance Using Phosphine Based Ligands

Figure 17.5 summarizes the proposed mechanism of the catalytic borylation of methane that was discussed above based on our DFT calculations. The cycle is based on an Ir(III)/Ir(V) redox couple that undergoes oxidative addition and reductive elimination reactions, as discussed above. At the initial stage of the catalytic cycle, the methane activation and the methylboryl formation occurs via oxidative addition and reductive elimination, respectively. Next, the catalyst regenerates by interacting with an equivalent of diboron reagent and forming the hydroborane. The catalyst carrying the 1,10-phenanthroline ligand was well-known for the C—H bonds activation of other hydrocarbons.

To test the theoretical predictions that phosphine-based ligand may improve catalytic performance, high-pressure and high-throughput screening techniques were utilized with numerous monodentate and bidentate phosphines. Gratifyingly, and after extensive screening through various conditions, including temperature, time, catalyst concentrations, borylating agent, pre-catalyst, and solvent, it was found that using dmpe at 150 °C and at various concentrations of pre-catalyst $[Ir(COD)(\mu\text{-}Cl)]_2$ over 2–16 hours greatly affected the catalytic

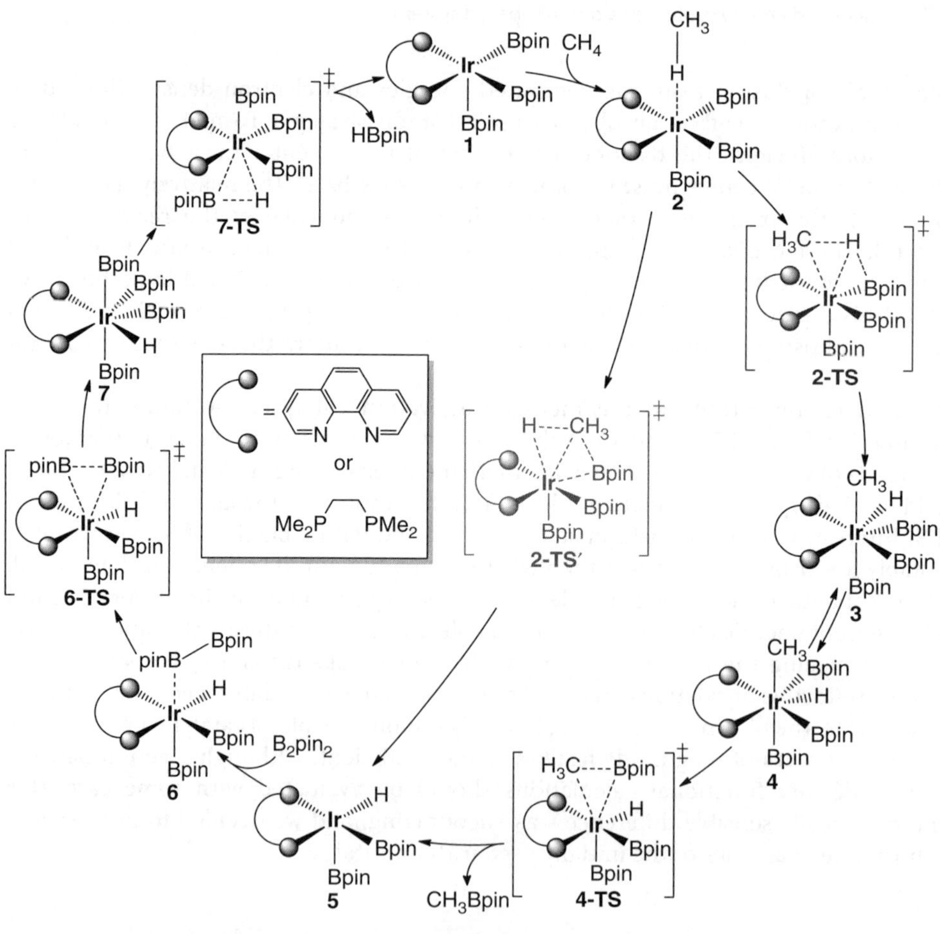

Figure 17.5 The proposed mechanism. (*See insert for color representation.*)

Table 17.2 Variations of catalyst loading and time using the ligand dmpe for the borylation of methane. The ligand dmpe ($Me_2PCH_2CH_2PMe_2$) was used in a 2 : 1 ratio relative to the Ir precatalyst $[(COD)Ir(\mu\text{-}Cl)]_2$ cyclohexane (CyH) under 500 psi of CH_4 [106].

$$\underset{500\ \text{psi}}{CH_4} + B_2pin_2 \xrightarrow[\text{CyH, X h}\ \ 150\,^\circ\text{C}]{[(COD)Ir(\mu\text{-}Cl)]_2\ \ \text{dmpe}} \underset{\mathbf{1}}{H_3C\text{-}Bpin} + \underset{\mathbf{2}}{H_2C(Bpin)_2} + HBpin + O(Bpin)_2$$

Entry	Loading/ mol%	Time h^{-1}	% yield 1	1 : 2	TON
1	25	16	9.4	9 : 1	<1
2	10	16	16	5 : 1	~1
3	5.0	16	23	5 : 1	4
4	1.0	16	25	5 : 1	25
5	0.5	16	52	3 : 1	104
6	25	6	17	7 : 1	<1
7	10	6	25	7 : 1	~2
8	5.0	6	27	6 : 1	5
9	1.0	6	21	8 : 1	21
10	0.5	6	28	6 : 1	56
11	10	2	22	3 : 1	22
12	5.0	2	27	4 : 1	5
13	1.0	2	24	4 : 1	24
14	0.5	2	16	2 : 1	32

reaction. Table 17.2 shows how catalyst concentration and time affected the yield of **1**, the TON, as well as the distribution ratio of products **1** : **2**. Varying catalyst concentrations from 0.5 to 25 mole percent (mol%) led to an optimum conversion of 52% yield corresponding to a TON to 104 with selectivity of 3 : 1 for products **1** versus **2** at a catalyst loading of 0.5 mol% (entry 5). Although an increase in the mol% of catalyst lowered the overall conversion, the selectivity in forming **1** over **2** increased to as high as 9 : 1 (entry 1) at 25 mol% catalyst loading. While lowering the pressure to 200 psi afforded lower conversions, increasing the pressure above 500 psi did not improve the overall yield of products **1** or **2**.

17.8 Conclusions and Outlook

The catalytic activation of methane remains a challenge that continues to inspire much work. Recently, some significant advances were made in designing catalytic systems that can activate methane under milder conditions with encouraging selectivities. It is clear that the great advances that organometallic and organic chemists have made in the recent decade in understanding and designing C—H activation processes are helpful for attempts at targeting the most difficult of the hydrocarbons. It is not a coincidence that the recent advances in the catalytic borylation of methane built on the successes of utilizing late metal catalysts that have been shown to activate alkylic hydrocarbons that are generally easier to activate. An interesting observation is the interactive and predictive use of computer models of the catalytic reaction. Whereas quantum chemical models have in the past been mainly used to rationalize experimental observations, the sophistication of the computational reaction modeling has now reached a level where computer-aided design of catalytic processes is possible. In this case, the computer models were utilized to derive a clear and intuitively understandable concept on principles of hard and soft acid/bases, which predicted that the phosphine ligands should render the iridium center a much more capable catalyst for the oxidative C–H bond addition step, which was identified as the most difficult step of the catalysis. Through the integration of theory and experiment, a very precise catalytic mechanism has been proposed, which should allow further optimizations. Indeed, recent studies have demonstrated that higher TON numbers and selectivities for the monoborylation of methane can be achieved through a combination of theory and experiment.

Acknowledgments

Financial support of this research was provided by the University of Pennsylvania (to D.J.M.). M.H.B. acknowledges support from the Institute for Basic Science (IBS-R10-D1) in Korea. We thank the Chemical Sciences, Geosciences, and BiosciencesDivision, Office of BES, Office of Science, U.S. DOE(DEFG02-07ER15893) for the financial support of this research (grant awarded to D.J.M. for the synthetic aspects of this project).

References

1 Gold, R. (2011). *Firms Plan to Export Gas*. New York: The Wall Street Journal.
2 Department of Energy, Office of Energy & Renewable Energy Hydrogen Production: Natural Gas Reforming https://energy.gov/eere/fuelcells/hydrogen-production-natural-gas-reforming (30 October 2017), 2017.
3 Hydrogen Production – Steam Methane Reforming (SMR), Hydrogen Fact Sheet, archived from the original on February 4, 2006, retrieved October 20, 2014.
4 Johnson, J. (2014). Methane's role in climate change. *Chem. Eng. News* 92: 10.
5 Patel, P. (2017). *Chem. Eng. News* 95: 19.
6 Ratnasamy, C. and Wagner, J.P. (2009). *Catal. Rev. Sci. Eng.* 51: 325.
7 Murkin, C. and Brightling, J. (2016). *Johnson Matthey Technol. Rev.* 60: 263.
8 Arpe, H.-J. (2009). *Industrial Organic Chemistry*, 5e. Weinheim: Wiley-VCH.
9 Schulz, H. (1999). *Appl. Catal., A* 186: 3.
10 Olah, G.A., Goeppert, A., Czaun, M., and Surya Prakash, G.K. (2013). *J. Am. Chem. Soc.* 135: 648.
11 Caballero, A. and Perez, P.J. (2013). *Chem. Soc. Rev.* 42: 8809.
12 U.S. Department of Energy (2008). *The Impact of Increased Use of Hydrogen on Petroleum Consumption and Carbon Dioxide Emissions*. Washington, DC: Office of Integrated Analysis and Forecasting Office of Coal, Nuclear, Electric and Alternate Fuels.
13 Stoltzenberg, D. (2004). *Fritz Haber: Chemist, Nobel Laureate, German, Jew*. Philadelphia, PA, USA: Chemical Heritage Press.
14 Ritter, S.K. (2008). *Chem. Eng. News* 86: https://cen.acs.org/articles/86/i33/Haber-Bosch-Reaction-Early-Chemical.html (20 August 2008).
15 Eyde, S. (1909). *J. R. Soc. Arts* 57: 568.
16 Caro, N. (1906). *Z. Angew. Chem.* 37: 1569.
17 Erisman, J.W., Sutton, M.A., Galloway, J. et al. (2008). *Nat. Geosci.* 1: 636.
18 Blanksby, S.J. and Ellison, G.B. (2003). *Acc. Chem. Res.* 36: 255.

19 Chupka, W.A. and Berkowitz, J. (1971). *J. Chem. Phys.* 54: 4256.
20 Berkowitz, J., Greene, J.P., Cho, H., and Ruscic, B. (1987). *J. Chem. Phys.* 86: 674.
21 Gunsalus, N.J., Koppaka, A., Park, S.H. et al. (2017). *Chem. Rev.* 117: 8521.
22 Gol'dshleger, N.F., Es'kova, V.V., Shilov, A.E., and Shteinmann, A.A. (1972). *Zh. Fiz. Khim.* 46: 1353.
23 Gol'dshleger, N.F., Tyabin, M.B., Shilov, A.E., and Shteinmann, A.A. (1969). *Zh. Fiz. Khim.* 43: 2174.
24 Fekl, U., Kaminsky, W., and Goldberg, K.I. (2001). *J. Am. Chem. Soc.* 123: 6423.
25 Vedernikov, A.N., Binfield, S.A., Zavalij, P.Y., and Khusnutdinova, J.R. (2006). *J. Am. Chem. Soc.* 128: 82.
26 Zimmermann, T., Soorholtz, M., Bilke, M., and Schüth, F. (2016). *J. Am. Chem. Soc.* 138: 12395.
27 Periana, R.A., Taube, D.J., Evitt, E.R. et al. (1993). *Science* 259: 340.
28 Dimroth, O. (1898). *Ber. Dtsch. Chem. Ges.* 31: 2154.
29 J. C. Snyder, A. V. Grosse, Reaction of methane with sulphur trioxide. U.S. Patent 2,493,038, Jan 3, 1950.
30 Lawton, T.J. and Rosenzweig, A.C. (2016). *J. Am. Chem. Soc.* 138: 9327.
31 Periana, R.A., Taube, D.J., Gamble, S. et al. (1998). *Science* 280: 560.
32 Paul, A. and Musgrave, C.B. (2007). *Organometallics* 26: 793.
33 Ahlquist, M., Periana, R.A., and Goddard, W.A. III, (2009). *Chem. Commun.* 17: 2373.
34 Mironov, O.A., Bischof, S.M., Konnick, M.M. et al. (2013). *J. Am. Chem. Soc.* 135: 14644.
35 Basickes, N., Hogan, T.E., and Sen, A. (1996). *J. Am. Chem. Soc.* 118: 13111.
36 Periana, R.A., Mironov, O., Taube, D. et al. (2013). *Science* 301: 814.
37 Olah, G.A., Klopman, G., and Schlosberg, R.H. (1969). *J. Am. Chem. Soc.* 91: 3261.
38 Periana, R.A., Mironov, O., Taube, D.J., and Gamble, S. (2002). *Chem. Commun.* 20: 2376.
39 Watson, P.L. (1983). *J. Am. Chem. Soc.* 105: 6491.
40 Sadow, A.D. and Tilley, T.D. (2003). *J. Am. Chem. Soc.* 125: 7971.
41 Fontaine, F.-G. and Tilley, T.D. (2005). *Organometallics* 24: 4340.
42 Sadow, A.D. and Tilley, T.D. (2003). *Angew. Chem. Int. Ed.* 42: 803.
43 Sadow, A.D. and Tilley, T.D. (2005). *J. Am. Chem. Soc.* 127: 643.
44 Fendrick, C.M. and Marks, T.J. (1986). *J. Am. Chem. Soc.* 3: 425.
45 Li, C., Dinoi, C., Coppel, Y., and Etienne, M. (2015). *J. Am. Chem. Soc.* 137: 12450.
46 Büchner, E. and Curtius, T. (1885). *Ber.* 18: 2371.
47 Davies, H.M.L. and Manning, J.R. (2008). *Nature* 451: 417.
48 Demonceau, A., Noels, N.F., Hubert, A.J., and Teyssié, P. (1981). *J. Chem. Soc., Chem. Commun.* 14: 688.
49 Caballero, A., Despagnet-Ayoub, E., Díaz-Requejo, M.M. et al. (2011). *Science* 332: 835.
50 Lovely, C.J., Browning, R.G., Badarinarayana, V., and Dias, H.V.R. (2005). *Tetrahedron Lett.* 46: 2453.
51 Rivilla, I., Gómez-Emeterio, B.P., Fructos, M.R. et al. (2011). *Organometallics* 30: 2855.
52 Fluores, J.A., Komine, N., Pal, K. et al. (2012). *ACS Catal.* 2: 2066.
53 Komine, N., Flores, J.A., Pal, K. et al. (2013). *Organometallics* 32: 3185.
54 Flores, J.A., Pal, K., Carroll, M.E. et al. (2014). *Organometallics* 33: 1544.
55 Doyle, M.P., Duffy, R., Ratnikov, M., and Zhou, L. (2010). *Chem. Rev.* 110: 704.
56 Gava, R., Olmos, A., Noverges, B. et al. (2015). *ACS Catal.* 5: 3726.
57 Soulivong, D., Norsic, S., Taoufik, M. et al. (2008). *J. Am. Chem. Soc.* 130: 5044.
58 Soulivong, D., Copéret, C., Thivolle-Cazat, J. et al. (2004). *Angew. Chem. Int. Ed.* 43: 5366.
59 Basset, J.-M., Copéret, C., Soulivong, D. et al. (2010). *Acc. Chem. Res.* 43: 323.
60 Amghizar, I., Vandewalle, L.A., Van Geem, K.M., and Marin, G.B. (2017). *Engineering* 3: 171.

61 Schrock, R.R. (1974). *J. Am. Chem. Soc.* 96: 6796.
62 Parkin, G., Bunel, E., Burger, B.J. et al. (1987). *J. Mol. Catal.* 41: 21.
63 Antonelli, D.M., Schaefer, W.P., Parkin, G., and Bercaw, J.E. (1993). *J. Organomet. Chem.* 462: 213.
64 Flores, J.A., Cavaliere, V.N., Buck, D. et al. (2011). *Chem. Sci.* 2: 1457.
65 Chirik, P.J., Zubris, D.L., Ackerman, L.J. et al. (2003). *Organometallics* 22: 172.
66 Klei, S.R., Golden, J.T., Burger, P., and Bergman, R.G. (2002). *J. Mol. Catal. A: Chem.* 189: 79.
67 Kletzin, H., Werner, H., Serhadli, O., and Ziegler, M.L. (1983). *Angew. Chem. Int. Ed.* 22: 46.
68 Hayes, J.C. and Cooper, N.J. (1982). *J. Am. Chem. Soc.* 104: 5570.
69 Chiu, K.W., Jones, R.A., Wilkinson, G. et al. (1981). *J. Chem. Soc., Dalton Trans.* 5: 1204.
70 Hoyano, J.K., McMaster, A.D., and Graham, W.A.G. (1983). *J. Am. Chem. Soc.* 105: 7190.
71 Wax, M.J., Stryker, J.M., Buchanan, J.M. et al. (1984). *J. Am. Chem. Soc.* 106: 1121.
72 Bernskoetter, W.H., Schauer, C.K., Goldberg, K.I., and Brookhart, M. (2009). *Science* 326: 553.
73 Arndtsen, B.A. and Bergman, R.G. (1995). *Science* 270: 1970.
74 Webster, C.E. and Hall, M.B. (2003). *Coord. Chem. Rev.* 238–239: 315.
75 Hinderling, C., Feichtinger, D., Plattner, D.A., and Chen, P. (1997). *J. Am. Chem. Soc.* 119: 10793.
76 Luecke, H.F. and Bergman, R.G. (1997). *J. Am. Chem. Soc.* 119: 11538.
77 Klei, S.R., Tilley, T.D., and Bergman, R.G. (2000). *J. Am. Chem. Soc.* 122: 1816.
78 Young, K.J.H., Oxgaard, J., Ess, D.H. et al. (2009). *Chem. Commun.* 22: 3270.
79 Mkhalid, I.A.I., Barnard, J.H., Marder, T.B. et al. (2010). *Chem. Rev.* 110: 890.
80 Hartwig, J.F. (2011). *Chem. Soc. Rev.* 40: 1992.
81 Molander, G.A. and Ellis, N. (2007). *Acc. Chem. Res.* 40: 275.
82 Boebel, T.A. and Hartwig, J.F. (2008). *J. Am. Chem. Soc.* 130: 7534.
83 Kallepalli, V.A., Shi, F., Paul, S. et al. (2009). *J. Org. Chem.* 74: 9199.
84 Paul, S., Chotana, G.A., Holmes, D. et al. (2006). *J. Am. Chem. Soc.* 128: 15552.
85 Nguyen, P., Blom, H.P., Westcott, S.A. et al. (1993). *J. Am. Chem. Soc.* 115: 9329.
86 Waltz, K.M., He, X., Muhoro, C., and Hartwig, J.F. (1995). *J. Am. Chem. Soc.* 117: 11357.
87 Waltz, K.M. and Hartwig, J.F. (1997). *Science* 277: 211.
88 Waltz, K.M., Muhoro, C., and Hartwig, J.F. (1999). *Organometallics* 18: 3383.
89 Chen, H. and Hartwig, J.F. (1999). *Angew. Chem. Int. Ed.* 38: 3391.
90 Chen, H., Schlecht, S., Semple, T.C., and Hartwig, J.F. (2000). *Science* 287: 1995.
91 Brotherton, R.J., McCloskey, A.L., Boone, J.L., and Manasevit, H.M. (1960). *J. Am. Chem. Soc.* 82: 6245.
92 Chotana, G.A., Vanchura, B.A. II,, Tse, M.K. et al. (2009). *Chem. Commun.* 38: 5731.
93 Murphy, J.M., Lawrence, J.D., Kawamura, K. et al. (2006). *J. Am. Chem. Soc.* 128: 13684.
94 Iverson, C.N. and Smith, M.R. III, (1999). *J. Am. Chem. Soc.* 121: 7696.
95 Raben, P.R. and Hartwig, J.F. (1996). *J. Am. Chem. Soc.* 118: 4648.
96 Hartwig, J.F., Cook, K.S., Hapke, M. et al. (2005). *J. Am. Chem. Soc.* 127: 2538.
97 Sakaki, S. and Kikuno, T. (1997). *Inorg. Chem.* 36: 226.
98 Cho, J.-Y., Iverson, C.N., and Smith, M.R. III, (2000). *J. Am. Chem. Soc.* 122: 12868.
99 Cho, J.-Y., Tse, M.K., Holmes, D. et al. (2002). *Science* 295: 305.
100 Ishiyama, T., Takagi, J., Ishida, K. et al. (2002). *J. Am. Chem. Soc.* 124: 390.
101 Preshlock, S.M., Ghaffari, B., Maligres, P.E. et al. (2013). *J. Am. Chem. Soc.* 135: 7572.
102 Murata, M., Odajima, H., Watanabe, S., and Masuda, Y. (2006). *Bull. Chem. Soc. Jpn.* 79: 1980.

103 Frey, G.D., Rentzsch, C.F., von Preysing, D. et al. (2006). *J. Organomet. Chem.* 691: 5725.
104 Tagata, T. and Nishida, M. (2004). *Adv. Synth. Catal.* 346: 1655.
105 Yinghuai, Z., Yan, K.C., Jizhong, L. et al. (2007). *J. Organomet. Chem.* 692: 4244.
106 Smith, K.T., Berritt, S., González-Moreiras, M. et al. (2016). *Science* 351: 1424.
107 Ghaffari, B., Vanchura, B.A. II,, Chotana, G.A. et al. (2015). *Organometallics* 34: 4732.
108 Cook, A.K., Schimler, S.D., Matzger, A.J., and Sanford, M.S. (2016). *Science* 351: 1421.
109 S. Ahn, D. Sorsche, S. Berritt, et al. (2018) *ACS Catal.* 8: 10021.
110 Vastine, B.A. and Hall, M.B. (2009). *Coord. Chem. Rev.* 253: 1202.

18

Alkane Carbonylation and Carbene Insertion Reactions

Ana M. Faisca Phillips and Armando J. L. Pombeiro

Universidade de Lisboa, Centro de Química Estrutural, Instituto Superior Técnico, Lisboa, Portugal

18.1 Introduction

The addition of carbon monoxide, or a surrogate, to an alkane, is an important carbon–carbon bond-forming reaction. Known as carbonylation, this reaction can be used to convert the saturated hydrocarbons found in cheap, abundant natural gas and petroleum, into value-added products [1–5]. However, despite its importance and the fact that many studies have been devoted to this subject over the years, there are still no practical general methods of synthesis run under mild conditions based on alkane carbonylation reactions. Important developments have been reported recently, but the carbonylation of alkanes is a difficult process to achieve. It shares the problems common to other reactions of alkanes, i.e. the challenge of achieving site selectivity for one C—H bond among others present in the molecule, the difficulty of monofunctionalization when the product obtained is more reactive than the starting material, and the fact that the new functional group may also react further. The carbonylation of alkane fragments in functionalized molecules can even be more complex, since it involves achieving selectivity at an unactivated C—H bond when other weaker or more acidic bonds near the functional groups are present. The major hurdles to overcome in this process are however the inertness of the C—H bond and the π-acidic nature of the C≡O triple bond [6]. The majority of carbonylation reactions are endergonic, i.e. the products are thermodynamically less stable than the starting materials, and when the alkane is placed under a CO atmosphere no reaction takes place, unless some form of C—H bond activation is used. The C—H bond is very strong. For instance, the bond dissociation energies for methane and ethane are above 100 kcal mol^{-1} and not much lower for a secondary site in a plain alkane, e.g. 98 kcal mol^{-1} for propane and butane, 97 kcal mol^{-1} for cyclohexane [7]. For the tertiary C—H bond of 2-methylpropane the dissociation energy is 96 kcal mol^{-1}. In contrast, the dissociation energy for an α-CH bond in tetrahydrofuran (THF) is 92 kcal mol^{-1} and in ethylamine it is 90 kcal mol^{-1}. Due to the inertness of hydrocarbons, the choice of solvent is also important in these reactions, since solvents can compete as substrates [8]. Hydrocarbons also have a very low polarity, and with a pKa value greater than 40, they are not acidic enough to be deprotonated by bases.

Alkane C—H bonds can be activated and form species, which can undergo carbonylation reactions in three distinct ways:

i. By reaction with superacids or superelectrophiles
ii. By reaction with metal complexes
iii. By reaction with free radicals

Alkane Functionalization, First Edition. Edited by Armando J. L. Pombeiro and M. Fátima C. Guedes da Silva.

The carbonylation products can be acids, esters if the reactions are carried out in the presence of alcohols, or amides if amines are present in the medium. Sometimes it is possible to stop the reaction at the aldehyde oxidation level. In this chapter, we review these important alkane functionalization reactions, as well as the related carbene insertion reactions in alkanes, from which additional products can be obtained: diesters, ketoesters, ketoamides, and others (Scheme 18.1). Only unactivated alkanes are included and not reactions involving the CH bonds of aromatic systems, alkenes, alkynes, or even those occurring at benzylic and allylic positions in a carbon chain.

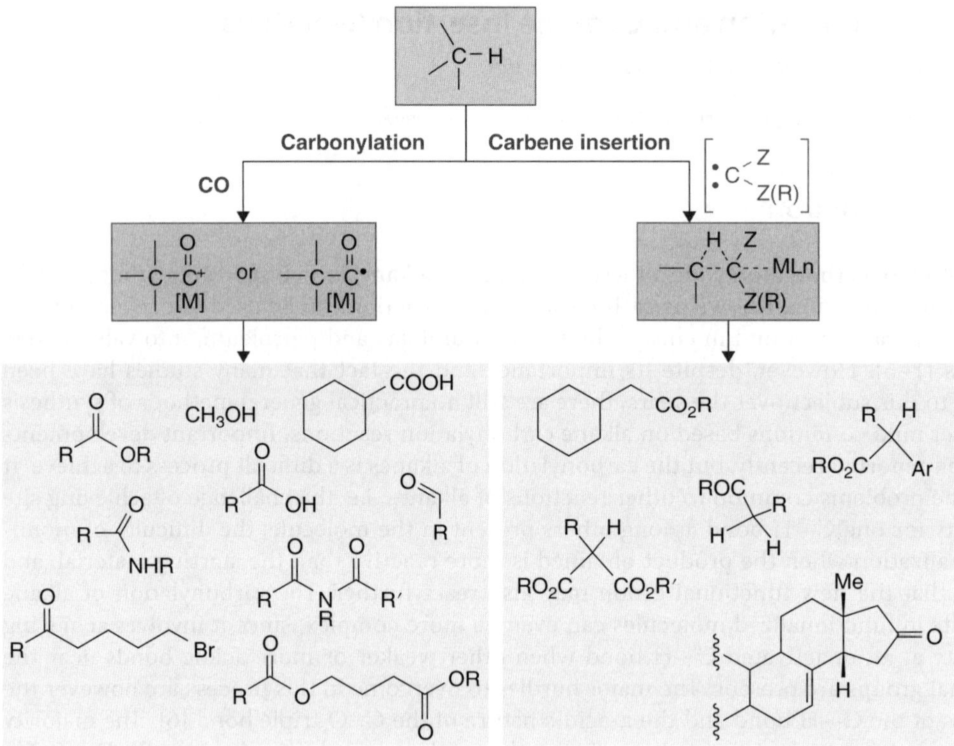

Scheme 18.1 General scheme for alkane functionalization via carbonylation and carbene insertion reactions.

18.2 Carbonylation via CH Activation with Superacids or Superelectrophiles

The reaction of alkanes with carbon monoxide in the presence of superacids has been known since the early 1930s [9]. By 1967, Paatz and Weisberger showed that the reaction of 1-methylcyclopentane (**1**) with CO catalyzed by HF-SbF_5 could be made to proceed regioselectively with respect to the carbonylated products, e.g. 1-methylcyclopentanecarboxylic acid (**2**) and cyclohexanecarboxylic acid (**3**), although one mole equivalent of open chain *iso*-hexane (**4**) was also obtained for every two moles of substrate converted (Scheme 18.2) [9]. The nature of the carbonylated products could be varied by a judicious selection of quenching substance. Hence, with water quenching, cyclohexane carboxylic acid **3** was the major product (Scheme 18.2), but ketones could be obtained if benzene was added at the end, although in this case a more complex mixture of products was produced. It was postulated

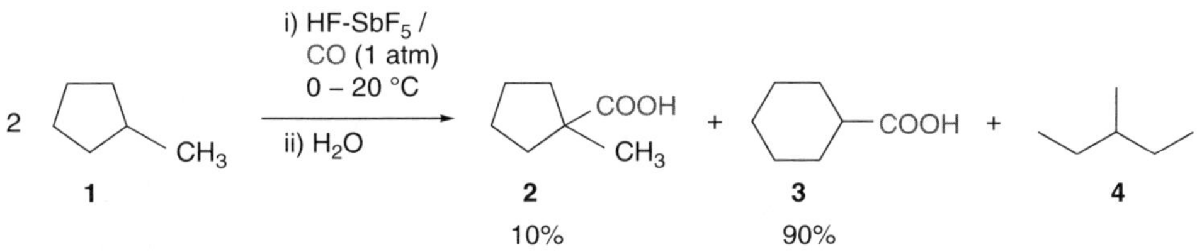

Scheme 18.2 The carbonylation of 1-methylcyclopentane catalyzed by HF-SbF_5 [9].

that the reaction proceeded via formation of the more stable tertiary carbocation **6**, which upon Wagner–Meerwein-type rearrangement yielded a cyclohexane carbonium ion **7**, which was carbonylated and subsequently hydrolyzed to yield carboxylic acid **3** (Scheme 18.3). The formation of the carbocation results from hydride abstraction from the alkane which, as described by Olah and Klumpp, is possible in superacidic media or with superelectrophiles, due to the σ-basicity of alkanes [10, 11]. In superacid solutions, 2-electron-3-center bonded intermediates like **5** can form, i.e. via pentacoordinated carbonium ion transition states, which allow this reaction to take place, leading to the concomitant formation of hydrogen gas.

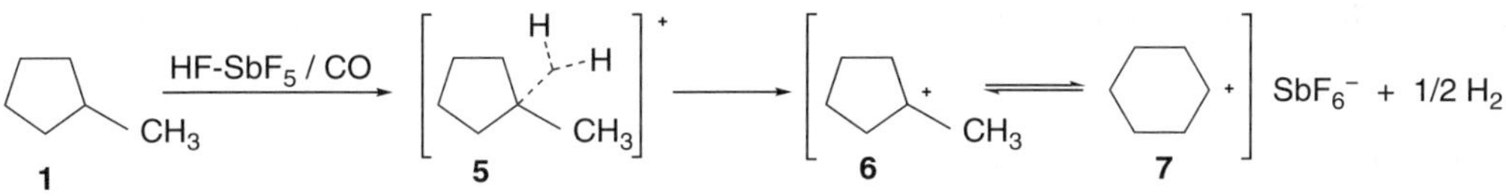

Scheme 18.3 The mechanism for carbonium ion generation during the carbonylation of 1-methylcyclopentane [9].

Olah and co-workers were the first to discover methods to prepare and to study long-lived persistent carbocations, which was possible in superacid solutions, and since then these hypercoordinated reaction intermediates have been observed spectroscopically by many research groups. His success with carbocation studies with superacids would earn G.A. Olah the Nobel prize in chemistry in 1994. The reactions of alkanes with reactive radicals, superacids, or superelectrophiles generally tend to produce branched substrates, since the C—H bond that is cleaved is the weakest one, i.e. the reaction takes place at the most substituted carbon atom [12].

In subsequent years other alkanes were shown to react with CO in superacidic media. In studies with adamantane (**8**), known to give a mixture of products, i.e. adamantane carboxylic acid (**11**), the alcohol **12** and the carboxaldehyde (**14**) in reactions with carbon monoxide in superacid, Olah and co-workers investigated the mechanisms that led to the formation of these species. Adamantane is a convenient system to study because its rigid framework does not allow the formation of olefins, backside nucleophilic, or electrophilic attack at the bridgehead position, and intramolecular hydrogen shifts are also not possible in the carbocation. It was postulated that the carboxylation reaction that yielded the carboxylic acid as the major product involved initial formation of the 1-adamantyl cation **9**. In superacidic media, a highly reactive protosolvated formyl cation $[HCO]^+$ can also be produced, and even the dicationic $[HCOH]^{2+}$, which may insert into the bridgehead tertiary C—H bond via a five-coordinate intermediate carbocation (Scheme 18.4) to produce **13** [13]. This insertion reaction was shown to account for the formation of 1-formyladamantane. The alternative route for the formation of the aldehyde would involve hydride abstraction from excess adamantane and carbocation formation. Deuteration experiments with 1,3,5,7-tetradeuterioadamantane followed by NMR spectroscopy showed that along with the usual 1-adamantanol and

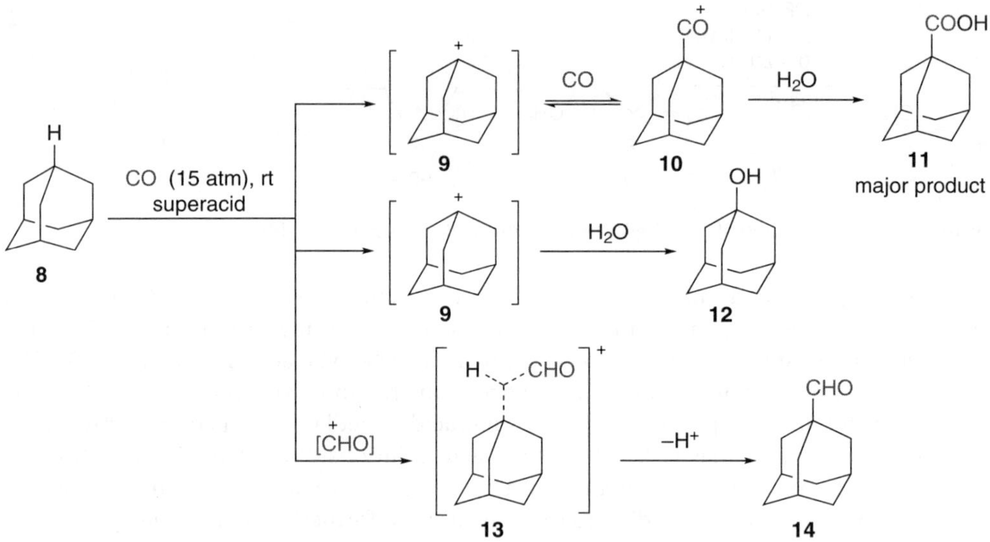

Scheme 18.4 The carbonylation of adamantane in superacid. Only some of the equilibria possible are shown [13].

1-adamantanecarboxylic acid, 3,5,7-trideuterio-1-adamantanecarboxaldehyde-H and 3,5,7-trideuterio-1-adamantanecarboxaldehyde-D were produced in a 94 : 6 ratio, which suggested that the direct σ-insertion of formyl cation was the major pathway leading to the formation of 1-adamantanecarboxaldehyde (**14**). Based on these results, Olah and co-workers reported in 2000 a method to formylate isoalkanes such as isobutane and isopentane using carbon monoxide in which ketones were the only products [14]. Carboxylic acids were either not detected or were produced in minimal amounts.

In order to achieve this selectivity, superacids weaker than HF-SbF_5 or the magic acid FSO_3H-SbF_5 were used, which do not oxidatively ionize isoalkanes readily to their corresponding cations, but are still strong enough to activate carbon monoxide to an electrophilic formylating agent. HF-BF_3 proved to be the best. The reaction of isobutane (**15**) yielded methyl isopropyl ketone (**16**) only (Scheme 18.5), as a result of carbonylation-rearrangement, whereas isopentane and 3-methylpentane yielded mixtures of ketones.

Superelectrophiles are also reactive enough to react with nucleophiles as weak as alkanes. Normally, they are not isolated as persistent stable salts, but they are generated *in situ* in highly acidic systems as needed [10]. They can be doubly electron-deficient species, stable in superacids, which are generated when a cationic electrophile reacts with a Brønsted acid or a Lewis acid, and they have a reactivity by far superior to that of the parent electrophiles. The superelectrophiles often used in reactions of alkanes are polyhalomethane/aluminum halide systems, generated by the reaction of halogenated methanes with an excess of aluminum halide, e.g. CCl_4-$AlCl_3$, CCl_4-nSbF_5, or CBr_4-$AlBr_3$. These aprotic Lewis acids were initially developed by Vol'pin et al. [15]. When these substances are mixed, e.g. tetrabromomethane (**17**) and aluminum tribromide (**18**), a charged electrophile like **19** is formed initially, which is subsequently complexed by another molecule of Lewis acid as in **20** (Scheme 18.6). This complexation decreases neighboring group participation into the electrophilic center, and generates the superelectrophile. These substances are so reactive that they can crack alkanes and cause rearrangements in the carbon skeleton under mild conditions [16]. However, they differ from the superacids in that the reactions are more selective and regioselective.

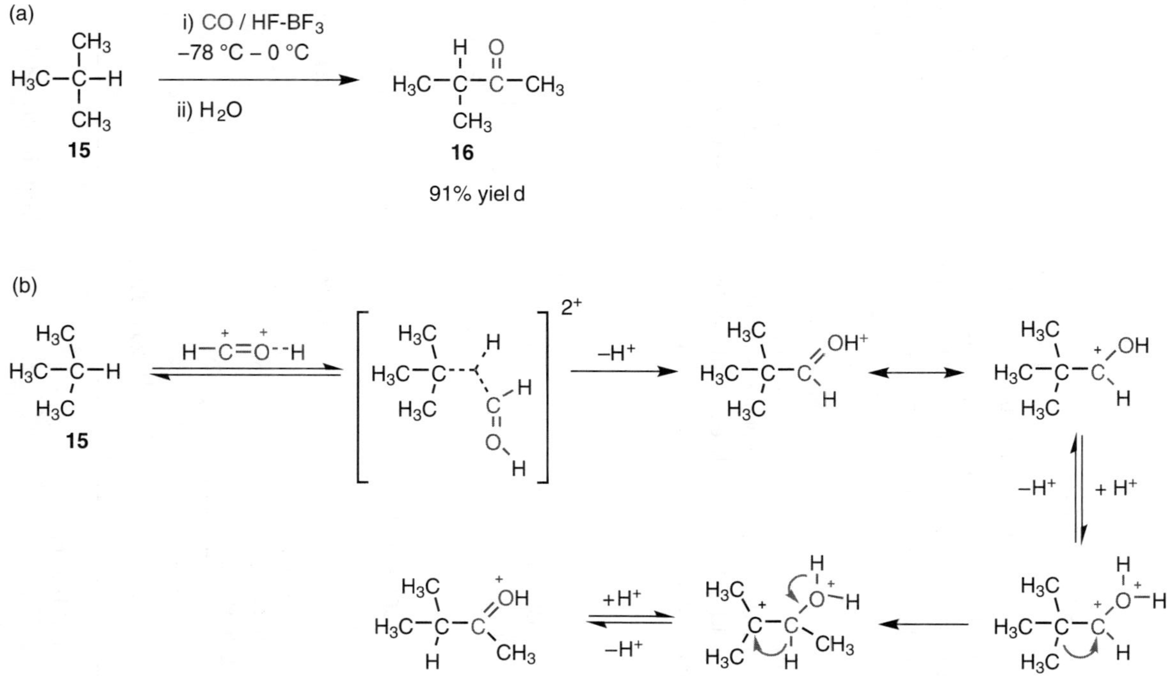

Scheme 18.5 Ketone synthesis by formylation of isoalkanes [14].

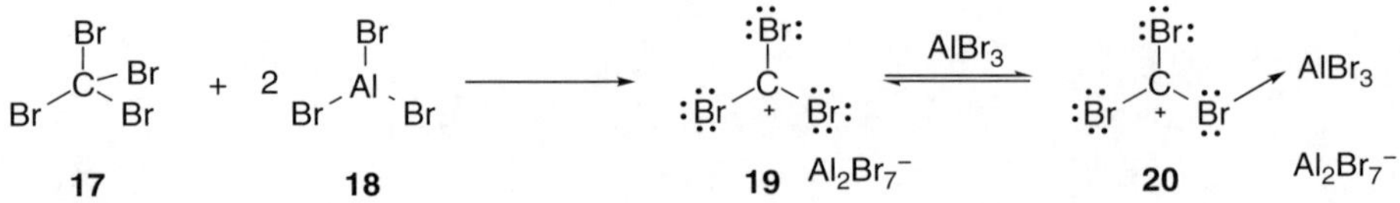

Scheme 18.6 Generation of a superelectrophile from tetrabromomethane and aluminum bromide [15].

The thermodynamically most stable carbocation is produced selectively in the reactions, which reacts with CO to form a stable acyl cation, RCO^+, which can react with nucleophiles.

Polyhalomethane-based superelectrophilic systems were the catalysts used in the first selective carbonylation of *n*-butane (**21**) and *n*-pentane (**22**) reported only in 1999, by Akhrem et al. (Scheme 18.7) [18]. Prior to that, mixtures of products were obtained in related reactions with superacids [19]. The reactions could be performed under mild conditions. The structure of the product obtained was dependent on the nature of the superelectrophile and the quencher used, i.e. with water quench the acids were obtained, whereas with *iso*-propanol as quencher the respective esters were produced. The selectivity observed in the reaction of *n*-butane could be switched according to the nature of the superelectrophile used. Hence catalysis by the more reactive acid $CBr_4{\cdot}2AlBr_3$ yielded preferentially the tertiary ester (**23**), whereas the less reactive acid $CCl_4{\cdot}2AlBr_3$ yielded preferentially the secondary product (**24**). In the presence of a larger excess of Lewis acid ($AlBr_3$), i.e. three equivalents, the percentage of **23** increased, presumably due to an increase in the decarboxylation of the less stable intermediate *tert*-$BuCO^+$ cation.

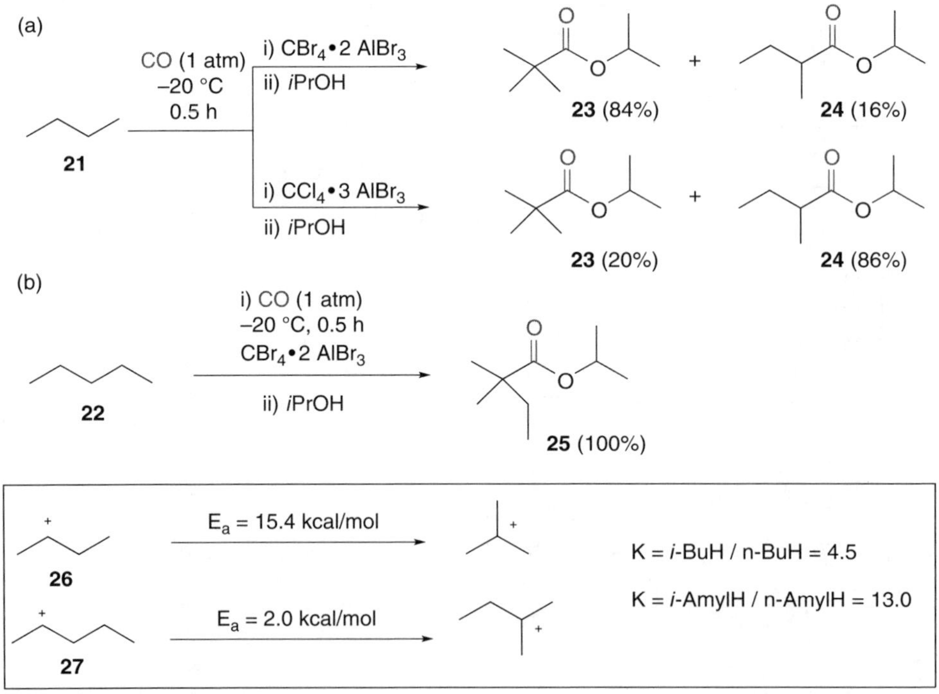

Scheme 18.7 The carbonylation of *n*-butane and *n*-pentane in superacidic media. The percentages are mol% of product based on the superacid (E_a = activation energy at 300 K) [17].

The high selectivity observed in these carbonylation reactions is due to the fast generation of butyl and amyl cations, which then isomerize into stable tertiary carbocations, the species that eventually react with CO. The carbocations are also stable to decarbonylation before quenching. The difference in selectivity between the two reactions, butane carbonylation and pentane carbonylation, is a consequence of the differences in the energies of activation for the isomerization of the carbocations, and it is also due to the difference in the equilibrium constants for the respective isomerizations. According to Hogeveen, not only is the activation energy for the isomerization of the pentyl cation (**27**) much lower than that of the butyl cation (**26**) (Scheme 18.7) [17], but the equilibrium constant K = *i*AmylH/*n*AmylH is also much higher [20]. The equilibrium for the amyl cation isomerization is so fast that upon alcohol quench only one product is observed, i.e. **25**. This equilibrium is not reached in the case of the butyl system.

The carbonylation of higher homologues, the C_6–C_{10} linear alkanes, under similar conditions, i.e. in the presence of $CBr_4{\cdot}2AlBr_3$ at −40 °C and 1 atm CO, was later shown by the same group to provide regioisomeric esters of tertiary carboxylic acids, $R^1C(Me)_2COOR$ and $R^2(Et)C(Me)COOR$, after the workup of reaction mixtures with ROH, instead of products of destructive carbonylation observed with superacids [21].

In the cyclic series, single reaction products could be obtained in a few cases, with or without rearrangement, as shown in Scheme 18.8, but the reactions of cyclooctane were very much dependent on the temperature, reaction time, and the substrate to Lewis acid ratio [22]. A large mixture of products was obtained, unless the conditions shown in Scheme 18.8 were used, as well as strictly the 1.2 : 1 ratio of substrate to Lewis acid. The carbonylation reaction of 1,3-dimethylcyclohexane also yielded a mixture of four products.

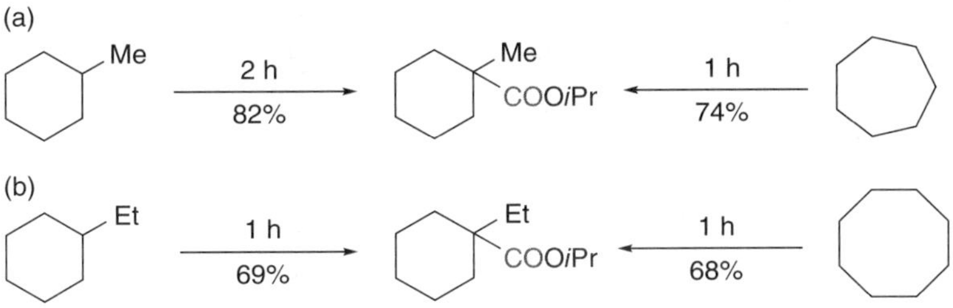

Scheme 18.8 The carbonylation of cyclic alkanes in a superacidic medium. Conditions: CO (1 atm), $CBr_4{\cdot}2AlBr_3$, −40 °C, CH_2Br_2 followed by quench with *i*PrOH [22].

One useful application of this chemistry is for the bifunctionalization of adamantane. The products, difficult to access by other routes, can be obtained selectively with superelectrophilic activation, because the bridging C—H bonds are always cleaved preferentially under these conditions in preference to the secondary C—H bonds. Monofunctionalization is easy to achieve, and COOMe and $CONH_2$ or $CONEt_2$ groups can be introduced, which can also be converted into a wide range of other functionalities [23–25]. Monofunctionalized adamantanes **28** can be subsequently reacted with carbon monoxide in the presence of a superelectrophile (in a 1 : 1.3–1.5 ratio of substrate to electrophile), and *in-situ* treatment of the resulting carbonylated product with a HY nucleophile (HY = alcohols, amines (diethylamine, pyrrolidine, piperidine, morpholine, and aniline), aromatics, and aromatic heterocycles (anisole, pyrrole, thiophene, and furan), and ethylene leads to 1,3-bifunctionalized adamantanes **29** (Scheme 18.9) [26]. These substances are useful substrates for the synthesis of new biologically active compounds, thermally and chemically stable polymers, novel crown ethers, peptide ionophores, and other valuable chemicals [22, 24]. This procedure allows the introduction of different carbonyl-containing functional groups into adamantane, contrary to the

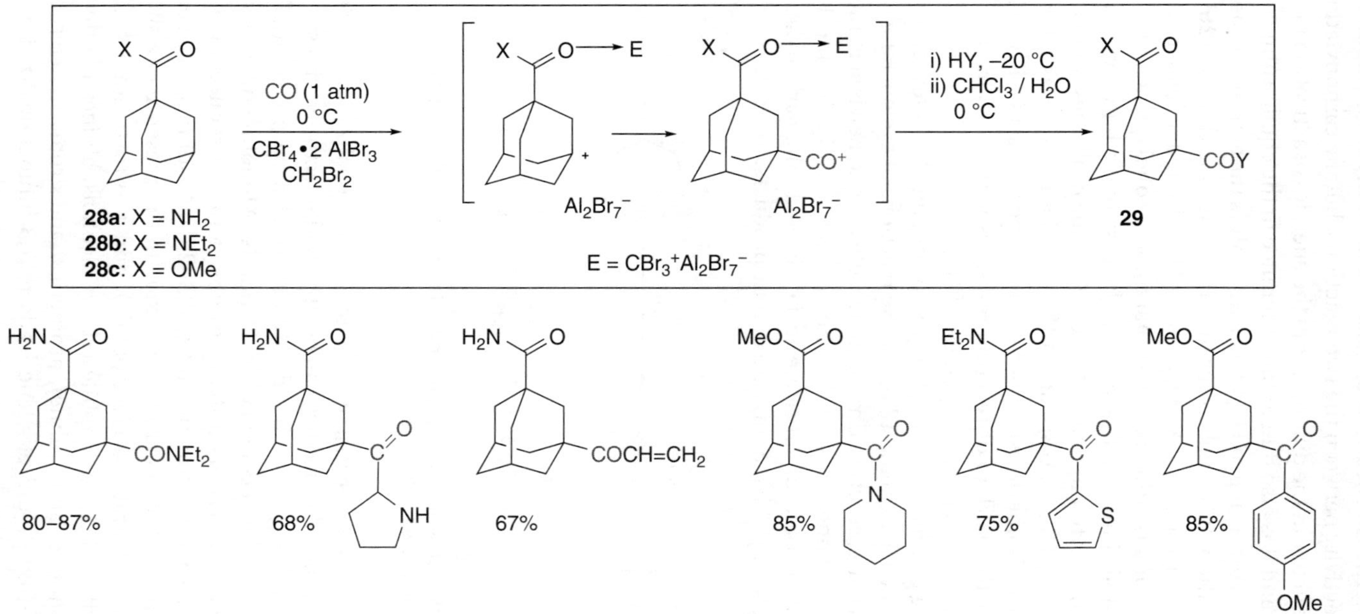

Scheme 18.9 Bifunctionalization of adamantane in superacidic media. The regioselectivity observed is possible because the bridging C—H bonds are cleaved preferentially to secondary C—H bonds under these conditions [26].

carbonylation of 1-bromoadamantane, in which 1,3-difunctionalized adamantanes containing identical substituents are obtained [27].

Alkane carbonylation with superacids or superelectrophiles has reached higher levels of complexity than those mentioned above. This is noteworthy, since although alkanes are abundant raw materials, the difficulties in functionalizing them in a selective manner and the harsh conditions required for these processes in many cases, have hampered the development of useful chemical routes to more complex substrates useful for organic synthesis. In more recent years, the use of superelectrophilic systems to produce acylium cations under mild conditions when alkanes are placed under a CO atmosphere, has led to the development of a few novel useful synthetic methods. Remote functionalization has also become possible [25].

In an earlier work, Akhrem et al. found that halogen substituted alkyl esters of the type $RCOO(CH_2)_4Br$ (**30**) could be produced in moderate to high yields when acylium salts generated from an alkane and CO in the presence of a superelectrophile like $CBr_4{\cdot}2AlBr_3$ in CH_2X_2 solution (X = Br, Cl), were quenched with THF at room temperature (Scheme 18.10) [28]. In some cases, warming the solution up to 50 °C was needed. This was possible since upon reaction with THF, oxonium salts (**31**) are generated, which are more stable to decarboxylation then the corresponding acylium cations. The new species then undergo a ring-opening reaction by the action of halide ion to yield the difunctionalized products. Only one regioisomer was obtained in these reactions, which were performed in most cases with equimolar ratios of alkane to superelectrophile, except for pentane and cyclopentane in which a large excess of alkane (10-fold) was present. Cyclic and bicyclic alkanes could be also functionalized in this way, as well as adamantane. In these reactions, the superelectrophile not only has the role of catalyst but is also a reagent, providing the halide ion necessary for the reaction.

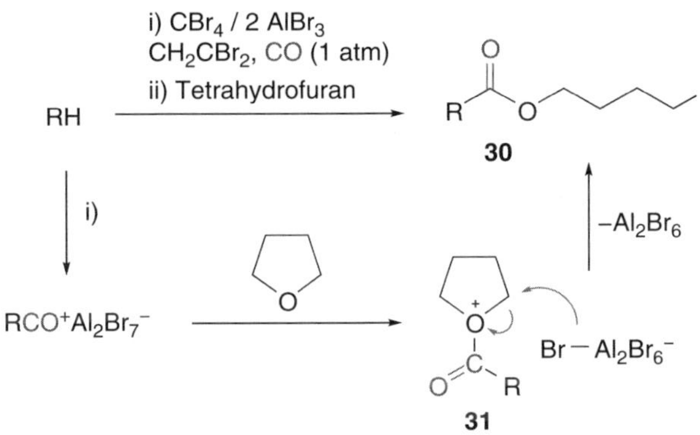

Scheme 18.10 Remote functionalization of alkanes leading to esters [28].

The remote functionalization of alkanes by intermolecular sp^3 C—H bond functionalization was reported by Akhrem and co-workers in 2010 [29], but up to this stage it was very rare [23, 25]. It has since been successful in a number of cases. This was the first example of an "alkane-like" functionalization of linear *n*-alkyl acetates with general formula $C_nH_{2n+1}OCOMe$. The functionalization took place at the end of the alkyl chain opposite to the acetate group with good selectivity and under mild conditions, when the substrates were placed under CO at atmospheric pressure and treated with an excess of the superelectrophilic system $CBr_4{\cdot}2AlBr_3$ and a nucleophile. The acetate group remained intact throughout the procedure, although the carbonyl group is a much stronger nucleophile than a sp^3 C—H bond. Even if CO-electrophile complexation also deactivates the molecule for further reactions at other sites, the fact that the interaction of the superelectrophile with the carbonyl group is reversible, whereas the C—H bond cleavage is irreversible, results in the reaction proceeding preferentially at the less nucleophilic center to yield bifunctional compounds. The reactions were highly regioselective and the products were obtained with 72–85% yield (Scheme 18.11). For the reactions of *n*-hexylacetate (**32**), products of degradative functionalization containing a different number of carbon atoms were not observed. The selectivity was dependent on temperature, and the addition of a CO molecule to the secondary $MeCOO(CH_2)_nCH^+Me$ cation appears to proceed more rapidly than its isomerization into the tertiary cation and the subsequent addition of CO to the sterically hindered cation. At −20 °C the non-isomerized alkyl products were obtained in the functionalization of *n*-hexyl acetate (**32**), but at 0 °C the neo-diester was obtained with 88% selectivity when *iso*-propanol was the nucleophile. Selectivity was also dependent on the nature of the nucleophile, being lower when the strong nucleophile piperidine was used. The site selectivity in these reactions seems to be due to the remoteness from the functional group, since C—H nucleophilicity decreases closer to the functional group as a result of its electron withdrawing properties and the stability of the carbocation generated. In the functionalization of *n*-octyl acetate (**33**) neo-products (**34**) were obtained, which suggests that the isomerization of a long-chained cation generated from an alkyl acetate is probably faster than that of shorter homologues [29]. The same group later extended the scope of the *n*-octyl acetate functionalization reaction to include a broader range of alcohols as nucleophiles (Scheme 18.12) [30].

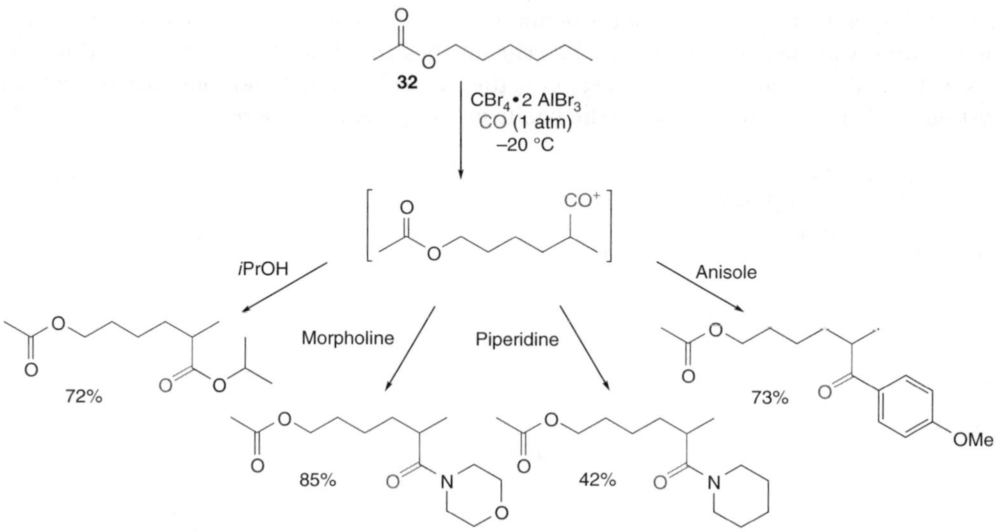

Scheme 18.11 Remote functionalization of long-chain acetates by carbonylation [29].

This procedure for the remote functionalization of alkanes is an interesting addition to organic chemistry since although the α-, β-, γ-, or δ-hydroxy acids and some of their derivatives have been well studied and their syntheses, properties, and applications in chemical transformations and practical use are well known, their long-chained homologues and, in particular, compounds with branched and neo-alkyl chains, have not been available by other methods.

The remote functionalization of ketones was also possible using a related method. The CO bond and the α- and β-CH bonds are the usual reactive centers of these molecules. The other C—H bonds in a ketone molecule are normally inert to functionalization by nucleophiles or electrophiles even in the presence of transition metals [25]. Carbonylation in the presence of superelectrophiles also allowed the remote sp^3 C—H functionalization of methyl *n*-alkyl ketones [31]. Hence, Akhrem and co-workers showed that when

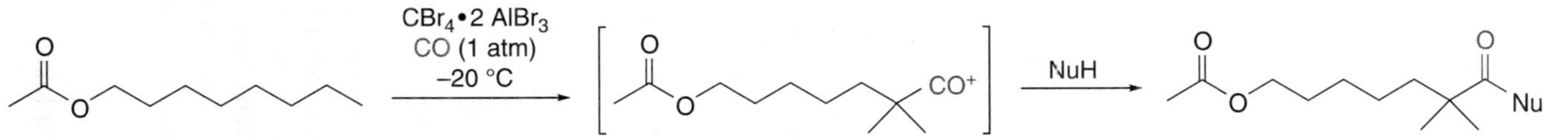

33 **34**

NuH = *i*PrOH, morpholine, anisole (75–80%)
NuH = CF_3CH_2OH, $HOCH_2C{\equiv}CH$, $HO(CH_2)_3Br$
$HOCH_2CH(Et)Bu$ (63–78%)

Scheme 18.12 Bifunctionalization of long chain acetates by carbonylation in superacidic media [30].

ketones of the type $C_nH_{2n+1}COMe$ (n = 6, 7, 8) were placed under an atmosphere of carbon monoxide at atmospheric pressure, in the presence of an excess of the superelectrophilic system $CBr_4{\cdot}2AlBr_3$, and a nucleophile was added to the carbonylated product at −20 °C, bifunctional products could be obtained in moderate to good yields. A trend similar to that obtained with the acetyl esters was observed, i.e. with octan-2-one linear products functionalized at C-7 were obtained, but with nonan-2-one (**35**) (Scheme 18.13) and decan-2-one, the neo-products were favored. A wide range of nucleophiles could be used to trap the intermediate cation, e.g. alcohols, furan, anisole, and thiophene, for **36** derived from nonan-2-one. When the nucleophiles were alcohols, high yields and conversions of 97–98% were obtained.

One year later, the Akhrem group also showed that the methodology used for the difunctionalization of *n*-alkyl acetyl esters and ketones could also be extended to methyl alkanoates [$C_nH_{2n+1}COOR$ (R = Me, Et; *n* = 7, 8, 9)], to provide the corresponding difunctionalized molecules in good to very high yields [32]. The successes obtained in the selective functionalizations mediated by the superelectrophilic system $CBr_4{\cdot}2AlBr_3$ achieved so far suggest that this carbonylative functionalization of sp^3 C—H bonds may provide a route for further developments in this field.

18.3 Carbonylation Reactions Catalyzed by Transition Metal Complexes

Although the use of transition metals makes the development of effective catalytic processes possible, there are still not many available for alkane carbonylation. A few factors account for this. Due to its π-acidic nature, when CO coordinates to a metal it makes it more electron deficient, and subsequent oxidative additions to the metal (if required) become more difficult [4]. In addition, in the presence of an atmosphere of carbon monoxide, there is a competitive inhibition toward coordination of inert C—H bonds by CO. When a metal center reacts with an alkane, a metal-alkyl bond may be formed, but the resulting complex may be either inert or simply revert to alkane instead of leading to further alkane functionalization [33]. In fact, the oxidative addition reaction of an alkane C—H bond to a metal is thermodynamically unfavored. In the methods available, oxidants are often required to re-oxidize the metal at the end of the reaction cycle to render the processes catalytic. Commonly used oxidants are organic compounds or inorganic salts, e.g. benzoquinone, peroxodisulfate, $CuCl_2$, silver salts, etc. In some cases, oxygen and air are also used. A number of metals have been applied successfully. In the first examples, rhodium was used, but palladium, vanadium, and copper have provided the best results.

One of the drives for research in this area for many years has been the quest for methods that allow the conversion of primary gaseous raw materials like methane and other low molecular weight alkanes directly into liquid fuels and also into functionalized substances which can be easily and safely transported [3, 34, 35]. Transition metal-catalyzed processes could help to reduce the greenhouse gas emissions that accompany the technological processes used nowadays for the conversion of petroleum and natural gas components into liquid fuels. Some seminal works in this field are highlighted below.

18.3.1 Carbonylation Mediated by Rhodium Complexes

The first example of a truly catalytic transition metal complex-catalyzed carbonylation of alkanes was reported by Sakakura and Tanaka in 1987. They found that the Vaska's type rhodium complex *trans*-[$RhCl(CO)(PMe_3)_2$] could promote the carbonylation of cyclohexane

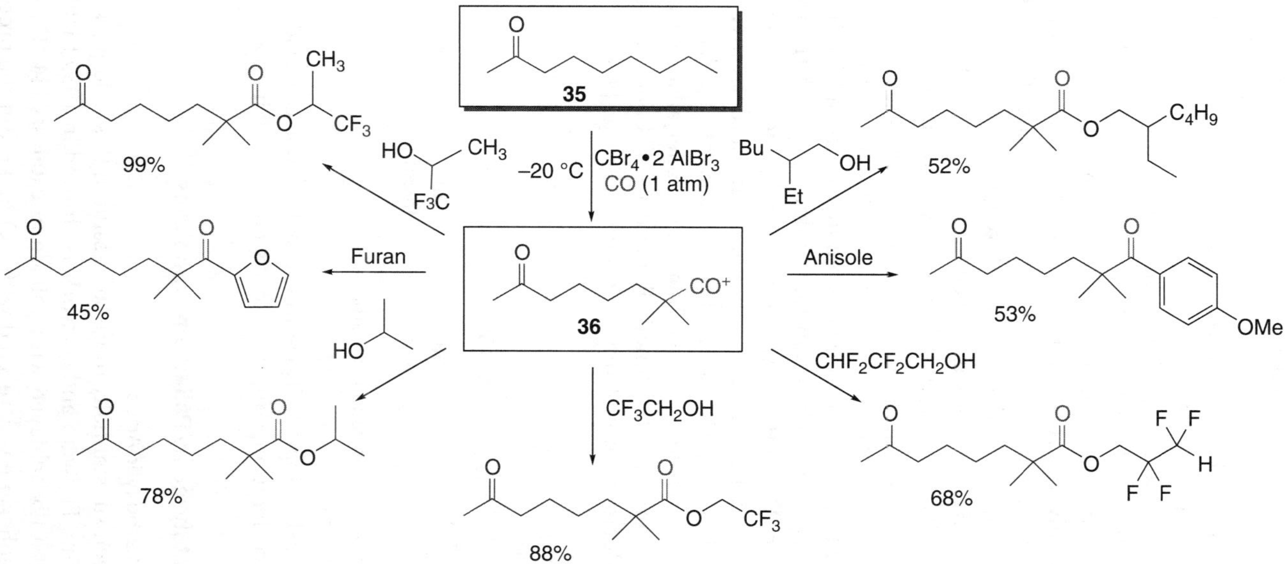

Scheme 18.13 Remote functionalization of long chain ketones [31].

and pentane even with 1 atm of carbon monoxide, at room temperature, to produce a mixture of products. Although this process was truly catalytic, since it required only 8 mol% of rhodium catalyst, it only proceeded in the presence of radiation [36]. The only previous report of alkane carbonylation was achieved with methane in the presence of [Cp_2TiCl_2] and $AlEt_3$, under very high pressure, but the process was not catalytic [37]. The regioselectivity at short wavelengths of radiation was contrary to that observed in radical reactions and reactions of superacids and superelectrophiles. Sakakura and Tanaka's work is described in more detail in Section 18.3.5, together with further work on photocatalysis.

In 1994 Sen and co-workers found that $RhCl_3$ could be used to catalyze the production of acetic acid (**38**) from methane (**37**) (Scheme 18.14) [38, 39]. Methanol (**39**) and formic acid (**40**) were obtained in much lower amounts. Despite the fact that the addition of carbon monoxide to methane is thermodynamically unfavorable, in the presence of oxygen and in aqueous media, acetic acid was produced (turnover number, TON = 28). If either CO or O_2 were omitted, no product was observed, suggesting the need for dioxygen as an oxidant. Isotopic labeling studies indicated that an Rh-CH_3 species was derived from CH_4 and not from CH_3OH or CH_3X (X = I or Cl), since when $^{13}CH_3OH$ was added to the reaction mixture, no $^{13}CH_3COOH$ was produced. Hence, the mechanism of the reaction is different from that of the Monsanto process for the conversion of methanol to acetic acid, which is believed to proceed via the formation of methyl iodide and water from the reaction between methanol and HI. In this new process, the nature of the solvent was found to be important, influencing both the rate of the reaction and the product distribution. When the solvent was a strong acid, e.g. a 6 : 1 (v/v) mixture of perfluorobutyric acid and water, the formation of methanol and its derivative $C_3F_7CO_2CH_3$ was favored. Addition of halide ions (e.g. HI, KI, HCl) also caused rate acceleration, although their specific role could not be determined. When ethane and higher alkanes were applied to this process, e.g. butane, isopentane, and 2,3-dimethylbutane, C—C bond cleavage products were observed, rather than products resulting from C—H bond cleavage [40].

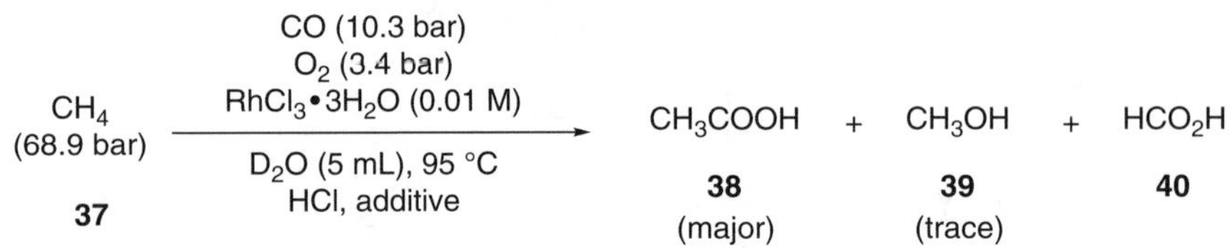

Scheme 18.14 Rhodium-catalyzed carbonylation of methane [38].

Chepaikin et al. also found, using $RhCl_3$ catalysis, that both acetic acid and trifluoroacetic acid (TFA) caused a rate acceleration, compared to systems in which no acid was used [41, 42].

18.3.2 Carbonylation Catalyzed by Palladium Complexes

18.3.2.1 Carbonylation to Carboxylic Acids

Palladium catalysis has been substantially more successful in this field [6, 43]. After the initial developments with rhodium, Fujiwara and co-workers showed in 1989 that the carbonylation of alkanes could be successfully achieved with palladium catalysts [44, 45]. No radiation was required in this case. Cyclohexane was reacted with CO in TFA in the presence of palladium acetate, to yield the corresponding carboxylic acid **3** (Scheme 18.15). Although the yield of **3** was low, this was the first report of transition metal-catalyzed alkane carbonylation that proceeded without the need for radiation as a form of activation. Potassium persulfate was used as an oxidant and it was found that methanol caused a rate acceleration. Cycloheptane also led to the

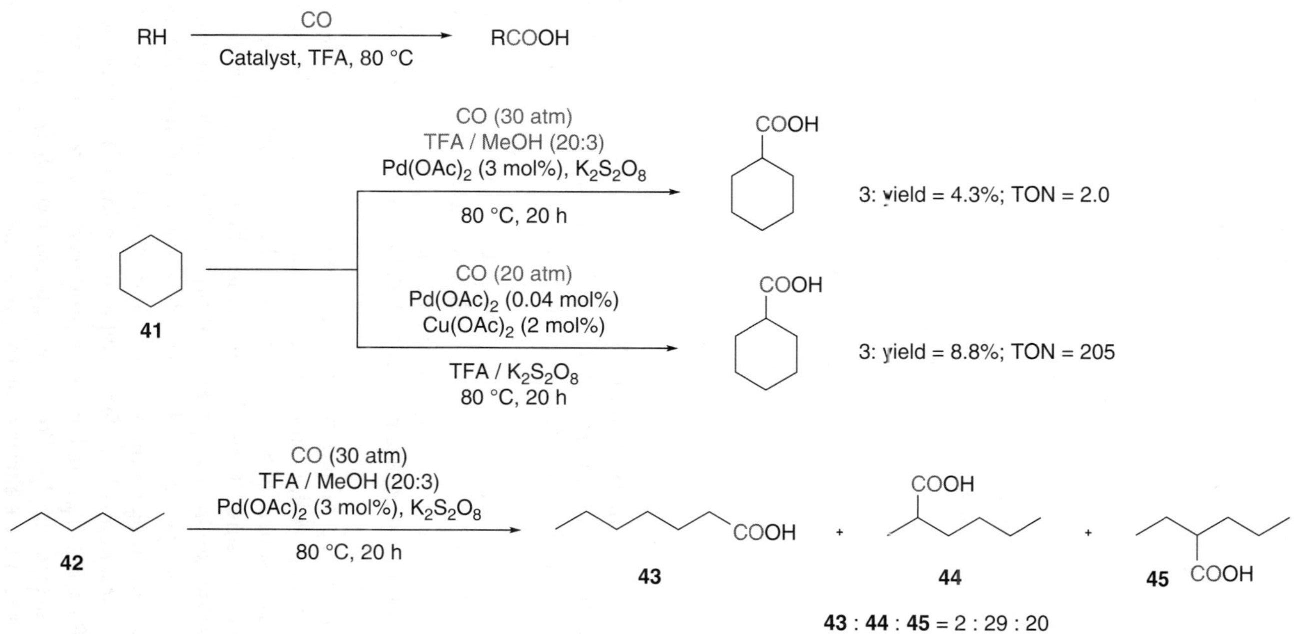

Scheme 18.15 The first examples of Pd-catalyzed carbonylation of alkanes [44].

monosubstituted acid, but with a straight chain alkane, *n*-hexane (**42**), a mixture of carboxylic acids (**43–45**) was obtained. Carbonylation took place preferentially at the secondary carbon atoms, contrary to what was observed with rhodium catalysis under radiation (Scheme 18.15). The authors postulated that reaction occurs via electrophilic attack of a cationic Pd(II) species on the alkane, to yield an alkyl Pd(II) complex, i.e. R-Pd-$OCOCF_3$, which undergoes CO insertion to form an acyl species (Scheme 18.16). Subsequent reductive elimination leads to the formation of a mixed anhydride, which in the presence of TFA yields the alkane carboxylic acid and trifluoroacetic anhydride [46].

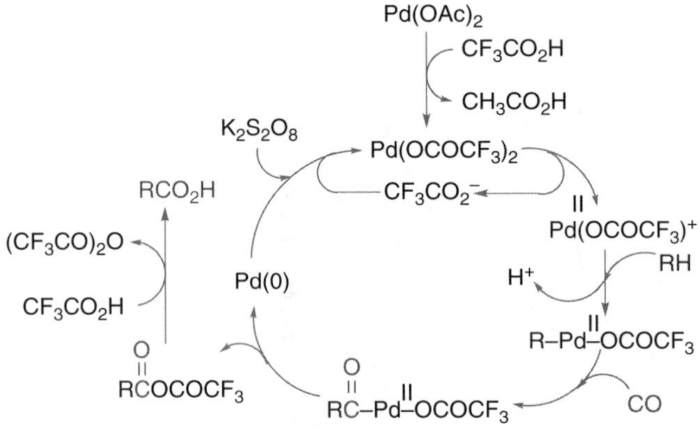

Scheme 18.16 Mechanism proposed for the Pd-catalyzed carbonylation of alkanes [46]. (*See insert for color representation.*)

The same group would later, in 1991, find that this procedure could be improved to produce cyclohexane carboxylic acid with a much higher TON, i.e. up to 200, if a co-catalyst was used, cuprous acetate, which in this case also permitted the use of a lower catalyst loading (Scheme 18.15) [47]. The increase in TON was attributed to an acceleration of the re-oxidation of Pd(0) to Pd(II) by $Cu(OAc)_2$. The carboxylation of methane with carbon monoxide was also possible, albeit in low yield [45]. In TFA at 80 °C, acetic acid was produced with TON of 8 when a Pd(II)-Cu(II) catalyst was used, in contrast with TON of 0.6 with Pd(II) alone, but 18 with Cu(II) alone, under an atmosphere of CO, at a pressure of 40 bar in the presence of peroxodisulfate ($K_2S_2O_8$) as oxidizing agent.

A distinct type of acetic acid production in a single step from methane was reported by Periana and co-workers in 2003 [5, 48]. It concerns the oxidative condensation of methane where this substrate behaves as the source of both the alkyl and the carbonyl groups, without requiring the use of CO, as proved by using ^{13}C-labeled methane. The reaction is catalyzed by $PdSO_4$ in 96% sulfuric acid. A mechanism involving CH activation by Pd(II) rather than acid catalysis was postulated (Scheme 18.17), since in control experiments performed in the absence of palladium, in deuterated methane containing methanol and acid mixtures, no acetic acid was produced. It was proposed that the reaction involves electrophilic activation of methane by Pd(II), resulting in the formation of a Pd-CH_3 intermediate. CO, produced *in situ* by the oxidation of methyl bisulfate or methanol, inserts into the Pd—C bond to form a Pd(II)-acyl intermediate, which subsequently undergoes reductive elimination to give acetic acid and Pd(0). Oxidation of Pd(0) by H_2SO_4 regenerates the active Pd(II) form.

Bell and co-workers subsequently studied this system, but adding Cu(II) so that O_2 could be used as the final oxidant. The reaction was performed in 96% sulfuric acid with a methane/oxygen atmosphere. Acetic acid and methanol were produced with 47% selectivity

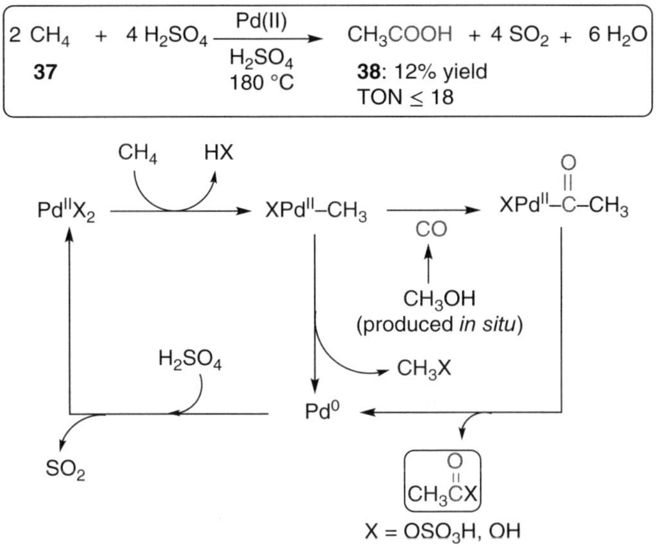

Scheme 18.17 Palladium-catalyzed oxidative condensation of methane and the mechanism proposed [48]. (*See insert for color representation.*)

after 4 hours at 180 °C. They also confirmed that CO is an intermediate and a byproduct of the reaction [49]. Additional experiments in the absence of the Cu cocatalyst provided similar results to those obtained by Periana with TON up to ~14. At a temperature of 160 °C, the selectivity for the formation of acetic acid went up to 82% [50].

Chempath and Bell also applied density functional theory to investigate the thermodynamics and activation barriers associated with the Pd(II) catalyzed reaction [51] (results published in 2006). Their results agreed essentially with Periana's mechanistic proposals and other works of that time [51]: electrophilic activation by a Pd species, in this case $Pd(\eta^2\text{-}HSO_4)(HSO_4)(H_2SO_4)$. Subsequent CO insertion into the P—C bond in Pd(Me)$(HSO_4)(H_2SO_4)_2$ yielding the acyl species $Pd^{II}(CH_3CO)(HSO_4)(H_2SO_4)_2$ was found to be thermodynamically favorable. However, the direct production of acetic acid by reductive elimination of the Pd^{II}—acyl intermediate was found to be energetically unfavorable. The authors proposed that oxidation of $Pd^{II}(CH_3CO)(HSO_4)(H_2SO_4)_2$ to the Pd^{IV} octahedral complex, $Pd^{IV}(HSO_4)_3(H_2SO_4)_2(CH_3CO)$, occurred instead. Reductive elimination of $CH_3CO(HSO_4)$ from this Pd^{IV} intermediate was determined to be much more facile and kinetically reasonable.

Among other cases where carboxylic acids are formed upon oxidation of alkanes, rather than upon carbonylation, the single-pot conversion of cyclohexane to adipic acid (HOOC$(CH_2)_4$COOH) (**47**) is noteworthy in view of its potential commercial significance. This has been achieved recently by Martins, Pombeiro, and co-workers under mild conditions by using the scorpionate complex $[FeCl_2\{HC(pz)_3\}]$, (pz = pyrazol-1-yl) (**46**) as catalyst (Scheme 18.18) [52]. Adipic acid could be obtained directly from cyclohexane in 96% yield, 98% selectivity, and TONs up to 1.8×10^3 using ozone in the presence of this catalyst in a solvent-free procedure, which also required no heating, no radiation, and which is also characteristically N_2O-free. This is an advantage over the method presently used for the industrial multi-step production of adipic acid, needed every year in large amounts as a precursor for the synthesis of Nylon-6,6 polyamide and also widely used to produce polyester and polyurethane resins. Undesirable N_2O is a byproduct of the manufacturing process produced in high amounts.

41

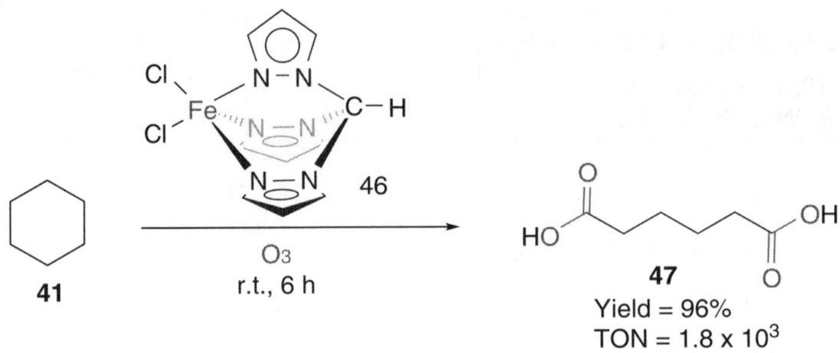

Scheme 18.18 One-pot synthesis of adipic acid from cyclohexane catalyzed by an Fe-scorpionate complex [52].

The scorpionate-catalyzed procedure provides an efficient, environmentally friendly, and much simpler (single-pot) alternative.

18.3.2.2 Other Carbonylation Reactions

Dyson and co-workers developed a Pd-catalyzed oxidative carbonylation of C(sp^3)-H bonds with carbon monoxide and amines in 2014, using $PdCl_2$ and diphosphine ligands [53]. A metal-catalyzed process like this provides an atom economical route to amides, which are often prepared from more functionalized substances like acetyl chlorides or anhydrides and even aldehydes. In addition, acid or base are not required in this method. The alkane is used in excess, helping to solubilize the reagents. Although the scope of the reaction was limited to aromatic amines and in most of the examples studied the functionalization of benzylic C—H bonds was involved, e.g. as in the reaction of aniline (**48**) with toluene (**49**) to provide amide **50**, it was shown that cyclohexane could also be used to afford the corresponding amide **51** in good yields (Scheme 18.19). Bisphosphines with a bite angle greater than 90° provided the best results, i.e. 4,5-bis(diphenylphosphino)-9,9-dimethylxanthene (xanthephos) (**L1**) and (*R*)-(−)-4,12-bis(diphenylphosphino)-[2.2]-paracyclophane [(*R*)-phanephos] (**L2**). The best oxidants were strong ones, i.e. no product was obtained with Ag_2O but di-*tert*-butylperoxide (DTBP) provided good yields. When the radical trap 2,2,6,6-tetramethylpiperidinoxyl (TEMPO) was added to the mixture, there was no reaction, which suggests that a radical mechanism is involved. A plausible one is shown in Scheme 18.19, where the N- and C-radicals are generated upon H-abstraction from aniline and cyclohexane, respectively, by DTBP [5].

In order to investigate the mechanism of the reaction, toluene-d_8 was used as a substrate and a significant isotopic effect was observed. The yield of amide after 2 hours was 7%, lower than the 18% observed with non-deuterated toluene, which suggests that C—H bond cleavage occurs during the rate-limiting step or prior to it. This result also indicates that the yield of the desired product is a function of the C(sp^3)-H bond dissociation energy.

Electron deficient amines provided better results, e.g. with 3,5-$(CF_3)_2C_6H_3NH_2$ as substrate, the yield of product was 78%, but aliphatic amines did not react.

The power of palladium catalysis was also recently illustrated by Lei et al. [54]. While studying the carbonylation of alkanes, they developed a radical oxidative process that uses alcohols *in situ* to produce alkyl carboxylates. The method was developed initially for the reaction of cyclohexane with benzyl alcohol (**52**) and optimization studies showed that in the presence of triphenylphosphine, 5 mol% $PdCl_2$ can be used to promote the reaction between equimolar amounts of the reagents, provided DTBP is present (Scheme 18.20). *tert*-Butyl hydroperoxide (TBHP) does not work. The pressure of CO was kept at 5 atm and ester **53a** was produced in 84% yield as determined by GLC. The process also works with CO at atmospheric pressure at the expense of a 10% decrease in yield. A large variety of aromatic alcohols could be

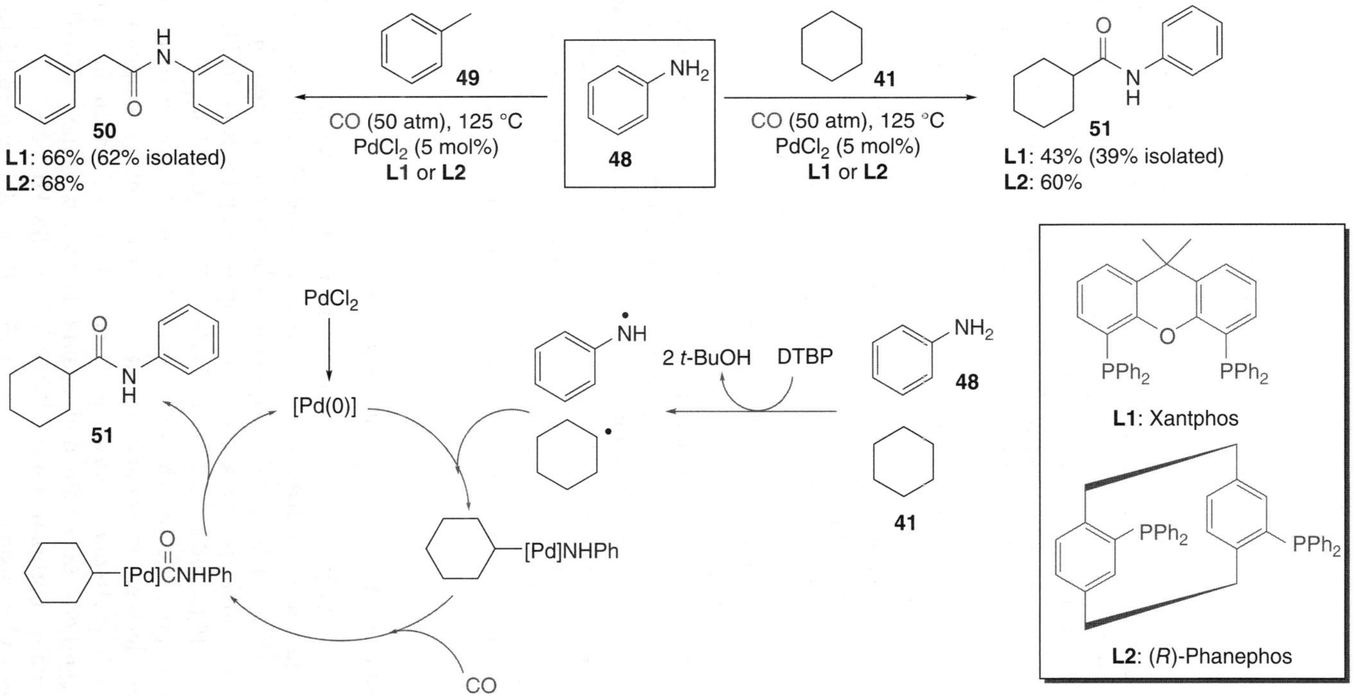

Scheme 18.19 The first examples of amide synthesis via Pd-catalyzed carbonylation of alkanes and proposec mechanism [53].

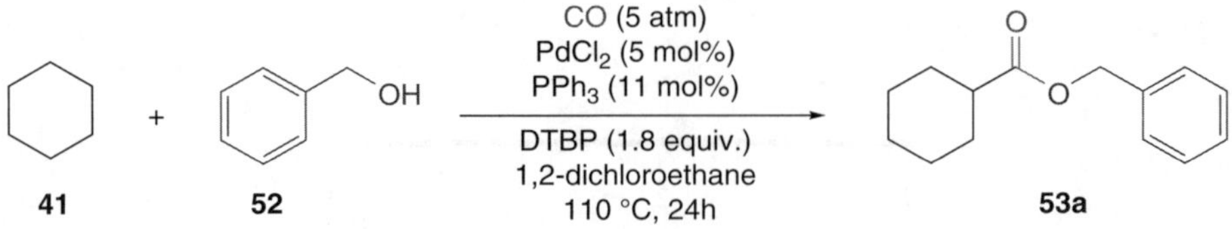

41 52 53a

Scheme 18.20 Synthesis of esters by Pd-catalyzed oxidative alkoxycarbonylation of alkanes [54].

used and esters were produced in yields varying from 58% to 87%. When α-chiral alcohols were applied, there was full retention of configuration and no racemization, e.g. **53e** was produced from (*R*)-1-phenylethanol (Scheme 18.21). In addition, the method could be extended to aliphatic alcohols and moderate to high yields (40–94%) of the desired alkyl carboxylates were produced. Generally, the tertiary alcohols gave lower yields, and no racemization was observed when enantiomerically pure substrates were used. Further, a variety of cyclic and even linear alkanes were also subjected to the reaction, providing the desired carboxylates in moderate to high yields (36–81%). This reaction could also be used to produce benzyl ester **53o** from ethane, a gas, in 61% yield (Scheme 18.21). Ethane is one of the constituents of natural gas, and this development suggests good prospects for further research in this area.

Some studies were performed to provide an insight into the mechanism of this alkane C—H activation process. Comparing the results obtained from a reaction performed with cyclohexane and cyclohexane-d_{12}, it was found that there was a significant isotopic kinetic effect, i.e. $k_H/k_D = 5.0$, which suggests that C—H bond cleavage occurs in the rate-determining step. In addition, when radical trapping experiments were performed with the addition of TEMPO to the reaction medium, no product was obtained, which suggests that a radical mechanism is involved in the process. The mechanism proposed is shown in Scheme 18.22. It involves H-abstraction from the alkane by DTBP, upon which the resulting alkyl radical reacts with CO coordinated to the metal. A metal-acyl species is produced, which is in equilibrium with an acyl radical. Coupling of this metal-acyl/radical species with the *t*-BuO• radical derived from the alcohol leads to the ester [53]. The reaction between cyclohexane, CO and benzyl alcohol could be also performed on a gram-scale without deterioration of the results. The product was obtained in 81% yield [53].

18.3.3 Carboxylation Catalyzed by Vanadium Complexes

Vanadium, with various easily interconvertible oxidation states, catalyzes many reactions, including the carboxylation of methane to produce acetic acid [55]. In 1999, Fujiwara and co-workers found that [VO(acac)$_2$] could catalyze the carboxylation of methane to acetic acid in a very effective manner in the presence of carbon monoxide and $K_2S_2O_8$ in TFA. Acetic acid was produced in 93% yield (based on CH_4) with TON = 27.5 at 80 °C [56].

However, even much more effective catalysts were developed by Pombeiro and co-workers, inspired on a biomolecule, amavadin, a natural vanadium(IV) complex present in *Amanita fungi*, including *Amanita muscaria* (Figure 18.1). Although its biological role is not yet known, synthetic amavadin and chemical models, namely $[V(HIDPA)_2]^{2-}$(**54**, HIDPA = *N*-hydroxyiminodipropionate) and $[V(HIDA)_2]^{2-}$(**55**, HIDA = *N*-hydroxyiminodiacetate), have catalase- and peroxidase-type activity [57]. They catalyze peroxidative oxidation, peroxidative halogenation, and carboxylation of alkanes and other hydrocarbons and the oxidation of alcohols and epoxidation of allylic alcohols, but under conditions that are very different from those that normally occur in biological media. The catalytic activities of amavadin

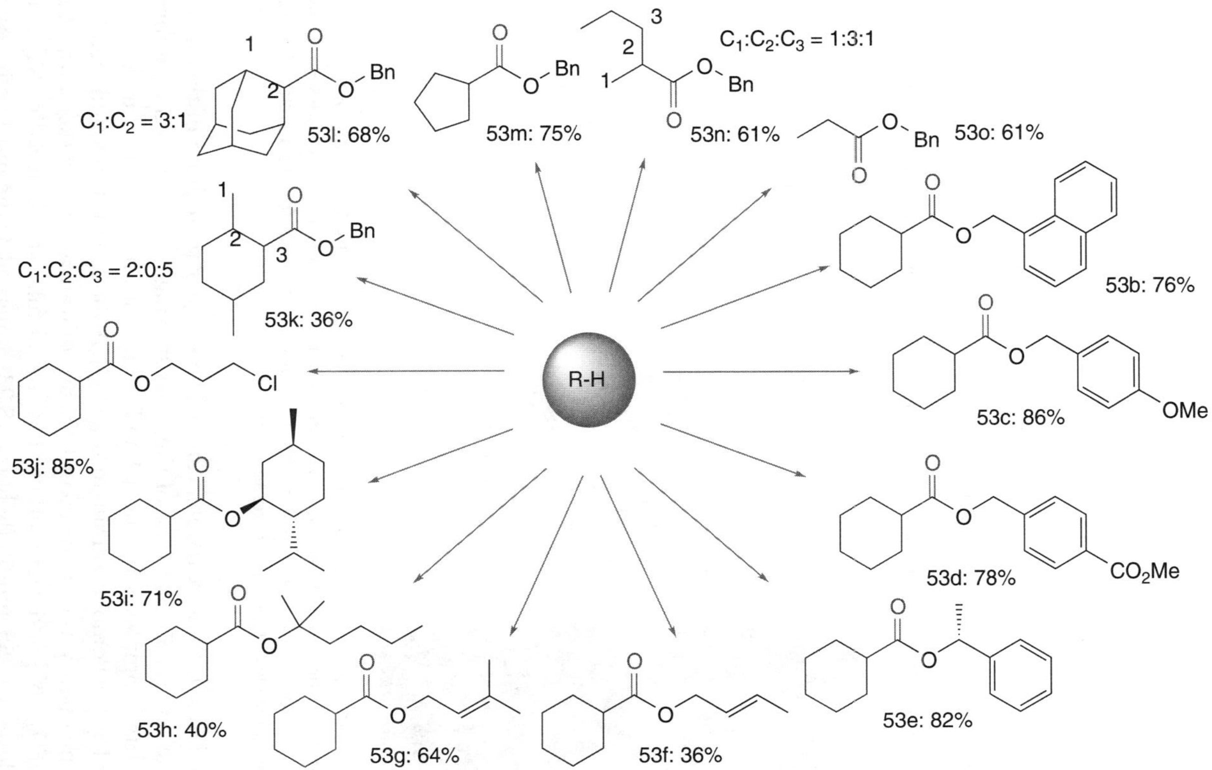

Scheme 18.21 The scope of the Pd-catalyzed alkoxycarbonylation reaction. For the synthesis of the examples shown, xantphos was used as ligand [54].

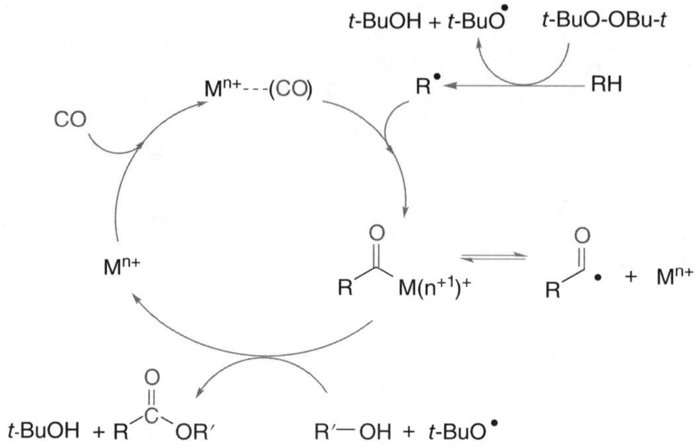

Scheme 18.22 Mechanistic proposal for the Pd-catalyzed alkoxycarbonylation of alkanes [54].

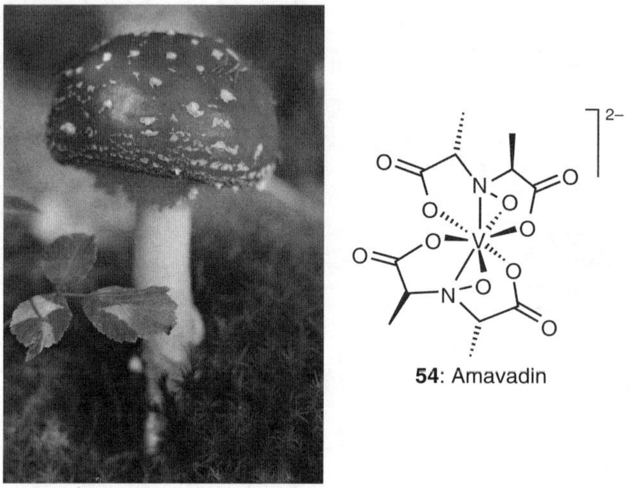

Figure 18.1 Amavadin (**54**), a non-oxido vanadium(IV) complex found in the fungus *Amanita muscaria*, seen in the figure, a poisonous mushroom with psychoactive and insecticidal properties. Amavadin catalyzes a few reactions of alkanes, including carboxylation and it has been used as a model for the development of other carboxylation catalysts. (*See insert for color representation.*)

and related complexes in these reactions can be even higher than those of the most efficient oxovanadium catalysts [1, 57, 58]. One of the most striking findings was the performance of amavadin and related complexes (in the +4 and +5 oxidation states), in the carboxylation of methane reported by the Pombeiro group in 2003 [59]. Although the transition metal-catalyzed carboxylation of methane was discovered in 1992 by Fujiwara and co-workers, with the system $CuSO_4/Pd(OCOEt)_2$, [60] and many other studies had followed, amavadin (**54**) and related vanadium complexes, e.g. **55** and **56** (Scheme 18.23), emerged as the most active catalysts for carboxylation of alkanes, including the gaseous ones. In fact, methane (**37**) could be converted to acetic acid (**38**) in TFA at 80 °C, in yields up to 55% and TONs up to 6×10^3, under an atmosphere of CO, at low pressures (typically below 10–15 atm) in the presence of an excess of peroxodisulfate ($K_2S_2O_8$) as oxidizing agent [59, 61]. The reaction also takes place in the absence of CO, in which case the methyl ester of TFA is produced, but in lower yield. The use

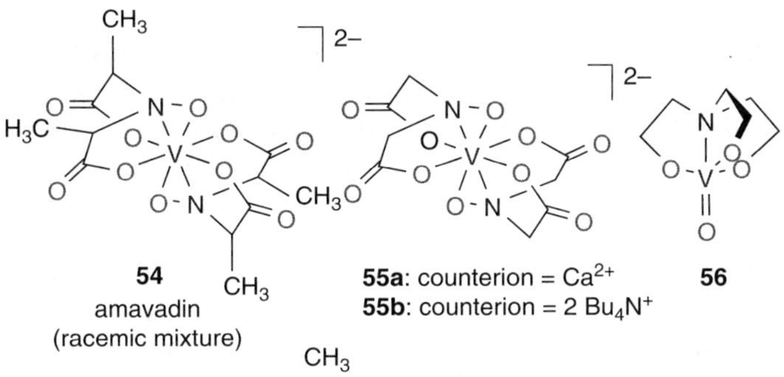

54 amavadin (racemic mixture)

55a: counterion = Ca^{2+}
55b: counterion = 2 Bu_4N^+

56

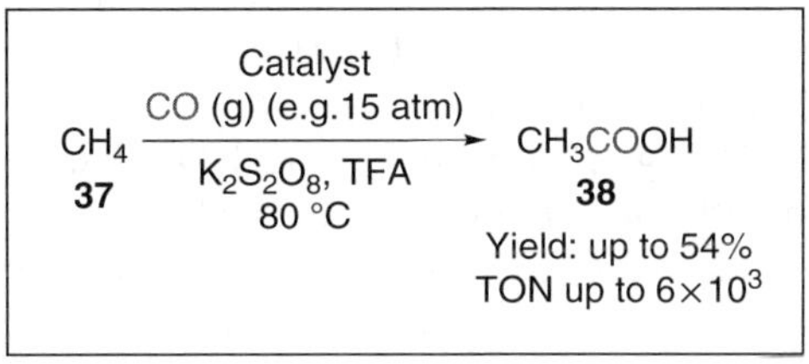

Scheme 18.23 Vanadium complexes (amavadin models) found to perform the best in the carboxylation of methane in the presence of carbon monoxide gas and potassium peroxodisulfate [59].

of ^{13}C-labeled methane confirms the alkane origin of the methyl group of the acetic acid produced. This process is more efficient than the important processes used for the industrial production of acetic acid, which proceed in three separate stages under harsh conditions (e.g. high temperatures and pressures) and in which the third stage, the carbonylation of methanol, requires an expensive catalyst containing either Rh or Ir (the Monsanto or Cativa processes). In contrast, the amavadin reaction is a single one-pot process from the alkane, proceeds under mild conditions, and has cheaper metal catalysts [55, 58].

The vanadium-catalyzed carboxylation reaction was dependent on the CO pressure, with the maximum activity occurring at pressures below 10 atm. This was presumably due to coordination of CO at higher pressures, leading to deactivation of the catalyst. Studies with radical traps suggested that the reaction proceeds via a radical mechanism (Scheme 18.24), as the catalytic activity was either dramatically reduced or suppressed in their presence. Further evidence came from the fact that the methyl radical was formed from H abstraction with peroxodisulfate, even in the absence of the vanadium catalyst. The assumption that a radical mechanism

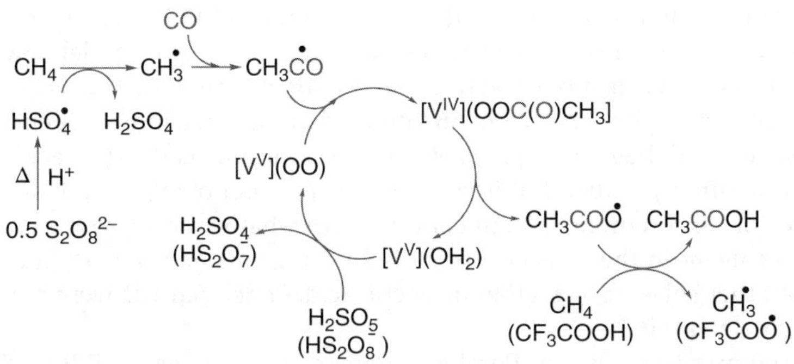

Scheme 18.24 Proposed mechanism for the vanadium-catalyzed carboxylation of methane [61]. (*See insert for color representation.*)

was involved was confirmed by DFT studies [61] and also by the fact that the reactions of other linear alkanes took place preferentially at a secondary carbon atom, rather than at a primary one, e.g. with *n*-pentane and *n*-hexane. In fact, in the carboxylation of pentane (**22**), three carboxylic acids (2-methylpentanoic acid (**62**), 2-ethylbutanoic acid (**63**), and hexanoic acid (**64**)) are formed in a 4 : 2 : 1 ratio, and the carboxylation of hexane (**42**) is regioselective, resulting in the production of 2-ethylpentanoic acid (**45**) with a selectivity of 95%. (Scheme 18.25) [64].

In the mechanism proposed for the vanadium-promoted methane carboxylation (Scheme 18.24) [61], a methyl radical ($CH_3^•$) is generated from methane upon hydrogen abstraction by a hydrosulfate radical ($HSO_4^•$). The hydrosulfate radical itself is generated from the oxidant, potassium peroxodisulfate, by metal-catalyzed thermal decomposition to $SO_4^{-•}$ followed by protonation in the acidic medium. The methyl radical undergoes facile exothermic carbonylation by CO to afford an acyl radical ($CH_3CO^•$), a process that has a low activation barrier (ΔH_s = 3.40 kcal mol^{-1}) [61]. Reaction of this radical with the active peroxido-vanadium(V) species $[V^V](OO)$, e.g. $[V(OO)\{N(CH_2CH_2O)_3\}]$ derived from the oxido-vanadium catalyst upon reaction with the oxidant leads to the formation of the acetate radical ($CH_3COO^•$). This radical abstracts hydrogen from the acid in the medium or from methane to yield the desired acetic acid and the catalyst, i.e. $[V^V(O)]$, is regenerated [61].

Other alkanes are also carboxylated catalytically with amavadin and related V complexes [59–64]. In the case of ethane, propionic acid is the predominant product, obtained with overall yields up to 93% and catalyst TONs up to 2.0×10^4 (Scheme 18.25) [62]. Amavadin and analogues (**54**, **55**) are the most active catalysts. Acetic acid is obtained as a minor product. The mechanisms of the reactions were studied with labeling experiments using ^{13}C-enriched ethane and carbon monoxide (^{13}CO), by radical experiments and by DFT calculations. The labeling experiments indicated that in the carboxylation of ethane to propionic acid the C—C bond remains intact, since when the reaction is performed with $^{13}CH_3{}^{13}CH_3$ and CO, double-labeled propionic acid ($^{13}CH_3{}^{13}CH_2COOH$) is obtained. The isotopic labeling studies also show that acetic acid (a minor product) can be produced in two ways: (i) by oxidation of ethane, the main route, since double-labeled acetic acid ($^{13}CH_3{}^{13}COOH$) is obtained; or (ii) by C—C bond rupture and carbonylation (minor route), since single-labeled acetic acid ($^{13}CH_3COOH$) is also observed but in a minor amount. In general, radical trapping experiments and the theoretical calculations support interpretations based on the involvement of radicals. Mechanistic proposals for both pathways for acetic acid were presented on the basis of DFT calculations [62] whereas the carboxylation of ethane to propionic acid is believed to follow a mechanism analogous to that of methane carboxylation to acetic acid (Scheme 18.24). In the carboxylation of propane, three products can be obtained: isobutyric (the main one), *n*-butyric, and acetic acids (the last as the minor one) (Scheme 18.25) [63].

The carboxylation of cyclic alkanes has also been studied with amavadin and analogues,which were shown to catalyze the reaction, the most active catalyst being the amavadin model **55a** (Scheme 18.26) [63, 64]. Cyclohexane (**41**) is more reactive than cyclopentane (**65**). The yields of the corresponding cycloalkane carboxylic acids **3** and **66** obtained in these reactions depend on the carbon monoxide pressure used, being as high as 50% for cyclohexane carboxylic acid. The TFA ester was obtained as a minor product, but it was the only product obtained in reactions performed in the absence of carbon monoxide, in contrast with what was observed with methane. The use of TFA as a solvent in these reactions is disadvantageous, particularly due to its cost, but attempts to use other solvents, e.g. ethanol, acetic acid, or acetonitrile were not very successful with vanadium catalysts [63].

In 2010, it was shown for the first time also by Pombeiro and co-workers that a different type of catalysts, dioxido- or chloridovanadium complexes containing C-based trispyrazolyl scorpionate ligands or pyrazole ligands, could also catalyze the carboxylation of methane

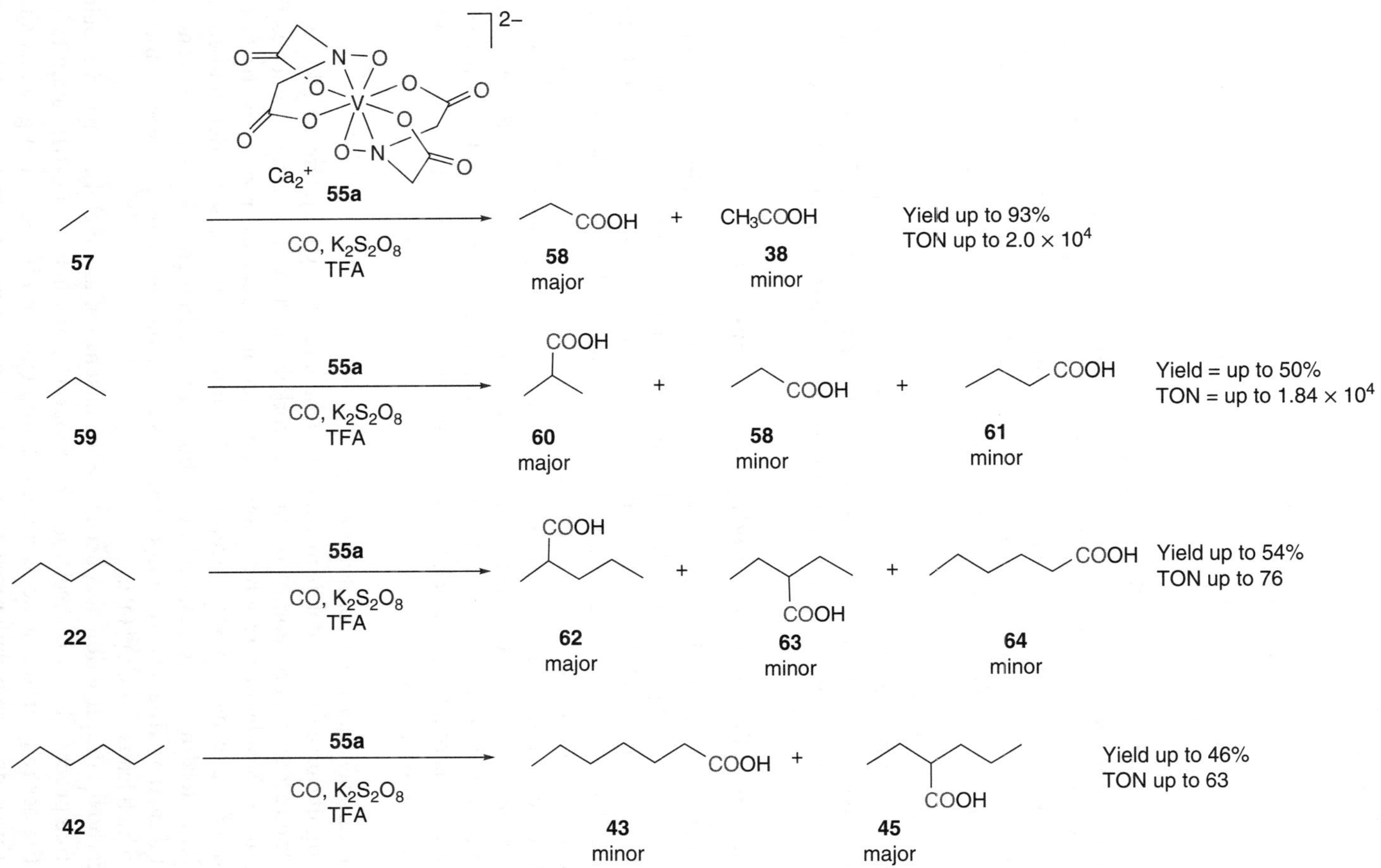

Scheme 18.25 The carboxylation of ethane [62], propane [63], pentane, and hexane [64] catalyzed by the amavadin model **55a**. The yields reported are relative to the alkane substrates [64].

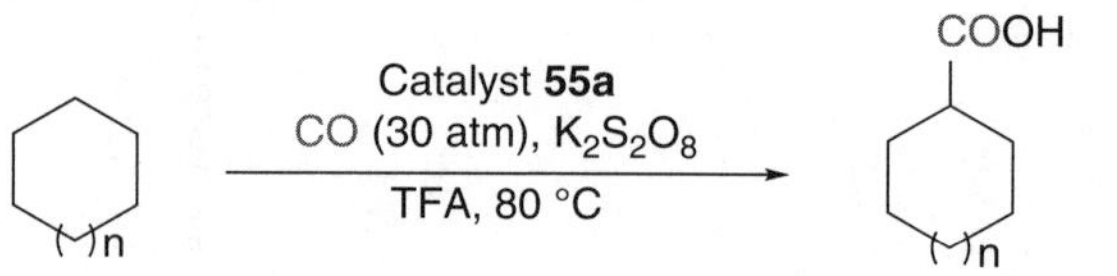

65: n = 0
41: n = 1

66: Yield up to 33%, TON = 50
3: Yield up to 50%, TON = 74

Scheme 18.26 The carboxylation of cycloalkanes [63, 64] catalyzed by the amavadin model **55a**. The yields reported are relative to the cycloalkane.

and ethane under mild conditions (Scheme 18.27) [65]. The more active catalysts toward methane carboxylation were the pyrazole derivative **67** and $[VCl_3\{SO_3C(pz)_3\}]$ (**68**), with the sulfonated and water-soluble tris(1-pyrazolyl)methanesulfonate ligand, with yields up to 40% (TONs up to 157). The activity decreased when the sulfonated scorpionate was changed for the analogous neutral hydrotris(1-pyrazolyl)methane as ligand. The maximum yield of acetic acid was obtained with a carbon monoxide pressure of 5 atm, and it dropped as the pressure was increased.

The peroxodisulfate/TFA system was extended by the same group to other metal catalysts, including group 5–7 metal oxides, e.g. rhenium oxides. The rhenium catalysts promote the conversion of ethane into propionic and acetic acids with overall yields up to 50% [66]. This was the first report of an application of rhenium complexes as catalysts for alkane carboxylation.

18.3.4 Carbonylation Reactions Catalyzed by Copper Complexes

The use of copper salts and complexes as catalysts for C—H bond activation has the advantages that they are generally inexpensive and have low toxicity. In 1989. Fujiwara and co-workers reported the first example of alkane carboxylation based on transition metal catalysis that did not require radiation [46, 67, 68]. Alkanes could be carboxylated to the corresponding acids with CO (20–40 atm) in TFA at 80 °C in the presence of a Pd(II)/Cu(II) catalytic system. $K_2S_2O_8$ was used as oxidant, and a radical mechanism was postulated [67]. The yields reported were not high and the system had the drawback of using TFA as solvent at 80 °C. These difficulties were overcome by the systems operating in water/acetonitrile solvent described below concerning hydroxycarboxylations of alkanes, which highlight the potential of this metal for future applications of alkanes in synthesis.

18.3.4.1 Hydrocarboxylation to Carboxylic Acids

Of particular significance to the establishment of a sustainable catalytic system for alkane carboxylation is the recognition by Pombeiro and co-workers in 2009 that water can act as the hydroxylating agent, besides being a solvent [69]. The reaction of alkanes proceeds with CO and water using $S_2O_8^{2-}$ as oxidant in water/acetonitrile media (Scheme 18.28), under mild conditions (30–60 °C), affording the corresponding carboxylic acids in high yields in the presence of a suitable metal catalyst. The involvement of water as a stoichiometric reagent was confirmed using $H_2{}^{18}O$, which leads to $RCO^{18}OH$.

The hydrocarboxylation reaction also occurs in the absence of any metal catalyst, i.e. being then a metal-free process, but it is accelerated by copper(II) or vanadium(V-III) catalysts and the yields of the desired carboxylic acids obtained are higher. Both liquid and gaseous alkanes undergo this hydrocarboxylation reaction, but the system is not effective for methane. The reactions take place preferentially at the secondary carbon atoms, leading to mixtures of mono-substituted carboxylic acids, which suggests that radical mechanisms are involved.

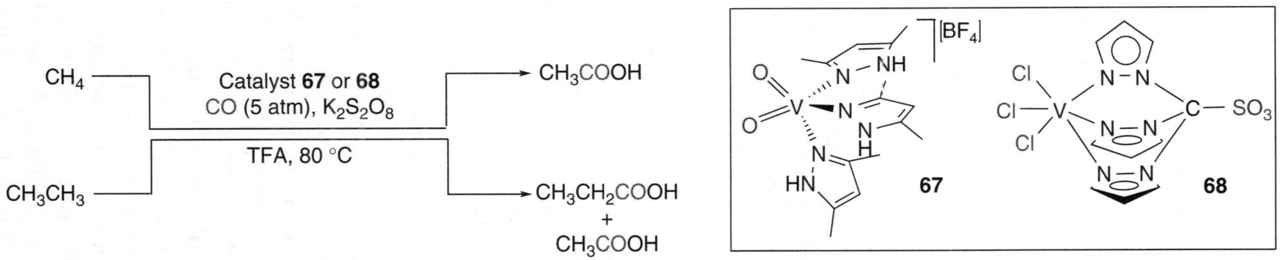

Scheme 18.27 The carboxylation of gaseous alkanes promoted by scorpionate and pyrazole dioxovanadium complexes [65].

$$RH + CO + H_2O + S_2O_8^{2-} \xrightarrow[\text{Catalyst}]{H_2O\,/\,MeCN} RCOOH + 2\,HSO_4^-$$

Scheme 18.28 The hydrocarboxylation of alkanes in a water-acetonitrile solution [69]. (*See insert for color representation.*)

The presence of acid is not required as in the reactions with vanadium catalysis, and the ratio of water to acetonitrile used has an important effect on the yield. A number of copper(II) catalysts promote the hydrocarboxylation of alkanes. The best results obtained so far were with water-soluble tetracopper(II) triethanolaminate $[O\subset Cu_4\{N(CH_2CH_2O)_3\}_4(BOH)_4][BF_4]$ (**69**), which catalyzes the reactions of both linear [69–71] and cyclic alkanes [69, 72] (Scheme 18.29) in yields up to 87% and very high selectivities. A ratio of acetonitrile to water of 2 : 1 or 1 : 1 provided the best results in most cases.

In the metal-catalyzed mechanism proposed (Scheme 18.30) [69] on the basis of labeling experiments and DFT calculations, the free alkyl radical ($R^•$) is formed by hydrogen abstraction from the alkane by a sulfate radical ($SO_4^{-•}$) resulting from thermal homolysis of $K_2S_2O_8$. Facile reaction of the alkyl radical with CO results in formation of an acyl radical ($RCO^•$), which is oxidized into the corresponding cation (RCO^+) by the action of the catalyst in its oxidized (M^n) form. This acyl cation reacts with one molecule of water forming the protonated acid, which upon deprotonation yields the carboxylic acid (RCOOH). The reduced catalyst (M^{n-1}) is reoxidized by $K_2S_2O_8$ for another catalytic cycle. In the absence of a metal catalyst, the hydrocarboxylation occurs less effectively upon oxidation of the acyl radical by $S_2O_8^{2-}$ (metal-free pathway) [69].

Other tetranuclear copper complexes were found to catalyze the hydrocarboxylation of alkanes, e.g. **75**, which contains hydrazone Schiff base ligands [74], and **76**, containing a macrocyclic disiloxane ligand [75], but were however less selective and provided lower yields of carboxylic acids (Scheme 18.31). Mono- and dinuclear copper complexes, aminopolyalcoholates i.e. **77** and **78**, displayed similar activities and were selective toward carboxylic acid production in favor of the corresponding ketone or alcohol, particularly in the case of cyclopentane and cyclohexane, less so for the larger C7 and C8 rings, providing yields up to 36%. With linear alkanes, mixtures of carboxylic acids were obtained, with secondary acids predominating [76].

Heterometallic coordination polymers, i.e. **79–82** are also good catalyst precursors for alkane hydrocarboxylation (Scheme 18.32) [73, 77–80]. Metal-organic framework (MOF) **79** is a 2D heterometallic Cu^{II}/Mg coordination polymer, $[Cu_2Mg_2(\mu\text{-}HTEA)_2(\mu_6\text{-}PMA)(H_2O)_6]_n\cdot 6nH_2O$, derived from triethanolamine (H_3TEA) and pyromellitic acid (H_4PMA) [77]. It is a very rare example of a Cu/Mg coordination network. It catalyzes the hydrocarboxylation of linear C_n (n = 5–9) alkanes into the corresponding C_{n+1} carboxylic acids in yields up to 25% and C_n (n = 5–8) cyclic alkanes into the corresponding acids in yields up to 44%. MOFs **80a** and **b** are Cu^{II}/Li materials, composed of similar $[Cu_4(\mu\text{-}HBES)_4(\mu\text{-carboxylate})]^-$ cores and $[\mu\text{-}Li(H_2O)_2]^+$ groups ($H_3BES = (HOCH_2CH_2)_2N(CH_2CH_2SO_3H)$), which are arranged into infinite 3D non-interpenetrating metal-organic networks [78]. In MOF **80c**, sodium replaces lithium [79]. MOF **80c** behaved similarly to **79,** reaching yields of carboxylic acids up to 37% (from linear alkanes) or up to 47% (from cycloalkanes), both being highly selective for hydrocarboxylation versus oxidation. The sodium analogue, MOF **80c**, is particularly effective in the hydrocarboxylation of gaseous alkanes (C_3H_8 and n-C_4H_{10}), converting them to the corresponding carboxylic acids in up to 95% yields based on the alkane. In the case of butane, a selectivity of 95% was observed (secondary versus primary acid).

The 1D coordination polymer $\{[Cu_2(Piv)_4(H_3tBuDea)](Piv)\}n$ (**81**) prepared through the self-assembly reaction of copper(II) nitrate with pivalic acid (HPiv) and N-*tert*-butyldiethanola

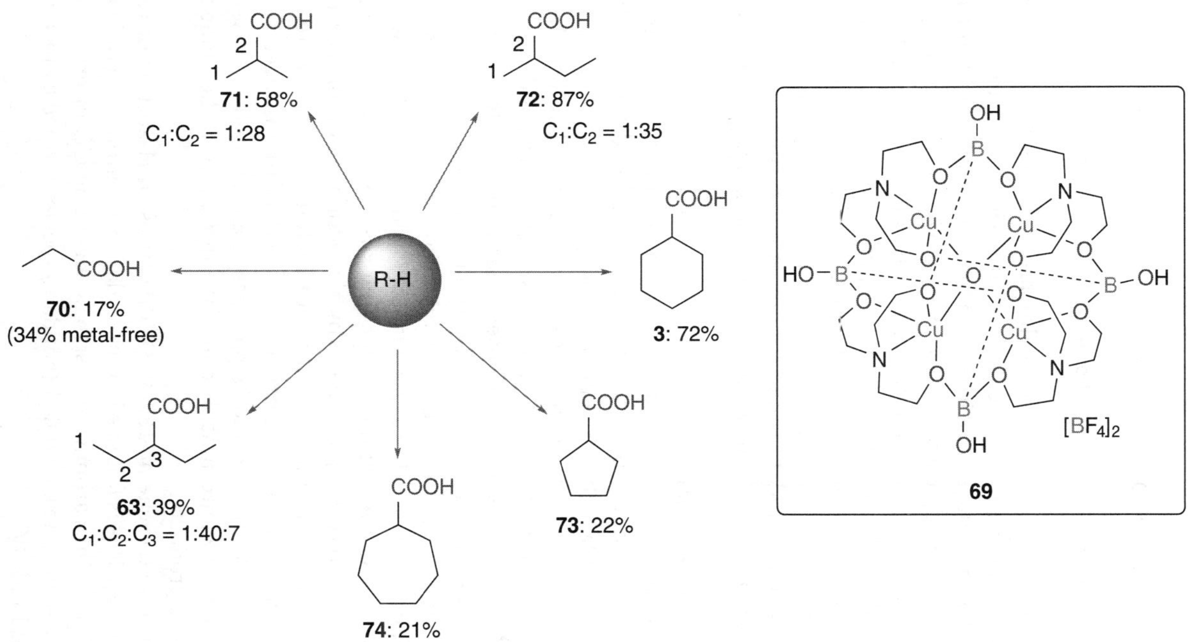

Scheme 18.29 The hydrocarboxylation of linear and cyclic alkanes in water-acetonitrile solution catalyzed by tetranuclear copper catalyst **69** [70–72].

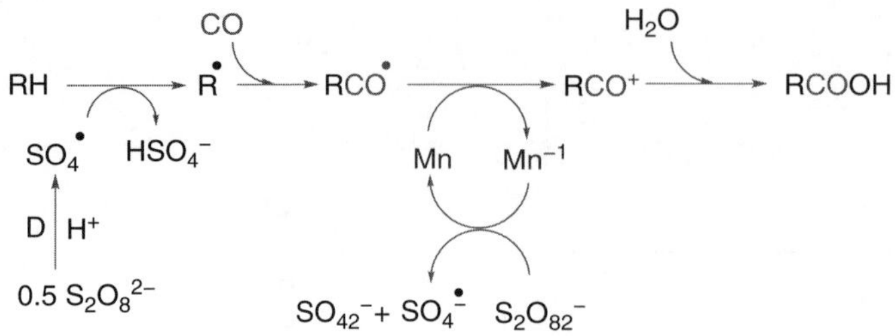

Scheme 18.30 The mechanism proposed for the hydrocarboxylation of alkanes in water-acetonitrile [72, 73].

mine (H_2tBuDea) was also used for the mild hydrocarboxylation of linear and cyclic C_5–C_8 alkanes into the corresponding carboxylic acids, providing yields of up to 26% [80].

Recently, it was also shown by the Pombeiro's team that the organic (acetonitrile) solvent can be replaced by an ionic liquid [81]. In fact, the hydrocarboxylation of cyclohexane was achieved with a Cu(II) MOF containing a flexible bifunctional terpyridine with a 4-phenoxycarboxylate unit as a catalyst (**82**) [81]. When the 1 : 1 ionic liquid ([BMPyr][NTf_2]):water system i.e. 1-butyl-1-methylpyrrolidinium bis(trifluoromethanesulfonyl)imide/water, was used in the presence of peroxodisulfate as oxidant, cyclohexane carboxylic acid was obtained with a maximum yield of 36%, at 60 °C and 20 atm CO. Comparable results could be obtained under the same conditions in the mixed solvent water/MeCN, although the reaction rate was lower.

18.3.4.2 Other Carbonylation Reactions

Recently, copper catalysis has been successfully used for different types of alkane carboxylation reactions. Wu and co-workers developed the first copper-catalyzed carbonylation reaction of cycloalkanes and amides (**83**), which was applied to the solvent-free synthesis of a wide range of imides (**84**) (Scheme 18.33) [82]. A screen for the best reagents identified the copper(I) bromide dimethyl sulfide complex [CuBr(Me_2S)] as the best metal source and 1,10-phenanthroline as the best ligand, out of several tested. In the absence of copper catalyst or ligand, only traces of the desired products were obtained. In this procedure, di-*tert*-butylperoxide (DTBP) was used as an oxidant, which was possible due to the relatively high stability of amides, which act in this process as weak nucleophiles, toward oxidants. A wide range of amides could be used, even lactams and also cycloalkanes with different ring sizes (Schemes 18.33 and 18.34). The yields varied from moderate to very high, under typical reaction conditions, i.e. p(CO) = 20 bar and 120 °C. Exceptions were the primary amides, for which no products were obtained under the standard reaction conditions, and amides in which one of the nitrogen substituents was aromatic, e.g. with *N*-phenylacetamide and *N*-phenylformamide. In these cases, only the corresponding amide was observed, presumably due to the instability of the imide formed, which decomposes to give these products.

In the mechanism proposed (Scheme 18.34), DTBP acts as an H-abstractor to form an alkyl radical from the alkane [82]. This is followed by an oxidative addition of this alkyl radical and of the *t*-BuO• radical to Cu(I). The carbonylation step occurs upon CO insertion into the Cu—C or Cu—N bond of an intermediate Cu^{III}(alkyl)(amidate) species formed by reaction with amide **85**, yielding the desired imide **86**.

Inspired by the results obtained, the Wu group developed subsequently a procedure to obtain amides using the carbonylation of cyclohexane in the presence of a variety of amines [83]. In this case, CuF_2 was identified as the best copper salt to promote this reaction. The reaction was run under solventless conditions, typically at p(CO) = 20 bar and 120 °C, and the presence of a Cu coordinating ligand was also necessary, in the absence of which no reaction took place. 1,10-Phenanthroline proved to be the best in this case too. A range of linear,

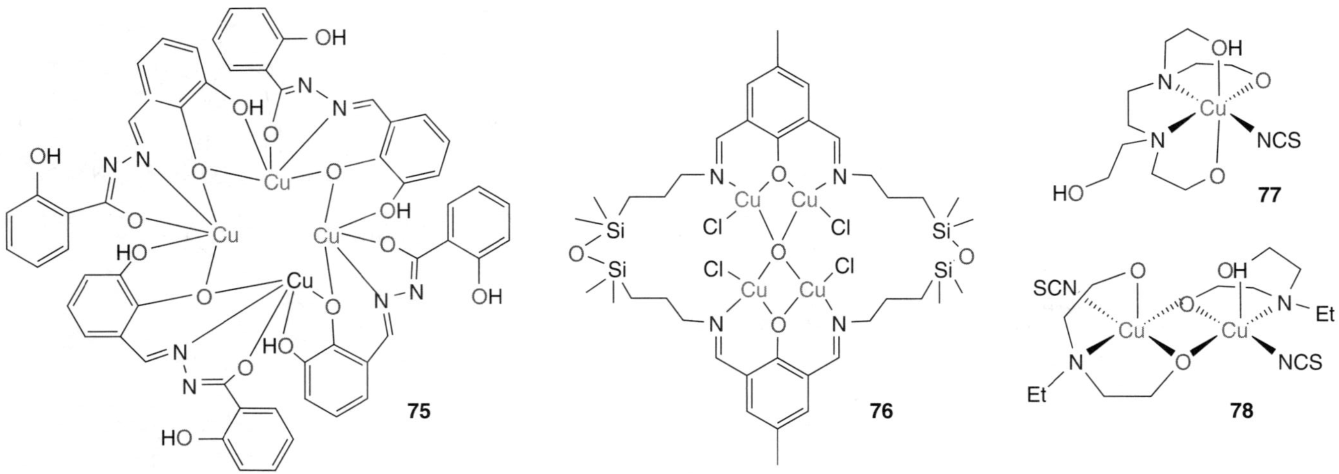

Scheme 18.31 Copper catalysts **75–78,** which catalyze the hydrocarboxylation of linear or cyclic alkanes [74–76].

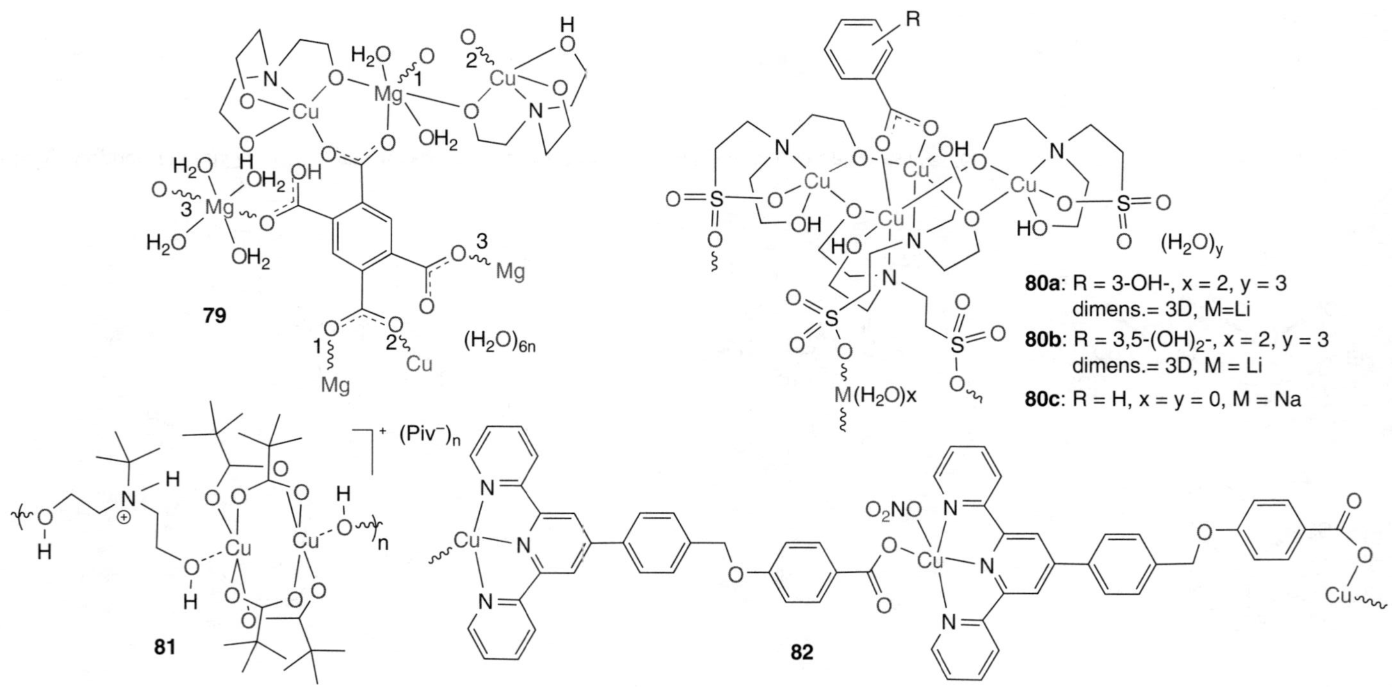

79

80a: R = 3-OH-, x = 2, y = 3 dimens.= 3D, M=Li

80b: R = 3,5-$(OH)_2$-, x = 2, y = 3 dimens.= 3D, M = Li

80c: R = H, x = y = 0, M = Na

81

82

Scheme 18.32 Metal-organic frameworks (MOFs) or coordination polymers (**79–82**), which catalyze the hydrocarboxylation of alkanes [73, 77–81].

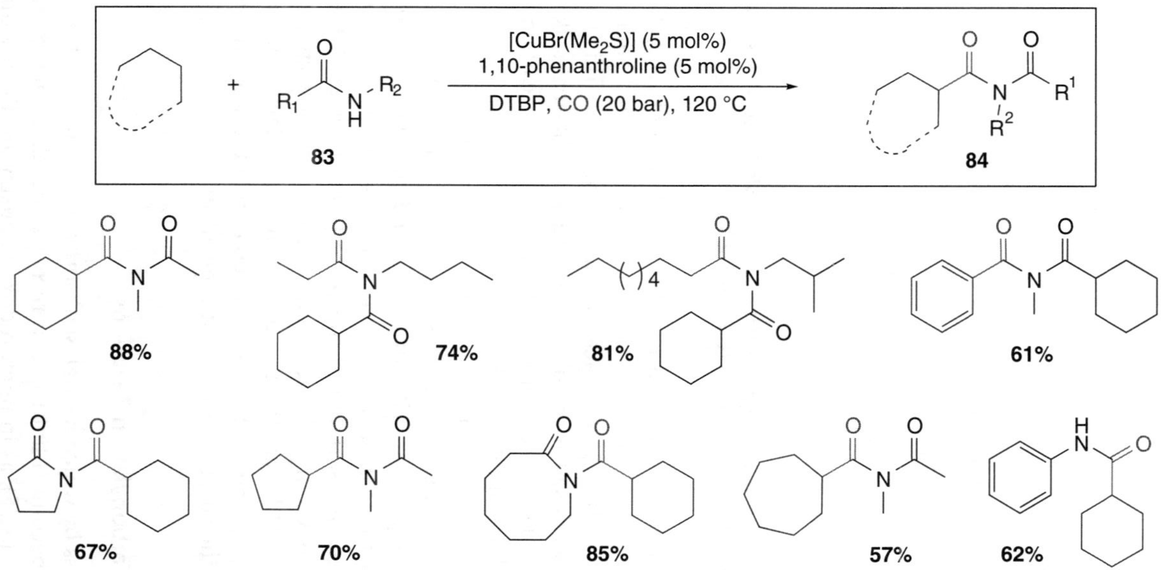

Scheme 18.33 Copper-catalyzed carbonylative synthesis of aliphatic imides from alkanes and primary amines [82].

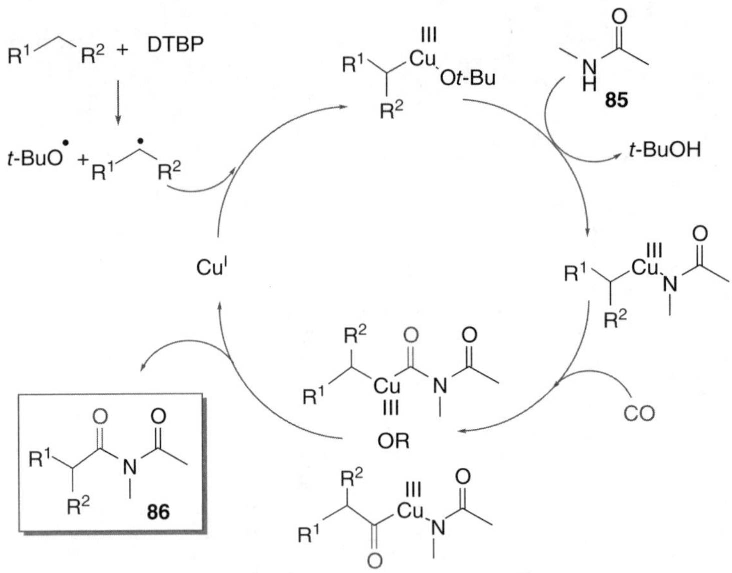

Scheme 18.34 Plausible mechanism for the copper-catalyzed carbonylative synthesis of aliphatic imides from alkanes and primary amines [82]. (*See insert for color representation.*)

aromatic, and benzylic primary amines **88** could be used, as well as cyclic and linear alkanes, to afford amides **89** (Scheme 18.35). Enantiomerically enriched amines may be used as substrates, since no racemization takes place during the reaction. This was observed when (*S*)-2-methylbenzylamine was used as the substrate. With straight-chain alkanes the selectivity is moderate, and reaction takes place preferentially at a secondary carbon atom (C-2). Some reaction occurred at C-3, but only about 5%.

In 2017, Wu and co-workers expanded the scope of the copper-1,10-phenanthroline catalysis to the carbonylative reactions of alkanes with alcohols [84]. In this case, copper chloride was found to be the best copper salt. Without catalyst and ligand only trace amounts of product were obtained. A wide range of alcohols (cyclic, linear, primary, secondary, tertiary) could be reacted in good to high yields, to afford the corresponding esters **90**, typically also with p(CO) = 20 bar and at 120 °C. In the case of linear alkanes the selectivity was poorer than with the analogous amides, although the reaction at the secondary centers was still the preferred one (Scheme 18.36). Interestingly, *in-situ* generation of the alcohol is also possible from paraformaldehyde, when the latter is reacted with the alkane.

18.3.5 Photocatalytic Carbonylation Reactions

Photocatalysis, the process by which a catalyst absorbs light and activates a reactant through a chemical reaction, has become a hot topic in recent years [85]. Substrates can be activated by single electron transfer (SET), as in photoredox chemistry, or directly through a hydrogen atom transfer (HAT) process. The most common catalysts are Ir or Ru complexes. Nevertheless, in alkane chemistry there are still very few applications of metal photocatalysis and even more particularly in alkane carbonylation reactions, although the success of the first alkane carbonylation mediated by transition metals ever reported, by Sakakura and Tanaka in 1987, relied on the concurrent use of radiation in conjunction with the rhodium catalyst *trans*-[RhCl(CO)$(PMe_3)_2$] (**91**) as described briefly in Section 18.3.1 [36]. Catalytic amounts

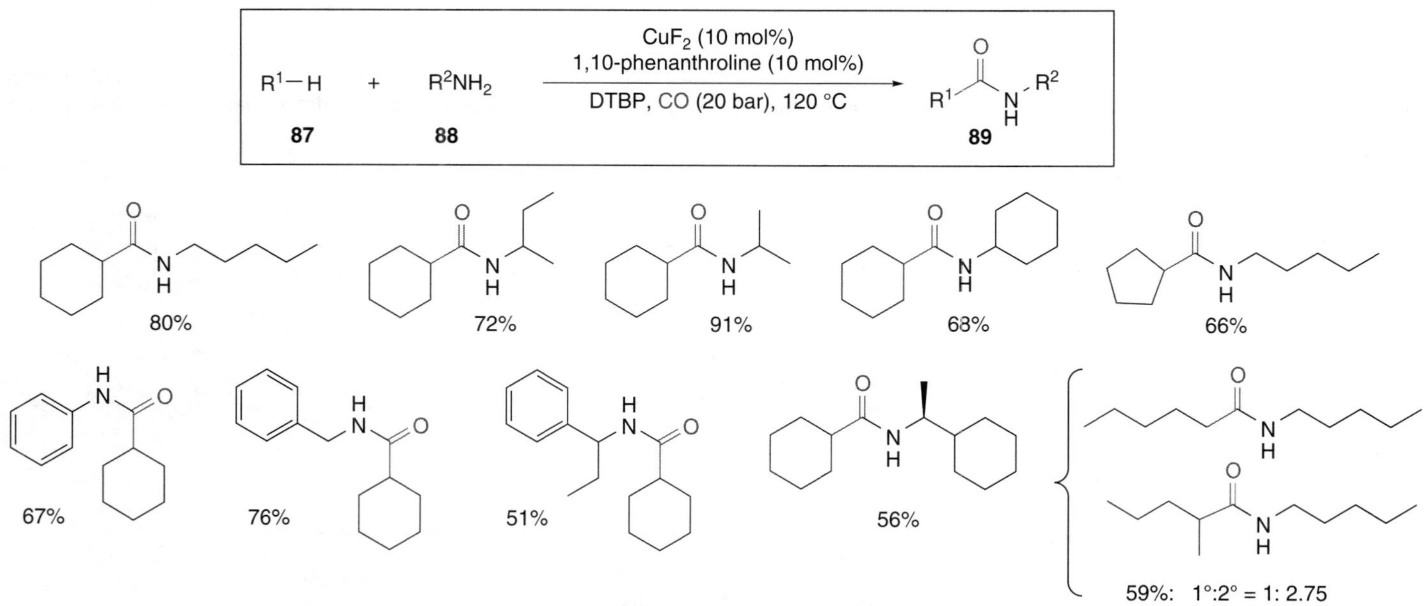

Scheme 18.35 Synthesis of aliphatic amides via copper-catalyzed carbonylation of alkanes [83].

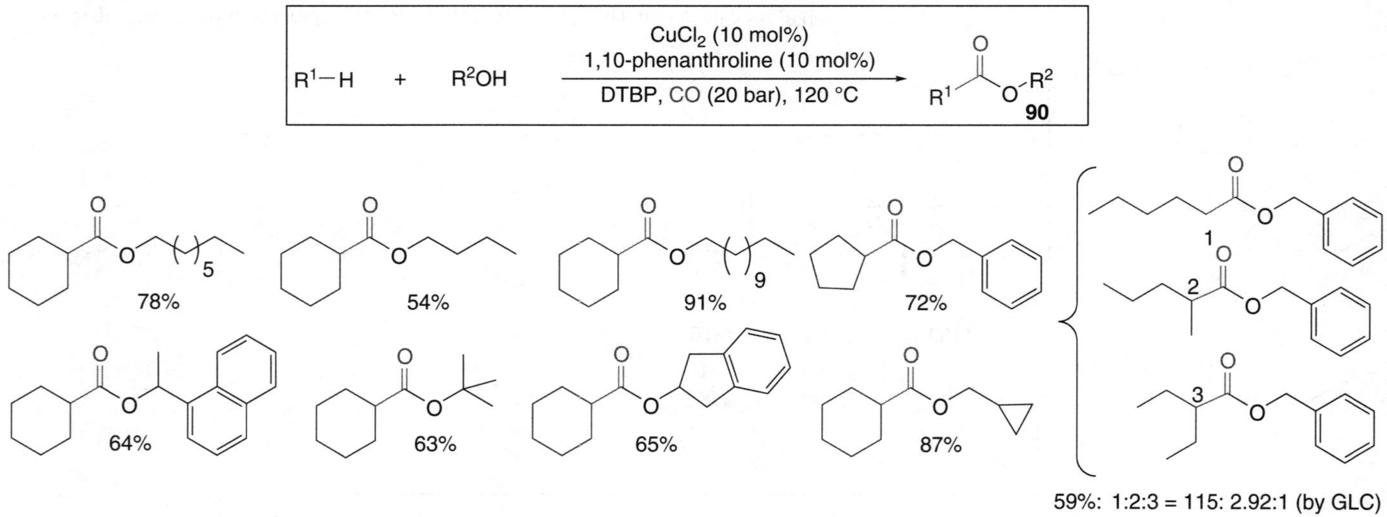

Scheme 18.36 Synthesis of esters by copper-catalyzed alkoxycarbonylation of alkanes with alcohols. The yields shown were those obtained after isolation by column chromatography [84].

of this complex promoted the carbonylation of cyclohexane and pentane with 1 atm of carbon monoxide, at room temperature, to produce a mixture of products. In the case of cyclohexane a complex mixture was obtained, consisting of cyclohexanecarboxaldehyde, cyclohexanemethanol, and cyclohexanecarboxylic acid. In the reaction of pentane (**22**) (Scheme 18.37), a mixture of products was also obtained, although the reaction was regioselective toward aldehyde formation, with the terminal aldehyde, hexanal (**92**), being produced preferentially, in 16% yield [86]. The branched aldehyde, 2-ethylbutanal (**93**) was present in <1% in the reaction mixture. The major product was, however, 1-butene (56%), which was also produced alongside with acetaldehyde (13%) and alcohols (15%). The process was truly catalytic. The reaction favored the formation of the least sterically hindered products, contrary to what was observed in radical reactions and in the reactions of superacids and superelectrophiles, i.e. a terminal, stronger C—H bond was cleaved preferentially, to produce terminal aldehydes (Scheme 18.37). This selectivity was observed mostly with short wavelength irradiation. At longer wavelengths, branched aldehydes were produced preferentially, suggesting that different reaction pathways may be involved [86].

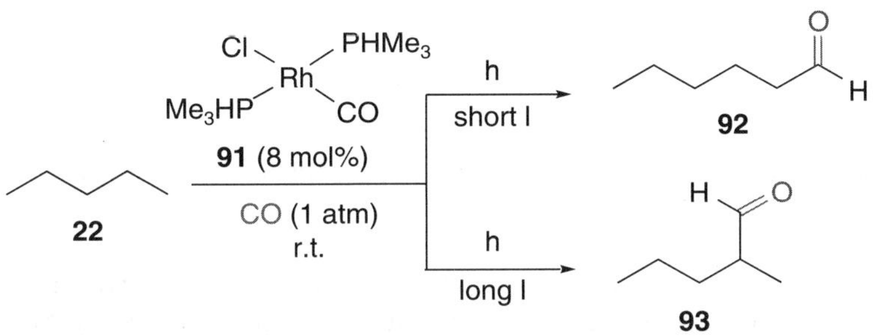

Scheme 18.37 The first transition metal catalyzed carbonylation of alkanes using substoichiometric quantities of the metal [36].

In this reaction, it was assumed that one of the effects of photocatalysis was to create the vacant coordination site for the alkane oxidative addition i.e. that the catalytically active species was $RhCl(PMe_3)_2$ (**94**) and that the oxidative addition was facilitated by the use of electron-rich phosphines as ligands. This assumption was based on the fact that in flash photolysis experiments with $RhCl(CO)(PPh_3)$, with light of wavelength >315 nm, the formation of $RhCl(PPh_3)_2$ was observed. The mechanism proposed for the reaction, consisting of oxidative addition of the C—H bond, photo-assisted CO dissociation, CO insertion into the carbon-metal bond and reductive elimination, is shown in Scheme 18.38.

Another pathway that could lead to aldehyde production would be dehydrogenation of the alkane, followed by hydroformylation of the resulting alkenes. However, this mechanism was ruled out, since C_5 aldehydes should have been produced by this mechanism from butenes and they were not detected. The major product obtained in this experiment, 1-butene, was in fact a secondary product formed together with acetaldehyde via a Norrish Type II photoreaction of hexanal. This was shown by a control experiment, since when the product of the carbonylation reaction, 1-hexanal, was irradiated for 16.5 hours in pentane in the absence of catalyst, these two substances were obtained as products. Similar results were obtained in the carbonylation of a longer-chain *n*-alkane, decane, i.e. the main products were 1-nonene and undecanal [86].

In addition to the metal-mediated insertion of CO into the C—H bond, rhodium catalyst **91** was subsequently shown to be also effective in promoting the dehydrogenation of alkanes to olefins with turnovers ranging up to several hundred [87].

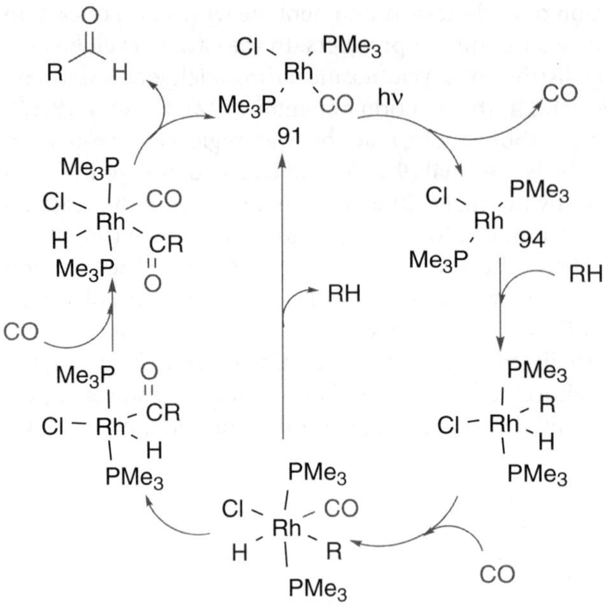

Scheme 18.38 A probable mechanism for the photocatalytic rhodium-catalyzed carbonylation of pentane. The steps shown are reversible [36].

A class of photocatalysts that activate alkanes via HAT processes are the polyoxometalates, and more specifically, decatungstate salts, e.g. tetrabutylammonium decatungstate [$(nBu_4N)W_{10}O_{32}$] (TBADT). This cluster can activate C—H bonds in a variety of compounds. It does not absorb in the visible region of the spectrum, but there is overlap with solar light emission. It abstracts a hydrogen atom from a C—H bond, leading to the formation of a nucleophilic carbon-centered radical, which reacts further. In the first example of this chemistry, reported in 1995 by Hill and Jaynes [88, 89], it was shown that cyclohexane could be converted to cyclohexane carbaldehyde in the presence of TBADT, light, and CO (1 atm), although the conversion was low (8% yield), mainly because the newly formed carboxaldehyde underwent decarbonylation as a secondary reaction. In 2011, the same photocatalyst was used in a much more efficient manner to promote the conversion of alkanes to unsymmetrical ketones [90].

Ryu and co-workers developed atom-economical three-component reactions between alkanes (**87**), carbon monoxide, and electron deficient alkenes (**95**), which provided the desired carbonylated products (**96**) in moderate to very high yields upon irradiation with a 500 W xenon lamp in the presence of 4 mol% TBADT, typically in MeCN, at p(CO) = 80 atm (Scheme 18.39) [90]. Initially, a substantial amount of reduced product was obtained, but the side reaction was effectively suppressed when the quartz liner of the reactor windows was changed to a Pyrex-glass liner. The pressure of CO was kept at 80 atm, because at low pressures the formation of noncarbonylated products competed. The mechanism proposed (Scheme 18.40), which was corroborated by deuterium-labeling experiments, involved H-abstraction from the alkane by the excited TBADT, to give the reduced catalyst and the alkyl radical ($R^•$), which upon carbonylation forms the acyl radical ($RCO^•$) that adds to the olefin. When the reaction was performed in CD_3CN, there was no incorporation of deuterium in the product, but when it was performed in $MeCN/D_2O$ there was, which showed that hydrogen was donated exclusively by the reduced photocatalyst, which thus regenerates the oxidized form, and not by the reaction medium,

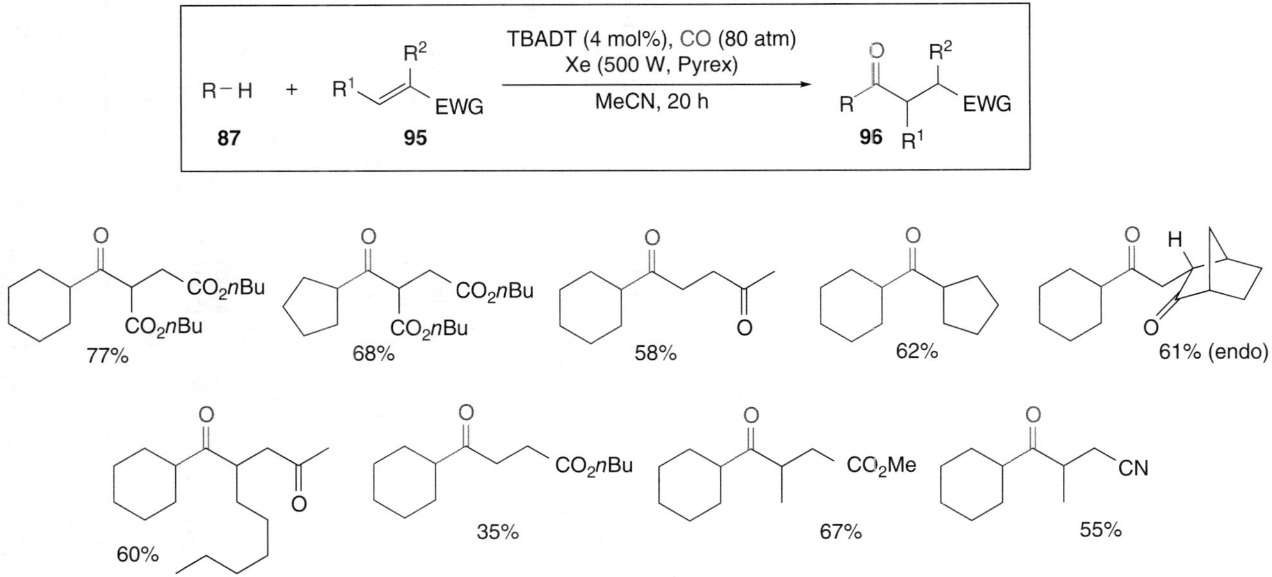

Scheme 18.39 Carbonylation of alkanes via polyoxotungstate photocatalysis [90].

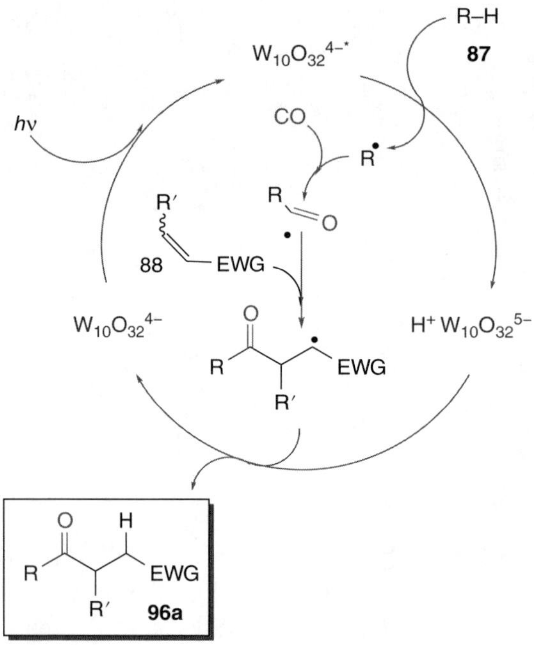

Scheme 18.40 Mechanistic proposal for the radical carbonylation of alkanes via polyoxotungstate photocatalysis [90].

i.e. the deuterium in the product came from $D^+W_{10}O_{32}^{-5}$, resulting from H/D exchange between the catalyst and D_2O.

In 2013, the Ryu and Fagnoni team reported another example of alkane activation by photocatalysis with TBADT, the three-component coupling between cyclic alkanes, CO, and diisopropyl azodicarboxylate (**97**) to give the corresponding acyl hydrazides (**98**) (Scheme 18.41) [91]. The C—C and C—N bond-forming reactions took place smoothly to provide the adducts in moderate to good yields. Since many acyl hydrazides are antioxidants, the method has potential applications in synthesis. At the same time, the direct addition of alkanes to azodicarboxylates in the absence of CO was also reported, and in that case the corresponding adducts were also obtained in good yields.

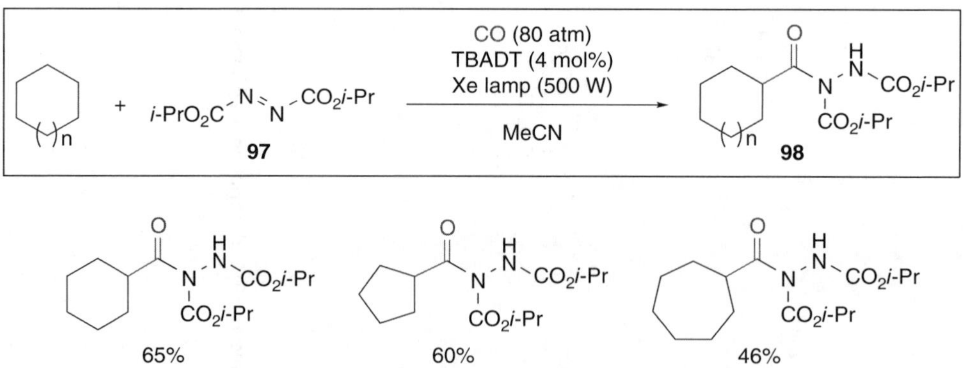

Scheme 18.41 Synthesis of acyl hydrazides via a photocatalytic three-component reaction [91].

18.4 Carbene Reactions

In the alkane carbonylation reactions covered in the previous section, there is formation of a C—C bond between an alkyl component derived from an alkane and a carbonyl group, which can proceed via addition of CO to the alkyl radical or via insertion into a metal-alkyl bond. Carbenes are reactive species of divalent carbon with electronic similarities to CO and can also be applied in alkane reactions with formation of a C—C bond, namely upon carbene insertion. In a carbene insertion reaction, a diazo compound, e.g. **99**, often reacts with a transition metal complex forming, upon N_2 loss, a transient electrophilic metallocarbene. The metallocarbene then reacts with a nucleophile present in the reaction medium, e.g. an alkane, in such a way that the carbene moiety is transferred from the metal center to the alkane, becoming inserted into one C—H σ-bond (Scheme 18.42). A new C—C bond and a new C—H bond are created as in **100** [92, 93]. The order of reactivity parallels that of the bond dissociation energies, i.e. tertiary > secondary > primary > methane.

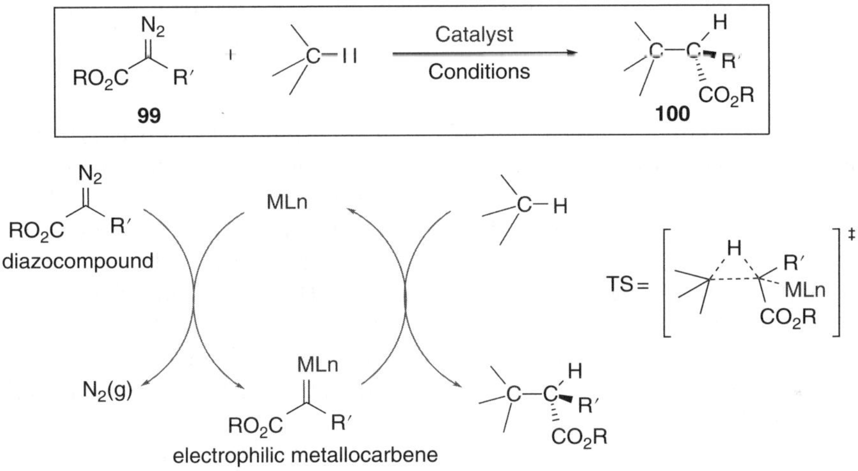

Scheme 18.42 Catalytic carbene insertion into a C—H bond of an alkane and a probable mechanism [92, 93].

A challenge in these reactions, as in the carbonylation of alkanes, is how to control regioselectivity. A useful feature is that there is retention of configuration at the carbon atom of the reacting C—H bond. The metals more commonly used for alkane C—H insertion reactions are the coinage metals (Cu, Ag, Au) and Rh, Ru, and Fe [94]. The more typical ligands are porphyrins for Rh(III), Fe or Ru, acetate and derivatives for Rh(II) and polypyrazolylborate or *N*-heterocyclic carbene (NHC) for coinage metals (Cu(I), Ag(I), or Au(I)) [94]. Theoretical studies by Nakamura et al. [95] confirmed Doyle's mechanistic proposals [96] that C—C and C—H bond formation occur at the same time that the metal-ligated carbene dissociates (Scheme 18.42), i.e. they are concerted, although they do not necessarily occur at the same rate. There is no direct interaction between the metal and the alkane C—H bond during the reaction [93]. A competitive reaction that may occur is the coupling of two carbene units, i.e. carbene dimerization, leading to the formation of olefins, as established by Pérez and co-workers [94]. Strategies to prevent this reaction often have to be used.

The carbenes used for C—H activation fall into three general groups: acceptor-substituted, i.e. one of the carbon atom substituents is an electron-withdrawing group, acceptor-acceptor substituted, in which both carbon atom substituents are electron-withdrawing and donor-acceptor substituted, in which one substituent is an electron withdrawing group and the other an electron donating group. The nature of the substituents is critical for the stability of the resulting metallocarbene and the latter must also be sufficiently electrophilic to react with a nucleophile as weak as an alkane. The more stable metallocarbenes, i.e. the less reactive ones, are the donor-acceptor substituted carbenes, and they are also the more chemoselective ones [97]. The reactivity follows the same order as the basicity of the compounds (Scheme 18.43). The metallocarbene patterns of substitution limit somewhat the nature of the products which can be obtained by carbene insertion reactions. However, diazoalkanes are usually not easy to prepare, they are unstable and even explosive. The presence of an aromatic ring that allows for conjugation increases their stability, but since they cannot be stored for very long, they are usually prepared *in situ*. The diazocarbonyl compounds, on the contrary, are more stable for longer periods of time and do not decompose easily in the absence of a catalyst. They are commonly used in the alkane insertion reactions.

The data in Scheme 18.44, from Davies et al., give an idea of the relative reactivities of several species, including alkanes, toward C—H bond insertion by aryldiazocarbenes and the regioselectivities possible [98]. The figures were obtained from competition experiments between two species, **101** and **102**, which reacted with diazocarbene **103** in the presence of $[Rh_2(S\text{-DOSP})_4]$. Although electron rich C—H bonds react preferentially, site selectivity is also dependent on steric factors [92].

Research on this topic is nowadays directed mainly at the development of novel catalysts, and the information available has been grouped according to the metal used to highlight some recent developments. The study of carbene insertion reactions into alkanes dates back to 1958. However, it was only in the early 1980s that the first results showing the reaction could have some synthetic applicability, attributed to Teyssie and his group, were reported. They concerned the carbon–hydrogen insertion reactions of ethyl diazoacetate (**106**: EDA) into alkanes, catalyzed by dirhodium(II) tetraacetate and related rhodium carboxylates [99, 100]. The selectivity and yields of the reactions were low, but the results already showed the influence that the ligand used could have on regioselectivity. $[Rh_2(9\text{-trp})_4]$, based on rhodium complexed to ligand **107**, was the catalyst that provided the highest yields (Scheme 18.45).

Rhodium was also shown to have better reactivity than copper (Scheme 18.46) [101]. In an application in total synthesis, virescenol B diacetate was transformed into diazoketone **108**. Intramolecular carbene insertion into a $C(sp^3)$—H bond allowed the construction of ring D of the tetracyclic steroid skeleton of **109.**

In 1982, Callot and Metz also showed that the reactions of *n*-alkanes (C_6–C_{12}) with diazoesters and rhodium(III) porphyrins gave products resulting primarily from attack at the primary C—H bonds [102]. The yields were also quite low (up to 20–25%). Presumably, in this case, the steric bulk of the ligands was enough to override any electronic effects. Recently, in 2008, Che and co-workers, using a bulky porphyrin ligand on catalyst **110**, also reported the same trend in the reactions of alkanes, e.g. the linear alkanes **42** and **111–112** to yield esters **113–115** (Scheme 18.47) [103].

Encouraged by the promising results obtained with the first catalyst, Che and co-workers prepared a chiral porphyrin, **117**, to see if an enantioselective reaction was possible. Good to high yields of products could be obtained in very high *ee*'s (Scheme 18.48) [103]. In the case of linear substrates, the main product was indeed that resulting from reaction at the primary C—H bond.

Coinage metal catalysts have also been used for alkane insertion reactions, but their activities, until 2008, were generally lower than those obtained with rhodium catalysis. In 2008, Etiénne and co-workers developed novel Ag- and Cu-based catalysts containing perfluorinated scorpionate ligands, prepared from the new tris(indazolyl)borate $[F_{21}{}^{-}Tp^{4Bo,3CF3}]Tl$ (**118**), of

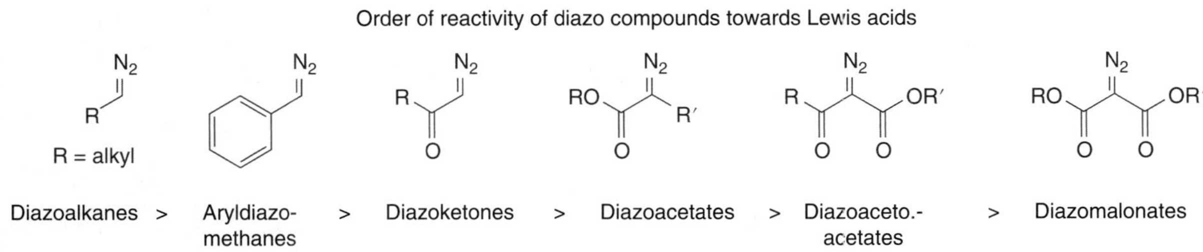

Scheme 18.43 The order of reactivity of some diazo compounds towards Lewis acids [97].

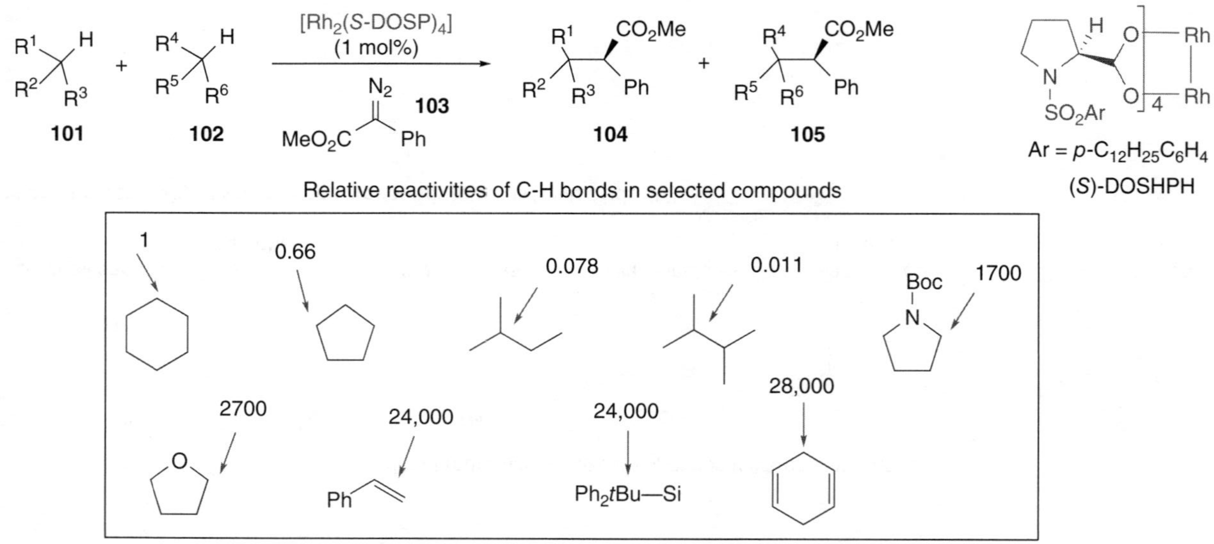

Scheme 18.44 Competition experiments in carbene insertion reactions of alkanes to elucidate the relative reactivities of a selection of alkanes [98].

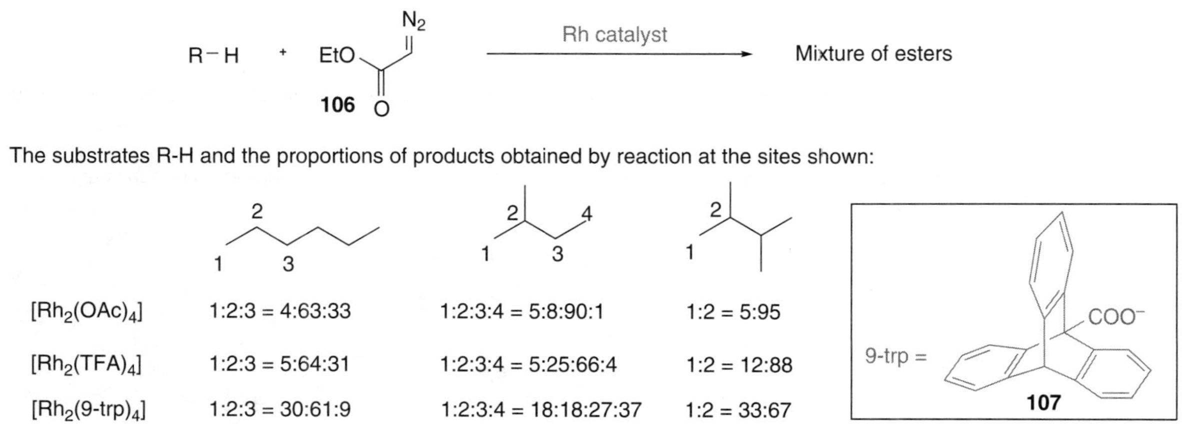

Scheme 18.45 Rhodium-catalyzed carbene insertion reactions of light alkanes [100].

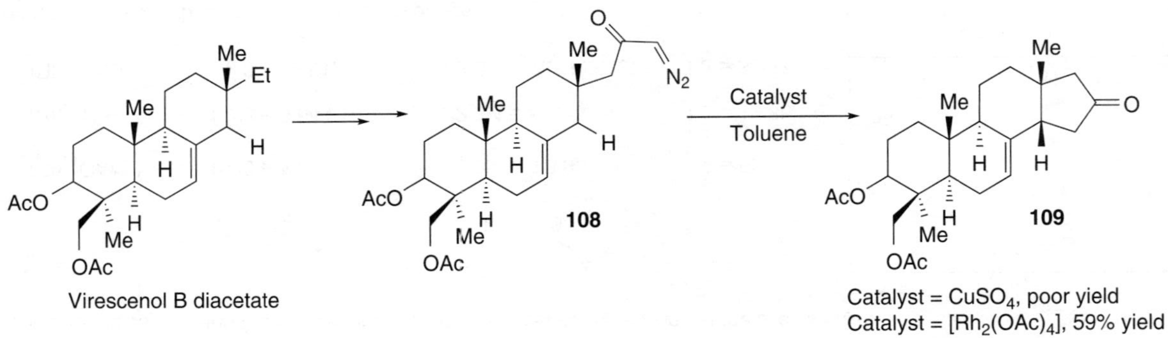

Scheme 18.46 Intramolecular carbene-insertion reaction of diazoketones into alkanes for the synthesis of steroids [101].

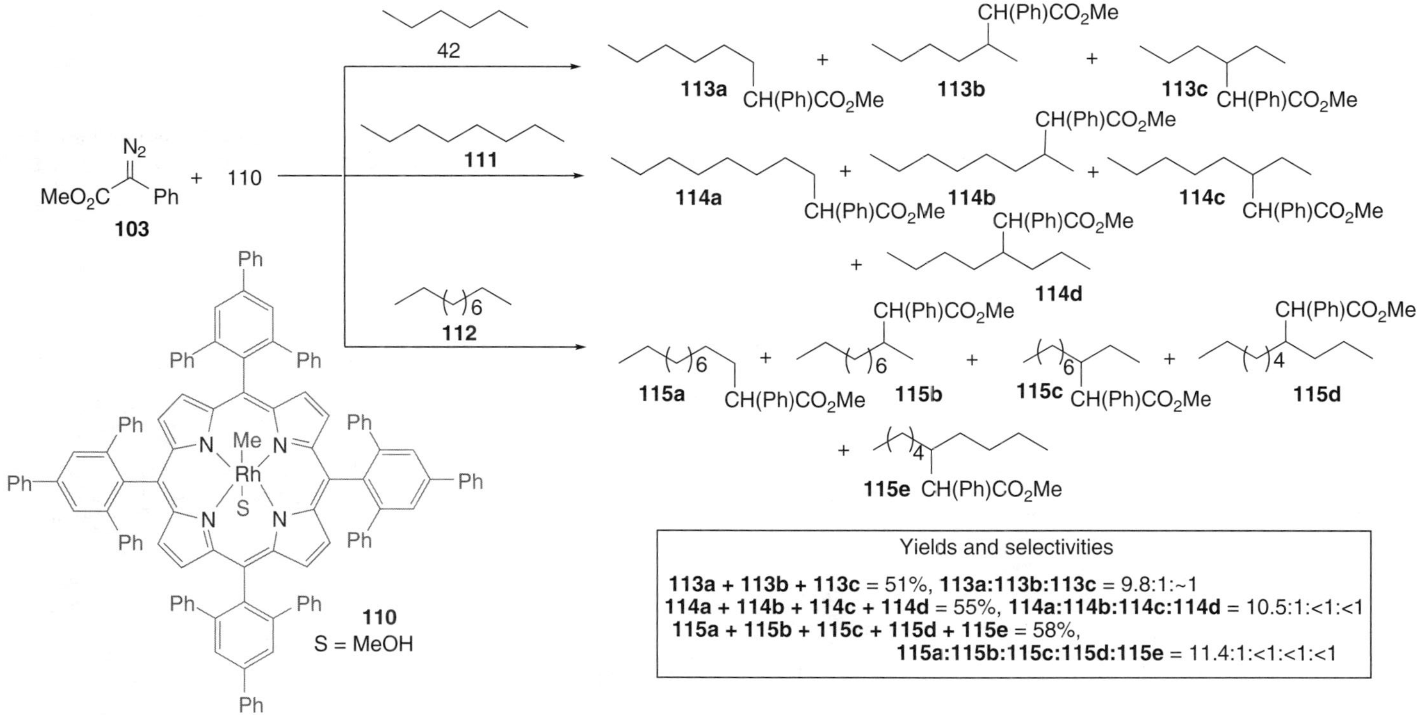

Scheme 18.47 Rhodium-catalyzed homologation of n-alkanes with diazoesters [103].

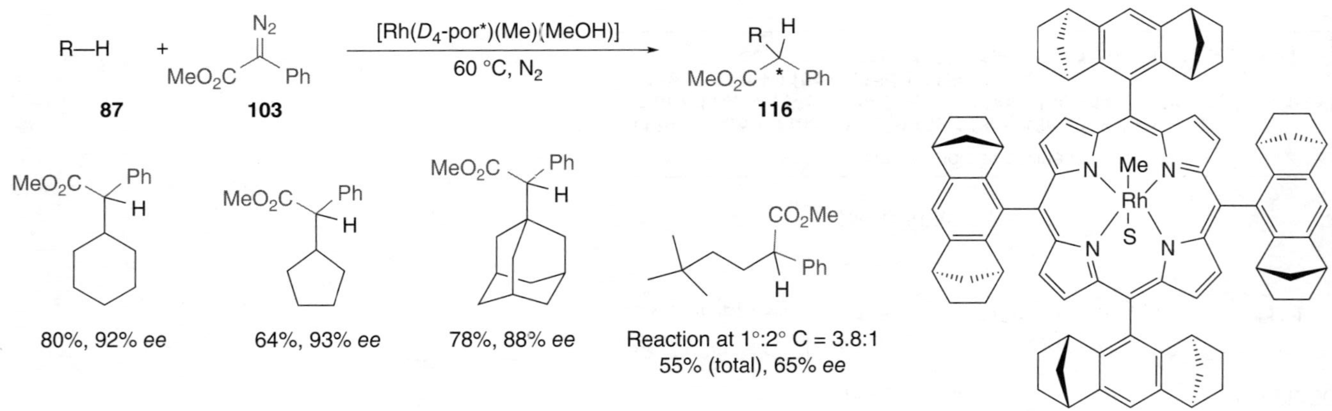

117: [Rh(D_4-por*)(Me)(MeOH)]

S = MeOH

Scheme 18.48 Enantioselective rhodium-catalyzed carbon-carbon bond formation by carbene insertion into alkanes. In the reactions with cyclic alkanes, 0.1 mol% catalyst was used, for the linear, 1.5 mol% [103].

which [F_{21}-$Tp^{4Bo,3CF3}$]ML (M = Cu, L = NCMe; M = Ag, L = Me_2CO) were the best. Taking into account that the metal centers in the new complexes should be electron deficient, the complexes were employed to promote the insertion of EDA into C(sp^3)—H bonds of alkanes (Scheme 18.49) [104]. The silver complex with the borate ligand from **118** in particular displayed excellent activity, and the products could be obtained in high yields (>90%) with low catalyst loadings (0.5%). TON values of 200 were obtained, which were 10 times higher than those previously reported for any other group 11 metal-based catalyst for this transformation. Interesting, to ensure success it was necessary to add EDA over 3 hours, because when it was added at once the desired products were not obtained.

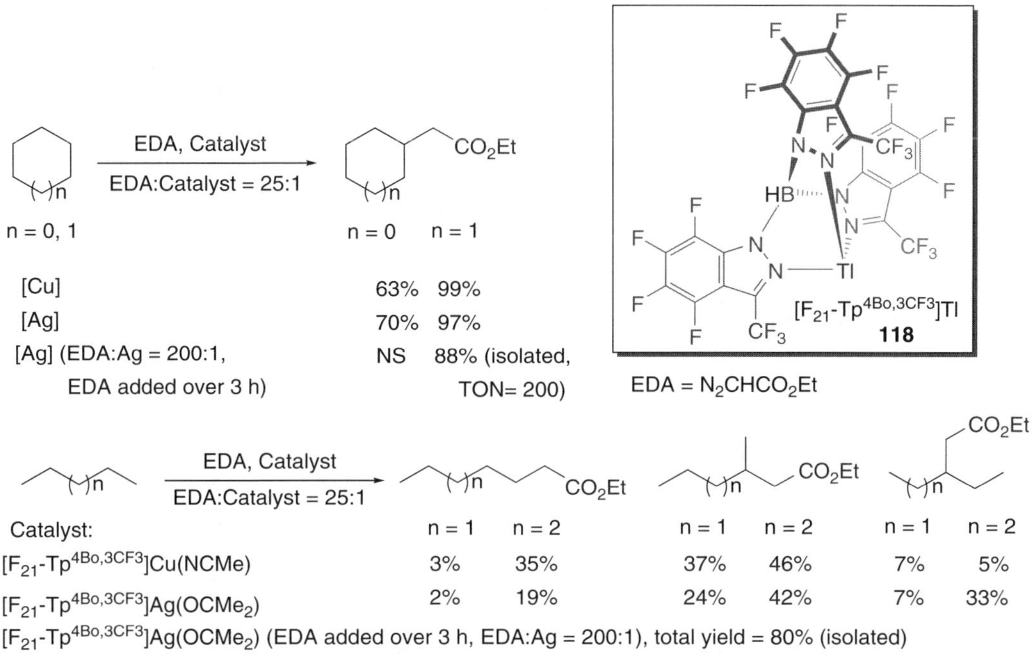

Scheme 18.49 Functionalization reactions of linear and cyclic alkanes by carbene insertion reactions promoted by transition metals complexed to a novel perfluorinated F21-Tp scorpionate ligand [104].

A preference for functionalization of secondary over primary C—H bonds was observed, particularly with the silver catalyst. These results are in agreement with theoretical studies performed by Pérez and co-workers in the past with homoscorpionate complexes [105], which showed that the energy barriers for the carbene insertion reaction follow the order tertiary > secondary and > primary and that for a given type of C—H bond the barriers for silver catalysts are lower than those for the copper catalysts. Later, the authors used the same silver catalyst to investigate reactions of methane with EDA. To homogenize methane, a solvent was required, which presented a problem since the C—H bonds of methane are less reactive than those of common solvents. A solution was found in the use of supercritical CO_2 ($scCO_2$) as reaction medium. In the presence of the silver perfluorinated catalysts, ethyl propionate could be obtained from methane in 19% yield (based on EDA) when the reaction was tried at 40 °C [106]. Similar procedures were later developed by Pérez and co-workers for the other gaseous alkanes, with the reactions following selectivity trends similar to those observed previously [107].

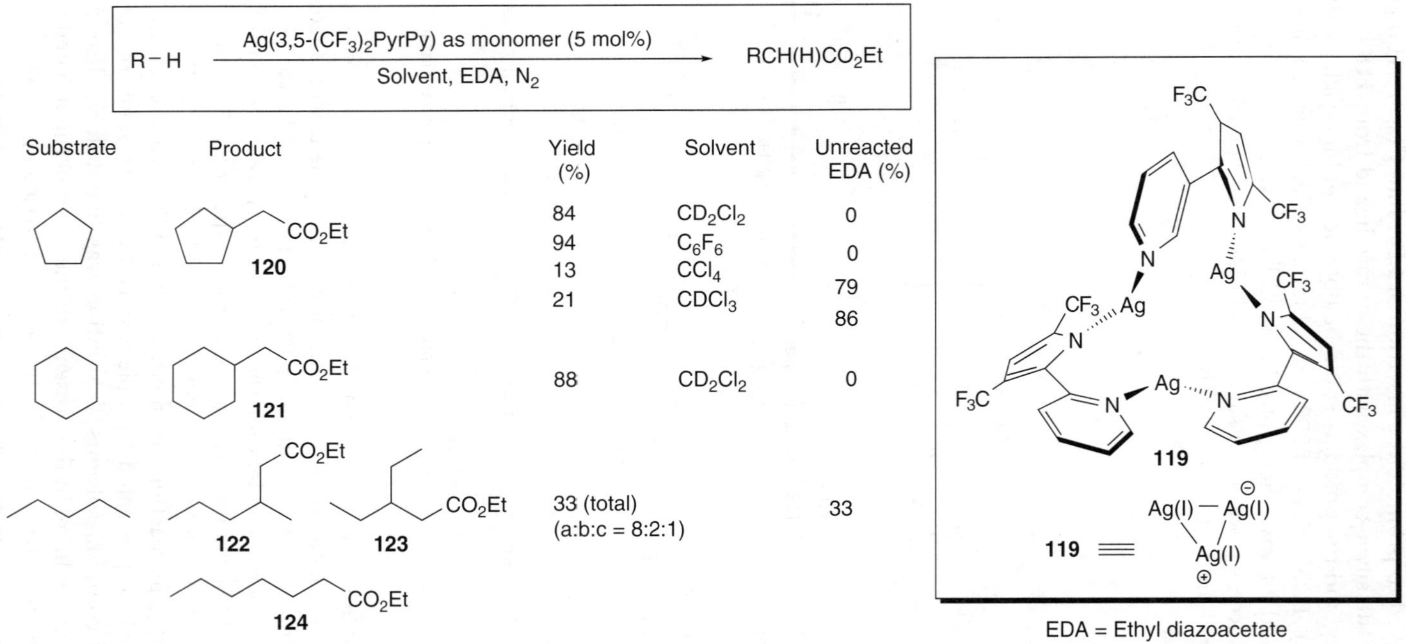

Scheme 18.50 Functionalization of alkanes by carbene insertion reactions promoted by the silver catalyst **119**. The reactions performed in the dark for three days [108].

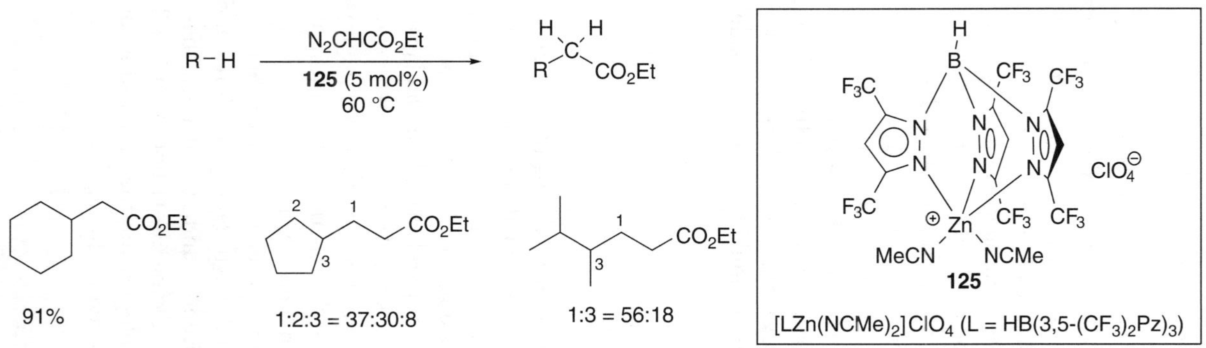

Scheme 18.51 Zinc-catalyzed carbene insertion of diazoesters into alkanes [109].

Caulton, Mindiola, and co-workers have used a different type of silver complexes to promote the insertion of EDA into the C—H bond of a series of alkanes at room temperature [108]. The trinuclear cluster $[Ag_3(\mu_2\text{-}3,5\text{-}(CF_3)_2PyrPy)_3]$ (**119**, $3,5\text{-}(CF_3)_2PyrPy$ = 2,2′-pyridylpyrrolide) has silver in three distinct environments, but the catalytic active species was assumed to be the transient silver carbene monomer $(3,5\text{-}(CF_3)_2PyrPy)Ag(CHCO_2Et)$. Deuterated solvents were used as reaction media, and the yields of product were found to be strongly dependent on the nature of the solvent used. In this case, insertion at secondary C—H bonds was also found to prevail (Scheme 18.50).

The cationic zinc adduct $[\{HB(3,5\text{-}(CF_3)_2Pz)_3\}Zn(NCMe)_2]ClO_4$ (**125**) was also found recently to catalyze the insertion of EDA into the C—H bonds of alkanes [109]. This zinc catalyst, with a scorpionate ligand, formed an intermediate that computational studies suggested was more a zinc(II)-bound carbocation than a zinc carbene and provided high yields of products (Scheme 18.51). Nevertheless, the species that is transferred to the alkane during the reaction is a neutral: $CHCO_2Et$ carbene moiety. Besides the use of this preprepared catalyst, a few experiments were conducted by mixing the ligand with metal ions *in situ*. *In-situ* generated $[\{HB(3,5\text{-}(CF_3)_2Pz)_3\}Zn]^+$species containing the bulkier $[B\{3,5\text{-}(CF_3)_2C_6H_3\}_4]^-$ anion gave the best results among the zinc catalysts used, i.e. a 98% yield of product with cyclohexane. The original ligand, $HB(3,5\text{-}(CF_3)_2Pz)_3$, when reacted *in situ* with silver ions in THF gave a more selective reaction with 2,3-dimethylbutane, i.e. a 75% to 20% yield of products was obtained resulting from reaction at the primary versus the tertiary C—H bond. These new zinc catalysts suggest that future research in this area should be fruitful, considering that zinc is an abundant and inexpensive metal.

18.5 Conclusions

The carbonylation and carbene insertion reactions of alkanes have attracted a high level of attention in recent years with the development of novel catalysts and new methodologies, which make it possible to achieve more selective and higher yielding transformations with these relatively inert substrates. It is now possible to convert linear and cyclic alkanes into carboxylic acids, ketones, esters, or amides with superelectrophiles, and even achieve certain difunctionalizations through reactions at remote sites. Transition metal catalysts allowed the conversion of alkanes directly into acids, esters, amides, and imides in high yields, and even into ketones by photochemical processes. Through the reactions of alkanes with carbenes it is now possible to obtain ketones or esters directly in high yields.

The use of lower temperatures and lower pressures is becoming more common and the novel one-pot transformations simplify synthetic procedures and product recovery, making processes more environmentally friendly and sustainable. These trends, together with the ability to activate and functionalize alkane C—H bonds with milder reagents, hopefully will allow the development of methodologies for the application of alkanes in target-oriented synthesis to be a reality of a not-so-distant future.

Acknowledgments

Financial support from Fundação para a Ciência e a Tecnologia (FCT), Portugal, for projects PTDC/QEQ-QIN/3967/2014 and UID/QUI/00100/2013, is gratefully acknowledged.

References

1 Pombeiro, A.J.L. (2014). Toward functionalization of alkanes under environmentally benign conditions. In: *Advances in Organometallic Chemistry and Catalysis: The Silver/Gold Jubilee ICOMC Celebratory Book* (ed. A.J.L. Pombeiro), 15–25. Hoboken), Chapter 2: Wiley.
2 Gadge, S.T., Gautam, P., and Bhanage, B.M. (2016). *Chem. Rec.* 16: 835–856.
3 Hartwig, J.F. (2016). *J. Am. Chem. Soc.* 138: 2–24.
4 Shul'pin, G.B. (2016). *Catalysts* 6: 1–40.
5 Gunsalus, N.J., Koppaka, A., Park, S.H. et al. (2017). *Chem. Rev.* 117: 8521–8573.
6 Liu, Q., Zhang, H., and Lei, A. (2011). *Angew. Chem. Int. Ed.* 50: 10788–10799.
7 Díaz-Requejo, M.M. and Pérez, P.J. (2008). *Chem. Rev.* 108: 3379–3394.
8 Shul'pin, G.B. (2013). *Dalton Trans.* 42: 12794–12818.
9 Paatz, R. and Weisgerber, G. (1967). *Chem. Ber.* 100: 984–990.
10 Olah, G.A. and Klumpp, D.A. (2008). *Superelectrophiles and their Chemistry*. Hoboken: Wiley.
11 Sommer, J. and Kala, J. (1993). *Acc. Chem. Res.* 26: 370–376.
12 Astruc, D. (2007). C-H activation and functionalization of alkanes and arenes. In: *Organometallic Chemistry and Catalysis*, 409–418. Berlin), Chapter 17: Springer-Verlag.
13 Farooq, O., Marcelli, M., Prakash, G.K.S., and Olah, G.A. (1988). *J. Am. Chem. Soc.* 110: 864–867.
14 Olah, G.A., Prakash, G.K.S., Mathew, T., and Marinez, E.R. (2000). *Angew. Chem. Int. Ed.* 39: 2547–2548.
15 Vol'pin, M., Akhrem, I., and Orlinkov, A. (1989). *New J. Chem.* 13: 771–790.
16 Olah, G.A., Rasul, G., Yudin, A.K. et al. (1996). *J. Am. Chem. Soc.* 118: 1446–1451.
17 Hogeveen, H. (1973). The reactivity of carbonium ions towards carbon monoxide. In: *Advances in Physical Organic Chemistry* (ed. V. Gold), 29–52. London), Chapter 10: Academic Press.
18 Akhrem, l., Orfinkov, A., Afanas'eva, L. et al. (1999). *Tetrahedron Lett.* 40: 5897–5900.
19 Akhrem, I.S. (2003). *Russ. Chem. B. Int. Ed.* 52: 2606–2627.
20 Zhorov, Y.M. (1985). *Thermodynamics of Chemical Processes*", Chemistry, 182. Moscow: Mir Publishers.
21 Akhrem, I.S., Afanas'eva, L.V., Vitt, S.V., and Petrovskii, P.V. (2002). *Mendeleev Commun.* 12: 180–182.
22 Akhrem, I.S., Afanas'eva, L.V., Petrovskii, P.V. et al. (2001). *Russ. Chem. B. Int. Ed.* 50: 2394–2400.
23 Akhrem, I.S. (2015). *J. Organomet. Chem.* 793: 54–77.
24 Bagrii, E.I. (1989). *Adamantany: poluchenie, svoistva, primenenie (Adamantanes: Production, Properties. Application)*. Moscow: Nauka (in Russian).
25 Akhrem, I.S. and Orlinkov, A.V. (2007). *Chem. Rev.* 107: 2037–2079.
26 Akhrem, I., Avetisyan, D., Afanas'eva, L. et al. (2010). *Tetrahedron Lett.* 51: 907–911.
27 Akhrem, I.S., Avetisyan, D.V., Afanas'eva, L.V. et al. (2011). *Mendeleev Commun.* 5: 259–261.
28 Akhrem, I.S., Avetisyan, D.V., Vitt, S.V., and Petrovskii, P.V. (2005). *Mandeleev Commun.* 15: 185–187.
29 Orlinkov, A.V., Kagramanov, N.D., Petrovskii, P.V., and Akhrem, I.S. (2010). *Tetrahedron Lett.* 51: 259–263.

30 Akhrem, I.S., Avetisyan, D.V., Afanas'eva, L.V. et al. (2011). *Mendeleev Commun.* 6: 323–325.

31 Akhrem, I.S., Afanas'eva, L.V., Kagramanov, N.D., and Petrovskii, P.V. (2012). *Tetrahedron Lett.* 53: 4221–4224.

32 Akhrem, I.S., Avetisyan, D.V., Churilova, I.M. et al. (2013). *Tetrahedron Lett.* 54: 6037–6040.

33 Crabtree, R.H. (2004). *J. Organomet. Chem.* 689: 4083–4091.

34 Fei, Q., Guarnieri, M.T., Tao, L. et al. (2014). *Biotechnol. Adv.* 32: 596–614.

35 Haynes, C.A. and Gonzalez, R. (2014). *Nat. Chem. Biol.* 10: 331–339.

36 Sakakura, T. and Tanaka, M. (1987). *J. Chem. Soc., Chem. Commun.* 758–759.

37 Enikolopyan, N.S., Menchikova, G.N., and Grigoryan, E.A. (1986). *Dokl. Akad. Nauk SSSR* 291: 110–113, Chem. Abstr. 1987, 107, 962932.

38 Lin, M. and Sen, A. (1994). *Nature* 368: 613–615.

39 Sen, A. (1998). *Acc. Chem. Res.* 31: 550–557.

40 Lin, M., Hogan, T.E., and Sen, A. (1996). *J. Am. Chem. Soc.* 118: 4574–4580.

41 Chepaikin, E.G., Boyko, G.N., Bezruchenko, A.P. et al. (1998). *J. Mol. Catal. A: Chem.* 129: 15–18.

42 Chepaikin, E.G., Bezruchenko, A.P., Boiko, G.N. et al. (2006). *Kinet. Catal.* 47: 12–19.

43 He, J., Wasa, M., Chan, K.S.L. et al. (2017). *Chem. Rev.* 117: 8754–8786.

44 Fujiwara, Y., Takaki, K., Watanabe, J. et al. (1989). *Chem. Lett.* 18: 1687–1688.

45 Jia, C., Kitamura, T., and Fujiwara, Y. (2001). *Acc. Chem. Res.* 34: 633–639.

46 Fujiwara, Y., Takaki, K., and Taniguchi, Y. (1996). *Synlett* 591–599.

47 Nakata, K., Watanabe, J., Takaki, K., and Fujiwara, Y. (1991). *Chem. Lett.* 20: 1437–1438.

48 Periana, R.A., Mironov, O., Taube, D. et al. (2003). *Science* 301: 814–818.

49 Zerella, M., Mukhopadhyay, S., and Bell, A.T. (2004). *Chem. Commun.* 1948–1949.

50 Zerella, M., Kahros, A., and Bell, A.T. (2006). *J. Catal.* 237: 111–117.

51 Chempath, S. and Bell, A.T. (2006). *J. Am. Chem. Soc.* 128: 4650–4657.

52 Ribeiro, A.P.C., Martins, L.M.D.R.S., and Pombeiro, A.J.L. (2017). *Green Chem.* 6: 1499–1501.

53 Liu, H., Laurenczy, G., Yan, N., and Dyson, P.J. (2014). *Chem. Commun.* 50: 341–343.

54 Lu, L., Shi, R., Liu, L. et al. (2016). *Chem. Eur. J.* 22: 14484–14488.

55 Sutradhar, M. and Pombeiro, A.J.L. (2014). *Coord. Chem. Rev.* 265: 89–124.

56 Taniguchi, Y., Hayashida, T., Shibasaki, H. et al. (1999). *Org. Lett.* 1: 557–560.

57 da Silva, J.A.L., da Silva, J.J.R.F., and Pombeiro, A.J.L. (2013). *Coord. Chem. Rev.* 257: 2388–2400.

58 Pombeiro, A.J.L. (2007). Vanadium-catalyzed alkane functionalization reactions under mild conditions. In: *Vanadium: The Versatile Metal*, ACS Symposium Series, vol. 974 (ed. K. Kustin, J.C. Pessoa and D.C. Crans), 51–60. Washington, DC), Chapter 4: American Chemical Society.

59 Reis, P.M., Silva, J.A.L., Palavra, A.F. et al. (2003). *Angew. Chem. Int. Ed.* 42: 821–823.

60 Nishiguchi, T., Nakata, K., and Fujiwara, Y. (1992). *Chem. Lett.* 1141–1142.

61 Kirillova, M.V., Kuznetsov, M.L., Reis, P.M. et al. (2007). *J. Am. Chem. Soc.* 129: 10531–10545.

62 Kirillova, M.V., Kuznetsov, M.L., da Silva, J.A.L. et al. (2008). *Chem. Eur. J.* 14: 1828–1842.

63 Kirillova, M.V., da Silva, J.A.L., da Silva, J.J.R.F. et al. (2007). *Adv. Synth. Catal.* 349: 1765–1774.

64 Reis, P.M., Silva, J.A.L., Palavra, A.F. et al. (2005). *J. Catal.* 235: 333–340.

65 Silva, T.F.S., Luzyanin, K.V., Kirillova, M.V. et al. (2010). *Adv. Synth. Catal.* 352: 171–187.

66 Kirillova, M.V., Kirillov, A.M., Reis, P.M. et al. (2007). *J. Catal.* 248: 130–136.
67 Taniguchi, Y., Yamaoka, Y., Nakata, K. et al. (1995). *Chem. Lett.* 345–346.
68 Nakata, K., Yamaoka, Y., Miyata, T. et al. (1994). *J. Organomet. Chem.* 473: 329–334.
69 Kirillova, M.V., Kirillov, A.M., Kuznetsov, M.L. et al. (2009). *Chem. Commun.* 2353–2355.
70 Kirillova, M.V., Kirillov, A.M., and Pombeiro, A.J.L. (2010). *Chem. Eur. J.* 16: 9485–9493.
71 Kirillova, M.V., Kirillov, A.M., and Pombeiro, A.J.L. (2011). *Appl. Catal., A* 401: 106–113.
72 Kirillova, M.V., Kirillov, A.M., and Pombeiro, A.J.L. (2009). *Adv. Synth. Catal.* 351: 2936–2948.
73 Kirillov, A.M., Kirillova, M.V., and Pombeiro, A.J.L. (2012). *Coord. Chem. Rev.* 256: 2741–2759.
74 Sutradhar, M., Kirillova, M.V., da Silva, M.F.C.G. et al. (2013). *Dalton Trans.* 42: 16578–16587.
75 Zaltariov, M.-F., Alexandru, M., Cazacu, M. et al. (2014). *Eur. J. Inorg. Chem.* 29: 4946–4956.
76 Kirillov, A.M., Kirillova, M.V., Shul'pina, L.S. et al. (2011). *J. Mol. Catal. A: Chem.* 350: 26–34.
77 Kirillov, A.M., Karabach, Y.Y., Kirillova, M.V. et al. (2012). *Cryst. Growth Des.* 12: 1069–1074.
78 Kirillova, M.V., Kirillov, A.M., Martins, A.N.C. et al. (2012). *Inorg. Chem.* 51: 5224–5234.
79 Kirillov, A.M., Coelho, J.A.S., Kirillova, M.V. et al. (2010). *Inorg. Chem.* 49: 6390–6392.
80 Nesterova, O.V., Kirillova, M.V., da Silva, M.F.C.G. et al. (2014). *Cryst. Eng. Comm.* 16: 775–783.
81 Paul, A., Ribeiro, A.P.C., Karmakar, A. et al. (2016). *Dalton Trans.* 45: 12779–12789.
82 Li, Y., Dong, K., Zhu, F. et al. (2016). *Angew. Chem. Int. Ed.* 55: 7227–7230.
83 Li, Y., Zhu, F., Wang, Z., and Wu, X.-F. (2016). *ACS Catal.* 6: 5561–5564.
84 Li, Y., Wang, C., Zhu, F. et al. (2017). *ChemSusChem* 10: 1341–1345.
85 Ravelli, D., Protti, S., and Fagnoni, M. (2016). *Acc. Chem. Res.* 49: 2232–2242.
86 Sakakura, T., Sodeyama, T., Sasaki, K. et al. (1990). *J. Am. Chem. Soc.* 112: 1221–1229.
87 Sakakura, T., Sodeyama, T., and Tanaka, M. (1989). *New J. Chem.* 13: 737–745.
88 Jaynes, B.S. and Hill, C.L. (1995). *J. Am. Chem. Soc.* 117: 4704–4705.
89 Hill, C.L. (1995). *Synlett* 2: 127–132.
90 Ryu, I., Tani, A., Fukuyama, T. et al. (2011). *Angew. Chem. Int. Ed.* 50: 1869–1872.
91 Ryu, I., Tani, A., Fukuyama, T. et al. (2013). *Org. Lett.* 15: 2554–2557.
92 Wang, B., Qiu, D., Zhang, Y., and Wang, J. (2016). *Beilstein J. Org. Chem.* 12: 796–804.
93 Doyle, M.P., Duffy, R., Ratnikov, M., and Zhou, L. (2010). *Chem. Rev.* 110: 704–724.
94 Caballero, A. and Pérez, P.J. (2015). *J. Organomet. Chem.* 793: 108–113.
95 Nakamura, E., Yoshikai, N., and Yamanaka, M. (2002). *J. Am. Chem. Soc.* 124: 7181–9192.
96 Doyle, M.P., Westrum, L.J., Wolthuis, W.N.E. et al. (1993). *J. Am. Chem. Soc.* 115: 958–964.
97 Davies, H.M.L. and Beckwith, R.E.J. (2003). *Chem. Rev.* 103: 2861–2904.
98 Davies, H.M.L., Hansen, T., and Churchill, M.R. (2000). *J. Am. Chem. Soc.* 122: 3063–3070.
99 Demonceau, A., Noels, A.F., Hubert, A.J., and Teyssie, P. (1981). *J. Chem. Soc., Chem. Commun.* 688–689.
100 Demonceau, A., Noels, A.F., Hubert, A.J., and Teyssie, P. (1984). *Bull. Soc. Chim. Belg.* 93: 945–948.
101 Wenkert, E., Davis, L.L., Mylari, B.L. et al. (1982). *J. Org. Chem.* 47: 3242–3247.
102 Callot, H.J. and Metz, F. (1982). *Tetrahedron Lett.* 23: 4321–4324.
103 Thu, H.-Y., Tong, G.S.-M., Huang, J.-S. et al. (2008). *Angew. Chem. Int. Ed.* 47: 9747–9751.
104 Despagnet-Ayoub, E., Jacob, K., Vendier, L. et al. (2008). *Organometallics* 27: 4779–4787.
105 Braga, A.A.C., Maseras, F., Urbano, J. et al. (2006). *Organometallics* 25: 5292–5300.

106 Caballero, A., Despagnet-Ayoub, E., Díaz-Requejo, M.M. et al. (2011). *Science* 332: 835–838.
107 Fuentes, M.A., Olmos, A., Muñoz, B.K. et al. (2014). *Chem. Eur. J.* 20: 11013–11018.
108 Flores, J.A., Komine, N., Pal, K. et al. (2012). *ACS Catal.* 2: 2066–2078.
109 Kulkarni, N.V., Dash, C., Jayaratna, N.B. et al. (2015). *Inorg. Chem.* 54: 11043–11045.

19

Catalytic Alkane Amidation and Related Reactions

Dmytro S. Nesterov, Luís M. T. Frija, Armando J. L. Pombeiro, and Maximilian N. Kopylovich

Universidade de Lisboa, Centro de Química Estrutural, Instituto Superior Técnico, Lisboa, Portugal

19.1 Introduction

The syntheses of many compounds of industrial and pharmaceutical importance can be significantly simplified if direct functionalization of C—H bonds is involved [1], thus encouraging a continuous search for novel synthetic methods of C—H bonds activation and modification. Of particular interest is the activation of sp^3 C—H bonds, which have the highest dissociation energy within the C—H bond family and, therefore, possess considerable chemical inertness [2, 3]. The ability of some natural enzymes, such as cytochromes P450 or methane monooxygenases [4, 5], to catalyze alkanes hydroxylation under mild conditions inspired chemists to mimic the activity of these enzymes. As a result, alkane oxidations with conversion of C—H into C—O bonds became the most studied catalytic processes in the field of C—H activation. Despite the renown achievements in alkane hydroxylation (upon formation of a C—O bond), the formation of C—N bonds from alkanes is much less studied, although the latter reaction is very promising for chemical synthesis since it provides a scope of functional groups of great synthetic value. This chapter covers recent advances in the direct nitrogenation of sp^3 C—H bonds of hydrocarbons with focus on alkanes, and in particular the amidation, imidation, amination, nitration, and azidation processes.

19.2 Amidation

19.2.1 Amidation of sp^3 C—H Bonds

The amide group is widespread in nature (e.g., in peptides) and in artificial compounds of industrial importance, such as polyamides (nylons) and many pharmaceuticals [6]. The reactivity of sp^3 C—H bonds in substrates containing non-sp^3 carbons at close proximity is usually higher than those of pure alkanes. Due to this reason, the amidation of sp^3 C—H bonds in activated hydrocarbons is much more studied and will be discussed in the first instance.

For a long time, the amidation of sp^3 C—H bonds was mostly limited to the nitrene insertion (reaction with, for instance, tosylimino phenyliodinane, TsN=IPh) [7] until in 1997 Katsuki and co-workers reported a catalytic system where peroxycarbamates were used to form a carbon–nitrogen bond [8a]. Copper triflate, $Cu(OTf)_2$, was applied as a catalyst, yielding 56% (Scheme 19.1a; $R^1 = (CO)OO^tBu$, $R^2 = p$-tolyl). A key point was the use of a peroxide to activate a C—H bond, while the necessity to synthesize N-(*p*-toluenesulfonyl)peroxycarbamate is an

Alkane Functionalization, First Edition. Edited by Armando J. L. Pombeiro and M. Fátima C. Guedes da Silva.

obvious limitation. Notably, the authors attempted to perform asymmetric amidation applying a chiral co-ligand, but the observed enantioselectivity was poor (3% *ee*).

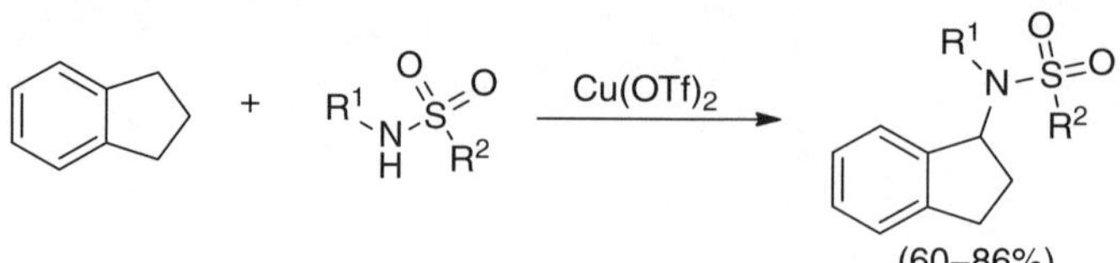

Scheme 19.1 Amidation of indane with sulfonamides [8].

A further development was performed by Pelletier and Powell in 2006, who succeeded in introducing a benzenesulfonamide into the indane substrate (Scheme 19.1b; R^1 = H, CH_3; R^2 = Ph, $C_6H_4OCH_3$, C_6H_4-Hal, thiophene) [8b]. Using the same copper triflate as a catalyst and employing ^{t}BuOOAc oxidant, the yield of 73% was achieved in the reaction with benzenesulfonamide. A crucial condition was a gradual addition of oxidant (during 6 hours) and catalyst (in two portions). Adamantane can also be amidated using this catalytic system, with 56% of overall yield and showing 95 : 5 regioselectivity.

Another related report extended the above approach toward amidation of primary benzylic C—H bonds for more than 20 examples catalyzed by $[Cu(MeCN)_4]PF_6$ and $Cu(OTf)_2$ (Scheme 19.2a; R^1 = H, CH_3, Hal; R^2 = H, CH_3; R^3 = H, OCH_3, Hal, Ph) [9]. In this case, more sophisticated peroxide was used to get yields up to 70% at room temperature. Simple ^{t}BuOOAc did not allow exceeding 50% yield (based on the sulfonamide), while the reaction required a large excess of substrate (25 eq. relatively to the sulfonamide). In spite of the low reactivity of C—H bonds under the studied conditions (substrate excess and 3 days' reaction time), the methodology can be extended to a broad range of substrates with no significant yield decrease. The mechanism was not investigated in detail, although the authors supposed that the formation of C—N bond may proceed via carbocation or Cu(III) intermediates [9].

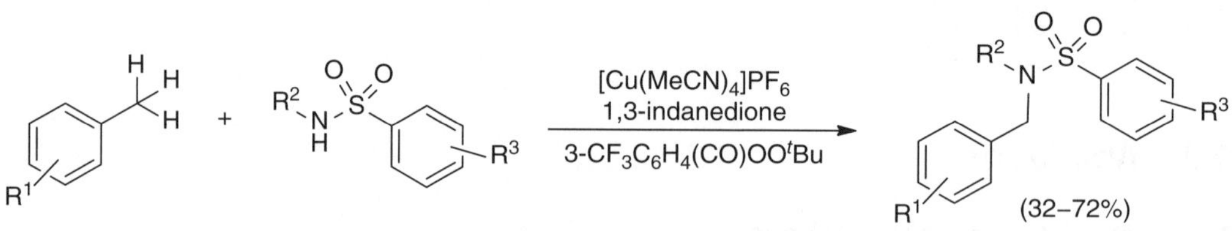

Scheme 19.2 Amidation of toluene and its derivatives [9, 10].

A system for toluene amidation (Scheme 19.2b; R^1, R^3 = H, CH_3, Hal; R^2 = H) was elaborated by replacing the copper catalyst ($Cu(OTf)_2$ or $[Cu(MeCN)_4]PF_6$) with a nickel one ($NiCl_2$ or $NiAc_2$), thus allowing yields up to 75% [10]. Depending on the conditions, the process can be turned toward production of imides (Ph–CH=N-SO_2–R), formed when the reaction proceeds under air rather than in N_2 atmosphere. Imide is proposed to be formed through oxidation of amide in air. Due to the use of ^{t}BuOO^{t}Bu di-*t*-butylperoxide (DTBP) oxidant, rather harsh conditions have been employed (120 °C for 24 hours). The reaction is solvent-free, where toluene serves as solvent and substrate. The possible mechanism involves participation of Ni(III) intermediates, while the main C—H bonds attacking species are ^{t}BuO• radicals (Scheme 19.3). Formation of the imide product is seen through the nickel-catalyzed oxidation of amide with air dioxygen.

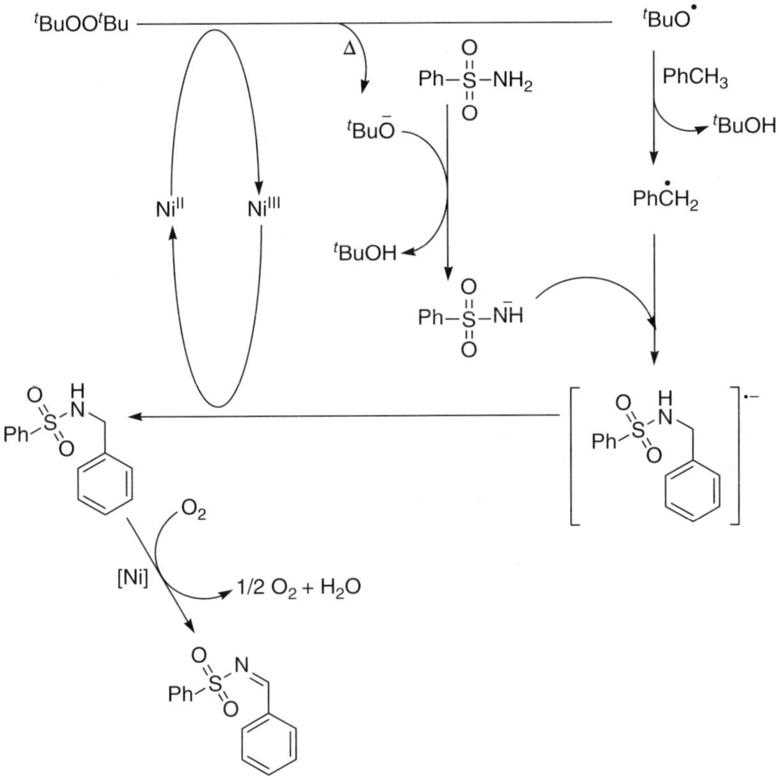

Scheme 19.3 Proposed mechanism for amidation and imidation of toluene with a nickel catalyst [10].

Chang and co-workers reported a catalytic system for amidation of sp^3 C—H bonds with sulfonyl and acyl azides (Scheme 19.4; $R^2 = CH_3$, Ph, C_6H_4, $C_6H_3(CF_3)_2$, $C_6H_3(NO_2)_2$) without using peroxides [11]. The amidation (with yield up to 90%) proceeds at a methyl carbon of a ketoxime and is promoted by a catalytic system involving [IrCp*Cl$_2$] (Cp* = η^5-C_5Me_5), AgNTf$_2$ (silver bis(trifluoromethanesulfonyl)imide) and AgAc (silver acetate) as an additive. The presence of the ketoxime group close to the amidated methyl is essential. The combination of the Ir/Ag two-component catalyst and the Ag additive is also crucial: no product was detected in the

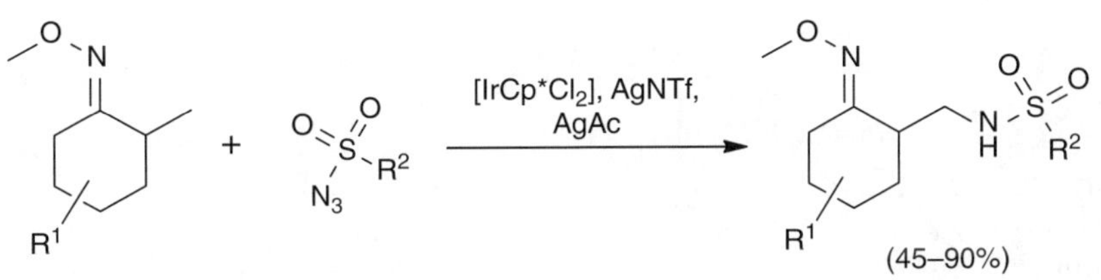

Scheme 19.4 Amidation of methyl groups of ketoximes with sulfonyl azides [11].

absence of additive or if iridium is replaced with other metal source (compounds of rhodium, ruthenium, and palladium were tested and revealed no activity). The authors proposed that silver oxidizes the iridium precatalyst to a catalytically active cationic species.

19.2.2 Amidation of C—H Bonds in Alkanes

Several amides (particularly, derivatives of benzamide) are simple amidation agents for alkanes, as revealed by Hartwig and co-workers in their comprehensive work (32 examples) in 2014 (Scheme 19.5) [12]. The reaction occurs in benzene at 100 °C using DTBP as oxidant and CuI/$(MeO)_2$Ph equimolar combination as the most efficient catalyst. The nonsubstituted benzamide gives the highest product yield (99%) among all the amides tested, with acetamide being less reactive (38% yield). Introduction of a methyl group (N-methylbenzamide) markedly decreases the activity to give a lower than 10% yield. Phthalimide leads to 20% yield, which can be increased up to 75% by replacing benzene solvent with 1,2-dichlorobenzene. Only a low discrimination is observed during the amidation of adamantane (ratio of tertiary to secondary products is 1.6 : 1). In contrast, 3-ehtylpentane and stereoisomers of 1,4-dimethylcyclohexane show reactivity toward secondary carbon only. 2,4-dimethylpentane, containing sterically hindered C—H bonds, shows exclusive preference toward amidation of primary carbon atom. The value of kinetic isotope effect measured for amidation of cyclohexane with benzamide is 2.9. The proposed reaction mechanism involves interaction of a copper complex with alkyl radical and is confirmed by isolated intermediates (Scheme 19.6) [12].

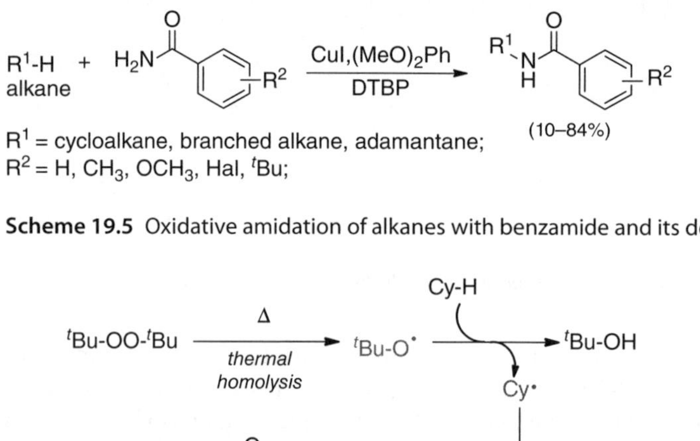

Scheme 19.5 Oxidative amidation of alkanes with benzamide and its derivatives [12].

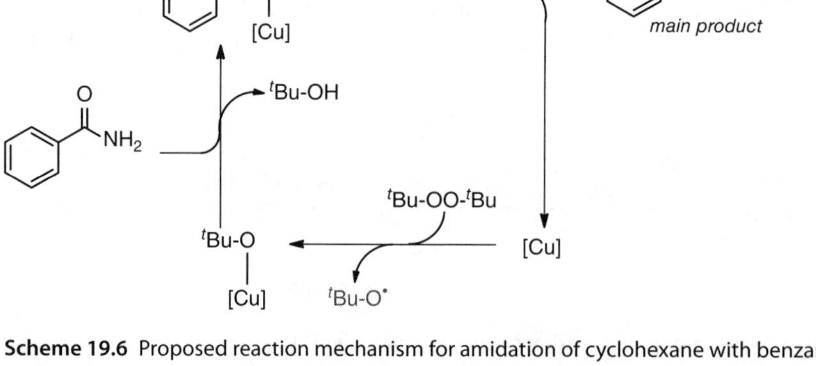

Scheme 19.6 Proposed reaction mechanism for amidation of cyclohexane with benzamide [12].

Similar reaction conditions were applied by Hoang et al. for the related amidation of cyclohexane with benzamide, using cyclohexane as solvent and a copper metal-organic framework heterogeneous catalyst based on the 2,5-dihydroxyterephthalic ligand [13]. During this study, yields up to 94% were achieved and the catalyst recyclability was attested.

Another related system of this type, reported by Zeng and Huang, concerns amidation of cyclohexane and toluene (at methyl group) with benzamide (and its) derivatives with CuCl as a catalyst [14]. The presence of a base, ^{t}BuOK, is essential to getting good yields of products (up to 84% for the reaction of cyclohexane and benzamide). The proposed reaction mechanism is similar to those reported by Hartwig (Scheme 19.6) [12].

Besides, a Ritter-type amidation reaction (a process where nitriles act as the nitrogen source) has been employed by Baran et al. in 2012 to functionalize alkanes mainly, along with other sp^3 C—H bonds [15]. The work represented eight examples of cyclic alkanes and seven examples of alcohol and ketone-containing substrates. The Selectfluor® reagent (F-TEDA-PF_6), commonly used in fluorination reactions, was used as oxidant and $CuBr_2$ as catalyst (Scheme 19.7). Although copper bromide itself can provide suitable yields of products, the efficiency of the catalytic system can be further improved by adding the Lewis acid $Zn(OTf)_2$. Thus, the highest yields of products of 90% and 51% are achieved for adamantane and cyclohexane amidation, respectively. The least reactive tested alkane is *cis*-decahydronaphthalene, which allows amidation with 21% yield, retention of stereoconfiguration of substrate and normalized 3° : 2° bond selectivity of 2.5 : 1. The respective *trans*-substrate is amidated at secondary positions with 36% of yield. A kinetic isotope effect of 3.5 was observed when an equimolar mixture of normal and deuterated cyclohexane was used. The proposed reaction mechanism involves hydrogen abstraction with formation of an alkyl radical, which undergoes oxidation to the carbon cation, followed by coupling with acetonitrile and nitrilium cation hydrolysis to give the final product (Scheme 19.8).

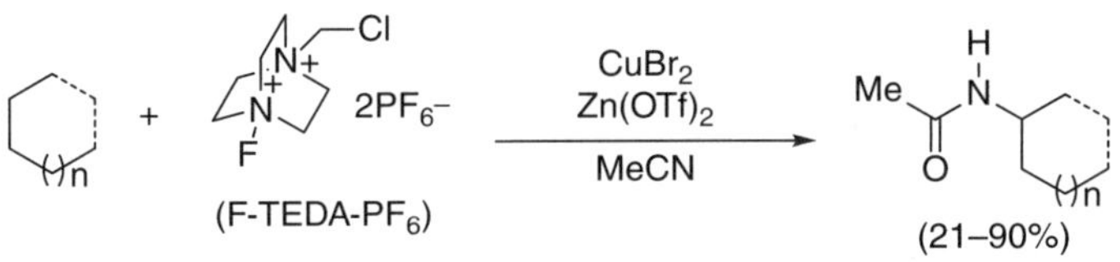

Scheme 19.7 Amidation of cycloalkanes with F-TEDA-PF_6/CH_3CN [16].

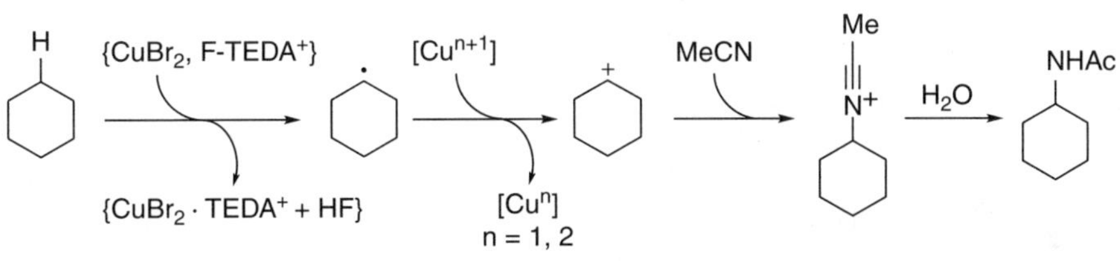

Scheme 19.8 Proposed reaction mechanism for amidation of cyclohexane with F-TEDA-PF_6/MeCN [15].

The Ritter-type amidation was further explored by Minakata and co-workers, who reported a metal-free organocatalytic amidation of tertiary sp^3 C—H bonds of alkanes (5 examples) and other substrates mainly containing the benzoyl group (16 examples) [16]. The screening for the best oxidant was performed using isoamyl benzoate as a model substrate (Scheme 19.9). With an I_2 additive, the best yields of product are achieved with $(NH_4)_2[Ce(NO_3)_6]$ (ceric ammonium nitrate, CAN, 53%), *m*-chloroperoxybenzoic acid (*m*-CPBA, 65%) and

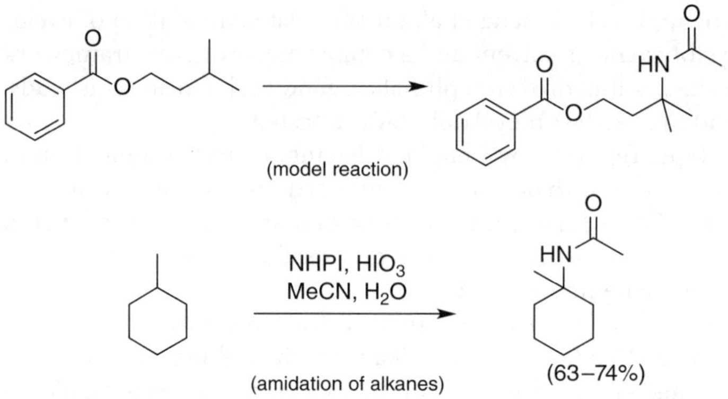

Scheme 19.9 Amidation of tertiary sp^3 C—H bonds of isoamyl benzoate and methylcyclohexane with the N-hydroxyphthalimide/HIO_3/MeCN system [16].

$KHSO_5 \cdot 0.5KHSO_4 \cdot 0.5K_2SO_4$ (Oxone®, 69%). However, the oxidants are inactive in the absence of iodine. The optimized reaction conditions involve the use of HIO_3 in the presence of N-hydroxyphthalimide (NHPI), giving 74% of yield in the model process. Acetonitrile can be replaced by EtCN (46% yield) or PhCN (52% yield), giving the respective amides. Alkanes (Scheme 19.9) can be amidated at moderate yields (up to 63% for methylcyclohexane, 61% for adamantane, and 47% for 2-methylpentane). Measurements of kinetic isotope effects in the course of oxidation of normal and deuterated methylcyclohexane reveals rather high 2.5 and 7.5 values for parallel and competitive oxidations, respectively. The proposed mechanism starts with formation of a phthalimide-N-oxyl radical, which abstracts one hydrogen atom from alkane to produce an alkyl radical (Scheme 19.10). The latter is trapped by iodine and then oxidized by HIO_3 to give iodine(III) species, then undergoing a Ritter-type amination to produce the final product (amide).

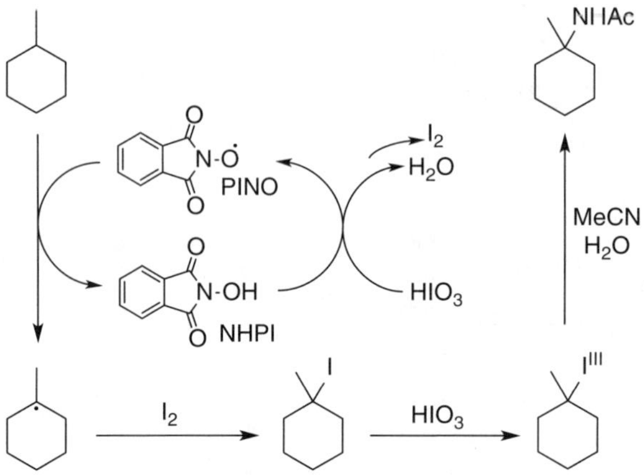

Scheme 19.10 Proposed reaction mechanism for amidation of methylcyclohexane with the N-hydroxyphthalimide/HIO_3/MeCN system [16].

19.3 Amination

The development of novel C—N bond forming methodologies, focused on the synthesis of new amine derivatives, is an intensively investigated field of the utmost importance [17]. Despite the fact that classical transformations range from nucleophilic displacement of a leaving group

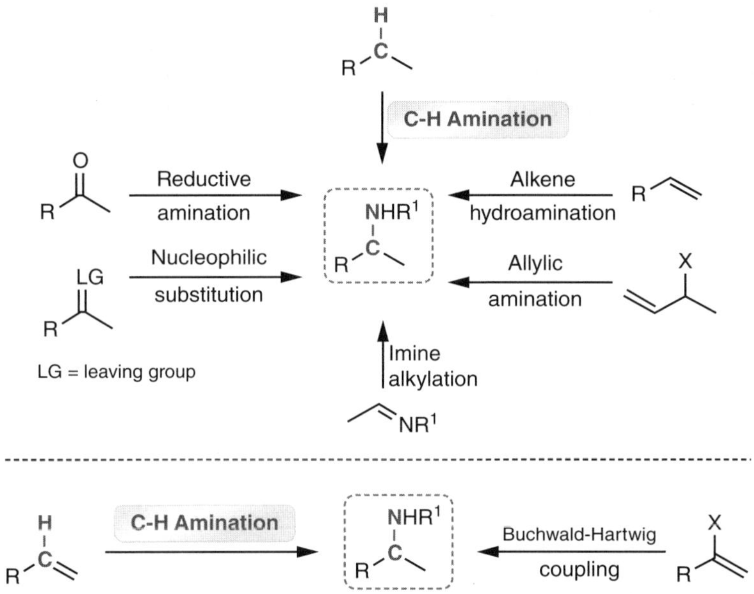

Scheme 19.11 General reactions involved in the formation of C—N bonds [17].

to reductive carbonyl amination or imine alkylation [18], recently discovered processes have mostly taken benefit from the introduction of transition metal catalysts [19], and modern amination methods include Buchwald-Hartwig C—N coupling [20], hydroamination [21] and diamination [22] of olefins, and allylic amination [23] (Scheme 19.11). Besides, the direct C—H functionalization processes for alkanes offer distinctive opportunities [24] complementary to those provided by the above-mentioned reactions, which are based on the transformation of a preinstalled functional group. This, however, requires finding suitable conditions for the generation of highly active but sufficiently mild species to induce selective C—H functionalization, a challenge given the high energy and the number of C—H bonds in alkanes. While adequate oxygenases for the selective C—H hydroxylation do exist in nature, the analogous direct nitrogen transfer has not yet been found in any biosynthetic pathway [25].

The primary examples of metal-mediated C—H amination date back to 1960s with the reports of Kwart and Kahn [26], Breslow and Sloan [27], and Turner et al. [28]. More attention from the organic community was, however, given to the studies involving iminoiodanes (e.g. PhI=NR) in the early 1980s [29]. A paper on the capacity of iron porphyrin or rhodium acetate to efficiently catalyze intramolecular C—H amination [29b], together with another one documenting the intermolecular version [30], opened the way for future research and demonstrated the potential of nitrene C—H insertion for the synthesis of nitrogen-containing products, a field that has considerably expanded in the last two decades with the development of smooth protocols for the generation of nitrenes.

The reaction of commercially available hypervalent iodine reagents with various NH_2-containing substrates has therefore led to the discovery of elegant and efficient methodologies that culminated with the total synthesis of tetrodotoxin [31]. These results have been summarized in several reviews that witnessed the growing impact of C—H amination in organic chemistry [32] and confirmed that intramolecular C—H amination now belongs to the toolbox that synthetic chemists have at their disposal for C—N bond formation.

Recent studies have allowed overcoming some, but not all, of the limitations of catalytic hypervalent iodine-mediated nitrene C—H insertions. Remarkably, novel aminating procedures involving C—H activation by a metal catalyst have emerged in parallel. These transformations, with the C—H bond being cleaved by a metal center (inner-sphere type mechanism) at the first step of the catalytic cycle, differ mechanistically from the nitrene-based methodologies (Scheme 19.12). The inner-sphere mechanism thus tends to give selectivities complementary to those induced by the outer-sphere one [17, 32f, 33]. Despite all the research covering C—H amination of a wide range of molecules, this section will emphasize the last developments made in the intermolecular amination of alkanes or related slightly activated hydrocarbons.

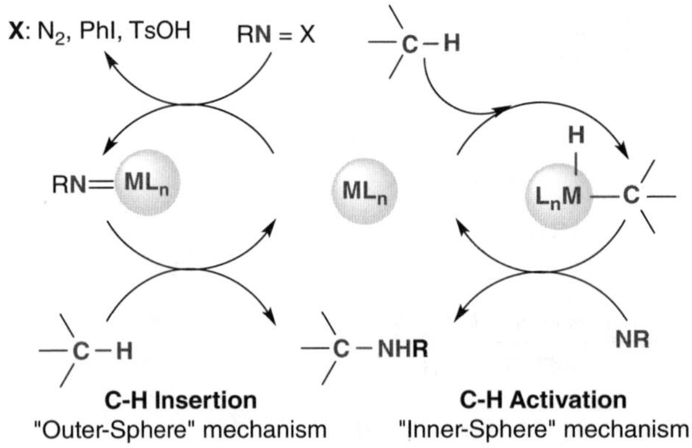

Scheme 19.12 General mechanisms of C—H amination [17].

Early in this century, Che et al. reported the great capability of ruthenium and manganese porphyrin complexes to catalyze the amination and amidation (with amine or amide, respectively) of benzylic and allylic C—H bonds in the presence of (diacetoxyiodo)benzene {$PhI(OAc)_2$} with yields up to 86% (Scheme 19.13) [34].

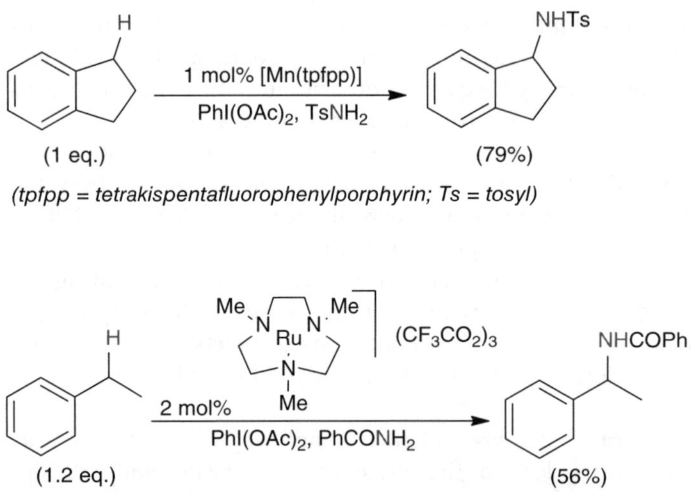

Scheme 19.13 Mn and Ru-catalyzed intermolecular C—H amination and amidation [34].

Likewise, C—H amination of cyclohexane showed to be chemoselective [34a]. Asymmetric transformations have also been described with an *ee* up to 56% [34c]. A stepwise mechanism has been proposed, with the formation of carboradical intermediates following *H*-abstraction by an imidometal species (Scheme 19.14) [34d]. Basically, these works demonstrated the possibility of generating *in situ* the somewhat capricious iminoiodanes [34a], a new procedure that paved way for the future intermolecular nitrene transfers from a variety of nitrogen precursors.

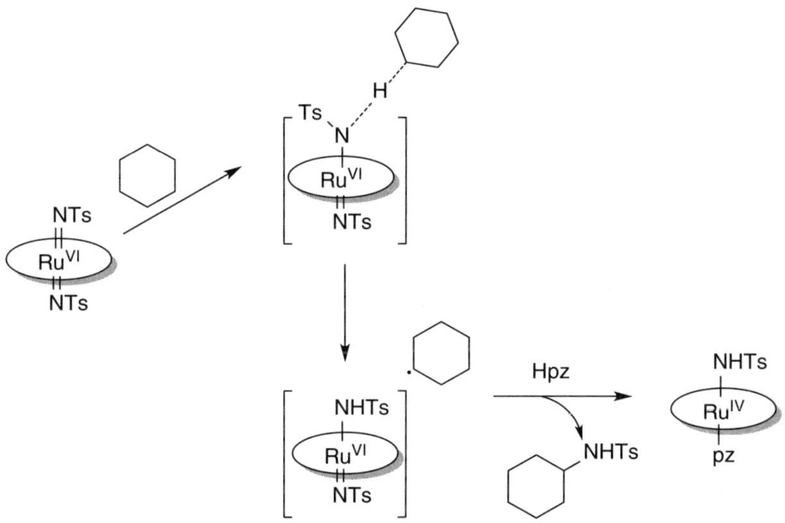

Scheme 19.14 Tosylamination mechanism of cyclohexane by a ruthenium imido complex.

Several transition metals have been methodically investigated in order to dispose a general method for intermolecular C—H amination. While cobalt [35], zinc(II) [36], or iron(II) [37] have been shown to catalyze such C—H functionalizations with limited scope and efficiencies, the results reported for silver(I), copper, and, particularly, rhodium(II) complexes are more remarkable. Thus, several dinuclear silver(I) complexes, initially found to catalyze intramolecular C—H amination of carbamates and sulfamates [38], have also proved to be active for the intermolecular amination of various substrates with PhI=NNs ([*N*-(*p*-nitrophenylsulfonyl)imino]phenyliodinane) or PhI=NTs ([*N*-(*p*-tolylsulfonyl)imino]phenyliodinane), and a series of benzylic derivatives and cyclic and aliphatic alkanes, though necessarily used in excess, can be aminated with significant yields (Scheme 19.15) [40]. The dinuclear core, assured by the stacking of two 4,7-disubstituted-1,10-phenanthroline ligands, has been found to be crucial for the catalytic activity and is probably one of the key factors for the insertion of a silver-bound nitrene.

An efficient catalytic intermolecular C—H amination involving a stoichiometric amount of substrate and the commercially available $[Cu(CH_3CN)_4]PF_6$ catalyst was also described nearly at the same time [39]. The reaction occurs with *N*-chloro-tosylamine sodium salts (TsNNaCl) or with tosylamine ($TsNH_2$) at primary, secondary, and tertiary benzylic positions, as well as at α-ethereal sites (Scheme 19.16), but chemoselective allylic C—H amination has proved impossible so far. Although these experiments highlight the potential of chloramine-T as a C—H aminating agent, the efficiency of the process remains lower than that involving hypervalent iodine reagents, as suggested by the study published within the same period and devoted to the copper-catalyzed amidation of cyclic ethers [41].

Important developments in the amination of alkanes have also been performed with rhodium catalysts through the design of new ligands and nitrene sources. Initial examples [42] as well as later asymmetric versions [43, 44] were based on the hypervalent iodine-mediated generation of nitrenes from sulfonamides. Nevertheless, the good yields and *ee's* (up to 94%)

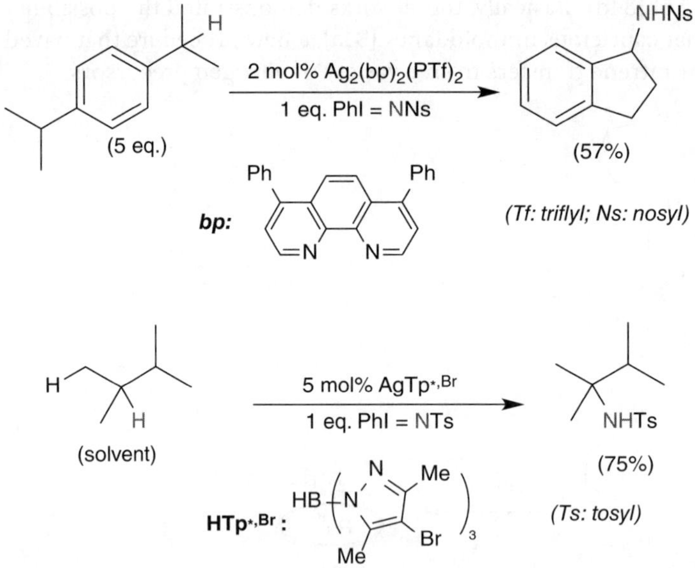

Scheme 19.15 Silver-catalyzed intermolecular C—H amination [40].

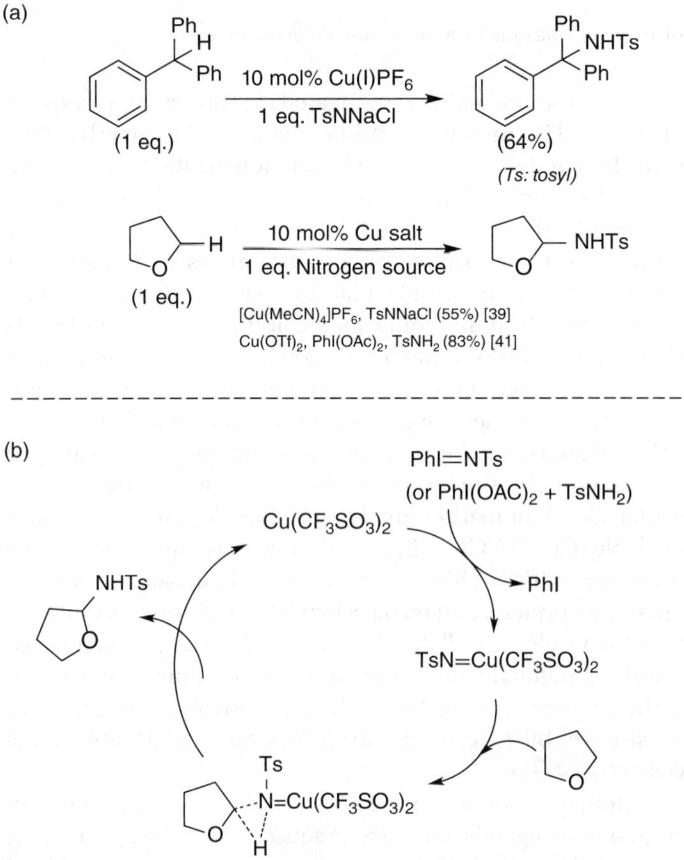

Scheme 19.16 (a) Two examples of copper-catalyzed intermolecular C—H amination reactions [39, 41]. (b) Mechanism for the copper-catalyzed amidation of tetrahydrofuran involving a hypervalent iodine reagent [41].

were achieved only with excess of substrate. Succeeding, it was observed that sulfamates can afford higher yields even with a stoichiometric amount of starting material [45, 46]. Thus, a combination of rhodium complex $[Rh_2(esp)_2]$, trichloroethylsulfamate and $PhI(OCO^tBu)_2$ [bis(*tert*-butylcarbonyloxy)iodobenzene] gives C—H aminated products with yields up to 74% (Scheme 19.17). The functionalization of secondary benzylic and tertiary positions is favored. In the case of products having multiple reactive sites, the regioselectivity is governed by a combination of electronic and steric factors, the more sterically accessible and/or electron-rich C—H bonds being preferentially aminated, as has also been observed in carbenoid chemistry [32a].

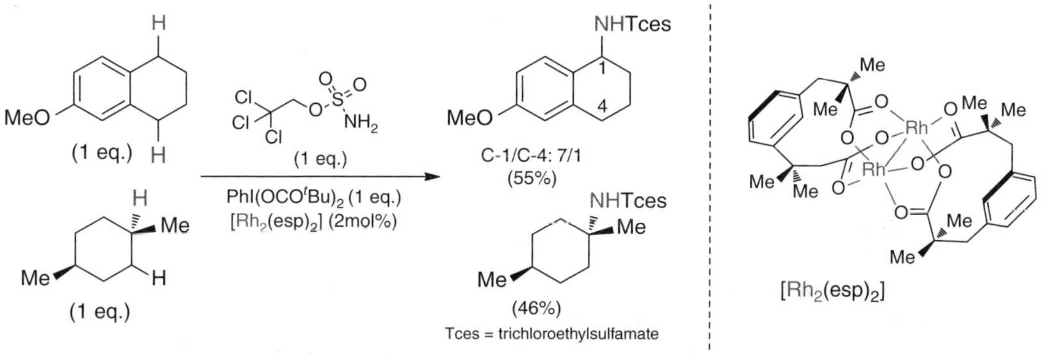

Scheme 19.17 Rhodium-catalyzed intermolecular C—H amination [45, 46]. (*See insert for color representation.*)

More recently, Suárez and Chiara have reported an elegant protocol for the intermolecular $C(sp^3)$—H amination of simple hydrocarbons using the shelf-stable nonafluorobutanesulfonyl azide ($CF_3(CF_2)_3SO_2N_3$) in the presence of a dirhodium(II) tetracarboxylate catalyst (Scheme 19.18) [48]. The amination products were obtained in moderate to good yields and could be further transformed to secondary amines via a simple two-step process.

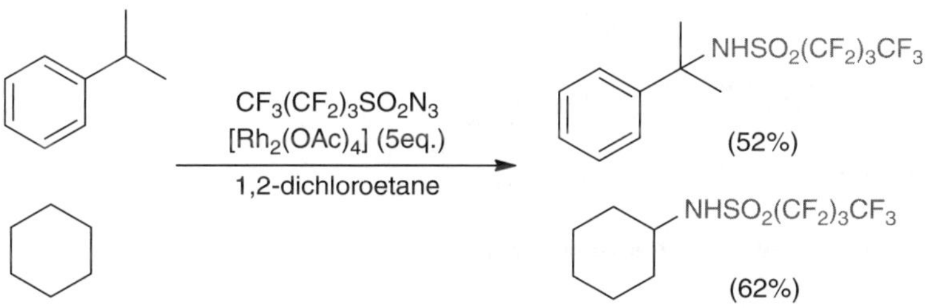

Scheme 19.18 Intermolecular amination of a benzylic C—H bond and a simple cyclic alkane [48].

As a final example of alkane amination, a simple method for direct metal-free C—H amination of unactivated hydrocarbons using easy-handling diacetoxy-λ^3-bromane and triflylamine or sulfamate esters (Scheme 19.19) can be highlighted [47]. This smooth protocol provides a direct route for the regioselective synthesis of alkylamines. Both cyclic and acyclic alkanes were efficiently transformed into N-triflylamines under metal-free conditions. Five-, seven-, and eight-membered cyclic alkanes similarly afforded good to high yields of N-triflylamines at 15 °C. These reactions showed marked preference for tertiary C—H bonds over secondary C—H bonds, while primary C—H insertion products were not observed.

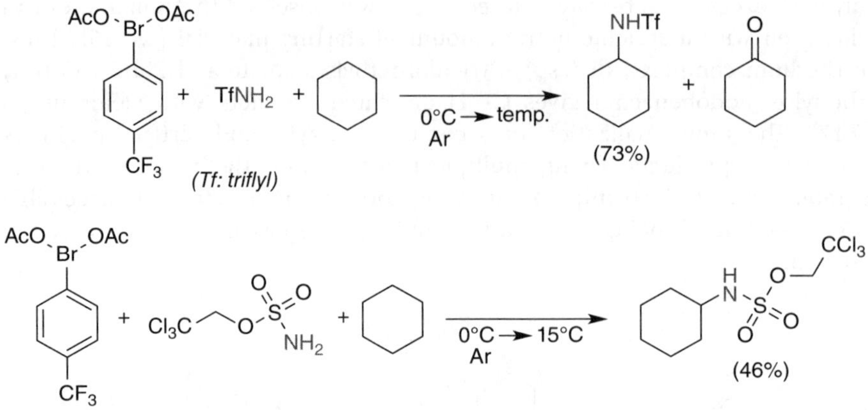

Scheme 19.19 Amination of cyclohexane with diacetoxybromane and triflylamine or 2,2,2-trichloroethylsulfonylamine [47].

19.4 Imidation

While the imidation of aromatic or benzylic C—H bonds is broadly studied [49], the respective processes for alkanes are much less developed.

In 2014, Cheng and co-workers reported that the imidation of activated sp^3 C—H bonds in diarylmethanes, particularly, diphenylmethane (used as solvent), with sulfoximines using DTBP oxidant and $FeBr_3$ as a catalyst proceeded with 85% yield (Scheme 19.20), while copper catalysts did not produce the desired products [51]. Other oxidants were also tested and revealed yields from 16 (for dioxygen) to 65% (for ^{t}BuOOH, *tert*-butyl hydroperoxide, TBHP). Considering the use of DTPB oxidant, which is thermally stable at 90 °C (temperature of the experiment), possibly the iron catalyst decomposes it, while the copper ones are inactive in this process.

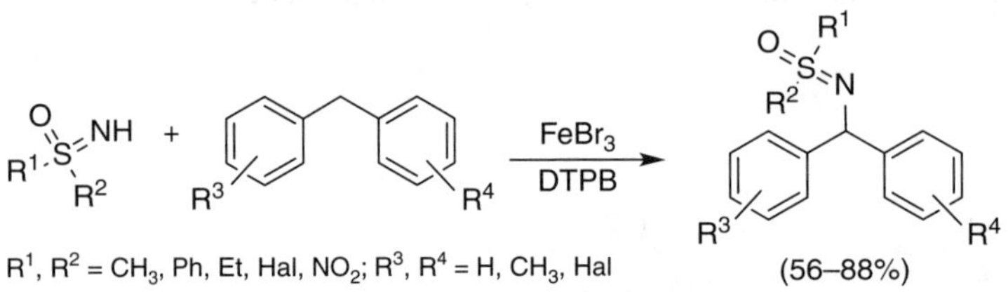

Scheme 19.20 Imidation of diphenylmethane and its derivatives [51].

Recently, an example of the C—H bond imidation with sulfoximines has been reported for a series of cycloalkanes (Scheme 19.21) [50]. Cyclohexane and cycloheptane react with sulfoximine derivatives in the presence of DTPB, giving the products with the corresponding cycloalkyl group (81% yield for cyclohexane), while cyclopentane and cyclooctane are less reactive (44% and 50% yields, respectively). Copper complexes are the most efficient catalysts for these reactions, while some iron and silver compounds ($FeCl_3$ and AgI) were also tested, showing lower yields (30–60%). It is proposed that the copper catalyst plays a key role in the reduction of ^{t}BuOO^{t}Bu to give ^{t}BuO• radical and ^{t}BuO$^-$ anion with formation of Cu(III) species which then oxidizes a radical anion intermediate (Scheme 19.22). A competitive reaction with an equimolar mixture of C_6H_{12} and C_6D_{12} led to the yields ratio of normal and deuterated products of 9 : 1, while the overall yield is only slightly hampered by air. The proposed reaction mechanism

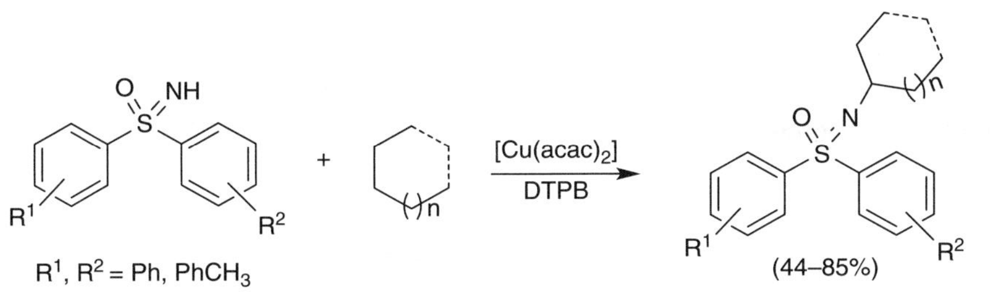

Scheme 19.21 Imidation of cycloalkanes [52].

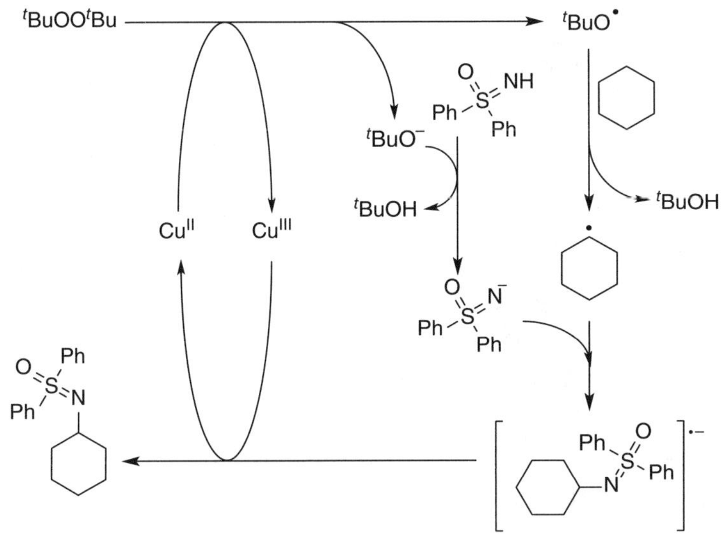

Scheme 19.22 Proposed reaction mechanism for imidation of cyclohexane with diphenylsulfoximine [50].

(Scheme 19.22) is similar to that depicted at Scheme 19.3, with sulphonamide being replaced with sulfoximine and the nickel catalyst with the copper one.

19.5 Carbamation

Kanai and co-workers in a 2015 study found that isocyanates were efficient agents for carbamation of alkanes and benzylic C—H bonds (Scheme 19.23) [52]. In this study, a complex catalyst comprising the copper compound $[Cu(MeCN)_4]BF_4$ and phenanthroline auxiliary ligand (neocuproine) showed the best product yield of 75% for carbamation of the cyclohexane substrate. In the absence of the auxiliary ligand, only 32% yield was obtained. Moreover, the salts of other metals (acetates of cobalt, iron, nickel, or silver) did not show catalytic activity. Almost no discrimination between tertiary and secondary C—H bonds was observed for the carbamation of adamantane (ratio of 3° and 2° products was 1.7 : 1). *n*-Hexane was carbamated with a low regioselectivity, the yields ratios being 1 : 3.2 : 1.4 for (1), (2), and (3) carbons, respectively (56% total yield). Interestingly, in spite of the high reactivity of primary positions of *n*-hexane and toluene (44% yield), ethylbenzene reacts exclusively at benzylic carbon with 65% yield. Kinetic isotope effect was measured for independent reactions with normal and deuterated cyclohexane and revealed the value of 2.9. A proposed reaction mechanism (Scheme 19.24)

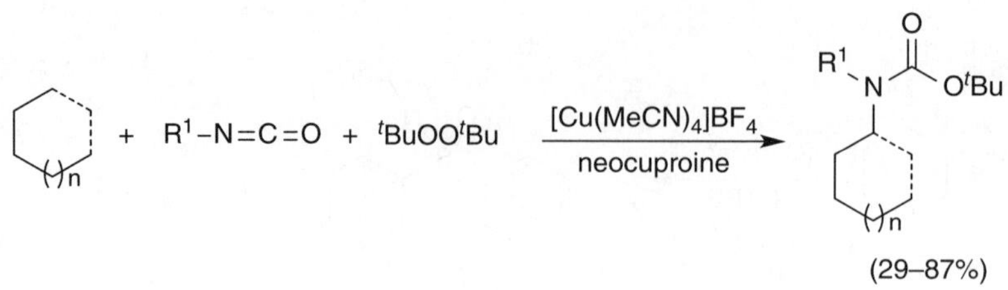

Scheme 19.23 Carbamation of alkanes with isocyanate and DTBP [52].

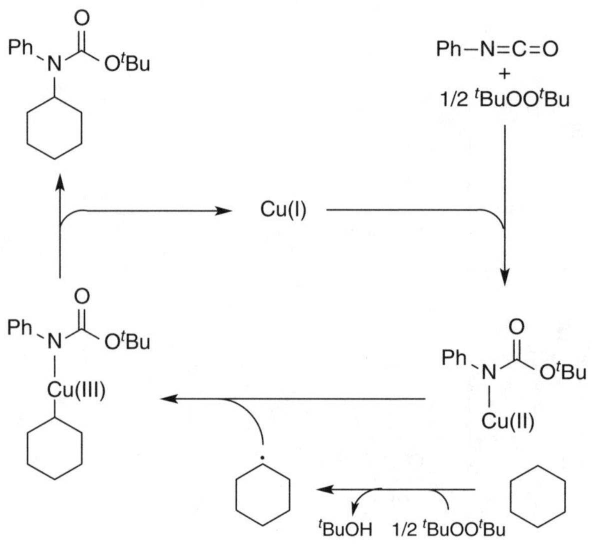

Scheme 19.24 Proposed mechanism for carbamation of cyclohexane with phenyl isocyanate [52].

involves coupling of ^{t}BuO• radical with the isocyanate to form a Cu(II)-carbamate intermediate, which adds the cyclohexyl radical to yield a Cu(III) intermediate from which the carbamate product is eliminated reductively.

19.6 Nitration and Nitrosation

Nitro- (R–NO_2) and nitroso- (R–N=O) compounds are versatile building blocks for organic synthesis, e.g. as precursors for many pharmaceuticals, pigments, and agrochemicals [53] or as starting materials in the production of industrially important intermediates such as cyclohexanone oxime [54] or ε-caprolactam [55]. Nitroalkanes are relatively stable and easy to handle, while the nitro group can be easily removed or transformed to amines, oximes, hydroxyl amines, etc., keeping other functional groups intact [56]. Hence, elaboration of selective, convenient and mild direct nitrations and nitrosations of available, and cheap alkanes are of significance for synthetic laboratory and industrial chemistry.

Although the production of nitroalkanes by direct nitration of aliphatic saturated hydrocarbons has long been known [57], forced reaction conditions (e.g. temperature of 250–400 °C) were used due to the difficulty of activation of C—H bonds and nitrating agents (generation of nitrogen dioxide NO_2 from nitric acid). Under such conditions free radical chain reactions occur, and since the bond dissociation energy of the C—C bonds of alkanes is lower than that of the C—H bonds, the homolysis of C—H bonds is accompanied by the C—C skeleton rupture [58]. Hence, the formation of undesirable oxidized products, lower alkanes, and nitroalkanes occurs [59]. For example, the nitration of propane with NO_2 at 300 °C produces nitropropanes and 26% of byproducts such as nitroethane and nitromethane, as a result of C—C bond cleavage [60]. Moreover, the acid waste produced in this process represents an operational and environmental problem.

Since the traditional nitration of saturated hydrocarbons proceeds via a free radical mechanism [61], the selectivity of this reaction can be improved by constructing a reactor with a high surface-to-volume ratio, which quenches the radical chain processes by wall effects [62]. Another approach to improving efficiency of the alkane nitration involves the elaboration of new catalysts, in particular heterogeneous ones [63].

Later, it was found that nitrogen dioxide can react with adamantanes preferably at the tertiary position in the presence of ozone at −78 °C to give 1-nitroadamantanes in a ~95% yield (so-named Kyodai-nitration) [64]. A related photoinduced nitration with N_2O_5 afforded a mixture of nitroadamantane and adamantane nitrite [65], while a laser-induced nitration of alkanes was also successfully employed [66]. Furthermore, nitroalkanes were obtained by treating alkanes with HNO_3 over metal nitrates [67]. However, the above systems require special handling and have a limited scope, operating only with reactive alkanes (e.g. adamantane).

Another method of nitration of some alkanes with superacid nitronium salts (e.g. nitronium hexafluorophosphate $NO_2^+PF_6^-$) is based on the electrophilic substitution of hydrogen by the nitro group, thus avoiding the free-radical pathway [68]. Such a nitration of adamantane allows 60–70% yields of 1-nitroadamantane [69], but the nitration of methane gave much lower yields (1–3%) [68]. Generally, tertiary C—H bonds show the highest reactivity, followed by C—C bonds, which in turn are more reactive than secondary C—H bonds. The possible mechanism involves a two-electron three-center bond in a carbocationic transition state formed by the nitronium ion attacking a two-electron covalent C—H or C—C σ-bond, thus leading to electron-pair sharing. Despite demonstrating a new route for nitration, this method was not able to provide high conversions and selectivities for most industrially important hydrocarbons.

At the beginning of the twenty-first century, Ishii et al. explored a more general and efficient catalytic nitration of light alkanes with nitrogen dioxide or nitric acid as nitrating agents and NHPI as the catalyst precursor [70]. In this method, phthalimide N-oxyl (PINO) radical, generated in situ by reaction of NHPI with dioxygen or nitric acid, serves as a carbon-radical producing catalyst by abstracting a hydrogen atom from a C—H bond in a hydrocarbon (Scheme 19.25). NHPI is regenerated with the formation of the alkyl radical, which is trapped by NO_2 to form nitroalkane and related byproducts (Scheme 19.25).

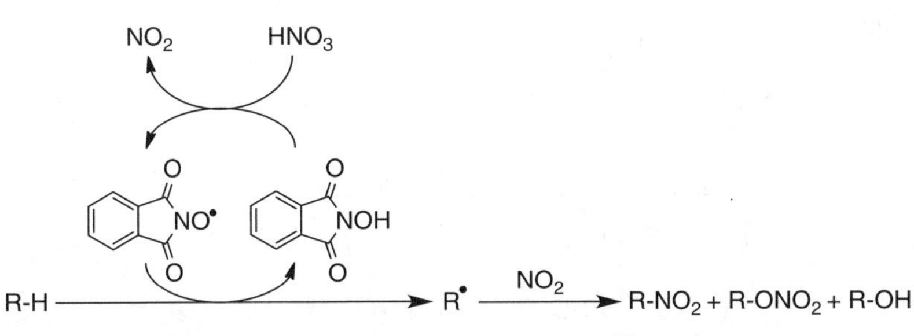

Scheme 19.25 General reaction pathway for the NHPI-catalyzed alkane nitration [70].

Similar processes occur upon nitration with gaseous nitrogen dioxide [71]. Nitrogen monoxide (NO) generated in the course of this reaction can be reoxidized by O_2 to NO_2, which is reused in the nitration. Therefore, the nitration of an alkane by a NO_2/air system produces a nitroalkane in higher yield than that by NO_2 alone [72]. Owing to the higher concentration of NO_2 than air, the formed alkyl radicals react with NO_2 rather than O_2 to give nitroalkanes in preference to oxygenated products. This allows nitration under aerobic conditions with 50–70% yields for the nitration of light alkanes at 70–100 °C after 14 hours. However, due to the potent oxygenating ability of NO_2, the nitration can be accompanied with oxygenation and, e.g. cyclooctane afforded nitrocyclooctane (50%) and cyclooctyl nitrate (2%) along with oxygenated products, such as cyclooctanone (17%) and cyclooctanol (7%) [72]. Apart from the above-mentioned drawbacks, a significant amount of NHPI is decomposed to phthalic acid and other byproducts in the hydrocarbon nitrations with nitric acid [72]. Therefore, at least 10 mol% of NHPI was required to achieve reasonable conversions while the catalyst was not recycled.

This issue was addressed [73], and recoverable and recyclable ionic liquid supported NHPI catalyst (Scheme 19.26) was prepared and successfully applied for the oxidation and nitration of alkanes with HNO_3. It was also demonstrated that nitrocyclohexane can be prepared from cyclohexane with ~27% cyclohexane conversion and ~88% selectivity to nitrocyclohexane using NHPI catalyst and *tert*-butyl nitrite ^{t}BuONO as the nitrating agent [74].

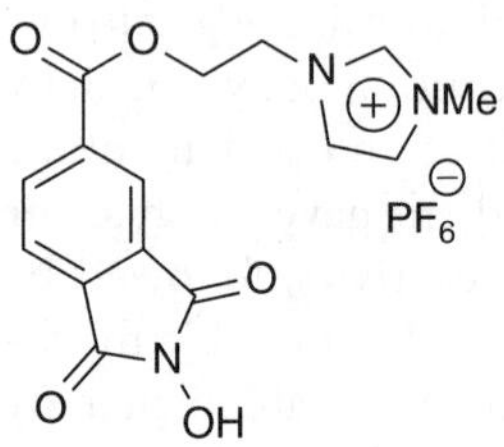

Scheme 19.26 Recyclable ionic liquid supported NHPI catalyst used for the nitration of alkanes [73].

The application of NHPI as a catalyst was further extended to the nitrosation of several cycloalkanes with *tert*-butyl nitrite (Scheme 19.27) [76]. Thus, cyclopentane was nitrosated to nitrosocyclopentane in ~50% yield with a small amount (2%) of cyclopentanone oxime formed. The nitrosating ^{t}BuONO is converted into *tert*-butyl alcohol and can be regenerated by reacting with NO_2 or sodium nitrite (Scheme 19.27). Interestingly, the reaction of cyclooctane leads to cyclooctanone oxime as a result of the nitrosocyclooctane isomerization; thus,

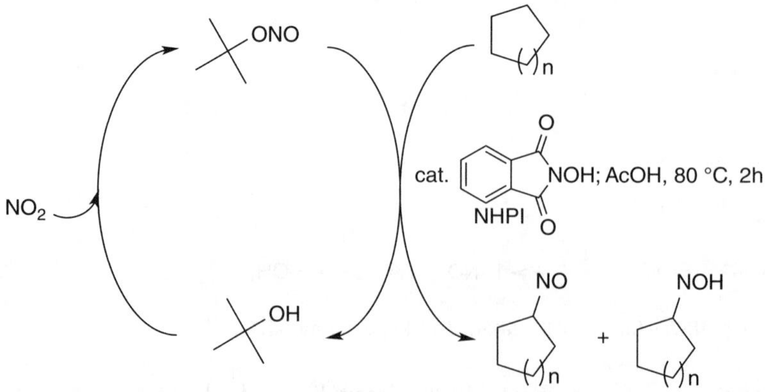

Scheme 19.27 NHPI-catalyzed nitrosation of cycloalkanes with *tert*-butyl nitrite [76].

direct transformation of cycloalkanes to cycloalkanone oximes can be achieved. A plausible pathway involves the formation of PINO and NO through the reaction of NHPI with *tert*-butyl nitrite. Then NO reacts with a small amount of the dissolved dioxygen to give NO_2, while the generated PINO abstracts a hydrogen atom from the cycloalkane to form the respective alkyl radical and regenerating NHPI. Subsequently, the alkyl radical is trapped by NO to produce nitrosocycloalkane.

The NHPI-based catalytic system was also applied for one-pot synthesis of lactams from cycloalkanes with *tert*-butyl nitrite as oxidant [75]. The synthetic strategy involved NHPI-catalyzed nitrosation of cycloalkanes followed by isomerization of nitrosocycloalkanes to oximes by treating them with amines and the Beckmann rearrangement to lactams using cyanuric chloride (Scheme 19.28). For example, treatment of cyclohexane with ^{t}BuONO followed by triethylamine and cyanuric chloride afforded ε-caprolactam in 74% yield within 5 hours and without salt formation [75b].

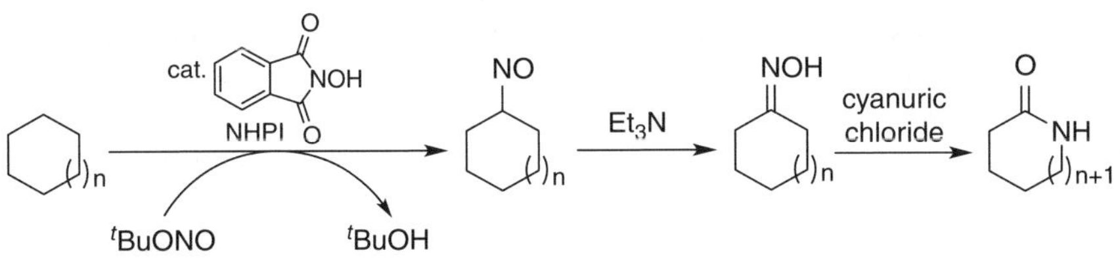

Scheme 19.28 One-pot NHPI-catalyzed synthesis of lactams from cycloalkanes and ^{t}BuONO [75].

Other catalytic systems, e.g. vanadium-substituted Keggin-type phosphomolybdate $[H_4PVMo_{11}O_{40}]$ [75b] or polyoxometalate $[VO(H_2O)_5]H[PMo_{12}O_{40}]$ [75a], were also applied for the nitration of alkanes by nitric acid. Notably, alkylbenzenes such as toluene and xylenes can be nitrated only at the alkyl side-chain C—H bonds while no nitration of the aromatic ring was observed. However, yields were not particularly high while significant amounts of oxygenated products and dinitro compounds were formed as well (e.g. 11% cyclooctanone and 14% dinitrocyclooctanes obtained along with 25% mononitrocyclooctane).

Co-production of nitrocyclohexane and adipic acid from vapor phase nitration–oxidation of cyclohexane with NO_2 and vanadium phosphorus oxide (VPO) catalysts has been lately developed (Scheme 19.29) [77]. A small amount of other oxygenated byproducts such as cyclohexanol and cyclohexanone were also formed in all cases. It can be noted that the reduction product NO can be recovered by oxidation with dioxygen and then reused. This work on co-production of nitrocyclohexane and adipic acid with total selectivity of ca. 87% at 22% conversion was extended to light-driven nitration–oxidation of cyclohexane with nitric acid in aqueous biphasic system under mild conditions [78]. The interesting feature of the biphasic systems involves the occurrence of the radical chain reaction in the organic phase, while the nitrocyclohexane and adipic acid products are transferred to the aqueous phase. This separates different stages and shifts the equilibrium, thus allowing higher yields and selectivities.

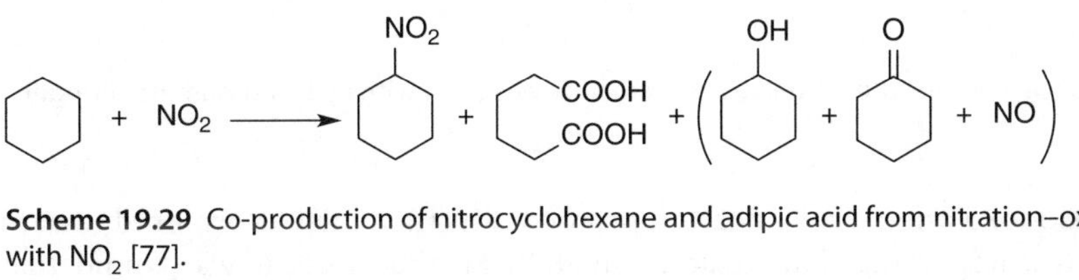

Scheme 19.29 Co-production of nitrocyclohexane and adipic acid from nitration–oxidation of cyclohexane with NO_2 [77].

19.7 Azidation

Due to its dipolar character and power of the azide moiety as a strong nucleophile, organic azides R–N_3 participate in many synthetic transformations to give various nitrogen-containing molecules [79]. For example, the organic azides are excellent substrates for 1,3-dipolar cycloaddition reactions (click chemistry) and are precursors of amines, nitrenes, and various azaheterocycles [79, 80]. Of particular importance is the ability of azides to provide one-nitrogen units as a result of the elimination of molecular nitrogen [81]. Syntheses of aryl- and alkylazides from activated substrates are well established [81, 82], but their direct synthesis from saturated hydrocarbons is in its infancy and only several examples are known.

Thus, unactivated alkanes can be transformed into alkyl azides using sodium azide and iodosobenzene (C_6H_5IO) or ^{t}BuOOH as oxidant and manganese porphyrins or iron complexes with tetradentate tripodal ligands as catalysts (Scheme 19.30) [84]. For example, with iodosobenzene as a stoichiometric oxidant and a Mn-porphyrin catalytic moiety, alkyl azides can be produced in yields >50% based on iodosobenzene together with alcohols in ca. 20% yields. Hence, a large amount of the corresponding alcohol byproduct is formed, while an excess of the alkane is required. Later, the system has been improved in terms of selectivity and scope of catalysts (e.g. manganese salen-type Jacobsen catalysts were applied as well) [83].

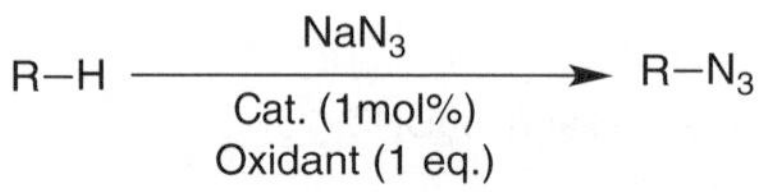

Scheme 19.30 Azidation of alkanes with sodium azide and iodosobenzene or ^{t}BuOOH as oxidant [84].

Another mild and selective protocol for the azidation of tertiary C—H bonds employs a cyclic hypervalent iodine reagent containing an azide unit, i.e. azidobenziodoxolone [85]. For example, *cis*-decalin can be transformed into the respective azido-containing product with 65% yield and a 6.3 : 1 ratio of diastereomers using a catalytic combination of iron(II) with PyBOX (2,6-Bis[(3a*R*,8a*S*)-(+)-8*H*-indeno[1,2-*d*]oxazolin-2-yl)pyridine) tridentate nitrogen ligands (Scheme 19.31) [87]. It was demonstrated that the azidobenziodoxolone-iron catalytic system is suitable for late-stage functionalization of various complex molecules [86, 87]. The site- and stereo-selectivities suggest that a tertiary alkyl radical intermediate is generated in the course of the reaction. Then formation of the C—N bond occurs in the iron-catalyzed process by reaction of the alkyl radical with an iron azide intermediate [86a].

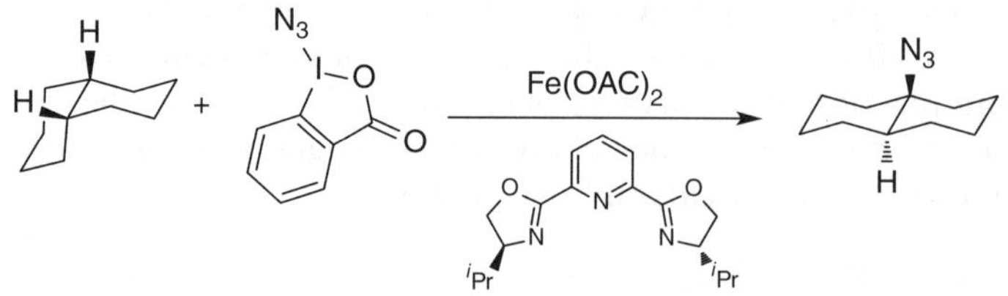

Scheme 19.31 Azidation of *cis*-decalin with a catalytic combination of iron(II) with PyBOX tridentate nitrogen ligand [87].

The found azidobenziodoxolone-based azidation protocol was then further explored toward new catalysts, substrates, and using light irradiation [88]. Furthermore, it was found that aliphatic tertiary C—H bonds can be selectively transformed into organic azides under transition-metal-free conditions in aqueous media using persulfate to induce the radical processes [89]. Finally, it should be noted that an iron-dependent halogenase can also be employed to functionalize unactivated aliphatic carbon centers toward their direct nitration and azidation [90].

19.8 Concluding Remarks

During the last decade, a number of novel protocols for mild nitrogen insertion into alkanes C—H bonds have been elaborated and perfected, and some of the most interesting examples, from our point of view, are listed in this chapter. The main types of C—N bond formation are amidation and related processes (amination, imidation, and carbamation), nitration, and azidation. Amidation and amination represent the largest family of reactions, being used for pure alkanes and for sp^3 C—H bonds in other substrates. The amidation, imidation, carbamation, and azidation processes require the presence of a strong oxidizing agent (such as a peroxide) and proceed via a free radical mechanism. In such cases, the N-containing group is coupled to a radical species formed from alkane. The roles of the metal catalyst concern stabilization of involved species and promotion of decomposition of the oxidant to form C—H attacking radicals. While nitration and azidation are typically limited to formation of NO_2 and N_3 groups instead of C—H bond, the amidation and amination processes are broader and allow the introduction of a wide range of functional groups, from simple acetamide to bulky substituted sulfoximines.

However, in spite of the achievements performed in this area, there is still lack of simple and more effective protocols, while selectivity over various C—H bonds of alkane substrates should also be improved. For instance, stereoselective functionalization is limited to rare examples of azidation or amination in the presence of chiral catalysts. Thus, the search for cheap but effective catalytic systems possessing higher regio- and stereoselectivities can be recognized as an important goal for further investigations in this field.

Acknowledgments

This work has been supported by the Foundation for Science and Technology (FCT), Portugal (projects UID/QUI/00100/2013 and PTDC/QEQ-QIN/3967/2014, "Investigador 2013" contract IF/01270/2013/CP1163/CT0007, and fellowships SFRH/BPD/99533/2014 and SFRH/BPD/99851/2014).

References

1 (a) Yamaguchi, J., Yamaguchi, A.D., and Itami, K. (2012). *Angew. Chem. Int. Edit.* 51: 8960–9009. (b) Gensch, T., Hopkinson, M.N., Glorius, F. et al. (2016). *Chem. Soc. Rev.* 45: 2900–2936. (c) Liu, C., Yuan, J., Gao, M. et al. (2015). *Chem. Rev.* 115: 12138–12204. (d) Pombeiro, A.J.L. (2014). Toward functionalization of alkanes under environmentally benign conditions. In: *Advances in Organometallic Chemistry and Catalysis* (ed. A.J.L. Pombeiro), 15–25. Wiley.

2 (a) Nesterov, D.S., Nesterova, O.V., and Pombeiro, A.J.L. (2018). *Coord. Chem. Rev.* 355: 199–222. (b) Hartwig, J.F. (2016). *J. Am. Chem. Soc.* 138: 2–24.

3 Hartwig, J.F. and Larsen, M.A. (2016). *ACS Central Sci.* 2: 281–292.

4 (a) Bordeaux, M., Galarneau, A., and Drone, J. (2012). *Angew. Chem. Int. Edit.* 51: 10712–10723. (b) Wang, V.C.-C., Maji, S., Chen, P.P.-Y. et al. (2017). *Chem. Rev.* 117: 8574–8621.
5 O'Reilly, E., Kohler, V., Flitsch, S.L., and Turner, N.J. (2011). *Chem. Commun.* 47: 2490–2501.
6 Wan, J.P. and Jing, Y.F. (2015). *Beilst. J. Org. Chem.* 11: 2209–2222.
7 Muller, P. and Fruit, C. (2003). *Chem. Rev.* 103: 2905–2919.
8 (a) Kohmura, Y., Kawasaki, K., and Katsuki, T. (1997). *Synlett* 1456–1458. (b) Pelletier G., and Powell, D.A. (2006). *Org. Lett.* 8: 6031–6034.
9 Powell, D.A. and Fan, H. (2010). *J. Org. Chem.* 75: 2726–2729.
10 Li, Z.L., Jin, L.K., and Cai, C. (2017). *Org. Biomol. Chem.* 15: 1317–1320.
11 Kang, T., Kim, Y., Lee, D. et al. (2014). *J. Am. Chem. Soc.* 136: 4141–4144.
12 Tran, B.L., Li, B.J., Driess, M. et al. (2014). *J. Am. Chem. Soc.* 136: 2555–2563.
13 Hoang, T.T., Nguyen, H.T.H., Le, T.T. et al. (2016). *Tetrahedron* 72: 8241–8251.
14 Zeng, H.T. and Huang, J.M. (2015). *Org. Lett.* 17: 4276–4279.
15 Michaudel, Q., Thevenet, D., and Baran, P.S. (2012). *J. Am. Chem. Soc.* 134: 2547–2550.
16 Kiyokawa, K., Takemoto, K., and Minakata, S. (2016). *Chem. Commun.* 52: 13082–13085.
17 Collet, F., Dodd, R.H., and Dauban, P. (2009). *Chem. Commun.* 5061–5074.
18 Trost, B.M. and Fleming, I. (eds.) (1991). *Comprehensive Organic Synthesis*, vol. 6. Oxford: Pergamon Press.
19 (a) Hartwig, J.F. (2008). *Nature* 455: 314. (b) Ricci, A. (ed.) (2008). *Amino Group Chemistry from Synthesis to the Life Sciences.* Weinheim: Wiley-VCH. (c) Ricci, A. (ed.) (2000). *Modern Amination Methods.* Weinheim: Wiley-VCH. (d) Kienle, M., Dubbaka, S.R., Brade, K. et al. (2007). *Eur. J. Org. Chem.* 4166.
20 (a) Surry, D.S. and Buchwald, S.L. (2008). *Angew. Chem. Int. Ed.* 47: 6338. (b) Hartwig, J.F. (2008). *Acc. Chem. Res.* 41: 1534.
21 Müller, T.E., Hultzsch, K.C., Yus, M. et al. (2008). *Chem. Rev.* 108: 3795.
22 Marcia de Figueiredo, R. (2009). *Angew. Chem. Int. Ed.* 48: 1190.
23 (a) Takemoto, Y. and Miyabe, H. (2007). Applications II: Transition Metal Compounds in Organic Synthesis 1. In: *Comprehensive Organometallic Chemistry III*, vol. 10 (ed. R.H. Crabtree and D.M.P. Mingos), 695. Oxford: Elsevier. (b) Johannsen, M. and Jorgensen, K.A. (1998). *Chem. Rev.* 98: 1689.
24 (a) Bergman, R.G. (2007). *Nature* 446: 391. (b) Godula, K. and Sames, D. (2006). *Science* 312: 67. (c) Dyker, G. (ed.) (2005). *Handbook of C–H Transformations.* Weinheim: Wiley-VCH. (d) Kakiuchi, F. and Chatani, N. (2003). *Adv. Synth. Catal.* 345: 1077. (e) Ritleng, V., Sirlin, C., and Pfeffer, M. (2002). *Chem. Rev.* 102: 1731. (f) Labinger, J.A. and Bercaw, J.E. (2002). *Nature* 417: 507.
25 Hili, R. and Yudin, A.K. (2006). *Nat. Chem. Biol.* 6: 284.
26 Kwart, H. and Kahn, A.A. (1967). *J. Am. Chem. Soc.* 89: 1951.
27 Breslow, D.S. and Sloan, M.F. (1968). *Tetrahedron Lett.* 9 (51): 5349.
28 Carr, D., Seden, T.P., and Turner, R.W. (1969). *Tetrahedron Lett.* 10 (6): 477.
29 (a) Breslow, R. and Gellman, S.H. (1982). *J. Chem. Soc., Chem. Commun.* 1400. (b) Breslow, R. and Gellman, S.H. (1983). *J. Am. Chem. Soc.* 105: 6728.
30 Mahy, J.-P., Bedi, G., Battioni, P. et al. (1989). *New J. Chem.* 13: 651.
31 Hinman, A. and Du Bois, J. (2003). *J. Am. Chem. Soc.* 125: 11510.
32 (a) Davies, H.M.L. and Manning, J.R. (2008). *Nature* 451: 417. (b) Díaz-Requejo, M.M. and Pérez, P.J. (2008). *Chem. Rev.* 108: 3379. (c) Dick, A.R. and Sanford, M.S. (2006). *Tetrahedron* 62: 2439. (d) Davies, H.M.L. (2006). *Angew. Chem. Int. Ed.* 45: 6422. (e) Davies, H.M.L. and Long, M.S. (2005). *Angew. Chem. Int. Ed.* 44: 3518. (f) Espino, C.G. and Du Bois, J.

(2005). Rhodium(II)-Catalyzed Oxidative Amination,. In: *Modern Rhodium-Catalyzed Organic Reactions*, vol. 17 (ed. P.A. Evans), 379. Weinheim: Wiley-VCH. (g) Halfen, J.A. (2005). *Curr. Org. Chem.* 9: 657. (h) Du Bois, J. (2005). *Chemtracts: Org. Chem.* 18: 1. (i) Dauban, P. and Dodd, R.H. (2003). *Synlett* 1571.

33 Crabtree, R.H. (2001). *J. Chem. Soc., Dalton Trans.* 2437.

34 (a) Yu, X.-Q., Huang, J.-S., Zhou, X.-G. et al. (2000). *Org. Lett.* 2: 2233. (b) Au, S.-M., Huang, J.-S., Che, C.-M. et al. (2000). *J. Org. Chem.* 65: 7858. (c) Liang, J.-L., Huang, J.-S., Yu, X.-Q. et al. (2002). *Chem. Eur. J.* 8: 1563. (d) Au, S.-M., Huang, J.-S., Yu, W.-Y. et al. (1999). *J. Am. Chem. Soc.* 121: 9120.

35 Harden, J.D., Ruppel, J.V., Gao, G.-Y. et al. (2007). *Chem. Commun.* 4644.

36 Kalita, B., Lamar, A.A., and Nicholas, K.M. (2008). *Chem. Commun.* 4291.

37 Wang, Z., Zhang, Y., Fu, H. et al. (2008). *Org. Lett.* 10: 1863.

38 Cui, Y. and He, C. (2004). *Angew. Chem. Int. Ed.* 43: 4210.

39 Bhuyan, R. and Nicholas, K.M. (2007). *Org. Lett.* 9: 3957.

40 Li, Z., Capretto, D.A., Rahaman, R. et al. (2007). *Angew. Chem. Int. Ed.* 46: 5184.

41 He, L., Yu, J., Zhang, J. et al. (2007). *Org. Lett.* 9: 2277.

42 Nägeli, I., Baud, C., Bernardinelli, G. et al. (1997). *Helv. Chim. Acta* 80: 1087.

43 Yamawaki, M., Tsutsui, H., Kitagaki, S. et al. (2002). *Tetrahedron Lett.* 43: 9561.

44 Reddy, R.P. and Davies, H.M.L. (2006). *Org. Lett.* 8: 5013.

45 Fruit, C. and Müller, P. (2004). *Tetrahedron: Asymmetry* 15: 1019.

46 Espino, C.G., Fiori, K.W., Kim, M. et al. (2004). *J. Am. Chem. Soc.* 126: 15378.

47 Miyamoto, K., Ota, T., Hoquec, M.M. et al. (2015). *Org. Biomol. Chem.* 13: 2129.

48 Suárez, J.R. and Chiara, J.L. (2013). *Chem. Commun.* 49: 9194.

49 Rit, R.K., Shankar, M., and Sahoo, A.K. (2017). *Org. Biomol. Chem.* 15: 1282–1293.

50 Teng, F., Sun, S., Jiang, Y. et al. (2015). *Chem. Commun.* 51: 5902–5905.

51 Cheng, Y., Dong, W.R., Wang, L. et al. (2014). *Org. Lett.* 16: 2000–2002.

52 Chikkade, P.K., Kuninobu, Y., and Kanai, M. (2015). *Chem. Sci.* 6: 3195–3200.

53 (a) Feuer, H. and Nielsen, A.T. (eds.) (1990). *Nitro Compounds: Recent Advances in Synthesis and Chemistry*. New York: VCH Publishing. (b) Ono, N. (ed.) (2001). *The Nitro Group in Organic Synthesis*. New York: Wiley-VCH.

54 Yan, Y., Liu, S., Hao, F. et al. (2014). *Catal. Commun.* 50: 9–12.

55 Liu, P., Zhang, H., Liu, S. et al. (2012). *Reac. Kinet. Mech. Cat.* 107: 383–391.

56 (a) Ballini, R., Gabrielli, S., Palmieri, A. et al. (2011). *Curr. Org. Chem.* 15: 1482–1506. (b) Huang, J.-P., Chen, Z.-Y., Yuan, J.-J. et al. (2016). *Asian J. Org. Chem.* 5: 951–960.

57 (a) Beilstein, F. and Kurbatov, A. (1880). *Chem. Ber.* 13: 1818–1821. (b) Konovalov, M. (1895). *Chem. Ber.* 28: 1852–1865. (c) Markovnikov, W. (1899). *Chem. Ber.* 32: 1441–1449.

58 (a) Olah, G.A., Malhotra, R., and Narahg, S.C. (eds.) (1989). *Nitration: Methods and Mechanism*. New York: VCH. (b) Albright, L.F. (1995). Nitration. In: *Kirk-Othmer Encyclopedia of Chemical Technology*, vol. 17 (ed. J.I. Kroschwite and M. Howe-Grant), 68–107. New York: Wiley.

59 (a) Hass, H.B., Hodge, E.B., and Vanderbilt, B.M. (1936). *Ind. Eng. Chem.* 28: 339–344. (b) Grundmann, C. (1943). *Die Chemie* 56: 159–163.

60 Hass, H.B., Dorsky, J., and Hodge, E.B. (1941). *Ind. Eng. Chem.* 33: 1138–1143.

61 Audley, G.J., Baulch, D.L., and Campbell, I.M. (1982). *J. Chem. Soc. Chem. Commun.* 18: 1053–1055.

62 Löwe, H., Wei, G., Jiang, M. et al. (2012). *Green Process. Synth.* 1: 439–448.

63 (a) Suzuki, S., Tahmori, K., and Ono, Y. (1986). *Chem. Lett.* 5: 747. (b) Suzuki, S., Tahmori, K., and Ono, Y. (1987). *J. Mol. Catal.* 43: 41. (c) Olah, G.A., Krishnamurthy, V.V., and Narang, S. (1982). *J. Org. Chem.* 47: 596. (d) Cornelis, A., Gerstmans, A., and Laszlo, P.

(1987). *Chem. Lett.* 11: 1389. (e) Nagi, N.H., Zubkov, E.A., and Shubin, V.G. (1990). *Izv. Akad. Nauk. SSSR, Ser. Khim.* 7: 1650. (f) Suzuki, S., Tahmori, K., and Ono, Y. (1987). *Chem. Lett.* 11: 2273.
64 (a) Suzuki, H. and Nonomiya, N. (1996). *Chem. Commun.* 1783–1784. (b) Suzuki, H. and Nonomiya, N. (1997). *J. Chem. Soc., Perkin Trans. 1* 2965–2971.
65 Tabushi, I., Kojo, S., and Yoshida, Z. (1971). *Chem. Lett.* 1431.
66 (a) Umstead, M.E., Lloyd, S.A., Fleming, J.W., and Lin, M.C. (1985). *Appl. Phys. B: Laser Opt.* 38: 219. (b) Stanley, A.E. and Godbey, S.E. (1989). *Appl. Spectrosc.* 43: 674.
67 Minkov, D., Dimov, D., and Luvchieva, D. (1991). *Neftechimia* 30: 697.
68 (a) Olah, G.A. and Lin, H.C. (1971). *J. Am. Chem. Soc.* 93: 1259–1261. (b) Olah, G.A., Ramaiah, P., and Prakash, G.K.S. (1997). *Proc. Natl. Acad. Sci. U.S.A.* 94: 11783–11785.
69 Olah, G.A., Ramaiah, P., Rao, B.C. et al. (1993). *J. Am. Chem. Soc.* 115: 7246.
70 (a) Sakaguchi, S., Nishiwaki, Y., Kitamura, T. et al. (2001). *Angew. Chem. Int. Ed.* 40: 222–224. (b) Isozaki, S., Nishiwaki, Y., Sakaguchi, S. et al. (2001). *Chem. Commun.* 1352–1352. (c) Nishiwaki, Y., Sakaguchi, S., and Ishii, Y. (2002). *J. Org. Chem.* 67: 5663–5668.
71 (a) Iwahama, T., Sakaguchi, S., and Ishii, Y. (1998). *Org. Process Res. Dev.* 2: 255–260. (b) Ishii, Y., Sakaguchi, S., and Iwahama, T. (2001). *Adv. Synth. Catal.* 343: 393–427.
72 (a) Yamaguchi, K., Shinachi, S., and Mizuno, N. (2004). *Chem. Commun.* 4: 424–425. (b) Shinachi, S., Yahiro, H., Yamaguchi, K. et al. (2004). *Chem. Eur. J.* 10: 6489–6496.
73 Koguchi, S. and Kitazume, T. (2006). *Tetrahedron Lett.* 47: 2797–2801.
74 Liu, S., You, K., Jian, J. et al. (2016). *Res. Chem. Intermed.* 42: 2211–2220.
75 (a) Hashimoto, M., Sakaguchi, S., and Ishii, Y. (2006). *Chem. Asian J.* 1: 712–716. (b) Ishii, Y., Sakaguchi, S., and Obora, Y. (2008). *J. Synth. Org. Chem. Jpn.* 66: 1066–1075.
76 Hirabayashi, T., Sakaguchi, S., and Ishii, Y. (2004). *Angew. Chem. Int. Ed.* 43: 1120–1123.
77 You, K., Jian, J., Xiao, H. et al. (2012). *Catal. Commun.* 28: 174–178.
78 Zhong, W., Mao, L., Yi, W. et al. (2016). *Res. Chem. Intermed.* 42: 461–470.
79 Bräse, S., Gil, C., Knepper, K. et al. (2005). *Angew. Chem. Int. Ed.* 44: 5188–5240.
80 Chiba, S. (2012). *Synlett* 23: 21–44.
81 Lang, S. and Murphy, J.A. (2006). *Chem. Soc. Rev.* 35: 146–156.
82 Goswami, M. and de Bruin, B. (2017). *Eur. J. Org. Chem.* 2017 (8): 1152–1176.
83 Huang, X., Bergsten, T.M., and Groves, J.T. (2015). *J. Am. Chem. Soc.* 137: 5300–5303.
84 (a) Hill, C.L., Smegal, J.A., and Henly, T.J. (1983). *J. Org. Chem.* 48: 3277–3281. (b) Kojima, T., Leising, R.A., Yan, S.P., and Que, L. (1993). *J. Am. Chem. Soc.* 115: 11328–11335.
85 Zhdankin, V.V., Krasutsky, A.P., Kuehl, C.J. et al. (1996). *J. Am. Chem. Soc.* 118: 5192–5197.
86 (a) Vita, M.V. and Waser, J. (2015). *Angew. Chem. Int. Ed.* 54: 5290–5292. (b) Huang, X. and Groves, J.T. (2016). *ACS Catal.* 6: 751–759.
87 Sharma, A. and Hartwig, J.F. (2015). *Nature* 517: 600–604.
88 (a) Wang, Y., Li, G.-X., Yang, G. et al. (2016). *Chem. Sci.* 7: 2679–2683. (b) Kamijo, S., Watanabe, M., Kamijo, K. et al. (2016). *Synthesis* 48: 115–121.
89 Li, X. and Shi, Z.-J. (2016). *Org. Chem. Front.* 3: 1326–1330.
90 Matthews, M.L., Chang, W.-C., Layne, A.P. et al. (2014). *Nat. Chem. Biol.* 10: 209–215.

Part IV

Dehydrogenation Reactions

20

Oxidative and Nonoxidative Routes to Alkane Dehydrogenation

Jay A. Labinger

California Institute of Technology, Beckman Institute, Pasadena, CA, USA

20.1 Introduction

Among the wide range of topics included under the topic of this volume – alkane functionalization – dehydrogenation (henceforth abbreviated DH) looks particularly attractive from the point of view of practical applicability, especially in the current environment. The recent huge uptick in the availability and accompanying economic advantage of natural gas – both shale gas and more traditional resources – provides a large source of light hydrocarbons, particularly ethane and propane [1]. Other potentially significant sources include light end products of Fischer-Tropsch synthesis (an increasingly important process) and various refinery off-streams. Light alkanes per se have very limited value, essentially just their energy content. In contrast, the corresponding alkenes are of considerably greater value, as monomers for a variety of homo- and copolymers as well as precursors to other large-scale chemicals.

Dehydrogenation is also a viable approach to functionalization of more complex organic molecules: even if the corresponding alkene is *not* the desired target structure, indirect functionalization, via one of the many ways we have for transforming alkenes, may well be a superior alternative, in terms of selectivity, mild conditions, etc., to direct conversion of the alkane to the functionalized product. An example is the synthesis of the macrolactam rhazinilam (Scheme 20.1), which involved generation of an alkene intermediate via C—H activation leading to (stoichiometric) dehydrogenation of an ethyl group [2].

Species arising from C—H activation of alkanes are highly prone to loss of a second hydrogen atom to generate an alkene. We know that β-H elimination is virtually ubiquitous in transition metal alkyls – so much so that commentators have felt it necessary to explain why *other* modes of functionalization can even compete [3]. Surface-bound alkyls in heterogeneous catalysts also undergo facile β-H elimination (the microscopic reverse of catalytic hydrogenation); even gas-phase alkyl radicals readily lose an adjacent H atom. Thus mechanistically, at least, dehydrogenation would appear to be easily achievable. Why is it not more widely practiced? Two reasons: thermodynamics and selectivity.

Nonoxidative dehydrogenation (NODH) is entropically favored but enthalpically disfavored; ΔG_0 remains positive at temperatures below 400 °C or so [4], depending on pressure (Figure 20.1). That pretty much rules out straightforward application of homogeneous catalysts, which will rarely be stable at anything approaching such temperatures, although as we shall see, a variety of "tricks" may be used to achieve non-negligible conversions in particular cases. Heterogeneous catalysts can easily survive such temperatures, but at those conditions

Alkane Functionalization, First Edition. Edited by Armando J. L. Pombeiro and M. Fátima C. Guedes da Silva.

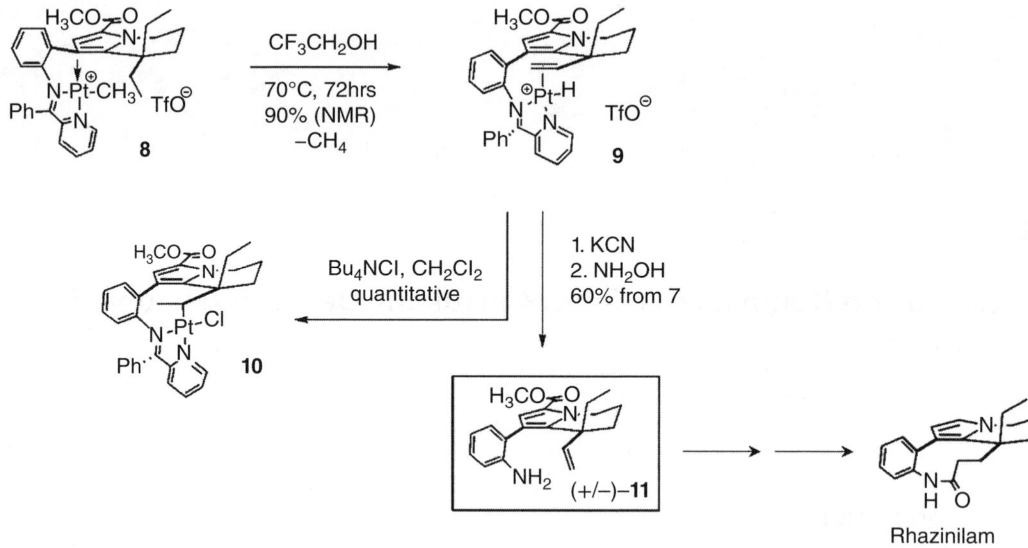

Scheme 20.1 Dehydrogenation-based route to a complex product. Source: Reprinted with permission from reference [2], copyright 2002, American Chemical Society.

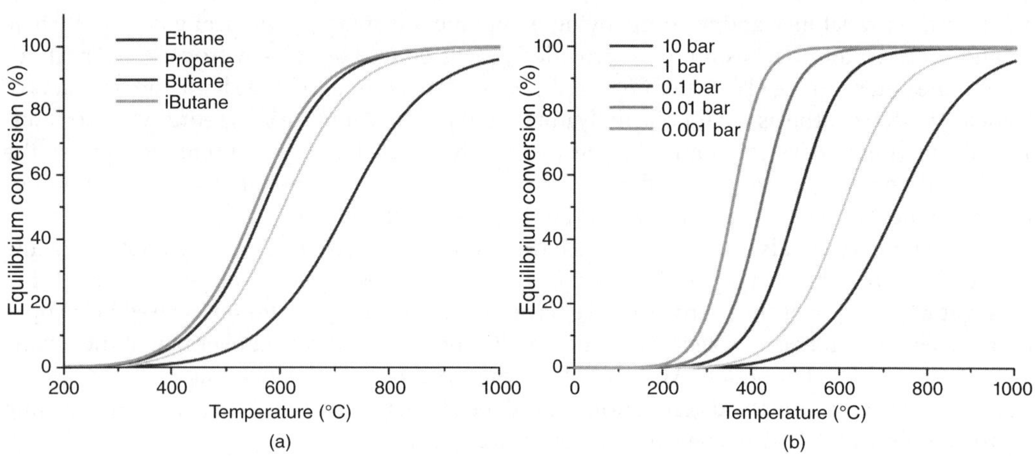

Figure 20.1 (a) Equilibrium conversion of ethane, propane, *n*-butane, and *i*-butane to the corresponding alkenes plus hydrogen as a function of temperature; the degree of conversion is greater for more substituted alkanes. (b) Equilibrium conversion of propane to propene plus hydrogen as a function of temperature and pressure; the degree of conversion is greater at lower pressures. Source: Reprinted with permission from reference [4], copyright 2014, American Chemical Society. (*See insert for color representation.*)

noncatalytic thermal reactions start becoming significant, leading to unwanted side reactions and reduced selectivity.

Oxidative dehydrogenation (ODH), in contrast, is always strongly favored thermodynamically; but *over*oxidation is even more so, and is also typically favored by kinetic considerations, so selectivity is the central problem. Formation of undesired secondary and side products – polyenes and aromatics (which actually may be desired in some cases), oxygenates, and especially CO and CO_2 – must be controlled somehow for any practically useful product yield to be achieved. In principle, that would seem to be more readily achievable with a homogeneous catalyst, where we might hope to have more control over mechanism, but so far at least there have not been many successful demonstrations.

This chapter examines these various modes of alkane DH, focusing on their comparative advantages and disadvantages, and tries to identify promising opportunities for future directions. It should be noted that the literature on heterogeneously catalyzed DH, both non-oxidative and oxidative, is massive, and that on homogeneously catalyzed NODH is starting to become comparably so (the Chapter 21 by Huang and Fang in the present volume [5] covers that topic in detail); hence I make no attempt at any sort of comprehensive coverage, but only highlight trends and generalizations, relying primarily on recent reviews. Reports on homogeneous catalysis of ODH are rather less common, so I will spend somewhat more time there.

I also largely restrict the discussion to alkanes. There is (again) a vast literature of dehydrogenative chemistry involving substrates with heteroatoms – alcohols to carbonyl compounds, amines to imines, etc. – reactions that are generally kinetically more facile and thermodynamically less disfavored than those of hydrocarbons, but I will only include examples that appear highly relevant to and/or cast some light upon alkane chemistry. For the broader picture of dehydrogenation and hydrogen transfer, there are several recent reviews, such as those by Dobereiner and Crabtree [6] and Beller et al. [7].

20.2 Heterogeneous Catalysis

The following three sections briefly summarizes the current state of the art for both NODH and ODH over heterogeneous catalysts and compares their practical applicability.

20.2.1 Heterogeneous NODH

Most of the information in this brief section, unless noted otherwise, is based on the excellent recent review by Weckhuysen et al. [4], which cites a number of other reviews on specific subtopics as well as extensively covering the primary literature.

As noted above, the increasing availability of light alkanes, especially from shale gas, makes dehydrogenation a more attractive route to high-volume alkenes such as ethene and propene. As yet only a small fraction (around 5%) of the large (on the order of 100 million tons per year!) demand for propene is met by propane dehydrogenation – the lion's share comes from cracking of higher hydrocarbons – but that fraction appears to be headed upward.

Several different process schemes have been implemented, based on one of two catalysts: chromia on alumina or platinum-tin on alumina. Both operate at temperatures well above 500 °C and pressures around or below 1 bar, required to achieve practically useful equilibrium conversions (see Figure 20.1); typical conversions are in the range of 10–30% with selectivities exceeding 90% (for propane to propene). The primary unwanted product is coke, which leads to deactivation by coating catalytic sites; hence it must be burned off to regenerate activity, either by frequent (minutes to several hours) cycles in fixed-bed reactors, or by operating in a recycling fluidized bed reactor.

A number of alternate catalysts based on other metal oxides (V, Mo, Ga, etc.) or carbon materials have been investigated extensively, and some lab-scale experiments have reported higher yields, approaching 50%; but none has yet been put into practice. For all of these, as well as those that have been put into practice, there of course has been (and is ongoing) a huge amount of work, both experimental and computational, aimed at identifying the precise nature of active sites, the roles played by different promoters and support materials, the factors controlling activity, selectivity, and deactivation processes, for example.

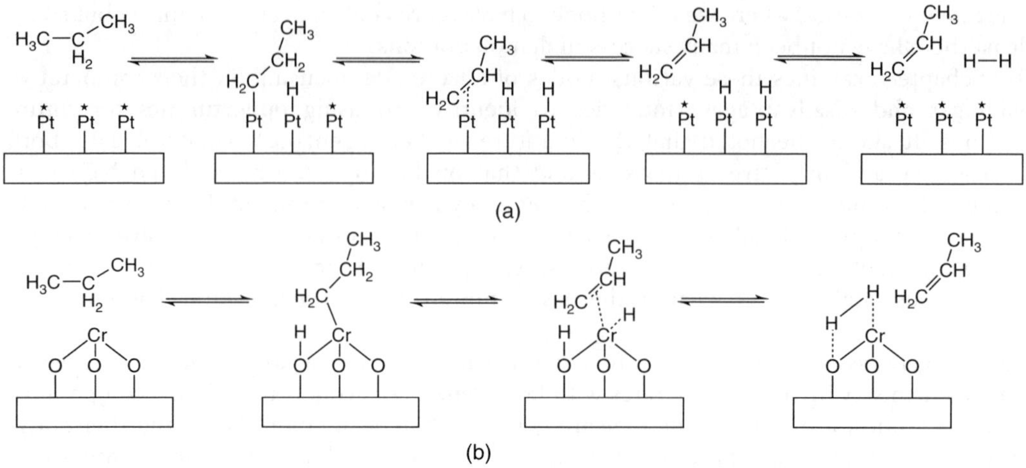

Figure 20.2 Probable mechanism of propane dehydrogenation over Pt (a) and CrO_x (b) based catalysts. Source: Reprinted with permission from reference [4], copyright 2014, American Chemical Society.

For both Pt- and CrO_x-based catalysts, the first step of the DH sequence is believed to be C—H activation at a metal center. By analogy to organometallic C—H activation, we might consider activation on a Pt surface as homolytic in nature – an oxidative addition-like step (but here involving two metal centers instead of one), whereas on CrO_x it is heterolytic, with the H going to an O center instead of another Cr – a sigma bond metathesis-like step. In either case, the resulting metal-bound alkyl group undergoes β-H elimination to liberate the products, alkene plus H_2 (Figure 20.2).

Although CrO_x, like most of the other MO_x systems studied, is potentially redox active, the mechanism of Figure 20.2 does not involve any oxidation state change, and indeed non-redox active MO_x-based catalysts have been shown to function for NODH as well. Recent work on Zn/SiO_2 reported propane DH with turnover frequencies (TOFs) as high as $2\,h^{-1}$, comparable to some of the better results for other MO_x catalysts (TOFs on Pt are considerably higher), with "catalytic dehydrogenation selectivity" >95% [8]. But that is somewhat misleading: to achieve such a high TOF, the reaction had to be carried out at 650 °C, around 100 °C higher than the comparison studies. At that temperature nonselective thermal reactions compete severely, so the >95% value is an estimate obtaining by subtracting the thermal contribution from the actual observed selectivity (not given in the paper; in a related study on a Fe/SiO_2 catalyst by the same group, which exhibited similar performance, the actual measured selectivity at 650 °C was around 70% [9]). A "true" DH selectivity of >95% was measured at 550 °C, where thermal reactions are negligible, but the TOF at that temperature was nearly an order of magnitude lower. The proposed mechanism, supported by DFT calculations, is very similar to that shown for CrO_x in Figure 20.2b.

20.2.2 Heterogeneous ODH

Several reviews in the last decade or so cover general aspects of the topic, although most are focused on particular substrates and/or catalysts: *n*-butane [10]; light alkanes [11]; ethane and propane [12]; ethane [13]; and propane over VO_x [14]. The following summary draws on all of them, as well as other references cited.

ODH, the reaction of alkane with O_2 to give alkene plus water, is strongly exothermic and hence not subject to the thermodynamic limitations of NODH: it can be carried out at any temperature where sufficient activity can be obtained, typically (but not always) well below those required for practical conversion levels with NODH. A wide variety of materials have been explored as catalysts; most of these are based on reducible metal oxides, especially V and Mo, but others involve nonreducible metal oxides (often in combination with a potentially redox-active promoter, such as Cl) and even nonoxidic materials.

The most commonly considered mechanism for ODH begins with hydrogen atom abstraction by a reactive oxygen center, producing as a first intermediate the equivalent of an alkyl radical, which may be either bound to the surface or released into the gas phase. The nature of the active oxygen species has been a subject of intense investigation; possibilities in any given case may include so-called lattice oxygen – oxide ions bound to metal centers in terminal or bridging mode – as well as surface-bound O_2-derived species (O^-, O_2^-, O_2^{2-}), defect sites, etc. In many cases it has been suggested that the availability of different sites is connected to selectivity, with, e.g. lattice sites promoting selective ODH and other types effecting nonselective side reactions. Participation of gas-phase reactions can also be significant, as the reaction of R· with O_2 is very rapid and leads predominantly to oxygenated products and/or CO_x.

Abstraction of a second H from the β position of the R· equivalent liberates alkene, followed by loss of water and regeneration of the active O species. If the latter is lattice oxygen, the overall sequence amounts to successive reduction and reoxidation of the metal – the so-called Mars-van Krevelen mechanism – as illustrated (in cartoon version) in Scheme 20.2 for a generic V-based catalyst. Other mechanistic patterns have also been considered.

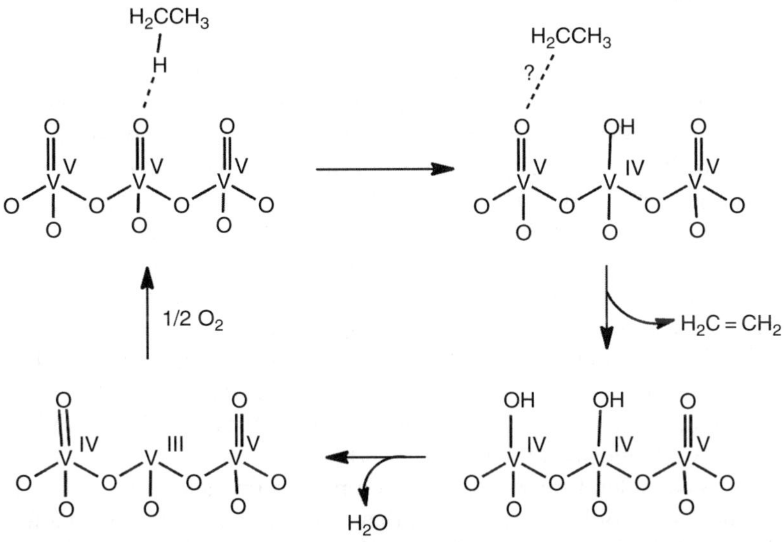

Scheme 20.2 Schematic Mars-van Krevelen route for ODH of ethane over a generic VO_x-based catalyst.

Of course, the thermodynamic advantage of ODH doesn't come for free: selectivity will always be considerably poorer, even without any complications from unwanted nonselective sites. This is because the alkene products will usually be as reactive as (in the best of cases), or considerably more reactive than, the alkane feed towards H atom abstraction, the rate of which is largely governed by C—H bond strength. For propane, for example, the dissociation energy of the secondary C—H bond is about 97 kcal mol^{-1}, vs. about 86 kcal mol^{-1} for the allylic C—H, so the relative rate of H abstraction for propene can be as much as an order of magnitude (or even more) higher than that for propane, depending on how "hot" the abstracting O center is.

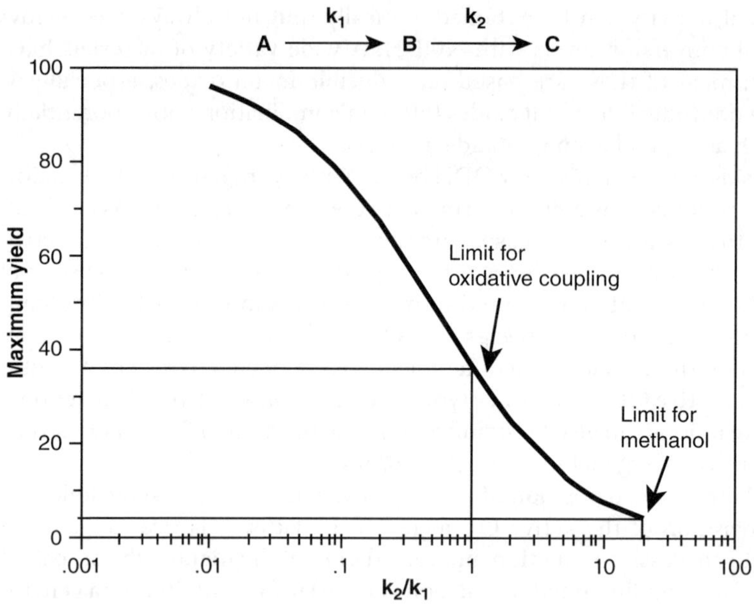

Figure 20.3 Maximum yield of desired product **B** as a function of the ratio of rate constants k_1 (for reactant feed **A** to **B**) and k_2 (for **B** to waste product **C**).

This places severe constraints on the maximum per-pass yield, which is given by conversion times selectivity: there will be an inverse relationship between the two, with selectivity decreasing as conversion increases. We can often achieve good selectivity at very low conversion, but for equal reactivity, once the product has built up to a concentration comparable to that of the reactant, it will be consumed by overoxidation as fast as it is formed; and that will happen even sooner if the product is more reactive. This is illustrated in Figure 20.3, which shows the maximum achievable yield as a function of the ratio of rate constants k_1 (for reactant A to desired product B) and k_2 (for B to waste product C): at equal rates the maximum yield is around 35%. Thus, for example, the oxidative coupling of methane to ethane should give a maximum yield in that neighborhood, as the C—H bond strengths of reactant and product are similar; while the direct oxidation of methane to methanol should give much lower yields – around 5% maximum – as the C—H bond strength of methanol is much smaller than that of methane. Both of those predictions have held up well, *independent of catalyst* [15].

For propane ODH, then, this consideration would predict a maximum yield in the range of 10–20%; and indeed, the majority of studies come out in that range, although there are a few scattered reports of yields approaching 30% or more [12]. Ethane might be thought to have a big advantage over propane, as ethene's olefinic C—H bonds are *stronger* than those of ethane, which by itself would predict $k_1 > k_2$. Indeed, there are some reports of higher (sometimes considerably) ethane ODH selectivities [12, 13]. However, there are other modes of reaction of C=C double bonds that work in the opposite direction, particularly the strength of adsorption: the more tightly the product alkene is held on the surface, the more likely it is to undergo secondary overoxidation rather than be liberated as product. The stronger binding of ethene appears to be a major part of the reason why ethane is not much (if at all) better than propane in ODH [16].

A completely different kind of material, boron nitride (BN), has recently been examined as a propane ODH catalyst [17]; the authors claim "extraordinary" selectivity to propene, up to 79% at 14% conversion (11% yield), exceeding the performance of "state of the art" catalysts.

They ascribe this superior performance to a different sort of active site – not lattice oxygen (which would not be at all likely for BN, of course) – instead they propose (supported by DFT calculations) that the active site consists of O—O terminated BN edges. It should be noted, though that (i) a top yield of 11% does not appear inconsistent with the C—H bond strength criteria outlined above; and (ii) a number of catalysts mentioned in the review articles cited above match or exceed that yield, so whether this really represents any sort of breakthrough remains to be seen.

20.2.3 Heterogeneous NODH vs. ODH

These two alternate approaches have rather obvious advantages and disadvantages with respect to one another. NODH generally operates at higher temperatures and, as an endothermic reaction, requires a source of heat, so it should be less energy efficient; also the overall process is more complex, in part due to the need to regenerate after coke deposition, and hence probably more expensive to construct and run. ODH is exothermic, so no heat supply would be needed. On the other side, ODH (at least to date) gives *much* lower selectivities/yields than NODH, requiring either operating at low conversion with a large recycle stream or accepting major losses to CO_x and other waste products; also NODH yields a valuable co-product, H_2.

This is considerably oversimplified, though: there are alternate possible reactor configurations and factors specific to individual process designs. For example, the H_2 produced by NODH can be used for heating the reactor, eliminating the need for an additional fuel source (but sacrificing the value of the H_2, of course). Or it can be operated in autothermal mode, where a sub-stoichiometric amount of O_2 is cofed with the alkane, hoping it will react selectively with H_2 rather than hydrocarbon and thus both provide the needed heat and shift the thermodynamics. ODH reactors can feature staged O_2 feeding or oxygen-transporting membranes, minimizing the effect of unselective gas-phase reactions.

A 2001 analysis compared three methods for propane DH, using literature data and a process simulation package [18]. The respective performances deduced from the literature were: 25% conversion, 88% selectivity for the existing UOP NODH process; 90% conversion, 17% selectivity for an autothermal variant over Pt/Al_2O_3; and 27% conversion, 32% selectivity for ODH over V_2O_5/TiO_2. Process modeling indicated that neither of the last two comes close to being competitive with existing NODH technology, although the autothermal process produces large amounts of ethene (more than propene!), so it could be of interest if ethene were in high demand. Somewhat unexpectedly, the ODH came out worst in terms of energy efficiency, largely as a consequence of a heavy requirement for propane/propene separation. Overall, this analysis implies that really *major* improvements in ODH selectivity/yield will be needed for it to become a viable route to light alkane conversion.

20.3 Homogeneous NODH

In light of the fact that this volume includes a separate chapter dedicated to this topic (Chapter 21) [5], as well as the existence of recent reviews on various aspects thereof [6, 19], I will confine myself to a very brief overview.

A variety of mechanistic modes for C—H activation of an alkane at a transition metal center – oxidative addition, sigma bond metathesis and variants thereof, electrophilic activation, etc. – lead to transition metal alkyls; many of these reactions take place under remarkably mild conditions [20]. From there, β-H elimination will very often be a facile next step; but subsequent liberation of alkene plus H_2 is far uphill thermodynamically, requiring some additional driving force for a net transformation to take place. Otherwise, the reaction may just stop at the alkene-hydride stage, where the ligand-metal bond energies provide stability. A number of different examples have been reported, such as that shown in Eq. (20.1) (which proceeds well for just about any linear or cyclic alkane tested) [21]. Sometimes further dehydrogenation is observed: formation of cyclopentadienyl complexes from reactions of cyclopentane or cyclopentene have been observed in several instances.

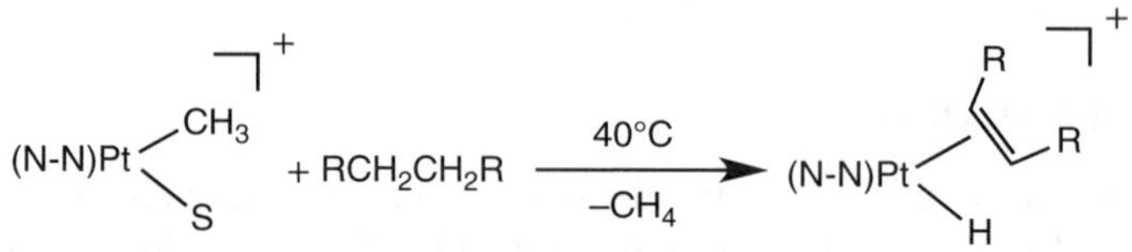

(20.1)

In many homogeneous NODH studies, the driving force is provided by a sacrificial olefin, most commonly *t*-butylethylene; the net reaction is thus H_2 transfer rather than simple DH. Early examples involved a cationic Ir complex (which led to bound ligand, not free alkene, as shown in Scheme 20.3) and the Rh(I) complex $RhCl(CO)(PMe_3)_2$; but the vast majority of work in this field has made use of "PCP" pincer complexes such as that shown as Eq. (20.2). A large number of such complexes, where the backbone arene, the substituents on the phosphines, and the linkers connecting them are all varied, have been tested for their ability to effect transfer hydrogenation, and the mechanistic details are well understood [5, 19].

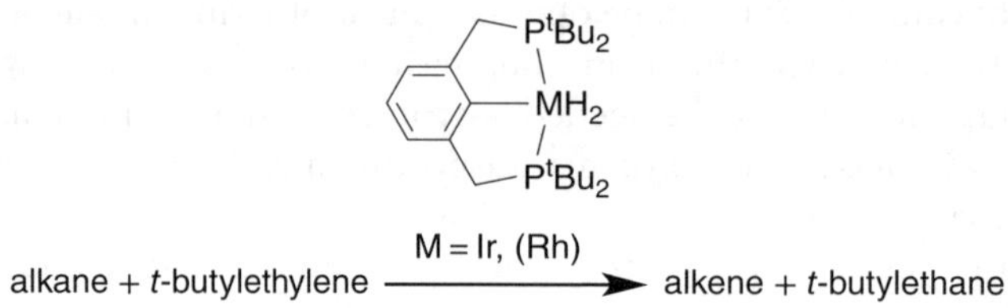

(20.2)

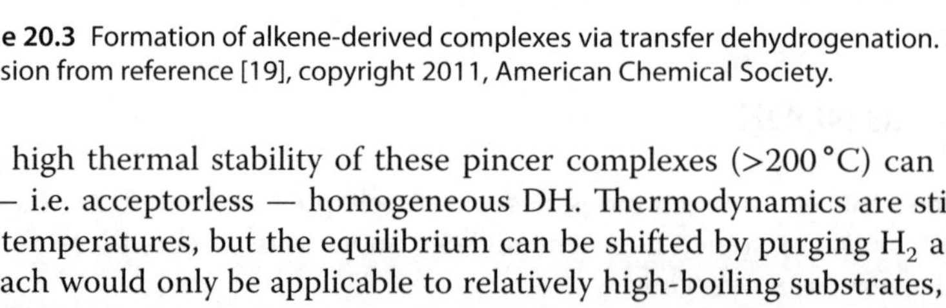

Scheme 20.3 Formation of alkene-derived complexes via transfer dehydrogenation. Source: Reprinted with permission from reference [19], copyright 2011, American Chemical Society.

The high thermal stability of these pincer complexes (>200 °C) can be exploited to effect true — i.e. acceptorless — homogeneous DH. Thermodynamics are still very unfavorable at these temperatures, but the equilibrium can be shifted by purging H_2 as it forms; clearly this approach would only be applicable to relatively high-boiling substrates, certainly not gaseous light alkanes.

Other means of overcoming thermodynamics have been devised as well. The reaction can be driven photochemically, an observation first reported in the 1980s for the Ir and Rh complexes cited above, and further studied more recently: up to 1500 turnovers were achieved in 7 hours for the DH of octane, using $RhCl(CO)(PMe_3)_2$ and near UV-blue light [22]. Unfortunately selectivity to the most desired isomer 1-octene is low (as is also the case for acceptorless thermal DH), and the economics of such a photochemical transformation are not clear.

Another approach is to couple the unfavorable step with one or more favorable ones in a "tandem" catalytic system whose overall thermodynamics are neutral or favorable. The combination of DH and alkene metathesis effects net *alkane* metathesis (Scheme 20.4), in which the extremely low equilibrium concentration of alkenes is compensated by the extremely high reactivity of alkene metathesis catalysts; this could provide a route, for example, of not-very-valuable hexane to much more valuable (as diesel or jet fuel) decane. Unfortunately, the reactions are not very selective: C_{10} could be obtained in up to 50% selectivity of all C_{7+} products, but only at relatively low conversion levels (and with a good deal of lighter byproducts as well) [23]. Alkane metathesis over a single catalyst – a surface-supported Ta—H species obtained by reaction of an organometallic Ta complex with silica – had been previously reported; the mechanism is thought to be quite similar, with the Ta centers able to effect *both* components of the tandem system, DH and alkene metathesis [24].

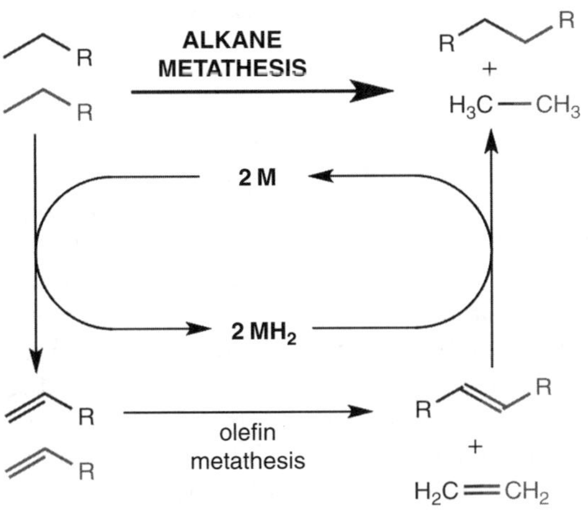

Scheme 20.4 Schematic route for alkane metathesis. Source: Reprinted with permission from reference [23], copyright 2012, American Chemical Society.

A different tandem approach couples DH with olefin oligomerization, in effect a homogeneously catalyzed alkylation (Scheme 20.5); it is favorable enthalpically but unfavorable entropically, with $\Delta G_o < 0$ up to around 250 °C. Partial proof of concept has been demonstrated as in Scheme 20.6 [25]; unfortunately, the C_{13} and C_{14} alkylation products are highly branched alkanes (a consequence of the particular dimerization catalyst used) that cannot undergo dehydrogenation to close the cycle [24].

$$C_nH_{2n+2} \underset{}{\overset{\text{dehydrogenate}}{\rightleftharpoons}} C_nH_{2n} + H_2$$

$$2\ C_nH_{2n} \xrightarrow{\text{dimerize}} C_{2n}H_{4n}$$

$$C_{2n}H_{4n} + H_2 \overset{\text{hydrogenate}}{\rightleftharpoons} C_{2n}H_{4n+2}$$

$$\text{Net: } C_nH_{2n+2} + C_nH_{2n} \longrightarrow C_{2n}H_{4n+2}$$

Scheme 20.5 Tandem alkane-alkene coupling scheme.

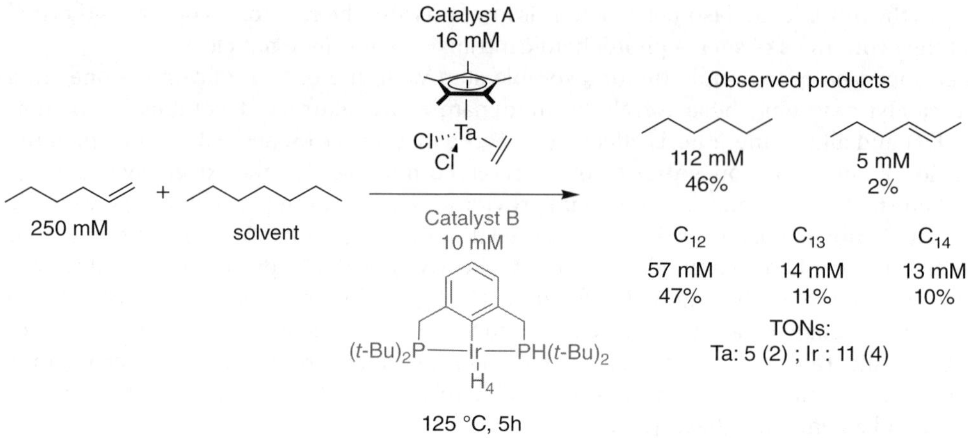

Scheme 20.6 Hexene-heptane coupling by a tandem Ir (DH)-Ta (dimerization) catalyst system.

20.4 Homogeneous ODH

From the preceding discussion, coupled with a good deal of work on alkane oxidation mediated by homogeneous C—H bond activation, this approach would appear to be *very* promising. Despite the fact that many organometallic species are unstable to O_2 and/or oxidizing conditions in general, there are a number of systems that can be exploited for alkane oxidation, including the original Shilov system of the 1970s, the Catalytica system of the 1990s, and many others [26]. There is no obvious inherent barrier to combining an alkane with O_2 and finding a homogeneous catalyst to induce them to react. The possibility of temporally separating the reactions – oxidizing the hydrocarbon by the metal, and reoxidizing the metal by O_2 or some other reagent, as in the Wacker process – offers additional possibilities for minimizing selectivity-degrading side reactions. Above all, the thermodynamics of ODH are favorable under the conditions most compatible with organometallic chemistry, unlike NODH. And yet there are only a few known examples of such chemistry – none at all for converting simple alkanes to the corresponding alkene.

The Shilov system provides one of the earliest examples. That system consists of heating hydrocarbons with a mixture of Pt(II) and Pt(IV) chlorides in water or aqueous CF_3COOH; whereas acyclic alkanes are oxidized to alcohols and alkyl chlorides; cyclohexane undergoes ODH to benzene [27]. In similar fashion, heating substituted cyclohexanes with a mixture of $Pd(OAc)_2$ and $Na_2Cr_2O_7$ afforded the corresponding substituted benzenes, along with further oxidation products thereof [28]. Note that both of these involve aromatization, not just simple ODH, and metal-based oxidants (Pt(IV), Cr(VI)) rather than O_2.

An example of ODH using O_2 – but not of an alkane – arose out of studies of hydroxy-bridged dimers of Pt(II) and Pd(II), which will activate allylic C—H bonds (among others). Both the Pd and Pt complexes behave identically toward indene, to give (indenyl)M(II) complexes as shown in Eq. (20.3). The corresponding reactions with cyclohexene, in contrast, differ dramatically: whereas the Pt complex gives the analogous (cyclohexenyl)Pt(II) complex, with Pd no stable organometallic complex is formed; instead the cyclohexene is catalytically disproportionated to benzene plus cyclohexane, along with deposition of Pd metal. Even more intriguingly, though, running the reaction under an O_2 atmosphere completely shuts down disproportionation and gives only catalytic ODH of cyclohexene to benzene, with essentially 100% selectivity [29]. In similar fashion, $Pd(CF_3CO_2)_2$, which had previously been known to bring about cyclohexene disproportionation under inert atmosphere [30], becomes an even more active catalyst for cyclohexene ODH under O_2, proceeding at a moderate rate even at room temperature (Eq. (20.4)) [31].

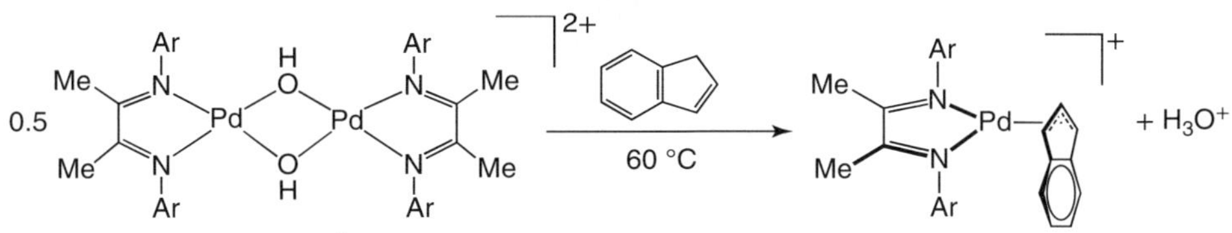

(1) (Ar = 3,5-$^tBu_2C_6H_3$)

(20.3)

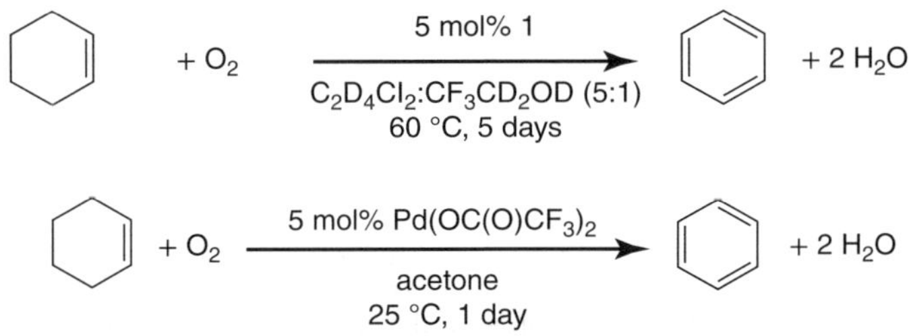

(20.4)

ODH by $Pd(CF_3CO_2)_2$ under such mild conditions could be extended only to a couple of cyclohexene derivatives: 1,2-dihydronaphthalene worked well at room temperature, but 1- and 4-methylcyclohexene required 60 °C, and at that higher temperature the presence of O_2 was not sufficient to completely inhibit formation of Pd metal and competing disproportionation [31]. More recently, Stahl has found that under still more stringent conditions – 110 °C with the addition of an oxidation-promoting quinone – a fairly wide range of substituted cyclohexenes can be effectively aromatized (Scheme 20.7) [32]. Another study examining a range of co-oxidants and solvents gave allylic oxidation and/or disproportionation as the primary products, with little or no ODH observed [33]. Stahl has also examined the Pd-catalyzed ODH of cyclohexanones to phenols; the reaction proceeds via cyclohexenones, which can be made the main product by suitably controlling reaction conditions, apparently a consequence of competing catalysis by molecular Pd(II) and soluble Pd nanoparticles [32].

A generic scheme for Pd(II)-catalyzed ODH is shown in Scheme 20.8. We originally explained the low-temperature ODH of cyclohexene by the sequence of Scheme 20.9 [31], where the initially formed (cyclohexenyl)Pd(II) complex (unlike its Pt analog) is unstable toward β-H elimination, leading to cyclohexadiene and a Pd—H species. The latter is subject to oxidation by O_2, as is known for a wide range of Pd systems, either by O_2 insertion into Pd—H (as in Scheme 20.9), as has been demonstrated for some related systems, or via preliminary deprotonation to Pd(0) and oxidation thereof (as in Scheme 20.8). A second cycle with the cyclohexadiene as substrate would give benzene. Stahl proposes the same mechanism [32].

But there are (at least) two problems with this interpretation. First, if it were correct, there is no obvious reason why linear alkenes should not exhibit similar behavior. It is of course true that the final product, benzene, is particularly stable; but how do the molecules "know" that such a stable endpoint is available at the end of the sequence (unless all steps are reversible, which seems most unlikely)? Generating a diene from an acyclic monoene should not be significantly less favored than cyclohexadiene from cyclohexene; nonetheless, no such chemistry is observed. Instead, acyclic alkenes react to generate Pd—H intermediates via a process resembling Wacker chemistry (trace amounts of dienes can also be detected,

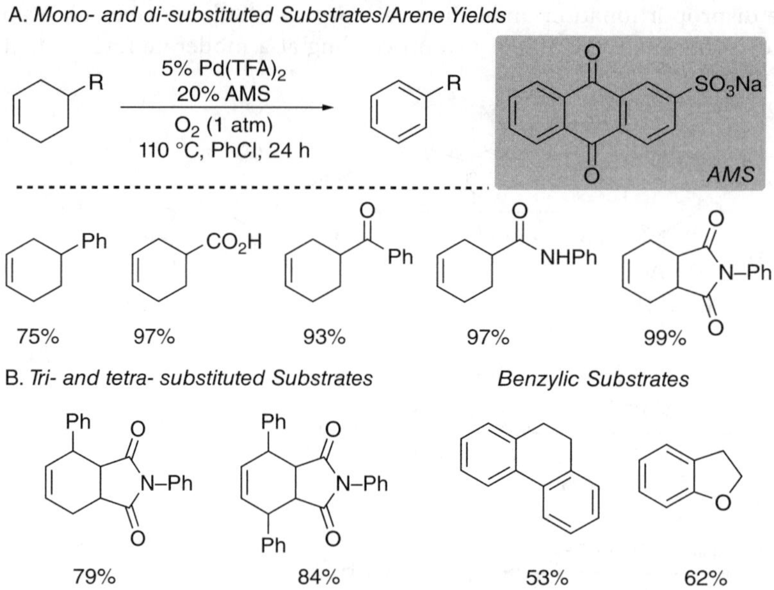

Scheme 20.7 ODH of substituted cyclohexenes. Source: Reprinted with permission from reference [32], copyright 2016, American Chemical Society.

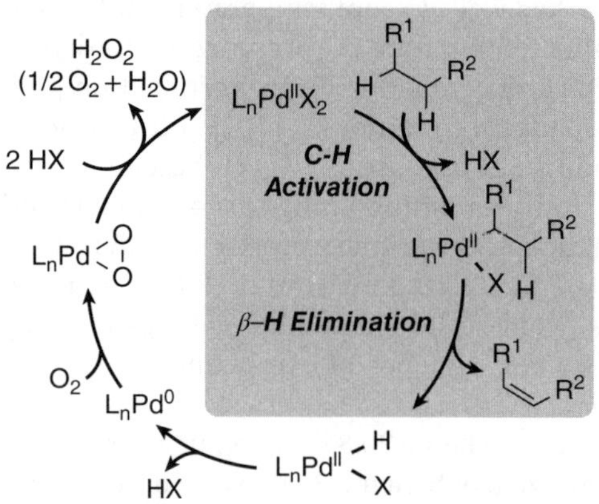

Scheme 20.8 Schematic route for Pd(II)-catalyzed alkane ODH. Source: Reprinted with permission from reference [32], copyright 2016, American Chemical Society.

implying that ODH is occurring, but at a very slow rate compared to other processes), which then catalyze alkene oligomerization [34]. In contrast, acyclic carbonyl compounds *do* undergo ODH to α,β-unsaturated carbonyls under conditions similar to those used for ODH of cyclohexanones [32].

There is an even stronger reason to be dubious about the mechanism of Scheme 20.9: the putative (cyclohexenyl)Pd(II) intermediate was independently synthesized and found to be completely inert to the conditions of catalytic cyclohexene ODH (Scheme 20.10) [35]! So there must be something special about the interaction of cyclohexene, and perhaps other such cyclic hydrocarbons as well, that allows this ODH chemistry to proceed for them but not acyclic analogs. Some considerable efforts toward achieving Pd-catalyzed ODH for linear alkenes

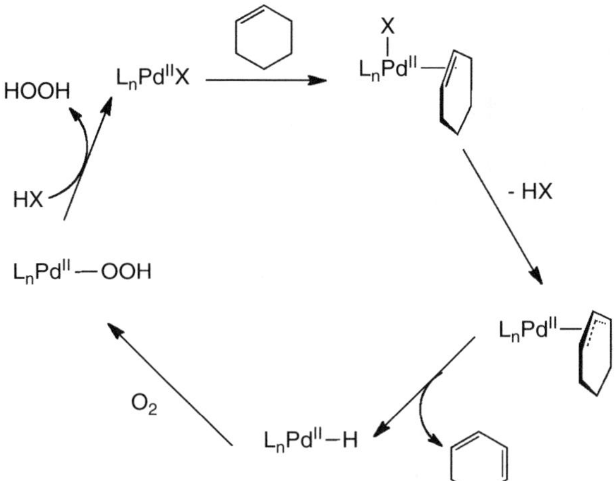

Scheme 20.9 Possible mechanism for Pd(II)-catalyzed ODH of cyclohexene to benzene.

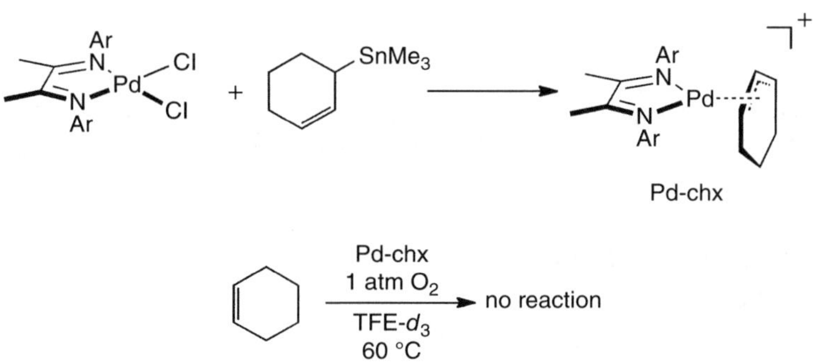

Scheme 20.10 Synthesis and non-reactivity of putative (cyclohexenyl)Pd(II) intermediate.

and alkanes, involving a wide range of ligands, metal complex structures, solvents, reaction conditions, etc., have as yet yielded no success [35].

An Ir-based C—H activator has been shown, in effect, to carry out all the steps of an ODH cycle for a *linear* alkene. The Ir(III) complex of Scheme 20.11 stoichiometrically dehydrogenates octane to a mixture of isomeric octenes and an Ir(III)-hydride; in contrast to NODH by pincer complexes of Ir(I) (see above), activation does not appear to proceed via C—H oxidative addition, but rather by some sort of sigma bond metathesis route that leaves the oxidation state of Ir unchanged [37]. Subsequently, it was discovered that the Ir(III) hydride could be reconverted to the starting species by reaction with O_2 at room temperature [36]. The sum of those two steps amounts to catalytic ODH of octane analogous to that of Scheme 20.8); unfortunately, it can't actually be run, as the reoxidation cannot be carried out at anything like the elevated temperature needed for C—H activation without destroying the system.

One last result is worth mentioning in this section: in the original report on the Catalytica system [38], ethane was found to react much more rapidly than methane, and to give a difunctionalized product, isethionic acid (ITA). Recently, a more careful examination has determined that the reaction proceeds via β-H elimination from a Pt-ethyl intermediate, followed by addition of H_2SO_4 or $H_2S_2O_7$ to C=C (Scheme 20.12). It is not clear whether (free) ethylene or ethyl sulfate is the primary product, as they are in rapid equilibrium, and in any case no ethylene is

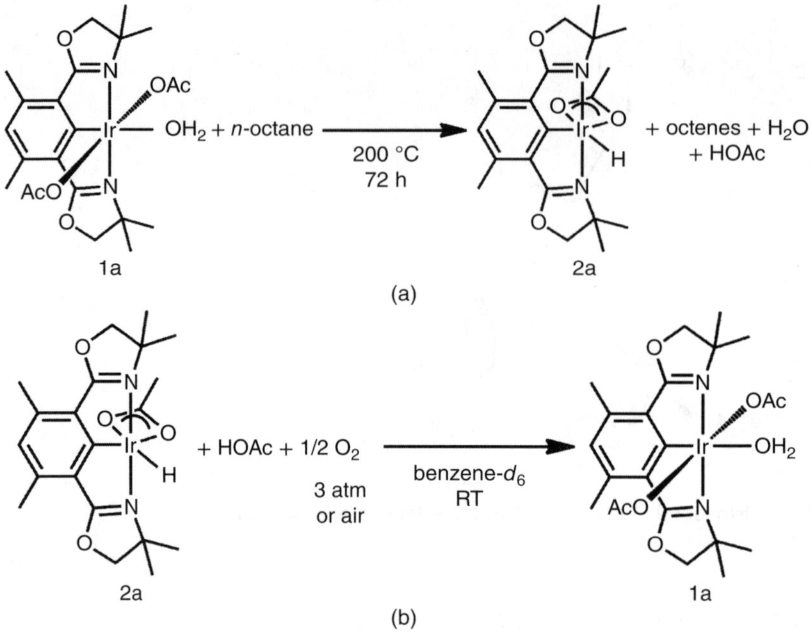

Scheme 20.11 (a) Stoichiometric alkane DH by Ir(III). (b) Aerobic regeneration of Ir(III) complex. Source: Reprinted with permission from reference [36], copyright 2014, American Chemical Society.

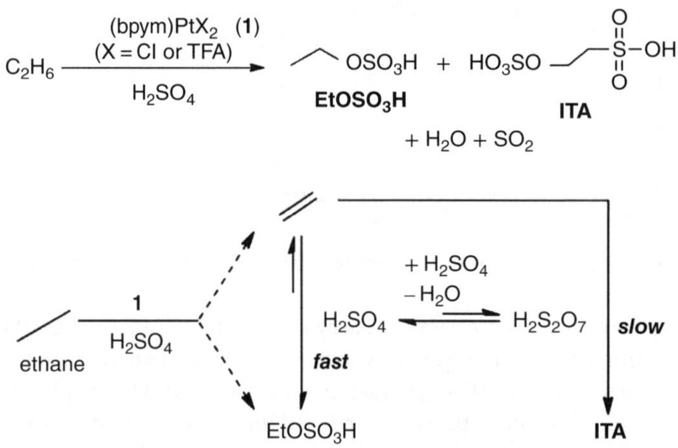

Scheme 20.12 Oxidative functionalization of ethane by the Catalytica system. Source: Reprinted with permission from reference [39], copyright 2014, American Chemical Society.

ever observed, so this chemistry does not amount to any achievable ODH, but such a step may well participate [39].

20.5 Summary and Prospects

A brief – and oversimplified – summary of the current state of catalytic dehydrogenation might run as follows. For conversion of light alkanes to the corresponding alkenes – very large-scale processes that are expected to increase in importance – heterogeneous NODH appears likely to be the method of choice for the foreseeable future. It is already competitive with other routes to light alkenes, such as cracking, despite the drawback of unfavorable thermodynamics, necessitating high temperatures that lead to extensive coking, in turn requiring complex (and hence relatively expensive) process designs.

Of course, the thermodynamics problem is much worse for homogeneous NODH, such that it is difficult to envision any large-scale application to alkene production. Processes that couple DH to another transformation, such as the tandem alkane metathesis and/or alkane-alkene couplings discussed earlier, might possibly be economically feasible, although selectivity is a major limitation in demonstrations so far; it is not clear how that could be overcome. Still, this looks like a promising area for further work.

Despite the attractiveness – and high level of interest – in replacing heterogeneous NODH with ODH, results to date are not all that encouraging. Assuming that the economic/engineering evaluations are correct, the best selectivities are just not good enough; and given that these appear to be a consequence of *inherent* relative reactivities of C—H and C=C bonds, it seems unlikely that there is a magic catalyst, waiting to be discovered, that would provide sufficient improvement. However, perhaps some combination of novel catalyst and process design (e.g. some sort of membrane reactor) might make ODH more competitive.

Smaller-scale DH of complex molecules offers considerable potential for applications in organic synthesis. Such reactions would almost surely require homogeneous catalysis, and indeed, some examples have already been demonstrated, such as the one in Scheme 20.1; Stahl's review on ODH of cyclo-C_6 species illustrates routes to a number of important phenol-based synthetic targets [32]. Homogeneous NODH approaches may also be promising, although subject to a couple of limitations. For most applications a sacrificial H_2 acceptor is required to drive the reaction, placing some economic constraints on any possible process; for acceptorless NODH, product double-bond isomerization is almost always as fast as or faster than alkene formation, so that regioselectivity will be problematic in most instances.

At present, homogeneous ODH appears (to me) to present the most intriguing opportunities. We have seen several examples of highly selective conversions under extremely mild conditions; granted, they all involve rather special substrates such as cyclohexene and derivatives thereof, but there doesn't appear to be any fundamental reason why analogous reactions of acyclic alkenes and even of alkanes should not be possible, given the vast number of known C—H activators that work under the same mild conditions. The Pd(II) complexes that catalyze cyclohexene ODH have not been shown to be reactive toward alkane C—H bonds, but some very closely related ones are; hence, it does not seem too far-fetched to imagine that some combination of ligand, solvent, and reaction conditions might make *all* the steps of a sequence such as that in Scheme 20.8 possible and compatible with one another. Similarly, a variant of the Ir(III) complex of Scheme 20.11 that lowers the temperature needed for C—H activation, while at the same time raising the temperature at which reoxidation can occur nondestructively, might permit the two reaction halves to operate simultaneously. Fruitful areas of research are certainly waiting to be explored in all of the DH approaches discussed here, with low-temperature, homogeneous ODH perhaps the most attractive of all.

Acknowledgment

Most of the author's own work discussed here, including the unpublished results, was supported by BP.

References

1 Alper, J. The Changing Landscape of Hydrocarbon Feedstocks for Chemical Production: Implications for Catalysis: Proceedings of a Workshop, The National Academies Press, 2016.
2 Johnson, J.A., Li, N., and Sames, D. (2002). *J. Am. Chem. Soc.* 124: 6900.
3 Eisenstein, O. and Crabtree, R.H. (2001). *New J. Chem.* 25: 665.
4 Sattler, J.J.H.B., Ruiz-Martinez, J., Santillan-Jimenez, E. et al. (2014). *Chem. Rev.* 114: 10613.
5 Fang, H., Liu, G., and Huang, Z. (2018). Dehydrogenation of alkanes using molecular catalysts. In: *Functionalization of Alkanes* (ed. A.J.L. Pombeiro) Chapter 21. Wiley.
6 Dobereiner, G.E. and Crabtree, R.H. (2010). *Chem. Rev.* 110: 681.
7 Werkmeister, S., Neumann, J., Junge, K. et al. (2015). *Chem. J. Eur.* 21: 12226.
8 Schweitzer, N.M., Hu, B., Das, U. et al. (2014). *ACS Catal.* 4: 1091.
9 Hu, B., Schweitzer, N.M., Zhang, G. et al. (2015). *ACS Catal.* 5: 3494.
10 Madeira, L.M. and Portela, M.F. (2002). *Catal. Rev.* 44: 247.
11 Grabowski, R. (2006). *Catal. Rev.* 48: 199.
12 Cavani, F., Ballarini, N., and Cericola, A. (2007). *Catal. Today* 127: 113.
13 Gärtner, C.A., van Veen, A.C., and Lercher, J.A. (2013). *ChemCatChem* 5: 3196.
14 Carrero, C.A., Schloegl, R., Wachs, I.E. et al. (2014). *ACS Catal.* 4: 3357.
15 Labinger, J.A. (2004). *J. Mol. Catal. A: Chem.* 220: 27.
16 Zboray, M., Bell, A.T., and Iglesia, E. (2009). *J. Phys. Chem. C* 113: 12380.
17 Grant, J.T., Carrero, C.A., Goeltl, F. et al. (2016). *Science* 354: 1570.
18 Wolf, D., Dropka, N., Smejkal, Q. et al. (2001). *Chem. Eng. Sci.* 56: 713.
19 Choi, J., Roy MacArthus, A.H., Brookhart, M. et al. (2011). *Chem. Rev.* 111: 1761.
20 Labinger, J.A. and Bercaw, J.E. (2002). *Nature* 417: 507.
21 Chen, G.S., Labinger, J.A., and Bercaw, J.E. (2007). *Proc. Natl. Acad. Sci. U.S.A.* 104: 6915.
22 Chowdhury, A.D., Weding, N., Julis, J. et al. (2014). *Angew. Chem. Int. Ed.* 53: 6477.
23 Haibach, M.C., Kundu, S., Brookhart, M. et al. (2012). *Acc. Chem. Res.* 45: 947.
24 Copéret, C. (2010). *Chem. Rev.* 110: 656.
25 Leitch, D.C., Lam, Y.C., Labinger, J.A. et al. (2013). *J. Am. Chem. Soc.* 135: 10302.
26 Labinger, J.A. (2014). Alkane functionalization via electrophilic activation. In: *Alkane C-H Activation by Single-Site Metal Catalysis* (ed. P.J. Pérez), 17–71. Dordrecht: Springer.
27 Shilov, A.E. and Shul'pin, G.B. (1987). *Russ. Chem. Rev.* 56: 442.
28 Kao, L.C. and Sen, A. (1991). *New J. Chem.* 15: 575.
29 Bercaw, J.E., Hazari, N., Labinger, J.A. et al. (2008). *Angew. Chem. Int. Ed.* 47: 9941.
30 Trost, B.M. and Metzner, P.J. (1980). *J. Am. Chem. Soc.* 102: 3572.
31 Bercaw, J.E., Hazari, N., and Labinger, J.A. (2008). *J. Org. Chem.* 73: 8654.
32 Iosub, A.V. and Stahl, S.S. (2016). *ACS Catal.* 6: 8201.
33 da Silva, M.J. and Berlini, L. (2016). *New J. Chem.* 40: 7996.
34 Winston, M.S., Oblad, P.F., Labinger, J.A. et al. (2012). *Angew. Chem. Int. Ed.* 51: 9822.
35 M.S. Winston, M.C. Lehman, J.A. Labinger, et al. unpublished results.
36 Allen, K.E., Heinekey, D.M., Goldman, A.S. et al. (2014). *Organometallics* 33: 1337.
37 Allen, K.E., Heinekey, D.M., Goldman, A.S. et al. (2013). *Organometallics* 32: 1579.
38 Periana, R.A., Taube, D.J., Gamble, S. et al. (1998). *Science* 280: 560.
39 Konnick, M.M., Bischof, S.M., Yousufuddin, M. et al. (2014). *J. Am. Chem. Soc.* 136: 10085.

21

Dehydrogenation of Alkanes Using Molecular Catalysts

Huaquan Fang, Guixia Liu, and Zheng Huang

Shanghai Institute of Organic Chemistry, State Key Laboratory of Organometallic Chemistry, Shanghai, China

21.1 Introduction

Olefins are widely used as building blocks for the production of a variety of commodity chemicals ranging from small molecules to large polymers. Their manufacture in industry, especially for light olefins, mainly relies on the steam cracking and fluid catalytic cracking of abundant and low-cost hydrocarbons (such as light diesel oil, naphtha, and other oil byproducts) by using heterogeneous dehydrogenation systems at very high temperature (500–900 °C) [1]. However, such harsh reaction conditions will result in low energy efficiency, unavoidable side reactions, and poor product selectivities. To date, the application of heterogeneous dehydrogenation systems is limited to the dehydrogenation of light alkanes, such as ethane, propane, and *n*-butane, and the conversion of ethylbenzene to styrene. Largely due to the development of organometallic chemistry, the homogeneous catalytic alkane dehydrogenation has emerged as an important research topic in the field of C—H activation. The alkane dehydrogenation reactions using molecular catalysts in general proceed under much milder reaction conditions (150–200 °C), and thus may offer better selectivity in comparison to the known heterogeneous system.

Direct dehydrogenation of alkanes (i.e. acceptorless dehydrogenation), which generates the alkene products and the hydrogen gas, is an atom-economical and environmentally benign reaction (Scheme 21.1). Although the removal of molecular hydrogen from alkanes is entropically favored, it is a thermodynamically unfavorable process in most cases. The high endothermicity of the process could be compensated by the use of a sacrificial hydrogen acceptors or by UV irradiation (Scheme 21.1). The focus of this chapter is on homogeneous dehydrogenation of alkanes mediated and/or catalyzed by various transition-metal molecular catalysts. This chapter is divided into two parts: (i) thermochemical dehydrogenation of alkanes; and (ii) photochemical dehydrogenation of alkanes. It is noteworthy that the oxidative dehydrogenation of alkanes, in which oxygen acts as an oxidant, is not covered in this chapter. The readers are directed to Chapter 20 by Dr. Labinger [2], which covers the topic of oxidative dehydrogenation of alkanes in detail. It should also be noted that reviews [3–6] and chapters [7–11] covering various aspects of alkane dehydrogenations have been published in recent years.

21.2 Thermochemical Dehydrogenation of Alkanes

The first stoichiometric transfer dehydrogenation of alkanes mediated by a cationic Ir(III) complex, $[IrH_2(acetone)_2(PPh_3)_2][BF_4]$, was reported by Crabtree in 1979 [12] (Eq. 21.1).

Alkane Functionalization, First Edition. Edited by Armando J. L. Pombeiro and M. Fátima C. Guedes da Silva.

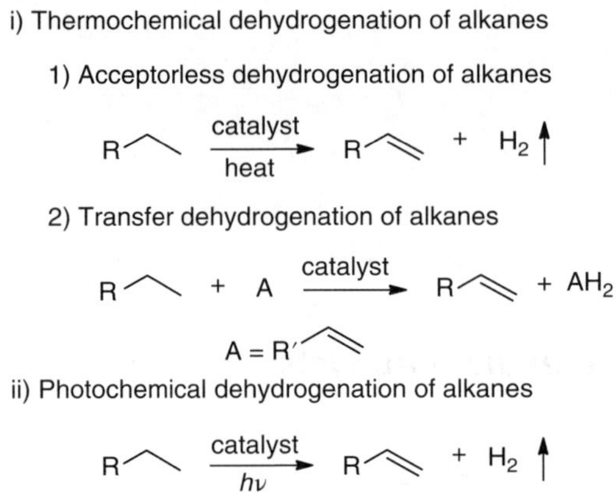

Scheme 21.1 Thermochemical and photochemical dehydrogenation of alkanes.

$[IrH_2(acetone)_2(PPh_3)_2][BF_4]$ reacted with cycloalkanes, such as cyclopentane and cyclooctane (COA), to generate the corresponding diene Ir(III) complexes in the presence of *tert*-butylethylene (TBE) as the hydrogen acceptor. The author found that TBE was indispensable for this transformation. Less bulky olefins, such as ethylene, styrene, and trimethylvinylsilane, inhibit the reaction due to the formation of stable iridium complexes. More bulky olefins, such as tetraphenylethylene, did not react with the cationic Ir(III) complex at all. Thus, TBE, bearing a bulky *t*Bu substituent and lacking allylic hydrogens, was recognized as an ideal hydrogen acceptor for the alkane dehydrogenation reaction. Subsequently, similar stoichiometric transfer dehydrogenation of cycloalkanes were reported by Felkin and co-workers by using $(PAr_3)_2ReH_7$ (Ar = *p*-F-Ph, Ph, or *p*-Me-Ph) [13, 14]. In 1991, Roddick also reported a stoichiometric dehydrogenation of cycloalkane using a ruthenium(II) (fluoroalkyl)phosphine hydride complex [15].

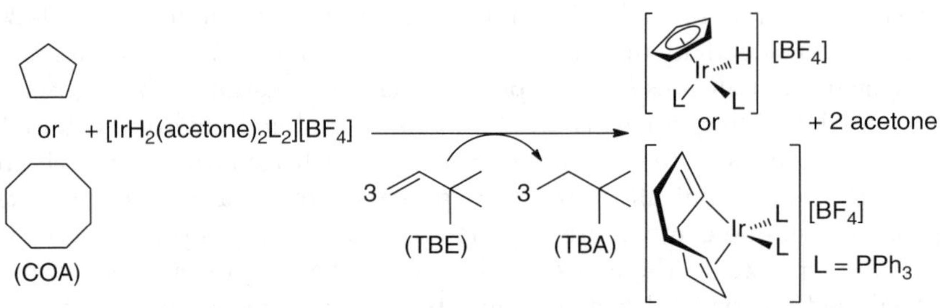

(21.1)

The first catalytic transfer dehydrogenation of alkanes was reported independently by the group of Felkin [16, 17] and Crabtree [18–20]. Felkin and co-workers found that the Re complexes $(PAr_3)_2ReH_7$ (Ar = *p*-F-Ph, Ph, or *p*-Me-Ph), catalyzed the transfer dehydrogenation of cycloalkanes [16]. Among the catalysts investigated, $[P(p\text{-F-}C_6H_4)_3]_2ReH_7$ exhibited the best results. For the transfer dehydrogenation of COA, a maximum of 1.6 turnover numbers (TONs) could be obtained at 30 °C after 10 minutes, and a maximum of 9 TONs was obtained at 80 °C under similar conditions. Later, the same group reported a variety of iridium and ruthenium polyhydrides for the transfer dehydrogenation of COA with TBE as the hydrogen acceptor [17]. The iridium pentahydride $(PiPr_3)_2IrH_5$ gave a maximum of 70 TONs at 150 °C after 3 hours, and the ruthenium tetrahydride $[P(p\text{-F}C_6H_4)_3]_3RuH_4$ afforded a maximum of 55 TONs at 150 °C after 10 days.

Crabtree et al. synthesized two new catalysts, $(PR_3)_2IrH_2(CF_3CO_2)$ (R = *p*-F-C_6H_4 or Cy), to catalyze the COA/TBE transfer dehydrogenation [19]. The (triarylphosphine)-based

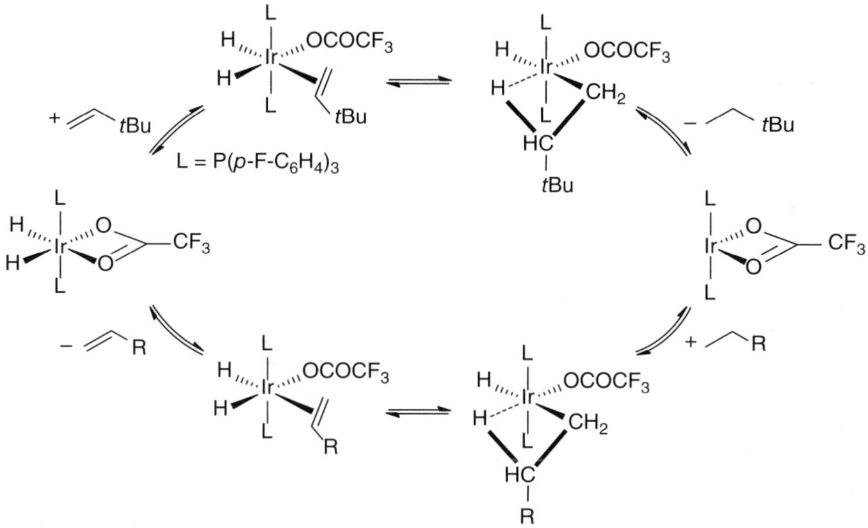

Scheme 21.2 Proposed mechanism of thermochemical dehydrogenation of COA with TBE as the hydrogen acceptor by $[P(p\text{-F-}C_6H_4)_3]_2IrH_2(CF_3CO_2)$.

complex, $[P(p\text{-F-}C_6H_4)_3]_2IrH_2(CF_3CO_2)$, gave a maximum of 34 TONs at 150 °C after 14 days. The proposed mechanism of this catalytic thermal dehydrogenation of alkanes was shown in Scheme 21.2 [20]. The facile opening of the trifluoroacetate chelate ring allows for the coordination of TBE. 1,2-Insertion of TBE into an Ir—H bond followed by reductive elimination liberates *tert*-butylethane (TBA) and generates a highly reactive 14e Ir(I) species, which is capable of oxidative addition of a C—H bond of alkane to give another Ir(H)(alkyl) intermediate. β—H elimination of the Ir(H)(alkyl) intermediate produces alkene and regenerates the starting Ir(III) dihydride complex. The author also developed catalytic systems employing $(AsPh_3)_2ReH_7$, $(disil)(PPh_3)_2ReH_5$ (disil = 1,2-bis(dimethylsilyl)benzene or 1,2-bis(dimethylsilyl)ethane), $[(triphos)ReH_5]$, and $[(triphos)WH_6]$ (triphos = $PhP(CH_2CH_2Ph_2)_2$) as the catalysts. However, limited TONs were achieved due to the rapid deactivation of catalysts via P—C cleavage [21–23]. C_6H_5F resulting from P—C cleavage of the phosphine ligand was detected, indicating the deactivation of the catalyst [20].

Subsequently, Jensen et al. explored a chloro iridium complex $(PiPr_3)_2IrClH_2$ as the catalyst for the COA dehydrogenation with TBE as the hydrogen acceptor at 150 °C [24]. Mechanistic studies suggested a pathway in which a 14e intermediate is not involved. While the reversible coordination and hydride migration to TBE is facile, the subsequent elimination of TBA may require the prior coordination of COA to the alkyl hydride intermediate.

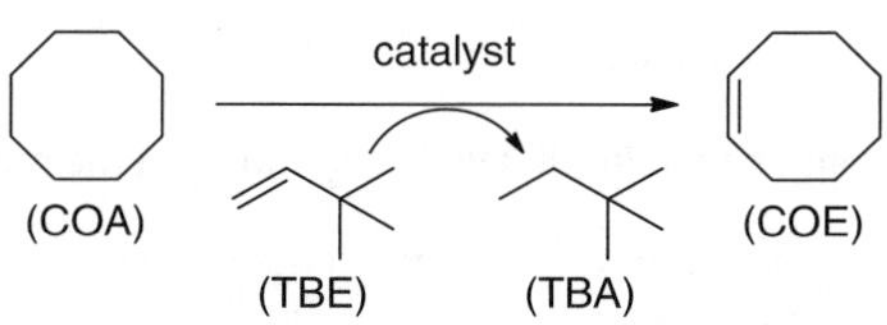

(21.2)

The enthalpy of dehydrogenation of COA is 23.3 kcal mol^{-1}, which is 5–7 kcal mol^{-1} lower than most alkanes. As described above, TBE is a particularly effective acceptor for alkane dehydrogenation reaction. Thus, the dehydrogenation of COA with TBE as the hydrogen acceptor, employed by both Felkin and Crabtree, has become a benchmark reaction used to evaluate the transfer dehydrogenation activity of new catalysts (Eq. 21.2).

In 1990, Saito et al. found that alkane dehydrogenation could be catalyzed by Wilkinson catalysts $RhCl(PR_3)_3$ (R = *p*-tolyl or Ph) or by its dimeric homologues under reflux conditions [25, 26] (Scheme 21.3). The highest TON (14) was observed for $RhCl[P(p\text{-tolyl})_3]_3$ at 180 °C after 48 hours. By replacing one phosphine ligand of the Wilkinson catalyst with ethylene or carbon monoxide, the groups of Tanaka and Goldman synthesized Rh complexes $RhCl(CH_2{=}CH_2)(PMe_3)_2$ and $RhCl(CO)(PMe_3)_2$ to catalyze alkane dehydrogenation, respectively [27, 28]. When a COA solution of the complex $RhCl(CH_2{=}CH_2)(PMe_3)_2$ was heated at 230 °C under ethylene pressure (30 atm), 32 equivalent of cyclooctene (COE) (relative to Rh catalyst) was observed [27]. Goldman and co-workers found $RhCl(CO)(PMe_3)_2$ could catalyze alkane dehydrogenation under thermal conditions with an added hydrogen atmosphere [28]. A COA solution of $RhCl(CO)(PMe_3)_2$ (0.4 mM) and norbornene (3.8 M) afforded 950 TONs under 1000 psi of H_2 at 100 °C for 15 minutes. Such high TON obtained are not limited by catalyst decomposition but, rather, by complete hydrogenation of the alkene hydrogen acceptor. The reactive fragment $RhCl(PMe_3)_2$ was generated by the reaction of $RhCl(CO)(PMe_3)_2$ with H_2 in the presence of added hydrogen acceptor, and a simplified mechanism of this process was presented (Scheme 21.4). Later, Goldman reported a series of rhodium complexes containing the $RhCl(L)_2$ moiety, $RhCl(PMe_3)_2(PR_3)$ (R = *i*Pr or Cy) and the dimer $[RhCl(PMe_3)_2]_2$. These complexes are significantly more efficient thermal dehydrogenation catalysts than $RhCl(CO)(PMe_3)_2$ under hydrogen atmosphere [29]. In the absence of H_2, such complexes also show dehydrogenation activity, but with much lower rates and total catalytic TONs.

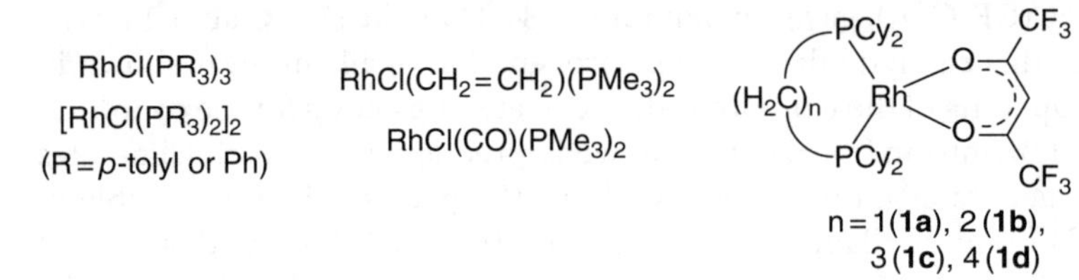

Scheme 21.3 Alkane dehydrogenation by Wilkinson catalyst and its analogues.

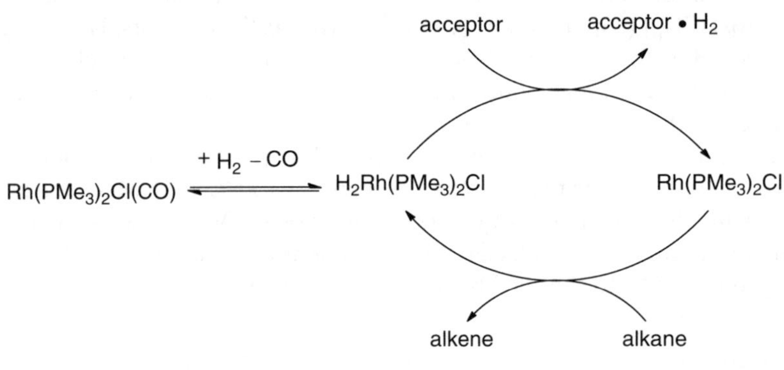

Scheme 21.4 Proposed mechanism of alkane dehydrogenation with $RhCl(CO)(PMe_3)_2$ and H_2 in the presence of an added hydrogen acceptor.

In 1997, Leitner et al. demonstrated that $[Cy_2P(CH_2)_nPCy_2]Rh(hfacac)_2$ **1** (n = 1~4, hfacac = $\kappa^2 O,O'$-hexafluoroacetylacetonate) and $[Cy_2P(CH_2)_nPCy_2]Ru(\eta^3\text{-}C_4H_7)_2$ **2** (n = 1~3) are active for the dehydrogenation of COA, while the catalytic process could only be achieved by bisallyl ruthenium complexes [30, 31]. Among these bisallyl ruthenium complexes investigated, $[Cy_2PCH_2PCy_2]Ru(\eta^3\text{-}C_4H_7)_2$ (**2a**) exhibited the highest catalytic activity and 5.2 TONs was observed for the dehydrogenation of COA under refluxing conditions. Saito et al. reported the $Ru_3(CO)_{12}$ catalyzed dehydrogenation of COA, which resulted in 6 TONs at 151 °C after 24 hours [32].

Catalyst deactivation is a serious problem in catalytic alkane dehydrogenation and is usually responsible for limited TONs. A major breakthrough was achieved by Kaska and Jensen in 1996 [33]. They synthesized a novel PCP pincer iridium complex (^{tBu4}PCP)IrH_2 **3**, which has long-term stability at high temperature (200 °C), to catalyze the transfer dehydrogenation of COA with TBE as the hydrogen acceptor. A turnover frequency (TOF) of 12 min^{-1} was observed for the dehydrogenation of COA at 200 °C. The catalytic activity is strongly inhibited by N_2 due to the formation of a stable dinitrogen complex. In addition, a high concentration of TBE could also inhibit the catalytic reaction [34]. Notably, these dehydrogenation reactions also occur in the absence of hydrogen acceptor. Refluxing cyclodecane with (^{tBu4}PCP)IrH_2 while passing a steam of Ar above the condenser gave 170 and 360 TONs after 4 and 24 hours, respectively [35].

The pioneering work by Kaska and Jensen stimulated extensive studies using various pincer iridium complexes for the dehydrogenation of alkanes. During the past two decades, numerous types of pincer iridium complexes have been synthesized and tested for the dehydrogenation of alkanes. Strategies for the development of novel pincer iridium complexes were based on the modification of the parent complex (^{tBu4}PCP)Ir$(H)_2$, by adjusting the steric and electronic properties of phosphorus atoms, introducing functional groups in the aromatic backbone, altering the linker between the aromatic backbone and two phosphorus atoms (Scheme 21.5), varying the aromatic backbone (Scheme 21.6), and replacing the side arms of the pincer ligand (Scheme 21.7).

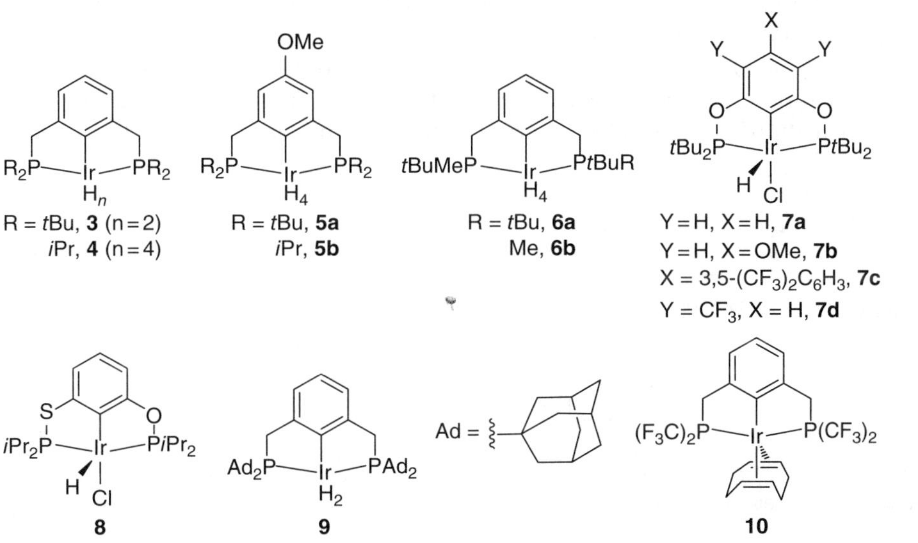

Scheme 21.5 Modification of the parent complex (^{tBu4}PCP)IrH_2 for alkane dehydrogenation.

By replacing the *t*Bu substituents on the phosphorus atoms with the less sterically hindered *i*Pr substituents, Goldman, Jensen, and co-workers synthesized a novel dihydrido iridium PCP pincer complex (^{iPr4}PCP)IrH_4 **4** to catalyze the dehydrogenation of alkanes [36–38]. (^{iPr4}PCP)IrH_4 **4** and (^{tBu4}PCP)IrH_2 **3** were found to be efficient and regioselective transfer dehydrogenation catalysts to convert *n*-octane into 1-octene, which initially constitute >90% of the olefin products. The less bulky (^{iPr4}PCP)Ir system exhibited a higher regioselectivity for the formation of α-olefin relative to the (^{tBu4}PCP)Ir system, while subsequent isomerization led ultimately to the mixture of internal olefins. For the transfer dehydrogenation of COA, 720 TONs was obtained at 200 °C for 1 hour [38]. (^{iPr4}PCP)IrH_4 has also been found extraordinarily active and robust catalyst for the acceptorless dehydrogenation of cyclic- and linear alkanes. Refluxing a CDA solution of complex **4**, afforded 460 and 863 TONs after 1 and 20 hours, respectively. For the acceptorless dehydrogenation of *n*-undecane, 35 TONs were obtained with a mixture of undecene isomers.

By adding a *p*-methoxy group on the aryl ring of the (PCP)Ir backbone, Goldman, Krogh-Jespersen, and co-workers synthesized two new complexes (*p*-OMe-^{tBu4}PCP)IrH_4 **5a** and (*p*-OMe-^{iPr4}PCP)IrH_4 **5b**, which showed improved activities for the dehydrogenation of alkanes compared with the parent complex [39]. For the acceptorless dehydrogenation of CDA, the *p*-methoxy substitution and the less crowed substitution on phosphorus significantly enhanced the dehydrogenation activity. For example, TONs of 360, 820, and 3050 were obtained for the acceptorless dehydrogenation of CDA at 250 °C when using **3**, **5a**, and **5b**, respectively. However, no significant improvement in activity was observed for the acceptorless dehydrogenation of *n*-octane and transfer dehydrogenation of COA or *n*-alkane. The same group also developed new Ir complexes **6a** and **6b** by altering the substitution on the phosphorus atom with a methyl group [40]. The less sterically hindered complex **6a** (^{tBu3Me}PCP)IrH_4 containing one methyl substituent exhibited a higher catalytic activity than (^{iPr4}PCP)IrH_4 for the dehydrogenation of alkanes, while the least crowded complex **6b** (^{tBu2Me2}PCP)IrH_4 with one methyl substituents displayed reduced productivity in the dehydrogenation reaction, probably due to the formation of catalytically inactive dinuclear species.

In 2004, Brookhart et al. reported a series of iridium bis(phosphinite)-based pincer complexes (*p*-X-POCOP)IrHCl **7a-c**. Upon *in situ* activation with NaO*t*Bu, these new complexes showed very high activities than the (PCP)Ir analogues for the transfer dehydrogenation of COA with TBE [41]. With a 3030:3030:1 ratio of COA:TBE:**7b**, TON up to 2537 was achieved. Subsequently, Wendt et al. synthesized complex **7d** with two CF_3 groups in the *meta*-position of the aryl backbone, which afforded 2398 TONs at 170 °C after 40 hours for the COA/TBE transfer dehydrogenation [42].

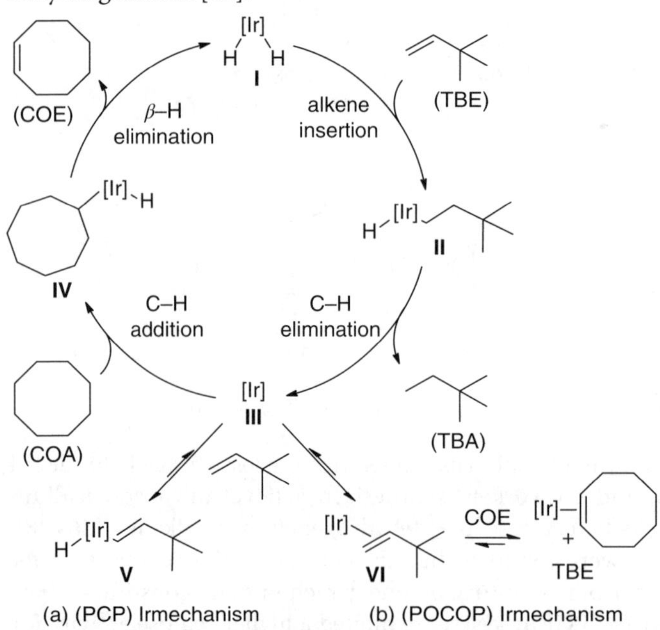

Scheme 21.6 Varing the aromatic backbone of the parent complex (^{tBu4}PCP)IrH_2 for alkane dehydrogenation.

The mechanism of COA/TBE transfer dehydrogenation catalyzed by (PCP)Ir and (POCOP)Ir systems were elucidated by the groups of Goldman [43] and Brookhart [44], respectively. Both catalytic systems underwent a similar dehydrogenation pathway (Scheme 21.8). That is, TBE coordinates to the Ir(III) dihydride complex **I**, followed by alkene insertion into the Ir—H bond to generate iridium *tert*-butyl ethyl hydride **II**. Reductive elimination of TBA forms the reactive three-coordinate 14e Ir(I) species **III**. This intermediate then undergoes oxidative addition of COA via C—H bond activation to generate iridium cyclooctyl hydride **IV**. Finally, β-H elimination produces COE and regenerates the Ir(III) dihydride complex **I**. A striking difference between (PCP)Ir and (POCOP)Ir for COA/TBE transfer dehydrogenation is their resting states in the presence of high concentrations of TBE. For (PCP)Ir system, the resting state is an Ir(III)(vinyl)(hydride) **V** formed by oxidative addition of TBE, while an Ir(I)(TBE) complex **VI** generated by simply coordination of TBE is the resting state for (POCOP)Ir system. Therefore, the (PCP)Ir system is more subject to TBE inhibition than the (POCOP)Ir system under a high concentration of the hydrogen acceptor.

Inspired by the great enhancement of dehydrogenation activity of Ir catalysts via subtle change of the linker from CH_2 group to an O atom, Huang et al. synthesized a *i*Pr-substituted PSCOP-type iridium pincer complex **8** supported by a hybrid phosphinothious/phosphinite pincer ligand. This Ir complex is one of the most active catalysts for the transfer dehydrogenation of cycloalkanes [45]. Upon activation with NaO*t*Bu, this system gave a maximum TON of 5901 after 15 hours at 200 °C with a COA/TBE/catalyst ratio of 6000 : 6000 : 1. For transfer dehydrogenation of *n*-octane, a maximum TON of 1211 was obtained after 4 hours at 200 °C with a *n*-octane/TBE/catalyst ratio of 3000 : 3000 : 1. Significantly, an appreciable activity (35 TONs after 8 hours) and high regioselectivity for α-olefin formation (100% of 1-octene within the first hour) were observed at 100 °C. Subsequently, a series of new PSCOP-type iridium pincer complexes containing substitutes other than the *i*Pr groups have been synthesized and the product selectivities in the transfer dehydrogenation of linear alkanes were investigated [46]. The results indicated that the less sterically hindered complex showed higher dehydrogenation activity than the more hindered complex, however, the latter offer higher regioselectivity for the formation of terminal olefin at the early stage of the reactions.

A highly stable adamantly substituted pincer-ligated iridium complex **9** was synthesized by the Goldman group, and the stability and catalytic activity for alkane dehydrogenation was conducted [47]. The results indicated that its dehydrogenation activity was comparable to that of $(^{tBu4}PCP)IrH_2$, while its stability was much more robust than (PCP)Ir and (POCOP)Ir systems.

It is generally recognized that pincer iridium complexes containing electron-rich alkylphosphine ligands were highly active for alkane dehydrogenation. However, Roddick and co-workers synthesized a pincer iridium complex $(^{CF3}PCP)Ir(COD)$ **10** (COD = 1,5-cyclooctadiene) ligated by a strongly π-accepting ligand to catalyze the dehydrogenation of alkane [48]. Unlike the classical donor pincer iridium systems, the catalytic activity of $(^{CF3}PCP)Ir$ system was remarkably insensitive to nitrogen and water. Another interesting feature of $(^{CF3}PCP)Ir$ system is that only modestly alkene inhibition was observed. Refluxing a 1 : 1 and 5 : 1 ratio of COA/TBE, TONs of 660 and 3020 were obtained at 200 °C after 58 and 64 hours, respectively. Compared to (PCP)Ir systems, low to moderate activities were obtained for transfer dehydrogenation of *n*-alkanes and acceptorless dehydrogenation of cycloalkanes.

The backbone of the (PCP)Ir systems were not limited to known benzene-based pincer ligands. By replacing the benzylic PCP pincer ligands with the more rigid polycyclic aromatic ligand backbone, Haenel, Kaska, Hall, and co-workers developed an anthraphos-supported iridium complex **11** to catalyze the dehydrogenation of alkanes [49]. The rigidity and chemical inertness of the anthracene ring renders the complex remarkably stable, while the catalytic activity was somewhat lower than the classical (PCP)Ir systems, likely due to the inflexibility of

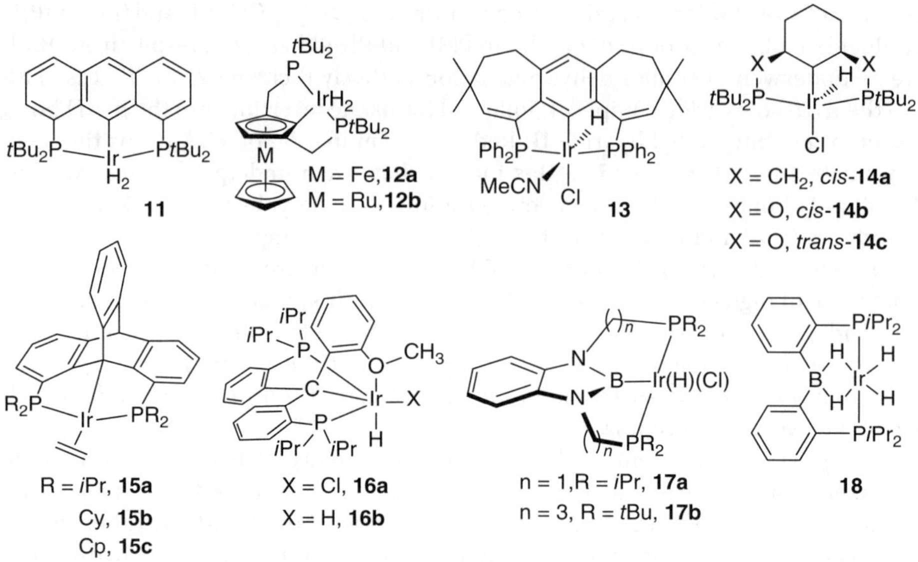

Scheme 21.7 Replacing the side arms of the pincer ligand of the parent complex (^{tBu4}PCP)IrH_2 for alkane dehydrogenation.

the anthracene backbone. Nevertheless, the extremely high thermal stability allows refluxing a CDA solution of the catalyst at 250 °C, giving 136 TONs after 148 hours.

Two sandwich nature of metallocene-based hydrido iridium pincer complexes **12a-b** were developed by the Koridze group [50], and high catalytic activity were achieved by both systems for COA/TBE transfer dehydrogenation, giving TONs of 3300 and 2571 after 8 hours at 180 °C, respectively. The authors supposed that both steric and electronic effects are responsible for the difference in catalytic activity between these two isostructural complexes.

More recently, an iridium complex **13** bearing a novel 7-6-7 fused-ring skeleton, which combines the robustness of the anthracene system and the flexibility of the (PCP)Ir system, has been synthesized by the Yamamoto group [51]. The complex was found to be highly efficient for the COA/TBE transfer dehydrogenation. Upon the activation of NaO*t*Bu, TONs of 4100 and 4820 were obtained at 200 and 230 °C with a COA/TBE/catalyst ratio of 10000 : 6750 : 1, respectively.

The majority of pincer ligands investigated contain an arene backbone, while iridium pincer complexes with aliphatic backbones have received much less attention, which might primarily be due to their instability, resulting from the presence of easily abstractable α- and β-hydrogens. The Wendt group synthesized a series of cyclohexane-based pincer complexes **14a–c** to catalyze the dehydrogenation of alkane [52, 53]. The thermal stability of aliphatic phosphinite complexes **14b** and **14c** were found to be more robust than that of aliphatic phosphine complex **14a**. Upon the activation of NaO*t*Bu, complex *trans*-**14c** was found to be highly active for the COA/TBE transfer dehydrogenation, and 3238 TONs were obtained after 32 hours at 120 °C and additional three hours at 180 °C with a 20000 : 5000 : 1 ratio of COA/TBE/catalyst. In contrast, complexes **14a** and **14b** displayed relatively lower activities (200 and 144 TONs, respectively) for the COA/TBE transfer dehydrogenation.

Brookhart and co-workers reported a variety of triptycene-based PC(sp^3)P-Ir complexes **15a-c**, which lack labile α- and β-hydrogens [54]. The complex **15a** bearing *i*Pr substituents on the phosphine atoms showed highest activity, giving 2820 and 6000 TONs at 200 °C for the transfer dehydrogenation of COA and *n*-octane with TBE as the hydrogen acceptor, respectively. The authors proposed that the lack of labile α- and β-hydrogens combined with the strong σ-donation of the metalated bridgehead sp^3 hybridized carbon is essential for the stability and robustness of these complexes. Iridium PC(sp^3)P type complexes **16a-b** with a hemilabile anisole tether were synthesized by the Iluc group [55]. They envisioned that the pincer ligands incorporated by a hemilabile tethered anisole would provide enhancements to catalytic activity and bond activation via metal-ligand cooperation. However, upon activation with NaO*t*Bu, low activity (<12 TONs) was obtained at 150 °C for the COA/TBE transfer dehydrogenation.

Most of the pincer complexes employ a carbon atom as the central coordinating atom. The replacement of the anionic carbon center with a heteroatom may generate a complex with unique features. Boryl ligands are one of the most donating and trans influencing donor ligands, however, there has been no reports on alkane dehydrogenation catalyzed by PBP-pincer complexes until recently. Yamashita and co-workers developed a series of PBP-pincer iridium complexes **17a-b** to catalyze alkane dehydrogenation [56]. Upon activation with LiTMP (TMP = 2,2,6,6-tetramethylpiperidide), complexes **17a** exhibited low activity for COA/TBE transfer dehydrogenation, giving 43 TONs at 220 °C. The authors attributed the low activity to the small P—Ir—P bond angles of the complex, thus disfavoring the reductive elimination of iridium (hydrido)alkyl intermediate. Hence, by elongating the linkers between the phosphine atoms and benzodiazaborole ring, Yamashita synthesized a PBP-pincer iridium complex **17b** to adjust the steric environment around the iridium center [57]. The dehydrogenation using **17b** resulted in a slight increase in the catalytic activity, giving 113 TONs at 160 °C for the COA/TBE transfer dehydrogenation. More recently, the Ozerov group synthesized diarylboryl/bis(phosphine)-based PBP pincer iridium complex **18** [58]. A maximum TON of 221 was achieved after 26 hours at 200 °C for the transfer dehydrogenation of COA with 1-hexene as the hydrogen acceptor.

The replacement of two flanking phosphine side-arms of the pincer ligand by other types of donors is another strategy for the ligand design. *N*-heterocyclic carbene (NHC) ligands, have been considered as valuable alternatives to the phosphine ligands because of their

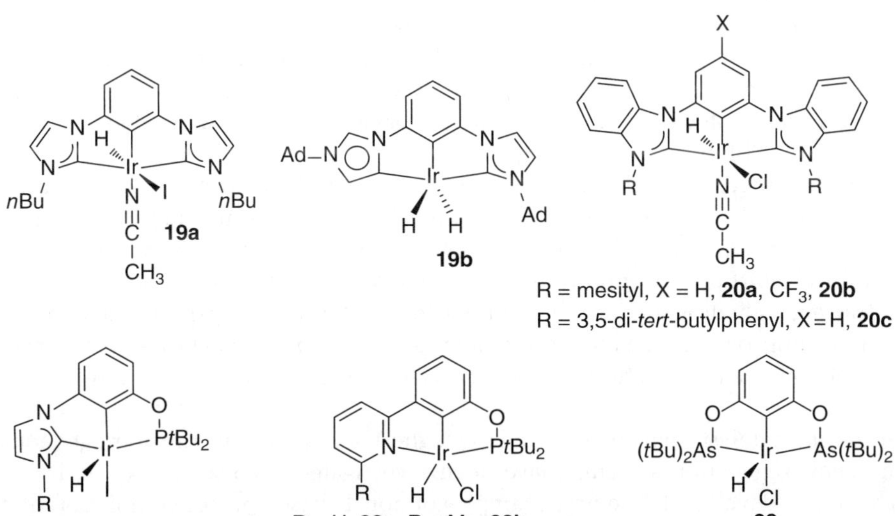

Scheme 21.8 Proposed mechanism of COA/TBE transfer dehydrogenation by (^{tBu4}PCP)IrH_2 and (^{iPr4}POCOP)IrH_2 systems.

strong σ-donating and weak π-accepting properties. By incorporating the NHC moiety into pincer-type ligands, Braunstein and co-workers synthesized novel CCC-pincer iridium complexes **19a-b** with one or two NHC ligands as the side arms of pincer ligands [59, 60]. Upon activation with NaO*t*Bu, complex **19a** was inactive for the COA/TBE transfer dehydrogenation. In contrast, the mixed NHC/abnormal NHC complex **19b**, which was expected to enhance the thermal stability of the resulting complexes due to the presence of adamantyl substituents on the NHC ligands, exhibited a detectable catalytic activity (4 TONs at 200 °C after 10 hours) for the COA/TBE transfer dehydrogenation. The authors attributed the low catalytic activity to the poor solubility of the catalyst in the nonpolar reaction medium.

The Chianese group synthesized a series of bis(NHC)-based CCC-pincer hydridochloride iridium complex **20a-c** [61–63]. Upon activation with NaO*t*Bu, TONs of 103, 84, and 35 were obtained for the acceptorless dehydrogenation of COA with precatalysts **20a**, **20b**, and **20c**, respectively. For the acceptorless dehydrogenation of *n*-undecane, complex **20c** exhibited the highest activity (96 TONs after 22 hours) under similar conditions, while complexes **20a** and **20b** showed a slightly lower activity (56 and 63 TONs after 22 hours, respectively).

Although the dehydrogenation activity of these bis(NHC) iridium complexes were shown to be relatively lower than that of (PCP)Ir systems, it provided a new strategy to design novel pincer iridium complexes for alkane dehydrogenation. By using monodentate phosphinite combined with NHC ligands in iridium complexes, the Braunstein group synthesized two hybrid P-NHC iridium pincer complexes **21a-b** [64]. The catalytic activity of these complexes were found to be very low for COA/TBE transfer dehydrogenation, affording 5 and 15 TONs at 200 °C after 10 hours with **21a** and **21b** as the precatalysts, respectively.

By replacing one phosphinite arm in the POCOP ligand with a pyridine group, Huang et al. reported new NCP pincer Ir complexes **22a-c** [65]. The catalytic activity of **22a** was comparable to that of the (PCP)Ir system. Upon activation with NaO*t*Bu, 466 TONs was obtained at 150 °C after 12 hours for the COA/TBE transfer dehydrogenation. However, the process with **22b** and **22c** gave a very low activity (6 TONs) under otherwise identical conditions. It is noteworthy that the dehydrogenation reaction was inhibited by high concentration of TBE. The (NCP)Ir complexes were also active for the transfer dehydrogenation of linear alkanes. Upon activation with NaO*t*Bu, TONs of 78, 28, and 19 were obtained at 150 °C after 1 hour with **22a**, **22b**, and **22c** as the precatalysts, respectively. To increase the thermal stability of the NCP complexes, Huang et al. synthesized a novel PNCP iridium complex **22d** supported by a tetradendate ligand [66]. However, upon activation with NaO*t*Bu, the active species generated showed low activity (16 TONs) for the COA/TBE transfer dehydrogenation.

An alteration of the pincer ligand by using arsenic instead of phosphorus as the chelating atoms was reported by the Jensen group [67]. They synthesized a new bis(arsenic)-based pincer iridium complex (^{tBu4}AsOCOAs)IrHCl, and the catalytic activity of complex **23** was explored. Upon activation with NaO*t*Bu, a maximum TON of 960 was obtained after 24 hours at 175 °C for COA/TBE transfer dehydrogenation. The auhors found the dehydrogenation activity of complex **23** could be improved as the reaction temperatures increase (125–175 °C). However, conducting the reaction at a temperature higher than 175 °C resulted in the decomposition of the complex.

While pincer iridium catalysts, which exhibited the highest activity among the homogeneous catalytic alkane dehydrogenation systems, have been investigated extensively, non-Ir based pincer catalysts are less developed. The first examples of non-Ir based pincer complexes were reported by Goldman and co-workers [68] (Scheme 21.9). They synthesized a series of rhodium complexes bearing tridentate pincer ligands, (η^3-PXP)Rh(L) (PXP = $(Me_2PCH_2Me_2Si)_2N$, **24**, $Me_2PCH_2(2,6\text{-}C_6H_3)CH_2PMe_2$, **25a**), which were found to catalyze the dehydrogenation of alkanes, albeit with low activities. Later, (^{tBu4}PCP)RhH_2 **25b** was used to catalyze the

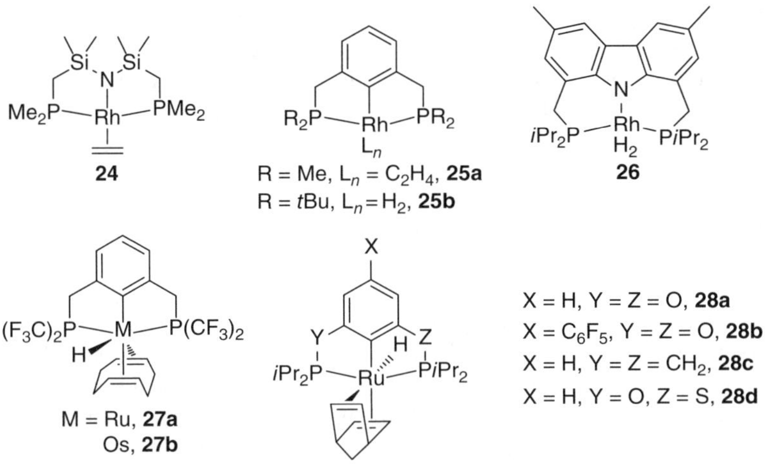

Scheme 21.9 Non-Ir based pincer complexes for alkane dehydrogenation.

dehydrogenation of alkanes, which exhibited lower dehydrogenation activity compared with the Ir analogues under the similar conditions.

An effective rhodium-based alkane transfer-dehydrogenation catalyst (carb-PNP)RhH_2 **26** with the bis-phosphine carbazolide pincer ligand was developed by the Brookhart group [69]. Heating a 1 : 1 COA:TBE solution of **26** (0.15 mol %) at 200 °C gave 340 TONs after 4 hours with a TOF 10 min^{-1}. However, (carb-PNP)RhH_2 was less effective for the transfer dehydrogenation of *n*-octane, giving 14 TONs after 1 hour at 200 °C with a *n*-octane/TBE/catalyst ratio of 341 : 341 : 1. In contrast, the Ir analogue was inactive for transfer dehydrogenation of alkane.

Roddick and co-workers reported the synthesis of (^{CF3}PCP)RuH(COD) **27a** and (^{CF3}PCP)OsH(COD) **27b** bearing the same perfluoroalkyl-substituted pincer ligand [70, 71]. Thermolysis of **27a** in 1 : 1 mixtures of COA and TBE at 200 °C resulted a maximum TON of ~350 after only 20 minutes with an initial TOF of 1950 h^{-1}. An initial TOF of 1520 h^{-1}, approximately 75% to that of **27a**, was observed for **27b** under the similar reaction conditions, however, the catalyst lifetime was significantly improved, affording 610 TONs after 8 hours. Acceptorless dehydrogenation of COA by **27a** and **27b** were also investigated. Refluxing a COA solution of **27a** (~140 °C at 590 Torr ambient pressure) for 1 hour produced 10 equivalents of COE with an initial rate of 14 h^{-1}, however, the (^{CF3}PCP)Os system is catalytically inactive under the similar conditions. Refluxing a higher boiling CDA solution of **27b** (~190 °C at 590 Torr ambient pressure) for 1 hour generated 125 total equivalents of cyclodecenes. Alkane dehydrogenation catalyzed by these Ru and Os-systems were also found to be uninhibited by the presence of nitrogen, water, even (for M = Ru) oxygen, highlighting the potential advantage of less electron rich dehydrogenation catalysts in alkane functionalization.

More recently, Huang and co-workers developed a series of hydrido ruthenium complexes (*p*-X-^{iPr4}PYCZP)Ru(H)(NBD) **28a-d** supported by isopropyl-substituted pincer ligands for alkane dehydrogenation [72]. The less-electron-rich complex **28b** with a pentafluorophenyl substituent on the aromatic backbone exhibited the highest activity for COA/TBE transfer dehydrogenation, with a TON of 322 after 24 hours at 200 °C. However, it showed relatively low to modest activity for acceptorless dehydrogenation of alkane. These Ru systems tolerated various added polar species (such as H_2O, ethyl acetate, Et_2O, and acetone) and impurities from the starting materials, highlighting a remarkable advance in the development of practical dehydrogenation catalysts. This work indicated that nonfluoroalkyl-pincer ruthenium complexes are also effective dehydrogenation catalysts, which reveals that this is an intrinsic property of the metal and its configuration, not the electron-poor fluoroalkyl-PCP ligands.

Several groups have also made progress in the field of homogeneous alkane dehydrogenation using Ti [73–75], Pt [76–82], Zr [73], and Hf [73] complexes. However, most of them were stoichiometric reactions or give limited TONs (<45) due to fast decomposition of the catalyst.

21.3 Photochemical Dehydrogenation of Alkanes

Although significant progress has been achieved for thermochemical dehydrogenation of alkanes, an excess of sacrificial alkene is generally required as the hydrogen acceptor to overcome the positive enthalpy. For the acceptorless dehydrogenation of alkane, the efficiency of the removal of the liberated H_2 from the system is the key to the success of the reaction. In fact, the catalytic dehydrogenation in the absence of an hydrogen acceptor reactions in general gives much lower TONs than the reactions with an hydrogen acceptor. Photochemical dehydrogenation of alkanes based on the irradiation of light, not requiring the use of an added hydrogen acceptor and high temperature, has been considered as an environmentally benign option for the alkane dehydrogenation (Scheme 21.10).

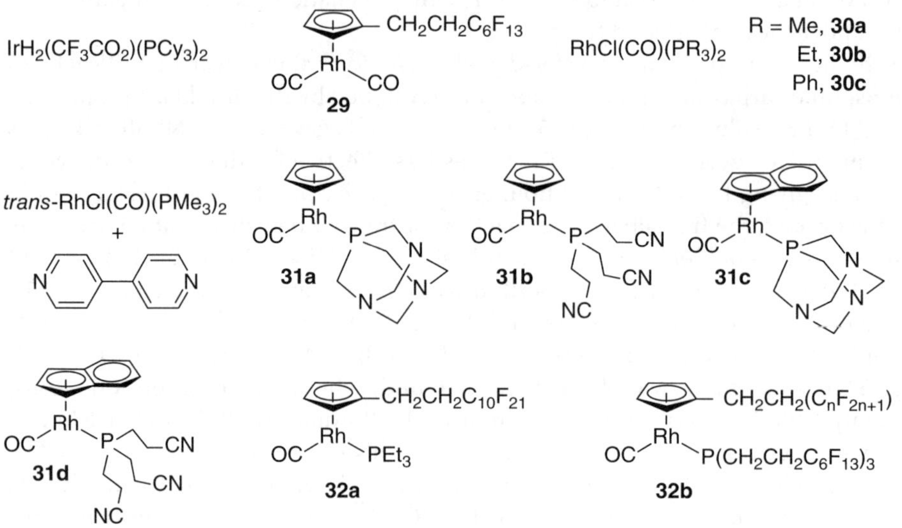

Scheme 21.10 Photochemical dehydrogenation of alkanes by noble metal complexes.

The first example of photochemical catalytic dehydrogenation of alkanes was discovered by the Crabtree group in 1985 [19]. They found that the less-effective thermal dehydrogenation catalyst (2 TONs at 150 °C), $(PCy_3)_2IrH_2(CF_3CO_2)$, was proven to be much more active under photochemical conditions, giving 28 TONs at 25 °C after 7 days and no P-C cleavage products were detected. The photochemical dehydrogenation reaction could also take place even in the absence of TBE, and 7 TONs was observed after 7 days. The proposed mechanism of this photochemical dehydrogenation of alkanes was similar to that of thermal dehydrogenation of alkanes. Photo-induced direct release of molecular hydrogen from $(PCy_3)_2IrH_2(CF_3CO_2)$ generates the unsaturated 14e Ir(I) complex [20] (Scheme 21.11).

Braun and co-workers also demonstrated that $[Ir(\eta^5\text{-}C_5H_4CH_2CH_2C_6F_{13})(CO)_2]$ **29** was an active catalyst for the photocatalytic dehydrogenation of cyclic alkanes, although the TONs were rather limited [83].

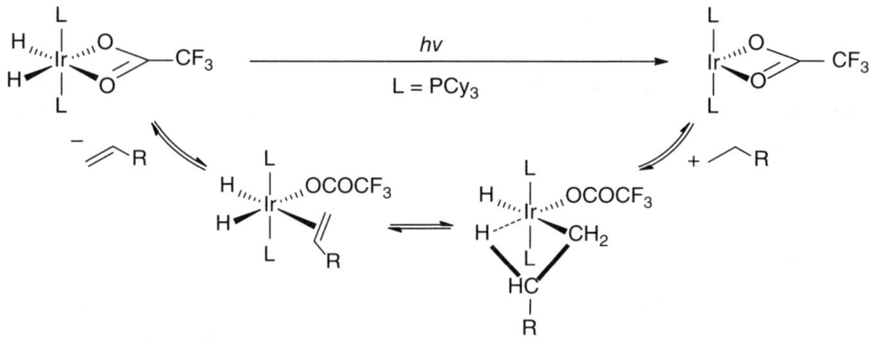

Scheme 21.11 Proposed mechanism of photochemical dehydrogenation of COA by $[PCy_3]_2IrH_2(CF_3CO_2)$.

Improved activities on photochemical dehydrogenation of alkanes were achieved by the groups of Saito [84], Tanaka [85], and Goldman [28, 29, 86] by using $RhCl(CO)(PMe_3)_2$ system. Saito and co-workers found the alkane dehydrogenations could proceed under photoirradiation conditions with $RhCl(CO)(PR_3)_2$ (R = Me, **30a**, Et, **30b**, or Ph, **30c**) as the catalyst and the PMe_3-based complex exhibited the highest activity. For example, TOF up to 795 h^{-1} was obtained from a *n*-heptane solution of $RhCl(CO)(PMe_3)_2$ at 92 °C [84]. The author also utilized $RhCl(CS)(PMe_3)_2$ to photocatalytic dehydrogenation of cyclohexane. However, the estimated TON was relatively low (3.8) [85]. The group of Tanaka utilized the $RhCl(CO)(PMe_3)_2$ system to catalytic photochemical dehydrogenation of alkane under nitrogen atmosphere in a closed system or under a slow stream of nitrogen [86]. The results indicated that the dehydrogenation reaction was accelerated when a slow stream of nitrogen was passed, and TON up to 930 was reached for acceptorless dehydrogenation of COA after 68.5 hours at ambient temperature.

A detailed investigation of photochemical dehydrogenation of alkanes catalyzed by *trans*-$RhCl(CO)(PMe_3)_2$ was reported by Goldman and co-workers [87]. The author demonstrated that *trans*-$RhCl(CO)(PMe_3)_2$ was to be the only significant photoactive species in solution, and a total TON of 5000 was achieved for the dehydrogenation of COA. Goldman proposed two hypothetical mechanisms, in which the order to release COE and H_2 from the 18e cyclooctene dihydride Rh(III) complex is different. The unfavorable energy for these dehydrogenation reactions could be compensated only by Rh-CO bond photolysis.

A practical and efficient catalytic system for the acceptorless dehydrogenation was developed by Beller and co-workers [88]. Under the optimal conditions for light transmittance, a combination of the well-known *trans*-$RhCl(CO)(PMe_3)_2$ complex with 4,4′-bipyridine exhibited an excellent activity for the acceptorless dehydrogenation of both cyclic and linear alkanes under at much lower temperatures than previously reported. For the acceptorless dehydrogenation of *n*-octane and COA, TONs of 1518 and 2883 were obtained under the optimized conditions after 7 hours, respectively. The author found that the light transmittance has a profound influence on the dehydrogenation reaction. High efficiency and increased product yield could be obtained by decreasing the glass thickness. This method allows for the challenging dehydrogenations, such as the dehydrogenation of potentially liquid organic hydrogen carriers (LOHC), on a practical scale. However, this catalytic system suffers from catalyst deactivation over time (after 7 hours) due to the formation of catalytically inactive Rh_2P nanoparticles. Some mechanistic studies were conducted by the Beller group [89], and they found the catalyst deactivation could be prevented by using either ethyl formate or CO_2 as a CO-releasing stabilizer. Up to 2182 and 5350 TONs were achieved for the acceptorless dehydrogenation of *n*-octane and COA under modified optimal conditions.

The Smoleński group synthesized novel rhodium(I) complexes **31a-d**, bearing aminophosphine PTA (PTA = 1,3,5-triaza-7-phosphaadamantane) and its analogue tris(2-cyanoethyl) phosphine, for photochemical dehydrogenation of alkanes [90]. All the complexes were found to be active catalysts for the photo-assisted dehydrogenation of alkanes without additional hydrogen acceptor, and the rates of dehydrogenation decreased in the order **31c** > **31a** > **31d** > **31b**. Complex **31c** bearing PTA and indenyl ligand exhibited the highest activity, giving TOFs of 26 and 56 for the photo-assisted dehydrogenation of *n*-octane and COA.

Braun and co-workers synthesized fluorinated cyclopentadienyl complexes $[Rh(\eta^5\text{-}C_5H_4CH_2CH_2C_{10}F_{21})(CO)(PEt_3)]$ **32a** and $[Rh(\eta^5\text{-}C_5H_4CH_2CH_2C_{10}F_{21})(CO)\{P(CH_2CH_2C_6F_{13})_3)\}]$ **32b** and their dehydrogenation activities were investigated [91]. In a C_7F_{14}/COA biphasic system, 16 TONs was obtained for the dehydrogenation of COA with complex **32b**.

Most catalytic alkane dehydrogenations involve noble metal catalysts, such as Rh and Ir. The development of catalysts based on earth-abundant and environmentally benign base metal catalysts has not been intensively studied. A stoichiometric photochemical example of *n*-pentane dehydrogenation with a bis(diphosphine) dihydride Fe(II) complex **33** was reported by Field and co-workers [92]. The reactive iron species is easily trapped by 1-pentene, resulting in the loss of the dehydrogenation activity (Scheme 21.12).

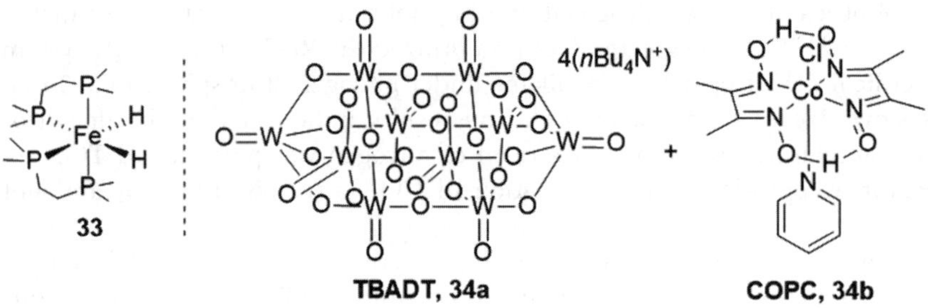

Scheme 21.12 Photochemical dehydrogenation of alkanes by base metal catalysts.

By combining tetra-*n*-butylammonium decatungstate (TBADT) **34a** and cobaloxime pyridine chloride (COPC) **34b**, Sorensen et al. demonstrated alkane dehydrogenation through cooperative base metal catalysis under near-UV irradiation at room temperature, giving 48 TONs after 48 hours for the acceptorless dehydrogenation of COA [93].

21.4 Summary and Outlook

Significant achievements have been made on the dehydrogenation of alkanes using molecular catalysts over the past four decades. Pioneered by Crabtree's report on a stoichiometric dehydrogenation of cycloalkanes by $[IrH_2(acetone)_2(PPh_3)_2][BF_4]$ in 1979, numerous homogeneous thermal dehydrogenation catalysts have been developed, however, the catalytic TONs were limited due to catalyst decomposition at such high temperature. A major breakthrough for homogeneous thermal alkane dehydrogenation was the use of bisphosphine pincer iridium complexes as catalysts by Kaska, Jensen, and Goldman. More stable and active iridium complexes for the dehydrogenation of cyclic and linear alkanes were subsequently developed via modification of the pincer ligand. The highest activity (up to 5901 TONs) for the transfer dehydrogenation of COA with TBE as the hydrogen acceptor was achieved using a (PSCOP)Ir system with a phosphinothious/phosphinite pincer ligand, and the highest activity for the *n*-octane/TBE transfer dehydrogenation (up to 6000 TONs) was demonstrated by Brookhart's (PC(sp^3)P)Ir system based on triptycene pincer ligand. However, both systems were found to be highly sensitive to oxygen, moisture, and even nitrogen, and the reactions were carried out at relatively high temperatures. Consequently, the development of air- and moisture-stable dehydrogenation catalysts with high activity and selectivity is of significant interest.

Since the report of the first example of photochemical dehydrogenation of alkanes by the Crabtree group with $(PCy_3)_2IrH_2(CF_3CO_2)$ as the catalyst, the groups of Saito, Tanaka, and Goldman have developed Vaska-type rhodium complex $[RhCl(CO)(PMe_3)_2]$ with enhanced dehydrogenation activity. By combining *trans*-$RhCl(CO)(PMe_3)_2$ complex with 4,4′-bipyridine and increasing the light transmittance, a further improvement in TONs was achieved by Beller.

Despite the significant advances for both thermochemical and photochemical dehydrogenation of alkanes, most catalytic systems are based on noble-metal catalysts, and thus developing base-metal dehydrogenation catalysts is highly desirable. The development of efficient catalysts for the dehydrogenation of functionalized alkanes and for the late stage modification of complex molecules would also be significant.

Acknowledgments

This work has been supported by the Ministry of Science and Technology of China (2015CB856600, 2016YFA0202900), the National Science Foundation of China (21432011, 21422209, 21421091, 21572255), and Chinese Academy of Science (XDB20000000).

References

1 Ren, T., Patel, M., and Blok, K. (2006). *Energy* 31: 425–451.
2 Labinger, J.A. (2018). Oxidative and non-oxidative routes to alkane dehydrogenation. In: *Functionalization of Alkanes* (ed. A.J.L. Pombeiro) Chapter 20. Wiley.
3 Choi, J., MacArthur, A.H.R., Brookhart, M. et al. (2011). *Chem. Rev.* 111: 1761–1779.
4 Zhang, Y., Yao, W., Fang, H. et al. (2015). *Sci. Bull.* 60: 1316–1331.
5 Werkmeister, S., Neumann, J., Junge, K. et al. (2015). *Chem. Eur. J.* 21: 12226–12250.
6 Daugulis, O., MacArthur, A.H.R., Rix, F.C. et al. (2016). *ACS Catal.* 6: 1518–1532.
7 Goldman, A.S., Renkema, K.B., Czerw, M. et al. (2004). Alkane Transfer-Dehydrogenation Catalyzed by a Pincer-Ligated Iridium Complex. In: *Activation and Functionalization of C—H Bonds*, vol. 885, 198–215. American Chemical Society.
8 Findlater, M., Choi, J., Goldman, A.S. et al. (2012). Alkane Dehydrogenation. In: *Alkane C-H Activation by Single-Site Metal Catalysis* (ed. P.J. Pérez), 113–141. Dordrecht: Springer Netherlands.
9 Yamaguchi, R. and Fujita, K.-I. (2014). Dehydrogenation of Alkanes Catalyzed By Transition Metal Complexes Bearing Pincer Ligands. In: *Ligand Platforms in Homogenous Catalytic Reactions with Metals*, 107–127. Wiley.
10 Kumar, A. and Goldman, A.S. (2016). Recent Advances in Alkane Dehydrogenation Catalyzed by Pincer Complexes. In: *The Privileged Pincer-Metal Platform: Coordination Chemistry & Applications* (ed. G. van Koten and R.A. Gossage), 307–334. Cham: Springer International Publishing.

11 Bézier, D. and Brookhart, M. (2016). Transfer Dehydrogenation of Alkanes and Related Reactions Using Iridium Pincer Complexes. In: *C—H Bond Activation and Catalytic Functionalization II* (ed. P.H. Dixneuf and H. Doucet), 189–207. Cham: Springer International Publishing.
12 Crabtree, R.H., Mihelcic, J.M., and Quirk, J.M. (1979). *J. Am. Chem. Soc.* 101: 7738–7740.
13 Baudry, D., Ephritikhine, M., and Felkin, H. (1980). *J. Chem. Soc., Chem. Commun.* 1243–1244.
14 Baudry, D., Ephritikhine, M., and Felkin, H. (1982). *J. Chem. Soc., Chem. Commun.* 606–607.
15 Koola, J.D. and Roddick, D.M. (1991). *J. Am. Chem. Soc.* 113: 1450–1451.
16 Baudry, D., Ephritikhine, M., Felkin, H. et al. (1983). *J. Chem. Soc., Chem. Commun.* 788–789.
17 Felkin, H., Fillebeen-Khan, T., Gault, Y. et al. (1984). *Tetrahedron Lett.* 25: 1279–1282.
18 Burk, M.J., Crabtree, R.H., Parnell, C.P. et al. (1984). *Organometallics* 3: 816–817.
19 Burk, M.J., Crabtree, R.H., and McGrath, D.V. (1985). *J. Chem. Soc., Chem. Commun.* 1829–1830.
20 Burk, M.J. and Crabtree, R.H. (1987). *J. Am. Chem. Soc.* 109: 8025–8032.
21 Aoki, T. and Crabtree, R.H. (1993). *Organometallics* 12: 294–298.
22 Loza, M.L., de Gala, S.R., and Crabtree, R.H. (1994). *Inorg. Chem.* 33: 5073–5078.
23 Loza, M.L. and Crabtree, R.H. (1995). *Inorg. Chim. Acta* 236: 63–66.
24 Belli, J. and Jensen, C.M. (1996). *Organometallics* 15: 1532–1534.
25 Fujii, T. and Saito, Y. (1990). *J. Chem. Soc., Chem. Commun.* 757–758.
26 Fujii, T., Higashino, Y., and Saito, Y. (1993). *J. Chem. Soc., Dalton Trans.* 517–520.
27 Toshiyasu, S., Fujio, A., and Masato, T. (1991). *Chem. Lett.* 20: 359–362.
28 Maguire, J.A. and Goldman, A.S. (1991). *J. Am. Chem. Soc.* 113: 6706–6708.
29 Maguire, J.A., Petrillo, A., and Goldman, A.S. (1992). *J. Am. Chem. Soc.* 114: 9492–9498.
30 Leitner, W. and Six, C. (1997). *Chem. Ber. Recl.* 130: 555–558.
31 Six, C., Gabor, B., Görls, H. et al. (1999). *Organometallics* 18: 3316–3326.
32 Yukawa, K., Kanaboshi, H., and Saito, Y. (1992). *Chem. Lett.* 21: 1177–1180.
33 Gupta, M., Hagen, C., Flesher, R.J. et al. (1996). *Chem. Commun.* 2083–2084.
34 Lee, D.W., Kaska, W.C., and Jensen, C.M. (1998). *Organometallics* 17: 1–3.
35 Xu, W.-W., Rosini, G.P., Krogh-Jespersen, K. et al. (1997). *Chem. Commun.* 2273–2274.
36 Liu, F., Pak, E.B., Singh, B. et al. (1999). *J. Am. Chem. Soc.* 121: 4086–4087.
37 Liu, F. and Goldman, A.S. (1999). *Chem. Commun.* 655–656.
38 Jensen, C.M. (1999). *Chem. Commun.* 2443–2449.
39 Zhu, K., Achord, P.D., Zhang, X. et al. (2004). *J. Am. Chem. Soc.* 126: 13044–13053.
40 Kundu, S., Choliy, Y., Zhuo, G. et al. (2009). *Organometallics* 28: 5432–5444.
41 Göttker-Schnetmann, I., White, P., and Brookhart, M. (2004). *J. Am. Chem. Soc.* 126: 1804–1811.
42 Kovalenko, O.O. and Wendt, O.F. (2016). *Dalton Trans.* 45: 15963–15969.
43 Renkema, K.B., Kissin, Y.V., and Goldman, A.S. (2003). *J. Am. Chem. Soc.* 125: 7770–7771.
44 Göttker-Schnetmann, I. and Brookhart, M. (2004). *J. Am. Chem. Soc.* 126: 9330–9338.
45 Yao, W., Zhang, Y., Jia, X. et al. (2014). *Angew. Chem. Int. Ed.* 53: 1390–1394.
46 Yao, W., Jia, X., Leng, X. et al. (2016). *Polyhedron* 116: 12–19.
47 Punji, B., Emge, T.J., and Goldman, A.S. (2010). *Organometallics* 29: 2702–2709.
48 Adams, J.J., Arulsamy, N., and Roddick, D.M. (2012). *Organometallics* 31: 1439–1447.
49 Haenel, M.W., Oevers, S., Angermund, K. et al. (2001). *Angew. Chem. Int. Ed.* 40: 3596–3600.
50 Kuklin, S.A., Sheloumov, A.M., Dolgushin, F.M. et al. (2006). *Organometallics* 25: 5466–5476.

51 Shi, Y., Suguri, T., Dohi, C. et al. (2013). *Chem. Eur. J.* 19: 10672–10689.
52 Polukeev, A.V., Gritcenko, R., Jonasson, K.J. et al. (2014). *Polyhedron* 84: 63–66.
53 Polukeev, A.V. and Wendt, O.F. (2017). *Organometallics* 36: 639–649.
54 Bézier, D. and Brookhart, M. (2014). *ACS Catal.* 4: 3411–3420.
55 Babbini, D.C. and Iluc, V.M. (2015). *Organometallics* 34: 3141–3151.
56 Tanoue, K. and Yamashita, M. (2015). *Organometallics* 34: 4011–4017.
57 Kwan, E.H., Kawai, Y.J., Kamakura, S. et al. (2016). *Dalton Trans.* 45: 15931–15941.
58 Shih, W.-C. and Ozerov, O.V. (2017). *Organometallics* 36: 228–233.
59 Raynal, M., Pattacini, R., Cazin, C.S.J. et al. (2009). *Organometallics* 28: 4028–4047.
60 Zuo, W. and Braunstein, P. (2012). *Organometallics* 31: 2606–2615.
61 Chianese, A.R., Mo, A., Lampland, N.L. et al. (2010). *Organometallics* 29: 3019–3026.
62 Chianese, A.R., Shaner, S.E., Tendler, J.A. et al. (2012). *Organometallics* 31: 7359–7367.
63 Chianese, A.R., Drance, M.J., Jensen, K.H. et al. (2014). *Organometallics* 33: 457–464.
64 Liu, X. and Braunstein, P. (2013). *Inorg. Chem.* 52: 7367–7379.
65 Jia, X., Zhang, L., Qin, C. et al. (2014). *Chem. Commun.* 50: 11056–11059.
66 Jia, X. and Huang, Z. (2015). *Sci. China Chem.* 58: 1340–1344.
67 Brayton, D.F., Beaumont, P.R., Fukushima, E.Y. et al. (2014). *Organometallics* 33: 5198–5202.
68 Wang, K., Goldman, M.E., Emge, T.J. et al. (1996). *J. Organomet. Chem.* 518: 55–68.
69 Bezier, D., Guan, C., Krogh-Jespersen, K. et al. (2016). *Chem. Sci.* 7: 2579–2586.
70 Gruver, B.C., Adams, J.J., Warner, S.J. et al. (2011). *Organometallics* 30: 5133–5140.
71 Gruver, B.C., Adams, J.J., Arulsamy, N. et al. (2013). *Organometallics* 32: 6468–6475.
72 Zhang, Y., Fang, H., Yao, W. et al. (2016). *Organometallics* 35: 181–188.
73 Alt, H.G. and Denner, C.E. (2009). *J. Organomet. Chem.* 694: 3270–3275.
74 Cavaliere, V.N., Crestani, M.G., Pinter, B. et al. (2011). *J. Am. Chem. Soc.* 133: 10700–10703.
75 Kamitani, M., Pinter, B., Searles, K. et al. (2015). *J. Am. Chem. Soc.* 137: 11872–11875.
76 Vedernikov, A.N. and Caulton, K.G. (2003). *Chem. Commun.* 358–359.
77 Fekl, U., Kaminsky, W., and Goldberg, K.I. (2003). *J. Am. Chem. Soc.* 125: 15286–15287.
78 Vedernikov, A.N., Huffman, J.C., and Caulton, K.G. (2003). *New J. Chem.* 27: 665–667.
79 Khaskin, E., Zavalij, P.Y., and Vedernikov, A.N. (2006). *J. Am. Chem. Soc.* 128: 13054–13055.
80 Kostelansky, C.N., MacDonald, M.G., White, P.S. et al. (2006). *Organometallics* 25: 2993–2998.
81 West, N.M., White, P.S., and Templeton, J.L. (2007). *J. Am. Chem. Soc.* 129: 12372–12373.
82 Khaskin, E., Lew, D.L., Pal, S. et al. (2009). *Chem. Commun.* 6270–6272.
83 Rabay, B., Braun, T., and Falkenhagen, J.P. (2013). *Dalton Trans.* 42: 8058–8065.
84 Nomura, K. and Saito, Y. (1988). *J. Chem. Soc., Chem. Commun.* 161–162.
85 Iwamoto, A., Itagaki, H., and Saito, Y. (1991). *J. Chem. Soc., Dalton Trans.* 1093–1097.
86 Sakakura, T., Sodeyama, T., Tokunaga, Y. et al. (1988). *Chem. Lett.* 17: 263–264.
87 Maguire, J.A., Boese, W.T., and Goldman, A.S. (1989). *J. Am. Chem. Soc.* 111: 7088–7093.
88 Chowdhury, A.D., Weding, N., Julis, J. et al. (2014). *Angew. Chem. Int. Ed.* 53: 6477–6481.
89 Chowdhury, A.D., Julis, J., Grabow, K. et al. (2015). *ChemSusChem* 8: 323–330.
90 Smoleński, P. (2011). *J. Organomet. Chem.* 696: 3867–3872.
91 Liebau, F., Braun, T., and Braun, B. (2015). *J. Fluorine Chem.* 180: 192–200.
92 Baker, M.V. and Field, L.D. (1987). *J. Am. Chem. Soc.* 109: 2825–2826.
93 West, J.G., Huang, D., and Sorensen, E.J. (2015). *Nat. Commun.* 6: 10093.

Part V

Unconventional Systems

22

Nontransition Metal Catalyzed Oxidation of Alkanes with Peroxides

Maxim L. Kuznetsov

Universidade de Lisboa, Centro de Química Estrutural, Instituto Superior Técnico, Lisboa, Portugal

22.1 Introduction

Hydrocarbons, in general, and alkanes, in particular, are the most important, abundant, and cheapest carbon raw material. Their functionalization leads to industrially important products such as alcohols, aldehydes, ketones, esters, carboxylic acids, amines, and epoxides [1]. The greatest problem standing on the way of the alkane functionalization is their high chemical inertness. Activation of alkanes requires the application of a catalyst and an appropriate oxidant. Among various oxidants, peroxides attract much attention. The O—O bond dissociation energy in peroxides is relatively low, being typically 36–49 kcal mol^{-1} for H_2O_2 and organic peroxides [2] and only 22–29 kcal mol^{-1} for $H_2S_2O_8$ and $S_2O_8{}^{2-}$ [3]. Such low values provide a high oxidative power of peroxides, in particular, in the presence of a corresponding catalyst. Additionally, hydrogen peroxide is also attractive from the economic and ecological points of view – the only product of the H_2O_2 reduction is water and the only byproduct is molecular oxygen.

Various complexes of transition metals are usually used as catalysts in the reactions of alkane oxidation. The chemistry of transition metal catalyzed alkane oxidation by peroxides started with the development of a simple system bearing a salt of iron(II) and hydrogen peroxide (Fenton reagent) [4]. The Fenton reagent is capable to oxidize various organic species including hydrocarbons. However, the efficiency of this catalytic system toward oxidation of alkanes is rather low. During past decades, a number of various complexes of Fe as well as other transition metals (for instance, V, Cu, Mn, Mo, Re, Ru, Os, and many others) have been developed as efficient catalysts for the oxidation of alkanes [1].

At the same time, the application of catalysts based on nontransition metal compounds in these reactions with peroxides is rare, and this field is at the very beginning stage of development. Nevertheless, recent experimental and theoretical publications demonstrated that some nontransition metal catalysts may be quite efficient in these processes. Furthermore, catalysts based on nontransition metals such as Al, Ca, and Mg are promising from the economic and ecological points of view being environmentally friendly, cheap, and accessible. In this chapter, reactions of the alkane oxidation with peroxides catalyzed by nontransition metal compounds are discussed. Publications dealing with combined catalytic systems bearing both nontransition metal and transition metal moieties are not included in this report since namely transition metal center is typically responsible for the catalytic activity in such systems.

Alkane Functionalization, First Edition. Edited by Armando J. L. Pombeiro and M. Fátima C. Guedes da Silva.

22.2 Homogeneous Hydroperoxidation of Alkanes

To date, the homogeneous hydroperoxidation of alkanes catalyzed by nontransition metal species was reported for Al, Be, Zn, Cd, and Bi (experimental and theoretical studies [5–8]) and Ga and In (theoretical study [9]).

22.2.1 Catalytic Experiments

The first report on the homogeneous oxidation of alkanes R–H with hydrogen peroxide catalyzed by nontransition metal compounds was published in 2008 by Shul'pin et al. [5]. It was found that simple aluminum nitrate $Al(NO_3)_3{\cdot}9H_2O$ efficiently catalyzes the oxidation of alkanes in acetonitrile-water solution in the presence of H_2O_2 at 70 °C. Later, water soluble salts (nitrates) of Be, Zn, Cd [7], and Bi [8] were also found to be catalytically active in this process (an activation of the allylic or benzylic C—H bonds with *tert*-BuOOH (TBHP) or H_2O_2 catalyzed by Bi(III) or Zn(II) complexes or salts was reported by Barrett et al. [10], by Salvador and Silvestre [11], and by Wu [12]).

The main reaction product of the R–H oxidation is the corresponding alkyl hydroperoxide ROOH as was indicated by the triphenylphosphine method developed by Shul'pin [1, 13]. Indeed, without an addition of PPh_3, chromatograms of the reaction mixture demonstrated the presence of both alcohols and ketones formed in a comparable amount upon thermal decomposition of ROOH in the gas chromatograph (GC) injector (Scheme 22.1). In experiments with the addition of an excess of PPh_3 to the reaction sample prior to GC analysis, chromatograms showed the almost exclusive formation of alcohols since under such conditions alkyl hydroperoxides ROOH are completely reduced into ROH. Comparing the intensities of peaks attributed to alcohols and ketones before and after addition of PPh_3, concentration of ROOH in the reaction mixture may be calculated.

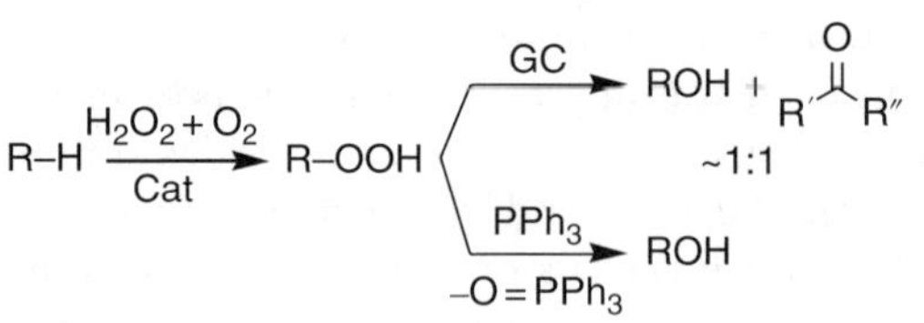

Scheme 22.1 Formation of alkyl hydroperoxides and their decomposition in GC and upon treatment with PPh_3.

In the oxidation of cyclooctane, the maximum yield of cyclooctyl hydroperoxide attains ca. 30% with the Al and Bi catalysts with the maximum turnover number (TON) of 110–150, while the Be, Zn, and Cd catalysts are less active (yields are 3–12%) (Table 22.1). In the case of the Bi catalyst, a saturation of the yield with time was not achieved even after 30 hours (Figure 22.1). The oxidation of linear (*n*-heptane, *n*-octane), branched (isooctane), and substituted cyclic (methylcyclohexane) alkanes with the Al and Bi systems demonstrates low relative selectivity. The normalized yield of each secondary alcohol in chromatograms is only 4–6 times higher than the yield of the primary alcohol and the yield of tertiary alcohols is 7.4–23 times higher compared to the primary ones (Table 22.2). Finally, for the oxidation of both *trans*- and *cis*-1,2-dimethylcyclohexane, the ratios of *trans*- and *cis*-1,2-dimethylhydroxycyclohexane are similar. These selectivity parameters are close to those obtained previously for reactions involving hydroxyl radicals $HO^{\bullet}$ as active oxidizing species, whereas the selectivity parameters for nonradical processes should be much higher [14].

Table 22.1 Maximum total yield of the oxygenate products based on alkane obtained in the oxidation of cyclooctane with the systems $[M(NO_3)_n]xH_2O/H_2O_2/CH_3CN + H_2O$.

Catalyst	Maximum total yield (%)	References
$Al(NO_3)_3{\bullet}9H_2O$	31	[5]
$Be(NO_3)_2{\bullet}4H_2O$	5.4[a)]	[7]
$Zn(NO_3)_2{\bullet}6H_2O$	3.0[a)]	[7]
$Cd(NO_3)_2{\bullet}4H_2O$	12.2[a)]	[7]
$Bi(NO_3)_3{\bullet}5H_2O$	32[b)]	[8]

a) Conditions were not optimized.
b) Nitric acid was added to the system to hamper the hydrolysis of the catalyst.

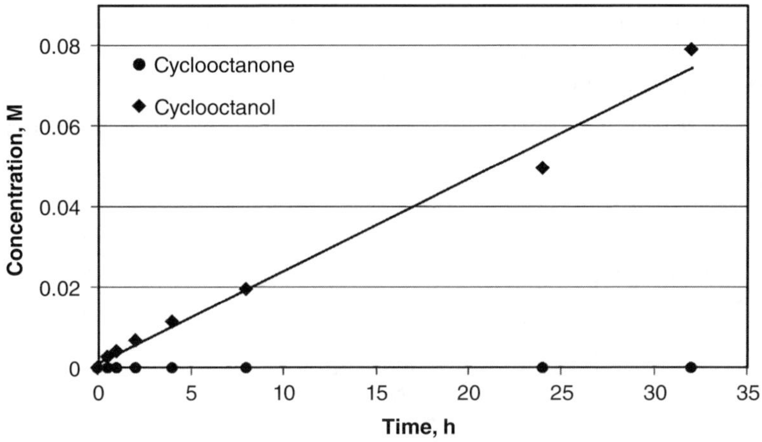

Figure 22.1 Accumulation of oxygenate (cyclooctanol and cyclooctanone) after reduction with PPh_3 with time in the cyclooctane (0.25 M) oxidation with H_2O_2 (1.2 M) catalyzed by $Bi(NO_3)_3$ (1.0×10^{-3} M) and HNO_3 (0.072 M) in acetonitrile containing water (2.4 M) at 75 °C.

Table 22.2 Selectivity parameters in the oxidation of linear, branched and substituted cyclic alkanes with the system $Al(NO_3)_3/H_2O_2/CH_3CN + H_2O$ at 70 °C [5].[a)]

	n-Heptane	*n*-Octane
C(1) : C(2) : C(3) : C(4)[b)]	1 : 5.3 : 5.4 : 4.9	1 : 5.6 : 5.6 : 5.0
	2,2,4-TMP	MCH
1° : 2° : 3°[c)]	1 : 4.3 : 7.4	1 : 6 : 23
	cis-DMCH	*trans*-DMCH
trans-/*cis-*[d)]	0.8	0.8

a) 2,2,4-TMP is isooctane, MCH is methylcyclohexane, DMCH is dimethylcyclohexane.
b) Relative normalized reactivities of the hydrogen atoms at carbons 1, 2, 3, and 4 of the alkane chain.
c) Relative normalized reactivities of the hydrogen atoms at primary, secondary, and tertiary carbons.
d) Ratio of the *trans*- and *cis*-isomers of dimethylhydroxycyclohexane.

The Zn and Cd complexes $[Zn_4O(dmpzc)_3]$, $[Cd(dmpzc)(DMF)(H_2O)]$ (dmpzc = 3,5-dimethylpyrazolate-4-carboxilate) as well as $Zn(NO_3)_2{\cdot}6H_2O$ and $Cd(MeCOO)_2{\cdot}2H_2O$ demonstrate low activity toward oxidation of cyclohexane with TBHP in CH_3CN in the presence of trifluoroacetic acid (TFA) at 50 °C, the total yield of the oxygenate products being 0–0.8% [15]. The oxidation of cyclohexane by H_2O_2 catalyzed by complex $Cd(IAA)_2(Phen)(H_2O)$ (IAA = indole acetic acid, Phen = 1,10-phenanthroline) in the presence of HNO_3 at 40 °C after 16 hours is characterized by TON of 2.4 [16].

22.2.2 Experimental Kinetic Studies

Experimental dependences of the initial reaction rate (W_0) vs. concentration of the substrate, catalyst, oxidant, additive, and solvent confirm results of the selectivity tests about the participation of the $HO^\bullet$ radical in the reactions of alkane oxidation with H_2O_2 catalyzed by the Al and Bi nitrates (Figure 22.2) [5, 8]. Indeed, the saturation of W_0 vs. the concentration of alkane at the excess of the latter (Figure 22.2a) demonstrates the competition of RH and acetonitrile solvent for the oxidizing species. Since CH_3CN is rather stable toward oxidation, the oxidizing species should be highly reactive such as $HO^\bullet$ radicals (note that the hydroperoxo radical $HOO^\bullet$ is not sufficiently "strong" to oxidize efficiently inert alkanes). Correspondingly, the global reaction mechanism may be represented by the set of Eqs. (22.1)–(22.5):

$$H_2O_2/M(NO_3)_3 \rightarrow\rightarrow HO^\bullet \quad W_i \tag{22.1}$$

$$HO^\bullet + MeCN \rightarrow \text{products} \quad k_2 \tag{22.2}$$

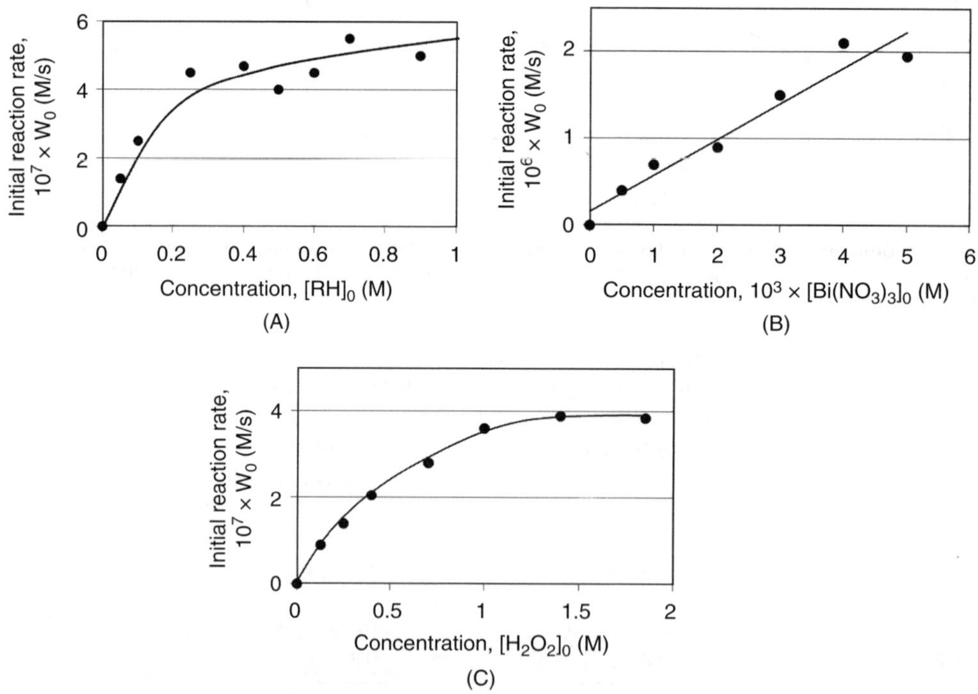

Figure 22.2 Dependences of the oxidation rate W_0 on initial concentrations of cyclooctane (RH) (A), $Bi(NO_3)_3$ (B), and H_2O_2 (C) in the oxidation of cyclooctane with H_2O_2 in acetonitrile containing water (2.4 M) catalyzed by $Bi(NO_3)_3$ and HNO_3 (0.072 M) at 60 °C. Conditions: $[Bi(NO_3)_3]_0 = 1.0 \times 10^{-3}$ M (A, C), $[RH]_0 = 0.25$ M (B, C), $[H_2O_2]_0 = 1.2$ M (A, B).

$$HO^{\bullet} + RH \rightarrow R^{\bullet} + H_2O \; k_3 \tag{22.3}$$

$$R^{\bullet} + O_2 \rightarrow ROO^{\bullet} \tag{22.4}$$

$$ROO^{\bullet} + H^+ + e^- \rightarrow ROOH \tag{22.5}$$

Hydrogen peroxide is catalytically decomposed to give the $HO^{\bullet}$ radical with the rate W_i. This radical reacts with alkane (constant k_3) abstracting the hydrogen and affording the alkyl radical $R^{\bullet}$ and with MeCN in a side reaction (constant k_2). The alkyl radical reacts with molecular oxygen producing the $ROO^{\bullet}$ radical and then final alkyl hydroperoxide ROOH. An application of the steady-state approximation relative to the hydroxyl radical gives the following equation for the RH oxidation rate:

$$W_0 = \left(\frac{d[ROOH]}{dt}\right)_0 = \frac{k_3[RH]_0 W_i}{k_3[RH]_0 + k_2[CH_3CN]_0}$$

or in a form which is more convenient for analysis:

$$\frac{[RH]_0}{W_0} = \frac{1}{W_i}\left([RH]_0 + \frac{k_2}{k_3}[CH_3CN]_0\right)$$

The calculated k_2/k_3 ratio (0.045 for the Al system and 0.008 for the Bi system) is typical for the reactions involving $HO^{\bullet}$ radicals [17].

It may be assumed that the generation of the $HO^{\bullet}$ radical starts with the formation of a complex of H_2O_2 with the catalyst "Cat•H_2O_2", which is decomposed to give $HO^{\bullet}$:

$$\text{"Cat"} + H_2O_2 \rightleftarrows \text{"Cat•}H_2O_2\text{"} \; K_6 \tag{22.6}$$

$$\text{"Cat•}H_2O_2\text{"} \rightarrow\rightarrow HO^{\bullet} \; k_7 \tag{22.7}$$

Taking into account that concentration of "Cat•H_2O_2" is quasi-stationary (i.e. $k_{-6} \gg k_7$) and that $[\text{"Cat"}]_0 \ll [H_2O_2]_0$, such a global mechanistic scheme leads to the following equation for the W_i rate:

$$W_i = \frac{k_7 K_6[\text{"Cat"}]_0[H_2O_2]}{1 + K_6[H_2O_2]}$$

The experimental dependence of W_0 on $[H_2O_2]$ (Figure 22.2c) fits well with this kinetic model. The calculated K_6 values are 1.4 and 1.3 M^{-1} for the Al and Bi catalysts, respectively, while the rate constants k_7 are 7.6×10^{-3} and 9×10^{-4} s^{-1}, correspondingly. Experimentally determined activation energy for the Bi catalyzed oxidation of cyclooctane with H_2O_2 is 20 ± 2.5 kcal mol^{-1}.

22.2.3 Theoretical Mechanistic Studies

As was discussed, in accord with experimental data, the oxidation of alkanes with H_2O_2 catalyzed by simple soluble Al and Bi salts involves the formation of $HO^{\bullet}$ radicals – the active oxidizing species directly reacting with alkanes. Formation of $HO^{\bullet}$ radicals is the rate-limiting step of the whole process. At the same time, the mechanism of the $HO^{\bullet}$ generation with nontransition metal catalysts was a mystery until recently.

Historically, the first mechanistic proposal for the $HO^{\bullet}$ radical generation was formulated by Haber and Weiss [18] for the Fenton system Fe(II) + H_2O_2 (another, nonradical mechanism associated with the Fenton reagent was proposed by Bray and Gorin, and it includes the formation of the ferryl species $[Fe^{(IV)}{=}O]^{2+}$, which then oxidizes alkanes [19]). In accord with this hypothesis, $HO^{\bullet}$ is formed as a result of the one electron oxidation of Fe^{2+} to Fe^{3+} and the homolytic O—O bond rupture in the peroxide molecule (reaction (22.8)). The initial Fe(II) species then may be restored upon reaction between Fe(III) and H_2O_2, also leading to the formation of the HOO• radical (reaction (22.9)) [20].

$$[Fe^{II}(H_2O)_6]^{2+} + H_2O_2 \rightarrow [Fe^{III}(H_2O)_5(OH)]^{2+} + HO^{\bullet} + H_2O \tag{22.8}$$

$$[Fe^{III}(H_2O)_5(OH)]^{2+} + H_2O_2 \rightarrow [Fe^{II}(H_2O)_6]^{3+} + HOO^{\bullet} \tag{22.9}$$

Later, a number of modifications of this radical mechanism were proposed, some of them involving two [21, 22] or even three [23] hydrogen peroxide molecules. When the metal of a catalyst exhibits its highest oxidation state, a modified mechanism with the reversal sequence of reaction steps 8 and 9 was proposed [17b]. This mechanism includes the initial reduction of the catalyst with the formation of the HOO• radical followed by the oxidation of the metal complex with another H_2O_2 molecule to give the HO• radical (Scheme 22.2) [24].

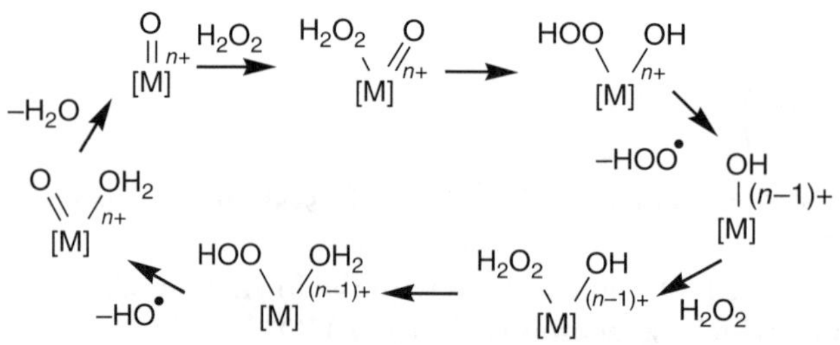

Scheme 22.2 Mechanism of HO• radical generation catalyzed by a transition metal complex with a metal in its highest oxidation state.

All these proposed mechanisms require a change of the oxidation state of the metal in a catalyst molecule. However, many nontransition metals and some transition metals have a single stable nonzero oxidation state (the SSOS metals); among them are Be, Zn, Cd, Al, Ga, In, Sc, Y, and some lanthanides. Therefore, the mechanism of the HO• generation with catalytic systems based on these metals should be of a fundamentally different type. Such a mechanism was proposed and investigated theoretically in detail in the recently published cycle of works [6–9]. Aqua complexes of the SSOS metals $[M(H_2O)_n]^{p+}$ (**1**) were considered as an initial form of the catalyst. Additionally, it was found [8] that the same mechanism is also feasible for Bi catalysts, despite this metal exhibiting several nonzero oxidation states.

At the first step of this mechanism, a water ligand in the aqua complex **1** is substituted with an hydrogen peroxide molecule to give the hydrogen peroxide adduct $[M(H_2O)_{(n-1)}(H_2O_2)]^{p+}$ (**2**) (Scheme 22.3). This step is usually slightly endergonic and occurs in a dissociative (for Zn, Al, Ga, Y, and Bi), an associative (for Sc and La), or an interchange I_a (for Be, Cd, and In) manner. All metal centers considered are Lewis acids. Hence, at the second step, the coordinated hydrogen peroxide molecule undergoes the hydrolysis yielding the hydroperoxo complex $[M(H_2O)_{(n-1)}(OOH)]^{(p-1)+}$ (**3**). The energy of hydrolysis depends on the metal center nature and varies from 0.6 kcal mol^{-1} (for Ga) to 13.4 kcal mol^{-1} (for Cd).

Complex **3** participates in the second substitution step, i.e. a water ligand in **3** is replaced for another H_2O_2 molecule to give intermediate $[M(H_2O)_{(n-m)}(H_2O_2)(OOH)]^{(p-1)+}$ (**4**) bearing two peroxide ligands. For Zn and Cd, this step is accompanied by the elimination of one additional water ligand from the coordination sphere of the metal. Intermediate **4** is a crucial species in the proposed mechanism. Calculations indicated that the HO—OH bond of the hydrogen peroxide in **4** is tremendously activated toward the homolytic cleavage. Indeed, the energy of

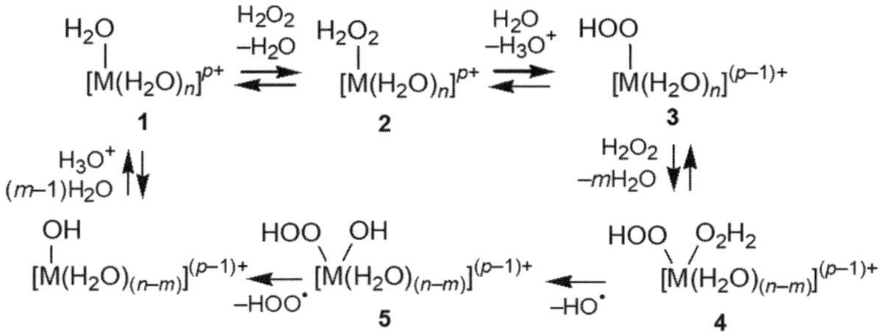

M = Al, Ga, In, Sc, Bi; $n = 5$; $m = 1$; $p = 3$
M = Y; $n = 7$; $m = 1$; $p = 3$
M = La; $n = 8$; $m = 1$; $p = 3$
M = Be; $n = 3$; $m = 1$; $p = 2$
M = Zn, Cd; $n = 5$; $m = 2$; $p = 2$

Scheme 22.3 Mechanism of the $HO^•$ radical generation proposed for the systems $[M(H_2O)_n]^{p+}/H_2O_2$ (M = Al, Ga, In, Sc, Y, La, Bi, Be, Zn, and Cd).

Table 22.3 DFT calculated energies of the HO—OH bond dissociation (in terms of ΔG in CH_3CN solution, in kcal mol^{-1}).[a)]

Species	ΔG_b	References
H_2O_2	39.4	[6]
cis-$[Al(H_2O)_4(OOH)(H_2O_2)]^{2+}$	6.1	[6]
cis-$[Ga(H_2O)_4(OOH)(H_2O_2)]^{2+}$	6.2	[9]
cis-$[In(H_2O)_4(OOH)(H_2O_2)]^{2+}$	7.5	[9]
trans-$[Sc(H_2O)_4(OOH)(H_2O_2)]^{2+}$	4.4	[9]
$[Y(H_2O)_6(OOH)(H_2O_2)]^{2+}$	10.3	[9]
$[La(H_2O)_7(OOH)(H_2O_2)]^{2+}$	15.3	[9]
$[Be(H_2O)_2(OOH)(H_2O_2)]^{+}$	−0.8	[7]
$[Zn(H_2O)_3(OOH)(H_2O_2)]^{+}$	11.2	[7]
$[Cd(H_2O)_3(OOH)(H_2O_2)]^{+}$	8.3	[7]
cis-$[Bi(H_2O)_4(OOH)(H_2O_2)]^{2+}$	5.5	[8]

a) Outer sphere water molecules in complexes are omitted.

the homolytic HO—OH bond dissociation in **4** is (−0.8)–(15.3) kcal mol^{-1} (in terms of Gibbs free energy) vs. 39.4 kcal mol^{-1} in free H_2O_2 (Table 22.3). As a result, the $HO^•$ radical is easily generated in a monomolecular step **4** → **5** + $HO^•$.

The reasons for such a great activation of H_2O_2 in **4** may be understood from the analysis of the electronic structure of **5**. For this complex, three resonance structures (**a**–**c**) may be drawn (Figure 22.3). In structure **a**, the unpaired electron is localized at the OH ligand, while, in structures **b** and **c**, this electron is localized at one of the oxygen atoms of the OOH ligand. The calculations indicated that the resonance structure **a** does not play any role in complex **5**,

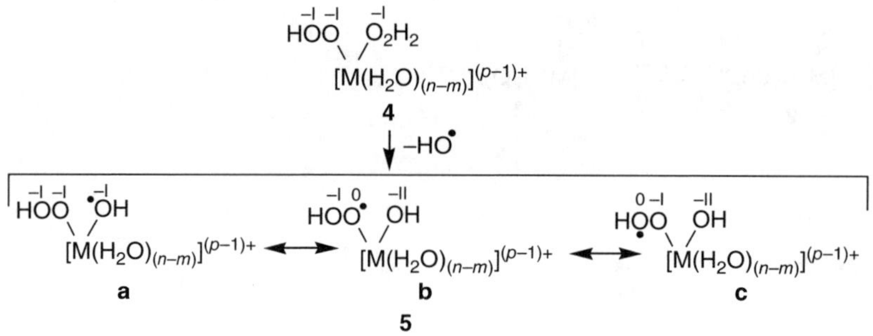

Figure 22.3 Resonance structures of complexes **5** and oxygen oxidation states (Roman numerals) in **4** and **5**.

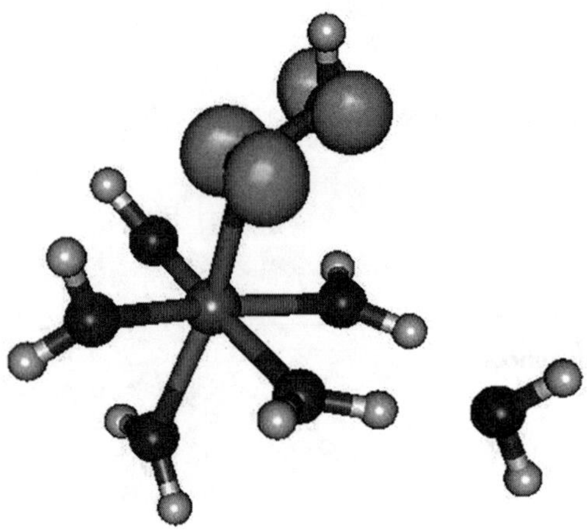

Figure 22.4 Spin density distribution in complex *cis*-$[Ga(H_2O)_4(OOH)(OH)](H_2O)^{2+}$. (*See insert for color representation*).

whereas the contribution ratio of resonances **b** and **c** is ca. 2 : 1 (Figure 22.4). Such a situation corresponds to the reduction of the OH ligand and oxidation of the OOH^- ligand upon the HO—OH bond cleavage in **4**. As a result, the OOH ligand in **5** becomes neutral. The facts that the OOH^- ligand is able to be oxidized rather easily and that the spin density in **5** is delocalized among two oxygen atoms of the OOH ligand provide a stabilization of **5** and a decrease the HO—OH bond dissociation energy in **4**. Thus, the OOH^- co-ligand in intermediate **4** plays a crucial role in the mechanism of the $HO^•$ radical generation. This role is similar to that of a transition metal in the Fenton or Fenton-like processes (stabilization of products of the HO—OH bond rupture). Similar stabilization due to delocalization of the spin density among the atoms of co-ligands was previously found for the titanium complexes $[Cp_2Ti(O)L]$ ($L = Cl^-$, OTf^-, Br^-, OEt_2, Et_3P) [25].

One of the crucial features of the proposed mechanism is the participation of two H_2O_2 molecules. The first H_2O_2 molecule is responsible for the formation of the redox active OOH^- ligand, while the second H_2O_2 molecule is directly decomposed after coordination to give the $HO^•$ radical without a change of the metal oxidation state.

The rate-limiting step of the whole reaction of the $HO^•$ generation is the second H_2O-for H_2O_2 substitution (for Be and Al) or the monomolecular cleavage of the HO—OH bond in **4** (for other metals). The calculated overall activation energies are given in Table 22.4. The catalytic activity of $[M(H_2O)_n]^{p+}$ is controlled by three main factors, i.e. (i) lability of the water ligands in the coordination sphere of the metal; (ii) acidity of the metal-bound ligands; and (iii) activation of the coordinated H_2O_2 toward the homolytic HO—OH bond cleavage.

Table 22.4 Calculated Gibbs free energies of activation (in kcal mol^{-1}) of the HO$^{\bullet}$ radical generation in the systems $[M(H_2O)_n]^{p+}/H_2O_2/CH_3CN + H_2O$ [6–9].

Catalyst	ΔG≠	References
$[Al(H_2O)_6]^{3+}$	25.6	[6]
$[Ga(H_2O)_6]^{3+}$	19.5	[9]
$[In(H_2O)_6]^{3+}$	23.5	[9]
$[Sc(H_2O)_6]^{3+}$	22.2	[9]
$[Y(H_2O)_8]^{3+}$	23.7	[9]
$[La(H_2O)_9]^{3+}$	25.2	[9]
$[Be(H_2O)_4]^{2+}$	26.2	[7]
$[Zn(H_2O)_6]^{2+}$	26.4	[7]
$[Cd(H_2O)_6]^{2+}$	27.3	[7]
$[Bi(H_2O)_6]^{3+}$	15.2	[8]

22.3 Hydroxylation of Methane with a Monooxygenase Zn Model

Recently, a theoretical investigation of the methane hydroxylation with a methane monooxygenase model bearing the peroxo Zn_2O_2 core was reported [26]. It was found the mechanism of the reaction is radical-rebound (Scheme 22.4), including the formation of a molecular complex **7** between the catalyst **6** and substrate, the radical hydrogen abstraction to give methyl radical and **8**, the rebound of $CH_3^{\bullet}$ to the hydroperoxo group with the formation of complex **9,** and the release of methanol and generation of the μ-O bridged species **10**. The rate limiting step is the hydrogen abstraction, and the calculated activation energy (in terms of the total energy with zero point energy (ZPE) correction) is 25.2 kcal mol^{-1}.

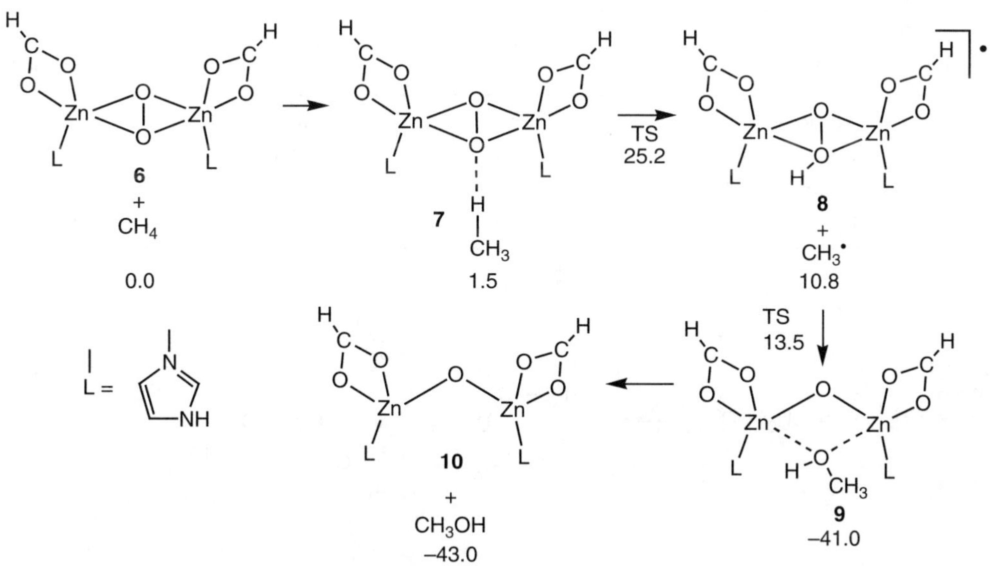

Scheme 22.4 Mechanism of the methane hydroxylation with the zinc peroxo complex **6** (relative ZPE corrected energies are given in kcal mol^{-1}).

22.4 Carboxylation of Alkanes

The system catalyst/$K_2S_2O_8$/TFA/CO was developed by Fujiwara et al. for the efficient carboxylation of alkanes to corresponding carboxylic acids. Typically, the catalyst is based on transition metal species (Pd, Yb, V, Mn, etc.) [27]. However, it was found that some catalytic systems based on unconventional for this process nontransition metal such as Ca are also very efficient. The optimization of reaction conditions indicated that the yield of acetic acid upon oxidation of methane with $K_2S_2O_8$ in TFA catalyzed by simple calcium chloride, $CaCl_2$, may reach 25.7% (based on methane) at 85 °C after 15 hours and it is almost quantitative (93.8%) after 140 hours [28]. The activation parameters were measured experimentally and the reaction was found to proceed with strongly negative activation entropy (−40.6 cal (mol K)$^{-1}$, Table 22.5). Other Ca compounds (CaO, $CaCO_3$, $Ca(OAc)_2$•$2H_2O$, $[Ca(acac)_2]$•$2H_2O$, $Ca(OH)_2$, $CaSO_4$•$2H_2O$) are not efficient as catalysts in the carboxylation of methane.

The application of radical traps (2,4,6-tri-*tert*-butylphenol, tetrachloro-*p*-benzoquinone and 2,2,6,6-tetramethylpiperidine) suppresses the formation of acetic acid. The ^{13}C labeled experiments indicated that when $^{13}CH_4$ or ^{13}CO were used, the exclusive formation of $^{13}CH_3COOH$ and $CH_3{}^{13}COOH$ was observed. Hence, it may be assumed that the mechanism is based on the formation of radicals $CH_3^{\bullet}$, which react with carbon monoxide to give $CH_3CO^{\bullet}$. The acetyl radical then may be converted into acetic acid via the mixed anhydride with TFA (Scheme 22.5).

The same catalytic system is also efficient for the oxidation of ethane, propane, and heavier alkanes (cyclopentane, cyclohexane, cycloheptane, and cyclooctane) under the same conditions [29]. Ethane oxidizes into propionic acid (major product) and acetic acid (minor product). In the case of carboxylation of propane, the main product is isobutyric acid, while *n*-butyric acid and isopropyl trifluoroacetate are minor products (Scheme 22.6). Propane

Table 22.5 Experimentally measured activation parameters for the carboxylation of alkanes with the system $CaCl_2/K_2S_2O_8$/TFA [28, 29].

Substrate	Product	ΔH≠ (kcal mol^{-1})	ΔS≠ (cal (mol K)$^{-1}$)	ΔG≠ (kcal mol^{-1})
CH_4	Acetic acid	16.2	−40.6	30.5
C_3H_8	Isobutiryc acid	30.6	2.5	29.7
C_3H_8	*n*-Butyric acid	32.2	6.8	29.6
C_3H_8	Isopropyl trifluoroacetate	36.0	16.0	30.3

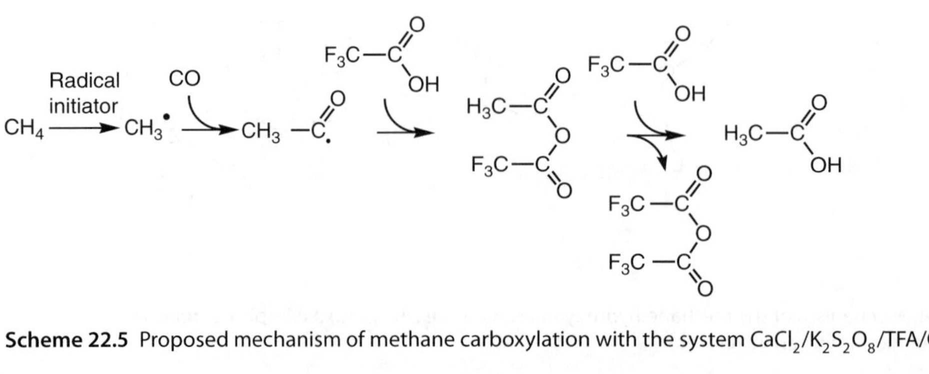

Scheme 22.5 Proposed mechanism of methane carboxylation with the system $CaCl_2/K_2S_2O_8$/TFA/CO.

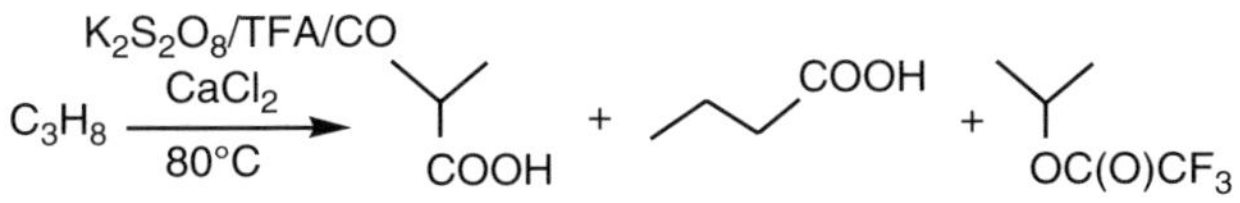

Scheme 22.6 Carboxylation of propane with the system $CaCl_2/K_2S_2O_8$/TFA/CO.

undergoes the reaction already at room temperature although in a low total yield of the products (2%). The catalytic system demonstrates a high efficiency toward oxidation of light alkanes (ethane and propane). At 80 °C, the product yields are up to 30% after 15 hours and 95% after 24 hours. The maximum TON obtained is 26.5 based on Ca. Interestingly, that this system is less efficient toward oxidation of cycloalkanes (total product yields 12–44%), while cyclooctane is not oxidized at these conditions. Some other Ca species (CaO, $CaCO_3$, and $Ca(OAc)_2{\bullet}2H_2O$) show lower but still significant catalytic activity toward carboxylation of propane.

Dependences of the product yield on various conditions were analyzed and such an analysis allowed the optimization of reaction conditions (Table 22.6). The highest yields

Table 22.6 Selected catalytic data on the alkane carboxylation with the systems [Ca]/$K_2S_2O_8$/TFA [28, 29].[a)]

Substrate	T (°C)	CO (atm)	TFA (ml)	Substrate (atm)	$CaCl_2$ (mmol)	Total product yield (%)
C_2H_6	80	30	5	1	0.5	88[b)]
C_3H_8	25	30	5	6	0.5	2
CH_4/C_3H_8	60	30	6/5	20/6	0.5	3/3
CH_4/C_3H_8	85/80	30	6/5	20/6	0.5	9/25
C_3H_8	80	30	5	6	0.5	20[c)] 16[d)] 12[e)] 1–4[f)]
CH_4/C_3H_8	85/80	1	6/5	20/6	0.5	1/19
CH_4/C_3H_8	85/80	40/30	3/2	20/6	0.5	8[g)]/13
C_3H_8	80	30	10	6	0.5	22
CH_4/C_3H_8	85/80	40/30	5	1	0.5	21[g)]/51 (95[h)])
CH_4/C_3H_8	85/80	40/30	5	20/6	0.05	0[g)]/6
CH_4/C_3H_8	85/80	40/30	5	20/6	1	6[g)]/17
cyclo-C_5H_{10}	80	30	5	1[i)]	0.5	44[h)]
cyclo-C_6H_{12}	80	30	5	1[i)]	0.5	33[h)]
cyclo-C_7H_{14}	80	30	5	1[i)]	0.5	12[h)]
cyclo-C_8H_{16}	80	30	5	1[i)]	0.5	—[h)]

a) Conditions: autoclave (25 ml), $K_2S_2O_8$ (4–5 mmol), 15 h.
b) Reaction time 48 h.
c) CaO catalyst.
d) $CaCO_3$ catalyst.
e) $Ca(OAc)_2{\bullet}2H_2O$ catalyst.
f) $Ca(acac)_2{\bullet}2H_2O$, $Ca(OH)_2$, $CaSO_4{\bullet}2H_2O$ catalysts.
g) $K_2S_2O_8$ 10 mmol.
h) Reaction time 24 h.
i) 1 mmol.

Table 22.7 Catalytic data on the alkane carboxylation with the system $Mg/K_2S_2O_8$/TFA [30].[a)]

Substrate	Conversion, %	Selectivity[b)]
CH_4	1.1	96 : 4
C_3H_8	11	96 : 4
cyclo-C_5C_{10}	39	66 : 22
cyclo-C_6H_{12}	43 (80)[c)]	82 : 14 (83 : 13)[c)]
cyclo-C_7H_{14}	48	31 : 59
Adamantane	18	80 : 13

a) Conditions: autoclave 25 ml, methane (30 atm), propane (7 atm), other alkanes (5 mmol), CO (50 atm), Mg (5 mmol), $K_2S_2O_8$ (5 mmol), TFA (3 ml), 80 °C, 30 h.
b) Ratio of RCOOH and $ROC(O)CF_3$.
c) Cyclohexane (1 mmol).

were observed at 75–80 °C, the further increase of temperature decreases yields of the acid products conceivably due to their thermal or catalytic decomposition. The yield of the acids also increases with the enhancement of concentrations of the catalyst and oxidant (up to 0.08 and 0.8 M, respectively) and of the pressure of substrate and carbon monoxide (up to 6 and 30 atm, respectively). Finally, the dependence of the product yield on the concentration of TFA demonstrates the presence of a maximum.

The experimentally measured activation parameters for the propane oxidation are given in Table 22.5. Although the authors proposed the same mechanism for the reactions with methane and propane, quite different kinetic may suggest that these processes occur via different pathways.

The same authors reported [30] that alkanes RH are efficiently carboxylated with $K_2S_2O_8$ and CO in the presence of magnesium powder in TFA to give the corresponding carboxylic acid RCOOH and alkyl trifluoroacetate $ROC(O)CF_3$. The selectivity of the reaction strongly depends on the substrate nature (Table 22.7). In contrast to the Ca-based systems described above, magnesium is a promoter rather than a catalyst in this reaction. The highest yields were obtained when Mg and $K_2S_2O_8$ were taken in equimolar amounts at 80 °C. Other Mg, as well as Zn, Ni, Al, and Ag additives are less efficient than the Mg powder. The IR spectra of the reaction mixture demonstrated the presence of $(CF_3COO)_2Mg$ and MgO; the latter is conceivably the active species in this reaction.

22.5 Heterogeneous Oxidation of Alkanes

Aluminum oxides (alumina) and various aluminosilicates and aluminophosphates are among the most popular supports used in heterogeneous catalysis for the oxidation of alkanes. The doping of Al_2O_3, aluminosilicates, and aluminophosphates by various transition metal ions often provides a significant catalytic activity of such systems [31]. However, these Al species can also exhibit the catalytic activity by themselves.

Without additives, aluminum oxide exhibits rather low catalytic activity toward the oxidation of cyclooctane with H_2O_2 (CH_3CN solvent, 80 °C) [32]. The maximum yield of oxygenates (cyclooctanol and cyclooctanone) does not exceed 2% even after 25 hours. Nevertheless, such yield is ca. four times higher than that in a blank experiment without the catalyst. The addition of TFA at a low concentration significantly increases the yield of the products up to 7.2% after 5 hours. Detailed experimental kinetic studies and the selectivity tests with *n*-heptane, *n*-octane, isooctane, methylcyclohexane, and dimethylcyclohexane indicated that the mechanism includes the generation of $HO^{\bullet}$ radicals that are catalytically formed from H_2O_2 with the rate of 4.4×10^{-6} M s^{-1}. The role of TFA in this process remains unclear.

The protonated zeolite H-ZSM-5 was found to be efficient for the partial oxidation of methane with H_2O_2 into formic acid at 100 °C and pressure of 2.6 MPa [33]. The maximum performance was achieved upon addition of PPh_3 as a methane oxidation promoter. The observed yield of HCOOH after 5 hours reaches 13% and the selectivity is 66.8%. The yield of the product is the highest at a low SiO_2/Al_2O_3 ratio (24%). It is interesting that catalytic systems based on some transition metal complexes (V, Fe, Ti) were found to be much less active in this reaction than H-ZSM-5. Silica supported tin catalysts are also effective in the reaction of methane oxidation to formaldehyde and methanol [34] and in the oxidation of the benzylic C—H bonds with H_2O_2 [35].

Bismuth oxide occluded in a zeolite ZSM-5 matrix was used as a catalyst for the oxidation of cyclohexane as well as benzene and toluene with H_2O_2 at 50–90 °C [36]. The conversion of the alkane substrate to cyclohexanone (selectivity 98–99%) reaches 25%. Bismuth containing MCM-41 is also active toward cyclohexane oxidation by molecular oxygen instead of H_2O_2. However, this reaction requires harsher conditions (temperature 140 °C) [37].

The mesoporous tin silicates were used as catalysts in the oxidation of cyclohexane in the presence of dilute H_2O_2 (25% aqueous) at 60–80 °C [38]. The reaction strongly depends on the solvent nature. When acetone, *tert*-butanol, or benzonitrile were used, the conversion was 1.7–6.5% after 12–16 hours, and the main observed products were cyclohexanone (47–83%) and cyclohexanol (17–50%). However, in the acetonitrile solution, the conversion reaches 98%, and the dehydrogenation becomes the main reaction channel leading to 1,4-cyclohexadiene as the main product (95.7%). The catalytic system is inactive in the methanol solution.

Cyclohexane was efficiently oxidized into observable cyclohexanol and cyclohexanone by zinc peroxide ZnO_2, which acts as an oxidizing agent (160 °C, 12 hours, conversion 60%) [39].

22.6 Final Remarks

Although nontransition metal-based catalytic systems remain unconventional for the oxidation of alkanes with peroxides, they demonstrate significant potential. In contrast to many transition metal complexes, catalysts based on nontransition metals such as Al, Mg, Ca, Ga, and Bi are attractive from economical and/or ecological points of view, as they are inexpensive and/or nontoxic. The area of the peroxidative oxidation of alkanes with nontransition metal compounds is still at the very beginning stage of its development. Nevertheless, the available experimental and theoretical data on this topic clearly indicate that even simple inorganic complexes or salts of many nontransition metals exhibit significant catalytic activity in these processes. Therefore, it is quite likely that the optimization of such catalytic systems could dramatically improve their efficiency, and further investigations in this exciting area are highly important.

Acknowledgments

The work has been partially supported by the Fundação para a Ciência e a Tecnologia (FCT), Portugal (project UID/QUI/00100/2013).

References

1 (a) Shilov, A.E. and Shul'pin, G.B. (1997). *Chem. Rev.* 97: 2879. (b) Shul'pin, G.B. (2009). *Mini-Rev. Org. Chem.* 6: 95. (c) Crabtree, R.H. (1995). *Chem. Rev.* 95: 987. (d) Jia, C., Kitamura, T., and Fujiwara, Y. (2001). *Acc. Chem. Res.* 34: 633. (e) Lane, B.S. and Burgess, K. (2003). *Chem. Rev.* 103: 2457. (f) Grigoropoulou, G., Clark, J.H., and Elings, J.A. (2003). *Green Chem.* 5: 1. (g) Brégeault, J.-M. (2003). *Dalton Trans.* 3289. (h) Labinger, J.A. (2004). *J. Mol. Catal. A* 220: 27. (i) Muzart, J. (2007). *J. Mol. Catal. A* 276: 62. (j) Díaz-Requejo, M.M. and Pérez, P.J. (2008). *Chem. Rev.* 108: 3379. (k) Crabtree, R.H. (2010). *Chem. Rev.* 110: 575. (l) da Silva, J.A.L., Fraústo da Silva, J.J.R., and Pombeiro, A.J.L. (2011). *Coord. Chem. Rev.* 255: 2232. (m) Shul'pin, G.B. (2013). *Dalton Trans.* 42: 12794.

2 (a) Luo, X., Fleming, P.R., and Rizzo, T.R. (1992). *J. Chem. Phys.* 96: 5659. (b) Bach, R.D., Ayala, P.Y., and Schlegel, H.B. (1996). *J. Am. Chem. Soc.* 118: 12758.

3 (a) Benson, S.W. (1978). *Chem. Rev.* 78: 23. (b) Brusa, M.A., Churio, M.S., Grela, M.A. et al. (2000). *Phys. Chem. Chem. Phys.* 2: 2383.

4 (a) Fenton, H.J.H. (1876). *Chem. News* 190. (b) Fenton, H.J.H. (1894). *J. Chem. Soc.* 65: 33, 899.

5 Mandelli, D., Chiacchio, K.C., Kozlov, Y.N. et al. (2008). *Tetrahedron Lett.* 49: 6693.

6 Kuznetsov, M.L., Kozlov, Y.N., Mandelli, D. et al. (2011). *Inorg. Chem.* 50: 3996.

7 Kuznetsov, M.L., Teixeira, F.A., Bokach, N.A. et al. (2014). *J. Catal.* 313: 135.

8 Rocha, B.G.M., Kuznetsov, M.L., Kozlov, Y.N. et al. (2015). *Catal. Sci. Technol.* 5: 2174.

9 Novikov, A.S., Kuznetsov, M.L., Pombeiro, A.J.L. et al. (2013). *ACS Catal.* 3: 1195.

10 (a) Bonvin, Y., Callens, E., Larrosa, I. et al. (2005). *Org. Lett.* 7: 4549. (b) Callens, E., Burton, A.J., White, A.J.P., and Barrett, A.G.M. (2008). *Tetrahedron Lett.* 49: 3709.

11 Salvador, J.A.R. and Silvestre, S.M. (2005). *Tetrahedron Lett.* 46: 2581.

12 Wu, X.-F. (2012). *Tetrahedron Lett.* 53: 6123.

13 (a) Shul'pin, G.B. (2002). *J. Mol. Catal. A* 189: 39. (b) Shul'pin, G.B. (2010). *Org. Biomol. Chem.* 8: 4217.

14 (a) Deno, N.C., Jedziniak, E.J., Messer, L.A. et al. (1977). *Tetrahedron* 33: 2503. (b) Shul'pin, G.B., Süss-Fink, G., and Shilov, A.E. (2001). *Tetrahedron Lett.* 42: 7253. (c) Shul'pin, G.B., Süss-Fink, G., and Shul'pina, L.S. (2001). *J. Mol. Catal. A* 170: 17. (d) Romakh, V.B., Therrien, B., Süss-Fink, G. et al. (2007). *Inorg. Chem.* 46: 1315. (e) Kozlov, Y.N., Nizova, G.V., and Shul'pin, G.B. (2008). *J. Phys. Org. Chem.* 21: 119. (f) Shul'pin, G.B., Matthes, M.G., Romakh, V.B. et al. (2008). *Tetrahedron* 64: 2143.

15 Tăbăcaru, A., Xhaferaj, N., Martins, L.M.D.R.S. et al. (2016). *Inorg. Chem.* 55: 5804.

16 Xing, N., Xu, L.T., Bai, F.Y. et al. (2014). *Inorg. Chim. Acta* 409: 360.

17 (a) Shul'pin, G.B., Nizova, G.V., Kozlov, Y.N. et al. (2004). *Adv. Synth. Catal.* 346: 317. (b) Shul'pin, G.B., Kozlov, Y.N., Nizova, G.V. et al. (2001). *J. Chem. Soc., Perkin Trans. 2* 1351. (c) Nesterov, D.S., Chygorin, E.N., Kokozay, V.N. et al. (2012). *Inorg. Chem.* 51: 9110.

18 Haber, F. and Weiss, J. (1932). *Naturwissemschaften* 20: 948.

19 Bray, W.C. and Gorin, M.H. (1932). *J. Am. Chem. Soc.* 54: 2124.

20 Barb, W.G., Baxendale, J.H., George, P. et al. (1951). *Trans. Faraday Soc.* 47: 591.

21 Kozlov, Y.N., Nadezhdin, A.D., and Purmal', A.P. (1973). *Kinet. Katal.* 14: 141.

22 Popivker, I., Zilbermann, I., Maimon, E. et al. (2013). *Dalton Trans.* 42: 16666.

23 Burg, A., Shusterman, I., Kornweitz, H. et al. (2014). *Dalton Trans.* 43: 9111.

24 (a) Kuznetsov, M.L. and Pombeiro, A.J.L. (2009). *Inorg. Chem.* 48: 307. (b) Khaliullin, R.Z., Bell, A.T., and Head-Gordon, M. (2005). *J. Phys. Chem. B* 109: 17984. (c) Kirillova, M.V., Kuznetsov, M.L., Romakh, V.B. et al. (2009). *J. Catal.* 267: 140. (d) Kirillova, M.V., Kuznetsov, M.L., Kozlov, Y.N. et al. (2011). *ACS Catal.* 1: 1511.

25 DiPasquale, A.G., Hrovat, D.A., and Mayer, J.M. (2006). *Organometallics* 25: 915.
26 Li, C.-Q., Yang, H.-Q., Xua, J. et al. (2012). *Org. Biomol. Chem.* 10: 3924.
27 (a) Nishiguchi, T., Nakata, K., Takaki, K. et al. (1992). *Chem. Lett.* 1141. (b) Asadullah, M., Taniguchi, Y., Kitamura, T. et al. (1998). *Appl. Organomet. Chem.* 12: 277. (c) Kirillova, M.V., Kuznetsov, M.L., Reis, P.M. et al. (2007). *J. Am. Chem. Soc.* 129: 10531. (d) Kirillova, M.V., Kuznetsov, M.L., da Silva, J.A.L. et al. (2008). *Chem. Eur. J.* 14: 1828.
28 Asadullah, M., Kitamura, T., and Fujiwara, Y. (2000). *Angew. Chem. Int. Ed.* 39: 2475.
29 (a) Asadullah, M., Kitamura, T., and Fujiwara, Y. (2000). *Catal. Lett.* 69: 37. (b) Asadullah, M., Kitamura, T., and Fujiwara, Y. (2000). *J. Catal.* 195: 180.
30 Asadullah, M., Kitamura, T., and Fujiwara, Y. (1999). *Appl. Organomet. Chem.* 13: 539.
31 (a) Thomas, J.M., Raja, R., Sankar, G. et al. (2001). *Acc. Chem. Res.* 34: 191. (b) Schuchardt, U., Cardoso, D., Sercheli, R. et al. (2001). *Appl. Catal. A* 211: 1. (c) Ricci, G.P., Rocha, Z.N., Nakagaki, S. et al. (2010). *J. Appl. Catal. A* 389: 147. (d) Borah, P. and Datta, A. (2010). *Appl. Catal. A* 376: 19. (e) Modén, B., Zhan, B.-Z., Dakka, J. et al. (2006). *J. Catal.* 239: 390. (f) Selvam, P. and Mohapatra, S.K. (2006). *J. Catal.* 238: 88. (g) Zahedi-Niaki, M.H., Kapoor, M.P., and Kaliaguine, S. (1998). *J. Catal.* 177: 231.
32 Mandelli, D., do Amaral, A.C.N., Kozlov, Y.N. et al. (2009). *Catal. Lett.* 132: 235.
33 Kalam, A., Rahman, L., Kurnashiro, M. et al. (2011). *Catal. Commun.* 12: 1198.
34 Eskendirov, I., Coville, N.J., Parmaliana, A. et al. (1997). *Natural Gas Conversion IV*, Studies in Surface and Catalysis, vol. 107 (ed. M. DePontes, R.L. Espinosa, C.P. Nicolaides, et al.), 301. New York: Elsevier Science.
35 Mal, N.K. and Ramaswamy, A.V. (1996). *Appl. Cat. A* 143: 75.
36 Dumitriu, D., Bârjega, R., Frunza, L. et al. (2003). *J. Catal.* 219: 337.
37 Qian, G., Ji, D., Lu, G. et al. (2005). *J. Catal.* 232: 378.
38 Samanta, S., Mal, N.K., Manna, A. et al. (2004). *Appl. Catal. A* 273: 157.
39 Lingampalli, S.R., Gupta, U., Gautam, U.K. et al. (2013). *ChemPlusChem* 78: 837.

23

Metal-Free Functionalization of Alkanes

Jincan Zhao[1] *and Jianlin Han*[2]

[1] *Hebei University, College of Chemistry & Environmental Science, Baoding, Hebei, People's Republic of China*
[2] *Nanjing University, School of Chemistry and Chemical Engineering, State Key Laboratory of Coordination Chemistry, Jiangsu Key Laboratory of Advanced Organic Materials, Nanjing, People's Republic of China*

23.1 Introduction

Hydrocarbons, especially alkanes, are the main feedstocks in the chemical industry from oil and natural gas. Therefore, direct and catalytic transformation of alkanes to various useful chemicals via sp^3 C—H bond functionalization is of considerable interest to chemical industries and remains a challenge to chemists [1–6].

The direct and selective functionalization of the sp^3 C—H bonds, which is a great challenge due to their high bond-dissociation energy (BDE) and low polarity, has attracted considerable attention in recent years [7–16]. Over the past decades, significant advances have been made in C(sp^3)—H bond activation adjacent to heteroatoms, double bonds, and phenyl or electron-withdrawing groups [17–24]. However, selective activation of the inactive C—H bonds of simple alkanes to generate C—X (X = C, N, O, S, S et al.) bonds is a more challenging target. Recently, transition-metal-catalyzed C(sp^3)—H bond activation has attracted much scientific attention [25–31]. C(sp^3)—H bond activation under metal-free condition may be more simple and green, as no toxic and expensive transition metals are needed for this transformation. With the development of synthetic protocol, recently, several related works have indicated that the C—H bond of alkane tends to be deprotonated in the presence of peroxides, and further functionalized via oxidative radical pathway. Therefore, we want to summarize these works and envisage the next stage of functionalization of simple alkanes under metal-free conditions.

23.2 The Formation of C—C Bond

23.2.1 Cross-Dehydrogenative-Coupling (CDC) Reactions

Cross-dehydrogenative-coupling (CDC) reactions, which combine two C—H bonds to form new C—C bonds, avoid the preparation of prefunctionalized materials and make syntheses simpler and more efficient. CDC strategy, looked at as green chemistry, has made significant contributions to organic chemical synthesis.

In 2009, Li et al. demonstrated a transition-metal-free system for the direct cross-coupling reactions of nitrogen heteroaromatics and alkanes (Scheme 23.1). Under the conditions of di-*tert*-butyl peroxide (DTBP), pyridine *N*-oxide derivatives react with alkanes to give the corresponding cross-coupling products (alkylated nitrogen heterocycles) in good chemical

Alkane Functionalization, First Edition. Edited by Armando J. L. Pombeiro and M. Fátima C. Guedes da Silva.

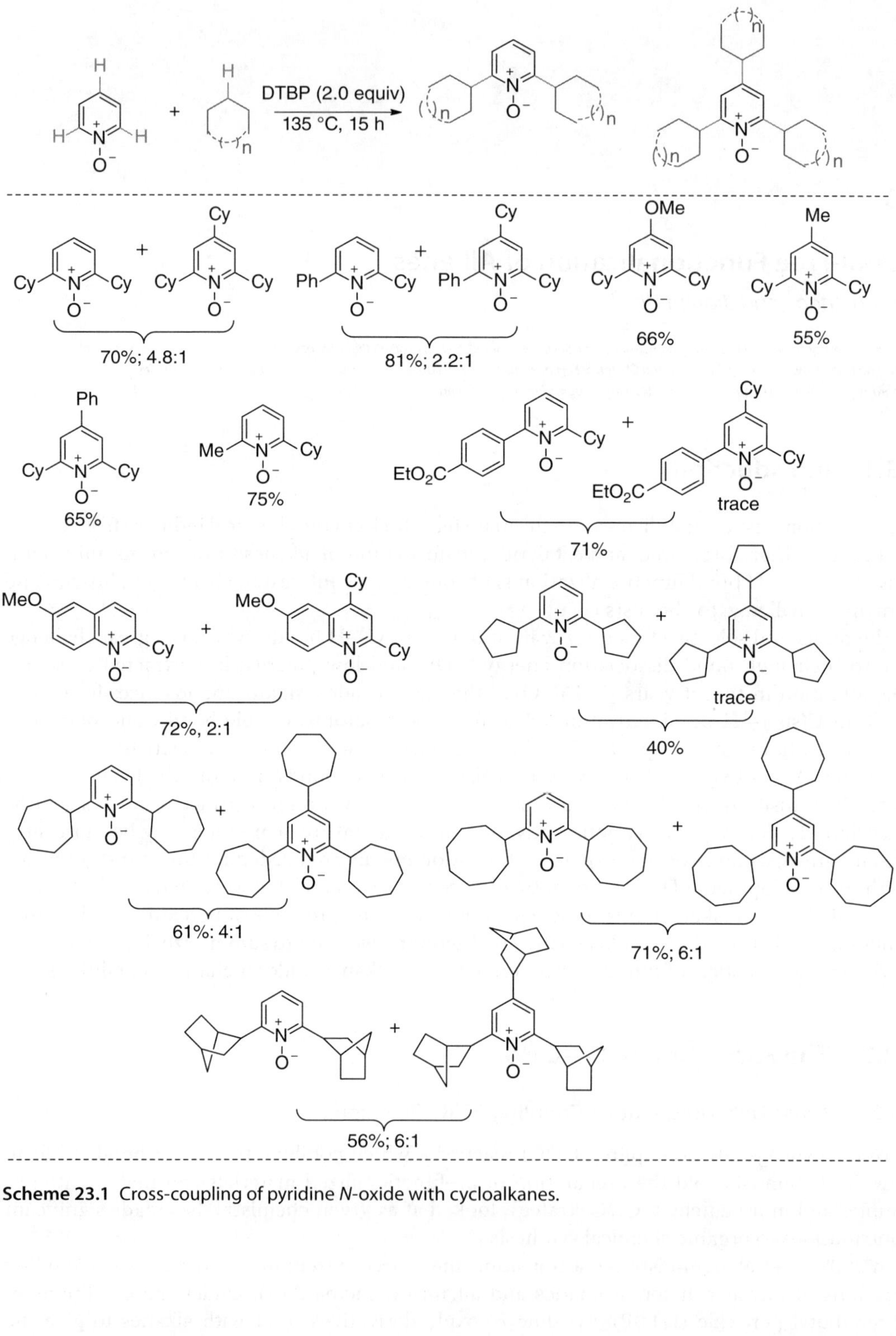

Scheme 23.1 Cross-coupling of pyridine *N*-oxide with cycloalkanes.

yields [32]. Reaction of activated substrate pyridine-*N*-oxide with cycloalkanes in the presence of 2 equiv of DTBP provided regioselective di- and trialkylated (2,6- and 2,4,6-) products. However, the use of relatively less activated pyridine substrate afforded only a trace of alkylated product even after employing excess (3 equiv.) of oxidant. They reasoned that the acidity of $C(sp^2)$—H is crucial for the success of this transformation.

In 2012, Guo et al. developed a new method of modifying purines (nucleosides), benzothiazole, and benzoxazole through $C(sp^3)$—H activation with alkanes (Scheme 23.2). The reaction exclusively occurred on C8 position to selectively give 8-cylcoakylpurines by C—C bond formation under metal-free conditions with the use of DTBP as oxidant. The C—C bond formation reaction presumably involves the formation of a cycloalkyl radical and subsequent radical addition-oxidation reactions with heteroaromatics [33].

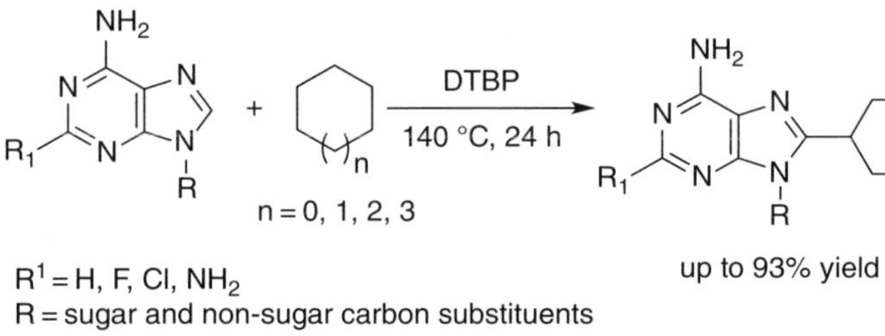

Scheme 23.2 Cross-coupling of heteroaromatics with cycloalkanes.

In 2013, Antonchick et al. developed an efficient and scalable method for the oxidative cross-coupling of heteroarenes with simple alkanes (Scheme 23.3). The desired products were smoothly formed under mild reaction conditions, within short reaction time, at ambient temperature. The formation of new $C(sp^3)$—$C(sp^2)$ bonds selectively occurred at the electron-poor position of arenes. Mechanism studies indicated that PIFA and NaN_3 played crucial roles in the generation of alkyl radical intermediates [34]. Later in 2014, the same group reported another example on the hypervalent iodine-mediated selective oxidative functionalization of aliphatic C—H bonds of alkanes with chromones and (thio)chromones. A wide range of alkanes, both cyclic and acyclic, has been found to react selectively and predictably, resulting in good yields [35].

In 2014, Cheng group developed a DTBP-promoted alkylation of α-amino carbonyl compounds by simple alkanes (Scheme 23.4). The procedure features the metal-free cleavage of inactive sp^3 C—H bonds. The reaction was applicable for the substrates, like α-amino ketones and α-amino esters, providing a facile strategy for the α-functionalization of these substrates. The radical pathway is also involved in this transformation [36].

In 2016, Kwong et al. demonstrated an effective metal-free method for a CDC reaction between indole derivatives and cycloalkanes (Scheme 23.5). With the aid of oxidant DTBP, a number of inactive cycloalkanes can directly couple with indole derivatives in acceptable yields. It is believed that a *tert*-butoxyl radical is generated by homolytic cleavage through thermal decomposition of DTBP under heating. Then, the *tert*-butoxyl radical undergoes a hydrogen abstraction of the proton from cycloalkanes, to afford an alkyl radical. The radical species further undergoes addition to indole derivative, followed by single electron transfer (SET) leading to the oxidative coupling final product [37].

In the same year, Yi et al. conducted a regiochemical investigation on the radical reaction of indoles and cycloalkanes through $C(sp^3)$—H activation under metal-free conditions in the presence of DTBP (Scheme 23.6). A wide range of indoles bearing substituents on different positions was functionalized directly with simple cycloalkanes in moderate to high regioselectivity.

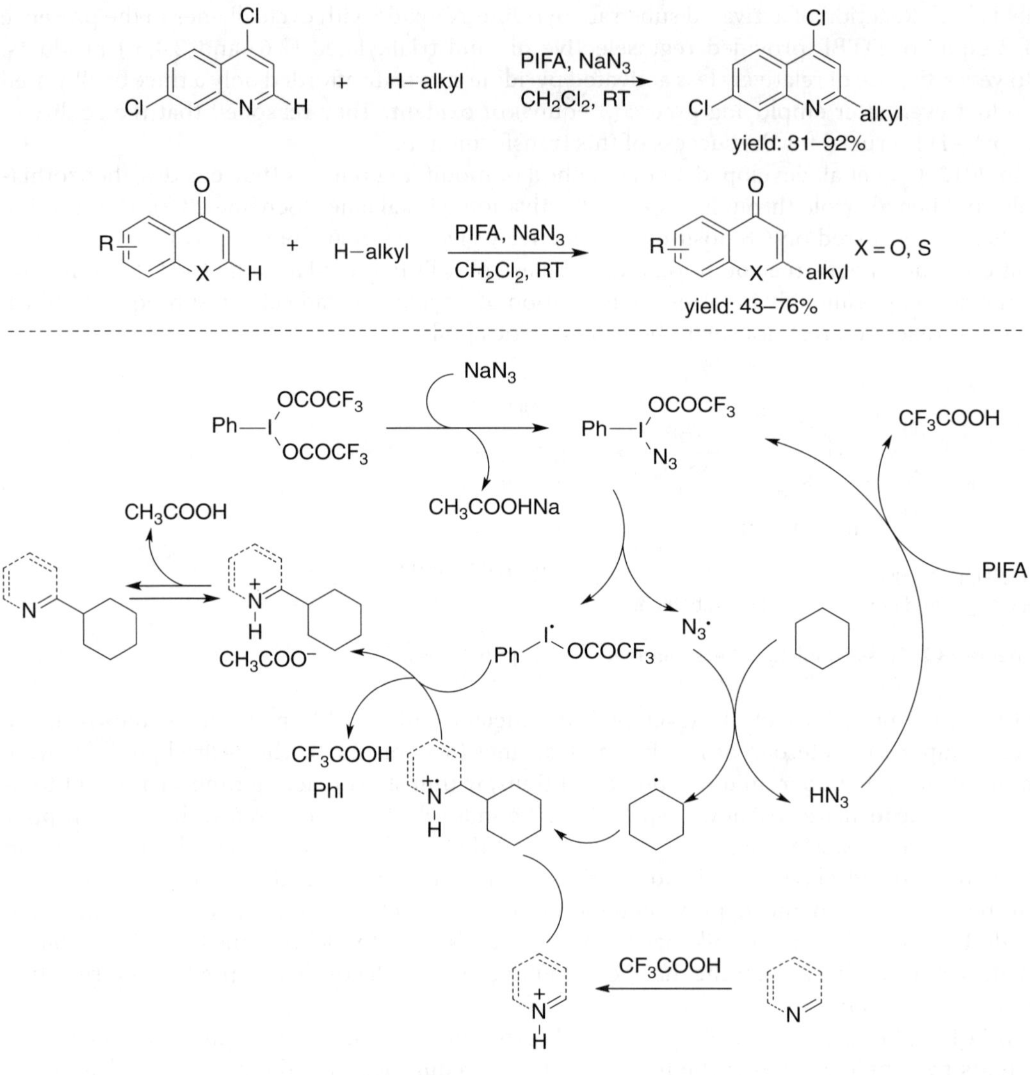

Scheme 23.3 Cross-coupling of simple alkanes with heteroarenes.

2-, 4-, and 7-position have been found to be mainly functionalized [38] under the optimized conditions.

23.2.2 Decarboxylative Cross-Coupling Reactions

Transition-metal-catalyzed decarboxylative cross-coupling reactions have recently emerged as a new and important method for organic transformations that have been found versatile applications in the construction of carbon–carbon and carbon–heteroatom bonds. The use of relatively cheap and stable carboxylic acids to replace organometallic reagents enables the decarboxylative cross-coupling reactions to proceed with excellent selectivities and functional group tolerance [39].

In 2015, Sun et al. developed a tunable method for the synthesis of alkenes and ketones via the oxidative alkylation of cinnamic acids with alkanes under transition metal-free conditions

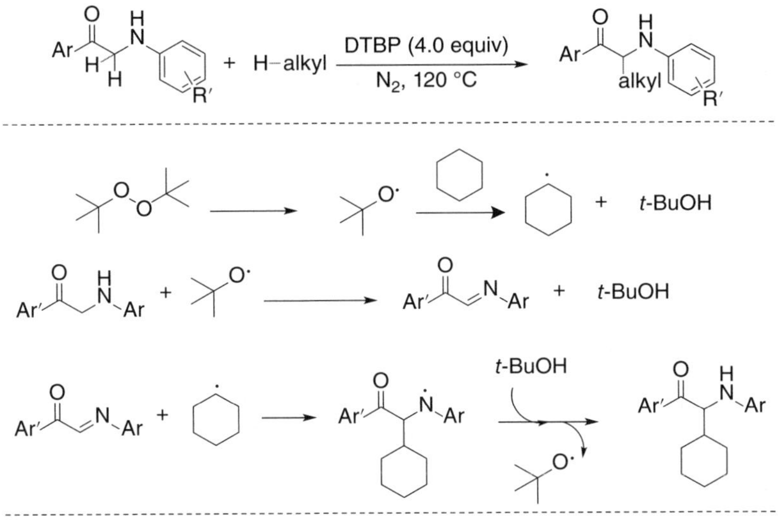

Scheme 23.4 DTBP-promoted α-alkylation of α-amino carbonyl compounds by simple alkanes.

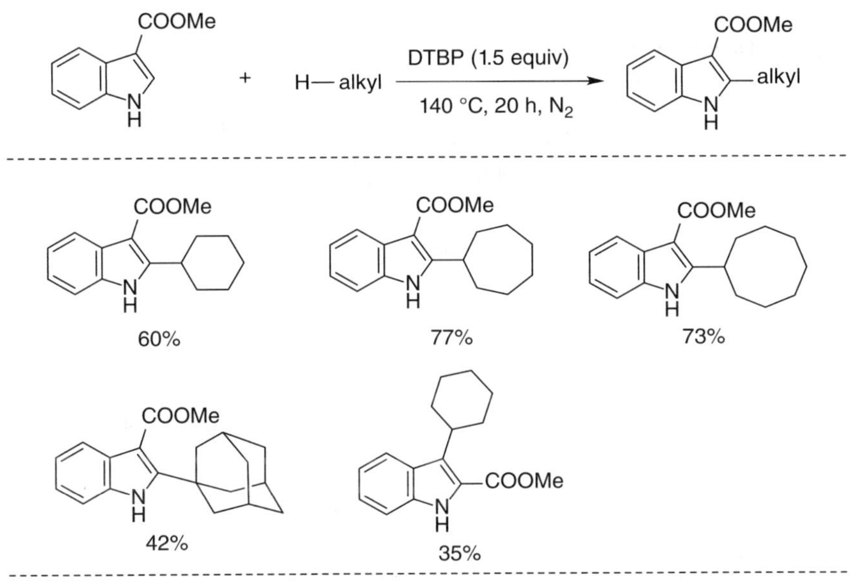

Scheme 23.5 Cross-coupling of indoles with cycloalkanes.

(Scheme 23.7). Peroxide DTBP and DTBP/*tert*-butyl hydroperoxide (TBHP) were proved to be efficient oxidants, respectively, for these transformations. Alkanes and various substituted cinnamic acids were well tolerated in the reactions and gave the corresponding products in moderate to good yields [40].

23.2.3 Conjugate Addition Reactions

In 2014, Han et al. developed an unprecedented directed oxidative sp^3 C—H functionalization and subsequent conjugate addition to chromones using DTBP as the oxidant without use of

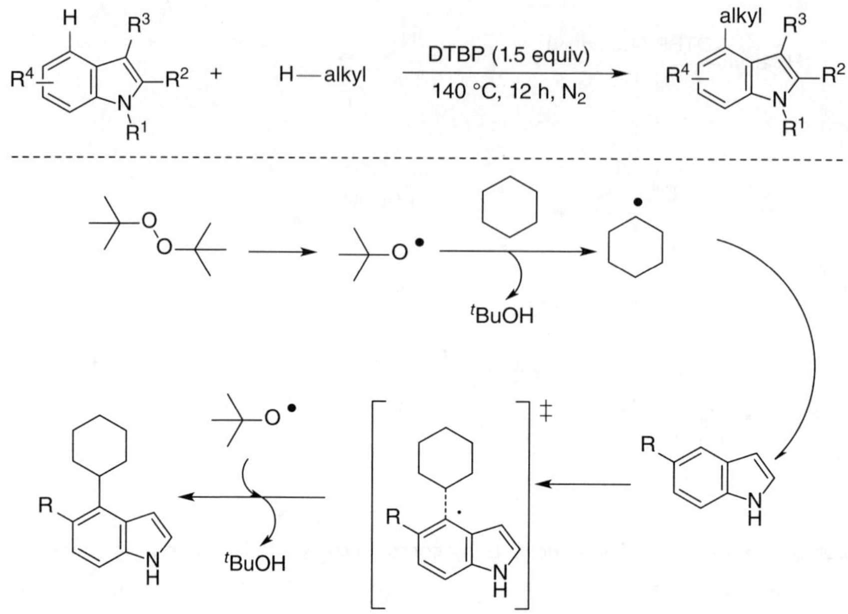

Scheme 23.6 Cross-coupling of indoles with cycloalkanes.

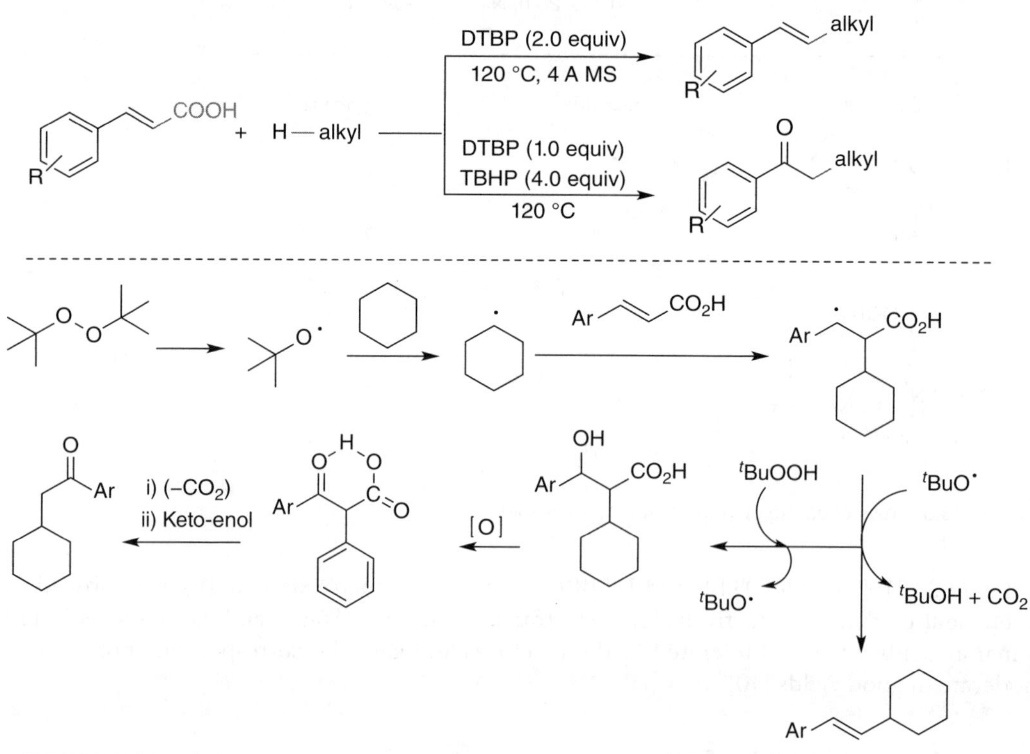

Scheme 23.7 Decarboxylative alkylation of cinnamic acids to form ketones and alkenes.

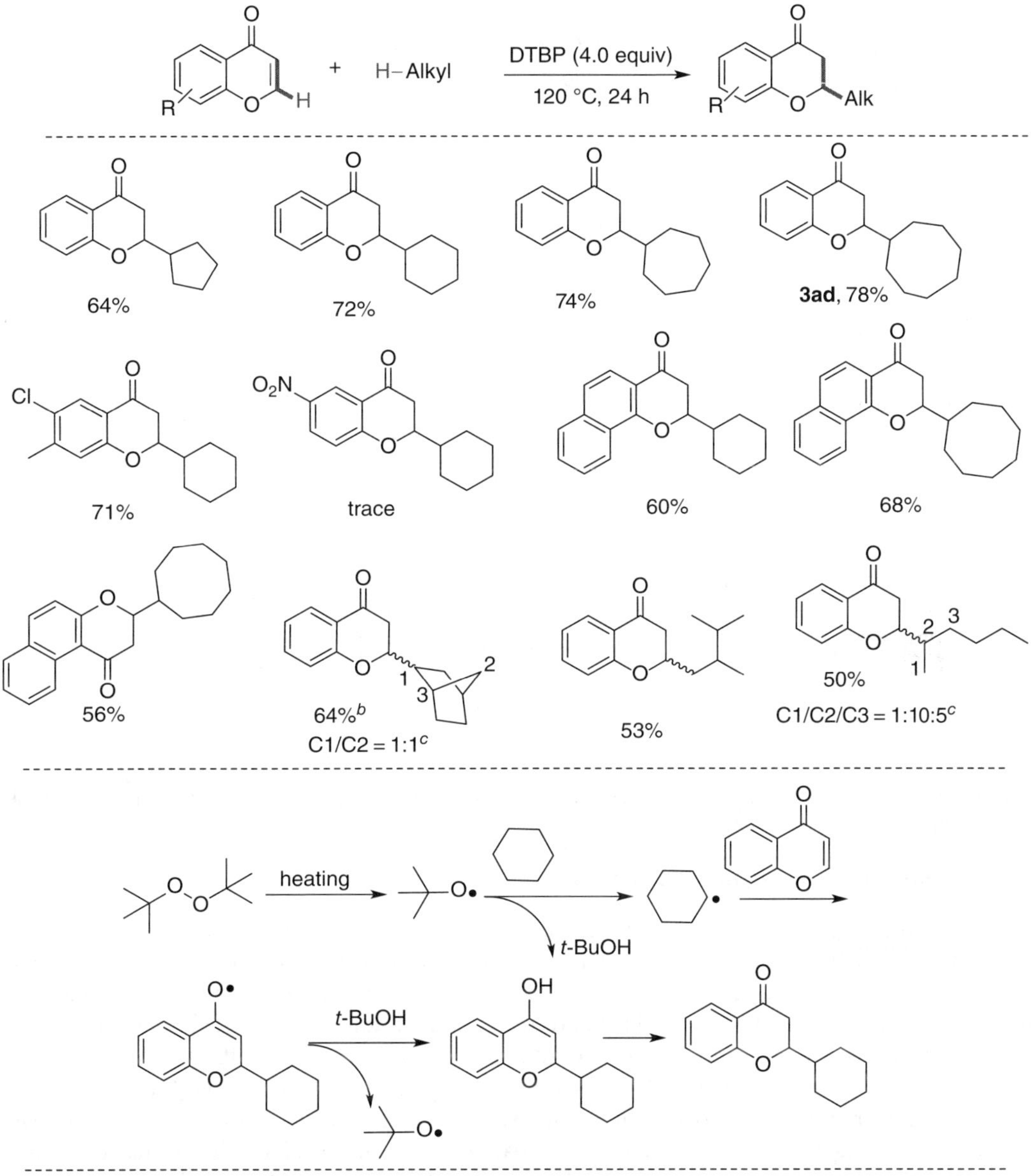

Scheme 23.8 Metal-free oxidative C(sp^3)—H bond functionalization of alkanes and conjugate addition to chromones.

any metal catalyst (Scheme 23.8). Different substituted chromones and various simple alkanes could be well tolerated, affording 2-alkylchromanones in good yields. The reaction did not work when chromones containing a nitro group was used as a substrate. Interestingly, these reactions of alkanes with high boiling point also resulted in moderate yields of the products. The author also carried out the reaction of open chain alkanes, and the corresponding products were also obtained but with poor regioselectivities [41].

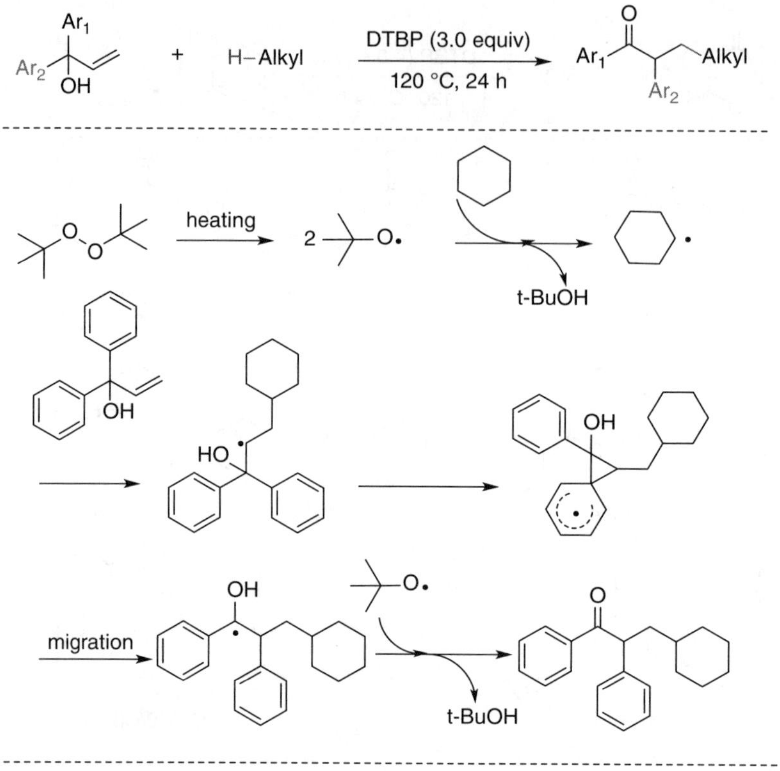

Scheme 23.9 Metal-free oxidative C(sp^3)—H bond functionalization of alkanes and alkylation-initiated radical 1,2-aryl migration in α,α-diaryl allylic alcohols.

23.2.4 Radical Rearrangement Reactions

In 2015, Han group also developed a DTBP-promoted radical alkylation reaction of α,α-diaryl allylic alcohols with simple alkanes under metal free conditions (Scheme 23.9). This cascade process involves new C(Ar)—C(sp^3) and C(sp^3)—C(sp^3) bond formation in one reaction through the sequence of C(sp^3)—H bond functionalization and 1,2-aryl migration. The reaction tolerates a wide range of simple alkane substrates to react with symmetrical and unsymmetrical α,α-diaryl allylic alcohols for direct preparation of α-aryl-β-alkylated carbonyl ketones in moderate to excellent yields. The chemoselective migration of the two different aryl groups was investigated in this reaction, and the results indicate that the more electron-deficient aryl group migrates preferentially. The authors indicated that the reaction might involve a radical ("neophyl") rearrangement process [42].

Then, the same group explored an unexpected radical cascade difunctionalization reaction of aryl alkynoates, which proceeded through cleavage of the sp^3 C—H bond of cycloalkanes, alkylation of alkynoates, 1,4-aryl migration, and decarboxylation (Scheme 23.10). This complex radical process tolerated a wide range of substrates, affording trisubstituted olefins with moderate chemical yields. As indicated in this work, substrates bearing different aromatic rings afforded the products as a 1 : 1 mixture of stereoisomers. These reactions could also afford the corresponding product bearing five- and seven-membered rings with up to 65% yield. The reactions also showed poor regioselectivity when the coupling partner cycloalkanes contain several potential reaction sites, and a mixture of regioisomers were obtained from these reactions [43].

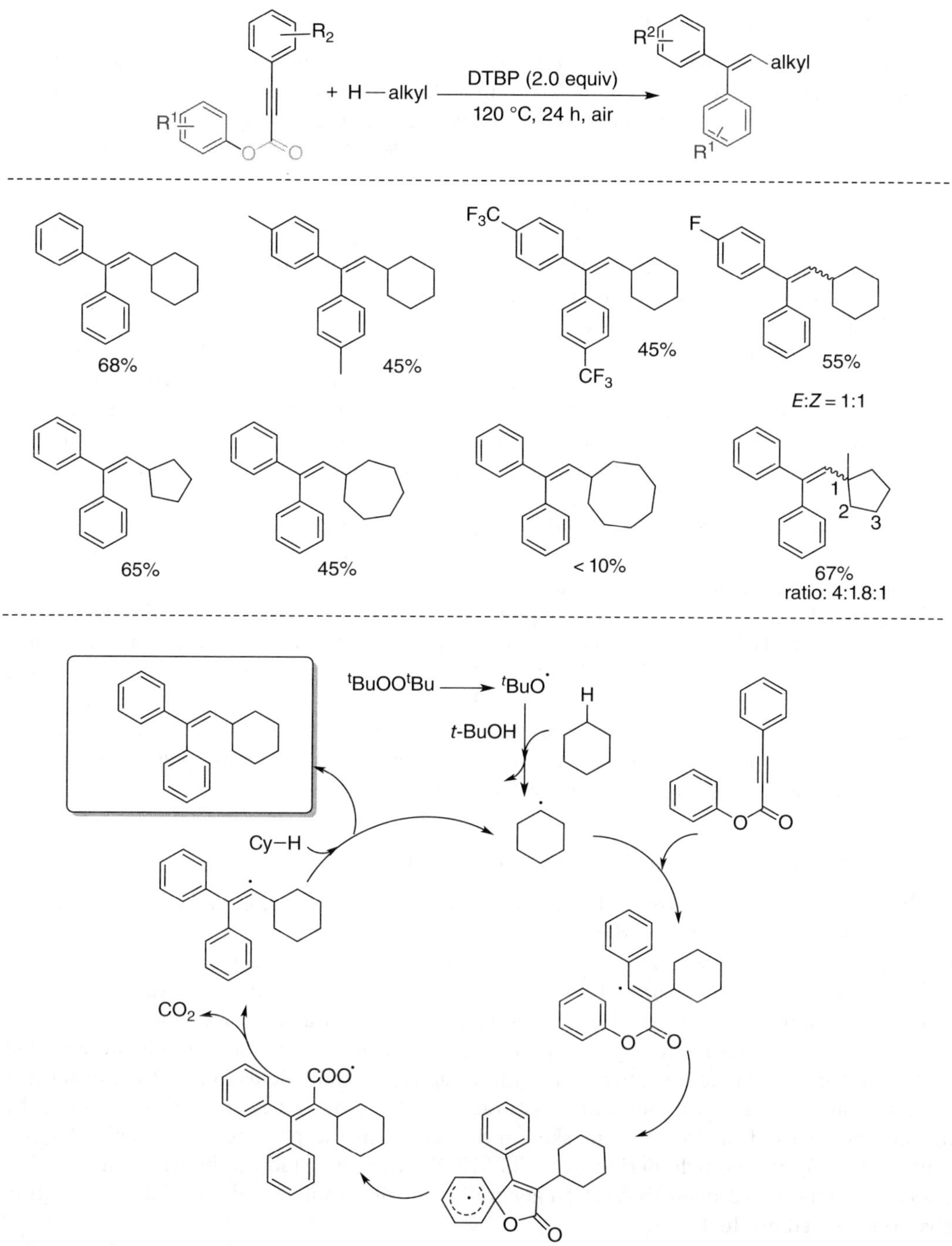

Scheme 23.10 Oxidative difunctionalization of alkynoates through alkylation and migration decarboxylative arylation.

23.2.5 Radical Cascade Reactions

Cascade reaction has been demonstrated as an efficient organic synthetic strategy, since it usually provides the desired complex product directly in one reaction, without multiple processes, no need for separation of intermediates and time-consuming.

In 2014, Cheng group developed the benzoyl peroxide (BPO)-promoted phenanthridinylation of simple alkanes with isonitrile under metal-free conditions (Scheme 23.11). This protocol shows good substrate generality, and a range of isocyanides containing electron-donating and electron-withdrawing substituents as well as various cycloalkanes could work very well affording the desired products in moderate chemical yields (30–81%). This procedure is featured by dual C—C bond formation via the sequence of the addition of an alkyl radical to isonitrile and radical aromatic cyclization [44].

In 2015, Zhu group we reported a metal-free cascade alkylation of alkenes via radical progress with *N*-phenyl-*N*-(phenyl-sulfonyl)-methacryl-amides as starting materials, in which activation of inert C(sp^3)—H bond, elimination of SO_2 and formation of C—C bond were achieved in sequence (Scheme 23.12). Different substituent groups in *N*-phenyl-*N*-(phenylsulfonyl)-methacrylamides were tolerated in this reaction, providing the alkylation products in moderate to good yields. Amides with electron-donating group on phenyl such as CH_3, OCH_3, *t*-Bu, gave desired products in high yields. However, amides with strong electron-withdrawing group, such as NO_2 on the *N*-phenyl (SO_2), could not work in this reaction and did not give any products [45].

One year later, the same group developed a tert-butyl peroxybenzoate (TBPB) promoted oxidative C(sp^3)—H bond functionalization of simple alkanes and alkylation-initiated radical dearomatizing spirocyclization of *N*-phenyl-cinnamamides, in which two C(sp^3)—C(sp^3) bonds formation were involved (Scheme 23.13). A series of simple alkanes substrates were examined in this reaction, and were found to react very well with different cinnamamides, directly generating alkyl substituted 1-azaspiro[4.5] decanes in moderate to good yields with excellent regioselectivities and diastereoselectivities [46].

23.2.6 Guided Desaturation of Unactivated Aliphatics

In 2012, Baran et al. developed directing group-assisted alkane desaturation using a portable desaturase (TzoCl) under a mild, operationally simple and metal-free conditions (Scheme 23.14). The desaturation reaction described in their work is applicable on simple substrates derived from saturated alcohols and amines, to give the corresponding olefin products without any overoxidation. Various complex natural products such as terpenes eudesmane and triterpene 18-β-glycyrrhetinic acid derivatives were successfully desaturated under the optimized reaction conditions with moderate to good yields (32–92%). The authors have demonstrated that the portable desaturase (TzoCl) plays a key role in this reaction by forming an aryl radical from the masked aryl triazene in the presence of a single-electron shuttle 2,2,6,6-tetramethylpiperidinoxyl (TEMPO). This aryl radical intermediate further undergoes remote hydrogen abstraction generating an alkyl radical and final dehydrogenation gives the desired product [47].

23.2.7 Difunctionalisation of 3-Substituted Coumarins

In 2015, Patel group developed C4 cycloalkylation-C3 peroxidation of 3-substituted coumarins in the presence of TBHP (Scheme 23.15). The metal free difunctionalization is associated

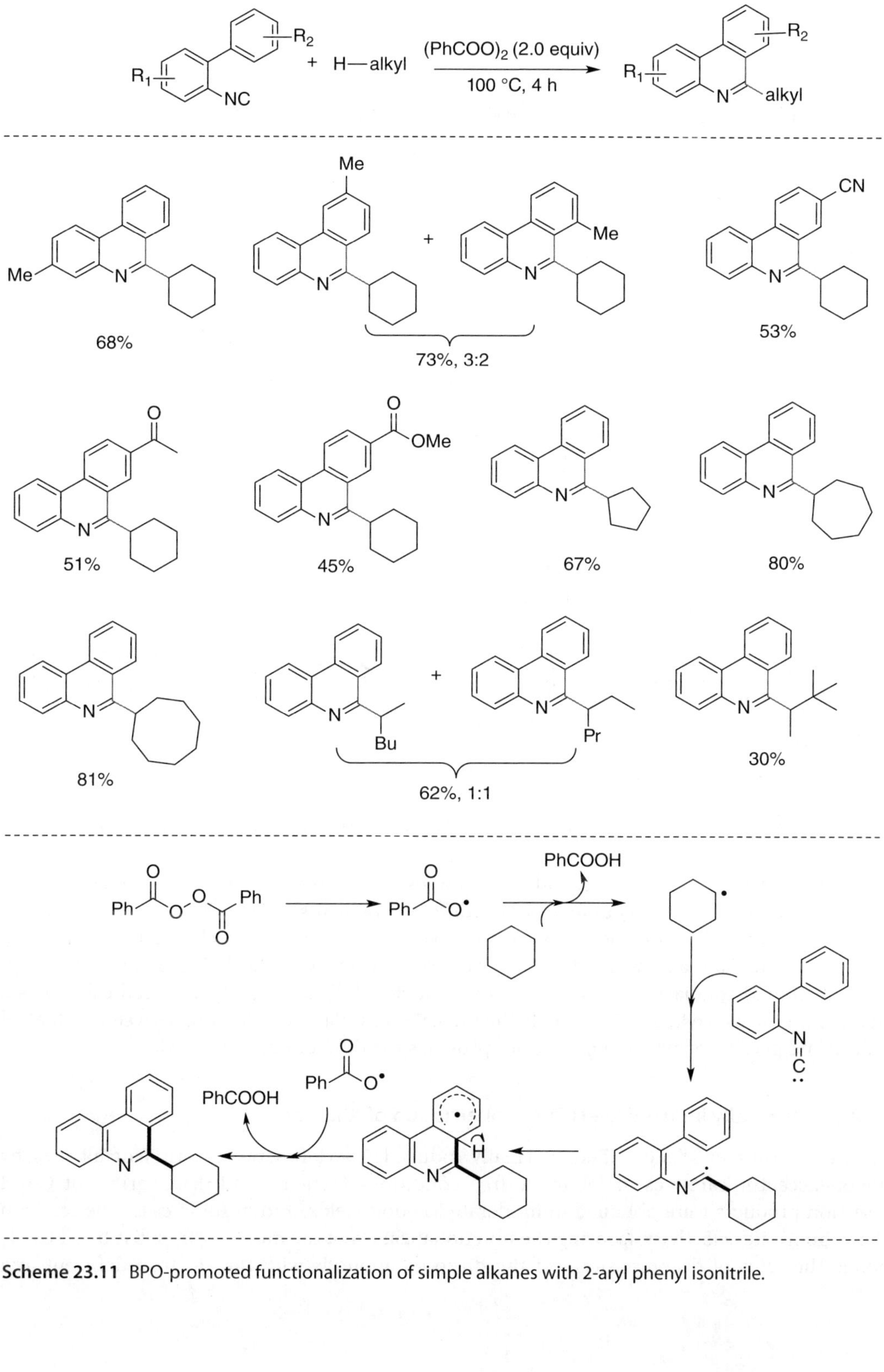

Scheme 23.11 BPO-promoted functionalization of simple alkanes with 2-aryl phenyl isonitrile.

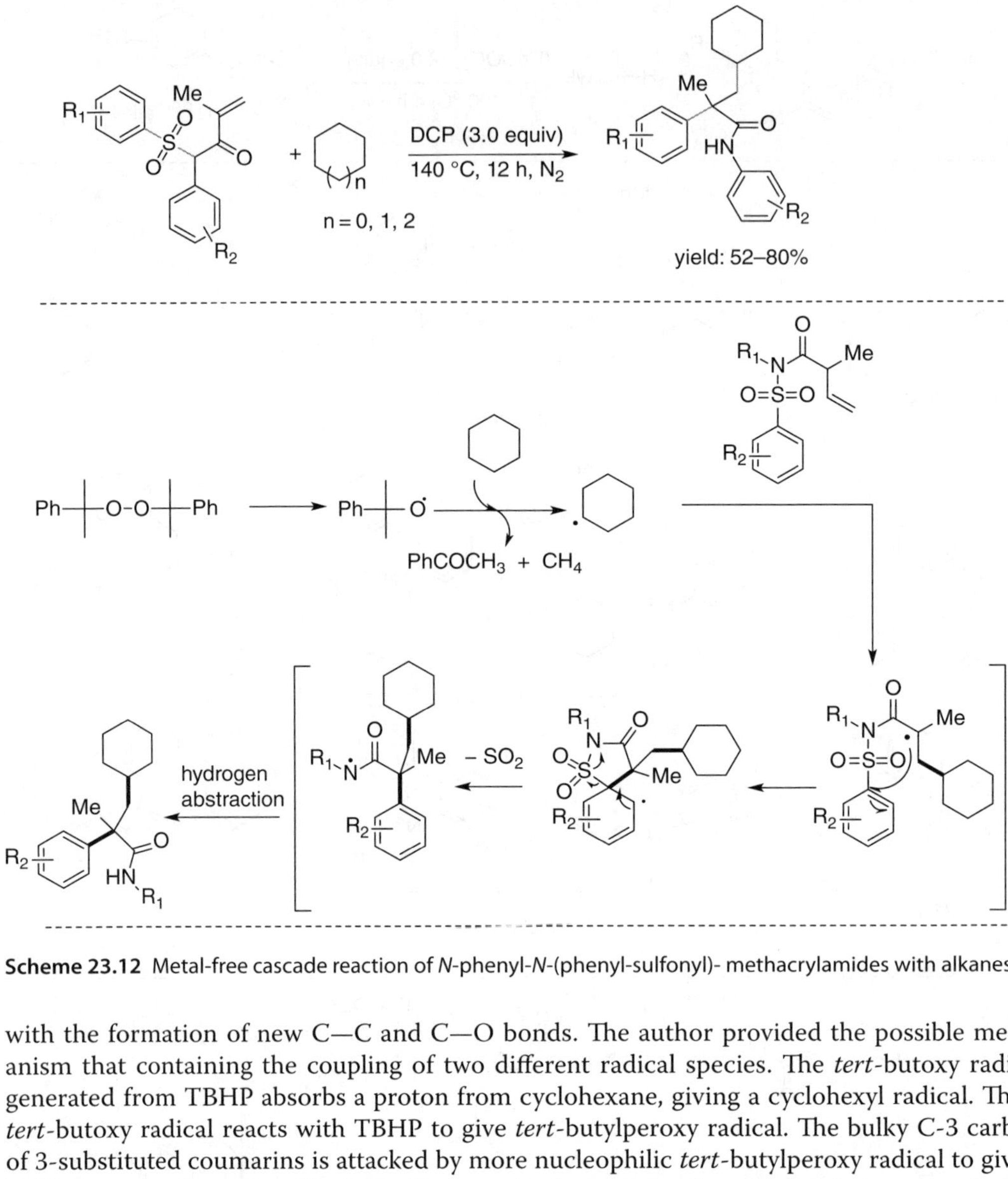

Scheme 23.12 Metal-free cascade reaction of *N*-phenyl-*N*-(phenyl-sulfonyl)- methacrylamides with alkanes.

with the formation of new C—C and C—O bonds. The author provided the possible mechanism that containing the coupling of two different radical species. The *tert*-butoxy radical generated from TBHP absorbs a proton from cyclohexane, giving a cyclohexyl radical. Then, *tert*-butoxy radical reacts with TBHP to give *tert*-butylperoxy radical. The bulky C-3 carbon of 3-substituted coumarins is attacked by more nucleophilic *tert*-butylperoxy radical to give a stable α-peroxy benzylic radical. Finally the radical cross-coupling reaction between cycloalkyl radical and peroxybenzylic radical species provides the difunctionalized product [48].

23.2.8 Thermally Induced C—H Functionalization of Alkanes

In 2017, Davies et al. described thermally induced C—H functionalization of alkanes by donor–acceptor carbenes under metal-free conditions (Scheme 23.16). A variety of C—H insertion products were obtained in moderate to good yields, and in some cases the reaction shows good site selectivity, favoring the functionalization of the more highly substituted C—H bond. The rates of decomposition of the diazo compounds with the electron-deficient aryl

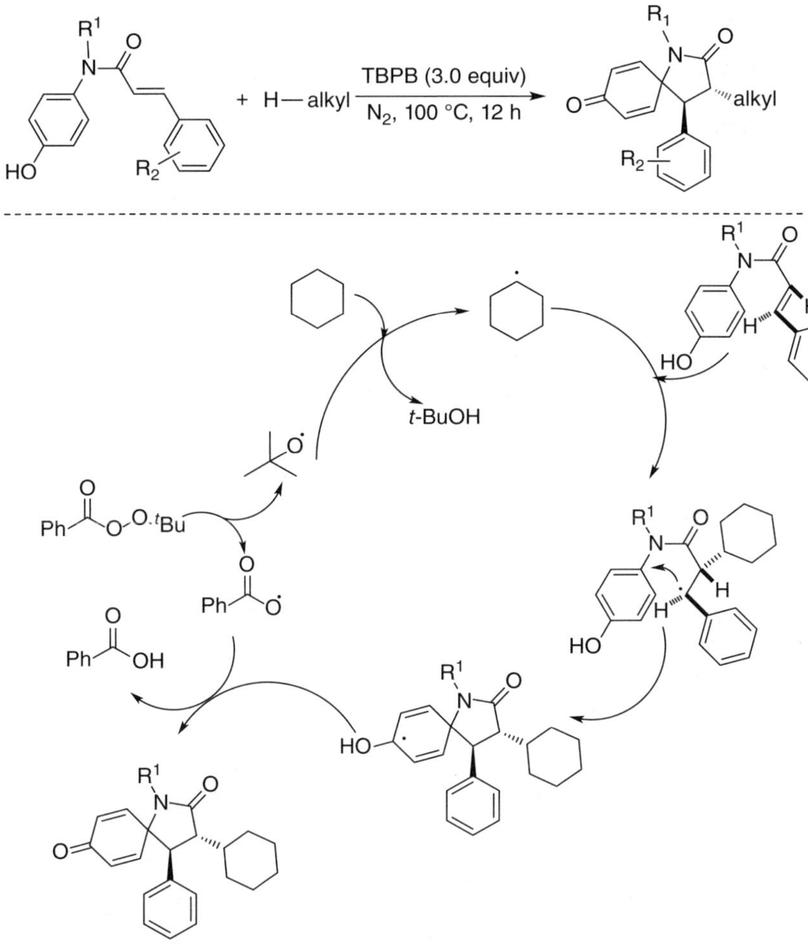

Scheme 23.13 Metal-free cyclization and dearomatization of *N*-(4-hydroxyphenyl)-*N*- alkylcinnamamides with alkanes.

groups were low, and the reactions of these substrates needed extended times to achieve full conversion (up to 96 hours) [49].

23.3 The Formation of C—N Bond

In 2009 Li et al. reported a direct amination of simple cycloalkanes with nitroarenes mediated by peroxides under metal-free conditions (Scheme 23.17). Various secondary arylamines were obtained efficiently from this system using cycloalkanes as starting materials. The reaction features several advantages, such as a wide range of substrate scope, aqueous conditions, and no need of transition-metal as catalyst. It is should be mentioned that peroxides act as nitro group reducing reagents in this novel amination reaction, although they are commonly used as radical initiators and oxidants [50].

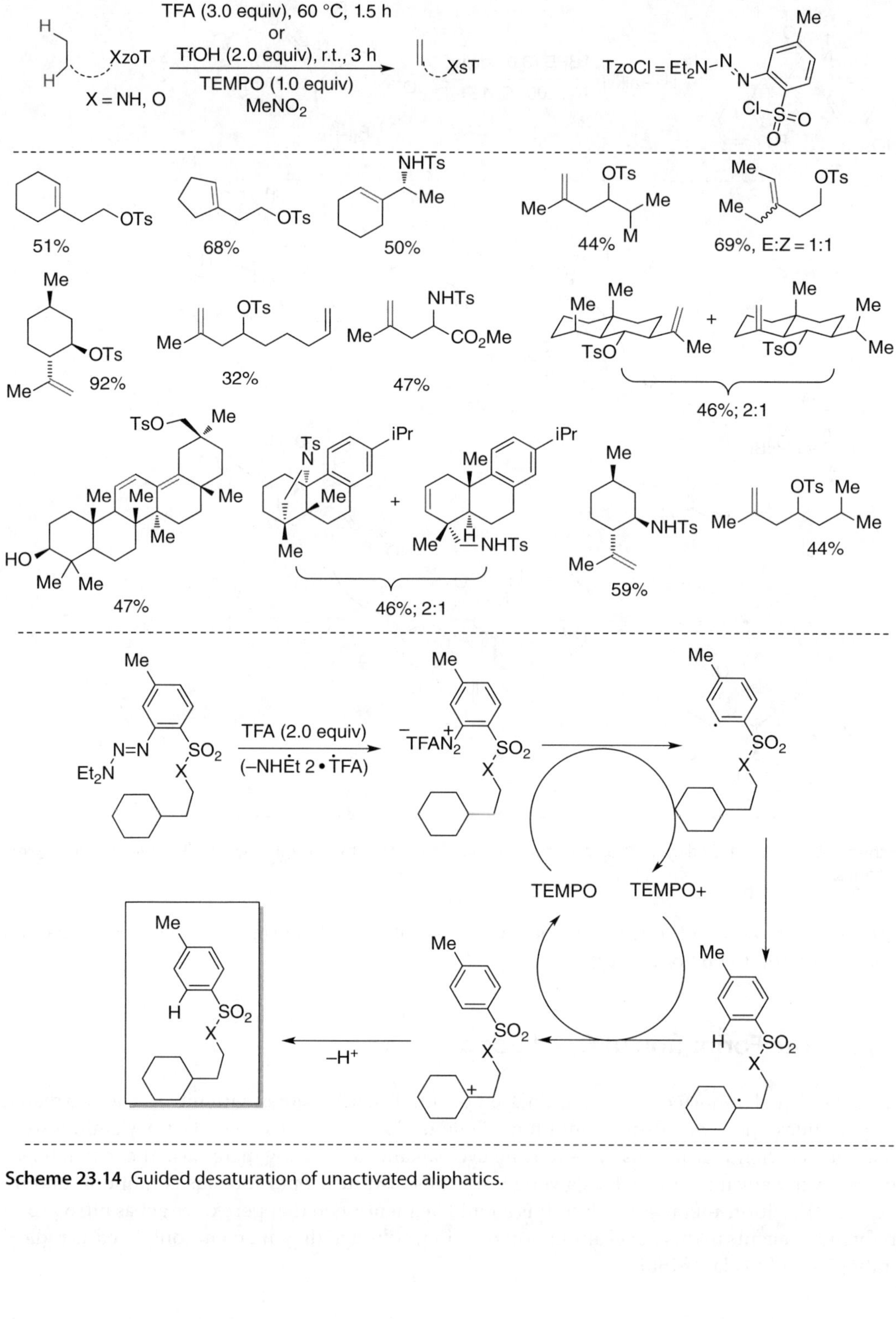

Scheme 23.14 Guided desaturation of unactivated aliphatics.

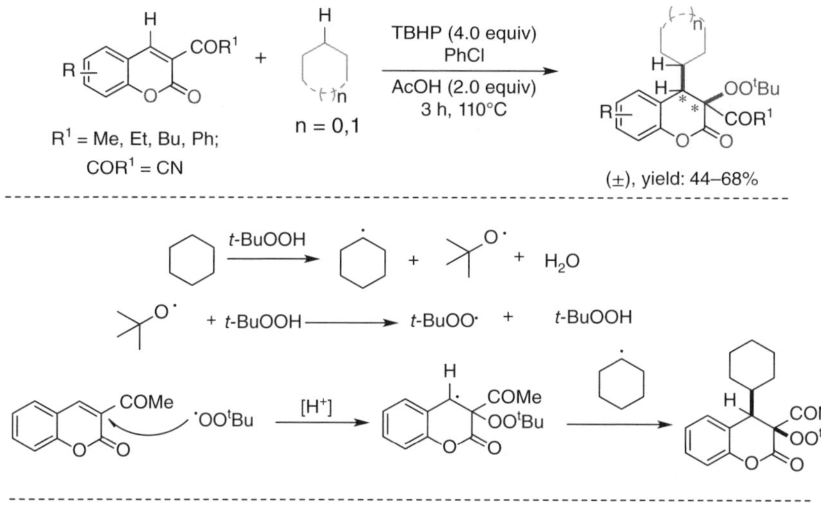

Scheme 23.15 C-4 cycloalkylation-C-3 peroxidation of 3-substitued coumarins in the presence of TBHP.

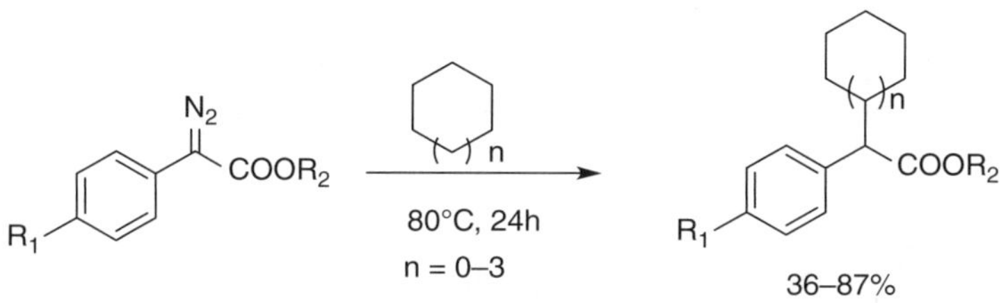

Scheme 23.16 Metal-free C—H functionalization of alkanes by aryldiazoacetates.

23.4 The Formation of C—S Bond and C—Se Bond

The formation of C—S, C—Se bonds represents an active research area in general organic chemistry, material science, and biological and pharmaceutical chemistry (Scheme 23.18). In 2014, Han et al. developed a concise thiolation of C(sp^3)—H bond of cycloalkanes with diaryl disulfides in the presence of oxidant of DTBP. This reaction was conducted under simple conditions without using any metal catalyst, giving the corresponding alkylaryl sulfides good to excellent chemical yields. The process has a broad scope and high compatibility with functional groups, and the substituents on different positions of aromatic rings were well tolerated. The limitation of this method is that 1,2-dibenzyldisulfane was not a suitable substrate for this thiolation system. This reaction provides a useful approach to cycloalkyl aryl sulfides from unactivated cycloalkanes [51].

At almost the same time, Sun et al. reported C—S and C—Se bond formation based on the direct oxidative cross-couplings of simple alkanes with disulfides or diselenides in the absence of a transition metal (Scheme 23.19). They also used DTBP as the oxidant, providing the corresponding products with good to excellent yields [52].

23.5 Final Comments

In summary, we have presented a review on the metal-free functionalization of simple alkanes. The direct and selective functionalization of the sp^3 C—H bonds is a great challenge due to

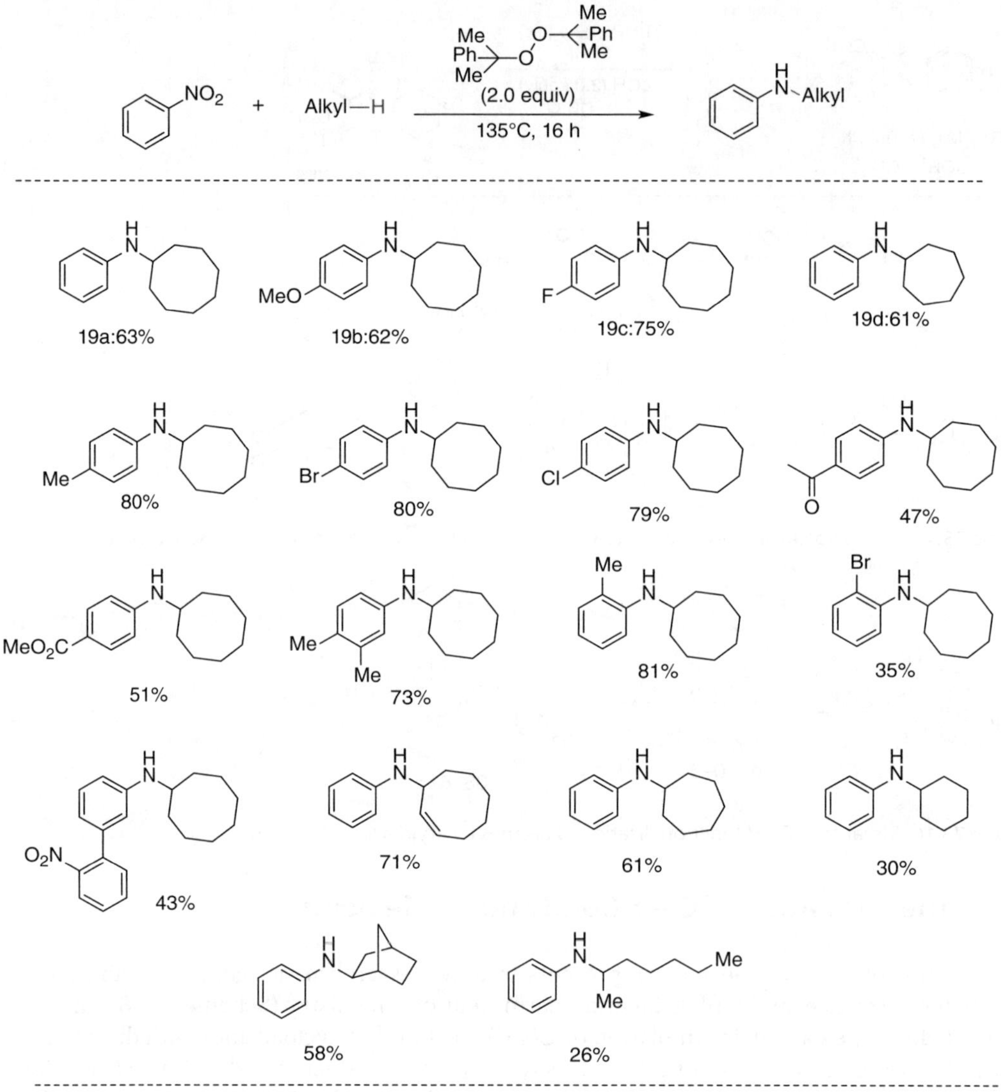

Scheme 23.17 An unusual peroxide-mediated amination of cycloalkanes with nitroarenes.

their high BDE and low polarity. However, the sp^3 C—H bond of alkanes is prone to be functionalized via a radical pathway under certain reaction conditions. In view of these points, simple alkanes can be elaborated on a broad range of substrates, leading to a series of alkylated products, which enormously promote the development of the metal-free functionalization of alkanes.

Although there have been major advancements in this area, significant limitations still remain in terms of practicability such as use of strong oxidants, high reaction temperature, and poor regioselectivity. We should pay more attention to the site-selective functionalization of long chain alkanes, fused polycyclic alkanes in the near future. Despite these challenges, further advances in this area that empower synthetic chemists with efficient tools are highly anticipated.

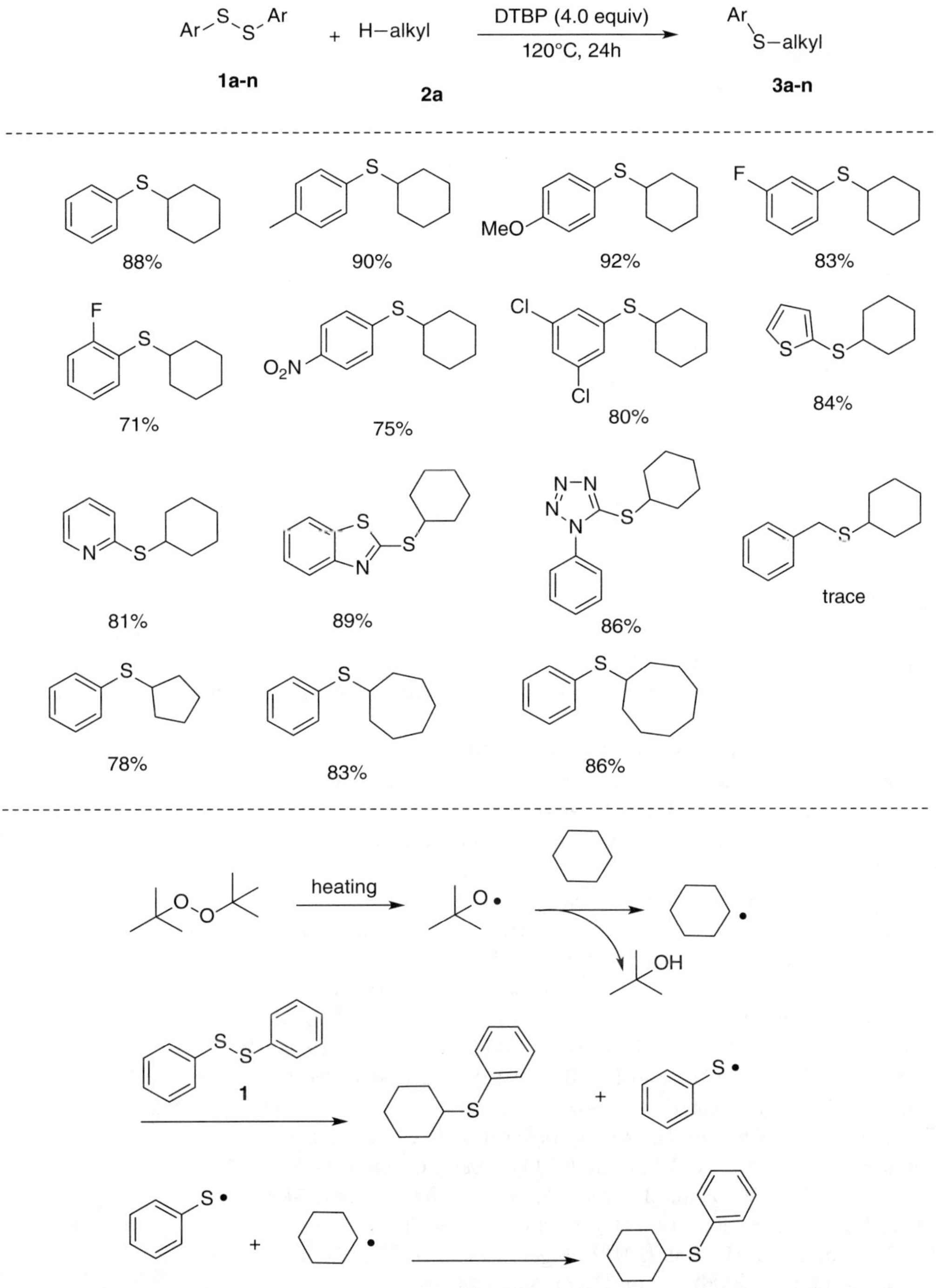

Scheme 23.18 Metal-free preparation of cycloalkyl aryl sulfides via DTBP-promoted oxidative C(sp^3)—H bond thiolation of cycloalkanes.

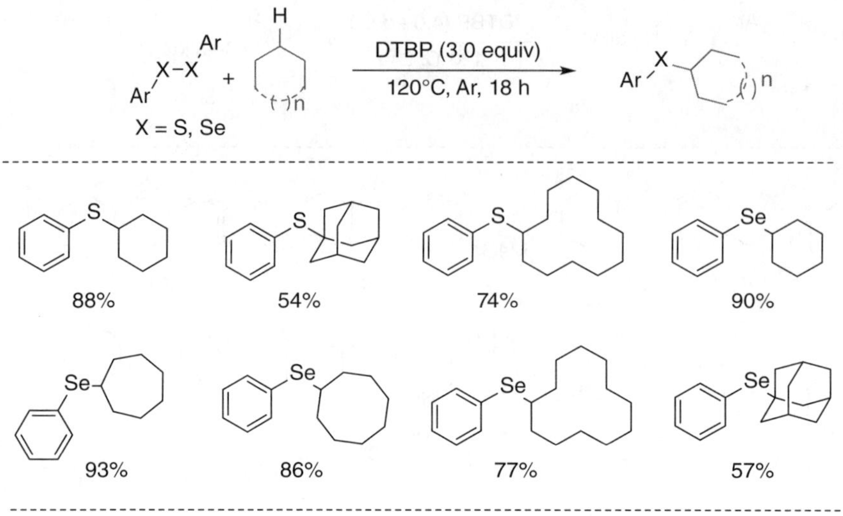

Scheme 23.19 Syntheses of sulfides and selenides through direct oxidative functionalization of C(sp^3)—H bond.

References

1 Shilov, E. and Shul'pin, G.B. (1997). *Chem. Rev.* 97: 2879.
2 Murai, S. (ed.) (1999). *Activation of Unreactive Bonds and Organic Synthesis*. New York: Springer.
3 Dyker, G. (1999). *Angew. Chem. Int. Ed.* 38: 1698.
4 Fujiwara, Y., Takaki, K., and Taniguchi, Y. (1996). *Synlett* 591.
5 Gupta, M., Hagen, C., Kaska, W.C. et al. (1997). *J. Am. Chem. Soc.* 119: 840.
6 Jia, C., Kitamura, T., and Fujiwara, Y. (2001). *Acc. Chem. Res.* 34: 633.
7 Godula, K. and Sames, D. (2006). *Science* 312: 67.
8 Bergman, R.G. (2007). *Nature* 446: 391.
9 Davies, H.M.L. and Morton, D. (2011). *Chem. Soc. Rev.* 40: 1857.
10 Engle, K.M., Mei, T.S., Wasa, M. et al. (2011). *Acc. Chem. Res.* 45: 788.
11 Liu, C., Zhang, H., Shi, W. et al. (2011). *Chem. Rev.* 111: 1780.
12 Newhouse, T. and Baran, P.S. (2011). *Angew. Chem. Int. Ed.* 50: 3362.
13 Zhang, S.Y., Zhang, F.-M., and Tu, Y.Q. (2011). *Chem. Soc. Rev.* 40: 1937.
14 Roizen, J.L., Harvey, M.E., and Du, B.J. (2012). *Acc. Chem. Res.* 45: 911.
15 White, M.C. (2012). *Science* 335: 807.
16 Rouquet, G. and Chatani, N. (2013). *Angew. Chem. Int. Ed.* 52: 11726.
17 Tang, R.Y., Xie, Y.X., Xie, Y.L. et al. (2011). *Chem. Commun.* 47: 12867.
18 Zhao, J.C., Zhou, W., Han, J.L. et al. (2013). *Tetrahedron Lett.* 54: 6507.
19 Zhao, J.C., Fang, H. et al. (2014). *J. Org. Chem.* 79: 3847.
20 Liu, D., Liu, C., Li, H. et al. (2013). *Angew. Chem. Int. Ed.* 52: 4453.
21 Li, Z. and Li, C.J. (2006). *J. Am. Chem. Soc.* 128: 56.
22 Li, Z., Cao, L., and Li, C.-J. (2007). *Angew. Chem. Int. Ed.* 46: 6505.
23 Zhou, S.-L., Guo, L.-N., Wang, S. et al. (2014). *Chem. Commun.* 50: 3589.
24 Li, Z. and Li, C.J. (2005). *J. Am. Chem. Soc.* 127: 3672.
25 Deng, G., Zhao, L., and Li, C.J. (2008). *Angew. Chem. Int. Ed.* 47: 6278.

26 Deng, G. and Li, C.J. (2009). *Org. Lett.* 11: 1171.
27 Zhu, Y. and Wei, Y. (2014). *Chem. Sci.* 5: 2379.
28 Cui, Z., Shang, X., Shao, X.F. et al. (2012). *Chem. Sci.* 3: 2853.
29 Zhao, J., Fang, H., Han, J. et al. (2014). *Org. Lett.* 16: 2530.
30 Banerjee, A., Sarkar, S., and Patel, B.K. (2017). *Org. Biomol. Chem.* 15: 505.
31 Teng, F. and Cheng, J. (2017). *Chin. J. Chem.* 35: 289.
32 Deng, G.J., Ueda, K., Yanagisawa, S. et al. (2009). *Chem. Eur. J.* 15: 333.
33 Xia, R., Niu, H.-Y., Qu, G.-R. et al. (2012). *Org. Lett.* 14: 5546.
34 Antonchick, A.P. and Burgmann, L. (2013). *Angew. Chem. Int. Ed.* 52: 3267.
35 Narayan, R. and Antonchick, A.P. (2014). *Chem. Eur. J.* 20: 4568.
36 Peng, H., Yu, J.-T., Jiang, Y. et al. (2014). *J. Org. Chem.* 79: 9847.
37 Yang, Q., Choy, P.Y., Wu, Y. et al. (2016). *Org. Biomol. Chem.* 14: 2608.
38 Xiu, J. and Yi, W. (2016). *Catal. Sci. Technol.* 6: 998.
39 Shang, R. and Liu, L. (2011). *Sci. China Chem.* 54: 1670.
40 Ji, J., Liu, P., and Sun, P. (2015). *Chem. Commun.* 51: 7546.
41 Zhao, J., Fang, H., Qian, P. et al. (2014). *Org. Lett.* 16: 5342.
42 Zhao, J., Fang, H., Song, R. et al. (2015). *Chem. Commun.* 51: 599.
43 Ni, S., Zhang, Y., Xie, C. et al. (2015). *Org. Lett.* 17: 5524.
44 Sha, W., Yu, J.-T., Jiang, Y. et al. (2014). *Chem. Commun.* 50: 9179.
45 Zhang, H., Pan, C., Jin, N. et al. (2015). *Chem. Commun.* 51: 1320.
46 Zhang, H., Gu, Z., Xu, P. et al. (2016). *Chem. Commun.* 52: 477.
47 Voica, A.-F., Mendoza, A., Gutekunst, W.R. et al. (2012). *Nat. Chem.* 4: 629.
48 Banerjee, A., Santra, S.K., Khatun, N. et al. (2015). *Chem. Commun.* 51: 15422.
49 Tortoreto, C., Rackl, D., and Davies, H.M.L. (2017). *Org. Lett.* 19: 770.
50 Deng, G., Chen, W., and Li, C.-J. (2009). *Adv. Synth. Catal.* 351: 353.
51 Zhao, J., Fang, H., Han, J. et al. (2014). *Adv. Synth. Catal.* 356: 2719.
52 Du, B., Jin, B., and Sun, P. (2014). *Org. Lett.* 16: 3032.

24

Alkane Functionalization under Unconventional Conditions: in Ionic Liquid, in Supercritical CO_2, and Microwave Assisted

Ana P. C. Ribeiro[1], *Elisabete C. B. A. Alegria*[1,2], *António Palavra*[1], *and Armando J. L. Pombeiro*[1]

[1] *Universidade de Lisboa, Centro de Química Estrutural, Instituto Superior Técnico, Lisboa, Portugal*
[2] *Instituto Superior de Engenharia de Lisboa, Instituto Politécnico de Lisboa, Chemical Engineering Department, Lisboa, Portugal*

24.1 Introduction

In recent years, the need for designing clean and green chemical processes has attracted much attention in a diversity of fields, including the selective oxidation of chemically inert alkanes to value-added products such as alcohols and carboxylic acids [1–5]. This constitutes an important and challenging field in industry since many of the current industrial processes use toxic oxidants and organic solvents, eventually producing wastes.

Some recent investigations have focused on the application of unconventional conditions toward that aim. They include the use of green solvents, namely ionic liquids (ILs) [6] and supercritical fluids (SCFs) [7], and new procedures involving microwave (MW) irradiation [8–10], among others (Figure 24.1), as illustrated below.

24.2 Alkane Oxidations in Ionic Liquids

24.2.1 Properties of Ionic Liquids

An IL is a salt with an organic cation and a suitable anion combination in the liquid state. The most common salts are those with *N*-alkylpyridinium, alkylphosphonium, alkylammonium, and *N*,*N*′-dialkylimidazolium cations with various organic or inorganic anions. Structures and abbreviations of some common IL cations and anions are illustrated in Figure 24.2.

ILs are quite unique due to their special properties, such as the large liquid state range, relatively high thermal stability, solvation behavior, high ionic conductivity, etc. [11–20] They have been widely studied in catalysis [21–25], chemical industry [26], electrochemistry [27, 28], biomass conversion [29–31], extraction [32, 33], and many other areas [34].

It is well established that besides electrostatic forces, hydrogen bonding is also very important in ILs, being commonly different from the conventional hydrogen bonds [35, 36]. Hydrogen bonding in ILs can be of the "double ionic" type between charged centers. A variety of distinct hydrogen bonds can be formed in view of the different types and numbers of hydrogen bond donor and acceptor sites that can be present [37–41]. Pointing out 1,3-dialkylimidazolium ILs as an example, they can be regarded as hydrogen-bonded polymeric supramolecules (highly ordered hydrogen bonded materials) with one cation surrounded by at least three anions and one anion surrounded by at least three cations (Scheme 24.1) [41]. The number of anions that

Alkane Functionalization, First Edition. Edited by Armando J. L. Pombeiro and M. Fátima C. Guedes da Silva.

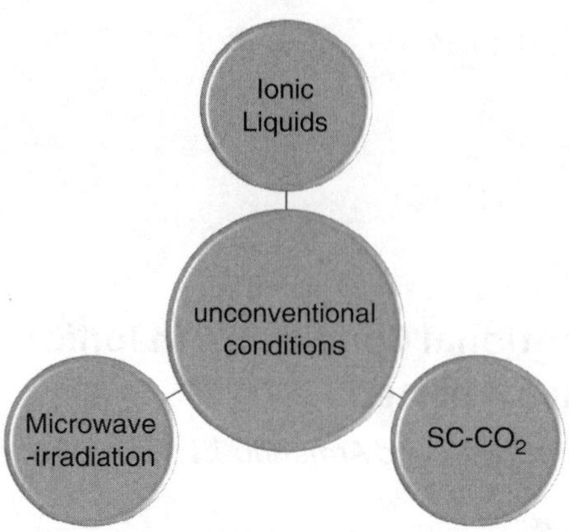

Figure 24.1 Unconventional conditions in alkane functionalization address herein.

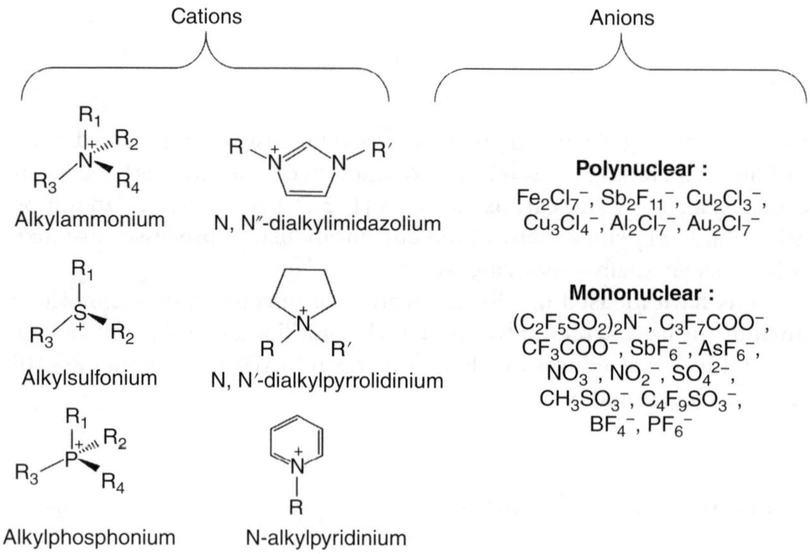

Figure 24.2 Common cations and anions of ILs.

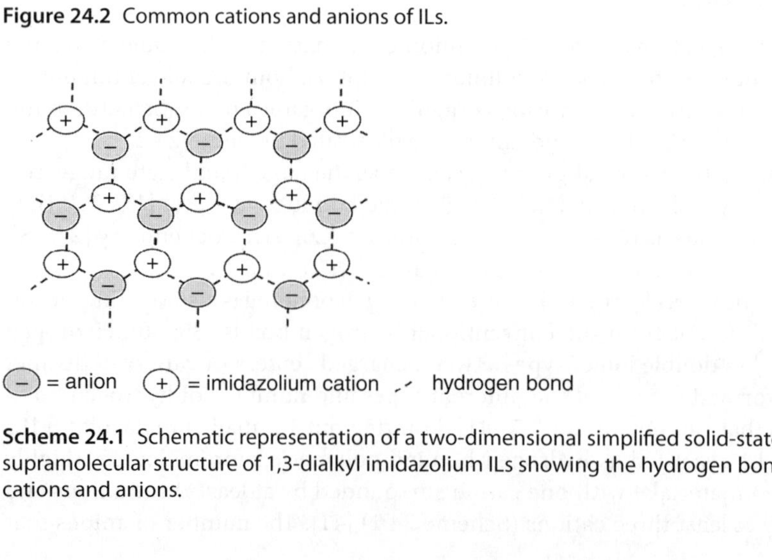

Scheme 24.1 Schematic representation of a two-dimensional simplified solid-state model of the polymeric supramolecular structure of 1,3-dialkyl imidazolium ILs showing the hydrogen bonds between imidazolium cations and anions.

surround the cation (and vice-versa) depends on the anion size and the *N*-alkyl group in the cation. A strong hydrogen bonding in the imidazolium ring is originated normally from the most acidic hydrogen (H2) of the imidazolium cation, followed by the other two hydrogens (H4 and H5) and/or the hydrogens of the *N*-alkyl groups [42]. Replacing the H2 hydrogen by a methyl group in a imidazolium IL weakens the hydrogen bonding as proved by spectroscopic evidence [43].

The three-dimensional arrangement of the imidazolium ILs is generally formed through chains of the imidazolium rings. In some cases, there are typical π-π stacking interactions among the imidazolium rings, and for 1-alkyl-3-methylimidazolium salts a relatively weak C-H...π interaction via the methyl group and the imidazolium ring-π system can also be found. This molecular arrangement can generate channels in which the anions are accommodated in chains. The structural pattern depends on the anion geometry, and the internal arrangements along the imidazolium columns vary with the type of the *N*-alkyl substituents [43]. There are also reports on the opposite view, i.e. that the hydrogen bonding in an IL (e.g. 1-butyl-3-methylimidazolium hexafluorophosphate, [bmim][PF_6]) is not so strong as expected and far from dominating the structure and dynamics [44]. When other molecules are introduced into a pure IL, the hydrogen-bonding network can be disrupted and ion pairs formed. In some cases, nanostructures with polar and nonpolar regions are generated [45–49], and even larger ionic and neutral aggregates can be formed when dissolved in solvents with relatively high dielectric constants, such as acetonitrile and dimethyl sulfoxide (Scheme 24.2) [50].

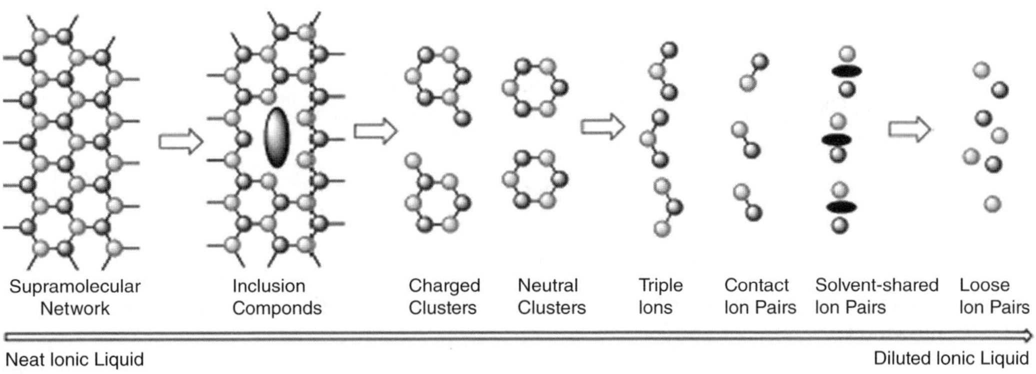

Scheme 24.2 Simplified 2D representation of the structure of a neat ionic liquid to its infinite dilution in the presence of other organic compounds. Most of these structures have been experimentally already observed (red spheres = anions, blue spheres = cations, black spots = solvent molecules; the lines represent the hydrogen bonds and/or other weaker interactions). Source: Reproduced with permission from ref. [50a]. (*See insert for color representation*).

The stability and reactivity of the aggregates are dependent on the nature of the anion and imidazolium substituents and the aggregates are more abundant in ILs containing strong coordinating anions, in particular those that can form charge transfer complexes [51–53].

Compared with imidazolium-based ILs, *N*-alkylpyridinium-, *N*-alkylpiperidinium-, alkylammonium-, and alkylphosphonium-based ILs show weaker hydrogen bonding interactions between the hydrogen atoms in the cation and anion. However, when the alkyls are replaced by hydrogen atoms, the ILs become protic and show much stronger hydrogen bonds. The key property that distinguishes protic ILs from other ILs concerns proton transfer, with the associated presence of proton-donor and -acceptor sites and the formation of a hydrogen bonding network. Experiments and calculations have proven that there are stronger hydrogen bonds in protic ILs than in aprotic ones [54, 55]. This difference can lead to an improvement in the catalytic activity. Some cases concerning the oxidation of alkanes will be discussed below.

24.2.2 Ionic Liquids in Catalysis

Applications of ILs to replace conventional solvents or to act as catalysts have been extensively reported [56–58]. A common disadvantage of the homogeneous catalytic systems concerns the difficulty of catalyst separation and recycling, and ILs have been used as a possible solution to overcome this problem, in a few cases even supported on solid surfaces [59]. ILs have become a good alternative as recyclable solvents and unconventional media in catalytic reactions [25, 60–69]. They can even be applied as *biphasic catalysts* or *immobilized catalysts* in heterogeneous systems without requiring an additional complex as a catalyst.

In 2005, Bianchini et al. [70] reported a C—H insertion reaction with H_2O_2 catalyzed by homogeneous methyltrioxorhenium [$MeReO_3$] (MTO), heterogeneous poly(4-vinylpyridine)/methyltrioxo rhenium (PVP/MTO), and microencapsulated polystyrene/methyltrioxorhenium (PS/MTO) systems in ILs. In some cases, a higher activity was observed in ILs than in molecular solvents under the same reaction conditions. Since then, a considerable work on the oxidation of alkanes using ILs has been performed. In particular, the partial oxidation of cyclohexane to cyclohexanol (CyOH) and cyclohexanone (Cy=O) (as shown in Scheme 24.3) has attracted the highest interest [71–73] and recent outcomes are summarized in Table 24.1.

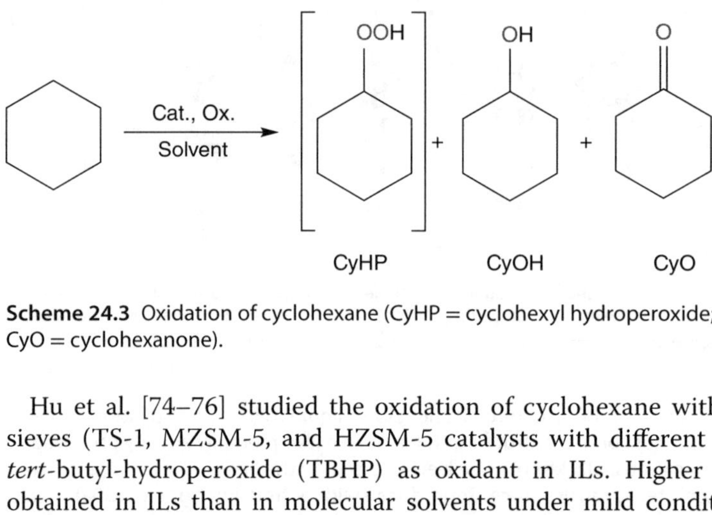

Scheme 24.3 Oxidation of cyclohexane (CyHP = cyclohexyl hydroperoxide; CyOH = cyclohexanol, CyO = cyclohexanone).

Hu et al. [74–76] studied the oxidation of cyclohexane with metal-containing molecular sieves (TS-1, MZSM-5, and HZSM-5 catalysts with different Si/Al ratios) as catalysts and *tert*-butyl-hydroperoxide (TBHP) as oxidant in ILs. Higher yields and selectivities were obtained in ILs than in molecular solvents under mild conditions. FeZSM-5 exhibited the highest catalytic activity among all the MZSM-5 catalysts investigated, and the conversion of cyclohexane and the selectivity of desired products could reach 20.9% and 98.2%, respectively. The catalyst/IL system could be successfully recycled by simple decantation without significant loss of activity.

Some of the authors of this chapter reported [77] the catalytic peroxidative oxidation of cyclohexane with H_2O_2 in an IL using the tetracopper(II) complex $[(CuL)_2(\mu_4\text{-}O,O',O'',O'''\text{-}CDC)]_2{\cdot}2H_2O$ [L = 2-(2-pyridylmethyleneamino)benzenesulfonate, CDC = cyclohexane-1, 4-dicarboxylate] as catalyst (Table 24.1, entries 18–20; Figure 24.3).

Significant improvement in the catalytic performances were achieved in [bmim][PF_6], with the catalyst leading to a product yield up to 36% (entry 20, Table 24.1) and a TON up to 529 (entry 18, Table 24.1) in a short reaction period (2 hours) [77]. Moreover, the

Table 24.1 Peroxidative oxidation of cyclohexane in an ionic liquid.

Entry	Catalyst	Solvent	Oxidant	Time	Conversion (%)	Yield(%)		Ref
						CyOH	Cy = O	
1[a)]	TS-1	Acetone	TBHP	24 h	3.62	1.7	1.1	[74]
2[a)]	TS	No solvent	TBHP	24 h	1.25	0.9	0.4	[74]
3[a)]	TS	[emim][BF_4]	H_2O_2	24 h	0.18	0.1	0.0	[74]
4[a)]	TS	[emim][BF_4]	TBHP	6 h	6.77	5.4	1.3	[74]
5[a)]	TS	[emim][BF_4]	TBHP	24 h	13.0	10.7	1.9	[74]
6[b)]	HZSM-5	No solvent	TBHP	12 h	0.98	0.3	0.68	[75]
7[b)]	HZSM-5	Acetone	TBHP	12 h	3.29	0	1.29	[75]
8[b)]	FeZSM-5	Acetone	TBHP	12 h	3.81	0.6	1.35	[75]
9[b)]	HZSM-5	[emim][BF_4]	TBHP	12 h	15.8	6.98	2.70	[75]
10[b)]	NiZSM-5	[emim][BF_4]	TBHP	12 h	15.9	8.2	2.84	[75]
11[b)]	CoZSM-5	[emim][BF_4]	TBHP	12 h	14.2	9.82	3.53	[75]
12[b)]	MnZSM-5	[emim][BF_4]	TBHP	12 h	15.5	10.3	3.06	[75]
13[b)]	FeZSM-5	[emim][BF_4]	TBHP	12 h	20.9	12.1	3.27	[75]
14[b)]	CuZSM-5	[emim][BF_4]	TBHP	12 h	9.5	6.15	3.31	[75]
15[c)]	Z1	[emim][BF_4]	TBHP	12 h	15.8	7.0	2.7	[76]
16[c)]	Z2	[emim][BF_4]	TBHP	12 h	12.8	5.9	1.9	[76]
17[c)]	Z3	[emim][BF_4]	TBHP	12 h	10.8	4.8	1.6	[76]
18[d)]	$[(CuL)_2(CDC)]_2{\cdot}2H_2O$	[bmim][PF_6]	H_2O_2	2 h	23.1	4.0	19.0	[77]
19[e)]	$[(CuL)_2(CDC)]_2{\cdot}2H_2O$	[bmim][PF_6]	H_2O_2	2 h	34.9	8.9	25.3	[77]
20[f)]	$[(CuL)_2(CDC)]_2{\cdot}2H_2O$	[bmim][PF_6]	H_2O_2	2 h	36.1	7.1	28.9	[77]
21[g)]	$[CuL'(NO_3)(H_2O)]{\cdot}H_2O$	[emim][BF_4]	H_2O_2	3 h	–	10.3	3.1	[78]
22[g)]	$[CuL'(NO_3)(H_2O)]{\cdot}H_2O$	[bmim][PF_6]	H_2O_2	3 h	–	6.4	4	[78]
23[h)]	$[FeCl_2\{HC(pz)_3\}]$ **1**	[bmim][BF_4]	H_2O_2	6 h	–	4.2	12.3	[79]
24[h)]	$[FeCl_2\{HC(pz)_3\}]$ **1**	[bmim][$N(CN)_2$]	H_2O_2	6 h	–	17.7	1.7	[79]
25[i)]	$[FeCl_2\{HC(pz)_3\}]$ **1**	[bmim][PF_6]	TBHP	6 h	–	8.3	12.6	[80]
26[i)]	$[FeCl_2\{HC(pz)_3\}]$**1**	[bmim][FAP]	TBHP	6 h	–	4.1	12.8	[80]
27[i)]	$[FeCl_2\{HOCH_2C(pz)_3\}]$ **2**	[bmim][PF_6]	TBHP	6 h	–	7.5	9.5	[80]
28[i)]	$[FeCl_2\{HOCH_2C(pz)_3\}]$ **2**	[bmim][FAP]	TBHP	6 h	–	2.5	8.6	[80]

TS- titanium silicate, HZSM-5 – Molecular sieves; Z1, Z2 and Z3 represent H-ZSM-5 catalysts with Si/Al ratios of 25, 38, and 50, respectively; L = 2-(2-pyridylmethyleneamino)benzenesulfonate; CDC = cyclohexane-1,4-dicarboxylate; L′ = (*E*)-N′-(2-oxido-5-methylbenzylidene)benzohydrazide; [bmim][FAP] = 1-butyl-3-methylimidazolium tris(pentafluoroethyl)trifluorophosphate.

a) 0.15 g of catalyst, 27.8 mmol of cyclohexane, 55.6 mmol of TBHP (80% in H_2O), 5 ml (6.25 g) of solvent, 90 °C.
b) 0.15 g of catalyst, 27.8 mmol of cyclohexane, 55.6 mmol of TBHP (85% in H_2O), 5 ml (6.25 g) of IL, 90 °C.
c) 0.15 g of catalyst, 27.8 mmol of cyclohexane, 55.6 mmol of TBHP (85% in H_2O), 5 ml (6.25 g) of IL, 90 °C.
d) Catalyst amount 2×10^{-5} mol l^{-1}, 5 mmol of cyclohexane, 10 mmol of H_2O_2 (50% in H_2O), 5 ml total volume, 50 °C.
e) Catalyst amount 2×10^{-4} mol l^{-1}, 5 mmol of cyclohexane, 10 mmol of H_2O_2 (50% in H_2O), 5 ml total volume, 50 °C.
f) Catalyst amount 2×10^{-3} mol l^{-1}, 5 mmol of cyclohexane, 10 mmol of H_2O_2 (50% in H_2O), 5 ml total volume, 50 °C.
g) Catalyst amount 1×10^{-5} mol l^{-1}, 5 mmol of cyclohexane, 10 mmol of H_2O_2 (50% in H_2O), 5 ml total volume, 40 °C.
h) Catalyst amount 1×10^{-5} mol l^{-1}, 5 mmol of cyclohexane, 10 mmol of H_2O_2 (50% in H_2O), 3 ml total volume, 50 °C.
i) Catalyst amount 1×10^{-5} mol l^{-1}, 5 mmol of cyclohexane, 10 mmol of TBHP (70% in H_2O), 3 ml total volume, 70 °C.

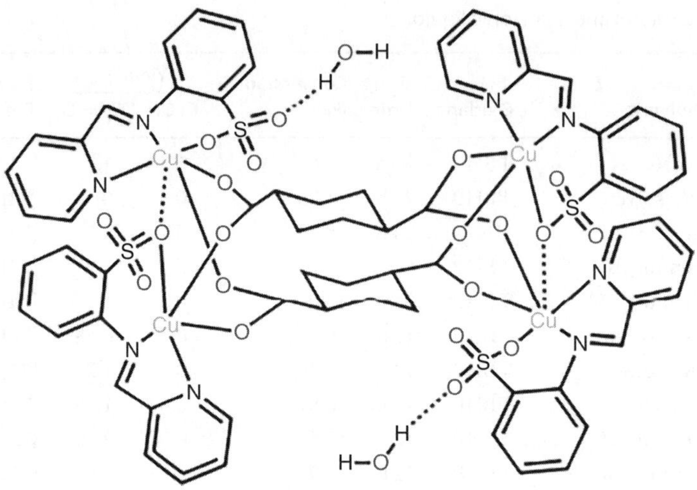

Figure 24.3 $[(CuL)_2(\mu_4\text{-}O,O',O'',O''''\text{-}CDC)]_2 \cdot 2H_2O$ [L = 2-(2-pyridylmethyleneamino) benzenesulfonate, CDC = cyclohexane-1,4-dicarboxylate]. Source: Reproduced with permission from ref. [77].

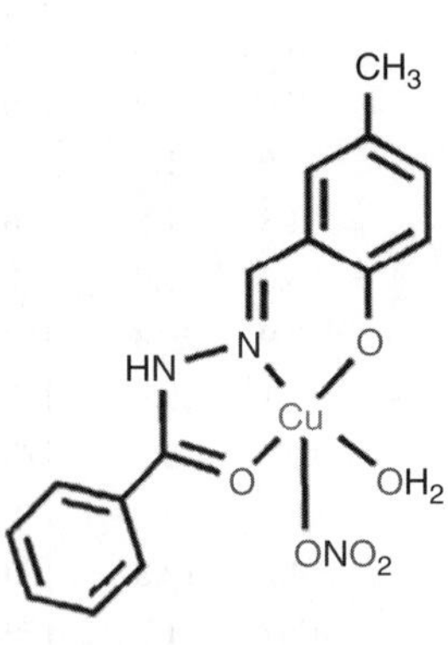

Figure 24.4 $[CuL'(NO_3)(H_2O)] \cdot H_2O$ [L′ = (*E*)-N′-(2-oxido-5-methylbenzylidene) benzohydrazide]. Source: Reproduced with permission from ref. [78].

copper(II) compound with a Schiff base type ligand $[CuL'(NO_3)(H_2O)] \cdot H_2O$ (L′ = (*E*)-N′-(2-oxido-5-methylbenzylidene)benzohydrazide) (Figure 24.4) exhibits also a high catalytic activity for alkane peroxidative oxidation in IL media (Table 24.1, entries 21–22) [78]. The use of IL as reaction media promotes the catalytic activity and allows catalyst recycling in contrast with a typical organic solvent.

Mncube and Bala [81] reported the use of simple metal salts in a triazolium IL as recyclable catalysts for the oxidation of alkanes using H_2O_2 as oxidant to obtain oxygenated products in an aqueous biphasic system. The IL is required for catalyst stability and enhancement of its activity, and the chemical constitution of the IL by virtue of the substituents around the triazole ring has an influence on the oxidation performance. For example, an IL containing a hydrophobic group was more active.

The catalytic activity of the iron(II) C-scorpionate complexes $[FeCl_2\{HC(pz)_3\}]$ **1** (pz = pyrazol-1-yl) and $[FeCl_2\{HOCH_2C(pz)_3\}]$ **2** (Figure 24.5) toward cyclohexane oxidation with TBHP was evaluated and compared by us [80] in different ILs (1-butyl-3-methylimidazolium hexafluorophosphate, $[bmim][PF_6]$, and 1-butyl-3-methylimidazolium tris(pentafluoroethyl)trifluorophosphate, [bmim][FAP]. Complexes **1** and **2** exhibited the same catalytic trend in these ILs. The use of $[bmim][PF_6]$ led to the maximum yields (20.9% for **1** and 17% for **2**, entries 25 and 27, respectively). When the solvent was [bmim][FAP], the maximum yields were 16.9% for **1** and 11.1% for **2** (entries 26 and 28, respectively). The use of such alternative solvents led to efficient and selective protocols for the oxidation of cyclohexane. Moreover, tuning the alcohol/ketone selectivity was possible by choosing the suitable solvent.

Related observations concern the use of **1** in $[bmim][N(CN)_2]$ (1-butyl-3-methylimidazolium dicyanamide) and $[bmim][BF_4]$, but a better overall yield (19.4%, entry 24) and a much higher selectivity toward cyclohexanol were achieved in the former IL than in the latter (entry 23).

The possibility of recycling and reuse of **1** in a considerable number of consecutive catalytic cycles with retention of its activity and selectivity is an advantage of using these ILs in comparison with an organic solvent (Figure 24.6) [79].

Figure 24.5 Iron(II) C-scorpionate complexes $[FeCl_2\{HC(pz)_3\}]$ **1** (pz = pyrazol-1-yl) and $[FeCl_2\{HOCH_2C(pz)_3\}]$ **2**. Source: Reproduced with permission from ref. [80].

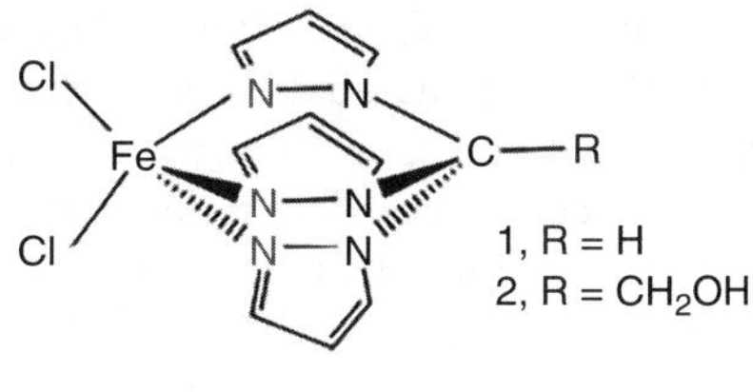

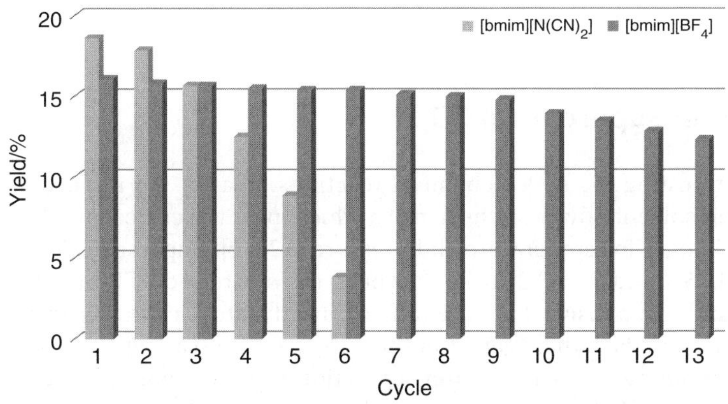

Figure 24.6 Effect of the catalyst recycling on the total (A + K) yield for the MW-assisted oxidation of cyclohexane with H_2O_2 catalyzed by $[FeCl_2\{HC(pz)_3\}]$ **1** in $[bmim][N(CN)_2]$ or $[bmim][BF_4]$. Reaction conditions are those of entry 24 of Table 24.1. Source: Reproduced with permission from ref. [79].

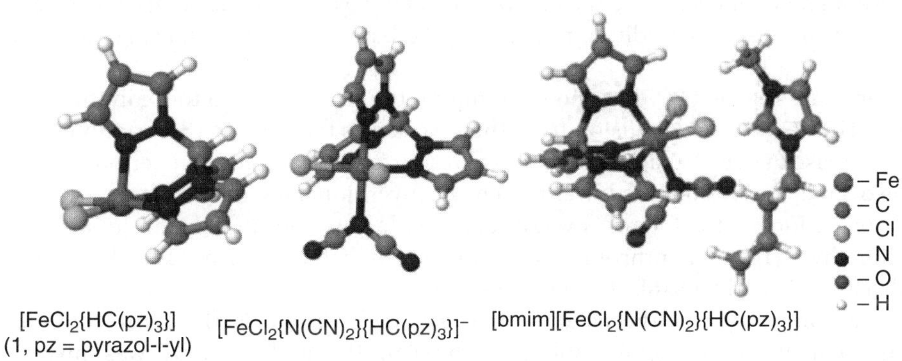

Figure 24.7 Calculated equilibrium structures for $[FeCl_2\{HC(pz)_3\}]$ **1** (pz = pyrazol-1-yl), $[FeCl_2\{N(CN)_2\}\{HC(pz)_3\}]^-$ and $[bmim][FeCl_2\{N(CN)_2\}\{HC(pz)_3\}]$. Source: Reproduced with permission from ref. [79]. (*See insert for color representation*).

The interaction of the scorpionate catalyst $[FeCl_2\{HC(pz)_3\}]$ **1** (Figure 24.5) with the IL $[bmim][N(CN)_2]$ was investigated using theoretical density functional theory (DFT) calculations [79]. The coordination polyhedron of the equilibrium structure of the Fe scorpionate complex (tetragonal pyramid with one of the donor N atoms being at the axial position) discloses one empty coordination position to which the IL anion $N(CN)_2^-$ can bind. Indeed, the equilibrium structure of the coordinatively saturated complex $[FeCl_2\{N(CN)_2\}\{HC(pz)_3\}]^-$ was found as a result of geometry optimization (Figure 24.7). The formation of this complex from $[FeCl_2\{HC(pz)_3\}]$ and $N(CN)_2^-$ is energetically favorable in terms of both gas-phase Gibbs free energy and SMD energy in solution (by −8.7 and −12.3 kcal mol^{-1}, respectively) [79].

The interaction of $[FeCl_2\{HC(pz)_3\}]$ with both the cation $[bmim]^+$ and the dicyanamide anion of the IL was also considered, and the equilibrium structure of the neutral associate

[bmim][$FeCl_2${N(CN)$_2$}{HC(pz)$_3$}] was obtained (Figure 24.6). The [bmim]$^+$ cation in this complex is situated between the Cl^- and $N(CN)^{2-}$ ligands, forming with them several C—H···Cl and C—H···N hydrogen bonds (Figure 24.6). Such a position of [bmim]$^+$ in the second coordination sphere corresponds to the most extended regions of the negative electrostatic potential of the complex. The formation of the [bmim][$FeCl_2${N(CN)$_2$}{HC(pz)$_3$}] associate from [$FeCl_2${HC(pz)$_3$}] and [bmim][N(CN)$_2$] (optimized as an ion pair) is highly exergonic (by −26.7 kcal mol^{-1}), indicating that the interaction of the catalyst with IL is efficient [79].

24.3 Alkane Oxidations in Supercritical CO_2

SCFs constitute particularly interesting media for chemical reactions [82a–c]. An important feature is the dependence of their solvent power on the density, which, near the critical point, is particularly sensitive to small changes in temperature and pressure [82d, e]. Supercritical carbon dioxide (SC-CO_2, T_c = 304.25 K and P_c = 7.39 MPa) has been the most used SCF, because it is nonflammable, nontoxic, and inexpensive, is in the fully oxidized state, easy to separate, and completely miscible with the oxidant dioxygen. Another important advantage of using SC-CO_2 is the possibility of combining the reaction and separation into only one integrated process. However, the apolar nature of SC-CO_2 makes it usually a bad solvent for metal complexes, which is a severe limitation for application in homogeneous systems [83]. Nevertheless, a selectivity similar to those obtained in organic media has been achieved in SC-CO_2 using lower catalyst concentrations [84].

Srinivas and Mukhopadhyay reported the first results on the partial oxidation of cyclohexane utilizing SC-CO_2 as the reaction medium in noncatalyzed batch experiments, using O_2 as the oxidant [85].

The reaction pressure reached 207 bar and the temperature was similar to those in the commercial process (410–433 K). A conclusion from their work was that raising the reaction pressure caused an increase in cyclohexane conversion and shortened the induction period. On the other hand, the overall conversion was lower than in a conventional liquid solvent.

The partial oxidation with O_2 of cyclohexane with *cis*-[Fe(dmp)$_2$(H_2O)$_2$](CF_3SO_3)$_2$ (dmp = 2,9-dimethyl-1,10-phenanthroline) in SC-CO_2, in batch runs at 343–373 K and 207 bar, was also investigated by Sahle-Demessie [86].

Very recently, a comparison of the catalytic performances in one of the most common solvents for the oxidation of cyclohexane (acetonitrile), ILs and SC-CO_2 was investigated for the C-scorpionate iron(II) catalysts [$FeCl_2${HC(pz)$_3$}] **1** (pz = pyrazol-1-yl) and [$FeCl_2${$HOCH_2C$(pz)$_3$}] **2** (Figure 24.5) [80]. TBHP was the chosen oxidant, and the obtained activity in the various solvents followed the order [bmim][PF_6] > [bmim][FAP] > SC-CO_2 > CH_3CN. In each medium, catalyst **1** leads to higher yield and TON values than **2**. Thus, the highest oxygenates yield (21%) was achieved with **1** in [bmim][PF_6]. In comparison, the organic molecular solvent CH_3CN led to much lower yields (up to 6%) and to lower selectivity toward cyclohexanone. The catalysts can be easily recycled in IL and SC-CO_2 media, preserving their activity up to three consecutive cycles.

Another solvent comparative study was recently undertaken using the new *cis*-dioxidomolybdenum(VI) complexes [$MoO_2(L^1)$] (**3**), [$MoO_2(L^2)$]·MeOH (**4**) and [$MoO_2(L^3)$] (**5**) which were prepared by reaction of [MoO_2(acac)$_2$] with the corresponding aroylhydrazone Schiff base [H_2L^1 = 2,3-dihydroxybenzylidene-2-hydroxybenzohydrazide, H_2L^2 = 2, 3-dihydroxybenzylidene-benzohydrazide, H_2L^3 = (3,5-di-*tert*-butyl-2-hydroxybenzylidene)-2-hydroxybenzohydrazide]. They were applied as catalysts in the oxidation of cyclohexane [87a].

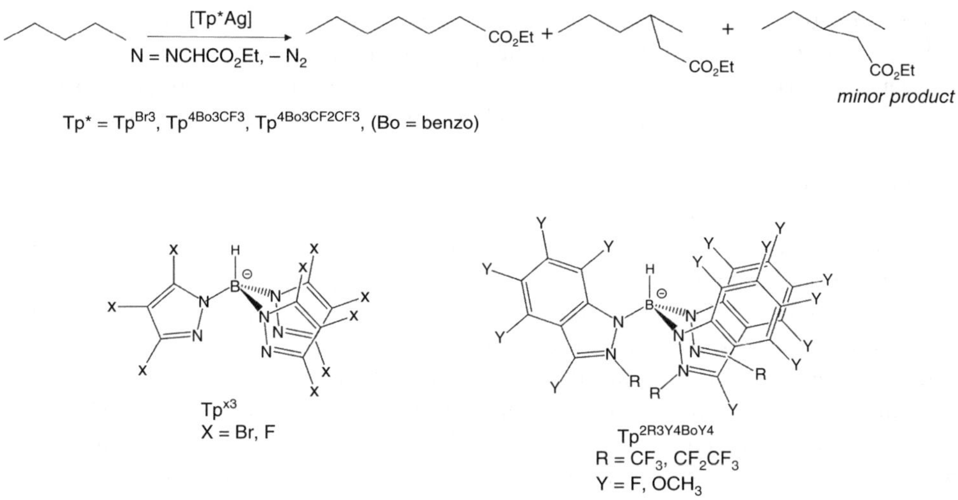

Figure 24.8 Functionalization of C—H bonds (illustrated for n-pentane) by carbene insertion from ethyl diazoacetate using Tp*Ag complexes as catalysts [87b].

For each medium, the least active catalyst is that with the highest steric hindrance. Contrary to the iron(II) C-scorpionate complexes, the selectivity is higher toward cyclohexanol than cyclohexanone.

C—C bond formation upon reaction of methane with ethyl diazoacetate NN=CHCOOEt (EDA) was investigated using SC-CO_2 as a solvent and [Tp*Ag] (Tp* = tris(bromo- or fluoro-substituted pyrazolyl)borate) as the catalyst (Figure 24.8, for *n*-pentane) [87b]. This approach is favorable due to the SC-CO_2 inertness toward intermediates, the complete miscibility with methane and a strong solvating ability with EDA, and the excellent transport properties. Although the catalyst is only sparingly soluble in CH_4/CO_2 mixtures, optimized conditions result in a 19% yield of ethyl propionate (based on the starting quantity of the diazoester) at 40 °C for 14 hours. A plausible mechanism for the C—H functionalization of methane and the other hydrocarbons involves a metal-catalyzed N_2 elimination from the diazo substrate followed by the CHCOOEt carbene transfer mediated by a highly electrophilic metal carbene intermediate [87b]. The reactivity of the silver complexes [Tp*Ag] toward the C—H bonds gives a product distribution that is governed, at least in part, by the C—H bond dissociation energy in accord with the observation that ethane functionalization is favored over methane by a factor of ca. 14 [87b].

24.4 Microwave Promoted Oxidations

The application of various radiation types (namely gamma, MW or sound-range waves, UV and visible light) in synthetic organic chemistry is attracting increasing attention due to possible improvements in efficiency, reaction rate, selectivity and energy saving, and the environmentally friendly nature of this alternative approach. The effect of MW irradiation on the initiation and stimulation of chemical reactions is known, but the mechanism of action has not yet been fully clarified and has been the object of some debate [8, 88].

The interest in the use of MW radiation as a source of energy in chemical reactions emerged in the 1980s. In recent years, there have been many reports on the interaction between MW and matter, including the effect of MW heating on catalytic reactions [8–10]. MW-assisted chemistry relies on the efficient interaction of molecules with the electromagnetic waves, generated by a *microwave dielectric effect*. Although most organic compounds have a considerably lower dielectric constant compared to water, they can achieve a high absorption and consequently a rapid heating due to their high ability to convert electromagnetic energy into heat.

Nonpolar liquids (hydrocarbons, CCl_4 and dimethyl ether) remain almost not heated by MW fields under certain conditions, which complicate the performance of processes in these media [88]. In these cases, it is common to add polar components (such as ILs) to increase the absorbance of a reaction medium [89]. In addition, the heat distribution that is observed in the MW heated object differs from that resulting from conventional heating [90] since the MW irradiation interacts directly with the polar molecules and can provide an efficient energy transfer directly to the reagents [8–10]. The simultaneous application of a catalyst and MW irradiation has shown in some cases a marked synergistic effect [8–10]. Over the past few years, it was recognized that transition metal-catalyzed bond transformations can be significantly enhanced by employing MW heating under sealed-vessel conditions, even without requiring an inert atmosphere. Promising studies relate to the use of MW fields in gas-phase catalytic processes which are of industrial interest, i.e. involving oxidative condensation, partial oxidation, hydrogenation, and dehydrogenation reactions.

In this section, we do not intend to discuss the mechanisms of irradiation action but focus on some recent applications of MW irradiation in alkane functionalization.

There are some examples of the successful application of MW-assisted chemistry to alkane functionalization. They can involve the use of benign reaction media, solvent-free conditions, and application of supported and reusable catalysts.

Bond et al. [91] pioneered MW-assisted catalysis in the activation of methane to C_2 hydrocarbons using sodium aluminate as catalyst. According to the authors, the use of MW radiation allows a decrease in the gas phase temperature and in the catalyst amount, and consequently hampers reactions of deep oxidation. Under MW activation conditions, the ethylene/ethane product ratio increased, while ethane was the main product under thermal activation. The CO formation also increased relatively to CO_2 [91].

Recently, a catalytic system based on the C-scorpionate iron complex $[FeCl_2\{HC(pz)_3\}]$ **1** (Figure 24.5) as catalyst, $[bmim][BF_4]$ (IL) as solvent and a low power MW irradiation was shown to provide a very selective, efficient, and reusable catalyst procedure for the mild oxidation of cyclohexane with hydrogen peroxide, yielding selectively the KA (cyclohexanol + cyclohexanone) oil mixture (Scheme 24.3) with 28% yield in 1 hours at 50 °C, through a radical mechanism [79]. Higher catalytic activities are obtained with the ILs in comparison to those with other solvents. The possibility to recycle and reuse the catalyst (up to 9 consecutive cycles) without loss of activity is also recognized [79]. The higher product yields and selectivities achieved by these systems under milder conditions in comparison with the industrial process (typically operating at ca. 150 °C, with ca. 6% yield to get a good selectivity [92] constitute promising advantages.

The advantage of using low-power MW irradiation over the conventional heating was also observed for the peroxidative (with aq. *tert*-butyl hydroperoxide, *t*-BuOOH, TBHP) oxidation of cyclohexane to cyclohexanol and cyclohexanone (Scheme 24.3) performed under solvent- and additive-free conditions catalyzed by the dinuclear vanadium(V) complexes $NH_4[(VO_2)_2(^{H}LH)]$, $NH_4[(VO_2)_2(^{t\text{-}Bu}LH)]$, $NH_4[(VO_2)_2(^{Cl}LH)]$, $[(VO_2)(VO)(^{H}LH)(CH_3O)]$, $[(VO_2)(VO)(^{t\text{-}Bu}LH)(C_2H_5O)]$, and $[(VO_2)(VO)(^{Cl}LH)(CH_3O)(CH_3OH/H_2O)]$ ($^{H}LH_4$ = 1,5-bis (2-hydroxybenzaldehyde)carbohydrazone, $^{t\text{-}Bu}LH_4$ = 1,5-bis(3,5-di-tert-butyl-2-hydroxybenzaldehyde)carbohydrazone, and $^{Cl}LH_4$ = 1,5-bis(3,5-dichloro-2-hydroxybenzaldehyde)carbohydrazone), via a radical mechanism [93]. Cyclohexanol and cyclohexanone were the only products obtained (high selectivity), after 1.5 hours under MW irradiation.

Environmentally benign MW-assisted neat oxidation of cyclohexane to a cyclohexanol and cyclohexanone mixture (KA oil) with TBHP (aq.), through a radical mechanism, in high yields and selectivity, was also successfully achieved in the presence of the dinuclear bis(μ-chlorido)-bridged cobalt(II) complex $[Co_2(HL_2)_4(\mu\text{-}Cl)_2][CoCl_4]$ (HL_2 is a silyl-containing Schiff base) [94] and copper(I) and copper(II) metallacycles $[Cu_2(\mu\text{-}N,N\text{-}3,5\text{-}(NO_2)_2pz)_2(PPh_3)_2]$ (pz = pyrazol-1-yl), and $[Cu_6(\mu\text{-}OH)_6(\mu\text{-}3,5\text{-}(CF_3)_2pz)_6]$ [95].

Following the same research line, the vanadium(IV) C-scorpionate complexes $[VO_xCl_{3\text{-}x}\{RC(pz)_3\}]$ [x = 0, R = SO_3 (**6**); x = 1, R = CH_2OH (7) or CH_2OSO_2Me (**8**)] supported on functionalized carbon nanotubes (CNTs) were applied in the MW-assisted neat oxidation of *o*-, *m*- and *p*-xylene with TBHP (aq. solution) to the corresponding toluic acids (major products), tolualdehydes and methyl benzyl alcohols (Scheme 24.4). Recycling of catalysts was possible up to five consecutive cycles without loss of activity [96].

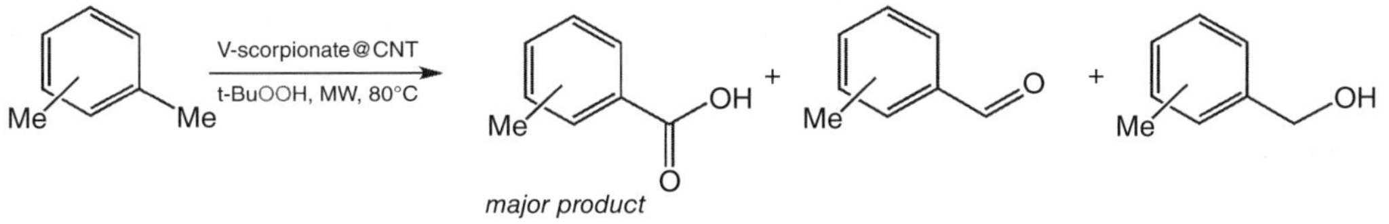

Scheme 24.4 Microwave-assisted peroxidative oxidation of *o*-, *m*- or *p*-xylenes to the corresponding methylbenzyl alcohols, tolualdehydes, and toluic acids. Source: Reproduced with permission from ref. [96].

A related solvent-free MW-assisted oxidation of toluene was undertaken successfully with TBHP in the presence of *keto* and polymeric *enol* aroylhydrazone Cu(II) complexes [97]. Benzaldehyde was formed as the major product (Scheme 24.5), whereas benzyl alcohol and benzoic acid were minor products. A promoting effect of MW-irradiation is observed, even under low power (5 W) relatively to the reactions performed under conventional heating. Under the same conditions, but using conventional heating (oil-bath) instead of MW, only 4.4% of product yield was obtained after 1 hours reaction at 50 °C, whereas 39% yield was achieved at 50 °C and under a very low MW power (5 W) with a very high selectivity (ca. 99%) toward benzaldehyde. The best catalytic results were achieved under solvent- and additive-free conditions [97].

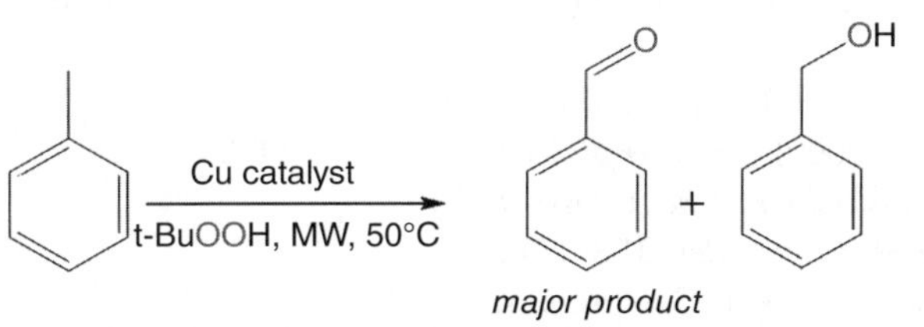

Scheme 24.5 Microwave-assisted peroxidative oxidation of toluene. Source: Reproduced with permission from ref. [97].

24.5 Final Comments

The functionalization of alkanes can benefit from the use of ILs or SC-CO_2 as unconventional solvents and of MW irradiation as an unconventional heating method. The advantages concern not only the development of environmentally friendly conditions but also achieving higher product yields and selectivities in shorter reaction times under milder conditions. Among these methods, the use of ILs has attracted great attention since they constitute a class of tunable and designable solvents that are also attractive for clean separation and green synthesis.

The selected examples discussed in this chapter show the potential of such approaches, but the field still remains understudied. Further research is required to propose general behaviors and to understand the differences relatively to the use of conventional solvents and heating.

Acknowledgments

Financial support from the Fundação para a Ciência e a Tecnologia (FCT), Portugal, for fellowship SFRH/BPD/90883/2012 to A.P.C.R., and for the PTDC/QEQ-QIN/3967/2014 and UID/QUI/00100/2013 projects, and from the Instituto Politécnico de Lisboa (MechSynCat Project-Concurso Anual de IDI&CA, 710044/2016) are gratefully acknowledged.

References

1 Sivaramakrishna, A., Suman, P., Goudi, E.V. et al. (2012). *Res. Rev. Mater. Sci. Chem.* 1: 75.
2 Nizova, G.V., Bolm, C., Ceccarelli, S. et al. (2002). *Adv. Synth. Catal.* 344: 899.
3 Periana, R.A., Bhalla, G., Tenn, W.J. et al. (2004). *J. Mol. Catal. A: Chem.* 220: 7.
4 Weaver, J.F., Hakanoglu, C., Antony, A. et al. (2014). *Chem. Soc. Rev.* 43: 7536.
5 Sun, M., Zhang, J., Putaj, P. et al. (2014). *Chem. Rev.* 114: 981.
6 Hardacre, C. and Parvulescu, V. (2014). *Catalysis in Ionic Liquids, from Catalyst Synthesis to Application*. Cambridge, UK: Royal Society of Chemistry.
7 (a) Jessop, P.G. and Leitner, W. (1999). *Chemical Synthesis Using Supercritical Fluids*. Wiley-VCH. (b) Leitner, W. (2002). *Acc. Chem. Res.* 35: 746.
8 Karabach, Y.Y., Kopylovich, M.N., Mahmudov, K.T. et al. (2014, Chapter 18). Microwave-assisted catalytic oxidation of alcohols to carbonyl compounds. In: *Advances in Organometallic Chemistry and Catalysis* (ed. A.J.L. Pombeiro), 233–246. Hoboken, USA: Wiley.
9 (a) de la Hoz, A. and Loupy, A. (eds.) (2012). *Microwaves in Organic Synthesis* 3rd. Weinheim: Wiley-VCH. (b) de la Hoz, A., Diaz-Ortiz, A., and Moreno, A. (2005). *Chem. Soc. Rev.* 34: 164.
10 Lidström, P., Tierney, J., Wathey, B. et al. (2001). *Tetrahedron* 57: 9225.
11 Seddon, K.R., Stark, A., and Torres, M.J. (2000). *Pure Appl. Chem.* 72 (12): 2275.
12 Endres, F. and Abedin, S.Z.E. (2006). *Phys. Chem. Chem. Phys.* 8: 2101.
13 Greaves, T.L. and Drummond, C.J. (2008). *Chem. Rev.* 108: 206.
14 Dong, K., Zhang, S.J., and Wang, J.J. (2016). *Chem. Commun.* 52: 6744.
15 Maton, C., Vos, N.D., and Stevens, C.V. (2013). *Chem. Soc. Rev.* 42: 5963.
16 Amarasekara, A.S. (2016). *Chem. Rev.* 116: 6133.
17 Podgoršek, A., Jacquemin, J., Pádua, A.A.H. et al. (2016). *Chem. Rev.* 116: 6075.
18 Thompson, D., Coleman, S., Diamond, D. et al. (2011). *Phys. Chem. Chem. Phys.* 13: 6156.
19 MacFarlane, D.R., Pringle, J.M., Johansson, K.M. et al. (2006). *Chem. Commun.* 37: 1905.

20 Ohno, H. and Fukumoto, K. (2007). *Acc. Chem. Res.* 40: 1122.
21 Dupont, J., de Souza, R.F., and Suarez, P.A.Z. (2002). *Chem. Rev.* 102: 3667.
22 Welton, T. (2004). *Coord. Chem. Rev.* 248: 2459.
23 Rantwijk, F.V. and Sheldon, R.A. (2007). *Chem. Rev.* 107: 2757.
24 Olivier-Bourbigou, H., Magna, L., and Morvan, D. (2010). *Appl. Catal. A* 373: 1.
25 Pârvulescu, V.I. and Hardacre, C. (2007). *Chem. Rev.* 107: 2615.
26 Plechkova, N.V. and Seddon, K.R. (2008). *Chem. Soc. Rev.* 37: 123.
27 Armand, M., Endres, F., MacFarlane, D.R. et al. (2009). *Nat. Mater.* 8: 621.
28 Liu, H.T., Liu, Y., and Li, J.H. (2010). *Phys. Chem. Chem. Phys.* 12: 1685.
29 Wang, H., Gurau, G., and Rogers, R.D. (2012). *Chem. Soc. Rev.* 41: 1519.
30 Hulsbosch, J., De Vos, D.E., Binnemans, K. et al. (2016). *ACS Sustainable Chem. Eng.* 4: 2917.
31 Farrán, A., Cai, C., Sandoval, M. et al. (2015). *Chem. Rev.* 115: 6811.
32 Zhao, H., Xia, S.Q., and Ma, P.S.S. (2005). *J. Chem. Technol. Biotechnol.* 80: 1089.
33 Sun, X.Q., Luo, H.M., and Dai, S. (2012). *Chem. Rev.* 112: 2100.
34 Zhang, S.J., Sun, J., Zhang, X.C. et al. (2014). *Chem. Soc. Rev.* 43: 7838.
35 Dong, K., Zhang, S.J., and Wang, J.J. (2016). *Chem. Commun.* 52: 6744.
36 Hunt, P.A., Ashworth, C.R., and Matthews, R.P. (2015). *Chem. Soc. Rev.* 44: 1257.
37 Oehlke, A., Hofmann, K., and Spange, S. (2006). *New J. Chem.* 30: 533.
38 Lungwitz, R. and Spange, S. (2008). *New J. Chem.* 32: 392.
39 Lungwitz, R., Friedrich, M., Linert, W. et al. (2008). *New J. Chem.* 32: 1493.
40 Lungwitz, R., Strehmel, V., and Spange, S. (2010). *New J. Chem.* 34: 1135.
41 Crowhurst, L., Mawdsley, P.R., Perez-Arlandis, J.M. et al. (2003). *Phys. Chem. Chem. Phys.* 5: 2790.
42 Kempter, V. and Kirchner, B. (2010). *J. Mol. Struct.* 972: 22.
43 Wulf, A., Fumino, K., and Ludwig, R. (2010). *Angew. Chem. Int. Ed.* 49: 449.
44 Zhao, W., Leroy, F., Heggen, B. et al. (2009). *J. Am. Chem. Soc.* 131: 15825.
45 Schroder, U., Wadhawan, J.D., Compton, R.G. et al. (2000). *New J. Chem.* 24: 1009.
46 Jiao, J.J., Dong, B., Zhang, H.N. et al. (2012). *J. Phys. Chem. B* 116: 958.
47 Jankowski, P., Dranka, M., and Żukowska, G.Z. (2015). *J. Phys. Chem. C* 119: 10247.
48 Pierola, I.F. and Agzenai, Y.J. (2012). *Phys. Chem. B* 116: 3973.
49 Dupont, J. (2004). *J. Braz. Chem. Soc.* 15: 341.
50 (a) Stassen, H.K., Ludwig, R., Wulf, A. et al. (2015). *Chem. Eur. J.* 21: 8324. (b) Qiao, Y., Ma, W., Theyssen, N. et al. (2017). *Chem. Rev.* 117: 6881.
51 Bester-Rogac, M., Fedotova, M.V., Kruchinin, S.E. et al. (2016). *Phys. Chem. Chem. Phys.* 18: 28594.
52 Rai, N. and Maginn, E.J. (2011). *J. Phys. Chem. Lett.* 2: 1439.
53 Cade, E.A., Petenuci, J., and Hoffmann, M.M. (2016). *ChemPhysChem* 17: 520.
54 Fumino, K., Wulf, A., and Ludwig, R. (2009). *Phys. Chem. Chem. Phys.* 11: 8790.
55 Hayes, R., Imberti, S., Warr, G.G. et al. (2013). *Angew. Chem. Int. Ed.* 52: 4623.
56 Welton, T. (2004). *Coord. Chem. Rev.* 248: 2459.
57 Scholten, J.D., Leal, B.C., and Dupont, J. (2012). *ACS Catal.* 2: 184.
58 Owens, G.S. and Abu-Omar, M.M. (2002). *ACS Symp. Ser.* 818: 321.
59 Hintermair, U., Zhao, G.Y., Santini, C.C. et al. (2007). *Chem. Commun.* (14): 1462.
60 Welton, T. (1999). *Chem. Rev.* 99: 2071.
61 Earle, M.J. and Seddon, K.R. (2000). *Pure Appl. Chem.* 72: 1391.

62 Rogers, R.D. and Seddon, K.R. (2003). *Science* 302: 792.
63 Rogers, R.D. and Seddon, K.R. (eds.) (2005). *Ionic Liquids IIIA: Fundamentals, Progress, Challenges, and Opportunities: Properties and Structure*, ACS Symp Ser., vol. 901. Washington, DC: American Chemical Society.
64 Rogers, R.D. and Seddon, K.R. (eds.) (2005). *Ionic Liquids IIIB: Fundamentals, Progress, Challenges, and Opportunities: Transformations and Processes*, ACS Symp Ser., vol. 902. Washington, DC: American Chemical Society.
65 Wilkes, J.S. (2004). *J. Mol. Catal. A: Chem.* 214: 11.
66 Armel, V., Pringle, J.M., Forsyth, M. et al. (2010). *Chem. Commun.* 46: 3146.
67 Hallett, J.P. and Welton, T. (2011). *Chem. Rev.* 111: 3508.
68 Lei, Z., Dai, C., Zhu, J. et al. (2014). *AlChE J.* 60: 3312.
69 Lei, Z., Dai, C., and Chen, B. (2014). *Chem. Rev.* 114: 1289.
70 Bianchini, G., Crucianelli, M., De Angelis, F. et al. (2005). *Tetrahedron Lett.* 46: 2427.
71 Ebadi, A., Safari, N., and Peyrovi, M.H. (2007). *Appl. Catal. A* 321: 135.
72 Guo, C.C., Huang, G., Zhang, X.B. et al. (2003). *Appl. Catal. A* 247: 261.
73 Tian, P., Liu, Z.M., Wu, Z.B. et al. (2004). *Catal. Today* 93-95: 735.
74 Wang, J., Zhao, H., Zhang, X. et al. (2008). *Chin. J. Chem. Eng.* 16: 373.
75 Wang, J., Zhao, F., Liu, R. et al. (2008). *J. Mol. Catal. A: Chem.* 279: 153.
76 Hu, Y., Wang, J., Zhao, R. et al. (2009). *Chin. J. Chem. Eng.* 17: 407.
77 Ribeiro, A.P.C., Martins, L.M.D.R.S., Hazra, S. et al. (2015). *C.R. Chim.* 18: 758.
78 Anbu, S., Paul, A., Ribeiro, A.P.C. et al. (2016). *Inorg. Chim. Acta* 450: 426.
79 Ribeiro, A.P.C., Martins, L.M.D.R.S., Kuznetsov, M.L. et al. (2017). *Organometallics* 36 (1): 192.
80 Ribeiro, A.P.C., Martins, L.M.D.R.S., Alegria, E.C.B.A. et al. (2017). *Catalysts* 7: 230.
81 Mncube, S.G. and Bala, M.D.J. (2016). *Mol. Liq.* 215: 396.
82 (a) Karmee, S.K., Greiner, L., Kraynov, A. et al. (2010). *Chem. Commun.* 46: 6705. (b) Vinod, C.P., Wilson, K., and Lee, A.F. (2011). *J. Chem. Technol. Biotechnol.* 86: 161. (c) Fernandes, R.R., Lasri, J., Guedes da Silva, M.F.C. et al. (2011). *Adv. Synth. Catal.* 353: 1153. (d) Mendes, R.L., Nobre, B.P., Cardoso, M.T. et al. (2003). *Inorg. Chim. Acta* 356: 328. (e) Bruno, T.J., Castro, C.A.N., Hamel, J.P.F. et al. (1993, Chapter 11). *Recovery Process for Biological Materials* (ed. J.F. Kennedy and J.M.S. Cabral), 303–356. New York: Wiley.
83 Harwardt, T., Francio, G., and Leitner, W. (2010). *Chem. Commun.* 46: 6669.
84 Olsen, M.H.N., Salomao, G.C., Drago, V. et al. (2005). *J. Supercrit. Fluids* 34: 119.
85 (a) Srinivas, P. and Mukhopadhyay, M. (1994). *Ind. Eng. Chem. Res.* 33: 3118. (b) Mukhopadhyay, M. and Srinivas, P. (1996). *Ind. Eng. Chem. Res.* 35: 4713. (c) Srinivas, P. and Mukhopadhyay, M. (1997). *Ind. Eng. Chem. Res.* 36: 2066.
86 Sahle-Demessie, E., Gonzalez, M.A., Enriquez, J. et al. (2000). *Ind. Eng. Chem. Res.* 39: 4858.
87 (a) Sutradhar, M., Ribeiro, A.P.C., Guedes da Silva, M.F.C. et al. (2017). *Mol. Catal.* http://dx.doi.org/10.1016/j.mcat.2017.10.026. (b) Caballero, A., Despagnet-Ayoub, E., Mar Díaz-Requejo, M. et al. (2011). *Sci. New Ser.* 332: 835.
88 Hoz, A., Diaz-Ortiz, A., and Moreno, A. (2005). *Chem. Soc. Rev.* 34: 164.
89 (a) Herrero, M.A., Kremsner, J.M., and Kappe, C.O. (2007). *J. Organomet. Chem.* 73: 36. (b) Obermayer, D. and Kappe, C.O. (2010). *Org. Biomol. Chem.* 8: 114. (c) Kappe, C.O. (2004). *Angew. Chem. Int. Ed.* 43: 6250.
90 Kustov, L.M. and Sinev, I.M. (2010). *Russ. J. Phys. Chem. A* 84: 1676.
91 Bond, G., Moyes, R.B., and Whan, D.A. (1993). *Catal. Today* 17: 427.

92 (a) Ribeiro, A.P.C., Martins, L.M.D.R.S., and Pombeiro, A.J.L. (2017). *Green Chem.* 19: 1499. (b) Schuchardt, U., Cardoso, D., Sercheli, R. et al. (2001). *Appl. Catal. A* 211: 1.
93 Dragancea, D., Talmaci, N., Shova, S. et al. (2016). *Inorg. Chem.* 55: 9187.
94 Zaltariov, M.-F., Vieru, V., Zalibera, M. et al. (2017). *Eur. J. Inorg. Chem.* 4324.
95 Galassi, R., Simon, O.C., Burini, A. et al. (2017). *Polyhedron* 134: 143.
96 Wang, J., Martins, L.M.D.R.S., Ribeiro, A.P.C. et al. (2017). *Chem. Asian J.* 12: 1915.
97 Sutradhar, M., Alegria, E.C.B.A., Barman, T.R. et al. (2017). *Mol. Catal.* 439: 224.

25

Noncovalent Interactions in Alkane Chemistry

Kamran T. Mahmudov[1,2], M. Fátima C. Guedes da Silva[1], Fedor I. Zubkov[3], and Armando J. L. Pombeiro[1]

[1] *Universidade de Lisboa, Centro de Química Estrutural, Instituto Superior Técnico, Lisboa, Portugal*
[2] *Baku State University, Department of Chemistry, Baku, Azerbaijan*
[3] *RUDN University, Organic Chemistry Department, Moscow, Russian Federation*

25.1 Introduction

Noncovalent interactions, such as hydrogen, halogen, chalcogen, pnicogen, tetrel, and icosagen bonds, as well as agostic, anagostic, π-π stacking, n-π^*, π-cation, and π-anion interactions, etc. [1–32], play a vital role in synthesis, catalysis, crystal engineering, adsorption of molecules on surfaces, protein folding, stacking of base pairs in DNA, etc. Due to their fundamental and practical importance, noncovalent interactions are studied very extensively using a broad range of experimental and theoretical methods. Unlike covalent bonds, noncovalent interactions are known to act at long distances of several angstroms; thus, these interactions are traditionally considered to be weak, although their strength covers a substantial range, starting from a few kJ mol^{-1} to several hundreds of kJ mol^{-1} depending on their nature/type. On account of orchestrated actions (multiplicity) and facile transformations, their overall influence on the course of a reaction can be decisive.

Alkanes or saturated hydrocarbons contain strong C—C and C—H single bonds. Their molecules have no empty orbitals of low energy or filled orbitals of high energy that could readily participate in a chemical reaction, and are generally inert to chemical transformations [33–36]. The low reactivity of alkane C—C and C—H bonds is often associated with their high bond energies (typically 340–400 and 370–440 kJ mol^{-1}, respectively) and with the very low acidity (difficult to measure directly, but estimated to have pK_a = 45–60) and basicity of the latter ones [37]. The structure of alkanes can be viewed as a combination of sp^3 hybridized orbitals with a tetrahedral orientation. Short, unbranched alkanes prefer linear conformations, while longer unbranched ones are folded. These features are important for their involvement in intra- or intermolecular noncovalent interactions. The relative stability of conformers in normal alkanes may be strongly influenced by intramolecular dispersion interactions, especially when some conformers possess compact structures.

Alkanes can form noncovalent interactions with organic or inorganic molecules leading to the corresponding associates (Tables 25.1 and 25.2). For instance, methane forms two hydrates that are associates in which the methane is trapped as a guest in an ice-like lattice host and the water molecules are held together by H-bonds [4]. Due to the hydrophobic character of alkanes, many types of noncovalent interactions are excluded but can possibly be used in the stabilization of radicals formed in their reactions. Moreover, chalcogen bond assisted nanotubes are able to

Alkane Functionalization, First Edition. Edited by Armando J. L. Pombeiro and M. Fátima C. Guedes da Silva.

Table 25.1 Number of acyclic alkane associates deposited in the CCDC (01.04.2017).

Alkanes	CH_4	C_2H_5	C_3H_8	C_4H_{10}	C_5H_{12}	C_6H_{14}	C_7H_{16}	C_8H_{18}	C_9H_{20}	$C_{10}H_{22}$
Organic	15	25	13	15	425	1207	132	34	5	1
Inorganic	31	3	8	21	2053	3981	341	66	3	9

Table 25.2 Number of cyclic alkane associates deposited in the CCDC (01.04.2017).

Alkanes	3-membered	4-membered	5-membered	6-membered	7-membered	8-membered
Organic	4	3	23	283	8	18
Inorganic	2	0	100	498	2	4

host *n*-hexane, as evidenced by the X-ray structure analysis [38], and this type of materials can be used in the transportation or storage of alkanes.

In nature, alkane C—H bond oxidation is catalyzed by both heme and nonheme metalloenzymes, where substrate positioning via specific noncovalent interactions and steric effects can promote site-selective C—H functionalization [39–41]. This has motivated a considerable interest in the study of noncovalent interactions of alkanes in metal complex systems. A main strategy employed to promote the reactivity of alkanes is based on the use of transition metal complexes for C—H functionalization [42–50]. The possibility of using transition metal complexes to activate alkane C—H bonds has been considered viable since Shilov first discovered that certain platinum compounds could be employed to convert alkanes into alkyl halides and acetates [48, 51]. Complexes in which an alkane acts as a ligand to a metal center using its C—H bonding electrons (intermolecular interaction), engaging in a three-centre two-electron, so-called σ-alkane complexes, are of considerable interest due to their role as key intermediates in catalytic C—H activation processes and the challenges involved with their synthesis and characterization [52]. σ-Alkane complexes are closely related to complexes that show *agostic* M···H—C interactions. In an *agostic* complex, the two electrons involved in a C—H bond intramolecularly (in contrast to σ-alkane complexes) enter into an empty d-orbital of a transition metal, resulting in a three-centre two-electron bond [9, 52], and the C—H and M···H bond distances (typically 1.8–2.3 Å, with the M···H—C angle falling in the 90–140° range) are 5–20% longer than expected for isolated metal hydride and hydrocarbons. When the M···H distances and M···H—C angles fall into the ranges of 2.3–2.9 Å and 110–170°, respectively, this type of noncovalent interaction is called *anagostic* [9], which is more electrostatic in character. Transition metal catalyzed transformation of alkanes, in some cases, are proposed [9, 52] to proceed via alkyl intermediates featuring *agostic* or *anagostic* interactions, which will also be illustrated in this chapter.

25.2 Noncovalent Organic Associates with Alkanes

The C—H groups can act as weak hydrogen bond donors with a strength that depends on the carbon hybridisation [C(sp)—H > C(sp^2)—H > C(sp^3)—H], as proved by vibrational spectroscopy [53]. In alkanes, the C(sp^3)—H moiety can behave as a H-bond donor toward strong nucleophiles leading to organic or inorganic associates depending on the nature of the latter ones. The homonuclear R—H· · ·H—R dihydrogen interactions or the C· · ·H—C interactions hold alkanes together in the solid state and in the lipid parts of cell membranes [54, 55]. Differences between weak hydrogen bonds and R—H· · ·H—R dihydrogen types of van der Waals interactions concern their directionality and physical nature [56]. Thus, van der Waals interactions relate to the polarizability of the interacting molecules, and are characterized further by a dependence on r^{-6} (r is the intermolecular distance). The weak C_{alkane}—H· · ·nucleophile hydrogen bonds exhibit preferred directions and can be explained, in part, through molecular orbital theory, whereas the R—H· · ·H—R dihydrogen van der Waals interactions can be ascribed to nondirectional induced dipole· · ·induced dipole interactions [54].

C_{alkane}—H· · ·π and R—H· · ·H—R interactions were observed in the associate of methane with an organic zeolite i.e. 1,2-dimethoxy-*p-tert*-butylcalix[4]dihydroquinone (Figure 25.1) [55]. This microporous organic zeolite is valuable for trapping a significant amount of methane at modest pressures, due to its outstanding porosity, and C_{alkane}—H· · ·π bonds are responsible for the host–guest interactions. According to the X-ray analysis, one methane molecule (green) interacts directly through a C_{alkane}—H· · ·π interaction with one of the hydroquinone rings of the host framework (C_{alkane}· · ·π 3.23 Å) and with another symmetry-equivalent methane molecule ($C_{methane}$· · ·$C_{methane}$ 3.68 Å) forming a ring of eight methane and eight calixarene molecules (Figure 25.1) [55]. The other methane molecules (light gray) are located in the middle of this ring with weak intermolecular CH_3—H· · ·H—CH_3 dihydrogen van der Waals forces, and do not involve any specific interaction with the wall of the channels, which explains the presence of positional disorder at this site (Figure 25.1) [55].

Noncovalent interactions of the types R—H· · ·H—R, O—H· · ·H—O and C_{alkane}—H· · ·O were observed in the organic associate of 17-(4-(2,2-dimethylcyclopropyl)butan-2-yl)-4,4,10, 13,14-pentamethyl-2,3,4,5,6,7,8,10,12,13,14,15,16,17-tetradecahydro-1H-cyclopenta[a] phenanthren-3-ol with ethane (Figure 25.2a) [57]. Intermolecular noncovalent distances between hydrogen atoms are 1.475 and 1.076 Å in C_2H_5—H· · ·H—C_2H_5 and O—H· · ·H—O, respectively, whereas the hydrogen bond distance between oxygen and hydrogen atoms is 1.716 Å for

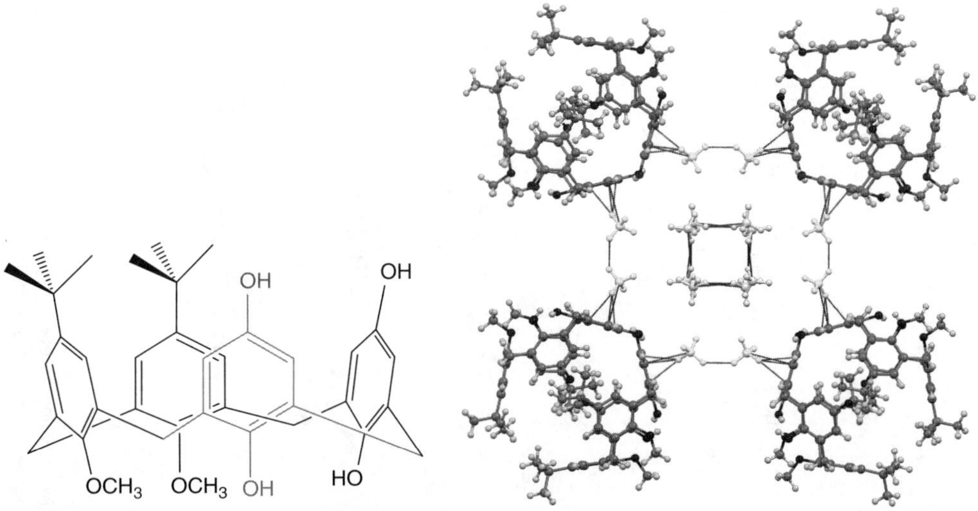

(a) (b)

Figure 25.1 1,2-Dimethoxy-*p-tert*-butylcalix[4]dihydroquinone (a) and its associate with methane (b) [55]. (*See insert for color representation.*)

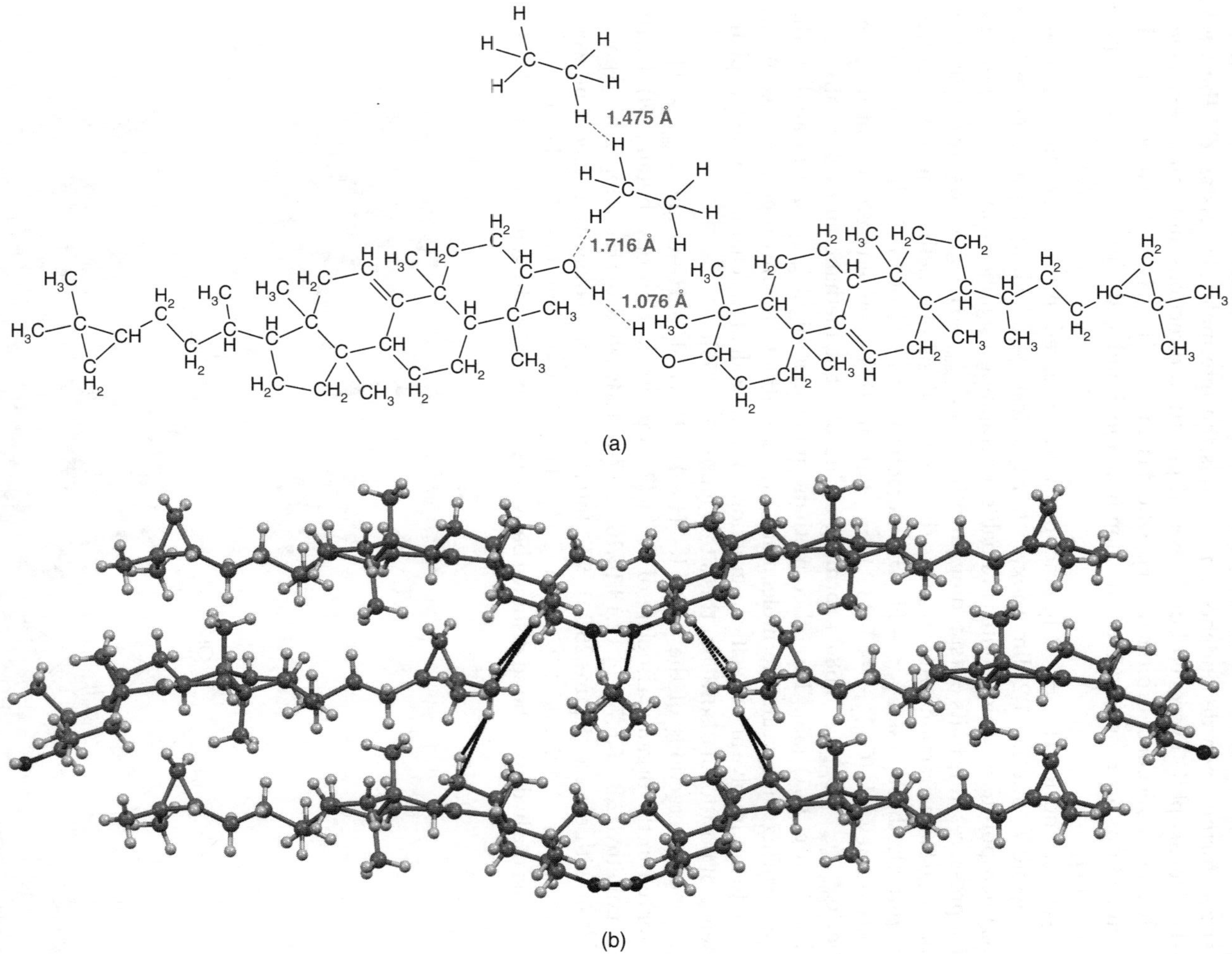

Figure 25.2 Noncovalent interactions in 17-(4-(2,2-dimethylcyclopropyl)butan-2-yl)-4,4,10,13,14-pentamethyl-2,3,4,5,6,7,8,10,12,13,14,15,16,17-tetradecahydro-1H-cyclo-penta[a]phenanthren-3-ol associate with ethane [57].

C_{ethane}—H· · ·O (Figure 25.2a). The van der Waals interactions support an ethane dimer that behaves as an H-bond donor toward an oxygen atom of a host molecule in the supramolecular channel, leading to a "strong" hydrogen bond (Figure 25.2b).

Almost all protons of butane behave as H-bond donors toward 3-benzoyl-3a-hydroxy-1-tosyl-1H-pyrazolo[5,1-a]isoindol-8(3aH)-one acceptors leading to C—H· · ·π and C—H· · ·O=C types of supramolecular synthons (Figure 25.3a) [58]. As expected, in this associate, the C—H· · ·O distances (2.621 Å) between oxygen and hydrogen atoms in the C—H· · ·O=C moiety are shorter than the H· · ·π ones in C—H· · ·π (2.928–3.967 Å) (Figure 25.3a), and the structure is stabilized by multiple intermolecular hydrogen bond interactions organized via C=O, S(=O)$_2$, π-systems (as H-bond acceptors), C—H and O—H (as H-bond donors) moieties.

In contrast to this example, the associate of (*Z*)-(2-(*bis*(perfluorophenyl)boryl)-2-(perfluoro-phenyl)-1-(trimethylsilyl)vinyl)(mesityl)((trimethylsilyl)ethynyl) phosphine with pentane is constructed via weak intermolecular dihydrogen van der Waals interactions (Figure 25.3b) [59].

The previously discussed types of noncovalent interactions are important in molecular recognition of flexible alkanes by synthetic hosts, which determine the host–guest structures [60]. Short unbranched alkanes, such as pentane, hexane, and heptane, prefer extended conformations in solution and in the solid state, as well as in a host cavity. The helically folded structures of linear alkanes require additional noncovalent interactions that compensate for the energetic cost of the gauche strains [60]. Thus, the nucleophilicity of H-bond acceptor sites of host molecules is crucial toward H-bond donor sites of alkanes in order to obtain "strong" noncovalent interactions promoting helically folded structures. Calix[5]arene offers several H-bond acceptor sites (five aromatic π-systems) and a wide bowl-shaped cavity, which is more attractive for capturing sizable alkane guests (Figure 25.4) [60]. Interaction of alkane guests, such as hexane, heptane, octane, nonane, decane, undecane, and dodecane, with the *p-tert*-butylcalix[5]arene host produce noncovalent host-guest complexes (Figure 25.4). X-ray

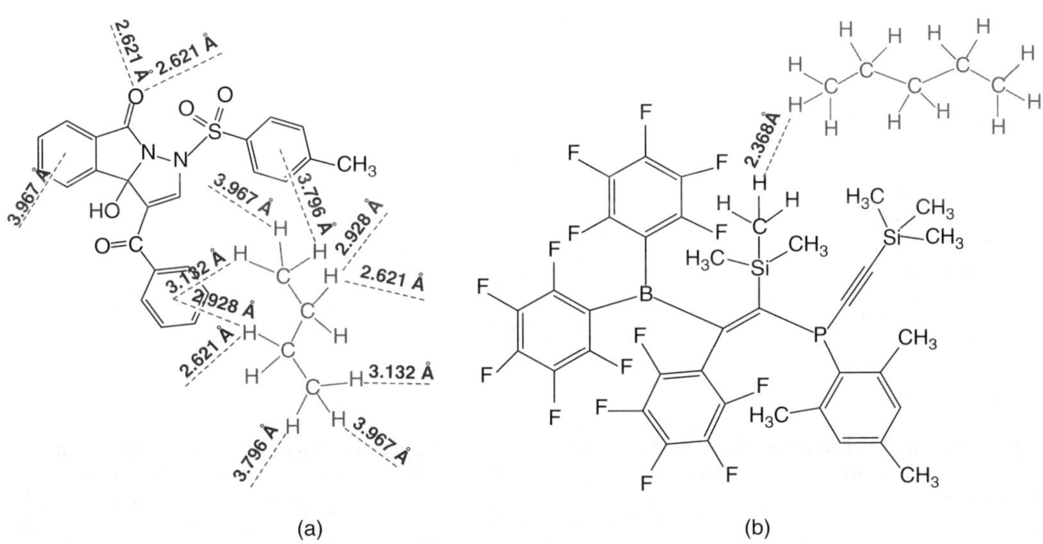

Figure 25.3 (a) Multiple hydrogen bonds between 3-benzoyl-3a-hydroxy-1-tosyl-1H-pyrazolo[5,1-a]isoindol-8(3aH)-one and butane [58]. (b) Dihydrogen van der Waals interactions in the associate of (*Z*)-(2-(*bis* (perfluorophenyl)boryl)-2-(perfluoro-phenyl)-1-(trimethylsilyl)vinyl)(mesityl)((tri-methylsilyl)ethynyl) phosphine with pentane [59]. (*See insert for color representation.*)

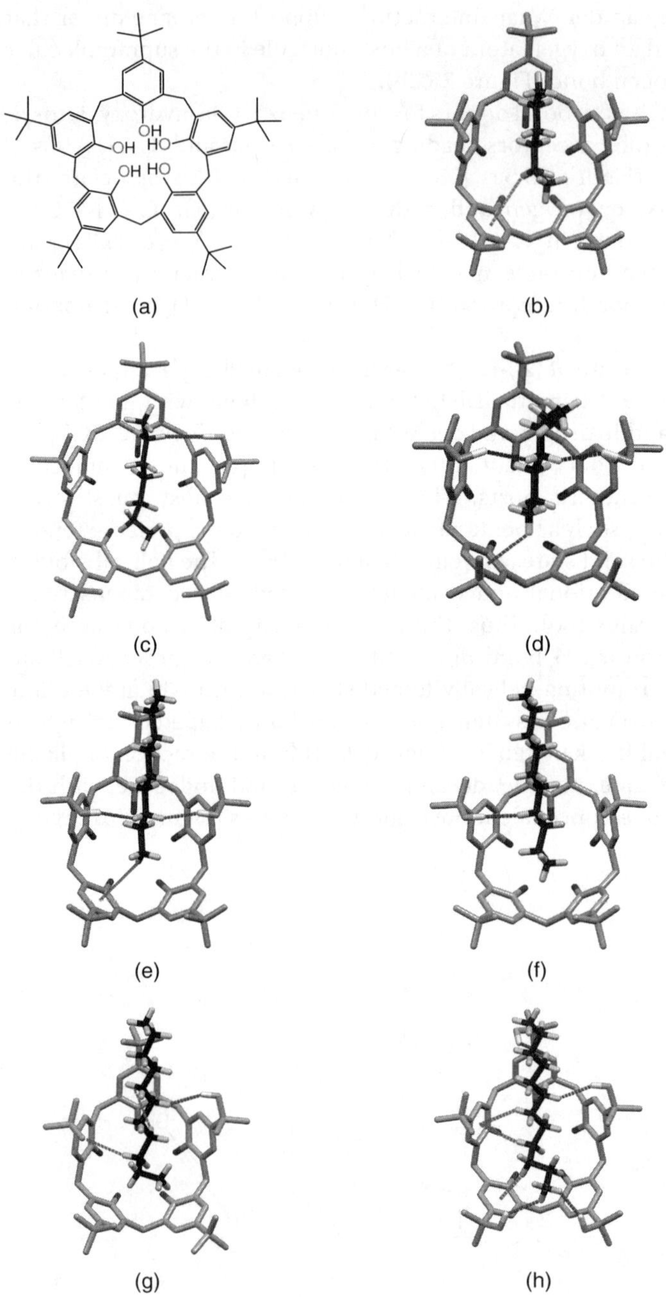

Figure 25.4 (a) *p-tert*-butylcalix[5]arene: (b) Host-guest complexes of *p-tert*-butylcalix[5]arene with hexane; (c) heptane; (d) octane; (e) nonane; (f) decane; (g) undecane; and (h) dodecane [60]. (*See insert for color representation.*)

diffraction analyses of these noncovalent compounds show that the alkyl chain termini of the guests adopt folded conformations and form helix-like structures in the hydrophobic cavity of the host (Figure 25.4), which are stabilized by C—H···π and dihydrogen van der Waals interactions. Moreover, the folded conformations of the alkyl chains involved in noncovalent interactions are bulkier than their extended conformations as indicated by DFT calculations. Correspondingly, the mutual induced-fit shape regulations between the host cavity and the flexible alkane guests play an important role in maximizing the host-guest interactions directed by the C(sp^3)—H···π noncovalent interactions, promoting particular conformations of the guests and a selective binding of the alkyl chains.

As in the above discussed acyclic alkanes, the cycloalkane hydrogens can also be involved in hydrogen bonding (as H-bond donors) and in intermolecular dihydrogen noncovalent interactions. Cycloalkanes may have many possible conformations depending on the size of the cycle. For instance, cyclopropane is the only cycloalkane that is planar, whereas cyclobutane and other higher cycloalkanes adopt nonplanar conformations in order to minimize the eclipsing of bonds on adjacent atoms. The angle strain in cyclobutane is less than in cyclopropane, whereas cyclopentane and higher cycloalkanes are virtually free of angle strain. With the exception of cyclopropane, all flexible cycloalkanes undergo rapid internal motion involving interconversion of nonplanar "puckered" conformations [61, 62]. Not only internal factors, but also external ones, such as intermolecular noncovalent interactions, are important in the stabilization of these conformers. For example, cyclooctane or cyclodecane has various theoretically possible low-energy conformations (Figure 25.5a), and one (or two) of them can be selected/stabilized by a host, following weak intermolecular C(guest)—H···π, C(guest)—H···O and dihydrogen van der Waals interactions in its cavity [63]. Thus, the guest-binding potential of the host (1*r*,3*r*,6*s*)-1,3,6,8-tetrakis(4-methoxy-2,6-dimethylphenyl)pyrene (TMDP) has been exploited to demonstrate the utility of this host system in the conformational locking of flexible cyclooctane in its chair–boat and boat–boat conformations (the asymmetric unit cell contains two molecules of cyclooctane) or cyclodecane in its boat–chair–boat conformation (Figure 25.5a–d) [63]. The structure with the lowest total energy for cyclooctane and cyclodecane is boat–chair and boat–chair–boat, respectively, which corresponds to the best combination of bond distances, bond angles, and conformation.

25.3 Noncovalent Inorganic Associates with Alkanes

Alkanes (or alkyl moieties of organometallics) may be involved in inter- or intramolecular noncovalent interactions with/in coordination compounds. A growing number of experimental facts has suggested that transition metals are capable of forming intermolecular noncovalent interactions with alkanes leading to σ-complexes, as well as intramolecular agostic or anagostic interactions with the C—H bond of appended alkyl moieties (Figure 25.6) [9, 52, 64–70]. All these noncovalent interactions lengthen the C—H bond and lower its bond stretching vibrational frequency, which is important in C—H functionalization in alkanes [65].

The first structurally characterized agostic compound [$TiEtCl_3(Me_2PCH_2CH_2PMe_2)$] was reported in 1980s (Figure 25.7) [71–73]. There is a direct interaction between the C—H moiety of the ethyl group and the titanium atom, which can be described in terms of an agostic interaction with distance of 2.29 Å. In the crystal structure of this complex a remarkably acute angle of 85.9(6)° at the β-carbon was found, whereas a normal angle close to 109° would have been expected in the absence of agostic interaction. Moreover, a similar but smaller distortion of an α-hydrogen was observed by neutron diffraction studies for the corresponding methyl-titanium agostic compound [73].

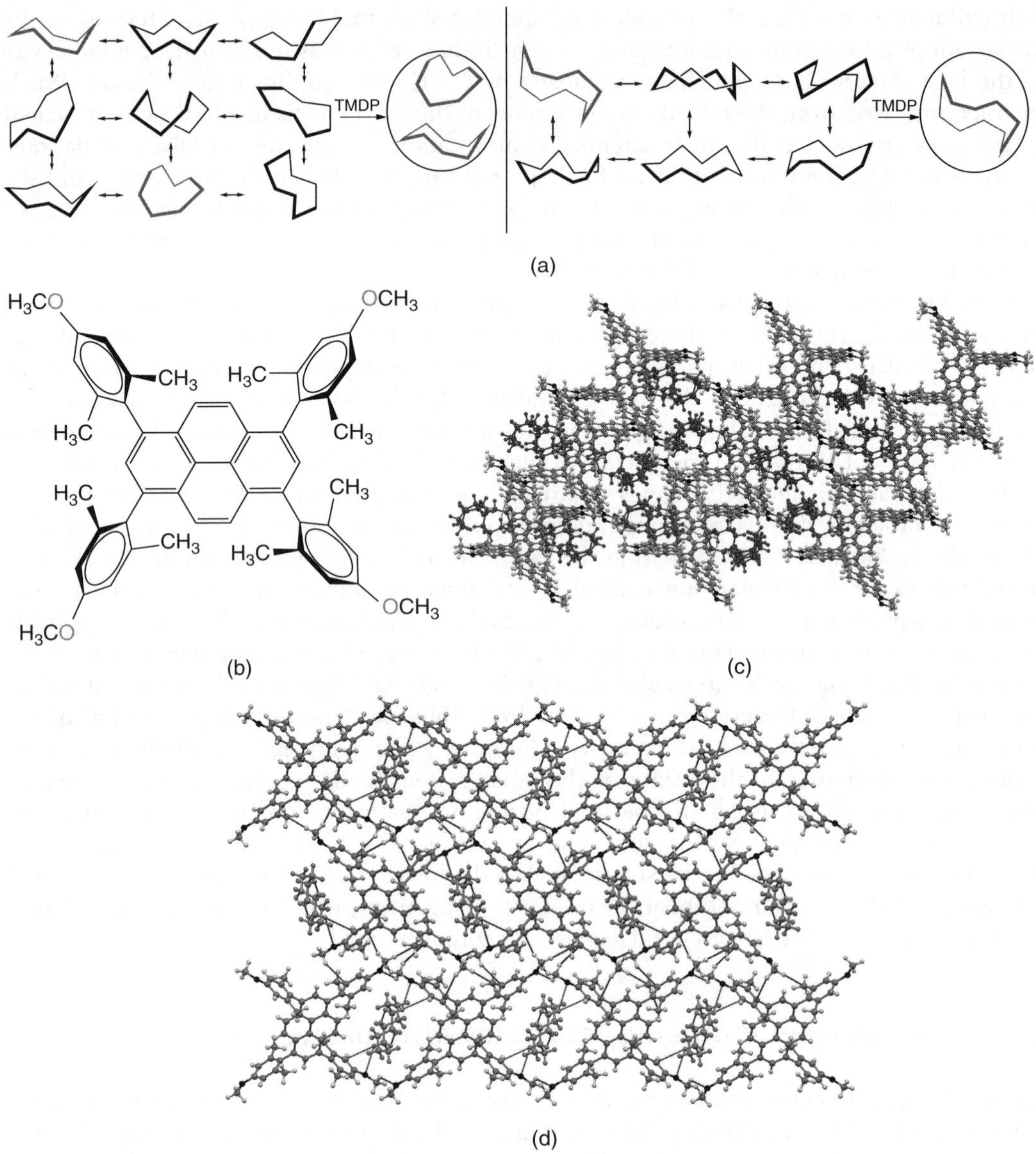

Figure 25.5 Various low-energy conformations of (a) cyclooctane (left) and cyclodecane (right); (b) highlighting (in dotted lines) those that are found in their inclusion compounds with **TMDP;** (c) Host-guest complexes of **TMDP** with cyclooctane; (d) or cyclodecane [63].

The σ-alkane complexes represent the intermolecular analogues of agostic or anagostic complexes in which a saturated C—H bond weakly interacts with the metal center [9]. In these complexes, the alkane can intermolecularly interact with the metal center in an end-on fashion through one, two, or three hydrogen atoms (k-*H*, k-*H,H* or k-*H,H,H*, respectively) (Figure 25.6a) [65]. The metal carbon distance decreases when the number of noncovalent interactions increases from k-*H* to k-*H,H,H*. Alternatively, the carbon atom of an alkane C—H moiety can also be involved in the intermolecular interaction with the metal center leading to the side-on η^2-*C,H* mode of interaction (Figure 25.6a).

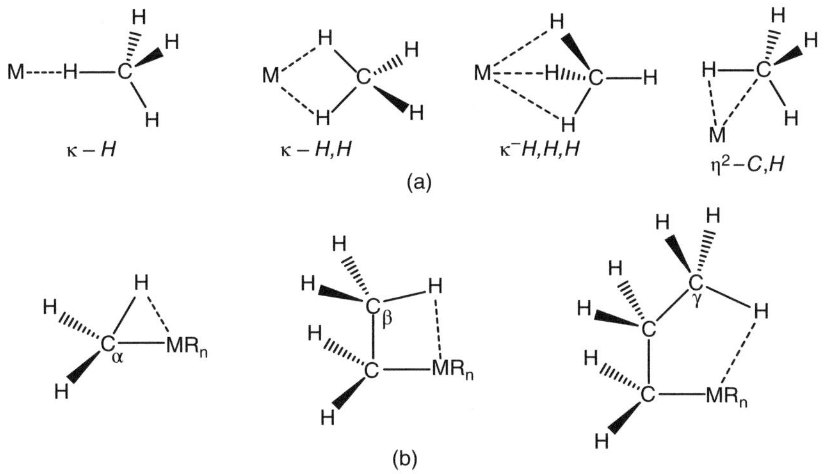

Figure 25.6 Possible modes of (a) intermolecular noncovalent methane C—H bond interaction with a metal center (σ-complexes); (b) α, β and γ agostic or anagostic intramolecular noncovalent interactions of an alkyl C—H bond with a metal center [64, 65].

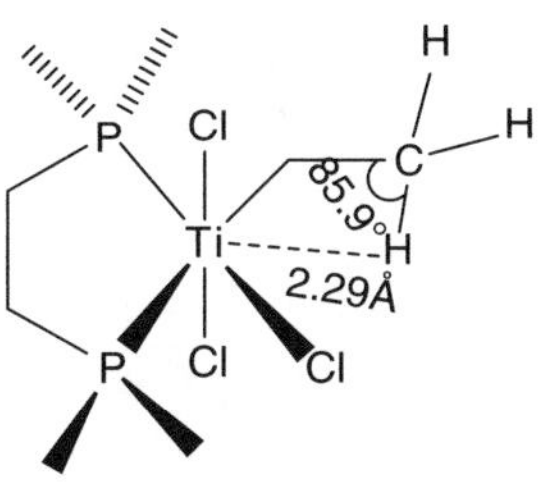

Figure 25.7 The first structurally characterized agostic compound [71–73].

The known σ-alkane complexes are unstable in solution, but a number of experimental facts suggest their existence. For instance, two isomers of the manganese propane or butane complexes derived from [$CpMn(CO)_3$] were generated photochemically at 130–136 K with the alkane as solvent and characterized by ^{1}H NMR and FTIR spectroscopies with *in situ* laser photolysis (Figure 25.8) [74]. D-labeling experiments using $DH_2CCH_2CH_3$ or $H_3CCD_2CH_3$ proved the η^2-*C,H* interaction, in which the methyl or methylene group interacts with the metal center.

The first crystallographic example of a noncovalent interaction of a metal center with an alkane was reported by Reed et al. in 1997, where *n*-heptane was found to be in close approach to the iron center in an Fe(II) porphyrin complex (Figure 25.9) [75]. In Fe(DAP)·(*n*-heptane), the hydrophobic pocket of a double A-framed porphyrin supported the heptane-iron associate through noncovalent interactions (Figure 25.9). The *n*-heptane molecule shows a positional disorder where its different orientations provide two different Fe· · ·C distances of 2.5 and 2.8 Å (Figure 25.9b), which are also proved by DFT calculations that propose a η^2-*C,H* noncovalent interaction between the *n*-heptane and the iron center.

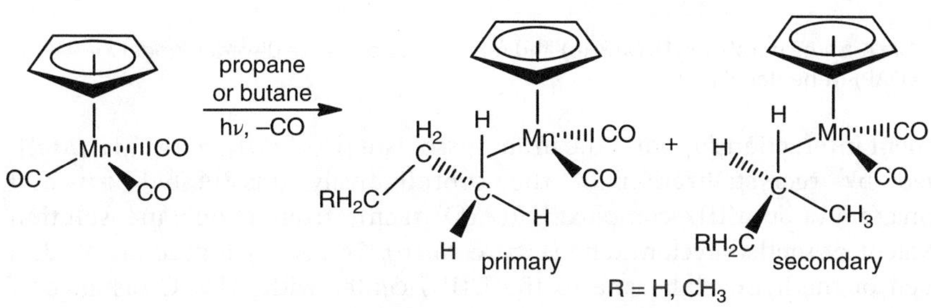

Figure 25.8 Photolysis of [$CpMn(CO)_3$] in liquid propane or butane [74].

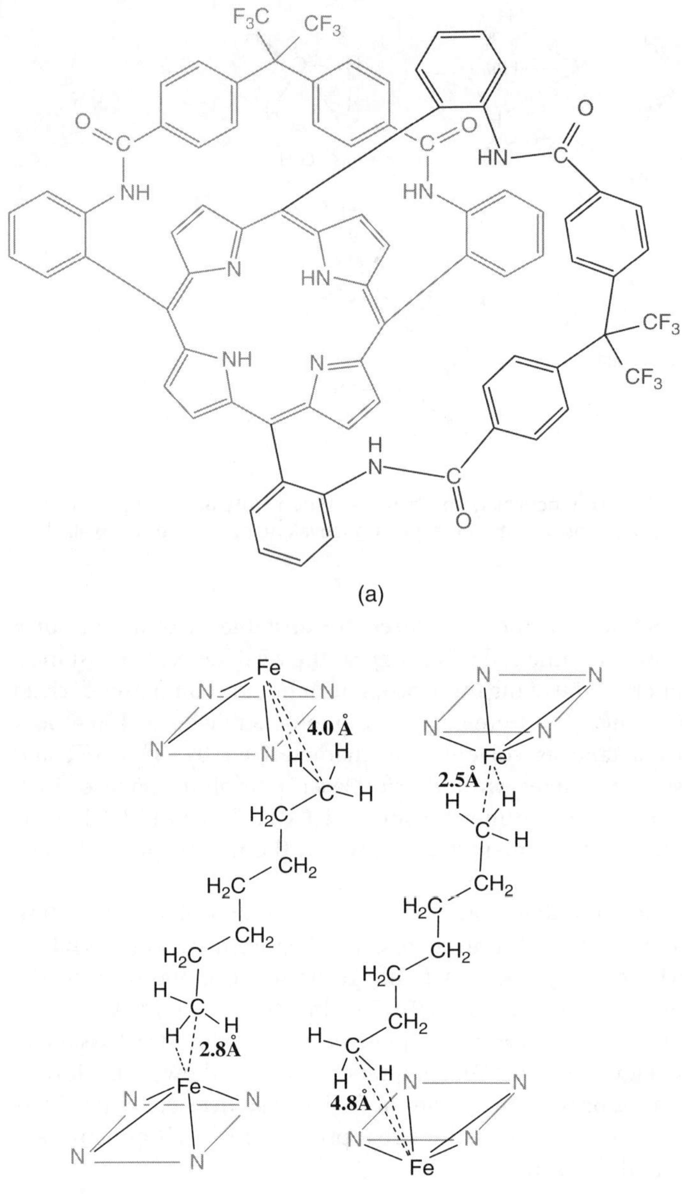

Figure 25.9 The (a) double A-frame porphyrin, H_2DAP; (b) and representation of the disorder model for n-heptane binding in Fe(DAP)·(n-heptane) [75].

A similar noncovalent interaction supports the uranyl associate $[U((ArO)_3tacn)(C_6H_{11}Me)]\cdot(C_6H_{11}Me)$ obtained by recrystallization of the coordinatively unsaturated *tris*-(aryl oxido)triazacyclononane uranium(III) complex $[U((ArO)_3tacn)]$ from a pentane solution containing 50 equivalent of methyl cyclohexane (Figure 25.10) [76]. X-ray diffraction studies confirm the approach of methyl cyclohexane to the U(III) center with a U···C distance of 3.864(7) Å, suggesting a noncovalent interaction between these atoms (the combined covalent radii of U and C is 3.34 Å). Moreover, C—H· · ·H—C and C—H· · ·C types of noncovalent

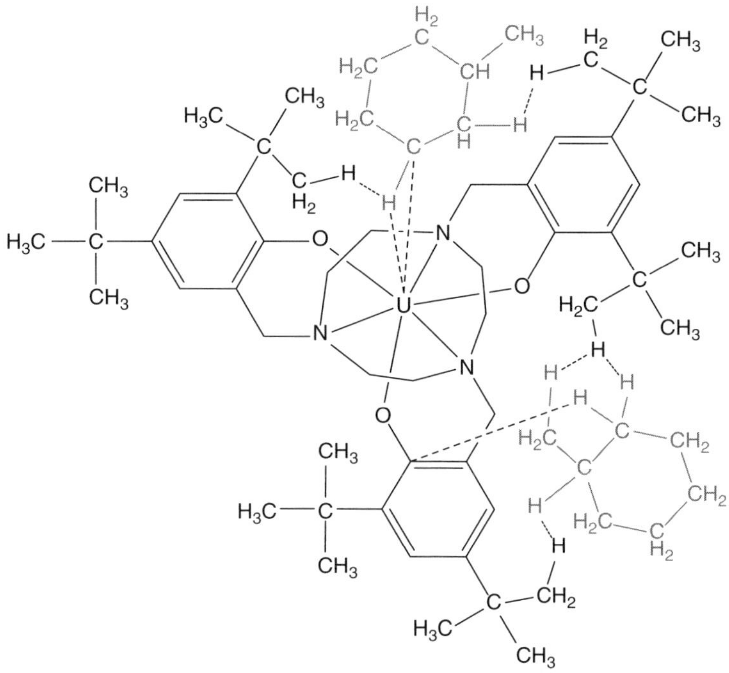

Figure 25.10 Noncovalent interactions in $[U((ArO)_3tacn)(C_6H_{11}Me)]\cdot(C_6H_{11}Me)$ [76]. (*See insert for color representation.*)

interactions between a solvated methyl cyclohexane and a $[U((ArO)_3tacn)\ (C_6H_{11}Me)]$ unit were also observed (Figure 25.10).

Many inorganic associates of alkanes (see Tables 25.1 and 25.2) were isolated in solid state when the alkanes were used as solvents. In those compounds, the alkane-metal interactions naturally depend on the nature of both components. For instance, cyclohexane C—H moieties can interact with the Bi(III) center and a ligand moiety of an hypervalent organobismuth(III) tetrafluoroborate complex [77] but, in the case of the associate of *trans*-$[Pt(Cl)\{BN(SiMe_3)_2CCPh\}(PMe_3)_2]$ with hexane, only alkane noncovalent interactions with a ligand were found [78] (Figure 25.11). In this adduct, a hexane molecule is involved in two noncovalent interactions with $C_{aromatic}$—H and —$SiMe_3$ moieties of the $BN(SiMe_3)_2CCPh$ ligand leading to the shortest C—HH—C and C—H· · ·C distances of 2.127 and 2.698 Å, respectively (Figure 25.11).

A particularly relevant field where noncovalent inorganic associates with alkanes can play an important role concerns the preparation of high-capacity gas storage metal organic frameworks (MOFs) [79, 80]. Understanding the sites of alkane adsorption within the MOFs requires the analysis of the types of noncovalent interactions occurring upon adsorption and can ultimately lead to the design of those materials for CH_4, C_2H_6, etc. uptake, separation, and transportation [81]. Noncovalent interactions between the alkane and the pore surface of MOFs can determine the orientation of the alkane molecule. This is especially important in cases where the adsorption behavior, which can include conformational changes, depends on the type and amount of alkane adsorbed [79–86]. Due to the disorder and thermal motion, it is difficult to achieve the experimental location of alkane molecules in MOFs, even by X-ray analysis at low temperature.

Methane plays a vital role in the chemical and petroleum industries, and to facilitate its use in modern technology it is desirable to find suitable porous materials to store it in a safe,

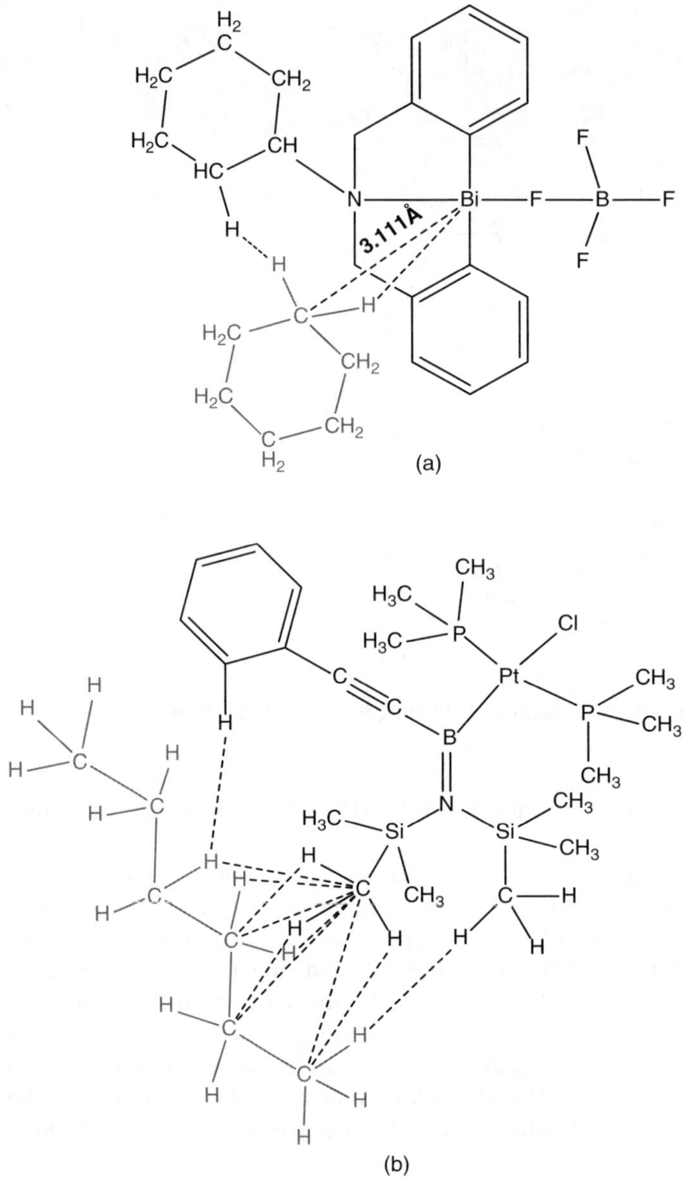

Figure 25.11 Inorganic associates of alkanes with hypervalent organobismuth(III) tetrafluoroborate (a) [77] and the platinum boryl complex *trans*-[Pt(Cl){BN(SiMe$_3$)$_2$CCPh}(PMe$_3$)$_2$] [78].

cheap, and convenient manner without the need for a massive pressure container. Up to date, various adsorbents, including zeolites and porous carbon materials, have been reported for methane uptake. MOFs are possible alternatives due to their large surface areas, well-defined pore sizes, and structural features (such as coordinatively unsaturated metals and versatile organic linkers) [79–94]. Zhou et al. have shown that methane adsorption occurs primarily on the open metal coordination sites of the MOFs M_2(dhtp) (M = Mg, Mn, Co, Ni, Zn; dhtp = 2,5-dihydroxyterephthalate) through enhanced noncovalent interactions [91] where methane is in dispersive contact with multiple MOF porous. The enhanced noncovalent sites are present exclusively in small cages and at the windows to these cages, whereas large cages with relatively flat pore surfaces bind very little methane. Hence, development of new MOFs for methane storage uses should promote the enrichment of coordinatively unsaturated metal sites, and the increase of the volume percentage of accessible small cages and channels, and minimize the fraction of large pores (e.g. by functionalization of organic linkers). Several types of noncovalent interactions can coexist, contributing to the total methane or other alkane storage capacities.

The first report on methane adsorption by a porous compound, $[Co_2(NO_3)_4(4,4'\text{-bpy})_3]$ having 52 $cm^3\ g^{-1}$, at 298 K and 3.0 MPa, dates back to as early as 1997, as pioneered by Kitagawa et al. [92]. To date, many MOFs based on Zn, Sc, Ni, Mn, Ca, etc. and carboxylate ligands, have already been constructed and applied for alkane adsorption [79, 82–86]. According to *in situ* crystallographic experiments, in those adsorption processes the multiplicity of

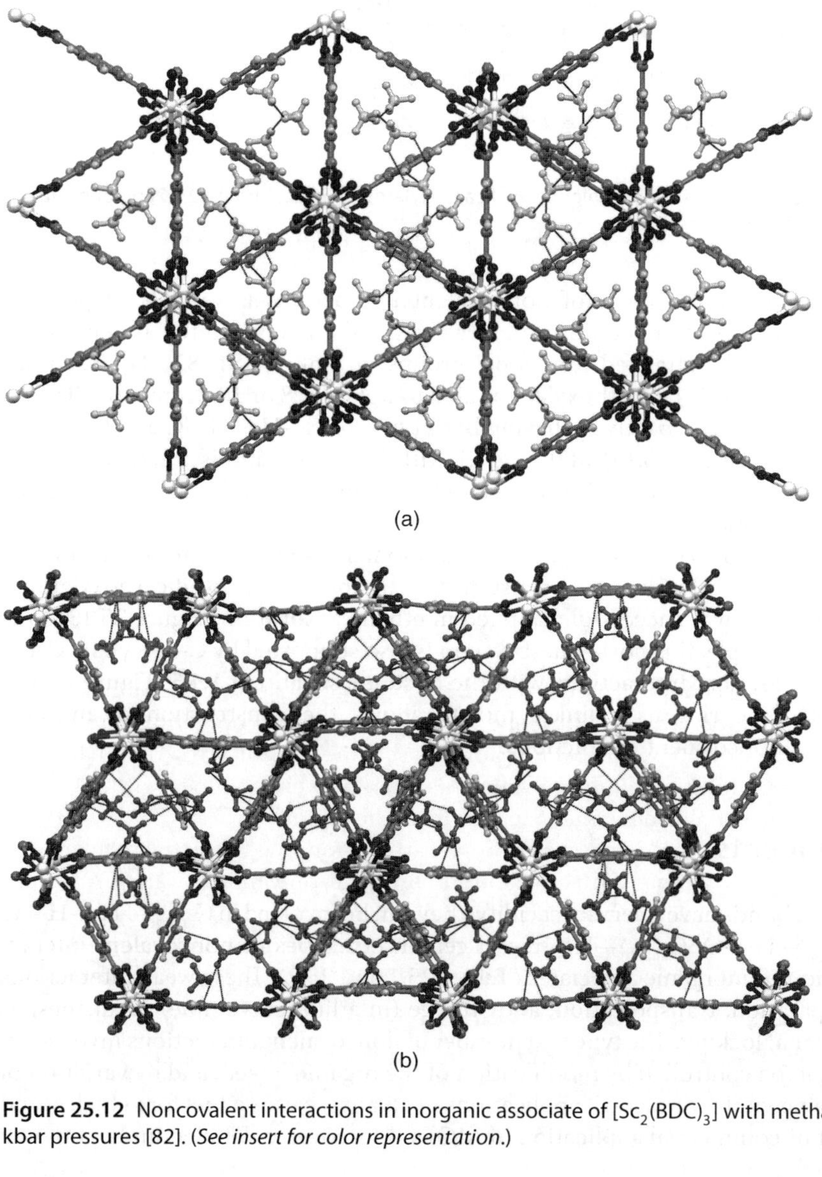

Figure 25.12 Noncovalent interactions in inorganic associate of $[Sc_2(BDC)_3]$ with methane at 3 (a) or 13 (b) kbar pressures [82]. (*See insert for color representation.*)

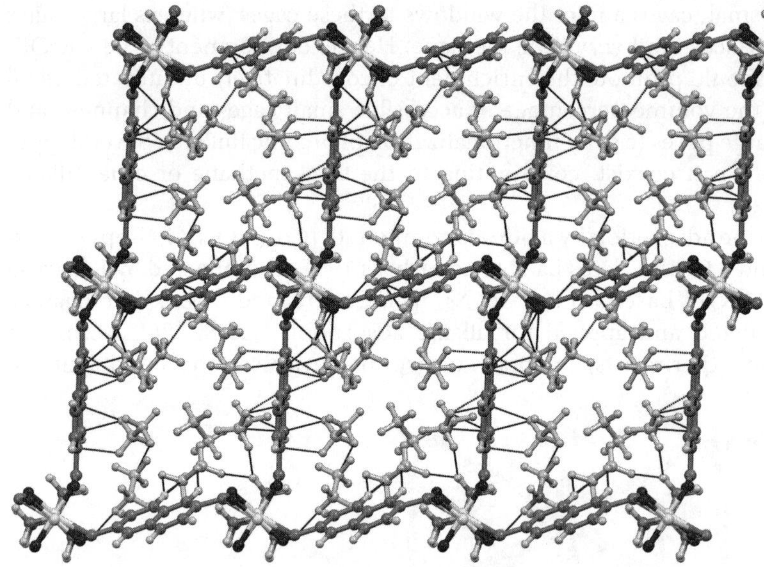

Figure 25.13 Inorganic associate of Ni_2(naphthalene-2,6-dicarboxylate)$_2$1,4-diazabicyclo[2.2.2] octane with ethane [84]. (*See insert for color representation.*)

alkane···alkane or alkane· · ·MOF types of noncovalent interactions is variable, depending on external factors (pressure, temperature, etc.), as well as on unsaturated (open) metal sites within pores and on the attached functional groups on linkers [81, 82]. For example, [$Sc_2(BDC)_3$] (BDC = benzene-1,4-dicarboxylate) is able to uptake 7.8 or 10.2 mmol of CH_4 per gram at 3 or 13 kbar pressure, respectively, having both types of C—H· · ·H—C and C—H· · ·π noncovalent interactions (Figure 25.12) [82]. The [$Sc_2(BDC)_3$]:CH_4 ratio is 1:2.27 and 1:2.88 at 3 or 13 kbar pressures, respectively, with the phase transitions from *Fddd* to *Fdd*2, and the shortest C—H· · ·H—C contact of 2.44 Å [82].

Not only for methane, but also for other alkanes, separation, transport, and storage are of economic importance for chemical industry. For instance, Ni_2(naphthalene-2,6-dicarboxylate)$_2$1, 4-diazabicyclo[2.2.2]octane was successfully applied in ethane adsorption (Figure 25.13) [84]. Similarly to the above example, this host–guest system is also supported by C—H· · ·H—C and C—H· · ·π types of noncovalent interactions with the host: guest ratio of 1:4.5 (Figure 25.13). The use of a naphthalene ring as a linker moiety allows the construction of multiple C_{ethane}—H· · ·$\pi_{naphthalene}$ noncovalent interactions.

25.4 Final Comments

Although alkane C—H bonds have a very low acidity, they can be involved in C_{alkane}—H· · ·H—C, C_{alkane}—H· · ·π, C_{alkane}—H· · ·O, C_{alkane}—H· · ·metal center, etc. types of noncovalent interactions in many organic and inorganic associates (Tables 25.1 and 25.2). These weak interactions are important in separation, transportation, and storage (in a limited volume) of alkanes, as well as in conformational locking. The type and number of noncovalent interactions involved in alkanes and MOFs can be controlled by modification of the organic linkers and by variation of the metal centers, being vital for practical applications of alkanes, e.g. as a fuel for vehicles, and for the development of commercial application of MOFs. Moreover, a detailed understanding of these interactions in alkane associates will be helpful for their functionalization and for driving alkane chemistry toward new frontiers.

Acknowledgments

This work has been partially supported by the Foundation for Science and Technology (FCT), Portugal [projects UID/QUI/00100/2013 and PTDC/QEQ-QIN/3967/2014]. This work also was supported by the RUDN University Program 5-100, Russian Federation.

References

1 J. D. van der Waals, Over de Continuïteit van den Gas-en Vloeistoftoestand (On the continuity of the gaseous and liquid state). PhD. Dissertation, University of Leiden, 1873.
2 Etter, M.C. (1990). *Acc. Chem. Res.* 23: 120.
3 Desiraju, G.R. (1995). *Angew. Chem. Int. Ed.* 34: 2311.
4 Jeffrey, G.A. (1997). *An Introduction to Hydrogen Bonding*. New York: Oxford University Press.
5 Desiraju, G.R. and Steiner, T. (1999). *The Weak Hydrogen Bond in Structural Chemistry and Biology*. New York: Oxford University Press Inc.
6 Steiner, T. (2002). *Angew. Chem. Int. Ed.* 41: 48.
7 Karshikoff, A. (2006). *Non-covalent Interactions in Proteins*. Singapore: Imperial College Press.
8 Desiraju, G.R. (2007). *Angew. Chem. Int. Ed.* 46: 8342.
9 Brookhart, M., Green, M.L.H., and Parkin, G. (2007). *Proc. Natl. Acad. Sci. U.S.A.* 104: 690.
10 Pihko, P.M. (ed.) (2009). *Hydrogen Bonding in Organic Synthesis*. Weinheim: Wiley-VCH.
11 Gilli, G. and Gilli, P. (2009). *The Nature of the Hydrogen Bond*. Oxford, USA: Oxford University Press.
12 Metrangolo, P. and Resnati, G. (2010). *Halogen Bonding: Fundamentals and Applications (Structure and Bonding)*. Heidelberg: Springer.
13 Hobza, P. and Müller-Dethlefs, K. (2010). *Non-covalent Interactions: Theory and Experiment*. UK: Royal Society of Chemistry.
14 Baev, A.K. (ed.) (2012). *Specific Intermolecular Interactions of Organic Compounds*. Heidelberg: Springer.
15 Scheiner, S. (2013). *Acc. Chem. Res.* 46: 280.
16 Alvarez, S. (2013). *Dalton Trans.* 42: 8617.
17 Baev, A.K. (ed.) (2014). *Specific Intermolecular Interactions of Nitrogenated and Bioorganic Compounds*. Heidelberg: Springer.
18 Cavallo, G., Metrangolo, P., Pilati, T. et al. (2014). *Cryst. Growth Des.* 14: 2697.
19 Giese, M., Albrecht, M., and Rissanen, K. (2015). *Chem. Rev.* 115: 8867.
20 Scheiner, S. (ed.) (2015). *Noncovalent Forces*. Dordrecht: Springer.
21 Gilday, L.C., Robinson, S.W., Barendt, T.A. et al. (2015). *Chem. Rev.* 115: 7188.
22 Singh, S.K. and Das, A. (2015). *Phys. Chem. Chem. Phys.* 17: 9596.
23 Maharramov, A.M., Mahmudov, K.T., Kopylovich, M.N., and Pombeiro, A.J.L. (eds.) (2016). *Non-covalent Interactions in the Synthesis and Design of New Compounds*. Hoboken: Wiley.
24 Cavallo, G., Metrangolo, P., Milani, R. et al. (2016). *Chem. Rev.* 116: 2478.
25 Mahadevi, A.S. and Sastry, G.N. (2016). *Chem. Rev.* 116: 2775.
26 Wang, H., Wang, W., and Jin, W.J. (2016). *Chem. Rev.* 116: 5072.

27 Kolář, M.H. and Hobza, P. (2016). *Chem. Rev.* 116: 5155.
28 Mahmudov, K.T. and Pombeiro, A.J.L. (2016). *Chem. Eur. J.* 22: 16356.
29 Crabtree, R.H. (2017). *Chem. Soc. Rev.* 46: 1720.
30 Davis, H.J. and Phipps, R.J. (2017). *Chem. Sci.* 8: 864.
31 Mahmudov, K.T., Kopylovich, M.N., Guedes da Silva, M.F.C. et al. (2017). *Coord. Chem. Rev.* 345: 54.
32 Mahmudov, K.T., Kopylovich, M.N., Guedes da Silva, M.F.C. et al. (2017). *Dalton Trans.* 46: 10121.
33 Crabtree, R.H. (1985). *Chem. Rev.* 85: 245.
34 Crabtree, R.H. (1995). *Chem. Rev.* 95: 987.
35 Crabtree, R.H. (2001). *J. Chem. Soc., Dalton Trans.* 2437.
36 Crabtree, R.H. (2004). *J. Organomet. Chem.* 689: 4083.
37 Arndtsen, B.A., Bergman, R.G., Mobley, T.A. et al. (1995). *Acc. Chem. Res.* 28: 154.
38 Werz, D.B., Gleiter, R., and Rominger, F. (2002). *J. Am. Chem. Soc.* 124: 10638.
39 Que, L. Jr., and Tolman, W.B. (2008). *Nature* 455: 333.
40 Shteinman, A.A. (2008). *Russ. Chem. Rev.* 77: 945.
41 Ortiz de Montellano, P.R. (2010). *Chem. Rev.* 110: 932.
42 Pérez, P.J. (2012). *Alkane C–H Activation by Single-Site Metal Catalysis.* Dordrecht: Springer.
43 Caballero, A. and Pérez, P.J. (2013). *Chem. Soc. Rev.* 42: 8809.
44 Hashiguchi, B.G., Bischof, S.M., Konnick, M.M. et al. (2012). *Acc. Chem. Res.* 45: 885. (other reviews in the same journal topical issue also include alkane reactions).
45 Pombeiro, A.J.L. (2014). Toward functionalization of alkanes under environmentally benign conditions. In: *Advances in Organometallic Chemistry and Catalysis* (ed. A.J.L. Pombeiro). Hoboken: Wiley, Ch. 2.
46 Kirillov, A.M., Kirillova, M.V., and Pombeiro, A.J.L. (2013). *Adv. Inorg. Chem.* 65: 1.
47 Kirillov, A.M., Kirillova, M.V., and Pombeiro, A.J.L. (2012). *Coord. Chem. Rev.* 256: 2741.
48 Shilov, A.E. and Shul'pin, G.B. (1997). *Chem. Rev.* 97: 2879.
49 Shul'pin, G.B. (2016). *Catalysts* 6: 50.
50 Shul'pin, G.B., Mini-rev.(2009). *Organomet. Chem.* 6: 95.
51 Gol'dshleger, N.F., Es'kova, V.V., Shilov, A.E. et al. (1972). *Zh. Fiz. Khim. (Engl. Transl.)* 46: 785.
52 Weller, A.S., Chadwick, F.M., and McKay, A.I. (2016). *Adv. Organomet. Chem.* 66: 223.
53 Allerhand, A. and von, Schleyer, P. (1963). *J. Am. Chem. Soc.* 85: 1715.
54 Echeverría, J., Aullón, G., Danovich, D. et al. (2011). *Nat. Chem.* 3: 323.
55 Tedesco, C., Erra, L., Brunelli, M. et al. (2010). *Chem. Eur. J.* 16: 2371.
56 Steiner, T. and Desiraju, G.R. (1998). *Chem. Commun.* 891.
57 Lin, Y.-L., Wang, W.-Y., Kuo, Y.-H. et al. (2001). *Chem. Pharm. Bull.* 49: 1098.
58 Zhou, Q.-F., Ge, F.-F., Chen, Q.-Q. et al. (2016). *RSC Adv.* 6: 1395.
59 Möbus, J., Bonnin, Q., Ueda, K. et al. (2012). *Angew. Chem. Int. Ed.* 51: 1954.
60 Haino, T. (2016). *Bull. Chem. Soc. Jpn.* 89: 220.
61 Cotton, F.A., Day, V.W., Frenz, B.A. et al. (1973). *J. Am. Chem. Soc.* 95: 4522.
62 Bocian, D.F., Pickett, H.M., Rounds, T.C. et al. (1975). *J. Am. Chem. Soc.* 97: 687.
63 Moorthy, J.N., Natarajan, P., and Venugopalan, P. (2009). *J. Organomet. Chem.* 74: 8566.
64 Zins, E.-L., Silvi, B., and Alikhani, M.E. (2015). *Phys. Chem. Chem. Phys.* 17: 9258.
65 Perutz, R.N. and Hall, C. (1996). *Chem. Rev.* 96: 3125.
66 Brookhart, M. and Green, M.L.H. (1983). *J. Organomet. Chem.* 250: 395.
67 Crabtree, R.H. (1993). *Angew. Chem., Int. Ed.* 32: 789.
68 Schneider, J.J. (1996). *Angew. Chem., Int. Ed.* 35: 1068.

69 Scherer, W. and McGrady, G.S. (2004). *Angew. Chem. Int. Ed.* 43: 1782.
70 Etienne, M. and Weller, A.S. (2014). *Chem. Soc. Rev.* 43: 242.
71 Dawoodi, Z., Green, M.L.H., Mtetwa, V.S.B. et al. (1986). *J. Chem. Soc., Dalton Trans.* 1629.
72 Dawoodi, Z., Green, M.L.H., Mtetwa, V.S.B. et al. (1982). *J. Chem. Soc., Chem. Commun.* 802.
73 Dawoodi, Z., Green, M.L.H., Mtetwa, V.S.B. et al. (1982). *J. Chem. Soc., Chem. Commun.* 1410.
74 Calladine, J.A., Duckett, S.B., George, M.W. et al. (2011). *J. Am. Chem. Soc.* 133: 2303.
75 Evans, D.R., Drovetskaya, T., Bau, R. et al. (1997). *J. Am. Chem. Soc.* 119: 3633.
76 Castro-Rodriguez, I., Nakai, H., Gantzel, P. et al. (2003). *J. Am. Chem. Soc.* 125: 15734.
77 Zhang, X., Qiu, R., Tan, N. et al. (2010). *Tetrahedron Lett.* 51: 153.
78 Braunschweig, H., Ye, Q., Radacki, K. et al. (2011). *Dalton Trans.* 40: 3666.
79 He, Y., Zhou, W., Krishna, R. et al. (2012). *Chem. Commun.* 48: 11813.
80 Mason, J.A., Veenstra, M., and Long, J.R. (2014). *Chem. Sci.* 5: 32.
81 Carrington, E.J., Vitórica-Yrezábal, I.J., and Brammer, L. (2014). *Acta Cryst.* B70: 404.
82 Sotelo, J., Woodall, C.H., Allan, D.R. et al. (2015). *Angew. Chem. Int. Ed.* 54: 13332.
83 Kim, H., Samsonenko, D.G., Das, S. et al. (2009). *Chem. Asian J.* 4: 886.
84 Bon, V., Klein, N., Senkovska, I. et al. (2015). *Phys. Chem. Chem. Phys.* 17: 17471.
85 Banerjee, D., Wang, H., Plonka, A.M. et al. (2016). *Chem. Eur. J.* 22: 11816.
86 Banerjee, D., Wang, H., Gong, Q. et al. (2016). *Chem. Sci.* 7: 759.
87 He, Y., Zhou, W., Qiand, G. et al. (2014). *Chem. Soc. Rev.* 43: 5657.
88 Peng, Y., Krungleviciute, V., Eryazici, I. et al. (2013). *J. Am. Chem. Soc.* 135: 11887.
89 Konstas, K., Osl, T., Yang, Y. et al. (2012). *J. Mater. Chem.* 22: 16698.
90 Kuppler, R.J., Timmons, D.J., Fang, Q.-R. et al. (2009). *Coord. Chem. Rev.* 253: 3042.
91 Wu, H., Zhou, W., and Yildirim, T. (2009). *J. Am. Chem. Soc.* 131: 4995.
92 Kondo, M., Yoshitomi, T., Seki, K. et al. (1997). *Angew. Chem. Int. Ed. Engl.* 36: 1725.
93 He, Y., Chen, F., Li, B. et al. (2018). *Coord. Chem. Rev.* 373: 167. 10.1016/j.ccr.2017.10.002.
94 Lin, R.-B., Xiang, S., Xing, H. et al. (2017, in press). *Coord. Chem. Rev.* doi: 10.1016/j.ccr.2017.09.027.

26

Alkane Complexation

Robert H. Crabtree

Yale University Yale Chemistry Department, 225 Prospect Street, New Haven, CT, USA
Energy Sciences Institute, 810 West Campus Drive, West Haven, CT, USA

26.1 Introduction and Early Work

Although alkanes R—H can react with metal complexes to give oxidative addition products, $L_nM(R)(H)$ (R = alkyl, L_nM = metal and ancillary ligands) [1], organometallic complexes of intact alkanes of the type $L_nM(H—R)$ were sought for many years without success. This failure is not surprising – alkanes are poor ligands, being both weak donors via their low energy C—H σ bonding electrons and weak back-bonding acceptors via their high energy C—H σ^* orbitals. Sufficient evidence for alkane complexation has been available since 1996, however, to permit Hall and Perutz to review the state of the field [2]. They noted that applying the standard η^n nomenclature is complicated by the ambiguity of whether to count the carbon as metal-bound or not. Is $L_nM(H—R)$ an η^1-*H* or an η^2-*C,H* ligand? One way to avoid this problem would be to retain the atom labels, *H* or *C,H*, after the η^n symbol. Alternatively, since the CH bond is the donor function, one could argue that both atoms are necessarily involved, and η^2 is the more appropriate label. Important and comprehensive recent reviews of alkane complexes are available [3, 4].

Binding as weakly as they do, special efforts are normally needed to prevent alkane dissociation from their complexes. Common strategies involve (i) working in matrix at low temperature, (ii) using host-guest effects to promote binding, and (iii) generating and trapping the alkane adjacent to the metal in a crystal lattice. When working in solution, great care must be taken to eliminate all potential competing ligands such as solvent and counterion, as well as avoiding cluster formation by which an unsaturated fragment L_nM prefers to react with itself rather than bind the alkane.

The matrix strategy was the first to be applied, and in important early work alkane complexes were identified by IR spectroscopy in a low-temperature matrix. For example, from 1972, $Cr(CO)_5(CH_4)$ and $Fe(CO)_4(CH_4)$ were detected in matrix at 12 K [5–8]. More recently, George et al. and Perutz et al. explored this strategy with regard to fast time-resolved infrared (TRIR) spectroscopy [9, 10]. The related $Cr(CO)_5(C_6H_{12})$ had also been detected in flash photolysis on the hexacarbonyl in cold cyclohexane solution, so working in matrix is not absolutely required for success [11]. Success was first achieved at ambient temperature in solution by Hermann et al. [12] in the case of $M(CO)_5$(cyclohexane) (M = Cr, Mo, W). The strength of the metal–alkane bond in $M(CO)_5$(heptane) was estimated by photoacoustic calorimetry at 10 ± 2.5 (Cr) and $13 \pm 3\,kcal\,mol^{-1}$ (W) [13]. Gas-phase experiments [14] on $M(CO)_5$ alkane adducts give similar results, with the advantage of altogether avoiding both solvent and matrix. These values

Alkane Functionalization, First Edition. Edited by Armando J. L. Pombeiro and M. Fátima C. Guedes da Silva.

can usefully be compared with the ones [15] for $M(CO)_5(H_2)$: 15 ± 1 kcal mol^{-1} for Cr and >16 (exptl.) or 19.1 kcal mol^{-1} (calc'd) for W. This 50% advantage for the H_2 over the CH complex may be related to the enhanced basicity for H_2, arising from H having a lower electronegativity than C, or to the lower steric hindrance for H_2 bringing the ligand closer to the metal. The strictly side-on structure of L_nM—(H_2) also allows good back-bonding overlap with both lobes of the σ^* orbital of the ligand, not just one as in L_nM—(H—R), where the ligand binds slantwise, hydrogen atoms first.

It now seems likely [2, 10] that transient alkane complexation occurs whenever alkanes are either formed, as in reductive eliminations of alkyl hydrides, or act as reactants, as in the influential early Shilov chemistry [16–18]. They have also often been proposed as transients on the basis of computational or experimental work [1]. This chapter, however, emphasizes alkane complexes that are stable enough at or near room temperature to be isolated or detected by standard organometallic methods in the solid state or in solution.

26.2 Comparison of C—H with H—H Complexation

Work with dihydrogen complexes [15, 19] provided early encouragement for work on alkanes because H_2 can, in some respects, be considered as the $n = 0$ member of the alkane series, $C_nH_{(n+2)}$. Dihydrogen complexation and reactivity when bound to metals therefore gave early warning of what might be expected for the $n = 1$ and higher members of the series. The two main reactivity patterns of H_2 with transition metals, beyond simple dissociation, are oxidative addition to give a dihydride (Scheme 26.1, Eq. (1a)) and proton loss to give a monohydride (Scheme 26.1, Eq. (1b)). This means that dihydrogen complexation underlies much of the reactivity of H_2 with these metals. From this body of results it is but a short step to alkane complexation and reactivity, where, once again, oxidative addition to give an alkyl hydride (Scheme 26.1, Eq. (2a)) and proton loss to give an alkyl complex (Scheme 26.1, Eq. (2b)) are both important pathways, often from transient alkane complexes that are not directly observed [20].

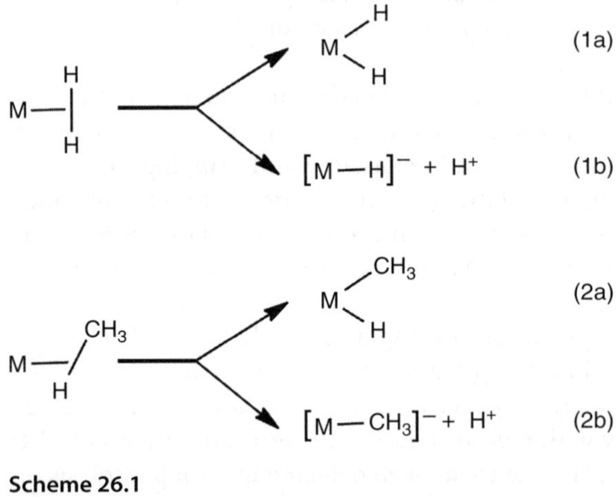

Scheme 26.1

Alkane loss to generate vacant sites at the metal probably occurs in the hydrogenation of alkene ligands present in catalyst precursors, as in the case of [(cod)IrLL′]BF_4 (cod = 1,5-cyclooctadiene. L = $P(C_6H_{11})_3$ or other P-donor. L′ = pyridine or other N-donor) [21, 22]. Recent work from Weller [23], discussed in Section 26.1, supports this proposal by trapping the alkane as a ligand in closely related hydrogenation reactions.

26.3 Sigma Complexation of C—H Bonds in Agostic Complexes

A very large number of agostic complexes are known in which CH bonds of an alkyl or aryl substituent of a ligand that is already stably bound, typically via a P- or N-donor, also coordinates to the metal [24]. In this type of binding, the CH bond benefits from the tight binding of the adjacent N- or P-donor, and from the chelate effect. The stability of the resulting M(H—C) interaction is such that hundreds of crystal structures have been obtained. Detailed analysis shows that the agostic C—H bond is stretched but not broken on binding, suggesting that the structure is an arrested state that lies on the trajectory for oxidative addition [25, 26]. Binding is not always side-on as in H_2 complexes but is aslant, as indicated in Scheme 26.1, Eq. (2). A linear C—H···M arrangement, on the other hand, would be more suggestive of a hydrogen bonding interaction, as has been discussed in detail elsewhere [27].

The existence of agostic complexes clearly indicated that alkane complexation should be possible, at least in principle. The problem is the weak binding energy of the C—H bond to the metal, as suggested both by calculation and by the low barrier for fluxional exchange between CH bonds, for example in an agostic CH_3 group.

26.4 Silane σ-Complexes

Another clue that alkane complexes should be accessible came from the existence of numerous sigma complexes [15, 28, 29] with element-hydrogen bonds other than H—H and C—H, for example with Si—H bonds. These do not require chelation for stability because the interaction energy is sufficient to permit them to bind strongly as monodentate ligands via an Si—H bond. The advantage of Si—H bonds over C—H bonds comes in part from the greater hydridic character of the H atom, or, more correctly, greater donor power of the Si—H bond itself. The weaker strength of the Si—H bond itself also plays a role because the Si—H σ^* orbital is now more accessible for back donation than in H—H or C—H, strengthening both σ and π silane-metal bonding components.

26.5 Transient Alkane Complexation

In Shilov Chemistry, aqueous Pt(II) salts bring about H/D isotope exchange between alkanes and a deuterated solvent, typically CH_3COOD/D_2O. Careful mechanistic work by Shilov showed that at a very early stage of the reaction, multiple H/D isotope exchanges had already occurred in a small fraction of the alkane molecules. Evidently, multiple H/D exchange was possible during a single contact of the alkane with the catalyst, as confirmed by detailed quantitative kinetics, which ruled out alkane dissociation and reassociation. For most alkanes, polydeuteration was most easily explained by a proposed dehydrogenation to form the alkene (e.g. C_2H_4 from C_2H_6. Scheme 26.2, Eqs. (3)–(5)). This alkene could be easily envisaged as binding to the

$$L_nPt + C_2H_6 \longrightarrow [L_nPtC_2H_5]^- + H^+ \qquad (3)$$

$$[L_nPtC_2H_5]^- \longrightarrow [L_nPtH(C_2H_4)]^- \qquad (4)$$

$$[L_nPtH(C_2H_4)]^- \xrightarrow{2D^+} L_nPt + C_2H_4D_2 + H^+ \qquad (5)$$

Scheme 26.2

metal and undergoing multiple reversible 1,2-migratory insertions/β-elimination with putative Pt—D bonds of the catalyst.

This explanation failed for CH_4, however, because although there is no C-1 alkene, multiple exchange occurred exactly as before. Shilov was forced to postulate carbene intermediates of type Pt=CH_2 as intermediates (Scheme 26.3, Eqs. (6) and (7)) with 1,1-migratory insertions/α-elimination with D^+ exchange. In fact, what appears to be the true explanation in all cases of polydeuteration involves the formation of alkane complexes that remain metal-bound as σ-complexes (Scheme 26.3, Eq. (8)) while multiple exchanges occur [29]. This was not the only case where transient alkane complexation was required to explain multiple isotope exchange, as shown for H/D exchange in t-BuCH=CH_2 with $[IrH_2(O_2CCF_3)(PAr_3)_2]$ as catalyst [30].

$$L_nPt + CH_4 \longrightarrow [L_nPtCH_3]^- + H^+ \quad (6)$$

$$[L_nPtCH_3]^- \longrightarrow [L_nPtH(CH_2)]^- \quad (7)$$

$$[L_nPtCH_3]^- \xrightarrow{nD^+} L_nPt(H\text{-}CD_n) + nH^+ \quad (8)$$

Scheme 26.3

In the case of $[(C_5Me_5)Os(dmpm)(CH_2R)H]$, the H/D exchange of Scheme 26.4 even proved to be fast on the NMR timescale, with the alkane complex being identified as the transient intermediate [31].

$$Cp^*(dmpm)Os(CH_3)H' \rightleftharpoons Cp^*(dmpm)Os(CH_2H')H \quad (9)$$

Scheme 26.4

The Bergman group had early indications from kinetics and isotope work for the presence of transient alkane complexes prior to CH oxidative addition to the photogenerated $Cp^*Ir(PMe_3)$ fragment [32]. Later work led to the direct detection of a transient alkane complex prior to oxidative addition of a C—H bond in the case of a CpRh(CO) transient formed by photoextrusion of CO from $CpRh(CO)_2$ in a solvent mixture of alkane in liquid Kr or Xe. The first formed product to be detected by time resolved IR spectroscopy was the rare gas complex CpRh(CO)(Xe or Kr), which gave way to the alkane complex, CpRh(CO)(H—R), which, in turn, subsequently underwent C—H oxidative addition to give the final alkyl hydride, CpRh(CO)(R)(H) [33]. The CH oxidative addition reaction could also be followed in the gas phase where irradiation of $CpRh(CO)_2$ gave unsolvated CpRh(CO), detected by transient absorption IR spectroscopy, followed by activation of C—H bonds, which occurred on almost every collision for alkanes of moderate size [34].

Another important early example was Geftakis & Ball's direct NMR detection of the $[CpRe(CO)_2(\text{cyclopentane})]$ alkane complex by photoextrusion of CO from $CpRe(CO)_3$ in cold cyclopentane as solvent [35]. This time, there was no subsequent oxidative addition, although the complex itself was not isolable because of the loss of alkane on warming. Estimates for the alkane–metal bond dissociation energy for these early examples ranged from 5 to 10 kcal mol^{-1} [33], not quite enough for isolation of the complex.

Outside a cryogenic matrix, methane had long evaded capture even in a transient complex, when Bernskoetter et al. detected one from the low temperature (−110 °C) protonation of the

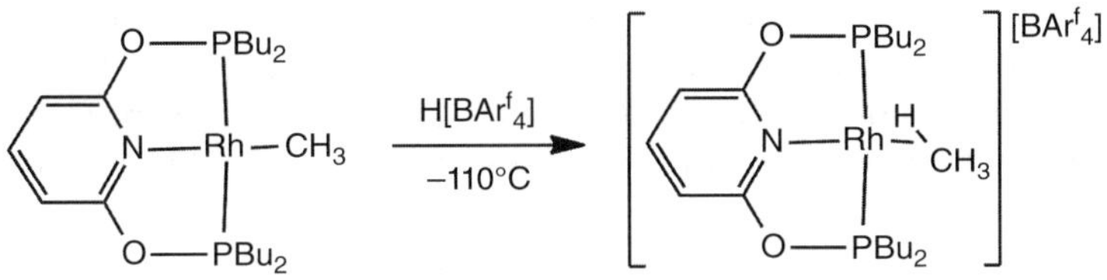

(10)

Scheme 26.5

Rh methyl shown in Scheme 26.5, where all four methane protons in the product were in fast exchange, implying fast tumbling of the CH_4 [36]. As always, both the counterion, $B(Ar^f)_4$, (Ar^f = 3,5-$(CF_3)_2C_6H_3$) and the solvent, $CDCl_2F$, chosen for such studies have to be highly "noncoordinating" to prevent displacement of the alkane from the metal. The activation energy for alkane dissociation, providing an estimate of the alkane binding energy, is 14.5 kcal mol^{-1} at −87 °.

By means of sophisticated kinetic studies, Jones [37] has shown how alkane complex intermediates can undergo fast chain walking by which a metal – the Tp′Rh(NCR) fragment in this case – may bind to the middle of an alkane chain where oxidative addition is prohibited but the metal can then move easily from methylene to methylene until it encounters a terminal methyl group, in which case oxidative addition becomes possible to give the *n*-alkyl hydride complex. Kinetic data suggest that the metal fragment coordinates more rapidly to a methylene over a methyl group by a factor of ~1.5, presumably as a result of the more basic character of the methylene C—H bonds. More recent work on C—H activation in the related Tp′Rh(PMe_3) series has elicited valuable information on M—C bond strengths in Tp′Rh(PMe_3)(R)(H) for a broad range of R groups [38].

The methane complex [CpRe$(CO)_2(CH_4)$] has become a standard in the field and the subject of many studies. TRIR spectroscopy shows it to be in equilibrium with the oxidative addition product, [CpRe$(CO)_2(CH_3)$(H)] [39]. DFT calculations on the methane complex by Cundari and coworkers predict Re· · ·H, 1.99, Re· · ·C, 2.73, and C—H, 1.14 Å, while, with a higher level of theory, Chan and Ball predict a Re· · ·H distance of 1.92 Å, a Re· · ·C distance of 2.6 Å, and a C—H bond distance of 1.15 Å along with a M—HCH_3 binding energy of 15 kcal mol^{-1} [40]. Gordon et al. [41] also looked at methane complexation in [CpRe$(CO)_2(CH_4)$] and concluded that electron donation from the methane to the metal is the most important contributor to the bonding.

26.6 Stable Alkane Complexes

Moving to the heart of the matter, a number of stable alkane complexes are now known. The general strategies that have proved useful are (i) host–guest assistance; (ii) hydrogenation of alkene or other π complexes within single crystals where the resulting alkane cannot escape; (iii) generation, often photochemical, of an unsaturated precursor in an alkane at low temperature; and (iv) generation of an alkane complex from an alkylmetal precursor.

In the first category, Reed et al. [42] found that the iron complex of the double A frame porphyrin (**1**) cocrystallized with *n*-heptane in such a way that each terminal methyl group was bound to the each of two adjacent iron atoms in the crystal. The terminus of the heptane occupied the cavities in the structure of the complex [Fe(**1**)(n-heptane)]. Disorder complicated the structural analysis but the shortest Fe—H distance was estimated by DFT computations to be 2.05 Å. Since H atom positions are poorly located in X-ray crystal structures, the M···C distance, d_{MC}, seems the best indicator of the degree of advancement of the metal alkane interaction. Of course we need to account for the different radii of the metals and for this purpose the metal covalent radius, r_{covM}, seems best adapted. The % expansion of the r_{cov} would then be $\{100(d_{MC} - r_{covC})/r_{covM}\}$. For the iron-heptane case of Structure **1**, the two independent d_{MC} values are 2.5 and 2.8 Å, corresponding to an expansion of r_{covM} by 32% and 55%.

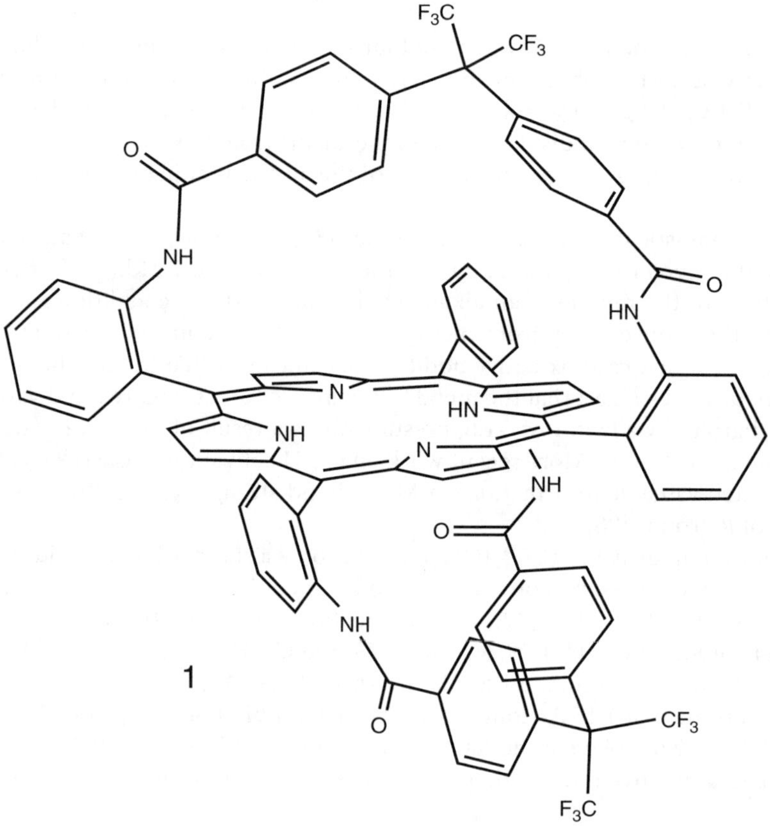

1

The K. Meyer group [43] cocrystallized the U(III) complex of ligand **2** with cyclohexane and related molecules to produce a series of crystalline adducts. The cyclohexane data was problematic, but good diffraction data came from the methyl cyclohexane, methyl cyclopentane, and neohexane adducts where the shortest U—H distances were in the range 3.19–3.36 Å. Given the covalent radii of U (1.96 Å) and H (0.31 Å), this indicates a weak interaction, as confirmed by DFT calculations that estimate the binding energy as ~2.7 kcal mol^{-1}. The important role of the host–guest effect of the crystal lattice in constraining the alkane from dissociation is therefore very clear in this low binding-energy case. Likewise, Andreychuk and Emslie report multiple short K···H—C contacts in the hydrophobic pocket of the potassium salt of a dimethylxanthene derivative [44]. The shortest K···C distance of 3.18 Å represents an expansion of r_{covK} as defined above of only 19%, so a short contact, at least in the crystal, does not require any special coordinating ability as in a transition metal ion. This is not an isolated example: in their discussion the same authors point to a number of similar cases of short K· · ·C distances in potassium salts of organic anions. The origin of the effect is thought to be cation-induced dipole electrostatic bonding.

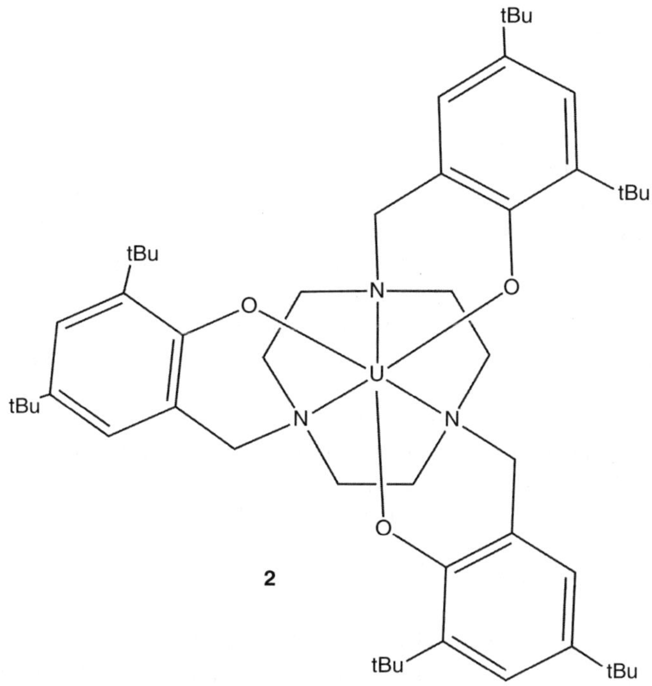

2

In the second strategy, due to Weller [23], the alkane is trapped within a crystal by a hydrogenation of the crystal itself, the resulting alkane being prevented from departure by the constraints of the crystal lattice as well as being stabilized by chelation. For example, in the case of [(norbornadiene)Rh(diphos)]$BAr^F{}_4$, the hydrogenation produces the norbornane complex, **3**. Since this is a single-crystal-to-single-crystal transition, the structure of the product can then be obtained by standard X-ray diffraction methods. Computed metrics in the agostic Rh—H—C unit of the product are Rh—H—C, 102.8°, C—H, 1.16 Å, Rh—H, 1.92 Å, and Rh—C, 2.45 Å, to be compared with the experimental Rh—C distance of 2.480(11) Å; the H not being well located, the experimental data involving this atom are not reported. Comparison with metric data on prior agostic systems [45] suggests that the strength of C—H to Rh binding is strong, so the stability of the complex depends on more than just crystal forces. Bond critical points were found along the Rh—H vectors, confirming the bonding interaction. Deuteration gave rise to a number of isotopomers whose formation was interpreted in terms of a number of fluxional processes occurring in the hydrogenation step. The isotopomers also differed from those obtained by carrying out the reaction in solution [46].

Other similar complexes were prepared and identified in the same way [47], notably an *n*-pentane complex bound at carbons 2 and 4 by hydrogenation of the analogous 1,3-pentadiene complex [48]. The solid-state NMR data are consistent with chain walking in the complex so that the 1,3 and 1,4 pentane isomers are accessed.

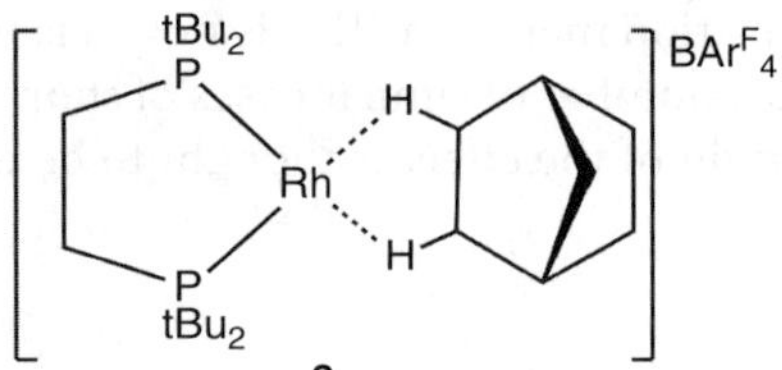

3

As an example of the third strategy, photochemical generation of an unsaturated species, George et al. showed that $CpRe(CO)_3$ could be photolyzed in hexane to form $CpRe(CO)_2$(hexane), which only decomposed at a rate of 40 s^{-1} at 298 K [49]. This led Geftakis and Ball [35] to look at the system at low temperatures. They showed that $CpRe(CO)_3$ could be photolyzed in cyclopentane at −80 °C to form $CpRe(CO)_2$(cyclopentane), which could be detected by NMR spectroscopy. Duckett, George, and Perutz collaborated on $TpRe(CO)_3$ (Tp = tris-(pyrazolyl)borate) and found that this could be photolyzed at 266 nm in cyclopentane at 200 K to give the expected $TpRe(CO)_2$(cyclopentane), detected by NMR spectroscopy. Fluxional behavior of all the alkane protons suggests that the metal can undergo both 1,1′ and 1,2 shifts around the ring. George and Perutz also used NMR and IR data to identify a similar Mn complex of n-butane and propane, having lifetimes of 5.5 and 22 minutes, from photolysis of $CpMn(CO)_3$ at 130–136 K [10]. The Mn complexes were systematically less well bound than the analogous Re complexes, the barriers for decay being 12.5 (Re) and 8 kcal mol^{-1} (Mn). A tungsten complex of pentane, (η^6-C_6Et_6)$W(CO)_2$(pentane) was made in a similar way. One new feature here is the selectivity of the binding exclusively to the terminal methyl group of the alkane [50]. In contrast, all three possible isomers were formed on photolysis of $CpRe(CO)_2(N_2)$ at 180 K in $tBuCH_2CH_3$ to give the alkane complexes [51]. Common to all the cases in this section is the monodentate character of the alkane, limited by the 16e configuration of the unsaturated precursor, showing that chelation is not required for this level of stability.

The fourth strategy is exemplified by low temperature protonation of the Rh(I) methyl complex, (PNP)RhMe, discussed previously. The resulting cation $[(PNP)RhMe]^-$ was shown to be more thermodynamically stable than the methyl hydride oxidative addition product. The complex is relatively kinetically stable at −110 °C in $CDCl_2F$ and kinetic data indicate a half life of 83 minutes at −87 °C, leading to an activation energy for dissociation of 14.5 kcal mol^{-1} [36]. Once again, monodentate alkane complexes prove to be stable when synthesized in this way.

26.7 Conclusion

We now see a greatly expanding portfolio of alkane complexes, both transient and stable, which puts the reality of the phenomenon beyond doubt. Computational work has identified many more cases of a transient complex in a reaction path where experiment proved powerless to see it. The hope for the future is that we can now use alkane complexation to promote reactivity and functionalization beyond simple isotope exchange.

Acknowledgment

I thank the US Department of Energy, Basic Energy Sciences, for funding us when we worked in this area, also the TomKat Charitable Trust for present funding in energy sciences.

References

1 Arndtsen, B.A., Bergman, R.G., Mobley, T.A. et al. (1995). *Acc. Chem. Res.* 28: 154. and references cited.
2 Hall, C. and Perutz, R. (1996). *Chem. Rev.* 96: 3125.
3 Cowan, A.J. and George, M.W. (2008). *Coord. Chem. Rev.* 252: 2504.
4 Weller, A.S., Chadwick, F.M., and Mckay, A.I. (2016). *Adv. Organometal. Chem.* 66: 223.
5 Graham, M.A., Turner, J.J., Poliakoff, M. et al. (1972). *J. Organomet. Chem.* 34: C34.
6 Poliakoff, M. and Turner, J.J. (1974). *J. Chem. Soc., Dalton Trans.* 2276.
7 Perutz, R.N. and Turner, J.J. (1975). *J. Am. Chem. Soc.* 97: 4791.
8 Dobson, G.R., Hodges, P.M., Healy, M.A. et al. (1987). *J. Amer. Chem. Soc.* 109: 4218.
9 Calladine, J.A., Vuong, K.Q., Sun, X.Z. et al. (2009). *Pure Appl. Chem.* 81: 1667.
10 Calladine, J.A., Duckett, S.B., George, M.W. et al. (2011). *J. Am. Chem. Soc.* 133: 2303. and refs cited.
11 Kelly, J.M., Hermann, H., and Von Gustorf, E.K. (1973). *Chem. Commun.* 105.
12 Hermann, H., Grevels, F.W., Henne, A. et al. (1982). *J. Phys. Chem.* 86: 5151.
13 Leu, G.L. and Burkey, T.J.J. (1995). *Coord. Chem.* 34: 87.
14 Brown, C.E., Ishikawa, Y., Hackett, P.A. et al. (1990). *J. Am. Chem. Soc.* 112: 2530.
15 *Metal Dihydrogen and σ-Bond Complexes*, by Gregory J. Kubas, Kluwer, New York, 2001.
16 *Activation of Saturated Hydrocarbons by Transition Metal Complexes*, by A. E. Shilov, Riedel, Dordrecht, 1984.
17 Lersch, M. and Tilset, M. (2005). *Chem. Rev.* 105: 2471.
18 Crabtree, R.H. (2015). *J. Organometal. Chem.* 793: 41–46. and refs. cited.
19 Crabtree, R.H. (2016). *Chem. Rev.* 116: 8750.
20 Labinger, J.A. and Bercaw, J.E. (2015). *J. Organometal. Chem.* 793: 47.
21 Crabtree, R.H. (1979). *Acc. Chem. Res.* 12: 331.
22 Yang, S., Che, W., Wu, W.L. et al. (2017). *Chem. Sci.* 8: 1977.
23 Pike, S.D., Thompson, A.L., Algarra, A.G. et al. (2012). *Science* 337: 1648.
24 Brookhart, M., Green, M.L.H., and Parkin, G. (2007). *Proc. Natl. Acad. U.S.A.* 104: 6908.
25 Crabtree, R.H., Holt, E.M., Lavin, M. et al. (1985). *Inorg. Chem.* 24: 1986.
26 Cundari, T.R. (1994). *J. Am. Chem. Soc.* 116: 340.
27 Yao, W., Eisenstein, O., and Crabtree, R.H. (1997). *Inorg. Chim. Acta* 254: 105–111.
28 Crabtree, R.H. (1993). *Angew. Chem. Int. Ed.* 32: 789.
29 Labinger, J.A. and Bercaw, J.E. (2002). *Nature* 417: 507.
30 Gérard, H., Eisenstein, O., Lee, D.-H. et al. (2001). *New J. Chem.* 25: 1121.
31 Gross, C.L. and Girolami, G.S. (1998). *J. Am. Chem. Soc.* 120: 6605.
32 Periana, R.A. and Bergman, R.G. (1986). *J. Am. Chem. Soc.* 108: 7332.
33 (a) B.H. Weiller, E.P. Wasserman, R.G. Bergman, et al., *J. Am. Chem. Soc.* 1989, 111, 8288–8290. (b) A.A. Bengali, R.H. Schultz, C.B. Moore, et al. *J. Am. Chem. Soc.* 1994, 116, 9585–9589.
34 Wasserman, E.P., Moore, C.B., and Bergman, R.G. (1992). *Science* 255: 315.
35 Geftakis, S. and Ball, G.E. (1998). *J. Am. Chem. Soc.* 120: 9953.
36 Bernskoetter, W.H., Schauer, C.K., Goldberg, K.I. et al. (2009). *Science* 326: 553.
37 Jones, W.D. (2005). *Inorg. Chem.* 44: 4475.
38 Jiao, Y.Z., Morris, J., Brennessel, W.W. et al. (2013). *J. Am. Chem. Soc.* 135: 16198.
39 Cowan, A.J., Poritus, P., Kawanami, H.K. et al. (2007). *Proc. Natl. Acad. Sci. U.S.A.* 104: 6933–6938.
40 Chan, B. and Ball, G.E. (2013). *J. Chem. Theory Comput.* 9: 2199.
41 Khaliullin, R.Z., Cobar, E.A., Lochan, R.C. et al. (2007). *J. Phys. Chem. A* 111: 8753.

42 Evans, D.R., Drovetskaya, T., Bau, R. et al. (1997). *J. Am. Chem. Soc.* 119: 3633.
43 Castro-Rodriguez, I., Nakai, H., Gantzel, P. et al. (2003). *J. Am. Chem. Soc.* 125: 15734.
44 Andreychuk, N.R. and Emslie, D.J.H. (2013). *Angew. Chem. Int. Ed.* 52: 1696.
45 Crabtree, R.H., Holt, E.M., Lavin, M. et al. (1985). *Inorg. Chem.* 24 (13): 1986–1992.
46 Chadwick, F.M., Krämer, T., Gutmann, T. et al. (2016). *J. Am. Chem. Soc.* 138: 13369.
47 Pike, S.D., Chadwick, F.M., Rees, N.H. et al. (2015). *J. Am. Chem. Soc.* 137: 820.
48 Chadwick, F.M., Rees, N.H., Weller, A.S. et al. (2016). *Angew. Chem. Int. Ed.* 55: 3677.
49 Sun, X.-Z., Grills, D.C., Nikiforov, S.M. et al. (1997). *J. Am. Chem. Soc.* 119: 7521.
50 Young, R.D., Lawes, D.J., Hill, A.F. et al. (2012). *J. Am. Chem. Soc.* 134: 8294.
51 Calladine, J.A., Torres, O., Anstey, M. et al. (2010). *Chem. Sci.* 1: 622.

27

Alkane-Related C—H Bond Activation and Functionalization of Aliphatic Amines

Christian Bruneau and Mathieu Achard

Univ Rennes, CNRS, ISCR (Institut des Sciences Chimiques de Rennes) - UMR 6226, Rennes, France

27.1 Introduction

Aliphatic nitrogen-containing heterocycles represent a wide variety of natural products, and many of them have revealed interesting biological activities for many kinds of diseases and have been used as drugs [1]. Pyrrolidine, piperidine, and tetrahydroisoquinoline are major building blocks in this family of commercial pharmaceuticals (Scheme 27.1) [2].

In nature, synthesis of cyclic amines and amides mainly involves iminium and enamine intermediates arising from the corresponding amino acids thanks to hydrogenases, carboxylases, and proteins. Pyrrolidine alkaloids arise from arginine or ornithine whereas lysine metabolites constitute the biosynthetic scaffolds of piperidine alkaloids such as lupine alkaloids [3]. Related biomimetic syntheses of these derivatives efficiently allowed the access to various valuable alkaloids [4].

Their syntheses include either a final cyclization step with formation of one or two carbon-nitrogen bonds, or the functionalization of the saturated heterocyclic core. The most straightforward methods rely on the selective activation of sp^3C—H bond at α (C2) or β (C3) carbon atoms and formation of carbon–carbon or carbon–heteroatom bonds. It must be noted that in many cases, these transformations including some enantioselective versions have been performed starting from *N*-protected amines, mainly as amide, sulfonamide, or carbamate [5].

In this chapter, we will focus on regioselective C—C bond formation at the α, β or remote position of unprotected secondary and tertiary cyclic amines. This review thus excludes the C—C bond formation at sp^3carbons of *N*-protected amines such as amides, sulfonamides, carbamates, where the protecting group can act via its electronic effect or its coordinating ability to direct C—H bond activation mainly encountered in palladium catalysis. We will show the different possibilities of selective functionalization at C(α) and C(β) positions using different methodologies with a special focus on metal-catalyzed transformations, namely metal-catalyzed redox processes, directed sp^3C—H activation, and C—C bond formation triggered by hydrogen transfer and hydrogen borrowing processes.

Alkane Functionalization, First Edition. Edited by Armando J. L. Pombeiro and M. Fátima C. Guedes da Silva.

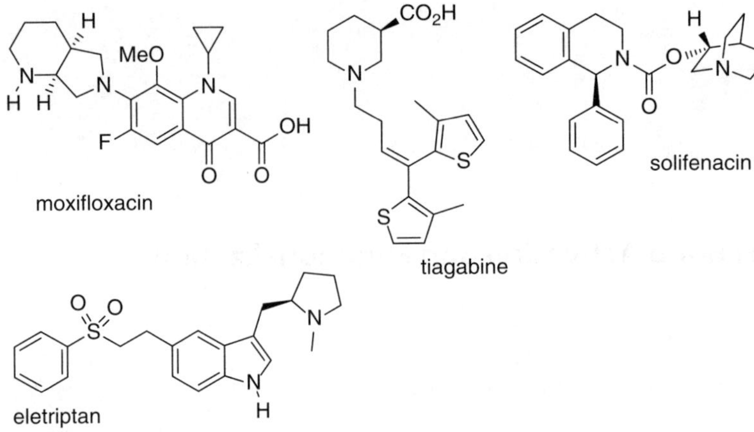

Scheme 27.1 Some biologically active aliphatic amines.

27.2 Metal-Free Carbon–Carbon Bond Formation from Aliphatic Amines

The formation of carbon–carbon bonds at the β-position of cyclic aliphatic amines upon reaction with aromatic aldehydes has been carried out in the presence of glacial acetic acid in refluxing toluene with continuous elimination of water. This protocol led to aromatization of the starting unsaturated heterocycle to give 3,5-disubstituted pyridine and 4-substituted isoquinoline from piperidine [6] and tetrahydroisoquinoline [7], respectively. Under more drastic thermal conditions, the N,C(β)-dialkylation of pyrrolidine into the corresponding disubstituted pyrrole could be achieved without acid catalyst at 180–200 °C [8a] and under microwave irradiation [8b] (Scheme 27.2).

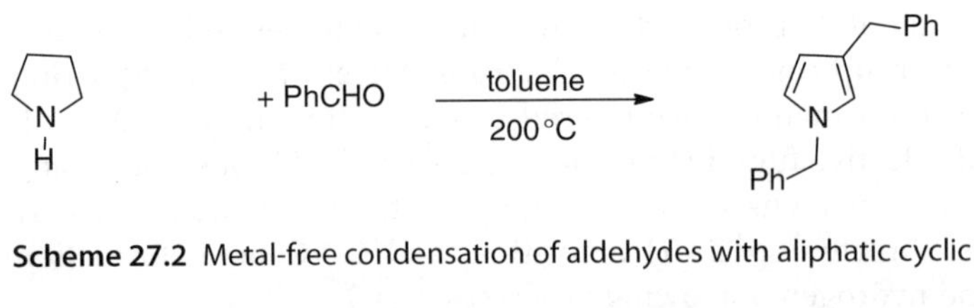

Scheme 27.2 Metal-free condensation of aldehydes with aliphatic cyclic amines.

More recently, D. Seidel has shown the high potential of iminium, azomethine, and enamine intermediates (Scheme 27.3) that are generated during the reaction of cyclic saturated amines with aldehydes in the presence of carboxylic acid by trapping them with nucleophiles, electrophiles, or via cycloadditions to generate a huge variety of heterocycles with formation of N—C bonds as well as C—C bonds at $sp^3C(\alpha)$ or $sp^3C(\beta)$ without aromatization (Scheme 27.4) [9].

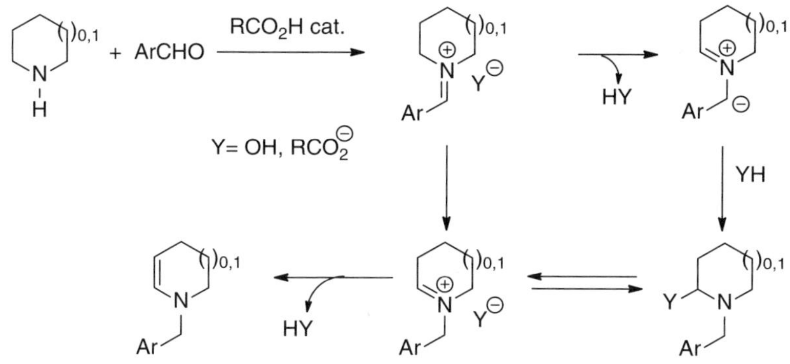

Scheme 27.3 Reactive intermediates leading to α- and β-functionalization, and α,β-difunctionalization of saturated amines.

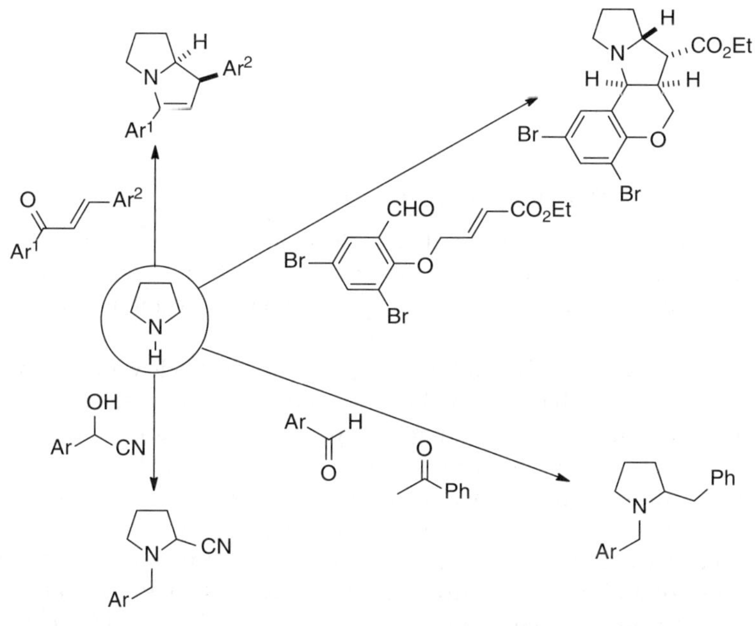

Scheme 27.4 Selected examples of redox-neutral C—C bond formation from pyrrolidine promoted by carboxylic acids.

It is noteworthy that a metal-free method for β-functionalization of cyclic and acyclic aliphatic amines has been developed by treatment of the amine with aldehyde without any additional nucleophile in refluxing xylene in the presence of 3,5-dinitrobenzoic acid. This method led to the formation of α,β-conjugated imines via an intermediate enamine [10].

27.3 Redox Cross Dehydrogenative Coupling (CDC)

The sp^{3}C—H bond oxidation of tertiary amines at the α-carbon of the nitrogen atom generates a reactive iminium salt intermediate that is prone to react with carbonucleophiles to create a new carbon–carbon bond (Scheme 27.5). The general mechanism involves electron transfers and deprotonation. The oxidation steps have been carried out using different approaches: (i) electrochemical oxidation; (ii) chemical oxidation; (iii) catalytic oxidation

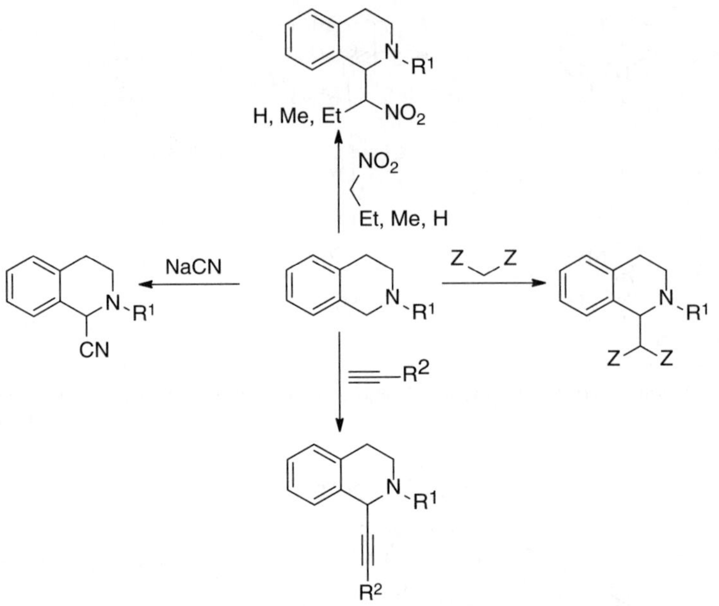

Scheme 27.5 Selected examples of α-functionalization of tetrahydroisoquinolines via CDC.

and (iv) photoredox catalysis. The bibliography on this topic mainly focuses on C(α)-H functionalization of tetrahydroisoquinoline (Scheme 27.5) and *N*-methylaniline derivatives. These substrates lead to C—C bond formation at the benzylic position and at the methyl group, respectively.

As early as 1969, the anodic cyanation of *N*-alkylanilines at a platinum electrode was demonstrated [11] and then further extended to tertiary pyrrolidine substrates [12]. This green process was then used for the elegant preparation of biologically active cyclic amines as targets [13, 14].

The regioselective α-functionalization of amines resulting from chemical oxidation without catalyst was shown in 1988 during the oxidative cyanation of cyclic aliphatic amines using ClO_2 as oxidant to generate α-cyanopyrrolidine and cyanopiperidine derivatives [15].

More recently, transition metal catalysts associated with various oxidants were used to perform the α-functionalization of amines. S.-I. Murahashi first showed that $RuCl_2(PPh_3)_3$ catalyzed the sp^3C—H bond oxidation of tertiary amine with *tert*-butyl hydroperoxide at the α-carbon of the nitrogen atom leading to cyclization products resulting from intramolecular nucleophilic attack of an alkene followed by reaction with a nucleophile (halide, hydroxide) (Scheme 27.6) [16].

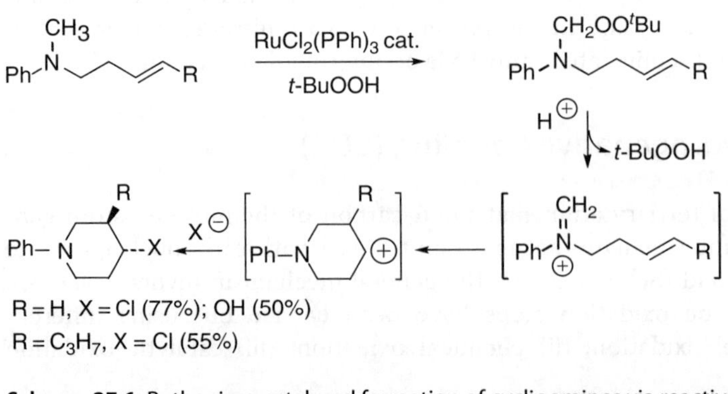

Scheme 27.6 Ruthenium-catalyzed formation of cyclic amines via reactive iminium intermediates.

α-Cyanation of tertiary amines, especially *N*-methylanilines, using $RuCl_3 \cdot nH_2O$, $K_2[RuCl_5(H_2O)]$, or $Ru_2(OAc)_4Cl$ as the ruthenium catalyst and oxygen in the presence of acetic acid as the oxidant were then achieved (Scheme 27.7) [17]. It was also shown that aqueous solution of hydrogen peroxide as well as HCN as the source of cyanide could be used to perform this cyanation reaction [18]. Slightly different catalytic steps have been proposed for the reaction using oxygen or hydrogen peroxide as an oxidant [18, 19].

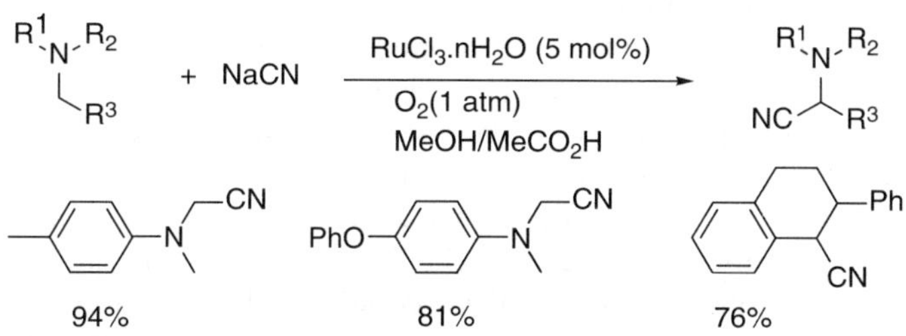

Scheme 27.7 Oxidative cyanation of tertiary amines with sodium cyanide with $RuCl_3$/oxygen.

Other carbonucleophiles such as nitromethane [20] and indoles [21] have been used under similar oxidative conditions to produce sp^3-sp^3 and sp^2-sp^3 carbon–carbon bonds, respectively. The oxidative coupling of tertiary amines with indoles as nonclassical nucleophiles has been successfully achieved with the ruthenium porphyrin catalyst [Ru(2,6-Cl_2tpp)CO] (tpp = tetraphenylporphyrinato) in excellent yields (Scheme 27.8) [21].

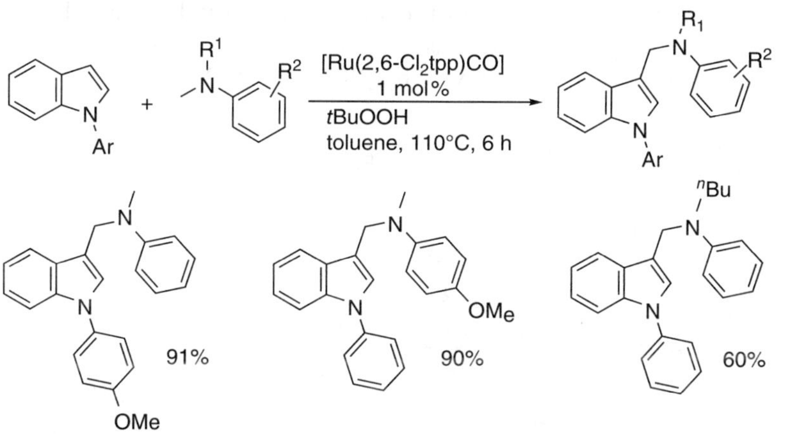

Scheme 27.8 Ruthenium-catalyzed alkylation of indoles with tertiary methylanilines.

Copper(I) and copper(II) catalysts associated in most cases with *tert*-butyl hydroperoxide as oxidant were then developed by the group of C.-J. Li [22]. In this case, copper catalysts promoted a single electron transfer process, which produced iminium cuprate intermediates [23]. Starting from terminal alkynes, α-alkynylation of sp^3C—H bonds of amines was first carried out with copper bromide as catalyst (5 mol%) at 100 °C under neat conditions with ^{t}BuOOH as oxidant and gave rise to a variety of tertiary propargylic amines [24], whereas the use of CuBr associated to NBS gave the same products but via a free radical process [25]. Enantioselective versions of these alkynylation reactions were achieved by adding enantiopure chelating ligands such as Pybox [26, 27] or Quinap when a dihydroisoquinolinium salt was directly used [27]. It is interesting to note that terminal propargylamines have been prepared in good to excellent yields using calcium carbide CaC_2 as a source of acetylenic fragment with 20 mol% of CuBr and ^{t}BuOOH or diethylazodicarboxylate (DEAD) as preferred oxidant [28].

The creation of sp^3-sp^2 carbon–carbon bonds has been illustrated by the coupling of tetrahydroisoquinolines with electron-rich aromatic substrates such as indoles and naphthols catalyzed by CuBr with ^{t}BuOOH as an oxidant [29, 22b]. Using the same catalytic system, *N,N*-dimethylanilines can also undergo the same type of coupling with indoles [30]. Electron-deficient olefins such as conjugated enones, enals, and acrylonitrile have been coupled at the C(α) position of amines via an Aza-Baylis-Hillmann-type cross dehydrogenative coupling (CDC) reaction in the presence of an additional basic cocatalyst such as DABCO [22b, c] or a chiral base in the asymmetric version [31]. The copper-catalyzed CDC starting from activated methylene or methyl-containing compounds is also a convenient route for the formation of sp^3C-sp^3C bonds with the appropriate oxidant selected with respect to the substrates and solvents among ^{t}BuOOH, DDQ, H_2O_2, or oxygen. The soft carbonucleophiles that are involved in these reactions originate from malonates [32, 33], malononitrile [32], nitroalkanes [33–35], and trifluoromethyltrimethylsilane [36]. The CDC starting from enolisable ketones or aldehydes is also possible under copper catalysis conditions but requires the help of organocatalysis provided by pyrrolidine/benzoic acid [30] or an equivalent organocatalyst [37]. It must be noted that $VO(acac)_2$/proline (20 mol% each) with ^{t}BuOOH as oxidant provides an efficient system for the α-alkylation of *N*-methylpyrrolidine with acetone [38]. Finally, it is worth mentioning that other metal salts such as $PtCl_2$ [39a] used in $MeNO_2/H_2O$ (10 : 1) as solvent without added oxidant, $FeCl_2$ with $(^tBuO)_2$ as oxidant [39b], V_2O_5 [39c], and gold complexes [39d] under oxygen have shown catalytic activity for CDC of aliphatic amines with nitromethane as pronucleophile.

More recently, many of these catalytic CDCs have been carried out using photoredox catalysis with transition metals [40]. The most common photocatalysts are based on tris(bipyridine) ruthenium(II) complexes such as $[Ru(bpy)_3]Cl_2$ and cyclometallated phenylpyridine iridium(III) complexes such as $[Ir(ppy)_2(bpy)]PF_6$ (Scheme 27.9).

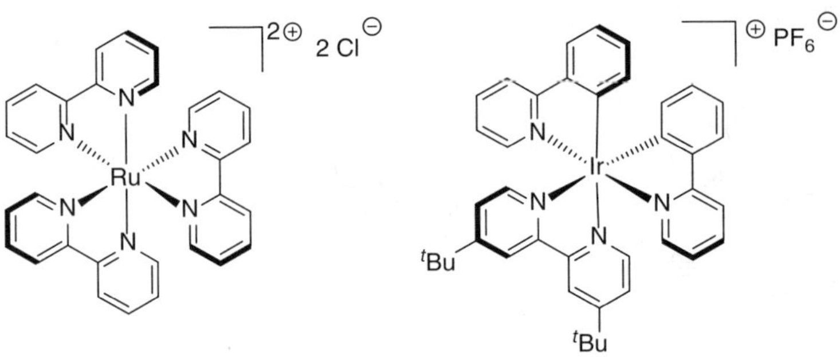

Scheme 27.9 Typical ruthenium and iridium photoredox catalysts.

These catalysts usually operate under visible light irradiation in the presence of air (or oxygen) and make the previous reactions possible. For instance, mechanistic (Scheme 27.10) [41, 42] and recycling [43] studies have been performed for the Aza-Henry reaction catalyzed by ruthenium photocatalysts. The pioneering works of Stephenson [41, 44], who revealed the possibility of generating reactive iminium intermediates notably from tetrahydroisoquinolines, provided impetus for the development of CDC to create carbon–carbon bonds under photocatalytic conditions. Coupling with cyanide [44, 45], malonates [44], enol silanes [44, 46], terminal alkynes [44, 47], allylic trimethylsilane [44], indole [44], and electron-deficient olefins [48], including asymmetric version [49] as well as simple ketones in the presence of a Lewis base such as proline [50], have been reported.

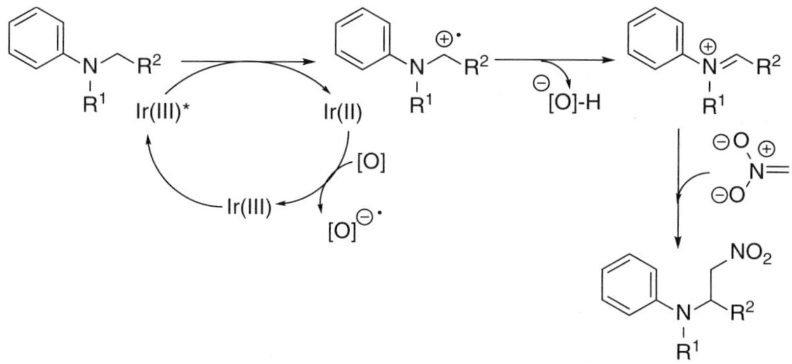

Scheme 27.10 Mechanism proposed for photoredox catalytic Aza-Henry reaction [41].

Based on similar elemental steps, three component reactions for the direct synthesis of α-amino amides from *N*-methylanilines, isocyanides, and alcohol or water have been proposed via mild aerobic oxidation in the presence of $[Ir(ppy)_2(bpy)]PF_6$ [51] and TiO_2 [52] as photocatalysts (Scheme 27.11).

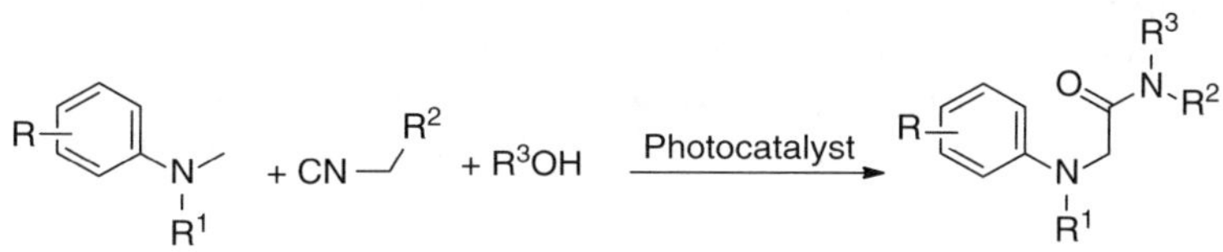

Scheme 27.11 Photoredox three component condensation.

Dual catalysis involving a photocatalytic system and another organometallic or organocatalyst have been explored as exemplified by the association with palladium catalyst for coupling with allylic substrates [53], with nickel catalyst for arylation [54] or carbene for acylation [55].

Besides photocatalysts based on metal complexes, organophotocatalysts – namely, EosinY [56], rose bengal [57], and dicyanopyrazine derivatives [58] – have also been used efficiently to achieved the same CDC of aliphatic amines.

27.4 Directed C—H Bond Activation

27.4.1 Rhodium-Catalyzed Acylation

Regioselective activation/functionalization of sp^3C—H bond of aliphatic amines have been achieved mainly with rhodium and ruthenium catalysts with the assistance of a pyridine as a chelating neighboring group [59]. This transient coordination provides an easy activation of a selected C—H bond and further functionalization by an external reagent, which interacts either via direct insertion into a C-metal or H-metal bond or through oxidative addition to the metal center followed by reductive elimination.

In 1997, S. Murai reported the deshydrogenative acylation of *N*-(2-pyridyl)piperazines upon reaction with CO/ethylene in the presence of $Rh_4(CO)_{12}$ as catalyst at 160 °C (Scheme 27.12) [60].

This reactivity was enhanced by the presence of electron-withdrawing groups on the pyridine and electron-donating groups at the distal piperazine nitrogen substituent. The reaction

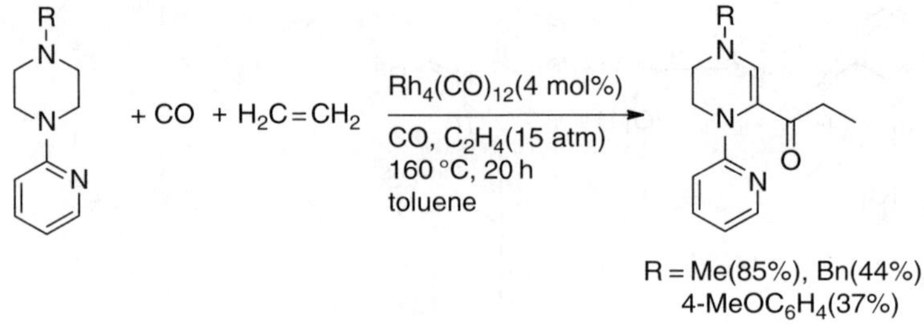

Scheme 27.12 Rhodium-catalyzed acylation of *N*-(2-pyridyl)piperazines.

was extended to other cyclic aliphatic amines such as *N*-(2-pyridyl)-pyrrolidine, -piperidine and -tetrahydroisoquinoline [61]. Starting from an amide with an acyl replacing the directing pyridyl group, the same transformation took place under similar conditions [62].

27.4.2 Ruthenium(0)-Catalyzed Alkylation

The trinuclear ruthenium(0) complex $Ru_3(CO)_{12}$ also catalyzed the alkylation at sp^3 carbons by terminal olefins with the assistance of pyridine chelation. In 1998, C.-H. Jun showed the insertion of an alkene into the sp^3C—H bond of a benzyl group using this ruthenium(0) catalyst [63]. The strongly chelating pyridine-containing benzylamine in the presence of a large excess of hex-1-ene at 130 °C with 10 mol% of $Ru_3(CO)_{12}$ led to the formal insertion of the alkene into one benzyl sp^3C—H bond and thus to the alkylation of a sp^3C—H bond to give the alkylated benzylic amine (Scheme 27.13).

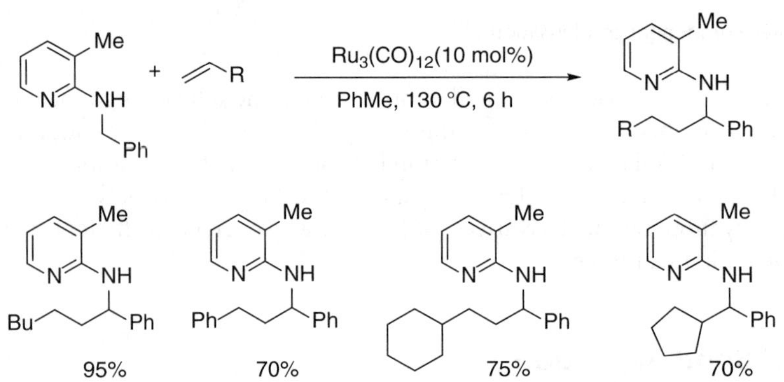

Scheme 27.13 Chelation-assisted ruthenium-catalyzed alkylation of benzyl amine derivatives.

The proposed mechanism (Scheme 27.14) involves the coordination of the pyridine that directs oxidative addition of one benzylic sp^3C—H bond to the Ru(0) center, followed by insertion of the alkene into the Ru—H bond, and reductive elimination of the formed alkyl ligands leading to sp^3C-sp^3C bond formation.

Starting from *ortho*-substituted benzylamines, the more electron-donating substituents led to faster sp^3C—H bond alkylation. This alkylation at the benzylic position of benzylamines was extended to alkylation of a variety of cyclic alkyl amines *N*-protected by a 2-pyridyl or 2-pyrimidyl group as chelating and directing group using the same catalyst $Ru_3(CO)_{12}$ but in the presence of carbon monoxide (Scheme 27.15) [64]. Under these conditions, α,α-dialkylated amines were formed as the major products. The monoalkylated derivatives were obtained only when one adjacent position of the amine *N*-atom was already substituted.

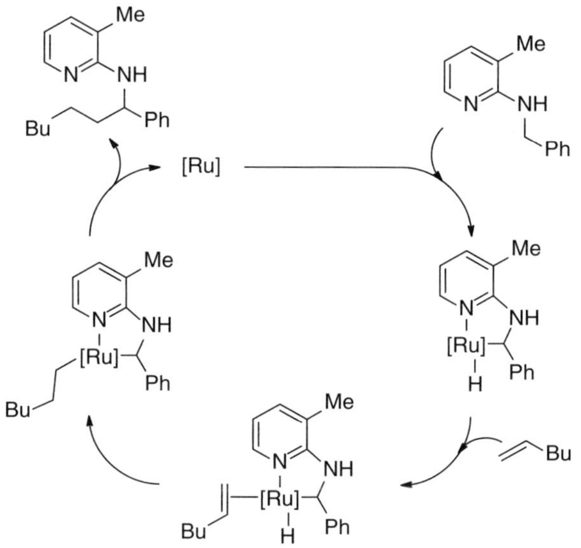

Scheme 27.14 Catalytic cycle for sp^{3}C—H alkylation of *N*-(2-pyridyl)benylamine with alkenes.

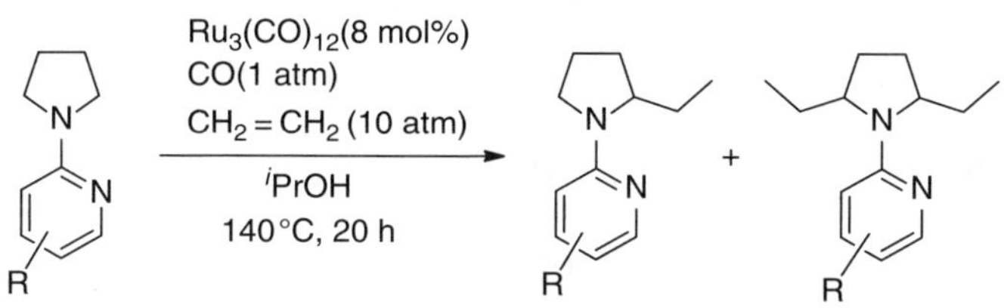

R = H	0%	92%
R = 3-Me	0%	76%
R = 5-Me	0%	93%
R = 6-Me	39%	11%
R= 5-CF_3	0%	94%

Scheme 27.15 $Ru_3(CO)_{12}$-catalyzed α-alkylation of cyclic amines with ethylene.

The nature of the solvent was crucial as only protic solvents were suitable for an efficient conversion and selective formation of the dialkylated products. Electronic properties and position of substituents on the pyridine ring played a noticeable role on reactivity (Scheme 27.15). The use of CO was not compulsory but allowed reaching better yields, likely avoiding the decomposition of the catalyst. For instance, the dialkylated pyrrolidine was obtained in 77% instead of 92% yield in isopropanol when carbon monoxide was replaced by nitrogen.

Acyclic tertiary *N*-(2-pyridyl)benzylamines were also ethylated with ethylene at the benzylic position. The alkylation was extended to a variety of alkenes such as 1- and 2-hexene, styrene, *tert*-butylethylene and cyclohexene (Scheme 27.16) [64]. Under typical conditions, a large excess of the alkene (10 equivalents) was reacted at 140 °C for 40–60 hours in the presence of 8 mol% of $Ru_3(CO)_{12}$ without carbon monoxide. From aliphatic olefins including cyclic ones, the reaction was less selective toward the formation of dialkylated amines as observed with ethylene. With 2-hexene, the alkylated products contained 1-hexyl substituents, showing that double-bond migration took place after insertion of the double bond into an initially generated Ru—H bond due to the steric congestion of the secondary alkyl group, as also noticed by C.-H. Jun [63].

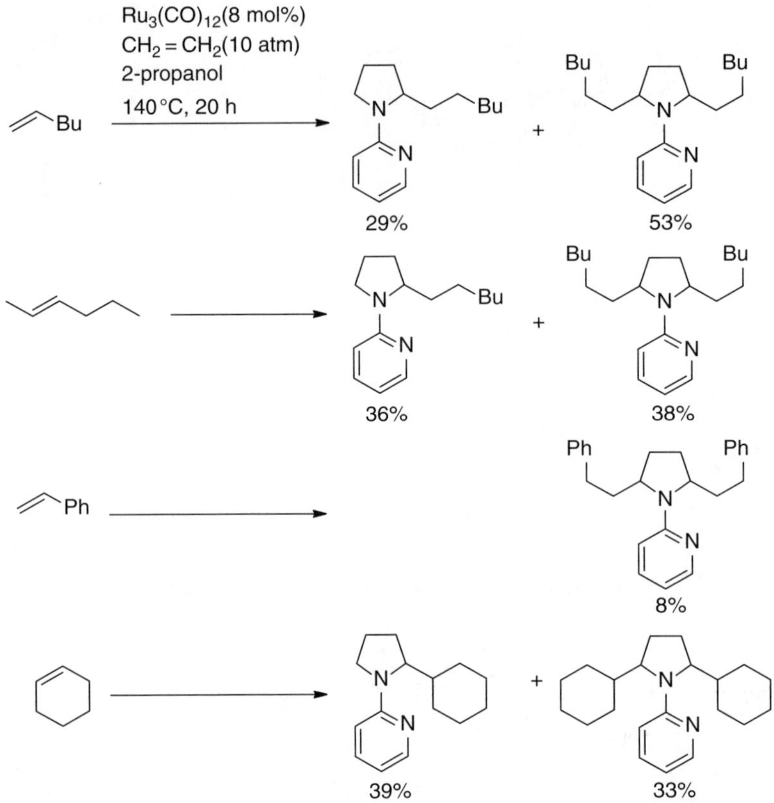

Scheme 27.16 $Ru_3(CO)_{12}$-catalyzed α-alkylation of *N*-(2-pyridyl)pyrrolidine with various alkenes.

On the other hand, functional olefins such as acrylonitrile and butyl vinyl ether were not reactive.

The proposed mechanism is very similar to the one described in Scheme 27.14. It involves initial coordination of the pyridine to direct sp^3C—H bond oxidative addition to the Ru(0) center followed by alkene insertion into a Ru—H bond and reductive elimination.

It has been shown that the presence of catalytic amounts of carboxylic acid such as *trans*-1,2-cyclohexyldicarboxylic or 3,4,5-trifluorobenzoic acid improved the ruthenium-catalyzed sp^3C—H bond alkylation of cyclic amines by terminal olefins and allowed reaching full conversion of less reactive *N*-pyridylpiperidine derivatives. Thus, in the absence of CO but in the presence of *trans*-1,2-cyclohexyldicarboxylic acid, *N*-pyridylpiperidine was efficiently hexylated into mono-**A** and dialkylated **B** piperidine derivatives, whereas in the absence of carboxylic acid under otherwise identical conditions, the conversion was lower, affording the monoalkylated **39** as the major product (Scheme 27.17) [65, 66].

An application of this alkylation reaction is illustrated in the preparation of an intermediate piperidine derivative, which after cleavage of the directing pyridyl group under heterogeneous rhodium hydrogenation catalysis conditions leads to *trans*-2-methyl-6-undecylpiperidine, (±)-solenopsin A (Scheme 27.18) [65]. Removal of pyridine directing group based on platinum catalysis has also been proposed for the same synthesis [67].

Recently, it was shown that the ruthenium(II) complex $RuCl_2(PPh_3)_3$ associated with catalytic amounts of BINAP and silver triflate enabled the directed alkylation of *N*-(2-pyridyl)pyrrolidine

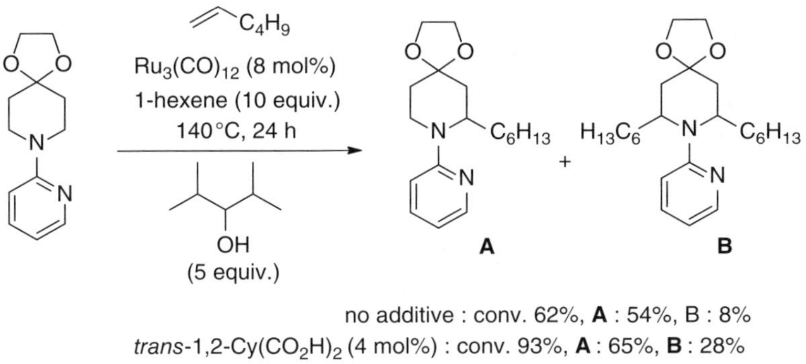

Scheme 27.17 Influence of carboxylic acid additive in $Ru_3(CO)_{12}$-catalyzed alkylation of cyclic amines.

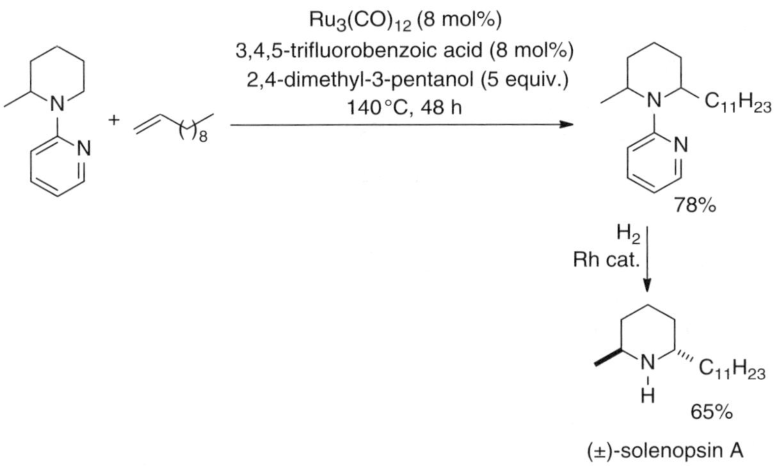

Scheme 27.18 Synthesis of (±)-solenopsin A.

with unactivated alkenes at the α-position of the cyclic saturated ring under milder and safer conditions (Scheme 27.19) [68]. The reaction operated with aliphatic terminal alkenes, including long-chain haloalkenes, ketones, ethers, and also with vinyl silanes and styrene derivatives. Removal of the pyridine directing group using platinum catalysis [67] has also been applied to prepare unprotected pyrrolidines.

27.4.3 Ruthenium(0)-Catalyzed Arylation

Arylation of cyclic amines with ruthenium(0) precursors, essentially $Ru_3(CO)_{12}$ have been carried out with arylboronates as a coupling partner and a nitrogen-containing directing group, thus leading to α-substituted amines. The first example, reported by D. Sames in 2006, described the arylation of pyrrolidines *N*-substituted by an amidine, a pyridine, or a pyrimidine group [68]. Starting from a mono 2-substituted pyrrolidine, among other mono- and binuclear ruthenium complexes, $Ru_3(CO)_{12}$ was the best catalyst and gave 5-arylated pyrrolidine derivatives at 150 °C in good yields with the *trans*-product as the major diastereomer in the presence of five equivalents of *tert*-butylmethylketone, which played a critical role in the reaction (Scheme 27.20) [69].

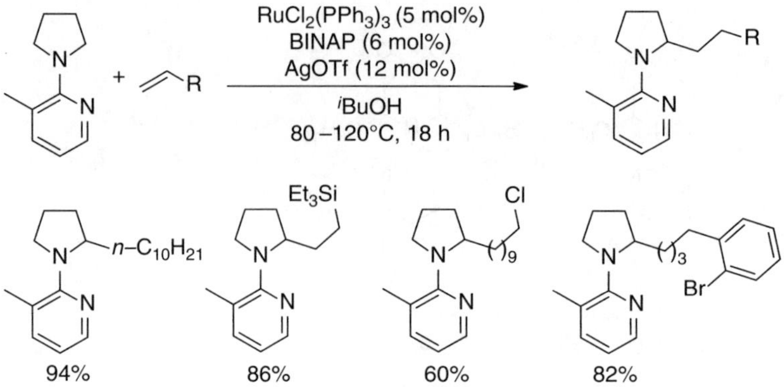

Scheme 27.19 Directed $RuCl_2(PPh_3)_3$-catalyzed α-alkylation of pyrrolidine derivatives.

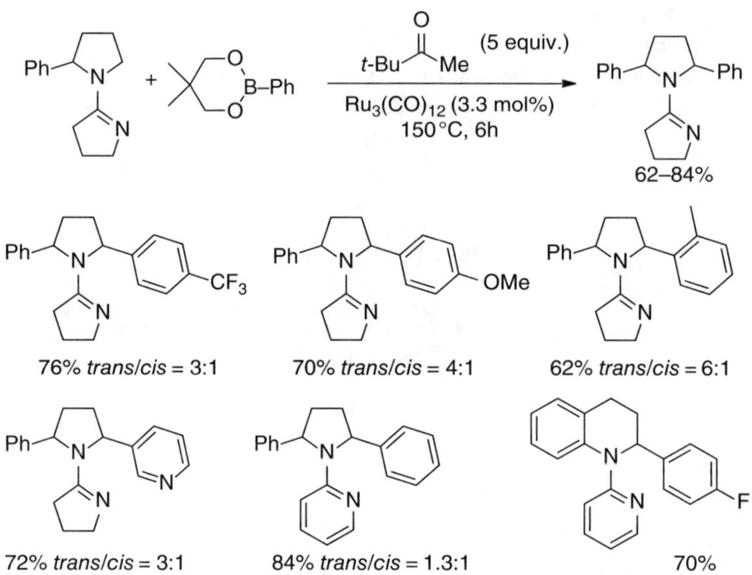

Scheme 27.20 $Ru_3(CO)_{12}$-catalyzed sp^3C—H bond arylation with arylboronates.

This catalytic reaction was compatible with a large variety of substituted arylboronates, including indole- and pyridine-containing boronate esters. This arylation was applied to pyrrolidines containing either a pyridine or a pyrimidine group as *N*-directing group but required higher catalyst loading (10 mol%). With these directing groups, the arylated products were obtained as a mixture of diastereoisomers with low stereoselectivity, which was attributed to a fast isomerization with respect to arylation under the reaction conditions.

The proposed mechanism (Scheme 27.21) involves the initial sp^3C—H bond oxidative addition to ruthenium(0), followed by aryl transmetallation and reductive elimination. It was suggested by D. Sames that the role of the ketone was to insert into the Ru—H bond to favor *trans*-arylation with the phenylboronate ester. Later on, it was shown by B.U.W. Maes that pinacolyl alcohol was formed via Ru-catalyzed hydrogenation of pinacolone rather than through a Ru-alkoxide intermediate [70].

The deprotection of the 2,5-diarylated amidines to give 2,5-disubstituted pyrrolidines could be performed by treatment with H_2NNH_2/CF_3CO_2H (Scheme 27.22) [69].

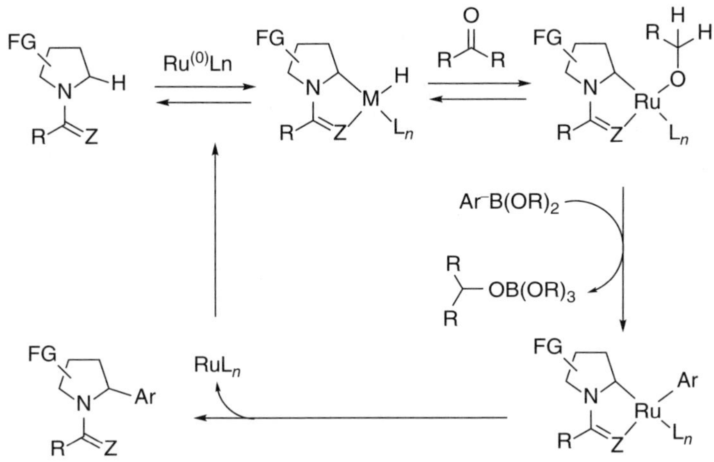

Scheme 27.21 A mechanistic proposal for sp^{3}C—H bond arylation with arylboronates and Ru(0) catalysts.

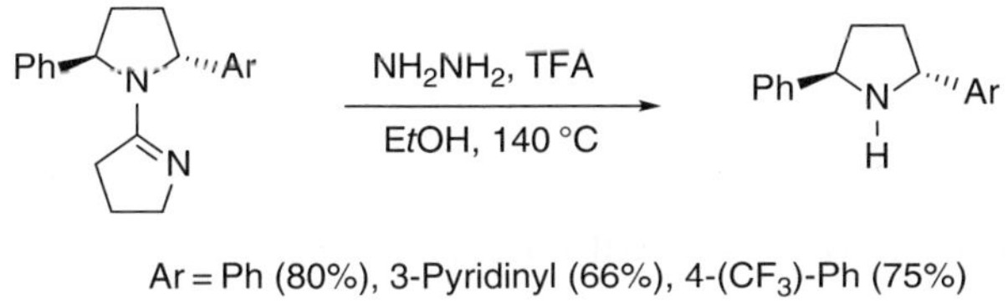

Scheme 27.22 Deprotection of 2,5-disubstituted amidines.

The more difficult direct arylation of the less-reactive *N*-(2-pyridyl)piperidines was efficiently achieved by B.U.W. Maes with another catalytic system also based on ruthenium(0) catalyst and using an alcohol as solvent (^{t}BuOH or Et_3COH), which was more effective than the ketone (Scheme 27.23) [70, 71].

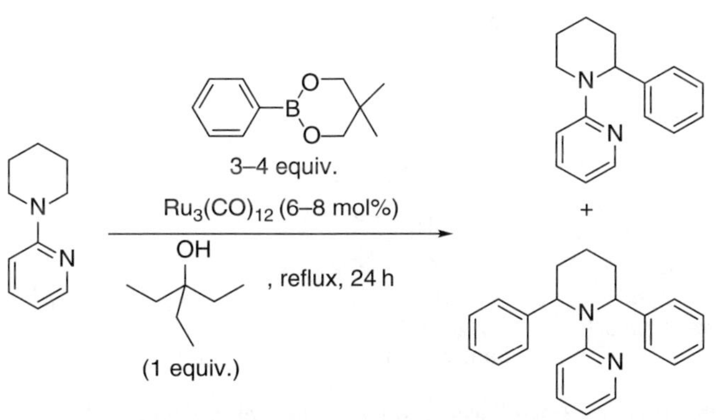

Scheme 27.23 sp^{3}C—H mono and diarylation from *N*-(2-pyridinyl)piperidine.

These observations led to the catalytic cycle proposed in Scheme 27.24. The innovation consisted in the direct transarylation on the Ru—H species and formation of the pinacolborane that is trapped by the tertiary alcohol with the formation of hydrogen.

The reaction usually required three equivalents of phenylboronic acid neopentylglycol ester with 6 mol% of $Ru_3(CO)_{12}$ in 3-ethylpentan-3-ol (one equivalent) at reflux for 24 hours

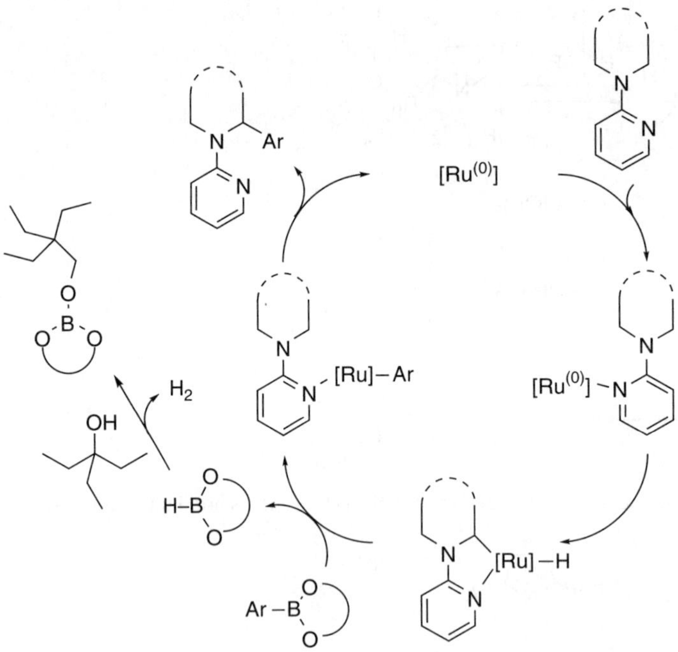

Scheme 27.24 Proposed catalytic cycle for ruthenium(0)-catalyzed arylation with arylboronates.

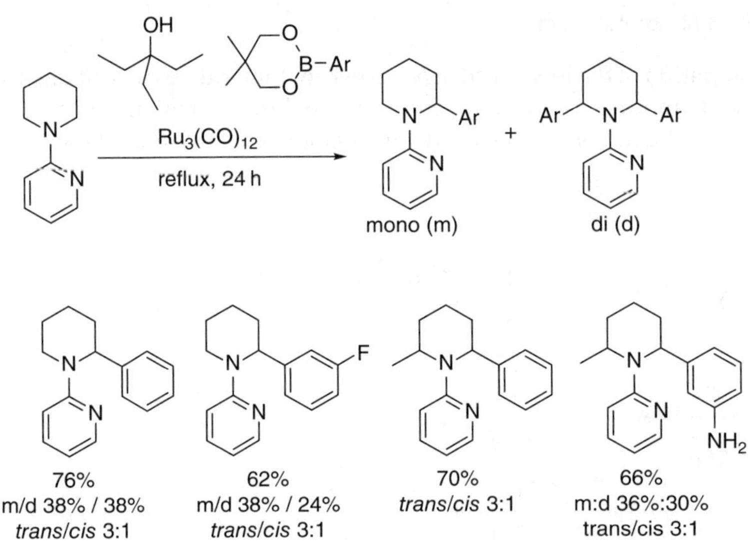

Scheme 27.25 sp^3C—H arylation of saturated cyclic *N*-(2-pyridinyl)piperidines with arylboronates.

(Scheme 27.25). In the presence of an excess of arylboronate, the mono- and di-arylated products were formed without selectivity, but in the case of 2,6-disubstitution, the *trans*-products were always formed as the major stereoisomers.

The reaction was then extended to α-(hetero)arylation of various saturated *N*-(2-pyridinyl) cyclic amines under similar catalytic conditions [71]. The pyridine directing group could be removed by sequential catalytic hydrogenation on Pd/C followed by reaction with

$H_2NNH_2/AcOH/^iPrOH$ [67, 70] or by quaternization by MeOTf followed by reduction with $NaBH_4$ [67].

The first selective monoarylation of *N*-(2-pyridyl)piperidine was made possible by introduction of a trifluoromethyl substituent at the 3-position of the pyridin-2-yl directing group (Scheme 27.26) [72]. The steric hindrance brought by this substituent favored the sole formation of monoarylated piperidines. A beneficial effect of metal salt additives to the $Ru_3(CO)_{12}$-based catalyst was observed on conversion of the substrates. $Cu(SO_4)\cdot 5H_2O$ provided the best results when the reactions were performed at 140 °C in *o*-xylene in the presence of 0.5 equiv. of 1,3-propanediol.

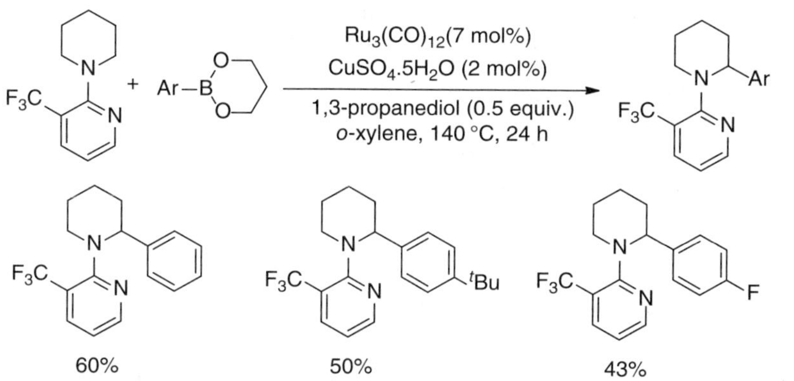

Scheme 27.26 Selected examples of monoarylation of *N*-(3-trifluoromethylpyridin-2-yl)piperidine with arylboronates in the presence of $Ru_3(CO)_{12}$as catalyst precursor.

Similarly, secondary benzylic amines featuring a 2-pyridyl substituent have been coupled with arylboronates at their sp^3C—H bond in the presence of $Ru_3(CO)_{12}$ and pinacolone as solvent. The reaction was not only directed by a pyridine substituent but also required substitution of this pyridyl group at the 3-position, preferentially by a methyl or a phenyl group (Scheme 27.27) [73, 74].

The arylation worked better with six-membered 2-aryl-1,3,2-dioxaborinane reagents in pinacolone as the solvent of choice, which worked as a hydrogen and boron scavenger simultaneously. Whereas alkoxy and halogen substituents were tolerated, the coordinating 4-CN and 4-NO_2 substituents on the aryl group led to complete inhibition of the arylation reaction. It is noteworthy that the formation of the carbon–carbon bond was also possible at the α-position of nitrogen starting from homobenzylic amines, as exemplified by the arylation by phenylboronate of 3-methyl-*N*-(2-phenylethyl)-2-pyridineamine in 39% yield, and with 3-methyl-2-(2-phenylethyl)pyridine in 75% yield. A mechanism related to Scheme 27.21 was proposed where the initial step is coordination of pyridine nitrogen to the ruthenium(0) center, followed by oxidative addition to give a 5-membered cyclometalated ruthenium hydride. Insertion of the carbonyl group of the ketone into the Ru—H bond leads to the formation of a ruthenium alkoxide, which facilitates the transmetalation with the arylboronate. Finally, reductive elimination delivers the arylated benzylamine derivative.

27.4.4 Ruthenium(II)-Catalyzed Arylation

The same type of α-arylation of benzylamine derivatives equipped with a 2-pyridyl directing group was carried out with arylhalides instead of arylboronates but with $[RuCl_2(p\text{-cymene})]_2$ as the ruthenium(II) catalyst precursor in the presence of a carbonate (K_2CO_3 or

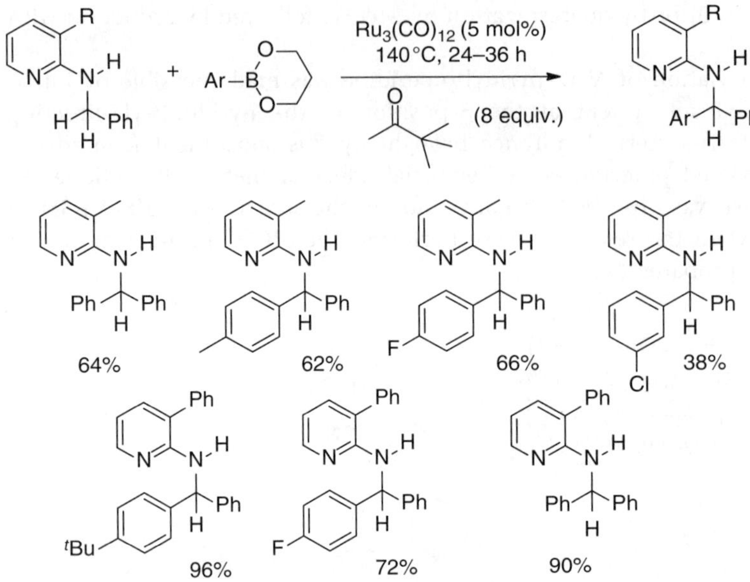

Scheme 27.27 Directed cross coupling at the benzylic position of benzyl amine.

Na_2CO_3) as a base and a catalytic amount of potassium pivalate [75, 76] or directly with Ru(carboxylate)$_2$(*p*-cymene) [77]. Both catalytic systems operated at 140 °C for 24 hours in toluene or *o*-xylene in the presence of 2.5–5 mol% of catalyst loading. The advantage of using the well-defined ruthenium(II)bis(carboxylate) complex is that it does not require an excess of carboxylate as it is the case with the in situ generated catalytic system.

In the case of well-defined catalysts, Ru(O_2CMes)$_2$(*p*-cymene) was more active than Ru(O_2C^tBu)$_2$(*p*-cymene), and Na_2CO_3 led to better results than K_2CO_3, KOAc, and K_3PO_4 as a base in *o*-xylene as solvent. Selected examples with the two catalytic systems are presented in Schemes 27.28 [76] and 27.29 [77].

A proposed mechanism is directly related to the concerted metalation deprotonation mechanism, which is operative in sp^2C-H bond activation with ruthenium(II) carboxylate catalytic

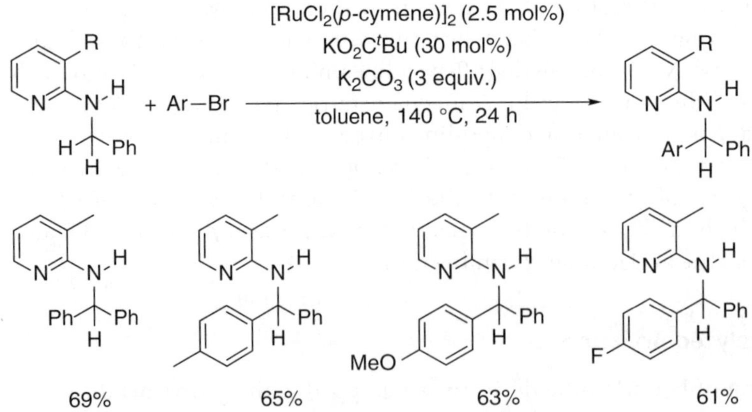

Scheme 27.28 sp^3C—H bond arylation with an *in-situ* generated ruthenium(II) catalyst.

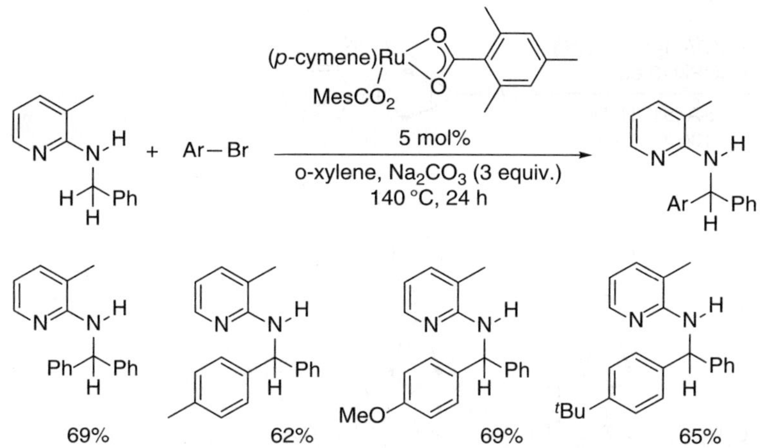

Scheme 27.29 sp^{3}C—H bond arylation with arylhalides in the presence of (ruthenium)bis(carboxylate) catalyst.

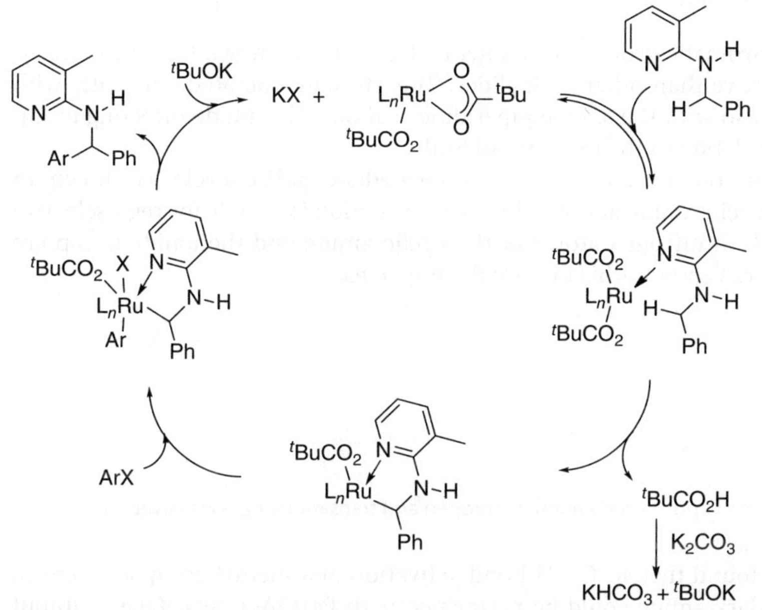

Scheme 27.30 Proposed CMD mechanism for the ruthenium(II)-catalyzed sp^{3}C-H bond activation.

systems (Scheme 27.30) [76]. A mechanism involving external deprotonation by a free carboxylate cannot be ruled out [78].

27.4.5 Palladium-Catalyzed Directed C—C Bond Formation

There are only few examples of the application of palladium catalysis in directed C—C bond formation via sp^{3}C—H bond arylation. It has successfully been achieved by M.S. Sandford, who used an amide derived from perfluorinated *p*-methylaniline as the directing group to functionalize a remote position of saturated cyclic amines (Scheme 27.31) [79]. Various arylated substrates with diverse functional groups (methoxy, fluoride, bromide, hydroxyl,

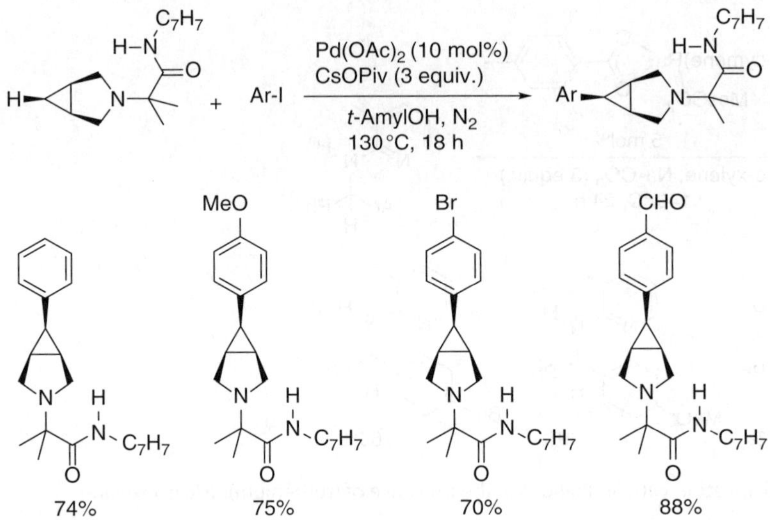

Scheme 27.31 Palladium-catalyzed arylation of alicylic amines.

aldehyde) and even heteroaromatic groups, were tolerated, and it was shown that the iodoaryl derivatives were more reactive than other arylhalides. This strategy operated efficiently with fused bicyclic amines but also with the simple piperidine scaffold. The amide directing group was easily removed after arylation in the presence of SmI_2.

The success is based on the formation of an intermediate palladacycle as shown in Scheme 27.32, where the acyclic amine adopts a boat conformation favorable for regioselective remote C—H activation. Both nitrogen atoms of the cyclic amine and the amide group are coordinated to the palladium center in this intermediate species.

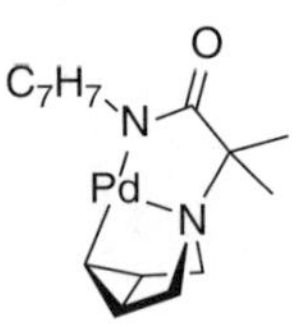

Scheme 27.32 Palladacycle formed upon coordination of nitrogen and transannular C—H activation.

M.J. Gaunt has recently found that sp^{3}C—H bond activation of a methyl group adjacent to nitrogen in a cyclic secondary amine could be performed with Pd(OAc)$_2$ as catalyst without any additional directing group but thanks to the sole coordination of the nitrogen atom [80]. Arylation of a variety of piperidine derivatives could thus be carried out with arylboronic acids and esters in the presence of two equivalents of Ag_2CO_3, benzoquinone, $NaHCO_3$ as a base and *N*-acetyl phenylalanine as a ligand (Scheme 27.33).

A postulated intermediate in this catalytic reaction is a four-membered palladacycle formed via a concerted metalation-deprotonation process leading to the C—H bond activation, which is rendered possible by the high steric hindrance around the amine motif. It is noteworthy that a similar type of activation of an adjacent methyl group has been proposed for the C—H carbonylation of alicyclic amines to lactams, which involves concomitant formation of a C—C and an N—C bond [81].

In the case of spirocyclic oxazolidines (*N*,*O*-ketals) presenting high steric constraints, direct arylation has been achieved with ArIOTf as an arylating agent (Scheme 27.34) [82]. In this case

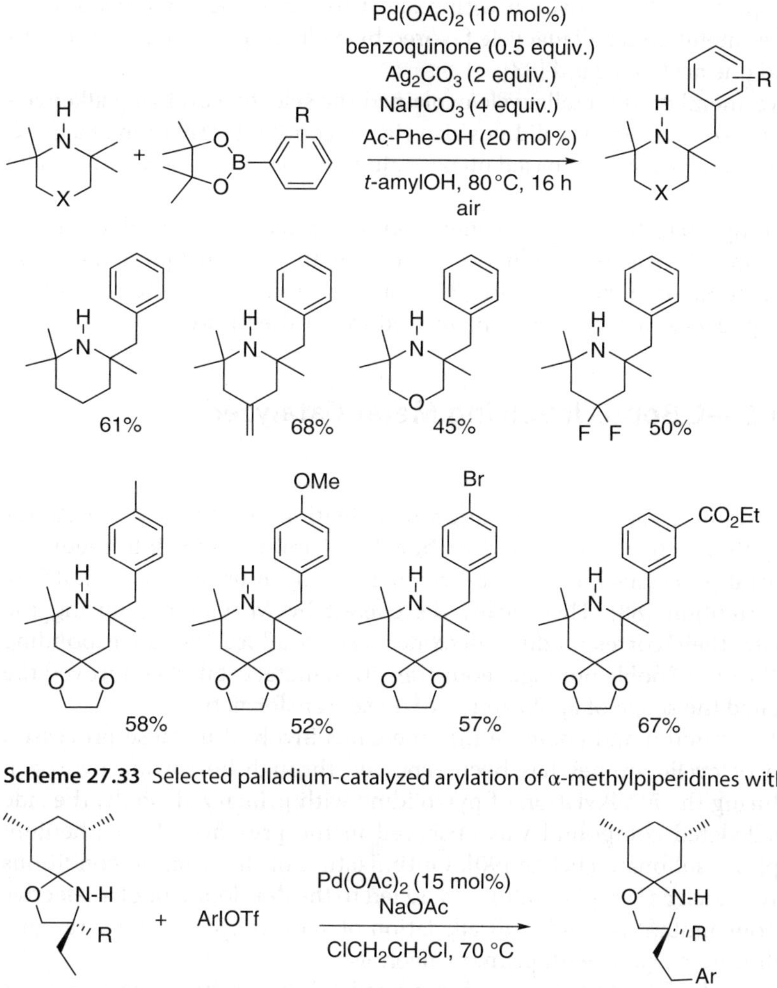

Scheme 27.33 Selected palladium-catalyzed arylation of α-methylpiperidines with arylboronates.

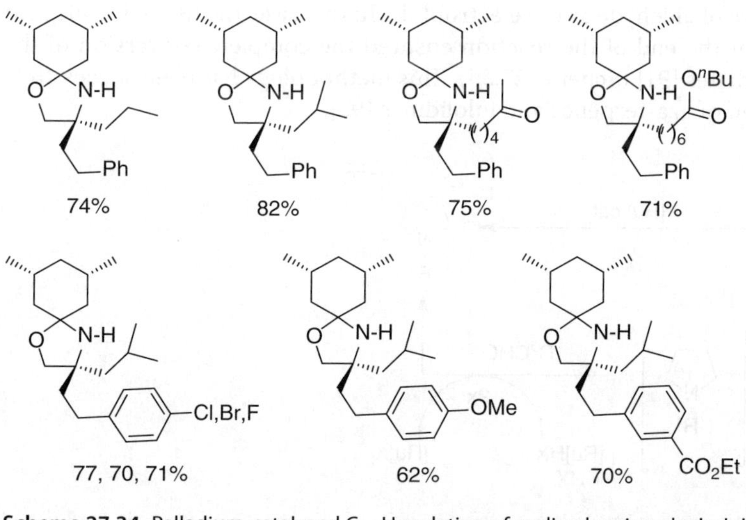

Scheme 27.34 Palladium-catalyzed C—H arylation of cyclized amino alcohol derivatives.

the C—H activation took place at the γ-position with respect to the nitrogen atom via intermediate formation of a five-membered palladacycle favored by hydrogen bonding interactions between the substrate and one acetate ligand [83].

Similar formation of five-membered metallacycle explained the selective cyclizing alkenylation of morpholinones to produce fused pyrrolidines also via concomitant formation of a C—C and an N—C bond with the catalytic system based on palladium acetate and *N*-acyl amino acid as ligand [84].

It also worth mentioning that the *in-situ* generation of transient removable directing groups upon reaction of primary aliphatic amines with 8-formylquinoline [85], 2-hydroxynicotinaldehyde [86], or glyoxylic acid [87] for the arylation at the γ-position of linear aliphatic chains appears as a very promising procedure for the future.

27.5 Formation of C—C Bonds Involving Metal-Catalyzed Hydrogen Transfer

Activation of alcohols or amines via dehydrogenation or oxidation constitute the methods of choice to generate highly reactive unsaturated carbonyl and imin-e/-ium intermediates. Among those, redox neutral processes known as borrowing hydrogen or hydrogen shuttling have attracted a lot of attention [88]. Historically discovered by M. Guerbet during the treatment of alcohols with their corresponding alkoxide base to afford the corresponding β-dimeric alcohols [89], the use of noble homogeneous transition metal catalysts improved the sustainability and broadened the scope of applicability of these transformations.

Taking advantage of the iminium and enamine intermediates involved in these processes, C(β)-H functionalization of cyclic amines has been reported through borrowing hydrogen methodology. In 2010, during the *N*-alkylation of pyrrolidine with primary alcohols, the side formation of an *N*,*C*-dialkylated compound was observed in the presence of a ruthenium complex featuring phosphine-sulfonate chelate [90]. Optimization of the reaction conditions using sulfonic acid additives such as camphorsulfonic acid led to the development of a selective catalytic system for the concomitant *N*- and *C*(β)-alkylation of various cyclic amines ranging from pyrrolidine, piperidine, to azepane with primary alcohols.

Using a modified strategy, the late stage *C*(β)-H functionalization of tertiary cyclic amines was achieved in the presence of aldehyde as the electrophile. In this case, the use of formic acid acting as hydrogen donor at the end of the reaction ensured the complete conversion of the alkylated enamine intermediates [91] (Scheme 27.35). This methodology has been successfully extended for the preparation of aza-terpenes and julolidines [92].

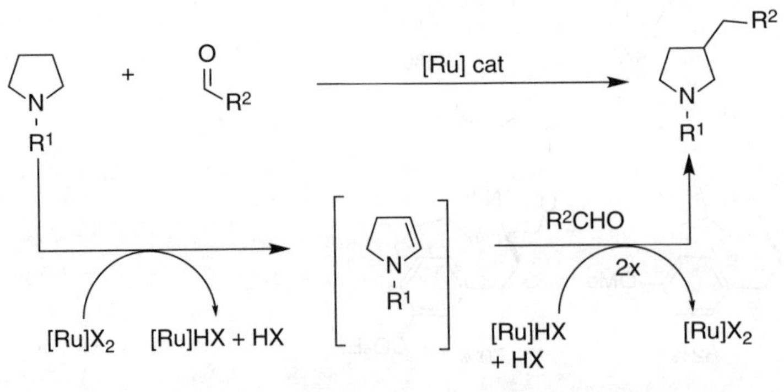

Scheme 27.35 Ruthenium-catalyzed β-functionalization of cyclic amines.

Since the seminal results on transition-metal-catalyzed *N*-alkylation through hydrogen auto-transfer, reports dealing on the preparation of cyclic amines from diols and primary amine or ammonia highlighted the versatility of these processes [93–95]. The use of a well-defined iridium complex featuring phosphine-sulfonate chelate enabled the tandem redox processes from pentanediols, anilines, and aldehydes for the two-step construction of various β-alkylated piperidines generating water as the only side product of the reaction (Scheme 27.36) [96].

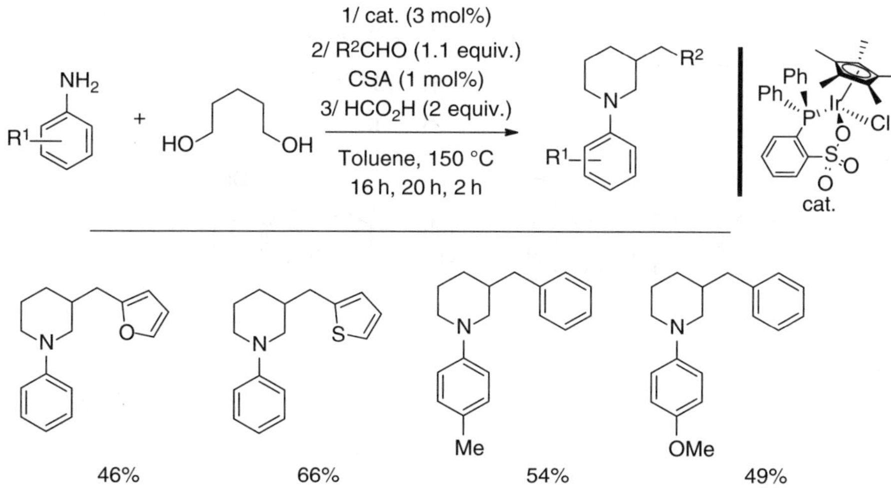

Scheme 27.36 Sequential formation and β-functionalization of piperidines.

Vicinal difunctionalization of tertiary cyclic amine was then achieved. Taking advantage of the enamine intermediate, the use of salicylaldehyde both acting as a source of electrophile and nucleophile enabled the formation of the corresponding cyclic hemiaminal ethers. By modification of the reaction conditions, including the nature of the catalyst associated with a phosphine-sulfonate ligand, the major formation of the alkylated lactams arising from a dehydrogenation process was also observed [97]. Although a metallo-catalyzed hydrogen transfer between the cyclic amine and the salicylaldehyde to generate *o*-quinone methide cannot totally be excluded, the authors proposed the involvement of α,β-unsaturated iminium intermediates **II** arising from the borrowing hydrogen pathway to account for the formation of the key cyclic hemiaminals (Scheme 27.37). The cyclic hemiaminal ethers, upon protonation, can lead either to the reduced β-alkylated amines through reduction or to the corresponding lactams via dehydrogenation.

27.6 Miscellaneous

27.6.1 Rhodium-Carbenoid Insertion into C(α)—H Bonds of Aliphatic Amines

Single and double sp^3C—H bond activation of the methyl groups of *N,N*-dimethylanilines has been carried out at room temperature with rhodium catalysts. Donor/acceptor-substituted carbenoids derived from aryldiazoacetates undergo selective insertion into the *N*-methyl groups to give β-amino esters or symmetrical diesters, depending on the aniline/diazoacetate ratio [98]. The use of (S)-DOSP as chiral ligand led to the formation of these products with good to excellent enantioselectivities (Scheme 27.38).

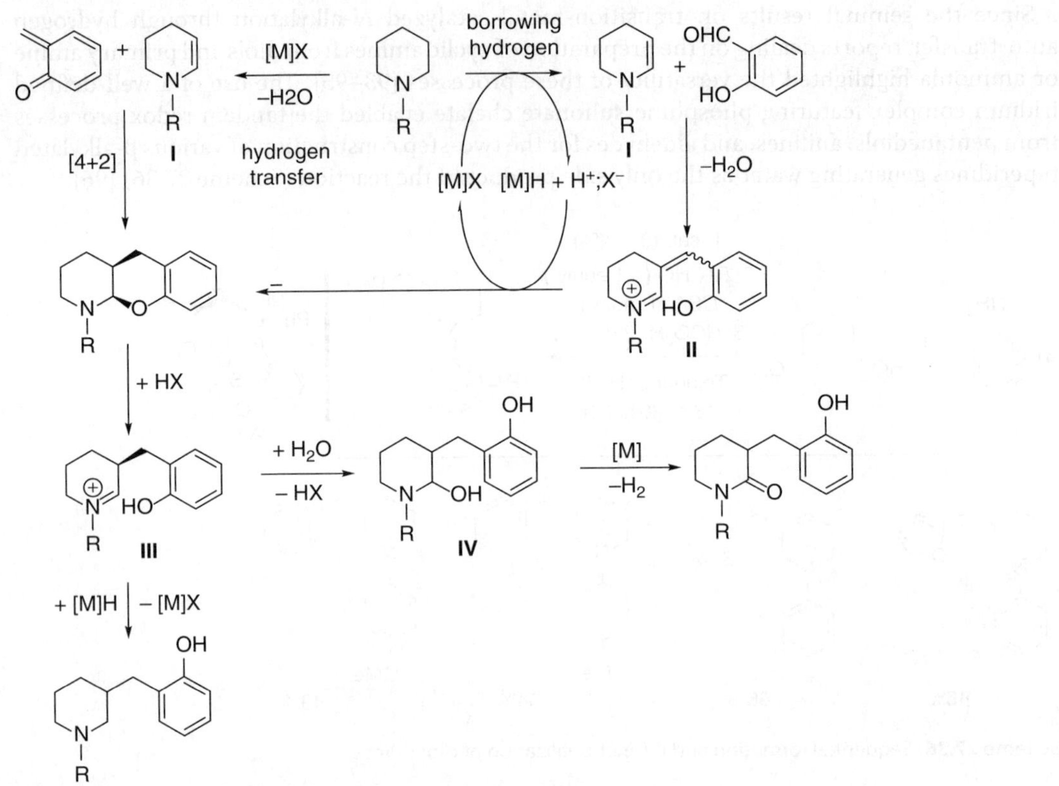

Scheme 27.37 Formation of various heterocycles from salicylaldehyde and piperidine.

Profit has been taken from this specific reactivity to perform rhodium-catalyzed C—H carbene insertion of brucine with different methylene site selectivities on the different cyclic structures, depending on the nature of the catalyst [99].

27.6.2 Formation of C(2)-Substituted Imines from Cyclic Secondary Amines and Alkenes

Using $RuH_2(CO)(PCy_3)_2$ as a catalyst, a dehydrogenative coupling of saturated secondary cyclic amines with an olefin such as ethylene, propene, or *tert*-butylethene operating under neutral conditions in THF at 70–120 °C has been reported (Scheme 27.39) [100]. The final products were imines, and significant amounts of cyclic 2-alkylamine were obtained with the bulky 3,3-dimethylbut-1-ene as olefin or the larger azepane as cyclic amine.

Imines were preferentially formed and the carbon–carbon bond was selectively formed at the C(α) position of the cyclic amine. With ethylene and propene, only imines were formed but with more bulky olefins and seven-membered cyclic amines, reduction to 2-alkyl cyclic amine took place. A mechanism was proposed, invoking nitrogen coordination and successive C(α)-H bond activation of amine and imine (Scheme 27.40).

27.6.3 Formation of C(β)-Substituted Enamines and Heteroaromatics Involving Hydrogen Transfer under Oxidative Conditions

Under oxidative conditions, the formation of reactive nucleophilic enamine intermediates can take place efficiently using $K_2PtCl_4/Sc(OTf)_3$ as a catalyst in the presence of oxygen at 60 °C

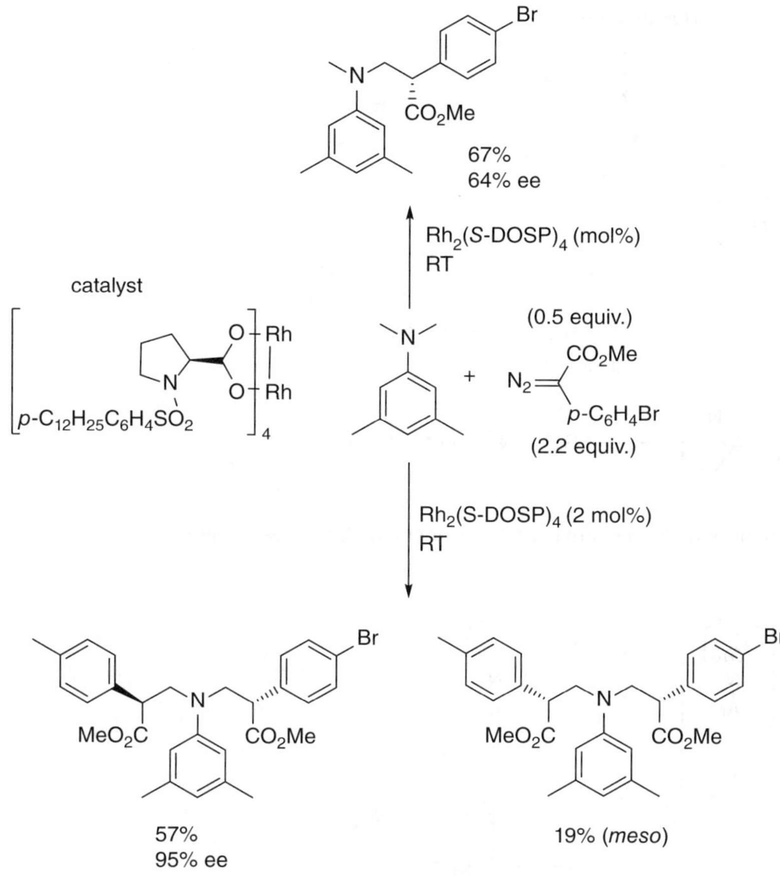

Scheme 27.38 C—H activation via rhodium-carbenoid insertion.

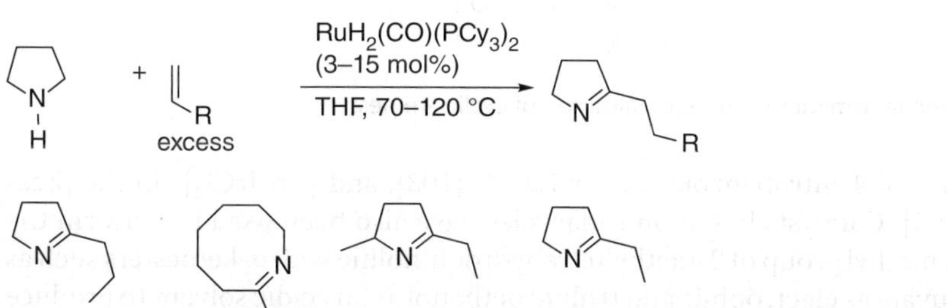

Scheme 27.39 Coupling of pyrrolidine with olefins catalyzed by ruthenium hydride.

[101] or $FeCl_3$/DMAP associated with *tert*-butyl peroxide as an oxidant [102]. These enamines further react with β-nitrostyrene in a Michael addition reaction to form a carbon–carbon bond selectively at the β-position of the final enamine (Scheme 27.41).

Finally, recent reports have described the formation of polycyclic aromatic amines substituted at the β-*endo* or *exo*-cyclic positions of secondary tetrahydroquinolines via intermediate formation of enamines. The creation of carbon–carbon bond (alkylation) has been shown starting from aldehydes as electrophiles with catalytic systems consisting in $[RuCl_2(p\text{-cymene})]_2$

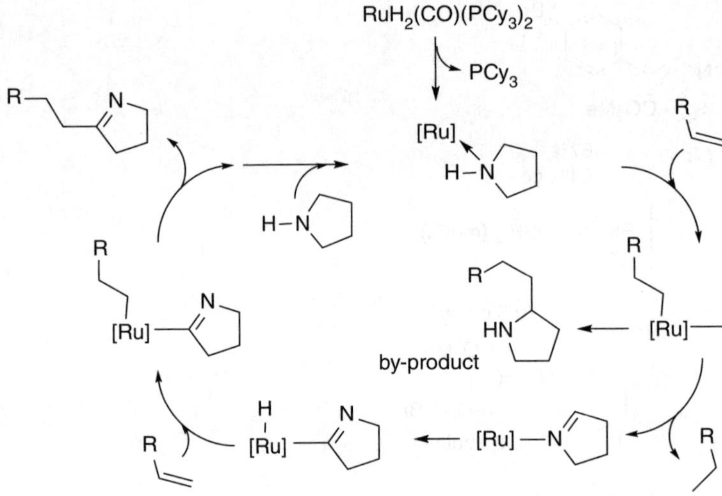

Scheme 27.40 Tentative mechanism for imine formation with ruthenium(II) hydride catalyst.

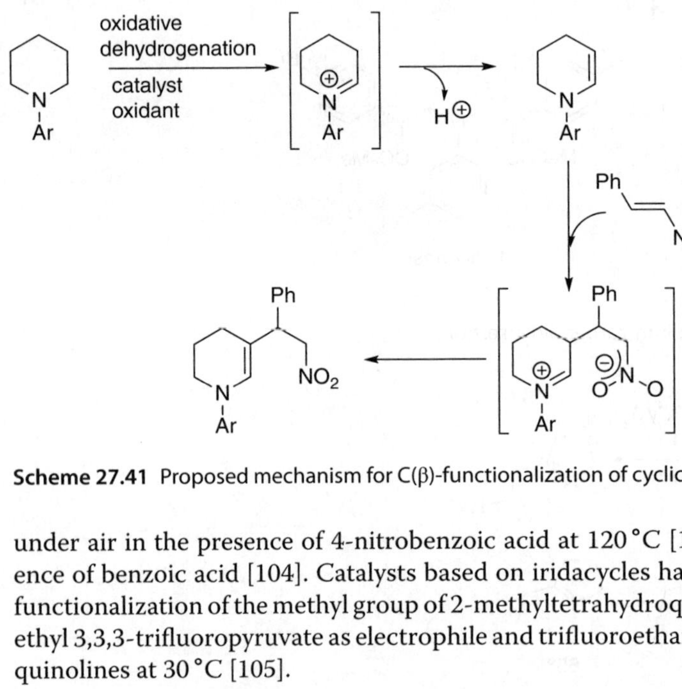

Scheme 27.41 Proposed mechanism for C(β)-functionalization of cyclic amines.

under air in the presence of 4-nitrobenzoic acid at 120 °C [103], and $[Cp^*IrCl_2]_2$ in the presence of benzoic acid [104]. Catalysts based on iridacycles have also been used to perform the functionalization of the methyl group of 2-methyltetrahydroquinoline with α-ketoesters such as ethyl 3,3,3-trifluoropyruvate as electrophile and trifluoroethanol as an acidic solvent to produce quinolines at 30 °C [105].

27.7 Final Comments and Conclusions

The functionalization of alicyclic amines can be achieved according to various types of mechanisms. The catalytic oxidation at the C(α) carbon leading to the formation of iminium intermediates followed by trapping with a nucleophile has been used extensively to form α-functionalized amines. Profit can be taken from the directed activation of the C(α)-H bond by transition metals leading to the formation of a metal-carbon bond to achieve coupling with arylboronates, arylhalides, and olefins to produce carbon–carbon bonds at this selected position. In most cases, rhodium, ruthenium catalysts have led to α-functionalization, but recent examples with palladium catalysts have shown the possibility of remote functionalization. Finally, functionalization at β-position of amines has been efficiently achieved when the intermediate formation of enamines able to react with external electrophiles, most often aldehydes, could be generated *in situ*.

It must be noted that most of these transformations have been possible only with cyclic amines, but the reactivity of linear ones is much less documented except in the special case of methylanilines.

Acknowledgments

The authors wish to thank CNRS and Université de Rennes1 for financial support, and the contribution of the European COST Action CA15106 (C–H Activation in Organic Synthesis, CHAOS).

References

1 Orru, R.V.A. and Ruijter, E. (eds.) (2010). *Synthesis of Heterocycles via Multicomponent Reactions I*, Topics in Heterocyclic Chemistry, vol. 23. Berlin, Heidelberg: Springer-Verlag.
2 (a) Baumann, M. and Baxendale, I.R. (2013). *Beilstein J. Org. Chem.* 9: 2265–2319. (b) Baumann, M., Baxendale, I.R., Ley, S.V., and Nikbin, N. (2011). *Beilstein J. Org. Chem.* 7: 442–495.
3 (a) Poupon, E. and Nay, B. (2011). *Biomimetic Organic Syntheses.* Wiley-VCH Verlag. (b) Poupon, E. and Gravel, E. (2015). *Chem. Eur. J.* 21: 10604–10615.
4 (a) Wypych, J.-C., Nguyen, T.M., Nuhant, P. et al. (2008). *Angew. Chem. Int. Ed.* 47: 5418–5421. (b) Brown, P.D., Willis, A.C., Sherburn, M.S. et al. (2013). *Angew. Chem. Int. Ed.* 52: 13273–13275.
5 Selected examples of sp^3C-H functionalization and C-C bond formation of *N*-protected aliphatic amines – as amide(a)Wang, C., Zhang, L., Chen, C. et al. (2015). *Chem. Sci.* 6: 4610–4614. (b) Spangler, J.E., Kobayashi, Y., Verma, P. et al. (2015). *J. Am. Chem. Soc.* 137: 11876–11879. (c) Pedroni, J., Boghi, M., Saget, T. et al. (2014). *Angew. Chem. Int. Ed.* 53: 9064–9067. (d) Guyonnet, M. and Baudoin, O. (2012). *Org. Lett.* 14: 398–401. (e) He, G., Zhao, Y., Zhang, S. et al. (2012). *J. Am. Chem. Soc.* 134: 3–6. as sulfonamide:; (f) Saget, T., Lemouzy, S.J., and Cramer, N. (2012). *Angew. Chem. Int. Ed.* 51: 2238–2242; (g) Saget, T. and Cramer, N. (2012). *Angew. Chem. Int. Ed.* 51: 12842–12845. (h) Donets, P.A., Saget, T., and Cramer, N. (2012). *Organometallics* 31: 8040–8046; as carbamate:; (i) Yang, L., Melot, R., Neuburger, M. et al. (2017). *Chem. Sci.* 8: 1344–1349. (j) Katayev, D., Larionov, E., Nakanishi, M. et al. (2014). *Chem. Eur. J.* 20: 15021–15030. (k) Larionov, E., Nakanishi, M., Katayev, D. et al. (2013). *Chem. Sci.* 4: 1995–2005. (l) Millet, A., Larini, P., Clot, E. et al. (2013). *Chem. Sci.* 4: 2241–2247. (m) Katayev, D., Nakanishi, M., Bügi, T., and Kündig, E.P. (2012). *Chem. Sci.* 3: 1422–1425. (n) Watanabe, T., Oishi, S., Fujii, N. et al. (2008). *Org. Lett.* 10: 1759–1762. (o) Anas, S., Cordi, A., and Kagan, H.B. (2011). *Chem. Commun.* 47: 11483–11485.
6 Poirier, R.H., Morin, R.D., McKim, A.M. et al. (1961). *J. Org. Chem.* 26: 4275–4278.
7 Burrows, E.P., Hutton, R.F., and Burrows, W.D. (1962). *J. Org. Chem.* 27: 316–317.

8 (a) Oda, M., Fukuchi, Y., Ito, S. et al. (2007). *Tetrahedron Lett.* 48: 9159–9162. (b) Polackova, V., Veverkova, E., Toma, S. et al. (2009). *Synth. Commun.* 39: 1871–1876.
9 (a) Ma, L., Paul, A., Breugst, M. et al. (2016). *Chem. Eur. J.* 22: 18179–18189. (b) Seidel, D. (2015). *Acc. Chem. Res.* 48: 317–328. (c) Kang, Y., Richers, M.T., Sawicki, C.H. et al. (2015). *Chem. Commun.* 51: 10648–10651. (d) Chen, W., Kang, Y., Wilde, R.G. et al. (2014). *Angew. Chem. Int. Ed.* 53: 5179–5182. (e) Mantelingu, K., Lin, Y., and Seidel, D. (2014). *Org. Lett.* 16: 5910–5913. (f) Ma, L., Chen, W., and Seidel, D. (2012). *J. Am. Chem. Soc.* 134: 15305–15308.
10 Mandal, S., Mahato, S., and Jana, C.K. (2015). *Org. Lett.* 17: 3762–3765.
11 Andreades, S. and Zahnow, E.W. (1969). *J. Am. Chem. Soc.* 91: 4181–4190.
12 Chiba, T. and Takata, Y. (1977). *J. Org. Chem.* 42: 2973–2977.
13 (a) Renaud, T., Hurvois, J.-P., and Uriac, P. (2001). *Eur. J. Org. Chem.* 987–996. (b) Malassene, R., Vanquelef, E., Toupet, L. et al. (2003). *Org. Biomol. Chem.* 1: 547–551. (c) Louafi, F., Hurvois, J.-P., Chibani, A., and Roisnel, T. (2010). *J. Org. Chem.* 75: 5721–5724.
14 Onomura, O. (2012). *Heterocycles* 85: 2111–2133.
15 (a) Chen, C.-K., Hortmann, A.G., and Marzabadi, M.R. (1988). *J. Am. Chem. Soc.* 110: 4829–4831. (b) Rao, G.A. and Periasamy, M. (2015). *Synlett* 2231–2236.
16 Murahashi, S.-I., Naota, T., and Yonemura, K. (1988). *J. Am. Chem. Soc.* 110: 8256–8282.
17 Murahashi, S.-I., Komiya, N., Terai, H. et al. (2003). *J. Am. Chem. Soc.* 125: 15312–15313.
18 Murahashi, S.-I., Nakae, T., Terai, H. et al. (2008). *J. Am. Chem. Soc.* 130: 11005–11012.
19 Murahashi, S.-I., Komiya, N., and Terai, H. (2005). *Angew. Chem. Int. Ed.* 44: 6931–6933.
20 Yu, A., Gu, Z., Chen, D. et al. (2009). *Catal. Commun.* 11: 162–166.
21 Wang, M.Z., Zhou, C.Y., Wong, M.K. et al. (2010). *Chem. Eur. J.* 16: 5723–5735.
22 (a) Li, C.-J. (2009). *Acc. Chem. Res.* 42: 335–344. (b) Li, Z., Bohle, D.S., and Li, C.-J. (2006). *Proc. Natl. Acad. Sci. U.S.A.* 103: 8928–8933. (c) Yeung, C.S. and Dong, V.M. (2011). *Chem. Rev.* 111: 1215–1292. (d) Zhang, C., Tang, C., and Jiao, N. (2012). *Chem. Soc. Rev.* 41: 3464–3484. (e) Scheuermann, C.J. (2010). *Chem. Asian J.* 5: 436–451. (f) Wendlandt, A.E., Suess, A.M., and Stahl, S.S. (2011). *Angew. Chem. Int. Ed.* 50: 11062–11087. (g) Jones, K.M. and Klussmann, M. (2012). *Synlett* 159–162. (h) Girard, S.A., Knauber, T., and Li, C.-J. (2014). *Angew. Chem. Int. Ed.* 53: 74–100. (i) Zhang, W., Wang, N.-X., and Xing, Y. (2015). *Synlett* 2088–2098. (j) Mitchell, E.A., Peschiulli, A., Lefevre, N. et al. (2012). *Chem. Eur. J.* 18: 10092–10142. (k) Doye, S. (2001). *Angew. Chem. Int. Ed.* 40: 3351–3353.
23 Boess, E., Sureshkumar, D., Sud, A. et al. (2011). *J. Am. Chem. Soc.* 133: 8106–8109.
24 Li, Z. and Li, C.-J. (2004). *J. Am. Chem. Soc.* 126: 11810–11811.
25 Niu, M., Yin, Z., Fu, H. et al. (2008). *J. Org. Chem.* 73: 3961–3963.
26 Li, Z. and Li, C.-J. (2004). *Org. Lett.* 6: 4997–4999.
27 Li, Z., MacLeod, P.D., and Li, C.-J. (2006). *Tetrahedron: Asymmetry* 17: 590–597.
28 Teong, S.P., Yu, D., Sum, Y.N. et al. (2016). *Green Chem.* 18: 3499–3502.
29 Li, Z. and Li, C.-J. (2005). *J. Am. Chem. Soc.* 127: 6968–6969.
30 Yang, F., Li, J., Xie, J. et al. (2010). *Org. Lett.* 12: 5214–5217.
31 Wang, G., Ma, Y., Wang, S. et al. (2012). *J. Am. Chem. Soc.* 134: 12334–12337.
32 Li, Z. and Li, C.-J. (2005). *Eur. J. Org. Chem.* 3173–3176.
33 Baslé, O. and Li, C.-J. (2007). *Green Chem.* 9: 1047–1050.
34 Li, Z. and Li, C.-J. (2005). *J. Am. Chem. Soc.* 127: 3672–3673.
35 Baslé, O., Borduas, N., Dubois, P. et al. (2010). *Chem. Eur. J.* 16: 8162–8166.
36 Mitsudera, H. and Li, C.-J. (2011). *Tetrahedron Lett.* 52: 1898–1900.
37 Zhang, J., Tiwari, B., Xing, C. et al. (2012). *Angew. Chem. Int. Ed.* 51: 3649–3652.
38 Sud, A., Sureskumar, D., and Klussmann, M. (2009). *Chem. Commun.* 3169–3171.

39 (a) Shu, X.-Z., Yang, Y.-F., Xia, X.-F. et al. (2010). *Org. Biomol. Chem.* 8: 4077–4079. (b) Rao Volla, C.M. and Vogel, P. (2009). *Org. Lett.* 11: 1701–1704. (c) Alagiri, K. and Prabhu, K.R. (2012). *Org. Biomol. Chem.* 10: 835–842. (d) Xie, J., Li, H., Zhou, J. et al. (2012). *Angew. Chem. Int. Ed.* 51: 1252–1255.

40 (a) Prier, C.K., Rankic, D.A., and MacMillan, D.W.C. (2013). *Chem. Rev.* 113: 5322–5363. (b) Xuan, J. and Xiao, W.-J. (2012). *Angew. Chem. Int. Ed.* 51: 6828–6838.

41 Condie, A.G., Gonzales-Gomez, J.C., and Stephenson, C.R.J. (2010). *J. Am. Chem. Soc.* 132: 1464–1465.

42 Bartling, H., Eisenhofer, A., König, B. et al. (2016). *J. Am. Chem. Soc.* 138: 11860–11871.

43 Wang, J.-L., Wang, C., deKrafft, K.E. et al. (2012). *ACS Catal.* 2: 417–424.

44 Freeman, D.B., Furst, L., Condie, A.G. et al. (2012). *Org. Lett.* 14: 94–97.

45 (a) Rueping, M., Zhu, S., and Koenigs, R.M. (2011). *Chem. Commun.* 47: 12709–12711. (b) Rueping, M., Zoller, J., Fabry, D.C. et al. (2012). *Chem. Eur. J.* 18: 3478–3481.

46 Zhao, G., Yang, C., Guo, L. et al. (2012). *Chem. Commun.* 48: 2337–2339.

47 Rueping, M., Koenigs, R.M., Poscharny, K. et al. (2012). *Chem. Eur. J.* 18: 5170–5174.

48 Zhu, S., Das, A., Bui, L. et al. (2013). *J. Am. Chem. Soc.* 135: 1823–1829.

49 Feng, Z.-J., Xuan, J., Xia, X.-D. et al. (2014). *Org. Biomol. Chem.* 12: 2037–2040.

50 Rueping, M., Vila, C., Koenigs, R.M. et al. (2011). *Chem. Commun.* 47: 2360–2362.

51 Rueping, M. and Vila, C. (2013). *Org. Lett.* 15: 2092–2095.

52 Vila, C. and Rueping, M. (2013). *Green Chem.* 15: 2056–2059.

53 Xuan, J., Zeng, T.-T., Feng, Z.-J. et al. (2015). *Angew. Chem. Int. Ed.* 54: 1625–1628.

54 Ahneman, D.T. and Doyle, A.G. (2016). *Chem. Sci.* 7: 7002–7006.

55 DiRocco, D.A. and Rovis, T. (2012). *J. Am. Chem. Soc.* 134: 8094–8097.

56 (a) Hari, D.P. and König, B. (2011). *Org. Lett.* 13: 3852–3855. (b) Liu, Q., Li, Y.-N., Zhang, H.-H. et al. (2012). *Chem. Eur. J.* 18: 620–627. (c) Cantillo, D., de Frutos, O., Rincon, J.A. et al. (2014). *Org. Lett.* 16: 896–899.

57 Pan, Y., Wang, S., Kee, C.W. et al. (2011). *Green Chem.* 13: 3341–3344.

58 Wei, G., Zhang, C., Bures, F. et al. (2016). *ACS Catal.* 6: 3708–3712.

59 Campos, K.R. (2007). *Chem. Soc. Rev.* 36: 1069–1084.

60 Ishii, Y., Chatani, N., Kakiuchi, F. et al. (1997). *Organometallics* 16: 3615–3622.

61 Chatani, N., Asaumi, T., Ikeda, T. et al. (2000). *J. Am. Chem. Soc.* 122: 12882–12883.

62 Ishii, Y., Chatani, N., Kakiuchi, F., and Murai, S. (1997). *Tetrahedron Lett.* 38: 7565–7568.

63 Jun, C.-H., Hwang, D.-C., and Na, S.-J. (1998). *Chem. Commun.* 1405–1406.

64 Chatani, N., Asaumi, T., Yorimitsu, S. et al. (2001). *J. Am. Chem. Soc.* 123: 10935–10941.

65 Bergman, S.D., Storr, T.E., Prokopcova, H. et al. (2012). *Chem. Eur. J.* 18: 10393–10398.

66 Kulago, A.A., Van Steijvoort, B.F., Mitchell, E.A. et al. (2014). *Adv. Synth. Catal.* 356: 1610–1618.

67 Smout, V., Peschiulli, A., Verbeeck, S. et al. (2013). *J. Org. Chem.* 78: 9803–9814.

68 Schinkel, M., Wang, L., Bielefeld, K. et al. (2014). *Org. Lett.* 16: 1876–1879.

69 Pastine, S.J., Gribkov, D.V., and Sames, D. (2006). *J. Am. Chem. Soc.* 128: 14220–14221.

70 Prokopcova, H., Bergman, S.D., Aelvoet, K. et al. (2010). *Chem. Eur. J.* 16: 13063–13067.

71 Peschiulli, A., Smout, V., Storr, T.E. et al. (2013). *Chem. Eur. J.* 19: 10378–10387.

72 Schwarz, M.C., Dastbaravardeh, N., Kirchner, K. et al. (2013). *Monatsh. Chem.* 144: 539–552.

73 Dastbaravardeh, N., Schnürch, M., and Mihovilovic, M.D. (2012). *Org. Lett.* 14: 1930–1933.

74 Dastbaravardeh, N., Kirchner, K., Schnürch, M. et al. (2013). *J. Org. Chem.* 78: 658–672.

75 Dastbaravardeh, N., Schnürch, M., and Mihovilovic, M.D. (2012). *Org. Lett.* 14: 3792–3795.

76 Dastbaravardeh, N., Schnürch, M., and Mihovilovic, M.D. (2013). *Eur. J. Org. Chem.* 2878–2890.

77 Kumar, N.Y.P., Jeyachandran, R., and Ackermann, L. (2013). *J. Org. Chem.* 78: 4145–4152.
78 Ferrer-Flegeau, E., Bruneau, C., Dixneuf, P.H. et al. (2011). *J. Am. Chem. Soc.* 133: 10161–10170.
79 Topczewski, J.J., Cabrera, P.J., Saper, N.I. et al. (2016). *Nature* 531: 220–224.
80 He, C. and Gaunt, M.J. (2015). *Angew. Chem. Int. Ed.* 54: 15840–15844.
81 McNally, A., Haffemayer, B., Collins, B.S.L. et al. (2014). *Nature* 510: 129–133.
82 Calleja, J., Pla, D., Gorman, T.W. et al. (2015). *Nat. Chem.* 7: 1009–1016.
83 (a) Zhang, Q., Yu, H., and Fu, Y. (2015). *Sci. China Chem.* 58: 1316–1322. (b) Zhang, Y., Qi, Z.-H., Ruan, G.-Y. et al. (2015). *RSC Adv.* 5: 71586–71592.
84 He, C. and Gaunt, M.J. (2017). *Chem. Sci.* 8: 3586–3592.
85 Xu, Y., Young, M.C., Wang, C. et al. (2016). *Angew. Chem. Int. Ed.* 55: 9084–9087.
86 Wu, Y., Chen, Y.-Q., Liu, T. et al. (2016). *J. Am. Chem. Soc.* 138: 14554–14557.
87 Liu, Y. and Ge, H. (2016). *Nat. Chem.* 9: 26–32.
88 (a) Dobereiner, G.E. and Crabtree, R.H. (2010). *Chem. Rev.* 110: 681–703. (b) Guillena, G., Ramón, D.J., and Yus, M. (2010). *Chem. Rev.* 110: 1611–1641. (c) Gunanathan, C. and Milstein, D. (2013). *Science* 341: 249–260. (d) Yang, Q., Wang, Q., and Yu, Z. (2015). *Chem. Soc. Rev.* 44: 2305–2329.
89 (a) Gabriels, D., Hernandez, W.Y., Sels, B. et al. (2015). *Catal. Sci. Technol.* 5: 3876–3902. (b) Guerbet, M. (1899). *C. R. Hebd. Séances Acad. Sci.* 128: 511–513.
90 Sundararaju, B., Tang, Z., Achard, M. et al. (2010). *Adv. Synth. Catal.* 352: 3141–3146.
91 Sundararaju, B., Achard, M., Sharma, G.V.M. et al. (2011). *J. Am. Chem. Soc.* 133: 10340–10343.
92 (a) Sahli, Z., Sundararaju, B., Achard, M. et al. (2013). *Green Chem.* 15: 775–779. (b) Labed, A., Jiang, F., Labed, I. et al. (2015). *ChemCatChem* 7: 1090–1096.
93 (a) Tsuji, Y., Huh, K.-T., Ohsugi, Y. et al. (1985). *J. Org. Chem.* 50: 1365–1370. (b) Huh, K.-T., Shim, S.C., and Doh, C.H. (1990). *Bull. Korean Chem. Soc.* 11: 45–49. (c) Marsella, J.A. (1991). *J. Organomet. Chem.* 407: 97–105. (d) Abbenhuis, R.A.T.M., Boersma, J., and van Koten, G. (1998). *J. Org. Chem.* 63: 4282–4290. (e) Hamid, M.H.S.A., Allen, C.L., Lamb, G.W. et al. (2009). *J. Am. Chem. Soc.* 131: 1766–1774. (f) Saidi, O., Blacker, A.J., Lamb, G.W. et al. (2010). *Org. Process Res. Dev.* 14: 1046–1049.
94 (a) Fujita, K.-I., Fujii, T., and Yamaguchi, R. (2004). *Org. Lett.* 6: 3525–3528. (b) Eary, C.T. and Clasen, D. (2006). *Tetrahedron Lett.* 47: 6899–6902. (c) Nordstrøm, L.U. and Madsen, R. (2007). *Chem. Commun.* 5034–5036.
95 Bähn, S., Imm, S., Neubert, L. et al. (2011). *ChemCatChem* 3: 1853–1864.
96 Yuan, K., Jiang, F., Sahli, Z. et al. (2012). *Angew. Chem. Int. Ed.* 51: 8876–8880.
97 Jiang, F., Achard, M., and Bruneau, C. (2015). *Chem. Eur. J.* 21: 14319–14323.
98 Davies, H.M.L. and Jin, Q. (2004). *Org. Lett.* 6: 1769–1772.
99 He, J., Hamann, L.G., Davies, H.M.L. et al. (2015). *Nat. Commun.* 6: 5943–5951.
100 Yi, C.S. and Yun, S.Y. (2004). *Organometallics* 23: 5392–5395.
101 (a) Xia, X.-F., Shu, X.-Z., Ji, K.-G. et al. (2010). *J. Org. Chem.* 75: 2893–2902. (b) Xia, X.-F., Shu, X.-Z., Ji, K.-G. et al. (2011). *J. Org. Chem.* 76: 342–345.
102 Takasu, N., Oisaki, K., and Kanai, M. (2013). *Org. Lett.* 15: 1918–1921.
103 Tan, Z., Jiang, H., and Zhang, M. (2016). *Org. Lett.* 18: 3174–3177.
104 Tan, Z., Jiang, H., and Zhang, M. (2016). *Chem. Commun.* 52: 9359–9362.
105 Talwar, D., Gonzalez-de-Castro, A., Li, H.Y. et al. (2015). *Angew. Chem. Int. Ed.* 54: 5223–5227.

Index

Alkane Functionalization, First Edition. Edited by Armando J. L. Pombeiro and M. Fátima C. Guedes da Silva.

b

C

g

h

j

k

l

m

p